SHOCK AND VIBRATION HANDBOOK

McGRAW-HILL HANDBOOKS

ABBOTT AND STETKA · National Electrical Code Handbook, 11th ed.
ALJIAN · Purchasing Handbook
AMERICAN INSTITUTE OF PHYSICS · American Institute of Physics Handbook, 2d ed.
AMERICAN SOCIETY OF MECHANICAL ENGINEERS · ASME Handbooks:
    Engineering Tables   Metals Engineering—Processes
    Metals Engineering—Design   Metals Properties
AMERICAN SOCIETY OF TOOL AND MANUFACTURING ENGINEERS:
    Die Design Handbook   Manufacturing Planning and Estimating
    Handbook of Fixture Design   Handbook
                                       Tool Engineers Handbook, 2d ed.
BEEMAN · Industrial Power Systems Handbook
BELL · Petroleum Transportation Handbook
BERRY, BOLLAY, AND BEERS · Handbook of Meteorology
BLATZ · Radiation Hygiene Handbook
BRADY · Materials Handbook, 9th ed.
BURINGTON · Handbook of Mathematical Tables and Formulas, 3d ed.
BURINGTON AND MAY · Handbook of Probability and Statistics with Tables
CARROLL · Industrial Instrument Servicing Handbook
COCKRELL · Industrial Electronics Handbook
CONDON AND ODISHAW · Handbook of Physics
CONSIDINE · Process Instruments and Controls Handbook
CROCKER · Piping Handbook, 4th ed.
CROFT AND CARR · American Electricians' Handbook, 8th ed.
DAVIS · Handbook of Applied Hydraulics, 2d ed.
DUDLEY · Gear Handbook
ETHERINGTON · Nuclear Engineering Handbook
FACTORY MUTUAL ENGINEERING DIVISION · Handbook of Industrial Loss Prevention
FINK · Television Engineering Handbook
FLÜGGE · Handbook of Engineering Mechanics
FRICK · Petroleum Production Handbook, 2 vols.
GUTHRIE · Petroleum Products Handbook
HARRIS · Handbook of Noise Control
HARRIS AND CREDE · Shock and Vibration Handbook, 3 vols.
HENNEY · Radio Engineering Handbook, 5th ed.
HUNTER · Handbook of Semiconductor Electronics, 2d ed.
HUSKEY AND KORN · Computer Handbook
JASIK · Antenna Engineering Handbook
JURAN · Quality Control Handbook, 2d ed.

KALLEN · Handbook of Instrumentation and Controls
KING AND BRATER · Handbook of Hydraulics, 5th ed.
KNOWLTON · Standard Handbook for Electrical Engineers, 9th ed.
KOELLE · Handbook of Astronautical Engineering
KORN AND KORN · Mathematical Handbook for Scientists and Engineers
KURTZ · The Lineman's Handbook, 3d ed.
LA LONDE AND JANES · Concrete Engineering Handbook
LANDEE, DAVIS, AND ALBRECHT · Electronic Designers' Handbook
LANGE · Handbook of Chemistry, 10th ed.
LAUGHNER AND HARGAN · Handbook of Fastening and Joining of Metal Parts
LE GRAND · The New American Machinist's Handbook
LIDDELL · Handbook of Nonferrous Metallurgy, 2d ed.
MAGILL, HOLDEN, AND ACKLEY · Air Pollution Handbook
MANAS · National Plumbing Code Handbook
MANTELL · Engineering Materials Handbook
MARKS AND BAUMEISTER · Mechanical Engineers' Handbook, 6th ed.
MARKUS · Handbook of Electronic Control Circuits
MARKUS AND ZELUFF · Handbook of Industrial Electronic Circuits
MARKUS AND ZELUFF · Handbook of Industrial Electronic Control Ciruits
MAYNARD · Industrial Engineering Handbook, 2d ed.
MEITES · Handbook of Analytical Chemistry
MERRITT · Building Construction Handbook
MOODY · Petroleum Exploration Handbook
MORROW · Maintenance Engineering Handbook
PERRY · Chemical Business Handbook
PERRY · Chemical Engineers' Handbook, 4th ed.
SHAND · Glass Engineering Handbook, 2d ed.
STANIAR · Plant Engineering Handbook, 2d ed.
STREETER · Handbook of Fluid Dynamics
STUBBS · Handbook of Heavy Construction
TERMAN · Radio Engineers' Handbook
TRUXAL · Control Engineers' Handbook
URQUHART · Civil Engineering Handbook, 4th ed.
WALKER · NAB Engineering Handbook, 5th ed.
WOODS · Highway Engineering Handbook
YODER, HENEMAN, TURNBULL, AND STONE · Handbook of Personnel Management and Labor Relations

SHOCK AND VIBRATION HANDBOOK

IN THREE VOLUMES

Edited by

CYRIL M. HARRIS

Columbia University

and

CHARLES E. CREDE

California Institute of Technology

VOLUME 1
BASIC THEORY AND MEASUREMENTS

McGRAW-HILL BOOK COMPANY

New York   Toronto   London

SHOCK AND VIBRATION HANDBOOK

Copyright © 1961 by McGraw-Hill, Inc. All Rights Reserved. Printed in the United States of America. This book, or parts thereof, may not be reproduced in any form without permission of the publishers.

Library of Congress Catalog Card Number: 60–16636

26798

PREFACE

A comparison of books in the field of vibration published during the past several years with corresponding books published a generation ago reveals a very significant contrast: Whereas the scope of earlier books included the entire field of vibration in reasonable detail, the scope of the more recent books includes only special aspects of the field. Furthermore, two important areas have not yet been treated in a comprehensive manner. There has not been published (1) an adequate treatment of mechanical shock—a topic generically related to vibration yet involving many different concepts and procedures, or (2) an appropriate discussion of instruments and methods of measurement, the development of which have progressed greatly in recent years. The "Shock and Vibration Handbook" recognizes, for the first time, the full scope of the field of shock and vibration by bringing together under one title classical vibration theory combined with modern applications of the theory to current engineering practice, including particularly the recently matured topics of mechanical shock and instrumentation for the measurement of shock and vibration.

This Handbook, the equivalent in content of at least eight textbooks of usual size, presents a unified treatment of the subject of shock and vibration. The 50 chapters were written by 72 authorities from industry, government laboratories, and universities. Each chapter covers the particular subject matter in the most comprehensive manner possible. In fact, on a number of specialized topics, some chapters contain more material than can be found collectively in all previously published books in the field. Others contain considerable material that has not been summarized previously in the literature. Duplication between chapters is avoided insofar as this is desirable—cross references to other chapters and to the technical literature being used frequently.

Whereas each chapter covers a particular topic, the chapters dealing with topics in the various categories are grouped together. The first group of chapters provides a theoretical basis for shock and vibration. The second group of chapters considers instrumentation and measurements. The next group of chapters deals with analysis and testing—concepts in the treatment of data obtained from measurements, and procedures for analyzing and testing systems subjected to vibration and shock. Then the important subject of methods of controlling shock and vibration is discussed in a group of chapters dealing with isolation, damping, and balancing. This is followed by chapters devoted to equipment design, packaging, and the effects of shock and vibration on man. The final group of chapters discusses the nature of environmental conditions existing in various classes of vehicles and in circumstances where shock and vibration are transmitted through the air or ground.

Although this Handbook is not intended primarily as a textbook, many teachers will find the classical and rigorous treatment suitable for classroom use; in particular, the extensive discussion of practical examples will be of value as a supplement to the usual classroom theory. The control of shock and vibration is of practical importance in many phases of engineering. The Handbook is particularly intended to find use as a working reference by engineers and scientists in the mechanical, aeronautical, electrical, refrigeration, air-conditioning, civil, acoustical, automotive, railroad, and chemical fields. Engineers in manufacturing, including plant maintenance, measurement and control, environmental testing, and packing and shipping, will find much of value, as will those engaged in development and design work.

Approximately four years have elapsed from initial detailed planning to publication. During this period many persons and organizations made contributions far too numerous to acknowledge here. To our contributing authors we owe much. They worked diligently with us toward the objective of making each chapter the definitive treatment in its field; this work was punctuated in many instances by repeated reviews through correspondence and personal conference, often leading to revised draft upon revised draft. Thanks to the cooperation of the contributing authors and many organizations, this Handbook contains considerable information not previously released for publication. Chapters dealing with commercially available equipment were sent to the manufacturers of such equipment to ensure authenticity; the cooperation of these manufacturers is acknowledged. Finally we wish to express our appreciation to the government agencies with whom some of our contributors are associated for clearing the material presented in chapters written by these contributors.

Cyril M. Harris and Charles E. Crede

CONTENTS

Volume 1. BASIC THEORY AND MEASUREMENTS

Preface   vii

1. **INTRODUCTION TO THE HANDBOOK**   1–1
   Cyril M. Harris, *Associate Professor of Electrical Engineering; Columbia University, New York 27, N.Y.*
   AND
   Charles E. Crede, *Professor of Mechanical Engineering; California Institute of Technology, Pasadena, Calif.*

2. **BASIC VIBRATION THEORY**   2–1
   Ralph E. Blake, *Staff Scientist; Lockheed Missiles and Space Company, Sunnyvale, Calif.*

3. **VIBRATION OF A RESILIENTLY SUPPORTED RIGID BODY**   3–1
   Harry Himelblau, Jr., *Specialist; Nortronics, A Division of Northrop Corporation, Hawthorne, Calif.*
   AND
   Sheldon Rubin, *Aerospace Corporation, Los Angeles 45, Calif.*

4. **NONLINEAR VIBRATION**   4–1
   H. Norman Abramson, *Director, Department of Mechanical Sciences; Southwest Research Institute, San Antonio 6, Texas*

5. **SELF-EXCITED VIBRATION**   5–1
   Robert S. Hahn, *Chief Research Engineer; The Heald Machine Company, Worcester, Mass.*

6. **DYNAMIC VIBRATION ABSORBERS AND AUXILIARY MASS DAMPERS**   6–1
   F. Everett Reed, *Partner; CONESCO, Arlington 74, Mass.*

7. **VIBRATION OF SYSTEMS HAVING DISTRIBUTED MASS AND ELASTICITY**   7–1
   William F. Stokey, *Associate Professor of Mechanical Engineering; Carnegie Institute of Technology, Pittsburgh 13, Pa.*

8. **TRANSIENT RESPONSE TO STEP AND PULSE FUNCTIONS**   8–1
   Robert S. Ayre, *Professor and Chairman, Department of Civil Engineering; Yale University, New Haven, Conn.*

CONTENTS

9. **EFFECTS OF IMPACT ON STRUCTURES** — 9-1
   William H. Hoppmann II, *Professor of Mechanics; The Rensselaer Polytechnic Institute, Troy, N.Y.*

10. **MECHANICAL IMPEDANCE AND MOBILITY** — 10-1
    Elmer L. Hixson, *Assistant Professor of Electrical Engineering, and Research Engineer in the Defense Research Laboratory; The University of Texas, Austin 12, Texas*

11. **STATISTICAL CONCEPTS IN VIBRATION** — 11-1
    John W. Miles, *Professor of Applied Mathematics; Institute of Advanced Studies, Australian National University, Canberra, Australia.*
    AND
    William T. Thomson, *Professor of Engineering; University of California, Los Angeles 24, Calif.*

12. **INTRODUCTION TO SHOCK AND VIBRATION MEASUREMENTS** — 12-1
    Wilson Bradley, Jr., *Executive Vice-president; Endevco Corporation, Pasadena, Calif.*
    AND
    Eldon E. Eller, *Engineer; Endevco Corporation, Pasadena, Calif.*

13. **SELF-CONTAINED VIBRATION-MEASURING INSTRUMENTS** — 13-1
    Fred Mintz, *Manager, Physical and Chemical Sciences Department; Lockheed Aircraft Corporation, Burbank, Calif.*
    AND
    John J. Dreher, *Group Engineer; Acoustics Research Group, Lockheed Aircraft Corporation, Burbank, Calif.*

14. **SPECIAL-PURPOSE AND MISCELLANEOUS SHOCK AND VIBRATION TRANSDUCERS** — 14-1
    Charles S. Duckwald, *Vibration Engineer; General Engineering Laboratory, General Electric Company, Schenectady, N.Y.*
    AND
    Burton S. Angell, *Machinery Apparatus Operations, General Electric Company, Schenectady, N.Y.*

15. **INDUCTIVE-TYPE PICKUPS** — 15-1
    R. R. Bouche, *Test Engineering Manager; Endevco Corporation, Pasadena, Calif.*

16. **PIEZOELECTRIC AND PIEZORESISTIVE PICKUPS**
    PART I. PICKUP CHARACTERISTICS — 16-1
    Abraham Dranetz, *Vice-president; Gulton Industries Inc., Metuchen, N.J.*
    AND
    Anthony W. Orlacchio, *General Manager, Instrumentation Division; Gulton Industries Inc., Metuchen, N.J.*

    PART II. PROPERTIES OF PIEZOELECTRIC AND PIEZORESISTIVE MATERIALS — 16-27
    Warren P. Mason, *Head, Mechanics Research; Bell Telephone Laboratories, Murray Hill, N.J.*

CONTENTS

17. STRAIN-GAGE INSTRUMENTATION 17–1
    Herbert R. Lissner, *Professor and Chairman, Engineering Mechanics Department; Wayne State University, Detroit 2, Mich.*
    AND
    C. C. Perry, *Associate Professor of Engineering Mechanics; Wayne State University, Detroit 2, Mich.*

18. CALIBRATION OF PICKUPS 18–1
    Samuel Levy, *Senior Mechanical Engineer, Mechanical Engineering Laboratory, General Electric Company, Schenectady 1, N.Y.*
    AND
    Russell H. Bickford, *Functional Leader Instrumentation; Structures Laboratory, Missile and Space Vehicle Department, General Electric Company, Philadelphia 4, Pa.*

19. SHOCK AND VIBRATION INSTRUMENTATION 19–1
    Robert W. Conrad, *Aeronautical Research Engineer; Goddard Space Flight Center, National Aeronautics and Space Administration, Greenbelt, Md.*

20. MEASUREMENT TECHNIQUES 20–1
    Richard D. Baxter, *Convair (Division of General Dynamics Corporation), San Diego, Calif.*
    AND
    John J. Beckman, *Convair (Division of General Dynamics Corporation), San Diego, Calif.*
    AND
    Harold A. Brown, *Convair (Division of General Dynamics Corporation), San Diego, Calif.*

Volume 2. DATA ANALYSIS, TESTING, AND METHODS OF CONTROL

21. INTRODUCTION TO DATA ANALYSIS AND TESTING 21–1
    Edward J. Lunney, *Chief, Test Laboratory; Research and Advanced Development Division, AVCO Corporation, Wilmington, Mass.*

22. CONCEPTS IN VIBRATION DATA ANALYSIS 22–1
    Allen J. Curtis, *Senior Staff Engineer; Hughes Aircraft Company, Culver City, Calif.*

23. CONCEPTS IN SHOCK DATA ANALYSIS 23–1
    Sheldon Rubin, *Aerospace Corporation, Los Angeles 45, Calif.*

24. SPECIFICATION OF LABORATORY TESTS 24–1
    Maurice Gertel, *Director, Shock and Vibration Division; MITRON Research and Development Corporation, Waltham 54, Mass.*

25. VIBRATION TESTING MACHINES 25–1
    Karl Unholtz, *Vice-president, Engineering; Unholtz-Dickie Corporation, Hamden 18, Conn.*

CONTENTS

26. **SHOCK TESTING MACHINES** — 26-1
    Irwin Vigness, *Head, Shock and Vibration Branch; U.S. Naval Research Laboratory, Washington 25, D.C.*

27. **SCALE-MODEL PRINCIPLES** — 27-1
    Donald E. Hudson, *Professor of Mechanical Engineering; California Institute of Technology, Pasadena, Calif.*

28. **NUMERICAL METHODS OF ANALYSIS** — 28-1
    Stephen H. Crandall, *Professor of Mechanical Engineering; Massachusetts Institute of Technology, Cambridge 39, Mass.*
    AND
    Robert B. McCalley, Jr., *Consulting Engineer; Knolls Atomic Power Laboratory, General Electric Company, Schenectady, N.Y.*

29. **ANALOG METHODS OF ANALYSIS** — 29-1
    Walter W. Soroka, *Professor of Mechanical Engineering; University of California, Berkeley 4, Calif.*

30. **THEORY OF VIBRATION ISOLATION** — 30-1
    Charles E. Crede, *Professor of Mechanical Engineering; California Institute of Technology, Pasadena, California*
    AND
    Jerome E. Ruzicka, *Manager, Contract Engineering; Barry Controls Division of Barry Wright Corporation, Watertown 72, Mass.*

31. **THEORY OF SHOCK ISOLATION** — 31-1
    R. E. Newton, *Professor and Chairman; Department of Mechanical Engineering, United States Naval Postgraduate School, Monterey, Calif.*

32. **APPLICATION AND DESIGN OF ISOLATORS** — 32-1
    Charles E. Crede, *Professor of Mechanical Engineering; California Institute of Technology, Pasadena, Calif.*

33. **AIR SUSPENSION AND SERVO-CONTROLLED ISOLATION SYSTEMS** — 33-1
    Richard D. Cavanaugh, *Manager of Technical Staff; Barry Controls Division of Barry Wright Corporation, Watertown 72, Mass.*

34. **MECHANICAL SPRINGS** — 34-1
    A. M. Wahl, *Advisory Engineer; Westinghouse Research Laboratories, Pittsburgh 35, Pa.*

35. **RUBBER SPRINGS** — 35-1
    William A. Frye, *Chief Test Engineer; Inland Manufacturing Division, General Motors Corporation, Dayton 1, Ohio*

36. **MATERIAL AND INTERFACE DAMPING** — 36-1
    Benjamin J. Lazan, *Professor and Head, Department of Aeronautical Engineering; University of Minnesota, Minneapolis 14, Minn.*
    AND
    Lawrence E. Goodman, *Professor of Mechanics; University of Minnesota, Minneapolis 14, Minn.*

37. **VIBRATION CONTROL BY APPLIED DAMPING TREATMENTS** — 37-1
    Richard N. Hamme, *Technical Director; Geiger and Hamme Laboratories, Ann Arbor, Mich.*

Volume 3. ENGINEERING DESIGN AND ENVIRONMENTAL CONDITIONS

38. TORSIONAL VIBRATION IN RECIPROCATING MACHINES   38–1
Frank M. Lewis, *Professor of Marine Engineering; Massachusetts Institute of Technology, Cambridge 39, Mass.*

39. BALANCING OF ROTATING MACHINERY
PART I. THEORY OF BALANCING   39–1
Douglas Muster, *Professor of Mechanical Engineering, University of Houston, Houston 4, Texas*
PART II. PRACTICE OF BALANCING   39–23
Werner I. Senger, *Vice-president in Charge of Engineering; Gisholt Machine Company, Madison 10, Wis.*

40. MACHINE-TOOL VIBRATION   40–1
Stephan A. Tobias, *Professor and Head, Department of Mechanical Engineering; University of Birmingham, Birmingham, England*

41. PACKAGING DESIGN   41–1
P. E. Franklin, *Senior Research Engineer; North American Aviation, Inc., Space and Information Systems Division, Downey, Calif.*
AND
M. T. Hatae, *Senior Research Engineer; North American Aviation, Inc., Space and Information Systems Division, Downey, Calif.*

42. THEORY OF EQUIPMENT DESIGN   42–1
Edward G. Fischer, *Advisory Engineer; Westinghouse Research Laboratories, Pittsburgh 35, Pa.*

43. PRACTICE OF EQUIPMENT DESIGN   43–1
Edward G. Fischer, *Advisory Engineer; Westinghouse Research Laboratories, Pittsburgh 35, Pa.*
AND
Harold M. Forkois, *Head, Mechanical Evaluation and Development Section; U.S. Naval Research Laboratory, Washington 25, D.C.*

44. EFFECTS OF SHOCK AND VIBRATION ON MAN   44–1
David E. Goldman, *Head, Biophysics Division; Naval Medical Research Institute, Bethesda 14, Md.*
AND
Henning E. von Gierke, *Chief, Bio-Acoustics Branch; Aerospace Medical Research Laboratories, Wright-Patterson Air Force Base, Ohio*

45. SHOCK AND VIBRATION IN ROAD AND RAIL VEHICLES
PART I. ROAD VEHICLES   45–1
Robert E. Engelhardt, *Manager, Environmental Research Section; Southwest Research Institute, San Antonio 6, Texas*
AND
Kenneth D. Mills, *Manager, Road Evaluation Section; Southwest Research Institute, San Antonio 6, Texas*
AND
Kurt Schneider, *Senior Research Engineer; Southwest Research Institute, San Antonio 6, Texas*

PART II. RAIL VEHICLES 45–21
Sergei G. Guins, *Assistant Director of Research; Chesapeake and Ohio Railway Co., Cleveland 13, Ohio*

46. SHOCK AND VIBRATION IN SHIPS 46–1
Francis F. Vane, *Supervisory Mechanical Engineer; Bureau of Ships, Department of the Navy, Washington 25, D.C.*

47. SHOCK AND VIBRATION IN AIRCRAFT AND MISSILES 47–1
H. A. Magrath, *Technical Director; Flight Dynamics Laboratory, Aeronautical Systems Division, Wright-Patterson Air Force Base, Ohio*
AND
O. R. Rogers, *Aeronautical Research Engineer; Flight Dynamics Laboratory, Aeronautical Systems Division, Wright-Patterson Air Force Base, Ohio*
AND
Charles K. Grimes, *Major, U.S.A.F.; Wright-Patterson Air Force Base, Ohio*

48. VIBRATIONS INDUCED BY ACOUSTIC WAVES 48–1
Harvey H. Hubbard, *Head, Acoustics Branch; National Aeronautics and Space Administration, Langley Research Center, Langley Field, Va.*
AND
John C. Houbolt, *Chief, Theoretical Mechanics Division; National Aeronautics and Space Administration, Langley Research Center, Langley Station, Hampton, Va.*

49. DESIGN OF BLAST-RESISTANT STRUCTURES 49–1
Nathan M. Newmark, *Professor and Head, Department of Civil Engineering; University of Illinois, Urbana, Ill.*
AND
Robert J. Hansen, *Professor of Structural Engineering; Massachusetts Institute of Technology, Cambridge 39, Mass.*

50. VIBRATION OF STRUCTURES INDUCED BY SEISMIC WAVES
PART I. EARTHQUAKES 50–1
George W. Housner, *Professor of Civil Engineering; California Institute of Technology, Pasadena, Calif.*

PART II. MAN-MADE GROUND MOTIONS 50–33
Donald E. Hudson, *Professor of Mechanical Engineering; California Institute of Technology, Pasadena, Calif.*

Index follows Chapter 50.

SHOCK AND VIBRATION HANDBOOK

1
INTRODUCTION TO THE HANDBOOK

Cyril M. Harris
Columbia University

Charles E. Crede
California Institute of Technology

CONCEPTS OF SHOCK AND VIBRATION

Vibration is a term that describes oscillation in a mechanical system. It is defined by the frequency (or frequencies) and amplitude. Either the motion of a physical object or structure or, alternatively, an oscillating force applied to a mechanical system is vibration in a generic sense. Conceptually, the time-history of vibration may be considered to be sinusoidal or simple harmonic in form. The frequency is defined in terms of cycles per unit time, and the magnitude in terms of amplitude (the maximum value of a sinusoidal quantity). The vibration encountered in practice does not have this regular pattern. It may be a combination of several sinusoidal quantities, each having a different frequency and amplitude. If each frequency component is an integral multiple of the lowest frequency, the vibration repeats itself after a determined interval of time and is called *periodic*. If there is no integral relation among the frequency components, there is no periodicity and the vibration is defined as *complex*.

Vibration may be described as *deterministic* or *random*. If it is deterministic, it follows an established pattern so that the value of the vibration at any designated future time is completely predictable from the past history. If the vibration is random, its future value is unpredictable except on the basis of probability. Random vibration is defined in statistical terms wherein the probability of occurrence of designated magnitudes and frequencies can be indicated. The analysis of random vibration involves certain physical concepts that are different from those applied to the analysis of deterministic vibration.

Vibration of a physical structure often is thought of in terms of a model consisting of a mass and a spring. The vibration of such a model, or system, may be "free" or "forced." In *free vibration*, there is no energy added to the system but rather the vibration is the continuing result of an initial disturbance. An *ideal system* may be considered undamped for mathematical purposes; in such a system the free vibration is assumed to continue indefinitely. In any *real system*, damping (i.e., energy dissipation) causes the amplitude of free vibration to decay continuously to a negligible value. Such free vibration sometimes is referred to as *transient vibration*. *Forced vibration*, in contrast to free vibration, continues under "steady-state" conditions because energy is supplied to the system continuously to compensate for that dissipated by damping in the system. In general, the frequency at which energy is supplied (i.e., the forcing frequency) appears in the vibration of the system. Forced vibration may be either deterministic or random. In either instance, the vibration of the system depends upon the relation of the excitation or forcing function to the properties of the system. This relationship is a prominent feature of the analytical aspects of vibration.

Shock is a somewhat loosely defined aspect of vibration wherein the excitation is nonperiodic, e.g., in the form of a pulse, a step, or transient vibration. The word "shock" implies a degree of suddenness and severity. These terms are relative rather than absolute measures of the characteristic; they are related to a popular notion of the characteristics of shock and are not necessary in a fundamental analysis of the applicable principles. From the analytical viewpoint, the important characteristic of shock is that the motion of the system upon which the shock acts includes both the frequency of the shock excitation and the natural frequency of the system. If the excitation is brief, the continuing motion of the system is free vibration at its own natural frequency.

The technology of shock and vibration embodies both theoretical and experimental facets prominently. Thus, methods of analysis and instruments for the measurement of shock and vibration are of primary significance. The results of analysis and measurement are used to evaluate shock and vibration environments, to devise testing procedures and testing machines, and to design and operate equipment and machinery. Shock and/or vibration may be either wanted or unwanted, depending upon circumstances. For example, vibration is involved in the primary mode of operation of such equipment as conveying and screening machines; the setting of rivets depends upon the application of impact or shock. More frequently, however, shock and vibration are unwanted. Then the objective is to eliminate or reduce their severity or, alternatively, to design equipment to withstand their influences. These procedures are embodied in the control of shock and vibration. Methods of control are emphasized throughout this Handbook.

CONTROL OF SHOCK AND VIBRATION

Methods of shock and vibration control may be grouped into three broad categories:

1. **Reduction at the Source**
    a. *Balancing of Moving Masses.* Where the vibration originates in rotating or reciprocating members, the magnitude of a vibratory force frequently can be reduced or possibly eliminated by balancing or counterbalancing. For example, during the manufacture of fans and blowers, it is common practice to rotate each rotor and to add or subtract material as necessary to achieve balance.
    b. *Balancing of Magnetic Forces.* Vibratory forces arising in magnetic effects of electrical machinery sometimes can be reduced by modification of the magnetic path. For example, the vibration originating in an electric motor can be reduced by skewing the slots in the armature laminations.
    c. *Control of Clearances.* Vibration and shock frequently result from impacts involved in operation of machinery. In some instances, the impacts result from inferior design or manufacture, such as excessive clearances in bearings, and can be reduced by closer attention to dimensions. In other instances, such as the movable armature of a relay, the shock can be decreased by employing a rubber bumper to cushion motion of the plunger at the limit of travel.
2. **Isolation**
    a. *Isolation of Source.* Where a machine creates significant shock or vibration during its normal operation, it may be supported upon isolators to protect other machinery and personnel from shock and vibration. For example, a forging hammer tends to create shock of a magnitude great enough to interfere with the operation of delicate apparatus in the vicinity of the hammer. This condition may be alleviated by mounting the forging hammer upon isolators.
    b. *Isolation of Sensitive Equipment.* Equipment often is required to operate in an environment characterized by severe shock or vibration. The equipment may be protected from these environmental influences by mounting it upon isolators. For example, equipment mounted in ships of the navy is subjected to shock of great severity during naval warfare and may be protected from damage by mounting it upon isolators.

3. **Reduction of the Response**
   a. Alteration of Natural Frequency. If the natural frequency of the structure of an equipment coincides with the frequency of the applied vibration, the vibration condition may be made much worse as a result of resonance. Under such circumstances, if the frequency of the excitation is substantially constant, it often is possible to alleviate the vibration by changing the natural frequency of such structure. For example, the vibration of a fan blade was reduced substantially by modifying a stiffener on the blade, thereby changing its natural frequency and avoiding resonance with the frequency of rotation of the blade. Similar results are attainable by modifying the mass rather than the stiffness.
   b. Energy Dissipation. If the vibration frequency is not constant or if the vibration involves a large number of frequencies, the desired reduction of vibration may not be attainable by altering the natural frequency of the responding system. It may be possible to achieve equivalent results by the dissipation of energy to eliminate the severe effects of resonance. For example, the housing of a washing machine may be made less susceptible to vibration by applying a coating of damping material on the inner face of the housing.
   c. Auxiliary Mass. Another method of reducing the vibration of the responding system is to attach an auxiliary mass to the system by a spring; with proper tuning the mass vibrates and reduces the vibration of the system to which it is attached. For example, the vibration of a textile-mill building subjected to the influence of several hundred looms was reduced by attaching large masses to a wall of the building by means of springs; then the masses vibrated with a relatively large motion and the vibration of the wall was reduced. The incorporation of damping in this auxiliary mass system may further increase its effectiveness.

CONTENT OF HANDBOOK

The chapters of this Handbook each deal with a discrete phase of the subject of shock and vibration. Frequent references are made from one chapter to another, to refer to basic theory in other chapters, to call attention to supplementary information, and to give illustrations and examples. Therefore, each chapter when read with other referenced chapters presents one complete facet of the subject of shock and vibration.

Chapters dealing with similar subject matter are grouped together. The first ten chapters following this introductory chapter deal with fundamental concepts of shock and vibration. Chapter 2 discusses the free and forced vibration of linear systems that can be defined by lumped parameters with similar types of coordinates. The properties of rigid bodies are discussed in Chap. 3, together with the vibration of resiliently supported rigid bodies wherein several modes of vibration are coupled. Nonlinear vibration is discussed in Chap. 4, and self-excited vibration in Chap. 5. Chapter 6 discusses two degree-of-freedom systems in detail—including both the basic theory and the application of such theory to dynamic absorbers and auxiliary mass dampers. The vibration of systems defined by distributed parameters, notably beams and plates, is discussed in Chap. 7. Chapters 8 and 9 relate to shock; Chap. 8 discusses the response of lumped parameter systems to step- and pulse-type excitation, and Chap. 9 discusses the effects of impact on structures. Chapter 10, entitled "Mechanical Impedance," discusses the concepts whereby the characteristics of a composite system may be determined from the characteristics of the component parts. Random vibration is discussed in Chap. 11 under the title of "Statistical Concepts in Vibration."

The second group of chapters deals with instrumentation for the measurement of shock and vibration. The general principles are discussed in Chap. 12. A wide range of instruments are described—from the compact and self-contained instruments discussed in Chap. 13 to elaborate instrumentation systems comprised of transducers and associated recording means. There are many types of transducers with a wide range of characteristics. The importance of this topic is indicated by the fact that four chapters (Chaps. 14 to 17) are devoted to discussions of the design, performance, and general characteristics of a large number of transducers. The calibration of instruments is discussed in Chap. 18. Chapter 19 describes the electrical circuits and associated equipment used with transducers for the measurement of shock and vibration, including filters, telemetering equipment, and recorders of various types. Field measurement techniques are discussed in Chap. 20.

Equipment intended for use under environmental conditions characterized by shock and vibration frequently is tested in the laboratory under simulated conditions. The importance of such testing is recognized by the group of chapters devoted to this subject. Chapter 21 introduces the concepts, and Chaps. 22 and 23 discuss the applicable concepts of data analysis. Chapter 22 is concerned with the analysis of data defining vibration conditions, and discusses the transformation of a time-history of a vibration measurement into more compact forms of data. Chapter 23 is an analogous presentation—treating the transformation of time-histories defining conditions of shock. Chapter 24 discusses the specification of laboratory tests for simulating the conditions, based upon data of the type described in Chaps. 22 and 23. Certain available hypotheses of equipment failure are used to specify the laboratory tests. Chapter 25 describes the construction and operation of vibration testing machines for laboratory use, and Chap. 26 is the analogous presentation dealing with shock testing machines.

The next group of chapters deals with computational methods. Dimensional analysis and models for vibration and shock studies are discussed in Chap. 27; the concepts which are presented are useful in both analytical and experimental work. Chapter 28 describes modern numerical methods in vibration analysis, dealing largely with the formulation of matrices for use with digital computers, and other numerical calculating methods. Chapter 29 discusses analog methods in vibration and shock analysis; it is based primarily upon the equivalence of the response of mechanical systems and electrical circuits. This type of analysis is used widely in shock and vibration work.

One of the most common techniques of shock and vibration control involves the concept of isolation. The theory of vibration isolation is discussed in detail in Chap. 30; an analogous presentation of shock isolation is included in Chap. 31. Chapter 30 includes treatment of vibration isolation when the vibration frequencies are so high that associated structures cannot be considered as rigid bodies, and when the vibration is random in nature. The more practical aspects of isolation are considered in Chap. 32, including the application and design of isolators. Chapter 33 discusses two particular classes of isolators—using pneumatic springs and involving servo control. Such principles often are applied to a single isolation system and are discussed in a single chapter. Chapters 34 and 35 describe the materials for and design of springs for use in isolators, metal springs being discussed in Chap. 34 and rubber springs being discussed in Chap. 35.

An important method of controlling shock and vibration involves the addition of damping or energy-dissipating means to structures that are susceptible to vibration. Chapter 36 discusses the general concepts of damping together with the application of such concepts to hysteresis and slip damping. The application of damping materials to structures is discussed in Chap. 37.

The latter chapters of the Handbook deal with the specific application of fundamentals of analysis, methods of measurement, and control techniques—where these are developed sufficiently to form a separate and discrete subject. Torsional vibration is discussed in Chap. 38, with particular application to internal-combustion engines. The balancing of rotating equipment is discussed in Chap. 39, and balancing machines are described. Chapter 40 describes the special vibration problems associated with the design and operation of machine tools. Among the most prominent occurrences of shock and vibration are those that arise during handling and shipping of merchandise. Packaging of equipment to protect against such shock and vibration is discussed in Chap. 41. Chapters 42 and 43 describe procedures for the design of equipment to withstand shock and vibration—the former considering primarily the theory of design and the latter considering practical aspects. A comprehensive discussion of the human aspect of shock and vibration is considered in Chap. 44 which describes the effect of shock and vibration on man.

Many problems in the field of shock and vibration are related to the environmental conditions existing in various types of vehicles. This is important not only for the design and satisfactory operation of the respective vehicles, but also in the protection and safety of passengers and cargo which are transported. For purposes of this Handbook, vehicles are grouped into those adapted for movement on land, on water, and in air. Chapter 45 discusses the sources of shock and vibration, and the characteristic environmental conditions, in road and rail vehicles. Chapter 46 presents similar material dealing with ships, and Chap. 47 considers manned aircraft and guided missiles.

The final three chapters consider important sources of excitation other than those of vehicular transportation. Chapter 48 presents information on sources of excitation and the response of structures as a result of vibration transmitted through the air, such as the vibration of an aircraft structure closely adjacent to the exhaust from a jet engine. Chapter 49 considers the design of structures adapted to withstand an air blast, such as that resulting from a bomb explosion or the shock wave induced by supersonic aircraft. The last chapter discusses the shock and vibration transmitted through the earth; the source of the excitation may be a natural earthquake, a blasting operation, the operation of machinery supported by the soil, or the travel of vehicles on nearby rights-of-way.

SYMBOLS

This section includes a list of symbols with their usual English units as used generally in the Handbook; metric units are given as alternates in the text. In special circumstances, some of the following symbols have different meanings in certain chapters but are defined in those chapters. Other symbols of special or limited application are defined in the respective chapters as they are used.

a	radius	in.
B	magnetic flux density	gauss
c	damping coefficient	lb-sec/in.
c	velocity of sound	in./sec
c_c	coefficient for critical damping	lb-sec/in.
C	capacitance	farads
D	diameter	in.
e	electrical voltage	volts
e	eccentricity	in.
E	energy	in.-lb
E	modulus of elasticity in tension and compression (Young's modulus)	lb/in.^{2}
f	frequency	cycles/sec (cps)
f_n	undamped natural frequency	cycles/sec (cps)
f_i	undamped natural frequencies in a multiple degree-of-freedom system, where $i = 1, 2, \ldots$	cycles/sec (cps)
f_d	damped natural frequency	cycles/sec (cps)
f_r	resonant frequency	cycles/sec (cps)
F	force	lb
f_f	Coulomb friction force	lb
g	acceleration of gravity	in./sec^2
G	modulus of elasticity in shear	lb/in.^{2}
h	height, depth	in.
H	magnetic field strength	oersteds
i	electric current	amperes
I_i	area or mass moment of inertia (subscript indicates axis)	in.^{4}
I_p	polar moment of inertia	in.^{4}
I_{ij}	area or mass product of inertia (subscripts indicate axes)	in.^{4}
$\mathcal{I}$	imaginary part of	
j	$\sqrt{-1}$	
J	inertia constant (weight moment of inertia)	lb-in.^{2}
J	impulse	lb-sec
k	linear stiffness	lb/in.
k_t	rotational (torsional) stiffness	lb-in./rad
l	length	in.
L	inductance	henrys
m	mass	lb-sec^2/in.

m_u	unbalanced mass	lb-sec²/in.
M	torque	lb.-in.
M	mutual inductance	henrys
$\mathfrak{M}$	mobility	in./lb-sec
n	number of coils, supports, etc.	
p	alternating pressure	lb/in.²
p	probability density	
P	probability distribution	
P	static pressure	lb/in.²
q	electric charge	coulombs
Q	resonance factor (also ratio of reactance to resistance)	
r	resistance	ohms
R	radius	in.
$\mathfrak{R}$	real part of	
s	arc length	in.
S	area of diaphragm, tube, etc.	in.²
t	thickness	in.
t	time	sec
T	transmissibility	
T	kinetic energy	in.-lb
v	linear velocity	in./sec
V	potential energy	in.-lb
w	width	in.
W	weight	lb
W	power	in.-lb/sec
W_e	spectral density of the excitation	(rms)²/unit freq.
W_r	spectral density of the response	(rms)²/unit freq.
x	linear displacement in direction of X axis	in.
y	linear displacement in direction of Y axis	in.
z	linear displacement in direction of Z axis	in.
Z	impedance	lb-sec/in.
α	rotational displacement about X axis	radians
β	rotational displacement about Y axis	radians
γ	rotational displacement about Z axis	radians
γ	shear strain	
γ	weight density	lb/in.³
δ	deflection	in.
δ_{st}	static deflection	in.
Δ	logarithmic decrement	
ϵ	tension or compression strain	
ζ	fraction of critical damping	
η	stiffness ratio	
θ	phase angle	radians
λ	wavelength	in.
μ	coefficient of friction	
μ	mass density	lb-sec²/in.⁴
ν	Poisson's ratio	
ρ	mass density	lb-sec²/in.⁴
ρ_i	radius of gyration (subscript indicates axis)	in.
σ	Poisson's ratio	
σ	normal stress	lb/in.²
σ	root-mean-square (rms) value	
τ	period	sec
τ	shear stress	lb/in.²
ϕ	magnetic flux	maxwell
ψ	phase angle	radians
ω	forcing frequency—angular	rad/sec
ω_n	undamped natural frequency—angular	rad/sec

ω_i	undamped natural frequencies—angular—in a multiple degree-of-freedom system, where $i = 1, 2, \ldots$	rad/sec
ω_d	damped natural frequency—angular.....................	rad/sec
ω_r	resonant frequency—angular...........................	rad/sec
Ω	rotational speed......................................	rad/sec
$\simeq$	approximately equal to	

CHARACTERISTICS OF HARMONIC MOTION

Harmonic functions are employed frequently in the analysis of shock and vibration. A body that experiences simple harmonic motion follows a displacement pattern defined by

$$x = x_0 \sin(2\pi ft) = x_0 \sin \omega t \tag{1.1}$$

where f is the *frequency* of the simple harmonic motion, $\omega = 2\pi f$ is the corresponding *angular frequency*, and x_0 is the *amplitude* of the displacement.

The velocity $\dot{x}$ and acceleration $\ddot{x}$ of the body are found by differentiating the displacement once and twice, respectively:

$$\dot{x} = x_0(2\pi f) \cos 2\pi ft = x_0 \omega \cos \omega t \tag{1.2}$$

$$\ddot{x} = -x_0(2\pi f)^2 \sin 2\pi ft = -x_0 \omega^2 \sin \omega t \tag{1.3}$$

The maximum absolute values of the displacement, velocity, and acceleration of a body undergoing harmonic motion occur when the trigonometric functions in Eqs. (1.1) to (1.3) are numerically equal to unity. These values are known, respectively, as displacement, velocity, and acceleration amplitudes; they are defined mathematically as follows:

$$x_0 = x_0 \qquad \dot{x}_0 = (2\pi f) x_0 \qquad \ddot{x}_0 = (2\pi f)^2 x_0 \tag{1.4}$$

It is common to express the displacement amplitude x_0 in inches when the English system of units is used and in centimeters or millimeters when the metric system is used. Accordingly, the velocity amplitude $\dot{x}_0$ is expressed in inches per second in the English system (centimeters per second or millimeters per second in the metric system). The acceleration amplitude $\ddot{x}_0$ usually is expressed as a dimensionless multiple of the gravitational acceleration g, where $g = 386$ in./sec^2 or 980 cm/sec^2. For example, an acceleration of $3,860$ in./sec^2 is written $10g$.

Table 1.1. Conversion Factors for Translational Velocity and Acceleration

Multiply Value in → or → By ↘ To obtain value in ↓	g-sec, g	ft/sec ft/sec^2	in./sec in./sec^2	cm/sec cm/sec^2	m/sec m/sec^2
g-sec, g	1	0.0311	0.00259	0.00102	0.102
ft/sec ft/sec^2	32.16	1	0.0833	0.0328	3.28
in./sec in./sec^2	386	12.0	1	0.3937	39.37
cm/sec cm/sec^2	980	30.48	2.540	1	100
m/sec m/sec^2	9.80	0.3048	0.0254	0.010	1

Factors for converting values of rectilinear velocity and acceleration to different units are given in Table 1.1; similar factors for angular velocity and acceleration are given in Table 1.2. Displacement, velocity, and acceleration amplitudes (expressed in the English system of units) as a function of frequency are shown graphically in Fig. 1.1 and by the nomograph of Fig. 1.2.

For certain purposes in analysis, it is convenient to express the amplitude in terms of the average value of the harmonic function, the root-mean-square (rms) value, or two times the amplitude (i.e., peak-to-peak value). These terms are defined mathematically in Chap. 22; numerical conversion factors are set forth in Table 1.3 for ready reference.

Table 1.2. Conversion Factors for Rotational Velocity and Acceleration

Multiply Value in → or → By ↘ To obtain value in ↓	rad/sec rad/sec^2	degree/sec degree/sec^2	rev/sec rev/sec^2	rev/min rev/min/sec
rad/sec rad/sec^2	1	0.01745	6.283	0.1047
degree/sec degree/sec^2	57.30	1	360	6.00
rev/sec rev/sec^2	0.1592	0.00278	1	0.0107
rev/min rev/min/sec	9.549	0.1667	60	1

Table 1.3. Conversion Factors for Simple Harmonic Motion

Multiply numerical value in terms of → By ↘ To obtain value in terms of ↓	Amplitude	Average value	Root-mean-square value (rms)	Peak-to-peak value
Amplitude	1	1.571	1.414	0.500
Average value	0.637	1	0.900	0.318
Root-mean-square value (rms)	0.707	1.111	1	0.354
Peak-to-peak value	2.000	3.142	2.828	1

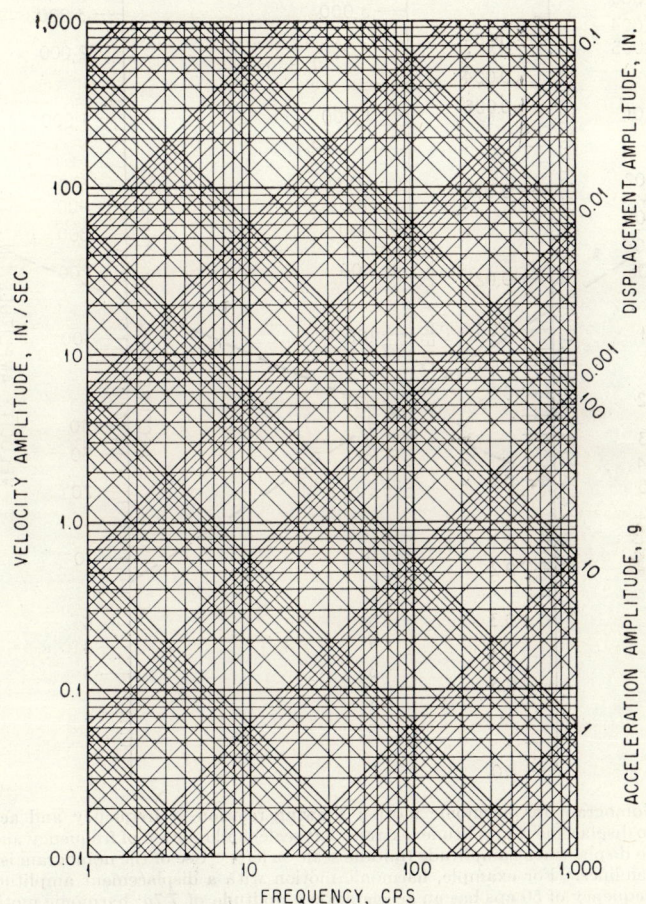

Fig. 1.1. Relation of frequency to the amplitudes of displacement, velocity, and acceleration in harmonic motion.

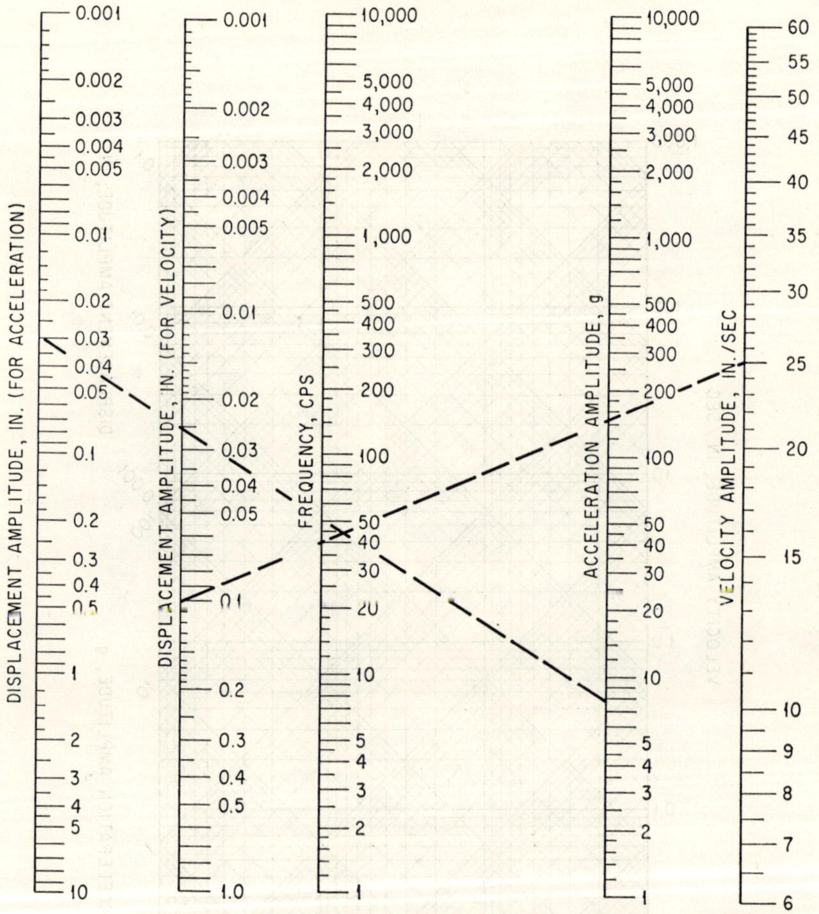

FIG. 1.2. Nomogram for harmonic motion showing relation of frequency and acceleration amplitude to displacement amplitude (extreme left scale) and relation of frequency and velocity amplitude to displacement amplitude (middle scale on left). Use of the nomogram is indicated by the broken lines. For example, harmonic motion with a displacement amplitude of 0.03 in. and a frequency of 50 cps has an acceleration amplitude of 7.7g; harmonic motion with a displacement amplitude of 0.10 in. and a frequency of 40 cps has a velocity amplitude of 25.1 in./sec.

APPENDIX 1.1
NATURAL FREQUENCIES OF COMMONLY USED SYSTEMS

The most important aspect of vibration analysis often is the calculation or measurement of the natural frequencies of mechanical systems. Natural frequencies are discussed prominently in many chapters of the Handbook. Appendix 1.1 includes in tabular form, convenient for ready reference, a compilation of frequently used expressions for the natural frequencies of common mechanical systems. The data for beams and plates are abstracted from Chap. 7.

NATURAL FREQUENCIES OF COMMONLY USED SYSTEMS 1-11

MASS-SPRING SYSTEMS IN TRANSLATION
(RIGID MASS AND MASSLESS SPRING)

k = SPRING STIFFNESS, LB/IN.
m = MASS, LB-SEC^2/IN.
ω_n = ANGULAR NATURAL FREQUENCY, RAD/SEC

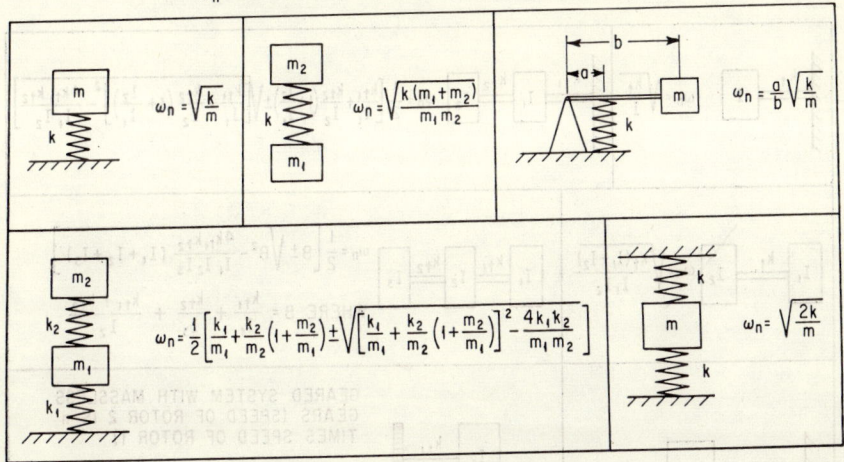

$$\omega_n = \sqrt{\frac{k}{m}}$$

$$\omega_n = \sqrt{\frac{k(m_1+m_2)}{m_1 m_2}}$$

$$\omega_n = \frac{a}{b}\sqrt{\frac{k}{m}}$$

$$\omega_n = \sqrt{\frac{1}{2}\left[\frac{k_1}{m_1}+\frac{k_2}{m_2}\left(1+\frac{m_2}{m_1}\right)\right]\pm\sqrt{\left[\frac{k_1}{m_1}+\frac{k_2}{m_2}\left(1+\frac{m_2}{m_1}\right)\right]^2-\frac{4k_1 k_2}{m_1 m_2}}}$$

$$\omega_n = \sqrt{\frac{2k}{m}}$$

SPRINGS IN COMBINATION
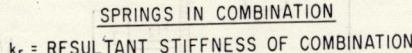
k_r = RESULTANT STIFFNESS OF COMBINATION

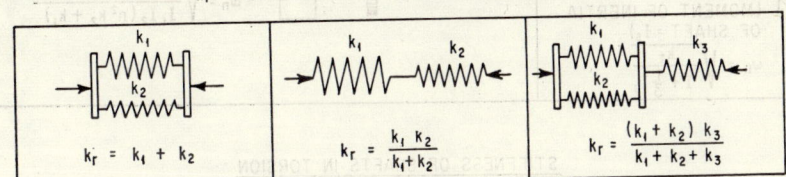

$k_r = k_1 + k_2$

$k_r = \dfrac{k_1 k_2}{k_1 + k_2}$

$k_r = \dfrac{(k_1 + k_2) k_3}{k_1 + k_2 + k_3}$

HELICAL SPRINGS

d = WIRE DIAMETER, IN.
D = MEAN COIL DIAMETER, IN.
n = NUMBER OF ACTIVE COILS
E = YOUNG'S MODULUS, LB/IN^2
G = MODULUS OF ELASTICITY IN SHEAR, LB/IN^2

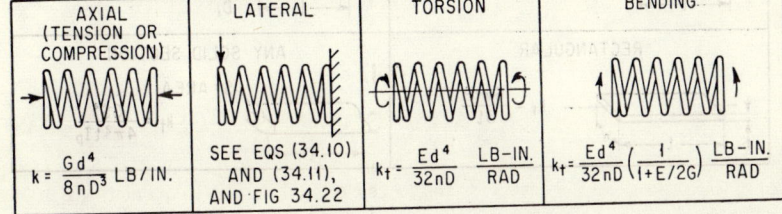

AXIAL (TENSION OR COMPRESSION)	LATERAL	TORSION	BENDING
$k = \dfrac{Gd^4}{8nD^3}$ LB/IN.	SEE EQS (34.10) AND (34.11), AND FIG 34.22	$k_t = \dfrac{Ed^4}{32nD}$ $\dfrac{\text{LB-IN.}}{\text{RAD}}$	$k_t = \dfrac{Ed^4}{32nD}\left(\dfrac{1}{1+E/2G}\right)$ $\dfrac{\text{LB-IN.}}{\text{RAD}}$

ROTOR–SHAFT SYSTEMS
(RIGID ROTOR AND MASSLESS SHAFT)

k_t = TORSIONAL STIFFNESS OF SHAFT, LB-IN./RAD
I = MASS MOMENT OF INERTIA OF ROTOR, LB-IN.-SEC^2
ω_n = ANGULAR NATURAL FREQUENCY, RAD/SEC

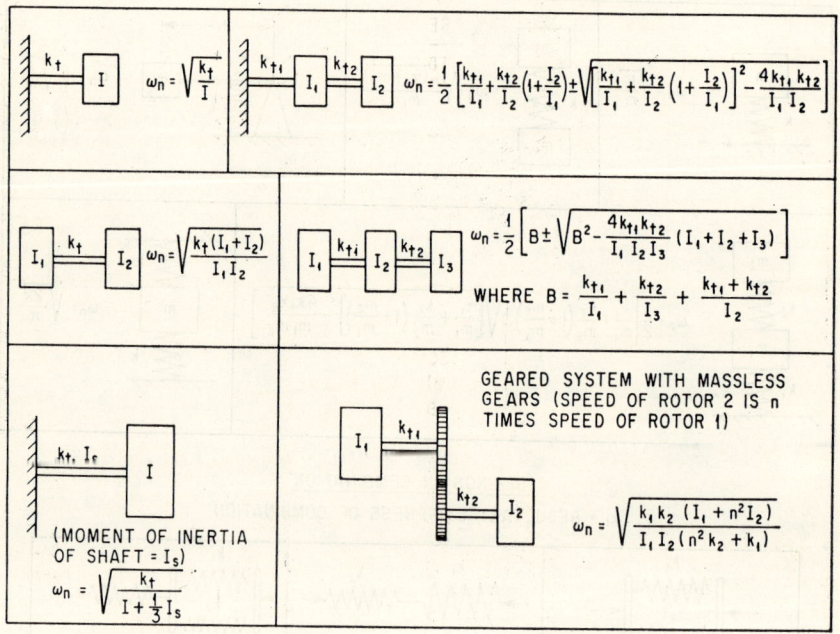

STIFFNESS OF SHAFTS IN TORSION

G = MODULUS OF ELASTICITY IN SHEAR, LB/IN.^{2}
ℓ = LENGTH OF SHAFT, IN.
I_p = POLAR MOMENT OF INERTIA OF SHAFT CROSS-SECTION, IN.^{4}

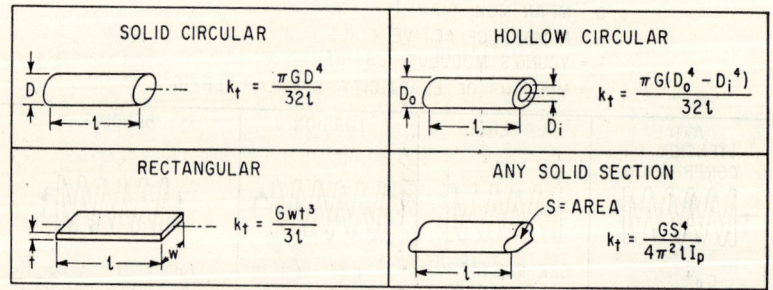

NATURAL FREQUENCIES OF COMMONLY USED SYSTEMS 1-13

MASSLESS BEAMS WITH CONCENTRATED MASS LOADS

- m = MASS OF LOAD, LB-SEC^2/IN.
- l = LENGTH OF BEAM, IN.
- I = AREA MOMENT OF INERTIA OF BEAM CROSS SECTION, IN.^{4}
- E = YOUNG'S MODULUS, LB/IN.^{2}
- ω_n = ANGULAR NATURAL FREQUENCY, RAD/SEC

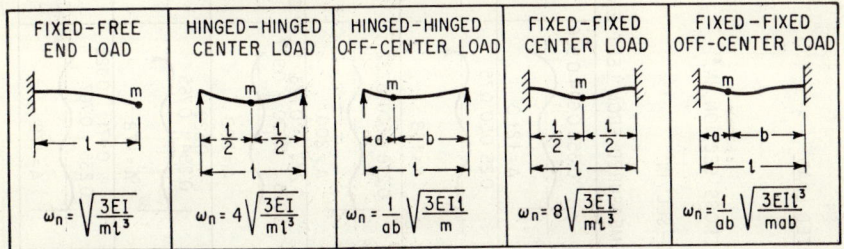

FIXED-FREE END LOAD	HINGED-HINGED CENTER LOAD	HINGED-HINGED OFF-CENTER LOAD	FIXED-FIXED CENTER LOAD	FIXED-FIXED OFF-CENTER LOAD
$\omega_n = \sqrt{\dfrac{3EI}{ml^3}}$	$\omega_n = 4\sqrt{\dfrac{3EI}{ml^3}}$	$\omega_n = \dfrac{l}{ab}\sqrt{\dfrac{3EIl}{m}}$	$\omega_n = 8\sqrt{\dfrac{3EI}{ml^3}}$	$\omega_n = \dfrac{1}{ab}\sqrt{\dfrac{3EIl^3}{mab}}$

MASSIVE SPRINGS (BEAMS) WITH CONCENTRATED MASS LOADS

- m = MASS OF LOAD, LB-SEC^2/IN.
- $m_s(m_b)$ = MASS OF SPRING (BEAM), LB-SEC^2/IN.
- k = STIFFNESS OF SPRING LB/IN.
- l = LENGTH OF BEAM, IN.
- I = AREA MOMENT OF INERTIA OF BEAM CROSS SECTION, IN.^{4}
- E = YOUNG'S MODULUS, LB/IN.^{2}
- ω_n = ANGULAR NATURAL FREQUENCY, RAD/SEC

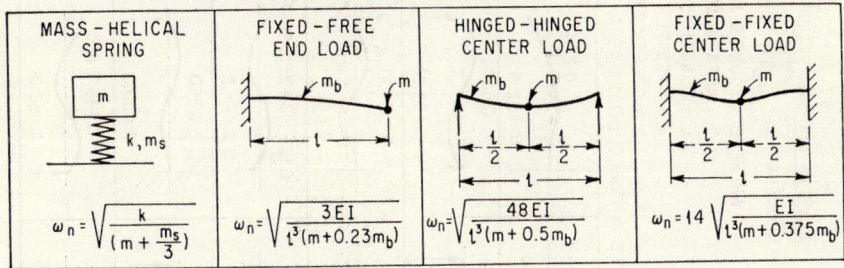

MASS-HELICAL SPRING	FIXED-FREE END LOAD	HINGED-HINGED CENTER LOAD	FIXED-FIXED CENTER LOAD
$\omega_n = \sqrt{\dfrac{k}{(m+\dfrac{m_s}{3})}}$	$\omega_n = \sqrt{\dfrac{3EI}{l^3(m+0.23m_b)}}$	$\omega_n = \sqrt{\dfrac{48EI}{l^3(m+0.5m_b)}}$	$\omega_n = 14\sqrt{\dfrac{EI}{l^3(m+0.375m_b)}}$

AREA MOMENT OF INERTIA OF BEAM SECTIONS
(WITH RESPECT TO AXIS a-a)

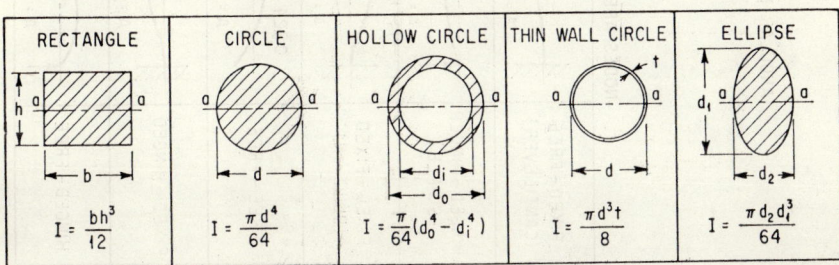

RECTANGLE	CIRCLE	HOLLOW CIRCLE	THIN WALL CIRCLE	ELLIPSE
$I = \dfrac{bh^3}{12}$	$I = \dfrac{\pi d^4}{64}$	$I = \dfrac{\pi}{64}(d_o^4 - d_i^4)$	$I = \dfrac{\pi d^3 t}{8}$	$I = \dfrac{\pi d_2 d_1^3}{64}$

BEAMS OF UNIFORM SECTION AND UNIFORMLY DISTRIBUTED LOAD

ANGULAR NATURAL FREQUENCY $\omega_n = A\sqrt{\dfrac{EI}{\mu \ell^4}}$ RAD/SEC

WHERE E = YOUNG'S MODULUS, LB/IN^2
  I = AREA MOMENT OF INERTIA OF BEAM CROSS SECTION, IN.^{4}
  ℓ = LENGTH OF BEAM, IN.
  μ = MASS PER UNIT LENGTH OF BEAM, LB-SEC^2/IN.^{2}
  A = COEFFICIENT FROM TABLE BELOW

NODES ARE INDICATED IN TABLE BELOW AS A PROPORTION OF LENGTH ℓ MEASURED FROM LEFT END

Beam Type	1st mode	2nd mode	3rd mode	4th mode	5th mode
FIXED–FREE (CANTILEVER)	$A=3.52$	$A=22.4$; node 0.774	$A=61.7$; nodes 0.500, 0.868	$A=121.0$; nodes 0.356, 0.644, 0.906	$A=200.0$; nodes 0.279, 0.500, 0.723, 0.926
HINGED–HINGED (SIMPLE)	$A=9.87$	$A=39.5$; node 0.500	$A=88.9$; nodes 0.333, 0.667	$A=158$; nodes 0.25, 0.50, 0.75	$A=247$; nodes 0.20, 0.40, 0.60, 0.80
FIXED–FIXED (BUILT–IN)	$A=22.4$	$A=61.7$; node 0.500	$A=121$; nodes 0.359, 0.641	$A=200$; nodes 0.278, 0.500, 0.722	$A=298$; nodes 0.227, 0.409, 0.591, 0.773
FREE–FREE	$A=22.4$; nodes 0.224, 0.776	$A=61.7$; nodes 0.132, 0.500, 0.868	$A=121$; nodes 0.094, 0.356, 0.644, 0.906	$A=200$; nodes 0.073, 0.277, 0.500, 0.723, 0.927	$A=298$; nodes 0.060, 0.227, 0.409, 0.591, 0.773, 0.940
FIXED–HINGED	$A=15.4$; node 0.736	$A=50.0$; nodes 0.560	$A=104$; nodes 0.384, 0.692	$A=178$; nodes 0.294, 0.529, 0.765	$A=272$; nodes 0.238, 0.429, 0.619, 0.810
HINGED–FREE	$A=15.4$	$A=50.0$; node 0.446, 0.853	$A=104$; nodes 0.308, 0.616, 0.898	$A=178$; nodes 0.235, 0.471, 0.707, 0.922	$A=272$; nodes 0.190, 0.381, 0.581, 0.763, 0.937

NATURAL FREQUENCIES OF COMMONLY USED SYSTEMS

NATURAL FREQUENCIES OF THIN FLAT PLATES OF UNIFORM THICKNESS

$$\omega_n = B \sqrt{\frac{E t^2}{\rho a^4 (1-\nu^2)}} \text{ RAD/SEC}$$

E = YOUNG'S MODULUS, LB/IN.^{2}
t = THICKNESS OF PLATE, IN.
ρ = MASS DENSITY, LB-SEC^2/IN.^{4}
a = DIAMETER OF CIRCULAR PLATE OR SIDE OF SQUARE PLATE, IN.
ν = POISSON'S RATIO

SHAPE OF PLATE	DIAGRAM	EDGE CONDITIONS	VALUE OF B FOR MODE:							
			1	2	3	4	5	6	7	8
CIRCULAR		CLAMPED AT EDGE	11.84	24.61	40.41	46.14	103.12			
CIRCULAR		FREE	6.09	10.53	14.19	23.80	40.88	44.68	61.38	69.44
CIRCULAR		CLAMPED AT CENTER	4.35	24.26	70.39	138.85				
CIRCULAR		SIMPLY SUPPORTED AT EDGE	5.90							
SQUARE		ONE EDGE CLAMPED – THREE EDGES FREE	1.01	2.47	6.20	7.94	9.01			
SQUARE		ALL EDGES CLAMPED	10.40	21.21	31.29	38.04	38.22	47.73		
SQUARE		TWO EDGES CLAMPED – TWO EDGES FREE	2.01	6.96	7.74	13.89	18.25			
SQUARE		ALL EDGES FREE	4.07	5.94	6.91	10.39	17.80	18.85		
SQUARE		ONE EDGE CLAMPED – THREE EDGES SIMPLY SUPPORTED	6.83	14.94	16.95	24.89	28.99	32.71		
SQUARE		TWO EDGES CLAMPED – TWO EDGES SIMPLY SUPPORTED	8.37	15.82	20.03	27.34	29.54	37.31		
SQUARE		ALL EDGES SIMPLY SUPPORTED	5.70	14.26	22.82	28.52	37.08	48.49		

MASSLESS CIRCULAR PLATE WITH CONCENTRATED CENTER MASS

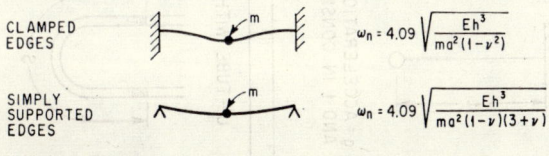

CLAMPED EDGES: $\omega_n = 4.09 \sqrt{\dfrac{E h^3}{m a^2 (1-\nu^2)}}$

SIMPLY SUPPORTED EDGES: $\omega_n = 4.09 \sqrt{\dfrac{E h^3}{m a^2 (1-\nu)(3+\nu)}}$

NATURAL FREQUENCIES OF MISCELLANEOUS SYSTEMS
(ω_n = ANGULAR NATURAL FREQUENCY, RAD/SEC)

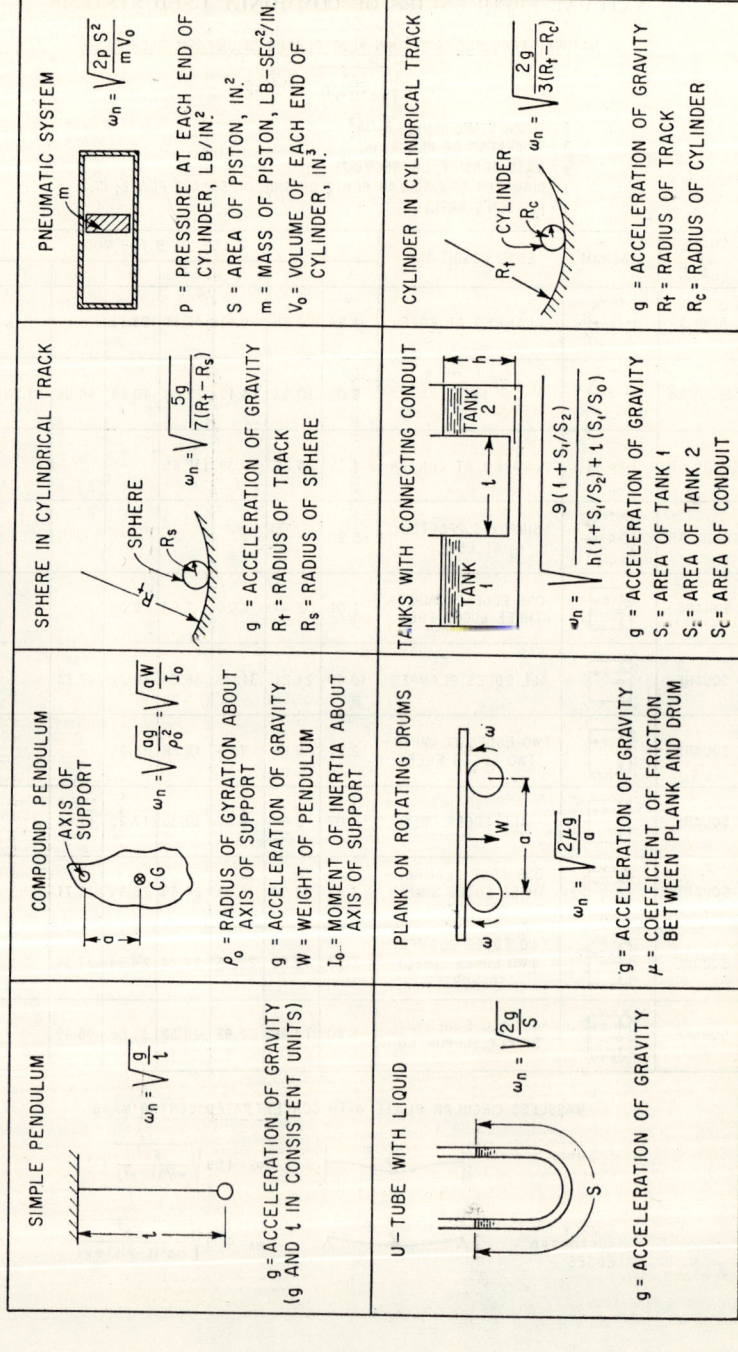

SIMPLE PENDULUM

$$\omega_n = \sqrt{\frac{g}{\ell}}$$

g = ACCELERATION OF GRAVITY
(g AND ℓ IN CONSISTENT UNITS)

COMPOUND PENDULUM

$$\omega_n = \sqrt{\frac{ag}{\rho_0^2}} = \sqrt{\frac{aW}{I_0}}$$

ρ_0 = RADIUS OF GYRATION ABOUT AXIS OF SUPPORT
g = ACCELERATION OF GRAVITY
W = WEIGHT OF PENDULUM
I_0 = MOMENT OF INERTIA ABOUT AXIS OF SUPPORT

SPHERE IN CYLINDRICAL TRACK

$$\omega_n = \sqrt{\frac{5g}{7(R_t - R_s)}}$$

g = ACCELERATION OF GRAVITY
R_t = RADIUS OF TRACK
R_s = RADIUS OF SPHERE

PNEUMATIC SYSTEM

$$\omega_n = \sqrt{\frac{2pS^2}{mV_0}}$$

p = PRESSURE AT EACH END OF CYLINDER, LB/IN.²
S = AREA OF PISTON, IN.²
m = MASS OF PISTON, LB-SEC²/IN.
V_0 = VOLUME OF EACH END OF CYLINDER, IN.³

U-TUBE WITH LIQUID

$$\omega_n = \sqrt{\frac{2g}{S}}$$

g = ACCELERATION OF GRAVITY

PLANK ON ROTATING DRUMS

$$\omega_n = \sqrt{\frac{2\mu g}{a}}$$

g = ACCELERATION OF GRAVITY
μ = COEFFICIENT OF FRICTION BETWEEN PLANK AND DRUM

TANKS WITH CONNECTING CONDUIT

$$\omega_n = \sqrt{\frac{g(1 + S_1/S_2)}{h(1 + S_1/S_2) + \ell(S_1/S_c)}}$$

g = ACCELERATION OF GRAVITY
S_1 = AREA OF TANK 1
S_2 = AREA OF TANK 2
S_c = AREA OF CONDUIT

CYLINDER IN CYLINDRICAL TRACK

$$\omega_n = \sqrt{\frac{2g}{3(R_t - R_c)}}$$

g = ACCELERATION OF GRAVITY
R_t = RADIUS OF TRACK
R_c = RADIUS OF CYLINDER

APPENDIX 1.2
TERMINOLOGY

For convenience, definitions of terms which are used frequently in the field of shock and vibration are assembled here. Certain definitions are reproduced with permission from *Acoustical Terminology, Including Mechanical Shock and Vibration (S1.1-1960)*, American Standards Association.† These definitions have been denoted by an asterisk (*). For terms which are not listed below, the reader is referred to the Index.

Acceleration.* Acceleration is a vector quantity that specifies the time rate of change of velocity.

Amplitude. Amplitude is the maximum value of a sinusoidal quantity.

Analog. If a first quantity or structural element is analogous to a second quantity or structural element belonging in another field of knowledge, the second quantity is called the analog of the first, and vice versa.

Analogy.* An analogy is a recognized relationship of consistent mutual similarity between the equations and structures appearing within two or more fields of knowledge, and an identification and association of the quantities and structural elements that play mutually similar roles in these equations and structures, for the purpose of facilitating transfer of knowledge of mathematical procedures of analysis and behavior of the structures between these fields.

Angular Frequency (Circular Frequency).* The angular frequency of a periodic quantity, in radians per unit time, is the frequency multiplied by 2π.

Angular Mechanical Impedance (Rotational Mechanical Impedance). Angular mechanical impedance is the impedance involving the ratio of torque to angular velocity. (See *Impedance*.)

Antinode (Loop).* An antinode is a point, line, or surface in a standing wave where some characteristic of the wave field has maximum amplitude.

Antiresonance.* For a system in forced oscillation, antiresonance exists at a point when any change, however small, in the frequency of excitation causes an increase in the response at this point.

Audio Frequency.* An audio frequency is any frequency corresponding to a normally audible sound wave.

Autocorrelation Coefficient. The autocorrelation coefficient of a signal is the ratio of the autocorrelation function to the mean-square value of the signal:

$$R(\tau) = \overline{x(t)x(t+\tau)}/[\overline{x(t)}]^2$$

Autocorrelation Function. The autocorrelation function of a signal is the average of the product of the value of the signal at time t with the value at time $t + \tau$:

$$R(\tau) = \overline{x(t)x(t+\tau)}$$

For a stationary random signal of infinite duration, the power spectral density (except for a constant factor) is the cosine Fourier transform of the autocorrelation function.

Auxiliary Mass Damper (Damped Vibration Absorber). An auxiliary mass damper is a system consisting of a mass, spring, and damper which tends to reduce vibration by the dissipation of energy in the damper as a result of relative motion between the mass and the structure to which the damper is attached.

Background Noise.* Background noise is the total of all sources of interference in a system used for the production, detection, measurement, or recording of a signal, independent of the presence of the signal.

Balancing. Balancing is a procedure for adjusting the mass distribution of a rotor so that vibration of the journals, or the forces on the bearings at once-per-revolution, are reduced or controlled. (See *Chap. 39, Part I*, for a complete list of definitions related to *balancing*.)

Band-pass Filter.* A band-pass filter is a wave filter that has a single transmission band extending from a lower cutoff frequency greater than zero to a finite upper cutoff frequency.

Bandwidth, Effective. (See *Effective Bandwidth*.)

Beats.* Beats are periodic variations that result from the superposition of two simple harmonic quantities of different frequencies f_1 and f_2. They involve the periodic increase and decrease of amplitude at the beat frequency $(f_1 - f_2)$.

---

† Copies of this standard, as well as others in the field of mechanical shock and vibration, are available from the American Standards Association, Inc., 10 East 40th St., New York 16, N.Y.

Broad-band Random Vibration. Broad-band random vibration is random vibration having its frequency components distributed over a broad frequency band. (See *Random Vibration*.)

Circular Frequency. (See *Angular Frequency*.)

Complex Angular Frequency. As applied to a function $a = Ae^{\sigma t} \sin(\omega t - \phi)$, where σ, ω, and ϕ are constant, the quantity $\omega_c = \sigma + j\omega$ is the complex angular frequency where j is an operator with rules of addition, multiplication, and division as suggested by the symbol $\sqrt{-1}$. If the signal decreases with time, σ must be negative.

Complex Function. A complex function is a function having real and imaginary parts.

Complex Vibration. Complex vibration is vibration whose components are sinusoids not harmonically related to one another. (See *Harmonic*.)

Compliance.* Compliance is the reciprocal of stiffness.

Continuous System (Distributed System).* A continuous system is one that is considered to have an infinite number of possible independent displacements. Its configuration is specified by a function of a continuous spatial variable or variables in contrast to a discrete or lumped parameter system which requires only a finite number of coordinates to specify its configuration.

Correlation Coefficient. The correlation coefficient of two variables is the ratio of the correlation function to the product of the averages of the variables:

$$\overline{x_1(t) \cdot x_2(t)} / \overline{x_1(t)} \cdot \overline{x_2(t)}$$

Correlation Function. The correlation function of two variables is the average value of their product:

$$\overline{x_1(t) \cdot x_2(t)}$$

Coulomb Damping (Dry Friction Damping).* Coulomb damping is the dissipation of energy that occurs when a particle in a vibrating system is resisted by a force whose magnitude is a constant independent of displacement and velocity, and whose direction is opposite to the direction of the velocity of the particle.

Coupled Modes.* Coupled modes are modes of vibration that are not independent but which influence one another because of energy transfer from one mode to the other. (See *Mode of Vibration*.)

Coupling Factor, Electromechanical.* The electromechanical coupling factor is a factor used to characterize the extent to which the electrical characteristics of a transducer are modified by a coupled mechanical system, and vice versa.

Critical Damping.* Critical damping is the minimum viscous damping that will allow a displaced system to return to its initial position without oscillation.

Critical Speed.* Critical speed is a speed of a rotating system that corresponds to a resonant frequency of the system.

Cycle.* A cycle is the complete sequence of values of a periodic quantity that occur during a period.

Damped Natural Frequency. The damped natural frequency is the frequency of free vibration of a damped linear system. The free vibration of a damped system may be considered periodic in the limited sense that the time interval between zero crossings in the same direction is constant, even though successive amplitudes decrease progressively. The frequency of the vibration is the reciprocal of this time interval.

Damping.* Damping is the dissipation of energy with time or distance.

Damping Ratio. (See *Fraction of Critical Damping*.)

Decibel (db). The decibel is a unit which denotes the magnitude of a quantity with respect to an arbitrarily established reference value of the quantity, in terms of the logarithm (to the base 10) of the ratio of the quantities. For example, in electrical transmission circuits a value of power may be expressed in terms of a power level in decibels; the power level is given by 10 times the logarithm (to the base 10) of the ratio of the actual power to a reference power (which corresponds to 0 db).

Degrees-of-freedom. The number of degrees-of-freedom of a mechanical system is equal to the minimum number of independent coordinates required to define completely the positions of all parts of the system at any instant of time. In general, it is equal to the number of independent displacements that are possible.

Deterministic Function. A deterministic function is one whose value at any time can be predicted from its value at any other time.

Displacement.* Displacement is a vector quantity that specifies the change of position of a body or particle and is usually measured from the mean position or position of rest. In general, it can be represented as a rotation vector or a translation vector, or both.

Distortion.* Distortion is an undesired change in waveform. Noise and certain desired changes in waveform, such as those resulting from modulation or detection, are not usually classed as distortion.

TERMINOLOGY

Distributed System. (See *Continuous System.*)

Driving Point Impedance. Driving point impedance is the impedance involving the ratio of force to velocity when both the force and velocity are measured at the same point and in the same direction. (See *Impedance.*)

Dry Friction Damping. (See *Coulomb Damping.*)

Duration of Shock Pulse.* The duration of a shock pulse is the time required for the acceleration of the pulse to rise from some stated fraction of the maximum amplitude and to decay to this value. (See *Shock Pulse.*)

Dynamic Vibration Absorber (Tuned Damper). A dynamic vibration absorber is an auxiliary mass-spring system which tends to neutralize vibration of a structure to which it is attached. The basic principle of operation is vibration out-of-phase with the vibration of such structure, thereby applying a counteracting force.

Effective Bandwidth.* The effective bandwidth of a specified transmission system is the bandwidth of an ideal system which (1) has uniform transmission in its pass band equal to the maximum transmission of the specified system and (2) transmits the same power as the specified system when the two systems are receiving equal input signals having a uniform distribution of energy at all frequencies.

Electromechanical Coupling Factor. (See *Coupling Factor, Electromechanical.*)

Electrostriction.* Electrostriction is the phenomenon wherein some dielectric materials experience an elastic strain when subjected to an electric field, this strain being independent of the polarity of the field.

Ensemble. A collection of signals. (Also see *Process.*)

Environment. (See *Natural Environments* and *Induced Environment.*)

Equivalent System.* An equivalent system is one that may be substituted for another system for the purpose of analysis. Many types of equivalence are common in vibration and shock technology: (1) equivalent stiffness; (2) equivalent damping; (3) torsional system equivalent to a translational system; (4) electrical or acoustical system equivalent to a mechanical system; etc.

Equivalent Viscous Damping.* Equivalent viscous damping is a value of viscous damping assumed for the purpose of analysis of a vibratory motion, such that the dissipation of energy per cycle at resonance is the same for either the assumed or actual damping force.

Ergodic Process. An ergodic process is a random process that is stationary and of such a nature that all possible time averages performed on one signal are independent of the signal chosen and hence are representative of the time averages of each of the other signals of the entire random process.

Excitation (Stimulus).* Excitation is an external force (or other input) applied to a system that causes the system to respond in some way.

Filter. A filter is a device for separating waves on the basis of their frequency. It introduces relatively small insertion loss to waves in one or more frequency bands and relatively large insertion loss to waves of other frequencies. (See *Insertion Loss.*)

Force Factor.* The force factor of an electromechanical transducer is: (*a*) the complex quotient of the force required to block the mechanical system divided by the corresponding current in the electric system; (*b*) the complex quotient of the resulting open-circuit voltage in the electric system divided by the velocity in the mechanical system. Force factors (*a*) and (*b*) have the same magnitude when consistent units are used and the transducer satisfies the principle of reciprocity. It is sometimes convenient in an electrostatic or piezoelectric transducer to use the ratios between force and charge or electric displacement, or between voltage and mechanical displacement.

Forced Vibration (Forced Oscillation).* The oscillation of a system is forced if the response is imposed by the excitation. If the excitation is periodic and continuing, the oscillation is steady-state.

Foundation (Support).* A foundation is a structure that supports the gravity load of a mechanical system. It may be fixed in space, or it may undergo a motion that provides excitation for the supported system.

Fraction of Critical Damping. The fraction of critical damping (damping ratio) for a system with viscous damping is the ratio of actual damping coefficient c to the critical damping coefficient c_c.

Free Vibration. Free vibration of a system is vibration that occurs in the absence of forced vibration.

Frequency.* The frequency of a function periodic in time is the reciprocal of the period. The unit is the cycle per unit time and must be specified. In many European countries the unit *cycle per second* is called *Hertz* (Hz).

Frequency, Angular. (See *Angular Frequency.*)

Fundamental Frequency.* (1) The fundamental frequency of a periodic quantity is the frequency of a sinusoidal quantity which has the same period as the periodic quantity. (2) The fundamental frequency of an oscillating system is the lowest natural frequency. The normal mode of vibration associated with this frequency is known as the fundamental mode.

Fundamental Mode of Vibration.* The fundamental mode of vibration of a system is the mode having the lowest natural frequency.

g. The quantity g is the acceleration produced by the force of gravity, which varies with the latitude and elevation of the point of observation. By international agreement, the value 980.665 cm/sec^2 = 386.087 in./sec^2 = 32.1739 ft/sec^2 has been chosen as the standard acceleration due to gravity.

Harmonic.* A harmonic is a sinusoidal quantity having a frequency that is an integral multiple of the frequency of a periodic quantity to which it is related.

Harmonic Motion. (See *Simple Harmonic Motion*.)

Harmonic Response. Harmonic response is the periodic response of a vibrating system exhibiting the characteristics of resonance at a frequency that is a multiple of the excitation frequency.

High-pass Filter.* A high-pass filter is a wave filter having a single transmission band extending from some critical or cutoff frequency, not zero, up to infinite frequency.

Image Impedances. The image impedances of a structure or device are the impedances that will simultaneously terminate all of its inputs and outputs in such a way that at each of its inputs and outputs the impedances in both directions are equal.

Impact.* An impact is a single collision of one mass in motion with a second mass which may be either in motion or at rest.

Impedance. Mechanical impedance is the ratio of a force-like quantity to a velocity-like quantity when the arguments of the real (or imaginary) parts of the quantities increase linearly with time. Examples of force-like quantities are: force, sound pressure, voltage, temperature. Examples of velocity-like quantities are: velocity, volume velocity, current, heat flow. *Impedance* is the reciprocal of *mobility*. (Also see *Angular Mechanical Impedance, Linear Mechanical Impedance, Driving Point Impedance,* and *Transfer Impedance*.)

Impulse.* Impulse is the product of a force and the time during which the force is applied; more specifically, the impulse is $\int_{t_1}^{t_2} F dt$ where the force F is time dependent and equal to zero before time t_1 and after time t_2.

Induced Environments. Induced environments are those conditions generated as a result of the operation of a structure or equipment.

Insertion Loss. The insertion loss, in decibels, resulting from insertion of an element in a transmission system is 10 times the logarithm to the base 10 of the ratio of the power delivered to that part of the system that will follow the element, before the insertion of the element, to the power delivered to that same part of the system after insertion of the element.

Isolation.* Isolation is a reduction in the capacity of a system to respond to an excitation, attained by the use of a resilient support. In steady-state forced vibration, isolation is expressed quantitatively as the complement of transmissibility.

Jerk.* Jerk is a vector that specifies the time rate of change of acceleration; jerk is the third derivative of displacement with respect to time.

Line Spectrum.* A line spectrum is a spectrum whose components occur at a number of discrete frequencies.

Linear Mechanical Impedance. Linear mechanical impedance is the impedance involving the ratio of force to linear velocity. (See *Impedance*.)

Linear System.* A system is linear if for every element in the system the response is proportional to the excitation. This definition implies that the dynamic properties of each element in the system can be represented by a set of linear differential equations with constant coefficients, and that for the system as a whole superposition holds.

Logarithmic Decrement.* The logarithmic decrement is the natural logarithm of the ratio of any two successive amplitudes of like sign, in the decay of a single-frequency oscillation.

Longitudinal Wave. A longitudinal wave in a medium is a wave in which the direction of displacement at each point of the medium is normal to the wave front.

Low-pass Filter.* A low-pass filter is a wave filter having a single transmission band extending from zero frequency up to some critical or cutoff frequency which is not infinite.

Magnetic Recorder.* A magnetic recorder is equipment incorporating an electromagnetic transducer and means for moving a ferromagnetic recording medium relative to the transducer for recording electric signals as magnetic variations in the medium.

Magnetostriction.* Magnetostriction is the phenomenon wherein ferromagnetic materials experience an elastic strain when subjected to an external magnetic field. Also, magnetostriction is the converse phenomenon in which mechanical stresses cause a change in the magnetic induction of a ferromagnetic material.

Mechanical Admittance. (See *Mobility*.)

TERMINOLOGY
1-21

Mechanical Impedance. (See *Impedance*.)

Mechanical Shock. Mechanical shock is a nonperiodic excitation (e.g., a motion of the foundation or an applied force) of a mechanical system that is characterized by suddenness and severity, and usually causes significant relative displacements in the system.

Mechanical System. A mechanical system is an aggregate of matter comprising a defined configuration of mass, stiffness, and damping.

Mobility (Mechanical Admittance). Mobility is the ratio of a velocity-like quantity to a force-like quantity when the arguments of the real (or imaginary) parts of the quantities increase linearly with time. *Mobility* is the reciprocal of *impedance*. The terms *Angular Mobility, Linear Mobility, Driving-point Mobility*, and *Transfer Mobility* are used in the same sense as corresponding impedances.

Modal Numbers.* When the normal modes of a system are related by a set of ordered integers, these integers are called modal numbers.

Mode of Vibration.* In a system undergoing vibration, a mode of vibration is a characteristic pattern assumed by the system in which the motion of every particle is simple harmonic with the same frequency. Two or more modes may exist concurrently in a multiple degree-of-freedom system.

Modulation. Modulation is the variation in the value of some parameter which characterizes a periodic oscillation. Thus, amplitude modulation of a sinusoidal oscillation is a variation in the amplitude of the sinusoidal oscillation.

Multiple Degree-of-freedom System.* A multiple degree-of-freedom system is one for which two or more coordinates are required to define completely the position of the system at any instant.

Narrow-band Random Vibration (Random Sine Wave). Narrow-band random vibration is random vibration having frequency components only within a narrow band. It has the appearance of a sine wave whose amplitude varies in an unpredictable manner. (See *Random Vibration*.)

Natural Environments. Natural environments are those conditions generated by the forces of nature and whose effects are experienced when the equipment or structure is at rest as well as when it is in operation.

Natural Frequency.* Natural frequency is the frequency of free vibration of a system. For a multiple degree-of-freedom system, the natural frequencies are the frequencies of the normal modes of vibration.

Noise.* Noise is any undesired signal. By extension, noise is any unwanted disturbance within a useful frequency band, such as undesired electric waves in a transmission channel or device.

Nominal Bandwidth.* The nominal bandwidth of a filter is the difference between the nominal upper and lower cutoff frequencies. The difference may be expressed (1) in cycles per second; (2) as a percentage of the pass-band center frequency; or (3) in octaves.

Nominal Pass-band Center Frequency.* The nominal passband center frequency is the geometric mean of the nominal cutoff frequencies.

Nominal Upper and Lower Cutoff Frequencies.* The nominal upper and lower cutoff frequencies of a filter pass-band are those frequencies above and below the frequency of maximum response of a filter at which the response to a sinusoidal signal is 3 db below the maximum response.

Nonlinear Damping.* Nonlinear damping is damping due to a damping force that is not proportional to velocity.

Normal Mode of Vibration. A normal mode of vibration is a mode of vibration that is uncoupled from (i.e., can exist independently of) other modes of vibration of a system. When vibration of the system is defined as an eigenvalue problem, the normal modes are the eigenvectors and the normal mode frequencies are the eigenvalues. The term "classical normal mode" is sometimes applied to the normal modes of a vibrating system characterized by vibration of each element of the system at the same frequency and phase. In general, classical normal modes exist only in systems having no damping or having particular types of damping.

Oscillation.* Oscillation is the variation, usually with time, of the magnitude of a quantity with respect to a specified reference when the magnitude is alternately greater and smaller than the reference.

Partial Node.* A partial node is the point, line, or surface in a standing-wave system where some characteristic of the wave field has a minimum amplitude differing from zero. The appropriate modifier should be used with the words "partial node" to signify the type that is intended; e.g., displacement partial node, velocity partial node, pressure partial node.

Peak-to-peak Value.* The peak-to-peak value of a vibrating quantity is the algebraic difference between the extremes of the quantity.

Period. The period of a periodic quantity is the smallest increment of the independent variable for which the function repeats itself.

Periodic Quantity.* A periodic quantity is an oscillating quantity whose values recur for certain increments of the independent variable.

Phase of a Periodic Quantity.* The phase of a periodic quantity, for a particular value of the independent variable, is the fractional part of a period through which the independent variable has advanced, measured from an arbitrary reference.

Pickup. (See *Transducer*.)

Piezoelectric (Crystal) (Ceramic) Transducer.* A piezoelectric transducer is a transducer that depends for its operation on the interaction between the electric charge and the deformation of certain asymmetric crystals having piezoelectric properties.

Piezoelectricity.* Piezoelectricity is the property exhibited by some asymmetrical crystalline materials which when subjected to strain in suitable directions develop electric polarization proportional to the strain. Inverse piezoelectricity is the effect in which mechanical strain is produced in certain asymmetrical crystalline materials when subjected to an external electric field; the strain is proportional to the electric field.

Power Spectral Density. Power spectral density is the limiting mean-square value (e.g., of acceleration, velocity, displacement, stress or other random variable) per unit bandwidth, i.e., the limit of the mean-square value in a given rectangular bandwidth divided by the bandwidth, as the bandwidth approaches zero.

Power Spectral Density Level. The spectrum level of a specified signal at a particular frequency is the level in decibels of that part of the signal contained within a band 1 cycle per second wide, centered at the particular frequency. Ordinarily this has significance only for a signal having a continuous distribution of components within the frequency range under consideration.

Process. A process is a collection of signals. The word *process* rather than the word *ensemble* ordinarily is used when it is desired to emphasize the properties the signals have or do not have as a group. Thus, one speaks of a stationary process rather than a stationary ensemble.

Pulse Rise Time.* The pulse rise time is the interval of time required for the leading edge of a pulse to rise from some specified small fraction to some specified larger fraction of the maximum value.

Q (Quality Factor). The quantity Q is a measure of the sharpness of resonance or frequency selectivity of a resonant vibratory system having a single degree of freedom, either mechanical or electrical. In a mechanical system, this quantity is equal to one-half the reciprocal of the damping ratio. It is commonly used only with reference to a lightly damped system, and is then approximately equal to the following:

(1) Transmissibility at resonance,
(2) π/logarithmic decrement,
(3) $2\pi W/\Delta W$ where W is the stored energy and ΔW the energy dissipation per cycle, and
(4) $f_r/\Delta f$ where f_r is the resonant frequency and Δf is the bandwidth between the half-power points.

Quasi-ergodic Process. A quasi-ergodic process is a random process which is not necessarily stationary but of such a nature that some time averages performed on a signal are independent of the signal chosen.

Quasi-periodic Signal. A signal consisting only of quasi-sinusoids.

Quasi-sinusoid. A function of the form $a = A \sin(2\pi ft - \phi)$ where either A or f, or both, is not a constant but may be expressed readily as a function of time. Ordinarily ϕ is considered constant.

Random Sine Wave. (See *Narrow-band Random Vibration*.)

Random Vibration. Random vibration is vibration whose instantaneous magnitude is not specified for any given instant of time. The instantaneous magnitudes of a random vibration are specified only by probability distribution functions giving the probable fraction of the total time that the magnitude (or some sequence of magnitudes) lies within a specified range. Random vibration contains no periodic or quasi-periodic constituents. If random vibration has instantaneous magnitudes that occur according to the Gaussian distribution, it is called "Gaussian random vibration."

Ratio of Critical Damping. (See *Fraction of Critical Damping*.)

Rayleigh Wave.* A Rayleigh wave is a surface wave associated with the free boundary of a solid, such that a surface particle describes an ellipse whose major axis is normal to the surface, and whose center is at the undisturbed surface. At maximum particle displacement away from the solid surface the motion of the particle is opposite to that of the wave.

Recording Channel.* The term "recording channel" refers to one of a number of independent recorders in a recording system or to independent recording tracks on a recording medium.

TERMINOLOGY

Recording System. A recording system is a combination of transducing devices and associated equipment suitable for storing signals in a form capable of subsequent reproduction.

Relaxation Time.* Relaxation time is the time taken by an exponentially decaying quantity to decrease in amplitude by a factor of $1/e = 0.3679$.

Re-recording.* Re-recording is the process of making a recording by reproducing a recorded signal source and recording this reproduction.

Resonance.* Resonance of a system in forced vibration exists when any change, however small, in the frequency of excitation causes a decrease in the response of the system.

Resonant Frequency. A resonant frequency is a frequency at which resonance exists.

Response.* The response of a device or system is the motion (or other output) resulting from an excitation (stimulus) under specified conditions.

Response Spectrum. (See *Shock Spectrum*.)

Rotational Mechanical Impedance. (See *Angular Mechanical Impedance*.)

Self-induced (Self-excited) Vibration.* The vibration of a mechanical system is self-induced if it results from conversion, within the system, of nonoscillatory excitation to oscillatory excitation.

Shake Table. (See *Vibration Machine*.)

Shear Wave (Rotational Wave).* A shear wave is a wave in an elastic medium which causes an element of the medium to change its shape without a change of volume.

Shock. (See *Mechanical Shock*.)

Shock Absorber.* A shock absorber is a device which dissipates energy to modify the response of a mechanical system to applied shock.

Shock Isolator * (Shock Mount). A shock isolator is a resilient support that tends to isolate a system from a shock motion.

Shock Machine.* A shock machine is a device for subjecting a system to controlled and reproducible mechanical shock.

Shock Motion. Shock motion is an excitation involving motion of a foundation. (See *Foundation* and *Mechanical Shock*.)

Shock Mount. (See *Shock Isolator*.)

Shock Pulse.* A shock pulse is a substantial disturbance characterized by a rise of acceleration from a constant value and decay of acceleration to the constant value in a short period of time. Shock pulses are normally displayed graphically as curves of acceleration as functions of time.

Shock-pulse Duration. (See *Duration of Shock Pulse*.)

Shock Spectrum (Response Spectrum). A shock spectrum is a plot of the maximum response experienced by a single degree-of-freedom system, as a function of its own natural frequency, in response to an applied shock. The response may be expressed in terms of acceleration, velocity or displacement.

Signal.* A signal is (1) a disturbance used to convey information; (2) the information to be conveyed over a communication system.

Simple Harmonic Motion.* A simple harmonic motion is a motion such that the displacement is a sinusoidal function of time; sometimes it is designated merely by the term *harmonic motion*.

Single Degree-of-freedom System.* A single degree-of-freedom system is one for which only one coordinate is required to define completely the configuration of the system at any instant.

Sinusoidal Motion. (See *Simple Harmonic Motion*.)

Snubber.* A snubber is a device used to increase the stiffness of an elastic system (usually by a large factor) whenever the displacement becomes larger than a specified value.

Spectrum. A spectrum is a definition of the magnitude of the frequency components that constitute a quantity.

Spectrum Density. (See *Power Spectral Density*.)

Standard Deviation. Standard deviation is the square root of the variance; i.e., the square root of the mean of the squares of the deviations from the mean value of a vibrating quantity.

Standing Wave.* A standing wave is a periodic wave having a fixed distribution in space which is the result of interference of progressive waves of the same frequency and kind. Such waves are characterized by the existence of nodes or partial nodes and antinodes that are fixed in space.

Stationary Process. A stationary process is an ensemble of signals such that an average of values over the ensemble at any given time is independent of time.

Stationary Signal. A stationary signal is a random signal of such nature that averages over samples of finite time intervals are independent of the time at which the sample occurs.

Steady-state Vibration.* Steady-state vibration exists in a system if the velocity of each particle is a continuing periodic quantity.

Stiffness.* Stiffness is the ratio of change of force (or torque) to the corresponding change in translational (or rotational) deflection of an elastic element.

Subharmonic.* A subharmonic is a sinusoidal quantity having a frequency that is an integral submultiple of the fundamental frequency of a periodic quantity to which it is related.

Subharmonic Response.* Subharmonic response is the periodic response of a mechanical system exhibiting the characteristic of resonance at a frequency that is a submultiple of the frequency of the periodic excitation.

Superharmonic Response. Superharmonic response is a term sometimes used to denote a particular type of harmonic response which dominates the total response of the system; it frequently occurs when the excitation frequency is a submultiple of the frequency of the fundamental resonance.

Transducer (Pickup). A transducer is a device which converts shock or vibratory motion into an optical, a mechanical, or most commonly to an electrical signal that is proportional to a parameter of the experienced motion.

Transfer Impedance. Transfer impedance between two points is the impedance involving the ratio of force to velocity when force is measured at one point and velocity at the other point. The term *transfer impedance* also is used to denote the ratio of force to velocity measured at the same point but in different directions. (See *Impedance*.)

Transient Vibration.* Transient vibration is temporarily sustained vibration of a mechanical system. It may consist of forced or free vibration or both.

Transmissibility.* Transmissibility is the nondimensional ratio of the response amplitude of a system in steady-state forced vibration to the excitation amplitude. The ratio may be one of forces, displacements, velocities, or accelerations.

Transmission Loss.* Transmission loss is the reduction in the magnitude of some characteristic of a signal, between two stated points in a transmission system.

Transverse Wave.* A transverse wave is a wave in which the direction of displacement at each point of the medium is parallel to the wave front.

Tuned Damper. (See *Dynamic Vibration Absorber*.)

Uncorrelated. Two signals or variables $a_1(t)$ and $a_2(t)$ are said to be uncorrelated if the average value of their product is zero: $\overline{a_1(t) \cdot a_2(t)} = 0$. If the correlation coefficient is equal to unity, the variables are said to be completely correlated. If the coefficient is less than unity but larger than zero, they are said to be partially correlated. (See *Correlation Coefficient*.)

Uncoupled Mode.* An uncoupled mode of vibration is a mode that can exist in a system concurrently with and independently of other modes.

Undamped Natural Frequency.* The undamped natural frequency of a mechanical system is the frequency of free vibration resulting from only elastic and inertial forces of the system.

Variance. Variance is the mean of the squares of the deviations from the mean value of a vibrating quantity.

Velocity. Velocity is a vector quantity that specifies the time rate of change of displacement with respect to a reference frame. If the reference frame is not inertial, the velocity is often designated "relative velocity."

Velocity Shock. Velocity shock is a particular type of shock motion characterized by a sudden velocity change of the foundation. (See *Foundation* and *Mechanical Shock*.)

Vibration.* Vibration is an oscillation wherein the quantity is a parameter that defines the motion of a mechanical system. (See *Oscillation*.)

Vibration Isolator.* A vibration isolator is a resilient support that tends to isolate a system from steady-state excitation.

Vibration Machine. A vibration machine is a device for subjecting a mechanical system to controlled and reproducible mechanical vibration.

Vibration Meter. A vibration meter is an apparatus for the measurement of displacement, velocity, or acceleration of a vibrating body.

Vibration Mount. (See *Vibration Isolator*.)

Vibration Pickup. (See *Transducer*.)

Viscous Damping.* Viscous damping is the dissipation of energy that occurs when a particle in a vibrating system is resisted by a force that has a magnitude proportional to the magnitude of the velocity of the particle and direction opposite to the direction of the particle.

Viscous Damping, Equivalent. (See *Equivalent Viscous Damping*.)

Wave.* A wave is a disturbance which is propagated in a medium in such a manner that at any point in the medium the quantity serving as measure of disturbance is a function of the time, while at any instant the displacement at a point is a function of the position of the point. Any physical quantity that has the same relationship to some independent variable (usually time) that a propagated disturbance has, at a particular instant, with respect to space, may be called a wave.

Wave Interference.* Wave interference is the phenomenon which results when waves of the same or nearly the same frequency are superposed; it is characterized by a spatial or temporal distribution of amplitude of some specified characteristic differing from that of the individual superposed waves.

Wavelength.* The wavelength of a periodic wave in an isotropic medium is the perpendicular distance between two wave fronts in which the displacements have a difference in phase of one complete period.

White Noise. White noise is a noise whose power spectral density is substantially independent of frequency over a specified range.

2

BASIC VIBRATION THEORY

Ralph E. Blake
Lockheed Aircraft Corporation

INTRODUCTION

This chapter presents the theory of free and forced steady-state vibration of single degree-of-freedom systems. Undamped systems and systems having viscous damping and structural damping are included. Multiple degree-of-freedom systems are discussed, including the normal-mode theory of linear elastic structures and Lagrange's equations.

ELEMENTARY PARTS OF VIBRATORY SYSTEMS

Vibratory systems are comprised of means for storing potential energy (spring), means for storing kinetic energy (mass or inertia), and means by which the energy is gradually lost (damper). The vibration of a system involves the alternating transfer of energy between its potential and kinetic forms. In a damped system, some energy is dissipated at each cycle of vibration and must be replaced from an external source if a steady vibration is to be maintained. Although a single physical structure may store both kinetic and potential energy, and may dissipate energy, this chapter considers only *lumped parameter systems* comprised of ideal springs, masses, and dampers wherein each element has only a single function. In translational motion, displacements are defined as linear distances; in rotational motion, displacements are defined as angular motions.

TRANSLATIONAL MOTION

SPRING. In the linear spring shown in Fig. 2.1, the change in the length of the spring is proportional to the force acting along its length:

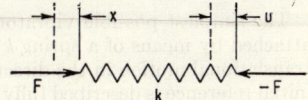

Fig. 2.1. Linear spring.

$$F = k(x - u) \quad (2.1)$$

The ideal spring is considered to have no mass; thus, the force acting on one end is equal and opposite to the force acting on the other end. The constant of proportionality k is the *spring constant or stiffness*.

MASS. A mass is a rigid body (Fig. 2.2) whose acceleration $\ddot{x}$ according to Newton's second law is proportional to the resultant F of all forces acting on the mass: †

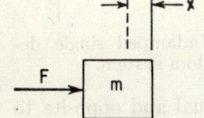

Fig. 2.2. Rigid mass.

$$F = m\ddot{x} \quad (2.2)$$

† It is common to use the word "mass" in a general sense to designate a rigid body. Mathematically, the mass of the rigid body is defined by m in Eq. (2.2).

2–1

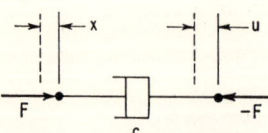

FIG. 2.3. Viscous damper.

DAMPER. In the viscous damper shown in Fig. 2.3, the applied force is proportional to the relative velocity of its connection points:

$$F = c(\dot{x} - \dot{u}) \tag{2.3}$$

The constant c is the *damping coefficient*, the characteristic parameter of the damper. The ideal damper is considered to have no mass; thus the force at one end is equal and opposite to the force at the other end. "Structural damping" is considered below and several other types of damping are considered in Chap. 30.

ROTATIONAL MOTION

The elements of a mechanical system which moves with pure rotation of the parts are wholly analogous to the elements of a system that moves with pure translation. The property of a rotational system which stores kinetic energy is inertia; stiffness and damping coefficients are defined with reference to angular displacement and angular velocity, respectively. The analogous quantities and equations are listed in Table 2.1.

Table 2.1. Analogous Quantities in Translational and Rotational Vibrating Systems

Translational quantity	*Rotational quantity*
Linear displacement x	Angular displacement α
Force F	Torque M
Spring constant k	Spring constant k_r
Damping constant c	Damping constant c_r
Mass m	Moment of inertia I
Spring law $F = k(x_1 - x_2)$	Spring law $M = k_r(\alpha_1 - \alpha_2)$
Damping law $F = c(\dot{x}_1 - \dot{x}_2)$	Damping law $M = c_r(\dot{\alpha}_1 - \dot{\alpha}_2)$
Inertia law $F = m\ddot{x}$	Inertia law $M = I\ddot{\alpha}$

Inasmuch as the mathematical equations for a rotational system can be written by analogy from the equations for a translational system, only the latter are discussed in detail. Whenever translational systems are discussed, it is understood that corresponding equations apply to the analogous rotational system, as indicated in Table 2.1.

SINGLE DEGREE-OF-FREEDOM SYSTEM

The simplest possible vibratory system is shown in Fig. 2.4; it consists of a mass m attached by means of a spring k to an immovable support. The mass is constrained to translational motion in the direction of the X axis so that its change of position from an initial reference is described fully by the value of a single quantity x. For this reason it is called a *single degree-of-freedom system*. If the mass m is displaced from its equilibrium position and then allowed to vibrate free from further external forces, it is said to have *free vibration*. The vibration also may be forced; i.e., a continuing force acts upon the mass or the foundation experiences a continuing motion. Free and forced vibration are discussed below.

FREE VIBRATION WITHOUT DAMPING [1,2,3]

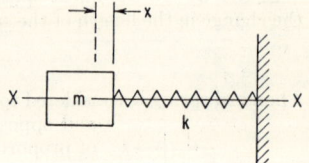

FIG. 2.4. Undamped single degree-of-freedom system.

Considering first the free vibration of the undamped system of Fig. 2.4, Newton's equation is written for the mass m. The force $m\ddot{x}$ exerted by the mass on the spring is equal and opposite to the force kx applied by the spring on the mass:

$$m\ddot{x} + kx = 0 \tag{2.4}$$

where $x = 0$ defines the equilibrium position of the mass.

SINGLE DEGREE-OF-FREEDOM SYSTEM

The solution of Eq. (2.4) is

$$x = A \sin \sqrt{\frac{k}{m}} t + B \cos \sqrt{\frac{k}{m}} t \qquad (2.5)$$

where the term $\sqrt{k/m}$ is the *angular natural frequency* defined by

$$\omega_n = \sqrt{\frac{k}{m}} \qquad \text{rad/sec} \qquad (2.6)$$

The sinusoidal oscillation of the mass repeats continuously, and the time interval to complete one cycle is the *period*:

$$\tau = \frac{2\pi}{\omega_n} \qquad (2.7)$$

The reciprocal of the period is the *natural frequency*:

$$f_n = \frac{1}{\tau} = \frac{\omega_n}{2\pi} = \frac{1}{2\pi} \sqrt{\frac{k}{m}} = \frac{1}{2\pi} \sqrt{\frac{kg}{W}} \qquad (2.8)$$

where $W = mg$ is the weight of the rigid body forming the mass of the system shown in Fig. 2.4. The relations of Eq. (2.8) are shown by the solid lines in Fig. 2.5.

INITIAL CONDITIONS. In Eq. (2.5), B is the value of x at time $t = 0$, and the value of A is equal to $\dot{x}/\omega_n$ at time $t = 0$. Thus, the conditions of displacement and velocity which exist at zero time determine the subsequent oscillation completely.

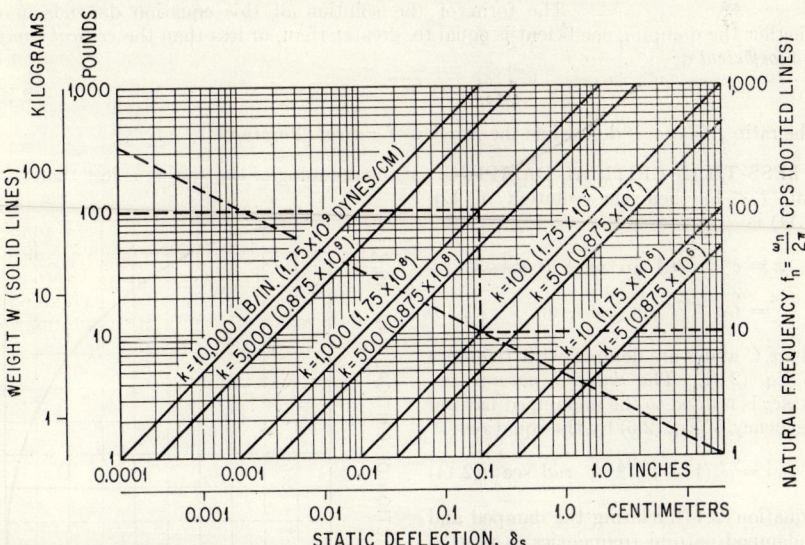

FIG. 2.5. Natural frequency relations for a single degree-of-freedom system. Relation of natural frequency to weight of supported body and stiffness of spring [Eq. (2.8)] is shown by solid lines. Relation of natural frequency to static deflection [Eq. (2.10)] is shown by diagonal-dashed line. Example: To find natural frequency of system with $W = 100$ lb and $k = 1,000$ lb/in., enter at $W = 100$ on left ordinate scale; follow the dashed line horizontally to solid line $k = 1,000$, then vertically down to diagonal-dashed line, and finally horizontally to read $f_n = 10$ cps from right ordinate scale.

PHASE ANGLE. Equation (2.5) for the displacement in oscillatory motion can be written, introducing the frequency relation of Eq. (2.6),

$$x = A \sin \omega_n t + B \cos \omega_n t = C \sin (\omega_n t + \theta) \tag{2.9}$$

where $C = (A^2 + B^2)^{1/2}$ and $\theta = \tan^{-1}(B/A)$. The angle θ is called the *phase angle*.

STATIC DEFLECTION. The static deflection of a simple mass-spring system is the deflection of spring k as a result of the gravity force of the mass, $\delta_{st} = mg/k$. (For example, the system of Fig. 2.4 would be oriented with the mass m vertically above the spring k.) Substituting this relation in Eq. (2.8),

$$f_n = \frac{1}{2\pi}\sqrt{\frac{g}{\delta_{st}}} \tag{2.10}$$

The relation of Eq. (2.10) is shown by the diagonal-dashed line in Fig. 2.5. This relation applies only when the system under consideration is both linear and elastic. For example, rubber springs tend to be nonlinear or exhibit a dynamic stiffness which differs from the static stiffness; hence, Eq. (2.10) is not applicable.

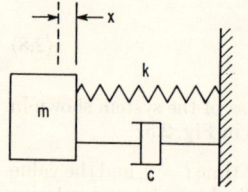

FIG. 2.6. Single degree-of-freedom system with viscous damper.

FREE VIBRATION WITH VISCOUS DAMPING [1,2,3]

Figure 2.6 shows a single degree-of-freedom system with a viscous damper. The differential equation of motion of mass m, corresponding to Eq. (2.4) for the undamped system, is

$$m\ddot{x} + c\dot{x} + kx = 0 \tag{2.11}$$

The form of the solution of this equation depends upon whether the damping coefficient is equal to, greater than, or less than the *critical damping coefficient* c_c:

$$c_c = 2\sqrt{km} = 2m\omega_n \tag{2.12}$$

The ratio $\zeta = c/c_c$ is defined as the *fraction of critical damping*.

LESS-THAN-CRITICAL-DAMPING. If the damping of the system is less than critical, $\zeta < 1$; then the solution of Eq. (2.11) is

$$x = e^{-ct/2m}(A \sin \omega_d t + B \cos \omega_d t)$$
$$= Ce^{-ct/2m} \sin (\omega_d t + \theta) \tag{2.13}$$

where C and θ are defined with reference to Eq. (2.9). The *damped natural frequency* is related to the undamped natural frequency of Eq. (2.6) by the equation

$$\omega_d = \omega_n(1 - \zeta^2)^{1/2} \quad \text{rad/sec} \tag{2.14}$$

Equation (2.14), relating the damped and undamped natural frequencies, is plotted in Fig. 2.7.

CRITICAL DAMPING. When $c = c_c$, there is no oscillation and the solution of Eq. (2.11) is

$$x = (A + Bt)e^{-ct/2m} \tag{2.15}$$

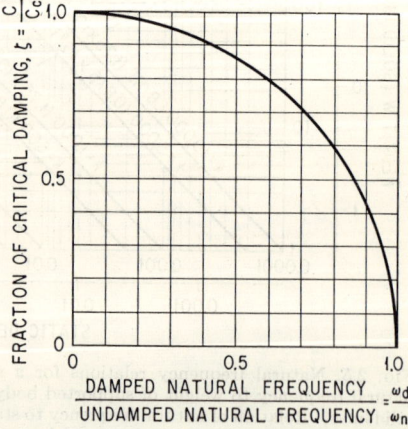

FIG. 2.7. Damped natural frequency as a function of undamped natural frequency and fraction of critical damping.

GREATER-THAN-CRITICAL-DAMPING. When $\zeta > 1$, the solution of Eq. (2.11) is

$$x = e^{-ct/2m}(Ae^{\omega_n\sqrt{\zeta^2-1}\,t} + Be^{-\omega_n\sqrt{\zeta^2-1}\,t}) \qquad (2.16)$$

This is a nonoscillatory motion; if the system is displaced from its equilibrium position, it tends to return gradually.

LOGARITHMIC DECREMENT. The degree of damping in a system having $\zeta < 1$ may be defined in terms of successive peak values in a record of a free oscillation. Substituting the expression for critical damping from Eq. (2.12), the expression for free vibration of a damped system, Eq. (2.13), becomes

$$x = Ce^{-\zeta\omega_n t} \sin(\omega_d t + \theta) \qquad (2.17)$$

Consider any two maxima (i.e., value of x when $dx/dt = 0$) separated by n cycles of oscillation, as shown in Fig. 2.8. Then the ratio of these maxima is

$$\frac{x_n}{x_0} = e^{-2\pi n\zeta/(1-\zeta^2)^{1/2}} \qquad (2.18)$$

Values of x_n/x_0 are plotted in Fig. 2.9 for several values of n over the range of ζ from 0.001 to 0.10.

The *logarithmic decrement* Δ is the natural logarithm of the ratio of the amplitudes of two successive cycles of the damped free vibration:

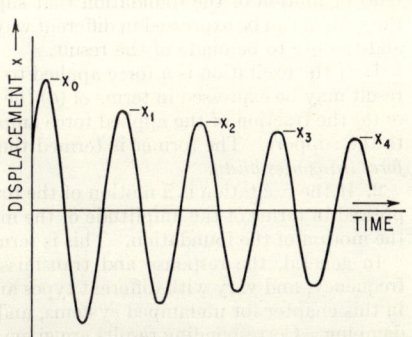

FIG. 2.8. Trace of damped free vibration showing amplitudes of displacement maxima.

$$\Delta = \log \frac{x_1}{x_2} \quad \text{or} \quad \frac{x_2}{x_1} = e^{-\Delta} \qquad (2.19)$$

A comparison of this relation with Eq. (2.18) when $n = 1$ gives the following expression for Δ:

$$\Delta = \frac{2\pi\zeta}{(1-\zeta^2)^{1/2}} \qquad (2.20)$$

The logarithmic decrement can be expressed in terms of the difference of successive amplitudes by writing Eq. (2.19) as follows:

$$\frac{x_1 - x_2}{x_1} = 1 - \frac{x_2}{x_1} = 1 - e^{-\Delta}$$

Writing $e^{-\Delta}$ in terms of its infinite series, the following expression is obtained which gives a good approximation for $\Delta < 0.2$:

$$\frac{x_1 - x_2}{x_1} = \Delta \qquad (2.21)$$

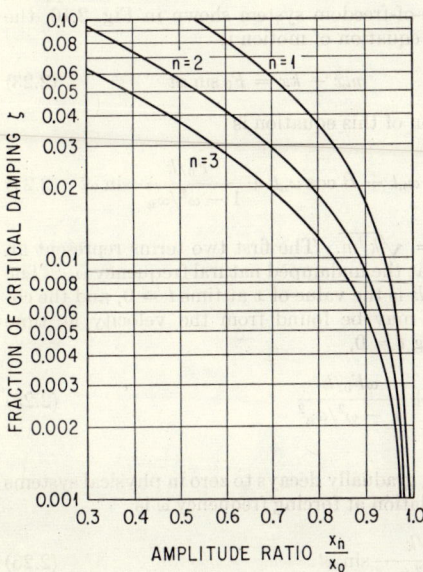

FIG. 2.9. Effect of damping upon the ratio of displacement maxima of a damped free vibration.

For small values of ζ (less than about 0.10), an approximate relation between the fraction of critical damping and the logarithmic decrement, from Eq. (2.20), is

$$\Delta \simeq 2\pi\zeta \qquad (2.22)$$

FORCED VIBRATION

Forced vibration in this chapter refers to the motion of the system which occurs in response to a continuing excitation whose magnitude varies sinusoidally with time. (See Chaps. 8 and 23 for a treatment of the response of a simple system to step, pulse, and transient vibration excitations.) The excitation may be, alternatively, force applied to the system (generally, the force is applied to the mass of a single degree-of-freedom system) or motion of the foundation that supports the system. The resulting response of the system can be expressed in different ways, depending upon the nature of the excitation and the use to be made of the result.

1. If the excitation is a force applied to the mass of the system shown in Fig. 2.4, the result may be expressed in terms of (a) the amplitude of the resulting motion of the mass or (b) the fraction of the applied force amplitude that is transmitted through the system to the support. The former is termed the *motion response* and the latter is termed the *force transmissibility*.

2. If the excitation is a motion of the foundation, the resulting response usually is expressed in terms of the amplitude of the motion of the mass relative to the amplitude of the motion of the foundation. This is termed the *motion transmissibility* for the system.

In general, the response and transmissibility relations are functions of the forcing frequency, and vary with different types and degrees of damping. Results are presented in this chapter for undamped systems, and for systems with either viscous or structural damping. Corresponding results are given in Chap. 30 for systems with Coulomb damping, and for systems with either viscous or Coulomb damping in series with a linear spring.

FORCED VIBRATION WITHOUT DAMPING

FORCE APPLIED TO MASS. When the sinusoidal force $F = F_0 \sin \omega t$ is applied to the mass of the undamped single degree-of-freedom system shown in Fig. 2.10, the differential equation of motion is

$$m\ddot{x} + kx = F_0 \sin \omega t \qquad (2.23)$$

The solution of this equation is

$$x = A \sin \omega_n t + B \cos \omega_n t + \frac{F_0/k}{1 - \omega^2/\omega_n^2} \sin \omega t \qquad (2.24)$$

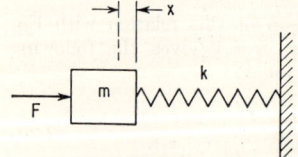

Fig. 2.10. Undamped single degree-of-freedom system excited in forced vibration by force acting on mass.

where $\omega_n = \sqrt{k/m}$. The first two terms represent an oscillation at the undamped natural frequency ω_n. The coefficient B is the value of x at time $t = 0$, and the coefficient A may be found from the velocity at time $t = 0$. Differentiating Eq. (2.24) and setting $t = 0$,

$$\dot{x}(0) = A\omega_n + \frac{\omega F_0/k}{1 - \omega^2/\omega_n^2} \qquad (2.25)$$

The value of A is found from Eq. (2.25).

The oscillation at the natural frequency ω_n gradually decays to zero in physical systems because of damping. The steady-state oscillation at forcing frequency ω is

$$x = \frac{F_0/k}{1 - \omega^2/\omega_n^2} \sin \omega t \qquad (2.26)$$

This oscillation exists after a condition of equilibrium has been established by decay of the oscillation at the natural frequency ω_n and persists as long as the force F is applied.

The force transmitted to the foundation is directly proportional to the spring deflection: $F_t = kx$. Substituting x from Eq. (2.26) and defining transmissibility $T = F_t/F$,

$$T = \frac{1}{1 - \omega^2/\omega_n^2} \qquad (2.27)$$

If the mass is initially at rest in the equilibrium position of the system (i.e., $x = 0$ and $\dot{x} = 0$) at time $t = 0$, the ensuing motion at time $t > 0$ is

$$x = \frac{F_0/k}{1 - \omega^2/\omega_n^2}\left(\sin \omega t - \frac{\omega}{\omega_n} \sin \omega_n t\right) \qquad (2.28)$$

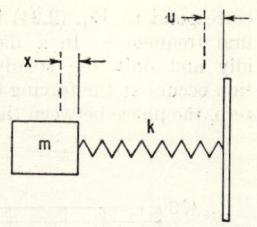

Fig. 2.11. Undamped single degree-of-freedom system excited in forced vibration by motion of foundation.

For large values of time, the second term disappears because of the damping inherent in any physical system, and Eq. (2.28) becomes identical to Eq. (2.26).

When the forcing frequency coincides with the natural frequency, $\omega = \omega_n$ and a condition of resonance exists. Then Eq. (2.28) is indeterminate and the expression for x may be written as

$$x = -\frac{F_0 \omega}{2k} t \cos \omega t \qquad (2.29)$$

According to Eq. (2.29), the amplitude x increases continuously with time, reaching an infinitely great value only after an infinitely great time.

MOTION OF FOUNDATION. The differential equation of motion for the system of Fig. 2.11 excited by a continuing motion $u = u_0 \sin \omega t$ of the foundation is

$$m\ddot{x} = -k(x - u_0 \sin \omega t)$$

The solution of this equation is

$$x = A_1 \sin \omega_n t + B_2 \cos \omega_n t + \frac{u_0}{1 - \omega^2/\omega_n^2} \sin \omega t$$

where $\omega_n = \sqrt{k/m}$ and the coefficients A_1, B_1 are determined by the velocity and displacement of the mass, respectively, at time $t = 0$. The terms representing oscillation at the natural frequency are damped out ultimately, and the ratio of amplitudes is defined in terms of transmissibility T:

$$\frac{x_0}{u_0} = T = \frac{1}{1 - \omega^2/\omega_n^2} \qquad (2.30)$$

where $x = x_0 \sin \omega t$. Thus, in the forced vibration of an undamped single degree-of-freedom system, the motion response, the force transmissibility, and the motion transmissibility are numerically equal.

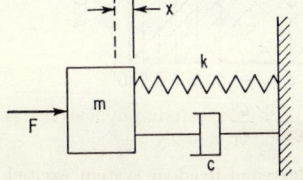

Fig. 2.12. Single degree-of-freedom system with viscous damping, excited in forced vibration by force acting on mass.

FORCED VIBRATION WITH VISCOUS DAMPING

FORCE APPLIED TO MASS. The differential equation of motion for the single degree-of-freedom system with viscous damping shown in Fig. 2.12, when the excitation is a force $F = F_0 \sin \omega t$ applied to the mass, is

$$m\ddot{x} + c\dot{x} + kx = F_0 \sin \omega t \qquad (2.31)$$

Equation (2.31) corresponds to Eq. (2.23) for forced vibration of an undamped system; its solution would correspond to Eq. (2.24) in that it includes terms representing oscillation at the natural frequency. In a damped system, however, these terms are damped out rapidly and only the steady-state solution usually is considered. The resulting motion occurs at the forcing frequency ω; when the damping coefficient c is greater than zero, the phase between the force and resulting motion is different than zero. Thus, the

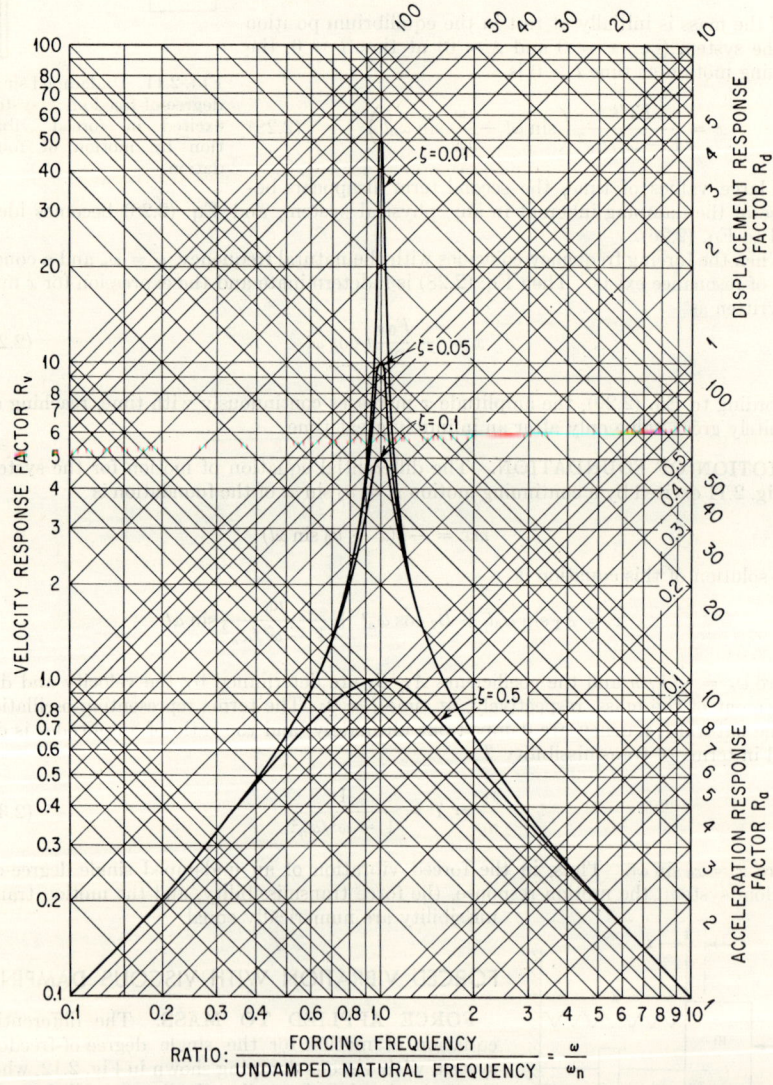

FIG. 2.13. Response factors for a viscous-damped single degree-of-freedom system excited in forced vibration by a force acting on the mass. The velocity response factor shown by horizontal lines is defined by Eq. (2.36); the displacement response factor shown by diagonal lines of positive slope is defined by Eq. (2.33); and the acceleration response factor shown by diagonal lines of negative slope is defined by Eq. (2.37).

response may be written

$$x = R \sin(\omega t - \theta) = A_1 \sin \omega t + B_1 \cos \omega t \tag{2.32}$$

Substituting this relation in Eq. (2.31), the following result is obtained:

$$\frac{x}{F_0/k} = \frac{\sin(\omega t - \theta)}{\sqrt{(1 - \omega^2/\omega_n^2)^2 + (2\zeta\omega/\omega_n)^2}} = R_d \sin(\omega t - \theta) \tag{2.33}$$

where
$$\theta = \tan^{-1}\left(\frac{2\zeta\omega/\omega_n}{1 - \omega^2/\omega_n^2}\right)$$

and R_d is a *dimensionless response factor* giving the ratio of the amplitude of the vibratory displacement to the spring displacement that would occur if the force F were applied statically. At very low frequencies R_d is approximately equal to 1; it rises to a peak near ω_n and approaches zero as ω becomes very large. The displacement response is defined at these frequency conditions as follows:

$$x \simeq \left(\frac{F_0}{k}\right) \sin \omega t \qquad [\omega \ll \omega_n]$$

$$x = \frac{F_0}{2k\zeta} \sin\left(\omega_n t + \frac{\pi}{2}\right) = -\frac{F_0 \cos \omega_n t}{c\omega_n} \qquad [\omega = \omega_n] \tag{2.34}$$

$$x \simeq \frac{\omega_n^2 F_0}{\omega^2 k} \sin(\omega t + \pi) = \frac{F_0}{m\omega^2} \sin \omega t \qquad [\omega \gg \omega_n]$$

For the above three frequency conditions, the vibrating system is sometimes described as *spring-controlled*, *damper-controlled*, and *mass-controlled*, respectively, depending on which element is primarily responsible for the system behavior.

Curves showing the dimensionless response factor R_d as a function of the frequency ratio ω/ω_n are plotted in Fig. 2.13 on the coordinate lines having a positive 45° slope. Curves of the phase angle θ are plotted in Fig. 2.14. A phase angle between 180 and 360°

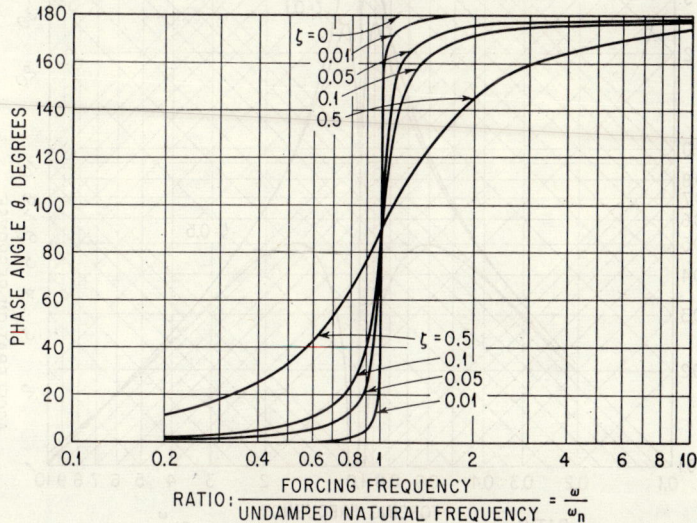

Fig. 2.14. Phase angle between the response displacement and the excitation force for a single degree-of-freedom system with viscous damping, excited by a force acting on the mass of the system.

2-10 BASIC VIBRATION THEORY

cannot exist in this case since this would mean that the damper is furnishing energy to the system rather than dissipating it.

An alternative form of Eqs. (2.33) and (2.34) is

$$\frac{x}{F_0/k} = \frac{(1 - \omega^2/\omega_n^2)\sin\omega t - 2\zeta(\omega/\omega_n)\cos\omega t}{(1 - \omega^2/\omega_n^2)^2 + (2\zeta\omega/\omega_n)^2} \quad (2.35)$$

$$= (R_d)_x \sin\omega t + (R_d)_R \cos\omega t$$

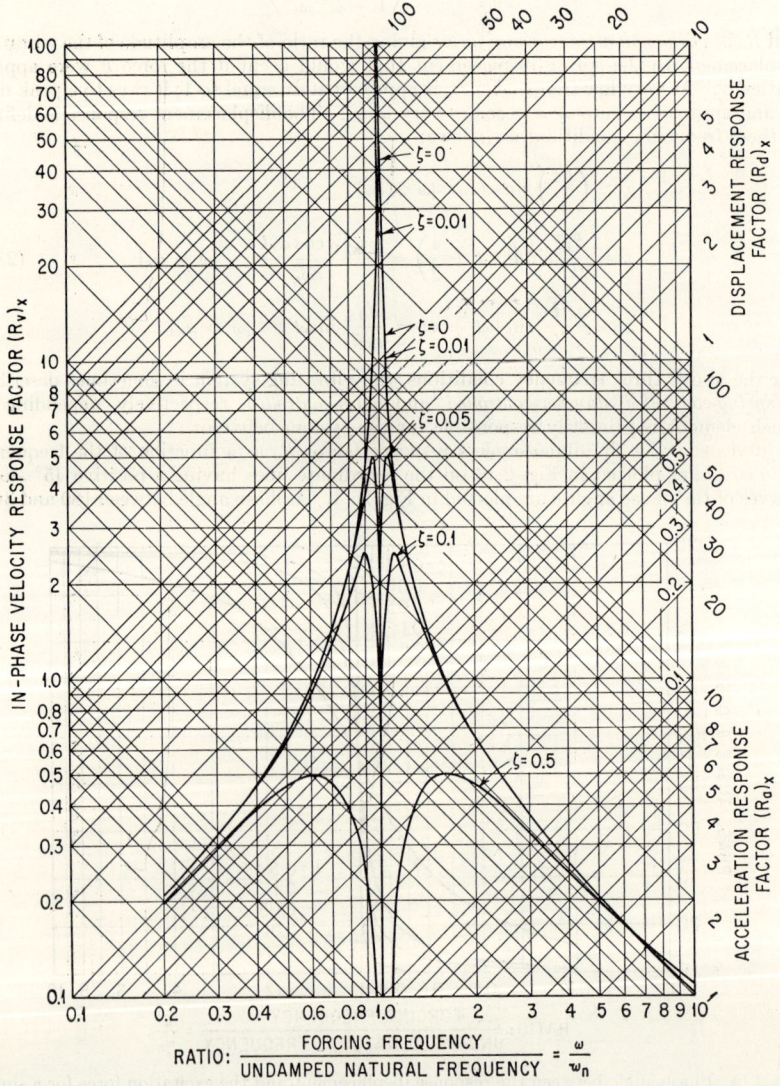

Fig. 2.15. In-phase component of response factor of a viscous-damped system in forced vibration. All values of the response factor for $\omega/\omega_n > 1$ are negative but are plotted without regard for sign. The fraction of critical damping is denoted by ζ.

SINGLE DEGREE-OF-FREEDOM SYSTEM

This shows the components of the response which are in phase [$(R_d)_x \sin \omega t$] and 90° out of phase [$(R_d)_R \cos \omega t$] with the force. Curves of $(R_d)_x$ and $(R_d)_R$ are plotted as a function of the frequency ratio ω/ω_n in Figs. 2.15 and 2.16.

Velocity and Acceleration Response. The shape of the response curves changes distinctly if velocity $\dot{x}$ or acceleration $\ddot{x}$ is plotted instead of displacement x. Differentiating Eq. (2.33),

$$\frac{\dot{x}}{F_0/\sqrt{km}} = \frac{\omega}{\omega_n} R_d \cos(\omega t - \theta) = R_v \cos(\omega t - \theta) \qquad (2.36)$$

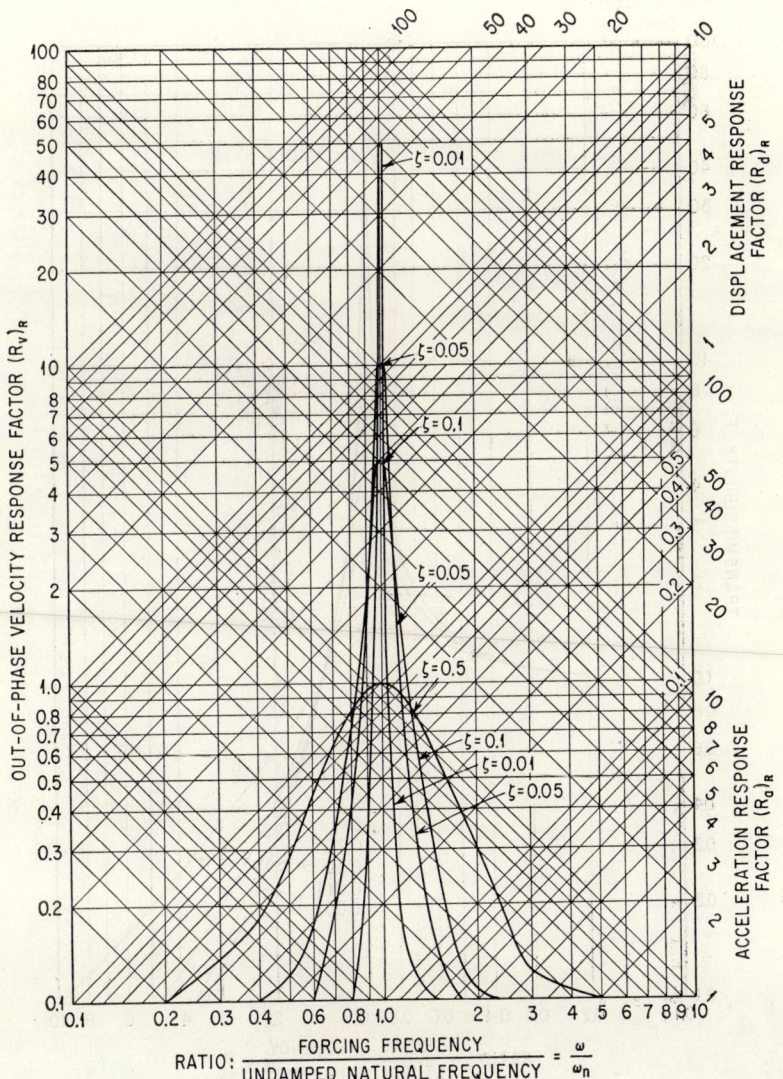

Fig. 2.16. Out-of-phase component of response factor of a viscous-damped system in forced vibration. The fraction of critical damping is denoted by ζ.

The acceleration response is obtained by differentiating Eq. (2.36):

$$\frac{\ddot{x}}{F_0/m} = -\frac{\omega^2}{\omega_n^2} R_d \sin(\omega t - \theta) = -R_a \sin(\omega t - \theta) \quad (2.37)$$

The velocity and acceleration response factors defined by Eqs. (2.36) and (2.37) are shown graphically in Fig. 2.13, the former to the horizontal coordinates and the latter to the coordinates having a negative 45° slope. Note that the velocity response factor approaches zero as $\omega \to 0$ and $\omega \to \infty$, whereas the acceleration response factor approaches 0 as $\omega \to 0$ and approaches unity as $\omega \to \infty$.

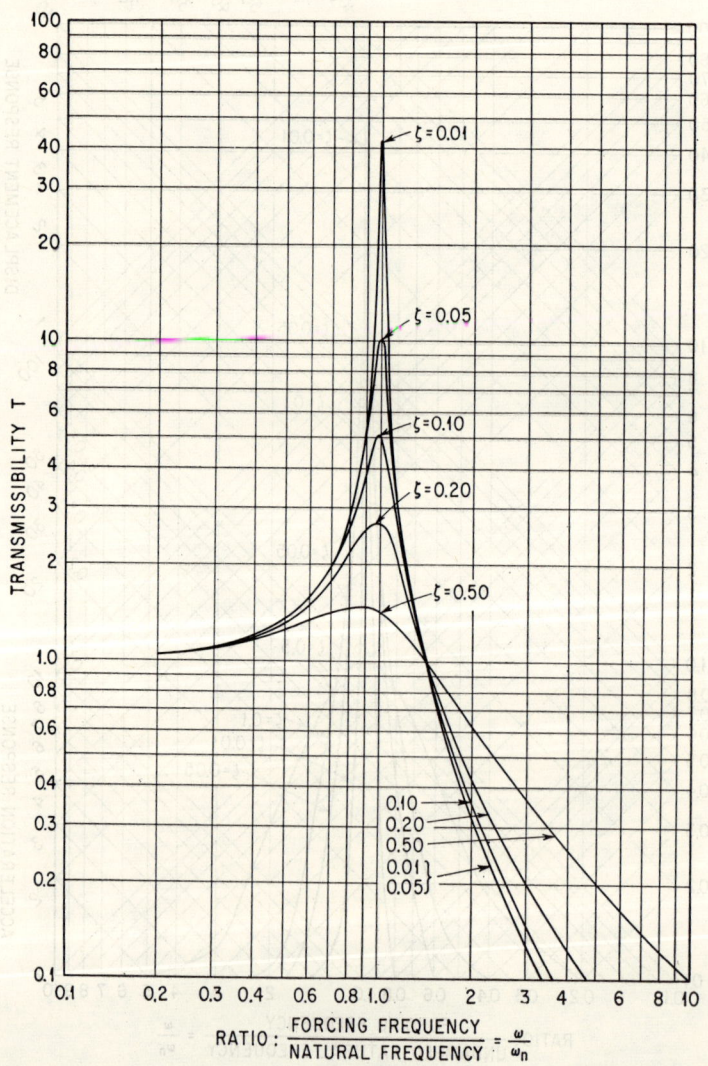

Fig. 2.17. Transmissibility of a viscous-damped system. Force transmissibility and motion transmissibility are identical numerically. The fraction of critical damping is denoted by ζ.

SINGLE DEGREE-OF-FREEDOM SYSTEM

Force Transmission. The force transmitted to the foundation of the system is

$$F_T = c\dot{x} + kx \tag{2.38}$$

Since the forces $c\dot{x}$ and kx are 90° out of phase, the magnitude of the transmitted force is

$$|F_T| = \sqrt{c^2\dot{x}^2 + k^2x^2} \tag{2.39}$$

The ratio of the transmitted force F_T to the applied force F_0 can be expressed in terms of transmissibility T:

$$\frac{F_T}{F_0} = T \sin(\omega t - \psi) \tag{2.40}$$

where

$$T = \sqrt{\frac{1 + (2\zeta\omega/\omega_n)^2}{(1 - \omega^2/\omega_n^2)^2 + (2\zeta\omega/\omega_n)^2}} \tag{2.41}$$

and

$$\psi = \tan^{-1} \frac{2\zeta(\omega/\omega_n)^3}{1 - \omega^2/\omega_n^2 + 4\zeta^2\omega^2/\omega_n^2}$$

The transmissibility T and phase angle ψ are shown in Figs. 2.17 and 2.18, respectively, as a function of the frequency ratio ω/ω_n and for several values of the fraction of critical damping ζ.

Hysteresis. When the viscous damped, single degree-of-freedom system shown in Fig. 2.12 undergoes vibration defined by

$$x = x_0 \sin \omega t \tag{2.42}$$

the net force exerted on the mass by the spring and damper is

$$F = kx_0 \sin \omega t + c\omega x_0 \cos \omega t \tag{2.43}$$

Equations (2.42) and (2.43) define the relation between F and x; this relation is the ellipse

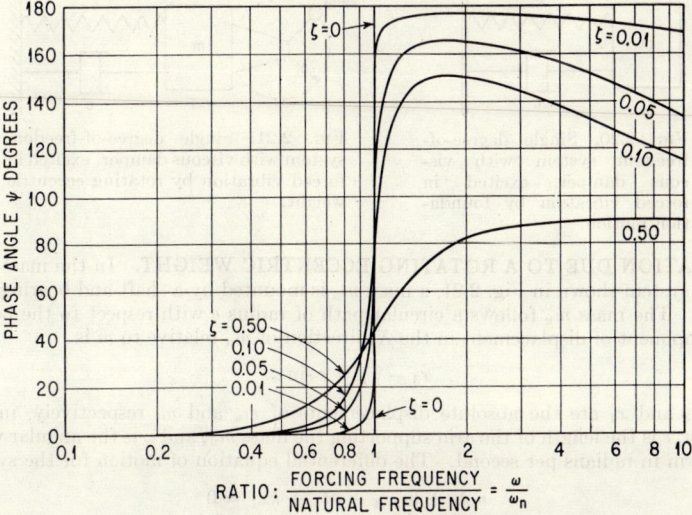

Fig. 2.18. Phase angle of force transmission (or motion transmission) of a viscous-damped system excited (1) by force acting on mass and (2) by motion of foundation. The fraction of critical damping is denoted by ζ.

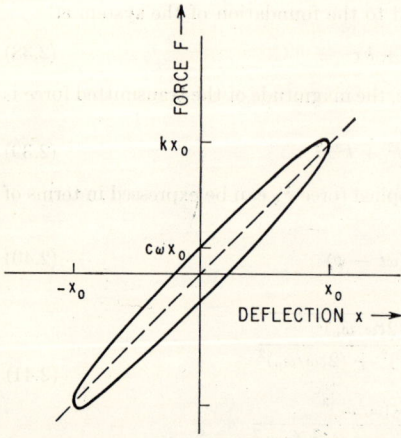

FIG. 2.19. Hysteresis curve for a spring and viscous damper in parallel.

shown in Fig. 2.19. The energy dissipated in one cycle of oscillation is

$$W = \int_{T}^{T+2\pi/\omega} F \frac{dx}{dt} dt = \pi c \omega x_0^2 \quad (2.44)$$

MOTION OF FOUNDATION. The excitation for the elastic system shown in Fig. 2.20 may be a motion $u(t)$ of the foundation. The differential equation of motion for the system is

$$m\ddot{x} + c(\dot{x} - \dot{u}) + k(x - u) = 0 \quad (2.45)$$

Consider the motion of the foundation to be a displacement that varies sinusoidally with time, $u = u_0 \sin \omega t$. A steady-state condition exists after the oscillations at the natural frequency ω_n are damped out, defined by the displacement x of mass m:

$$x = Tu_0 \sin(\omega t - \psi) \quad (2.46)$$

where T and ψ are defined in connection with Eq. (2.40) and are shown graphically in Figs. 2.17 and 2.18, respectively. Thus, the motion transmissibility T in Eq. (2.46) is identical numerically to the force transmissibility T in Eq. (2.40). The motion of the foundation and of the mass m may be expressed in any consistent units, such as displacement, velocity, or acceleration, and the same expression for T applies in each case.

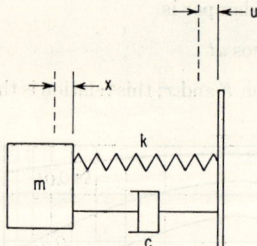

FIG. 2.20. Single degree-of-freedom system with viscous damper, excited in forced vibration by foundation motion.

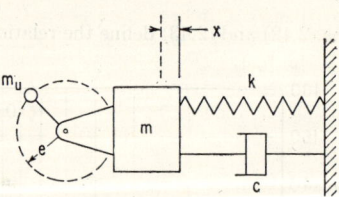

FIG. 2.21. Single degree-of-freedom system with viscous damper, excited in forced vibration by rotating eccentric weight.

VIBRATION DUE TO A ROTATING ECCENTRIC WEIGHT. In the mass-spring-damper system shown in Fig. 2.21, a mass m_u is mounted by a shaft and bearings to the mass m. The mass m_u follows a circular path of radius e with respect to the bearings. The component of displacement in the X direction of m_u relative to m is

$$x_3 - x_1 = e \sin \omega t \quad (2.47)$$

where x_3 and x_1 are the absolute displacements of m_u and m, respectively, in the X direction; e is the length of the arm supporting the mass m_u; and ω is the angular velocity of the arm in radians per second. The differential equation of motion for the system is

$$m\ddot{x}_1 + m_u \ddot{x}_3 + c\dot{x}_1 + kx_1 = 0 \quad (2.48)$$

Differentiating Eq. (2.47) with respect to time, solving for $\ddot{x}_3$, and substituting in Eq. (2.48):

$$(m + m_u)\ddot{x}_1 + c\dot{x}_1 + kx_1 = m_u e \omega^2 \sin \omega t \quad (2.49)$$

SINGLE DEGREE-OF-FREEDOM SYSTEM

Equation (2.49) is of the same form as Eq. (2.31); thus, the response relations of Eqs. (2.33), (2.36), and (2.37) apply by substituting $(m + m_u)$ for m and $m_u e \omega^2$ for F_0. The resulting displacement, velocity, and acceleration responses are

$$\frac{x_1}{m_u e \omega^2} = R_d \sin(\omega t - \theta)$$

$$\frac{\dot{x}_1 \sqrt{km}}{m_u e \omega^2} = R_v \cos(\omega t - \theta) \qquad (2.50)$$

$$\frac{\ddot{x}_1 m}{m_u e \omega^2} = -R_a \sin(\omega t - \theta)$$

RESONANT FREQUENCIES. The peak values of the displacement, velocity, and acceleration response of a system undergoing forced, steady-state vibration occur at slightly different forcing frequencies. Since a *resonant frequency* is defined as the frequency for which the response is a maximum, a simple system has three resonant frequencies if defined only generally. The natural frequency is different from any of the resonant frequencies. The relations among the several resonant frequencies, the damped natural frequency, and the undamped natural frequency ω_n are:

Displacement resonant frequency: $\omega_n(1 - 2\zeta^2)^{1/2}$
Velocity resonant frequency: ω_n
Acceleration resonant frequency: $\omega_n/(1 - 2\zeta^2)^{1/2}$
Damped natural frequency: $\omega_n(1 - \zeta^2)^{1/2}$

For the degree of damping usually embodied in physical systems, the difference among the three resonant frequencies is negligible.

RESONANCE, BANDWIDTH, AND THE QUALITY FACTOR Q. Damping in a system can be determined by noting the maximum response, i.e., the response at the resonant frequency as indicated by the maximum value of R_v in Eq. (2.36). This is defined by the factor Q sometimes used in electrical engineering terminology and defined with respect to mechanical vibration as

$$Q = (R_v)_{\max}$$

The factor Q is defined by

$$Q = \frac{1}{2\zeta}$$

The damping in a system is also indicated by the sharpness or width of the response curve in the vicinity of a resonant frequency ω_n. Designating the width as a frequency increment $\Delta\omega$ measured at the "half-power point" (i.e., at a value of R equal to $R_{\max}/\sqrt{2}$), as illustrated in Fig. 2.22, the damping of the system is defined to a good approximation by

$$\frac{\Delta\omega}{\omega_n} = \frac{1}{Q} = 2\zeta \qquad (2.51)$$

for values of ζ less than 0.1.

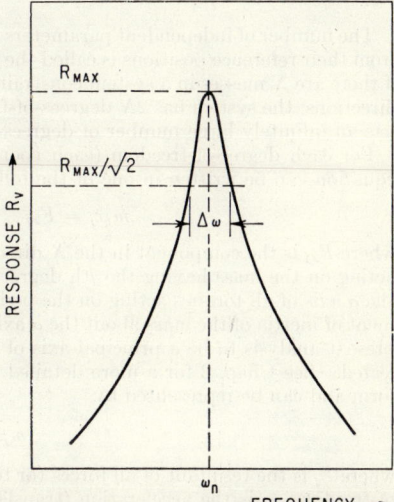

Fig. 2.22. Response curve showing bandwidth at "half-power point."

STRUCTURAL DAMPING. The energy dissipated by the damper is known as *hysteresis loss;* as indicated by Eq. (2.44), it is proportional to the forcing frequency ω. However, the hysteresis loss of many engineering structures has been found to be independent of frequency. To provide a better model for defining the *structural damping* experienced during vibration, an arbitrary damping term $k\mathfrak{g} = c\omega$ is introduced. In effect, this defines the damping force as being equal to the viscous damping force at some frequency, depending upon the value of $\mathfrak{g}$, but being invariant with frequency. The relation of the damping force F to the displacement x is defined by an ellipse similar to Fig. 2.19, and the displacement response of the system is described by an expression corresponding to Eq. (2.33) as follows:

$$\frac{x}{F_0/k} = R_g \sin(\omega t - \theta) = \frac{\sin(\omega t - \theta)}{\sqrt{(1 - \omega^2/\omega_n^2)^2 + \mathfrak{g}^2}} \tag{2.52}$$

where $\mathfrak{g} = 2\zeta\omega/\omega_n$. The resonant frequency is ω_n, and the value of R_g at resonance is $1/\mathfrak{g} = Q$.

The equations for the hysteresis ellipse for structural damping are

$$\begin{aligned} F &= kx_0(\sin \omega t + \mathfrak{g} \cos \omega t) \\ x &= x_0 \sin \omega t \end{aligned} \tag{2.53}$$

UNDAMPED MULTIPLE DEGREE-OF-FREEDOM SYSTEMS

An elastic system sometimes cannot be described adequately by a model having only one mass but rather must be represented by a system of two or more masses considered to be *point masses* or *particles* having no rotational inertia. If a group of particles is bound together by essentially rigid connections, it behaves as a rigid body having both mass (significant for translational motion) and moment of inertia (significant for rotational motion). There is no limit to the number of masses that may be used to represent a system. For example, each mass in a model representing a beam may be an infinitely thin slice representing a cross section of the beam; a differential equation is required to treat this continuous distribution of mass.

DEGREES-OF-FREEDOM

The number of independent parameters required to define the distance of all the masses from their reference positions is called the number of *degrees-of-freedom N*. For example, if there are N masses in a system constrained to move only in translation in the X and Y directions, the system has $2N$ degrees-of-freedom. A continuous system such as a beam has an infinitely large number of degrees-of-freedom.

For each degree-of-freedom (each coordinate of motion of each mass) a differential equation can be written in one of the following alternative forms:

$$m_j \ddot{x}_j = F_{xj} \qquad I_k \ddot{\alpha}_k = M_{\alpha k} \tag{2.54}$$

where F_{xj} is the component in the X direction of all external, spring, and damper forces acting on the mass having the jth degree-of-freedom, and $M_{\alpha k}$ is the component about the α axis of all torques acting on the body having the kth degree-of-freedom. The moment of inertia of the mass about the α axis is designated by I_k. (This is assumed for the present analysis to be a principal axis of inertia, and product of inertia terms are neglected. See Chap. 3 for a more detailed discussion.) Equations (2.54) are identical in form and can be represented by

$$m_j \ddot{x}_j = F_j \tag{2.55}$$

where F_j is the resultant of all forces (or torques) acting on the system in the jth degree-of-freedom, $\ddot{x}_j$ is the acceleration (translational or rotational) of the system in the jth degree-of-freedom, and m_j is the mass (or moment of inertia) in the jth degree-of-freedom. Thus, the terms defining the motion of the system (displacement, velocity, and acceleration) and the deflections of structures may be either translational or rotational, depending upon the type of coordinate. Similarly, the "force" acting on a system may be either a force or a torque, depending upon the type of coordinate. For example, if a system has n bodies each free to move in three translational modes and three rotational modes, there would be $6n$ equations of the form of Eq. (2.55), one for each degree-of-freedom.

UNDAMPED MULTIPLE DEGREE-OF-FREEDOM SYSTEMS

DEFINING A SYSTEM AND ITS EXCITATION

The first step in analyzing any physical structure is to represent it by a mathematical model which will have essentially the same dynamic behavior. A suitable number and distribution of masses, springs, and dampers must be chosen, and the input forces or foundation motions must be defined. The model should have sufficient degrees-of-freedom to determine the modes which will have significant response to the exciting force or motion.

The properties of a system that must be known are the natural frequencies ω_n, the normal mode shapes D_{jn}, the damping of the respective modes, and the mass distribution m_j. The detailed distributions of stiffness and damping of a system are not used directly but rather appear indirectly as the properties of the respective modes. The characteristic properties of the modes may be determined experimentally as well as analytically.

STIFFNESS COEFFICIENTS

The spring system of a structure of N degrees-of-freedom can be defined completely by a set of N^2 *stiffness coefficients*.[5] A stiffness coefficient K_{jk} is the change in spring force acting on the jth degree-of-freedom when only the kth degree-of-freedom is slowly displaced a unit amount in the negative direction. This definition is a generalization of the linear, elastic spring defined by Eq. (2.1). Stiffness coefficients have the characteristic of reciprocity, i.e., $K_{jk} = K_{kj}$. The number of independent stiffness coefficients is $(N^2 + N)/2$.

The total elastic force acting on the jth degree-of-freedom is the sum of the effects of the displacements in all of the degrees-of-freedom:

$$F_{el} = -\sum_{k=1}^{N} K_{jk} x_k \tag{2.56}$$

Inserting the spring force F_{el} from Eq. (2.56) in Eq. (2.55) together with the external forces F_j results in the n equations:

$$m_j \ddot{x}_j = F_j - \sum_k K_{jk} x_k \tag{2.56a}$$

FREE VIBRATION

When the external forces are zero, the preceding equations become

$$m_j \ddot{x}_j + \sum_k K_{jk} x_k = 0 \tag{2.57}$$

Solutions of Eq. (2.57) have the form

$$x_j = D_j \sin(\omega t + \theta) \tag{2.58}$$

Substituting Eq. (2.58) in Eq. (2.57),

$$m_j \omega^2 D_j = \sum_k K_{jk} D_k \tag{2.59}$$

This is a set of n linear algebraic equations with n unknown values of D. A solution of these equations for values of D other than zero can be obtained only if the determinant of the coefficients of the D's is zero:

$$\begin{vmatrix} (m_1\omega^2 - K_{11}) & -K_{12} & \cdot & \cdot & -K_{in} \\ -K_{21} & (m_2\omega^2 - K_{22}) & \cdot & \cdot & \cdot \\ \cdot & \cdot & \cdot & \cdot & \cdot \\ \cdot & \cdot & \cdot & \cdot & \cdot \\ -K_{ni} & \cdot & \cdot & \cdot & (m_n\omega^2 - K_{nn}) \end{vmatrix} = 0 \tag{2.60}$$

Equation (2.60) is an algebraic equation of the nth degree in ω^2; it is called the *frequency equation* since it defines n values of ω which satisfy Eq. (2.57). The roots are all real; some may be equal, and others may be zero. These values of frequency determined from Eq. (2.60) are the frequencies at which the system can oscillate in the absence of external forces. These frequencies are the *natural frequencies* ω_n of the system. Depending upon the initial conditions under which vibration of the system is initiated, the oscillations may occur at any or all of the natural frequencies and at any amplitude.

Example 2.1. Consider the three degree-of-freedom system shown in Fig. 2.23; it consists of three equal masses m and a foundation connected in series by three equal springs k. The absolute displacements of the masses are x_1, x_2, and x_3. The stiffness coefficients (see section entitled *Stiffness Coefficients*) are thus $K_{11} = 2k$, $K_{22} = 2k$, $K_{33} = k$, $K_{12} = K_{21} = -k$, $K_{23} =$

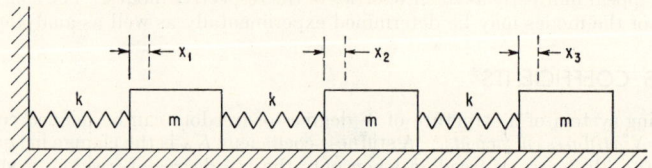

Fig. 2.23. Undamped three degree-of-freedom system on foundation.

$K_{32} = -k$, and $K_{13} = K_{31} = 0$. The frequency equation is given by the determinant, Eq. (2.60),

$$\begin{vmatrix} (m\omega^2 - 2k) & k & 0 \\ k & (m\omega^2 - 2k) & k \\ 0 & k & (m\omega^2 - k) \end{vmatrix} = 0$$

The determinant expands to the following polynomial:

$$\left(\frac{m\omega^2}{k}\right)^3 - 5\left(\frac{m\omega^2}{k}\right)^2 + 6\left(\frac{m\omega^2}{k}\right) - 1 = 0$$

Solving for ω,

$$\omega = 0.445\sqrt{\frac{k}{m}}, \quad 1.25\sqrt{\frac{k}{m}}, \quad 1.80\sqrt{\frac{k}{m}}$$

NORMAL MODES OF VIBRATION. A structure vibrating at only one of its natural frequencies ω_n does so with a characteristic pattern of amplitude distribution called a *normal mode of vibration*.[3,4,5] A normal mode is defined by a set of values of D_{jn} [see Eq. (2.58)] which satisfy Eq. (2.59) when $\omega = \omega_n$:

$$\omega_n^2 m_j D_{jn} = \sum_k K_{jn} D_{kn} \tag{2.61}$$

A set of values of D_{jn} which form a normal mode is independent of the absolute values of D_{jn} but depends only on their relative values. To define a mode shape by a unique set of numbers, any arbitrary *normalizing condition* which is desired can be used. A condition often used is to set $D_{1n} = 1$ but $\sum_j m_j D_{jn}^2 = 1$ and $\sum_j m_j D_{jn}^2 = \sum_j m_j$ also may be found convenient.

ORTHOGONALITY OF NORMAL MODES. The usefulness of normal modes in dealing with multiple degree-of-freedom systems is due largely to the orthogonality of the normal modes. It can be shown[3,4,5] that the set of inertia forces $\omega_n^2 m_j D_{jn}$ for one mode does no work on the set of deflections D_{jm} of another mode of the structure:

$$\sum_j m_j D_{jm} D_{jn} = 0 \quad [m \neq n] \tag{2.62}$$

This is the *orthogonality condition*.

UNDAMPED MULTIPLE DEGREE-OF-FREEDOM SYSTEMS 2–19

NORMAL MODES AND GENERALIZED COORDINATES. Any set of N deflections x_j can be expressed as the sum of normal mode amplitudes:

$$x_j = \sum_{n=1}^{N} q_n D_{jn} \qquad (2.63)$$

The numerical values of the D_{jn}'s are fixed by some normalizing condition, and a set of values of the N variables q_n can be found to match any set of x_j's. The N values of q_n constitute a set of *generalized coordinates* which can be used to define the position coordinates x_j of all parts of the structure. The q's are also known as the amplitudes of the normal modes, and are functions of time. Equation (2.63) may be differentiated to obtain

$$\ddot{x}_j = \sum_{n=1}^{N} \ddot{q}_n D_{jn} \qquad (2.64)$$

Any quantity which is distributed over the j coordinates can be represented by a linear transformation similar to Eq. (2.63). It is convenient now to introduce the parameter γ_n relating D_{jn} and F_j/m_j as follows:

$$\frac{F_j}{m_j} = \sum_n \gamma_n D_{jn} \qquad (2.65)$$

where F_j may be zero for certain values of n.

FORCED MOTION

Substituting the expressions in generalized coordinates, Eqs. (2.63) to (2.65), in the basic equation of motion, Eq. (2.56a),

$$m_j \sum_n \ddot{q}_n D_{jn} + \sum_k k_{jk} \sum_n q_n D_{kn} - m_j \sum_n \gamma_n D_{jn} = 0 \qquad (2.66)$$

The center term in Eq. (2.66) may be simplified by applying Eq. (2.61) and the equation rewritten as follows:

$$\sum_n (\ddot{q}_n + \omega_n^2 q_n - \gamma_n) m_j D_{jn} = 0 \qquad (2.67)$$

Multiplying Eqs. (2.67) by D_{jm} and taking the sum over j (i.e., adding all the equations together),

$$\sum_n (\ddot{q}_n + \omega_n^2 q_n - \gamma_n) \sum_j m_j D_{jn} D_{jm} = 0$$

All terms of the sum over n are zero, except for the term for which $m = n$, according to the orthogonality condition of Eq. (2.62). Then since $\sum_j m_j D_{jn}^2$ is not zero, it follows that

$$\ddot{q}_n + \omega_n^2 q_n - \gamma_n = 0$$

for every value of n from 1 to N.

An expression for γ_n may be found by using the orthogonality condition again. Multiplying Eq. (2.65) by $m_j D_{jm}$ and taking the sum taken over j,

$$\sum_j F_j D_{jm} = \sum_n \gamma_n \sum_j m_j D_{jn} D_{jm} \qquad (2.68)$$

All the terms of the sum over n are zero except when $n = m$, according to Eq. (2.62), and Eq. (2.68) reduces to

$$\gamma_n = \frac{\sum_j F_j D_{jn}}{\sum_j m_j D_{jn}^2} \qquad (2.69)$$

Then the differential equation for the response of any generalized coordinate to the externally applied forces F_j is

$$\ddot{q}_n + \omega_n^2 q_n = \gamma_n = \frac{\sum_j F_j D_{jn}}{\sum_j m_j D_{jn}^2} \quad (2.70)$$

where $\Sigma F_j D_{jn}$ is the generalized force, i.e., the total work done by all external forces during a small displacement δq_n divided by δq_n, and $\Sigma m_j D_{jn}^2$ is the generalized mass.

Thus the amplitude q_n of each normal mode is governed by its own equation, independent of the other normal modes, and responds as a simple mass-spring system. Equation (2.70) is a generalized form of Eq. (2.23).

The forces F_j may be any functions of time. Any equation for the response of an undamped mass-spring system applies to each mode of a complex structure by substituting:

The *generalized coordinate* q_n for x

The *generalized force* $\sum_j F_j D_{jn}$ for F (2.71)

The *generalized mass* $\sum_j m_j D_{jn}$ for m

The *mode natural frequency* ω_n for ω_n

RESPONSE TO SINUSOIDAL FORCES. If a system is subjected to one or more sinusoidal forces $F_j = F_{0j} \sin \omega t$, the response is found from Eq. (2.26) by noting that $k = m\omega_n^2$ [Eq. (2.6)] and then substituting from Eq. (2.71):

$$q_n = \frac{\sum_j F_{0j} D_{jn}}{\omega_n^2 \sum_j m_j D_{jn}^2} \frac{\sin \omega t}{(1 - \omega^2/\omega_n^2)} \quad (2.72)$$

Then the displacement of the kth degree-of-freedom, from Eq. (2.63), is

$$x_k = \sum_{n=1}^{N} \frac{D_{kn} \sum_j F_{0j} D_{jn} \sin \omega t}{\omega_n^2 \sum_j m_j D_{jn}^2 (1 - \omega^2/\omega_n^2)} \quad (2.73)$$

This is the general equation for the response to sinusoidal forces of an undamped system of N degrees-of-freedom. The application of the equation to systems free in space or attached to immovable foundations is discussed below.

Example 2.2. Consider the system shown in Fig. 2.24; it consists of 3 equal masses m connected in series by 2 equal springs k. The system is free in space and a force $F \sin \omega t$ acts on the first mass. Absolute displacements of the masses are x_1, x_2, and x_3. Determine the ac-

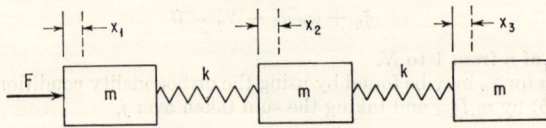

FIG. 2.24. Undamped three degree-of-freedom system acted on by sinusoidal force.

celeration $\ddot{x}_3$. The stiffness coefficients (see section entitled *Stiffness Coefficients*) are $K_{11} = K_{33} = k$, $K_{22} = 2k$, $K_{12} = K_{21} = -k$, $K_{13} = K_{31} = 0$, and $K_{23} = K_{32} = -k$. Substituting in Eq. (2.60), the frequency equation is

$$\begin{vmatrix} (m\omega^2 - k) & k & 0 \\ k & (m\omega^2 - 2k) & k \\ 0 & k & (m\omega^2 - k) \end{vmatrix} = 0$$

UNDAMPED MULTIPLE DEGREE-OF-FREEDOM SYSTEMS   2–21

The roots are $\omega_1 = 0$, $\omega_2 = \sqrt{k/m}$, and $\omega_3 = \sqrt{3k/m}$. The zero value for one of the natural frequencies indicates that the entire system translates without deflection of the springs. The mode shapes are now determined by substituting from Eq. (2.58) in Eq. (2.57), noting that $\ddot{x} = -D\omega^2$, and writing Eq. (2.59) for each of the three masses in each of the oscillatory modes 2 and 3:

$$mD_{21}\left(\frac{k}{m}\right) = K_{11}D_{21} + K_{21}D_{22} + K_{31}D_{23}$$

$$mD_{22}\left(\frac{k}{m}\right) = K_{12}D_{21} + K_{22}D_{22} + K_{32}D_{23}$$

$$mD_{23}\left(\frac{k}{m}\right) = K_{13}D_{21} + K_{23}D_{22} + K_{33}D_{23}$$

$$mD_{31}\left(\frac{3k}{m}\right) = K_{11}D_{31} + K_{21}D_{32} + K_{31}D_{33}$$

$$mD_{32}\left(\frac{3k}{m}\right) = K_{12}D_{31} + K_{22}D_{32} + K_{32}D_{33}$$

$$mD_{33}\left(\frac{3k}{m}\right) = K_{13}D_{31} + K_{23}D_{32} + K_{33}D_{33}$$

where the first subscript on the D's indicates the mode number (according to ω_1 and ω_2 above) and the second subscript indicates the displacement amplitude of the particular mass. The values of the stiffness coefficients K are calculated above. The mode shapes are defined by the relative displacements of the masses. Thus, assigning values of unit displacement to the first mass (i.e., $D_{21} = D_{31} = 1$), the above equations may be solved simultaneously for the D's:

$$D_{21} = 1 \qquad D_{22} = 0 \qquad D_{23} = -1$$
$$D_{31} = 1 \qquad D_{32} = -2 \qquad D_{33} = 1$$

Substituting these values of D in Eq. (2.71), the generalized masses are determined: $M_2 = 2m$, $M_3 = 6m$.

Equation (2.73) then can be used to write the expression for acceleration $\ddot{x}_3$:

$$\ddot{x}_3 = \left[\frac{1}{3m} + \frac{(\omega^2/\omega_2{}^2)(-1)(+1)}{2m(1-\omega^2/\omega_2{}^2)} + \frac{(\omega^2/\omega_3{}^2)(+1)(+1)}{6m(1-\omega^2/\omega_3{}^2)}\right] F_1 \sin \omega t$$

FREE AND FIXED SYSTEMS. For a structure which is free in space, there are six "normal modes" corresponding to $\omega_n = 0$. These represent motion of the structure without relative motion of its parts; this is rigid body motion with six degrees-of-freedom.

The rigid body modes all may be described by equations of the form

$$D_{jm} = a_{jm} D_m \qquad [m = 1, 2, \ldots, 6]$$

where D_m is a motion of the rigid body in the m coordinate and a is the displacement of the jth degree-of-freedom when D_m is moved a unit amount. The geometry of the structure determines the nature of a_{jm}. For example, if D_m is a rotation about the Z axis, $a_{jm} = 0$ for all modes of motion in which j represents rotation about the X or Y axis and $a_{jm} = 0$ if j represents translation parallel to the Z axis. If D_{jm} is a translational mode of motion parallel to X or Y, it is necessary that a_{jm} be proportional to the distance r_j of m_j from the Z axis and to the sine of the angle between r_j and the jth direction. The above relations may be applied to an elastic body. Such a body moves as a rigid body in the gross sense in that all particles of the body move together generally but may experience relative vibratory motion. The orthogonality condition applied to the relation between any rigid body mode D_{jm} and any oscillatory mode D_{jn} yields

$$\sum_j m_j D_{jn} D_{jm} = \sum_j m_j a_{jm} D_{jn} = 0 \qquad \begin{bmatrix} m \leq 6 \\ n > 6 \end{bmatrix} \qquad (2.74)$$

These relations are used in computations of oscillatory modes, and show that normal modes of vibration involve no net translation or rotation of a body.

A system attached to a fixed foundation may be considered as a system free in space in which one or more "foundation" masse- or moments of inertia are infinite. Motion of the system as a rigid body is determined entirely by the motion of the foundation. The amplitude of an oscillatory mode representing motion of the foundation is zero; i.e., $M_j D_{jn}^2 = 0$ for the infinite mass. However, Eq. (2.73) applies equally well regardless of the size of the masses.

FOUNDATION MOTION. If a system is small relative to its foundation, it may be assumed to have no effect on the motion of the foundation. Consider a foundation of large but unknown mass m_0 having a motion $x_0 \sin \omega t$, the consequence of some unknown force

$$F_0 \sin \omega t = -m_0 x_0 \omega^2 \sin \omega t \tag{2.75}$$

acting on m_0 in the x_0 direction. Equation (2.73) is applicable to this case upon substituting

$$-m_0 x_0 \omega^2 D_{0n} = \sum_j F_{0j} D_{jn} \tag{2.76}$$

where D_{0n} is the amplitude of the foundation (the 0 degree-of-freedom) in the nth mode. The oscillatory modes of the system are subject to Eqs. (2.74):

$$\sum_{j=0} m_j a_{jm} D_{jn} = 0$$

Separating the 0th degree-of-freedom from the other degrees-of-freedom:

$$\sum_{j=0} m_j a_{jm} D_{jn} = m_0 a_{0m} D_{0n} + \sum_{j=1} m_j a_{jm} D_{jn}$$

If m_0 approaches infinity as a limit, D_{0n} approaches zero and motion of the system as a rigid body is identical with the motion of the foundation. Thus, a_{0m} approaches unity for motion in which $m = 0$, and approaches zero for motion in which $m \neq 0$. In the limit:

$$\lim_{m_0 \to \infty} m_0 D_{0n} = -\sum_j m_j a_{j0} D_{jn} \tag{2.77}$$

Substituting this result in Eq. (2.76),

$$\lim_{m_0 \to \infty} \sum_j F_{0j} D_{jn} = x_0 \omega^2 \sum_j m_j a_{j0} D_{jn} \tag{2.78}$$

The generalized mass in Eq. (2.73) includes the term $m_0 D_{0n}^2$, but this becomes zero as m_0 becomes infinite.

The equation for response of a system to motion of its foundation is obtained by substituting Eq. (2.78) in Eq. (2.73):

$$x_k = \sum_{n=1}^{N} \frac{\omega^2}{\omega_n^2} D_{kn} \frac{\sum_j m_j a_{j0} D_{jn} x_0 \sin \omega t}{\sum_j m_j D_{jn}^2 (1 - \omega^2/\omega_n^2)} + x_0 \sin \omega t \tag{2.79}$$

DAMPED MULTIPLE DEGREE-OF-FREEDOM SYSTEMS

Consider a set of masses interconnected by a network of springs and acted upon by external forces, with a network of dampers acting in parallel with the springs. The viscous dampers produce forces on the masses which are determined in a manner analogous to that used to determine spring forces and summarized by Eq. (2.56). The damping force acting on the jth degree-of-freedom is

$$(F_d)_j = -\sum_k C_{jk} \dot{x}_k \tag{2.80}$$

where C_{jk} is the resultant force on the jth degree-of-freedom due to a unit velocity of the kth degree-of-freedom.

In general, the distribution of damper sizes in a system need not be related to the spring or mass sizes. Thus, the dampers may couple the normal modes together, allowing motion of one mode to affect that of another. Then the equations of response are not easily separable [6] into independent normal mode equations. However, there are two types of damping distribution which do not couple the normal modes.[6,7,8] These are known as *uniform viscous damping* and *uniform mass damping*.

UNIFORM VISCOUS DAMPING

Uniform damping is an appropriate model for systems in which the damping effect is an inherent property of the spring material. Each spring is considered to have a damper acting in parallel with it, and the ratio of damping coefficient to stiffness coefficient is the same for each spring of the system. Thus, for all values of j and k,

$$\frac{C_{jk}}{k_{jk}} = 2\mathcal{G} \qquad (2.81)$$

where $\mathcal{G}$ is a constant.

Substituting from Eq. (2.81) in Eq. (2.80),

$$-(F_d)_j = \sum_k C_{jk}\dot{x}_k = 2\mathcal{G} \sum_k k_{jk}\dot{x}_k \qquad (2.82)$$

Since the damping forces are "external" forces with respect to the mass-spring system, the forces $(F_d)_j$ can be added to the external forces in Eq. (2.70) to form the equation of motion:

$$\ddot{q}_n + \omega_n^2 q = \frac{\sum_j (F_d)_j D_{jn} + \sum_j F_j D_{jn}}{\sum_j m_j D_{jn}^2} \qquad (2.83)$$

Combining Eqs. (2.61), (2.63), and (2.82), the summation involving $(F_d)_j$ in Eq. (2.83) may be written as follows:

$$\sum_j (F_d)_j D_{jn} = -2\mathcal{G}\omega_n^2 \dot{q}_n \sum_j m_j D_{jn}^2 \qquad (2.84)$$

Substituting Eq. (2.84) in Eq. (2.83),

$$\ddot{q}_n + 2\mathcal{G}\omega_n^2 \dot{q}_n + \omega_n^2 q_n = \gamma_n \qquad (2.85)$$

Comparison of Eq. (2.85) with Eq. (2.31) shows that each mode of the system responds as a simple damped oscillator.

The damping term $2\mathcal{G}\omega_n^2$ in Eq. (2.85) corresponds to $2\zeta\omega_n$ in Eq. (2.31) for a simple system. Thus, $\mathcal{G}\omega_n$ may be considered the critical damping ratio of each mode. Note that the effective damping for a particular mode varies directly as the natural frequency of the mode.

FREE VIBRATION. If a system with uniform viscous damping is disturbed from its equilibrium position and released at time $t = 0$ to vibrate freely, the applicable equation of motion is obtained from Eq. (2.85) by substituting $2\zeta\omega$ for $2\mathcal{G}\omega_n^2$ and letting $\gamma_n = 0$:

$$\ddot{q}_n + 2\zeta\omega_n \dot{q}_n + \omega_n^2 q_n = 0 \qquad (2.86)$$

The solution of Eq. (2.86) for less than critical damping is

$$x_j(t) = \sum_n D_{jn} e^{-\zeta\omega_n t}(A_n \sin \omega_d t + B_n \cos \omega_d t) \qquad (2.87)$$

where $\omega_d = \omega_n(1 - \zeta^2)^{1/2}$.

The values of A and B are determined by the displacement $x_j(0)$ and velocity $\dot{x}_j(0)$ at time $t = 0$:

$$x_j(0) = \sum_n B_n D_{jn}$$

$$\dot{x}_j(0) = \sum_n (A_n \omega_{dn} - B_n \zeta \omega_n) D_{jn}$$

Applying the orthogonality relation of Eq. (2.62) in the manner used to derive Eq. (2.69),

$$B_n = \frac{\sum_j x_j(0) m_j D_{jn}}{\sum_j m_j D_{jn}^2}$$

$$A_n \omega_{dn} - B_n \zeta \omega_{dn} = \frac{\sum_j \dot{x}_j(0) m_j D_{jn}}{\sum_j m_j D_{jn}^2}$$

(2.88)

Thus each mode undergoes a decaying oscillation at the damped natural frequency for the particular mode, and the amplitude of each mode decays from its initial value, which is determined by the initial displacements and velocities.

UNIFORM STRUCTURAL DAMPING

To avoid the dependence of viscous damping upon frequency, as indicated by Eq. (2.85), the uniform viscous damping factor $\mathfrak{g}$ is replaced by $\mathfrak{g}/\omega$ for uniform structural damping. This corresponds to the structural damping parameter $\mathfrak{g}$ in Eqs. (2.52) and (2.53) for sinusoidal vibration of a simple system. Thus, Eq. (2.85) for the response of a mode to a sinusoidal force of frequency ω is

$$\ddot{q}_n + \frac{2\mathfrak{g}}{\omega} \omega_n^2 \dot{q}_n + \omega_n^2 q_n = \gamma_n \qquad (2.89)$$

The amplification factor at resonance ($Q = 1/\mathfrak{g}$) has the same value in all modes.

UNIFORM MASS DAMPING

If the damping force on each mass is proportional to the magnitude of the mass,

$$(F_d)_j = -B m_j \dot{x}_j \qquad (2.90)$$

where B is a constant. For example, Eq. (2.90) would apply to a uniform beam immersed in a viscous fluid.

Substituting as $\dot{x}_j$ in Eq. (2.90) the derivative of Eq. (2.63),

$$\Sigma (F_d)_j D_{jn} = -B \sum_j m_j D_{jn} \sum_m \dot{q}_m D_{jm} \qquad (2.91)$$

Because of the orthogonality condition, Eq. (2.62):

$$\Sigma (F_d)_j D_{jn} = -B \dot{q}_n \sum_j m_j D_{jn}^2$$

Substituting from Eq. (2.91) in Eq. (2.83), the differential equation for the system is

$$\ddot{q}_n + B \dot{q}_n + \omega_n^2 q_n = \gamma_n \qquad (2.92)$$

where the damping term B corresponds to $2\zeta\omega$ for a simple oscillator, Eq. (2.31). Then $B/2\omega_n$ represents the fraction of critical damping for each mode, a quantity which diminishes with increasing frequency.

GENERAL EQUATION FOR FORCED VIBRATION

All the equations for response of a linear system to a sinusoidal excitation may be regarded as special cases of the following general equation:

$$x_k = \sum_{n=1}^{N} \frac{D_{kn}}{\omega_n^2} \frac{F_n}{m_n} R_n \sin(\omega t - \theta_n) \qquad (2.93)$$

where x_k = displacement of structure in kth degree-of-freedom
N = number of degrees-of-freedom, including those of the foundation
D_{kn} = amplitude of kth degree-of-freedom in nth normal mode
F_n = generalized force for nth mode
m_n = generalized mass for nth mode
R_n = response factor, a function of the frequency ratio ω/ω_n (Fig. 2.13)
θ_n = phase angle (Fig. 2.14)

Equation (2.93) is of sufficient generality to cover a wide variety of cases, including excitation by external forces or foundation motion, viscous or structural damping, rotational and translational degrees-of-freedom, and from one to an infinite number of degrees-of-freedom.

LAGRANGIAN EQUATIONS

The differential equations of motion for a vibrating system sometimes are derived more conveniently in terms of kinetic and potential energies of the system than by the application of Newton's laws of motion in a form requiring the determination of the forces acting on each mass of the system. The formulation of the equations in terms of the energies, known as Lagrangian equations,[3,4,5] is expressed as follows:

$$\frac{d}{dt}\frac{\partial T}{\partial \dot{q}_n} - \frac{\partial T}{\partial q_n} + \frac{\partial V}{\partial q_n} = F_n \qquad (2.94)$$

where T = total kinetic energy of system
V = total potential energy of system
q_n = generalized coordinate—a displacement
$\dot{q}_n$ = velocity at generalized coordinate q_n
F_n = generalized force, the portion of the total forces not related to the potential energy of the system (gravity and spring forces appear in the potential energy expressions and are not included here)

The method of applying Eq. (2.94) is to select a number of independent coordinates (generalized coordinates) equal to the number of degrees-of-freedom, and to write expressions for total kinetic energy T and total potential energy V. Differentiation of these expressions successively with respect to each of the chosen coordinates leads to a number of equations similar to Eq. (2.94), one for each coordinate (degree-of-freedom). These are the applicable differential equations and may be solved by any suitable method.

Example 2.3. Consider free vibration of the three degree-of-freedom system shown in Fig. 2.23; it consists of three equal masses m connected in tandem by equal springs k. Take as coordinates the three absolute displacements x_1, x_2, and x_3. The kinetic energy of the system is

$$T = \tfrac{1}{2}m(\dot{x}_1^2 + \dot{x}_2^2 + \dot{x}_3^2)$$

The potential energy of the system is

$$V = \frac{k}{2}[x_1^2 + (x_1 - x_2)^2 + (x_2 - x_3)^2] = \frac{k}{2}(2x_1^2 + 2x_2^2 + x_3^2 - 2x_1 x_2 - 2x_2 x_3)$$

Differentiating the expression for the kinetic energy successively with respect to the velocities,

$$\frac{\partial T}{\partial \dot{x}_1} = m\dot{x}_1 \qquad \frac{\partial T}{\partial \dot{x}_2} = m\dot{x}_2 \qquad \frac{\partial T}{\partial \dot{x}_3} = m\dot{x}_3$$

The kinetic energy is not a function of displacement; therefore, the second term in Eq. (2.94) is zero. The partial derivatives with respect to the displacement coordinates are

$$\frac{\partial V}{\partial x_1} = 2kx_1 - kx_2 \qquad \frac{\partial V}{\partial x_2} = 2kx_2 - kx_1 - kx_3 \qquad \frac{\partial V}{\partial x_3} = kx_3 - kx_2$$

In free vibration, the generalized force term in Eq. (2.93) is zero. Then, substituting the derivatives of the kinetic and potential energies from above into Eq. (2.94),

$$m\ddot{x}_1 + 2kx_1 - kx_2 = 0$$
$$m\ddot{x}_2 + 2kx_2 - kx_1 - kx_3 = 0$$
$$m\ddot{x}_3 + kx_3 - kx_2 = 0$$

The natural frequencies of the system may be determined by placing the preceding set of simultaneous equations in determinant form, in accordance with Eq. (2.60):

$$\begin{vmatrix} (m\omega^2 - 2k) & k & 0 \\ k & (m\omega^2 - 2k) & k \\ 0 & k & (m\omega^2 - k) \end{vmatrix} = 0$$

The natural frequencies are equal to the values of ω that satisfy the preceding determinant equation.

Example 2.4. Consider the compound pendulum of mass m shown in Fig. 2.25, having its center-of-gravity located a distance l from the axis of rotation. The moment of inertia is I

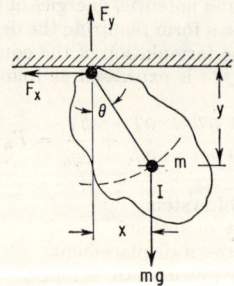

FIG. 2.25. Forces and motions of a compound pendulum.

about an axis through the center-of-gravity. The position of the mass is defined by three coordinates, x and y to define the location of the center-of-gravity, and θ to define the angle of rotation.

The *equations of constraint* are $y = l(1 - \cos\theta)$; $x = l\sin\theta$. Each equation of constraint reduces the number of degrees-of-freedom by one; thus the pendulum is a one degree-of-freedom system whose position is defined uniquely by θ alone.

The kinetic energy of the pendulum is

$$T = \tfrac{1}{2}(I + ml^2)\dot{\theta}^2$$

The potential energy is

$$V = mgl(1 - \cos\theta)$$

Then

$$\frac{\partial T}{\partial \dot{\theta}} = (I + ml^2)\dot{\theta} \qquad \frac{d}{dt}\left(\frac{\partial T}{\partial \dot{\theta}}\right) = (I + ml^2)\ddot{\theta}$$

$$\frac{\partial T}{\partial \theta} = 0 \qquad \frac{\partial V}{\partial \theta} = mgl\sin\theta$$

Substituting these expressions in Eq. (2.94), the differential equation for the pendulum is

$$(I + ml^2)\ddot{\theta} + mgl\sin\theta = 0$$

Example 2.5. Consider oscillation of the water in the U tube shown in Fig. 2.26. If the displacements of the water levels in the arms of a uniform-diameter U tube are h_1 and h_2, then conservation of matter requires that $h_1 = -h_2$. The kinetic energy of the water flowing in the tube with velocity $\dot{h}_1$ is

$$T = \tfrac{1}{2}\rho S l \dot{h}_1{}^2$$

where ρ is the water density, S is the cross-section area of the tube, and l is the developed length of the water column. The potential energy (difference in potential energy between arms of tube) is

$$V = S\rho g h_1{}^2$$

Taking h_1 as the generalized coordinate, differentiating the expressions for energy, and substituting in Eq. (2.94),

$$S\rho l \ddot{h}_1 + 2\rho g S h_1 = 0$$

Dividing through by $\rho S l$,

$$\ddot{h}_1 + \frac{2g}{l} h_1 = 0$$

This is the differential equation for a simple oscillating system of natural frequency ω_n, where

$$\omega_n = \sqrt{\frac{2g}{l}}$$

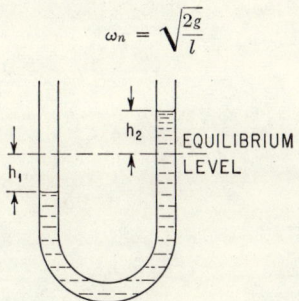

Fig. 2.26. Water column in a U tube.

REFERENCES

1. Timoshenko, S.: "Vibration Problems in Engineering," D. Van Nostrand Company, Inc., Princeton, N.J., 1937.
2. Hansen, H. M., and P. F. Chenea: "Mechanics of Vibration," John Wiley & Sons, Inc., New York, 1952.
3. Kármán, T. V., and M. A. Biot: "Mathematical Methods in Engineering," McGraw-Hill Book Company, Inc., New York, 1940.
4. Slater, J. C., and N. H. Frank: "Mechanics," McGraw-Hill Book Company, Inc., New York, 1947.
5. Pipes, L. A.: "Applied Mathematics for Engineers and Physicists," McGraw-Hill Book Company, Inc., New York, 1958.
6. Foss, K. A.: "Coordinates Which Uncouple the Equations of Motion of Damped Linear Dynamic Systems," *ASME Applied Mechanics Paper* 57-A-86, December, 1957.
7. Rayleigh: "The Theory of Sound," vol. I, Macmillan & Co., Ltd., London, 1894.
8. Crumb, S. F.: "A Study of the Effects of Damping on Normal Modes of Electrical and Mechanical Systems," California Institute of Technology, Pasadena, Calif., 1955.

Example 2.5. Consider oscillation of the water in the U-tube shown in Fig. 2.26. If the displacements of the water levels in the arms of a uniform-diameter U tube are h_1 and h_2, then conservation of matter requires that $h_1 = -h_2 = h$. The kinetic energy of the water flowing in the tube with velocity h is

$$T = \tfrac{1}{2} m \dot{h}^2$$

where m is the water density δ times cross-sectional area of the tube, and l is the developed length of the water columns. The potential energy difference in potential energy between arms of the tube is

$$V = \delta g h^2$$

Taking h as the generalized coordinate, differentiating the expressions for energy, and substituting in Eq. (2.64),

$$\frac{d}{dt}(m\dot{h}) + 2 \delta g h = 0$$

Dividing through by $m\delta$,

$$\ddot{h} + \frac{2g}{l} h = 0$$

This is the differential equation for simple oscillation at natural frequency ω, where

$$\omega = \sqrt{\frac{2g}{l}}$$

Fig. 2.26. Water oscillation in a U-tube.

REFERENCES

1. Timoshenko, S., "Vibration Problems in Engineering," D. Van Nostrand Company, Inc., Princeton, N. J., 1955.
2. Hansen, H. M., and P. F. Chenea, "Mechanics of Vibration," John Wiley & Sons, Inc., New York, 1952.
3. Den Hartog, J. P., "Mechanical Vibrations," McGraw-Hill Book Company, Inc., New York, 1956.
4. Jacobsen, L. S., and R. S. Ayre, "Engineering Vibrations," McGraw-Hill Book Company, Inc., New York, 1958.
5. Pipes, L. A., "Applied Mathematics for Engineers and Physicists," McGraw-Hill Book Company, Inc., New York, 1958.
6. Franke, F. A., "Exponential Wave Function in the Discussion of Motion of Damped Linear Dynamic Systems," AIEE Annual Meeting, Paper 57-A-68, December, 1957.
7. Rayleigh, "The Theory of Sound," Vol. I, Macmillan & Co., Ltd., London, 1937.
8. Coupland, P. L., "A Study of the Influence of Dampers on Natural Modes of Electrical and Mechanical Systems," Cambridge University of Technology, England, c. 1957.

3
VIBRATION OF A RESILIENTLY SUPPORTED RIGID BODY

Harry Himelblau, Jr.
North American Aviation, Inc.*

Sheldon Rubin
Hughes Aircraft Company

INTRODUCTION

SCOPE

This chapter discusses the vibration of a rigid body on resilient supporting elements, including (1) methods of determining the inertial properties of a rigid body, (2) discussion of the dynamic properties of resilient elements, and (3) motion of a single rigid body on resilient supporting elements for various dynamic excitations and degrees of symmetry.

The general equations of motion for a rigid body on linear massless resilient supports are given; these equations are general in that they include any configuration of the rigid body and any configuration and location of the supports. They involve six simultaneous equations with numerous terms, for which a general solution is impracticable without the use of high-speed automatic computing equipment. Various degrees of simplification are introduced by assuming certain symmetry, and results useful for engineering purposes are presented. Several topics are considered: (1) determination of undamped natural frequencies and discussion of coupling of modes of vibration; (2) forced vibration where the excitation is a vibratory motion of the foundation; (3) forced vibration where the excitation is a vibratory force or moment generated within the body; and (4) free vibration caused by an instantaneous change in velocity of the system (velocity shock). Results are presented mathematically and, where feasible, graphically.

SYSTEM OF COORDINATES

The motion of the rigid body is referred to a fixed "inertial" frame of reference. The inertial frame is represented by a system of cartesian coordinates $\overline{X}, \overline{Y}, \overline{Z}$. A similar system of coordinates X, Y, Z fixed in the body has its origin at the center-of-mass. The two sets of coordinates are coincident when the body is in equilibrium under the action of gravity alone. The motions of the body are described by giving the displacement of the body axes relative to the inertial axes. The translational displacements of the center-of-mass of the body are x_c, y_c, z_c in the $\overline{X}, \overline{Y}, \overline{Z}$ directions, respectively. The rotational displacements of the body are characterized by the angles of rotation α, β, γ of

---
*Now at Nortronics Div., Northrop Corp.

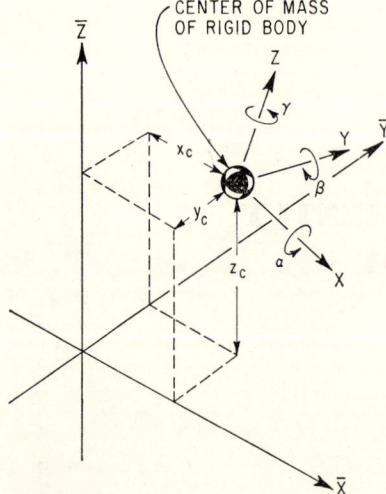

FIG. 3.1. System of coordinates for the motion of a rigid body consisting of a fixed inertial set of reference axes ($\bar{X}$, $\bar{Y}$, $\bar{Z}$), and a set of axes (X, Y, Z) fixed in the moving body with its origin at the center-of-mass. The axes $\bar{X}$, $\bar{Y}$, $\bar{Z}$ and X, Y, Z are coincident when the body is in equilibrium under the action of gravity alone. The displacement of the center-of-mass is given by the translational displacements x_c, y_c, z_c and the rotational displacements α, β, γ as shown. A positive rotation about an axis is one which advances a right-handed screw in the positive direction of the axis.

the body axes about the $\bar{X}$, $\bar{Y}$, $\bar{Z}$ axes, respectively. These displacements are shown graphically in Fig. 3.1.

Only small translations and rotations are considered. Hence, the rotations are commutative (i.e., the resulting position is independent of the order of the component rotations) and the angles of rotation about the body axes are equal to those about the inertial axes. Therefore, the displacements of a point b in the body (with the coordinates b_x, b_y, b_z in the X, Y, Z directions, respectively) are the sums of the components of the center-of-mass displacement in the directions of the $\bar{X}$, $\bar{Y}$, $\bar{Z}$ axes plus the tangential components of the rotational displacement of the body:

$$x_b = x_c + b_z\beta - b_y\gamma$$
$$y_b = y_c - b_z\alpha + b_x\gamma \qquad (3.1)$$
$$z_b = z_c - b_x\beta + b_y\alpha$$

EQUATIONS OF SMALL MOTION OF A RIGID BODY

The equations of motion for the translation of a rigid body are

$$m\ddot{x}_c = \mathbf{F}_x \qquad m\ddot{y}_c = \mathbf{F}_y \qquad m\ddot{z}_c = \mathbf{F}_z \qquad (3.2)$$

where m is the mass of the body, $\mathbf{F}_x$, $\mathbf{F}_y$, $\mathbf{F}_z$ are the summation of all forces acting on the body, and $\ddot{x}_c$, $\ddot{y}_c$, $\ddot{z}_c$ are the accelerations of the center-of-mass of the body in the $\bar{X}$, $\bar{Y}$, $\bar{Z}$ directions, respectively. The motion of the center-of-mass of a rigid body is the same as the motion of a particle having a mass equal to the total mass of the body and acted upon by the resultant external force.

The equations of motion for the rotation of a rigid body are

$$I_{xx}\ddot{\alpha} - I_{xy}\ddot{\beta} - I_{xz}\ddot{\gamma} = \mathbf{M}_x$$
$$-I_{xy}\ddot{\alpha} + I_{yy}\ddot{\beta} - I_{yz}\ddot{\gamma} = \mathbf{M}_y \qquad (3.3)$$
$$-I_{xz}\ddot{\alpha} - I_{yz}\ddot{\beta} + I_{zz}\ddot{\gamma} = \mathbf{M}_z$$

where $\ddot{\alpha}$, $\ddot{\beta}$, $\ddot{\gamma}$ are the rotational accelerations about the X, Y, Z axes, as shown in Fig. 3.1; $\mathbf{M}_x$, $\mathbf{M}_y$, $\mathbf{M}_z$ are the summation of torques acting on the rigid body about the axes X, Y, Z, respectively; and $I_{xx} \ldots$, $I_{xy} \ldots$ are the moments and products of inertia of the rigid body as defined below.

INERTIAL PROPERTIES OF A RIGID BODY

The properties of a rigid body that are significant in dynamics and vibration are the mass, the position of the center-of-mass (or center-of-gravity), the moments of inertia, the products of inertia, and the directions of the principal inertial axes. This section discusses the properties of a rigid body, together with computational and experimental methods for determining the properties.

MASS

COMPUTATION OF MASS. The mass of a body is computed by integrating the product of mass density $\rho(V)$ and elemental volume dV over the body:

$$m = \int_V \rho(V)\,dV \tag{3.4}$$

If the body is made up of a number of elements, each having constant or an average density, the mass is

$$m = \rho_1 V_1 + \rho_2 V_2 + \cdots + \rho_n V_n \tag{3.5}$$

where ρ_1 is the density of the element V_1, etc. Densities of various materials may be found in handbooks containing properties of materials.[1]

If a rigid body has a common geometrical shape, or if it is an assembly of subbodies having common geometrical shapes, the volume may be found from compilations of formulas. Typical formulas are included in Tables 3.1 and 3.2. Tables of areas of plane sections as well as volumes of solid bodies are useful.

If the volume of an element of the body is not given in such a table, the integration of Eq. (3.4) may be carried out analytically, graphically, or numerically. A graphical approach may be used if the shape is so complicated that the analytical expression for its boundaries is not available or is not readily integrable. This is accomplished by graphically dividing the body into smaller parts, each of whose boundaries may be altered slightly (without change to the area) in such a manner that the volume is readily calculable or measurable.

The weight W of a body of mass m is a function of the acceleration of gravity g at the particular location of the body in space:

$$W = mg \tag{3.6}$$

Unless otherwise stated, it is understood that the weight of a body is given for an average value of the acceleration of gravity on the surface of the earth. For engineering purposes, $g = 32.2$ ft/sec^2 or 386 in./sec^2 is usually used.

EXPERIMENTAL DETERMINATION OF MASS. Although Newton's second law of motion, $F = m\ddot{x}$, may be used to measure mass, this usually is not convenient. The mass of a body is most easily measured by performing a static measurement of the weight of the body and converting the result to mass. This is done by use of the value of the acceleration of gravity at the measurement location [Eq. (3.6)].

CENTER-OF-MASS

COMPUTATION OF CENTER-OF-MASS. The center-of-mass (or center-of-gravity) is that point located by the vector

$$\mathbf{r}_c = \frac{1}{m} \int_m \mathbf{r}(m)\,dm \tag{3.7}$$

where $\mathbf{r}(m)$ is the radius vector of the element of mass dm. The center-of-mass of a body in a cartesian coordinate system X, Y, Z is located at

$$X_c = \frac{1}{m} \int_V X(V)\rho(V)\,dV$$

$$Y_c = \frac{1}{m} \int_V Y(V)\rho(V)\,dV \tag{3.8}$$

$$Z_c = \frac{1}{m} \int_V Z(V)\rho(V)\,dV$$

where $X(V)$, $Y(V)$, $Z(V)$ are the X, Y, Z coordinates of the element of volume dV and m is the mass of the body.

If the body can be divided into elements whose centers-of-mass are known, the center-of-mass of the entire body having a mass m is located by equations of the following type:

$$X_c = \frac{1}{m}(X_{c1}m_1 + X_{c2}m_2 + \cdots + X_{cn}m_n), \text{ etc.} \tag{3.9}$$

where X_{c1} is the X coordinate of the center-of-mass of element m_1. Tables (see Tables 3.1 and 3.2) which specify the location of centers of area and volume (called centroids) for simple sections and solid bodies often are an aid in dividing the body into the submasses indicated in the above equation. The centroid and center-of-mass of an element are coincident when the density of the material is uniform throughout the element.

EXPERIMENTAL DETERMINATION OF CENTER-OF-MASS. The location of the center-of-mass is normally measured indirectly by locating the center-of-gravity of the body, and may be found in various ways. Theoretically, if the body is suspended by a flexible wire attached successively at different points on the body, all lines represented by the wire in its various positions when extended inwardly into the body intersect at the center-of-gravity. Two such lines determine the center-of-gravity, but more may be used as a check. There are practical limitations to this method in that the point of intersection often is difficult to designate.

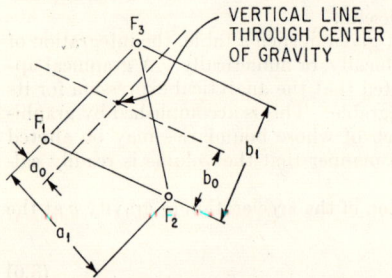

FIG. 3.2. Three-scale method of locating the center-of-gravity of a body. The vertical forces F_1, F_2, F_3 at the scales result from the weight of the body. The vertical line located by the distances a_0 and b_0 [see Eqs. (3.10)] passes through the center-of-gravity of the body.

Other techniques are based on the balancing of the body on point or line supports. A point support locates the center-of-gravity along a vertical line through the point; a line support locates it in a vertical plane through the line. The intersection of such lines or planes determined with the body in various positions locates the center-of-gravity. The greatest difficulty with this technique is the maintenance of the stability of the body while it is balanced, particularly where the height of the body is great relative to a horizontal dimension. If a perfect point or edge support is used, the equilibrium position is inherently unstable. It is only if the support has width that some degree of stability can be achieved, but then a resulting error in the location of the line or plane containing the center-of-gravity can be expected.

Another method of locating the center-of-gravity is to place the body in a stable position on three scales. From static moments the vector weight of the body is the resultant of the measured forces at the scales as shown in Fig. 3.2. The vertical line through the center-of-gravity is located by the distances a_0 and b_0:

$$a_0 = \frac{F_2}{F_1 + F_2 + F_3}a_1$$

$$b_0 = \frac{F_3}{F_1 + F_2 + F_3}b_1 \tag{3.10}$$

This method cannot be used with more than three scales.

MOMENT AND PRODUCT OF INERTIA

COMPUTATION OF MOMENT AND PRODUCT OF INERTIA.
The moments of inertia of a rigid body with respect to the orthogonal axes X, Y, Z fixed in the body are

$$I_{xx} = \int_m (Y^2 + Z^2)\, dm \qquad I_{yy} = \int_m (X^2 + Z^2)\, dm \qquad I_{zz} = \int_m (X^2 + Y^2)\, dm \qquad (3.11)$$

where dm is the infinitesimal element of mass located at the coordinate distances X, Y, Z; and the integration is taken over the mass of the body. Similarly, the products of inertia are

$$I_{xy} = \int_m XY\, dm \qquad I_{xz} = \int_m XZ\, dm \qquad I_{yz} = \int_m YZ\, dm \qquad (3.12)$$

It is conventional in rigid body mechanics to take the center of coordinates at the center-of-mass of the body. Unless otherwise specified, this location is assumed and the moments of inertia and products of inertia refer to axes through the center-of-mass of the body. For a unique set of axes, the products of inertia vanish. These axes inherently originate at the center-of-mass and are called the principal inertial axes of the body. The moments of inertia about these axes are called the principal moments of inertia.

The moments of inertia of a rigid body can be defined in terms of radii of gyration as follows:

$$I_{xx} = m\rho_x^2 \qquad I_{yy} = m\rho_y^2 \qquad I_{zz} = m\rho_z^2 \qquad (3.13)$$

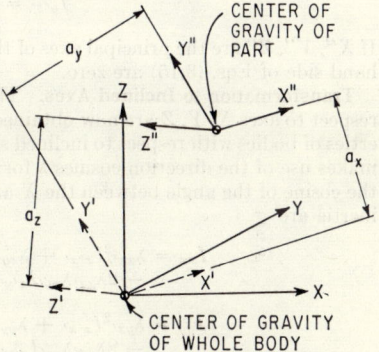

Fig. 3.3. Axes required for moment and product of inertia transformations. Moments and products of inertia with respect to the axes X'', Y'', Z'' are transferred to the mutually parallel axes X', Y', Z' by Eqs. (3.14) and (3.15), and then to the inclined axes X, Y, Z by Eqs. (3.16) and (3.17).

where I_{xx}, ... are the moments of inertia of the body as defined by Eqs. (3.11), m is the mass of the body, and ρ_x, ... are the radii of gyration. The radius of gyration has the dimension of length, and often leads to convenient expressions in dynamics of rigid bodies when distances are normalized to an appropriate radius of gyration. Solid bodies of various shapes have characteristic radii of gyration which sometimes are useful intuitively in evaluating dynamic conditions.

Unless the body has a very simple shape, it is laborious to evaluate the integrals of Eqs. (3.11) and (3.12). The problem is made easier by subdividing the body into parts for which simplified calculations are possible. The moments and products of inertia of the body are found by first determining the moments and products of inertia for the individual parts with respect to appropriate reference axes chosen in the parts, and then summing the contributions of the parts. This is done by selecting axes through the centers-of-mass of the parts, and then determining the moments and products of inertia of the parts relative to these axes. Then the moments and products of inertia are transferred to the axes chosen through the center-of-mass of the whole body, and the transferred quantities summed. In general, the transfer involves two sets of nonparallel coordinates whose centers are displaced. Two transformations are required as follows.

Transformation to Parallel Axes. Referring to Fig. 3.3, suppose that X, Y, Z is a convenient set of axes for the moment of inertia of the whole body with its origin at the center-of-mass. The moments and products of inertia for a part of the body are $I_{x''x''}$,

$I_{y''y''}$, $I_{z''z''}$, $I_{x''y''}$, $I_{x''z''}$, and $I_{y''z''}$, taken with respect to a set of axes X'', Y'', Z'' fixed in the part and having their center at the center-of-mass of the part. The axes X', Y', Z' are chosen parallel to X'', Y'', Z'' with their origin at the center-of-mass of the body. The perpendicular distance between the X'' and X' axes is a_x; between Y'' and Y' is a_y; between Z'' and Z' is a_z. The moments and products of inertia of the part of mass m_n with respect to the X', Y', Z' axes are

$$I_{x'x'} = I_{x''x''} + m_n a_x^2$$
$$I_{y'y'} = I_{y''y''} + m_n a_y^2 \tag{3.14}$$
$$I_{z'z'} = I_{z''z''} + m_n a_z^2$$

The corresponding products of inertia are

$$I_{x'y'} = I_{x''y''} + m_n a_x a_y$$
$$I_{x'z'} = I_{x''z''} + m_n a_x a_z \tag{3.15}$$
$$I_{y'z'} = I_{y''z''} + m_n a_y a_z$$

If X'', Y'', Z'' are the principal axes of the part, the product of inertia terms on the right-hand side of Eqs. (3.15) are zero.

Transformation to Inclined Axes. The desired moments and products of inertia with respect to axes X, Y, Z are now obtained by a transformation theorem relating the properties of bodies with respect to inclined sets of axes whose centers coincide. This theorem makes use of the direction cosines λ for the respective sets of axes. For example, $\lambda_{xx'}$ is the cosine of the angle between the X and X' axes. The expressions for the moments of inertia are

$$\begin{aligned}
I_{xx} &= \lambda_{xx'}^2 I_{x'x'} + \lambda_{xy'}^2 I_{y'y'} + \lambda_{xz'}^2 I_{z'z'} \\
&\quad - 2\lambda_{xx'}\lambda_{xy'} I_{x'y'} - 2\lambda_{xx'}\lambda_{xz'} I_{x'z'} - 2\lambda_{xy'}\lambda_{xz'} I_{y'z'} \\
I_{yy} &= \lambda_{yx'}^2 I_{x'x'} + \lambda_{yy'}^2 I_{y'y'} + \lambda_{yz'}^2 I_{z'z'} \\
&\quad - 2\lambda_{yx'}\lambda_{yy'} I_{x'y'} - 2\lambda_{yx'}\lambda_{yz'} I_{x'z'} - 2\lambda_{yy'}\lambda_{yz'} I_{y'z'} \\
I_{zz} &= \lambda_{zx'}^2 I_{x'x'} + \lambda_{zy'}^2 I_{y'y'} + \lambda_{zz'}^2 I_{z'z'} \\
&\quad - 2\lambda_{zx'}\lambda_{zy'} I_{x'y'} - 2\lambda_{zx'}\lambda_{zz'} I_{x'z'} - 2\lambda_{zy'}\lambda_{zz'} I_{y'z'}
\end{aligned} \tag{3.16}$$

The corresponding products of inertia are

$$\begin{aligned}
-I_{xy} &= \lambda_{xx'}\lambda_{yx'} I_{x'x'} + \lambda_{xy'}\lambda_{yy'} I_{y'y'} + \lambda_{xz'}\lambda_{yz'} I_{z'z'} \\
&\quad - (\lambda_{xx'}\lambda_{yy'} + \lambda_{xy'}\lambda_{yx'}) I_{x'y'} - (\lambda_{xy'}\lambda_{yz'} + \lambda_{xz'}\lambda_{yy'}) I_{y'z'} \\
&\quad - (\lambda_{xz'}\lambda_{yx'} + \lambda_{xx'}\lambda_{yz'}) I_{x'z'} \\
-I_{xz} &= \lambda_{xx'}\lambda_{zx'} I_{x'x'} + \lambda_{xy'}\lambda_{zy'} I_{y'y'} + \lambda_{xz'}\lambda_{zz'} I_{z'z'} \\
&\quad - (\lambda_{xx'}\lambda_{zy'} + \lambda_{xy'}\lambda_{zx'}) I_{x'y'} - (\lambda_{xy'}\lambda_{zz'} + \lambda_{xz'}\lambda_{zy'}) I_{y'z'} \\
&\quad - (\lambda_{xx'}\lambda_{zz'} + \lambda_{xz'}\lambda_{zx'}) I_{x'z'} \\
-I_{yz} &= \lambda_{yx'}\lambda_{zx'} I_{x'x'} + \lambda_{yy'}\lambda_{zy'} I_{y'y'} + \lambda_{yz'}\lambda_{zz'} I_{z'z'} \\
&\quad - (\lambda_{yx'}\lambda_{zy'} + \lambda_{yy'}\lambda_{zx'}) I_{x'y'} - (\lambda_{yy'}\lambda_{zz'} + \lambda_{yz'}\lambda_{zy'}) I_{y'z'} \\
&\quad - (\lambda_{yz'}\lambda_{zx'} + \lambda_{yx'}\lambda_{zz'}) I_{x'z'}
\end{aligned} \tag{3.17}$$

EXPERIMENTAL DETERMINATION OF MOMENTS OF INERTIA. The moment of inertia of a body about a given axis may be found experimentally by suspending the body as a pendulum so that rotational oscillations about that axis can occur. The period of free oscillation is then measured, and is used with the geometry of the pendulum to calculate the moment of inertia.

Two types of pendulums are useful: the compound pendulum and the torsional pendulum. When using the compound pendulum, the body is supported from two overhead points by wires, illustrated in Fig. 3.4. The distance l is measured between the axis of support O–O and a parallel axis C–C through the center-of-gravity of the body. The moment of inertia about C–C is given by

$$I_{cc} = ml^2 \left[\left(\frac{\tau_0}{2\pi}\right)^2 \left(\frac{g}{l}\right) - 1 \right] \quad (3.18)$$

where τ_0 is the period of oscillation in seconds, l is the pendulum length in inches, g is the gravitational acceleration in in./sec^2, and m is the mass in lb-sec^2/in., yielding a moment of inertia in lb-in.-sec^2.

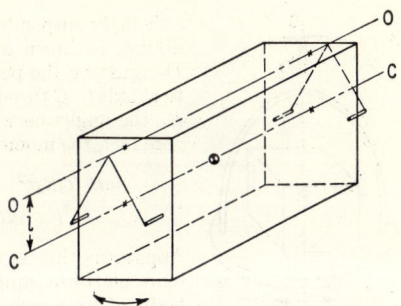

Fig. 3.4. Compound pendulum method of determining moment of inertia. The period of oscillation of the test body about the horizontal axis O–O and the perpendicular distance l between the axis O–O and the parallel axis C–C through the center-of-gravity of the test body give I_{cc} by Eq. (3.18).

The accuracy of the above method is dependent upon the accuracy with which the distance l is known. Since the center-of-gravity often is an inaccessible point, a direct measurement of l may not be practicable. However, a change in l can be measured quite readily. If the experiment is repeated with a different support axis O'–O', the length l becomes $l + \Delta l$ and the period of oscillation becomes τ_0'. Then, the distance l can be written in terms of Δl, and the two periods τ_0, τ_0':

$$l = \Delta l \left[\frac{(\tau_0'^2/4\pi^2)(g/\Delta l) - 1}{[(\tau_0^2 - \tau_0'^2)/4\pi^2][g/\Delta l] - 1} \right] \quad (3.19)$$

This value of l can be substituted into Eq. (3.18) to compute I_{cc}.

Note that accuracy is not achieved if l is much larger than the radius of gyration ρ_c of the body about the axis C–C ($I_{cc} = m\rho_c^2$). If l is large, then $(\tau_0/2\pi)^2 \simeq l/g$ and the expression in brackets in Eq. (3.18) is very small; thus, it is sensitive to small errors in the measurement of both τ_0 and l. Consequently, it is highly desirable that the distance l be chosen as small as convenient, preferably with the axis O–O passing through the body.

A torsional pendulum may be constructed with the test body suspended by a single torsional spring (in practice, a rod or wire) of known stiffness, or by three flexible wires. A solid body supported by a single torsional spring is shown in Fig. 3.5. From the known torsional stiffness k_t and the measured period of torsional oscillation τ, the moment of inertia of the body about the vertical torsional axis is

$$I_{cc} = \frac{k_t \tau^2}{4\pi^2} \quad (3.20)$$

A platform may be constructed below the torsional spring to carry the bodies to be measured, as shown in Fig. 3.6. By repeating the experiment with two different bodies placed on the platform, it becomes unnecessary to measure the torsional stiffness k_t. If a body with a *known* moment of inertia I_1 is placed on the platform and an oscillation period τ_1 results, the moment of inertia I_2 of a body which produces a period τ_2 is given by

$$I_2 = I_1 \left[\frac{(\tau_2/\tau_0)^2 - 1}{(\tau_1/\tau_0)^2 - 1} \right] \quad (3.21)$$

where τ_0 is the period of the pendulum comprised of platform alone.

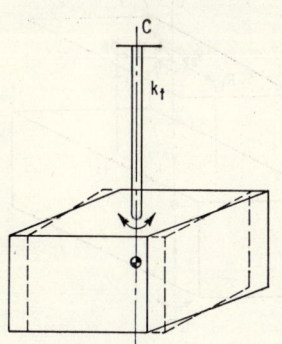

Fig. 3.5. Torsional pendulum method of determining moment of inertia. The period of torsional oscillation of the test body about the vertical axis C–C passing through the center-of-gravity and the torsional spring constant k_t give I_{cc} by Eq. (3.20).

VIBRATION OF A RESILIENTLY SUPPORTED RIGID BODY

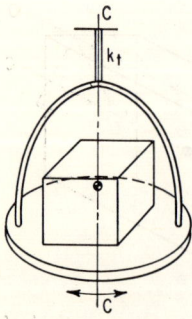

FIG. 3.6. A variation of the torsional pendulum method shown in Fig. 3.5 wherein a light platform is used to carry the test body. The moment of inertia I_{cc} is given by Eq. (3.20).

A body suspended by three flexible wires, called a trifilar pendulum, as shown in Fig. 3.7, offers some utilitarian advantages. Designating the perpendicular distances of the wires to the vertical axis C-C through the center-of-gravity of the body by R_1, R_2, R_3, the angles between wires by ϕ_1, ϕ_2, ϕ_3, and the length of each wire by l, the moment of inertia about axis C-C is

$$I_{cc} = \frac{mgR_1R_2R_3\tau^2}{4\pi^2 l} \frac{R_1\sin\phi_1 + R_2\sin\phi_2 + R_3\sin\phi_3}{R_2R_3\sin\phi_1 + R_1R_3\sin\phi_2 + R_1R_2\sin\phi_3} \quad (3.22)$$

Apparatus that is more convenient for repeated use embodies a light platform supported by three equally spaced wires. The body whose moment of inertia is to be measured is placed on the platform with its center-of-gravity equidistant from the wires. Thus $R_1 = R_2 = R_3 = R$ and $\phi_1 = \phi_2 = \phi_3 = 120°$. Substituting these relations in Eq. (3.22), the moment of inertia about the vertical axis C-C is

$$I_{cc} = \frac{mgR^2\tau^2}{4\pi^2 l} \quad (3.23)$$

where the mass m is the sum of the masses of the test body and the platform. The moment of inertia of the platform is subtracted from the test result to obtain the moment of inertia of the body being measured. It becomes unnecessary to know the distances R and l in Eq. (3.23) if the period of oscillation is measured with the platform empty, with the body being measured on the platform, and with a second body of known mass m_1 and known moment of inertia I_1 on the platform. Then the desired moment of inertia I_2 is

$$I_2 = I_1 \left[\frac{[1 + (m_2/m_0)][\tau_2/\tau_0]^2 - 1}{[1 + (m_1/m_0)][\tau_1/\tau_0]^2 - 1} \right] \quad (3.24)$$

where m_0 is the mass of the unloaded platform, m_2 is the mass of the body being measured, τ_0 is the period of oscillation with the platform unloaded, τ_1 is the period when loaded with known body of mass m_1, and τ_2 is the period when loaded with the unknown body of mass m_2.

EXPERIMENTAL DETERMINATION OF PRODUCT OF INERTIA. The experimental determination of a product of inertia usually requires the measurement of moments of inertia. (An exception is the balancing machine technique described later.) If possible, symmetry of the body is used to locate directions of principal inertial axes, thereby simplifying the relationship between the moments of inertia as known and the products of inertia to be found. Several alternative procedures are described below, depending on the number of principal inertia axes whose directions are known. Knowledge of two principal axes implies a knowledge of all three since they are mutually perpendicular.

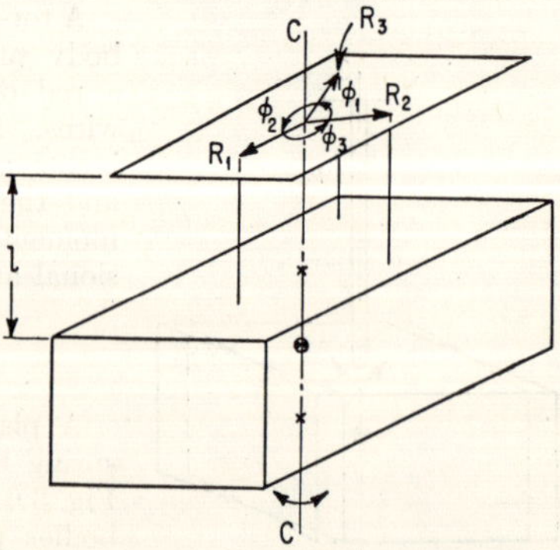

If the directions of all three principal axes (X', Y', Z') are known and it is desirable to use another set of axes (X, Y, Z), Eqs. (3.16) and (3.17) may be simplified because the products of inertia with respect to the principal directions are zero. First, the three principal moments of inertia $(I_{x'x'}, I_{y'y'}, I_{z'z'})$ are measured by one of the above techniques; then

FIG. 3.7. Trifilar pendulum method of determining moment of inertia. The period of torsional oscillation of the test body about the vertical axis C-C passing through the center-of-gravity and the geometry of the pendulum give I_{cc} by Eq. (3.22); with a simpler geometry, I_{cc} is given by Eq. (3.23).

the moments of inertia with respect to the X, Y, Z axes are

$$I_{xx} = \lambda_{xx'}{}^2 I_{x'x'} + \lambda_{xy'}{}^2 I_{y'y'} + \lambda_{xz'}{}^2 I_{z'z'}$$
$$I_{yy} = \lambda_{yx'}{}^2 I_{x'x'} + \lambda_{yy'}{}^2 I_{y'y'} + \lambda_{yz'}{}^2 I_{z'z'} \quad (3.25)$$
$$I_{zz} = \lambda_{zx'}{}^2 I_{x'x'} + \lambda_{zy'}{}^2 I_{y'y'} + \lambda_{zz'}{}^2 I_{z'z'}$$

The products of inertia with respect to the X, Y, Z axes are

$$-I_{xy} = \lambda_{xx'}\lambda_{yx'} I_{x'x'} + \lambda_{xy'}\lambda_{yy'} I_{y'y'} + \lambda_{xz'}\lambda_{yz'} I_{z'z'}$$
$$-I_{xz} = \lambda_{xx'}\lambda_{zx'} I_{x'x'} + \lambda_{xy'}\lambda_{zy'} I_{y'y'} + \lambda_{xz'}\lambda_{zz'} I_{z'z'} \quad (3.26)$$
$$-I_{yz} = \lambda_{yx'}\lambda_{zx'} I_{x'x'} + \lambda_{yy'}\lambda_{zy'} I_{y'y'} + \lambda_{yz'}\lambda_{zz'} I_{z'z'}$$

The direction of one principal axis Z may be known from symmetry. The axis through the center-of-gravity perpendicular to the plane of symmetry is a principal axis. The product of inertia with respect to X and Y axes, located in the plane of symmetry, is determined by first establishing another axis X' at a counterclockwise angle θ from X, as shown in Fig. 3.8. If the three moments of inertia $I_{xx}, I_{x'x'}$, and I_{yy} are measured by any applicable means, the product of inertia I_{xy} is

$$I_{xy} = \frac{I_{xx}\cos^2\theta + I_{yy}\sin^2\theta - I_{x'x'}}{\sin 2\theta} \quad (3.27)$$

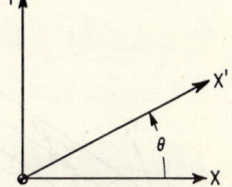

Fig. 3.8. Axes required for determining the product of inertia with respect to the axes X and Y when Z is a principal axis of inertia. The moments of inertia about the axes X, Y, and X', where X' is in the plane of X and Y at a counterclockwise angle θ from X, give I_{xy} by Eq. (3.27).

where $0 < \theta < \pi$. For optimum accuracy, θ should be approximately $\pi/4$ or $3\pi/4$. Since the third axis Z is a principal axis, I_{xz} and I_{yz} are zero.

Another method is illustrated in Fig. 3.9. The plane of the X and Z axes is a plane of symmetry, or the Y axis is otherwise known to be a principal axis of inertia. For determining I_{xz}, the body is suspended by a cable so that the Y axis is horizontal and the Z axis is vertical. Torsional stiffness about the Z axis is provided by four springs acting in the Y direction at the points shown. The body is oscillated about the Z axis with various positions of the springs so that the angle θ can be varied. The spring stiffnesses and locations must be such that there is no net force in the Y direction due to a rotation about the Z axis. In general, there is coupling between rotations about the X and Z axes, with the result that oscillations about both axes occur as a result of an initial rotational displacement about the Z axis. At some particular value of $\theta = \theta_0$, the two rotations are uncoupled; i.e., oscillation about the Z axis does not cause oscillation about the X axis. Then

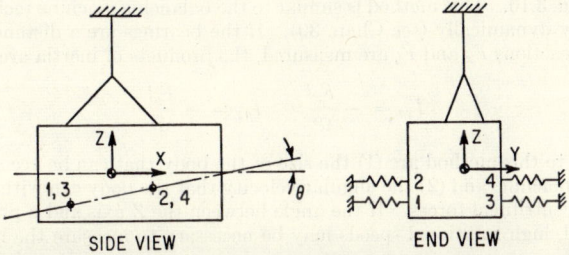

Fig. 3.9. Method of determining the product of inertia with respect to the axes X and Z when Y is a principal axis of inertia. The test body is oscillated about the vertical Z axis with torsional stiffness provided by the four springs acting in the Y direction at the points shown. There should be no net force on the test body in the Y direction due to a rotation about the Z axis. The angle θ is varied until, at some value of $\theta = \theta_0$, oscillations about X and Z are uncoupled. The angle θ_0 and the moment of inertia about the Z axis give I_{xz} by Eq. (3.28).

$$I_{xz} = I_{zz} \tan \theta_0 \qquad (3.28)$$

The moment of inertia I_{zz} can be determined by one of the methods described under *Experimental Determination of Moments of Inertia*.

When the moments and product of inertia with respect to a pair of axes X and Z in a principal plane of inertia XZ are known, the orientation of a principal axis P is given by

$$\theta_p = \tfrac{1}{2} \tan^{-1}\left(\frac{2I_{xz}}{I_{zz} - I_{xx}}\right) \qquad (3.29)$$

where θ_p is the counterclockwise angle from the X axis to the P axis. The second principal axis in this plane is at $\theta_p + 90°$.

Consider the determination of products of inertia when the directions of all principal axes of inertia are unknown. In one method, the moments of inertia about two independent sets of three mutually perpendicular axes are measured, and the direction cosines between these sets of axes are known from the positions of the axes. The values for the six moments of inertia and the nine direction cosines are then substituted into Eqs. (3.16) and (3.17). The result is six linear equations in the six unknown products of inertia, from which the values of the desired products of inertia may be found by simultaneous solution of the equations. This method leads to experimental errors of relatively large magnitude because each product of inertia is, in general, a function of all six moments of inertia, each of which contains an experimental error.

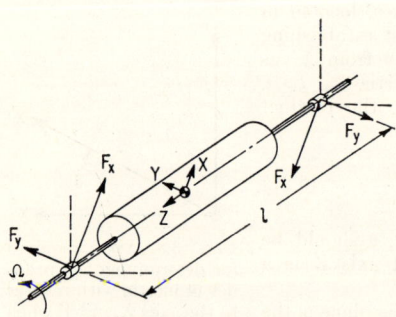

Fig. 3.10. Balancing machine technique for determining products of inertia. The test body is rotated about the Z axis with angular velocity Ω. The dynamic reactions F_x and F_y measured at the bearings, which are a distance l apart, give I_{xz} and I_{yz} by Eq. (3.30).

An alternative method is based upon the knowledge that one of the principal moments of inertia of a body is the largest and another is the smallest that can be obtained for any axis through the center-of-gravity. A trial-and-error procedure can be used to locate the orientation of the axis through the center-of-gravity having the maximum and/or minimum moment of inertia. After one or both are located, the moments and products of inertia for any set of axes are found by the techniques previously discussed.

The products of inertia of a body also may be determined by rotating the body at a constant angular velocity Ω about an axis passing through the center of gravity, as illustrated in Fig. 3.10. This method is similar to the balancing machine technique used to balance a body dynamically (see Chap. 39). If the bearings are a distance l apart and the dynamic reactions F_x and F_y are measured, the products of inertia are

$$I_{xz} = -\frac{F_x l}{\Omega^2} \qquad I_{yz} = -\frac{F_y l}{\Omega^2} \qquad (3.30)$$

Limitations to this method are (1) the size of the body that can be accommodated by the balancing machine and (2) the angular velocity that the body can withstand without damage from centrifugal forces. If the angle between the Z axis and a principal axis of inertia is small, high rotational speeds may be necessary to measure the reaction forces accurately.

PROPERTIES OF RESILIENT SUPPORTS

A resilient support is considered to be a three-dimensional element having two terminals or end connections. When the end connections are moved one relative to the other in any direction, the element resists such motion. In this chapter, the element is considered to be massless; the force that resists relative motion across the element is considered to consist of a spring force that is directly proportional to the relative displacement (deflection across the element) and a damping force that is directly proportional to the relative velocity (velocity across the element). Such an element is defined as a *linear resilient support*. Nonlinear elements are discussed in Chap. 4; elements with mass are discussed in Chap. 30; and nonlinear damping is discussed in Chaps. 2 and 30.

In a single degree-of-freedom system or in a system having constraints on the paths of motion of elements of the system (Chap. 2), the resilient element is constrained to deflect in a given direction and the properties of the element are defined with respect to the force opposing motion in this direction. In the absence of such constraints, the application of a force to a resilient element generally causes a motion in a different direction. The *principal elastic axes* of a resilient element are those axes for which the element, when unconstrained, experiences a deflection colineal with the direction of the applied force. Any axis of symmetry is a principal elastic axis.

In rigid body dynamics, the rigid body sometimes vibrates in modes that are coupled by the properties of the resilient elements as well as by their location. For example, if the body experiences a static displacement x in the direction of the X axis only, a resilient element opposes this motion by exerting a force $k_{xx}x$ on the body in the direction of the X axis where one subscript on the spring constant k indicates the direction of the force exerted by the element and the other subscript indicates the direction of the deflection. If the X direction is not a principal elastic direction of the element and the body experiences a static displacement x in the X direction, the body is acted upon by a force $k_{yx}x$ in the Y direction if no displacement y is permitted. The stiffnesses have reciprocal properties; i.e., $k_{xy} = k_{yx}$. In general, the stiffnesses in the directions of the coordinate axes can be expressed in terms of (1) principal stiffnesses and (2) the angles between the coordinate axes and the principal elastic axes of the element. (See Chap. 30 for a detailed discussion of a biaxial stiffness element.) Therefore, the stiffness of a resilient element can be represented pictorially by the combination of three mutually perpendicular, idealized springs oriented along the principal elastic directions of the resilient element. Each spring has a stiffness equal to the principal stiffness represented.

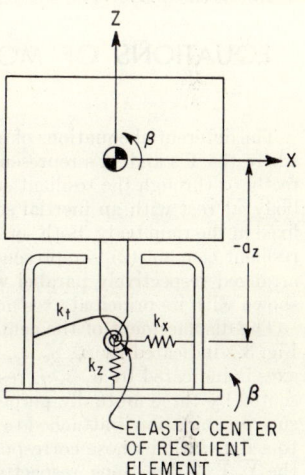

Fig. 3.11. Pictorial representation of the properties of an undamped resilient element in the XZ plane including a torsional spring k_t. An analysis of the motion of the supported body in the XZ plane shows that the torsional spring can be neglected if $k_t \ll a_z^2 k_x$.

A resilient element is assumed to have damping properties such that each spring representing a value of principal stiffness is paralleled by an idealized viscous damper, each damper representing a value of principal damping. Hence, coupling through damping exists in a manner similar to coupling through stiffness. Consequently, the viscous damping coefficient c is analogous to the spring coefficient k; i.e., the force exerted by the damping of the resilient element in response to a velocity $\dot{x}$ is $c_{xx}\dot{x}$ in the direction of the X axis and $c_{yx}\dot{x}$ in the direction of the Y axis if $\dot{y}$ is zero. Reciprocity exists; i.e., $c_{xy} = c_{yx}$.

The point of intersection of the principal elastic axes of a resilient element is designated as the *elastic center of the resilient element*. The elastic center is important since it defines the theoretical point location of the resilient element for use in the equations of motion of a resiliently supported rigid body. For example, the torque on the rigid body about the Y axis due to a force $k_{xx}x$ transmitted by a resilient element in the X direction is $k_{xx}a_z x$, where a_z is the Z coordinate of the elastic center of the resilient element.

In general, it is assumed that a resilient element is attached to the rigid body by means of "ball joints"; i.e., the resilient element is incapable of applying a couple to the body. If this assumption is not made, a resilient element would be represented not only by translational springs and dampers along the principal elastic axes but also by torsional springs and dampers resisting rotation about the principal elastic directions.

Figure 3.11 shows that the torsional elements usually can be neglected. The torque which acts on the rigid body due to a rotation β of the body and a rotation **β** of the support is $(k_t + a_z^2 k_x)(\beta - \boldsymbol{\beta})$, where k_t is the torsional spring constant in the β direction. The torsional stiffness k_t usually is much smaller than $a_z^2 k_x$ and can be neglected. Treatment of the general case indicates that if the torsional stiffnesses of the resilient element are small compared with the product of the translational stiffnesses times the square of distances from the elastic center of the resilient element to the center-of-gravity of the rigid body, the torsional stiffnesses have a negligible effect on the vibrational behavior of the body. The treatment of torsional dampers is completely analogous.

EQUATIONS OF MOTION FOR A RESILIENTLY SUPPORTED RIGID BODY

The differential equations of motion for the rigid body are given by Eqs. (3.2) and (3.3) where the **F**'s and **M**'s represent the forces and moments acting on the body, either directly or through the resilient supporting elements. Figure 3.12 shows a view of a rigid body at rest with an inertial set of axes $\bar{X}, \bar{Y}, \bar{Z}$ and a coincident set of axes X, Y, Z fixed in the rigid body, both sets of axes passing through the center-of-mass. A typical resilient element (2) is represented by parallel spring and viscous damper combinations arranged respectively parallel with the $\bar{X}, \bar{Y}, \bar{Z}$ axes. Another resilient element (1) is shown with its principal axes not parallel with $\bar{X}, \bar{Y}, \bar{Z}$.

The displacement of the center-of-gravity of the body in the $\bar{X}, \bar{Y}, \bar{Z}$ directions is in Fig. 3.1 indicated by x_c, y_c, z_c, respectively; and rotation of the rigid body about these axes is indicated by α, β, γ, respectively. In Fig. 3.12, each resilient element is represented by three mutually perpendicular spring-damper combinations. One end of each such combination is attached to the rigid body; the other end is considered to be attached to a foundation whose corresponding translational displacement is defined by u, v, w in the $\bar{X}, \bar{Y}, \bar{Z}$ directions, respectively, and whose rotational displacement about these axes is defined by **α, β, γ**, respectively. The point of attachment of each of the idealized resilient elements is located at the coordinate distances a_x, a_y, a_z of the elastic center of the resilient element.

Consider the rigid body to experience a translational displacement x_c of its center-of-gravity and no other displacement, and neglect the effects of the viscous dampers. The force developed by a resilient element has the effect of a force $-k_{xx}(x_c - u)$ in the X direction, a moment $k_{xx}(x_c - u)a_y$ in the γ coordinate (about the Z axis) and a moment $-k_{xx}(x_c - u)a_z$ in the β coordinate (about the Y axis). The coupling stiffness causes a force $-k_{xy}(x_c - u)$ in the Y direction and a force $-k_{xz}(x_c - u)$ in the Z direction. These forces have the moments: $k_{xy}(x_c - u)a_z$ in the α coordinate; $-k_{xy}(x_c - u)a_x$ in the γ coordinate; $k_{xz}(x_c - u)a_x$ in the β coordinate; and $-k_{xz}(x_c - u)a_y$ in the α coordinate. By considering in a similar manner the forces and moments developed by a resilient element for successive displacements of the rigid body in the three translational and three rotational coordinates, and summing over the number of resilient elements, the equations of motion are written as follows:[5,6]

$$m\ddot{x}_c + \Sigma k_{xx}(x_c - u) + \Sigma k_{xy}(y_c - v) + \Sigma k_{xz}(z_c - w)$$
$$+ \Sigma(k_{xz}a_y - k_{xy}a_z)(\alpha - \boldsymbol{\alpha}) + \Sigma(k_{xx}a_z - k_{xz}a_x)(\beta - \boldsymbol{\beta})$$
$$+ \Sigma(k_{xy}a_x - k_{xx}a_y)(\gamma - \boldsymbol{\gamma}) = F_x \qquad (3.31a)$$

$$I_{xx}\ddot{\alpha} - I_{xy}\ddot{\beta} - I_{xz}\ddot{\gamma} + \Sigma(k_{xz}a_y - k_{xy}a_z)(x_c - u)$$
$$+ \Sigma(k_{yz}a_y - k_{yy}a_z)(y_c - v) + \Sigma(k_{zz}a_y - k_{yz}a_z)(z_c - w)$$
$$+ \Sigma(k_{yy}a_z^2 + k_{zz}a_y^2 - 2k_{yz}a_y a_z)(\alpha - \boldsymbol{\alpha})$$
$$+ \Sigma(k_{xz}a_y a_z + k_{yz}a_x a_z - k_{zz}a_x a_y - k_{xy}a_z^2)(\beta - \boldsymbol{\beta})$$
$$+ \Sigma(k_{xy}a_y a_z + k_{yz}a_x a_y - k_{yy}a_x a_z - k_{xz}a_y^2)(\gamma - \boldsymbol{\gamma}) = M_x \qquad (3.31b)$$

EQUATIONS OF MOTION FOR A RESILIENTLY SUPPORTED RIGID BODY

$$m\ddot{y}_c + \Sigma k_{xy}(x_c - u) + \Sigma k_{yy}(y_c - v) + \Sigma k_{yz}(z_c - w)$$
$$+ \Sigma(k_{yz}a_y - k_{yy}a_z)(\alpha - \mathbf{a}) + \Sigma(k_{xy}a_z - k_{yz}a_x)(\beta - \boldsymbol{\beta})$$
$$+ \Sigma(k_{yy}a_x - k_{xy}a_y)(\gamma - \boldsymbol{\gamma}) = F_y \quad (3.31c)$$

$$I_{yy}\ddot{\beta} - I_{xy}\ddot{\alpha} - I_{yz}\ddot{\gamma} + \Sigma(k_{xx}a_z - k_{xz}a_x)(x_c - u)$$
$$+ \Sigma(k_{xy}a_z - k_{yz}a_x)(y_c - v) + \Sigma(k_{xz}a_z - k_{zz}a_x)(z_c - w)$$
$$+ \Sigma(k_{xz}a_y a_z + k_{yz}a_x a_z - k_{zz}a_x a_y - k_{xy}a_z{}^2)(\alpha - \mathbf{a})$$
$$+ \Sigma(k_{xx}a_z{}^2 + k_{zz}a_x{}^2 - 2k_{xz}a_x a_z)(\beta - \boldsymbol{\beta})$$
$$+ \Sigma(k_{xy}a_x a_z + k_{xz}a_x a_y - k_{xx}a_y a_z - k_{yz}a_x{}^2)(\gamma - \boldsymbol{\gamma}) = M_y \quad (3.31d)$$

$$m\ddot{z}_c + \Sigma k_{xz}(x_c - u) + \Sigma k_{yz}(y_c - v) + \Sigma k_{zz}(z_c - w)$$
$$+ \Sigma(k_{zz}a_y - k_{yz}a_z)(\alpha - \mathbf{a}) + \Sigma(k_{xz}a_z - k_{zz}a_x)(\beta - \boldsymbol{\beta})$$
$$+ \Sigma(k_{yz}a_x - k_{xz}a_y)(\gamma - \boldsymbol{\gamma}) = F_z \quad (3.31e)$$

$$I_{zz}\ddot{\gamma} - I_{xz}\ddot{\alpha} - I_{yz}\ddot{\beta} + \Sigma(k_{xy}a_x - k_{xx}a_y)(x_c - u)$$
$$+ \Sigma(k_{yy}a_x - k_{xy}a_y)(y_c - v) + \Sigma(k_{yz}a_x - k_{xz}a_y)(z_c - w)$$
$$+ \Sigma(k_{xy}a_y a_z + k_{yz}a_x a_y - k_{yy}a_x a_z - k_{xz}a_y{}^2)(\alpha - \mathbf{a})$$
$$+ \Sigma(k_{xy}a_x a_z + k_{xz}a_x a_y - k_{xx}a_y a_z - k_{yz}a_x{}^2)(\beta - \boldsymbol{\beta})$$
$$+ \Sigma(k_{xx}a_y{}^2 + k_{yy}a_x{}^2 - 2k_{xy}a_x a_y)(\gamma - \boldsymbol{\gamma}) = M_z \quad (3.31f)$$

where the moments and products of inertia are defined by Eqs. (3.11) and (3.12) and the stiffness coefficients are defined as follows:

$$\begin{aligned}
k_{xx} &= k_p\lambda_{xp}{}^2 + k_q\lambda_{xq}{}^2 + k_r\lambda_{xr}{}^2 \\
k_{yy} &= k_p\lambda_{yp}{}^2 + k_q\lambda_{yq}{}^2 + k_r\lambda_{yr}{}^2 \\
k_{zz} &= k_p\lambda_{zp}{}^2 + k_q\lambda_{zq}{}^2 + k_r\lambda_{zr}{}^2 \\
k_{xy} &= k_p\lambda_{xp}\lambda_{yp} + k_q\lambda_{xq}\lambda_{yq} + k_r\lambda_{xr}\lambda_{yr} \\
k_{xz} &= k_p\lambda_{xp}\lambda_{zp} + k_q\lambda_{xq}\lambda_{zq} + k_r\lambda_{xr}\lambda_{zr} \\
k_{yz} &= k_p\lambda_{yp}\lambda_{zp} + k_q\lambda_{yq}\lambda_{zq} + k_r\lambda_{yr}\lambda_{zr}
\end{aligned} \quad (3.32)$$

where the λ's are the cosines of the angles between the principal elastic axes of the resilient supporting elements and the coordinate axes. For example, λ_{xp} is the cosine of the angle between the X axis and the P axis of principal stiffness.

The equations of motion, Eqs. (3.31), do not include forces applied to the rigid body by damping forces from the resilient elements. To include damping, appropriate damping terms analogous to the corresponding stiffness terms are added to each equation. For example, Eq. (3.31a) would become

$$m\ddot{x}_c + \Sigma c_{xx}(\dot{x}_c - \dot{u}) + \Sigma k_{xx}(x_c - u) + \cdots$$
$$+ \Sigma(c_{xz}a_y - c_{xy}a_z)(\dot{\alpha} - \dot{\mathbf{a}}) + \Sigma(k_{xz}a_y - k_{xy}a_z)(\alpha - \mathbf{a}) + \cdots = F_x \quad (3.31a')$$

where
$$c_{xx} = c_p\lambda_{xp}{}^2 + c_q\lambda_{xq}{}^2 + c_r\lambda_{xr}{}^2$$
$$c_{xy} = c_p\lambda_{xp}\lambda_{yp} + c_q\lambda_{xq}\lambda_{yq} + c_r\lambda_{xr}\lambda_{yr}$$

The number of degrees-of-freedom of a vibrational system is the minimum number of coordinates necessary to define completely the positions of the mass elements of the system in space. The system of Fig. 3.12 requires a minimum of six coordinates $(x_c, y_c, z_c, \alpha, \beta, \gamma)$ to define the position of the rigid body in space; thus, the system is said to vibrate in six

3-14   VIBRATION OF A RESILIENTLY SUPPORTED RIGID BODY

degrees-of-freedom. Equations (3.31) may be solved simultaneously for the three components x_c, y_c, z_c of the center-of-gravity displacement and the three components α, β, γ of the rotational displacement of the rigid body. In most practical instances, the equations are simplified considerably by one or more of the following simplifying conditions:

1. The reference axes X, Y, Z are selected to coincide with the principal inertial axes of the body; then
$$I_{xy} = I_{xz} = I_{yz} = 0 \tag{3.33}$$

2. The resilient supporting elements are so arranged that one or more planes of symmetry exist; i.e., motion parallel to the plane of symmetry has no tendency to excite motion perpendicular to it, or rotation about an axis lying in the plane does not

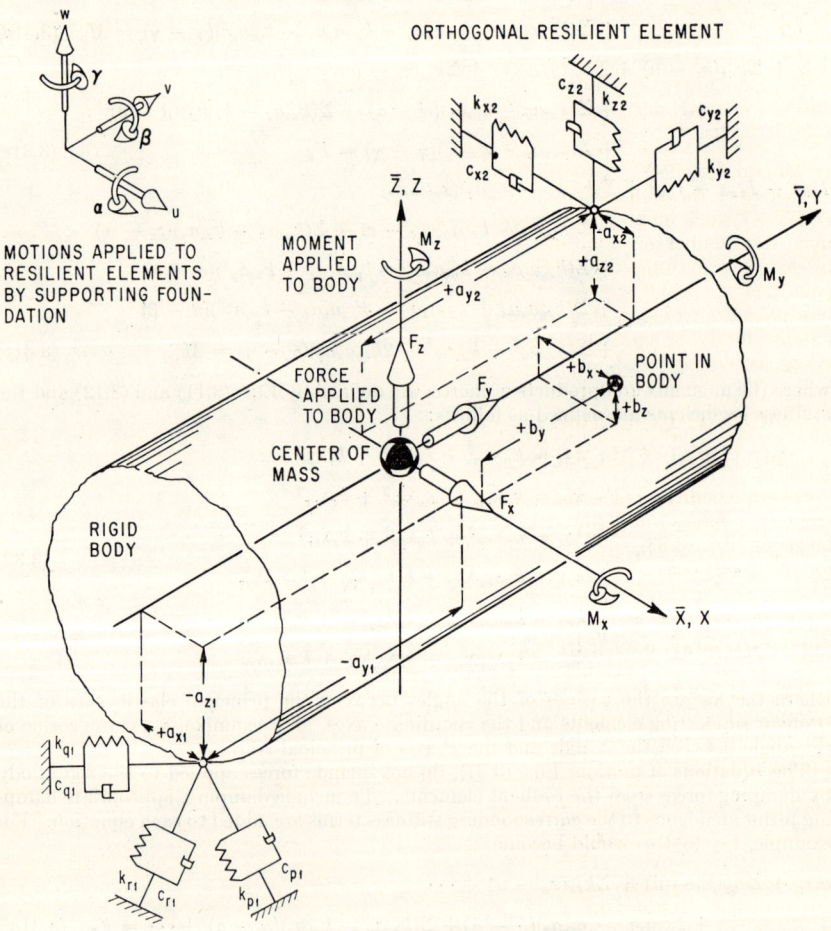

FIG. 3.12. Rigid body at rest supported by resilient elements, with inertial axes $\overline{X}$, $\overline{Y}$, $\overline{Z}$ and coincident reference axes X, Y, Z passing through the center-of-mass. The forces F_x, F_y, F_z and the moments M_x, M_y, M_z are applied directly to the body; the translations u, v, w and rotations **α**, **β**, **γ** in and about the X, Y, Z axes, respectively, are applied to the resilient elements located at the coordinates a_x, a_y, a_z. The principal directions of resilient element (2) are parallel to the $\overline{X}$, $\overline{Y}$, $\overline{Z}$ axes (orthogonal) and those of resilient element (1) are not parallel to the $\overline{X}$, $\overline{Y}$, $\overline{Z}$ axes (inclined).

excite motion parallel to the plane. For example, in Eq. (3.31a), motion in the XY plane does not tend to excite motion in the XZ or YZ plane if Σk_{xz}, $\Sigma(k_{xz}a_y - k_{xy}a_z)$, and $\Sigma(k_{xx}a_z - k_{xz}a_x)$ are zero.

3. The principal elastic axes P, Q, R of all resilient supporting elements are orthogonal with the reference axes X, Y, Z of the body, respectively. Then, in Eqs. (3.32),

$$k_{xx} = k_p = k_x \qquad k_{yy} = k_q = k_y \qquad k_{zz} = k_r = k_z$$
$$k_{xy} = k_{xz} = k_{yz} = 0 \tag{3.34}$$

where k_x, k_y, k_z are defined for use when orthogonality exists. The supports are then called *orthogonal supports*.

4. The forces F_x, F_y, F_z and moments M_x, M_y, M_z are applied directly to the body and there are no motions ($u = v = w = \alpha = \beta = \gamma = 0$) of the foundation; or alternatively, the forces and moments are zero and excitation results from motion of the foundation.

In general, the effect of these simplifications is to reduce the numbers of terms in the equations and, in some instances, to reduce the number of equations that must be solved simultaneously. Simultaneous equations indicate coupled modes; i.e., motion cannot exist in one coupled mode independently of motion in other modes which are coupled to it.

MODAL COUPLING AND NATURAL FREQUENCIES

Several conditions of symmetry resulting from zero values for the product of inertia terms in Eq. (3.33) are discussed in the following sections.

ONE PLANE OF SYMMETRY WITH ORTHOGONAL RESILIENT SUPPORTS

When the YZ plane of the rigid body system in Fig. 3.12 is a plane of symmetry, the following terms in the equations of motion are zero:

$$\Sigma k_{yy}a_x = \Sigma k_{zz}a_x = \Sigma k_{yy}a_xa_z = \Sigma k_{zz}a_xa_y = 0$$
$$\tag{3.35}$$

Introducing the further simplification that the principal elastic axes of the resilient elements are parallel with the reference axes, Eqs. (3.34) apply. Then the motions in the three coordinates y_c, z_c, α are coupled but are independent of motion in any of the other coordinates; furthermore, the other three coordinates x_c, β, γ also are coupled. For example, Fig. 3.13 illustrates a resiliently supported rigid body, wherein

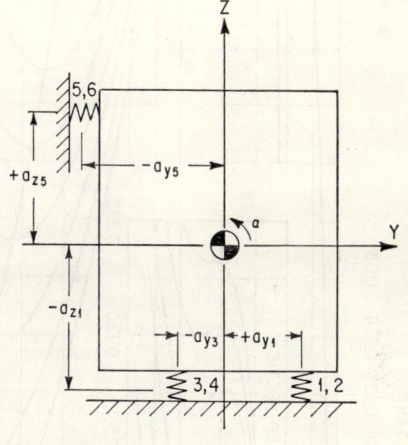

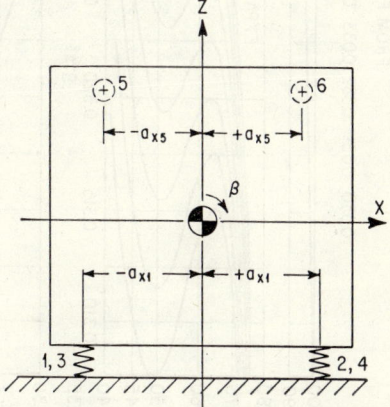

Fig. 3.13. Example of a rigid body on orthogonal resilient supporting elements with one plane of symmetry. The YZ plane is a plane of symmetry since each resilient element has properties identical to those of its mirror image in the YZ plane; i.e., $k_{x1} = k_{x2}$, $k_{x3} = k_{x4}$, $k_{x5} = k_{x6}$, etc. The conditions satisfied are Eqs. (3.33) to (3.35).

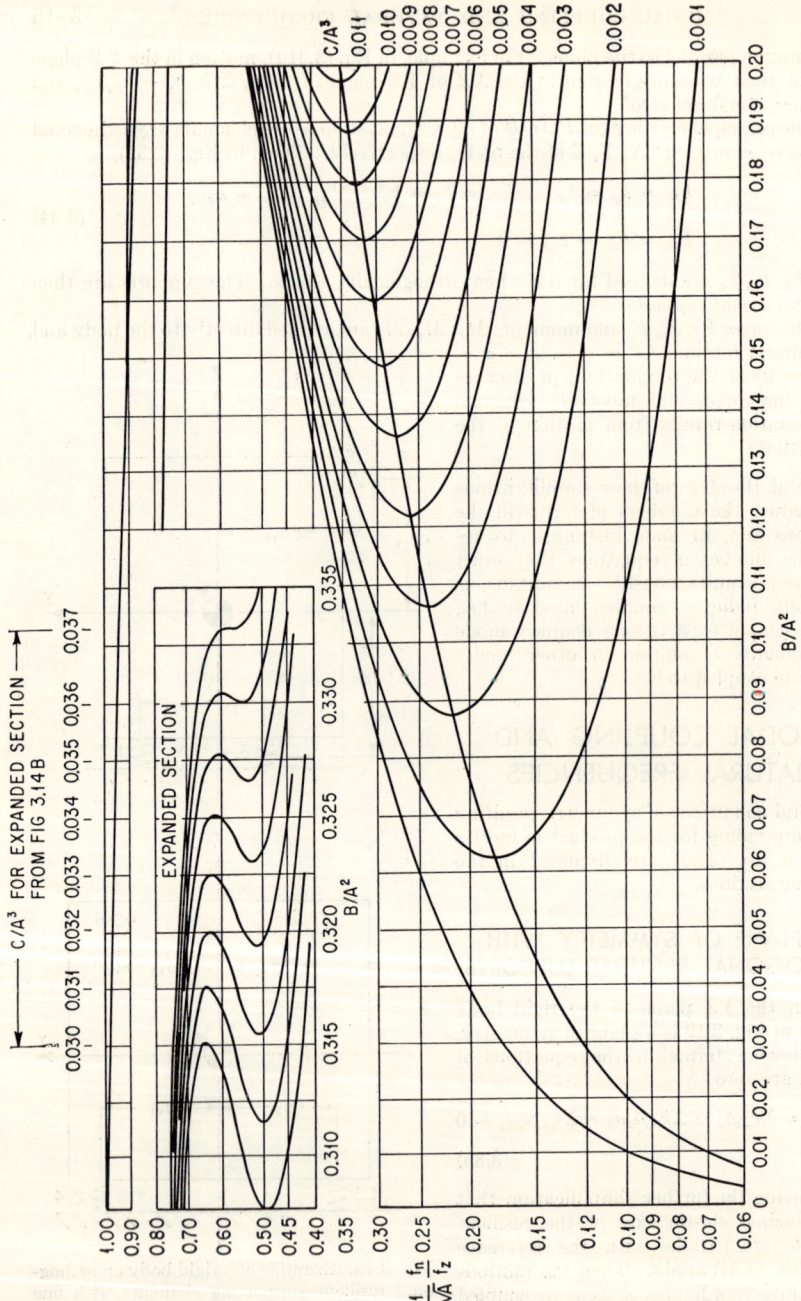

FIG. 3.14A. Graphical method of determining solutions of the cubic Eq. (3.36). Calculate A, B, C for the appropriate set of coupled coordinates, enter the abscissa at B/A^2 (values less than 0.2 on Fig. 3.14A, values greater than 0.2 on Fig. 3.14B), and read three values of $(f_n/f_z)/\sqrt{A}$ from the curve having the appropriate value of C/A^3.

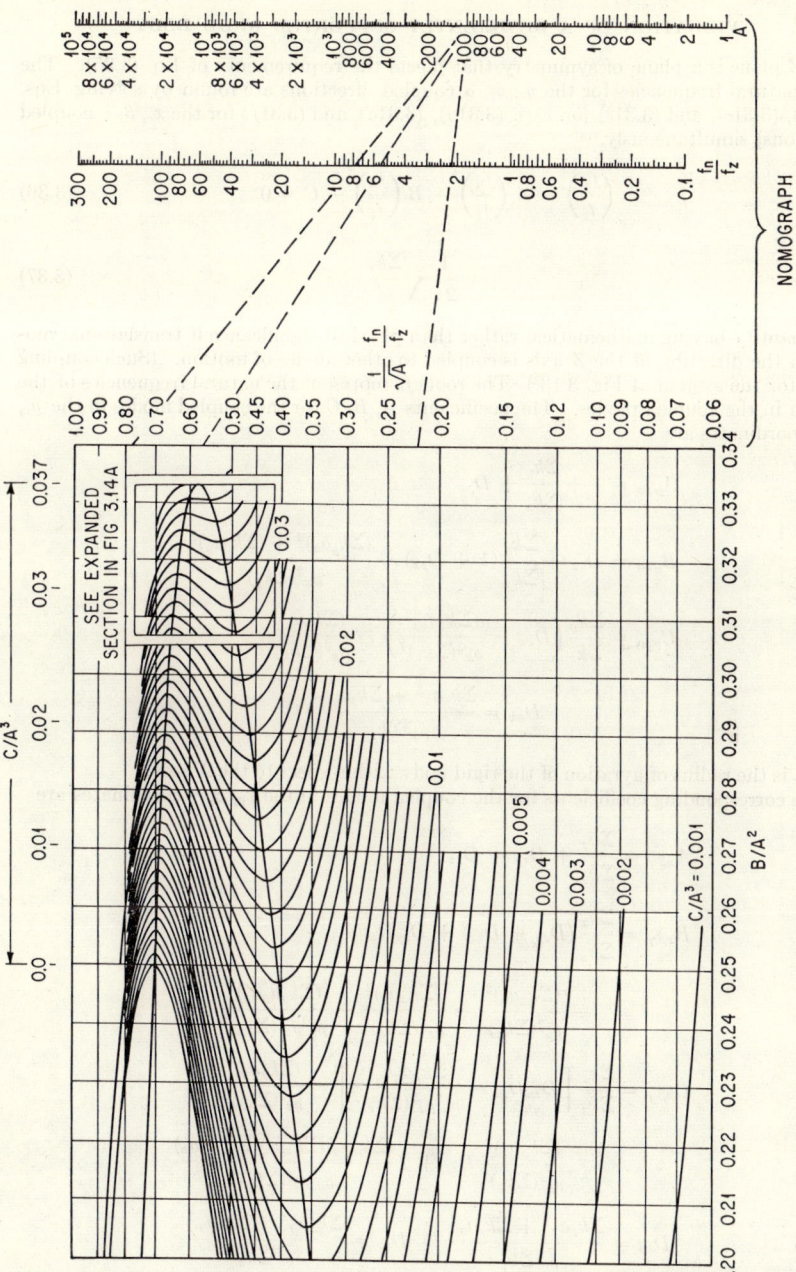

FIG. 3.14B. Using the above nomograph with values of $(f_n/f_z)/\sqrt{A}$ (see Fig. 3.14A) on the left scale of the nomograph to the value of A on the right scale, as indicated by the dotted lines. The three roots f_n/f_z of Eq. (3.36) are given by the intercept of these dotted lines with the center scale of the nomograph. (*After F. F. Vane.*)

3–17

the YZ plane is a plane of symmetry that meets the requirements of Eq. (3.35). The three natural frequencies for the y_c, z_c, α coupled directions are found by solving Eqs. (3.31b), (3.31c), and (3.31e) [or Eqs. (3.31a), (3.31d), and (3.31f) for the x_c, β, γ coupled directions] simultaneously.[7,8]

$$\left(\frac{f_n}{f_z}\right)^6 - A\left(\frac{f_n}{f_z}\right)^4 + B\left(\frac{f_n}{f_z}\right)^2 - C = 0 \tag{3.36}$$

where
$$f_z = \frac{1}{2\pi}\sqrt{\frac{\Sigma k_z}{m}} \tag{3.37}$$

is a quantity having mathematical rather than physical significance if translational motion in the direction of the Z axis is coupled to other modes of motion. (Such coupling exists for the system of Fig. 3.13.) The roots f_n represent the natural frequencies of the system in the coupled modes. The coefficients A, B, C for the coupled modes in the y_c, z_c, α coordinates are

$$A_{yz\alpha} = 1 + \frac{\Sigma k_y}{\Sigma k_z} + D_{zx}$$

$$B_{yz\alpha} = D_{zx} + \frac{\Sigma k_y}{\Sigma k_z}(1 + D_{zx}) - \frac{(\Sigma k_y a_z)^2 + (\Sigma k_z a_y)^2}{\rho_x^2 (\Sigma k_z)^2}$$

$$C_{yz\alpha} = \frac{\Sigma k_y}{\Sigma k_z}\left(D_{zx} - \frac{(\Sigma k_z a_y)^2}{\rho_x^2(\Sigma k_z)^2}\right) - \frac{(\Sigma k_y a_z)^2}{\rho_x^2(\Sigma k_z)^2}$$

where
$$D_{zx} = \frac{\Sigma k_y a_z^2 + \Sigma k_z a_y^2}{\rho_x^2 \Sigma k_z}$$

and ρ_x is the radius of gyration of the rigid body with respect to the X axis.

The corresponding coefficients for the coupled modes in the x_c, β, γ coordinates are

$$A_{x\beta\gamma} = \frac{\Sigma k_x}{\Sigma k_z} + D_{zy} + D_{zz}$$

$$B_{x\beta\gamma} = \frac{\Sigma k_x}{\Sigma k_z}(D_{zy} + D_{zz}) + D_{zy}D_{zz}$$
$$- \frac{(\Sigma k_x a_z)^2}{\rho_y^2(\Sigma k_z)^2} - \frac{(\Sigma k_x a_y)^2}{\rho_z^2(\Sigma k_z)^2} - \frac{(\Sigma k_x a_y a_z)^2}{\rho_y^2 \rho_z^2(\Sigma k_z)^2}$$

$$C_{x\beta\gamma} = \frac{\Sigma k_x}{\Sigma k_z}\left[D_{zy}D_{zz} - \frac{(\Sigma k_x a_y a_z)^2}{\rho_y^2 \rho_z^2 (\Sigma k_z)^2}\right] - \frac{(\Sigma k_x a_y)^2}{\rho_z^2(\Sigma k_z)^2}D_{zy}$$
$$- \frac{(\Sigma k_x a_z)^2}{\rho_y^2(\Sigma k_z)^2}D_{zz} + 2\frac{(\Sigma k_x a_y)(\Sigma k_x a_z)(\Sigma k_x a_y a_z)}{\rho_y^2 \rho_z^2(\Sigma k_z)^3}$$

where
$$D_{zy} = \frac{\Sigma k_x a_z^2 + \Sigma k_z a_x^2}{\rho_y^2 \Sigma k_z} \qquad D_{zz} = \frac{\Sigma k_x a_y^2 + \Sigma k_y a_x^2}{\rho_z^2 \Sigma k_z}$$

and ρ_y, ρ_z are the radii of gyration of the rigid body with respect to the Y, Z axes.

The roots of the cubic equation Eq. (3.36) may be found graphically from Fig. 3.14.[7] The coefficients A, B, C are first calculated from the above relations for the appropriate set of coupled coordinates. Figure 3.14 is entered on the abscissa scale at the appropriate value for the quotient B/A^2. Small values of B/A^2 are in Fig. 3.14A, and large values in Fig. 3.14B. The quotient C/A^3 is the parameter for the family of curves. Upon selecting the appropriate curve, three values of $(f_n/f_z)/\sqrt{A}$ are read from the ordinate and transferred to the left scale of the nomograph in Fig. 3.14B. Diagonal lines are drawn for each root to the value of A on the right scale, as indicated by dotted lines, and the roots f_n/f_z of the equation are indicated by the intercept of these dotted lines with the center scale of the nomograph.

The coefficients A, B, C can be simplified if all resilient elements have equal stiffness in the same direction. The stiffness coefficients always appear to equal powers in numerator and denominator, and lead to dimensionless ratios of stiffness. For n resilient elements, typical terms reduce as follows:

$$\frac{\Sigma k_y}{\Sigma k_z} = \frac{k_y}{k_z} \qquad \frac{\Sigma k_z a_y^2}{\rho_x^2 \Sigma k_z} = \frac{\Sigma a_y^2}{n \rho_x^2}$$

$$\frac{(\Sigma k_x a_y a_z)^2}{\rho_y^2 \rho_z^2 (\Sigma k_z)^2} = \left(\frac{k_x}{n k_z} \frac{\Sigma a_y a_z}{\rho_y \rho_z}\right)^2, \text{ etc.}$$

TWO PLANES OF SYMMETRY WITH ORTHOGONAL RESILIENT SUPPORTS

Two planes of symmetry may be achieved if, in addition to the conditions of Eqs. (3.33) to (3.35), the following terms of Eqs. (3.31) are zero:

$$\Sigma k_{xx} a_y = \Sigma k_{zz} a_y = \Sigma k_{xx} a_y a_z = 0 \quad (3.38)$$

Under these conditions, Eqs. (3.31) separate into two independent equations, Eqs. (3.31e) and (3.31f), and two sets each consisting of two coupled equations [Eqs. (3.31a) and (3.31d); Eqs. (3.31b) and (3.31c)]. The planes of symmetry are the XZ and YZ planes. For example, a common system is illustrated in Fig. 3.15, where four identical resilient supporting elements are located symmetrically about the Z axis in a plane not containing the center-of-gravity.[5,9,10] Coupling exists between translation in the X direction and rotation about the Y axis (x_c, β), as well as between translation in the Y direction and rotation about the X axis (y_c, α). Translation in the Z direction (z_c) and rotation about the Z axis (γ) are each independent of all other modes.

The natural frequency in the Z direction is found by solving Eq. (3.31e) to obtain Eq. (3.37), where $\Sigma k_{zz} = 4k_z$. The rotational natural frequency f_γ about the Z axis is found by solving Eq. (3.31f); it can be expressed with respect to the natural frequency in the direction of the Z axis:

$$\frac{f_\gamma}{f_z} = \sqrt{\frac{k_x}{k_z}\left(\frac{a_y}{\rho_z}\right)^2 + \frac{k_y}{k_z}\left(\frac{a_x}{\rho_z}\right)^2} \quad (3.39)$$

where ρ_z is the radius of gyration with respect to the Z axis.

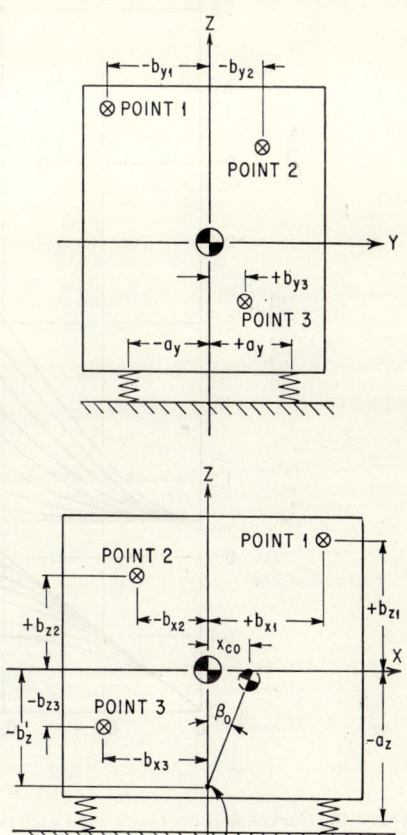

FIG. 3.15. Example of a rigid body on orthogonal resilient supporting elements with two planes of symmetry. The XZ and YZ planes are planes of symmetry since the four resilient supporting elements are identical and are located symmetrically about the Z axis. The conditions satisfied are Eqs. (3.33), (3.34), (3.35), and (3.38). At any single frequency, coupled vibration in the x_c, β direction due to X vibration of the foundation is equivalent to a pure rotation of the rigid body with respect to an axis of rotation as shown. Points 1, 2, and 3 refer to the example of Fig. 3.26.

3-20 VIBRATION OF A RESILIENTLY SUPPORTED RIGID BODY

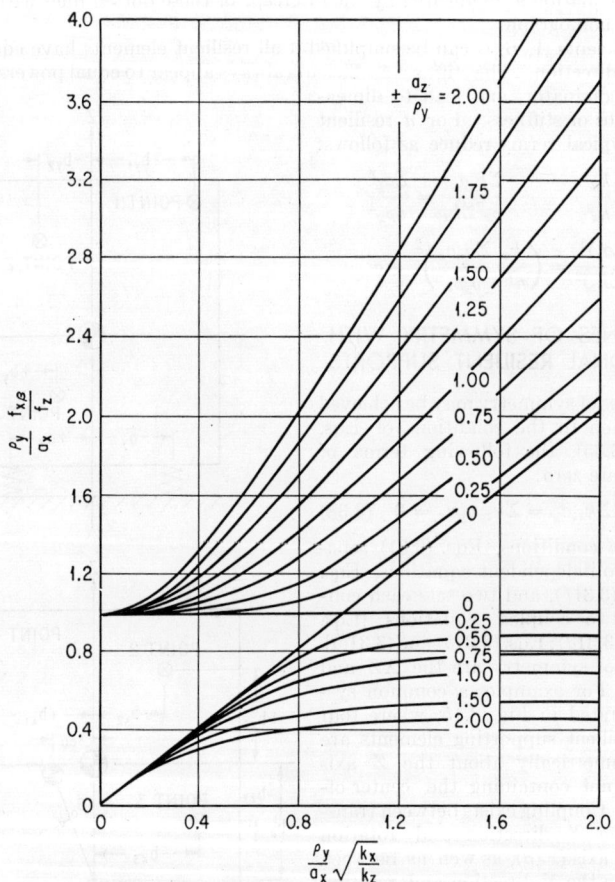

FIG. 3.16. Curves showing the ratio of each of the two coupled natural frequencies $f_{x\beta}$ to the decoupled natural frequency f_z, for motion in the XZ plane of symmetry for the system in Fig. 3.15 [see Eq. (3.40)]. Calculate the abscissa $(\rho_y/a_x)/\sqrt{k_x/k_z}$ and the parameter a_z/ρ_y, where a_x, a_z are indicated in Fig. 3.15; k_x, k_z are the stiffnesses of the resilient supporting elements in the X, Z directions, respectively; and ρ_y is the radius of gyration of the body about the Y axis. The two values read from the ordinate when divided by ρ_y/a_x give the natural frequency ratios $f_{x\beta}/f_z$. (*After C. E. Crede.*[10])

MODAL COUPLING AND NATURAL FREQUENCIES   3-21

The natural frequencies in the coupled x_c, β modes are found by solving Eqs. (3.31a) and (3.31d) simultaneously; the roots yield the following expression for natural frequency:

$$\frac{f_{x\beta}^2}{f_z^2} = \frac{1}{2}\left\{\frac{k_x}{k_z}\left(1 + \frac{a_z^2}{\rho_y^2}\right) + \frac{a_x^2}{\rho_y^2} \pm \sqrt{\left[\frac{k_x}{k_z}\left(1 + \frac{a_z^2}{\rho_y^2}\right) + \frac{a_x^2}{\rho_y^2}\right]^2 - 4\frac{k_x}{k_z}\frac{a_x^2}{\rho_y^2}}\right\} \quad (3.40)$$

Figure 3.16 provides a convenient graphical method for determining the two coupled natural frequencies $f_{x\beta}$. An expression similar to Eq. (3.40) is obtained for $f_{y\alpha}^2/f_z^2$ by

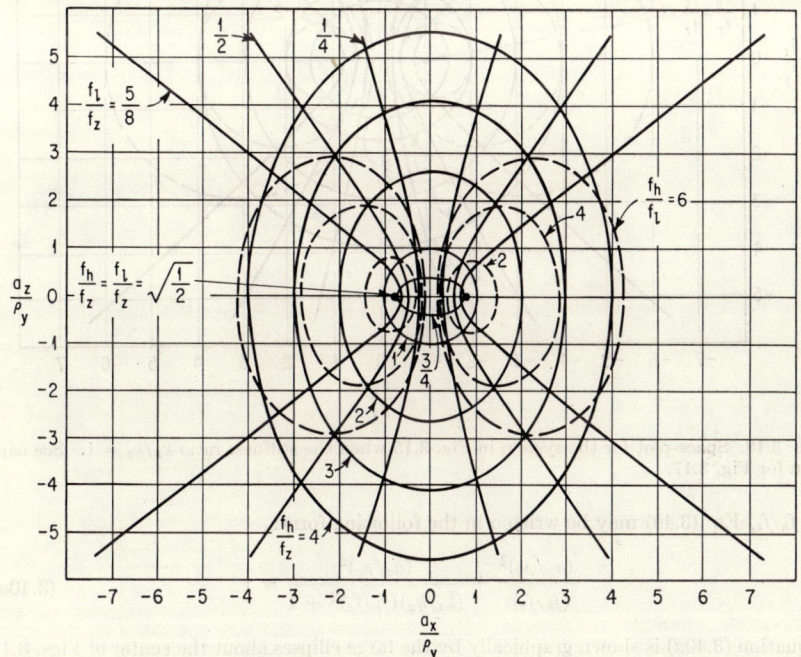

FIG. 3.17. Space-plot for the system in Fig. 3.15 when the stiffness ratio $k_x/k_z = 0.5$, obtained from Eqs. (3.40a) to (3.40c). With all dimensions divided by the radius of gyration ρ_y about the Y axis, superimpose the outline of the rigid body in the XZ plane on the plot; the center-of-gravity of the body is located at the coordinate center of the plot. The elastic centers of the resilient supporting elements give the natural frequency ratios f_l/f_z, f_h/f_z and f_h/f_l for x_c, β coupled motion, each ratio being read from one of the three families of curves as indicated on the plot. Replacing k_x, ρ_y, a_x with k_y, ρ_x, a_y, respectively, allows the plot to be applied to motions in the YZ plane.

solving Eqs. (3.31b) and (3.31d) simultaneously. By replacing ρ_y, a_x, k_x, $f_{x\beta}$ with ρ_x, a_y, k_y, $f_{y\alpha}$, respectively, Fig. 3.16 also can be used to determine the two values of $f_{y\alpha}$.

It may be desirable to select resilient element locations a_x, a_y, a_z which will produce coupled natural frequencies in specified frequency ranges, with resilient elements having specified stiffness ratios k_x/k_z, k_y/k_z. For this purpose it is convenient to plot solutions of Eq. (3.40) in the form shown in Figs. 3.17 to 3.19. These plots are termed *space-plots* and their use is illustrated in Example 3.1.[11,12]

The space-plots are derived as follows: In general, the two roots of Eq. (3.40) are numerically different, one usually being greater than unity and the other less than unity. Designating the root associated with the positive sign before the radical (higher value)

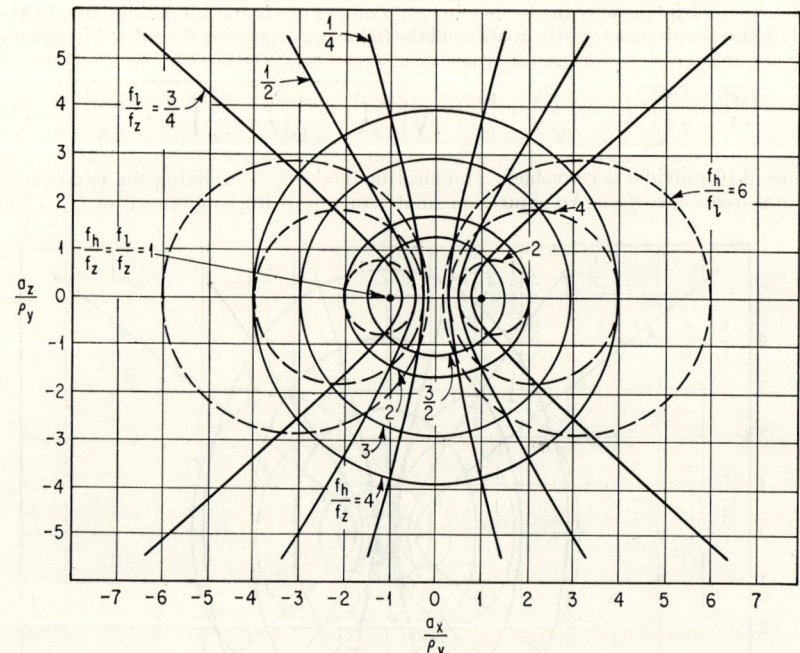

Fig. 3.18. Space-plot for the system in Fig. 3.15 when the stiffness ratio $k_x/k_z = 1$. See caption for Fig. 3.17.

as f_h/f_z, Eq. (3.40) may be written in the following form:

$$\frac{(a_x/\rho_y)^2}{(f_h/f_z)^2} + \frac{(a_z/\rho_y)^2}{(k_z/k_x)(f_h/f_z)^2 - 1} = 1 \tag{3.40a}$$

Equation (3.40a) is shown graphically by the large ellipses about the center of Figs. 3.17 to 3.19, for stiffness ratios k_x/k_z of $\frac{1}{2}$, 1, and 2, respectively. A particular type of resilient element tends to have a constant stiffness ratio k_x/k_z; thus, Figs. 3.17 to 3.19 may be used by cut-and-try methods to find the coordinates a_x, a_z of such elements to attain a desired value of f_h.

Designating the root of Eq. (3.40) associated with the negative sign (lower value) by f_l, Eq. (3.40) may be written as follows:

$$\frac{(a_x/\rho_y)^2}{(f_l/f_z)^2} - \frac{(a_z/\rho_y)^2}{1 - (k_z/k_x)(f_l/f_z)^2} = 1 \tag{3.40b}$$

Equation (3.40b) is shown graphically by the family of hyperbolas on each side of the center in Figs. 3.17 to 3.19, for values of the stiffness ratio k_x/k_z of $\frac{1}{2}$, 1, and 2.

The two roots f_h/f_z and f_l/f_z of Eq. (3.40) may be expressed as the ratio of one to the other. This relationship is given parametrically as follows:

$$\left[\frac{2\dfrac{a_x}{\rho_y} \pm \sqrt{\dfrac{k_x}{k_z}}\left(\dfrac{f_h}{f_l} + \dfrac{f_l}{f_h}\right)}{\sqrt{\dfrac{k_x}{k_z}}\left(\dfrac{f_h}{f_l} - \dfrac{f_l}{f_h}\right)}\right]^2 + \left[\frac{2\dfrac{a_z}{\rho_y}}{\dfrac{f_h}{f_l} - \dfrac{f_l}{f_h}}\right]^2 = 1 \tag{3.40c}$$

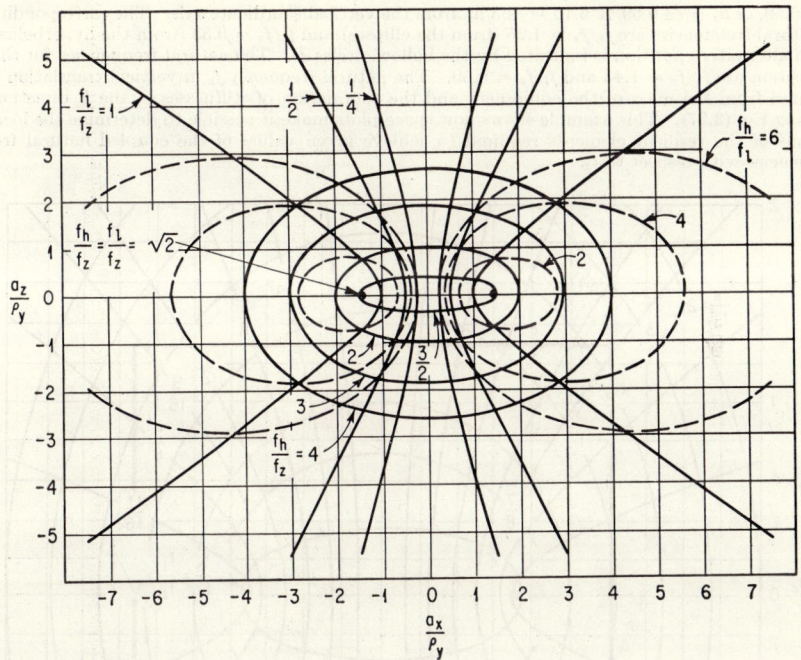

Fig. 3.19. Space-plot for the system in Fig. 3.15 when the stiffness ratio $k_x/k_z = 2$. See caption for Fig. 3.17.

Equation (3.40c) is shown graphically by the smaller ellipses (shown dotted) displaced from the vertical center line in Figs. 3.17 to 3.19.

Example 3.1. A rigid body is symmetrical with respect to the XZ plane; its width in the X direction is 13 in. and its height in the Z direction is 12 in. The center-of-gravity is 5½ in. from the lower side and 6¾ in. from the right side. The radius of gyration about the Y axis through the center-of-gravity is 5.10 in. Use a space-plot to evaluate the effects of the location for attachment of resilient supporting elements having the characteristic stiffness ratio $k_x/k_z = $ ½.

Superimpose the outline of the body on the space-plot of Fig. 3.20, with its center-of-gravity at the coordinate center of the plot. (Figure 3.20 is an enlargement of the central portion of Fig. 3.17.) All dimensions are divided by the radius of gyration ρ_y. Thus, the four corners of the body are located at coordinate distances as follows:

Upper right corner:
$$\frac{a_z}{\rho_y} = \frac{+6.50}{5.10} = +1.28 \qquad \frac{a_x}{\rho_y} = \frac{+6.75}{5.10} = +1.32$$

Upper left corner:
$$\frac{a_z}{\rho_y} = \frac{+6.50}{5.10} = +1.28 \qquad \frac{a_x}{\rho_y} = \frac{-6.25}{5.10} = -1.23$$

Lower right corner:
$$\frac{a_z}{\rho_y} = \frac{-5.50}{5.10} = -1.08 \qquad \frac{a_x}{\rho_y} = \frac{+6.75}{5.10} = +1.32$$

Lower left corner:
$$\frac{a_z}{\rho_y} = \frac{-5.50}{5.10} = -1.08 \qquad \frac{a_x}{\rho_y} = \frac{-6.25}{5.10} = -1.23$$

The resilient supports are shown in heavy outline at A in Fig. 3.20, with their elastic centers indicated by the solid dots. The horizontal coordinates of the resilient supports are $a_x/\rho_y =$

±0.59, or $a_x = \pm 0.59 \times 5.10 = \pm 3$ in. from the vertical coordinate axis. The corresponding natural frequencies are $f_h/f_z = 1.25$ (from the ellipses) and $f_l/f_z = 0.33$ (from the hyperbolas). An alternative position is indicated by the hollow circles B. The natural frequencies for this position are $f_h/f_z = 1.43$ and $f_l/f_z = 0.50$. The natural frequency f_z in vertical translation is found from the mass of the equipment and the summation of stiffnesses in the Z direction, using Eq. (3.37). This example shows how space-plots make it possible to determine the locations of the resilient elements required to achieve given values of the coupled natural frequencies with respect to f_z.

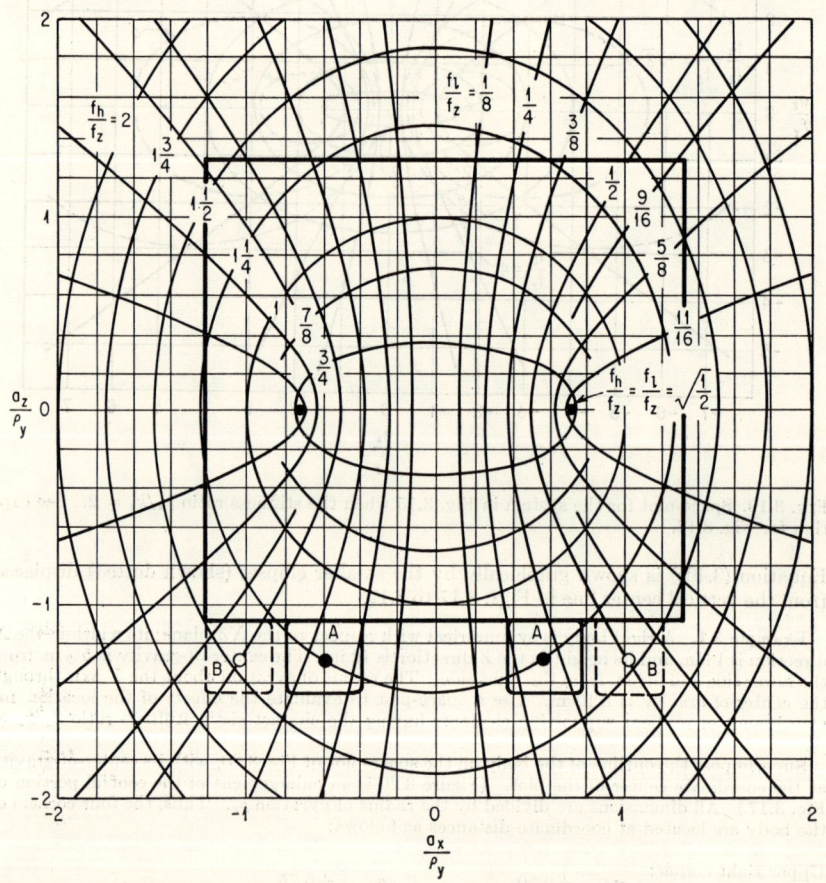

FIG. 3.20. Enlargement of the central portion of Fig. 3.17 with the outline of the rigid body discussed in Example 3.1.

THREE PLANES OF SYMMETRY WITH ORTHOGONAL RESILIENT SUPPORTS

A system with three planes of symmetry is defined by six independent equations of motion. A system having this property is sometimes called a *center-of-gravity system*. The equations are derived from Eqs. (3.31) by substituting, in addition to the conditions of Eqs. (3.33), (3.34), (3.35), and (3.38), the following condition:

$$\Sigma k_{xx}a_z = \Sigma k_{yy}a_z = 0 \qquad (3.41)$$

The resulting six independent equations define six uncoupled modes of vibration, three in translation and three in rotation. The natural frequencies are:

Translation along X axis:
$$f_x = \frac{1}{2\pi}\sqrt{\frac{\Sigma k_x}{m}}$$

Translation along Y axis:
$$f_y = \frac{1}{2\pi}\sqrt{\frac{\Sigma k_y}{m}}$$

Translation along Z axis:
$$f_z = \frac{1}{2\pi}\sqrt{\frac{\Sigma k_z}{m}}$$

Rotation about X axis:
$$f_\alpha = \frac{1}{2\pi}\sqrt{\frac{\Sigma(k_y a_z^2 + k_z a_y^2)}{m\rho_x^2}} \quad (3.42)$$

Rotation about Y axis:
$$f_\beta = \frac{1}{2\pi}\sqrt{\frac{\Sigma(k_x a_z^2 + k_z a_x^2)}{m\rho_y^2}}$$

Rotation about Z axis:
$$f_\gamma = \frac{1}{2\pi}\sqrt{\frac{\Sigma(k_x a_y^2 + k_y a_x^2)}{m\rho_z^2}}$$

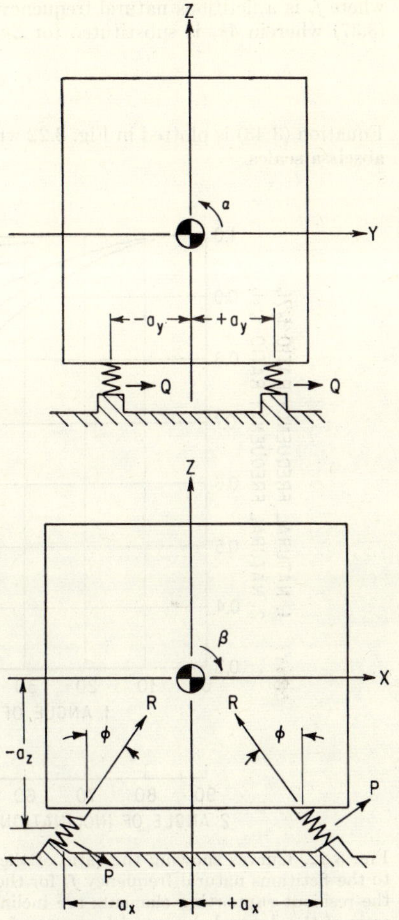

TWO PLANES OF SYMMETRY WITH RESILIENT SUPPORTS INCLINED IN ONE PLANE ONLY

When the principal elastic axes of the resilient supporting elements are inclined with respect to the X, Y, Z axes, the stiffness coefficients k_{xy}, k_{xz}, k_{yz} are nonzero. This introduces elastic coupling which must be considered in evaluating the equations of motion. Two planes of symmetry may be achieved by meeting the conditions of Eqs. (3.33), (3.35), and (3.38). For example, consider the rigid body supported by four identical resilient supporting elements located symmetrically about the Z axis, as shown in Fig. 3.21. The XZ and the YZ planes are planes of symmetry and the resilient elements are inclined toward the YZ plane so that one of their principal elastic axes R is inclined at the angle ϕ with the Z direction as shown; hence $k_{yy} = k_q$, and $k_{xy} = k_{yz} = 0$.

Because of symmetry, translational motion z_c in the Z direction and rotation γ about the Z axis are each decoupled from the other modes. The pairs of translational and rotational modes in the x_c, β and y_c, α coordinates are coupled. The natural frequency in the Z direction is

Fig. 3.21. Example of a rigid body on resilient supporting elements inclined toward the YZ plane. The resilient supporting elements are identical and are located symmetrically about the Z axis, making XZ and YZ planes of symmetry. The principal stiffnesses in the XZ plane are k_p and k_r. The conditions satisfied are Eqs. (3.33), (3.35), and (3.38).

$$\frac{f_z}{f_r} = \sqrt{\frac{k_p}{k_r}\sin^2\phi + \cos^2\phi} \quad (3.43)$$

where f_r is a fictitious natural frequency used for convenience only; it is related to Eq (3.37) wherein $4k_r$ is substituted for Σk_z:

$$f_r = \frac{1}{2\pi}\sqrt{\frac{4k_r}{m}}$$

Equation (3.43) is plotted in Fig. 3.22 where the angle ϕ is indicated by the upper of the abscissa scales.

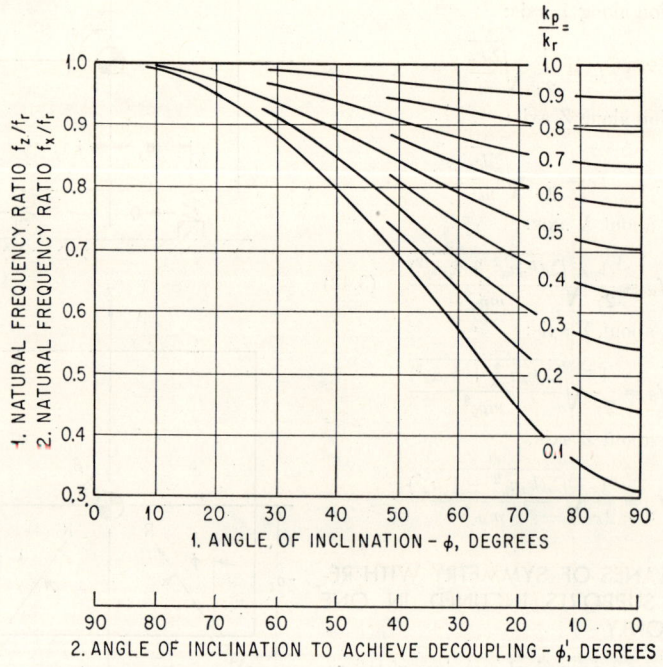

Fig. 3.22. Curves showing the ratio of the decoupled natural frequency f_z of translation z_c to the fictitious natural frequency f_r for the system shown in Fig. 3.21 [see Eq. (3.43)] when the resilient supporting elements are inclined at the angle ϕ. The curves also indicate the ratio of the decoupled natural frequency f_x of translation x_c to f_r when ϕ has a value ϕ' (use lower abscissa scale) which decouples x_c, β motions [see Eqs. (3.47) and (3.48)]. (*After C. E. Crede.*[13])

The rotational natural frequency about the Z axis is obtained from

$$\frac{f_\gamma}{f_r} = \sqrt{\left(\frac{k_p}{k_r}\cos^2\phi + \sin^2\phi\right)\left(\frac{a_y}{\rho_z}\right)^2 + \frac{k_q}{k_r}\left(\frac{a_x}{\rho_z}\right)^2} \qquad (3.44)$$

For the x_c, β coupled mode, the two natural frequencies are

$$\frac{f_{x\beta}}{f_r} = \frac{1}{2}\left[A \pm \sqrt{A^2 - 4\frac{k_p}{k_r}\left(\frac{a_x}{a_y}\right)^2}\right] \qquad (3.45)$$

where $A = \left(\frac{k_p}{k_r}\cos^2\phi + \sin^2\phi\right)\left[1 + \left(\frac{a_z}{\rho_y}\right)^2\right] + \left(\frac{k_p}{k_r}\sin^2\phi + \cos^2\phi\right)\left(\frac{a_x}{\rho_y}\right)^2$

$$+ 2\left(1 - \frac{k_p}{k_r}\right)\left|\frac{a_x}{\rho_y}\right|\sin\phi\cos\phi$$

For the y_c, α coupled mode, the natural frequencies are

$$\frac{f_{y\alpha}}{f_r} = \frac{1}{2}\left[B \pm \sqrt{B^2 - 4\frac{k_q}{k_r}\left(\frac{k_p}{k_r}\sin^2\phi + \cos^2\phi\right)\left(\frac{a_y}{\rho_x}\right)^2}\right] \quad (3.46)$$

where
$$B = \frac{k_q}{k_r}\left[1 + \left(\frac{a_z}{\rho_x}\right)^2\right] + \left(\frac{k_p}{k_r}\sin^2\phi + \cos^2\phi\right)\left(\frac{a_y}{\rho_x}\right)^2$$

DECOUPLING OF MODES IN A PLANE USING INCLINED RESILIENT SUPPORTS

The angle ϕ of inclination of principal elastic axes (see Fig. 3.21) can be varied to produce changes in the amount of coupling between the x_c and β coordinates. Decoupling

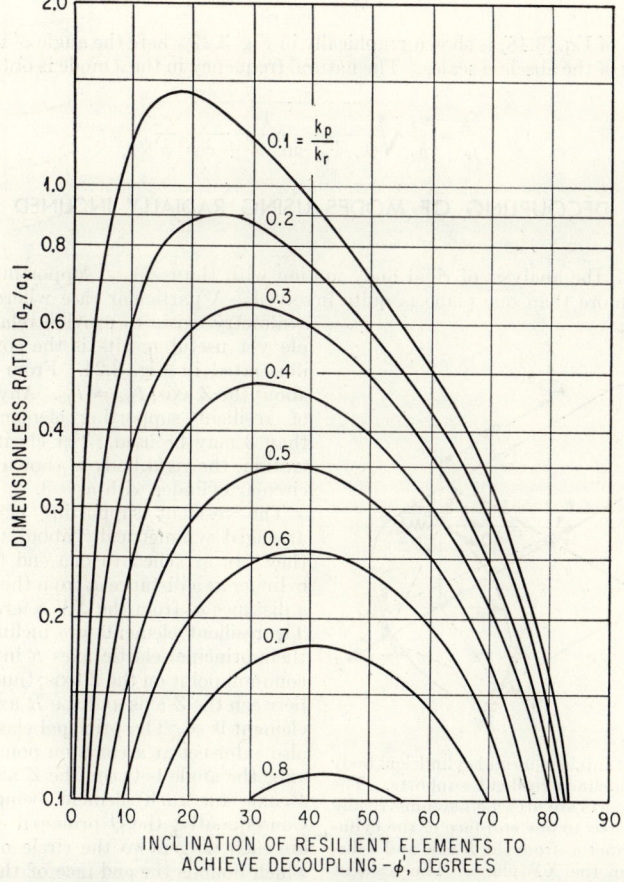

Fig. 3.23. Curves showing the angle of inclination ϕ' of the resilient elements which achieves decoupling of the x_c, β motions in Fig. 3.21 [see Eq. (3.47)]. Calculate the ordinate $|a_z/a_x|$ and with the stiffness ratio k_p/k_r determine two values of ϕ' for which decoupling is possible. Decoupling is not possible for a particular value of k_p/k_r if $|a_z/a_y|$ has a value greater than the maximum ordinate of the k_p/k_r curve.

of the x_c and β coordinates is effected if [13]

$$\left|\frac{a_z}{a_x}\right| = \frac{[1 - (k_p/k_r)]\cot\phi'}{1 + (k_p/k_r)\cot^2\phi'} \qquad (3.47)$$

where ϕ' is the value of the angle of inclination ϕ required to achieve decoupling. When Eq. (3.47) is satisfied, the configuration is sometimes called an "equivalent center-of-gravity system" in the YZ plane since all modes of motion in that plane are decoupled. Figure 3.23 is a graphical presentation of Eq. (3.47). There may be two values of ϕ' that decouple the x_c and β modes for any combination of stiffness and location for the resilient supporting elements.

The decoupled natural frequency for translation in the X direction is obtained from

$$\frac{f_x}{f_r} = \sqrt{\frac{k_p}{k_r}\cos^2\phi' + \sin^2\phi'} \qquad (3.48)$$

The relation of Eq. (3.48) is shown graphically in Fig. 3.22 where the angle ϕ' is indicated by the lower of the abscissa scales. The natural frequency in the β mode is obtained from

$$\frac{f_\beta}{f_r} = \frac{a_x}{\rho_y}\sqrt{\frac{1}{(k_r/k_p)\sin^2\phi' + \cos^2\phi'}} \qquad (3.49)$$

COMPLETE DECOUPLING OF MODES USING RADIALLY INCLINED RESILIENT SUPPORTS

In general, the analysis of rigid body motion with the resilient supporting elements inclined in more than one plane is quite involved. A particular case where sufficient symmetry exists to provide relatively simple yet useful results is the configuration illustrated in Fig. 3.24. From symmetry about the Z axis, $I_{xx} = I_{yy}$. Any number n of resilient supporting elements greater than 3 may be used. For clarity of illustration, the rigid body is shown as a right circular cylinder with $n = 3$.

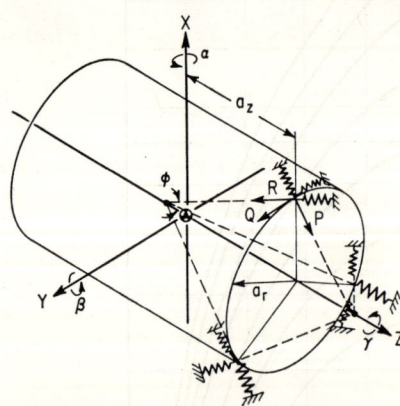

FIG. 3.24. Example of a rigid cylindrical body on radially inclined resilient supports. The resilient supports are attached symmetrically about the Z axis to one end face of the cylinder at a distance a_r from the Z axis and a distance a_z from the XY plane. The resilient elements are inclined so that their principal elastic axes R and P intersect the Z axis at common points. The angle between the R axes and the Z axis is ϕ; and the angle between the P axis and Z axis is $90° - \phi$. The Q principal elastic axes are each tangent to the circle of radius a_r.

The resilient supporting elements are arranged symmetrically about the Z axis; they are attached to one end face of the cylinder at a distance a_r from the Z axis and a distance a_z from the XY reference plane. The resilient elements are inclined so that their principal elastic axes R intersect at a common point on the Z axis; thus, the angle between the Z axis and the R axis for each element is ϕ. The principal elastic axes P also intersect at a common point on the Z axis, the angle between the Z axis and the P axis for each element being $90° - \phi$. Consequently, the Q principal elastic axes are each tangent to the circle of radius a_r which bounds the end face of the cylinder.

The use of such a configuration permits decoupling of all six modes of vibration of the rigid body. This complete decoupling is achieved if the angle of inclination ϕ has the value ϕ' which satisfies the following equation:

$$\left|\frac{a_z}{a_r}\right| = \frac{(1/2)[1 - (k_p/k_r)] \sin 2\phi'}{(k_q/k_r) + (k_p/k_r) + [1 - (k_p/k_r)] \sin^2 \phi'} \quad (3.50)$$

Since complete decoupling is effected, the system may be termed an "equivalent center-of-gravity system."[14,15] The natural frequencies of the six decoupled modes are

$$\frac{f_x}{f_r} = \frac{f_y}{f_r} = \sqrt{\frac{1}{2}\left(\frac{k_p}{k_r}\cos^2\phi' + \sin^2\phi' + \frac{k_q}{k_r}\right)} \quad (3.51)$$

$$\frac{f_\alpha}{f_r} = \frac{f_\beta}{f_r} = \left\{\frac{a_r}{2\rho_x}\left[\frac{k_p}{k_r}\sin\phi'\left(\frac{a_r}{\rho_x}\sin\phi' + \frac{a_z}{\rho_x}\cos\phi'\right) + \cos\phi'\left(\frac{a_r}{\rho_x}\cos\phi' - \frac{a_z}{\rho_x}\sin\phi'\right)\right]\right\}^{1/2} \quad (3.52)$$

$$\frac{f_\gamma}{f_r} = \sqrt{\frac{k_q}{k_r}\frac{a_r}{\rho_z}} \quad (3.53)$$

The frequency ratio f_z/f_r is given by Eq. (3.43) or Fig. 3.22. The fictitious natural frequency f_r is given by

$$f_r = \frac{1}{2\pi}\sqrt{\frac{nk_r}{m}}$$

FORCED VIBRATION

Forced vibration results from a continuing excitation that varies sinusoidally with time. The excitation may be a vibratory displacement of the foundation for the resiliently supported rigid body (*foundation-induced vibration*), or a force or moment applied to or generated within the rigid body (*body-induced vibration*). These two forms of excitation are considered separately.

FOUNDATION-INDUCED SINUSOIDAL VIBRATION

This section includes an analysis of foundation-induced vibration for two different systems, each having two planes of symmetry. In one system, the principal elastic axes of the resilient elements are parallel to the X, Y, Z axes; in the other system, the principal elastic axes are inclined with respect to two of the axes but in a plane parallel to one of the reference planes. The excitation is translational movement of the foundation in its own plane, without rotation. No forces or moments are applied directly to the rigid body; i.e., in the equations of motion [Eqs. (3.31)], the following terms are equal to zero:

$$F_x = F_y = F_z = M_x = M_y = M_z = \alpha = \beta = \gamma = 0 \quad (3.54)$$

TWO PLANES OF SYMMETRY WITH ORTHOGONAL RESILIENT SUPPORTS
The system is shown in Fig. 3.15. The excitation is a motion of the foundation in the direction of the X axis defined by $u = u_0 \sin \omega t$. (Alternatively, the excitation may be the displacement $v = v_0 \sin \omega t$ in the direction of the Y axis, and analogous results are obtained.) The resulting motion of the resiliently supported rigid body involves translation x_c and rotation β simultaneously. The conditions of symmetry are defined by Eqs. (3.33), (3.34), (3.35), and (3.38); these conditions decouple Eqs. (3.31) so that only Eqs. (3.31a) and (3.31d), and Eqs. (3.31b) and (3.31c), remain coupled. Upon substituting $u = u_0 \sin \omega t$ as the excitation, the response in the coupled modes is of a form

$x_c = x_{c0} \sin \omega t$, $\beta = \beta_0 \sin \omega t$ where x_{c0} and β_0 are related to u_0 as follows:[16]

$$\frac{x_{c0}}{u_0} = \frac{\frac{k_x}{k_z}\left[\left(\frac{a_x}{\rho_y}\right)^2 - \left(\frac{f}{f_z}\right)^2\right]}{\left(\frac{f}{f_z}\right)^4 - \left[\frac{k_x}{k_z} + \frac{k_x}{k_z}\left(\frac{a_z}{\rho_y}\right)^2 + \left(\frac{a_x}{\rho_y}\right)^2\right]\left(\frac{f}{f_z}\right)^2 + \frac{k_x}{k_z}\left(\frac{a_x}{\rho_y}\right)^2} \quad (3.55)$$

$$\frac{\beta_0}{u_0/\rho_y} = \frac{-\frac{k_x}{k_z}\frac{a_z}{\rho_y}\left(\frac{f}{f_z}\right)^2}{\left(\frac{f}{f_z}\right)^4 - \left[\frac{k_x}{k_z} + \frac{k_x}{k_z}\left(\frac{a_z}{\rho_y}\right)^2 + \left(\frac{a_x}{\rho_y}\right)^2\right]\left(\frac{f}{f_z}\right)^2 + \frac{k_x}{k_z}\left(\frac{a_x}{\rho_y}\right)^2} \quad (3.56)$$

where $f_z = \frac{1}{2\pi}\sqrt{4k_z/m}$ in accordance with Eq. (3.37). A similar set of equations apply for vibration in the coupled y_c, α coordinates. There is no response of the system in the z_c or γ modes since there is no net excitation in these directions; that is, $\mathbf{F}_z$ and $\mathbf{M}_z$ are zero.

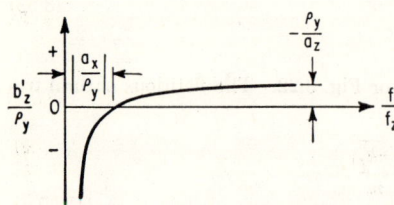

FIG. 3.25. Curve showing the position of the axis of pure rotation of the rigid body in Fig. 3.15 as a function of the frequency ratio f/f_z when the excitation is sinusoidal motion of the foundation in the X direction [see Eq. (3.57)]. The axis of rotation is parallel to the Y axis and in the XZ plane, and its coordinate along the Z axis is designated by b_z'.

As indicated by Eqs. (3.1), the displacement at any point in a rigid body is the sum of the displacement at the center-of-gravity and the displacements resulting from motion of the body in rotation about axes through the center-of-gravity. Equations (3.55) and (3.56) together with analogous equations for y_{c0}, α_0 provide the basis for calculating these displacements. Care should be taken with phase angles, particularly if two or more excitations u, v, w exist concurrently.

At any single frequency, coupled vibration in the x_c,β modes is equivalent to a pure rotation of the rigid body with respect to an axis parallel to the Y axis, in the YZ plane and displaced from the center-of-gravity of the body (see Fig. 3.15). As a result, the rigid body has zero displacement x in the horizontal plane containing this axis. Therefore, the Z coordinate of this axis b_z' satisfies $x_{c0} + b_z'\beta_0 = 0$, which is obtained from the first of Eqs. (3.1) by setting $x_b = 0$ (γ_0 motion is not considered). Substituting Eqs. (3.55) and (3.56) for x_{c0} and β_0, respectively, the axis of rotation is located at

$$\frac{b_z'}{\rho_y} = \frac{(a_x/\rho_y)^2 - (f/f_z)^2}{(a_z/\rho_y)(f/f_z)^2} \quad (3.57)$$

Figure 3.25 shows the relation of Eq. (3.57) graphically. At high values of frequency f/f_z, the axis does not change position significantly with frequency; b_z'/ρ_y approaches a positive value as f/f_z becomes large since a_z is negative (see Fig. 3.15).

When the resilient supporting elements have damping as well as elastic properties, the solution of the equations of motion [see Eq. (3.31a)] becomes too laborious for general use. Responses of systems with damping have been obtained for several typical cases using a digital computer.[17] Figures 3.26A, B, and C show the response at three points in the body of the system shown in Fig. 3.15, with the excitation $u = u_0 \sin \omega t$. The weight of the body is 45 lb; each of the four resilient supporting elements has a stiffness $k_z = 1,050$ lb/in. and stiffness ratios $k_x/k_z = k_y/k_z = \frac{1}{2}$. The critical damping coefficients in the X, Y, Z directions are taken as $c_{cx} = 2\sqrt{4k_xm}$, $c_{cy} = 2\sqrt{4k_ym}$, $c_{cz} = 2\sqrt{4k_zm}$, respectively, where the expression for c_{cz} follows from the single degree-of-freedom case defined by Eq. (2.12). The fractions of critical damping are $c_x/c_{cx} = c_y/c_{cy} = c_z/c_{cz} = c/c_c$, the parameter of the curves in Figs.

3.26A, B, and C. Coordinates locating the resilient elements are $a_x = \pm 5.25$ in., $a_y = \pm 3.50$ in., and $a_z = -6.50$ in. The radii of gyration with respect to the X, Y, Z axes are $\rho_x = 4.40$ in., $\rho_y = 5.10$ in., and $\rho_z = 4.60$ in.

Natural frequencies calculated from Eqs. (3.37) and (3.40) are $f_z = 30.0$ cps; $f_{x\beta} = 43.7$ cps, 15.0 cps; and $f_{y\alpha} = 43.2$ cps, 11.7 cps. The fraction of critical damping c/c_c varies between 0 and 0.25. Certain characteristic features of the response curves in Figs. 3.26A, B, and C are:

1. The relatively small response at the frequency of 24.2 cps in Fig. 3.26C occurs because point 3 lies near the axis of rotation of the rigid body at that frequency. Point 2 lies near the axis of rotation at higher frequencies and the response becomes correspondingly low, as shown in Fig. 3.26B. The position of the axis of rotation changes rapidly for small changes of fre-

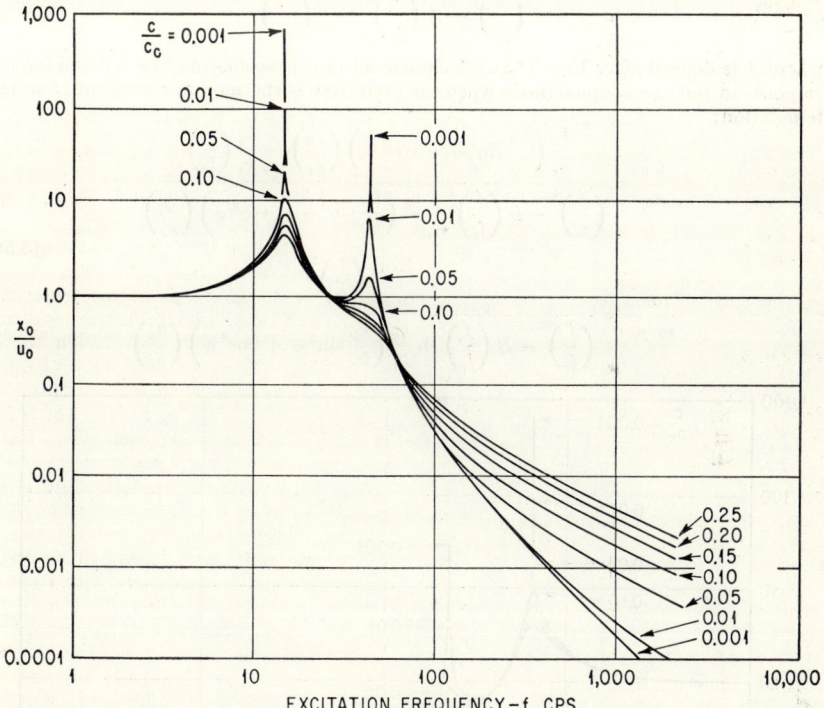

Fig. 3.26A. Response curves with damping in the resilient supports at the point 1 in the system shown in Fig. 3.15. The response is the ratio of the amplitude at the center-of-gravity of the rigid body in the X direction to the amplitude of the foundation in the X direction (x_0/u_0). The fraction of critical damping c/c_c is the same in the X, Y, Z directions.

quency in the low- and intermediate-frequency range (indicated by the sharp dip in the curves for small damping in Fig. 3.26C) and varies asymptotically toward a final position as the forcing frequency increases (see Fig. 3.25).

2. The effect of damping on the magnitude of the response at the higher and lower natural frequencies in coupled modes is illustrated. When the fraction of critical damping is between 0.01 and 0.10, the response at the lower of the coupled natural frequencies is approximately 10 times as great as the response at the higher of the coupled natural frequencies. With greater damping ($c/c_c \geq 0.15$), the effect of resonance in the vicinity of the higher coupled natural frequency becomes so slight as to be hardly discernible.

TWO PLANES OF SYMMETRY WITH RESILIENT SUPPORTS INCLINED IN ONE PLANE ONLY. The system is shown in Fig. 3.21, and the excitation is $u = u_0 \sin \omega t$. The conditions of symmetry are defined by Eqs. (3.33), (3.35), and (3.38).

3-32   VIBRATION OF A RESILIENTLY SUPPORTED RIGID BODY

The response is entirely in the x_c, β coupled mode with the following amplitudes:

$$\frac{x_{c0}}{u_0} = \frac{\frac{k_p}{k_r}\left(\frac{a_x}{\rho_y}\right)^2 - \left(\frac{k_p}{k_r}\cos^2\phi + \sin^2\phi\right)\left(\frac{f}{f_r}\right)^2}{\left(\frac{f}{f_r}\right)^4 - A\left(\frac{f}{f_r}\right)^2 + \frac{k_p}{k_r}\left(\frac{a_x}{\rho_y}\right)^2}$$

$$\frac{\beta_0}{u_0/\rho_y} = \frac{-\left[\left(\frac{k_p}{k_r}\cos^2\phi + \sin^2\phi\right)\left(\frac{a_z}{\rho_y}\right) + \left(1 - \frac{k_p}{k_r}\right)\left|\frac{a_x}{\rho_y}\right|\sin\phi\cos\phi\right]\left(\frac{f}{f_r}\right)^2}{\left(\frac{f}{f_r}\right)^4 - A\left(\frac{f}{f_r}\right)^2 + \frac{k_p}{k_r}\left(\frac{a_x}{\rho_y}\right)^2}$$
(3.58)

where A is defined after Eq. (3.45). A similar set of expressions may be written for the response in the y_c, α coupled mode when the excitation is the motion $v = v_0 \sin \omega t$ of the foundation:

$$\frac{y_{c0}}{v_0} = \frac{\frac{k_q}{k_r}\left(\frac{k_p}{k_r}\sin^2\phi + \cos^2\phi\right)\left(\frac{a_y}{\rho_x}\right)^2 - \frac{k_q}{k_r}\left(\frac{f}{f_r}\right)^2}{\left(\frac{f}{f_r}\right)^4 - B\left(\frac{f}{f_r}\right)^2 + \frac{k_q}{k_r}\left(\frac{k_p}{k_r}\sin^2\phi + \cos^2\phi\right)\left(\frac{a_y}{\rho_x}\right)}$$

$$\frac{\alpha_0}{v_0/\rho_x} = \frac{\frac{k_q}{k_r}\frac{a_z}{\rho_x}\left(\frac{f}{f_r}\right)^2}{\left(\frac{f}{f_r}\right)^4 - B\left(\frac{f}{f_r}\right)^2 + \frac{k_q}{k_r}\left(\frac{k_p}{k_r}\sin^2\phi + \cos^2\phi\right)\left(\frac{a_y}{\rho_x}\right)}$$
(3.59)

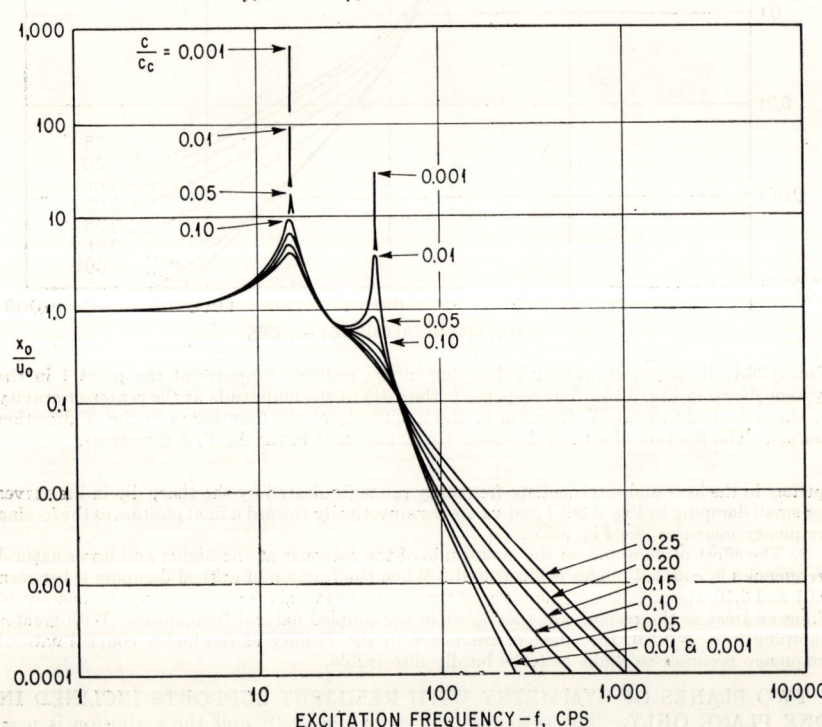

FIG. 3.26B. Response curves at point 2 in the system shown in Fig. 3.15. See caption for Fig. 3.26A.

where B is defined after Eq. (3.46). No motion occurs in the z_c or γ mode since the quantities F_z and M_z are zero in Eqs. (3.31e) and (3.31f).

Response curves for the system shown in Fig. 3.21 when damping is included are qualitatively similar to those shown in Figs. 3.26.[18] The significant advantage in the use of inclined resilient supports is the additional versatility gained from the ability to vary the angle of inclination ϕ which directly affects the degree of coupling in the x_c,β coupled mode. For example, a change in the angle ϕ produces a change in the position of the axis of pure rotation of the rigid body. In a manner similar to that used to derive Eq. (3.57), Eqs. (3.58) yield the following expression defining the location of the axis of rotation:

$$\frac{b_z'}{\rho_y} = \frac{\dfrac{k_p}{k_r}\left(\dfrac{a_x}{\rho_y}\right)^2 - \left(\dfrac{k_p}{k_r}\cos^2\phi + \sin^2\phi\right)\left(\dfrac{f}{f_r}\right)^2}{\left[\left(\dfrac{k_p}{k_r}\cos^2\phi + \sin^2\phi\right)\dfrac{a_z}{\rho_y} + \left(1 - \dfrac{k_p}{k_r}\right)\left(\dfrac{a_x}{\rho_y}\right)\sin\phi\cos\phi\right]\left(\dfrac{f}{f_r}\right)^2} \quad (3.60)$$

BODY-INDUCED SINUSOIDAL VIBRATION

This section includes the analysis of a resiliently supported rigid body wherein the excitation consists of forces and moments applied directly to the rigid body (or originating within the body). The system has two planes of symmetry with orthogonal resilient supports; the modal coupling and natural frequencies for such a system are considered above. Two types of excitation are considered: (1) a force rotating about an axis parallel to one of the principal inertial axes and (2) an oscillatory moment acting about one of the

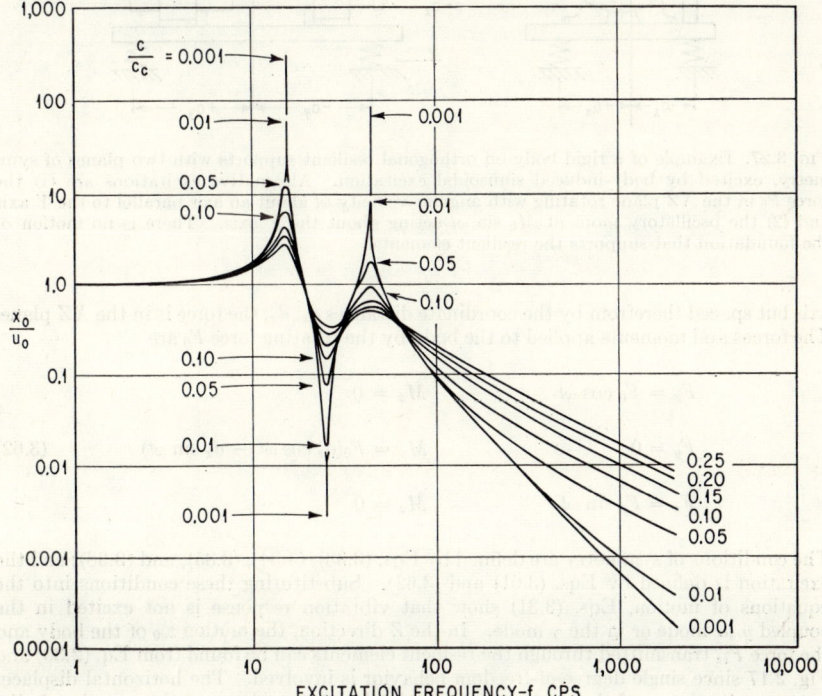

FIG. 3.26C. Response curves at point 3 in the system shown in Fig. 3.15. See caption for Fig. 3.26A.

3-34 VIBRATION OF A RESILIENTLY SUPPORTED RIGID BODY

principal inertial axes.[19,20] There is no motion of the foundation that supports the resilient elements; thus, the following terms in Eqs. (3.31) are equal to zero:

$$u = v = w = \alpha = \beta = \gamma = 0 \tag{3.61}$$

TWO PLANES OF SYMMETRY WITH ORTHOGONAL RESILIENT ELEMENTS EXCITED BY A ROTATING FORCE. The system excited by the rotating force is illustrated in Fig. 3.27. The force F_0 rotates at frequency ω about an axis parallel to the Y

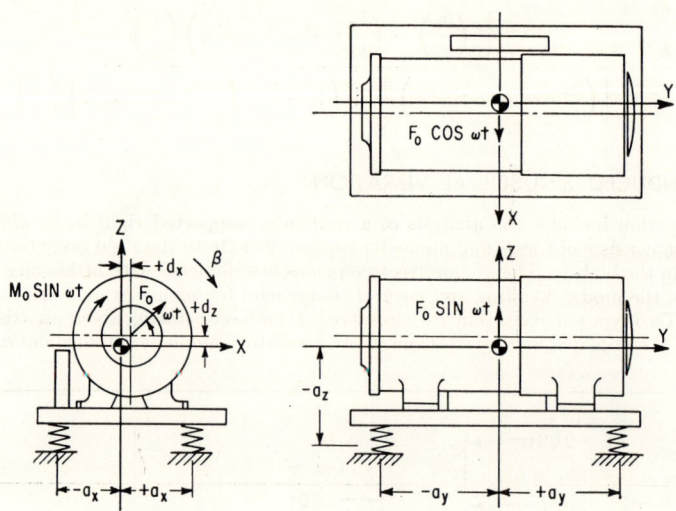

Fig. 3.27. Example of a rigid body on orthogonal resilient supports with two planes of symmetry, excited by body-induced sinusoidal excitation. Alternative excitations are (1) the force F_0 in the XZ plane rotating with angular velocity ωt about an axis parallel to the Y axis and (2) the oscillatory moment $M_0 \sin \omega t$ acting about the Y axis. There is no motion of the foundation that supports the resilient elements.

axis but spaced therefrom by the coordinate distances d_x, d_z; the force is in the XZ plane. The forces and moments applied to the body by the rotating force F_0 are

$$F_x = F_0 \cos \omega t \qquad M_x = 0$$

$$F_y = 0 \qquad M_y = F_0(d_z \cos \omega t - d_x \sin \omega t) \tag{3.62}$$

$$F_z = F_0 \sin \omega t \qquad M_z = 0$$

The conditions of symmetry are defined by Eqs. (3.33), (3.34), (3.35), and (3.38); and the excitation is defined by Eqs. (3.61) and (3.62). Substituting these conditions into the equations of motion, Eqs. (3.31) show that vibration response is not excited in the coupled y_c, α mode or in the γ mode. In the Z direction, the motion z_{c0} of the body and the force F_{tz} transmitted through the resilient elements can be found from Eq. (2.30) and Fig. 2.17 since single degree-of-freedom behavior is involved. The horizontal displacement amplitude x_{c0} of the center-of-gravity in the X direction and the rotational displacement amplitude β_0 about the Y axis are given by

FORCED VIBRATION 3-35

$$\frac{x_{c0}}{F_0/4k_x} = \frac{k_x}{k_z} \frac{\sqrt{\left[\frac{k_x}{k_z}\frac{a_z}{\rho_y}\left(\frac{a_z}{\rho_y}-\frac{d_z}{\rho_y}\right) + \left(\frac{a_x}{\rho_y}\right)^2 - \left(\frac{f}{f_z}\right)^2\right]^2 + \left[\frac{k_x}{k_z}\frac{d_x}{\rho_y}\frac{a_z}{\rho_y}\right]^2}}{\left(\frac{f}{f_z}\right)^4 - \left[\frac{k_x}{k_z} + \frac{k_x}{k_z}\left(\frac{a_z}{\rho_y}\right)^2 + \left(\frac{a_x}{\rho_y}\right)^2\right]\left(\frac{f}{f_z}\right)^2 + \frac{k_x}{k_z}\left(\frac{a_x}{\rho_y}\right)^2}$$

(3.63)

$$\frac{\beta_0}{F_0/4k_x\rho_y} = \frac{k_x}{k_z} \frac{\sqrt{\left[\frac{k_x}{k_z}\left(\frac{a_z}{\rho_y}-\frac{d_z}{\rho_y}\right) + \frac{d_z}{\rho_y}\left(\frac{f}{f_z}\right)^2\right]^2 + \left[\frac{d_x}{\rho_y}\left(\frac{k_x}{k_z} - \frac{f^2}{f_z^2}\right)\right]^2}}{\left(\frac{f}{f_z}\right)^4 - \left[\frac{k_x}{k_z} + \frac{k_x}{k_z}\left(\frac{a_z}{\rho_y}\right)^2 + \left(\frac{a_x}{\rho_y}\right)^2\right]\left(\frac{f}{f_z}\right)^2 + \frac{k_x}{k_z}\left(\frac{a_x}{\rho_y}\right)^2}$$

where a_x, a_z are location coordinates of the resilient supports, and

$$f_z = \frac{1}{2\pi}\sqrt{\frac{4k_z}{m}} \qquad (3.64)$$

The amplitude of the oscillating force F_{tx} in the X direction and the amplitude of the oscillating moment M_{ty} about the Y axis which are transmitted to the foundation by the combination of resilient elements are

$$F_{tx} = 4k_x\sqrt{x_{c0}^2 - 2a_z x_{c0}\beta_0 \cos(\phi_x - \phi_\beta) + a_z^2\beta_0^2}$$

$$M_{ty} = 4k_z a_z^2 \beta_0$$

(3.65)

where F_{tx} is the sum of the forces transmitted by the individual resilient elements and M_{ty} is a moment formed by forces in the Z direction of opposite sign at opposite resilient supports. The angles ϕ_x and ϕ_β are defined by

$$\tan \phi_x = \frac{\frac{k_x}{k_z}\frac{a_z}{\rho_y}\left(\frac{a_z}{\rho_y}-\frac{d_z}{\rho_y}\right) + \left(\frac{a_x}{\rho_y}\right)^2 - \left(\frac{f}{f_z}\right)^2}{\frac{k_x}{k_z}\frac{a_z}{\rho_y}\frac{d_x}{\rho_y}} \qquad [0° \le \phi_x \le 360°]$$

$$\tan \phi_\beta = \frac{\frac{k_x}{k_z}\left(\frac{a_z}{\rho_y}-\frac{d_z}{\rho_y}\right) + \frac{d_z}{\rho_y}\left(\frac{f}{f_z}\right)^2}{\frac{d_x}{\rho_y}\left[\frac{k_x}{k_z} - \left(\frac{f}{f_z}\right)^2\right]} \qquad [0° \le \phi_\beta \le 360°]$$

To obtain the correct value of $(\phi_x - \phi_\beta)$ in Eq. (3.65), the signs of the numerator and denominator in each tangent term must be inspected to determine the proper quadrant for ϕ_x and ϕ_β.

Example 3.2. Consider an electric motor which has an unbalanced rotor, creating a centrifugal force. The motor weighs 3,750 lb, and has a radius of gyration $\rho_y = 9.10$ in. The distances $d_x = d_y = d_z = 0$, that is, the axis of rotation is the Y principal axis and the center-of-gravity of the rotor is in the XZ plane. The resilient supports each have a stiffness ratio of $k_x/k_z = 1.16$, and are located at $a_z = -14.75$ in., $a_x = \pm 12.00$ in. The resulting displacement amplitudes of the center-of-gravity, expressed dimensionlessly, are shown in Fig. 3.28; the force and moment amplitudes transmitted to the foundation, expressed dimensionlessly, are shown in Fig. 3.29. The displacements of the center-of-gravity of the body are dimensionalized with respect to the displacements at zero frequency:

$$z_{c0}(0) = \frac{F_0}{4k_z}$$

$$x_{c0}(0) = \frac{F_0}{4k_x}\left[1 + \frac{k_z}{k_x}\left(\frac{a_z}{a_x}\right)^2\right] \tag{3.66}$$

$$\beta_0(0) = \frac{F_0}{4k_z a_z}\left(\frac{a_z}{a_x}\right)^2$$

At excitation frequencies greater than the higher natural frequency of the x_c,β coupled motion, the displacements, forces, and moment all continuously decrease as the frequency increases.

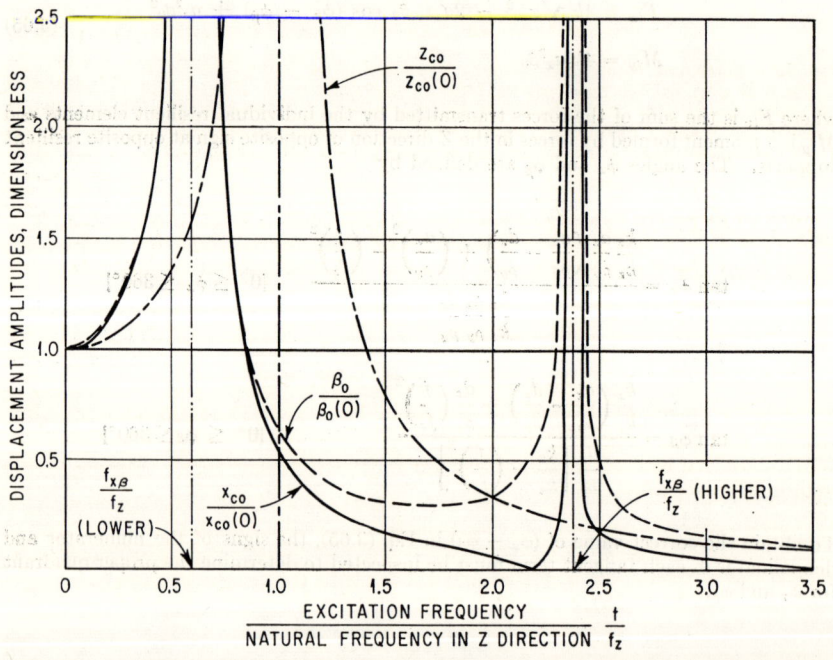

Fig. 3.28. Response curves for the system shown in Fig. 3.27 when excited by a rotating force F_0 acting about the Y axis. The parameters of the system are $k_x/k_z = 1.16$, $a_x/\rho_y = \pm 1.32$, $a_z/\rho_y = -1.62$. Only x_c, z_c, β displacements of the body are excited [see Eqs. (3.63)]. The displacements are expressed dimensionlessly by employing the displacements at zero frequency [see Eqs. (3.66)].

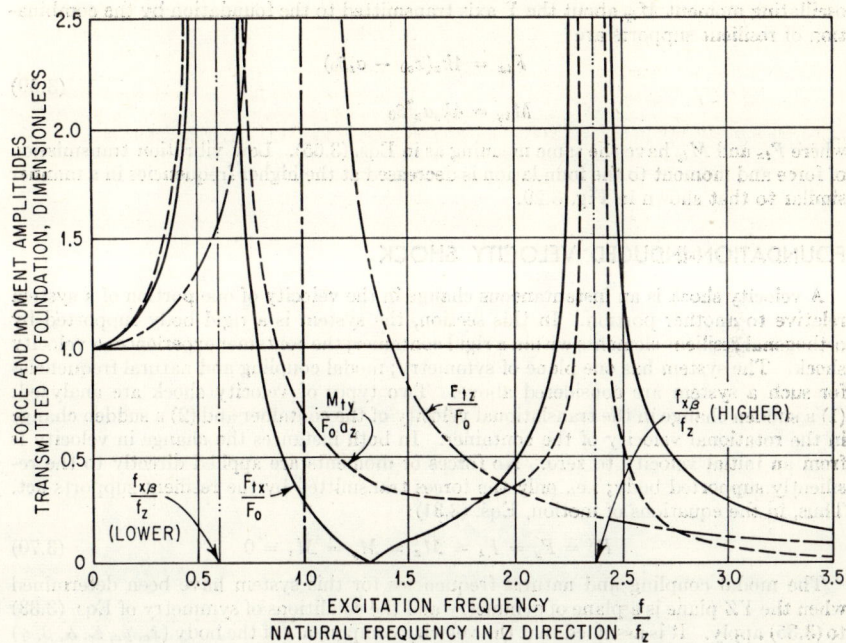

FIG. 3.29. Force and moment amplitudes transmitted to the foundation for the system shown in Fig. 3.27 when excited by a rotating force F_0 acting about the Y axis. The parameters of the system are $k_x/k_z = 1.16$, $a_x/\rho_y = \pm 1.32$, $a_z/\rho_y = -1.62$. The amplitudes of the oscillating forces in the X and Z directions transmitted to the foundation are F_{tx} and F_{tz}, respectively. The amplitude of the total oscillating moment about the Y axis transmitted to the foundation is M_{ty}.

TWO PLANES OF SYMMETRY WITH ORTHOGONAL RESILIENT ELEMENTS EXCITED BY AN OSCILLATING MOMENT. Consider the oscillatory moment M_0 acting about the Y axis with forcing frequency ω. The resulting applied forces and moments acting on the body are

$$M_y = M_0 \sin \omega t \tag{3.67}$$
$$F_x = F_y = F_z = M_x = M_z = 0$$

Substituting conditions of symmetry defined by Eqs. (3.33), (3.34), (3.35), and (3.38), and the excitation defined by Eqs. (3.61) and (3.67), the equations of motion [Eqs. (3.31)] show that oscillations are excited only in the x_c, β coupled mode. Solving for the resulting displacements,

$$\frac{x_{c0}}{M_0/4k_x\rho_y} = \frac{\left(\frac{k_x}{k_z}\right)^2 \frac{a_z}{\rho_y}}{\left(\frac{f}{f_z}\right)^4 - \left[\frac{k_x}{k_z} + \frac{k_x}{k_z}\left(\frac{a_z}{\rho_y}\right)^2 + \left(\frac{a_x}{\rho_y}\right)^2\right]\left(\frac{f}{f_z}\right)^2 + \frac{k_x}{k_z}\left(\frac{a_x}{\rho_y}\right)^2} \tag{3.68}$$

$$\frac{\beta_0}{M_0/4k_x\rho_y{}^2} = \frac{\frac{k_x}{k_z}\left[\frac{k_x}{k_z} - \left(\frac{f}{f_z}\right)^2\right]}{\left(\frac{f}{f_z}\right)^4 - \left[\frac{k_x}{k_z} + \frac{k_x}{k_z}\left(\frac{a_z}{\rho_y}\right)^2 + \left(\frac{a_x}{\rho_y}\right)^2\right]\left(\frac{f}{f_z}\right)^2 + \frac{k_x}{k_z}\left(\frac{a_x}{\rho_y}\right)^2}$$

The amplitude of the oscillating force F_{tx} in the X direction and the amplitude of the oscillating moment M_{ty} about the Y axis transmitted to the foundation by the combination of resilient supports are

$$F_{tx} = 4k_x(x_{c0} - a_z\beta_0)$$
$$M_{ty} = 4k_za_x^2\beta_0 \qquad (3.69)$$

where F_{tx} and M_{ty} have the same meaning as in Eqs. (3.65). Low vibration transmission of force and moment to the foundation is decreased at the higher frequencies in a manner similar to that shown in Fig. 3.29.

FOUNDATION-INDUCED VELOCITY SHOCK

A velocity shock is an instantaneous change in the velocity of one portion of a system relative to another portion. In this section, the system is a rigid body supported by orthogonal resilient elements within a rigid container; the container experiences a velocity shock. The system has one plane of symmetry; modal coupling and natural frequencies for such a system are considered above. Two types of velocity shock are analyzed: (1) a sudden change in the translational velocity of the container and (2) a sudden change in the rotational velocity of the container. In both instances the change in velocity is from an initial velocity to zero. No forces or moments are applied directly to the resiliently supported body; i.e., only the forces transmitted by the resilient supports act. Thus, in the equations of motion, Eqs. (3.31):

$$F_x = F_y = F_z = M_x = M_y = M_z = 0 \qquad (3.70)$$

The modal coupling and natural frequencies for this system have been determined when the YZ plane is a plane of symmetry and the conditions of symmetry of Eqs. (3.33) to (3.35) apply. It is assumed that the velocity components of the body ($\dot{x}_c, \dot{y}_c, \dot{z}_c, \dot{\alpha}, \dot{\beta}, \dot{\gamma}$) and the velocity components of the supporting container ($\dot{u}, \dot{v}, \dot{w}, \dot{\alpha}, \dot{\beta}, \dot{\gamma}$) are respectively equal at time $t < 0$. At $t = 0$, all velocity components of the supporting container are brought to zero instantaneously. To determine the subsequent motion of the resiliently supported body, the natural frequencies f_n in the coupled modes of response are first calculated using Eq. (3.36). Then the response motion of the resiliently supported body to the two types of velocity shock can be found by the analyses which follow.[21]

ONE PLANE OF SYMMETRY WITH ORTHOGONAL RESILIENT SUPPORTS EXCITED BY A TRANSLATIONAL VELOCITY SHOCK. Figure 3.30 shows a rigid body supported within a rigid container by resilient supports in such a manner that the YZ plane is a plane of symmetry. The entire system moves with constant velocity $\dot{v}_0$ and without relative motion. At time $t = 0$, the container impacts inelastically against the rigid wall shown at the right. The following initial conditions of displacement and velocity apply at the instant of impact ($t = 0$):

$$\dot{y}_c(0) = \dot{v}_0$$
$$x_c(0) = y_c(0) = z_c(0) = \alpha(0) = \beta(0) = \gamma(0) = 0 \qquad (3.71)$$
$$\dot{x}_c(0) = \dot{z}_c(0) = \dot{\alpha}(0) = \dot{\beta}(0) = \dot{\gamma}(0) = 0$$

As a result of the impact, the velocity of the supported rigid body tends to continue and is responsible for excitation of the system in the coupled mode of the y_c, z_c, α motions. The maximum displacements of the center-of-gravity of the supported body are

$$\frac{y_{cm}}{2\pi\dot{v}_0/f_z} = \frac{1}{B}\sum_{n=1}^{3}\left(\frac{|A_n|}{f_n/f_z}\right)$$
$$\frac{z_{cm}}{2\pi\dot{v}_0/f_z} = \frac{1}{B}\sum_{n=1}^{3}\left(\frac{|M_nA_n|}{f_n/f_z}\right) \qquad (3.72)$$
$$\frac{\alpha_m}{2\pi\dot{v}_0/\rho_xf_z} = \frac{1}{B}\sum_{n=1}^{3}\left(\frac{|N_nA_n|}{f_n/f_z}\right)$$

FORCED VIBRATION 3-39

The maximum accelerations of the center-of-gravity of the supported body are

$$\frac{\ddot{y}_{cm}}{2\pi f_z \dot{v}_0} = \frac{1}{B} \sum_{n=1}^{3} \left(|A_n| \frac{f_n}{f_z} \right)$$

$$\frac{\ddot{z}_{cm}}{2\pi f_z \dot{v}_0} = \frac{1}{B} \sum_{n=1}^{3} \left(|M_n A_n| \frac{f_n}{f_z} \right) \quad (3.73)$$

$$\frac{\ddot{\alpha}_m}{2\pi f_z \dot{v}_0/\rho_x} = \frac{1}{B} \sum_{n=1}^{3} \left(|N_n A_n| \frac{f_n}{f_z} \right)$$

where the subscript m denotes maximum value and

$$M_n = \frac{1}{1 - (f_n/f_z)^2} \left[\frac{\Sigma k_y}{\Sigma k_z} - \left(\frac{f_n}{f_z}\right)^2 \right] \frac{\Sigma k_z a_y}{\Sigma k_y a_z}$$

$$N_n = \left[\frac{\Sigma k_y}{\Sigma k_z} - \left(\frac{f_n}{f_z}\right)^2 \right] \frac{\rho_x \Sigma k_z}{\Sigma k_y a_z} \quad (3.74)$$

$$A_n = M_{n+1} N_{n+2} - M_{n+2} N_{n+1}$$

$$B = \left| \sum_{n=1}^{3} M_n (N_{n+1} - N_{n+2}) \right|$$

The fictitious natural frequency f_z is defined for mathematical purposes by Eq. (3.37). The numerical values of the subscript numbers n, $n+1$, $n+2$ denote the three natural frequencies in the coupled mode of the y_c, z_c, α motions determined from Eq. (3.36). These natural frequencies are arbitrarily assigned the values $n = 1, 2, 3$. When $n+1$

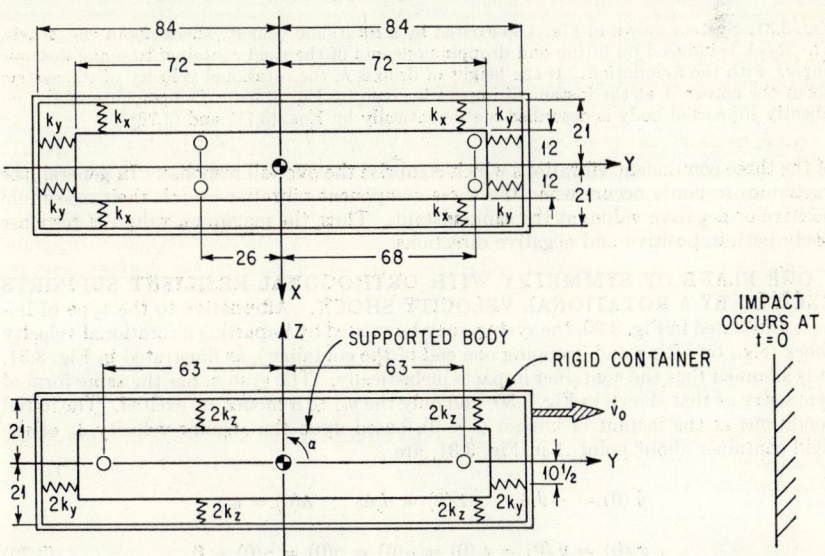

Fig. 3.30. Example of a rigid body supported within a rigid container by resilient elements with YZ a plane of symmetry. Excitation is by a translational velocity shock in the Y direction. Prior to impact the entire system moves with constant velocity $\dot{v}_0$ and without relative motion. The rigid container impacts inelastically against the wall shown at the right, and y_c, z_c, α motions of the internally supported body result, as described mathematically by Eqs. (3.72) and (3.73).

3-40 VIBRATION OF A RESILIENTLY SUPPORTED RIGID BODY

or $n + 2$ equals 4, use 1 instead; when $n + 2$ equals 5, use 2 instead. Maximum displacements and accelerations may be calculated for other points in the supported rigid body by using Eqs. (3.1) except that each of the terms must be made numerically additive. For example, the maximum value of the y displacement at the point b having the Z coordinate b_z is

$$y_{bm} = y_{cm} + |b_z|\alpha_m \qquad (3.75)$$

since $\gamma = 0$.

Since the system is assumed undamped, the response of the suspended body in terms of displacement or acceleration consists of a superposition of three sinusoidal components at the three natural frequencies in the coupled y_c, z_c, α mode. The absolute values of terms appear in Eq. (3.75) because the maximum response is the sum of the amplitudes

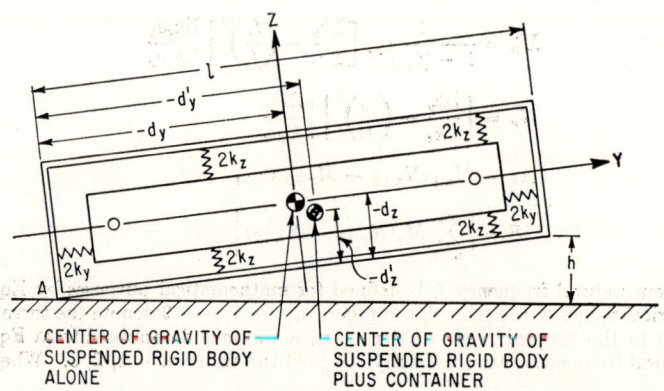

FIG. 3.31. System shown in Fig. 3.30 excited by a rotational velocity shock about the X axis. The shock is induced by lifting and dropping one end of the rigid container to make inelastic impact with the foundation. If the height of drop is h, the rotational velocity of the system about the corner A at the instant of impact is given by Eq. (3.79). The response of the resiliently supported body is described mathematically by Eqs. (3.77) and (3.78).

of the three component vibrations which comprise the over-all response. In general, the maximum response occurs when the three component vibrations reach their maximum positive or negative values at the same instant. Thus, the maximum values of response apply both in positive and negative directions.

ONE PLANE OF SYMMETRY WITH ORTHOGONAL RESILIENT SUPPORTS EXCITED BY A ROTATIONAL VELOCITY SHOCK. Alternative to the type of impact illustrated in Fig. 3.30, the system may be excited by imparting a rotational velocity shock (e.g., by lifting and dropping one end of the container), as illustrated in Fig. 3.31. It is assumed that the container impacts inelastically. The system has the same form of symmetry as that shown in Fig. 3.30, and only the y_c, z_c, α modes are excited. The initial conditions at the instant of impact ($t = 0$), based upon the angular velocity $\dot\alpha_0$ of the rigid container about point A in Fig. 3.31, are

$$\dot{y}_c(0) = -d_z\dot\alpha_0 \qquad \dot{z}_c(0) = d_y\dot\alpha_0 \qquad \dot\alpha(0) = \dot\alpha_0$$

$$x_c(0) = y_c(0) = z_c(0) = \alpha(0) = \beta(0) = \gamma(0) = 0 \qquad (3.76)$$

$$\dot{x}_c(0) = \dot\beta(0) = \dot\gamma(0) = 0$$

Note that d_y and d_z are negative quantities. The initial conditions in Eqs. (3.76) are based on the assumption that motion of the rigid body relative to the container during the fall is negligible compared to that which occurs after the impact. The maximum displacements of the center-of-gravity of the supported body are

$$\frac{y_{cm}}{2\pi\rho_x\dot{\alpha}_0/f_z} = \frac{1}{B}\sum_{n=1}^{3}\left[\left|\frac{d_z}{\rho_x}A_n + \frac{d_y}{\rho_x}(N_{n+1} - N_{n+2}) + (M_{n+2} - M_{n+1})\right|\frac{f_z}{f_n}\right]$$

$$\frac{z_{cm}}{2\pi\rho_x\dot{\alpha}_0/f_z} = \frac{1}{B}\sum_{n=1}^{3}\left[\left|M_n\left(\frac{d_z}{\rho_x}A_n + \frac{d_y}{\rho_x}(N_{n+1} - N_{n+2}) + (M_{n+2} - M_{n+1})\right)\right|\frac{f_z}{f_n}\right] \quad (3.77)$$

$$\frac{\alpha_m}{2\pi\dot{\alpha}_0/f_z} = \frac{1}{B}\sum_{n=1}^{3}\left[\left|N_n\left(\frac{d_z}{\rho_x}A_n + \frac{d_y}{\rho_x}(N_{n+1} - N_{n+2}) + (M_{n+2} - M_{n+1})\right)\right|\frac{f_z}{f_n}\right]$$

The maximum accelerations of the center-of-gravity of the supported body are

$$\frac{\ddot{y}_{cm}}{2\pi\rho_x f_z\dot{\alpha}_0} = \frac{1}{B}\sum_{n=1}^{3}\left[\left|\frac{d_z}{\rho_x}A_n + \frac{d_y}{\rho_x}(N_{n+1} - N_{n+2}) + (M_{n+2} - M_{n+1})\right|\frac{f_n}{f_z}\right]$$

$$\frac{\ddot{z}_{cm}}{2\pi\rho_x f_z\dot{\alpha}_0} = \frac{1}{B}\sum_{n=1}^{3}\left[\left|M_n\left(\frac{d_z}{\rho_x}A_n + \frac{d_y}{\rho_x}(N_{n+1} - N_{n+2}) + (M_{n+2} - M_{n+1})\right)\right|\frac{f_n}{f_z}\right] \quad (3.78)$$

$$\frac{\ddot{\alpha}_m}{2\pi f_z\dot{\alpha}_0} = \frac{1}{B}\sum_{n=1}^{3}\left[\left|N_n\left(\frac{d_z}{\rho_x}A_n + \frac{d_y}{\rho_x}(N_{n+1} - N_{n+2}) + (M_{n+2} - M_{n+1})\right)\right|\frac{f_n}{f_z}\right]$$

where d_z and d_y are the Z and Y coordinates, respectively, of the edges of the container, as shown in Fig. 3.31, and the other quantities are the same as those appearing in Eqs. (3.72) and (3.74). The maximum response at any point in the suspended body can be found in the manner of Eq. (3.75).

The rotational velocity $\dot{\alpha}_0$ of the container about the corner A in Fig. 3.31 may be induced by lifting the opposite end to a height h and dropping it. The resulting velocity $\dot{\alpha}_0$ is

$$\dot{\alpha}_0 = \left\{\frac{2g}{\rho_A{}^2}\left[\frac{h}{l}d_y' + d_z'\sqrt{1 - \left(\frac{h}{l}\right)^2} - d_z'\right]\right\}^{1/2} \quad (3.79)$$

where g is the acceleration of gravity, ρ_A is the radius of gyration of the rigid body plus container about the corner A, h is the initial elevation of the raised end of the container, l is the length of the container, and d_y' and d_z' are the Y and Z coordinates, respectively, of the edges of the container with respect to the center-of-gravity of the assembly of rigid body plus container (see Fig. 3.31).

Example 3.3. The rigid body shown in Fig. 3.31 weighs 1,500 lb and has a radius of gyration $\rho_x = 42$ in. with respect to the X axis. The resilient supporting elements apply forces parallel to their longitudinal axes *only*. Each element with its longitudinal axis in the X or Y direction has a stiffness of $k_x = k_y = 500$ lb/in. Each element whose longitudinal axis extends in the Z direction has a stiffness $k_z = 1,000$ lb/in. The resilient elements are positioned as shown in Fig. 3.30, and $l = 168$ in., $d_y = d_y' = -84$ in., $d_z = d_z' = -21$ in., $\rho_A = 308$ in. The rotational velocity shock results from a height of drop $h = 36$ in.

The fictitious natural frequency f_z is obtained from Eq. (3.37), yielding $f_z = 7.22$ cps. From Eq. (3.36) or Fig. 3.14, the natural frequencies in the y_c, z_c, α mode are $f_1 = 3.58$ cps, $f_2 = 6.02$ cps, and $f_3 = 9.75$ cps. From Eqs. (3.74), it is determined that $M_1 \simeq 0$, $M_2 = 11.7$, $M_3 = -15.3$, $N_1 = -0.1$, $N_2 = 7.1$, $N_3 = 25.1$, $A_1 = 402$, $A_2 = 2$, $A_3 = 1$, $B = 405$. Sample calculations for M_1 and A_1 are

$$M_1 = \frac{1}{1 - (3.58/7.22)^2}\left[\frac{4(500)}{8(1,000)} - \left(\frac{3.58}{7.22}\right)^2\right]\frac{4(1,000)(68 - 26)}{4(500)(-10.5)} = -0.04$$

$$A_1 = M_2N_3 - M_3N_2 = (11.7)(25.1) - (-15.3)(7.1) = 402$$

From Eq. (3.79), $\dot{\alpha}_0 = 0.38$ rad/sec. Then Eqs. (3.78) give the maximum acceleration of the center-of-gravity in the Y direction of the supported body as follows:

$$\ddot{y}_{cm} = \frac{2\pi\rho_x f_z \ddot{\alpha}_0}{B} \begin{bmatrix} \left|\dfrac{d_z}{\rho_x} A_1 + \dfrac{d_y}{\rho_x}(N_2 - N_3) + (M_3 - M_2)\right| \dfrac{f_1}{f_z} \\ + \left|\dfrac{d_z}{\rho_x} A_2 + \dfrac{d_y}{\rho_x}(N_3 - N_1) + (M_1 - M_3)\right| \dfrac{f_2}{f_z} \\ + \left|\dfrac{d_z}{\rho_x} A_3 + \dfrac{d_y}{\rho_x}(N_1 - N_2) + (M_2 - M_1)\right| \dfrac{f_3}{f_z} \end{bmatrix}$$

$$= \frac{724 \text{ in./sec}^2}{405} \begin{bmatrix} \left|\dfrac{-21}{42}(402) + \dfrac{-84}{42}(7.1 - 25.1) + (-15.3 - 11.7)\right| \dfrac{3.58}{7.22} \\ + \left|\dfrac{-21}{42}(2) + \dfrac{-84}{42}(25.1 + 0.1) + (0 + 15.3)\right| \dfrac{6.02}{7.22} \\ + \left|\dfrac{-21}{42}(1) + \dfrac{-84}{42}(-0.1 - 7.1) + (11.7 - 0)\right| \dfrac{9.75}{7.22} \end{bmatrix}$$

$= 286 \text{ in./sec}^2 = 0.74g$

In a similar manner:

$$z_{cm} = 1{,}580 \text{ in./sec}^2 = 4.09g$$
$$\ddot{\alpha}_m = 45.9 \text{ rad/sec}^2$$

REFERENCES

1. Marks, L. S., and T. Baumeister: "Mechanical Engineer's Handbook," 6th ed., p. **6**-6, McGraw-Hill Book Company, Inc., New York, 1958.
2. Housner, G. W., and D. E. Hudson: "Applied Mechanics-Dynamics," 2d ed., chap. 7, D. Van Nostrand Company, Inc., Princeton, N. J., 1959.
3. Boucher, R. W., D. A. Rich, H. L. Crane, and C. E. Matheny: *NACA Tech. Note* 3084, 1954.
4. Rubin, S.: *SAE Preprint* 197, 1957.
5. Macduff, J. N.: *Product Eng.*, **17**:106, 154 (1946).
6. Vane, F. F.: "A Guide for the Selection and Application of Resilient Mountings to Shipboard Equipment—Revised," *David W. Taylor Model Basin Rept.* 880, February, 1958, p. 98.
7. Ref. 6, p. 50.
8. Crede, C. E.: "Vibration and Shock Isolation," p. 68, John Wiley & Sons, Inc., New York, 1951.
9. Ref. 6, pp. 37–49.
10. Ref. 8, pp. 53–58.
11. Lewis, R. C., and K. Unholtz: *Trans. ASME*, **69**:813 (1947).
12. Klein, E., R. S. Ayre, and I. Vigness: "Fundamentals of Guided Missile Packaging," *Dept. Defense (U.S.) Rept.* RD 219/3, July, 1955, appendix 8, pp. 49–52.
13. Ref. 8, p. 73.
14. Taylor, E. S., and K. A. Browne: *J. Aeronaut. Sci.*, **6**:43 (1938).
15. Browne, K. A.: *Trans. SAE*, **44**:185 (1939).
16. Ref. 8, p. 50.
17. Himelblau, H.: "A Reliable Approach to Protecting Fragile Equipment from Aircraft Vibration," *North American Aviation, Inc., Rept.* NA-56-1030, 1957, pp. 16, 86.
18. Ref. 17, pp. 22, 95.
19. Ref. 8, pp. 43, 61.
20. Himelblau, H.: *Product Eng.*, **23**:151 (1952).
21. Ref. 12, chap. 11, November, 1955.
22. Ref. 2, appendix IV.

Table 3.1. Properties of Plane Sections
(After G. W. Housner and D. E. Hudson.[22])

The dimensions X_c, Y_c are the X, Y coordinates of the centroid, A is the area, $I_x\ldots$ is the area moment of inertia with respect to the $X\ldots$ axis, $\rho_x\ldots$ is the radius of gyration with respect to the $X\ldots$ axis; uniform solid cylindrical bodies of length l in the Z direction having the various plane sections as their cross sections have mass moment and product of inertia values about the Z axis equal to ρl times the values given in the table, where ρ is the mass density of the body; the radii of gyration are unchanged.

	Plane section	Area and centroid	Area moment of inertia	Square of radius of gyration	Area product of inertia
1	Right triangle	$A = \tfrac{1}{2}bh$ $X_c = \tfrac{2}{3}b$ $Y_c = \tfrac{1}{3}h$	$I_{x_c} = \dfrac{bh^3}{36}$ $I_{y_c} = \dfrac{b^3h}{36}$	$\rho_{x_c}^2 = \tfrac{1}{18}h^2$ $\rho_{y_c}^2 = \tfrac{1}{18}b^2$	$I_{x_c y_c} = \dfrac{A}{36}hb = \dfrac{h^2 b^2}{72}$
2		$A = \tfrac{1}{2}bh$ $X_c = \tfrac{1}{3}b$ $Y_c = \tfrac{1}{3}h$	$I_{x_c} = \dfrac{bh^3}{36}$ $I_{y_c} = \dfrac{b^3h}{36}$	$\rho_{x_c}^2 = \tfrac{1}{18}h^2$ $\rho_{y_c}^2 = \tfrac{1}{18}b^2$	$I_{x_c y_c} = -\dfrac{A}{36}hb = -\dfrac{h^2 b^2}{72}$
3	Triangle	$A = \tfrac{1}{2}bh$ $X_c = \tfrac{1}{3}(a+b)$ $Y_c = \tfrac{1}{3}h$	$I_{x_c} = \dfrac{bh^3}{36}$ $I_{y_c} = \dfrac{bh}{36}(b^2 - ab + a^2)$	$\rho_{x_c}^2 = \tfrac{1}{18}h^2$ $\rho_{y_c}^2 = \tfrac{1}{18}(b^2 - ab + a^2)$	$I_{x_c y_c} = \dfrac{Ah}{36}(2a - b) = \dfrac{bh^2}{72}(2a - b)$

Table 3.1 Properties of Plane Sections (Continued)

	Plane Section	Area and centroid	Area moment of inertia	Square of radius of gyration	Area product of inertia
4	Square	$A = a^2$ $X_c = \frac{1}{2}a$ $Y_c = \frac{1}{2}a$	$I_{x_c} = I_{y_c} = \frac{a^4}{12}$	$\rho_{x_c}^2 = \rho_{y_c}^2 = \frac{1}{12}a^2$	$I_{x_c y_c} = 0$
5	Rectangle	$A = bh$ $X_c = \frac{1}{2}b$ $Y_c = \frac{1}{2}h$	$I_{x_c} = \frac{bh^3}{12}$ $I_{y_c} = \frac{b^3 h}{12}$	$\rho_{x_c}^2 = \frac{1}{12}h^2$ $\rho_{y_c}^2 = \frac{1}{12}h^2$	$I_{x_c y_c} = 0$
6	Parallelogram	$A = ab \sin \theta$ $X_c = \frac{1}{2}(b + a \cos \theta)$ $Y_c = \frac{1}{2}(a \sin \theta)$	$I_{x_c} = \frac{a^3 b}{12} \sin^3 \theta$ $I_{y_c} = \frac{ab}{12} \sin \theta \, (b^2 + a^2 \cos^2 \theta)$	$\rho_{x_c}^2 = \frac{1}{12}(a \sin \theta)^2$ $\rho_{y_c}^2 = \frac{1}{12}(b^2 + a^2 \cos^2 \theta)$	$I_{x_c y_c} = \frac{a^3 b}{12} \sin^2 \theta \cos \theta$
7	Trapezoid	$A = \frac{1}{2} h (a + b)$ $Y_c = \frac{1}{3} h \left(\frac{2a + b}{a + b} \right)$	$I_{x_c} = \frac{h^3 (a^2 + 4ab + b^2)}{36(a + b)}$	$\rho_{x_c}^2 = \frac{h^2 (a^2 + 4ab + b^2)}{18(a + b)^2}$	

3-44

	Shape	Properties		
8	Circle	$A = \pi a^2$ $X_c = a$ $Y_c = a$	$I_{x_c} = I_{y_c} = \tfrac{1}{4}\pi a^4$	$\rho_{x_c}^2 = \rho_{y_c}^2 = \tfrac{1}{4}a^2$	$I_{x_c y_c} = 0$
9	Annulus	$A = \pi(a^2 - b^2)$ $X_c = a$ $Y_c = a$	$I_{x_c} = I_{y_c} = \dfrac{\pi}{4}(a^4 - b^4)$	$\rho_{x_c}^2 = \rho_{y_c}^2 = \tfrac{1}{4}(a^2 + b^2)$	$I_{x_c y_c} = 0$
10	Semicircle	$A = \tfrac{1}{2}\pi a^2$ $X_c = a$ $Y_c = \dfrac{4a}{3\pi}$	$I_{x_c} = \dfrac{a^4(9\pi^2 - 64)}{72\pi}$ $I_{y_c} = \tfrac{1}{8}\pi a^4$	$\rho_{x_c}^2 = \dfrac{a^2(9\pi^2 - 64)}{36\pi^2}$ $\rho_{y_c}^2 = \tfrac{1}{4}a^2$	$I_{x_c y_c} = 0$
11	Circular sector	$A = a^2 \theta$ $X_c = \dfrac{2a \sin\theta}{3\,\theta}$ $Y_c = 0$	$I_x = \tfrac{1}{4}a^4(\theta - \sin\theta\cos\theta)$ $I_y = \tfrac{1}{4}a^4(\theta + \sin\theta\cos\theta)$	$\rho_x^2 = \tfrac{1}{4}a^2\left(\dfrac{\theta - \sin\theta\cos\theta}{\theta}\right)$ $\rho_y^2 = \tfrac{1}{4}a^2\left(\dfrac{\theta + \sin\theta\cos\theta}{\theta}\right)$	$I_{x_c y_c} = 0$ $I_{xy} = 0$

Table 3.1. Properties of Plane Sections (Continued)

	Plane section	Area and centroid	Area moment of inertia	Square of radius of gyration	Area product of inertia
12	Circular segment	$A = a^2(\theta - \frac{1}{2}\sin 2\theta)$ $X_c = \frac{2a}{3}\left(\frac{\sin^3\theta}{\theta - \sin\theta\cos\theta}\right)$ $Y_c = 0$	$I_x = \frac{Aa^2}{4}\left[1 - \frac{2\sin^3\theta\cos\theta}{3(\theta - \sin\theta\cos\theta)}\right]$ $I_y = \frac{Aa^2}{4}\left[1 + \frac{2\sin^3\theta\cos\theta}{\theta - \sin\theta\cos\theta}\right]$	$\rho_x^2 = \frac{a^2}{4}\left[1 - \frac{2\sin^3\theta\cos\theta}{3(\theta - \sin\theta\cos\theta)}\right]$ $\rho_y^2 = \frac{a^2}{4}\left[1 + \frac{2\sin^3\theta\cos\theta}{\theta - \sin\theta\cos\theta}\right]$	$I_{x_c y_c} = 0$ $I_{xy} = 0$
13	Ellipse	$A = \pi ab$ $X_c = a$ $Y_c = b$	$I_{x_c} = \frac{\pi}{4}ab^3$ $I_{y_c} = \frac{\pi}{4}a^3b$	$\rho_{x_c}^2 = \frac{1}{4}b^2$ $\rho_{y_c}^2 = \frac{1}{4}a^2$	$I_{x_c y_c} = 0$
14	Semiellipse	$A = \frac{1}{2}\pi ab$ $X_c = a$ $Y_c = \frac{4b}{3\pi}$	$I_{x_c} = \frac{ab^3}{72\pi}(9\pi^2 - 64)$ $I_{y_c} = \frac{\pi}{8}a^3b$	$\rho_{x_c}^2 = \frac{b^2}{36\pi^2}(9\pi^2 - 64)$ $\rho_{y_c}^2 = \frac{1}{4}a^2$	$I_{x_c y_c} = 0$
15	Parabola	$A = \frac{4}{3}ab$ $X_c = \frac{3}{5}a$ $Y_c = 0$	$I_{x_c} = \frac{4}{15}ab^3$ $I_{y_c} = \frac{16}{175}a^3b$	$\rho_{x_c}^2 = \frac{1}{5}b^2$ $\rho_{y_c}^2 = \frac{12}{175}a^2$	$I_{x_c y_c} = 0$
16	Semiparabola	$A = \frac{2}{3}ab$ $X_c = \frac{3}{5}a$ $Y_c = \frac{3}{8}b$	$I_x = \frac{2}{15}ab^3$ $I_y = \frac{2}{7}a^3b$	$\rho_x^2 = \frac{1}{5}b^2$ $\rho_y^2 = \frac{3}{7}a^2$	$I_{xy} = \frac{A}{4}ab = \frac{1}{6}a^2b^2$

3–46

17 $Y \mid \underset{Y=\frac{h}{b^n}X^n}{}$![nth-degree parabola with C, h, b, X]	$A = \dfrac{bh}{n+1}$ $X_c = \dfrac{n+1}{n+2}b$ $Y_c = \dfrac{h}{2}\left(\dfrac{n+1}{2n+1}\right)$	$I_x = \dfrac{bh^3}{3(3n+1)}$ $I_y = \dfrac{hb^3}{n+3}$	$\rho_x^2 = \dfrac{h^2(n+1)}{3(3n+1)}$ $\rho_y^2 = \dfrac{n+1}{n+3}b^2$
18 $Y=\frac{h}{b^{\frac{1}{n}}}X^{\frac{1}{n}}$![nth-degree parabola]	$A = \dfrac{n}{n+1}bh$ $X_c = \dfrac{n+1}{2n+1}b$ $Y_c = \dfrac{n+1}{2(n+2)}h$	$I_x = \dfrac{n}{3(n+3)}bh^3$ $I_y = \dfrac{n}{3n+1}hb^3$	$\rho_x^2 = \dfrac{n+1}{3(n+3)}h^2$ $\rho_y^2 = \dfrac{n+1}{3n+1}b^2$

Table 3.2. Properties of Homogeneous Solid Bodies
(After G. W. Housner and D. E. Hudson.[22])

The dimensions X_c, Y_c, Z_c are the X, Y, Z coordinates of the centroid, S is the cross-sectional area of the thin rod or hoop in cases 1 to 3, V is the volume, I_x... is the mass moment of inertia with respect to the X... axis, ρ_x... is the radius of gyration with respect to the X... axis, ρ is the mass density of the body.

	Solid body	Volume and centroid	Mass moment of inertia	Radius of gyration squared	Mass product of inertia
1	Thin rod	$V = Sl$ $X_c = \frac{1}{2}l$ $Y_c = 0$ $Z_c = 0$	$I_{x_c} = 0$ $I_{y_c} = I_{z_c} = \frac{\rho V}{12}l^2$	$\rho_{x_c}^2 = 0$ $\rho_{y_c}^2 = \rho_{z_c}^2 = \frac{1}{12}l^2$	$I_{x_c y_c}$, etc. $= 0$
2	Thin circular rod	$V = 2SR\theta$ $X_c = \dfrac{R\sin\theta}{\theta}$ $Y_c = 0$ $Z_c = 0$	$I_x = I_{x_c}$ $= \dfrac{\rho V R^2(\theta - \sin\theta\cos\theta)}{2\theta}$ $I_y = \dfrac{\rho V R^2(\theta + \sin\theta\cos\theta)}{2\theta}$ $I_z = \rho V R^2$	$\rho_x^2 = \rho_{x_c}^2 = \dfrac{R^2(\theta - \sin\theta\cos\theta)}{2\theta}$ $r_y^2 = \dfrac{R^2(\theta + \sin\theta\cos\theta)}{2\theta}$ $\rho_z^2 = R^2$	$I_{x_c y_c}$, etc. $= 0$ I_{xy}, etc. $= 0$
3		$V = 2\pi SR$ $X_c = R$ $Y_c = R$ $Z_c = 0$	$I_{x_c} = I_{y_c} = \dfrac{\rho V}{2}R^2$ $I_{z_c} = \rho V R^2$	$\rho_{x_c}^2 = \rho_{y_c}^2 = \frac{1}{2}R^2$ $\rho_{z_c}^2 = R^2$	$I_{x_c y_c}$, etc. $= 0$

4 Cube	$V = a^3$ $X_c = \frac{1}{2}a$ $Y_c = \frac{1}{2}a$ $Z_c = \frac{1}{2}a$	$I_{x_c} = I_{y_c} = I_{z_c} = \frac{1}{6}\rho V a^2$	$\rho_{x_c}^2 = \rho_{y_c}^2 = \rho_{z_c}^2 = \frac{1}{6}a^2$	$I_{x_cy_c},$ etc. $= 0$
5 Rectangular prism	$V = abc$ $X_c = \frac{1}{2}a$ $Y_c = \frac{1}{2}b$ $Z_c = \frac{1}{2}c$	$I_{x_c} = \frac{1}{12}\rho V(b^2 + c^2)$	$\rho_{x_c}^2 = \frac{1}{12}(b^2 + c^2)$	$I_{x_cy_c},$ etc. $= 0$
6 Right rectangular pyramid	$V = \frac{1}{3}abh$ $X_c = 0$ $Y_c = \frac{1}{4}h$ $Z_c = 0$	$I_{x_c} = \frac{1}{80}\rho V(4b^2 + 3h^2)$ $I_{y_c} = \frac{1}{20}\rho V(a^2 + b^2)$	$\rho_{x_c}^2 = \frac{1}{80}(4b^2 + 3h^2)$ $\rho_{y_c}^2 = \frac{1}{20}(a^2 + b^2)$	$I_{x_cy_c},$ etc. $= 0$

3-49

Table 3.2. Properties of Homogeneous Solid Bodies (Continued)

	Solid body	Volume and centroid	Mass moment of inertia	Radius of gyration squared	Mass product of inertia
7	Right circular cone	$V = \tfrac{1}{3}\pi R^2 h$ $X_c = 0$ $Y_c = \tfrac{1}{4}h$ $Z_c = 0$	$I_{x_c} = I_{z_c} = \dfrac{3\rho V}{80}(4R^2 + h^2)$ $I_{y_c} = \tfrac{3}{10}\rho V R^2$	$\rho_{x_c}^2 = \rho_{z_c}^2 = \tfrac{3}{80}(4R^2 + h^2)$ $\rho_{y_c}^2 = \tfrac{3}{10}R^2$	$I_{x_c y_c}$, etc. $= 0$
8	Right circular cylinder	$V = \pi R^2 h$ $X_c = 0$ $Y_c = \tfrac{1}{2}h$ $Z_c = 0$	$I_{x_c} = I_{z_c} = \tfrac{1}{12}\rho V(3R^2 + h^2)$ $I_{y_c} = \tfrac{1}{2}\rho V R^2$	$\rho_{x_c}^2 = \rho_{z_c}^2 = \tfrac{1}{12}(3R^2 + h^2)$ $\rho_{y_c}^2 = \tfrac{1}{2}R^2$	$I_{x_c y_c}$, etc. $= 0$
9	Hollow right circular cylinder	$V = \pi h(R_1^2 - R_2^2)$ $X_c = 0$ $Y_c = \tfrac{1}{2}h$ $Z_c = 0$	$I_{x_c} = I_{z_c}$ $\quad = \tfrac{1}{12}\rho V(3R_1^2 + 3R_2^2 + h^2)$ $I_{y_c} = \tfrac{1}{2}\rho V(R_1^2 + R_2^2)$	$\rho_{x_c}^2 = \rho_{z_c}^2 = \tfrac{1}{12}(3R_1^2 + 3R_2^2 + h^2)$ $\rho_{y_c}^2 = \tfrac{1}{2}(R_1^2 + R_2^2)$	$I_{x_c y_c}$, etc. $= 0$

10	Sphere	$V = \frac{4}{3}\pi R^3$ $X_c = 0$ $Y_c = 0$ $Z_c = 0$	$I_{x_c} = \frac{2}{5}\rho V R^2$ $I_{y_c} = \frac{2}{5}\rho V R^2$ $I_{z_c} = \frac{2}{5}\rho V R^2$	$\rho_{x_c}^2 = \frac{2}{5}R^2$ $\rho_{y_c}^2 = \frac{2}{5}R^2$ $\rho_{z_c}^2 = \frac{2}{5}R^2$ $I_{x_c y_c}$, etc. $= 0$
11	Hollow sphere	$V = \frac{4}{3}\pi(R_1^3 - R_2^3)$ $X_c = 0$ $Y_c = 0$ $Z_c = 0$	$I_x = I_y = I_z$ $= \frac{2}{5}\rho V \dfrac{R_1^5 - R_2^5}{R_1^3 - R_2^3}$	$\rho_x^2 = \rho_y^2 = \rho_z^2$ $= \dfrac{2}{5}\dfrac{R_1^5 - R_2^5}{R_1^3 - R_2^3}$ I_{xy}, etc. $= 0$
12	Hemisphere	$V = \frac{2}{3}\pi R^3$ $X_c = 0$ $Y_c = \frac{3}{8}R$ $Z_c = 0$	$I_x = I_y = I_z = \frac{2}{5}\rho V R^2$	$\rho_x^2 = \rho_y^2 = \rho_z^2 = \frac{2}{5}R^2$ $I_{x_c y_c}$, etc. $= 0$ I_{xy}, etc. $= 0$

3-51

Table 3.2. Properties of Homogeneous Solid Bodies (Continued)

	Solid body	Volume and centroid	Mass moment of inertia	Radius of gyration squared	Mass product of inertia
13	Ellipsoid	$V = \frac{4}{3}\pi abc$ $X_c = 0$ $Y_c = 0$ $Z_c = 0$	$I_x = \frac{1}{5}\rho V(b^2 + c^2)$ $I_y = \frac{1}{5}\rho V(a^2 + c^2)$ $I_z = \frac{1}{5}\rho V(a^2 + b^2)$	$\rho_x^2 = \frac{1}{5}(b^2 + c^2)$ $\rho_y^2 = \frac{1}{5}(a^2 + c^2)$ $\rho_z^2 = \frac{1}{5}(a^2 + b^2)$	I_{xy}, etc. $= 0$
14	Paraboloid of revolution	$V = \frac{1}{2}\pi R^2 h$ $X_c = \frac{2}{3}h$ $Y_c = 0$ $Z_c = 0$	$I_{x_c} = \frac{1}{3}\rho V R^2$ $I_{y_c} = I_{z_c} = \frac{1}{18}\rho V(3R^2 + h^2)$	$\rho_{x_c}^2 = \frac{1}{3}R^2$ $\rho_{y_c}^2 = \rho_{z_c}^2 = \frac{1}{18}(3R^2 + h^2)$	$I_{x_c y_c}$, etc. $= 0$
15	Elliptic paraboloid	$V = \frac{1}{2}\pi abc$ $X_c = \frac{2}{3}a$ $Y_c = 0$ $Z_c = 0$	$I_{x_c} = \frac{1}{6}\rho V(b^2 + c^2)$ $I_{y_c} = \frac{1}{18}\rho V(3c^2 + a^2)$ $I_{z_c} = \frac{1}{18}\rho V(3b^2 + a^2)$	$\rho_{x_c}^2 = \frac{1}{6}(b^2 + c^2)$ $\rho_{y_c}^2 = \frac{1}{18}(3c^2 + a^2)$ $\rho_{z_c}^2 = \frac{1}{18}(3b^2 + a^2)$	$I_{x_c y_c}$, etc. $= 0$

4
NONLINEAR VIBRATION

H. Norman Abramson
Southwest Research Institute

INTRODUCTION

A vast body of scientific knowledge has been developed over a long period of time devoted to a description of natural phenomena. In the field of mechanics, rapid progress in the past two centuries has occurred, due in large measure to the ability of investigators to represent physical laws in terms of rather simple equations. In many cases the governing equations were not so simple; therefore, certain assumptions, more or less consistent with the physical situation, were employed to reduce the equations to types more easily soluble. Thus, the process of linearization has become an intrinsic part of the rational analysis of physical problems. An analysis based on linearized equations, then, may be thought of as an analysis of a corresponding but idealized problem.

In many instances the linear analysis is insufficient to describe the behavior of the physical system adequately. In fact, one of the most fascinating features of a study of nonlinear problems is the occurrence of new and totally unsuspected phenomena; i.e., new in the sense that the phenomena are not predicted, or even hinted at, by the linear theory. On the other hand, certain phenomena observed physically are unexplainable except by giving due consideration to nonlinearities present in the system.

The branch of mechanics that has been subjected to the most intensive attack from the nonlinear viewpoint is the theory of vibration of mechanical and electrical systems. Other branches of mechanics, such as incompressible and compressible fluid flow, elasticity, plasticity, wave propagation, etc., also have been studied as nonlinear problems, but the greatest progress has been made in treating vibration of nonlinear systems. The systems treated in this chapter are systems with a finite number of degrees-of-freedom which can be defined by a finite number of simultaneous ordinary differential equations; on the other hand, the mechanics of continua involves partial differential equations. Nonlinear ordinary differential equations are easier to handle than nonlinear partial differential equations. Interesting surveys of the entire realm of nonlinear mechanics are given in Refs. 1 and 2.

This chapter provides information concerning features of nonlinear vibration theory likely to be encountered in practice, and methods of nonlinear vibration analysis which find ready application.

EXAMPLES OF SYSTEMS POSSESSING NONLINEAR CHARACTERISTICS

SIMPLE PENDULUM

As a first example of a system possessing nonlinear characteristics, consider a simple pendulum of length l having a bob of mass m, as shown in Fig. 4.1. The well-known differential equation governing free vibration is

$$ml^2\ddot{\theta} + mgl\theta = 0 \qquad (4.1)$$

This equation holds only for small oscillations about the position of equilibrium since the actual restoring moment is characterized by the quantity $\sin \theta$. Equation (4.1) thus employs the assumption $\sin \theta \simeq \theta$. The exact, but nonlinear, equation of motion is

$$ml^2\ddot{\theta} + mgl \sin \theta = 0 \qquad (4.2)$$

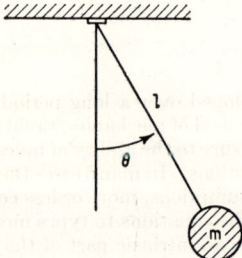 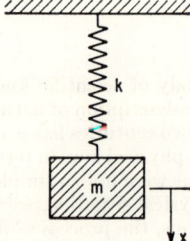

Fig. 4.1. Simple pendulum.   Fig. 4.2. Simple spring-mass system.

SIMPLE SPRING-MASS SYSTEM

A simple spring-mass system, as shown in Fig. 4.2, is characterized by the equation

$$m\ddot{x} + kx = 0$$

This equation is based on the assumption that the elastic spring obeys Hooke's law; i.e., the characteristic curve of restoring force versus displacement is a straight line. How-

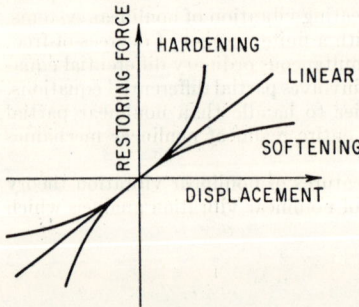

Fig. 4.3. Restoring force characteristic curves for linear, hardening, and softening vibration systems.

ever, many materials do not exhibit such a linear characteristic. Further, in the case of a simple coil spring, a deviation from linearity occurs at large compression as the coils begin to close up, or conversely, when the extension becomes so great that the coils begin to lose their individual identity. In either case, the spring exhibits a characteristic such that the restoring force increases more rapidly than the displacement. Such a characteristic is called *hardening*. In a similar manner, certain systems (e.g., a simple pendulum) exhibit a *softening* characteristic. Both types of characteristic are shown in Fig. 4.3. A simple system with either softening or hardening restoring force may be described approximately by an

EXAMPLES OF SYSTEMS POSSESSING NONLINEAR CHARACTERISTICS

equation of the form

$$m\ddot{x} + k(x \pm \mu^2 x^3) = 0$$

where the upper sign refers to the hardening characteristic and the lower to the softening characteristic.

It is possible for a system with only linear components to exhibit nonlinear characteristics, by snubber action for example, as shown in Fig. 4.4. A system undergoing vibration of small amplitude also may exhibit nonlinear characteristics; for example, in the pendulum shown in Fig. 4.5, the length depends on the amplitude.

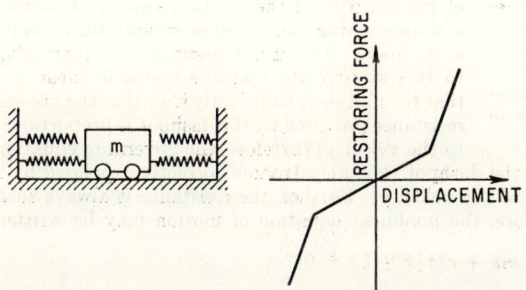
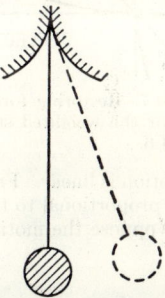

FIG. 4.4. Nonlinear mechanical system with snubber action showing piecewise linear restoring force characteristic curve.

FIG. 4.5. Pendulum with nonlinear characteristic resulting from dependence of length on vibration amplitude.

STRETCHED STRING WITH CONCENTRATED MASS

The large amplitude vibration of a stretched string with a concentrated mass, as shown in Fig. 4.6, offers another example of a nonlinear system. The governing nonlinear differential equation is, approximately,

$$m\ddot{w} + F_0 \left(\frac{l}{ab}\right) w + (SE - F_0) \left(\frac{a^3 + b^3}{2a^3 b^3}\right) w^3 = 0$$

where F_0 is the initial tension, S is the cross-sectional area, and E is the elastic modulus of the string. Consider now the special case of $a = b$ and denote the unstretched length

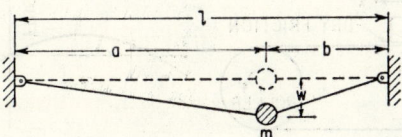

FIG. 4.6. Vibration of a weighted string as an example of a nonlinear system.

of the half string by l_0. Then the initial tension and the restoring force become

$$F_0 = SE\left(\frac{a - l_0}{l_0}\right)$$

$$F_r \simeq SE\left[2\left(\frac{a}{l_0} - 1\right)\left(\frac{w}{a}\right) + \left(2 - \frac{a}{l_0}\right)\left(\frac{w}{a}\right)^3\right]$$

NONLINEAR VIBRATION

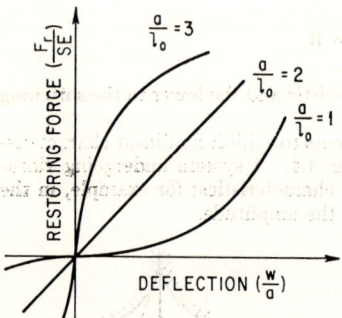

FIG. 4.7. Restoring force characteristics for the weighted string shown in Fig. 4.6.

An interesting feature of this system is that it exhibits a wide variety of either hardening or softening characteristics, depending upon the value of a/l_0, as shown in Fig. 4.7.

SYSTEM WITH VISCOUS DAMPING

The foregoing examples all involve nonlinearities in the elastic components, either as a result of appreciable amplitudes of vibration or as a result of peculiarities of the elastic element. Consider a simple spring-mass system which also includes a dashpot. The usual assumptions pertaining to this system are that the spring is linear and that the motion is sufficiently slow that the viscous resistance provided by the dashpot is proportional to the velocity; therefore, the governing equation of motion is linear. Frequently, the dashpot resistance is more correctly expressed by a term proportional to the square of the velocity. Further, the resistance is always such as to oppose the motion; therefore, the nonlinear equation of motion may be written

$$m\ddot{x} + c|\dot{x}|\dot{x} + kx = 0$$

BELT FRICTION SYSTEM

The system shown in Fig. 4.8 involves a nonlinearity depending upon the dry friction between the mass and the moving belt. The belt has a constant speed v_0, and the applicable equation of motion is

$$m\ddot{x} + F(\dot{x}) + kx = 0$$

where the friction force $F(\dot{x})$ is shown in Fig. 4.9. For large values of displacement, the damping term is positive, has positive slope, and removes energy from the system; for small values of displacement, the damping term is negative, has negative slope, and actually puts energy into the system. Even though there is no external stimulus, the system can have an oscillatory solution, and thus corresponds to a nonlinear *self-excited* system.

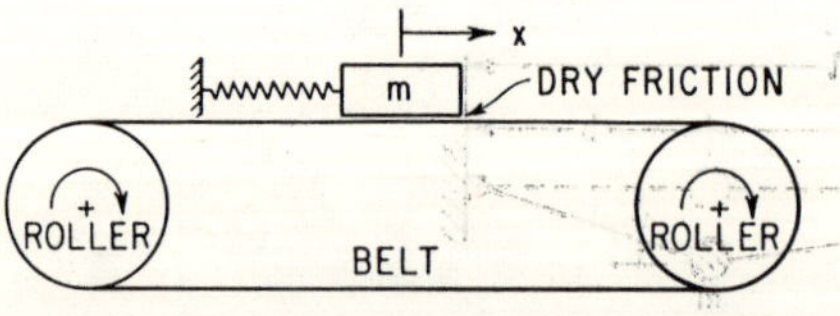

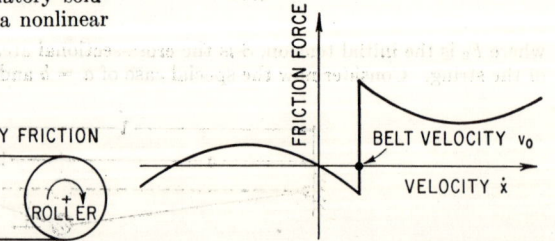

FIG. 4.8. Belt friction system which exhibits self-excited vibration.

FIG. 4.9. Damping force characteristic curve for the belt friction system shown in Fig. 4.8.

Many other examples of nonlinear systems are given in the references of this chapter, particularly Refs. 3 to 5.

DESCRIPTION OF NONLINEAR PHENOMENA

This section describes briefly, largely in nonmathematical terms, certain of the more important features of nonlinear vibration. Further details and methods of analysis are given later.

FREE VIBRATION

In so far as the free vibration of a system is concerned, one distinguishing feature between linear and nonlinear behavior is the dependence of the period of the motion in nonlinear vibration on the amplitude. For example, the simple pendulum of Fig. 4.1 may be analyzed on the basis of the linearized equation of motion, Eq. (4.1), from which it is found that the period of the vibration is given by the constant value $\tau_0 = 2\pi/\omega_n$. An analysis on the basis of the nonlinear equation of motion, Eq. (4.2), leads to an expression for the period of the form

$$\frac{\tau}{\tau_0} = 1 + \tfrac{1}{4}(U)^2 + \tfrac{9}{64}(U)^4 + \tfrac{25}{256}(U)^6 + \cdots \tag{4.3}$$

where U is related to the amplitude of the vibration Θ by the relation $U = \sin(\Theta/2)$. The linear solution thus corresponds to the first term of Eq. (4.3). The dependence of

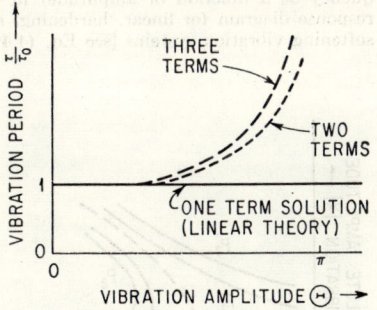

FIG. 4.10. Period of free vibration of a simple pendulum according to Eq. (4.3) and showing the effect of nonlinear terms.

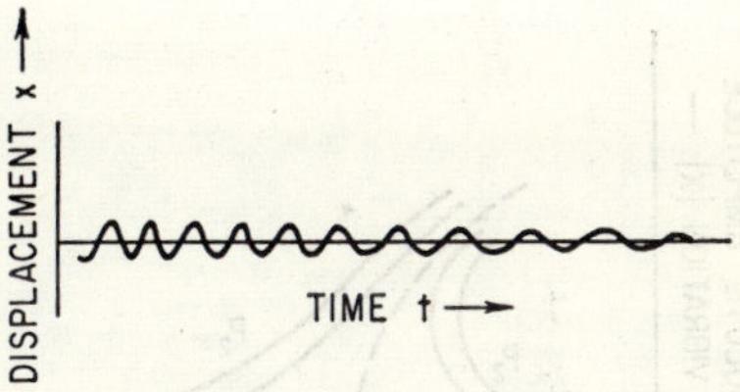

FIG. 4.11. Deflection time-history for free damped vibration of the nonlinear system described by Duffing's equation [Eq. (4.16)].

the period of vibration on amplitude is shown in Fig. 4.10. Systems in which the period of vibration is independent of the amplitude are called *isochronous*, while those in which the period τ is dependent on the amplitude are called *nonisochronous*.

The dependence of period on amplitude also may be seen from the vibration trace shown in Fig. 4.11, which corresponds to a solution of the equation

$$m\ddot{x} + c\dot{x} + k(x + \mu^2 x^3) = 0$$

CHARACTER OF THE RESPONSE CURVES FOR FORCED VIBRATION

Representations of vibration behavior in the form of curves of response amplitude versus exciting frequency are called *response curves*. The response curves for an undamped linear system acted on by a harmonic exciting force of amplitude p and frequency ω may be derived from the equation of motion

$$\ddot{x} + \omega_n^2 x = \frac{p}{m} \cos \omega t \tag{4.4}$$

The solution has the form shown in Fig. 4.12. The vertical line at $\omega = \omega_n$ corresponds not only to resonance, but also to free vibration ($p = 0$); the amplitude in this instance is determined by the initial conditions of the motion. In a nonlinear system the character of the motion is dependent upon the amplitude. This requires that the natural frequency likewise be amplitude-dependent; hence, it follows that the free vibration curve $p = 0$ for nonlinear systems cannot be a straight line. Figure 4.13 shows free vibration

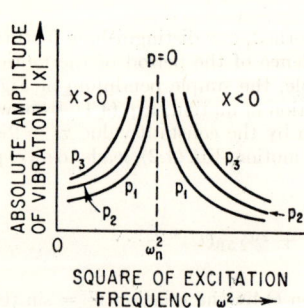
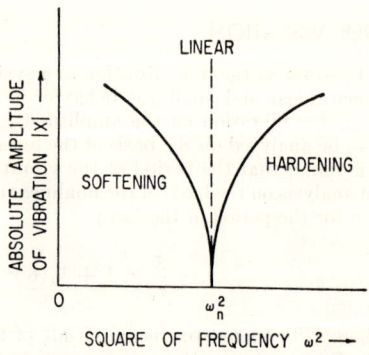

Fig. 4.12. Family of response curves for the undamped linear system defined by Eq. (4.4).

Fig. 4.13. Free vibration curves (natural frequency as a function of amplitude) in the response diagram for linear, hardening, and softening vibration systems [see Eq. (4.49)].

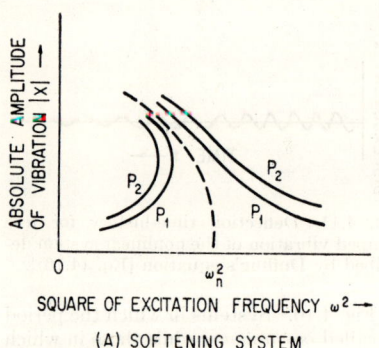
(A) SOFTENING SYSTEM

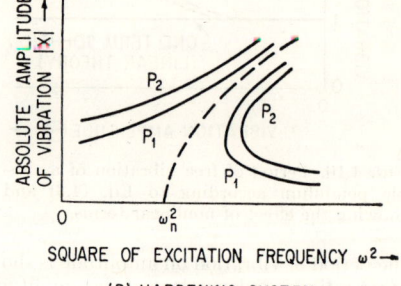
(B) HARDENING SYSTEM

Fig. 4.14. Response curves for undamped nonlinear systems with hardening and softening restoring force characteristics [see Eq. (4.50)].

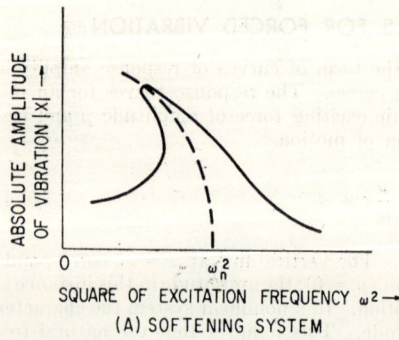
(A) SOFTENING SYSTEM

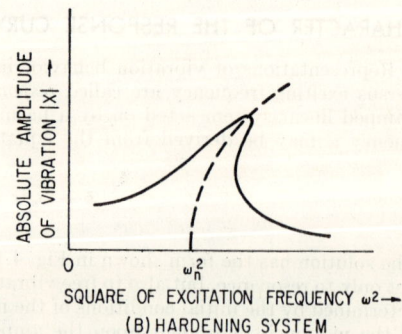
(B) HARDENING SYSTEM

Fig. 4.15. Response curves for damped nonlinear systems with hardening and softening restoring force characteristics [see Eq. (4.52)].

curves (i.e., natural frequency as a function of amplitude) for hardening and softening systems.

Figures 4.12 and 4.13 suggest that the forced vibration response curves for systems with nonlinear restoring forces have the general form of those of a linear system, but are "swept over" to the right or left, depending on whether the system is hardening or softening. These are shown in Fig. 4.14. The principal effect of damping in forced vibration of a nonlinear system is to limit the amplitude at resonance, as shown in Fig. 4.15.

STABILITY

Consider a hardening system whose response curve is shown in Fig. 4.15B. Suppose that the exciting frequency starts at a low value, and increases continuously at a slow rate. The amplitude of the vibration also increases, but only up to a point. In particular, at

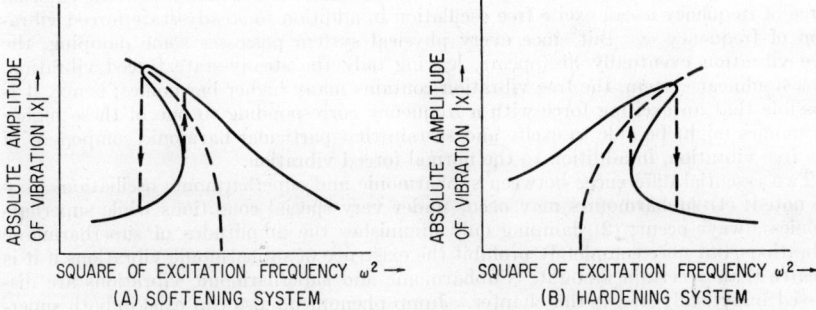

FIG. 4.16. Jump phenomenon in hardening and softening systems.

the point of vertical tangency of the response curve, a slight increase in frequency requires that the system perform in an unusual manner; i.e., that it "jump" down in amplitude to the lower branch of the response curve. This experiment may be repeated by starting with a large value of exciting frequency, but requiring that the forcing frequency be continuously reduced. A similar situation again is encountered; the system must jump up in amplitude in order to meet the conditions of the experiment. This *jump phenomenon* is shown in Fig. 4.16 for both the hardening and softening systems. The jump is not instantaneous in time, but requires a few cycles of vibration to establish a steady-state vibration at the new amplitude.

There is a portion of the response curve which is "unattainable"; it is not possible to obtain that particular amplitude by a suitable choice of forcing frequency. Thus, for certain values of ω there appear to be three possible amplitudes of vibration but only the upper and lower can actually exist. If by some means it were possible to initiate a steady-state vibration with just the proper amplitude and frequency to correspond to the middle branch, the condition would be unstable; at the slightest disturbance the motion would jump to either of the other two states of motion. The direction of the jump depends on the direction of the disturbance. Thus, of the three possible states of motion, one in phase and two out-of-phase with the exciting force, the one having the larger amplitude of the two out-of-phase motions is

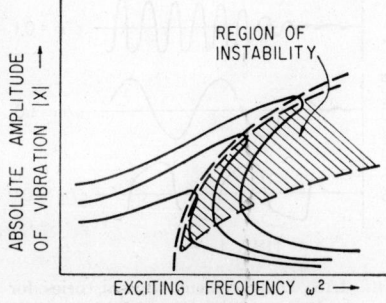

FIG. 4.17. Instability region defined by the loci of vertical tangents of the damped response curves for the hardening system.

unstable. This region of instability in the response diagram is defined by the loci of vertical tangents to the response curves, and is shown for a hardening system in Fig. 4.17.

SUBHARMONIC AND SUPERHARMONIC VIBRATIONS

In the preceding discussion, only the harmonic solutions of nonlinear differential equations were considered; i.e., the solutions described were those whose frequency is the same as that of the exciting force $F \cos \omega t$. However, permanent oscillations whose frequency is a fraction ($\frac{1}{2}$, $\frac{1}{3}$, ..., $1/n$) of that of the exciting force also can occur in nonlinear systems. The term *subharmonic vibration* usually is applied to this phenomenon, although the term *frequency demultiplication* is also used. In a similar manner, *superharmonic vibration* of frequency equal to some multiple (2, 3, ..., n) of the exciting frequency also can occur.

Although subharmonic and superharmonic oscillations occur in nonlinear systems, it is not a simple matter to give a plausible physical explanation for their occurrence. In a linear system, if the frequency of the free vibration is ω_n, then a periodic external force of frequency ω can excite free oscillation in addition to steady-state forced vibration of frequency ω. But since every physical system possesses some damping, the free vibration eventually disappears, leaving only the steady-state forced vibration. In a nonlinear system, the free vibration contains many higher harmonics; hence, it is possible that an exciting force with a frequency corresponding to one of these higher harmonics might be able to excite and sustain that particular harmonic component of the free vibration, in addition to the normal forced vibration.

Two essential differences between subharmonic and superharmonic oscillations may be noted: (1) subharmonics may occur under very special conditions while superharmonics always occur; (2) damping only diminishes the amplitudes of superharmonic vibrations, but may completely prohibit the existence of subharmonic vibrations if it is greater than a certain amount. Subharmonic and superharmonic vibrations are discussed analytically later in this chapter. Jump phenomena also can exist in both superharmonic and subharmonic vibrations.

OTHER PHENOMENA

SELF-EXCITED VIBRATION.* Consider the nonlinear equation of motion

$$m\ddot{x} + c(x^2 - 1)\dot{x} + kx = 0$$

This is known as Van der Pol's equation and may be written alternatively

$$\ddot{x} - \epsilon(1 - x^2)\dot{x} + \kappa^2 x = 0 \tag{4.5}$$

The principal feature of this self-excited system resides in the damping term; for small displacements the damping is negative, and for large displacements the damping is positive. Thus, even an infinitesimal disturbance causes the system to oscillate; however, when the displacement becomes sufficiently large, the damping becomes positive and limits further increase in amplitude. This is shown in Fig. 4.18. Such systems, which start in a spontaneous manner, often are called "soft" systems in contrast with "hard" systems which exhibit sustained oscillations only if a shock in excess of a certain level is applied. Note that stability questions arise here (which are different from those discussed earlier in connection with jump phenomena) concerning the existence of one or more limiting amplitudes, such as the one noted above in the Van der Pol oscillator.

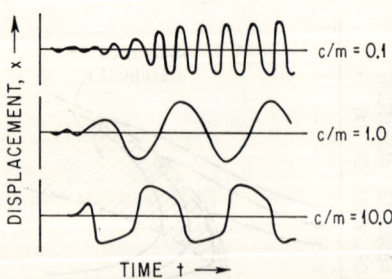

FIG. 4.18. Displacement time-histories for Van der Pol's equation [Eq. (4.5)] for various values of the damping.

* A general treatment of self-excited vibration is given in Chap. 5.

RELAXATION OSCILLATIONS. As shown in Fig. 4.18, the motion of the Van der Pol oscillator is very nearly harmonic for $c/m = 0.1$ while the motion is made up of relatively sudden transitions between deflections of opposite sign for $c/m = 10.0$. The period of the harmonic motion for $c/m = 0.1$ is determined essentially by the linear spring stiffness k and the mass m; the period of the motion corresponding to $c/m = 10.0$ is very much larger and depends also on c. Thus, it is possible to obtain an undamped periodic oscillation in a damped system as a result of the particular behavior of the damping term. Such oscillations are often called *relaxation oscillations*.

ASYNCHRONOUS EXCITATION AND QUENCHING. In linear systems, the principle of superposition is valid, and there is no interaction between different oscillations. Moreover, the mathematical existence of a periodic solution always indicates the existence of a periodic phenomenon. In nonlinear systems, there is an interaction between oscillations; the mathematical existence of a periodic solution is only a necessary condition for the existence of corresponding physical phenomena. When supplemented by the condition of stability, the conditions become both necessary and sufficient for the appearance of the physical oscillation. Therefore, it is conceivable that under these conditions the appearance of one oscillation may either create or destroy the stability condition for another oscillation. In the first case, the other oscillation appears (*asynchronous excitation*), and in the second case, disappears (*asynchronous quenching*). The term asynchronous is used to indicate that there is no relation between the frequencies of these two oscillations.

ENTRAINMENT OF FREQUENCY. According to linear theory, if two frequencies ω_1 and ω_2 are caused to beat in a system, the period of beating increases indefinitely as ω_2 approaches ω_1. In nonlinear systems, the beats disappear as ω_2 reaches certain values. Thus, the frequency ω_1 falls in synchronism with, or is entrained by, the frequency ω_2 within a certain range of values. This is called *entrainment of frequency*, and the band of frequencies in which entrainment occurs is called the zone of entrainment or the interval of synchronization. In this region, the frequencies ω_1 and ω_2 combine and only vibration at a single frequency ensues.

EXACT SOLUTIONS

It is possible to obtain exact solutions for only a relatively few second-order nonlinear differential equations. In this section, some of the more important of these exact solutions are listed. They are exact in the sense that the solution is given either in closed form or in an expression that can be evaluated numerically to any desired degree of accuracy. The examples given here are fairly general in nature; more specialized examples are given in Ref. 4.

FREE VIBRATION

Consider the free vibration of an undamped system with a general restoring force $f(x)$ as governed by the differential equation

$$\ddot{x} + \kappa^2 f(x) = 0$$

This can be rewritten as

$$\frac{d(\dot{x}^2)}{dx} + 2\kappa^2 f(x) = 0 \tag{4.6}$$

and integrated to yield

$$\dot{x}^2 = 2\kappa^2 \int_x^X f(\xi)\, d\xi$$

where ξ is an integration variable and X is the value of the displacement when $\dot{x} = 0$. Thus

$$|\dot{x}| = \kappa\sqrt{2}\sqrt{\int_x^X f(\xi)\, d\xi}$$

NONLINEAR VIBRATION

This may be integrated again to yield

$$t - t_0 = \frac{1}{\kappa\sqrt{2}} \int_0^x \frac{d\zeta}{\sqrt{\int_\zeta^X f(\xi)\, d\xi}} \tag{4.7}$$

where ζ is an integration variable and t_0 corresponds to the time when $x = 0$. The displacement-time relation may be obtained by inverting this result. Considering the restoring force term to be an odd function, i.e.,

$$f(-x) = -f(x)$$

and considering Eq. (4.7) to apply to the time from zero displacement to maximum displacement, the period τ of the vibration is

$$\tau = \frac{4}{\kappa\sqrt{2}} \int_0^X \frac{d\zeta}{\sqrt{\int_\zeta^X f(\xi)\, d\xi}} \tag{4.8}$$

Exact solutions can be obtained in all cases where the integrals in Eq. (4.8) can be expressed explicitly in terms of X.

CASE 1. PURE POWERS OF DISPLACEMENT. Consider the restoring force function

$$f(x) = x^n$$

Equation (4.8) then becomes

$$\tau = \frac{4}{\kappa} \sqrt{\frac{n+1}{2}} \int_0^X \frac{d\zeta}{\sqrt{X^{n+1} - \zeta^{n+1}}}$$

Setting $u = \zeta/X$,

$$\tau = \frac{4}{\kappa\sqrt{X^{n-1}}} \left(\sqrt{\frac{n+1}{2}} \int_0^1 \frac{du}{\sqrt{1 - u^{n+1}}} \right)$$

The expression within the parentheses depends only on the parameter n and is denoted by $\psi(n)$. Thus

$$\tau = \frac{4}{\kappa\sqrt{X^{n-1}}} \psi(n) \tag{4.9}$$

The factor $\psi(n)$ may be evaluated numerically to any desired degree of accuracy, and is tabulated in Table 4.1.

CASE 2. POLYNOMIALS OF DISPLACEMENT. Consider the binomial restoring force

$$f(x) = x^n + \mu x^m \qquad [m > n \geq 0]$$

Introducing this expression into Eq. (4.8) and performing the integrations:[5]

$$\tau = \frac{4}{\kappa\sqrt{X^{n+1}}} \left(\sqrt{\frac{n+1}{2}} \int_0^1 \frac{du}{\sqrt{(1 + \bar{\mu}) - (u^{n+1} + \bar{\mu} u^{m+1})}} \right) \tag{4.10}$$

where

$$\bar{\mu} = \mu X^{m-n} \left(\frac{n+1}{m+1} \right) \tag{4.11}$$

For particular values of n, m, and $\bar{\mu}$, the expression within the parentheses can be evaluated to any desired degree of accuracy by numerical methods. The extension of this method to higher-order polynomials can be made quite readily.

EXACT SOLUTIONS

CASE 3. HARMONIC FUNCTION OF DISPLACEMENT. Consider now the problem of the simple pendulum which has a restoring force of the form

$$f(x) = \sin x$$

Introducing this relation into Eq. (4.7):

$$t - t_0 = \frac{1}{2\kappa} \int_0^x \frac{d\zeta}{\sqrt{\sin^2 \frac{X}{2} - \sin^2 \frac{\zeta}{2}}}$$

If $x = X$ and $t_0 = 0$, this integral can be reduced to the standard form of the complete elliptic integral of the first kind:

$$\widehat{K}(\alpha) = \int_0^{\pi/2} \frac{dv}{\sqrt{1 - \sin^2 \alpha \sin^2 v}} \tag{4.12}$$

Thus, the period of vibration is

$$\tau = \frac{1}{\kappa} \widehat{K}\left(\frac{X}{2}\right) \tag{4.13}$$

The displacement-time function can be obtained by inversion and leads to the inverse elliptic functions. Replacing $\sin \alpha$ by U in Eq. (4.12), expanding by the binomial theorem, and then integrating yields Eq. (4.3).

CASE 4. VELOCITY SQUARED DAMPING. As indicated by Eq. (4.6), the introduction of any other function of $\dot{x}^2$ does not complicate the problem. Thus, the differential equation *

$$\ddot{x} \pm \frac{\delta}{2} \dot{x}^2 + \kappa^2 f(x) = 0$$

can be reduced to

$$\frac{d(\dot{x}^2)}{dx} \pm \delta \dot{x}^2 = 2\kappa^2 f(x)$$

Integrating the above equation,

$$\dot{x}^2 = 2\kappa^2 e^{\mp \delta x} \int_x^X e^{\pm \delta \xi} f(\xi)\, d\xi$$

Integrating again,

$$t = \int_{x_0}^x \frac{d\eta}{\dot{x}(\eta)}$$

where η is an integration variable.

FORCED VIBRATION

Exact solutions for forced vibration of nonlinear systems are virtually nonexistent, except as the system can be represented in a stepwise linear manner. For example, consider a system with a stepwise linear symmetrical restoring force characteristic, as shown in Fig. 4.4. Denote the lower of the two stiffnesses by k_1, the upper by k_2, and the dis-

---

* The $\pm$ sign is employed here, and elsewhere in this chapter, to account for the proper direction of the resisting force. Consequently, reference frequently is made to upper or lower sign rather than to plus or minus.

placement at which the change in stiffness occurs by x_1. Thus, the problem reduces to the solution of two linear differential equations:

$$m\ddot{x}' + k_1 x' = \pm P \sin \omega t \qquad [x_1 \geq x' \geq 0] \qquad (4.14a)$$

$$m\ddot{x}'' + (k_1 - k_2)x_1 + k_2 x'' = \pm P \sin \omega t \qquad [x'' \geq x_1] \qquad (4.14b)$$

where the upper sign refers to in-phase exciting force and the lower sign to out-of-phase exciting force. The appropriate boundary conditions are

$$\begin{aligned} x'(t=0) &= 0 \\ x'(t=t_1) &= x''(t=t_1) = x_1 \\ \dot{x}'(t=t_1) &= \dot{x}''(t=t_1) \\ \dot{x}''\left(t = \frac{\pi}{2\omega}\right) &= 0 \end{aligned} \qquad (4.15)$$

The solutions of Eqs. (4.14) are

$$x' = \frac{\pm P/k_1}{1 - \omega^2/\omega_1^2} \sin \omega t + A_1 \cos \omega_1 t + B_1 \sin \omega_1 t$$

$$x'' = \frac{\pm P/k_2}{1 - \omega^2/\omega_2^2} \sin \omega t + A_2 \cos \omega_2 t + B_2 \sin \omega_2 t + \left(1 - \frac{k_1}{k_2}\right) x_1$$

where $\omega_1^2 = k_1/m$, $\omega_2^2 = k_2/m$, and the constants A_1, A_2, B_1, B_2 may be evaluated from the boundary conditions, Eq. (4.15).

This analysis also applies to the case of free vibration by setting $P = 0$. By assigning various values to k_1 and k_2, a wide variety of specific problems may be treated; a collection of such solutions is given in Refs. 7 and 8. It is not necessary to restrict the restoring forces to odd functions.

APPROXIMATE METHODS OF NONLINEAR VIBRATION ANALYSIS

Any second-order ordinary differential equation can be integrated numerically in a stepwise manner to yield a time-history of the motion. Known methods of stepwise integration [9] may be employed for this purpose and result in a solution of any desired degree of accuracy, depending on the size of the steps. Numerical methods apply only to specific equations, and are not useful for general studies of the behavior of nonlinear vibrating systems.

A large number of approximate analytical methods of nonlinear vibration analysis exist, each of which may or may not possess advantages for certain classes of problems. Some of these are restricted techniques which may work well with some types of equations, but not with others. The methods which are outlined below are among the better known and possess certain advantages as to ranges of applicability.[4-21]

DUFFING'S METHOD

Consider the nonlinear differential equation

$$\ddot{x} + \kappa^2(x \pm \mu^2 x^3) = p \cos \omega t \qquad (4.16)$$

where the $\pm$ sign indicates either a hardening or softening system.* As a first approximation to a harmonic solution, assume that

$$x_1 = A \cos \omega t \qquad (4.17)$$

* Equation (4.16) is known as Duffing's equation.

and rewrite Eq. (4.16) to obtain an equation for the second approximation:

$$\ddot{x}_2 = -(\kappa^2 A \pm \tfrac{3}{4}\kappa^2\mu^2 A^3 - p)\cos\omega t - \tfrac{1}{4}\kappa^2\mu^2 A^3 \cos\omega t$$

This equation may now be integrated to yield

$$x_2 = \frac{1}{\omega^2}(\kappa^2 A \pm \tfrac{3}{4}\kappa^2\mu^2 A^3 - p)\cos\omega t + \tfrac{1}{36}\kappa^2\mu^2 A^3 \cos 3\omega t \qquad (4.18)$$

where the constants of integration have been taken as zero to insure periodicity of the solution.

This may be regarded as an iteration procedure by reinserting each successive approximation into Eq. (4.16) and obtaining a new approximation. For this iteration procedure to be convergent, the nonlinearity must be small; i.e., κ^2, μ^2, A, and p must be small quantities. This restricts the study to motions in the neighborhood of linear vibration (but not near $\omega = \kappa$, since A would then be large); thus, Eq. (4.17) must represent a reasonable first approximation. It follows that the coefficient of the cos ωt term in Eq. (4.18) must be a good second approximation, and should not be far different from the first approximation.[22] Since this procedure furnishes the exact result in the linear case, it might be expected to yield good results for the "slightly nonlinear" case. Thus, a relation between frequency and amplitude is found by equating the coefficients of the first and second approximations:

$$\omega^2 = \kappa^2(1 \pm \tfrac{3}{4}\mu^2 A^2) - \frac{p}{A} \qquad (4.19)$$

This relation describes the response curves, as shown in Fig. 4.14.

The above method applies equally well when linear velocity damping is included. Further details concerning this method, and an analysis of its applicability when only μ^2 and p are considered small, may be found in Ref. 10.

RAUSCHER'S METHOD [23]

Duffing's method considered above is based on the idea of starting the iteration procedure from the linear vibration. More rapid convergence might be expected if the approximations were to begin with free nonlinear vibration; Rauscher's method [23] is based on this idea.

Consider a system with general restoring force described by the differential equation

$$\omega^2 x'' + \kappa^2 f(x) = p \cos \omega t \qquad (4.20)$$

where primes denote differentiation with respect to ωt, and $f(x)$ is an odd function. Assume that the conditions at time $t = 0$ are $x(0) = A$, $x'(0) = 0$. Start with the free nonlinear vibration as a first approximation, i.e., with the solution of the equation

$$\omega_0^2 x'' + \kappa^2 f(x) = 0 \qquad (4.21)$$

such that $x = x_0(\phi)$ (where $\omega t = \phi$) has the period 2π and $x_0(0) = A$, $x_0'(0) = 0$. Equation (4.21) may be solved exactly in the form of quadratures according to Eq. (4.7):

$$\phi = \phi_0(x) = \frac{\omega_0}{\kappa\sqrt{2}} \int_A^x \frac{d\zeta}{\sqrt{\int_\zeta^A f(\xi)\,d\xi}} \qquad (4.22)$$

Since $f(x)$ is an odd function and noting that ωt varies from 0 to $\pi/2$ as x varies from 0 to A,

$$\frac{1}{\omega_0} = \frac{\sqrt{2}}{\kappa\pi} \int_0^A \frac{d\zeta}{\sqrt{\int_\zeta^A f(\xi)\,d\xi}} \qquad (4.23)$$

With ω_0 and ϕ_0 determined by Eqs. (4.23) and (4.22), respectively, the next approximation may be found from the equation

$$\omega_1^2 x'' + \kappa^2 f(x) - p \cos \phi_0 = 0 \tag{4.24}$$

In the original differential equation, Eq. (4.20), ωt is replaced by its first approximation ϕ_0 and ω_0 (now known) is replaced by its second approximation ω_1, thus giving Eq. (4.24). This equation is again of a type which may be integrated explicitly; therefore, the next approximation ω_1 and ϕ_1 may be determined. In those cases where $f(x)$ is a complicated function, the integrals may be evaluated graphically.[23]

This method involves reducing nonautonomous systems to autonomous ones * by an iteration procedure in which the solution of the free vibration problem is used to replace the time function in the original equation which is then solved again for $t(x)$. The method is accurate and frequently two iterations will suffice.

THE PERTURBATION METHOD

In one of the most common methods of nonlinear vibration analysis, the desired quantities are developed in powers of some parameter which is considered small; then the coefficients of the resulting power series are determined in a stepwise manner. The method is straightforward, although it becomes cumbersome for actual computations if many terms in the perturbation series are required to achieve a desired degree of accuracy.

Consider Duffing's equation, Eq. (4.16), in the form

$$\omega^2 x'' + \kappa^2(x + \mu^2 x^3) - p \cos \phi = 0 \tag{4.25}$$

where $\phi = \omega t$ and primes denote differentiation with respect to ϕ. The conditions at time $t = 0$ are $x(0) = A$ and $x'(0) = 0$, corresponding to harmonic solutions of period $2\pi/\omega$. Assume that μ^2 and p are small quantities, and define $\kappa^2 \mu^2 \equiv \epsilon$, $p \equiv \epsilon p_0$. The displacement $x(\phi)$ and the frequency ω may now be expanded in terms of the small quantity ϵ:

$$\begin{aligned} x(\phi) &= x_0(\phi) + \epsilon x_1(\phi) + \epsilon^2 x_2(\phi) + \cdots \\ \omega &= \omega_0 + \epsilon \omega_1 + \epsilon^2 \omega_2 + \cdots \end{aligned} \tag{4.26}$$

The initial conditions are taken as $x_i(0) = x_i'(0) = 0$ [$i = 1, 2, \ldots$].

Introducing Eq. (4.26) into Eq. (4.25) and collecting terms of zero order in ϵ gives the linear differential equation

$$\omega_0^2 x_0'' + \kappa^2 x_0 = 0$$

Introducing the initial conditions into the solution of this linear equation gives $x_0 = A \cos \omega t$ and $\omega_0 = \kappa$. Collecting terms of the first order in ϵ,

$$\omega_0^2 x_1'' + \kappa^2 x_1 - (2\omega_0 \omega_1 A - \tfrac{3}{4} A^3 + p_0) \cos \phi + \tfrac{1}{4} A^3 \cos 3\phi = 0 \tag{4.27}$$

The solution of this differential equation has a nonharmonic term of the form $\phi \cos \phi$, but since only harmonic solutions are desired, the coefficient of this term is made to vanish so that

$$\omega_1 = \frac{1}{2\kappa} \left(\tfrac{3}{4} A^2 - \frac{p_0}{A} \right)$$

Using this result and the appropriate initial conditions, the solution of Eq. (4.27) is

$$x_1 = \frac{A^3}{32\kappa^2} (\cos 3\phi - \cos \phi)$$

* An autonomous system is one in which the time *does not* appear explicitly, while a nonautonomous system is one in which the time *does* appear explicitly.

APPROXIMATE METHODS OF NONLINEAR VIBRATION ANALYSIS 4–15

To the first order in ϵ, the solution of Duffing's equation, Eq. (4.25), is

$$x = A \cos \omega t + \epsilon \frac{A^3}{32\kappa^2}(\cos 3\omega t - \cos \omega t)$$

$$\omega = \kappa + \frac{\epsilon}{2\kappa}\left(\tfrac{3}{4}A^2 - \frac{p_0}{A}\right)$$

This agrees with the results obtained previously [Eqs. (4.18) and (4.19)]. The analysis may be carried beyond this point, if desired, by application of the same general procedures.

As a further example of the perturbation method, consider the self-excited system described by Van der Pol's equation

$$\ddot{x} - \epsilon(1 - x^2)\dot{x} + \kappa^2 x = 0 \qquad (4.5)$$

where the initial conditions are $x(0) = 0$, $\dot{x}(0) = A\kappa_0$. Assume that

$$x = x_0 + \epsilon x_1 + \epsilon^2 x_2 + \cdots$$

$$\kappa^2 = \kappa_0^2 + \epsilon \kappa_1^2 + \epsilon^2 \kappa_2^2 + \cdots$$

Inserting these series into Eq. (4.5) and equating coefficients of like terms, the result to the order ϵ^2 is

$$x = \left(2 - \frac{29\epsilon^2}{96\kappa_0^2}\right)\sin \kappa_0 t + \frac{\epsilon}{4\kappa_0}\cos \kappa_0 t + \frac{\epsilon}{4\kappa_0}\left(\frac{3\epsilon}{4\kappa_0}\sin 3\kappa_0 t - \cos 3\kappa_0 t\right) - \frac{5\epsilon^2}{124\kappa_0^2}\sin 5\kappa_0 t \qquad (4.28)$$

An application of the perturbation method which employs operational calculus to solve the resulting linear differential equations is given in Ref. 21, while other applications of the method are given in Refs. 24 and 25. An application to the problem of subharmonic response is outlined in Ref. 1.

THE METHOD OF KRYLOFF AND BOGOLIUBOFF [11]

Consider the general autonomous differential equation

$$\ddot{x} + F(x,\dot{x}) = 0$$

which can be rewritten in the form

$$\ddot{x} + \kappa^2 x + \epsilon f(x,\dot{x}) = 0 \qquad [\epsilon \ll 1] \qquad (4.29)$$

For the corresponding linear problem ($\epsilon \equiv 0$), the solution is

$$x = A \sin(\kappa t + \theta) \qquad (4.30)$$

where A and θ are constants.

The procedure employed often is used in the theory of ordinary linear differential equations, and known variously as the method of variation of parameters or the method of Lagrange. In the application of this procedure to a nonlinear equation of the form of Eq. (4.29), assume the solution to be of the form of Eq. (4.30) but with A and θ as time-dependent functions rather than constants. This procedure, however, introduces an excessive variability into the solution; consequently, an additional restriction may be introduced. The assumed solution, of the form of Eq. (4.30), is differentiated once considering A and θ as time-dependent functions; this is made equal to the corresponding relation from the linear theory (A and θ constant) so that the additional restriction

$$\dot{A}(t)\sin[\kappa t + \theta(t)] + \dot{\theta}(t)A(t)\cos[\kappa t + \theta(t)] = 0 \qquad (4.31)$$

is placed on the solution. The second derivative of the assumed solution is now formed and these relations are introduced into the differential equation, Eq. (4.29). Combining this result with Eq. (4.31),

$$\dot{A}(t) = -\left(\frac{\epsilon}{\kappa}\right) f[A(t) \sin \Phi, A(t)\kappa \cos \Phi] \cos \Phi$$

$$\dot{\theta}(t) = \frac{\epsilon}{\kappa A(t)} f[A(t) \sin \Phi, A(t)\kappa \cos \Phi] \sin \Phi$$

where
$$\Phi = \kappa t + \theta(t).$$

Thus, the second-order differential equation, Eq. (4.29), has been transformed into two first-order differential equations for $A(t)$ and $\theta(t)$.

The expressions for $\dot{A}(t)$ and $\dot{\theta}(t)$ may now be expanded in Fourier series:

$$\dot{A}(t) = -\left(\frac{\epsilon}{\kappa}\right) \left\{ K_0(A) + \sum_{n=1}^{r} [K_n(A) \cos n\Phi + L_n(A) \sin n\Phi] \right\}$$

$$\dot{\theta}(t) = \frac{\epsilon}{\kappa A} \left\{ P_0(A) + \sum_{n=1}^{r} [P_n(A) \cos n\Phi + Q_n(A) \sin n\Phi] \right\}$$
(4.32)

where
$$K_0(A) = \frac{1}{2\pi} \int_0^{2\pi} f[A \sin \Phi, A\kappa \cos \Phi] \cos \Phi \, d\Phi$$

$$P_0(A) = \frac{1}{2\pi} \int_0^{2\pi} f[A \sin \Phi, A\kappa \cos \Phi] \sin \Phi \, d\Phi$$

It is apparent that A and θ are periodic functions of time of period $2\pi/\kappa$; therefore, during one cycle, the variation of $\dot{A}$ and $\dot{\theta}$ is small because of the presence of the small parameter ϵ in Eqs. (4.32). Hence, the average values of $\dot{A}$ and $\dot{\theta}$ are considered. Since the motion is over a single cycle, and since the terms under the summation signs are of the same period and consequently vanish, then approximately:

$$\dot{A} \simeq -\left(\frac{\epsilon}{\kappa}\right) K_0(A)$$

$$\dot{\theta} \simeq \frac{\epsilon}{\kappa A} P_0(A)$$

$$\dot{\Phi} \simeq \kappa + \frac{\epsilon}{\kappa A} P_0(A)$$

For example, consider Rayleigh's equation

$$\ddot{x} - (\alpha - \beta \dot{x}^2)\dot{x} + \kappa^2 x = 0 \quad (4.33)$$

By application of the above procedures:

$$\dot{A} = -\left(\frac{1}{\kappa}\right) K_0(A) = -\frac{1}{\kappa} \left[\frac{1}{2\pi} \int_0^{2\pi} (-\alpha + \beta A^2 \kappa^2 \cos^2 \Phi) A\kappa \cos^2 \Phi \, d\Phi \right]$$

$$= \frac{A}{2} (\alpha - \tfrac{3}{4}\beta A^2 \kappa^2) \quad (4.34)$$

Equation (4.34) may be integrated directly:

$$t = 2 \int_{A_0}^{A} \frac{dA}{A(\alpha - \gamma A^2)} = \frac{1}{\alpha} \ln \frac{A^2}{\alpha - \gamma A^2}$$

Solving for A,

$$A = \frac{\alpha}{\gamma}\left[\frac{1}{1+\left(\frac{\alpha}{\gamma A_0^2}-1\right)e^{-\alpha t}}\right]^{1/2} \qquad (4.35)$$

where $\gamma = \frac{3}{8}\beta^2\kappa^2$

The application of the method to Van der Pol's equation, Eq. (4.5), is easily accomplished and leads to a solution in the first approximation of the form

$$x = 2\cos o + \theta) \qquad (4.36)$$

This may be compared with the perturbation solution given by Eq. (4.28).
The method may be applied equally well to nonautonomous systems.[5, 19, 20]

THE RITZ METHOD

In addition to methods of nonlinear vibration analysis stemming from the idea of small nonlinearities and from extensions of methods applicable to linear equations, other methods are based on such ideas as satisfying the equation at certain points of the motion [13] or satisfying the equation in the average. The Ritz method is an example of the latter method, and is quite powerful for general studies.

One method of determining such "average" solutions is to multiply the differential equation by some "weight function" $\psi_n(t)$ and then integrate the product over a period of the motion. If the differential equation is denoted by E, this procedure leads to

$$\int_0^{2\pi} E \cdot \psi_n(t)\, dt = 0 \qquad (4.37)$$

A second method of obtaining such "average" solutions can be derived from the calculus of variations by seeking functions that minimize a certain integral:

$$I = \int_{t_0}^{t_1} F(\dot{x},x,t)\, dt = \text{minimum}$$

Consider a function of the form

$$\tilde{x}(t) = a_1\psi_1(t) + a_2\psi_2(t) + \cdots + a_n\psi_n(t)$$

where the $\psi_k(t)$ are prescribed functions. If $\tilde{x}$ is now introduced for x, then

$$I = I(a_1, a_2, \ldots, a_n)$$

and a necessary condition for I to be a minimum is

$$\frac{\partial I}{\partial a_1} = 0, \quad \frac{\partial I}{\partial a_2} = 0, \ldots, \quad \frac{\partial I}{\partial a_n} = 0 \qquad (4.38)$$

This gives n equations of the form

$$\frac{\partial I}{\partial a_k} = \int_{t_0}^{t_1}\left(\frac{\partial F}{\partial \tilde{x}}\psi_k + \frac{\partial F}{\partial \dot{\tilde{x}}}\dot{\psi}_k\right)dt = 0 \qquad (4.39)$$

for determining the n unknown coefficients. Integrating Eq. (4.39),

$$\frac{\partial I}{\partial a_k} = \left[\frac{\partial F}{\partial \dot{\tilde{x}}}\psi_k\right]_{t_0}^{t_1} + \int_{t_0}^{t_1}\left[\frac{\partial F}{\partial \tilde{x}} - \frac{d}{dt}\left(\frac{\partial F}{\partial \dot{\tilde{x}}}\right)\right]\psi_k\, dt = 0$$

The first term is zero because ψ_k must satisfy the boundary conditions; the expression in brackets under the integral in the second term is Euler's equation. The conditions given

4–18 NONLINEAR VIBRATION

in Eqs. (4.38) then reduce to

$$\int_{t_0}^{t} E(\tilde{x})\psi_k \, dt = 0 \qquad [k = 1, 2, \ldots, n] \qquad (4.40)$$

This is the same as Eq. (4.37); thus, it is not necessary to "know" the variational problem, but only the differential equation. The conditions given in Eqs. (4.40) then yield average solutions based on variational concepts.

EXAMPLES. As a first example of the application of the Ritz method, consider the equation

$$\ddot{x} + \kappa^2 x^n = 0$$

for which an exact solution was given earlier in this chapter [Eq. (4.9)]. Assume a single-term solution of the form

$$\tilde{x} = A \cos \omega t$$

The Ritz procedure, defined by Eq. (4.40), gives

$$\int_0^{2\pi} (-\omega^2 A \cos^2 \omega t + \kappa^2 A^n \cos^{n+1} \omega t) \, d(\omega t) = 0$$

from which

$$\frac{\omega^2}{\kappa^2} = \frac{4}{\pi} A^{n-1} \int_0^{\pi/2} \cos^{n+1} \omega t \, d(\omega t) = A^{n-1} \varphi(n) \qquad (4.41)$$

The comparable exact solution obtained previously by introducing in Eq. (4.9) the quantity $2\pi/\omega$ for the period τ is

$$\frac{\omega^2}{\kappa^2} = \left[\frac{\pi^2/4}{\psi^2(n)}\right] X^{n-1} = \Phi(n) X^{n-1} \qquad (4.42)$$

Values of $\varphi(n)$ from the approximate analysis and $\Phi(n)$ from the exact analysis are compared directly in Table 4.1, affording an appraisal of the accuracy of the method.

Table 4.1. Values of the Functions $\psi(n)$, $\Phi(n)$, $\varphi(n)$ *

n	$\psi(n)$	$\Phi(n)$	$\varphi(n)$
0	1.4142	1.2337	1.2732
1	1.5708	1.0000	1.0000
2	1.7157	0.8373	0.8488
3	1.8541	0.7185	0.7500
4	1.9818	0.6282	0.6791
5	2.1035	0.5577	0.6250
6	2.2186	0.5013	0.5820
7	2.3282	0.4552	0.5469

* The mathematical expressions for $\psi(n)$, $\Phi(n)$, and $\varphi(n)$ and the equations to which they refer are:

$$\psi(n) = \sqrt{\frac{n+1}{2}} \int_0^1 \frac{du}{\sqrt{1 - u^{n+1}}} \qquad \text{[Eq. (4.9)]}$$

$$\Phi(n) = \frac{\pi^2/4}{\psi^2(n)} \qquad \text{[Eq. (4.42)]}$$

$$\varphi(n) = \frac{4}{\pi} \int_0^{\pi/2} \cos^{n+1} \sigma \, d\sigma \qquad \text{[Eq. (4.41)]}$$

GENERAL EQUATIONS FOR RESPONSE CURVES 4–19

Consider now the nonautonomous system described by Duffing's equation

$$E \equiv \ddot{x} + \kappa^2(x + \mu^2 x^3) - p \cos \omega t = 0$$

Assuming

$$\bar{x} = A \cos \phi, \qquad \phi = \omega t$$

the Ritz condition, Eq. (4.40), leads to

$$\int_0^{2\pi} \{[(1-\eta^2)A - s] \cos \phi + \mu^2 A^3 \cos^3 \phi\} \cos \phi \, d\phi$$

from which the amplitude-frequency relation is

$$(1-\eta^2)A + \tfrac{3}{4}\mu^2 A^3 = \pm s \tag{4.43}$$

where

$$s = \frac{p}{\kappa^2}, \qquad \eta^2 = \frac{\omega^2}{\kappa^2} \tag{4.44}$$

The upper sign indicates vibration in phase with the exciting force. Equation (4.43) describes the response curves shown in Fig. 4.14A, and corresponds to Eq. (4.19) obtained by Duffing's method.

Application of the Ritz method to Van der Pol's equation, Eq. (4.5), leads to the identical result given by Eq. (4.36).

GENERAL EQUATIONS FOR RESPONSE CURVES

The Ritz method has been applied extensively in studies of nonlinear differential equations.[6,8] Some of the general equations for response curves thereby obtained are given here, both as a further example of the application of the method and as a collection of useful relations.

SYSTEM WITH LINEAR DAMPING AND GENERAL RESTORING FORCES

Consider a system with general elastic restoring force (an odd function) and described by the equation of motion

$$a\ddot{x} + b\dot{x} + cf(x) - P \cos \omega t = 0$$

A solution may be assumed in the form

$$\bar{x} = A \cos(\omega t - \theta) = B \cos \phi + C \sin \phi \tag{4.45}$$

where $\phi = \omega t$, $B = A \cos \theta$, $C = A \sin \theta$. Introducing Eq. (4.45) according to the Ritz conditions, and recalling that $f(x)$ is to be an odd function,

$$-a\omega^2 A \cos \theta + b\omega A \sin \theta + cAF(A) \cos \theta = P$$
$$-a\omega^2 A \sin \theta - b\omega A \cos \theta + cAF(A) \sin \theta = 0 \tag{4.46}$$

where

$$F(A) = \frac{1}{\pi A} \int_0^{2\pi} f(A \cos \sigma) \cos \sigma \, d\sigma$$

and σ is simply an integration variable.

Some algebraic manipulations with Eqs. (4.46) give independent equations for the two unknowns A and θ:

$$[F(A) - \eta^2]^2 + 4D^2\eta^2 = \left(\frac{s}{A}\right)^2 \tag{4.47}$$

$$\tan \theta = \frac{2D\eta}{F(A) - \eta^2} \tag{4.48}$$

where η^2 and s are defined according to Eq. (4.44) and

$$\kappa^2 = \frac{c}{a} \qquad p = \frac{P}{a} \qquad D = \frac{b}{2\sqrt{ac}}$$

Equation (4.47) describes response curves of the form shown in Fig. 4.15, and Eq. (4.48) gives the corresponding phase angle relationships. These two equations also yield other special relations which describe various curves in the response diagram:

Undamped free vibration curve (Fig. 4.13),

$$\eta^2 = F(A) \tag{4.49}$$

Undamped response curves (Fig. 4.14),

$$\eta^2 = F(A) \mp \frac{s}{A} \tag{4.50}$$

Locus of vertical tangents of undamped response curves (Fig. 4.17),

$$\eta^2 = F(A) + A\frac{\partial F(A)}{\partial A} \tag{4.51}$$

Damped response curves (Fig. 4.15),

$$\eta^2 = [F(A) - 2D^2] \mp \sqrt{\left(\frac{s}{A}\right)^2 - 4D^2[F(A) - D^2]} \tag{4.52}$$

Locus of vertical tangents of damped response curves (Fig. 4.17),

$$[F(A) - \eta^2]\left[F(A) + A\frac{\partial F(A)}{\partial A} - \eta^2\right] = -4D^2\eta^2 \tag{4.53}$$

The maximum amplitude of vibration is of interest. The amplitude at the point at which a response curve crosses the free vibration curve is termed the *resonance amplitude*, and is determined in the nonlinear case by solving Eqs. (4.49) and (4.52) simultaneously. This leads to [27]

$$2D\eta = \frac{s}{A} \qquad \theta = \frac{\pi}{2} \tag{4.54}$$

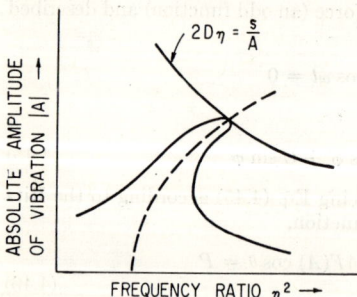

FIG. 4.19. Determination of the resonant amplitude in accordance with Eq. (4.54).

The first of these two equations defines a hyperbola in the response diagram, describing the locus of crossing points, as shown in Fig. 4.19; hence, the intersection of this curve with the free vibration curve gives the resonance amplitude. The phase angle at resonance has the value $\pi/2$, as in the linear case. This result is of great help in computing response curves since the effect of damping (except for very large values) is negligible except in the neighborhood of resonance. Therefore, one may compute only the undamped curves (which is not difficult) and the hyperbola (which does not contain the nonlinearity); then, the effect of damping may be sketched in from knowledge of the crossing point.

SYSTEM WITH GENERAL DAMPING AND GENERAL RESTORING FORCES

The preceding analysis may be extended to include the more general differential equation

$$E \equiv \ddot{x} + 2D\kappa g(\dot{x}) + \kappa^2 f(x) - p\cos\omega t = 0$$

By procedures similar to those employed above:

$$[F(A) - \eta^2]^2 + 4D^2S^2(A) = \left(\frac{s}{A}\right)^2 \qquad (4.55)$$

$$\tan \theta = \frac{2DS(A)}{F(A) - \eta^2} \qquad (4.56)$$

where
$$S(A) = \frac{1}{\pi \kappa A} \int_0^{2\pi} g(\omega A \sin \sigma) \sin \sigma \, d\sigma$$

In the case of linear velocity damping, $S(A) = \eta$, and Eqs. (4.55) and (4.56) reduce to Eqs. (4.47) and (4.48). The results for various types of damping forces are:

Coulomb damping: $\qquad g(\dot{x}) = \pm \nu_o \qquad S(A) = \dfrac{4}{\pi} \dfrac{\nu_o}{\kappa A}$

Linear velocity damping: $\qquad g(\dot{x}) = \nu_1 \dot{x} \qquad S(A) = \nu_1 \eta$

Velocity squared damping: $\qquad g(\dot{x}) = \nu_2 \dot{x}|\dot{x}| \qquad S(A) = \dfrac{8}{3\pi} \nu_2 \eta (A\omega)$

nth-power velocity damping: $\qquad g(\dot{x}) = \nu_n \dot{x}|\dot{x}|^{n-1} \qquad S(A) = \nu_n \eta (A\omega)^{n-1} \varphi(n)$

where $\varphi(n)$ is defined in Eq. (4.41) and values are given in Table 4.1.

The locus of resonance amplitudes or crossing points is now given by [27]

$$2DS(A) = \frac{s}{A} \qquad \theta = \frac{\pi}{2}$$

GRAPHICAL METHODS OF INTEGRATION

INTRODUCTION

Graphical methods may be employed in the analysis of nonlinear vibration and often prove to be of great value both for general studies of the behavior of a particular system and for actual integration of the equation of motion.

A single degree-of-freedom system requires two parameters to describe completely the state of the motion. When these two parameters are used as coordinate axes, the graphical representation of the motion is called a *phase-plane* representation. In dealing with ordinary dynamical problems, these parameters frequently are taken as the displacement and velocity. First consider an undamped linear system having the equation of motion

$$\ddot{x} + \omega_n^2 x = 0 \qquad (4.57)$$

and the solution
$$x = A \cos \omega_n t + B \sin \omega_n t$$
$$y = \frac{\dot{x}}{\omega_n} = -A \sin \omega_n t + B \cos \omega_n t \qquad (4.58)$$

Eliminating time as a variable between Eqs. (4.58):

$$x^2 + y^2 = c^2$$

Thus, the phase-plane representation is a family of concentric circles with centers at the origin. Such curves are called *trajectories*. The necessary and sufficient conditions in the phase-plane for periodic motions are (1) closed trajectories and (2) paths described in finite time.

Now, suppose that the solution of Eq. (4.57) is not known. By introducing $y = \dot{x}/\omega_n$,

$$\frac{dx}{dt} = \omega_n y \qquad \frac{dy}{dt} = -\omega_n x$$

Therefore

$$\frac{dy}{dx} = -\frac{x}{y} \qquad (4.59)$$

Thus, the path in the phase-plane is described by a simple first-order differential equation. This process of eliminating the time always can be done in principle, but frequently the problem is too difficult. Since the time is to be eliminated, *only autonomous systems can be treated by phase-plane methods*. When an equation of the type of Eq. (4.59) can

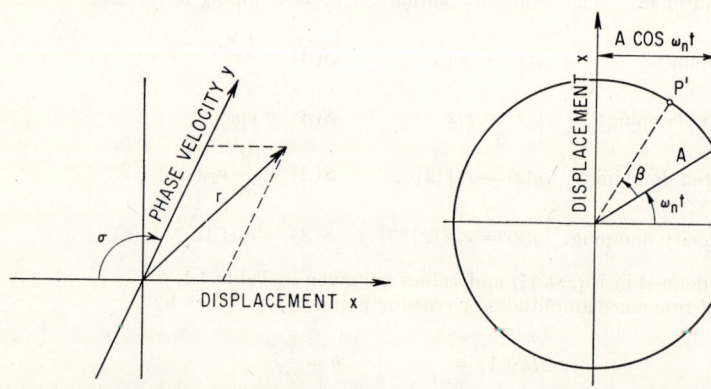

FIG. 4.20. Phase-plane using oblique coordinates which results in a logarithmic spiral trajectory for a linear system with viscous damping [Eq. (4.60)].

FIG. 4.21. Phase-plane solution for a linear undamped vibrating system.

be found, a direct solution of the problem follows since slopes of the trajectories can be sketched in the phase-plane and the trajectories determined by connecting the tangents; this is known as the method of *isoclines*. It sometimes happens that $dx = 0$, $dy = 0$ simultaneously so that there is no knowledge of the direction of the motion; such points in the phase-plane are called *singular* points. In the present example, the origin constitutes a singular point.

Consider now a damped linear system having the equation of motion

$$\ddot{x} + 2\zeta\omega_n\dot{x} + \omega_n^2 x = 0$$

and the solution

$$x = Ce^{-\delta t} \cos \Phi \qquad (4.60)$$

$$y = \frac{\dot{x}}{\omega_n} = Ce^{-\delta t} \cos(\Phi + \sigma)$$

where

$$\delta = \zeta\omega_n = -\omega_n \cos \sigma \qquad \left[\sigma > \frac{\pi}{2}\right]$$

$$\Phi = \nu t + \theta$$

$$\nu = \omega_n\sqrt{1 - \zeta^2} = \omega_n \sin \sigma$$

Equations (4.60) indicate that the trajectories in the phase-plane are some form of spiral (one of the simplest known of which is the logarithmic spiral). By referring to the oblique coordinate system shown in Fig. 4.20, and recalling that $\sin \sigma$ is a constant and

$$r^2 = x^2 + y^2 - 2xy \cos \sigma$$

Eqs. (4.60) reduce to
$$r = Ce^{-\delta t}\sin\sigma$$
This is a form of a logarithmic spiral.

The trajectories also could be found in a rectangular coordinate system, by the method of isoclines, without knowledge of the solution [Eq. (4.60)]. The governing differential equation is

$$\frac{dy}{dx} = -\frac{2\zeta y + x}{y} \qquad (4.61)$$

The resulting trajectories can be sketched in the phase-plane. On the other hand, Eq. (4.61) also can be integrated analytically by use of the substitution $z = y/x$ and separation of the variables:

$$y^2 + 2\zeta xy + x^2 = C\exp\left(\frac{2\zeta}{\sqrt{1-\zeta^2}}\tan^{-1}\frac{x\zeta + y}{x\sqrt{1-\zeta^2}}\right)$$

This is a spiral of the form of Eq. (4.60).

The method of isoclines is extremely useful in studying the behavior of solutions in the neighborhood of singular points and for the related questions of stability of solutions. In this sense, phase-plane methods may be thought of as topological methods.[5, 10, 12, 19] However, it is desirable also to study the over-all solutions, rather than solutions in the neighborhood of special points, and preferably by some straightforward method of graphical integration. Such integration methods are given in the following sections of this chapter, while topological studies, with special reference to questions of stability, are treated in Chap. 5.

PHASE-PLANE INTEGRATION OF STEPWISE LINEAR SYSTEMS

Consider the undamped linear system described by Eq. (4.57). The known solution $x = A\sin\omega_n t$, $\dot{x} = A\omega_n\cos\omega_n t$ may be shown graphically in the phase-plane representation of Fig. 4.21. The point P moves with constant angular velocity ω_n, and the deflection increases to P' in the time β/ω_n.

If the system has a nonlinear restoring force composed of straight lines (as in Fig. 4.4), the motion within the region represented by any one linear segment can be described as above. For example, consider a system with the force-deflection characteristic shown at the top of Fig. 4.22. If the motion starts with initial velocity q_6 and zero initial deflection, the motion is described by a circular arc with center at 0 and angular velocity $\omega_{n_1} = \sqrt{\dfrac{1}{m}\tan\alpha_1}$, from q_6 to q_5. At the point q_5, it is seen that $\dot{x}_A/\omega_{n_1} = \overline{x_A q_5}$ and $\dot{x}_A/\omega_{n_2} = \overline{x_A q_4}$. Therefore

$$\frac{\overline{x_A q_4}}{\overline{x_A q_5}} = \sqrt{\frac{\tan\alpha_2}{\tan\alpha_1}}$$

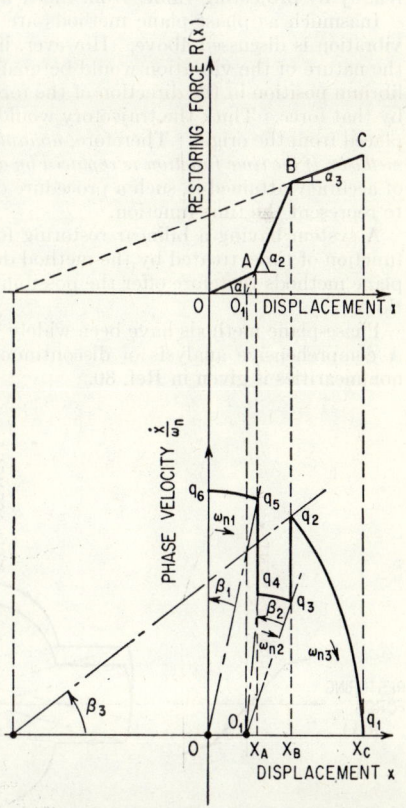

Fig. 4.22. Phase-plane solution for the stepwise linear restoring force characteristic curve shown at the top. The motion starts with zero displacement but finite velocity.

In this example, $\tan \alpha_1 < \tan \alpha_2$ so that $\overline{x_A q_4} < \overline{x_A q_5}$. The circular arc from q_4 to q_3 corresponds to the segment AB of the restoring force characteristic with center at the intersection 0_1 of the segment (extended) with the X axis. The radius of this circle is $\overline{0_1 q_4}$, where

$$q_4 = \frac{\dot{x}_B}{\omega_{n_2}} \quad \text{and} \quad \omega_{n_2} = \sqrt{\frac{1}{m} \tan \alpha_2}$$

The total time required to go from q_6 to q_1 is

$$t = \frac{\beta_1}{\omega_{n_1}} + \frac{\beta_2}{\omega_{n_2}} + \frac{\beta_3}{\omega_{n_3}}$$

For a symmetrical system this is one quarter of the period.

If the force-deflection characteristic of a nonlinear system is a smooth curve, it may be approximated by straight line segments and treated as above. It should be noted that the time required to complete one cycle is strongly influenced by the nature of the curve in regions where the velocity is low; therefore, linear approximations near the equilibrium position do not greatly affect the period.

The time-history of the motion (i.e., the x,t representation) may be obtained quite readily by projecting values from the X axis to an x,t plane.

Inasmuch as phase-plane methods are restricted to autonomous systems, only free vibration is discussed above. However, if a constant force were to act on the system, the nature of the vibration would be unaffected, except for a displacement of the equilibrium position in the direction of the force and equal to the static deflection produced by that force. Thus, the trajectory would remain a circular arc but with its center displaced from the origin. Therefore, *nonautonomous systems may be treated by phase-plane methods, if the time function is replaced by a series of stepwise constant values*. The degree of accuracy attained in such a procedure depends only on the number of steps assumed to represent the time function.

A system having a bilinear restoring force and acted upon by an external stepwise function of time, treated by the method described above, is shown in Fig. 4.23. Phase-plane methods therefore offer the possibility of treating transient as well as free vibrations.

Phase-plane methods have been widely used for the analysis of control mechanisms.[29] A comprehensive analysis of discontinuous-type systems possessing various types of nonlinearities is given in Ref. 30.

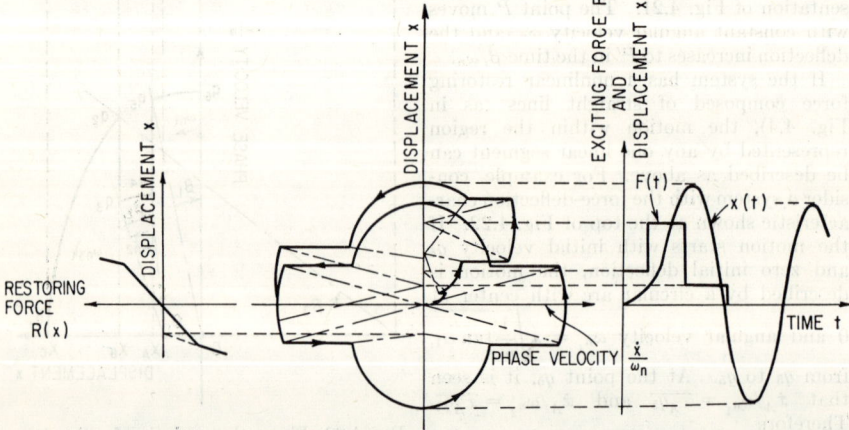

Fig. 4.23. Phase-plane solution for transient motion. The bilinear restoring force characteristic curve is shown at the left, and the exciting force $F(t)$ and the resulting motion of the system $X(t)$ are shown at the right. (*After Evaldson et al.*[28])

PHASE-PLANE INTEGRATION OF AUTONOMOUS SYSTEMS WITH NONLINEAR DAMPING

Consider the differential equation

$$\ddot{x} + g(\dot{x}) + \kappa^2 x = 0$$

Introducing $y = \dot{x}/\kappa$, the following isoclinic equation is obtained:

$$\frac{dy}{dx} = -\frac{g(y) + x}{y} \qquad (4.62)$$

For points of zero slope in the phase-plane, the numerator of Eq. (4.62) must vanish; therefore, the condition for zero slope is

$$x_0 = -g(y)$$

Points of infinite slope correspond to the X axis. Singular points occur where the x_0 curve intersects the X axis.

To construct the trajectory, the slope at any point P_i must be determined first. This is done as illustrated in Fig. 4.24: A line is drawn parallel to the X axis through P_i. The intersection of this line with the x_0 curve determines a point S_i on the X axis. With S_i as the center, a circular arc of short length is drawn through P_i; the tangent to this arc is the required slope. The termination of this short arc may be taken as the point P_{i+1}, etc. The accuracy of the construction is dependent on the lengths of the arcs. This construction is known as Liénard's method.[5,10]

As an example of Liénard's method, consider Rayleigh's equation, Eq. (4.33), in the form

$$\ddot{x} + \epsilon\left(\frac{\dot{x}^3}{3} - \dot{x}\right) + x = 0$$

The corresponding isoclinic equation is

$$\frac{dy}{dx} = \frac{\epsilon(y - y^3/3) - x}{y}$$

The x_0 curve is given by

$$x_0 = \epsilon\left(y - \frac{y^3}{3}\right) \qquad (4.63)$$

This is illustrated in Fig. 4.25.

A little experimentation shows that if a point P_1 is taken near the origin, the slope is such as to take the trajectory away from the origin (as compared with the undamped

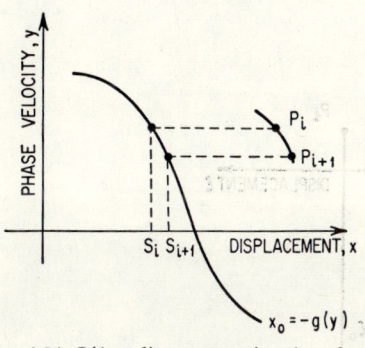

Fig. 4.24. Liénard's construction for phase-plane integration of autonomous systems with nonlinear damping.

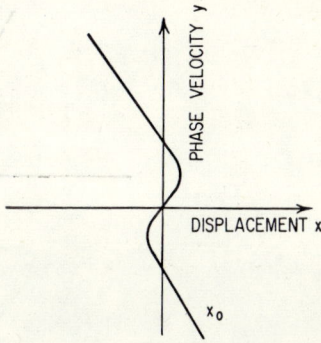

Fig. 4.25. Curve of x_0 for Rayleigh's equation [Eq. (4.33)] as given by Eq. (4.63).

vibration); by the same reasoning, a point P_2 far from the origin tends to take the trajectory toward the origin (again as compared with the undamped vibration). Therefore, there is some neutral curve, describing a periodic motion, toward which the trajectories tend; this neutral curve is called a *limit cycle* and is illustrated in Fig. 4.26. Such a limit cycle is obtained when x_0 has a different sign for different parts of the Y axis.

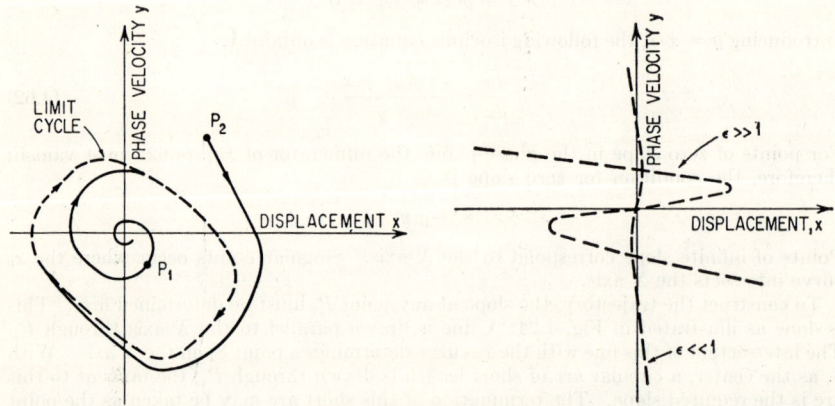

FIG. 4.26. Limit cycle for Rayleigh's equation [Eq. (4.33)].

FIG. 4.27. Curves of x_0 for extreme values of ϵ in Rayleigh's equation [Eq. (4.33)]. See Fig. 4.25 for a solution with a moderate value of ϵ.

For extreme values of ϵ, the x_0 curves would appear as shown in Fig. 4.27. For $\epsilon \gg 1$, introduce the notation $\xi = x/\epsilon$; then

$$\frac{dy}{d\xi} = \frac{\epsilon}{y}\left(y - \frac{y^3}{3} - \xi\right)$$

This leads to a trajectory as shown in Fig. 4.28. This type of motion is known as a *relaxation oscillation*. Note from Fig. 4.28 that for this case of large ϵ the slope changes quickly from horizontal to vertical. Hence, for a motion starting at some point P_i, a vertical

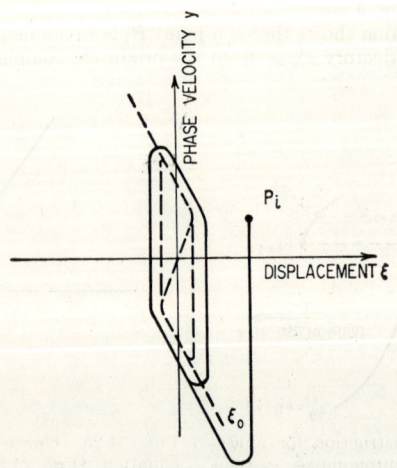

FIG. 4.28. Relaxation oscillations of Rayleigh's equation [Eq. (4.33)].

trajectory is followed until it intersects the ξ_0 curve; then, the trajectory turns and follows the ξ_0 curve until it enters the vertical field at the lower knee in the curve. The trajectory then moves straight up until it intersects ξ_0 again after which it swings right and down again. A few circuits bring the trajectory into the limit cycle.

There is a possibility that more than one limit cycle may exist. If the x_0 curve crosses the X axis more than three times, it can be shown that at least two limit cycles may exist. Further discussion of limit cycles is given in Chap. 5.

GENERALIZED PHASE-PLANE ANALYSIS

The following method of integrating second-order differential equations by phase-plane techniques has general application.[31, 32] Consider the general equation

$$\ddot{x} + F(x,\dot{x},t) = 0 \quad (4.64)$$

Equation (4.64) can be converted to the form

$$\ddot{x} + \kappa^2 x = g(x,\dot{x},t)$$

by adding $\kappa^2 x$ to both sides where

$$\kappa^2 x - F(x,\dot{x},t) = g(x,\dot{x},t)$$

Let

$$g(x_0,\dot{x}_0,t_0) = -\kappa^2 \Delta_0$$

where κ is chosen arbitrarily. At some point P_0 on the trajectory,

$$\ddot{x} + \kappa^2(x + \Delta_0) = 0$$

and

$$\frac{dy}{dx} = -\frac{x + \Delta_0}{y}$$

Referring to Fig. 4.29,

$$dt = \frac{1}{\kappa}\frac{dx}{y} = \frac{1}{\kappa}d\theta$$

Therefore, the time may be obtained by integration of the angular displacements. Thus, at a nearby point P_1 on the trajectory:

$$x_1 = x_0 + dx$$
$$y_1 = y_0 + dy$$
$$t_1 = t_0 + dt$$

Now, compute Δ_1 for the new center, and repeat the process.

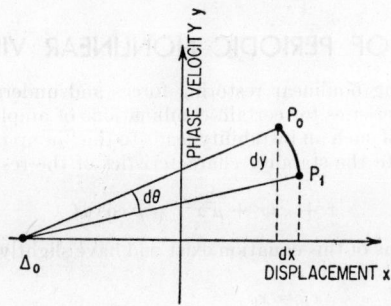

FIG. 4.29. Method of construction employed in the generalized phase-plane analysis.

This method has been applied to a very wide variety of linear and nonlinear equations.[31,33] For example, Fig. 4.30 shows the solution of Bessel's equation

$$\ddot{x} + \frac{1}{t}\dot{x} + \left(p^2 - \frac{n^2}{t^2}\right)x = 0$$

of order zero. The angle (or time) projection of x yields $J_0(pt)$, while the $\dot{x}/p$ projection yields $J_1(pt)$; that is, the Bessel functions of the zeroth and first order of the first kind. Bessel functions of the second kind also can be obtained.

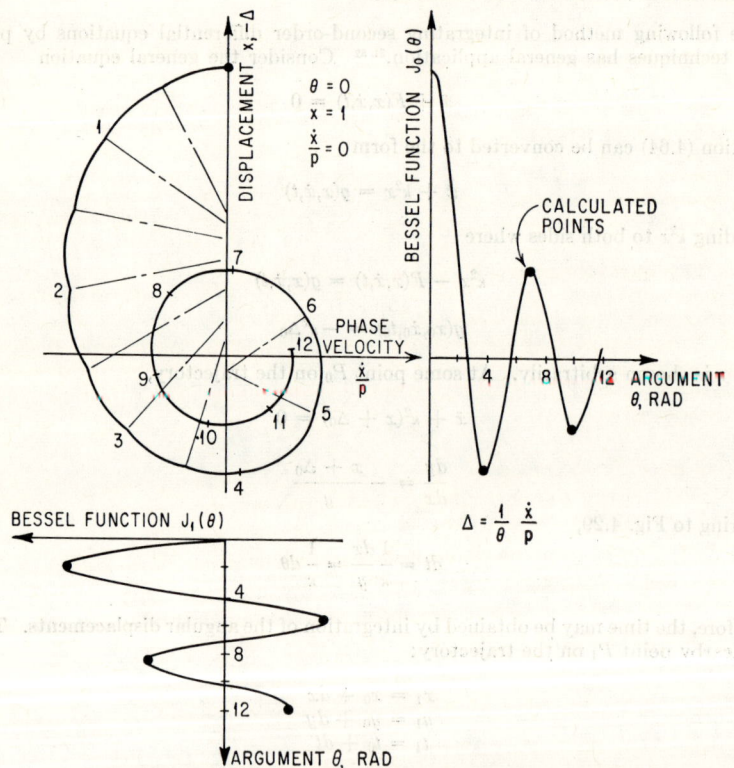

Fig. 4.30. Generalized phase-plane solution of Bessel's equation. (*Jacobsen*.[31])

STABILITY OF PERIODIC NONLINEAR VIBRATION

Certain systems having nonlinear restoring forces and undergoing forced vibration exhibit unstable characteristics for certain combinations of amplitude and exciting frequency. The existence of such an instability leads to the "jump phenomenon" shown in Fig. 4.16. To investigate the stability characteristics of the response curves, consider Duffing's equation

$$\ddot{x} + \kappa^2(x + \mu^2 x^3) = p \cos \omega t \qquad (4.65)$$

Assume that two solutions of this equation exist and have slightly different initial conditions:

$$x_1 = x_0$$

$$x_2 = x_0 + \delta \qquad [\delta \ll x_0]$$

Introducing the second of these into Eq. (4.65) and employing the condition that x_0 is also a solution,

$$\ddot{\delta} + \kappa^2(1 + 3\mu^2 x_0^2)\delta = 0 \tag{4.66}$$

Now an expression for x_0 must be obtained; assuming a one-term approximation of the form $x_0 = A \cos \omega t$, Eq. (4.66) becomes

$$\frac{d^2\delta}{d\varphi^2} + (\lambda + \gamma \cos \varphi)\delta = 0 \tag{4.67}$$

where
$$\kappa^2(1 + \tfrac{3}{2}\mu^2 A^2) = 4\omega^2\lambda$$
$$\tag{4.68}$$
and
$$\tfrac{3}{2}\kappa^2\mu^2 A^2 = 4\omega^2\gamma \qquad 2\omega t = \varphi$$

Equation (4.67) is known as Mathieu's equation.

Mathieu's equation has appeared in this analysis as a variational equation characterizing small deviations from the given periodic motion whose stability is to be investigated; thus, the stability of the solutions of Mathieu's equation must be studied. A given periodic motion is stable if *all* solutions of the variational equation associated with it tend toward zero for all positive time and unstable if there is at least one solution which does not tend toward zero. The stability characteristics of Eq. (4.67) often are represented in a chart as shown in Fig. 4.31.[10]

From the response diagram of Duffing's equation, the out-of-phase motion having the larger amplitude appears to be unstable. This portion of the response diagram (Fig. 4.17) corresponds to unstable motion in the Mathieu stability chart (Fig. 4.31), and the locus of vertical tangents of the response curves (considering undamped vibration for simplicity) corresponds exactly to the boundaries between stable and unstable regions in the stability chart. Thus, the region of interest in the response diagram is described by the free vibration

$$\omega^2 = \kappa^2(1 + \tfrac{3}{4}\mu^2 A^2) \tag{4.69}$$

and the locus of vertical tangents

$$\tfrac{3}{2}\kappa^2\mu^2 A^2 + \frac{p}{A} = 0 \tag{4.70}$$

The corresponding curves in the stability chart are taken as those for small positive values of γ and λ which have the approximate equations

$$\gamma = \tfrac{1}{2} - 2\lambda \tag{4.71}$$

$$\gamma = -\tfrac{1}{2} + 2\lambda \tag{4.72}$$

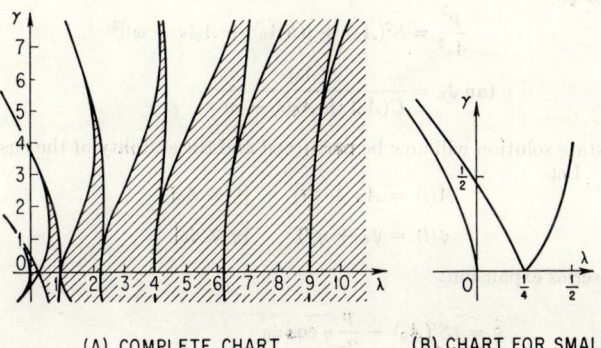

(A) COMPLETE CHART (B) CHART FOR SMALL λ AND γ

FIG. 4.31. Stability chart for Mathieu's equation [Eq. (4.67)].

Now, if Eq. (4.69) is introduced into Eqs. (4.68), the resulting equations expanded by the binomial theorem (assuming μ^2 small), and Eq. (4.72) introduced, the result is an identity. Therefore, the free vibration-response curve maps onto the curve of positive slope in the stability chart. The locus of vertical tangents to the response curves maps into the curve of negative slope in the stability chart; this may be seen from the identity obtained by introducing the equations obtained above by binomial expansion into Eq. (4.71) and then employing Eq. (4.70).

In any given case, it can be determined whether a motion is stable or unstable on the basis of the values of γ and λ, according to the location of the point in the stability chart.

The question of stability of response also can be resolved by means of a "stability criterion" developed from the Kryloff-Bogoliuboff procedures.[34] The differential equation of motion is considered in the form

$$\ddot{x} + \kappa^2 x + f(x,\dot{x}) = p \cos \omega t$$

Proceeding in the manner of the Kryloff-Bogoliuboff procedure described earlier,

$$\dot{A} = \frac{1}{\kappa} f(x,\dot{x}) \sin(\kappa t + \theta) - \frac{p}{\kappa} \cos \omega t \sin(\kappa t + \theta)$$

$$\dot{\theta} = \frac{1}{\kappa} f(x,\dot{x}) \cos(\kappa t + \theta) - \frac{p}{A\kappa} \cos \omega t \cos(\kappa t + \theta)$$

Expanding the last terms of these equations, the result contains motions of frequency κ, $\kappa + \omega$, and $\kappa - \omega$. The motion over a long interval of time is of interest, and the motions of frequencies $\kappa + \omega$ and $\kappa - \omega$ may be averaged out; this is accomplished by integrating over the period $2\pi/\omega$:

$$\dot{A} = S(A) - \frac{p}{2\kappa} \sin(\Phi - \omega t)$$

$$\dot{\theta} = \frac{C(A)}{A} - \frac{p}{2\kappa A} \cos(\Phi - \omega t)$$

where

$$S(A) = \frac{1}{2\pi\kappa} \int_0^{2\pi} f(A \cos \Phi, -A\kappa \sin \Phi) \sin \Phi \, d\Phi$$

$$C(A) = \frac{1}{2\pi\kappa} \int_0^{2\pi} f(A \cos \Phi, -A\kappa \sin \Phi) \cos \Phi \, d\Phi$$

The steady-state solution may be determined by employing the conditions $A = A_0$, $\psi = \Phi - \omega t = \psi_0$:

$$\frac{p^2}{4\kappa^2} = S^2(A_0) + [C(A_0) + A_0(\kappa - \omega)]^2$$

$$\tan \psi_0 = \frac{S(A_0)}{C(A_0) + A_0(\kappa - \omega)}$$

This steady-state solution will now be perturbed and the stability of the ensuing motion investigated. Let

$$A(t) = A_0 + \xi(t) \qquad [\xi \ll A_0]$$

$$\psi(t) = \psi_0 + \eta(t) \qquad [\eta \ll \psi_0]$$

By Taylor's series expansion:

$$\dot{\xi} = \xi S'(A_0) - \frac{p}{2\kappa} \eta \cos \psi_0$$

$$\dot{\eta} = \frac{\xi}{A_0}[(\kappa - \omega) + C'(A_0)] + \frac{p}{2\kappa A_0} \eta \sin \psi_0$$

where primes indicate differentiation with respect to A. These two differential equations are satisfied by the solutions

$$\xi = \mathcal{A}e^{zt} \qquad \eta = \mathcal{B}e^{zt}$$

where $\mathcal{A}$ and $\mathcal{B}$ are arbitrary constants and

$$z = \frac{1}{2A_0}\left\{[S(A_0) + A_0 S'(A_0)] \pm \sqrt{[S(A_0) + A_0 S'(A_0)]^2 - 4A_0 \bar{p}\frac{d\bar{p}}{dA_0}}\right\}$$

and $\bar{p} = p/2\kappa$.

For stability, the real parts of z must be negative; hence, the following criteria can be established:[34]

$$[S(A_0) + A_0 S'(A_0)] < 0, \frac{d\bar{p}}{dA_0} > 0, \text{ ensures stability}$$

$$[S(A_0) + A_0 S'(A_0)] < 0, \frac{d\bar{p}}{dA_0} < 0, \text{ ensures instability}$$

$$[S(A_0) + A_0 S'(A_0)] > 0, \frac{d\bar{p}}{dA_0} \gtreqless 0, \text{ ensures instability}$$

$$[S(A_0) + A_0 S'(A_0)] = 0, \frac{d\bar{p}}{dA_0} > 0, \text{ ensures stability}$$

These criteria can be interpreted in terms of response curves by reference to Fig. 4.14. For systems of this type, $[S(A_0) + A_0 S'(A_0)] < 0$; when $\frac{d\bar{p}}{dA_0} > 0$, $\bar{p}$ increases as A_0 also increases. This does not hold for the middle branch of the response curves, thus confirming the earlier results. Other analyses of stability are found in Refs. 35 to 38.

SUPERHARMONIC AND SUBHARMONIC VIBRATIONS

INTRODUCTION

Steady-state response of a nonlinear system also may occur at some multiple or submultiple of the forcing frequency. The response depends on both the frequency and the amplitude of the excitation, and may vary with time as well as with different initial conditions. Such vibration can be analyzed by several of the methods discussed above: The Kryloff and Bogoliuboff method,[39] the Ritz method,[40] a method employing Fourier series with either a perturbation or an iteration procedure used to determine the coefficients,[10] etc. Other studies are given in Refs. 41 to 43, as well as in many of the other references of this chapter.

SUPERHARMONIC RESPONSE

Consider forced vibration of a system defined by the differential equation

$$\ddot{x} + 2\delta\kappa\dot{x} + \kappa^2(x + \mu^2 x^3) = p\cos\omega t \qquad (4.73)$$

The solution of Eq. (4.73) may be of the form

$$x = A_1 \cos \omega t + A_3 \cos 3\omega t \qquad (4.74)$$

Typical values for the coefficients A_1 and A_3 are illustrated graphically in Fig. 4.32 [2] as a function of the frequency ratio ω/ω_n, where ω_n is the frequency of free vibration obtained from Eq. (4.73) by letting $\mu^2 = 0$.

If the term with frequency 3ω (i.e., the third harmonic) in Eq. (4.74) is neglected, the response can be described by plotting the coefficient A_1 against frequency, as shown by the broken lines a in Fig. 4.32. Such response is designated harmonic response; the resonance condition (at $\omega/\omega_n = 1$) is designated *harmonic resonance* and the peak is bent to the right as a consequence of the assumed hardening type of nonlinearity. When the third harmonic is included, the response is described by the two coefficients A_1 and A_3,

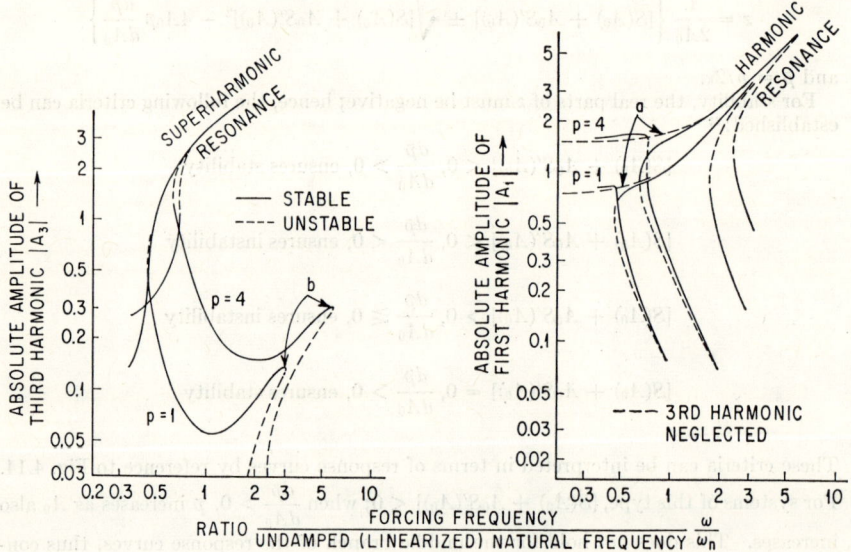

FIG. 4.32. Response curves for system defined by Eq. (4.73) with harmonic response. (*After Clauser*.[2])

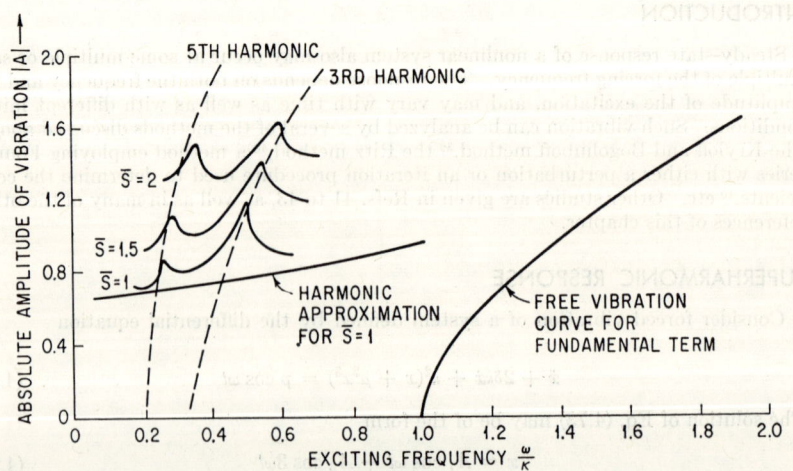

FIG. 4.33. Response curves for system with harmonic response in the third and fifth harmonics, where $\bar{s} = \dfrac{p}{\kappa^2}\sqrt{\dfrac{3}{4}\mu^2}$ and $\delta = 0.01$. (*After Atkinson*.[43])

SUPERHARMONIC AND SUBHARMONIC VIBRATIONS

plotted separately in Fig. 4.32. The latter is characterized by the following features:

1. A peak in the region $\omega/\omega_n = 1$; this represents a vibration at a frequency of approximately $3\omega_n$ superimposed upon the resonance of the harmonic response whose frequency is ω_n. This peak (b in Fig. 4.32) is designated the *third-order component of the harmonic resonance.*

2. A peak in the region $\omega/\omega_n = \frac{1}{3}$; this represents vibration at a frequency of approximately ω_n when the forcing frequency is $\omega_n/3$. This region of the response is designated *superharmonic resonance.** It is accompanied by a decrease in A_1, as indicated in Fig. 4.32; thus, the value of A_3 at the frequency of the superharmonic resonance may exceed the value of A_1.

If terms of frequency $5\omega, 7\omega \ldots$ with corresponding coefficients $A_5, A_7, \ldots$ had been included in Fig. 4.32, the same general pattern would have been repeated at the frequency ratios $\omega/\omega_n = \frac{1}{5}, \frac{1}{7} \ldots$. The vibration amplitude of the response may then be expressed as $A = A_1 + A_3 + \cdots$ and is illustrated graphically as a sequence of superharmonic resonances superposed on the approximate harmonic response, as shown in Fig. 4.33.[43] The jump phenomenon described with reference to Fig. 4.16 also may occur with superharmonic resonance. If the damping is very small, the superharmonic resonances may be of greater magnitude than the first-order component of the harmonic resonance because the nonlinearity provides a mechanism whereby energy from the excitation may be transferred to the higher harmonics.†

SUBHARMONIC RESPONSE

The subharmonic response of the system described by Eq. (4.73) may be studied by assuming a solution of the form

$$x = A_{1/3} \cos \tfrac{1}{3}\omega t + A_1 \cos \omega t \qquad (4.75)$$

Typical values for the coefficients $A_{1/3}$ and A_1 are illustrated graphically in Fig. 4.34 as a function of the frequency ratio ω/ω_n, where ω_n is again the frequency of the free vibration obtained from Eq. (4.73) by letting $\mu^2 = 0$.

If the term with frequency $(\frac{1}{3})\omega$ (i.e., the one-third harmonic) in Eq. (4.75) is neglected, the response can be described by the coefficient A_1 plotted against frequency, as shown by the lines c in Fig. 4.34.[2] Such response is designated harmonic response, and the resonance condition is designated harmonic resonance. When the one-third harmonic is included, the response is defined by the coefficients $A_{1/3}$ and A_1, plotted separately in Fig. 4.34. The response is characterized by a significant vibration $A_{1/3} \sin (\frac{1}{3})\omega t$ in the region of $\omega/\omega_n = 3$; physically, this means that the system vibrates at the frequency $\omega = \omega_n$ when the forcing frequency is $3\omega_n$. This is the *subharmonic resonance*. The harmonic response of the system is modified by the subharmonic resonance and is characterized by the solid (stable condition) and dotted (unstable condition) lines D in Fig. 4.34. The subharmonic response offers an alternate mode of vibration rather than a modification of the fundamental response. Since the subharmonic response is unstable at the junction with the fundamental response, the subharmonic cannot simply arise but must be induced by a disturbance of sufficient magnitude. Again, jump phenomena and subharmonics of higher orders are possible.

Consider the undamped system defined by

$$\omega^2 x'' + \kappa^2(x + \mu^2 x^3) = p \cos \omega t \qquad (4.76)$$

Introducing an assumed solution of the form of Eq. (4.75):

$$\left(\kappa^2 - \frac{\omega^2}{9}\right) A_{1/3} + \tfrac{3}{4}\kappa^2\mu^2(A_{1/3}^3 + A_{1/3}^2 A_1 + 2A_{1/3}A_1^2) = 0$$

$$(\kappa^2 - \omega^2)A_1 + \tfrac{1}{4}\kappa^2\mu^2(A_{1/3}^3 + 6A_{1/3}^2 A_1 + 3A_1^3) = p \qquad (4.77)$$

---

* The term *ultraharmonic resonance* sometimes is used in place of the term superharmonic resonance.

† Even as well as odd harmonics may appear in the superharmonic response.[49, 43]

NONLINEAR VIBRATION

These equations can be studied by an iteration procedure starting from the linear forced vibration.[10] Since $A_{1/3}$ is presumed to exist, Eqs. (4.77) can be rewritten

$$\omega^2 = 9\kappa^2 + \tfrac{27}{4}\kappa^2\mu^2(A_{1/3}^2 + A_{1/3}A_1 + 2A_1^2)$$

$$-8\kappa^2 A_1 = p + (\omega^2 - 9\kappa^2)A_1 - \tfrac{1}{4}\kappa^2\mu^2(A_{1/3}^3 + 6A_{1/3}^2 A_1 + 3A_1^3)$$

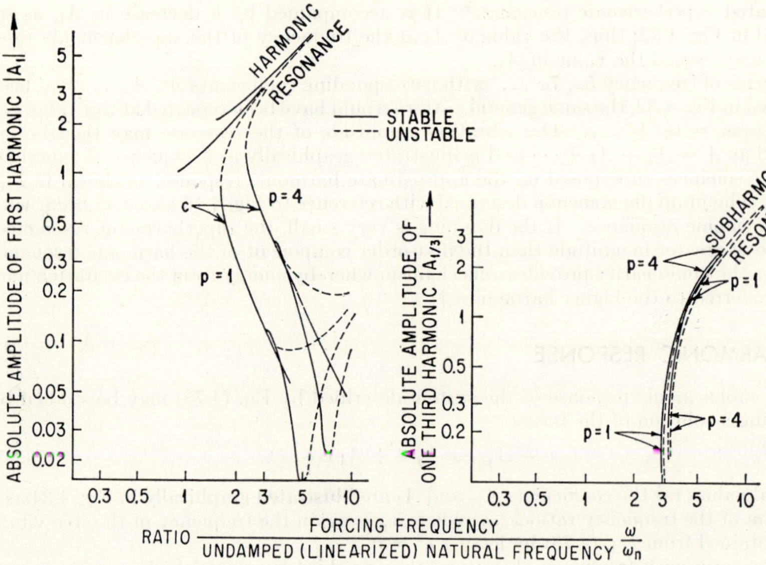

FIG. 4.34. Response curves for system defined by Eq. (4.74) with subharmonic response. (*After Clauser*.[2])

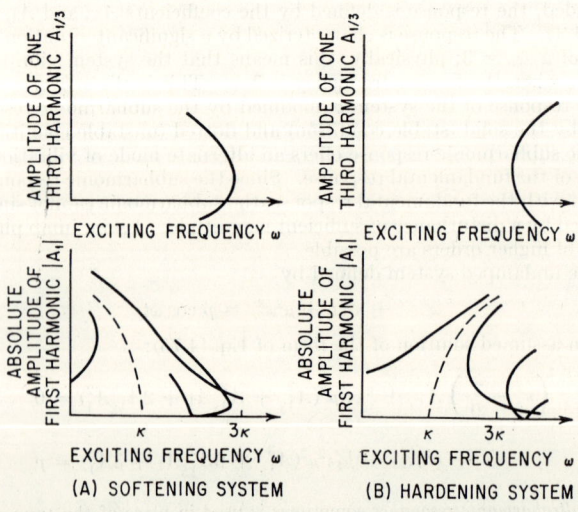

(A) SOFTENING SYSTEM

(B) HARDENING SYSTEM

FIG. 4.35. Response curves for hardening and softening systems with subharmonic response, defined by Eq. (4.76). (*After Stoker*.[10])

Beginning the iteration with $\mu = 0$ (linear forced vibration) yields

$$\omega = 3\kappa \qquad A_1 = -\frac{p}{8\kappa}$$

The next step becomes

$$\omega^2 = 9\kappa + {}^{27}\!/_{\!4}\kappa^2\mu^2\left(A_{1/3}^2 - \frac{p}{8\kappa}A_{1/3} + \frac{p^2}{32\kappa^2}\right)$$

$$A_1 = -\frac{p}{8\kappa} + {}^1\!/_{\!3\,2}\mu^2\left(A_{1/3}^3 + \frac{21}{8}\frac{p}{\kappa}A_{1/3}^2 - \frac{27}{64}\frac{p^2}{\kappa^2}A_{1/3} + \frac{51}{512}\frac{p^3}{\kappa^3}\right)$$

This solution applies to both the hardening system $(x + \mu^2 x^3)$ and the softening system $(x - \mu^2 x^3)$.

In the response diagram showing $A_{1/3}$ as a function of ω, ω^2 has a minimum for $+\mu^2$ and a maximum for $-\mu^2$ corresponding to

$$A_{1/3} = \frac{p}{16\kappa}$$

$$\omega^2 = 9[\kappa^2 + ({}^{21}\!/_{\!1024})\mu^2 p^2]$$

Thus, the subharmonic vibration can exist only for

$$\omega < 9[\kappa^2 + ({}^{21}\!/_{\!1024})\mu^2 p^2] \qquad \text{(hardening system)}$$
$$\omega > 9[\kappa^2 + ({}^{21}\!/_{\!1024})\mu^2 p^2] \qquad \text{(softening system)}$$

Response curves for both systems are shown in Fig. 4.35.

This analysis can be extended to include linear damping [10]; then the subharmonic of order $\frac{1}{3}$ cannot occur unless the damping is very small.

SYSTEMS OF MORE THAN A SINGLE DEGREE-OF-FREEDOM

Interest in systems of more than one degree-of-freedom arises from the problem of the dynamic vibration absorber. The earliest studies of nonlinear two degree-of-freedom systems were those of vibration absorbers having nonlinear elements.[44-46]

The analysis of multiple degree-of-freedom systems can be carried out by various of the methods described earlier in this chapter; thus, a stepwise linear system is treated in Ref. 47, and more general systems are treated in Refs. 47 to 51. The extension of phase-plane methods to such systems is given in Refs. 29, 30, and 52.

All the essential features of nonlinear vibration of single degree-of-freedom systems described earlier occur in the multiple degree-of-freedom systems as well. An analysis which considers subharmonic vibration in such systems by an iteration procedure is given in Ref. 53 and is completely analogous to that given here for the single degree-of-freedom system, with analogous results.

REFERENCES

Additional citations to the literature are given in Refs. 4, 29, 32, 54, and 55, and in past and current issues of the journal *Applied Mechanics Reviews*.

1. von Kármán, T.: *Bull. Am. Math. Soc.*, **46**: 615 (1940).
2. Clauser, F. H.: *J. Aeronaut. Sci.*, **23**:411 (1956).
3. Davis, S. A.: *Product Eng.*, **25**:181 (1954).
4. McLachlan, N. W.: "Ordinary Nonlinear Differential Equations," Oxford University Press, New York, 1956.
5. Minorsky, N.: "Introduction to Nonlinear Mechanics," Edwards Bros., Ann Arbor, Mich., 1947.
6. Klotter, K.: *Proc. 1st U.S. Natl. Congr. Appl. Mechs.*, 1951.

7. Klotter, K.: *Stanford Univ., Rept.* 17, Contract N6 ONR-251-II, 1951.
8. Klotter, K.: *Proc. Symposium on Nonlinear Circuit Analysis*, 1953.
9. Levy, H., and E. A. Baggott: "Numerical Solutions of Differential Equations," Dover Publications, New York, 1950.
10. Stoker, J. J.: "Nonlinear Vibrations," Interscience Publishers, Inc., New York, 1950.
11. Kryloff, N., and N. Bogoliuboff: "Introduction to Nonlinear Mechanics," Princeton University Press, Princeton, N.J., 1943.
12. Andronow, A. A., and C. E. Chaikin: "Theory of Oscillations," Princeton University Press, Princeton, N.J., 1949.
13. Rudenberg, R.: *ZAMM*, **3**:454 (1923).
14. Brock, J. E.: *J. Appl. Mechanics*, **18** (1951).
15. Roberson, R. E.: *J. Appl. Mechanics*, **20**:237 (1953).
16. Wylie, C. R.: *J. Franklin Inst.*, **236**:273 (1943).
17. Young, D.: *Proc. 1st Midwest. Conf. Solid Mechanics*, 1953.
18. Fifer, S.: *J. Appl. Phys.*, **22**:1421 (1951).
19. Minorsky, N.: "Advances in Applied Mechanics," Vol. I, Academic Press, Inc., New York, 1948.
20. Bellin, A. I.: "Advances in Applied Mechanics," Vol. III, Academic Press, Inc., New York, 1953.
21. Pipes, L. A.: *J. Appl. Phys.*, **13**:117 (1942).
22. Duffing, G.: "Erzwungene Schwingungen bei veranderlicher Eigenfrequenz," F. Vieweg u Sohn, Brunswick, 1918.
23. Rauscher, M.: *J. Appl. Mechanics*, **5**:169 (1938).
24. Minorsky, N.: *J. Franklin Inst.*, **248**:205 (1949).
25. Carrier, G. F.: *Quart. Appl. Math.*, **3**:157 (1945).
26. Schwesinger, G.: *J. Appl. Mechanics*, **17**:202 (1950).
27. Abramson, H. N.: *Product Eng.*, **25**:179 (1954).
28. Evaldson, R. L., R. S. Ayre, and L. S. Jacobsen: *J. Franklin Inst.*, **248**:473 (1949).
29. Ku, Y. H.: "Analysis and Control of Nonlinear Systems," The Ronald Press Company, New York, 1958.
30. Flügge-Lotz, I.: "Discontinuous Automatic Control," Princeton University Press, Princeton, N.J., 1953.
31. Jacobsen, L. S.: *J. Appl. Mechanics*, **19**:543 (1952).
32. Jacobsen, L. S., and R. S. Ayre: "Engineering Vibrations," McGraw-Hill Book Company, Inc., New York, 1958.
33. Bishop, R. E. D.: *Proc. Inst. Mech. Engrs.*, **168**:299 (1954).
34. Klotter, K., and E. Pinney: *J. Appl. Mechanics*, **20**:9 (1953).
35. John, F.: "Studies in Nonlinear Vibration Theory," New York University, 1946.
36. Rosenberg, R. M.: *Proc. 2nd Natl. Congr. Appl. Mechanics*, 1954.
37. Young, D., and P. N. Hess: *Proc. 2nd Natl. Congr. Appl. Mechanics*, 1954.
38. Hayashi, C.: "Forced Oscillations in Nonlinear Systems," Nippon Pub. Co., Osaka, Japan, 1953.
39. Caughey, T. K.: *J. Appl. Mechanics*, **21**:327 (1954).
40. Burgess, J. C.: *Stanford Univ., Rept.* 27, Contract N6 ONR-251-II, 1954.
41. Wu, M. H. L.: *Proc. 1st Natl. Congr. Appl. Mechanics*, 1951.
42. Rosenberg, R. M.: *Proc. 2nd Midwest. Conf. Solid Mechanics*, 1955.
43. Atkinson, C. P.: *J. Appl. Mechanics*, **24**:520 (1957).
44. Roberson, R. E.: *J. Franklin Inst.*, **254**:205 (1952).
45. Pipes, L. A.: *J. Appl. Mechanics*, **20**:515 (1953).
46. Arnold, F. R.: *J. Appl. Mechanics*, **22**:487 (1955).
47. Soroka, W. W.: *J. Appl. Mechanics*, **17**:185 (1950).
48. Sethna, P. R.: *Proc. 2nd Natl. Congr. Appl. Mechanics*, 1954.
49. Huang, T. C.: *J. Appl. Mechanics*, **22**:107 (1955).
50. Klotter, K.: *Trans. IRE on Circuit Theory*, **CT-1** (4):13 (1954).
51. Stoker, J. J.: *Proc. 2nd Natl. Congr. Appl. Mechanics*, 1954.
52. Ku, Y. H.: *J. Franklin Inst.*, **259**:115 (1955).
53. Huang, T. C.: *Proc. 2nd Natl. Congr. Appl. Mechanics*, 1954.
54. Minorsky, N.: *Appl. Mechanics Revs.*, **4**:266 (1951).
55. Klotter, K.: *Appl. Mechanics Revs.*, **10**:495 (1957).

5

SELF-EXCITED VIBRATION

R. S. Hahn
The Heald Machine Company

INTRODUCTION

GENERAL NATURE

Self-excited systems begin to vibrate of their own accord spontaneously, the amplitude increasing until some nonlinear effect limits any further increase. The energy supplying these vibrations is obtained from a uniform source of power associated with the system which, due to some mechanism inherent in the system, gives rise to oscillating forces. The nature of self-excited vibration compared to forced vibration is:[1]

In self-excited vibration the alternating force that sustains the motion is created or controlled by the motion itself; when the motion stops, the alternating force disappears.

In a forced vibration the sustaining alternating force exists independent of the motion and persists when the vibratory motion is stopped.

The occurrence of self-excited vibration in a physical system is intimately associated with the stability of equilibrium positions of the system. If the system is disturbed from a position of equilibrium, forces generally appear which cause the system to move either toward the equilibrium position or away from it. In the latter case the equilibrium position is said to be unstable; then the system may either oscillate with increasing amplitude or monotonically recede from the equilibrium position until nonlinear or limiting restraints appear. The equilibrium position is said to be stable if the disturbed system approaches the equilibrium position either in a damped oscillatory fashion or asymptotically.

The forces which appear as the system is displaced from its equilibrium position may depend on the displacement or the velocity, or both. If displacement-dependent forces appear and cause the system to move away from the equilibrium position, the system is said to be statically unstable. For example, an inverted pendulum is statically unstable. Velocity-dependent forces which cause the system to recede from a statically stable equilibrium position lead to dynamic instability.

The theory of self-excited vibration has important applications in mechanical systems. Several examples illustrate the importance of dynamical stability. Shortly after the official opening of the Tacoma Narrows Bridge at Puget Sound, Wash., the main span underwent a self-excited oscillation which resulted in the destruction of the bridge. The source of energy feeding this oscillation was a steady transversely blowing wind. Automatic control systems sometimes experience self-excited oscillations which cannot be tolerated. Rotating shafts sometimes whirl as a result of hysteresis or oil-film forces in the bearings. Machine tools frequently chatter, making the workpiece unfit for use. There are many other examples of undesirable self-excited oscillations. In general, self-excited oscillations are unwanted; however, there are important applications in electrical engineering and acoustics where self-excitation is desired.

STABILITY CRITERIA

Probably the most important engineering problem in the analysis of self-excited systems is that of determining the conditions under which the system is stable. These limiting conditions are known as stability criteria. The calculation of the frequencies and the final steady-state amplitude or limit cycle of self-excited vibration generally is much more difficult than the calculation of stability criteria because nonlinear effects enter into the former. The calculation of limit cycles is discussed in Chap. 4 on nonlinear vibration. Most of this chapter is concerned with the determination of the stability of equilibrium.

LINEAR SYSTEMS WITH CONSTANT COEFFICIENTS

CONCEPT OF NEGATIVE DAMPING

A simple mathematical representation of a self-excited vibration may be found in the concept of negative damping. Consider the differential equation for a damped, free vibration:

$$m\ddot{x} + c\dot{x} + kx = 0 \qquad (5.1)$$

This is generally solved by assuming a solution of the form

$$x = Ce^{st}$$

Substitution of this solution into Eq. (5.1) yields the characteristic (algebraic) equation

$$s^2 + \frac{c}{m}s + \frac{k}{m} = 0 \qquad (5.2)$$

If $c < 2\sqrt{mk}$, the roots are complex:

$$s_{1,2} = -\frac{c}{2m} \pm iq$$

where $\quad q = \sqrt{\frac{k}{m} - \left(\frac{c}{2m}\right)^2}$

The solution takes the form

$$x = e^{-ct/2m}(A \cos qt + B \sin qt) \qquad (5.3)$$

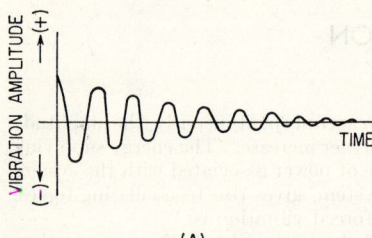
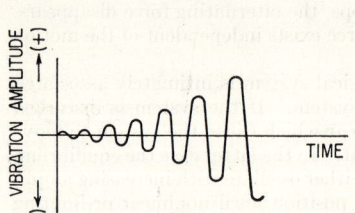

Fig. 5.1. (A) Illustration showing a decaying vibration (stable) corresponding to negative real parts of the complex roots. (B) Increasing vibration corresponding to positive real parts of the complex roots (unstable).

This represents a decaying oscillation because the exponential factor is negative, as illustrated in Fig. 5.1A. If $c < 0$, the exponential factor has a positive exponent and the vibration appears as shown in Fig. 5.1B. The system, initially at rest, begins to oscillate spontaneously with ever-increasing amplitude. Then, in any physical system, some nonlinear effect enters and Eq. (5.1) fails to represent the system realistically. Equation (5.4) defines a nonlinear system with negative damping at small amplitudes but with large positive damping at larger amplitudes, thereby limiting the amplitude to finite values:

$$m\ddot{x} + (-c + ax^2)\dot{x} + kx = 0 \qquad (5.4)$$

Thus, the fundamental criterion of stability in linear systems is that the roots of the characteristic equation have negative real parts, thereby producing decaying amplitudes.

DIRECT METHODS OF ESTABLISHING STABILITY CRITERIA

The presence of roots with positive real parts, indicating unstable operation, can be determined by several methods. For quadratic and cubic equations, the roots can be calculated directly, using the quadratic formula or Cardan's formulas.[2] For higher-degree equations with numerical coefficients, Graffe's root-squaring method may be used.[3]

In many instances it is sufficient to find the limiting values of the parameters in the characteristic equation which give zero real parts of the roots. Then, boundaries can be drawn which separate stable and unstable operating conditions.

For example, consider the characteristic equation which arises in certain metal-cutting vibration:

$$ms^2 + cs + k - re^{-sl} = 0$$

By placing $s = u + iv$, a pair of equations for the real and imaginary parts are obtained. In these equations u is set equal to zero, thereby giving two conditions which determine the boundaries between stable and unstable operating conditions.

ROUTH-HURWITZ STABILITY CRITERION

In many cases it is desirable to determine the presence of roots with positive real parts from a consideration of the coefficients of the characteristic equation, without solving the equation itself. Consider a system which has the following characteristic equation:

$$a_0 s^n + a_1 s^{n-1} + a_2 s^{n-2} \cdots \pm a_{n-1} s + a_n = 0 \qquad (5.5)$$

where the a_i are real numbers.

The Routh-Hurwitz criterion[4] states: *The number of roots with positive real parts is equal to the number of sign changes in the first column of Routh's array of the coefficients of his subsidiary functions.*

The array of coefficients of the subsidiary functions is formed by first arranging the coefficients of Eq. (5.5) as shown below in the first two rows; then additional rows are formed using the cross-multiplication scheme as illustrated:

$$a_0, \qquad a_2, \qquad a_4, \qquad a_6, \cdots$$

$$a_1, \qquad a_3, \qquad a_5, \qquad a_7, \cdots$$

$$\frac{a_1 a_2 - a_0 a_3}{a_1}, \quad \frac{a_1 a_4 - a_0 a_5}{a_1}, \quad \frac{a_1 a_6 - a_0 a_7}{a_1}, \quad \cdots$$

$$\frac{a_3 \left(\dfrac{a_1 a_2 - a_0 a_3}{a_1}\right) - a_1 \left(\dfrac{a_1 a_4 - a_0 a_5}{a_1}\right)}{\dfrac{a_1 a_2 - a_0 a_3}{a_1}}, \quad \frac{a_5 \left(\dfrac{a_1 a_2 - a_0 a_3}{a_1}\right) - a_1 \left(\dfrac{a_1 a_6 - a_0 a_7}{a_1}\right)}{\dfrac{a_1 a_2 - a_0 a_3}{a_1}}$$

. . . .

The process is continued until the last row has only a single term.

The subsidiary functions are

$$f_1(s) = a_0 s^n + a_2 s^{n-2} + a_4 s^{n-4} + \cdots$$

$$f_2(s) = a_1 s^{n-1} + a_3 s^{n-3} + a_5 s^{n-5} + \cdots$$

$$f_3(s) = \frac{a_1 a_2 - a_0 a_3}{a_1} s^{n-2} + \frac{a_1 a_4 - a_0 a_5}{a_1} s^{n-4} + \cdots$$

. . . .

If the terms of the first column of the array of coefficients all have the same sign, no positive real roots and no complex roots with positive real parts exist; hence, the system is stable.

The presence of equal roots but with opposite sign is indicated by the complete vanishing of a subsidiary function. To complete the array in this case, differentiate the last subsidiary function that did not vanish, enter the coefficients in the array, and proceed using the cross-multiplication scheme.

Example 5.1. Determine how many roots of the following equation have positive real parts: $s^{10} + s^9 - s^8 - 2s^7 + s^6 + 3s^5 + s^4 - 2s^3 - s^2 + s + 1 = 0$. Forming the array of coefficients and writing at the left-hand side the highest power of each subsidiary function:

$$s^{10}: \quad +1, -1, +1, +1, -1, +1$$
$$s^9: \quad +1, -2, +3, -2, +1$$
$$s^8: \quad +1, -2, +3, -2, +1$$

Note that the coefficients for s^8 are identical with those for s^9; hence, the next subsidiary function will vanish completely. The s^8 subsidiary function is

$$s^8 - 2s^6 + 3s^4 - 2s^2 + 1 = 0$$

Differentiating with respect to s and dividing by 4 to reduce the equation:

$$2s^7 - 3s^5 + 3s^3 - s = 0$$

The fourth row in the array then becomes

$$s^7: \quad +2, -3, +3, -1$$

The array may now be completed:

$$s^{10}: \quad +1, -1, +1, +1, -1, +1$$
$$s^9: \quad +1, -2, +3, -2, +1$$
$$s^8: \quad +1, -2, +3, -2, +1$$
$$s^7: \quad +2, -3, +3, -1$$
$$s^6: \quad -1, +3, -3, +2 \quad \text{(after multiplying by 2)}$$
$$s^5: \quad +1, -1, +1 \quad \text{(after dividing by 3)}$$
$$s^4: \quad +1, -1, +1$$
$$s^3: \quad +2, -1 \quad \text{(after dividing the differentiated coefficients by 2)}$$
$$s^2: \quad -1, +2 \quad \text{(after multiplying by 2)}$$
$$s: \quad +3$$
$$s^0: \quad +2$$

Since there are four changes of sign in the first column, there are four roots having positive real parts. Note that the original equation may be written:

$$(s^4 - s^2 + 1)^2 (s^2 + s + 1) = 0$$

Four roots of this equation are real and positive.

THE NYQUIST STABILITY CRITERION

The Nyquist stability criterion used in the analysis of feedback control systems provides another method of determining whether the roots of the characteristic equation have positive real parts. It involves the plotting of closed figures and determining whether or not they encircle the origin. The Nyquist method may be applied to evaluate the criteria that a system have not only no roots in the positive half plane but no roots with real parts greater than some arbitrary negative value. The positive half plane is that portion of the complex plane lying to the right of the axis of imaginary numbers and extending to $+\infty$ in the direction of the axis of real numbers. Both the existence of stability and the degree of stability can be assessed. This method can be extended to study the stability of systems with time lags.

LINEAR SYSTEMS WITH CONSTANT COEFFICIENTS 5-5

The Nyquist method has its foundation in complex-variable theory. Consider the characteristic equation

$$w(s) = s^n + a_1 s^{n-1} + a_2 s^{n-2} + \cdots + a^n = 0 \tag{5.6}$$

In general, the roots of this equation may be real or complex and of the form $s = x + iy$. Therefore the root s may be regarded as a complex variable, and $w(s)$ a function of a complex variable. The variable s may be represented as a point in the complex plane, Fig. 5.2A; the real part is plotted as abscissa and the imaginary part as ordinate. The function w also is a complex variable and may be plotted similarly, as shown in Fig. 5.2B. As a point representing the variable s describes a contour in the s plane of Fig. 5.2A, a corresponding point describes a related contour in the w plane of Fig. 5.2B.

The following important theorem [5] from complex-variable theory forms the basis of the Nyquist criterion:

Theorem. If a function $f(s)$ is analytic, except for possible poles, within and on a given contour, the number of times the plot of $f(s)$ encircles the origin in the $f(s)$ plane in the positive direction (i.e., counterclockwise), while s itself moves around the prescribed contour in the s plane once in a clockwise direction, is equal to the number of poles of $f(s)$ lying within the contour, diminished by the number of zeros of $f(s)$ within the contour, when each zero and pole is counted in accordance with its multiplicity.

The function $f(s)$ is said to be analytic if it satisfies the Cauchy-Riemann equations

$$\frac{\partial u}{\partial x} = \frac{\partial v}{\partial y} \qquad \frac{\partial u}{\partial y} = -\frac{\partial v}{\partial x} \tag{5.7}$$

where u and v are the real and imaginary parts of the complex variable $f(s)$ and x and y are the real and imaginary parts of s.[6,7] Values of s which make $f(s)$ infinite are called poles; values of s which make $f(s)$ zero are called zeros. The function $f(s) = a/(s - s_0)^2$ has a pole of multiplicity 2 at the point s_0. The function $f(s) = a(s - s_0)^2$ has a zero of multiplicity 2 at the point s_0.

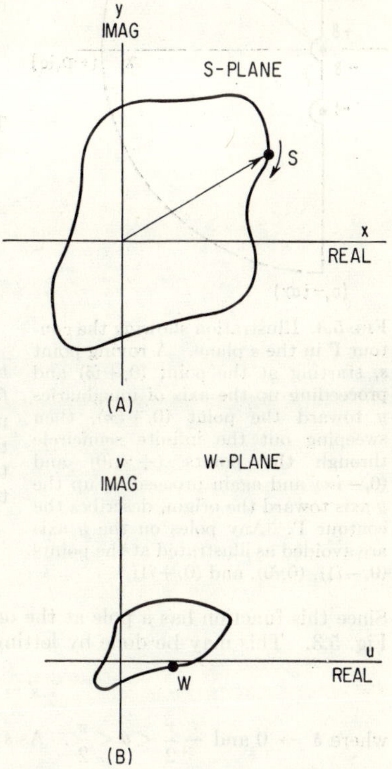

FIG. 5.2. (A) Illustration of a closed contour in the complex s plane where $s = x + iy$. (B) Illustration of the mapping of a contour in the s plane into the w plane where $w = u(x,y) + iv(x,y)$.

In order to examine the complete positive half-plane, the point s is usually considered to travel around the contour Γ shown in Fig. 5.3. If there are any poles on the imaginary axis or at the origin, they must be excluded as shown. As the point s moves toward infinity along the imaginary axis and around the infinite semicircle, $f(s)$ must approach either zero or a constant.

Associating the characteristic equation [Eq. (5.6)] with $1/f(s)$ so that the roots of $w(s)$ produce poles of $f(s)$:

$$f(s) = \frac{1}{w(s)} = \frac{1}{(s - r_1)(s - r_2)(s - r_3) \cdots (s - r_n)} \tag{5.8}$$

where $r_1, \ldots, r_n$ are the roots of Eq. (5.6). As s travels around the contour Γ in Fig. 5.3 in a clockwise sense, counterclockwise encirclement of the origin in the $f(s)$ plane indi-

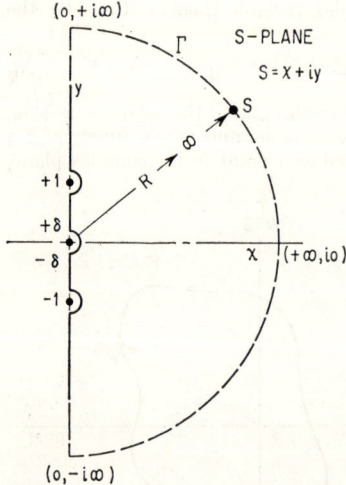

FIG. 5.3. Illustration showing the contour Γ in the s plane. A roving point s, starting at the point $(0, +i\delta)$ and proceeding up the axis of imaginaries y toward the point $(0, +i\infty)$, then sweeping out the infinite semicircle through the points $(+\infty, i0)$ and $(0, -i\infty)$ and again proceeding up the y axis toward the origin, describes the contour Γ. Any poles on the y axis are avoided as illustrated at the points $(0, -i1)$, $(0, i0)$, and $(0, +i1)$.

cates the presence of roots in the positive half-plane because $f(s)$ in this case has no zeros. Occasionally, $f(s)$ may contain a polynomial in s in the numerator, and thereby possess zeros. The presence of zeros in the positive half-plane must be ascertained in order to determine the correct number of poles.

For example, consider the following characteristic equation [8] arising in feedback control theory:

$$f(s) = \frac{s(1 + T_1 s)(1 + T_3 s)}{s(1 + T_1 s)(1 + T_3 s) + K(1 + T_2 s)}$$

This equation can be written:

$$f(s) = \frac{1}{1 + \dfrac{K(1 + T_2 s)}{s(1 + T_1 s)(1 + T_3 s)}}$$

$$= \frac{1}{w(s)} \equiv \frac{1}{1 + q(s)} \tag{5.9}$$

The number of poles enclosed within the contour Γ equals the net number of times the vector $f(s)$ rotates in the $f(s)$ plane; this is equal to the net number of revolutions of the $w(s)$ vector in the $w(s)$ plane, and equal to the number of times the plot of $q(s)$ encircles the point $(-1, +i0)$ in the $q(s)$ plane. From Eq. (5.9),

$$q(s) = \frac{K(1 + T_2 s)}{s(1 + T_1 s)(1 + T_3 s)}$$

Since this function has a pole at the origin, the origin must be avoided, as indicated in Fig. 5.3. This may be done by letting

$$s = \delta e^{i\theta} = x + iy$$

where $\delta \to 0$ and $-\dfrac{\pi}{2} < \theta < \dfrac{\pi}{2}$. As s proceeds from $-i\infty$ in Fig. 5.3 up the imaginary axis until $y = -\delta$,

$$q(-iy) = \frac{K(1 - iT_2 y)}{-iy(1 - iT_1 y)(1 - iT_3 y)}$$

plots from the origin and traverses the upper left-hand dotted curve in Fig. 5.4, approaching $(0, +i\infty)$. When $s = \delta e^{i\theta}$, the plot of $q(s)$ sweeps out a large semicircle in the first and fourth quadrant defined by

$$q(\delta e^{i\theta}) = \frac{K}{\delta e^{i\theta}}$$

The dotted portion of the semicircle in the first quadrant of Fig. 5.4 corresponds to negative values of θ, while the solid portion in the fourth quadrant corresponds to positive values of θ. Then as s proceeds from $(0, +i\delta)$ toward $(0, +i\infty)$ in Fig. 5.3, $q(iy)$ generates the solid portion in the third quadrant of Fig. 5.4. Since the point $(-1, +i0)$ is not encircled, the system is stable.

The plot in Fig. 5.5 illustrates a stable condition (A) and an unstable condition (B). For curve B, the vector $\mathbf{R}$ makes two clockwise revolutions about point $(-1, +i0)$, whereas for curve A it makes no net rotations. The sense of rotation is reversed from that given in the theorem because of the reciprocal relationship of $w(s)$ to $f(s)$ in Eq. (5.9); i.e., zeros and poles of $f(s)$ correspond, respectively, to poles and zeros of $w(s)$.

A further application of Nyquist's criterion is given in the next section in connection with systems with retarded actions or time lags.

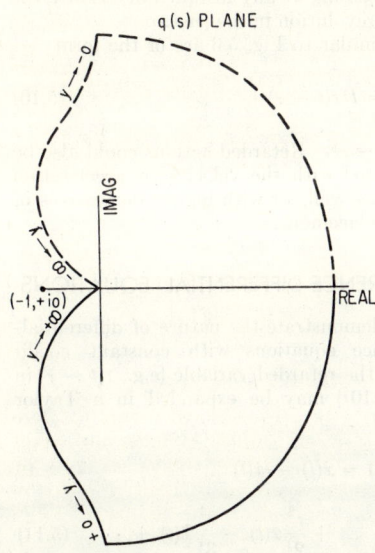

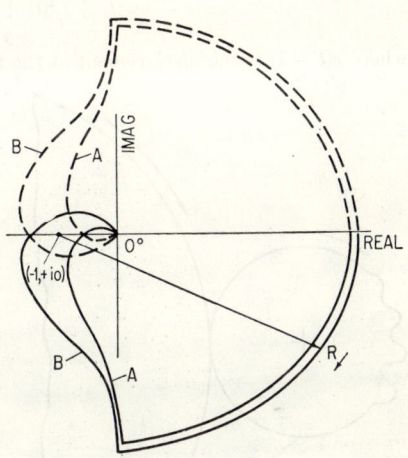

Fig. 5.4. Nyquist plot of

$$q(s) = \frac{k(1 + T_2 s)}{s(1 + T_1 s)(1 + T_3 s)}$$

As the roving point s in Fig. 5.3 proceeds from $(0, -i\infty)$ up the imaginary axis to $(0, -i\delta)$, $q(s)$ traces out the dotted portion of the curve in the upper left-hand quadrant; as the roving point s travels along the small semicircle around the origin, $q(s)$ traces out the large semicircle in the first and fourth quadrants, the dotted portion corresponding to negative values of y and the solid portion to positive values; as the roving point s proceeds from $(0, +i\delta)$ up the imaginary axis, $q(s)$ traces out the solid portion in the third quadrant; as the roving point s travels along the infinite semicircular portion of Γ, $q(s)$ describes an infinitesimal contour around the origin. (*After Chestnut and Mayer.*[3])

Fig. 5.5. Nyquist plot of

$$q(s) = \frac{10}{s(1 + s/4)(1 + s/16)}$$

[excluding point $(-1, i0)$ and illustrating a stable condition, curve A] and

$$q(s) = \frac{100}{s(1 + s/4)(1 + s/16)}$$

[encircling point $(-1, i0)$ and illustrating an unstable condition, curve B]. (*After Chestnut and Mayer.*[3])

LINEAR SYSTEMS WITH TIME LAG

INTRODUCTION

In many mechanical and electrical systems, there are time lags or retarded actions which are the essential feature in producing self-excited oscillations. For example, a time lag in a feedback control system may arise from the time required to transmit a hydraulic pressure signal through a long length of pipe.

Another example is the grinding wheel A operating on a workpiece B, as shown in

Fig. 5.6. Approximately one-half revolution has elapsed since the system underwent a small transient disturbance (shown exaggerated). This undulating profile ground into the work produces a pulsating force as it engages the wheel again. Consequently, the elastic system consisting of the wheel and work supporting structures is affected not only by the inertia, damping, and spring forces existing at any instant but also by the motion of the system which existed one period of revolution in the past.

The differential equations describing systems similar to Fig. 5.6 are of the form

$$m\ddot{x}(t) + c\dot{x}(t) + kx(t) = Dx(t - \tau) \tag{5.10}$$

where $x(t - \tau)$ is the displacement at the time $(t - \tau)$. Retarded actions could also be associated with the velocity or acceleration terms as well, or with higher derivatives of the displacement.

DIFFERENCE-DIFFERENTIAL EQUATIONS

To demonstrate the nature of differential-difference equations with constant coefficients, the retarded variable [e.g., $x(t - \tau)$ in Eq. (5.10)] may be expanded in a Taylor series:[9]

$$x(t - \tau) = x(t) - \tau\dot{x}(t)$$
$$+ \frac{\tau^2}{2!}\ddot{x}(t) - \frac{\tau^3}{3!}\dddot{x}(t) + \cdots \tag{5.11}$$

Substituting Eq. (5.11) into Eq. (5.10),

$$\frac{D}{m}\left[\cdots \frac{(-1)^{n+1}}{n!}\frac{x^n}{x} \cdots - \frac{\tau^2}{2!}\frac{\ddot{x}}{x} + \frac{\tau\dot{x}}{x} - 1\right]$$

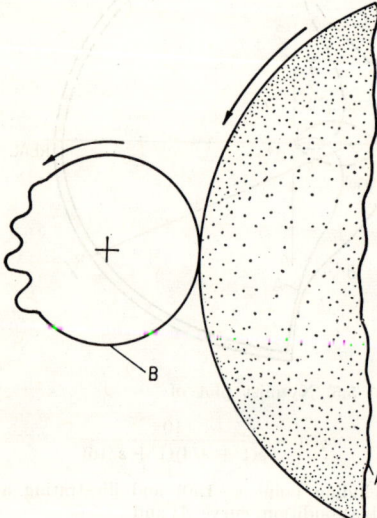

Fig. 5.6. Illustration of time lag in a mechanical system composed of a grinding wheel A and a workpiece B. The transient wave ground into the workpiece is reinjected into the system after one rotation of the work.

$$\times \left[x + \ddot{x} + \frac{c}{m}\dot{x} + \Omega^2 x\right] = 0 \tag{5.12}$$

This is a linear differential equation with constant coefficients and of infinite order; therefore, it possesses an infinite number of roots. Stability is attained by determining the parameters so that no roots have positive real parts. To reduce Eq. (5.12) to a bounded form, let $x = x_0 e^{zt}$. Then $\dot{x} = zx_0 e^{zt} = zx$ and $\dot{x}/x = z$. Also, $\ddot{x} = z^2 x_0 e^{zt} = z^2 x$ and $\ddot{x}/x = z^2$. Thus, the infinite series in the bracket is equal to $-e^{-\tau z}$ and Eq. (5.12) becomes

$$\ddot{x} + \frac{c}{m}\dot{x} + \Omega^2 x - \frac{D}{m}e^{-\tau z}x = 0 \tag{5.13}$$

The substitution of $x = x_0 e^{zt}$ gives the characteristic equation:

$$z^2 + \frac{c}{m}z + \Omega^2 - \frac{D}{m}e^{-\tau z} = 0$$

The exponential term involving z appears with the term z^0; exponential terms can appear with either the z or z^2 term, or both.

NYQUIST STABILITY CRITERION APPLIED TO RETARDED SYSTEMS

A convenient method of studying the stability of systems with time lag utilizes the Laplace transform method in conjunction with the general Nyquist criterion.[10] Consider a system with velocity time lag described by

$$I \frac{d^2f(t)}{dt^2} + R \frac{df(t)}{dt} + Kf(t) = -P \frac{df(t-\tau)}{dt} \qquad (5.14)$$

Taking the Laplace transform of both sides,

$$(s^2 I + sR + K + Pse^{-\tau s})F(s) = sIF_0 + I\dot{F}_0 + RF_0 \qquad (5.15)$$

where $F(s)$ is the transform of $f(t)$; $I, R, K, P,$ and τ are constants; and F_0 and $\dot{F}_0$ denote the values of $f(t)$ and $\dfrac{df(t)}{dt}$ at $t = 0$. Solving Eq. (5.15) for $F(s)$,

$$F(s) = \frac{L(s)}{Y(s) + Pse^{-s\tau}} = \frac{L(s)}{Y(s)} \cdot \frac{1}{1 + P \dfrac{se^{-s\tau}}{s^2 I + sR + K}}$$

where

$$L(s) \equiv sIF_0 + I\dot{F}_0 + RF_0$$
$$Y(s) \equiv s^2 I + sR + K$$

By Mellins's inversion theorem,[11, 12]

$$f(t) = \frac{1}{2\pi i} \int_\Gamma \frac{L(s)}{Y(s)} \cdot \frac{1}{1 + P \dfrac{se^{-s\tau}}{s^2 I + sR + K}} e^{st} \, ds \qquad (5.16)$$

where Γ is a contour enclosing all the roots of the denominator in the s plane. For a study of stability, it is necessary to consider only the presence of roots with positive real parts. Therefore, the contour can be reduced to that shown in Fig. 5.3 and described in the preceding section. The Nyquist stability theorem is applicable. Since the integral of Eq. (5.16) has no zeros (i.e., $L(s) \neq 0$ for any $s \geq 0$), the presence of roots in the positive half-plane is indicated by poles of the integrand or zeros of the denominator. The parameter $Y(s)$ has no zeros in the right half-plane for I, R, and K positive; therefore, it is necessary to investigate only the zeros of

$$w(s) = 1 + P \frac{se^{-s\tau}}{s^2 I + sR + K}$$

as described under the heading of Nyquist's criterion. As the representative point in the s plane traverses the contour Γ in Fig. 5.3, the corresponding point in the w plane traverses the contour shown in Fig. 5.7. Stability is ensured if the point $(-1, i0)$ is not encircled.

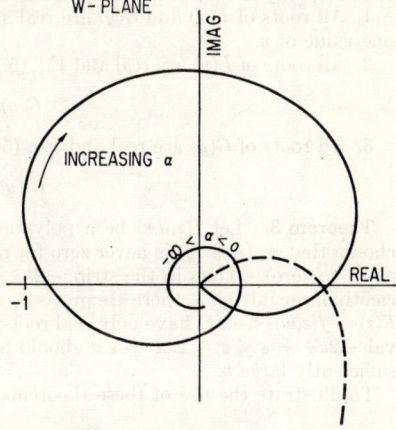

FIG. 5.7. Nyquist plot in the w plane of

$$1 + P \frac{se^{-s\tau}}{s^2 I + sR + K}$$

for $s = i\alpha$. Stability is ensured if the point $(-1, i0)$ is not encircled.

ZEROS OF EXPONENTIAL POLYNOMIALS

There are several theorems [13] which are helpful for investigating the roots of characteristic equations of the form

$$P(z, e^z) = 0 \quad \text{or} \quad Q(z, \cos z, \sin z) = 0$$

The concept of a leading term is necessary in discussing these theorems. If the polynomial $P(z,w)$ in the two variables z and w has coefficients which in general may be complex, and if it possesses a term $a_{rs}z^r w^s$ such that $r \geq m$ and $s \geq n$, where $a_{mn}z^m w^n$ is any other term, the term $a_{rs}z^r w^s$ is designated the leading term. For example, in the polynomial $(z^2 + pz + q)e^z$, the term $z^2 e^z$ is the leading term. However, not every polynomial has a leading term. For example, the polynomial

$$(z^2 + pz + q)e^z + rz^n$$

has the leading term $z^2 e^z$ when $n \leq 2$. If $n > 2$, the polynomial has no leading term.

Similarly, for $Q(z, \cos z, \sin z)$, if a term $z^r \phi_{rs}(\cos z, \sin z)$ exists such that $r \geq m$, $s \geq n$ for any other term $z^m \phi_{mn}(\cos z, \sin z)$, then $z^r \phi_{rs}(\cos z, \sin z)$ is said to be the leading term. Then $\phi_{rs}(\cos z, \sin z)$ denotes a homogeneous polynomial of degree s.

Theorem 1. If the polynomial $P(z,w)$ has no leading term, the function $P(z,e^z)$ has an infinity of zeros with arbitrarily large real parts. Conversely, if $P(z,w)$ has a leading term, the real parts of the zeros of $P(z,e^z)$ are uniformly bounded above.

For example, consider the characteristic equation

$$2\lambda e^\lambda - e^{2\lambda} + 1 = 0$$

Since this equation has no leading term, there exist an infinite number of roots with arbitrarily large real parts. A theorem similar to Theorem 1 applies to $Q(z, \cos z, \sin z)$.

Theorem 2. Let $H(z) = h(z, e^z)$, where $h(z,w)$ is a polynomial with a leading term. Write $H(iy)$, where y is real, as the sum of its real and imaginary parts, $H(iy) = F(y) + iG(y)$. If all the zeros of the function $H(z)$ lie to the left of the imaginary axis, then the zeros of the functions $F(y)$ and $G(y)$ are real, interlaced [i.e., the zeros of $F(y)$ are interspersed between those of $G(y)$, and vice versa], and

$$G'(y)F(y) - G(y)F'(y) > 0 \tag{5.17}$$

for every value of y. Conversely, for the roots of the function $H(z)$ to lie to the left of the imaginary axis, it is sufficient that at least one of the following conditions be satisfied:

1. All roots of $F(y)$ and $G(y)$ are real and interlaced, and Eq. (5.17) holds for at least one value of y.
2. All roots of $F(y)$ are real and Eq. (5.17) holds for each of them; i.e.,

$$G(y)F'(y) < 0 \tag{5.18}$$

3. All roots of $G(y)$ are real and Eq. (5.17) holds for each of them; i.e.,

$$G'(y)F(y) > 0 \tag{5.19}$$

Theorem 3. Let $f(z,u,v)$ be a polynomial with leading term $z^r \phi_{rs}(u,v)$. Let ϵ be so chosen that $\phi_{rs}(\epsilon + iy)$ is never zero for real values of y (note that ϵ often may be taken equal to zero). Then in the strip $-2k\pi + \epsilon \leq \Re[z] \leq 2\pi k + \epsilon$, beginning with a sufficiently large integer k, there are precisely $4sk + r$ roots. Thus, in order that the function $F(z) = f(z, \cos z, \sin z)$ have only real roots, it is necessary and sufficient that in the interval $-2k\pi + \epsilon \leq z \leq 2k\pi + \epsilon$ it should have exactly $4sk + r$ real roots, beginning with sufficiently large k.

To illustrate the use of these theorems, consider the characteristic equation

$$H(z) = pe^z + q - ze^z = 0 \tag{5.20}$$

Since this equation has a leading term (ze^z), according to Theorem 1, the real parts of the zeros are uniformly bounded above. Then for real values of y:

$$H(iy) = F(y) + iG(y)$$

$$= (p \cos y + y \sin y + q) + i(p \sin y - y \cos y)$$

According to Theorem 2, all roots of $G(y)$ must be real and
$$G'(y)F(y) > 0 \quad (5.19)$$
In this case
$$G(y) = -y \cos y + p \sin y = 0 \quad (5.21)$$
Theorem 3 gives the necessary and sufficient condition that the roots of this equation be real. The leading term is $y \cos y$. In this case $r = 1$, $s = 1$; therefore, in the interval $-2k\pi + \epsilon \le y \le 2k\pi + \epsilon$, there must be exactly $4k + 1$ real roots. Equation (5.21) can be rewritten as follows:

$$\tan y = \frac{1}{p} y \quad (5.22)$$

Equation (5.22) is plotted in Fig. 5.8; it is clear that in order to have $4k + 1$ real roots, p must lie between 0 and 1; i.e., $0 < p < 1$.

Introducing the expressions for $F(y)$, $G(y)$, and $G'(y)$ into the inequality Eq. (5.19) of Theorem 2,

$$-p(1-p)\cos^2 y$$
$$+ (2p-1)y \sin y \cos y + y^2 \sin^2 y$$
$$+ qy \sin y - q(1-p)\cos y > 0 \quad (5.23)$$

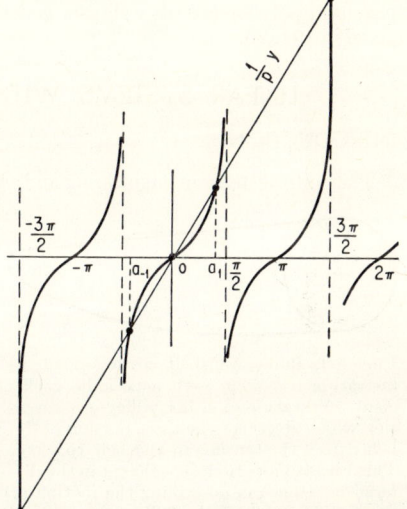

FIG. 5.8. Plot of $y/p = \tan y$ showing roots between $-\pi/2$ and $\pi/2$ when $p \le 1$.

This is the sufficient condition that all roots of Eq. (5.20) lie to the left of the imaginary axis. Inserting the root $y = a_0 = 0$,

$$(p-1)(p+q) > 0$$

Since $p < 1$, $(p+q) < 0$. Designating the roots of Eq. (5.22) by
$$\ldots, a_{-2}, a_{-1}, a_0, a_1, a_2, \ldots$$
as illustrated in Fig. 5.8, and inserting these roots into Eq. (5.23),

$$G'(a)F(a) = \frac{1}{\sqrt{a^2 + p^2}} (\sqrt{a^2 + p^2} + q)(a^2 + p^2 - p) > 0 \quad (5.24)$$

This applies for any positive or negative odd root. For any positive or negative even root:

$$G'(a)F(a) = \frac{1}{\sqrt{a^2 + p^2}} (\sqrt{a^2 + p^2} - q)(a^2 + p^2 - p) > 0 \quad (5.25)$$

Since $a_1 = p \tan a_1$, it is easy to show that $(a_1^2 + p^2 - p) > 0$. This also applies to $1/\sqrt{a_1^2 + p^2}$. Therefore, from Eq. (5.24),

$$\sqrt{a_1^2 + p^2} + q > 0$$

Hence, the roots of Eq. (5.20) have negative real parts if and only if $p < 1$ and $p < -q < \sqrt{a_1^2 + p^2}$, where a_1 is the root of Eq. (5.22) and $0 < a_1 < \pi$. If $p = 0$, it is assumed that $a_1 = \pi/2$.

By the use of the above theorems, necessary and sufficient conditions that the roots lie to the left of the imaginary axis have been worked out [13] for the following exponential polynomials:

$$(z^2 + pz + q)e^z + r = 0$$
$$(z^2 + pz + q)e^z + rz = 0$$
$$(z^2 + pz + q)e^z + rz^2 = 0$$
$$z^2 e^z + pz + q = 0$$

LINEAR SYSTEMS WITH PERIODIC COEFFICIENTS

INTRODUCTION

There are systems in engineering and physics which are described by linear differential equations having periodic coefficients,

FIG. 5.9. Pulley-and-belt arrangement illustrating a system with a periodic coefficient. Vibration of either pulley in a direction which stretches the tight portion of the belt causes the tension in the belt to vary. This leads to a periodic coefficient in the differential equation describing the motion of the tight portion of the belt.

$$\frac{d^2y}{dz^2} + p(z)\frac{dy}{dz} + q(z)y = 0 \quad (5.26)$$

where $p(z)$ and $q(z)$ are periodic in z. These systems also may exhibit self-excited vibrations, but the stability of the system cannot be evaluated by finding the roots of a characteristic equation.

For example, consider the belt-and-pulley system illustrated in Fig. 5.9. Assume that the larger pulley is attempting to vibrate horizontally. This produces a periodically varying tension in the stretched section of the belt. Considering the belt as a stretched string with tension F and mass density ρ,

$$\frac{\partial^2 \xi}{\partial t^2} - \frac{F}{\rho}\frac{\partial^2 \xi}{\partial x^2} = 0$$

where ξ is the lateral displacement of an element of the belt. Assuming

$$\xi(x,t) = f(t)\sin\frac{n\pi x}{l} \quad \text{and} \quad F = F_0(1 - 2\gamma \cos 2\omega t)$$

the differential equation becomes

$$\frac{d^2f}{dt^2} + \left(\frac{n\pi}{l}\right)^2 \frac{F_0}{\rho}(1 - 2\gamma \cos 2\omega t)f = 0$$

This may be written

$$\frac{d^2f}{dz^2} + (a - 2q \cos 2z)f = 0 \quad (5.27)$$

where $\quad z = \omega t \quad a = \dfrac{\left(\dfrac{n\pi}{l}\right)^2 F_0}{\omega^2 \rho} \quad q = \dfrac{\gamma\left(\dfrac{n\pi}{l}\right)^2 F_0}{\omega^2 \rho} = \gamma a$

Equation (5.27), known as Mathieu's equation, describes the vibration of the stretched section of the belt.

A system that possesses periodically varying coefficients is a shaft whose bending rigidity differs about the two axes 1–1 and 2–2, as shown in Fig. 5.10. The shaft may develop vibration of large amplitude when it is run at submultiples of the usual critical speed. The differential equations describing the motion of the center-of-gravity of the disc carried by the shaft are

$$\ddot{y} + \left(\frac{\omega_1^2 + \omega_2^2}{2} - \frac{\omega_1^2 - \omega_2^2}{2}\cos 2\Omega t\right)y = -\frac{\omega_1^2 - \omega_2^2}{2}\sin 2\Omega t\, x - g$$

$$\ddot{x} + \left(\frac{\omega_1^2 + \omega_2^2}{2} + \frac{\omega_1^2 - \omega_2^2}{2}\cos 2\Omega t\right)x = -\frac{\omega_1^2 - \omega_2^2}{2}\sin 2\Omega t\, y$$

where ω_1 and ω_2 are the natural circular frequencies in the first bending mode about the axes 1-1 and 2-2 and Ω is the rotational speed. If the shaft is operated in the vicinity of the critical speed $\Omega = \sqrt{\dfrac{\omega_1{}^2 + \omega_2{}^2}{2}}$ or at one-half, one-fourth, one-sixth, etc., of the critical speed, violent vibration is likely to occur. If the motion is restricted to the x coordinate only, a Mathieu equation results.

An equation somewhat more general than Mathieu's equation is Hill's equation:

$$\frac{d^2y}{dz^2} + [a - 2q\psi(z)]y = 0 \qquad (5.28)$$

where $\psi(z)$ is any even function of z.

GENERAL THEORY

The following theory applies to any linear differential equation of the second order with single-valued periodic coefficients, such as Eq. (5.26).[15] If $y_1(z)$ and $y_2(z)$ are any two linearly independent fundamental solutions, the complete solution is

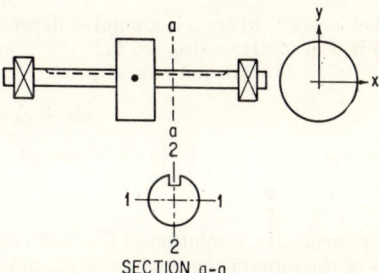

FIG. 5.10. Shaft system possessing unequal rigidities, leading to a pair of coupled inhomogeneous Mathieu equations.

$$y(z) = Ay_1(z) + By_2(z) \qquad (5.29)$$

where A and B are arbitrary constants. If the period of the coefficients is τ, then $y_1(z + \tau)$ and $y_2(z + \tau)$ are also solutions and each can be expressed in terms of the fundamental solutions $y_1(z)$ and $y_2(z)$:

$$\begin{aligned} y_1(z + \tau) &= \alpha_1 y_1(z) + \alpha_2 y_2(z) \\ y_2(z + \tau) &= \beta_1 y_1(z) + \beta_2 y_2(z) \end{aligned} \qquad (5.30)$$

From Eq. (5.29),

$$y(z + \tau) = Ay_1(z + \tau) + By_2(z + \tau)$$

Substituting in this equation from Eq. (5.30),

$$y(z + \tau) = (A\alpha_1 + B\beta_1)y_1(z) + (A\alpha_2 + B\beta_2)y_2(z)$$

If $y(z + \tau) = \phi y(z)$, where ϕ is a constant, then

$$(A\alpha_1 + B\beta_1)y_1(z) + (A\alpha_2 + B\beta_2)y_2(z) = \phi A y_1(z) + \phi B y_2(z)$$

Equating the coefficients of $y_1(z)$ and $y_2(z)$ to zero,

$$A(\alpha_1 - \phi) + B\beta_1 = 0$$
$$A\alpha_2 + B(\beta_2 - \phi) = 0$$

Since $A \neq B \neq 0$,

$$\begin{vmatrix} (\alpha_1 - \phi), & \beta_1 \\ \alpha_2, & (\beta_2 - \phi) \end{vmatrix} = 0$$

Writing the determinant as a quadratic in ϕ,

$$\phi^2 - (\alpha_1 + \beta_2)\phi + \alpha_1\beta_2 - \alpha_2\beta_1 = 0 \qquad (5.31)$$

Hence, ϕ can be determined if α_1, α_2, β_1, and β_2 are known. The parameters α_1, α_2, β_1, and β_2 are determined from Eqs. (5.30) and the following initial conditions:

$$y_1(0) = 1 \qquad y_1'(0) = 0 \qquad y_2(0) = 0 \qquad y_2'(0) = 1$$

It is assumed that $y_1(z)$ is an even function of z and $y_2(z)$ an odd function. Thus

$$\alpha_1 = y_1(\tau) \qquad \alpha_2 = y_1'(\tau)$$
$$\beta_1 = y_2(\tau) \qquad \beta_2 = y_2'(\tau)$$

Let $\phi \equiv e^{\mu\tau}$, where μ is a number depending on the parameters in Eq. (5.26), or on a and q in Eq. (5.27). Also, let $\Phi(z) \equiv e^{-\mu z}y(z)$. Then $\Phi(z + \tau) = e^{-\mu(z+\tau)}y(z+\tau)$. Since $y(z+\tau) = \phi y(z)$ and $\phi = e^{\mu\tau}$,

$$y(z+\tau) = e^{-\mu z}y(z) = \Phi(z)$$

Hence, $\Phi(z)$ is periodic in z with period τ and

$$y(z) = e^{\mu z}\Phi(z) \tag{5.32}$$

Consequently, a solution of Eq. (5.26) or (5.27), which itself may or may not be periodic, is of the form of the product of $e^{\mu z}$ and a periodic function of z, where $\mu = \alpha + i\beta$.

In the particular case of Mathieu's and Hill's equations where the coefficient is even in z,

$$y(-z) = e^{-\mu z}\Phi(-z) \tag{5.33}$$

is also a solution, independent of Eq. (5.32), provided either $\alpha \neq 0$ or, when $\alpha = 0$, β is nonintegral. Therefore, the complete solution of Mathieu's or Hill's equation is

$$y(z) = Ae^{\mu z}\Phi(z) + Be^{-\mu z}\Phi(-z)$$

where A and B are arbitrary constants.

Since $\Phi(z)$ is periodic, it may be expressed as

$$\Phi(z) = \sum_{r=-\infty}^{\infty} c_{2r}e^{2rzi} \qquad \Phi(-z) = \sum_{r=-\infty}^{\infty} c_{2r}e^{-2rzi}$$

Hence

$$y(z) = Ae^{\mu z}\sum_{r=-\infty}^{\infty} c_{2r}e^{2rzi} + Be^{-\mu z}\sum_{r=-\infty}^{\infty} c_{2r}e^{-2rzi} \tag{5.34}$$

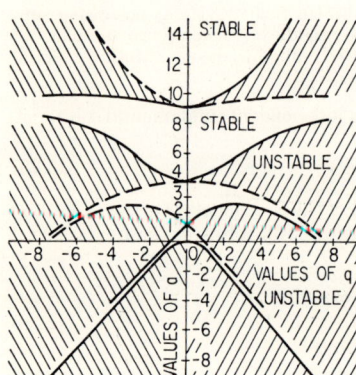

Fig. 5.11. Mathieu stability diagram showing the values of a and q which yield stable, unstable, and periodic solutions of Mathieu's equation, Eq. (5.27). If the point (a,q) lies in the uncrosshatched regions (stable), bounded solutions are obtained. If the point (a,q) lies in the crosshatched regions (unstable), the solutions are unbounded. If the point (a,q) lies on either the solid or dotted boundary curves, the solutions are periodic and are known as Mathieu functions of integral order.

MATHIEU'S STABILITY CRITERIA

The stability of the solutions to Mathieu's equation is dependent on the value of μ in Eq. (5.34), which in turn is a function of the parameters a and q in Eq. (5.27). If μ is a positive real number, the first term in Eq. (5.34) tends to infinity as $z \to \infty$; therefore, the complete solution is unstable. If μ is negative, the second term is unbounded and the complete solution also is unstable; if $\mu = \alpha + i\beta$ and $\alpha \neq 0$, the solution is unstable.

When $\mu = i\beta$ and β is nonintegral, Eq. (5.34) gives the stable solution:

$$y(z) = A\sum_{r=-\infty}^{\infty} c_{2r}e^{(2r+\beta)zi} + B\sum_{r=-\infty}^{\infty} c_{2r}e^{-(2r+\beta)zi} \tag{5.35}$$

If β is a rational fraction p/s, where p and s are prime numbers, both terms in Eq. (5.35) are periodic with period $2\pi s$. When β is irrational, the solution is oscillatory, bounded but not periodic.

The regions of the a,q plane corresponding to stable and unstable solutions of Eq.

(5.27) are mapped as shown in Fig. 5.11. If (a,q) lies in a stable area, Eq. (5.27) has bounded solutions. These solutions are referred to as Mathieu functions of fractional order. The limiting periodic solutions corresponding to the boundary curves of Fig. 5.11 are Mathieu functions of integral order. Figure 5.12 is an extended version of Fig. 5.11. Further details on Mathieu functions are given in Ref. 14.

For example, consider a long pin-jointed column which sustains a steady axial force P_0, as shown in Fig. 5.13. Because of out of balance in associated rotating machinery, an alternating component of force $-2\gamma P_0 \cos 2\omega t$ is superimposed. The equation of lateral motion [14] is

$$EI \frac{\partial^4 y}{\partial x^4} - P_0(1 - 2\gamma \cos 2\omega t) \frac{\partial^2 y}{\partial x^2} + m_l \frac{\partial^2 y}{\partial t^2} = 0 \quad (5.36)$$

where y = lateral displacement at point x
m_l = mass per unit length
2γ = strength of alternating force component relative to sustained force

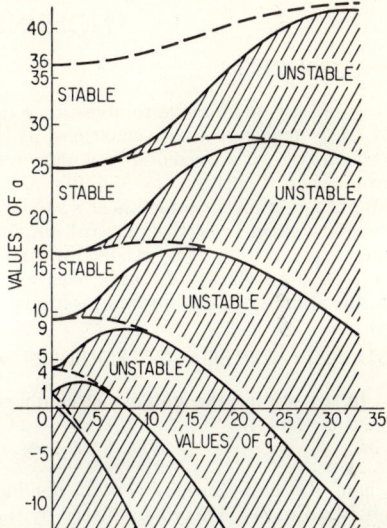

FIG. 5.12. Extension in the first quadrant of the Mathieu stability diagram of Fig. 5.11.

The boundary conditions are $y = 0$, $\partial^2 y / \partial x^2 = 0$ at $x = 0$ and $x = l$. These conditions are satisfied by assuming

$$y = f(t) \sin \frac{n\pi x}{l}$$

Substituting the assumed solution in Eq. (5.36),

$$\frac{d^2 f}{dz^2} + (a - 2q \cos 2z) f = 0$$

where
$$z = \omega t$$

$$a = \frac{\left(\frac{n\pi}{l}\right)^2 \left[EI \left(\frac{n\pi}{l}\right)^2 + P_0\right]}{m_l \omega^2}$$

$$q = \frac{\gamma P_0 \left(\frac{n\pi}{l}\right)^2}{m_l \omega^2}$$

For a given value of n, the stability is determined by computing values of a and q, and observing their location in Fig. 5.11.

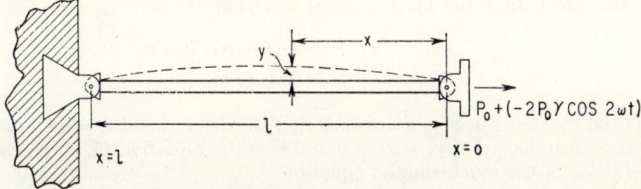

FIG. 5.13. Long column with pinned ends illustrating application of Figs. 5.11 and 5.12 for the determination of stability. A periodic force is superimposed upon a constant axial pull. (After McLachlan.[21])

NONLINEAR SYSTEMS

INTRODUCTION

It often is permissible to investigate the stability of equilibrium positions of real systems by means of linear equations, as discussed in the preceding sections. However, certain systems are essentially nonlinear and neglect of nonlinear terms is not permissible even when discussing the stability of equilibrium positions. Furthermore, since the amplitude of an unstable linear system tends to infinity, nonlinear terms must be considered (which limit the physical amplitude to finite values) in discussing the steady amplitude or limit cycle attained by the unstable system. This leads to questions concerning the stability of limit cycles, i.e., the stability of periodic motions, in contrast to the stability of equilibrium positions.

STABILITY CRITERIA OF EQUILIBRIUM—SINGULAR POINTS

In this section the stability of equilibrium positions is discussed for differential equations of the type

$$\ddot{x} + \phi(\dot{x}) + f(x) = 0 \tag{5.37}$$

where $\phi(\dot{x})$ and $f(x)$ may contain nonlinear terms. Since the time does not occur explicitly, Eq. (5.37) may be replaced by the two first-order equations

$$\frac{dv}{dt} = -f(x) - \phi(v)$$

$$\frac{dx}{dt} = v \tag{5.38}$$

Equations (5.38) are equivalent to

$$\frac{dv}{dx} = \frac{-f(x) - \phi(v)}{v} \tag{5.39}$$

Solutions or integral curves of Eq. (5.39) may be plotted in the phase-plane (x,v), as shown in Chap. 4.

A system is in a state of equilibrium when it is at rest $(dx/dt = 0)$ and when no force acts upon it $(dv/dt = 0)$. Applying this criterion to Eq. (5.39),

$$\frac{dv}{dx} = \frac{0}{0}$$

Consequently, the positions of equilibrium of linear as well as nonlinear systems correspond to singular points of Eq. (5.39), and questions concerning the stability of equilibrium are discussed in terms of the motion of the system in the neighborhood of these singular points.

It has been shown that the more general equation [16,17] is

$$\frac{dv}{dx} = \frac{ax + bv + P_2(x,v)}{cx + dv + Q_2(x,v)} \tag{5.40}$$

in which (1) the constants a, b, c, d are such that the determinant $\Delta = ad - bc \neq 0$ and (2) P_2 and Q_2 vanish like $x^2 + v^2$ as $x \to 0$ and $v \to 0$. Equation (5.40) has as its only singularities those of the much simpler equation

$$\frac{dv}{dx} = \frac{ax + bv}{cx + dv} \tag{5.41}$$

Equation (5.41) is equivalent to

$$\frac{dv}{dt} = ax + bv \qquad \frac{dx}{dt} = cx + dv \qquad (5.42)$$

Accordingly, the stability of equilibrium of complicated nonlinear systems represented by Eq. (5.40) can be determined by a study of Eq. (5.41).

Before studying the singularities of Eq. (5.41), consider the following special cases:
1. When $a = d = 0$ in Eqs. (5.41) and (5.42),

$$\frac{dv}{dx} = \frac{b}{c}\frac{v}{x}$$

Integrating the preceding equation,

$$v = v_0 \left(\frac{x}{x_0}\right)^{b/c} \qquad (5.43)$$

where $v = v_0$ when $x = x_0$.

If $b/c = 1$, Eq. (5.43) represents a family of straight lines through the origin of the phase-plane, as shown in Fig. 5.14A. If $b = c < 0$, the equivalent pair of Eqs. (5.42) show that the representative point in the phase-plane tends toward the origin as $t \to +\infty$; hence, the equilibrium is said to be stable. If $b = c > 0$, the representative point recedes from the origin (the singular point) as $t \to +\infty$ and the equilibrium position is

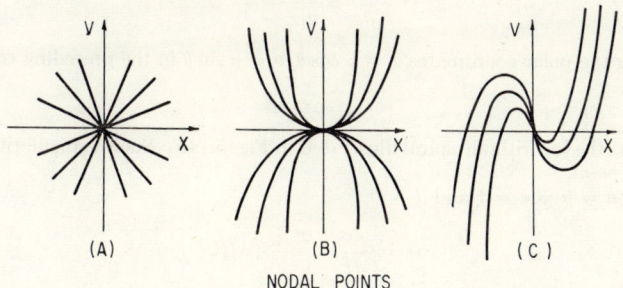

NODAL POINTS

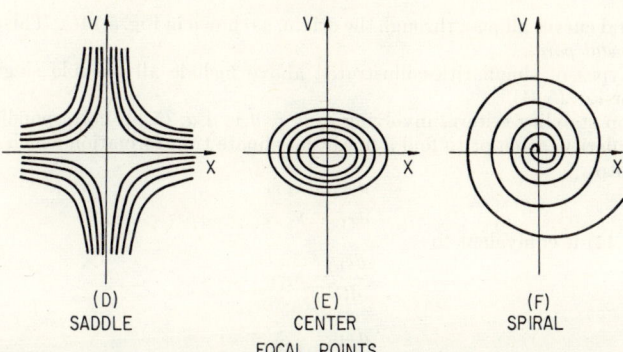

FOCAL POINTS

Fig. 5.14. Illustration showing the six possible types of singular points. The velocity of the representative point is plotted along the ordinate axis and the displacement is plotted along the abscissa. The center focal point (E) represents simple harmonic motion. The spiral focal point (F) represents a decaying vibration if the representative point spirals inwardly as time increases and growing vibration if the representative point spirals outwardly as time increases.

unstable. If $0 < b/c < 1$, the integral curves defined by Eq. (5.43) form a family of curves passing through the origin tangent to the v axis. If $b/c > 1$, the integral curves pass through the origin tangent to the X axis, as shown in Fig. 5.14B. In all these cases the singularity at the origin is called a *nodal point* or *node*; it is stable if the representative point approaches the *nodal point* for $t \to +\infty$ and unstable if the representative point recedes from the *nodal point* as $t \to +\infty$. If $b/c < 0$, the integral curves no longer go through the origin but instead are hyperbolas asymptotic to the x,v axes, as shown in Fig. 5.14D. In this case the origin is called a *saddle point* and is unstable.

2. When $b = c = 0$ in Eq. (5.41),

$$\frac{dv}{dx} = \frac{a}{d}\frac{x}{v} = k\frac{x}{v}$$

Upon integration of the preceding equation,

$$v^2 - kx^2 = v_0^2 - kx_0^2$$

where $v = v_0$ when $x = x_0$. If $k > 0$, the integral curves are hyperbolas and the system is unstable. If $k < 0$, the integral curves are ellipses with the origin as the center. This singularity is called a *center focal point*; it is shown in Fig. 5.14E and represents a stable position of equilibrium.

3. When $a = 1$, $d = -1$, and $b = c = a_1$ in Eq. (5.41),

$$\frac{dv}{dx} = \frac{x + a_1 v}{a_1 x - v}$$

Introducing the polar coordinates $x = \rho \cos \theta$, $v = \rho \sin \theta$ in the preceding equation and integrating,

$$\rho = Ce^{a_1 \theta}$$

This plots as the logarithmic spiral illustrated in Fig. 5.14F. Such a singularity is called a *spiral focal point*.

4. When $a = b = c = 1$ and $d = 0$,

$$\frac{dv}{dx} = \frac{x + v}{x}$$

Integrating,

$$v = x\frac{v_0}{x_0} + x \ln \left|\frac{x}{x_0}\right|$$

These integral curves all pass through the origin, as shown in Fig. 5.14C. This singularity is called a *nodal point*.

The six types of singularities illustrated above include all possible singularities of Eq. (5.40) or Eq. (5.41).

To develop stability criteria involving a, b, c, d in Eq. (5.40) corresponding to each type of singularity, attempt to find a linear coordinate transformation which transforms Eq. (5.41) into

$$\frac{dv_1}{dx_1} = \frac{\lambda_2}{\lambda_1}\frac{v_1}{x_1} \qquad (5.44)$$

Equation (5.44) is equivalent to

$$\frac{dv_1}{dt} = \lambda_2 v_1$$
$$\frac{dx_1}{dt} = \lambda_1 x_1 \qquad (5.45)$$

where λ_1 and λ_2 are constants. Inasmuch as the integral curves are not altered by a coordinate transformation, they are referred to a new coordinate system:

$$x_1 = \alpha x + \beta v \qquad v_1 = \gamma x + \delta v \qquad (5.46)$$

NONLINEAR SYSTEMS 5–19

where $\alpha\delta - \beta\gamma \neq 0$. With this transformation, Eqs. (5.45) and (5.42) combine to give

$$\gamma(cx + dv) + \delta(ax + bv) = \lambda_2(\gamma x + \delta v)$$
$$\alpha(cx + dv) + \beta(ax + bv) = \lambda_1(\alpha x + \beta v)$$
(5.47)

In general, the ratio of x and v is not constant; hence

$$\gamma(\lambda_2 - c) - \delta a = 0$$
$$-\gamma d + \delta(\lambda_2 - b) = 0$$
$$\alpha(\lambda_1 - c) - \beta a = 0$$
$$-\alpha d + \beta(\lambda_1 - b) = 0$$
(5.48)

Equations (5.48) must be satisfied for values of α, β, γ, and δ. The two pairs of equations require, after eliminating the ratio δ/γ or β/α, that λ_1 and λ_2 satisfy the equation

$$\lambda^2 - (b + c)\lambda - (ad - bc) = 0 \quad (5.49)$$

This is called the characteristic equation and has for a discriminant

$$D = (b - c)^2 + 4ad \quad (5.50)$$

The roots of Eq. (5.49) are

$$\lambda_{1,2} = \frac{b + c}{2} \pm \frac{1}{2}\sqrt{(b - c)^2 + 4ad}$$

From Eqs. (5.48),

$$\frac{\lambda_2 - c}{a} = \frac{\delta}{\gamma} \qquad \frac{\lambda_1 - c}{a} = \frac{\beta}{\alpha} \qquad \frac{\lambda_2 - b}{d} = \frac{\gamma}{\delta} \qquad \frac{\lambda_1 - b}{d} = \frac{\alpha}{\beta}$$

Therefore, in order that the condition $\alpha\delta - \beta\gamma \neq 0$ be associated with Eqs. (5.46), it is necessary that $\lambda_1 \neq \lambda_2$.

If λ_1 and λ_2 are real [i.e., $(b - c)^2 + 4ad > 0$] as well as unequal, Eq. (5.44) takes the form of case 1 and the singularity is either a saddle point ($\lambda_2/\lambda_1 < 0$; i.e., $\Delta = ad - bc > 0$) or a nodal point ($\lambda_2/\lambda_1 > 0$; i.e., $\Delta < 0$). The nodal point is stable if λ_1 and λ_2 are both negative. It is unstable if λ_1 and λ_2 are both positive. This is indicated by Eqs. (5.45).

If λ_1 and λ_2 are conjugate complex, x_1 and v_1 also are conjugate complex:

$$v_1 = p + iq \qquad x_1 = p - iq \quad (5.51)$$

Solving for p and q, and substituting from Eqs. (5.46),

$$p = \frac{x_1 + v_1}{2} = \left(\frac{\alpha + \gamma}{2}\right)x + \frac{\beta + \delta}{2}v$$
$$q = \frac{i}{2}(x_1 - v_1) = \frac{i}{2}[(\alpha - \gamma)x + (\beta - \delta)v]$$
(5.52)

Thus $(\alpha + \gamma)$ and $(\beta + \delta)$ must be real numbers, while $(\alpha - \gamma)$ and $(\beta - \delta)$ must be pure imaginary numbers. Accordingly, α and γ as well as β and δ must be conjugate complex; the transformation of Eqs. (5.46) shows that x_1 and v_1 also are conjugate complex. Then Eqs. (5.52) can be written in the following form, where α_1, β_1, γ_1, and δ_1 are real:

$$p = \alpha_1 x + \beta_1 v$$
$$q = \gamma_1 x + \delta_1 v$$
(5.53)

5-20 SELF-EXCITED VIBRATION

Differentiating Eqs. (5.52) with respect to t and eliminating t,

$$\frac{dp}{dq} = \frac{\dot{x}_1 + \dot{v}_1}{i(\dot{x}_1 - \dot{v}_1)} = \frac{\lambda_1 x_1 + \lambda_2 v_1}{i(\lambda_1 x_1 - \lambda_2 v_1)}$$

Substituting from Eqs. (5.51) and rearranging,

$$\frac{dp}{dq} = \frac{Ap + Bq}{Bp - Aq} \tag{5.54}$$

where A and B are real and $A = \lambda_1 + \lambda_2$, $B = \lambda_1 - \lambda_2$. This equation is of the same form discussed under case 3 and represents spirals in the p, q plane. However, if $B =$

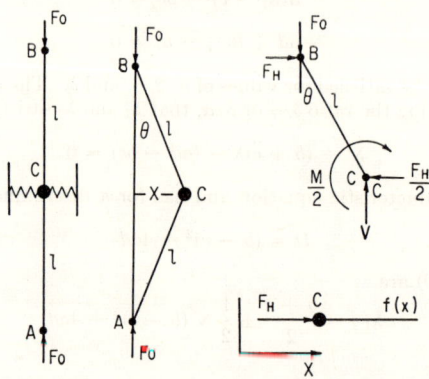

Fig. 5.15. Lumped mass and spring analog of an elastic pin-ended column illustrating elastic instability (buckling). If $2F_0/l > \alpha + 2k_r/l^2$, where α and k_1 describe the stiffness of the column, the column will buckle. (*After Stoker.*[22])

$\lambda_1 + \lambda_2 = 0$, the integral curves are circles concentric about the origin. Using Eqs. (5.52) to transform back into the x, v plane, the circles in general become ellipses when $\lambda_1 + \lambda_2 = 0$.

If $\lambda_1 = \lambda_2 = \lambda$, a real number, Eqs. (5.48) can be satisfied identically, i.e., when $\lambda = b = c$ and $a = d = 0$. Then Eq. (5.41) becomes

$$\frac{dv}{dx} = \frac{v}{x}$$

Thus, the integral curves are straight lines through the origin, as in case 1. When Eqs. (5.48) are not satisfied identically, nonzero values of α and β can be found which transform Eq. (5.41) into an equation of the form

$$\frac{dv_1}{dx_1} = \frac{a_1 x_1 + b_1 v_1}{x_1} \tag{5.55}$$

Equation (5.55) has the characteristic equation

$$\begin{vmatrix} 1 - \lambda, & a_1 \\ 0, & b_1 - \lambda \end{vmatrix} = 0$$

Since the roots are assumed equal, it follows that $b_1 = 1$. Another transformation, $v_2 = v_1/a$ and $x_2 = x_1$, transforms Eq. (5.55) into

$$\frac{dv_2}{dx_2} = \frac{x_2 + v_2}{x_2}$$

This corresponds to case 4; the singularity is a nodal point, as shown in Fig. 5.14C.

NONLINEAR SYSTEMS

The stability criteria relating to equilibrium positions discussed above can be summarized as follows, where the singularity is stable or unstable according to whether a point on the integral curve moves toward or away from the singularity as t increases:

Case 1: $(b - c)^2 + 4ad > 0$ $\begin{cases} \text{Node if } ad - bc < 0 \\ \text{Saddle if } ad - bc > 0 \end{cases}$ $\begin{cases} \text{Stable if } b + c < 0 \\ \text{Unstable if } b + c > 0 \end{cases}$

Case 2: $(b - c)^2 + 4ad < 0$ $\begin{cases} \text{Center if } b + c = 0 \\ \text{Spiral if } b + c \neq 0 \end{cases}$ $\begin{cases} \text{Stable if } b + c < 0 \\ \text{Unstable if } b + c > 0 \end{cases}$

Case 3: $(b - c)^2 + 4ad = 0$ $\quad\quad$ Node $\quad\quad$ $\begin{cases} \text{Stable if } b + c < 0 \\ \text{Unstable if } b + c > 0 \end{cases}$

To illustrate the above stability criteria, consider the simplified version of the long elastic column shown in Fig. 5.15.[17] The rods are pivoted at A, B, and C; a torsion spring (not shown) is assumed to act on each rod at C, producing a moment M in proportion to the angle through which the rods rotate at C. Equilibrium of a rod requires

$$F_0 = V$$

$$-\frac{M}{2} + Vl \sin\theta - \frac{F_H l}{2} \cos\theta = 0 \tag{5.56}$$

where F_0 is the force applied at the end of the rod, V is the vertical component of the force at hinge C, and F_H is the corresponding horizontal component of the force at the hinge.

The equation of motion of the mass m in horizontal motion is

$$m\ddot{x} = F_H - f(x)$$

The restoring moment M and the transverse spring force $f(x)$ are given by

$$M = 2k_r \theta$$

$$f(x) = \alpha x + \beta x^3$$

where k_r, α, and β are positive constants. Using Eqs. (5.56), the equation of motion becomes

$$m\ddot{x} - 2\frac{F_0 x - k_r \sin^{-1}(x/l)}{\sqrt{l^2 - x^2}} + (\alpha x + \beta x^3) = 0$$

Assuming $x \ll l$, expanding $\sin^{-1}(x/l)$ in a power series, and neglecting higher powers of x/l,

$$m\ddot{x} + \left(\alpha + \frac{2k_r}{l^2} - \frac{2F_0}{l}\right)x + \left(\beta + \frac{4k_r}{3l^4} - \frac{F_0}{l^3}\right)x^3 = 0$$

Writing the preceding equation in the form of Eq. (5.40),

$$\frac{dv}{dx} = \frac{1}{m} \frac{ax - \left(\beta + \frac{4k_r}{3l^4} - \frac{F_0}{l^3}\right)x^3}{v}$$

Thus
$$a = -\left(\alpha + \frac{2k_1}{l^2} - \frac{2P}{l}\right)$$
$$b = c = 0$$
$$d = +1$$

From the criteria summarized above, if $ad > 0$ (i.e., if $2F_0/l > \alpha + 2k_r/l^2$), the singular point is a saddle point, case 1. If $ad < 0$ (i.e., $2F_0/l < \alpha + 2k_r/l^2$), the singularity is a center, case 2.

LIMIT CYCLES

Where the singular points are unstable, the representative point in the phase-plane leaves the singularity as time increases, and may recede toward infinity or may tend to a closed integral curve known as a limit cycle. The latter condition results in a periodic motion. Moreover, a system may possess a number of limit cycles, as illustrated in Fig. 5.16.

These limit cycles or periodic motions may or may not be stable; i.e., if the system is started at a point slightly off a limit cycle, the ensuing motion may either tend toward or diverge away from the limit cycle. Thus, the limit cycle may be said to be stable or unstable.

It can be shown [18, 19] that in a succession of concentric limit cycles considered from the center outward, an outwardly stable limit cycle is followed by one that is inwardly un-

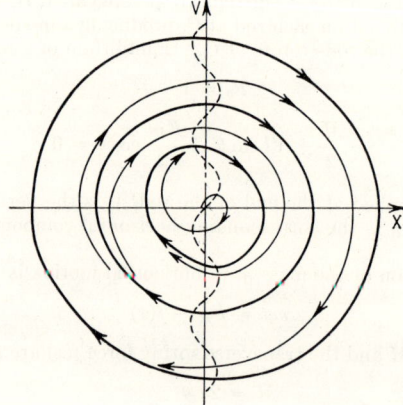

FIG. 5.16. Illustration showing a succession of stable and unstable limit cycles. (*After Stoker*.[23])

stable; a cycle that is outwardly unstable is followed by one that is inwardly stable. The point singularity at the center is considered a degenerate limit cycle possessing only the outward stability (or instability).

STABILITY OF PERIODIC MOTIONS

A basic method for investigating the stability of periodic solutions is the perturbation method illustrated below. However, other related methods have been used [17, 19] such as Van der Pol's method, the method of Andronow and Witt, and the first approximation of Kryloff and Bogoliuboff.

Consider the equations

$$\frac{dx}{dt} = P(x,y) \qquad \frac{dy}{dt} = Q(x,y) \qquad (5.57)$$

and assume that a periodic solution

$$x_1 = \phi(t) \qquad y_1 = \psi(t) \qquad (5.58)$$

of period τ is known. A solution, starting from initial conditions slightly displaced from the periodic solution, can be represented as

$$x = x_1 + \xi(t) \qquad y = y_1 + \eta(t) \qquad (5.59)$$

where ξ and η are considered to be small, so that higher powers may be neglected. If ξ

and η tend to zero as $t \to \infty$, the periodic solution x_1, y_1 is said to be stable. Substituting Eqs. (5.59) into Eqs. (5.57), after expanding $P(x,y)$ and $Q(x,y)$ in a Taylor series about x_1, y_1, and neglecting terms containing ξ^2, η^2, and higher powers,

$$\frac{d\xi}{dt} = \frac{\partial P(x_1 y_1)}{\partial x}\xi + \frac{\partial P(x_1 y_1)}{\partial y}\eta$$

$$\frac{d\eta}{dt} = \frac{\partial Q(x_1 y_1)}{\partial x}\xi + \frac{\partial Q(x_1 y_1)}{\partial y}\eta$$

(5.60)

where $\dfrac{\partial P}{\partial x}$, $\dfrac{\partial P}{\partial y}$, $\dfrac{\partial Q}{\partial x}$, and $\dfrac{\partial Q}{\partial y}$ are periodic functions of t with the common period τ. The system of Eqs. (5.60) possesses solutions of the form:[19]

$$\xi_1 = e^{h_1 t} f_{11}(t) \qquad \eta_1 = e^{h_2 t} f_{12}(t)$$
$$\xi_2 = e^{h_1 t} f_{21}(t) \qquad \eta_2 = e^{h_2 t} f_{22}(t)$$

where $f_{11}(t), \ldots, f_{22}(t)$ are periodic with period τ. The constant characteristic exponents h_1 and h_2, real or complex, are determined only to integral multiples of $2\pi i \tau$. Furthermore, for periodic solutions of Eqs. (5.60), it may be shown that one of the exponents vanishes and the remaining (nonvanishing) exponent is

$$h = \frac{1}{\tau}\int_0^\tau \left[\frac{\partial P(x_1 y_1)}{\partial x} + \frac{\partial Q(x_1 y_1)}{\partial y}\right] dt \tag{5.61}$$

The condition that the periodic solutions, Eqs. (5.58), are stable is that $h < 0$. If $h > 0$, the periodic solutions are unstable. If $h = 0$, no conclusion can be drawn.

The perturbation method also may be applied to second-order differential equations such as the Duffing equation [see Eqs. (4.16) and (4.65)]:

$$\ddot{x} + \alpha x + \beta x^3 = F \cos \omega t \tag{5.62}$$

Let $x_1(t)$ be a periodic solution and $x = x_1(t) + \xi(t)$ be a perturbed solution where ξ is very small. If this perturbed solution is inserted in Eq. (5.62) and powers of ξ neglected, the following variational equation results:

$$\ddot{\xi} + (\alpha + 3\beta x_1^2)\xi = 0 \tag{5.63}$$

Since x_1 is periodic, Eq. (5.63) is Hill's equation and the theory in the preceding section applies.

The periodic solution x_1 is said to be stable if all associated solutions of the variational equation are bounded for all positive values of t, and unstable if the variational equation has an unbounded solution.

If $x_1 = A \cos \omega t$, Eq. (5.63) reduces to a Mathieu equation, whose solutions are discussed in the preceding section, where conditions are given for ensuring boundedness of all solutions. If x_1 is a more general periodic function than $A \cos \omega t$, the general theory of Hill's equation must be applied.[14]

For a further treatment of stability of nonlinear systems, see Chap. 4.

REFERENCES

1. Den Hartog, J. P.: "Mechanical Vibrations," 4th ed., p. 346, McGraw-Hill Book Company, Inc., New York, 1956.
2. Rosenbach, J. B., and E. A. Whitman: "College Algebra," p. 265, Ginn & Company, Boston, 1939.
3. Doherty, R. E., and E. G. Keller: "Mathematics of Modern Engineering," pp. 98–130, John Wiley & Sons, Inc., New York, 1936.
4. Routh, E. J.: "Advanced Dynamics of a System of Rigid Bodies," p. 226, Dover Publications, New York, 1955.

5. Bode, H. W.: "Network Analysis and Feedback Amplifier Design," p. 149, D. Van Nostrand Company, Inc., Princeton, N.J., 1945.
6. McLachlan, N. W.: "Complex Variable Theory and Transform Calculus," p. 9, Cambridge University Press, New York, 1953.
7. Phillips, E. G.: "Functions of a Complex Variable," p. 12, Interscience Publishers, Inc., New York, 1953.
8. Chestnut, H., and R. W. Mayer: "Servomechanisms and Regulating System Design," vol. I, p. 146, John Wiley & Sons, Inc., New York, 1951.
9. Minorsky, N.: *J. Appl. Mechanics*, **64**:A65 (1942).
10. Ansoff, H. I.: *J. Appl. Mechanics*, **71**:158 (1949).
11. Gardner, M. F., and J. L. Barnes: "Transients in Linear Systems," p. 123, John Wiley & Sons, Inc., New York, 1942.
12. Ref. 6, p. 135.
13. Bellman, R., and J. M. Danskin, Jr.: "The Stability of Differential-Difference Equations," Publication P-381, The Rand Corporation, Santa Monica, Calif., 1953.
14. McLachlan, N. W.: "Theory and Applications of Mathieu Functions," p. 40, Oxford University Press, New York, 1947.
15. Floquet, G.: *Ann. l'école normale supérieure*, **12**:47 (1883).
16. Poincaré, H.: "Sur les courbes définies par une équation différentielle," Oeuvres, vol. 1, Gauthier-Villars, Paris, 1892.
17. Stoker, J. J.: "Non-linear Vibrations," p. 37, Interscience Publishers, Inc., New York, 1950.
18. Bendixson, I.: *Acta Mathematica*, **24**:1–88 (1901).
19. Minorsky, N.: "Introduction to Non-linear Mechanics," p. 78, J. W. Edwards, Publisher, Inc., Ann Arbor, Mich., 1947.
20. Pinney, E.: "Ordinary Difference-Differential Equations," p. 71, University of California Press, Berkeley, Calif., 1958.
21. Ref. 14, p. 292.
22. Ref. 17, p. 54.
23. Ref. 17, p. 133.

6

DYNAMIC VIBRATION ABSORBERS AND AUXILIARY MASS DAMPERS

F. Everett Reed
CONESCO

INTRODUCTION

Auxiliary masses are frequently attached to vibrating systems by springs and damping devices to assist in controlling the amplitude of vibration of the system. Depending upon the application, these auxiliary mass systems fall into two distinct classes.

1. If the primary system is excited by a force or displacement that has a constant frequency, or in some cases by an exciting force that is a constant multiple of a rotational speed, then it is possible to modify the vibration pattern and to reduce its amplitude significantly by the use of an auxiliary mass on a spring tuned to the frequency of the excitation. When the auxiliary mass system has as little damping as possible, it is called a *dynamic absorber*.

2. If it is impossible to incorporate damping into a structure that vibrates excessively, it may be possible to provide the damping in an auxiliary system attached to the structure. When used in this manner, the auxiliary mass system is one form of a damper. (Other forms may be incorporated as an integral part of the system.) The names *damped absorber* or *auxiliary mass damper* are given to this type of system.

It is sometimes useful to analyze the auxiliary mass system in terms of its electrical analog. Various types of analogs are described in Chap. 29.

FORMS OF DYNAMIC ABSORBERS AND AUXILIARY MASS DAMPERS

In its simplest form, as applied to a single degree-of-freedom system, the character of the auxiliary mass system is the same as that of the primary system. Thus a torsional system has a torsionally connected auxiliary mass; a linear system has a linear-spring connected mass; and a pendulum has an auxiliary pendulum. Examples of undamped auxiliary mass systems attached to single degree-of-freedom systems are shown in Figs. 6.1 and 6.2; examples of damped auxiliary mass systems are shown in Figs. 6.3 and 6.4. With multiple degree-of-freedom systems the attachment of the auxiliary masses is not as conventional as with the single degree-of-freedom system. For example, consider the two degree-of-freedom system shown in Fig. 6.5 ↑ consist-

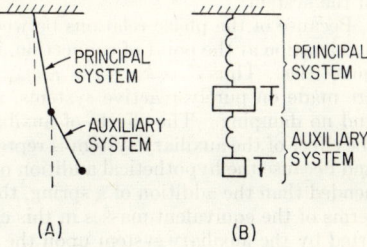

FIG. 6.1. Dynamic vibration absorbers in pendulum form (*A*) and linear form (*B*).

6–2 DYNAMIC VIBRATION ABSORBERS AND AUXILIARY MASS DAMPERS

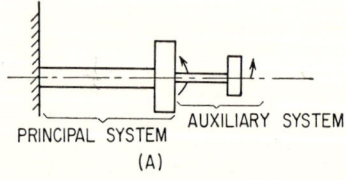

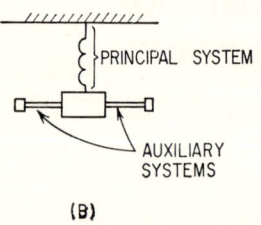

FIG. 6.2. Typical dynamic vibration absorbers. The principal and auxiliary systems vibrate in torsion in the arrangement at (A); the auxiliary system is in the form of masses and beams at (B).

ing of two weights m_1 and m_2 on a rigid, massless bar. A dynamic absorber of the type shown in Fig. 6.5B is effective for the vertical translational motion; however, if the auxiliary masses are on cantilever beams mounted on the rigid bar, as shown in Fig. 6.5C, the absorber can be made effective for both vertical translational motion and rotational motion about an axis normal to the page.

WAYS OF EXPRESSING THE EFFECTS OF AUXILIARY MASS SYSTEMS

Suppose a linear auxiliary mass system, consisting of one or more masses, springs, and dampers, is attached to a vibrating primary system. The reaction back on the primary system is proportional to the amplitude of motion at the point of attachment. It is a function of the frequency of excitation and of the masses, spring stiffnesses, and damping constants of the auxiliary mass system. If there is no damping in the auxiliary mass system, the reaction forces are either in phase or 180° out of phase with the displacement and the acceleration at the point of attachment. However, where there is damping in the auxiliary system, the reaction has a component that is 90° out of phase with the acceleration and the displacement.

Since the reaction is proportional to the amplitude of motion, it is possible to express the properties of the auxiliary mass system in terms of the motion at the point of attachment. This can be done in three ways: (1) the ratio of the reaction force to the displacement at the point of attachment, (2) the ratio of the reaction force to the velocity at the point of attachment, or (3) the ratio of the reaction force to the acceleration at the point of attachment. The first ratio can be considered equivalent to a spring whose stiffness changes with frequency. The second ratio can be considered equivalent to a damper; at any frequency it is equal in magnitude to the force-displacement ratio divided by the angular frequency. The phase angle between the force and the velocity is 90° from the phase angle between the force and the displacement. This force-velocity ratio is called the *mechanical impedance Z* of the auxiliary system. The third ratio corresponds to a mass and is designated *equivalent mass* m_{eq}. The equivalent mass of a system is $-1/\omega^2$ that of the equivalent spring k_{eq} of the system.

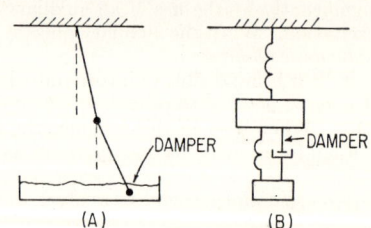

FIG. 6.3. Damped auxiliary mass systems corresponding to the undamped vibration absorbers shown in Fig. 6.1.

Because of the phase relations between the force and the displacement, velocity, and acceleration at the point of connection, it is customary to represent the ratios as complex quantities. Thus $Z = k_{eq}/j\omega = j\omega m_{eq}$. Most dynamic analyses of mechanical systems are made on purely reactive systems, i.e., systems having masses and stiffnesses only, and no damping. The effects of auxiliary mass systems are most easily understood if the effect of the auxiliary system is represented as a reactive subsystem. For this reason, and because the hypothetical addition of a mass to a system is often more easily comprehended than the addition of a spring, the effects of auxiliary mass systems are treated in terms of the equivalent masses in this chapter, i.e., in terms of the ratio of the force exerted by the auxiliary system upon the primary system to the acceleration at the point of attachment of the auxiliary system.

INFLUENCE OF A SIMPLE AUXILIARY MASS SYSTEM

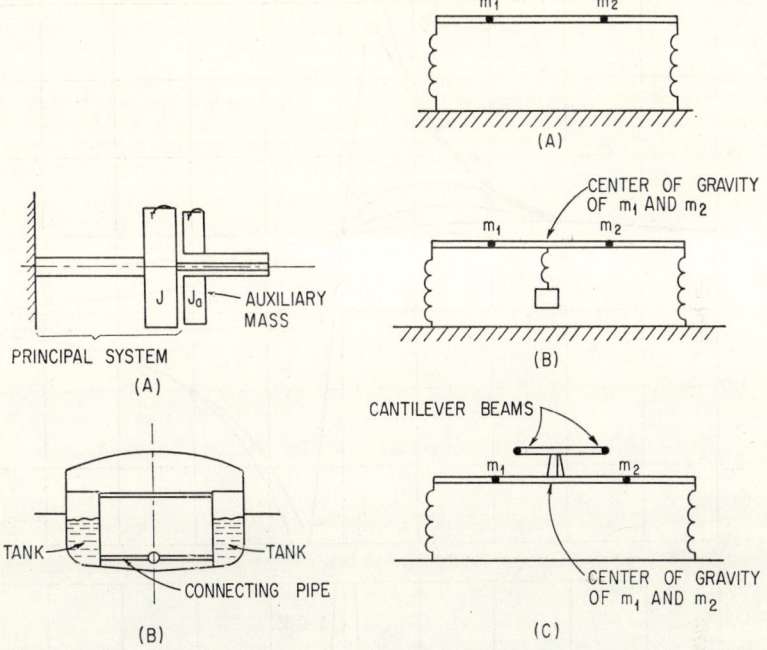

FIG. 6.4. Typical damped auxiliary mass systems. In the torsional system at (A), damping is provided by relative motion of the flywheels J, J_a. In the antiroll tanks for ships shown at (B), water flows from one tank to the other and damping is provided by a constriction in the connecting pipe.

FIG. 6.5. Application of a dynamic absorber to reduce the vibration of the spring-mounted bar at (A) in both vertical translational and rotational modes. The linear mass-spring system at (B) is effective for only translational motion, whereas the cantilever beams at (C) are effective for rotational as well as translational motion.

THE INFLUENCE OF A SIMPLE AUXILIARY MASS SYSTEM UPON A VIBRATING SYSTEM

The magnitude of the equivalent mass of a simple auxiliary mass system, consisting of a mass m_a, spring k_a, and viscous damper c_a, can be determined readily by evaluating the forces exerted by such a system upon a foundation vibrating at a frequency $f = \omega/2\pi$. The system with its assumed constants and displacements is shown in Fig. 6.6A. The spring and damping forces acting on m are shown in Fig. 6.6B, and the equation of motion is

$$(-k_a x_r - c_a j\omega x_r)e^{j\omega t} = -m_a(x_0 + x_r)\omega^2 e^{j\omega t}$$

Solving for x_r,

$$x_r = \frac{m_a \omega^2 x_0}{-m_a \omega^2 + jc_a \omega + k_a} \quad (6.1)$$

The force acting on the foundation is

$$Fe^{j\omega t} = (k_a + jc_a \omega)x_r e^{j\omega t}$$

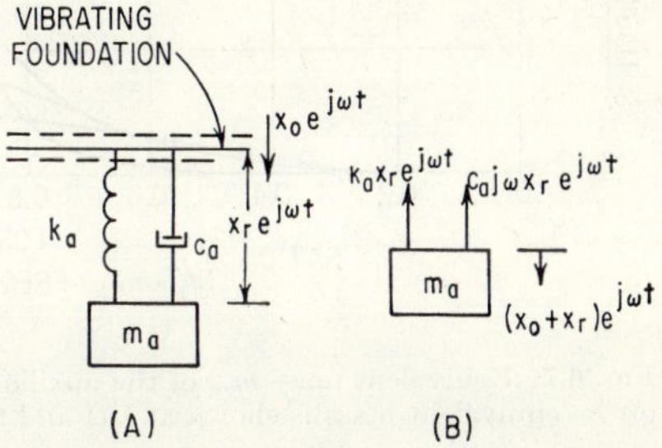

FIG. 6.6. Auxiliary mass damper. The arrangement of the damper is shown at (A), and the forces acting on the mass are indicated at (B).

6–4 DYNAMIC VIBRATION ABSORBERS AND AUXILIARY MASS DAMPERS

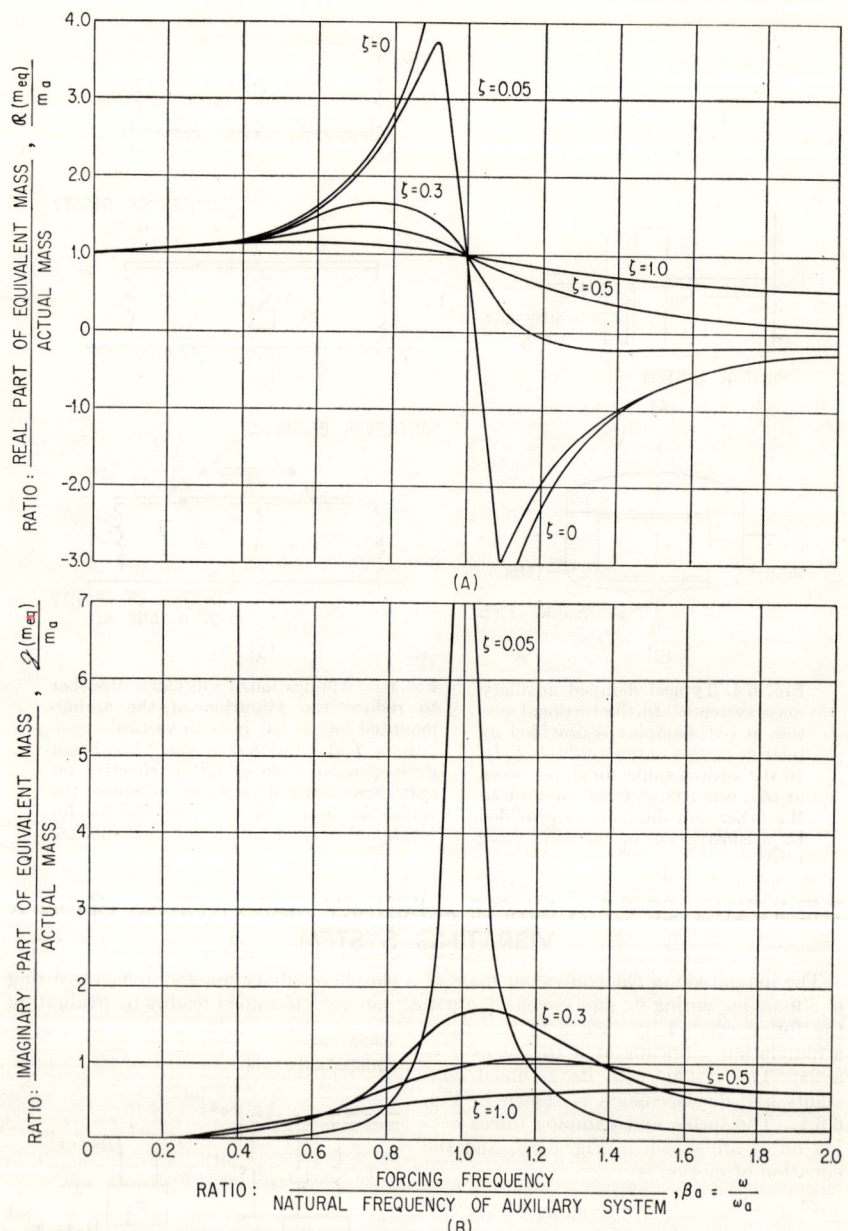

Fig. 6.7. Equivalent mass m_{eq} of the auxiliary-mass system shown in Fig. 6.6. The real part of the equivalent mass is shown at (A) and the imaginary part at (B).

Eliminating x_r from the preceding equations,

$$F = \frac{(k_a + jc_a\omega)m_a\omega^2}{-m_a\omega^2 + jc_a\omega + k_a} x_0 \tag{6.2}$$

Since the force exerted by an equivalent mass m_{eq} rigidly attached to the moving foundation is $F = m_{eq}\omega^2 x_0$:

$$m_{eq} = \frac{(k_a + jc_a\omega)}{k_a + jc_a\omega - m_a\omega^2} m \tag{6.3}$$

Equation (6.3) can be written in terms of nondimensional quantities:

$$m_{eq} = \frac{1 + 2\zeta\beta_a j}{(1 - \beta_a^2) + 2\zeta\beta_a j} m_a \tag{6.4}$$

where $\beta_a = \dfrac{\omega}{\omega_a}$, a tuning parameter

$\omega_a^2 = \dfrac{k_a}{m_a}$, the natural frequency of the auxiliary system

$\zeta = \dfrac{c_a}{c_{ca}}$, a damping parameter

$c_{ca} = 2\sqrt{k_a m_a}$, critical damping of the auxiliary system

Equation (6.4) can be divided into the following real and imaginary components:

$$m_{eq} = \frac{(1-\beta_a^2) + (2\zeta\beta_a)^2}{(1-\beta_a^2)^2 + (2\zeta\beta_a)^2} m_a - \frac{2\zeta\beta_a^3}{(1-\beta_a^2)^2 + (2\zeta\beta_a)^2} jm_a \tag{6.5}$$

The real and imaginary parts of m_{eq} are shown in Fig. 6.7A and Fig. 6.7B, respectively. If there is no damping, $\zeta = 0$ and

$$m_{eq} = \frac{1}{1 - \beta_a^2} m_a \tag{6.6}$$

If $\beta_a = 1$ in Eq. (6.6), m_{eq} becomes infinite and a finite force produces no displacement. Thus, the auxiliary mass enforces a point of no motion (i.e., a node) at its point of attachment.

This concept can be applied to reduce the amplitude of the forced vibration of a single degree-of-freedom system by attaching a damped absorber.[1,2] A sketch of the system with a damped auxiliary mass system is shown in Fig. 6.8A. In the equivalent system shown in Fig. 6.8B, there is no force acting on the mass m but instead the support is given a motion $ue^{j\omega t}$. The equations for the system of Fig. 6.8B are similar to those for the system of Fig. 6.8A with the value ku substituted for F. The amplitude of forced vibration of a single degree-of-freedom system, Eq. (2.24), is

$$x_0 = \frac{F/k}{1 - m\omega^2/k}$$

The effect of the auxiliary mass system is to increase the mass m of the primary system by the equivalent mass of the auxiliary system as

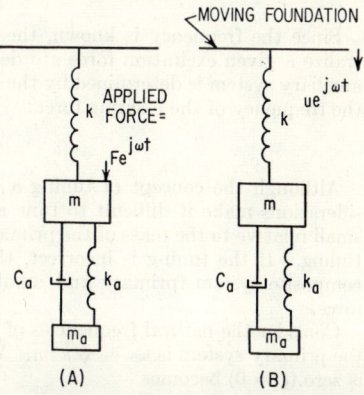

FIG. 6.8. Schematic diagram of auxiliary mass m_a coupled by a spring k_a and viscous damper c_a to a primary system k, m. The primary system is excited by the force F at (A), or alternatively by the foundation motion u at (B).

6–6 DYNAMIC VIBRATION ABSORBERS AND AUXILIARY MASS DAMPERS

given by Eq. (6.4):

$$x_0 = \frac{F/k}{1 - \frac{\omega^2}{k}\left[m + m_a \frac{(1 + 2\zeta\beta_a j)}{(1 - \beta_a^2) + 2\zeta\beta_a j}\right]}$$

Substituting $\mu = m_a/m$, the mass ratio, $\delta_{st} = F/k$, the static deflection of the spring of the primary system, and $\beta = \sqrt{m\omega^2/k}$, the ratio of the forcing frequency to the natural frequency of the primary system, and writing in dimensionless form,

$$\frac{x_0}{\delta_{st}} = \frac{(1 - \beta_a^2) + 2\zeta\beta_a j}{(1 - \beta_a^2) + 2\zeta\beta_a j - \beta^2[(1 - \beta_a^2) + 2\zeta\beta_a j + \mu(1 + 2\zeta\beta_a j)]}$$

The amplitude of motion of the primary mass, without regard for phase, is

$$\frac{x_0}{\delta_{st}} = \left\{\frac{(1 - \beta_a^2)^2 + (2\zeta\beta_a)^2}{[(1 - \beta_a^2)(1 - \beta^2) - \beta^2\mu]^2 + (2\zeta\beta_a)^2[1 - \beta^2 - \beta^2\mu]^2}\right\}^{1/2} \quad (6.7)$$

If $\zeta = 0$ (no damping), then

$$\frac{x_0}{\delta_{st}} = \frac{1 - \beta_a^2}{(1 - \beta_a^2)(1 - \beta^2) - \beta^2\mu} \quad (6.8)$$

If $\beta_a = 1$, $x_0 = 0$; that is, the vibration of the primary system is eliminated entirely when the auxiliary system is undamped and is tuned to the forcing frequency.

THE DYNAMIC ABSORBER

If the auxiliary mass system has no damping and is tuned to the forcing frequency, it acts as a dynamic absorber and enforces a node at its point of attachment. The auxiliary mass must be sufficiently large so that it will not have an excessive amplitude.[3] For a dynamic absorber attached to the primary system at the point where the excitation is introduced, the required mass of the auxiliary body is easily determined. Since the primary mass is motionless, the force exerted by the absorber, when the amplitude of motion of the auxiliary mass is u_0, is equal and of opposite sign to the exciting force F. Hence

$$F = m_a\omega^2 u_0 \quad (6.9)$$

Since the frequency is known, the mass and amplitude of motion necessary to neutralize a given excitation force are determined by Eq. (6.9). The spring stiffness in the auxiliary system is determined by the requirement that the auxiliary system be tuned to the frequency of the exciting force:

$$k_a = m_a\omega^2 \quad (6.10)$$

Although the concept of tuning a dynamic absorber appears simple, practical considerations make it difficult to tune any system exactly. When the auxiliary mass is small relative to the mass of the primary system, its effectiveness depends upon accurate tuning. If the tuning is incorrect, the addition of the auxiliary mass may bring the composite system (primary and auxiliary systems) into resonance with the exciting force.

Consider the natural frequencies of the composite system. The natural frequency of the primary system is $\omega_0 = \sqrt{k/m}$. With this relation, Eq. (6.8) in which the damping is zero ($\zeta = 0$) becomes

$$\frac{x_0}{\delta_{st}} = \frac{1 - \omega^2/\omega_a^2}{(1 - \omega^2/\omega_a^2)(1 - \omega^2/\omega_0^2) - (\omega^2/\omega_0^2)\mu}$$

At resonance the denominator is zero and ω is designated ω_n:

$$(\omega_n^2 - \omega_a^2)(\omega_n^2 - \omega_0^2) - \omega_n^2\omega_a^2\mu = 0 \quad (6.11)$$

The natural frequencies are found from the roots ω_n^2 of Eq. (6.11):

$$\omega_n^2 = \frac{\omega_a^2(1+\mu) + \omega_0^2}{2}$$
$$\pm \sqrt{\left[\frac{\omega_a^2(1+\mu) - \omega_0^2}{2}\right]^2 + \omega_a^2\omega_0^2\mu} \quad (6.12)$$

This last relation may be represented by Mohr's circle, Fig. 6.9.

Since the absorber is nominally tuned to the frequency of the excitation, the root ω_{n2}^2 that is closer to the forcing frequency is of interest. The ratio ω_{n2}/ω_a is a measure of the sensitivity of the tuning required to avoid resonance. This is given as a function of μ for various ratios of ω_0/ω_a in Fig. 6.10. Dynamic absorbers are most generally used when the primary system without the absorber is nearly in resonance with the excitation. If the natural frequency of the primary system is less than the forcing frequency, it is preferable to tune the dynamic absorber to a frequency slightly lower than the

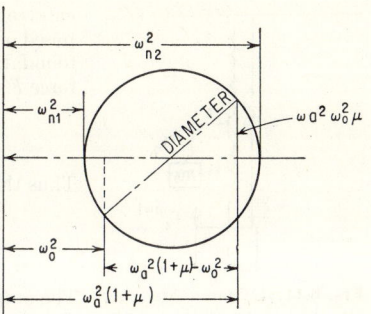

Fig. 6.9. Representation of the natural frequencies ω_n of the composite system by Mohr's circle. The circle is constructed on the diameter located by the natural frequencies ω_0, ω_a of the primary and auxiliary systems, respectively. The natural frequencies of the composite system are indicated by the intercept of the circle with the horizontal axis.

forcing frequency to avoid the resonance that lies above the natural frequency of the primary system. Likewise if the natural frequency of the primary system is above the forcing frequency, it is well to tune the damper to a frequency slightly greater than the forcing frequency. Figure 6.10 shows that the tuning for a primary system with high natural frequency is more sensitive than that for a primary system with low natural frequency. Mohr's circle of Fig. 6.9 provides a useful graphical representation.

Where the natural frequency of the composite system is nearly equal to the tuned frequency of the absorber, the amplitude of motion of the primary mass at resonance is much smaller than that of the absorber. Consequently, the motion of the primary mass does not become large even at resonance; but the motion of the absorber, unless limited by damping, may become so large that failure occurs.

The use of the dynamic absorber is not restricted to single degree-of-freedom systems or to locations in simple systems where the exciting forces act. However, dynamic absorbers are most effective if located where the excitation force acts. For example, consider a dynamic absorber that is attached to the spring in the simple system shown in Fig. 6.11. When the absorber is tuned so that $\sqrt{k_a/m_a} = \omega$, the equivalent mass is infinite at its point of attachment and enforces a node at point A. If the stiffness of the spring between A and the mass m is k_1, then the force F' exerted by the absorber to

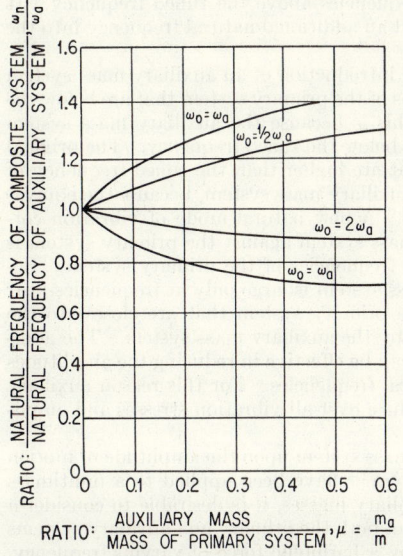

Fig. 6.10. Curves showing effect of mass ratio m_a/m on the natural frequencies ω_n of the composite system, for several ratios of the natural frequency ω_a of the auxiliary system to the natural frequency ω_0 of the primary system.

6-8 DYNAMIC VIBRATION ABSORBERS AND AUXILIARY MASS DAMPERS

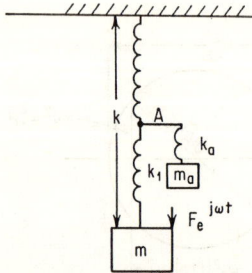

FIG. 6.11. Dynamic absorber attached to the spring of the primary system. The analysis shows that this is not as effective as if it were attached to the rigid body on which the force acts.

enforce the node is equal to that exerted by a system composed of the mass m and the spring k_1 attached to a fixed foundation at A and acted upon by the force $Fe^{j\omega t}$. The force F' is

$$F' = \frac{F}{1 - (m\omega^2/k_1)}$$

Thus the amplitude of motion of the auxiliary mass is

$$u_0 = \frac{F}{1 - (m\omega^2/k_1)} \times \frac{1}{m_a \omega^2} \quad (6.13)$$

The amplitude of motion of the primary mass is

$$x = \frac{F}{k_1}\left(1 - \frac{m\omega^2}{k_1}\right)^{-1} \quad (6.14)$$

Hence, an absorber attached to the spring is not as effective as one attached to the body where the force is acting. It is possible for the primary system to come into resonance about the new node at A.

AUXILIARY MASS DAMPERS

In general, the dynamic absorber is effective only for a system that is subjected to a constant frequency excitation. In the special case of a pendulum absorber (discussed later in this chapter), it is effective for an excitation that is a constant multiple of a rotating shaft speed. When excited at frequencies other than the frequency to which it is tuned, the absorber acts as an attached mass of positive value at frequencies below the tuned frequency and of negative value at frequencies above the tuned frequency. It introduces an additional degree-of-freedom and an additional natural frequency into the primary system.

In a multiple degree-of-freedom system, the introduction of an auxiliary mass system tends to lower those original natural frequencies of the primary system that are below the tuned frequency of the auxiliary system. This is because the auxiliary mass system adds a positive equivalent mass at frequencies below the tuned frequency. The original natural frequencies of the primary system that are higher than the tuned frequency of the auxiliary system are raised by adding the auxiliary mass system, because the equivalent mass of the auxiliary system is negative. A new natural mode of vibration corresponding to the vibration of the auxiliary mass system against the primary system is injected between the displaced initial natural frequencies of the primary system. Because the equivalent mass of the auxiliary mass system is large only at frequencies near the tuned frequency, those frequencies of the primary system that are closest to the tuned frequency are most strongly influenced by the auxiliary mass system. The addition of damping in the auxiliary mass system can be effective in reducing the amplitudes of motion of the primary system at the natural frequencies. For this reason auxiliary mass dampers are used quite commonly to reduce over-all vibration stresses and amplitudes.

Studies of the effects of a damped auxiliary mass system upon the amplitude of motion of an undamped, single degree-of-freedom system [1-5] have been applied to a multimass system.[6,7] In analyzing dampers utilizing auxiliary masses, it is desirable to consider a composite system in which the characteristics of both the primary and auxiliary systems are fixed. This composite system is excited by a harmonic force of varying frequency. It is desirable to express the tuned frequency of the auxiliary mass system in terms of the natural frequency of the primary system rather than the ratio β_a of the excitation frequency ω to the tuned frequency ω_a of the auxiliary system. Defining a new ratio α,

$$\alpha = \frac{\omega_a}{\omega_0} = \frac{\beta}{\beta_a}$$

INFLUENCE OF A SIMPLE AUXILIARY MASS SYSTEM

Then Eq. (6.7) becomes

$$\frac{x_0}{\delta_{st}} = \left\{ \frac{(\alpha^2 - \beta^2)^2 + (2\zeta\alpha\beta)^2}{[(\alpha^2 - \beta^2)(1 - \beta^2) - \alpha^2\beta^2\mu]^2 + (2\zeta\alpha\beta)^2(1 - \beta^2 - \beta^2\mu)^2} \right\}^{1/2} \quad (6.15)$$

This equation is plotted in Fig. 6.12. Note that all curves pass through two points A, B on the graph, independent of the damping parameter ζ. These points are known as *fixed points*. Their locations are independent of the value of ζ if the ratio of the coeffi-

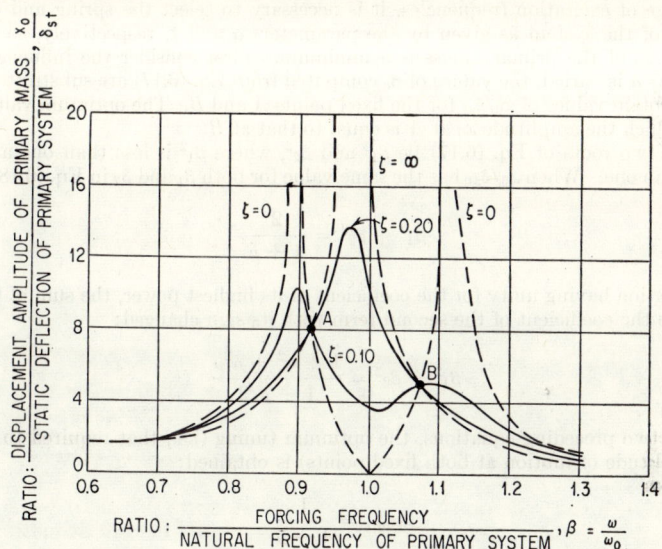

FIG. 6.12. Curves for auxiliary mass damper showing amplitude of vibration of mass of primary system, as given by Eq. (6.15), as a function of the ratio of forcing frequency ω to natural frequency of primary system $\omega = \sqrt{k/m}$. The mass ratio $m_a/m = 0.05$, and the natural frequency ω_a of the auxiliary mass system is equal to the natural frequency ω_0 of the primary system. Curves are included for several values of damping in the auxiliary system.

cient of ζ^2 to the term independent of ζ is the same in both numerator and denominator of Eq. (6.15):

$$\frac{(2\alpha\beta)^2}{(\alpha^2 - \beta^2)^2} = \frac{2\alpha\beta(1 - \beta^2 - \beta^2\mu)^2}{[(\alpha^2 - \beta^2)(1 - \beta^2) - \alpha^2\beta^2\mu]^2} \quad (6.16)$$

This equation is satisfied if

$$(2\alpha\beta)^2 = 0$$

$$\frac{1}{\alpha^2 - \beta^2} + \frac{(1 - \beta^2 - \beta^2\mu)}{(\alpha^2 - \beta^2)(1 - \beta^2) - \alpha^2\beta^2\mu} = 0$$

$$\frac{1}{\alpha^2 - \beta^2} - \frac{(1 - \beta^2 - \beta^2\mu)}{(\alpha^2 - \beta^2)(1 - \beta^2) - \alpha^2\beta^2\mu} = 0$$

The first two solutions are trivial. The third yields the equation

$$\beta^4\left(1 + \frac{\mu}{2}\right) - \beta^2(1 + \alpha^2 + \alpha^2\mu) + \alpha^2 = 0 \quad (6.17)$$

The solution of this equation gives two values of β, designated β_c, one corresponding to each fixed point.

6-10 DYNAMIC VIBRATION ABSORBERS AND AUXILIARY MASS DAMPERS

The amplitude of motion at each fixed point may be found by substituting each value of β_c given by Eq. (6.17) into Eq. (6.15). Since the amplitude is independent of ζ, the value that gives the simplest calculation (namely, $\zeta = \infty$) can be used for the calculation:

$$\left.\frac{x_0}{\delta_{st}}\right|_c = \left[\frac{1}{(1 - \beta_c^2 - \beta_c^2 \mu)^2}\right]^{1/2} \tag{6.18}$$

For the auxiliary mass damper to be most effective in limiting the value of x_0/δ_{st} over a full range of excitation frequencies, it is necessary to select the spring and damping constants of the system as given by the parameters α and ζ, respectively, so that the amplitude x_0 of the primary mass is a minimum. First consider the influence of the ratio α. As α is varied, the values of β_c computed from Eq. (6.17) are substituted in Eq. (6.18) to obtain values of x_0/δ_{st} for the fixed points A and B. The optimum value of α is that for which the amplitude x_0 at A is equal to that at B.

Let the two roots of Eq. (6.17) be β_1^2 and β_2^2, where β_1^2 is less than one and β_2^2 is greater than one. When x_0/δ_{st} has the same value for both β_1 and β_2 in Eq. (6.18),

$$\beta_1^2 + \beta_2^2 = \frac{2}{1 + \mu}$$

In an equation having unity for the coefficient of its highest power, the sum of the roots is equal to the coefficient of the second term with its sign changed:

$$\beta_1^2 + \beta_2^2 = \frac{1 + \alpha^2 + \alpha^2 \mu}{1 + \mu/2}$$

From the two preceding equations, the optimum tuning (i.e., that required to give the same amplitude of motion at both fixed points) is obtained:

$$\alpha_{opt} = \frac{1}{1 + \mu} \tag{6.19}$$

where α is defined by the equation preceding Eq. (6.15).

If the effect of the damping is considered, it is possible to choose a value of the damping parameter ζ that will make the fixed points nearly the points of greatest amplitude of the motion. Consider Fig. 6.13, which represents the curves defining the motion of a single degree-of-freedom system to which an ideally tuned damped vibration absorber is attached (Fig. 6.8). The solid curves (1) represent the response of a system fitted with an undamped absorber. Curve 2 represents infinite damping of the auxiliary system. Curves 3 have horizontal tangents at the fixed points A and B, respectively. Since it is difficult to determine the required damping from maxima at the fixed points, the assumption is made that an optimum damping gives the same value of x_0/δ_{st} at a convenient point between A and B as at these fixed points. First find the values of β at A and B. This is done by solving Eq. (6.17) with the values of α as determined by Eq. (6.19) substituted:

$$\beta^4 - \frac{2\beta^2}{1 + \mu} + \frac{2}{(2 + \mu)(1 + \mu)^2} = 0$$

Solving for β to obtain the abscissas at the fixed points,

$$\beta^2 = \frac{1}{1 + \mu}\left(1 \pm \sqrt{\frac{\mu}{2 + \mu}}\right) \tag{6.20}$$

A convenient value for β lying between the two fixed points A and B is defined by

$$\beta_l^2 = \frac{1}{1 + \mu} \tag{6.21}$$

The frequency corresponding to this frequency ratio β_l is the natural frequency of the composite system when the damping is infinite; it is called the locked frequency.[7] The value of x_0/δ_{st} at the fixed points is found by substituting Eq. (6.20) into Eq. (6.18):

$$\frac{x_0}{\delta_{st}} \text{ at fixed point} = \sqrt{1 + \frac{2}{\mu}} \quad (6.22)$$

An approximate value for the maximum damping is obtained by solving for the value of ζ in Eq. (6.15) that gives a value of $x_0/\delta_{st} = \sqrt{1 + 2/\mu}$ when β_l^2 (the locked frequency)

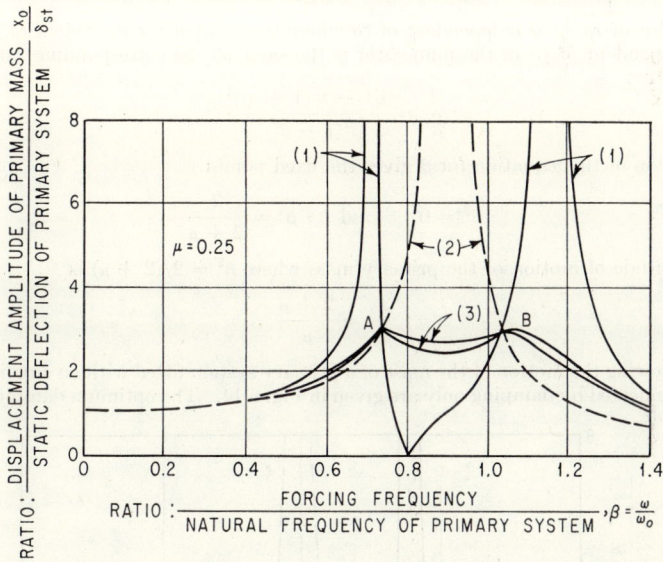

Fig. 6.13. Curves similar to Fig. 6.12 but with optimum tuning. Curves 1 apply to an undamped absorber, curve 2 represents infinite damping in the auxiliary system, and curves 3 have horizontal tangents at the fixed points A and B.

is given by Eq. (6.21) and α has the optimum value given by Eq. (6.19). This gives the following value for the optimum damping parameter:

$$\zeta_{opt} = \sqrt{\frac{\mu}{2(1+\mu)}} \quad (6.23)$$

It is possible to find the value of ζ that makes the fixed point A a maximum on the x_0/δ_{st} vs. β plot, Fig. 6.13, and also to find the value of ζ that makes the point B a maximum. The average of the two values so obtained indicates optimum damping: [4]

$$\zeta_{opt} = \sqrt{\frac{3\mu}{8(1+\mu)^3}} \quad (6.24)$$

OPTIMUM DAMPING FOR AN AUXILIARY MASS ABSORBER CONNECTED TO THE PRIMARY SYSTEM WITH DAMPING ONLY. In general, the most effective damping is obtained where the auxiliary mass damping system includes a spring in its connection to the primary system. However, such a design requires a calculation of the optimum stiffness of the spring. Sometimes it is more expedient to add an oversize mass, coupled only by damping to the primary system, than it is to compute the optimum system. However, if use is made of such a simplified damper by taking it from a list

6–12 DYNAMIC VIBRATION ABSORBERS AND AUXILIARY MASS DAMPERS

of standard dampers and applying it with a minimum of calculations, the stock dampers should be as efficient as the application will permit.

In computing the optimum damping characteristic for an auxiliary mass absorber, attached to a single degree-of-freedom system by damping only, from the relations that have been developed, note in Eq. (6.4) that $\zeta = \infty$ and $\beta_a = \infty$ when $k = 0$. Then $\alpha = \beta/\beta_a = 0$. However, the product $\zeta\alpha = \zeta\beta/\beta_a$ is finite; thus, substituting $\alpha = 0$ but retaining the product $\zeta\alpha$ in Eq. (6.15),

$$\frac{x_0}{\delta_{st}} = \sqrt{\frac{\beta^2 + 4(\zeta\alpha)^2}{\beta^2(1 - \beta^2)^2 + 4(\zeta\alpha)^2[1 - \beta^2(1 + \mu)]^2}} \tag{6.25}$$

The value of x_0/δ_{st} is independent of $\zeta\alpha$ where the ratio of the coefficient of $\zeta\alpha$ to the term independent of $\zeta\alpha$ in the numerator is the same as the corresponding ratio in the denominator:

$$\frac{4}{\beta^2} = \frac{4[1 - \beta^2(1 + \mu)]^2}{\beta^2(1 - \beta^2)^2}$$

The solution of this equation for β gives the fixed points

$$\beta^2 = 0 \quad \text{and} \quad \beta^2 = \frac{2}{2 + \mu} \tag{6.26}$$

The amplitude of motion of the primary mass where $\beta^2 = 2/(2 + \mu)$ is

$$\frac{x_0}{\delta_{st}} = \frac{2 + \mu}{\mu} \tag{6.27}$$

Curves showing the motion of the mass of a primary system fitted with an auxiliary mass system connected by damping only are given in Fig. 6.14. The optimum damping is that

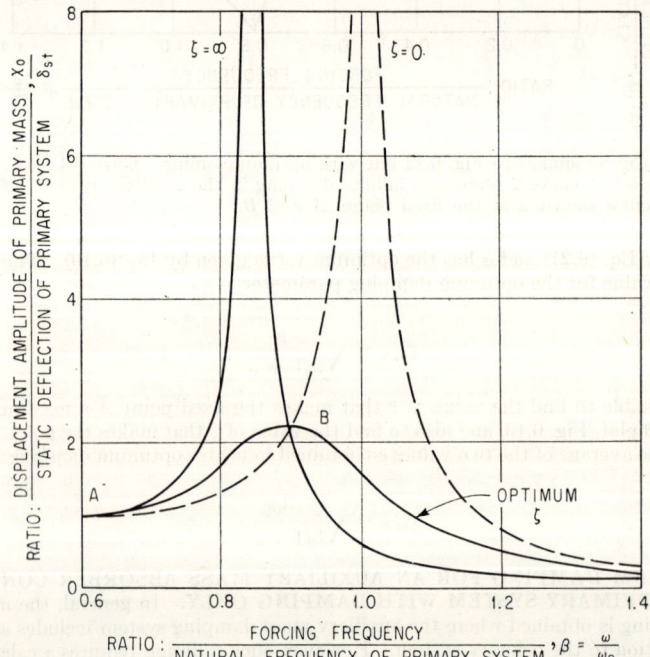

Fig. 6.14. Curves similar to Fig. 6.12 for system having auxiliary mass coupled by damping only. Several values of damping are included.

INFLUENCE OF A SIMPLE AUXILIARY MASS SYSTEM

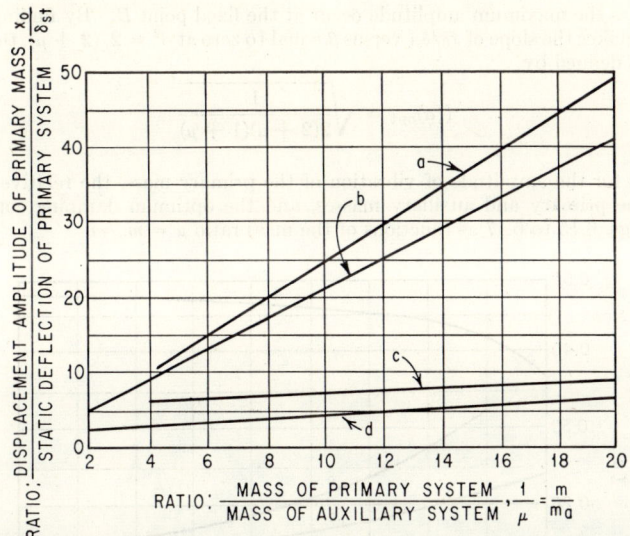

FIG. 6.15. Displacement amplitude of the primary mass as a function of the size of the auxiliary mass: (a) auxiliary system coupled only by Coulomb friction ($\alpha = 0$) with optimum damping; (b) auxiliary system coupled only by viscous damping ($\alpha = 0$) of optimum value; (c) auxiliary system coupled by spring and damper tuned to frequency of primary system ($\alpha = 1$) with optimum damping; (d) auxiliary system coupled by spring and damper with optimum tuning [$\alpha = 1/(1 + \mu)$] and optimum damping.

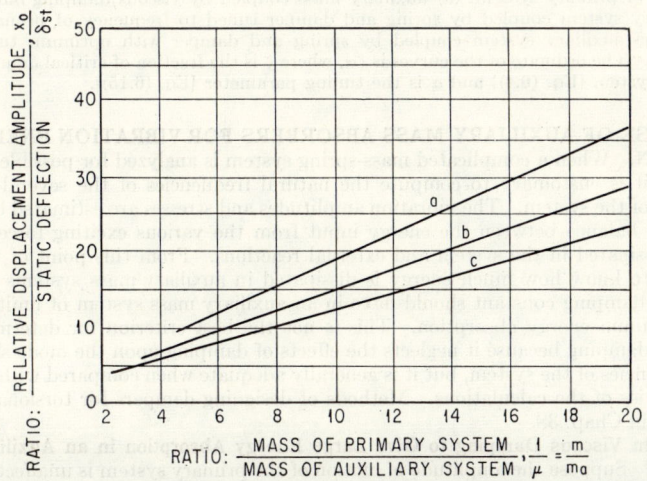

FIG. 6.16. Relative displacement amplitude between the primary mass and the auxiliary mass as a function of the size of the auxiliary mass: (a) auxiliary system coupled by spring and damper with optimum tuning [$\alpha = 1/(1 + \mu)$] and optimum damping; (b) auxiliary system coupled only by viscous damping ($\alpha = 0$) of optimum value; (c) auxiliary system coupled by spring and damper tuned to frequency of primary system ($\alpha = 1$) with optimum damping.

6-14 DYNAMIC VIBRATION ABSORBERS AND AUXILIARY MASS DAMPERS

which makes the maximum amplitude occur at the fixed point B. By finding the value of $\zeta\alpha$ that makes the slope of x_0/δ_{st} versus β equal to zero at $\beta^2 = 2/(2 + \mu)$, the optimum damping is defined by

$$(\zeta\alpha)_{opt} = \sqrt{\frac{1}{2(2 + \mu)(1 + \mu)}} \qquad (6.28)$$

The values for the amplitude of vibration of the primary mass, the relative amplitude between the primary and auxiliary masses, and the optimum damping constants are given in Figs. 6.15 to 6.17 as functions of the mass ratio $\mu = m_a/m$.

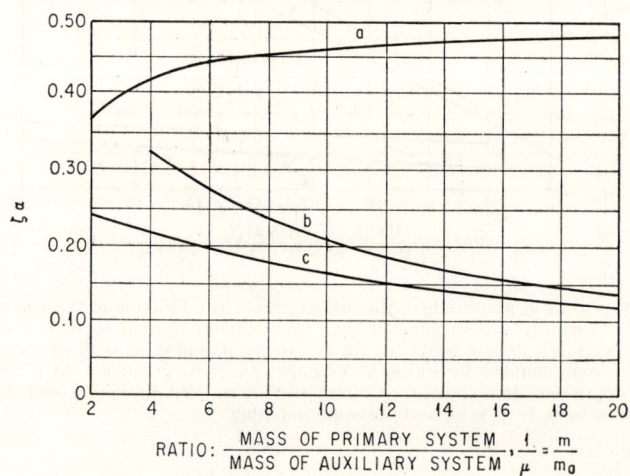

FIG. 6.17. Curves showing damping required in auxiliary mass systems to minimize vibration amplitude of primary system: (a) auxiliary mass coupled by viscous damping only ($\alpha = 0$); (b) auxiliary system coupled by spring and damper tuned to frequency of primary system ($\alpha = 1$); (c) auxiliary system coupled by spring and damper with optimum tuning [$\alpha = 1/(1 + \mu)$]. The ordinate of the curves is $\zeta\alpha$, where ζ is the fraction of critical damping in the auxiliary system [Eq. (6.4)] and α is the tuning parameter [Eq. (6.15)].

THE USE OF AUXILIARY MASS ABSORBERS FOR VIBRATION ENERGY DISSIPATION. When a complicated mass-spring system is analyzed for possible vibration troubles, it is customary to compute the natural frequencies of the several modes of vibration of the system. The vibration amplitudes and stresses are estimated by making an energy balance between the energy input from the various exciting forces and the energy dissipated in the system and external reactions. From this point of view, it is desirable to know how much energy is dissipated in auxiliary mass systems and what value the damping constant should have in an auxiliary mass system of limited size to give maximum energy absorption. This is not the best criterion for determining the optimum damping because it neglects the effects of damping upon the mode shapes and the frequencies of the system, but it is generally adequate when compared with the other uncertainties of the calculations. Methods of designing dampers for torsional systems are given in Chap. 38.

Optimum Viscous Damping to Give Large Energy Absorption in an Auxiliary Mass Absorber.[8] Suppose the amplitude of motion of the primary system is unaffected by the auxiliary mass system which is attached to it. All energy absorption occurs in the damping element of the auxiliary mass system and is obtained by integrating the differential work done in the damper over a vibration cycle. The force exerted by damping is $c\dot{x}_r$, where x_r is the relative motion and the increment of work is $c\dot{x}_r \, dx_r = c\dot{x}_r^2 \, dt$. If $x_r =$

$x_{r0} \cos \omega t$, the work done over a cycle is

$$V = \oint c\omega^2 x_{r0}^2 \sin^2 \omega t \, dt = \pi c x_{r0}^2 \omega \qquad (6.29)$$

For a damper attached to a support moving in harmonic motion of amplitude x_0, the relative motion x_r is given by Eq. (6.1). The amplitude of relative motion is

$$x_{r0} = \frac{m\omega^2 x_0}{\sqrt{(k - m\omega^2)^2 + c^2\omega^2}} = \frac{\beta_a^2 x_0}{\sqrt{(1 - \beta_a^2)^2 + (2\zeta\beta_a)^2}}$$

Substituting the above value of x_{r0} in Eq. (6.29) and integrating,

$$V = \frac{\pi c \omega x_0^2 m \omega^4}{(k - m\omega^2)^2 + c^2\omega^2} = \frac{\pi x_0^2 m \omega^2 (2\zeta\beta_a)\beta_a^2}{(1 - \beta_a^2)^2 + (2\zeta\beta_a)^2} \qquad (6.30)$$

Equation (6.30) can be used to find the tuning and the damping that gives the maximum energy dissipation when the amplitude of the forcing motion remains constant. Placing $\partial V/\partial \beta_a = 0$, the optimum value of β_a for given values of ζ is found from

$$(\beta_a)_{\text{opt}}^2 = (2\zeta^2 - 1) \pm 2\sqrt{1 - \zeta^2 + \zeta^4} \qquad (6.31)$$

Placing $\partial V/\partial \zeta = 0$, the optimum value of ζ for a given value of β_a is

$$\zeta_{\text{opt}} = \frac{1 - \beta_a^2}{2\beta_a} \qquad (6.32)$$

Where $k = 0$, the optimum damping is determined most conveniently by setting $\partial V/\partial c = 0$, using the dimensional form of Eq. (6.30), and determining c for maximum energy absorption:

$$c_{\text{opt}} = m\omega^2 \qquad (6.33)$$

AUXILIARY MASS DAMPER USING COULOMB FRICTION DAMPING.[9]

Dampers relying on Coulomb friction (i.e., friction whose force is constant) have been widely used. A damper relying on dry friction and connected to its primary system with a spring is too complicated to be analyzed or to be adjusted by experiment. For this reason, a damper with Coulomb friction has been used with only friction damping connecting the seismic mass (usually in a torsional application) to the primary system.[1, 2, 9] Because the motion is irregular, it is necessary to use energy methods of analysis. The analysis given here applies to the case of linear vibration. By analogy, the application to torsional or other vibration can be made easily (see Table 2.1 for analogous parameters).

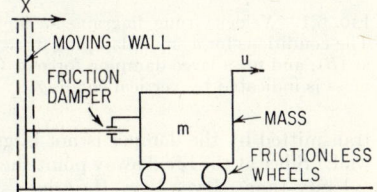

Fig. 6.18. Schematic diagram of auxiliary mass absorber with Coulomb friction damping.

Consider the system shown in Fig. 6.18. It consists of a mass resting on wheels that provide no resistance to motion and connected through a friction damper to a wall that is moving sinusoidally. The friction damper consists of two friction facings that are held on opposite sides of a plate by a spring that can be adjusted to give a desired clamping force. The maximum force that can be transmitted through each interface of the damper is the product of the normal force and the coefficient of friction; the maximum total force for the damper is the summation over the number of interfaces.

Consider the velocity diagrams shown in Fig. 6.19A, B, and C. In these diagrams the velocity of the moving wall, $\dot{x} = x_0 \omega \sin \omega t$, is shown by curve 1; the velocity $\dot{u}$ of the mass is shown by curve 2. The force exerted by the damper when slipping occurs is F_s.

6-16 DYNAMIC VIBRATION ABSORBERS AND AUXILIARY MASS DAMPERS

When $F_s \geq m\ddot{u}$, the mass moves sinusoidally with the wall. When $F_s < m\ddot{u}$, slipping occurs in the damper and the mass is accelerated at a constant rate. Since a constant acceleration produces a uniform change in velocity, the velocity of the mass when the damper is slipping is shown by straight lines. The relative velocity between the wall and the mass is shown by the vertical shading.

Figure 6.19A applies to a damper with a low friction force. The damper slips continuously. In Fig. 6.19B the velocities resulting from a larger friction force are shown. Slipping disappears for certain portions of the cycle. Where the wall and the mass have the same velocity, their accelerations also are equal. Slipping occurs when the force

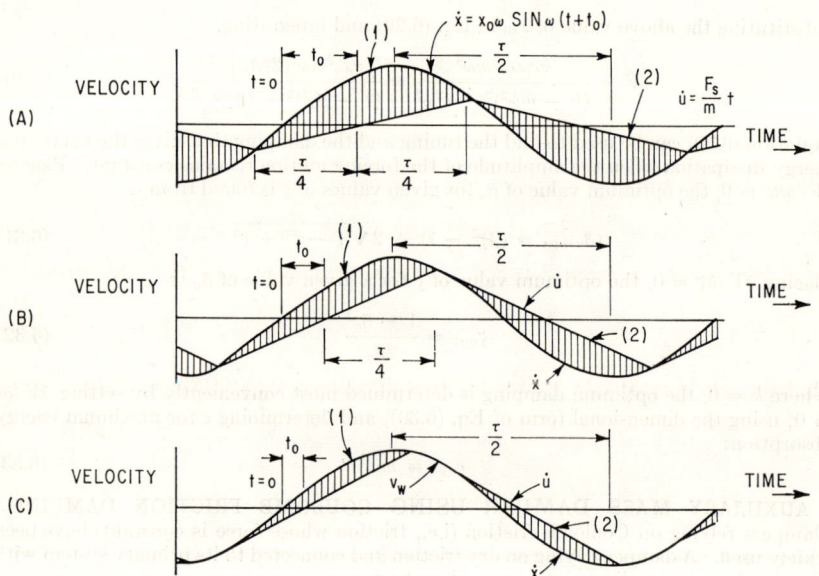

FIG. 6.19. Velocity-time diagrams for motion of wall (curve 1) and mass (curve 2) of Fig. 6.18. The conditions for a small damping force are shown at (A), for an intermediate damping force at (B), and for a large damping force at (C). The relative velocity between the wall and the mass is indicated by vertical shading.

transmitted by the damper is not large enough to keep the mass accelerating with the wall. Since at the breakaway point the accelerations of the wall and mass are equal, their velocity-time curves have the same slope; i.e., the curves are tangent at this point. In Fig. 6.19C, the damping force is so large that the mass follows the wall for a considerable portion of the cycle, and slips only where its acceleration becomes greater than the value of F_s/m. A slight increase in the clamping force or in the coefficient of friction locks the mass to the wall; then there is no relative motion and no damping.

Because of the nature of the damping force, the damping provided by the friction damper can be computed most practically in terms of energy. If the friction force exerted through the damper is F_s, the energy dissipated by the damper is the product of the friction force and the total relative motion between the mass and the moving wall. The time reference is taken at the moment when the auxiliary mass m has a zero velocity and is being accelerated to a positive velocity, Fig. 6.19A. Let the period of the vibratory motion of the wall be $\tau = 2\pi/\omega$, where ω is the angular frequency of the wall motion. By symmetry, the points of no slippage in the damper occur at times $-\tau/4$, $\tau/4$, and $3\tau/4$. Let the time when the velocity of the wall is zero be $-t_0$; then the velocity of the wall $\dot{x}$ is

$$\dot{x} = +x_0\omega \sin \omega(t + t_0)$$

NONLINEARITY IN SPRING OF AUXILIARY MASS DAMPER   6-17

The velocity $\dot{u}$ of the mass for $-\tau/4 < t < \tau/4$ is

$$\dot{u} = \ddot{u}t = \frac{F_s}{m}t$$

The velocities of the wall and the mass are equal at time $t = \tau/4$:

$$x_0\omega \sin \omega \left(\frac{\tau}{4} + t_0\right) = \frac{F_s}{m}\frac{\tau}{4}$$

Since $\omega\tau/4 = \pi/2$, $\sin \omega(\tau/4 + t_0) = \cos \omega t_0$. Therefore

$$\cos \omega t_0 = \frac{F_s}{m}\frac{\pi}{2x_0\omega^2}$$

The relative velocity between the moving wall and the mass is $\dot{x} - \dot{u}$, and the total relative motion is the integral of the relative velocity over a cycle. Note that the area between the two curves for the second half of the cycle is the same as for the first. Hence, the work V per cycle is

$$V = 2\int_{-\tau/4}^{\tau/4} F_s(\dot{x} - \dot{u})\,dt = 4F_s x_0 \sqrt{1 - \left(\frac{F_s\pi}{2mx_0\omega^2}\right)^2} \quad (6.34)$$

Optimum damping occurs when the work per cycle is a maximum. It can be determined by setting the derivative of V with respect to F_s in Eq. (6.34) equal to zero and solving for F_s:

$$(F_s)_{opt} = \frac{\sqrt{2}}{\pi}m\omega^2 x_0 \quad (6.35)$$

Energy absorption per cycle with optimum damping is, from Eq. (6.34),

$$V_{opt} = \frac{4}{\pi}m\omega^2 x_0^2 \quad (6.36)$$

A comparison of the effectiveness of the Coulomb friction damper with other types is given in Fig. 6.15.

EFFECT OF NONLINEARITY IN THE SPRING OF AN AUXILIARY MASS DAMPER

It is possible to extend the range of frequency over which a dynamic absorber is effective by using a nonlinear spring.[10-12] When a nonlinear spring is used, the natural frequency of the absorber is a function of the amplitude of vibration; it increases or decreases, depending upon whether the spring stiffness increases or decreases with deflection. Figure 6.20A shows a typical response curve for a system with increasing spring stiffness; Fig. 6.20B illustrates types of systems having increasing spring stiffness and shows typical force-deflection curves. Figure 6.21A shows a typical response curve for a system of decreasing spring stiffness; Fig. 6.21B illustrates types of systems having decreasing stiffnesses and shows typical force-deflection curves.

To compute the equivalent mass at a given frequency when a nonlinear spring is used, it is necessary to use a trial-and-error procedure. By the methods given in Chap. 4, compute the natural frequency of the auxiliary mass system, assuming the point of attachment fixed, as a function of the amplitude of motion of the auxiliary mass. This will

6–18 DYNAMIC VIBRATION ABSORBERS AND AUXILIARY MASS DAMPERS

result in a curve similar to the dotted curves in Figs. 6.20A and 6.21A. At the given frequency, compute β_a in Eq. (6.4) in terms of the tuned frequency of the absorber at zero amplitude. (The tuned frequency will change with amplitude because the spring constant changes.) With this value of β_a compute the equivalent mass from Eq. (6.6). With this mass in the system, compute the amplitude of motion x_0 of the primary mass to which the auxiliary system is attached [Eq. (6.7)] and the amplitude of the relative motion $x_{r0} = \nu^2(1 - \nu^2)x_0$. Using this value of x_{r0}, ascertain the corresponding value of

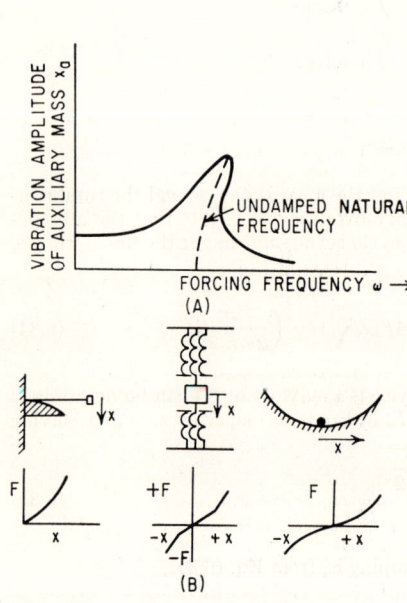
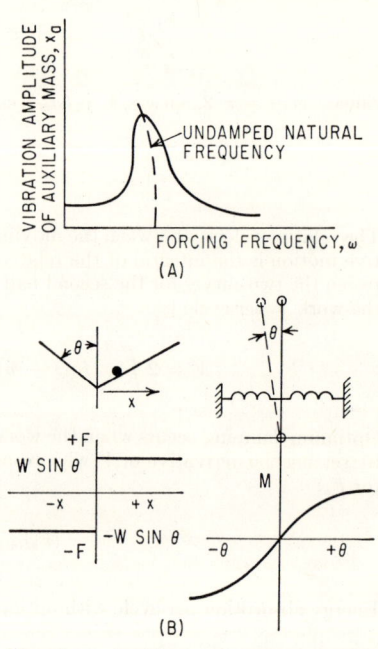

FIG. 6.20. Auxiliary mass damper with nonlinear spring having stiffness that increases as deflection increases. The response to forced vibration and the natural frequency are shown at (A). Several arrangements of nonlinear systems with the corresponding force-deflection curves are shown at (B).

FIG. 6.21. Auxiliary mass damper with nonlinear spring having stiffness that decreases as deflection increases. The response to forced vibration and the natural frequency are shown at (A). Two arrangements of nonlinear systems with the corresponding force-deflection curves are shown at (B).

resonant frequency of the system from the computed curve, and compute the new value of β_a. Repeat the process until the value of β_a remains unchanged upon repeated calculation.

A dynamic absorber having a nonlinear characteristic can be used to introduce nonlinearity into a resonant system. This can be useful in the case where a machine passes through a resonance rapidly as the speed is increased but slowly as the speed is decreased. In bringing this machine up to speed, there is a natural frequency that comes into strong resonance, giving a critical speed. A strongly nonlinear dynamic absorber tuned at low amplitudes to the optimum frequency for the damped absorber can be used to reduce the effects of the critical speed. Two resonant peaks will be introduced, as shown on curve 1 of Fig. 6.13. By making the dynamic absorber nonlinear, so that the stiffness becomes greater as the amplitude of vibration is increased, the peaks are bent over to provide the response curve shown in Fig. 6.22. In starting, the machine is accelerated through the two critical speeds so fast that a resonance is unable to build up. In coasting to a stop, there would be ample time for significant amplitudes to build up if the nonlinearity did not exist. Because of the nonlinearity, the amplitude of vibration as a function of speed (since β is proportional to speed) follows the path A, B, C, D, E, F, G and never reaches the extreme amplitudes H_1 and H_2.

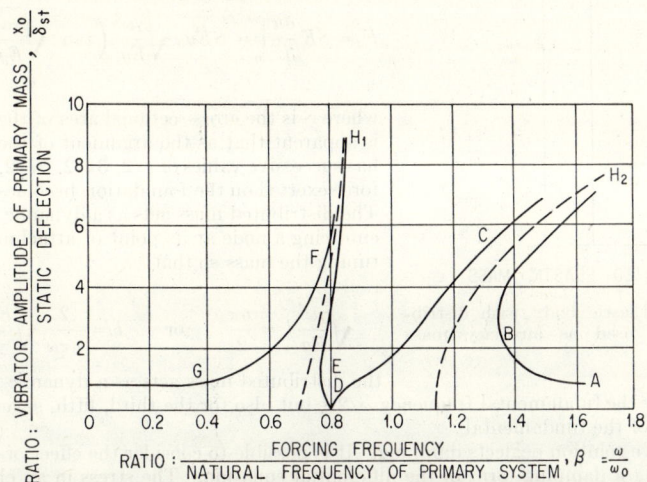

Fig. 6.22. Motion of the primary mass, as a function of forcing frequency, in a system having a nonlinear dynamic absorber whose natural frequency increases with amplitude. The mass of the absorber is 0.25 times the mass of the primary system ($\mu = 0.25$).

MULTIMASS ABSORBERS

In general, only one mass is used in a dynamic absorber. However, it is possible to provide a dynamic absorber that is effective for two or more frequencies by attaching an auxiliary mass system that resonates at the frequencies that are objectionable. The principle that would make such a dynamic absorber effective is utilized in the design of the elastic system of a ship's propulsion plant driven by independent high-pressure and low-pressure turbines. By making the frequencies of the two branches about the reduction gear identical, the gear becomes a node for one of the resonant modes. Then it is impossible to excite the mode of vibration where one turbine branch vibrates against the other as a result of excitation transmitted by the propeller shaft to that node.

DISTRIBUTED MASS ABSORBERS

It is possible to use distributed masses as vibration dampers. Consider an undamped rod of distributed mass and elasticity attached to a foundation that vibrates the rod axially, as shown in Fig. 6.23. The differential equation for the motion of this rod is derived in Chap. 7. The values of the constants are set by the boundary conditions:

$$\text{Stress} = E \frac{\partial u}{\partial x} = 0 \quad \text{where } x = l$$

$$u = u_0 \cos t \quad \text{where } x = 0$$

(6.37)

The solution of the equation of motion is

$$u = u_0 \cos \omega t \left(\cos \sqrt{\frac{\gamma \omega^2}{Eg}} x + \tan \sqrt{\frac{\gamma \omega^2}{Eg}} l \sin \sqrt{\frac{\gamma \omega^2}{Eg}} x \right)$$

(6.38)

6-20 DYNAMIC VIBRATION ABSORBERS AND AUXILIARY MASS DAMPERS

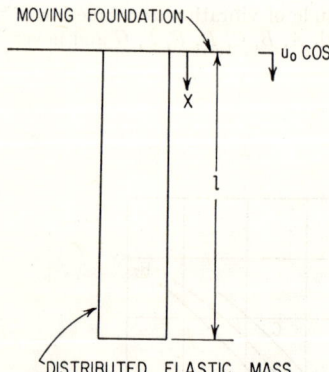

FIG. 6.23. Elastic body with distributed mass used as auxiliary mass damper.

where E is the modulus of elasticity and γ is the weight density of the material. When $x = 0$, the force F on the foundation is

$$F = SE \left. \frac{\partial u}{\partial x} \right|_0 = SEu_0 \sqrt{\frac{\gamma \omega^2}{Eg}} \left(\tan \sqrt{\frac{\gamma \omega^2}{Eg}} l \right) \tag{6.39}$$

where S is the cross-sectional area of the bar. It is apparent that as the argument of the tangent has successive values of $\pi/2, 3\pi/2, 5\pi/2, \ldots$, the force exerted on the foundation becomes infinite. The distributed mass acts as a dynamic absorber enforcing a node at its point of attachment. By tuning the mass so that

$$\sqrt{\frac{\gamma \omega^2}{Eg}} l = \frac{n\pi}{2} \quad \text{or} \quad l = \frac{2}{n\pi\omega} \sqrt{\frac{Eg}{\gamma}} \tag{6.40}$$

the distributed mass acts as a dynamic absorber for not only the fundamental frequency $\omega/2\pi$, but also for the third, fifth, seventh, ... harmonics of the fundamental.

The above solution neglects damping. It is possible to consider the effect of damping by including a damping term in the differential equation. The stress in an element is assumed to be the sum of a deformation stress and a stress related to the velocity of strain:

$$\sigma = E\epsilon + \mu \frac{d\epsilon}{dt} \tag{6.41}$$

where $\epsilon = \partial u/\partial x$ is the strain. Then the differential equation becomes

$$E \frac{\partial^2 u}{\partial x^2} + \mu \frac{\partial^3 u}{\partial x^2 \partial t} = \frac{\gamma}{g} \frac{\partial^2 u}{\partial t^2} \tag{6.42}$$

Since the absorber is excited by a foundation moving with a frequency $f = \omega/2\pi$, u may be expressed as $\mathcal{R} u_1 e^{j\omega t}$ and the partial differential equation can be written as the ordinary linear differential equation

$$E \frac{d^2 u_1}{dx^2} + j\omega\mu \frac{d^2 u_1}{dx^2} + \frac{\gamma \omega^2 u_1}{g} = 0$$

This equation may be written

$$\left(1 + \frac{\mu \omega j}{E}\right) \frac{d^2 u_1}{dx^2} + \frac{\gamma \omega^2}{Eg} u_1 = 0 \tag{6.43}$$

Since Eq. (6.43) is a second-order linear differential equation, the solution may be written

$$u = A_1 e^{\beta_1 x} + A_2 e^{\beta_2 x} \tag{6.44}$$

where β_1 and β_2 are the two roots of the equation

$$\beta^2 = \frac{-(\gamma/Eg)\omega^2}{1 + (j\mu\omega/E)} \tag{6.45}$$

DISTRIBUTED MASS ABSORBERS

For small values of μ, by a binomial expansion of the denominator,

$$\pm \beta = \frac{1}{2}\sqrt{\frac{\gamma}{Eg}}\frac{\mu\omega^2}{E} + j\sqrt{\frac{\gamma}{Eg}}\omega \tag{6.46}$$

where μ is defined by Eq. (6.41).

The boundary conditions to be met by the damper are:

At $x = 0$: $\quad u = u_0 \quad$ therefore, $A_1 + A_2 = u_0$

At $x = l$: $\quad \sigma = (E + j\omega\mu)\dfrac{\partial u}{\partial x} = 0 \quad$ therefore, $A_1 e^{\beta l} - A_2 e^{-\beta l} = 0$ (6.47)

Solving Eqs. (6.47) for A_1 and A_2 and substituting the result in (6.44),

$$u = u_0 \frac{\cosh \beta(l-x)}{\cosh \beta l} \tag{6.48}$$

Substituting from Eq. (6.48) in Eq. (6.39), the force exerted on the foundation by the damper is

$$F_{(x=0)} = S\sigma_{(x=0)} = -Su_0(E + j\omega\mu)\frac{\sinh \beta l}{\cosh \beta l} \tag{6.49}$$

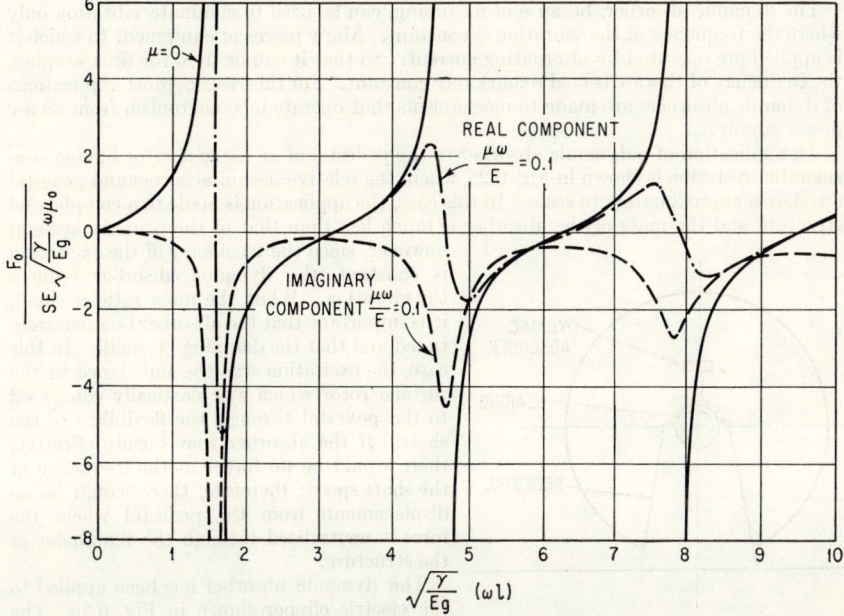

Fig. 6.24. Real and imaginary components of the force applied to a vibrating body by the distributed mass damper shown in Fig. 6.23. These relations are given mathematically by Eq. (6.50), and the terms are defined in connection with Eq. (6.38). The curves are for a value of the damping coefficient $\mu = 0.1$, where μ is defined by Eq. (6.41).

where S is the cross-section area of the bar. When the complex value of β as given in Eq. (6.46) is substituted in Eq. (6.49), the following value for the dynamic force exerted on the foundation is obtained:

$$\frac{F_{(x=0)}}{SE\sqrt{\frac{\gamma}{Eg}}\omega u_0} = \frac{\left(1 + \frac{\mu^2\omega^2}{2E^2}\right)\sin 2\sqrt{\frac{\gamma}{Eg}}\omega l + \frac{\mu\omega}{2E}\sinh\frac{\mu\omega}{E}\sqrt{\frac{\gamma}{Eg}}\omega l}{\cos\sqrt{\frac{\gamma}{Eg}}\omega l + \cosh\frac{\mu\omega}{E}\sqrt{\frac{\gamma}{Eg}}\omega l}$$

$$+ j\frac{\frac{\mu\omega}{2E}\sin 2\sqrt{\frac{\gamma}{Eg}}\omega l - \left(1 + \frac{\mu^2\omega^2}{2E^2}\right)\sinh\frac{\mu\omega}{E}\sqrt{\frac{\gamma}{Eg}}\omega l}{\cos\sqrt{\frac{\gamma}{Eg}}\omega l + \cosh\frac{\mu\omega}{E}\sqrt{\frac{\gamma}{Eg}}\omega l} \quad (6.50)$$

A plot of the real and imaginary values of $F_{(x=0)}/SE\sqrt{\frac{\gamma}{Eg}}\omega u_0$ is given in Fig. 6.24 for zero damping and for a damping coefficient $\mu\omega/E = 0.1$ as a function of a tuning parameter $\sqrt{\gamma/Eg}\,(\omega l)$. Damping decreases the effectiveness of the distributed mass damper substantially, particularly for the higher modes.

PRACTICAL APPLICATIONS OF AUXILIARY MASS DAMPERS AND ABSORBERS TO SINGLE DEGREE-OF-FREEDOM SYSTEMS

THE DYNAMIC ABSORBER

The dynamic absorber, because of its tuning, can be used to eliminate vibration only where the frequency of the vibration is constant. Many pieces of equipment to which it is applied are operated by alternating current. So that it can be used for time keeping, the frequency of this a-c is held remarkably constant. For this reason, most applications of dynamic absorbers are made to mechanisms that operate in synchronism from an a-c power supply.

An application of a dynamic absorber to the pedestal of an a-c generator having considerable vibration is shown in Fig. 6.25, where the relative sizes of absorber and pedestal are shown approximately to scale. In this case, the application is made to a complicated structure and the mass of the absorber is much less than that of the primary system; however, since the frequency of the excitation is constant, the dynamic absorber reduces the vibration. When the mass ratio is small, it is important that the absorber be accurately tuned and that the damping be small. In this case, the excitation was the unbalance in the turbine rotor which was elastically connec ed to the pedestal through the flexibility of the shaft. If the absorber were ideally effective, there would be no forces at the frequency of the shaft speed; therefore, there would be no displacements from the pedestal where the force is neutralized through the remainder of the structure.

The dynamic absorber has been applied to the electric clipper shown in Fig. 6.26. The structure consisting of the cutter blade and its driving mechanism is actuated by the magnetic field at a frequency of 120 cps, as a

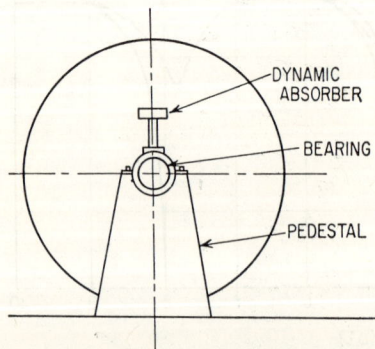

FIG. 6.25. Application of a dynamic absorber to the bearing pedestal of an a-c generator.

result of the 60-cps a-c power supply. The forces and torques required to move the blade are balanced by reactions on the housing, causing it to vibrate. The dynamic absorber tuned to a frequency of 120 cps enforces a node at the location of its mass. Since this is approximately the center-of-gravity of the assembly of the cutter and its driving mechanism, the absorber effectively neutralizes the unbalanced force. The

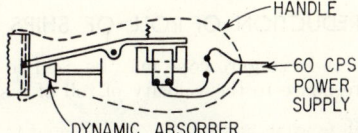

FIG. 6.26. Application of a dynamic absorber to a hair clipper.

moment caused by the rotation of the moving parts is still unbalanced. A second very small dynamic absorber placed in the handle of the clipper could enforce a node at the handle and substantially eliminate all vibration. The design of these absorbers is simple after the unbalanced forces and torques generated by the cutter mechanism are computed. The sum of the inertia forces generated by the two absorbers, $m_1 x_1 \omega^2 + m_2 x_2 \omega^2$ (where m_1 and x_1 are the mass and amplitude of motion of the first absorber, m_2 and x_2 are the corresponding values for the second absorber, and $\omega = 240\pi$), must equal the unbalanced force generated by the clipper mechanism. The torque generated by the two absorbers must balance the torque of the mechanism. Since the value of ω^2 is known, the values of $m_1 x_1$ and $m_2 x_2$ can be determined. Weights that fit into the available space with adequate room to move are chosen, and a spring is designed of such stiffness that the natural frequency is 120 cps.

Because of the desirable balancing properties of the simple dynamic absorber and the constancy of frequency of a-c power, it might be expected that devices operating at a frequency of 120 cps would be used more widely. However, their application is limited because the frequency of vibration is too high to allow large amplitudes of motion.

REDUCTION OF ROLL OF SHIPS BY AUXILIARY TANKS

One interesting although now obsolete application of auxiliary mass absorbers is found in the auxiliary tanks used to reduce the rolling of ships, [1, 13] as shown in Fig. 6.27. When a ship is heeled, the restoring moment $k_r \phi$ acting on it is proportional to the angle of heel (or roll). This restoring moment acts to return the ship (and the water that moves with it) to its equilibrium position. If I_s represents the polar moment of inertia of the ship and its entrained water, the differential equation for the rolling motion of the ship is

$$I_s \ddot{\phi} + k_r \phi = M_s \qquad (6.51)$$

where M_s represents the rolling moments exerted on the ship, usually by waves.

To reduce rolling of the ship, auxiliary wing tanks connected by pipes are used. The water flowing from one tank to another has a natural frequency that is determined by the length and cross-sectional area of the tube connecting the tanks. The damping is controlled by restricting the flow of water, either with a valve S in the line that allows air to flow between the tanks (Fig. 6.27) or with a valve V in the water line. Since the tanks occupy valuable space, the mass ratio of the water in the

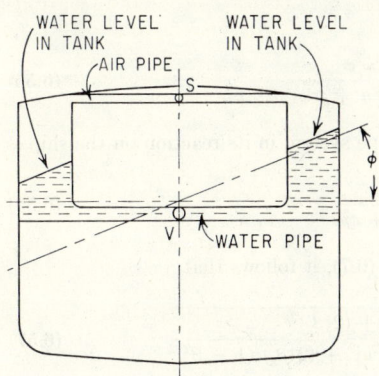

FIG. 6.27. Cross section of ship equipped with antiroll tanks. The flow of water from one tank to the other tends to counteract rolling of the ship.

tanks to the ship is small. Fortunately, the excitation from waves generally is not large relative to the restoring moments, and roll becomes objectionable only because the normal damping of a ship in rolling motion is not very large. The use of antirolling tanks in the German luxury liners *Bremen* and *Europa* reduced the maximum roll from 15 to 5°.

6-24 DYNAMIC VIBRATION ABSORBERS AND AUXILIARY MASS DAMPERS

REDUCTION OF ROLL OF SHIPS BY GYROSCOPES

A large gyroscope may be used to reduce roll in ships, as shown in Fig. 6.28.[1,14] In response to the velocity of roll of a ship, the gyroscope precesses in the plane of symmetry of the ship. By braking this precession, energy can be dissipated and the roll reduced. The torque exerted by the gyroscope is proportional to the rate of change of the angular momentum about an axis perpendicular to the torque. Letting I represent the polar moment of inertia of the gyroscope about its spin axis and $\dot{\theta}$ the angular velocity of precession of the gyroscope, then the equation of motion of the ship is

$$I_s\ddot{\phi} + k_r\phi + I\Omega\dot{\theta} = M_s \qquad (6.52)$$

Assume that the gyroscope has (1) a moment of inertia about the precession axis of I_g, (2) a weight of W, and (3) that its center-of-gravity is below the gimbal axis (as it must be for the gyro to come to equilibrium in a working position) a distance a, as shown in Fig. 6.28. Then the equation of motion of the gyroscope is

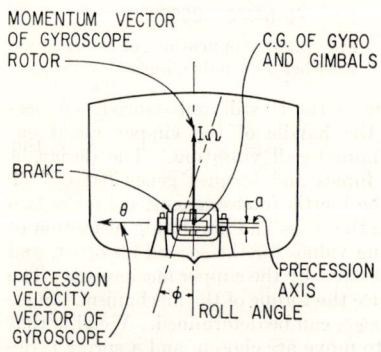

FIG. 6.28. Application of a gyroscope to a ship to reduce roll.

$$I_g\ddot{\theta} + Wa\theta + c\dot{\theta} - I\Omega\dot{\phi} = 0 \qquad (6.53)$$

where Ω is the spin velocity of the gyroscope. From Eq. (6.53), for a roll frequency of ω, the angle of precession of the gyroscope is

$$\theta = \frac{jI\Omega\omega\phi}{-I_g\omega^2 + Wa - jc\omega} \qquad (6.54)$$

The torque exerted on the ship is

$$I\Omega\dot{\theta} = \frac{-(I\Omega)^2\omega^2\phi}{-I_g\omega^2 + Wa + jc\omega} \qquad (6.55)$$

The equivalent moment of inertia of the gyroscope system in its reaction on the ship is

$$\frac{I\Omega^2}{-I_g\omega^2 + Wa + cj\omega} \qquad (6.56)$$

By analogy with the steps of Eqs. (6.2) through (6.7), it follows that

$$\frac{\phi}{\phi_{st}} = \sqrt{\frac{(1-\beta_g^2)^2 + (2\zeta\beta_g)^2}{[(1-\beta_g^2)(1-\beta^2) - \beta^2\mu]^2 + (2\zeta\beta_g)^2(1-\beta^2)^2}} \qquad (6.57)$$

where the parameters are defined in terms of ship and gyro constants as follows:

$$\beta_g = \frac{\omega}{\sqrt{Wa/I_g}} \qquad \beta = \frac{\omega}{\sqrt{k_r/I_s}} \qquad \zeta = \frac{c}{2\sqrt{WaI_g}} \qquad \mu = \frac{(I\Omega)^2}{WaI_s} \qquad \psi_{st} = \frac{M_s}{k_r}$$

Because $I\Omega$ can be made large by using a large gyro rotor and spinning it at a high speed, and Wa can be made small by choice of a design, the value of μ can be made quite large even though I_s is large. In one experimental ship, $\mu = 20$ was obtained. Even with this large value of μ, the precession angle of the gyroscope would become very large for optimum damping. Therefore it is necessary to use much more damping than optimum.

Gyro stabilizers were used on the Italian ship *Conte di Savoia;* they are sometimes installed on yachts.

Both antirolling tanks and gyro stabilizers are more effective if they are active rather than passive. Activated dampers are considered below.

AUXILIARY MASS DAMPERS APPLIED TO ROTATING MACHINERY

An important industrial use of auxiliary mass systems is to neutralize the unbalance of centrifugal machinery. A common application is the balance ring in the spin dryer of home washing machines. The operation of such a balancer is dependent upon the basket of the washer rotating at a speed greater than the natural frequency of its support. The balance ring is attached to the washing machine basket concentric with its axis of rotation, as shown in Fig. 6.29.

Consider the washing machine basket shown in Fig. 6.29. When its center-of-gravity does not coincide with its axis of rotation and it is rotating at a speed lower than its critical speed (corresponding to the natural frequency in rocking motion about the spherical seat), the centrifugal force tends to pull the rotational axis in the direction of the unbalance. This effect increases with an increase in rotational speed until the critical speed is reached. At this speed the amplitude would become infinite if it were not for the damping in the system. Above the critical speed, the phase position of the axis of rotation relative to the center-of-gravity shifts so that the basket tends to rotate about its center-of-gravity with the flexibly supported bearing moving in a circle about an axis through the center-of-gravity. The relative positions of the bearing center and the center-of-gravity are shown in Fig. 6.30A and B.

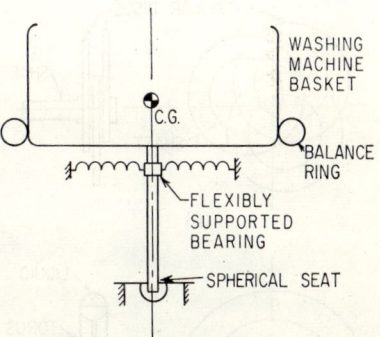

Fig. 6.29. Schematic diagram showing location of balance ring on basket of a spin dryer.

Since the balance ring is circular with a smooth inner surface, any weights or fluids contained in the ring can be acted upon only by forces directed radially. When the ring is rotated about a vertical axis, the weights or fluids will move within the ring in such a manner as to be concentrated on the side farthest from the axis of rotation. If this concentration occurs below the natural frequency (Fig. 6.30A), the weights tend to move further from the axis and the resultant center-of-gravity is displaced so as to give a greater eccentricity. The points A and G rotate about the axis O at the frequency ω. The initial eccentricity of the center-of-gravity of the washer basket and its load from the axis of rotation is represented by e, and ρ is the elastic displacement of this center of rotation due to the centrifugal force. Where the off-center rotating weight is W, the unbalanced force is $(W/g)(\rho + e)\omega^2$ [where $\rho = e/(1 - \beta^2)$ and $\beta^2 = \omega^2/\omega_n^2 < 1$] and acts in the direction from A to G.

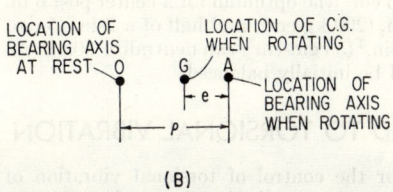

Fig. 6.30. Diagram in plane normal to axis of rotation of spin dryer in Fig. 6.29. Relative positions of axes when rotating speed is less than natural frequency are shown at (A); corresponding diagram for rotation speed greater than natural frequency is shown at (B).

If the displacement of the weights or fluids in the balance ring occurs above the natural frequency, the center-of-gravity tends to

6–26 DYNAMIC VIBRATION ABSORBERS AND AUXILIARY MASS DAMPERS

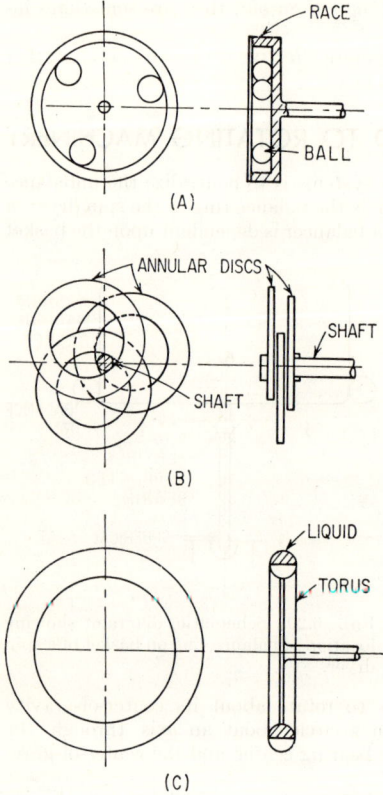

FIG. 6.31. Examples of balancing means for rotating machinery: (A) spheres (or cylinders) in a race; (B) annular discs rotating on shaft; (C) damping fluid in torus.

move closer to the dynamic location of the axis. The action in this case is shown in Fig. 6.30B. Then the points A and G rotate about O at the frequency ω. The unbalanced force is $(W/g)(\rho + e)\omega^2$ [where $\rho = e/(1 - \beta^2)$ and $\beta^2 = \omega^2/\omega_n^2 > 1$]. This gives a negative force that acts in a direction from G to A. Thus the eccentricity is brought toward zero and the rotor is automatically balanced. Because it is necessary to pass through the critical speed in bringing the rotor up to speed and in stopping it, it is desirable to heavily damp the balancing elements, either fluid or weights.

In practical applications, the balancing elements can take several forms. The earliest form consisted of two or more spheres or cylinders free to move in a race concentric with the axis of the rotor, as shown in Fig. 6.31A. A later modification consists of three annular discs that rotate about an enlarged shaft concentric with the axis, as indicated in Fig. 6.31B. These are contained in a sealed compartment with oil for lubrication and damping. A fluid type of damper is shown in Fig. 6.31C, the fluid usually being a high-density viscous material. With proper damping, mercury would be excellent, but it is too expensive. Therefore a more viscous, high-density halogenated fluid is used.

The balancers must be of sufficient weight and operate at such a radius that the product of their weight and the maximum eccentricity they can attain is equivalent to the unbalanced moment of the load. This requirement makes the use of the spheres or cylinders difficult because they cannot be made large; it makes the annular plates large because they are limited in the amount of eccentricity that can be obtained.

In a cylindrical volume 24 in. (61 cm) in diameter and 2 in. (5 cm) thick, seven spheres 2 in. (5 cm) in diameter can neutralize 98.6 lb-in. (114 kg-cm) of unbalance; three cylinders 4 in. (10 cm) in diameter by 2 in. (5 cm) thick can neutralize 255 lb-in. (295 kg-cm); three annular discs, each 5⁄8 in. (1.6 cm) thick with an outside diameter of 19.55 in. (50 cm) and an inside diameter of 10.45 in. (26.5 cm) [the optimum for a center post 6 in. (15.2 cm) in diameter], can neutralize 250 lb-in. (290 kg-cm); and half of a 2-in. (5-cm) diameter torus filled with fluid of density 0.2 lb-in.[3] (55 gm-cm³) can neutralize 609 lb-in. (700 kg-cm). Only the fluid-filled torus would be initially balanced.

AUXILIARY MASS DAMPERS APPLIED TO TORSIONAL VIBRATION

Dampers and absorbers are used widely for the control of torsional vibration of internal-combustion engines. The most common absorber is the viscous-damped, untuned auxiliary mass unit shown in Fig. 6.32. The device is comprised of a cylindrical housing carrying an inertia mass that is free to rotate. There is a preset clearance between the housing and the inertia mass that is filled with a silicone oil of proper viscosity. Silicone oil is used because of its high viscosity index; i.e., its viscosity changes relatively little with temperature. With the inertia mass and the damping medium contained, the housing is seal-welded to provide a leakproof and simple absorber. However, the silicone

AUXILIARY MASS DAMPERS APPLIED TO TORSIONAL VIBRATION 6–27

oil has poor boundary lubricating properties and if decomposed by a local hot spot (such as might be caused by a reduced clearance at some particular spot), the decomposed damping fluid is abrasive.

Because of the simplicity of this untuned damper, it is commonly used in preference to the more effective tuned absorber. However, it is possible to use the same construction methods for a tuned damper, as shown in Fig. 6.33. It is also possible to mount the standard damper with the housing for the unsprung inertia mass attached to the main

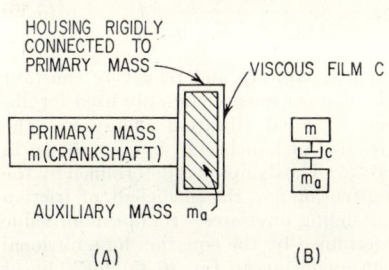

Fig. 6.32. Untuned auxiliary mass damper with viscous damping. The application to a torsional system is shown at (A), and the linear analog at (B).

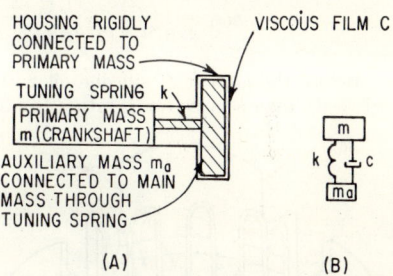

Fig. 6.33. Tuned auxiliary mass damper with viscous damping. The application to a torsional system is shown at (A), and the linear analog at (B).

mass by a spring, as shown in Fig. 6.34. If the viscosity of the oil and the dimensions of the masses and the clearance spaces are known, the damping effects of the dampers shown in Figs. 6.32 and 6.34 can be computed directly in terms of the equations previously developed. The damper in Fig. 6.34 can be analyzed by treating the spring and housing as additional elements in the main system and the untuned mass as a viscous damped auxiliary mass. If the inertia of the housing is negligible, the inertia mass is

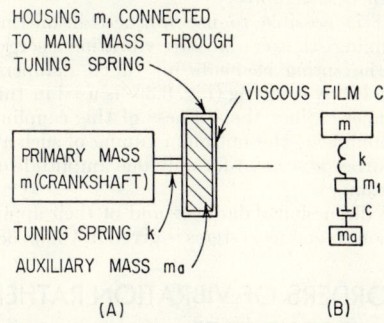

Fig. 6.34. Auxiliary mass damper with viscous damping and spring-mounted housing. The application to a torsional system is shown at (A), and the linear analog at (B).

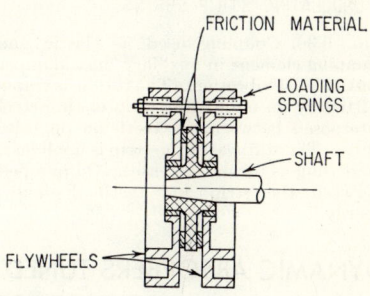

Fig. 6.35. Schematic cross section through Lanchester damper.

effectively connected to the main mass through a spring and a dashpot in series. The two elements in series can be represented by a complex spring constant equal to

$$\frac{1}{(1/jc\omega) + (1/k)} = \frac{kcj\omega}{k + cj\omega}$$

Where there is no damping in parallel with the spring, Eq. (6.3) becomes

$$m_{eq} = km/(k - m\omega^2)$$

6–28 DYNAMIC VIBRATION ABSORBERS AND AUXILIARY MASS DAMPERS

Substituting the complex value of the spring constant, the effective mass is

$$m_{\text{eq}} = \frac{ckj\omega}{k + cj\omega} \left[\frac{m}{-m\omega^2 + cjk\omega/(k + cj\omega)} \right] \quad (6.58)$$

In terms of the nondimensional parameters defined in Eq. (6.4):

$$m_{\text{eq}} = \frac{(2\zeta\beta_a)^2(1 - \beta_a^2)}{\beta_a^4 - (2\zeta\beta_a)^2(1 - \beta_a^2)} m + \frac{-2\zeta\beta_a^3 m}{\beta_a^4 - (2\zeta\beta_a)^2(1 - \beta_a^2)} j \quad (6.59)$$

Before the advent of silicone oil with its chemical stability and relatively constant viscosity over service temperature conditions, the damper most commonly used for absorbing torsional vibration energy was the dry friction or Lanchester damper shown in Fig. 6.35. The damping is determined by the spring tension and the coefficient of friction at the sliding interfaces. Its optimum value is determined by the equation for a torsional system analogous to Eq. (6.35) for a linear system:

$$(T_s)_{\text{opt}} = \frac{\sqrt{2}}{\pi} I\omega^2 \theta_0 \quad (6.60)$$

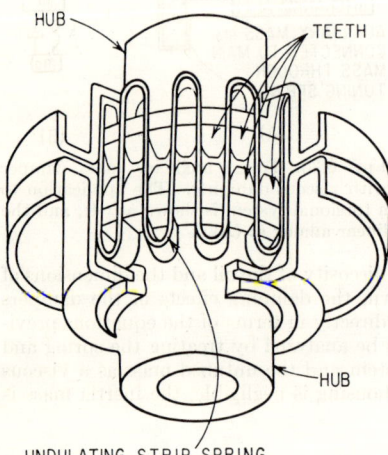

FIG. 6.36. Coupling used as elastic and damping element in auxiliary mass damper for torsional vibration. The torque is transmitted by an undulating strip of thin steel interposed between the teeth on opposite hubs. The stiffness of the strip is nonlinear, increasing as torque increases. Oil pumped between the strip and teeth dissipates energy.

where T_s is the slipping torque, I is the moment of inertia of the flywheels, and θ_0 is the amplitude of angular motion of the primary system. The dry-friction-based Lanchester damper requires frequent adjustment, as the braking material wears, to maintain a constant braking force.

It is possible to use torque-transmitting couplings that can absorb vibration energy, as the spring elements for tuned dampers. The Bibby coupling (Fig. 6.36) is used in this manner. Since the stiffness of this coupling is nonlinear, the optimum tuning of such an absorber is secured for only one amplitude of motion.

A discussion of dampers and of their application to engine systems is given in Chap. 38.

DYNAMIC ABSORBERS TUNED TO ORDERS OF VIBRATION RATHER THAN CONSTANT FREQUENCIES

In the torsional vibration of rotating machinery, it is generally found that exciting torques and forces occur at the same frequency as the rotational speed, or at multiples of this frequency. The ratio of the frequency of vibration to the rotational speed is called the *order of the vibration q*. Thus a power plant driving a four-bladed propeller may have a torsional vibration whose frequency is four times the rotational speed of the drive shaft; sometimes it may have a second torsional vibration whose frequency is eight times the rotational speed. These are called the fourth-order and eighth-order torsional vibrations.

If a dynamic absorber in the form of a pendulum acting in a centrifugal field is used, then its natural frequency increases linearly with speed. Therefore it can be used to neutralize an order of vibration.[15-19]

Consider a pendulum of length l and of mass m attached at a distance R from the center

DYNAMIC ABSORBERS TUNED TO ORDERS OF VIBRATION

of a rotating shaft, as shown in Fig. 6.37. Since the pendulum is excited by torsional vibration in the shaft, let the radius R be rotating at a constant speed Ω with a superposed vibration $\theta = \theta_0 \cos q\Omega t$, where q represents the order of the vibration. Then the angle of R with respect to any desired reference is $\Omega t + \theta_0 \cos q\Omega t$. The angle of the pendulum with respect to the radius R is defined as $\psi = \psi_0 \cos q\Omega t$, as shown by Fig. 6.37.

The acceleration acting on the mass m at

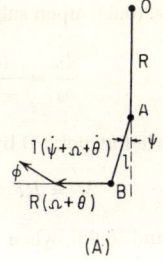

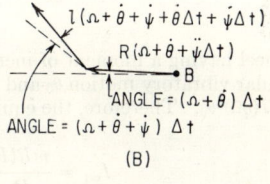

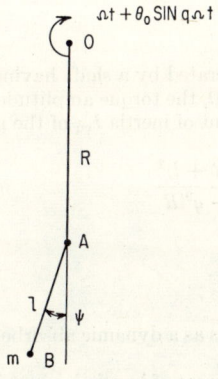

Fig. 6.37. Schematic diagram of pendulum absorber.

Fig. 6.38. Velocity vectors for the pendulum absorber: (A) velocities at time t; (B) velocities at time $t + \Delta t$; (C) change in velocities during time increment Δt.

position B is most easily ascertained by considering the change in velocity during a short increment of time Δt. The components of velocity of the mass m at time t are shown graphically in Fig. 6.38A; at time $t + \Delta t$, the corresponding velocities are shown in Fig. 6.38B. The change in velocity during the time interval Δt is shown in Fig. 6.38C. Since the acceleration is the change in velocity per unit of time, the accelerations along and perpendicular to l are:

Acceleration along l:

$$\frac{-l(\Omega + \dot\theta + \dot\psi)^2\,\Delta t - R(\Omega + \dot\theta)^2\,\Delta t \cos\psi + R\ddot\theta\,\Delta t \sin\psi}{\Delta t} \tag{6.61}$$

Acceleration perpendicular to l:

$$\frac{l(\ddot\theta + \ddot\psi)\,\Delta t + R(\Omega + \dot\theta)^2\,\Delta t \sin\psi + R\ddot\theta\,\Delta t \cos\psi}{\Delta t} \tag{6.62}$$

Only the force $-F$, directed along the pendulum, acts on the mass m. Therefore the equations of motion are

$$\begin{aligned}-F &= -ml(\Omega + \dot\theta + \dot\psi)^2 - mR(\Omega + \dot\theta)^2 \cos\psi + R\ddot\theta \sin\psi \\ 0 &= ml(\ddot\theta + \ddot\psi) + mR(\Omega + \dot\theta)^2 \sin\psi + mR\ddot\theta \cos\dot\psi\end{aligned} \tag{6.63}$$

Assuming that ψ and θ are small, Eqs. (6.63) simplify to

$$\begin{aligned}Ft &= m(R + l)\Omega^2 \\ l(\ddot\theta + \ddot\psi) &+ R\Omega^2\psi + R\ddot\theta = 0\end{aligned} \tag{6.64}$$

6-30 DYNAMIC VIBRATION ABSORBERS AND AUXILIARY MASS DAMPERS

The second of Eqs. (6.64) upon substitution of $\theta = \theta_0 \cos q\Omega t$ and $\psi = \psi_0 \cos q\Omega t$ yields

$$\frac{\psi_0}{\theta_0} = \frac{(q\Omega)^2(l+R)}{-(q\Omega)^2 l + \Omega^2 R} = \frac{q^2(l+R)}{R - q^2 l} \tag{6.65}$$

The torque M exerted at point 0 by the force F is

$$M = RF \sin \psi = RF\psi \qquad \text{when } \psi \text{ is small}$$

From Eqs. (6.64) and (6.65), when ψ is small,

$$M = \frac{mq^2 R(R+l)^2 \Omega^2}{R - q^2 l} \tag{6.66}$$

If a flywheel having a moment of inertia I is accelerated by a shaft having an amplitude of angular vibratory motion θ_0 and a frequency $q\Omega$, the torque amplitude exerted on the shaft is $I(q\Omega)^2 \theta_0$. Therefore, the equivalent moment of inertia I_{eq} of the pendulum is

$$I_{\text{eq}} = \frac{mR(R+l)^2}{R - q^2 l} = \frac{m(R+l)^2}{1 - q^2 l R} \tag{6.67}$$

When

$$\frac{R}{l} = q^2 \tag{6.68}$$

the equivalent inertia is infinite and the pendulum acts as a dynamic absorber by enforcing a node at its point of attachment.

Where the pendulum is damped, the equivalent moment of inertia is given by an equation analogous to Eqs. (6.4) and (6.5):

$$I_{\text{eq}} = \frac{1 + 2\zeta \nu j}{(1 - \nu^2) + 2\zeta \nu j} m(R+l)^2$$

$$= m(R+l)^2 \left[\frac{1 - \nu^2 + (2\zeta\nu)^2}{(1-\nu^2)^2 + (2\zeta\nu)^2} - \frac{2\zeta\nu^3 j}{(1-\nu^2)^2 + (2\zeta\nu)^2} \right] \tag{6.69}$$

where $\nu^2 = q^2 l/R$ and $\zeta = (c/2m\Omega)\sqrt{l/R}$.

When the pendulum is attached to a single degree-of-freedom system as is shown in Fig. 6.39, the amplitude of motion θ_a of the flywheel of inertia I is given, by analogy to Eq. (6.7), as

$$\frac{\theta_a}{\theta_{\text{st}}} = \sqrt{\frac{(1-\nu^2)^2 + (2\zeta\nu)^2}{[(1-\nu^2)(1-\beta_p^2) - \beta_p^2 \mu]^2 + (2\zeta\nu)^2[1 - \beta_p^2 - \beta_p^2 \mu_p]^2}} \tag{6.70}$$

where

$$2\zeta\nu = \frac{cql}{mR}$$

$$\mu_p = \frac{m(R+l)^2}{I}$$

$$\beta_p = \frac{q}{k_r I}$$

$$\theta_{\text{st}} = \frac{m_0}{k_r}$$

DYNAMIC ABSORBERS TUNED TO ORDERS OF VIBRATION

The pendulum tends to detune when the amplitude of motion of the pendulum is large, thereby introducing harmonics of the torque that it neutralizes.[17] Suppose the shaft rotates at a constant speed Ω, i.e., $\theta_0 = 0$, and consider the torque exerted on the shaft as m moves through a large amplitude ψ_0 about its equilibrium position. Equations (6.63) become

$$F = ml(\Omega + \dot\psi)^2 + mR\Omega^2 \cos\psi$$
$$l\ddot\psi + R\Omega^2 \sin\psi = 0 \tag{6.71}$$

A solution for the second of Eqs. (6.71) is

$$\dot\psi = \sqrt{\frac{2\Omega^2 R}{l}} \sqrt{\cos\psi - \cos\psi_0} \tag{6.72}$$

The solution of Eq. (6.72) involves elliptic integrals and is given approximately by

$$\psi = \psi_0 \sin \omega t$$

where
$$\omega = \sqrt{\frac{R}{l}} \frac{\pi/2}{F(\psi_0/2,\, \pi/2)} \Omega$$

and $F(\psi_0/2, \pi/2)$ is an elliptic function of the first kind whose value may be obtained from tables.

Since $\omega/\Omega = q$ (the order of the disturbance), the tuning of the damper will be changed for large angles and becomes

$$q^2 = \frac{R}{l} \left(\frac{\pi/2}{F(\psi_0/2,\, \pi/2)} \right)^2 \tag{6.73}$$

The value of $q^2 l/R = \nu^2$ used in Eqs. (6.69) and (6.70) is given in Fig. 6.40 as a function of the amplitude of the pendulum.

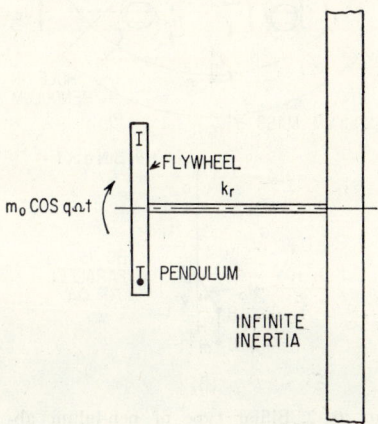

FIG. 6.39. Application of pendulum absorber to a rotational single degree-of-freedom system.

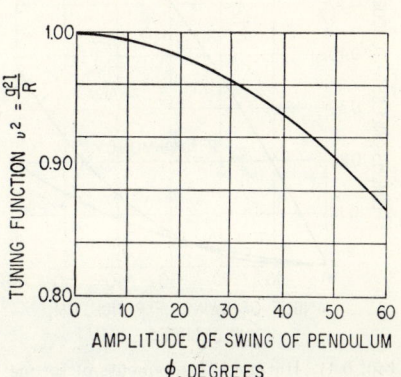

FIG. 6.40. Tuning function for a pendulum absorber used in Eqs. (6.69) and (6.70).

6–32 DYNAMIC VIBRATION ABSORBERS AND AUXILIARY MASS DAMPERS

Since the force exerted by the mass m is directed along the rod connecting it to the pivot A (Fig. 6.37), the reactive torque on the shaft is

$$M = FR \sin \psi$$

$$= mR^2\Omega^2 \left[\frac{l}{R}\left(1 + \frac{\dot\psi}{\Omega}\right)^2 \sin \psi + \sin \psi \cos \psi \right]$$

$$= mR^2\Omega^2(A_1 \sin q\Omega t + A_2 \sin 2q\Omega t + A_3 \sin 3q\Omega t + \cdots) \quad (6.74)$$

The values of the fundamental torque corresponding to the tuned frequency and to the second and third harmonics of this tuned frequency are given in Fig. 6.41 as a function of the angle of swing of the pendulum, for a typical installation. In this case, the pendulum is tuned to the $4\frac{1}{2}$ order of vibration. (The $4\frac{1}{2}$ order of vibration is one whose frequency is $4\frac{1}{2}$ times the rotational frequency and 9 times the fundamental frequency. The latter is called the half order and occurs at half of the rotational frequency. This is common in four-cycle engines.)

Two types of pendulum absorber are used. The one most commonly used is shown in Fig. 6.42. The counterweight, which also is used to balance rotating forces in the engine, is suspended from a hub carried by the crankshaft by pins that act through holes with clearance, Fig. 6.42A. By suspending the pendulum from two pins, the pendulum when oscillating does not rotate but rather moves as shown in Fig. 6.42B. Since it is not subjected to angular acceleration, it may be treated as a particle located at its center-

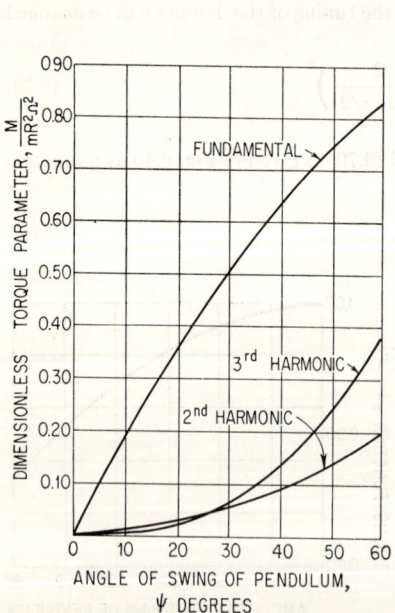

Fig. 6.41. Harmonic components of torque generated by a pendulum absorber as a function of its angle of swing. The torque is expressed by the parameters used in Eq. (6.74).

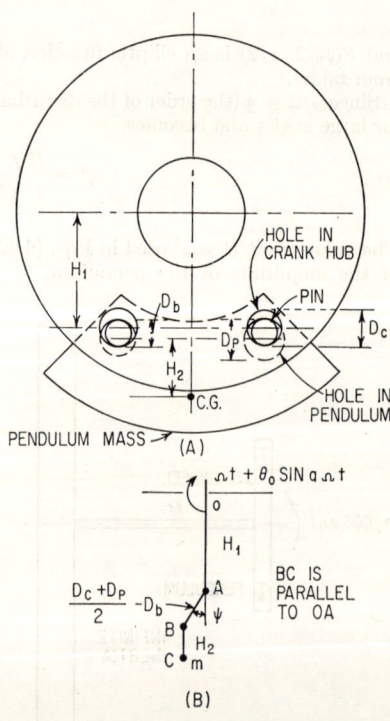

Fig. 6.42. Bifilar type of pendulum absorber. The mechanical arrangement is shown at (A), and a schematic diagram at (B).

of-gravity. Referring to Fig. 6.42A and B, the expression for acceleration [Eqs. (6.61) and (6.62)] and the equations of motion [Eqs. (6.63)] apply if

$$R = H_1 + H_2$$
$$l = \frac{D_c + D_p}{2} - D_b$$
(6.75)

where H_1 = distance from center of rotation to center of holes in crank hub

H_2 = distance from center of holes in pendulum to center-of-gravity of pendulum

D_c = diameter of hole in crank hub

D_p = diameter of hole in pendulum

D_b = diameter of pin

In practice, difficulty arises from the wear of the holes and the pin. Moreover, the motion on the pins generally is small and the loads due to centrifugal forces are large so that fretting is a problem. Because the radius of motion of the pendulum is short, only a small amount of wear can be tolerated. Hardened pins and bushings are used to reduce the wear.

The pendulum is most easily designed if it is recognized that the inertia torques generated by the pendulum must neutralize the forcing torques. Thus

$$m\omega^2 l \psi_0 R = M \qquad (6.76)$$

The radii l and R are set by the design of the crank and the order of vibration to be neutralized. The original motion ψ_0 is generally limited to a small angle, approximately 20°.

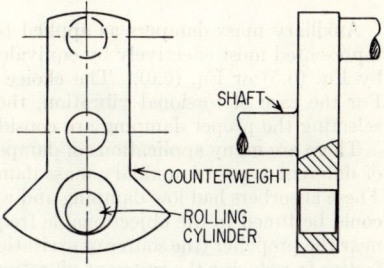

FIG. 6.43. Roller type of pendulum absorber.

It is probable that the most stringent condition is at the lowest operating speed, although the absorber may be required only to avoid difficulty at some particular critical speed. Knowing the excitation M, it is possible to compute the required mass of the pendulum weight.

A second type of pendulum absorber is a cylinder that rolls in a hole in a counterweight, as shown in Fig. 6.43. In this type, the radius of the pendulum corresponds to the difference in the radii of the hole and of the cylinder. It is found, by observing tests and checking the tuning of actual systems using cylindrical pendulums, that the weight rotates with a uniform angular velocity. Therefore the tuning is independent of the moments of inertia of the cylinder. It is common to allow a larger amplitude of motion with the absorber of Fig. 6.43 than with the absorber of Fig. 6.40.

Applications of pendulum absorbers to torsional-vibration problems are given in Chap. 38.

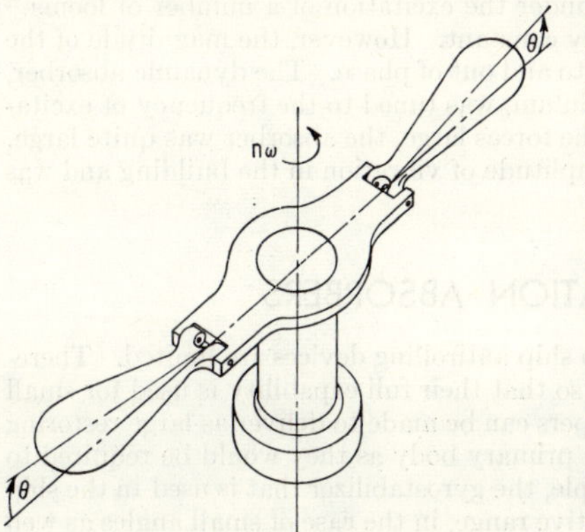

PENDULUM ABSORBER FOR LINEAR VIBRATION

FIG. 6.44. Application of pendulum absorbers to counteract linear vibration.

The principle of the pendulum absorber can be applied to linear vibration as well as to torsional vibration. To neutralize linear vibration, pendulums are rotated about an axis parallel to the direction of vibration, as shown in Fig. 6.44. This can be accomplished with an absorber mounted on the moving body. Two or more pendulums are used so that centrifugal forces are balanced. Free rotational movement of each pendulum in the plane of the axis allows the axial forces to be neutralized. The pendulum assembly must rotate about the axis at some submultiple of the frequency of vibration. The size of the absorber is determined by the condition that the components of the inertia forces of the weights in the axial direction $[\Sigma m\omega^2 r\theta]$ must balance the exciting forces. This device can be applied where the vibration is generated by the action of rotating members, but the magnitude of the vibratory forces is uncertain. A discussion of this absorber, including the influence of moments of inertia and damping of the pendulum, together with some applications to the elimination of vibration in special locations on a ship, is given in Ref. 20.

APPLICATIONS OF DAMPERS TO MULTIPLE DEGREE-OF-FREEDOM SYSTEMS

Auxiliary mass dampers as applied to systems of several degrees-of-freedom can be represented most effectively by equivalent masses or moments of inertia, as determined by Eq. (6.5) or Eq. (6.6). The choice of proper damping constants is more difficult. For the case of torsional vibration, the practical problems of designing dampers and selecting the proper damping are considered in Chap. 38.

There are many applications of dampers to vibrating structures that illustrate the use of different types of auxiliary mass damper. One such application has been to ships.[21] These absorbers had low damping and were designed to be filled with water so that they could be tuned to the objectionable frequencies. In one case, the absorber was located near the propeller (the source of excitation) and when properly tuned was found to be effective in reducing the resonant vibration of the ship. In another case, the absorber was located on an upper deck but was not as effective. It enforced a node at its point of attachment but, because of the flexibility between the upper deck and the bottom of the ship, there was appreciable motion in the vicinity of the propeller and vibratory energy was fed to the ship's structure. To operate properly, the absorbers must be closely tuned and the propeller speed closely maintained. Because the natural frequencies of the ship vary with the types of loading, it is not sufficient to install a fixed frequency absorber that is effective at only one natural frequency of the hull, corresponding to a particular loading condition.

An auxiliary mass absorber has been applied to the reduction of vibration in a heavy building that vibrated at a low frequency under the excitation of a number of looms.[22] The frequency of the looms was substantially constant. However, the magnitude of the excitation was variable as the looms came into and out of phase. The dynamic absorber, consisting of a heavy weight hung as a pendulum, was tuned to the frequency of excitation. Because the frequency was low and the forces large, the absorber was quite large. However, it was effective in reducing the amplitude of vibration in the building and was relatively simple to construct.

ACTIVATED VIBRATION ABSORBERS

The cost and space that can be allotted to ship antirolling devices are limited. Therefore it is desirable to activate the absorbers so that their full capability is used for small amplitudes as well as large. Activated dampers can be made to deliver as large restoring forces for small amplitudes of motion of the primary body as they would be required to deliver if the motions were large. For example, the gyrostabilizer that is used in the ship is precessed by a motor through its full effective range, in the case of small angles as well as large. Thus, it introduces a restoring torque that is much larger than would be introduced by the normal damped precession.[14] In the same manner the water in antiroll tanks is always pumped to the tank where it will introduce the maximum torque to counteract the roll. By pumping, much larger quantities of water can be transferred and larger damping moments obtained than can be obtained by controlled gravity flows.

Neither gyros nor antirolling tanks are used for reducing the roll of the most modern ships. The device now used is a retractable fin that is introduced into the water where the beam of the ship is large. The fin is actuated to tilt in such manner as to introduce forces resisting the rolling. For example, both the *Queen Elizabeth* and the *Queen Mary* use antirolling fins.

Activated vibration absorbers are essentially servomechanisms designed to maintain some desired steady state. Steam and gas turbine speed governors, wicket gate controls for frequency regulation in water turbines, and temperature control equipment can be considered as special forms of activated vibration absorbers.[23]

THE USE OF AUXILIARY MASS DEVICES TO REDUCE TRANSIENT AND SELF-EXCITED VIBRATIONS

Where the vibration is self-excited or caused by repeated impact, it is necessary to have sufficient damping to prevent a serious build-up of vibration amplitude. This damping, which need not always be large, may be provided by a loosely coupled auxiliary mass. A simple application of this type is the ring fitted to the interior of a gear, as shown in Fig. 6.45. By fitting this ring with the proper small clearance so that relative motion occurs between it and the gear, it is possible to obtain enough energy dissipation to damp the high-frequency, low-energy vibration that causes the gear to ring. The rubbery coatings applied to large, thin-metal panels such as automobile doors to give them a solid rather than a "tinny" sound depend for their effectiveness on a proper balance of mass, elasticity, and damping (see Chap. 37).

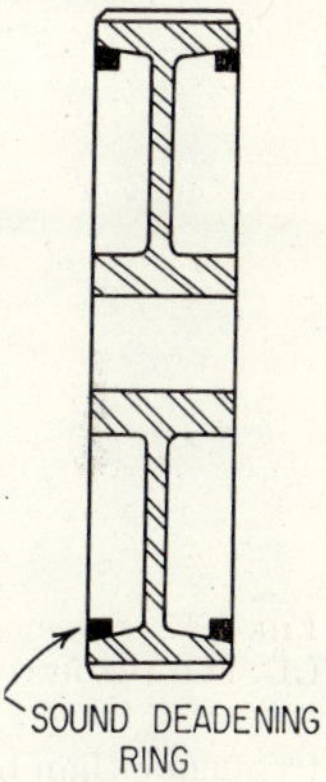

Another application where auxiliary mass dampers are useful is in the prevention of fatigue failures in turbines. At the high-pressure end of an impulse turbine, steam or hot gas is admitted through only a few nozzles. Consequently, as the blade passes the nozzle it is given an impulse by the steam and set into vibration at its natural frequency. It is a characteristic of alloy steels that they have very little internal damping at high operating temperature. For this reason the free vibration persists with only a slightly diminished amplitude until the blade again is subjected to the steam impulse. Some of these second impulses will be out of phase with the motion of the blade and will reduce its amplitude; however, successive impulses may increase the amplitude on subsequent passes until failure occurs. Damping can be increased by placing a number of loose wires in a cylindrical hole cut in the blade in a radial direction. The damping of a number of these wires has been computed in terms of the geometry of the application (number of wires, density of wires, size of the hole, radius of the blade, rotational speed, etc.) and the amplitude of vibration.[24] These computations show reasonable agreement with experimental results.

SOUND DEADENING RING

Fig. 6.45. Application of auxiliary mass damper to deaden noise in gear.

An auxiliary mass has been used to damp the cutting tool chatter set up in a boring bar.[25] Because of the characteristics of the metal-cutting process or of some coupling between motions of the tool parallel and perpendicular to the work face, it is sometimes found that a self-excited vibration is initiated at the natural frequency of the cutter system. Since the self-excitation energy is low, the vibration usually is initiated only if the damping is small. Chatter of the tool is most common in long, poorly supported tools, such as boring bars (see Chap. 40). To eliminate this chatter, a loose auxiliary mass is incor-

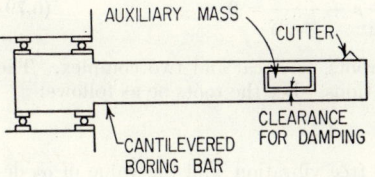

Fig. 6.46. Application of auxiliary mass damper to reduce chatter in boring bar.

6-36 DYNAMIC VIBRATION ABSORBERS AND AUXILIARY MASS DAMPERS

porated in the boring bar, as shown in Fig. 6.46. This may be air-damped or fluid-damped. Since the excitation is at the natural frequency of the tool, the damping should be such that the tool vibrates with a minimum amplitude at this frequency. The damping requirement can be estimated by substituting $\beta = 1$ in Eq. (6.25),

$$\frac{x_0}{\delta_{st}} = \sqrt{\frac{1 + 4(\zeta\alpha)^2}{4(\zeta\alpha)^2\mu^2}} \tag{6.77}$$

The optimum value of the parameter $(\zeta\alpha)$ is infinity. Thus when the frequency of excitation is constant, a greater reduction in amplitude can be obtained by a shift in natural

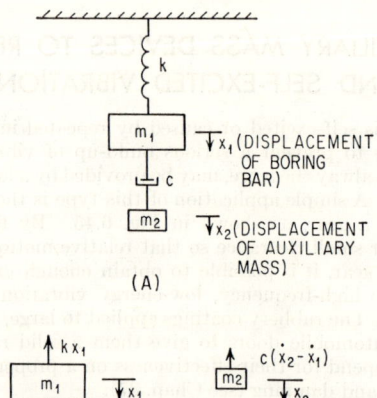

Fig. 6.47. Schematic diagram of damper shown in Fig. 6.46. The arrangement is shown at (A), and the forces acting on the boring bar and auxiliary mass are shown at (B).

frequency than by damping. However, such a shift cannot be attained because the frequency of the excitation always coincides with the natural frequency of the complete system. Instead, a better technique is to determine the damping that gives the maximum decrement of the free vibration.

Let the boring bar and damper be represented by a single degree-of-freedom system with a damper mass coupled to the main mass by viscous damping, as shown in Fig. 6.47A. The forces acting on the masses are shown in Fig. 6.47B. The equations of motion are

$$-kx_1 - c\dot{x}_1 + c\dot{x}_2 = m_1\ddot{x}_1$$
$$c\dot{x}_1 - c\dot{x}_2 = m_2\ddot{x}_2 \tag{6.78}$$

Substituting $x = Ae^{st}$, the resulting frequency equation is

$$s^3 + \frac{c(m_1 + m_2)}{m_1m_2}s^2 + \frac{k}{m_1}s + \frac{kc}{m_1m_2} = 0 \tag{6.79}$$

Where chatter occurs, this equation has three roots, one real and two complex. The complex roots correspond to decaying free vibrations. Let the roots be as follows:

$$\alpha_1, \quad \alpha_2 + j\beta, \quad \alpha_2 - j\beta$$

The value of β determines the frequency of the free vibration, and the value of α_2 determines the decrement (rate of decrease of amplitude) of the free vibration. The decrement α_2 is of primary interest; it is most easily found from the conditions that when the coefficient of s^3 is unity, (1) the sum of the roots is equal to the negative of the coefficient of s^2, (2) the sum of the products of the roots taken two at a time is the negative of the coefficient of s, and (3) the product of the roots is the negative of the constant term. The equations thus obtained are

$$\alpha_1 + 2\alpha_2 = -\frac{c(1+\mu)}{\mu m_1} \tag{6.80}$$

$$2\alpha_1\alpha_2 + \alpha_2^2 + \beta^2 = -\Omega_n^2 \tag{6.81}$$

$$\alpha_1(\alpha_2^2 + \beta^2) = -\Omega_n^2 \frac{c}{m_1\mu} \tag{6.82}$$

where $\omega_n^2 = k/m_1$ and $\mu = m_2/m_1$. It is not practical to find the optimum damping by solving these equations for α_2 and then setting the derivative of α_2 with respect to c equal to zero. However, it is possible to find the optimum damping by the following process. Eliminate $(\alpha_2^2 + \beta^2)$ between Eqs. (6.81) and (6.82) to obtain

$$2\alpha_1^2\alpha_2 = \omega_n^2 \left(\frac{c}{\mu m_1} - \alpha_1\right) \tag{6.83}$$

Substituting the value of α_1 from Eq. (6.80) in Eq. (6.83),

$$2\alpha_2 \left[2\alpha_2 + \frac{c(1+\mu)}{\mu m_1}\right]^2 = \frac{c\omega_n^2}{\mu m_1} + \omega_n^2 \left[2\alpha_2 + \frac{c(1+\mu)}{\mu m_1}\right] \tag{6.84}$$

To find the damping that gives the maximum decrement, differentiate with respect to c and set $d\alpha_2/dc = 0$:

$$2\alpha_2 \left[2\alpha_2 + \frac{c(1+\mu)}{\mu m_1}\right] = \tfrac{1}{2}\omega_n^2 \frac{2+\mu}{1+\mu} \tag{6.85}$$

Solving Eqs. (6.84) and (6.85) simultaneously,

$$c_{\text{opt}} = \frac{\mu^2 m_1 \omega_n}{2(1+\mu)^{3/2}} \tag{6.86}$$

$$(\alpha_2)_{\text{opt}} = -\frac{(2+\mu)\omega_n}{4(1+\mu)^{1/2}} \tag{6.87}$$

These values may be obtained by proper choice of clearance between the auxiliary mass and the hole in which it is located. Air damping is preferable to oil because it

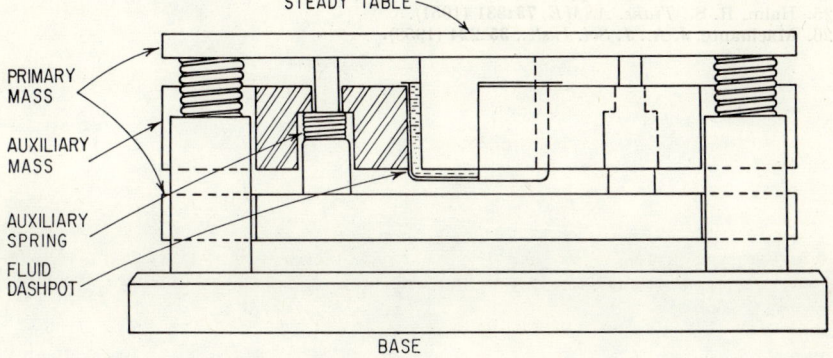

Fig. 6.48. Application of auxiliary mass to spring-mounted table to reduce vibration of table. (*Macinante*.[26])

requires less clearance. Therefore the plug is not immobilized by the centrifugal forces that, with the rotating boring bar, become larger as the clearance is increased.

In precision measurements, it is necessary to isolate the instruments from effects of shock and vibration in the earth and to damp any oscillations that might be generated in the measuring instruments. A heavy spring-mounted table fitted with a heavy auxiliary mass that is attached to the table by a spring and submerged in an oil bath (Fig. 6.48) has proved to be effective.[26] In this example the table has a top surface of 13½ in. (34 cm) by 13½ in. (34 cm) and a height of 6 in. (15 cm). Each auxiliary mass weighs about 70 lb (32 kg). The springs for both the primary table and the auxiliary system are designed to give a natural frequency between 2 and 4 cps in both the horizontal and vertical directions. By trying different fluids in the bath, suitable damping may be obtained experimentally.

REFERENCES

1. Timoshenko, S.: "Vibration Problems in Engineering," p. 240, D. Van Nostrand Company, Inc., Princeton, N.J., 1937.
2. Den Hartog, J. P.: "Mechanical Vibrations," chap. III, McGraw-Hill Book Company, Inc., New York, 1956.
3. Ormondroyd, J., and J. P. Den Hartog: *Trans. ASME*, **50**:A9 (1928).
4. Brock, J. E.: *J. Appl. Mechanics*, **13**(4):A-284 (1946).
5. Brock, J. E.: *J. Appl. Mechanics*, **16**(1):86 (1949).
6. Saver, F. M., and C. F. Garland: *J. Appl. Mechanics*, **16**(2):109 (1949).
7. Lewis, F. M.: *J. Appl. Mechanics*, **22**(3):377 (1955).
8. Georgian, J. C.: *Trans. ASME*, **16**:389 (1949).
9. Den Hartog, J. P., and J. Ormondroyd: *Trans. ASME*, **52**:133 (1930).
10. Roberson, R. E.: *J. Franklin Inst.*, **254**:205 (1952).
11. Pipes, L. A.: *J. Appl. Mechanics*, **20**:515 (1953).
12. Arnold, F. R.: *J. Appl. Mechanics*, **22**:487 (1955).
13. Hort, W.: "Technische Schwingungslehre," 2d ed., Springer-Verlag, Berlin, 1922.
14. Sperry, E. E.: *Trans. SNAME*, **30**:201 (1912).
15. Solomon, B.: *Proc. 4th Intern. Congr. Appl. Mechanics, Cambridge, England*, 1934.
16. Taylor, E. S.: *Trans. SAE*, **44**:81 (1936).
17. Den Hartog, J. P.: "Stephen Timoshenko 60th Anniversary Volume," The Macmillan Company, New York, 1939.
18. Porter, F. P.: "Evaluation of Effects of Torsional Vibration," SAE War Engineering Board, SAE, New York, 1945, p. 269.
19. Crossley, F. R. E.: *J. Appl. Mechanics*, **20**(1):41 (1953).
20. Reed, F. E.: *J. Appl. Mechanics*, **16**:190 (1949).
21. Constanti, M.: *Trans. Inst. of Naval Arch.*, **80**:181 (1938).
22. Crede, C. E.: *Trans. ASME*, **69**:937 (1947).
23. Brown, G. S., and D. P. Campbell: "Principles of Servomechanisms," John Wiley & Sons, Inc., New York, 1948.
24. DiTaranto, R. A.: *J. Appl. Mechanics*, **25**(1):21 (1958)
25. Hahn, R. S.: *Trans. ASME*, **73**:331 (1951).
26. Macinante, J. A.: *J. Sci. Instr.*, **35**:224 (1958).

7

VIBRATION OF SYSTEMS HAVING DISTRIBUTED MASS AND ELASTICITY

William F. Stokey
Carnegie Institute of Technology

INTRODUCTION

Preceding chapters consider the vibration of lumped parameter systems; i.e., systems that are idealized as rigid masses joined by massless springs and dampers. Many engineering problems are solved by analyses based on ideal models of an actual system, giving answers that are useful though approximate. In general, more accurate results are obtained by increasing the number of masses, springs, and dampers; i.e., by increasing the number of degrees-of-freedom. As the number of degrees-of-freedom is increased without limit, the concept of the system with distributed mass and elasticity is formed. This chapter discusses the free and forced vibration of such systems. Types of systems include rods vibrating in torsional modes and in tension-compression modes, and beams and plates vibrating in flexural modes. Particular attention is given to the calculation of the natural frequencies of such systems for further use in other analyses. Numerous charts and tables are included to define in readily available form the natural frequencies of systems commonly encountered in engineering practice.

FREE VIBRATION

DEGREES-OF-FREEDOM. Systems for which the mass and elastic parts are lumped are characterized by a finite number of degrees-of-freedom. In physical systems, all elastic members have mass, and all masses have some elasticity; thus, all real systems have distributed parameters. In making an analysis, it is often assumed that real systems have their parameters lumped. For example, in the analysis of a system consisting of a mass and a spring, it is commonly assumed that the mass of the spring is negligible so that its only effect is to exert a force between the mass and the support to which the spring is attached, and that the mass is perfectly rigid so that it does not deform and exert any elastic force. The effect of the mass of the spring on the motion of the system may be considered in an approximate way, while still maintaining the assumption of one degree-of-freedom, by assuming that the spring moves so that the deflection of each of its elements can be described by a single parameter. A commonly used assumption is that the deflection of each section of the spring is proportional to its distance from the support, so that if the deflection of the mass is given, the deflection of any part of the spring is defined. For the exact solution of the problem, even though the mass is considered to be perfectly rigid, it is necessary to consider that the deformation of the spring can occur in any manner consistent with the requirements of physical continuity.

Systems with distributed parameters are characterized by having an infinite number of degrees-of-freedom. For example, if an initially straight beam deflects laterally, it may

7–1

be necessary to give the deflection of each section along the beam in order to define completely the configuration. For vibrating systems, the coordinates usually are defined in such a way that the deflections of the various parts of the system from the equilibrium position are given.

NATURAL FREQUENCIES AND NORMAL MODES OF VIBRATION. The number of natural frequencies of vibration of any system is equal to the number of degrees-of-freedom; thus, any system having distributed parameters has an infinite number of natural frequencies. At a given time, such a system usually vibrates with appreciable amplitude at only a limited number of frequencies, often at only one. With each natural frequency is associated a shape, called the normal or natural mode, which is assumed by the system during free vibration at the frequency. For example, when a uniform beam with simply supported or hinged ends vibrates laterally at its lowest or fundamental natural frequency, it assumes the shape of a half sine wave; this is a normal mode of vibration. When vibrating in this manner, the beam behaves as a system with a single degree-of-freedom, since its configuration at any time can be defined by giving the deflection of the center of the beam. When any linear system, i.e., one in which the elastic restoring force is proportional to the deflection, executes free vibration in a single natural mode, each element of the system except those at the supports and nodes executes simple harmonic motion about its equilibrium position. All possible free vibration of any linear system is made up of superposed vibrations in the normal modes at the corresponding natural frequencies. The total motion at any point of the system is the sum of the motions resulting from the vibration in the respective modes.

There are always nodal points, lines, or surfaces, i.e., points which do not move, in each of the normal modes of vibration of any system. For the fundamental mode, which corresponds to the lowest natural frequency, the supported or fixed points of the system usually are the only nodal points; for other modes, there are additional nodes. In the modes of vibration corresponding to the higher natural frequencies of some systems, the nodes often assume complicated patterns. In certain problems involving forced vibrations, it may be necessary to know what the nodal patterns are, since a particular mode usually will not be excited by a force acting at a nodal point. Nodal lines are shown in some of the tables.

METHODS OF SOLUTION. The complete solution of the problem of free vibration of any system would require the determination of all the natural frequencies and of the mode shape associated with each. In practice, it often is necessary to know only a few of the natural frequencies, and sometimes only one. Usually the lowest frequencies are the most important. The exact mode shape is of secondary importance in many problems. This is fortunate, since some procedures for finding natural frequencies involve assuming a mode shape from which an approximation to the natural frequency can be found.

Classical Method. The fundamental method of solving any vibration problem is to set up one or more equations of motion by the application of Newton's second law of motion. For a system having a finite number of degrees-of-freedom, this procedure gives one or more ordinary differential equations. For systems having distributed parameters partial differential equations are obtained. Exact solutions of the equations are possible for only a relatively few configurations. For most problems other means of solution must be employed.

Rayleigh's and Ritz's Methods. For many elastic bodies, Rayleigh's method is useful in finding an approximation to the fundamental natural frequency. While it is possible to use the method to estimate some of the higher natural frequencies, the accuracy often is poor; thus, the method is most useful for finding the fundamental frequency. When any elastic system without damping vibrates in its fundamental normal mode, each part of the system executes simple harmonic motion about its equilibrium position. For example, in lateral vibration of a beam the motion can be expressed as $y = X(x) \sin \omega_n t$ where X is a function only of the distance along the length of the beam. For lateral vibration of a plate, the motion can be expressed as $w = W(x,y) \sin \omega_n t$ where x and y are the coordinates in the plane of the plate. The equations show that when the deflection from equilibrium is a maximum, all parts of the body are motionless. At that time all

FREE VIBRATION

the energy associated with the vibration is in the form of elastic strain energy. When the body is passing through its equilibrium position, none of the vibrational energy is in the form of strain energy so that all of it is in the form of kinetic energy. For conservation of energy, the strain energy in the position of maximum deflection must equal the kinetic energy when passing through the equilibrium position. Rayleigh's method of finding the natural frequency is to compute these maximum energies, equate them, and solve for the frequency. When the kinetic-energy term is evaluated, the frequency always appears as a factor. Formulas for finding the strain and kinetic energies of rods, beams, and plates are given in Table 7.1.

If the deflection of the body during vibration is known exactly, Rayleigh's method gives the true natural frequency. Usually the exact deflection is not known, since its determination involves the solution of the vibration problem by the classical method. If the classical solution is available, the natural frequency is included in it, and nothing is gained by applying Rayleigh's method. In many problems for which the classical solution is not available, a good approximation to the deflection can be assumed on the basis of physical reasoning. If the strain and kinetic energies are computed using such an assumed shape, an approximate value for the natural frequency is found. The correctness of the approximate frequency depends on how well the assumed shape approximates the true shape.

In selecting a function to represent the shape of a beam or a plate, it is desirable to satisfy as many of the boundary conditions as possible. For a beam or plate supported at a boundary, the assumed function must be zero at that boundary; if the boundary is built in, the first derivative of the function must be zero. For a free boundary, if the conditions associated with bending moment and shear can be satisfied, better accuracy

Table 7.1. Strain and Kinetic Energies of Uniform Rods, Beams, and Plates

Member	Strain energy V	Kinetic energy T General	Maximum *
Rod in tension or compression	$\dfrac{SE}{2}\int_0^l\left(\dfrac{\partial u}{\partial x}\right)^2 dx$	$\dfrac{S\gamma}{2g}\int_0^l\left(\dfrac{\partial u}{\partial t}\right)^2 dx$	$\dfrac{S\gamma\omega_n^2}{2g}\int_0^l V^2\, dx$
Rod in torsion	$\dfrac{GI_p}{2}\int_0^l\left(\dfrac{\partial \phi}{\partial x}\right)^2 dx$	$\dfrac{I_p\gamma}{2g}\int_0^l\left(\dfrac{\partial \phi}{\partial t}\right)^2 dx$	$\dfrac{I_p\gamma\omega_n^2}{2g}\int_0^l \Phi^2\, dx$
Beam in bending	$\dfrac{EI}{2}\int_0^l\left(\dfrac{\partial^2 y}{\partial x^2}\right)^2 dx$	$\dfrac{S\gamma}{2g}\int_0^l\left(\dfrac{\partial y}{\partial t}\right)^2 dx$	$\dfrac{S\gamma\omega_n^2}{2g}\int_0^l Y^2\, dx$
Rectangular plate in bending [1]	$\dfrac{D}{2}\int_S\int\left\{\left(\dfrac{d^2w}{dx^2}+\dfrac{d^2w}{dy^2}\right)^2 - 2(1-\mu)\left[\dfrac{\partial^2w}{\partial x^2}\dfrac{\partial^2w}{\partial y^2} - \left(\dfrac{\partial^2w}{\partial x\,\partial y}\right)^2\right]\right\} dx\,dy$	$\dfrac{\gamma h}{2g}\int_S\int\left(\dfrac{\partial w}{\partial t}\right)^2 dx\,dy$	$\dfrac{\gamma h\omega_n^2}{2g}\int_S\int W^2\, dx\,dy$
Circular plate (deflection symmetrical about center) [1]	$\pi D\int_0^a\left\{\left(\dfrac{\partial^2 w}{\partial r^2}+\dfrac{1}{r}\dfrac{\partial w}{\partial r}\right)^2 - 2(1-\mu)\dfrac{\partial^2 w}{\partial r^2}\dfrac{1}{r}\dfrac{\partial w}{\partial r}\right\} r\,dr$	$\dfrac{\pi\gamma h}{g}\int_0^a\left(\dfrac{\partial w}{\partial t}\right)^2 r\,dr$	$\dfrac{\pi\gamma h\omega_n^2}{g}\int_0^a W^2 r\,dr$

u = longitudinal deflection of cross section of rod
ϕ = angle of twist of cross section of rod
y = lateral deflection of beam
w = lateral deflection of plate
    Capitals denote values at extreme deflection for simple harmonic motion.
l = length of rod or beam
a = radius of circular plate
h = thickness of beam or plate

S = area of cross section
I_p = polar moment of inertia
I = moment of inertia of beam
γ = weight density
E = modulus of elasticity
G = modulus of rigidity
μ = Poisson's ratio
$D = \dfrac{Eh^3}{12(1-\mu^2)}$

* This is the maximum kinetic energy in simple harmonic motion.

7-4 VIBRATION OF SYSTEMS HAVING DISTRIBUTED MASS AND ELASTICITY

usually results. It can be shown [2] that the frequency that is found by using any shape except the correct shape always is higher than the actual frequency. Therefore, if more than one calculation is made, using different assumed shapes, the lowest computed frequency is closest to the actual frequency of the system.

In many problems for which a classical solution would be possible, the work involved is excessive. Often a satisfactory answer to such a problem can be obtained by the application of Rayleigh's method. In this chapter several examples are worked using both the classical method and Rayleigh's method. In all, Rayleigh's method gives a good approximation to the correct result with relatively little work. Many other examples of solutions to problems by Rayleigh's method are in the literature.[3,4,5]

Ritz's method is a refinement of Rayleigh's method. A better approximation to the fundamental natural frequency can be obtained by its use, and approximations to higher natural frequencies can be found. In using Ritz's method, the deflections which are assumed in computing the energies are expressed as functions with one or more undetermined parameters; these parameters are adjusted to make the computed frequency a minimum. Ritz's method has been used extensively for the determination of the natural frequencies of plates of various shapes, and is discussed in the section on the lateral vibrations of plates.

Lumped Parameters. A procedure that is useful in many problems for finding approximations to both the natural frequencies and the mode shapes is to reduce the system with distributed parameters to one having a finite number of degrees-of-freedom. This is done by lumping the parameters for each small region into an equivalent mass and elastic element. Several formalized procedures for doing this and for analyzing the re-

Table 7.2. Approximate Formulas for Natural Frequencies of Systems Having Both Concentrated and Distributed Mass

TYPE OF SYSTEM	NATURAL FREQUENCY $\omega_n = 2\pi f_n$	STIFFNESS
SPRING WITH MASS ATTACHED (k, m, M)	$\sqrt{\dfrac{k}{M + m/3}}$	$k = \dfrac{Gd^4}{8nD^3}$ D = COIL DIA d = WIRE DIA n = NUMBER OF TURNS
CIRCULAR ROD, WITH DISC ATTACHED, IN TORSION (k_r, I_s, I)	$\sqrt{\dfrac{k_r}{I + I_s/3}}$	$k_r = \dfrac{G\pi D^4}{32\,l}$ D = ROD DIAMETER l = ROD LENGTH
UNIFORM SIMPLY SUPPORTED BEAM WITH MASS IN CENTER ($m/2$, M, $m/2$)	$\sqrt{\dfrac{k}{M + m/2}}$	$k = \dfrac{48EI}{l^3}$ l = BEAM LENGTH I = MOMENT OF INERTIA
UNIFORM CANTILEVER BEAM WITH MASS ON END (m, M)	$\sqrt{\dfrac{k}{M + 0.23m}}$	$k = \dfrac{3EI}{l^3}$ l = BEAM LENGTH I = MOMENT OF INERTIA

FREE VIBRATION

sulting systems are described in Chap. 28. If a system consists of a rigid mass supported by a single flexible member whose mass is not negligible, the elastic part of the system sometimes can be treated as an equivalent spring; i.e., some of its mass is lumped with the rigid mass. Formulas for several systems of this kind are given in Table 7.2.

ORTHOGONALITY. It is shown in Chap. 2 that the normal modes of vibration of a system having a finite number of degrees-of-freedom are orthogonal to each other. For a system of masses and springs having n degrees-of-freedom, if the coordinate system is selected in such a way that X_1 represents the amplitude of motion of the first mass, X_2 that of the second mass, etc., the orthogonality relations are expressed by $(n-1)$ equations as follows:

$$m_1 X_1^a X_1^b + m_2 X_2^a X_2^b + \cdots = \sum_{i=1}^{n} m_i X_i^a X_i^b = 0 \qquad [a \neq b]$$

where X_1^a represents the amplitude of the first mass when vibrating only in the ath mode, X_1^b the amplitude of the first mass when vibrating only in the bth mode, etc.

For a body such as a uniform beam whose parameters are distributed only lengthwise; i.e., in the X direction, the orthogonality between two normal modes is expressed by

$$\int_0^l \rho \phi_a(x) \phi_b(x)\, dx = 0 \qquad [a \neq b] \tag{7.1}$$

where $\phi_a(x)$ represents the deflection in the ath normal mode, $\phi_b(x)$ the deflection in the bth normal mode, and ρ the density.

For a system, such as a uniform plate, in which the parameters are distributed in two dimensions, the orthogonality condition is

$$\int_A \int \rho \phi_a(x,y) \phi_b(x,y)\, dx\, dy = 0 \qquad [a \neq b] \tag{7.2}$$

LONGITUDINAL AND TORSIONAL VIBRATIONS OF UNIFORM CIRCULAR RODS

Equations of Motion. A circular rod having a uniform cross section can execute longitudinal, torsional, or lateral vibrations, either individually or in any combination. The equations of motion for longitudinal and torsional vibrations are similar in form, and the solutions are discussed together. The lateral vibration of a beam having a uniform cross section is considered separately.

In analyzing the longitudinal vibration of a rod, only the motion of the rod in the longitudinal direction is considered. There is some lateral motion because longitudinal stresses induce lateral strains; however, if the rod is fairly long compared to its diameter, this motion has a minor effect.

Consider a uniform circular rod, Fig. 7.1A. The element of length dx, which is formed by passing two parallel planes A–A and B–B normal to the axis of the rod, is shown in Fig. 7.1B. When the rod executes only longitudinal vibration, the force acting on the

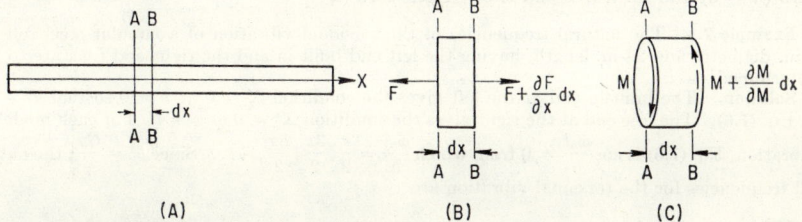

Fig. 7.1. (A) Rod executing longitudinal or torsional vibration. (B) Forces acting on element during longitudinal vibration. (C) Moments acting on element during torsional vibration.

7–6 VIBRATION OF SYSTEMS HAVING DISTRIBUTED MASS AND ELASTICITY

face A–A is F, and that on face B–B is $F + (\partial F/\partial x)\, dx$. The net force acting to the right must equal the product of the mass of the element $(\gamma/g)S\, dx$ and its acceleration $\partial^2 u/\partial t^2$, where γ is the weight density, S the area of the cross section, and u the longitudinal displacement of the element during the vibration:

$$\left(F + \frac{\partial F}{\partial x} dx\right) - F = \frac{\partial F}{\partial x} dx = \left(\frac{\gamma}{g}\right) S\, dx\, \frac{\partial^2 u}{\partial t^2} \quad \text{or} \quad \frac{\partial F}{\partial x} = \frac{\gamma S}{g} \frac{\partial^2 u}{\partial t^2} \qquad (7.3)$$

This equation is solved by expressing the force F in terms of the displacement. The elastic strain at any section is $\partial u/\partial x$, and the stress is $E\partial u/\partial x$. The force F is the product of the stress and the area, or $F = ES\, \partial u/\partial x$, and $\partial F/\partial x = ES\, \partial^2 u/\partial x^2$. Equation (7.3) becomes $Eu'' = \gamma/g\ddot{u}$, where $u'' = \partial^2 u/\partial x^2$ and $\ddot{u} = \partial^2 u/\partial t^2$. Substituting $a^2 = Eg/\gamma$,

$$a^2 u'' = \ddot{u} \qquad (7.4)$$

The equation governing the torsional vibration of the circular rod is derived by equating the net torque acting on the element, Fig. 7.1C, to the product of the moment of inertia J and the angular acceleration $\ddot{\phi}$, ϕ being the angular displacement of the section. The torque on the section A–A is M and that on section B–B is $M + (\partial M/\partial x)\, dx$. By an analysis similar to that for the longitudinal vibration, letting $b^2 = Gg/\gamma$,

$$b^2 \phi'' = \ddot{\phi} \qquad (7.5)$$

Solution of Equations of Motion. Since Eqs. (7.4) and (7.5) are of the same form, the solutions are the same except for the meaning of a and b. The solution of Eq. (7.5) is of the form $\phi = X(x)T(t)$ in which X is a function of x only and T is a function of t only. Substituting this in Eq. (7.5) gives $b^2 X''T = X\ddot{T}$. By separating the variables,[6]

$$T = A \cos(\omega_n t + \theta)$$

$$X = C \sin \frac{\omega_n x}{b} + D \cos \frac{\omega_n x}{b} \qquad (7.6)$$

The natural frequency ω_n can have infinitely many values, so that the complete solution of Eq. (7.5) is, combining the constants,

$$\phi = \sum_{n=1}^{n=\infty} \left(C_n \sin \frac{\omega_n x}{b} + D_n \cos \frac{\omega_n x}{b}\right) \cos(\omega_n t + \theta_n) \qquad (7.7)$$

The constants C_n and D_n are determined by the end conditions of the rod and by the initial conditions of the vibration. For a built-in or clamped end of a rod in torsion, $\phi = 0$ and $X = 0$ because the angular deflection must be zero. The torque at any section of the shaft is given by $M = (GI_p)\phi'$, where GI_p is the torsional rigidity of the shaft; thus, for a free end, $\phi' = 0$ and $X' = 0$. For the longitudinal vibration of a rod, the boundary conditions are essentially the same; i.e., for a built-in end the displacement is zero ($u = 0$) and for a free end the stress is zero ($u' = 0$).

Example 7.1. The natural frequencies of the torsional vibration of a circular steel rod of 2-in. diameter and 24-in. length, having the left end built in and the right end free, are to be determined.

Solution. The built-in end at the left gives the condition $X = 0$ at $x = 0$ so that $D = 0$ in Eq. (7.6). The free end at the right gives the condition $X' = 0$ at $x = l$. For each mode of vibration, Eq. (7.6) is $\cos \frac{\omega_n l}{b} = 0$ from which $\frac{\omega_n l}{b} = \frac{\pi}{2}, \frac{3\pi}{2}, \frac{5\pi}{2}, \ldots$. Since $b^2 = \frac{Gg}{\gamma}$, the natural frequencies for the torsional vibration are

$$\omega_n = \frac{\pi}{2l}\sqrt{\frac{Gg}{\gamma}},\ \frac{3\pi}{2l}\sqrt{\frac{Gg}{\gamma}},\ \frac{5\pi}{2l}\sqrt{\frac{Gg}{\gamma}},\ \ldots \quad \text{rad/sec}$$

For steel, $G = 11.5 \times 10^6$ lb/in.^{2} and $\gamma = 0.28$ lb/in.^{3} The fundamental natural frequency is

$$\omega_n = \frac{\pi}{2(24)} \sqrt{\frac{(11.5 \times 10^6)(386)}{0.28}} = 8{,}240 \text{ rad/sec} = 1{,}311 \text{ cps}$$

The remaining frequencies are 3, 5, 7, etc., times ω_n.

Since Eq. (7.4), which governs longitudinal vibration of the bar, is of the same form as Eq. (7.5), which governs torsional vibration, the solution for longitudinal vibration is the same as Eq. (7.7) with u substituted for ϕ and $a = \sqrt{Eg/\gamma}$ substituted for b. The natural frequencies of a uniform rod having one end built in and one end free are obtained by substituting a for b in the frequency equations found above in Example 7.1:

$$\omega_n = \frac{\pi}{2l}\sqrt{\frac{Eg}{\gamma}},\ \frac{3\pi}{2l}\sqrt{\frac{Eg}{\gamma}},\ \frac{5\pi}{2l}\sqrt{\frac{Eg}{\gamma}},\ \ldots$$

The frequencies of the longitudinal vibration are independent of the lateral dimensions of the bar, so that these results apply to uniform noncircular bars. Equation (7.5) for torsional vibration is valid only for circular cross sections.

TORSIONAL VIBRATIONS OF CIRCULAR RODS WITH DISCS ATTACHED. An important type of system is that in which a rod which may twist has mounted on it one or more rigid discs or members that can be considered as the equivalents of discs. Many systems can be approximated by such configurations. If the moment of inertia of the rod is small compared to the moments of inertia of the discs, the mass of the rod may be neglected and the system considered to have a finite number of degrees-of-freedom. Then the methods described in Chaps. 2 and 38 are applicable. Even if the moment of inertia of the rod is not negligible, it usually may be lumped with the moment of inertia of the disc. For a shaft having a single disc attached, the formula in Table 7.2 gives a close approximation to the true frequency.

The exact solution of the problem requires that the effect of the distributed mass of the rod be considered. Usually it can be assumed that the discs are rigid enough that their elasticity can be neglected; only such systems are considered. Equation (7.5) and its solution, Eq. (7.7), apply to the shaft where the constants are determined by the end conditions. If there are more than two discs, the section of shaft between each pair of discs must be considered separately; there are two constants for each section. The constants are determined from the following conditions:

1. For a disc at an end of the shaft, the torque of the shaft at the disc is equal to the product of the moment of inertia of the disc and its angular acceleration.
2. Where a disc is between two sections of shaft, the angular deflection at the end of each section adjoining the disc is the same; the difference between the torques in the two sections is equal to the product of the moment of inertia of the disc and its angular acceleration.

Example 7.2. The fundamental frequency of vibration of the system shown in Fig. 7.2 is to be calculated and the result compared with the frequency obtained by considering that each half of the system is a simple shaft-disc system with the end of the shaft fixed. The system consists of a steel shaft 24 in. long and 4 in. in diameter having attached to it at each end a rigid steel disc 12 in. in diameter and 2 in. thick. For the approximation, add one-third of the moment of inertia of half the shaft to that of the disc (Table 7.2). (Because of symmetry, the center of the shaft is a nodal point; i.e., it does not move. Thus, each half of the system can be considered as a rod-disc system.)

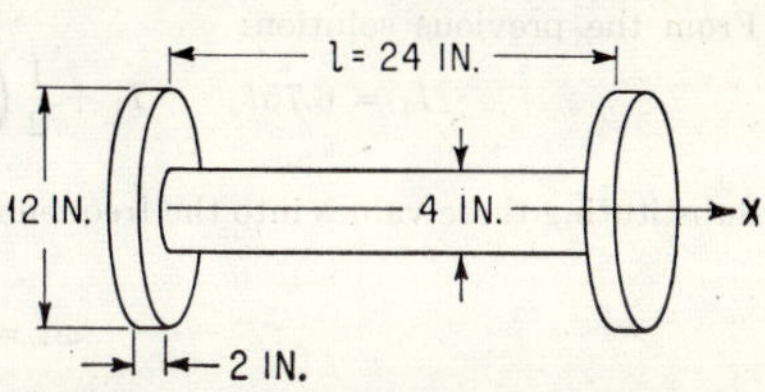

Fig. 7.2. Rod with disc attached at each end.

Exact Solution. The boundary conditions are: at $x = 0$, $M = GI_p\phi' = I_1\ddot{\phi}$; at $x = l$, $M = GI\phi' = -I_2\ddot{\phi}$, where I_1 and I_2 are the moments of inertia of the discs. The signs are opposite for the two boundary conditions because, if the shaft is twisted in a certain direction, it will tend to accelerate the disc at the left end in one direction and the disc at the right end in the other. In the present example, $I_1 = I_2$; however, the solution is carried out in general terms.

7-8 VIBRATION OF SYSTEMS HAVING DISTRIBUTED MASS AND ELASTICITY

Using Eq. (7.7), the following is obtained for each value of n:

$$\phi' = \frac{\omega_n}{b}\left(C \cos \frac{\omega_n x}{b} - D \sin \frac{\omega_n x}{b}\right) \cos(\omega_n t + \theta)$$

$$\ddot{\phi} = \omega_n^2 \left(C \sin \frac{\omega_n x}{b} + D \cos \frac{\omega_n x}{b}\right)[-\cos(\omega_n t + \theta)]$$

The boundary conditions give the following:

$$GI_p \frac{\omega_n}{b} C = -\omega_n^2 D I_1 \quad \text{or} \quad C = -\frac{b\omega_n I_1}{GI_p} D$$

$$\frac{\omega_n}{b} GI_p \left(C \cos \frac{\omega_n l}{b} - D \sin \frac{\omega_n l}{b}\right) = \omega_n^2 I_2 \left(C \sin \frac{\omega_n l}{b} + D \cos \frac{\omega_n l}{b}\right)$$

These two equations can be combined to give

$$-\frac{\omega_n}{b} GI_p \left(\frac{b\omega_n I_1}{GI_p} \cos \frac{\omega_n l}{b} + \sin \frac{\omega_n l}{b}\right) = \omega_n^2 I_2 \left(-\frac{b\omega_n I_1}{GI_p} \sin \frac{\omega_n l}{b} + \cos \frac{\omega_n l}{b}\right)$$

The preceding equation can be reduced to

$$\tan \alpha_n = \frac{(c+d)\alpha_n}{cd\alpha_n^2 - 1} \tag{7.8}$$

where $\alpha_n = (\omega_n l)/b$, $c = I_1/I_s$, $d = I_2/I_s$, and I_s is the polar moment of inertia of the shaft as a rigid body. There is a value for X in Eq. (7.6) corresponding to each root of Eq. (7.8) so that Eq. (7.7) becomes

$$\theta = \sum_{n=1}^{n=\infty} A_n \left(\cos \frac{\omega_n x}{b} - c\alpha_n \sin \frac{\omega_n x}{b}\right) \cos(\omega_n t + \theta_n)$$

For a circular disc or shaft, $I = \frac{1}{2}mr^2$ where m is the total mass; thus $c = d = \frac{D^4}{d^4}\frac{h}{l} = 6.75$. Equation (7.8) becomes $(45.56\alpha_n^2 - 1) \tan \alpha_n = 13.5\alpha_n$, the lowest root of which is $\alpha_n = 0.538$. The natural frequency is $\omega_n = 0.538\sqrt{Gg/\gamma l^2}$ rad/sec.

Approximate Solution. From Table 7.2, the approximate formula is

$$\omega_n = \left(\frac{k_r}{I + I_s/3}\right)^{1/2} \quad \text{where } k_r = \frac{\pi d^4}{32}\frac{G}{l}$$

For the present problem where the center of the shaft is a node, the values of moment of inertia I_s and torsional spring constant for half the shaft must be used:

$$\tfrac{1}{2}I_s = \frac{\pi d^4}{32}\frac{\gamma}{g}\frac{l}{2} \quad \text{and} \quad k_r = 2\left[\frac{\pi d^4}{32}\frac{G}{l}\right]$$

From the previous solution:

$$I_1 = 6.75 I_s \qquad I_1 + \frac{1}{2}\left(\frac{I_s}{3}\right) = \frac{\pi d^4}{32}\frac{\gamma}{g}\frac{l}{2}[2(6.75) + 0.333]$$

Substituting these values into the frequency equation and simplifying gives

$$\omega_n = 0.538 \sqrt{\frac{Gg}{\gamma l^2}}$$

In this example, the approximate solution is correct to at least three significant figures. For larger values of I_s/I, poorer accuracy can be expected.

For steel, $G = 11.5 \times 10^6$ lb/in.^{2} and $\gamma = 0.28$ lb/in.^{3}; thus

$$\omega_n = 0.538 \sqrt{\frac{(11.5 \times 10^6)(386)}{(0.28)(24)^2}} = 0.538 \times 5{,}245 = 2{,}822 \text{ rad/sec} = 449 \text{ cps}$$

FREE VIBRATION

LONGITUDINAL VIBRATION OF A ROD WITH MASS ATTACHED. The natural frequencies of the longitudinal vibration of a uniform rod having rigid masses attached to it can be solved in a manner similar to that used for a rod in torsion with discs attached. Equation (7.4) applies to this system; its solution is the same as Eq. (7.7) with a substituted for b. For each value of n,

$$u = \left(C_n \sin \frac{\omega_n x}{a} + D_n \cos \frac{\omega_n x}{a} \right) \cos(\omega_n t + \theta)$$

In Fig. 7.3, the rod of length l is fixed at $x = 0$ and has a mass m_2 attached at $x = l$. The boundary conditions are: at $x = 0$, $u = 0$ and at $x = l$, $SEu' = -m_2\ddot{u}$. The latter expresses the condition that the force in the bar equals the product of the mass and its acceleration at the end with the mass attached. The sign is negative because the force is tensile or positive when the acceleration of the mass is negative. From the first boundary condition, $D_n = 0$. The second boundary condition gives

$$\frac{\omega_n SE}{a} C_n \cos \frac{\omega_n l}{a} = m_2 \omega_n^2 C_n \sin \frac{\omega_n l}{a}$$

from which

$$\frac{SEl}{m_2 a^2} = \frac{\omega_n l}{a} \tan \frac{\omega_n l}{a}$$

Since $a^2 = Eg/\gamma$, this can be written

$$\frac{m_1}{m_2} = \frac{\omega_n l}{a} \tan \frac{\omega_n l}{a}$$

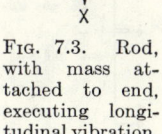

Fig. 7.3. Rod, with mass attached to end, executing longitudinal vibration.

where m_1 is the mass of the rod. This equation can be applied to a simple mass-spring system by using the relation that the constant k of a spring is equivalent to SE/l for the rod, so that $l/a = (m_1/k)^{1/2}$, where m_1 is the mass of the spring:

$$\frac{m_1}{m_2} = \omega_n \sqrt{\frac{m_1}{k}} \tan \omega_n \sqrt{\frac{m_1}{k}} \qquad (7.9)$$

Rayleigh's Method. An accurate approximation to the fundamental natural frequency of this system can be found by using Rayleigh's method. The motion of the mass can be expressed as $u_m = u_0 \sin \omega t$. If it is assumed that the deflection u at each section of the rod is proportional to its distance from the fixed end, $u = u_0(x/l) \sin \omega_n t$. Using this relation in the appropriate equation from Table 7.1, the strain energy V of the rod at maximum deflection is

$$V = \frac{SE}{2} \int_0^l \left(\frac{\partial u}{\partial x} \right)^2 dx = \frac{SE}{2} \int_0^l \left(\frac{u_0}{l} \right)^2 dx = \frac{SEu_0^2}{2l}$$

The maximum kinetic energy T of the rod is

$$T = \frac{S\gamma}{2g} \int_0^l V^2_{\max} dx = \frac{S\gamma}{2g} \int_0^l \left(\omega_n u_0 \frac{x}{l} \right)^2 dx = \frac{S\gamma}{2g} \omega_n^2 u_0^2 \frac{l}{3}$$

The maximum kinetic energy of the mass is $T_m = m_2 \omega_n^2 u_0^2/2$. Equating the total maximum kinetic energy $T + T_m$ to the maximum strain energy V gives

$$\omega_n = \left(\frac{SE}{l(m_2 + m_1/3)} \right)^{1/2}$$

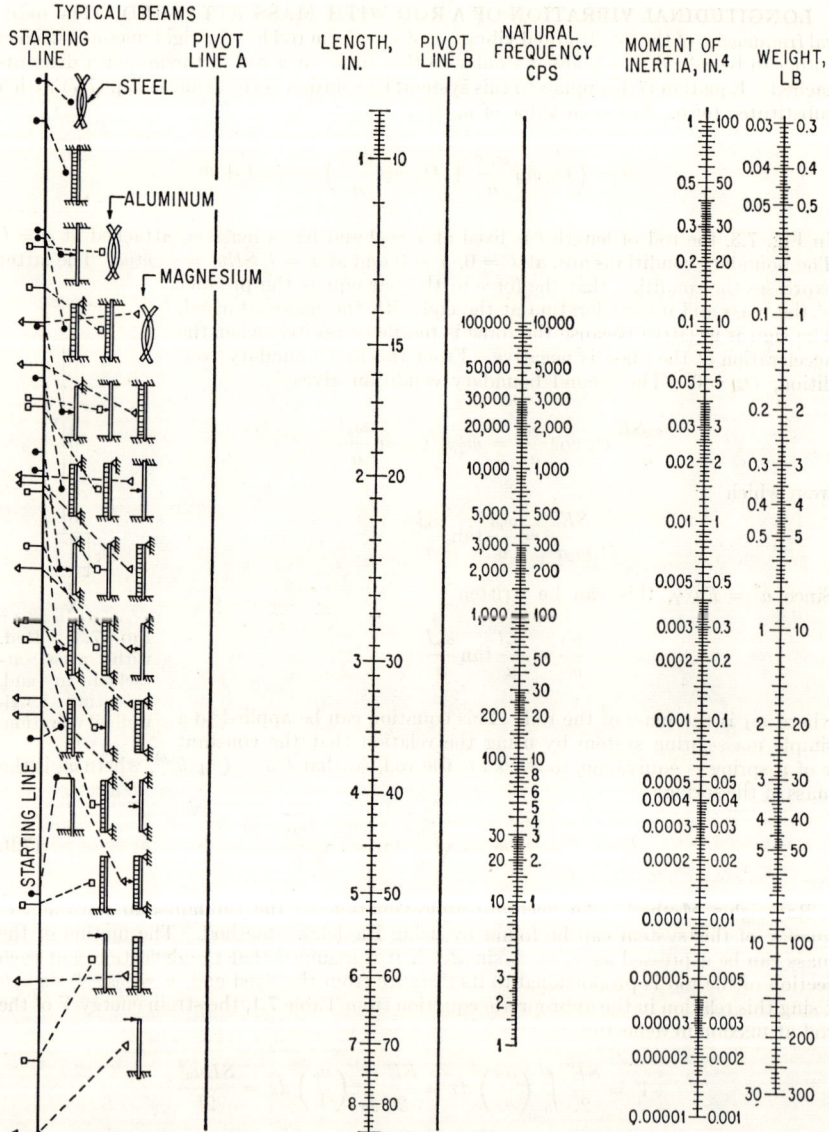

FIG. 7.4. Nomograph for determining fundamental natural frequencies of beams. From the point on the starting line which corresponds to the loading and support conditions for the beam, a straight line is drawn to the proper point on the length line. (If the length appears on the left side of this line, subsequent readings on all lines are made to the left; and if the length appears to the right, subsequent readings are made to the right.) From the intersection of this line with pivot line A, a straight line is drawn to the moment of inertia line; from the intersection of this line with pivot line B, a straight line is drawn to the weight line. (For concentrated loads, the weight is that of the load; for uniformly distributed loads, the weight is the total load on the beam, including the weight of the beam.) The natural frequency is read where the last line crosses the natural frequency line. (*J. J. Kerley.*[7])

where $m_1 = S\gamma l/g$ is the mass of the rod. Letting $SE/l = k$,

$$\omega_n = \sqrt{\frac{k}{M + m/3}} \qquad (7.10)$$

This formula is included in Table 7.2. The other formulas in that table are also based on analyses by the Rayleigh method.

Example 7.3. The natural frequency of a simple mass-spring system for which the weight of the spring is equal to the weight of the mass is to be calculated and compared to the result obtained by using Eq. (7.10).

Solution. For $m_1/m_2 = 1$, the lowest root of Eq. (7.9) is $\omega_n\sqrt{m/k} = 0.860$. When $m_2 = m_1$,

$$\omega_n = 0.860 \sqrt{\frac{k}{m_2}}$$

Using the approximate equation,

$$\omega_n = \sqrt{\frac{k}{m_2(1 + \frac{1}{3})}} = 0.866 \sqrt{\frac{k}{m_2}}$$

LATERAL VIBRATION OF STRAIGHT BEAMS

Natural Frequencies from Nomograph. For many practical purposes the natural frequencies of uniform beams of steel, aluminum, and magnesium can be determined with sufficient accuracy by the use of the nomograph, Fig. 7.4. This nomograph applies to many conditions of support and several types of load. Figure 7.4A indicates the procedure for using the nomograph.

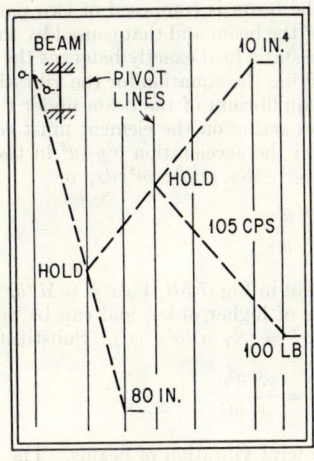
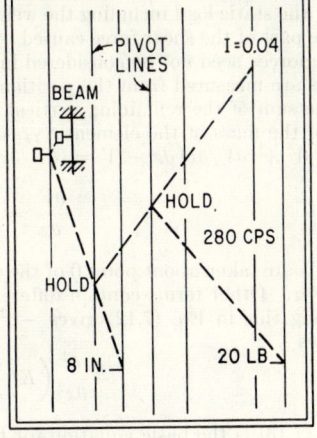

GIVEN:
1. E = 30×10⁶ (STEEL) PSI
2. BEAM: DOUBLE END FIXITY
3. I = 10 IN.⁴
4. l = 80 IN.
5. W = 100 LB

GIVEN:
1. E = 10.5×10⁶ (ALUM) PSI
2. I = 0.04 IN.⁴
3. l = 8 IN.
4. W = 20 LB
5. DOUBLE END FIXITY

FIG. 7.4A. Example of use of Fig. 7.4. The natural frequency of the steel beam is 105 cps and that of the aluminum beam is 280 cps. (*J. J. Kerley.*[7])

7–12 VIBRATION OF SYSTEMS HAVING DISTRIBUTED MASS AND ELASTICITY

Classical Solution. In the derivation of the necessary equation, use is made of the relation

$$EI \frac{d^2 y}{dx^2} = M \tag{7.11}$$

This equation relates the curvature of the beam to the bending moment at each section of the beam. This equation is based upon the assumptions that the material is homogeneous, isotropic and obeys Hooke's law, and that the beam is straight and of uniform cross section. The equation is valid for small deflections only, and for beams that are long compared to cross-sectional dimensions since the effects of shear deflection are neg-

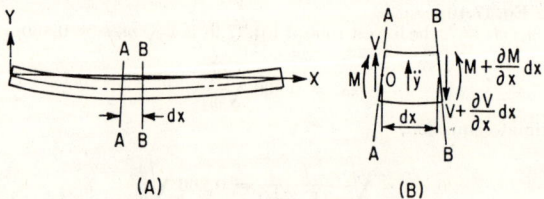

FIG. 7.5. (A) Beam executing lateral vibration. (B) Element of beam showing shear forces and bending moments.

lected. The effects of shear deflection and rotation of the cross sections are considered later.

The equation of motion for lateral vibration of the beam shown in Fig. 7.5A is found by considering the forces acting on the element, Fig. 7.5B, which is formed by passing two parallel planes A–A and B–B through the beam normal to the longitudinal axis. The vertical elastic shear force acting on section A–A is V, and that on section B–B is $V + (\partial V/\partial x)\, dx$. Shear forces acting as shown are considered to be positive. The total vertical elastic shear force at each section of the beam is composed of two parts: that caused by the static load including the weight of the beam and that caused by the vibration. The part of the shear force caused by the static load exactly balances the load, so that these forces need not be considered in deriving the equation for the vibration if all deflections are measured from the position of equilibrium of the beam under the static load. The sum of the remaining vertical forces acting on the element must equal the product of the mass of the element $S\gamma/g\, dx$ and the acceleration $\partial^2 y/\partial t^2$ in the lateral direction: $V + (\partial V/\partial x)\, dx - V = (\partial V/\partial x)\, dx = -(S\gamma/g)(\partial^2 y/\partial t^2)\, dx$, or

$$\frac{\partial V}{\partial x} = -\frac{\gamma S}{g}\frac{\partial^2 y}{\partial t^2} \tag{7.12}$$

If moments are taken about point 0 of the element in Fig. 7.5B, $V\, dx = (\partial M/\partial x)\, dx$, and $V = \partial M/\partial x$. Other terms contain differentials of higher order, and can be neglected. Substituting this in Eq. (7.12) gives $-\partial^2 M/\partial x^2 = (S\gamma/g)(\partial^2 y/\partial t^2)$. Substituting Eq. (7.11) gives

$$-\frac{\partial^2}{\partial x^2}\left(EI \frac{\partial^2 y}{\partial x^2}\right) = \frac{\gamma S}{g}\frac{\partial^2 y}{\partial t^2} \tag{7.13}$$

Equation (7.13) is the basic equation for the lateral vibration of beams. The solution of this equation, if EI is constant, is of the form $y = X(x)\,[\cos(\omega_n t + \theta)]$, in which X is a function of x only. Substituting

$$\kappa^4 = \frac{\omega_n^2 \gamma S}{EIg} \tag{7.14}$$

and dividing Eq. (7.13) by $\cos(\omega_n t + \theta)$:

$$\frac{d^4 X}{dx^4} = \kappa^4 X \tag{7.15}$$

where X is any function whose fourth derivative is equal to a constant multiplied by the function itself. The following functions satisfy the required conditions and represent the solution of the equation:

$$X = A_1 \sin \kappa x + A_2 \cos \kappa x + A_3 \sinh \kappa x + A_4 \cosh \kappa x$$

The solution can also be expressed in terms of exponential functions, but the trigonometric and hyperbolic functions usually are more convenient to use.

For beams having various support conditions, the constants A_1, A_2, A_3, and A_4 are found from the end conditions. In finding the solutions, it is convenient to write the equation in the following form in which two of the constants are zero for each of the usual boundary conditions:

$$X = A(\cos \kappa x + \cosh \kappa x) + B(\cos \kappa x - \cosh \kappa x)$$
$$+ C(\sin \kappa x + \sinh \kappa x) + D(\sin \kappa x - \sinh \kappa x) \quad (7.16)$$

In applying the end conditions, the following relations are used where primes indicate successive derivatives with respect to x:

The deflection is proportional to X and is zero at any rigid support.
The slope is proportional to X' and is zero at any built-in end.
The moment is proportional to X'' and is zero at any free or hinged end.
The shear is proportional to X''' and is zero at any free end.

The required derivatives are:

$$X' = \kappa[A(-\sin \kappa x + \sinh \kappa x) + B(-\sin \kappa x - \sinh \kappa x) \\ + C(\cos \kappa x + \cosh \kappa x) + D(\cos \kappa x - \cosh \kappa x)]$$

$$X'' = \kappa^2[A(-\cos \kappa x + \cosh \kappa x) + B(-\cos \kappa x - \cosh \kappa x) \\ + C(-\sin \kappa x + \sinh \kappa x) + D(-\sin \kappa x - \sinh \kappa x)]$$

$$X''' = \kappa^3[A(\sin \kappa x + \sinh \kappa x) + B(\sin \kappa x - \sinh \kappa x) \\ + C(-\cos \kappa x + \cosh \kappa x) + D(-\cos \kappa x - \cosh \kappa x)]$$

For the usual end conditions, two of the constants are zero, and there remain two equations containing two constants. These can be combined to give an equation which contains only the frequency as an unknown. Using the frequency, one of the unknown constants can be found in terms of the other. There always is one undetermined constant, which can be evaluated only if the amplitude of the vibration is known.

Example 7.4. The natural frequencies and modes of vibration of the rectangular steel beam shown in Fig. 7.6 are to be determined and the fundamental frequency compared with that obtained from Fig. 7.4. The beam is 24 in. long, 2 in. wide, and ¼ in. thick, with the left end built in and the right end free.

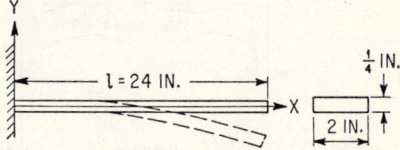

Fig. 7.6. First mode of vibration of beam with left end clamped and right end free.

Solution. The boundary conditions are: at $x = 0$, $X = 0$ and $X' = 0$; at $x = l$, $X'' = 0$ and $X''' = 0$. The first condition requires that $A = 0$ since the other constants are multiplied by zero at $x = 0$. The second condition requires that $C = 0$. From the third and fourth conditions, the following equations are obtained:

$$0 = B(-\cos \kappa l - \cosh \kappa l) + D(-\sin \kappa l - \sinh \kappa l)$$
$$0 = B(\sin \kappa l - \sinh \kappa l) + D(-\cos \kappa l - \cosh \kappa l)$$

Solving each of these for the ratio D/B and equating, or making use of the mathematical condition that for a solution the determinant of the two equations must vanish, the following equation results:

$$\frac{D}{B} = -\frac{\cos \kappa l + \cosh \kappa l}{\sin \kappa l + \sinh \kappa l} = \frac{\sin \kappa l - \sinh \kappa l}{\cos \kappa l + \cosh \kappa l} \quad (7.17)$$

7-14 VIBRATION OF SYSTEMS HAVING DISTRIBUTED MASS AND ELASTICITY

Table 7.3. Natural Frequencies and Normal Modes of Uniform Beams

SUPPORTS	MODE n	(A) SHAPE AND NODES (NUMBERS GIVE LOCATION OF NODES IN FRACTION OF LENGTH FROM LEFT END)	(B) BOUNDARY CONDITIONS EQ (7.16)	(C) FREQUENCY EQUATION	(D) CONSTANTS EQ (7.16)	(E) κl EQ (7.14) $\omega_n = \kappa^2 \sqrt{\frac{EIg}{A\gamma}}$	(F) R RATIO OF NON-ZERO CONSTANTS COLUMN (D)
HINGED-HINGED	1		$x=0 \begin{cases} X=0 \\ X''=0 \end{cases}$	$\sin \kappa l = 0$	$A=0$	3.1416	1.0000
	2	0.50			$B=0$	6.283	1.0000
	3	0.333 0.667			$\frac{C}{D}=1$	9.425	1.0000
	4	0.25 0.50 0.75	$x=l \begin{cases} X=0 \\ X''=0 \end{cases}$			12.566	1.0000
	$n>4$					$\approx n\pi$	1.0000
CLAMPED-CLAMPED	1		$x=0 \begin{cases} X=0 \\ X'=0 \end{cases}$	$(\cos \kappa l)(\cosh \kappa l) = 1$	$A=0$	4.730	-0.9825
	2	0.50			$C=0$	7.853	-1.0008
	3	0.359 0.641			$\frac{D}{B}=R$	10.996	$-1.0000-$
	4	0.278 0.50 0.722	$x=l \begin{cases} X=0 \\ X'=0 \end{cases}$			14.137	$-1.0000+$
	$n>4$					$\approx \frac{(2n+1)\pi}{2}$	$-1.0000-$
CLAMPED-HINGED	1		$x=0 \begin{cases} X=0 \\ X'=0 \end{cases}$	$\tan \kappa l = \tanh \kappa l$	$A=0$	3.927	-1.0008
	2	0.558			$C=0$	7.069	$-1.0000+$
	3	0.386 0.692			$\frac{D}{B}=R$	10.210	-1.0000
	4	0.294 0.529 0.765	$x=l \begin{cases} X=0 \\ X''=0 \end{cases}$			13.352	-1.0000
	$n>4$					$\approx \frac{(4n+1)\pi}{4}$	-1.0000
CLAMPED-FREE	1		$x=0 \begin{cases} X=0 \\ X'=0 \end{cases}$	$(\cos \kappa l)(\cosh \kappa l) = -1$	$A=0$	1.875	-0.7341
	2	0.783			$C=0$	4.694	-1.0185
	3	0.504 0.868			$\frac{D}{B}=R$	7.855	-0.9992
	4	0.358 0.644 0.906	$x=l \begin{cases} X''=0 \\ X'''=0 \end{cases}$			10.996	$-1.0000+$
	$n>4$					$\approx \frac{(2n-1)\pi}{2}$	$-1.0000-$
FREE-FREE	1	0.224 0.776	$x=0 \begin{cases} X''=0 \\ X'''=0 \end{cases}$	$(\cos \kappa l)(\cosh \kappa l) = 1$	$B=0$	0 (REPRESENTS TRANSLATION)	
	2	0.132 0.50 0.868			$D=0$	4.730	-0.9825
	3	0.094 0.356 0.644 0.906			$\frac{C}{A}=R$	7.853	-1.0008
	4	0.0734 0.277 0.50 0.723 0.927	$x=l \begin{cases} X''=0 \\ X'''=0 \end{cases}$			10.996	$-1.0000-$
	5					14.137	$-1.0000+$
	$n>5$					$\approx \frac{(2n-1)\pi}{2}$	$-1.0000-$

Equation (7.17) reduces to $\cos \kappa l \cosh \kappa l = -1$. The values of κl which satisfy this equation can be found by consulting tables of hyperbolic and trigonometric functions. The first five are: $\kappa_1 l = 1.875$, $\kappa_2 l = 4.694$, $\kappa_3 l = 7.855$, $\kappa_4 l = 10.996$, $\kappa_5 l = 14.137$. The corresponding frequencies of vibration are found by substituting the length of the beam to find each κ, and then solving Eq. (7.14) for ω_n:

$$\omega_n = \kappa_n^2 \sqrt{\frac{EIg}{\gamma S}}$$

FREE VIBRATION

For the rectangular section, $I = bh^3/12 = 1/384$ in.^{4} and $S = bh = 0.5$ in.^{2} For steel, $E = 30 \times 10^6$ lb/in.^{2} and $\gamma = 0.28$ lb/in.^{3} Using these values,

$$\omega_1 = \frac{(1.875)^2}{(24)^2} \sqrt{\frac{(30 \times 10^6)(386)}{(0.28)(384)(0.5)}} = 89.6 \text{ rad/sec} = 14.26 \text{ cps}$$

The remaining frequencies can be found by using the other values of κ. Using Fig. 7.4, the fundamental frequency is found to be about 12 cps.

To find the mode shapes, the ratio D/B is found by substituting the appropriate values of κl in Eq. (7.17). For the first mode:

$$\cosh 1.875 = 3.33710 \qquad \sinh 1.875 = 3.18373$$
$$\cos 1.875 = -0.29953 \qquad \sin 1.875 = 0.95409$$

Therefore, $D/B = -0.73410$. The equation for the first mode of vibration becomes

$$y = B_1[(\cos \kappa x - \cosh \kappa x) - 0.73410 (\sin \kappa x - \sinh \kappa x)] \cos (\omega_1 t + \theta_1)$$

in which B_1 is determined by the amplitude of vibration in the first mode. A similar equation can be obtained for each of the modes of vibration; all possible free vibration of the beam can be expressed by taking the sum of these equations.

Frequencies and Shapes of Beams. Table 7.3 gives the information necessary for finding the natural frequencies and normal modes of vibration of uniform beams having various boundary conditions. The various constants in the table were determined by computations similar to those used in Example 7.4. The table includes (1) diagrams showing the modal shapes including node locations; (2) the boundary conditions; (3) the frequency equation that results from using the boundary conditions in Eq. (7.16); (4) the constants that become zero in Eq. (7.16); (5) the values of κl from which the natural frequencies can be computed by using Eq. (7.14); and (6) the ratio of the nonzero constants in Eq. (7.16). By the use of the constants in this table, the equation of motion for any normal mode can be written. There always is a constant which is determined by the amplitude of vibration.

Values of characteristic functions representing the deflections of beams, at fifty equal intervals, for the first five modes of vibration have been tabulated.[8] Functions are given for beams having various boundary conditions, and the first three derivatives of the functions are also tabulated.

Rayleigh's Method. This method is useful for finding approximate values of the fundamental natural frequencies of beams. In applying Rayleigh's method, a suitable function is assumed for the deflection, and the maximum strain and kinetic energies are calculated, using the equations in Table 7.1. These energies are equated and solved for the frequency. The function used to represent the shape must satisfy the boundary conditions associated with deflection and slope at the supports. Best accuracy is obtained if other boundary conditions are also satisfied. The equation for the static deflection of the beam under a uniform load is a suitable function, although a simpler function often gives satisfactory results with less numerical work.

Example 7.5. The fundamental natural frequency of the cantilever beam in Example 7.4 is to be calculated using Rayleigh's method.

Solution. The assumed deflection $Y = (a/3l^4)[x^4 - 4x^3l + 6x^2l^2]$ is the static deflection of a cantilever beam under uniform load and having the deflection $Y = a$ at $x = l$. This deflection satisfies the conditions that the deflection Y and the slope Y' be zero at $x = 0$. Also, at $x = l$, Y'' which is proportional to the moment and Y''' which is proportional to the shear are zero. The second derivative of the function is $Y'' = (4a/l^4)[x^2 - 2xl + l^2]$. Using this in the expression from Table 7.1, the maximum strain energy is

$$V = \frac{EI}{2} \int_0^l \left(\frac{d^2Y}{dx^2}\right)^2 dx = \frac{8}{5} \frac{EIa^2}{l^3}$$

The maximum kinetic energy is

$$T = \frac{\omega_n^2 \gamma S}{2g} \int_0^l Y^2 dx = \frac{52}{405} \frac{\omega_n^2 \gamma S l a^2}{g}$$

7-16 VIBRATION OF SYSTEMS HAVING DISTRIBUTED MASS AND ELASTICITY

Equating the two energies and solving for the frequency,

$$\omega_n = \sqrt{\frac{162}{13} \times \frac{EIg}{\gamma Sl^4}} = \frac{3.530}{l^2}\sqrt{\frac{EIg}{\gamma S}}$$

The exact frequency as found in Example 4 is $(3.516/l^2)\sqrt{EIg/\gamma S}$; thus, Rayleigh's method gives good accuracy in this example.

If the deflection is assumed to be $Y = a[1 - \cos(\pi x/2l)]$, the calculated frequency is $(3.66/l^2)\sqrt{EIg/\gamma S}$. This is less accurate, but the calculations are considerably shorter. With this function, the same boundary conditions at $x = 0$ are satisfied; however, at $x = l$, $Y'' = 0$ but Y''' does not equal zero. Thus, the condition of zero shear at the free end is not satisfied. The trigonometric function would not be expected to give as good accuracy as the static deflection relation used in the example, although for most practical purposes the result would be satisfactory.

EFFECTS OF ROTARY MOTION AND SHEARING FORCE. In the preceding analysis of the lateral vibration of beams it has been assumed that each element of the beam moves only in the lateral direction. If each plane section that is initially normal to the axis of the beam remains plane and normal to the axis, as assumed in simple beam theory, then each section rotates slightly in addition to its lateral motion when the beam deflects.[9] When a beam vibrates, there must be forces to cause this rotation, and for a complete analysis these forces must be considered. The effect of this rotation is small except when the curvature of the beam is large relative to its thickness; this is true either for a beam that is short relative to its thickness or for a long beam vibrating in a higher mode so that the nodal points are close together.

Another factor that affects the lateral vibration of a beam is the lateral shear force. In Eq. (7.11) only the deflection associated with the bending stress in the beam is included. In any beam except one subject only to pure bending, a deflection due to the shear stress in the beam occurs. The exact solution of the beam vibration problem requires that this deflection be considered. The analysis of beam vibration including both the effects of rotation of the cross section and the shear deflection is called the "Timoshenko beam theory." The following equation governs such vibration:[10]

$$a^2 \frac{\partial^4 y}{\partial x^4} + \frac{\partial^2 y}{\partial t^2} - \rho^2\left(1 + \frac{E}{\varkappa G}\right)\frac{\partial^4 y}{\partial x^2 \partial t^2} + \rho^2 \frac{\gamma}{g \varkappa G}\frac{\partial^4 y}{\partial t^4} = 0 \qquad (7.18)$$

where $a^2 = EIg/S\gamma$, E = modulus of elasticity, G = modulus of rigidity, and $\rho = \sqrt{I/S}$, the radius of gyration; $\varkappa = F_s/GS\beta$, F_s being the total lateral shear force at any section and β the angle which a cross section makes with the axis of the beam because of shear deformation. Under the assumptions made in the usual elementary beam theory, $\varkappa$ is $\frac{2}{3}$ for a beam with a rectangular cross section and $\frac{3}{4}$ for a circular beam. More refined analysis shows[11] that, for the present purposes, $\varkappa = \frac{5}{6}$ and $\frac{9}{10}$ are more accurate values for rectangular and circular cross sections, respectively. Using a solution of the form $y = C \sin(n\pi x/l) \cos \omega_n t$, which satisfies the necessary end conditions, the following frequency equation is obtained for beams with both ends simply supported:

$$a^2 \frac{n^4 \pi^4}{l^4} - \omega_n^2 - \omega_n^2 \frac{n^2 \pi^2 \rho^2}{l^2} - \omega_n^2 \frac{n^2 \pi^2 \rho^2}{l^2}\frac{E}{\varkappa G} + \frac{\rho^2 \gamma}{g \varkappa G}\omega_n^4 = 0 \qquad (7.18a)$$

If it is assumed that $nr/l \ll 1$, Eq. (7.18a) reduces to

$$\omega_n = \frac{a\pi^2}{(l/n)^2}\left[1 - \frac{\pi^2 n^2}{2}\left(\frac{\rho}{l}\right)^2\left(1 + \frac{E}{\varkappa G}\right)\right] \qquad (7.18b)$$

When $nr/l < 0.08$, the approximate equation gives less than 5 per cent error in the frequency.[11]

Values of the ratio of ω_n to the natural frequency uncorrected for the effects of rotation and shear have been plotted,[11] using Eq. (7.18a) for three values of $E/\varkappa G$, and are shown in Fig. 7.7.

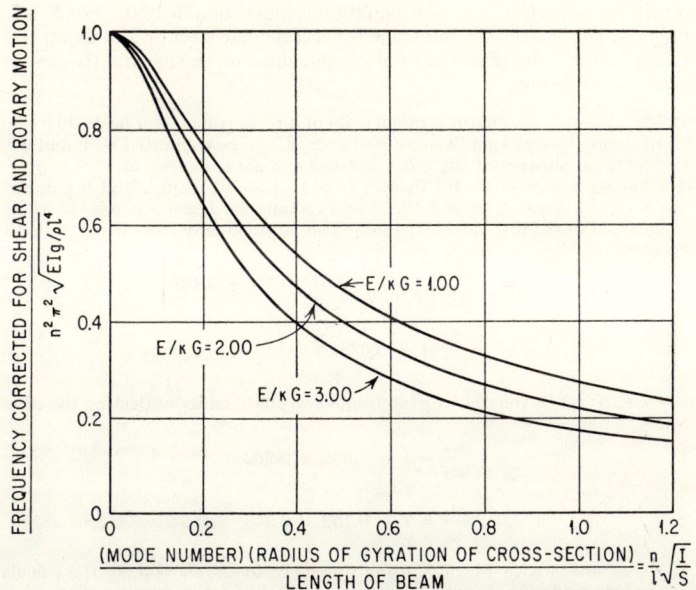

FIG. 7.7. Influence of shear force and rotary motion on natural frequencies of simply supported beams. The curves relate the corrected frequency to that given by Eq. (7.14). (*J. G. Sutherland and L. E. Goodman.*[11])

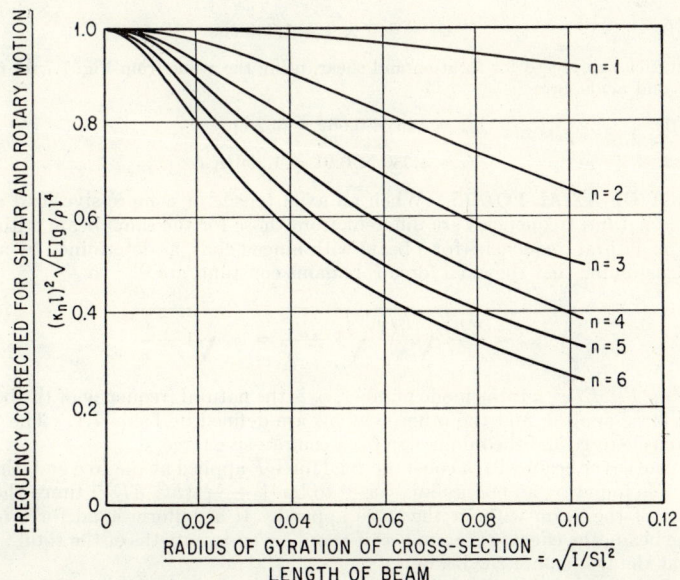

FIG. 7.8. Influence of shear force and rotary motion on natural frequencies of uniform cantilever beams ($E/\kappa G = 3.20$). The curves relate the corrected frequency to that given by Eq. (7.14). (*J. G. Sutherland and L. E. Goodman.*[11])

7-18 VIBRATION OF SYSTEMS HAVING DISTRIBUTED MASS AND ELASTICITY

For a cantilever beam the frequency equation is quite complicated. For $E/\varkappa G = 3.20$, corresponding approximately to the value for rectangular steel or aluminum beams, the curves in Fig. 7.8 show the effects of rotation and shear on the natural frequencies of the first six modes of vibration.

Example 7.6. The first two natural frequencies of a rectangular steel beam 40 in. long, 2 in. wide, and 6 in. thick, having simply supported ends, are to be computed with and without including the effects of rotation of the cross sections and shear deflection.

Solution. For steel $E = 30 \times 10^6$ lb/in.^{2}, $G = 11.5 \times 10^6$ lb/in.^{2}, and for a rectangular cross section $\varkappa = 5/6$; thus $E/\varkappa G = 3.13$. For a rectangular beam $\rho = h/\sqrt{12}$ where h is the thickness; thus $\rho/l = 6/(40\sqrt{12}) = 0.0433$. The approximate frequency equation, Eq. (7.18b), becomes

$$\omega_n = \frac{a\pi^2}{(l/n)^2}\left[1 - \frac{\pi^2}{2}(0.0433n)^2(1 + 3.13)\right]$$

$$= \frac{a\pi^2}{(l/n)^2}(1 - 0.038n^2)$$

Letting $\omega_0 = a\pi^2/(l/n)^2$ be the uncorrected frequency obtained by neglecting the effect of n in Eq. (7.18b):

For $n = 1$: $\qquad \dfrac{\omega_n}{\omega_0} = 1 - 0.038 = 0.962$

For $n = 2$: $\qquad \dfrac{\omega_n}{\omega_0} = 1 - 0.152 = 0.848$

Comparing these results with Fig. 7.7, using the curve for $E/\varkappa G = 3.00$, the calculated frequency for the first mode agrees with the curve as closely as the curve can be read. For the second mode, the curve gives $\omega_n/\omega_0 = 0.91$; therefore the approximate equation for the second mode is not very accurate. The uncorrected frequencies are, since $I/S = \rho^2 = h^2/12$:

For $n = 1$: $\qquad \omega_0 = \dfrac{\pi^2}{l^2}\sqrt{\dfrac{EIg}{S\gamma}} = \dfrac{\pi^2}{(40)^2}\sqrt{\dfrac{(30 \times 10^6)(36)386}{(12)(0.28)}} = 2{,}170$ rad/sec $= 345$ cps

For $n = 2$: $\qquad \omega_0 = 345 \times 4 = 1{,}380$ cps

The frequencies corrected for rotation and shear, using the value from Fig. 7.7 for correction of the second mode, are:

For $n = 1$: $\qquad f_n = 345 \times 0.962 = 332$ cps

For $n = 2$: $\qquad f_n = 1{,}380 \times 0.91 = 1{,}256$ cps

EFFECT OF AXIAL LOADS. When an axial tensile or compressive load acts on a beam, the natural frequencies are different from those for the same beam without such load. The natural frequencies for a beam with hinged ends, as determined by an energy analysis, assuming that the axial force F remains constant, are [12]

$$\omega_n = \frac{\pi^2 n^2}{l^2}\sqrt{\frac{EIg}{S\gamma}}\sqrt{1 \pm \frac{\alpha^2}{n^2}} = \omega_0\sqrt{1 \pm \frac{\alpha^2}{n^2}}$$

where $\alpha^2 = Fl^2/EI\pi^2$, n is the mode number, ω_0 is the natural frequency of the beam with no axial force applied, and the other symbols are defined in Table 7.1. The plus sign is for a tensile force and the minus sign for a compressive force.

For a cantilever beam with a constant axial force F applied at the free end, the natural frequency is found by an energy analysis [13] to be $[1 + \tfrac{5}{14}(Fl^2/EI)]^{1/2}$ times the natural frequency of the beam without the force applied. If a uniform axial force is applied along the beam, the effect is the same as if about seven-twentieths of the total force were applied at the free end of the beam.

If the amplitude of vibration is large, an axial force may be induced in the beam by the supports. For example, if both ends of a beam are hinged but the supports are rigid enough so that they cannot move axially, a tensile force is induced as the beam deflects. The force is not proportional to the deflection; therefore, the vibration is of the type characteristic of nonlinear systems in which the natural frequency depends on the amplitude of vibration. The natural frequency of a beam having immovable hinged ends is given in the following table where the axial force is zero at zero deflection of the beam [14] and where x_0 is the amplitude of vibration, I the moment of inertia, and S the area of the cross section; ω_0 is the natural frequency of the unrestrained bar.

$\dfrac{x_0}{\sqrt{I/S}}$	0	0.1	0.2	0.4	0.6	0.8
$\dfrac{\omega_n}{\omega_0}$	1	1.0008	1.0038	1.015	1.038	1.058
$\dfrac{x_0}{\sqrt{I/S}}$	1.0	1.5	2	3	4	5
$\dfrac{\omega_n}{\omega_0}$	1.089	1.190	1.316	1.626	1.976	2.35

BEAMS HAVING VARIABLE CROSS SECTIONS. The natural frequencies for beams of several shapes having cross sections that can be expressed as functions of the distance along the beam have been calculated.[15] The results are shown in Table 7.4. In the analysis, Eq. (7.13) was used, with EI considered to be variable.

Rayleigh's method or Ritz's method can be used to find approximate values for the frequencies of such beams. The frequency equation becomes, using the equations in Table 7.1, and letting $Y(x)$ be the assumed deflection,

$$\omega_n{}^2 = \frac{Eg}{\gamma} \frac{\int_0^l I \left(\dfrac{d^2 Y}{dx^2}\right)^2 dx}{\int_0^l S Y^2 \, dx}$$

where $I = I(x)$ is the moment of inertia of the cross section and $S = S(x)$ is the area of the cross section. Examples of the calculations are in the literature.[18] If the values of $I(x)$ and $S(x)$ cannot be defined analytically, the beam may be divided into two or more sections, for each of which I and S can be approximated by an equation. The strain and kinetic energies of each section may be computed separately, using an appropriate function for the deflection, and the total energies for the beam found by adding the values for the individual sections.

CONTINUOUS BEAMS ON MULTIPLE SUPPORTS. In finding the natural frequencies of a beam on multiple supports, the section between each pair of supports is considered as a separate beam with its origin at the left support of the section. Equation (7.16) applies to each section. Since the deflection is zero at the origin of each section, $A = 0$ and the equation reduces to

$$X = B(\cos \kappa x - \cosh \kappa x) + C(\sin \kappa x + \sinh \kappa x) + D(\sin \kappa x - \sinh \kappa x)$$

There is one such equation for each section, and the necessary end conditions are as follows:

1. At each end of the beam the usual boundary conditions are applicable, depending on the type of support.
2. At each intermediate support the deflection is zero. Since the beam is continuous, the slope and the moment just to the left and to the right of the support are the same.

General equations can be developed for finding the frequency for any number of spans.[19,20] Table 7.5 gives constants for finding the natural frequencies of uniform continuous beams on uniformly spaced supports for several combinations of end supports.

Table 7.4. Natural Frequencies of Variable-section Steel Beams
(J. N. Macduff and R. P. Felgar.[16,17])

BEAM STRUCTURE	$\dfrac{b}{b_0}$	$\dfrac{h}{h_0}$	$(f_n l^2/\rho)/10^4$ $n = 1$	$n = 2$	$n = 3$
(cantilever, constant b, tapered h)	1	$\dfrac{x}{l}$	17.09	48.89	96.57
(cantilever, tapered b and h)	$\dfrac{x}{l}$	$\dfrac{x}{l}$	26.08	68.08	123.64
(cantilever, b varies as sqrt)	$\left(\dfrac{x}{l}\right)^{\frac{1}{2}}$	$\dfrac{x}{l}$	22.30	58.18	109.90
(cantilever, exponential)	$e^{x/l}$	1	15.23	77.78	206.07
(free-free, constant b, double-tapered h)	1	$\dfrac{x}{l}$	21.21* 35.05†	56.97	
(free-free, tapered b and h)	$\dfrac{x}{l}$	$\dfrac{x}{l}$	32.73* 49.50†	76.57	
(free-free, b sqrt, h tapered)	$\left(\dfrac{x}{l}\right)^{\frac{1}{2}}$	$\dfrac{x}{l}$	25.66* 42.02†	66.06	

* SYMMETRIC † ANTISYMMETRIC

f_n = natural frequency, cps
$\rho = \sqrt{I/S}$ = radius of gyration, in.
h = depth of beam, in.
l = beam length, in.
n = mode number
b = width of beam, in.

For materials other than steel: $f_n = f_{n_s}\sqrt{\dfrac{E\gamma_s}{E_s\gamma}}$

E = modulus of elasticity, lb/in.^{2}
γ = density, lb/in.^{3}
Terms with subscripts refer to steel
Terms without subscripts refer to other material

Table 7.5. Natural Frequencies of Continuous Uniform Steel* Beams
(J. N. Macduff and R. P. Felgar.[16,17])

Beam structure	$(f_n l^2/\rho)/10^4$					
	N	$n=1$	$n=2$	$n=3$	$n=4$	$n=5$
	Extreme Ends Simply Supported					
	1	31.73	126.94	285.61	507.76	793.37
	2	31.73	49.59	126.94	160.66	285.61
	3	31.73	40.52	59.56	126.94	143.98
	4	31.73	37.02	49.59	63.99	126.94
	5	31.73	34.99	44.19	55.29	66.72
	6	31.73	34.32	40.52	49.59	59.56
	7	31.73	33.67	38.40	45.70	53.63
	8	31.73	33.02	37.02	42.70	49.59
	9	31.73	33.02	35.66	40.52	46.46
	10	31.73	33.02	34.99	39.10	44.19
	11	31.73	32.37	34.32	37.70	41.97
	12	31.73	32.37	34.32	37.02	40.52
	Extreme Ends Clamped					
	1	72.36	198.34	388.75	642.63	959.98
	2	49.59	72.36	160.66	198.34	335.20
	3	40.52	59.56	72.36	143.98	178.25
	4	37.02	49.59	63.99	72.36	137.30
	5	34.99	44.19	55.29	66.72	72.36
	6	34.32	40.52	49.59	59.56	67.65
	7	33.67	38.40	45.70	53.63	62.20
	8	33.02	37.02	42.70	49.59	56.98
	9	33.02	35.66	40.52	46.46	52.81
	10	33.02	34.99	39.10	44.19	49.59
	11	32.37	34.32	37.70	41.97	47.23
	12	32.37	34.32	37.02	40.52	44.94
	Extreme Ends Clamped-Supported					
	1	49.59	160.66	335.2	573.21	874.69
	2	37.02	63.99	137.30	185.85	301.05
	3	34.32	49.59	67.65	132.07	160.66
	4	33.02	42.70	56.98	69.51	129.49
	5	33.02	39.10	49.59	61.31	70.45
	6	32.37	37.02	44.94	54.46	63.99
	7	32.37	35.66	41.97	49.59	57.84
	8	32.37	34.99	39.81	45.70	53.63
	9	31.73	34.32	38.40	43.44	49.59
	10	31.73	33.67	37.02	41.24	46.46
	11	31.73	33.67	36.33	39.81	44.19
	12	31.73	33.02	35.66	39.10	42.70

*For materials other than steel, use equation at bottom of Table 7.4.

f_n = natural frequency, cps
$\rho = \sqrt{I/S}$ = radius of gyration, in.
l = span length, in.
n = mode number
N = number of spans

BEAMS WITH PARTLY CLAMPED ENDS. For a beam in which the slope at each end is proportional to the moment, the following empirical equation gives the natural frequency:[21]

$$f_n = f_0 \left[n + \frac{1}{2}\left(\frac{\beta_L}{5n + \beta_L}\right)\right] \left[n + \frac{1}{2}\left(\frac{\beta_R}{5n + \beta_R}\right)\right]$$

where f_0 is the frequency of the same beam with simply supported ends and n is the mode number. The parameters $\beta_L = k_L l/EI$ and $\beta_R = k_R l/EI$ are coefficients in which k_L and k_R are stiffnesses of the supports as given by $k_L = M_L/\theta_L$, where M_L is the moment and θ_L the angle at the left end, and $k_R = M_R/\theta_R$, where M_R is the moment and θ_R the angle at the right end. The error is less than 2 per cent except for bars having one end completely or nearly clamped ($\beta > 10$) and the other end completely or nearly hinged ($\beta < 0.9$).

LATERAL VIBRATION OF BEAMS WITH MASSES ATTACHED

The use of Fig. 7.4 is a convenient method of estimating the natural frequencies of beams with added loads.

Exact Solution. If the masses attached to the beam are considered to be rigid so that they exert no elastic forces, and if it is assumed that the attachment is such that the bending of the beam is not restrained, Eqs. (7.13) and (7.16) apply. The section of the beam between each two masses, and between each support and the adjacent mass, must be considered individually. The constants in Eq. (7.16) are different for each section. There are $4N$ constants, N being the number of sections into which the beam is divided. Each support supplies two boundary conditions. Additional conditions are provided by:

1. The deflection at the location of each mass is the same for both sections adjacent to the mass.
2. The slope at each mass is the same for each section adjacent thereto.
3. The change in the lateral elastic shear force in the beam, at the location of each mass, is equal to the product of the mass and its acceleration $\ddot{y}$.
4. The change of moment in the beam, at each mass, is equal to the product of the moment of inertia of the mass and its angular acceleration $(\partial^2/\partial t^2)(\partial y/\partial x)$.

Setting up the necessary equations is not difficult, but their solution is a lengthy process for all but the simplest configurations. Even the solution of the problem of a beam with hinged ends supporting a mass with negligible moment of inertia located anywhere except at the center of the beam is fairly long. If the mass is at the center of the beam, the solution is relatively simple because of symmetry, and is illustrated to show how the result compares with that obtained by Rayleigh's method.

Rayleigh's Method. Rayleigh's method offers a practical method of obtaining a fairly accurate solution of the problem, even when more than one mass is added. In carrying out the solution, the kinetic energy of the masses is added to that of the beam. The strain and kinetic energies of a uniform beam are given in Table 7.1. The kinetic energy of the ith mass is $(m_i/2)\omega_n^2 Y^2(x_i)$, where $Y(x_i)$ is the value of the amplitude at the location of mass. Equating the maximum strain energy to the total maximum kinetic energy of the beam and masses, the frequency equation becomes

$$\omega_n^2 = \frac{EI \int_0^l (Y'')^2\, dx}{\frac{\gamma S}{g}\int_0^l Y^2\, dx + \sum_{i=1}^n m_i Y^2(x_i)} \qquad (7.19)$$

where $Y(x)$ is the maximum deflection. If $Y(x)$ were known exactly, this equation would give the correct frequency; however, since Y is not known, a shape must be assumed. This may be either the mode shape of the unloaded beam or a polynomial that satisfies the necessary boundary conditions, such as the equation for the static deflection under a load.

FREE VIBRATION 7-23

Beam as Spring. A method for obtaining the natural frequency of a beam with a single mass mounted on it is to consider the beam to act as a spring, the stiffness of which is found by using simple beam theory. The equation $\omega_n = \sqrt{k/m}$ is used. Best accuracy is obtained by considering m to be made up of the attached mass plus some portion of the mass of the beam. The fraction of the beam mass to be used depends on the type of beam. The equations for simply supported and cantilevered beams with masses attached are given in Table 7.2.

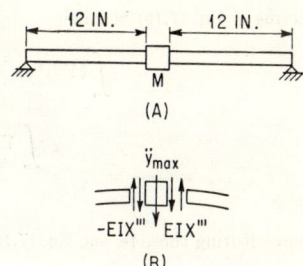

FIG. 7.9. (A) Beam having simply supported ends with mass attached at center. (B) Forces exerted on mass, at extreme deflection, by shear stresses in beam.

Example 7.7. The fundamental natural frequencies of a beam with hinged ends 24 in. long, 2 in. wide, and ¼ in. thick having a mass m attached at the center (Fig. 7.9) are to be calculated by each of the three methods, and the results compared for ratios of mass to beam mass of 1, 5, and 25. The result is to be compared with the frequency from Fig. 7.4.

Exact Solution. Because of symmetry, only the section of the beam to the left of the mass has to be considered in carrying out the exact solution. The boundary conditions for the left end are: at $x = 0$, $X = 0$, and $X'' = 0$. The shear force just to the left of the mass is negative at maximum deflection (Fig. 7.9B) and is $F_s = -EIX'''$; to the right of the mass, because of symmetry, the shear force has the same magnitude with opposite sign. The difference between the shear forces on the two sides of the mass must equal the product of the mass and its acceleration. For the condition of maximum deflection,

$$2EIX''' = m\ddot{y}_{max} \tag{7.20}$$

where X''' and $\ddot{y}_{max}$ must be evaluated at $x = l/2$. Because of symmetry the slope at the center is zero. Using the solution $y = X \cos \omega_n t$ and $\ddot{y}_{max} = -\omega_n^2 X$, Eq. (7.20) becomes

$$2EIX''' = -m\omega_n^2 X \tag{7.21}$$

The first boundary condition makes $A = 0$ in Eq. (7.16) and the second condition makes $B = 0$. For simplicity, the part of the equation that remains is written

$$X = C \sin \kappa x + D \sinh \kappa x \tag{7.22}$$

Using this in Eq. (7.20) gives

$$2EI\left(-C\kappa^3 \cos \frac{\kappa l}{2} + D\kappa^3 \cosh \frac{\kappa l}{2}\right) = -m\omega_n^2 \left(C \sin \frac{\kappa l}{2} + D \sinh \frac{\kappa l}{2}\right) \tag{7.23}$$

The slope at the center is zero. Differentiating Eq. (7.22) and substituting $x = l/2$,

$$\kappa \left(C \cos \frac{\kappa l}{2} + D \cosh \frac{\kappa l}{2}\right) = 0 \tag{7.24}$$

Solving Eqs. (7.23) and (7.24) for the ratio C/D and equating, the following frequency equation is obtained:

$$2 \frac{m_b}{m} = \frac{\kappa l}{2} \left(\tan \frac{\kappa l}{2} - \tanh \frac{\kappa l}{2}\right)$$

where $m_b = \gamma Sl/g$ is the total mass of the beam. The lowest roots for the specified ratios m/m_b are as follows:

m/m_b	1	5	25
$\kappa l/2$	1.1916	0.8599	0.5857

The corresponding natural frequencies are found from Eq. (7.14) and are tabulated, with the results obtained by the other methods, at the end of the example.

Solution by Rayleigh's Method. For the solution by Rayleigh's method it is assumed that $Y = B \sin (\pi x/l)$. This is the fundamental mode for the unloaded beam (Table 7.3). The

7-24 VIBRATION OF SYSTEMS HAVING DISTRIBUTED MASS AND ELASTICITY

terms in Eq. (7.19) become

$$\int_0^l (Y'')^2 \, dx = B^2 \left(\frac{\pi}{l}\right)^4 \int_0^l \sin^2 \frac{\pi x}{l} \, dx = B^2 \frac{l}{2} \left(\frac{\pi}{l}\right)^4$$

$$\int_0^l Y^2 \, dx = B^2 \int_0^l \sin^2 \frac{\pi x}{l} \, dx = B^2 \frac{l}{2}$$

$$Y^2(x_1) = B^2$$

Substituting these terms, Eq. (7.19) becomes

$$\omega_n = \sqrt{\frac{EIB^2 \frac{l}{2} \left(\frac{\pi}{l}\right)^4}{\frac{S\gamma B^2 l}{2g} + mB^2}} = \frac{\pi^2}{\sqrt{1 + 2m/m_b}} \sqrt{\frac{EIg}{S\gamma l^4}}$$

The frequencies for the specified values of m/m_b are tabulated at the end of the example. Note that if $m = 0$, the frequency is exactly correct, as can be seen from Table 7.3. This is to be expected since, if no mass is added, the assumed shape is the true shape.

Lumped Parameter Solution. Using the appropriate equation from Table 7.2, the natural frequency is

$$\omega_n = \sqrt{\frac{48EI}{l^3(m + 0.5m_b)}}$$

Since $m_b = \gamma Sl/g$, this becomes

$$\omega_n = \sqrt{\frac{48}{(m/m_b) + 0.5}} \sqrt{\frac{EIg}{S\gamma l^4}}$$

Comparison of Results. The results for each method can be expressed as a coefficient α multiplied by $\sqrt{EIg/S\gamma l^4}$. The values of α for the specified values by m/m_b for the three methods of solution are:

m/m_b	1	5	25
Exact.....	5.680	2.957	1.372
Rayleigh ..	5.698	2.976	1.382
Spring	5.657	2.954	1.372

The results obtained by all the methods agree closely. For large values of m/m_b the third method gives very accurate results.

Numerical Calculations. For steel, $E = 30 \times 10^6$ lb/in.^{2}, $\gamma = 0.28$ lb/in.^{3}; for a rectangular beam, $I = bh^3/12 = 1/384$ in.^{4} and $S = bh = \frac{1}{2}$ in.^{2} The fundamental frequency using the value of α for the exact solution when $m/m_b = 1$ is

$$\omega_1 = \frac{\alpha}{l^2} \sqrt{\frac{EIg}{S\gamma}} = \frac{5.680}{576} \sqrt{\frac{(30 \times 10^6)(386)}{(0.5)(384)(0.28)}} = 145 \text{ rad/sec} = 23 \text{ cps}$$

Other frequencies can be found by using the other values of α. Nearly the same result is obtained by using Fig. 7.4, if half the mass of the beam is added to the additional mass.

LATERAL VIBRATION OF PLATES

General Theory of Bending of Rectangular Plates. For small deflections of an initially flat plate of uniform thickness (Fig. 7.10) made of homogeneous isotropic material and subjected to normal and shear forces in the plane of the plate, the following equation relates the lateral deflection w to the lateral loading:[22]

$$D\nabla^4 w = D\left(\frac{\partial^4 w}{\partial x^4} + 2\frac{\partial^4 w}{\partial x^2 \partial y^2} + \frac{\partial^4 w}{\partial y^4}\right) = P + N_x \frac{\partial^2 w}{\partial x^2} + 2N_{xy} \frac{\partial^2 w}{\partial x \, \partial y} + N_y \frac{\partial^2 w}{\partial y^2} \quad (7.25)$$

where $D = Eh^3/12(1 - \mu^2)$ is the plate stiffness, h being the plate thickness and μ Poisson's ratio. The parameter P is the loading intensity, N_x the normal loading in the X direction per unit of length, N_y the normal loading in the Y direction, and N_{xy} the shear load parallel to the plate surface in the X and Y directions.

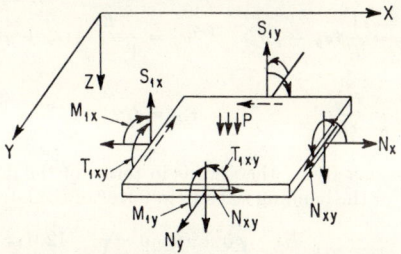

FIG. 7.10. Element of plate showing bending moments, normal forces, and shear forces.

The bending moments and shearing forces are related to the deflection w by the following equations:[23]

$$M_{1x} = -D\left(\frac{\partial^2 w}{\partial x^2} + \mu \frac{\partial^2 w}{\partial y^2}\right) \qquad M_{1y} = -D\left(\frac{\partial^2 w}{\partial y^2} + \mu \frac{\partial^2 w}{\partial x^2}\right)$$

$$T_{1xy} = D(1 - \mu)\frac{\partial^2 w}{\partial x \, \partial y} \qquad (7.26)$$

$$S_{1x} = -D\left(\frac{\partial^3 w}{\partial x^3} + \frac{\partial^3 w}{\partial x \, \partial y^2}\right) \qquad S_{1y} = -D\left(\frac{\partial^3 w}{\partial y^3} + \frac{\partial^3 w}{\partial x^2 \, \partial y}\right)$$

As shown in Fig. 7.10, M_{1x} and M_{1y} are the bending moments per unit of length on the faces normal to the X and Y directions, respectively, T_{1xy} is the twisting or warping moment on these faces, and S_{1x}, S_{1y} are the shearing forces per unit of length normal to the plate surface.

The boundary conditions that must be satisfied by an edge parallel to the X axis, for example, are as follows:

Built-in edge:
$$w = 0 \qquad \frac{\partial w}{\partial y} = 0$$

Simply supported edge:
$$w = 0 \qquad M_{1y} = -D\left(\frac{\partial^2 w}{\partial y^2} + \mu \frac{\partial^2 w}{\partial x^2}\right) = 0$$

Free edge:
$$M_{1y} = -D\left(\frac{\partial^2 w}{\partial y^2} + \mu \frac{\partial^2 w}{\partial x^2}\right) = 0 \qquad T_{1xy} = 0 \qquad S_{1y} = 0$$

which together give
$$\frac{\partial}{\partial y}\left[\frac{\partial^2 w}{\partial y^2} + (2 - \mu)\frac{\partial^2 w}{\partial x^2}\right] = 0$$

Similar equations can be written for other edges. The strains caused by the bending of the plate are

$$\epsilon_x = -z\frac{\partial^2 w}{\partial x^2} \qquad \epsilon_y = -z\frac{\partial^2 w}{\partial y^2} \qquad \gamma_{xy} = 2z\frac{\partial^2 w}{\partial x \, \partial y} \qquad (7.27)$$

where z is the distance from the center plane of the plate.

7-26 VIBRATION OF SYSTEMS HAVING DISTRIBUTED MASS AND ELASTICITY

Hooke's law may be expressed by the following equations:

$$\epsilon_x = \frac{1}{E}(\sigma_x - \mu\sigma_y) \qquad \sigma_x = \frac{E}{1-\mu^2}(\epsilon_x + \mu\epsilon_y)$$

$$\epsilon_y = \frac{1}{E}(\sigma_y - \mu\sigma_x) \qquad \sigma_y = \frac{E}{1-\mu^2}(\epsilon_y + \mu\epsilon_x) \qquad (7.28)$$

$$\gamma_{xy} = \frac{\tau_{xy}}{G} \qquad \tau_{xy} = G\gamma_{xy}$$

Substituting the expressions giving the strains in terms of the deflections, the following equations are obtained for the bending stresses in terms of the lateral deflection:

$$\sigma_x = -\frac{Ez}{1-\mu^2}\left(\frac{\partial^2 w}{\partial x^2} + \mu\frac{\partial^2 w}{\partial y^2}\right) = \frac{12M_{1x}}{h^3}z$$

$$\sigma_y = -\frac{Ez}{1-\mu^2}\left(\frac{\partial^2 w}{\partial y^2} + \mu\frac{\partial^2 w}{\partial x^2}\right) = \frac{12M_{1y}}{h^3}z \qquad (7.29)$$

$$\tau_{xy} = 2G\frac{\partial^2 w}{\partial x\, \partial y}z = \frac{12T_{1xy}}{h^3}z$$

Table 7.6 gives values of maximum deflection and bending moment at several points in plates which have various shapes and conditions of support, and which are subjected to uniform lateral pressure. The results are all based on the assumption that the deflections are small and that there are no loads in the plane of the plate. The bending stresses are found by the use of Eqs. (7.29). Bending moments and deflections for many other types of load are in the literature.[22]

The stresses caused by loads in the plane of the plate are found by assuming that the stress is uniform through the plate thickness. The total stress at any point in the plate is the sum of the stresses caused by bending and by the loading in the plane of the plate.

For plates in which the lateral deflection is large compared to the plate thickness but small compared to the other dimensions, Eq. (7.25) is valid. However, additional equations must be introduced because the forces N_x, N_y, and N_{xy} depend not only on the initial loading of the plate but also upon the stretching of the plate due to the bending. The equations of equilibrium for the X and Y directions in the plane of the plate are

$$\frac{\partial N_x}{\partial x} + \frac{\partial N_{xy}}{\partial y} = 0 \qquad \frac{\partial N_{xy}}{\partial x} + \frac{\partial N_y}{\partial y} = 0 \qquad (7.30)$$

It can be shown[27] that the strain components are given by

$$\epsilon_x = \frac{\partial u}{\partial x} + \frac{1}{2}\left(\frac{\partial w}{\partial x}\right)^2 \qquad \epsilon_y = \frac{\partial v}{\partial y} + \frac{1}{2}\left(\frac{\partial w}{\partial y}\right)^2$$

$$\gamma_{xy} = \frac{\partial u}{\partial y} + \frac{\partial v}{\partial x} + \frac{\partial w}{\partial x}\frac{\partial w}{\partial y} \qquad (7.31)$$

where u is the displacement in the X direction and v is the displacement in the Y direction. By differentiating and combining these expressions, the following relation is obtained:

$$\frac{\partial^2 \epsilon_x}{\partial y^2} + \frac{\partial^2 \epsilon_y}{\partial x^2} - \frac{\partial^2 \gamma_{xy}}{\partial x\, \partial y} = \left(\frac{\partial^2 w}{\partial x\, \partial y}\right)^2 - \frac{\partial^2 w}{\partial x^2}\frac{\partial^2 w}{\partial y^2} \qquad (7.32)$$

If it is assumed that the stresses caused by the forces in the plane of the plate are uni-

FREE VIBRATION

Table 7.6. Maximum Deflection and Bending Moments in Uniformly Loaded Plates under Static Conditions

RECTANGULAR PLATES

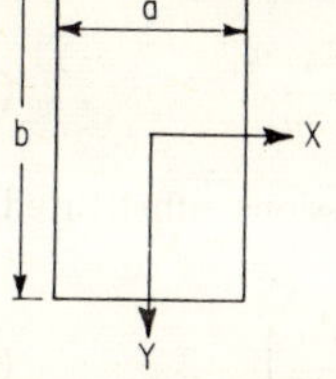

$\alpha = w_{MAX}/(Pa^4/Eh^3)$
$\beta = M_{lx}/Pa^2$
$\gamma = M_{ly}/Pa^2$
w = LATERAL DEFLECTION

P = UNIFORM PRESSURE
h = PLATE THICKNESS
E = MODULUS OF ELASTICITY
μ = POISSON'S RATIO

SIMPLY SUPPORTED EDGES ($\mu = 0.3$)[24]

b/a	1	1.2	1.4	1.6	1.8	2.0	3.0	∞
$(\alpha)_{x=0,\,y=0}$	0.044	0.062	0.077	0.091	0.102	0.111	0.134	0.142
$(\beta)_{x=0,\,y=0}$	0.048	0.063	0.075	0.086	0.095	0.102	0.119	0.125
$(\gamma)_{x=0,\,y=0}$	0.048	0.050	0.051	0.049	0.048	0.046	0.040	0.038

BUILT-IN EDGES ($\mu = 0.3$)[25]

	1	1.2	1.4	1.6	1.8	2.0
$(\alpha)_{x=0,\,y=0}$	0.014	0.019	0.023	0.025	0.027	0.028
$(\beta)_{x=a/2,\,y=0}$	-0.051	-0.064	-0.073	-0.078	-0.081	-0.083
$(\gamma)_{x=0,\,y=b/2}$	-0.051	-0.055	-0.057	-0.057	-0.057	-0.057
$(\beta)_{x=0,\,y=0}$		0.030	0.035	0.038	0.040	
$(\gamma)_{x=0,\,y=0}$		0.023	0.021	0.019	0.017	

CIRCULAR PLATES[26]

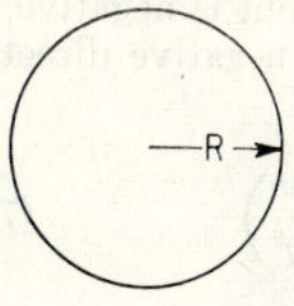

M_r = RADIAL MOMENT
M_t = TANGENTIAL MOMENT
$D = \dfrac{Eh^3}{12(1-\mu^2)}$

	SIMPLY SUPPORTED EDGES	BUILT-IN EDGES	
	CENTER	CENTER	EDGE
$w/(PR^4/D)$	$\dfrac{5+\mu}{64(1+\mu)}$	$\dfrac{1}{64}$	0
M_r/PR^2	$\dfrac{3+\mu}{16}$	$\dfrac{1+\mu}{16}$	$-\dfrac{1}{8}$
M_t/PR^2	$\dfrac{3+\mu}{16}$	$\dfrac{1+\mu}{16}$	$-\dfrac{\mu}{8}$

7–28 VIBRATION OF SYSTEMS HAVING DISTRIBUTED MASS AND ELASTICITY

formly distributed through the thickness, Hooke's law, Eqs. (7.28), can be expressed:

$$\epsilon_x = \frac{1}{hE}(N_x - \mu N_y) \qquad \epsilon_y = \frac{1}{hE}(N_y - \mu N_x) \qquad \gamma_{xy} = \frac{1}{hG} N_{xy} \qquad (7.33)$$

The equilibrium equations are satisfied by a stress function ϕ which is defined as follows:

$$N_x = h \frac{\partial^2 \phi}{\partial y^2} \qquad N_y = h \frac{\partial^2 \phi}{\partial x^2} \qquad N_{xy} = -h \frac{\partial^2 \phi}{\partial x \, \partial y} \qquad (7.34)$$

If these are substituted into Eqs. (7.33) and the resulting expressions substituted into Eq. (7.32), the following equation is obtained:

$$\frac{\partial^4 \phi}{\partial x^4} + 2\frac{\partial^4 \phi}{\partial x^2 \, \partial y^2} + \frac{\partial^4 \phi}{\partial y^4} = E\left[\left(\frac{\partial^2 w}{\partial x \, \partial y}\right)^2 - \frac{\partial^2 w}{\partial x^2}\frac{\partial^2 w}{\partial y^2}\right] \qquad (7.35)$$

A second equation is obtained by substituting Eqs. (7.34) in Eq. (7.25):

$$D\nabla^4 w = P + h\left(\frac{\partial^2 \phi}{\partial y^2}\frac{\partial^2 w}{\partial x^2} - 2\frac{\partial^2 \phi}{\partial x \, \partial y}\frac{\partial^2 w}{\partial x \, \partial y} + \frac{\partial^2 \phi}{\partial x^2}\frac{\partial^2 w}{\partial y^2}\right) \qquad (7.36)$$

Equations (7.35) and (7.36), with the boundary conditions, determine ϕ and w, from which the stresses can be computed. General solutions to this set of equations are not known, but some approximate solutions can be found in the literature.[28]

Free Lateral Vibrations of Rectangular Plates. In Eq. (7.25), the terms on the left are equal to the sum of the rates of change of the forces per unit of length in the X and Y directions where such forces are exerted by shear stresses caused by bending normal to the plane of the plate. For a rectangular element with dimensions dx and dy, the net force exerted normal to the plane of the plate by these stresses is $D\nabla^4 w \, dx \, dy$. The last three terms on the right-hand side of Eq. (7.25) give the net force normal to the plane of the plate, per unit of length, which is caused by the forces acting in the plane of the plate. The net force caused by these forces on an element with dimensions dx and dy is $(N_x \, \partial^2 w / \partial x^2 + 2N_{xy} \, \partial^2 w / \partial x \, \partial y + N_y \, \partial^2 w / \partial y^2) \, dx \, dy$. As in the corresponding beam problem, the forces in a vibrating plate consist of two parts: (1) that which balances the static load P including the weight of the plate and (2) that which is induced by the vibration. The first part is always in equilibrium with the load and together with the load can be omitted from the equation of motion if the deflection is taken from the position of static equilibrium. The force exerted normal to the plane of the plate by the bending stresses must equal the sum of the force exerted normal to the plate by the loads acting in the plane of the plate; i.e., the product of the mass of the element $(\gamma h/g) \, dx \, dy$ and its acceleration $\ddot{w}$. The term involving the acceleration of the element is negative, because when the bending force is positive the acceleration is in the negative direction. The equation of motion is

$$D\nabla^4 w = -\frac{\gamma}{g} h\ddot{w} + \left(N_x \frac{\partial^2 w}{\partial x^2} + 2N_{xy}\frac{\partial^2 w}{\partial x \, \partial y} + N_y \frac{\partial^2 w}{\partial y^2}\right) \qquad (7.37)$$

This equation is valid only if the magnitudes of the forces in the plane of the plate are constant during the vibration. For many problems these forces are negligible and the term in parentheses can be omitted.

When a system vibrates in a natural mode, all parts execute simple harmonic motion about the equilibrium position; therefore, the solution of Eq. (7.37) can be written as $w = AW(x,y) \cos(\omega_n t + \theta)$ in which W is a function of x and y only. Substituting this in Eq. (7.37) and dividing through by $A \cos(\omega_n t + \theta)$ gives

$$D\nabla^4 W = \frac{\gamma h \omega_n^2}{g} W + \left(N_x \frac{\partial^2 W}{\partial x^2} + 2N_{xy}\frac{\partial^2 W}{\partial x \, \partial y} + N_y \frac{\partial^2 W}{\partial y^2}\right) \qquad (7.38)$$

The function W must satisfy Eq. (7.38) as well as the necessary boundary conditions.

FREE VIBRATION

The solution of the problem of the lateral vibration of a rectangular plate with all edges simply supported is relatively simple; in general, other combinations of edge conditions require the use of other methods of solution. These are discussed later.

Example 7.8. The natural frequencies and normal modes of small vibration of a rectangular plate of length a, width b, and thickness h are to be calculated. All edges are hinged and subjected to unchanging normal forces N_x and N_y.

Solution. The following equation, in which m and n may be any integers, satisfies the necessary boundary conditions:

$$W = A \sin \frac{m\pi x}{a} \sin \frac{n\pi y}{b} \tag{7.39}$$

Substituting the necessary derivatives into Eq. (7.38),

$$D\left[\left(\frac{m}{a}\right)^4 + 2\left(\frac{m}{a}\right)^2\left(\frac{n}{b}\right)^2 + \left(\frac{n}{b}\right)^4\right]\pi^4 \sin \frac{m\pi x}{a} \sin \frac{n\pi y}{b}$$

$$= \frac{\gamma h \omega_n^2}{g} \sin \frac{m\pi x}{a} \sin \frac{n\pi y}{b} - \pi^2 \left[N_x \left(\frac{m}{a}\right)^2 + N_y \left(\frac{n}{b}\right)^2\right] \sin \frac{m\pi x}{a} \sin \frac{n\pi y}{b}$$

Solving for ω_n^2,

$$\omega_n^2 = \frac{g}{\gamma h}\left\{\pi^4 D\left[\left(\frac{m}{a}\right)^2 + \left(\frac{n}{b}\right)^2\right]^2 + \pi^2\left[N_x\left(\frac{m}{a}\right)^2 + N_y\left(\frac{n}{b}\right)^2\right]\right\} \tag{7.40}$$

By using integral values of m and n, the various frequencies are obtained from Eq. (7.40) and the corresponding normal modes from Eq. (7.39). For each mode, m and n represent the number of half sine waves in the X and Y directions, respectively. In each mode there are $m - 1$ evenly spaced nodal lines parallel to the Y axis, and $n - 1$ parallel to the X axis.

Rayleigh's and Ritz's Methods. The modes of vibration of a rectangular plate with all edges simply supported are such that the deflection of each section of the plate parallel to an edge is of the same form as the deflection of a beam with both ends simply supported. In general, this does not hold true for other combinations of edge conditions. For example, the vibration of a rectangular plate with all edges built in does not occur in such a way that each section parallel to an edge has the same shape as does a beam with both ends built in. A function that is made up using the mode shapes of beams with built-in ends obviously satisfies the conditions of zero deflection and slope at all edges, but it cannot be made to satisfy Eq. (7.38).

The mode shapes of beams give logical functions with which to formulate shapes for determining the natural frequencies, for plates having various edge conditions, by the Rayleigh or Ritz methods. By using a single mode function in Rayleigh's method an approximate frequency can be determined. This can be improved by using more than one of the modal shapes and using Ritz's method as discussed below.

The strain energy of bending and the kinetic energy for plates are given in Table 7.1. Finding the maximum values of the energies, equating them, and solving for ω_n^2 gives the following frequency equation:

$$\omega_n^2 = \frac{V_{\max}}{\frac{\gamma h}{2g} \int_A \int W^2 \, dx \, dy} \tag{7.41}$$

where V is the strain energy.

In applying the Rayleigh method, a function W is assumed that satisfies the necessary boundary conditions of the plate. An example of the calculations is given in the section on circular plates. If the shape assumed is exactly the correct one, Eq. (7.41) gives the exact frequency. In general, the correct shape is not known and a frequency greater than the natural frequency is obtained. The Ritz method involves assuming W to be of the form $W = a_1 W_1(x,y) + a_2 W_2(x,y) + \ldots$ in which $W_1, W_2, \ldots$ all satisfy the boundary conditions, and $a_1, a_2, \ldots$ are adjusted to give a minimum frequency. By the use of this method, solutions for plates having many combinations of edge conditions have been obtained. Detailed calculations for many problems are in the literature; several examples are cited in the following sections.

Table 7.7. Natural Frequencies and Nodal Lines of Square Plates with Various Edge Conditions (*After D. Young.*[29])

	1ST MODE	2ND MODE	3RD MODE	4TH MODE	5TH MODE	6TH MODE
$\omega_n / \sqrt{Dg/\gamma h a^4}$	3.494	8.547	21.44	27.46	31.17	
NODAL LINES						
$\omega_n / \sqrt{Dg/\gamma h a^4}$	35.99	73.41	108.27	131.64	132.25	165.15
NODAL LINES						
$\omega_n / \sqrt{Dg/\gamma h a^4}$	6.958	24.08	26.80	48.05	63.14	
NODAL LINES						

$\omega_n = 2\pi f_n$
$D = Eh^3/12(1-\mu^2)$
γ = WEIGHT DENSITY
h = PLATE THICKNESS
a = PLATE LENGTH

SQUARE, RECTANGULAR, AND SKEW RECTANGULAR PLATES. Tables of the functions necessary for the determination of the natural frequencies of rectangular plates by the use of the Ritz method are available,[29] these having been derived by using the modal shapes of beams having end conditions corresponding to the edge conditions of the plates. Information is included from which the complete shapes of the vibrational modes can be determined. Frequencies and nodal patterns for several modes of vibration of square plates having three sets of boundary conditions are shown in Table 7.7. By the use of functions which represent the natural modes of beams, the frequencies and nodal patterns for rectangular and skew cantilever plates have been determined [30] and are shown in Table 7.8. Comparison of calculated frequencies with experimentally determined values shows good agreement. Natural frequencies of rectangular plates having other boundary conditions are given in Table 7.9.

TRIANGULAR AND TRAPEZOIDAL PLATES. Nodal patterns and natural frequencies for triangular plates have been determined [32] by the use of functions derived from the mode shapes of beams, and are shown in Table 7.10. Certain of these have been compared with experimental values and the agreement is excellent. Natural frequencies and nodal patterns have been determined experimentally for six modes of vibration of a number of cantilevered triangular plates [33] and for the first six modes of cantilevered trapezoidal plates derived by trimming the tips of triangular plates parallel to the clamped edge.[34] These triangular and trapezoidal shapes approximate the shapes of various delta wings for aircraft and of fins for missiles.

CIRCULAR PLATES. The solution of the problem of small lateral vibration of circular plates is obtained by transforming Eq. (7.38) to polar coordinates and finding the solution that satisfies the necessary boundary conditions of the resulting equation.

Table 7.8. Natural Frequencies and Nodal Lines of Cantilevered Rectangular and Skew Rectangular Plates ($\mu = 0.3$)* (M. V. Barton.[30])

MODE \ a/b	1/2	1	2	5
FIRST	3.508	3.494	3.472	3.450
SECOND	5.372	8.547	14.93	34.73
THIRD	21.96	21.44	21.61	21.52
FOURTH	10.26	27.46	94.49	563.9
FIFTH	24.85	31.17	48.71	105.9

MODE	FIRST	SECOND	FIRST	SECOND	FIRST	SECOND
$\omega_n / \sqrt{Dg/\gamma h a^4}$	3.601	8.872	3.961	10.190	4.824	13.75
NODAL LINES	15°	15°	30°	30°	45°	45°

* For terminology, see Table 7.7.

Table 7.9. Natural Frequencies of Rectangular Plates (R. F. S. Hearmon.[31])

Plate with all edges simply supported (s,s,s,s):

b/a	1.0	1.5	2.0	2.5	3.0	∞
$\omega_n / \sqrt{Dg/\gamma h a^4}$	19.74	14.26	12.34	11.45	10.97	9.87

Plate with one edge clamped, three simply supported (c,s,s,s):

b/a	1.0	1.5	2.0	2.5	3.0	∞
$\omega_n / \sqrt{Dg/\gamma h a^4}$	23.65	18.90	17.33	16.63	16.26	15.43
a/b	1.0	1.5	2.0	2.5	3.0	∞
$\omega_n / \sqrt{Dg/\gamma h a^4}$	23.65	15.57	12.92	11.75	11.14	9.87

Plate with two opposite edges clamped (c,s,c,s):

b/a	1.0	1.5	2.0	2.5	3.0	∞
$\omega_n / \sqrt{Dg/\gamma h a^4}$	28.95	25.05	23.82	23.27	22.99	22.37
a/b	1.0	1.5	2.0	2.5	3.0	∞
$\omega_n / \sqrt{Dg/\gamma h a^4}$	28.95	17.37	13.69	12.13	11.36	9.87

Plate with all edges clamped (c,c,c,c):

b/a	1.0	1.5	2.0	2.5	3.0	∞
$\omega_n / \sqrt{Dg/\gamma h a^4}$	35.98	27.00	24.57	23.77	23.19	22.37

s DENOTES SIMPLY SUPPORTED EDGE
c DENOTES BUILT-IN OR CLAMPED EDGE
a = LENGTH OF PLATE
b = WIDTH OF PLATE
FOR OTHER TERMINOLOGY SEE TABLE 7.7

Omitting the terms involving forces in the plane of the plate,[35]

$$\left(\frac{\partial^2}{\partial r^2} + \frac{1}{r}\frac{\partial}{\partial r} + \frac{1}{r}\frac{\partial^2}{\partial \theta^2}\right)\left(\frac{\partial^2 W}{\partial r^2} + \frac{1}{r}\frac{\partial W}{\partial r} + \frac{1}{r}\frac{\partial^2 W}{\partial \theta^2}\right) = \varkappa^4 W \quad (7.42)$$

where

$$\varkappa^4 = \frac{\gamma h \omega_n^2}{gD}$$

The solution of Eq. (7.42) is [35]

$$W = A \cos(n\theta - \beta)[J_n(\varkappa r) + \lambda J_n(i\varkappa r)] \quad (7.43)$$

where J_n is a Bessel function of the first kind. When $\cos(n\theta - \beta) = 0$, a mode having a nodal system of n diameters, symmetrically distributed, is obtained. The term in

FREE VIBRATION

Table 7.10. Natural Frequencies and Nodal Lines of Triangular Plates (B. W. Anderson.[32])

MODE	k	2	4	8	14
FIRST		7.194	7.122	7.080	7.068
SECOND		30.803	30.718	30.654	30.638
THIRD		61.131	90.105	157.70	265.98
FOURTH		148.8	259.4	493.4	853.6
	k	2	4	7	
FIRST		5.887	6.617	6.897	
SECOND		25.40	28.80	30.28	

a = LENGTH OF TRIANGLE
k = RATIO OF LENGTH TO WIDTH OF TRIANGLE
FOR OTHER TERMINOLOGY SEE TABLE 7.7

brackets represents modes having concentric nodal circles. The values of $\varkappa$ and λ are determined by the boundary conditions, which are, for radially symmetrical vibration:

Simply supported edge:
$$W = 0 \qquad M_{1r} = D\left(\frac{d^2W}{dr^2} + \frac{\mu}{a}\frac{dW}{dr}\right) = 0$$

Fixed edge:
$$W = 0 \qquad \frac{dW}{dr} = 0$$

Free edge:
$$M_{1r} = D\left(\frac{d^2W}{dr^2} + \frac{\mu}{a}\frac{dW}{dr}\right) = 0 \qquad \frac{d}{dr}\left(\frac{d^2W}{dr^2} + \frac{1}{r}\frac{dW}{dr}\right) = 0$$

Example 7.9. The steel diaphragm of a radio earphone has an unsupported diameter of 2.0 in. and is 0.008 in. thick. Assuming that the edge is fixed, the lowest three frequencies for the free vibration in which only nodal circles occur are to be calculated, using the exact method and the Rayleigh and Ritz methods.

Exact Solution. In this example $n = 0$, which makes $\cos(n\theta - \beta) = 1$; thus, Eq. (7.43) becomes
$$W = A[J_0(\varkappa r) + \lambda I_0(\varkappa r)]$$
where $J_0(i\varkappa r) = I_0(\varkappa r)$ and I_0 is a modified Bessel function of the first kind.

At the boundary where $r = a$,
$$\frac{\partial W}{\partial r} = A\varkappa[-J_1(\varkappa a) + \lambda I_1(\varkappa a)] = 0 \qquad -J_1(\varkappa a) + \lambda I_1(\varkappa a) = 0$$

The deflection at $r = a$ is also zero:

$$J_0(\varkappa a) + \lambda I_0(\varkappa a) = 0$$

The frequency equation becomes

$$\lambda = \frac{J_1(\varkappa a)}{I_1(\varkappa a)} = -\frac{J_0(\varkappa a)}{I_0(\varkappa a)}$$

The first three roots of the frequency equation are: $\varkappa a = 3.196, 6.306, 9.44$. The corresponding natural frequencies are, from Eq. (7.42),

$$\omega_n = \frac{10.21}{a^2}\sqrt{\frac{Dg}{\gamma h}}, \quad \frac{39.77}{a^2}\sqrt{\frac{Dg}{\gamma h}}, \quad \frac{88.9}{a^2}\sqrt{\frac{Dg}{\gamma h}}$$

For steel, $E = 30 \times 10^6$ lb/in.^{2}, $\gamma = 0.28$ lb/in.^{3}, and $\mu = 0.28$. Hence

$$D = \frac{Eh^3}{12(1-\mu^2)} = \frac{30 \times 10^6 (0.008)^3}{12(1-0.078)} = 1.38 \text{ lb-in.}$$

Thus, the lowest natural frequency is

$$\omega_1 = 10.21\sqrt{\frac{(1.38)(386)}{(0.28)(0.008)}} = 4,960 \text{ rad/sec} = 790 \text{ cps}$$

The second frequency is 3,070 cps, and the third is 6,880 cps.

Solution by Rayleigh's Method. The equations for strain and kinetic energies are given in Table 7.1. The strain energy for a plate with clamped edges becomes

$$V = \pi D \int_0^a \left(\frac{\partial^2 W}{\partial r^2} + \frac{1}{r}\frac{\partial W}{\partial r}\right)^2 r\, dr$$

The maximum kinetic energy is

$$T = \frac{\omega_n^2 \pi \gamma h}{g}\int_0^a W^2 r\, dr$$

An expression of the form $W = a_1[1 - (r/a)^2]^2$, which satisfies the conditions of zero deflection and slope at the boundary, is used. The first two derivatives are $\partial W/\partial r = a_1(-4r/a^2 + 4r^3/a^4)$ and $\partial^2 W/\partial r^2 = a_1(-4/a^2 + 12r^2/a^4)$. Using these values in the equations for strain and kinetic energy, $V = 32\pi D a_1^2/3a^2$ and $T = \omega_n^2 \pi \gamma h a^2 a_1^2/10g$. Equating these values and solving for the frequency,

$$\omega_n = \sqrt{\frac{320\, Dg}{3a^4\, \gamma h}} = \frac{10.33}{a^2}\sqrt{\frac{Dg}{\gamma h}}$$

This is somewhat higher than the exact frequency.

Solution by Ritz's Method. Using an expression for the deflection of the form

$$W = a_1[1 - (r/a)^2]^2 + a_2[1 - (r/a)^2]^3$$

and applying the Ritz method, the following values are obtained for the first two frequencies:

$$\omega_1 = \frac{10.21}{a^2}\sqrt{\frac{Dg}{\gamma h}} \qquad \omega_2 = \frac{43.04}{a^2}\sqrt{\frac{Dg}{\gamma h}}$$

The details of the calculations giving this result are in the literature.[36] The first frequency agrees with the exact answer to four significant figures, while the second frequency is somewhat high. A closer approximation to the second frequency and approximations of the higher frequencies could be obtained by using additional terms in the deflection equation.

The frequencies of modes having n nodal diameters are:[36]

$n = 1$: $\qquad\omega_1 = \dfrac{21.22}{a^2}\sqrt{\dfrac{Dg}{\gamma h}}$

$n = 2$: $\qquad\omega_2 = \dfrac{34.84}{a^2}\sqrt{\dfrac{Dg}{\gamma h}}$

For a plate with its center fixed and edge free, and having m nodal circles, the frequencies are:[37]

m	0	1	2	3
$\omega_n a^2 / \sqrt{\dfrac{Dg}{\gamma h}}$	3.75	20.91	60.68	119.7

STRETCHING OF MIDDLE PLANE. In the usual analysis of plates, it is assumed that the deflection of the plate is so small that there is no stretching of the middle plane. If such stretching occurs, it affects the natural frequency. Whether it occurs depends

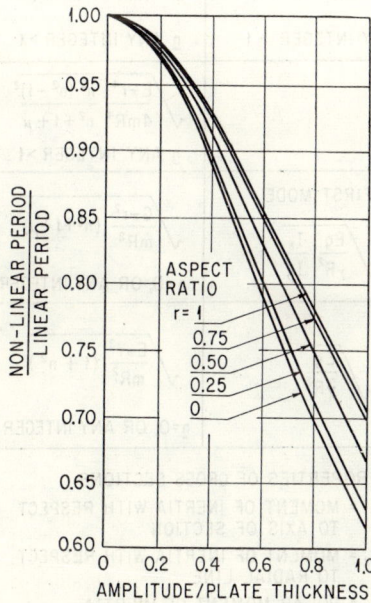

Fig. 7.11. Influence of amplitude on period of vibration of uniform rectangular plates with immovable hinged edges. The aspect ratio r is the ratio of width to length of the plate. (*H. Chu and G. Herrmann.*[38])

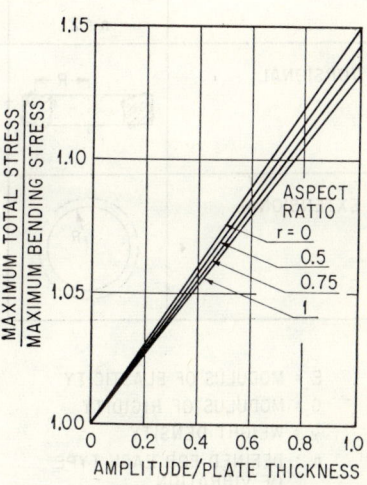

Fig. 7.12. Influence of amplitude on maximum total stress in rectangular plates with immovable hinged edges. The aspect ratio r is the ratio of width to length of the plate. (*H. Chu and G. Herrmann.*[38])

on the conditions of support of the plate, the amplitude of vibration, and possibly other conditions. In a plate with its edges built in, a relatively small deflection causes a significant stretching. The effect of stretching is not proportional to the deflection; thus, the elastic restoring force is not a linear function of deflection. The natural frequency is not independent of amplitude, but becomes higher with increasing amplitudes. If a plate is subjected to a pressure on one side, so that the vibration occurs about a deflected position, the effect of stretching may be appreciable. The effect of stretching in rectangular plates with immovable hinged supports has been discussed.[38] The effect of the amplitude on the natural frequency is shown in Fig. 7.11; the effect on the total stress in the plate is shown in Fig. 7.12. The natural frequency increases rapidly as the amplitude of vibration increases.

ROTATIONAL MOTION AND SHEARING FORCES. In the foregoing analysis, only the motion of each element of the plate in the direction normal to the plane of the plate is considered. There is also rotation of each element, and there is a deflection associated with the lateral shearing forces in the plate. The effects of these factors becomes

Table 7.11. Natural Frequencies of Complete Circular Rings Whose Thickness in Radial Direction Is Small Compared to Radius

TYPE OF VIBRATION	SHAPE OF LOWEST MODE	RECTANGULAR CROSS SECTION ω_n	CIRCULAR CROSS SECTION ω_n
FLEXURAL IN PLANE OF RING WITH n COMPLETE WAVELENGTH IN CIRCUMFERENCE	$n=2$	$\sqrt{\dfrac{Eg}{\gamma}\dfrac{I}{AR^4}\dfrac{n^2(n^2-1)^2}{n^2+1}}$ $\underline{n}$ ANY INTEGER >1	$\sqrt{\dfrac{E\pi r^4}{4mR^4}\dfrac{n^2(n^2-1)^2}{n^2+1}}$ $\underline{n}$ ANY INTEGER >1
FLEXURAL NORMAL TO PLANE OF RING	$n=2$		$\sqrt{\dfrac{E\pi r^4}{4mR^4}\dfrac{n^2(n^2-1)^2}{n^2+1+\mu}}$ $\underline{n}$ ANY INTEGER >1
TORSIONAL		FIRST MODE $\sqrt{\dfrac{Eg}{\gamma R^2}\dfrac{I_x}{I_p}}$	$\sqrt{\dfrac{G\pi r^2}{mR^2}}(n^2+1+\mu)$ $n=0$, OR ANY INTEGER
EXTENSIONAL		$\sqrt{\dfrac{Eg}{\gamma R^2}}$	$\sqrt{\dfrac{E\pi r^2}{mR^2}}(1+n^2)$ $\underline{n}=0$, OR ANY INTEGER

E = MODULUS OF ELASTICITY
G = MODULUS OF RIGIDITY
γ = WEIGHT DENSITY
n : DEFINED FOR EACH TYPE OF VIBRATION
R = RADIUS OF RING
μ = POISSON'S RATIO

PROPERTIES OF CROSS SECTIONS
I = MOMENT OF INERTIA WITH RESPECT TO AXIS OF SECTION
I_x = MOMENT OF INERTIA WITH RESPECT TO RADIAL LINE
I_p = POLAR MOMENT OF INERTIA
A = AREA
r = RADIUS
m = MASS PER UNIT OF LENGTH

significant if the curvature of the plate is large relative to its thickness, i.e., for a plate in which the thickness is large compared to the lateral dimensions or when the plate is vibrating in a mode for which the nodal lines are close together. These effects have been analyzed for rectangular plates [39] and for circular plates.[40]

COMPLETE CIRCULAR RINGS. Equations have been derived [41, 42] for the natural frequencies of complete circular rings for which the radius is large compared to the thickness of the ring in the radial direction. Such rings can execute several types of free vibration, which are shown in Table 7.11 with the formulas for the natural frequencies.

FORCED VIBRATION

CLASSICAL SOLUTION. The classical method of analyzing the forced vibration that results when an elastic system is subjected to a fluctuating load is to set up the equation of motion by the application of Newton's second law. During the vibration, each element of the system is subjected to elastic forces corresponding to those experienced during free vibration; in addition, some of the elements are subjected to the disturbing

FORCED VIBRATION 7–37

force. The equation which governs the forced vibration of a system can be obtained by adding the disturbing force to the equation for free vibration. For example, in Eq. (7.13) for the free vibration of a uniform beam, the term on the left is due to the elastic forces in the beam. If a force $F(x,t)$ is applied to the beam, the equation of motion is obtained by adding this force to Eq. (7.13), which becomes, after rearranging terms,

$$EI \frac{\partial^4 y}{\partial x^4} + \frac{\gamma S}{g} \frac{\partial^2 y}{\partial t^2} = F(x,t) \qquad (7.44)$$

where EI is a constant. The solution of this equation gives the motion that results from the force F. For example, consider the motion of a beam with hinged ends subjected to a sinusoidally varying force acting at its center. The solution is obtained by representing the concentrated force at the center by its Fourier series:

$$EIy'''' + \frac{\gamma S}{g} \ddot{y} = \frac{2F}{l}\left[\sin\frac{\pi x}{l} - \sin\frac{3\pi x}{l} + \sin\frac{5\pi x}{l} \cdots \right] \sin \omega t$$

$$= \frac{2F}{l} \sum_{n=1}^{n=\infty} \left(\sin\frac{n\pi}{2} \sin\frac{n\pi x}{l}\right) \sin \omega t \qquad (7.45)$$

where $\sin(n\pi/2)$, which appears in each term of the series, makes the nth term positive, negative, or zero. The solution of Eq. (7.45) is

$$y = \sum_{n=1}^{n=\infty}\left[A_n \sin\frac{n\pi x}{l} \sin \omega_n t + B_n \sin\frac{n\pi x}{l} \cos \omega_n t \right.$$

$$\left. + \sin\frac{n\pi}{2} \frac{2Fg/S\gamma l}{(n\pi/l)^4(EIg/S\gamma) - \omega^2} \sin\frac{n\pi x}{l} \sin \omega t \right] \qquad (7.46)$$

The first two terms of Eq. (7.46) are the values of y which make the left side of Eq. (7.45) equal to zero. They are obtained in exactly the same way as in the solution of the free-vibration problem and represent the free vibration of the beam. The constants are determined by the initial conditions; in any real beam, damping causes the free vibration to die out. The third term of Eq. (7.46) is the value of y which makes the left-hand side of Eq. (7.45) equal the right-hand side; this can be verified by substitution. The third term represents the forced vibration. From Table 7.3, $\kappa_n l = n\pi$ for a beam with hinged ends; then from Eq. (7.14), $\omega_n^2 = n^4\pi^4 EIg/S\gamma l^4$. The term representing the forced vibration in Eq. (7.46) can be written, after rearranging terms,

$$y = \frac{2Fg}{S\gamma l} \sum_{n=1}^{n=\infty} \frac{\sin(n\pi/2)}{\omega_n^2[1-(\omega/\omega_n)^2]} \sin\frac{n\pi x}{l} \sin \omega t \qquad (7.47)$$

From Table 7.3 and Eq. (7.16), it is evident that this deflection curve has the same shape as the nth normal mode of vibration of the beam since, for free vibration of a beam with hinged ends, $X_n = 2C \sin \kappa x = \sin(n\pi x/l)$.

The equation for the deflection of a beam under a distributed static load $F(x)$ can be obtained by replacing $-(\gamma S/g)\ddot{y}$ with F in Eq. (7.12); then Eq. (7.13) becomes

$$y_s'''' = \frac{F(x)}{EI} \qquad (7.48)$$

where EI is a constant. For a static loading $F(x) = 2F/l \sin\frac{n\pi}{2} \sin\frac{n\pi x}{l}$ corresponding to the nth term of the Fourier series in Eq. (7.45), Eq. (7.48) becomes $y_{sn}'''' = 2F/EIl \sin\frac{n\pi}{2} \sin\frac{n\pi x}{l}$. The solution of this equation is

$$y_{sn} = \frac{2F}{EIl}\left(\frac{l}{n\pi}\right)^4 \sin\frac{n\pi}{2} \sin\frac{n\pi x}{l}$$

Using the relation $\omega_n^2 = n^4\pi^4 EIg/S\gamma l^4$, this can be written

$$y_{sn} = \frac{2Fg}{\omega_n^2 S\gamma l}\sin\frac{n\pi x}{l}\sin\frac{n\pi}{2}$$

Thus, the nth term of Eq. (7.47) can be written

$$y_n = y_{sn}\frac{1}{1-(\omega/\omega_n)^2}\sin\omega t$$

Thus, the amplitude of the forced vibration is equal to the static deflection under the Fourier component of the load multiplied by the "amplification factor" $1/[1-(\omega/\omega_n)^2]$. This is the same as the relation that exists, for a system having a single degree-of-freedom, between the static deflection under a load F and the amplitude under a fluctuating load $F\sin\omega t$. Therefore, in so far as each mode alone is concerned, the beam behaves as a system having a single degree-of-freedom. If the beam is subjected to a force fluctuating at a single frequency, the amplification factor is small except when the frequency of the forcing force is near the natural frequency of a mode. For all even values of n, $\sin(n\pi/2) = 0$; thus, the even-numbered modes are not excited by a force acting at the center, which is a node for those modes. The distribution of the static load that causes the same pattern of deflection as the beam assumes during each mode of vibration has the same form as the deflection of the beam. This result applies to other beams since a comparison of Eqs. (7.15) and (7.48) shows that if a static load $F = (\omega_n^2 \gamma S/g)y$ is applied to any beam it will cause the same deflection as occurs during the free vibration in the nth mode.

The results for the simply supported beam are typical of those which are obtained for all systems having distributed mass and elasticity. Vibration of such a system at resonance is excited by a force which fluctuates at the natural frequency of a mode, since nearly any such force has a component of the shape necessary to excite the vibration. Even if the force acts at a nodal point of the mode, vibration may be excited because of coupling between the modes. The forced vibration of plates is discussed in Chap. 49.

METHOD OF WORK. Another method of analyzing forced vibration is by the use of the theorem of virtual work and D'Alembert's principle. The theorem of virtual work states that when any elastic body is in equilibrium, the total work done by all external forces during any virtual displacement equals the increase in the elastic energy stored in the body. A virtual displacement is an arbitrary small displacement that is compatible with the geometry of the body, and which satisfies the boundary conditions.

In applying the principle of work to forced vibration of elastic bodies, the problem is made into one of equilibrium by the application of D'Alembert's principle. This permits a problem in dynamics to be considered as one of statics by adding to the equation of static equilibrium an "inertia force" which, for each part of the body, is equal to the product of the mass and the acceleration. Using this principle, the theorem of virtual work can be expressed in the following equation:

$$\Delta V = \Delta(F_I + F_E) \tag{7.49}$$

in which V is the elastic strain energy in the body, F_I is the inertia force, F_E is the external disturbing force, and Δ indicates the change of the quantity when the body undergoes a virtual displacement. The various quantities can be found separately.

For example, consider the motion of a uniform beam having hinged ends with a sinusoidally varying force acting at the center, and compare the result with the solution obtained by the classical method. All possible motions of any beam can be represented by a series of the form

$$y = q_1 X_1 + q_2 X_2 + q_3 X_3 + \cdots = \sum_{n=1}^{n=\infty} q_n X_n \tag{7.50}$$

in which the X's are functions representing displacements in the normal modes of vibration and the q's are coefficients which are functions of time. The determination of the

FORCED VIBRATION 7-39

values of q_n is the problem to be solved. For a beam having hinged ends, Eq. (7.50) becomes

$$y = \sum_{n=1}^{n=\infty} q_n \sin \frac{n\pi x}{l} \qquad (7.51)$$

This is evident by using the values of $\kappa_n l$ from Table 7.3 in Eq. (7.16). A virtual displacement, being any arbitrary small displacement, can be assumed to be

$$\Delta y = \Delta q_m X_m = \Delta q_m \sin \frac{m\pi x}{l}$$

The elastic strain energy of bending of the beam is

$$V = \frac{EI}{2} \int_0^l \left(\frac{\partial^2 y}{\partial x^2}\right)^2 dx = \frac{EI}{2} \sum_{n=1}^{n=\infty} q_n^2 \int_0^l \left[\frac{\partial^2}{\partial x^2}\left(\sin \frac{n\pi x}{l}\right)\right]^2 dx$$

$$= \frac{EI}{2} \sum_{n=1}^{n=\infty} q_n^2 \left(\frac{n\pi}{l}\right)^4 \int_0^l \left(\sin \frac{n\pi x}{l}\right)^2 dx = \frac{EI}{2} \sum_{n=1}^{n=\infty} q_n^2 \left(\frac{n\pi}{l}\right)^4 \frac{l}{2}$$

For the virtual displacement, the change of elastic energy is

$$\Delta V = \frac{\partial V}{\partial q_m} \Delta q_m = \frac{EI}{2l^3} (n\pi)^4 q_m \Delta q_m = \frac{EI}{2l^3} (\kappa_n l)^4 q_m \Delta q_m$$

The value of the "inertia force" at each section is

$$F_I = -\frac{\gamma S}{g} \ddot{y} = -\frac{\gamma S}{g} \sum_{n=1}^{n=\infty} \frac{d^2 q_n}{dt^2} \sin \frac{n\pi x}{l}$$

The work done by this force during the virtual displacement Δy is

$$\Delta F_I = F_I \Delta y = -\frac{\gamma S}{g} \sum_{n=1}^{n=\infty} \frac{d^2 q_n}{dt^2} \Delta q_m \int_0^l \sin \frac{n\pi x}{l} \sin \frac{m\pi x}{l} dx$$

$$= -\frac{\gamma S l}{2g} \frac{d^2 q_m}{dt^2} \Delta q_m$$

The orthogonality relation of Eq. (7.1) is used here, making the integral vanish when $n = m$. For a disturbing force F_E, the work done during the virtual displacement is

$$\Delta F_E = F_E \Delta y = F(X_m)_{x=c} \Delta q_m$$

in which $(X_m)_{x=c}$ is the value of X_m at the point of application of the load. Substituting the terms into Eq. (7.49),

$$\frac{\gamma S l}{2g} \ddot{q}_m + \frac{EI}{2l^3} (\kappa_m l)^4 q_m = F(X_m)_{x=c}$$

Rearranging terms and letting $EI/S\gamma = a^2$,

$$\ddot{q}_m + \kappa_m^4 a^2 q_m = \frac{2g}{\gamma S l} F(X_m)_{x=c} \qquad (7.52)$$

If F_E is a force which varies sinusoidally with time at any point $x = c$,

$$F(X_m)_{x=c} = \bar{F} \sin \frac{m\pi c}{l} \sin \omega t$$

and Eq. (7.52) becomes

$$\ddot{q}_m + \kappa_m^4 a^2 q_m = \frac{2g\bar{F}}{\gamma S l} \sin \frac{m\pi c}{l} \sin \omega t$$

The solution of this equation is

$$q_m = A_m \sin \kappa_m^2 at + B_m \cos \kappa_m^2 at + \frac{2\bar{F}g}{\gamma Sl} \frac{\sin(m\pi c/l)}{\kappa_m^4 a^2 - \omega^2} \sin \omega t$$

Since $\kappa_m^2 a = \omega_m$,

$$q_m = A_m \sin \omega_m t + B_m \cos \omega_m t + \frac{2\bar{F}g}{\gamma Sl} \frac{\sin(m\pi c/l)}{\omega_m^2 - \omega^2} \sin \omega t$$

when the force acts at the center $c/l = \frac{1}{2}$. Substituting the corresponding values of q in Eq. (7.51), the solution is identical to Eq. (7.46), which was obtained by the classical method.

VIBRATION RESULTING FROM MOTION OF SUPPORT. When the supports of an elastic body are vibrated by some external force, forced vibration may be induced in the body.[43] For example, consider the motion that results in a uniform beam, Fig. 7.13,

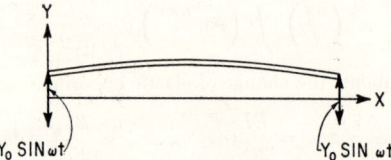

Fig. 7.13. Simply supported beam undergoing sinusoidal motion induced by sinusoidal motion of the supports.

when the supports are moved through a sinusoidally varying displacement $(y)_{x=0,l} = Y_0 \sin \omega t$. Although Eq. (7.13) was developed for the free vibration of beams, it is applicable to the present problem because there is no force acting on any section of the beam except the elastic force associated with the bending of the beam. If a solution of the form $y = X(x) \sin \omega t$ is assumed and substituted into Eq. (7.13):

$$X'''' = \frac{\omega^2 \gamma S}{EIg} X \qquad (7.53)$$

This equation is the same as Eq. (7.15) except that the natural frequency ω_n^2 is replaced by the forcing frequency ω^2. The solution of Eq. (7.53) is the same except that κ is replaced by $\kappa' = (\omega^2 \gamma S/EIg)^{1/4}$:

$$X = A_1 \sin \kappa' x + A_2 \cos \kappa' x + A_3 \sinh \kappa' x + A_4 \cosh \kappa' x \qquad (7.54)$$

The solution of the problem is completed by finding the constants, which are determined by the boundary conditions. Certain boundary conditions are associated with the supports of the beam and are the same as occur in the solution of the problem of free vibration. Additional conditions are supplied by the displacement through which the supports are forced. For example, if the supports of a beam having hinged ends are moved sinusoidally, the boundary conditions are: at $x = 0$ and $x = l$, $X'' = 0$, since the moment exerted by a hinged end is zero, and $X = Y_0$, since the amplitude of vibration is prescribed at each end. By the use of these boundary conditions, Eq. (7.54) becomes

$$X = \frac{Y_0}{2}\left[\tan\frac{\kappa' l}{2}\sin \kappa' x + \cos \kappa' x - \tanh\frac{\kappa' l}{2}\sinh \kappa' x + \cosh \kappa' x\right] \qquad (7.55)$$

The motion is defined by $y = X \sin \omega t$. For all values of κ', each of the coefficients except the first in Eq. (7.55) is finite. The tangent term becomes infinite if $\kappa' l = n\pi$, for odd values of n. The condition for the amplitude to become infinite is $\omega = \omega_n$ because $\kappa'/\kappa = \omega^2/\omega_n^2$ and, for natural vibration of a beam with hinged ends, $\kappa_n l = n\pi$. Thus, if the supports of an elastic body are vibrated at a frequency close to a natural frequency of the system, vibration at resonance occurs.

DAMPING. The effect of damping on forced vibration can be discussed only qualitatively. Damping usually decreases the amplitude of vibration, as it does in systems having a single degree-of-freedom. In some systems, it may cause coupling between modes, so that motion in a mode of vibration that normally would not be excited by a certain disturbing force may be induced.

REFERENCES

1. Timoshenko, S.: "Vibration Problems in Engineering," 3d ed., pp. 442, 448, D. Van Nostrand Company, Inc., Princeton, N.J., 1955.
2. Den Hartog, J. P.: "Mechanical Vibrations," 4th ed., p. 161, McGraw-Hill Book Company, Inc., New York, 1956.
3. Ref. 2, p. 152.
4. Hansen, H. M., and P. F. Chenea: "Mechanics of Vibration," p. 274, John Wiley & Sons, Inc., New York, 1952.
5. Jacobsen, L. S., and R. S. Ayre: "Engineering Vibrations," p. 73, McGraw-Hill Book Company, Inc., New York, 1958.
6. Ref. 4, p. 256.
7. Kerley, J. J.: *Prod. Eng., Design Digest Issue*, Mid-October, 1957, p. F34.
8. Young, D., and R. P. Felgar: "Tables of Characteristic Functions Representing Normal Modes of Vibration of a Beam," *Univ. Texas Bur. Eng. Research Bull.* 44, July 1, 1949.
9. Rayleigh, Lord: "The Theory of Sound," 2d rev. ed., vol. 1, p. 293; reprinted by Dover Publications, New York, 1945.
10. Timoshenko, S.: *Phil. Mag.* (ser. 6), **41**:744 (1921); **43**:125 (1922).
11. Sutherland, J. G., and L. E. Goodman: "Vibrations of Prismatic Bars Including Rotatory Inertia and Shear Corrections," Department of Civil Engineering, University of Illinois, Urbana, Ill., April 15, 1951.
12. Ref. 1, p. 374.
13. Ref. 1, 2d ed., p. 366.
14. Woinowsky-Krieger, S.: *J. Appl. Mechanics*, **17**:35 (1950).
15. Cranch, E. T., and A. A. Adler: *J. Appl. Mechanics*, **23**:103 (1956).
16. Macduff, J. N., and R. P. Felgar: *Trans. ASME*, **79**:1459 (1957).
17. Macduff, J. N., and R. P. Felgar: *Machine Design*, **29**(3):109 (1957).
18. Ref. 1, p. 386.
19. Darnley, E. R.: *Phil. Mag.*, **41**:81 (1921).
20. Smith, D. M.: *Engineering*, **120**:808 (1925).
21. Newmark, N. M., and A. S. Veletsos: *J. Appl. Mechanics*, **19**:563 (1952).
22. Timoshenko, S.: "Theory of Plates and Shells," p. 301, McGraw-Hill Book Company, Inc., New York, 1940.
23. Ref. 22, p. 88.
24. Ref. 22, p. 133.
25. Evans, T. H.: *J. Appl. Mechanics*, **6**:A-7 (1939).
26. Ref. 22, p. 58.
27. Ref. 22, p. 304.
28. Ref. 22, p. 344.
29. Young, D.: *J. Appl. Mechanics*, **17**:448 (1950).
30. Barton, M. V.: *J. Appl. Mechanics*, **18**:129 (1951).
31. Hearmon, R. F. S.: *J. Appl. Mechanics*, **19**:402 (1952).
32. Anderson, B. W.: *J. Appl. Mechanics*, **21**:365 (1954).
33. Gustafson, P. N., W. F. Stokey, and C. F. Zorowski: *J. Aeronaut. Sci.*, **20**:331 (1953).
34. Gustafson, P. N., W. F. Stokey, and C. F. Zorowski: *J. Aeronaut. Sci.*, **21**:621 (1954).
35. Ref. 9, p. 359.
36. Ref. 1, p. 449.
37. Southwell, R. V.: *Proc. Roy. Soc. (London)*, **A101**:133 (1922).
38. Chu, Hu-Nan, and G. Herrmann: *J. Appl. Mechanics*, **23**:532 (1956).
39. Mindlin, R. D., A. Schacknow, and H. Deresiewicz: *J. Appl. Mechanics*, **23**:430 (1956).
40. Deresiewicz, H., and R. D. Mindlin: *J. Appl. Mechanics*, **22**:86 (1955).
41. Love, A. E. H.: "A Treatise on the Mathematical Theory of Elasticity," 4th ed., p. 451, reprinted by Dover Publications, New York, 1944.
42. Ref. 1, p. 425.
43. Mindlin, R. D., and L. E. Goodman: *J. Appl. Mechanics*, **17**:377 (1950).

8
TRANSIENT RESPONSE TO STEP AND PULSE FUNCTIONS*

Robert S. Ayre
Yale University

INTRODUCTION

In analyses involving shock and transient vibration, it is essential in most instances to begin with the time-history of a quantity that describes a motion, usually displacement, velocity, or acceleration. The method of reducing the time-history depends upon the purpose for which the reduced data will be used. When the purpose is to compare shock motions, to design equipment to withstand shock, or to formulate a laboratory test as means to simulate an environmental condition, the *response spectrum* is found to be a useful concept. This concept in data reduction is discussed in Chap. 23, and its application to the simulation of environmental conditions is discussed in Chap. 24.

This chapter deals briefly with methods of analysis for obtaining the response spectrum from the time-history, and includes in graphical form certain significant spectra for various regular step- and pulse-type excitations. The usual concept of the response spectrum is based upon the single degree-of-freedom system, usually considered linear and undamped, although useful information sometimes can be obtained by introducing nonlinearity or damping. The single degree-of-freedom system is considered to be subjected to the shock or transient vibration, and its response determined.

The *response spectrum* is a graphical presentation of a selected quantity in the response taken with reference to a quantity in the excitation. It is plotted as a function of a dimensionless parameter that includes the natural period of the responding system and a significant period of the excitation. The excitation may be defined in terms of various physical quantities, and the response spectrum likewise may depict various characteristics of the response.

LINEAR, UNDAMPED, SINGLE DEGREE-OF-FREEDOM SYSTEMS

DIFFERENTIAL EQUATION OF MOTION

It is assumed that the system is linear and undamped. The excitation, which is a known function of time alone, may be a force function $F(t)$ acting directly on the mass of the system (Fig. 8.1A) or it may be a ground motion, i.e., foundation or base motion, acting on the spring anchorage. The ground motion may be expressed as a ground dis-

---

* Chapter 8 is based on Chaps. 3 and 4 of "Engineering Vibrations," by L. S. Jacobsen and R. S. Ayre, McGraw-Hill Book Company, Inc., 1958, and on the report, "A Comparative Study of Pulse and Step-type Loads on a Simple Vibratory System," by L. S. Jacobsen and R. S. Ayre, U.S. Navy Contract N6-ori-154, T. O. 1, Stanford University, 1952.

8-2   TRANSIENT RESPONSE TO STEP AND PULSE FUNCTIONS

placement function $u(t)$ (Fig. 8.1B). In many cases, however, it is more useful to express it as a ground acceleration function $\ddot{u}(t)$ (Fig. 8.1C).

The differential equation of motion, written in terms of each of the types of excitation, is given in Eqs. (8.1a), (8.1b), and (8.1c).

$$m\ddot{x} = -kx + F(t) \quad \text{or} \quad \frac{m\ddot{x}}{k} + x = \frac{F(t)}{k} \tag{8.1a}$$

$$m\ddot{x} = -k[x - u(t)] \quad \text{or} \quad \frac{m\ddot{x}}{k} + x = u(t) \tag{8.1b}$$

$$m[\ddot{\delta}_x + \ddot{u}(t)] = -k\delta_x \quad \text{or} \quad \frac{m\ddot{\delta}_x}{k} + \delta_x = -\frac{m\ddot{u}(t)}{k} \tag{8.1c}$$

where x is the displacement (absolute displacement) of the mass relative to a *fixed reference* and δ_x is the displacement relative to a *moving anchorage* or ground. These displacements are related to the ground displacement by $x = u + \delta_x$. Similarly, the accelerations are related by $\ddot{x} = \ddot{u} + \ddot{\delta}_x$.

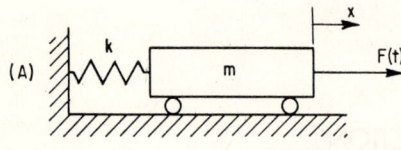

Furthermore, if Eq. (8.1b) is differentiated twice with respect to time, a differential equation is obtained in which ground acceleration $\ddot{u}(t)$ is the excitation and the absolute acceleration $\ddot{x}$ of the mass m is the variable. The equation is

$$\frac{m}{k}\frac{d^2\ddot{x}}{dt^2} + \ddot{x} = \ddot{u}(t) \tag{8.1d}$$

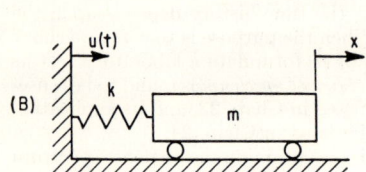

If Eq. (8.1d) is treated as a second-order equation in $\ddot{x}$ as the dependent variable, it is of the same general form as Eqs. (8.1a), (8.1b) and (8.1c).

Occasionally, the excitation is known in terms of ground velocity $\dot{u}(t)$. Differentiating Eq. (8.1b) once with respect to time, the following second-order equation in $\dot{x}$ is obtained:

$$\frac{m}{k}\frac{d^2\dot{x}}{dt^2} + \dot{x} = \dot{u}(t) \tag{8.1e}$$

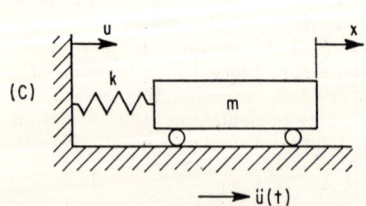

The analogy represented by Eqs. (8.1b), (8.1d), and (8.1e) may be extended further since it is generally possible to differentiate Eq. (8.1b) any number of times n:

$$\frac{m}{k}\frac{d^2}{dt^2}\left(\frac{d^n x}{dt^n}\right) + \left(\frac{d^n x}{dt^n}\right) = \left(\frac{d^n u}{dt^n}\right)(t) \tag{8.1f}$$

Fig. 8.1. Simple oscillator acted upon by known excitation functions of time: (A) force $F(t)$, (B) ground displacement $u(t)$, (C) ground acceleration $\ddot{u}(t)$.

This is of the same general form as the preceding equations if it is considered to be a second-order equation in $(d^n x/dt^n)$ as the response variable, with $(d^n u/dt^n)(t)$, a known function of time, as the excitation.

ALTERNATE FORMS OF THE EXCITATION AND OF THE RESPONSE

The foregoing equations are alike, mathematically, and a solution in terms of one of them may be applied to any of the others by making simple substitutions. Therefore,

LINEAR, UNDAMPED, SINGLE DEGREE-OF-FREEDOM SYSTEMS

the equations may be expressed in the single general form:

$$\frac{m}{k}\ddot{\nu} + \nu = \xi(t) \tag{8.2}$$

where ν and ξ are the *response* and the *excitation*, respectively, at time t.

A general notation (ν and ξ) is desirable in the presentation of response functions and response spectra for general use. However, in the discussion of examples of solution, it sometimes is preferable to use more specific notations. Both types of notation are used in this chapter. For ready reference, the alternate forms of the excitation and the response are given in Table 8.1 where $\omega_n^2 = k/m$.

Table 8.1. Alternate Forms of Excitation and Response in Eq. (8.2)

Excitation $\xi(t)$		Response ν	
Force	$\dfrac{F(t)}{k}$	Absolute displacement	x
Ground displacement	$u(t)$	Absolute displacement	x
Ground acceleration	$\dfrac{-\ddot{u}(t)}{\omega_n^2}$	Relative displacement	δ_x
Ground acceleration	$\ddot{u}(t)$	Absolute acceleration	$\ddot{x}$
Ground velocity	$\dot{u}(t)$	Absolute velocity	$\dot{x}$
nth derivative of ground displacement	$\dfrac{d^n u}{dt^n}(t)$	nth derivative of absolute displacement	$\dfrac{d^n x}{dt^n}$

METHODS OF SOLUTION OF THE DIFFERENTIAL EQUATION

A brief review of four methods of solution is given in the following sections.

CLASSICAL SOLUTION. The complete solution of the linear differential equation of motion consists of the sum of the *particular integral* x_1 and the *complementary function* x_2, that is, $x = x_1 + x_2$. Since the differential equation is of second order, two constants of integration are involved. They appear in the complementary function and are evaluated from a knowledge of the initial conditions.

Example 8.1. Versed-sine Force Pulse. In this case the differential equation of motion, applicable for the duration of the pulse, is

$$\frac{m\ddot{x}}{k} + x = \frac{F_p}{k}\frac{1}{2}\left(1 - \cos\frac{2\pi t}{\tau}\right) \quad [0 \leq t \leq \tau] \tag{8.3a}$$

where, in terms of the general notation, the excitation function $\xi(t)$ is

$$\xi(t) \equiv \frac{F(t)}{k} = \frac{F_p}{k}\frac{1}{2}\left(1 - \cos\frac{2\pi t}{\tau}\right)$$

and the response ν is displacement x. The maximum value of the pulse excitation force is F_p. The particular integral (particular solution) for Eq. (8.3a) is of the form

$$x_1 = M + N\cos\frac{2\pi t}{\tau} \tag{8.3b}$$

By substitution of the particular solution into the differential equation, the required values of the coefficients M and N are found.

The complementary function is

$$x_2 = A\cos\omega_n t + B\sin\omega_n t \tag{8.3c}$$

where A and B are the constants of integration. Combining x_2 and the explicit form of x_1 gives the complete solution:

$$x = x_1 + x_2 = \frac{F_p/2k}{1 - \tau^2/T^2}\left(1 - \frac{\tau^2}{T^2} + \frac{\tau^2}{T^2}\cos\frac{2\pi t}{\tau}\right) + A\cos\omega_n t + B\sin\omega_n t \quad (8.3d)$$

If it is assumed that the system is initially at rest, $x = 0$ and $\dot{x} = 0$ at $t = 0$, and the constants of integration are

$$A = -\frac{F_p/2k}{1 - \tau^2/T^2} \quad \text{and} \quad B = 0 \quad (8.3e)$$

The complete solution takes the following form:

$$\nu \equiv x = \frac{F_p/2k}{1 - \tau^2/T^2}\left(1 - \frac{\tau^2}{T^2} + \frac{\tau^2}{T^2}\cos\frac{2\pi t}{\tau} - \cos\omega_n t\right) \quad (8.3f)$$

If other starting conditions had been assumed, A and B would have been different from the values given by Eqs. (8.3e). It may be shown that if the starting conditions are general, namely, $x = x_0$ and $\dot{x} = \dot{x}_0$ at $t = 0$, it is necessary to superimpose on the complete solution already found, Eq. (8.3f), only the following additional terms:

$$x_0 \cos\omega_n t + \frac{\dot{x}_0}{\omega_n}\sin\omega_n t \quad (8.3g)$$

For values of time equal to or greater than τ, the differential equation is

$$m\ddot{x} + kx = 0 \quad [\tau \leq t] \quad (8.4a)$$

and the complete solution is given by the complementary function alone. However, the constants of integration must be redetermined from the known conditions of the system at time $t = \tau$. The solution is

$$\nu \equiv x = \frac{F_p}{k}\frac{\sin(\pi\tau/T)}{1 - \tau^2/T^2}\sin\omega_n\left(t - \frac{\tau}{2}\right) \quad [\tau \leq t] \quad (8.4b)$$

The additional terms given by expressions (8.3g) may be superimposed on this solution if the conditions at time $t = 0$ are general.

DUHAMEL'S INTEGRAL. The use of Duhamel's integral (convolution integral or superposition integral) is a well-known approach to the solution of transient vibration problems in linear systems. Its development [7,33] is based on the *superposition* of the responses of the system to a sequence of impulses. A general excitation function is shown in Fig. 8.2, where $F(t)$ is a known force function of time, the variable of integration is t_v between the limits of integration 0 and t and the elemental impulse is $F(t_v)\,dt_v$. It may be shown that the complete solution of the differential equation is

$$x = \left(x_0 - \frac{1}{m\omega_n}\int_0^t F(t_v)\sin\omega_n t_v\,dt_v\right)\cos\omega_n t + \left(\frac{\dot{x}_0}{\omega_n} + \frac{1}{m\omega_n}\int_0^t F(t_v)\cos\omega_n t_v\,dt_v\right)\sin\omega_n t$$
$$(8.5)$$

where x_0 and $\dot{x}_0$ are the initial conditions of the system at zero time.

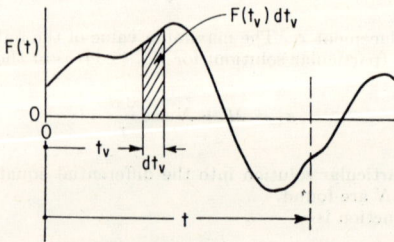

Fig. 8.2. General excitation and the elemental impulse.

Example 8.2. Half-cycle Sine, Ground Displacement Pulse. Consider the following excitation:

$$\xi(t) \equiv u(t) = \begin{cases} u_p \sin \dfrac{\pi t}{\tau} & [0 \leq t \leq \tau] \\ 0 & [\tau \leq t] \end{cases}$$

The maximum value of the excitation displacement is u_p. Assume that the system is initially at rest, so that $x_0 = \dot{x}_0 = 0$. Expressing the excitation function in terms of the variable of integration t_v, Eq. (8.5) may be rewritten for this particular case in the following form:

$$x = \frac{ku_p}{m\omega_n}\left(-\cos\omega_n t \int_0^t \sin\frac{\pi t_v}{\tau}\sin\omega_n t_v\, dt_v + \sin\omega_n t \int_0^t \sin\frac{\pi t_v}{\tau}\cos\omega_n t_v\, dt_v\right) \quad (8.6a)$$

Equation (8.6a) may be reduced, by evaluation of the integrals, to

$$\nu \equiv x = \frac{u_p}{1 - T^2/4\tau^2}\left(\sin\frac{\pi t}{\tau} - \frac{T}{2\tau}\sin\omega_n t\right) \quad [0 \leq t \leq \tau] \quad (8.6b)$$

where $T = 2\pi/\omega_n$ is the natural period of the responding system.

For the second era of time, where $\tau \leq t$, it is convenient to choose a new time variable $t' = t - \tau$. Noting that $u(t) = 0$ for $\tau \leq t$, and that for continuity in the system response the initial conditions for the second era must equal the closing conditions for the first era, it is found from Eq. (8.5) that the response for the second era is

$$x = x_\tau \cos\omega_n t' + \frac{\dot{x}_\tau}{\omega_n}\sin\omega_n t' \quad (8.7a)$$

where x_τ and $\dot{x}_\tau$ are the displacement and velocity of the system at time $t = \tau$ and hence at $t' = 0$. Equation (8.7a) may be rewritten in the following form:

$$\nu \equiv x = u_p \frac{(T/\tau)\cos(\pi\tau/T)}{(T^2/4\tau^2) - 1}\sin\omega_n\left(t - \frac{\tau}{2}\right) \quad [\tau \leq t] \quad (8.7b)$$

PHASE-PLANE GRAPHICAL METHOD. Several numerical and graphical methods,[18,23,27,37] all related in general but differing considerably in the details of procedure, are available for the solution of linear transient vibration problems. Of these methods, the phase-plane graphical method [25] is one of the most useful. The procedure is basically

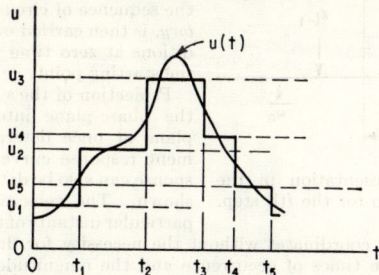

FIG. 8.3. General excitation approximated by a sequence of finite rectangular steps.

very simple, it gives a clear physical picture of the response of the system, and it may be applied readily to some classes of *nonlinear* systems.[3,5,6,8,13,15,21,22]

In Fig. 8.3 a general excitation in terms of ground displacement is represented, approximately, by a sequence of finite steps. The ith step has the total height u_i, where u_i is constant for the duration of the step. The differential equation of motion and its complete solution, applying for the duration of the step, are

$$\frac{m\ddot{x}}{k} + x = u_i \quad [t_{i-1} \leq t \leq t_i] \quad (8.8a)$$

$$x - u_i = (x_{i-1} - u_i)\cos\omega_n(t - t_{i-1}) + \frac{\dot{x}_{i-1}}{\omega_n}\sin\omega_n(t - t_{i-1}) \quad (8.8b)$$

where x_{i-1} and $\dot{x}_{i-1}$ are the displacement and velocity of the system at time t_{i-1}; consequently, they are the initial conditions for the ith step. The system velocity (divided by ω_n) during the ith step is

$$\frac{\dot{x}}{\omega_n} = -(x_{i-1} - u_i)\sin\omega_n(t - t_{i-1}) + \frac{\dot{x}_{i-1}}{\omega_n}\cos\omega_n(t - t_{i-1}) \qquad (8.8c)$$

Squaring Eqs. (8.8b) and (8.8c) and adding them,

$$\left(\frac{\dot{x}}{\omega_n}\right)^2 + (x - u_i)^2 = \left(\frac{\dot{x}_{i-1}}{\omega_n}\right)^2 + (x_{i-1} - u_i)^2 \qquad (8.8d)$$

This is the equation of a circle in a rectangular system of coordinates $\dot{x}/\omega_n$, x. The center is at 0, u_i; and the radius is

$$R_i = \left[\left(\frac{\dot{x}_{i-1}}{\omega_n}\right)^2 + (x_{i-1} - u_i)^2\right]^{1/2} \qquad (8.8e)$$

The solution for Eq. (8.8a) for the ith step may be shown, as in Fig. 8.4, to be the arc of the circle of radius R_i and center 0, u_i, subtended by the angle $\omega_n(t_i - t_{i-1})$ and starting at the point $\dot{x}_{i-1}/\omega_n$, x_{i-1}. Time is positive in the counterclockwise direction.

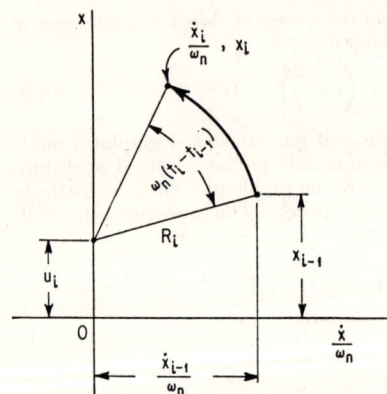

FIG. 8.4. Graphical representation in the phase-plane of the solution for the ith step.

Example 8.3. Application to a General Pulse Excitation. Figure 8.5 shows an application of the method for the general excitation $u(t)$ represented by seven steps in the time-displacement plane. Upon choice of the step heights u_i and durations $(t_i - t_{i-1})$, the arc-center locations can be projected onto the X axis in the phase-plane and the arc angles $\omega_n(t_i - t_{i-1})$ can be computed. The graphical construction of the sequence of circular arcs, the *phase trajectory*, is then carried out, using the system conditions at zero time (in this example, 0,0) as the starting point.

Projection of the system displacements from the phase-plane into the time-displacement plane at once determines the time-displacement response curve. The time-velocity response can also be determined by projection as shown. The velocities and displacements at particular instants of time can be found directly from the phase trajectory coordinates without the necessity for drawing the time-response curves. Furthermore, the times of occurrence and the magnitudes of all the maxima also can be obtained directly from the phase trajectory.

Good accuracy is obtainable by using reasonable care in the graphical construction and in the choice of the steps representing the excitation. Usually, the time intervals should not be longer than about one-fourth the natural period of the system.[22]

THE LAPLACE TRANSFORMATION. The Laplace transformation provides a powerful tool for the solution of linear differential equations. The following discussion of the technique of its application is limited to the differential equation of the type applying to the undamped linear oscillator. Application to the linear oscillator with viscous damping is illustrated in a later part of this chapter.

Definitions. The Laplace transform $F(s)$ of a known function $f(t)$, where $t > 0$, is defined by

$$F(s) = \int_0^\infty e^{-st} f(t)\, dt \qquad (8.9a)$$

where s is a complex variable. The transformation is abbreviated as

$$F(s) = \mathcal{L}[f(t)] \tag{8.9b}$$

The limitations on the function $f(t)$ are not discussed here. For the conditions for existence of $\mathcal{L}[f(t)]$, for complete accounts of the technique of application, and for extensive tables of function-transform pairs, the references should be consulted.[16, 17, 34, 36]

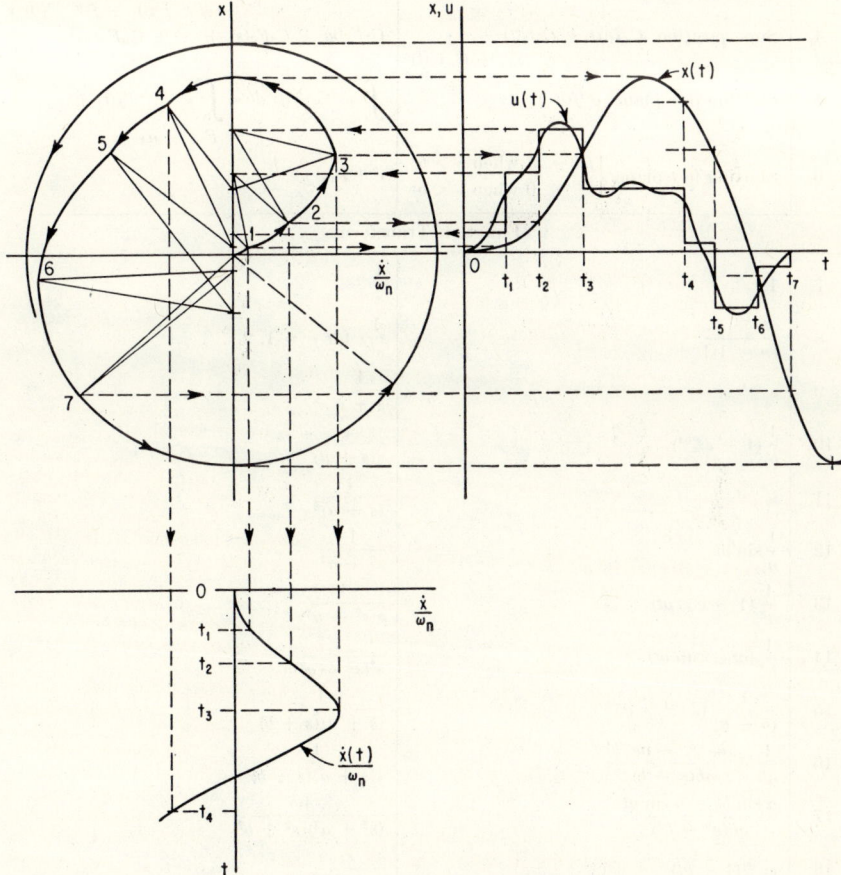

FIG. 8.5. Example of phase-plane graphical solution.[2]

General Steps in Solution of the Differential Equation. In the solution of a differential equation by Laplace transformation, the first step is to transform the differential equation, in the variable t, into an algebraic equation in the complex variable s. Then, the algebraic equation is solved, and the solution of the differential equation is determined by an *inverse* transformation of the solution of the algebraic equation. The process of inverse Laplace transformation is symbolized by

$$\mathcal{L}^{-1}[F(s)] = f(t) \tag{8.10}$$

Tables of Function-transform Pairs. The processes symbolized by Eqs. (8.9b) and (8.10) are facilitated by the use of tables of function-transform pairs. Table 8.2 is a brief

Table 8.2. Pairs of Functions $f(t)$ and Laplace Transforms $F(s)$

	$f(t)$	$F(s)$
	Operation Transforms	
1	Definition, $f(t)$	$F(s) = \int_0^\infty e^{-st} f(t)\, dt$
2	First derivative, $f'(t)$	$sF(s) - f(0)$
3	nth derivative, $f^{(n)}(t)$	$s^n F(s) - s^{n-1} f(0) - s^{n-2} f'(0) - \cdots - sf^{(n-2)}(0) - f^{(n-1)}(0)$ †
4	Superposition, $C_1 f_1(t) + C_2 f_2(t) + \cdots + C_n f_n(t)$	$C_1 F_1(s) + C_2 F_2(s) + \cdots + C_n F_n(s)$
5	Shifting in s plane, $e^{at} f(t)$	$\int_0^\infty e^{-st} e^{at} f(t)\, dt = \int_0^\infty e^{-(s-a)t} f(t)\, dt = F(s-a)$
6	Shifting in t plane $\begin{cases} f(t-b) \text{ when } t > b, \\ 0 \text{ when } t < b \end{cases}$	$e^{-bs} F(s)$
	Function Transforms	
7	1	$\dfrac{1}{s}$
8	$\dfrac{t^{n-1}}{(n-1)!}$	$\dfrac{1}{s^n}$, for $n = 1, 2, \ldots$
9	e^{-at}	$\dfrac{1}{s+a}$
10	$\dfrac{1}{a}(1 - e^{-at})$	$\dfrac{1}{s(s+a)}$
11	te^{-at}	$\dfrac{1}{(s+a)^2}$
12	$\dfrac{1}{a} \sin at$	$\dfrac{1}{s^2 + a^2}$
13	$\dfrac{1}{a^2}(1 - \cos at)$	$\dfrac{1}{s(s^2 + a^2)}$
14	$\dfrac{1}{a^3}(at - \sin at)$	$\dfrac{1}{s^2(s^2 + a^2)}$
15	$\dfrac{1}{(b-a)}(e^{-at} - e^{-bt})$	$\dfrac{1}{(s+a)(s+b)}$
16	$\dfrac{1}{ab} + \dfrac{be^{-at} - ae^{-bt}}{ab(a-b)}$	$\dfrac{1}{s(s+a)(s+b)}$
17	$\dfrac{a \sin bt - b \sin at}{ab(a^2 - b^2)}$	$\dfrac{1}{(s^2 + a^2)(s^2 + b^2)}$
18	$e^{-at}(1 - at)$	$\dfrac{s}{(s+a)^2}$
19	$\cos at$	$\dfrac{s}{s^2 + a^2}$
20	Rectangular pulse	$\dfrac{1 - e^{-s\tau}}{s}$
21	Sine pulse	$\dfrac{\pi/\tau}{s^2 + \pi^2/\tau^2}(1 + e^{-s\tau})$

† $f(t)$ and its derivatives through $f^{(n-1)}(t)$ must be continuous.

LINEAR, UNDAMPED, SINGLE DEGREE-OF-FREEDOM SYSTEMS

example. Transforms for general operations, such as differentiation, are included as well as transforms of explicit functions.

In general, the transforms of the explicit functions can be obtained by carrying out the integration indicated by the definition of the Laplace transformation. For example:

For $f(t) = 1$:

$$F(s) = \int_0^\infty e^{-st}\, dt = -\frac{1}{s} e^{-st} \Big]_0^\infty = \frac{1}{s}$$

Transformation of the Differential Equation. The differential equation for the undamped linear oscillator is given in general form by

$$\frac{1}{\omega_n^2} \ddot{\nu} + \nu = \xi(t) \tag{8.11}$$

Applying the operational transforms (items 1 and 3, Table 8.2), Eq. (8.11) is transformed to

$$\frac{1}{\omega_n^2} s^2 F_r(s) - \frac{1}{\omega_n^2} sf(0) - \frac{1}{\omega_n^2} f'(0) + F_r(s) = F_e(s) \tag{8.12a}$$

where $F_r(s)$ = the transform of the unknown response $\nu(t)$, sometimes called the *response transform*

$s^2 F_r(s) - sf(0) - f'(0)$ = the transform of the second derivative of $\nu(t)$

$f(0)$ and $f'(0)$ = the known *initial values* of ν and $\dot{\nu}$, i.e., ν_0 and $\dot{\nu}_0$

$F_e(s)$ = the transform of the known excitation function $\xi(t)$, written $F_e(s) = \mathcal{L}[\xi(t)]$, sometimes called the *driving transform*

It should be noted that the initial conditions of the system are explicit in Eq. (8.12a).

The Subsidiary Equation. Solving Eq. (8.12a) for $F_r(s)$,

$$F_r(s) = \frac{sf(0) + f'(0) + \omega_n^2 F_e(s)}{s^2 + \omega_n^2} \tag{8.12b}$$

This is known as the *subsidiary equation* of the differential equation. The first two terms of the transform derive from the initial conditions of the system, and the third term derives from the excitation.

Inverse Transformation. In order to determine the response function $\nu(t)$, which is the solution of the differential equation, an inverse transformation is performed on the subsidiary equation. The entire operation, applied explicitly to the solution of Eq. (8.11), may be abbreviated as follows:

$$\nu(t) = \mathcal{L}^{-1}[F_r(s)] = \mathcal{L}^{-1}\left[\frac{s\nu_0 + \dot{\nu}_0 + \omega_n^2 \mathcal{L}[\xi(t)]}{s^2 + \omega_n^2}\right] \tag{8.13}$$

Example 8.4. Rectangular Step Excitation. In this case $\xi(t) = \xi_c$ for $0 \leq t$ (Fig. 8.6A). The Laplace transform $F_e(s)$ of the excitation is, from item 7 of Table 8.2,

$$\mathcal{L}[\xi_c] = \xi_c \mathcal{L}[1] = \xi_c \frac{1}{s}$$

Assume that the starting conditions are general, that is, $\nu = \nu_0$ and $\dot{\nu} = \dot{\nu}_0$ at $t = 0$. Substituting the transform and the starting conditions into Eq. (8.13), the following is obtained:

$$\nu(t) = \mathcal{L}^{-1}\left[\frac{s\nu_0 + \dot{\nu}_0 + \omega_n^2 \xi_c(1/s)}{s^2 + \omega_n^2}\right] \tag{8.14a}$$

The foregoing may be rewritten as three separate inverse transforms:

$$\nu(t) = \nu_0 \mathcal{L}^{-1}\left[\frac{s}{s^2 + \omega_n^2}\right] + \dot{\nu}_0 \mathcal{L}^{-1}\left[\frac{1}{s^2 + \omega_n^2}\right] + \xi_c \omega_n^2 \mathcal{L}^{-1}\left[\frac{1}{s(s^2 + \omega_n^2)}\right] \tag{8.14b}$$

8-10 TRANSIENT RESPONSE TO STEP AND PULSE FUNCTIONS

The inverse transforms in Eq. (8.14b) are evaluated by use of items 19, 12 and 13, respectively, in Table 8.2. Thus, the time-response function is given explicitly by

$$v(t) = v_0 \cos \omega_n t + \frac{\dot{v}_0}{\omega_n} \sin \omega_n t + \xi_c(1 - \cos \omega_n t) \qquad (8.14c)$$

The first two terms are the same as the starting condition response terms given by expressions (8.20a). The third term agrees with the response function shown by Eq. (8.22), derived for the case of a start from rest.

Example 8.5. Rectangular Pulse Excitation. The excitation function, Fig. 8.6B, is given by

$$\xi(t) = \begin{cases} \xi_p & \text{for } 0 \leq t \leq \tau \\ 0 & \text{for } \tau \leq t \end{cases}$$

For simplicity, assume a start from rest, i.e., $v_0 = 0$ and $\dot{v}_0 = 0$ when $t = 0$.

During the first time interval, $0 \leq t \leq \tau$, the response function is of the same form as Eq. (8.14c) except that, with the assumed start from rest, the first two terms are zero.

During the second interval, $\tau \leq t$, the transform of the excitation is obtained by applying the delayed-function transform (item 6, Table 8.2) and the transform for the rectangular step function (item 7) with the following result:

$$F_e(s) = \mathcal{L}[\xi(t)] = \xi_p \left(\frac{1}{s} - \frac{e^{-s\tau}}{s} \right)$$

This is the transform of an excitation consisting of a rectangular step of height $-\xi_p$ starting at time $t = \tau$, superimposed on the rectangular step of height $+\xi_p$ starting at time $t = 0$.

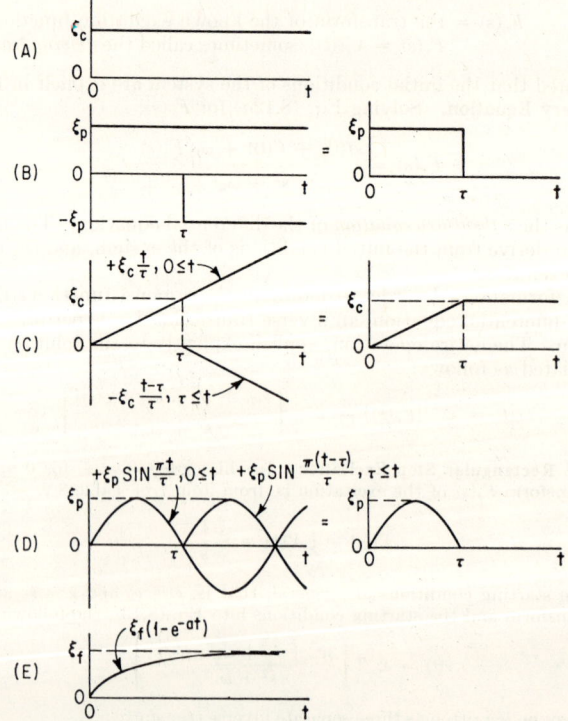

FIG. 8.6. Excitation functions in examples of use of the Laplace transform: (A) rectangular step, (B) rectangular pulse, (C) step with constant-slope front, (D) sine pulse, and (E) step with exponential asymptotic rise.

LINEAR, UNDAMPED, SINGLE DEGREE-OF-FREEDOM SYSTEMS

Substituting for $\mathcal{L}[\xi(t)]$ in Eq. (8.13),

$$v(t) = \xi_p \omega_n^2 \left\{ \mathcal{L}^{-1}\left[\frac{1}{s(s^2 + \omega_n^2)}\right] - \mathcal{L}^{-1}\left[\frac{e^{-s\tau}}{s(s^2 + \omega_n^2)}\right] \right\} \quad (8.15a)$$

The first inverse transform in Eq. (8.15a) is the same as the third one in Eq. (8.14b) and is evaluated by use of item 13 in Table 8.2. However, the second inverse transform requires the use of items 6 and 13. The function-transform pair given by item 6 indicates that when $t < b$ the inverse transform in question is zero, and when $t > b$ the inverse transform is evaluated by replacing t by $t - b$ (in this particular case, by $t - \tau$). The result is as follows:

$$v(t) = \xi_p \omega_n^2 \left\{ \frac{1}{\omega_n^2}(1 - \cos \omega_n t) - \frac{1}{\omega_n^2}[1 - \cos \omega_n(t - \tau)] \right\} \quad (8.15b)$$

$$= 2\xi_p \sin \frac{\pi\tau}{T} \sin \omega_n \left(t - \frac{\tau}{2}\right) \quad [\tau \leq t]$$

Theorem on the Transform of Functions Shifted in the Original (t) Plane. In Example 8.5, use is made of the theorem on the transform of functions shifted in the original plane. The theorem (item 6 in Table 8.2) is known variously as the second shifting theorem, the theorem on the transform of delayed functions, and the time-displacement theorem. In determining the transform of the excitation, the theorem provides for shifting, i.e., displacing the excitation or a component of the excitation in the positive direction along the time axis. This suggests the term *delayed function*. Examples of the shifting of component parts of the excitation appear in Fig. 8.6B, 8.6C, and 8.6D. Use of the theorem also is necessary in determining, by means of inverse transformation, the response following the delay in the excitation. Further illustration of the use of the theorem is shown by the next two examples.

Example 8.6. Step Function with Constant-slope Front. The excitation function (Fig. 8.6C) is expressed as follows:

$$\xi(t) = \begin{cases} \xi_c \dfrac{t}{\tau} & [0 \leq t \leq \tau] \\ \xi_c & [\tau \leq t] \end{cases}$$

Assume that $v_0 = 0$ and $\dot{v}_0 = 0$.

The driving transforms for the first and second time intervals are

$$\mathcal{L}[\xi(t)] = \begin{cases} \xi_c \dfrac{1}{\tau} \dfrac{1}{s^2} & [0 \leq t \leq \tau] \\ \xi_c \dfrac{1}{\tau}\left(\dfrac{1}{s^2} - \dfrac{e^{-s\tau}}{s^2}\right) & [\tau \leq t] \end{cases}$$

The transform for the second interval is the transform of a *negative* constant slope excitation, $-\xi_c(t-\tau)/\tau$, starting at $t = \tau$, superimposed on the transform for the positive constant slope excitation, $+\xi_c t/\tau$, starting at $t = 0$.

Substituting the transforms and starting conditions into Eq. (8.13), the responses for the two time eras, in terms of the transformations, are

$$v(t) = \begin{cases} \xi_c \dfrac{\omega_n^2}{\tau} \mathcal{L}^{-1}\left[\dfrac{1}{s^2(s^2 + \omega_n^2)}\right] & [0 \leq t \leq \tau] \\ \xi_c \dfrac{\omega_n^2}{\tau}\left\{\mathcal{L}^{-1}\left[\dfrac{1}{s^2(s^2 + \omega_n^2)}\right] - \mathcal{L}^{-1}\left[\dfrac{e^{-s\tau}}{s^2(s^2 + \omega_n^2)}\right]\right\} & [\tau \leq t] \end{cases} \quad (8.16a)$$

Evaluation of the inverse transforms by reference to Table 8.2 [item 14 for the first of Eqs. (8.16a), items 6 and 14 for the second] leads to the following:

$$v(t) = \begin{cases} \xi_c \dfrac{\omega_n^2}{\tau} \dfrac{1}{\omega_n^3}(\omega_n t - \sin \omega_n t) & [0 \leq t \leq \tau] \\ \xi_c \dfrac{\omega_n^2}{\tau}\left\{\dfrac{1}{\omega_n^3}(\omega_n t - \sin \omega_n t) - \dfrac{1}{\omega_n^3}[\omega_n(t - \tau) - \sin \omega_n(t - \tau)]\right\} & [\tau \leq t] \end{cases}$$

TRANSIENT RESPONSE TO STEP AND PULSE FUNCTIONS

Simplifying,

$$\nu(t) = \begin{cases} \xi_c \dfrac{1}{\omega_n \tau}(\omega_n t - \sin \omega_n t) & [0 \leq t \leq \tau] \\ \xi_c \left[1 + \dfrac{2}{\omega_n \tau} \sin \dfrac{\omega_n \tau}{2} \cos \omega_n \left(t - \dfrac{\tau}{2}\right)\right] & [\tau \leq t] \end{cases} \quad (8.16b)$$

Example 8.7. Half-cycle Sine Pulse. The excitation function (Fig. 8.6D) is

$$\xi(t) = \begin{cases} \xi_p \sin \dfrac{\pi t}{\tau} & [0 \leq t \leq \tau] \\ 0 & [\tau \leq t] \end{cases}$$

Let the system start from rest. The driving transforms are

$$\mathcal{L}[\xi(t)] = \begin{cases} \xi_p \dfrac{\pi}{\tau} \dfrac{1}{s^2 + \pi^2/\tau^2} & [0 \leq t \leq \tau] \\ \xi_p \dfrac{\pi}{\tau} \left(\dfrac{1}{s^2 + \pi^2/\tau^2} + \dfrac{e^{-s\tau}}{s^2 + \pi^2/\tau^2}\right) & [\tau \leq t] \end{cases}$$

The driving transform for the second interval is the transform of a sine wave of *positive* amplitude ξ_p and frequency π/τ starting at time $t = \tau$, superimposed on the transform of a sine wave of the same amplitude and frequency starting at time $t = 0$.

By substitution of the driving transforms and the starting conditions into Eq. (8.13), the following are found:

$$\nu(t) = \begin{cases} \xi_p \dfrac{\pi}{\tau} \omega_n^2 \mathcal{L}^{-1} \left[\dfrac{1}{s^2 + \pi^2/\tau^2} \cdot \dfrac{1}{s^2 + \omega_n^2}\right] & [0 \leq t \leq \tau] \\ \xi_p \dfrac{\pi}{\tau} \omega_n^2 \left\{ \mathcal{L}^{-1} \left[\dfrac{1}{s^2 + \pi^2/\tau^2} \cdot \dfrac{1}{s^2 + \omega_n^2}\right] + \mathcal{L}^{-1} \left[\dfrac{e^{-s\tau}}{s^2 + \pi^2/\tau^2} \cdot \dfrac{1}{s^2 + \omega_n^2}\right] \right\} & [\tau \leq t] \end{cases} \quad (8.17a)$$

Determining the inverse transforms from Table 8.2 [item 17 for the first of Eqs. (8.17a), items 6 and 17 for the second]:

$$\nu(t) = \begin{cases} \xi_p \dfrac{\pi}{\tau} \omega_n^2 \dfrac{\omega_n \sin(\pi t/\tau) - (\pi/\tau) \sin \omega_n t}{(\pi \omega_n/\tau)(\omega_n^2 - \pi^2/\tau^2)} & [0 \leq t \leq \tau] \\ \xi_p \dfrac{\pi}{\tau} \omega_n^2 \left[\dfrac{\omega_n \sin(\pi t/\tau) - (\pi/\tau) \sin \omega_n t}{(\pi \omega_n/\tau)(\omega_n^2 - \pi^2/\tau^2)} \right. \\ \left. \qquad + \dfrac{\omega_n \sin[\pi(t-\tau)/\tau] - (\pi/\tau) \sin \omega_n (t-\tau)}{(\pi \omega_n/\tau)(\omega_n^2 - \pi^2/\tau^2)}\right] & [\tau \leq t] \end{cases}$$

Simplifying,

$$\nu(t) = \begin{cases} \xi_p \dfrac{1}{1 - T^2/4\tau^2} \left(\sin \dfrac{\pi t}{\tau} - \dfrac{T}{2\tau} \sin \omega_n t\right) & [0 \leq t \leq \tau] \\ \xi_p \dfrac{(T/\tau) \cos(\pi \tau/T)}{(T^2/4\tau^2) - 1} \sin \omega_n \left(t - \dfrac{\tau}{2}\right) & [\tau \leq t] \end{cases} \quad (8.17b)$$

where $T = 2\pi/\omega_n$ is the natural period of the responding system. Equations (8.17b) are equivalent to Eqs. (8.6b) and (8.7b) derived previously by the use of Duhamel's integral.

Example 8.8. Exponential Asymptotic Step. The excitation function (Fig. 8.6E) is

$$\xi(t) = \xi_f(1 - e^{-at}) \quad [0 \leq t]$$

Assume that the system starts from rest. The driving transform is

$$\mathcal{L}[\xi(t)] = \xi_f \left(\dfrac{1}{s} - \dfrac{1}{s+a}\right) = \xi_f a \dfrac{1}{s(s+a)}$$

LINEAR, UNDAMPED, SINGLE DEGREE-OF-FREEDOM SYSTEMS 8–13

It is found by Eq. (8.13) that

$$\nu(t) = \xi_f a \omega_n^2 \mathcal{L}^{-1}\left[\frac{1}{s(s+a)(s^2+\omega_n^2)}\right] \quad [0 \le t] \quad (8.18a)$$

It frequently happens that the inverse transform is not readily found in an available table of transforms. Using the above case as an example, the function of s in Eq. (8.18a) is first *expanded in partial fractions*; then the inverse transforms are sought, thus:

$$\frac{1}{s(s+a)(s^2+\omega_n^2)} = \frac{\kappa_1}{s} + \frac{\kappa_2}{s+a} + \frac{\kappa_3}{s+j\omega_n} + \frac{\kappa_4}{s-j\omega_n} \quad (8.18b)$$

where $j = \sqrt{-1}$

$$\kappa_1 = \left[\frac{1}{(s+a)(s+j\omega_n)(s-j\omega_n)}\right]_{s=0} = \frac{1}{a\omega_n^2}$$

$$\kappa_2 = \left[\frac{1}{s(s+j\omega_n)(s-j\omega_n)}\right]_{s=-a} = \frac{1}{-a(a^2+\omega_n^2)}$$

$$\kappa_3 = \left[\frac{1}{s(s+a)(s-j\omega_n)}\right]_{s=-j\omega_n} = \frac{1}{-2\omega_n^2(a-j\omega_n)}$$

$$\kappa_4 = \left[\frac{1}{s(s+a)(s+j\omega_n)}\right]_{s=+j\omega_n} = \frac{1}{-2\omega_n^2(a+j\omega_n)}$$

Consequently, Eq. (8.18a) may be rewritten in the following expanded form:

$$\nu(t) = \xi_f \left\{ \mathcal{L}^{-1}\left[\frac{1}{s}\right] - \frac{\omega_n^2}{a^2+\omega_n^2}\mathcal{L}^{-1}\left[\frac{1}{s+a}\right] - \frac{a}{2(a-j\omega_n)}\mathcal{L}^{-1}\left[\frac{1}{s+j\omega_n}\right] - \frac{a}{2(a+j\omega_n)}\mathcal{L}^{-1}\left[\frac{1}{s-j\omega_n}\right] \right\} \quad (8.18c)$$

The inverse transforms may now be found readily (items 7 and 9, Table 8.2):

$$\nu(t) = \xi_f\left[1 - \frac{\omega_n^2}{a^2+\omega_n^2}e^{-at} - \frac{a}{2(a-j\omega_n)}e^{-j\omega_n t} - \frac{a}{2(a+j\omega_n)}e^{j\omega_n t}\right]$$

Rewriting,

$$\nu(t) = \xi_f\left[1 - \frac{\omega_n^2 e^{-at} + a^2 \tfrac{1}{2}(e^{j\omega_n t}+e^{-j\omega_n t}) - aj\omega_n \tfrac{1}{2}(e^{j\omega_n t}-e^{-j\omega_n t})}{a^2+\omega_n^2}\right]$$

Making use of the relations, $\cos z = (\tfrac{1}{2})(e^{jz}+e^{-jz})$ and $\sin z = -j(\tfrac{1}{2})(e^{jz}-e^{-jz})$, the equation for $\nu(t)$ may be expressed as follows:

$$\nu(t) = \xi_f\left[1 - \frac{(a/\omega_n)[\sin \omega_n t + (a/\omega_n)\cos \omega_n t] + e^{-at}}{1+a^2/\omega_n^2}\right] \quad (8.18d)$$

Partial Fraction Expansion of $F(s)$. The partial fraction expansion of $F_r(s)$, illustrated for a particular case in Eq. (8.18b), is a necessary part of the technique of solution. In general $F_r(s)$, expressed by the subsidiary equation (8.12b) and involved in the inverse transformation, Eqs. (8.10) and (8.13), is a quotient of two polynomials in s, thus

$$F_r(s) = \frac{A(s)}{B(s)} \quad (8.19)$$

The purpose of the expansion of $F_r(s)$ is to divide it into simple parts, the inverse transforms of which may be determined readily. The general procedure of the expansion is to factor $B(s)$ and then to rewrite $F_r(s)$ in partial fractions.[16, 17, 34, 36]

INITIAL CONDITIONS OF THE SYSTEM

In all the solutions for response presented in this chapter, unless otherwise stated, it is assumed that the initial conditions (ν_0 and $\dot{\nu}_0$) of the system are both zero. Other starting conditions may be accounted for merely by superimposing on the time-response functions given the additional terms

$$\nu_0 \cos \omega_n t + \frac{\dot{\nu}_0}{\omega_n} \sin \omega_n t \qquad (8.20a)$$

These terms are the complete solution of the homogeneous differential equation, $m\ddot{\nu}/k + \nu = 0$. They represent the free vibration resulting from the initial conditions.

The two terms in Eq. (8.20a) may be expressed by either one of the following combined forms:

$$\sqrt{\nu_0^2 + \left(\frac{\dot{\nu}_0}{\omega_n}\right)^2} \sin (\omega_n t + \theta_1) \qquad \text{where } \tan \theta_1 = \frac{\nu_0 \omega_n}{\dot{\nu}_0} \qquad (8.20b)$$

$$\sqrt{\nu_0^2 + \left(\frac{\dot{\nu}_0}{\omega_n}\right)^2} \cos (\omega_n t - \theta_2) \qquad \text{where } \tan \theta_2 = \frac{\dot{\nu}_0}{\nu_0 \omega_n} \qquad (8.20c)$$

where $\sqrt{\nu_0^2 + \left(\frac{\dot{\nu}_0}{\omega_n}\right)^2}$ is the resultant amplitude and θ_1 or θ_2 is the phase angle of the initial-condition free vibration.

PRINCIPLE OF SUPERPOSITION

When the system is linear, the *principle of superposition* may be employed. Any number of component excitation functions may be superimposed to obtain a prescribed total excitation function, and the corresponding component response functions may be superimposed to arrive at the total response function. However, the superposition must be carried out on a time basis and with complete regard for algebraic sign. The superposition of maximum component responses, disregarding time, may lead to completely erroneous results. For example, the response functions given by Eqs. (8.31) to (8.34) are defined completely with regard to time and algebraic sign, and may be superimposed for any combination of the excitation functions from which they have been derived.

COMPILATION OF RESPONSE FUNCTIONS AND RESPONSE SPECTRA; SINGLE DEGREE-OF-FREEDOM, LINEAR UNDAMPED SYSTEMS

STEP-TYPE EXCITATION FUNCTIONS

CONSTANT-FORCE EXCITATION (SIMPLE STEP IN FORCE). The excitation is a constant force applied to the mass at zero time, $\xi(t) \equiv F(t)/k = F_c/k$. Substituting this excitation for $F(t)/k$ in Eq. (8.1a) and solving for the absolute displacement x,

$$x = \frac{F_c}{k} (1 - \cos \omega_n t) \qquad (8.21a)$$

CONSTANT-DISPLACEMENT EXCITATION (SIMPLE STEP IN DISPLACEMENT). The excitation is a constant displacement of the ground which occurs at zero time, $\xi(t) \equiv u(t) = u_c$. Substituting for $u(t)$ in Eq. (8.1b) and solving for the absolute displacement x,

$$x = u_c(1 - \cos \omega_n t) \tag{8.21b}$$

CONSTANT-ACCELERATION EXCITATION (SIMPLE STEP IN ACCELERATION). The excitation is an instantaneous change in the ground acceleration at zero time, from zero to a constant value $\ddot{u}(t) = \ddot{u}_c$. The excitation is thus

$$\xi(t) \equiv -m\ddot{u}_c/k = -\ddot{u}_c/\omega_n^2.$$

Substituting in Eq. (8.1c) and solving for the *relative* displacement δ_x,

$$\delta_x = \frac{-\ddot{u}_c}{\omega_n}(1 - \cos \omega_n t) \tag{8.21c}$$

When the excitation is defined by a function of acceleration $\ddot{u}(t)$, it is often convenient to express the response in terms of the absolute acceleration $\ddot{x}$ of the system. The force acting on the mass in Fig. 8.1C is $-k\,\delta_x$; the acceleration $\ddot{x}$ is thus $-k\,\delta_x/m$ or $-\delta_x\omega_n^2$. Substituting $\delta_x = -\ddot{x}/\omega_n^2$ in Eq. (8.21c),

$$\ddot{x} = \ddot{u}_c(1 - \cos \omega_n t) \tag{8.21d}$$

The same result is obtained by letting $\xi(t) \equiv \ddot{u}(t) = \ddot{u}_c$ in Eq. (8.1d) and solving for $\ddot{x}$. Equation (8.21d) is similar to Eq. (8.21b) with acceleration instead of displacement on both sides of the equation. This analogy generally applies in step- and pulse-type excitations.

The absolute displacement of the mass can be obtained by integrating Eq. (8.21d) twice with respect to time, taking as initial conditions $x = \dot{x} = 0$ when $t = 0$,

$$x = \frac{\ddot{u}_c}{\omega_n^2}\left[\frac{\omega_n^2 t^2}{2} - (1 - \cos \omega_n t)\right] \tag{8.21e}$$

Equation (8.21e) also may be obtained from the relation $x = u + \delta_x$, noting that in this case $u(t) = \ddot{u}_c t^2/2$.

CONSTANT-VELOCITY EXCITATION (SIMPLE STEP IN VELOCITY). This excitation, when expressed in terms of ground or spring anchorage motion, is equivalent to prescribing, at zero time, an instantaneous change in the ground velocity from zero to a constant value $\dot{u}_c$. The excitation is $\xi(t) \equiv u(t) = \dot{u}_c t$, and the solution for the differential equation of Eq. (8.1b) is

$$x = \frac{\dot{u}_c}{\omega_n}(\omega_n t - \sin \omega_n t) \tag{8.21f}$$

For the velocity of the mass,

$$\dot{x} = \dot{u}_c(1 - \cos \omega_n t) \tag{8.21g}$$

The result of Eq. (8.21g) could have been obtained directly by letting $\xi(t) \equiv \dot{u}(t) = \dot{u}_c$ in Eq. (8.1e) and solving for the *velocity* response $\dot{x}$.

GENERAL STEP EXCITATION. A comparison of Eqs. (8.21a), (8.21b), (8.21c), (8.21d), and (8.21g) with Table 8.1 reveals that the response ν and the excitation ξ are related in a common manner. This may be expressed as follows:

$$\nu = \xi_c(1 - \cos \omega_n t) \tag{8.22}$$

where ξ_c indicates a constant value of the excitation. The excitation and response of the system are shown in Fig. 8.7.

ABSOLUTE DISPLACEMENT RESPONSE TO VELOCITY-STEP AND ACCELERATION-STEP EXCITATIONS. The absolute displacement responses to the velocity-step and the acceleration-step excitations are given by Eqs. (8.21f) and (8.21e) and are shown in Figs. 8.8 and 8.9, respectively. The comparative effects of displacement-

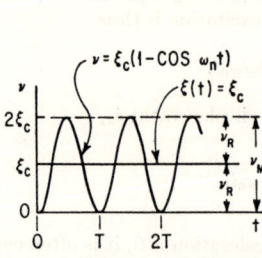

Fig. 8.7. Time response to a simple step excitation (general notation).

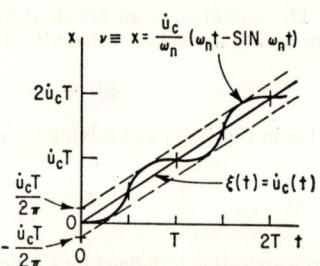

Fig. 8.8. Time-displacement response to a constant-velocity excitation (simple step in velocity).

step, velocity-step, and acceleration-step excitations, in terms of *absolute displacement* response, may be seen by comparing Figs. 8.7 to 8.9.

In the case of the velocity-step excitation, the *velocity* of the system is always positive, except at $t = 0, T, 2T, \ldots$, when it is zero. Similarly, an acceleration-step excitation results in system *acceleration* that is always positive, except at $t = 0, T, 2T, \ldots$, when it is zero. The natural period of the responding system is $T = 2\pi/\omega_n$.

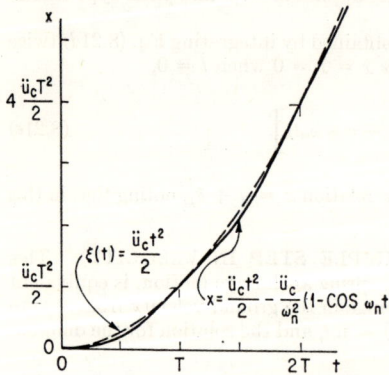

Fig. 8.9. Time-displacement response to a constant-acceleration excitation (simple step in acceleration).

RESPONSE MAXIMA. In the response of a system to step or pulse excitation, the maximum value of the response often is of considerable physical significance. Several kinds of maxima are important. One of these is the *residual response amplitude*, which is the amplitude of the free vibration about the final position of the excitation as a base. This is designated ν_R, and for the response given by Eq. (8.22):

$$\nu_R = \pm \xi_c \qquad (8.22a)$$

Another maximum is the *maximax response*, which is the greatest of the maxima of ν attained at *any time* during the response. In general, it is of the same sign as the excitation. For the response given by Eq. (8.22), the maximax response ν_M is

$$\nu_M = 2\xi_c \qquad (8.22b)$$

ASYMPTOTIC STEP. In the exponential function $\xi(t) = \xi_f(1 - e^{-at})$, the maximum value ξ_f of the excitation is approached asymptotically. This excitation may be defined alternatively by $\xi(t) = (F_f/k)(1 - e^{-at})$; $u_f(1 - e^{-at})$; $(-\ddot{u}_f/\omega_n^2)(1 - e^{-at})$; etc. (see Table 8.1). Substituting the excitation $\xi(t) = \xi_f(1 - e^{-at})$ in Eq. (8.2), the response ν is

$$\nu = \xi_f \left[1 - \frac{(a/\omega_n)[\sin \omega_n t + (a/\omega_n) \cos \omega_n t] + e^{-at}}{1 + a^2/\omega_n^2} \right] \qquad (8.23a)$$

The excitation and the response of the system are shown in Fig. 8.10. For large values of the exponent at, the motion is nearly simple harmonic. The residual amplitude, relative to the final position of equilibrium, approaches the following value asymptotically.

$$\nu_R \to \xi_f \frac{1}{\sqrt{1 + \omega_n^2/a^2}} \qquad (8.23b)$$

The maximax response $\nu_M = \nu_R + \xi_f$ is plotted against ω_n/a to give the response spectrum in Fig. 8.11.

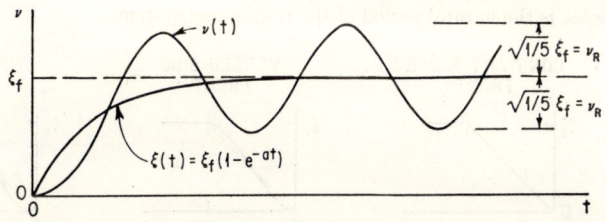

FIG. 8.10. Time response to an exponentially asymptotic step for the particular case $\omega_n/a = 2$.

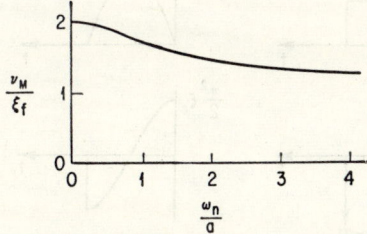

FIG. 8.11. Spectrum for maximax response resulting from exponentially asymptotic step excitation.

STEP-TYPE FUNCTIONS HAVING FINITE RISE TIME. Many step-type excitation functions rise to the constant maximum value ξ_c of the excitation in a finite length of time τ, called the *rise time*. Three such functions and their first three time derivatives are shown in Fig. 8.12. The step having a *cycloidal* front is the only one of the three that does not include an infinite third derivative; i.e., if the step is a ground displacement, it does not have an infinite rate of change of ground acceleration (infinite "jerk").

The excitation functions and the expressions for maximax response are given by the following equations:

Constant-slope front:

$$\xi(t) = \begin{cases} \xi_c \dfrac{t}{\tau} & [0 \le t \le \tau] \\ \xi_c & [\tau \le t] \end{cases} \qquad (8.24a)$$

$$\frac{\nu_M}{\xi_c} = 1 + \left| \frac{T}{\pi\tau} \sin \frac{\pi\tau}{T} \right| \qquad (8.24b)$$

Versed-sine front:

$$\xi(t) = \begin{cases} \dfrac{\xi_c}{2} \left(1 - \cos \dfrac{\pi t}{\tau}\right) & [0 \le t \le \tau] \\ \xi_c & [\tau \le t] \end{cases} \qquad (8.25a)$$

$$\frac{\nu_M}{\xi_c} = 1 + \left| \frac{1}{(4\tau^2/T^2) - 1} \cos \frac{\pi\tau}{T} \right| \qquad (8.25b)$$

Cycloidal front:

$$\xi(t) = \begin{cases} \dfrac{\xi_c}{2\pi}\left(\dfrac{2\pi t}{\tau} - \sin\dfrac{2\pi t}{\tau}\right) & [0 \leq t \leq \tau] \\ \xi_c & [\tau \leq t] \end{cases} \quad (8.26a)$$

$$\frac{\nu_M}{\xi_c} = 1 + \left|\frac{T}{\pi\tau(1 - \tau^2/T^2)} \sin\frac{\pi\tau}{T}\right| \quad (8.26b)$$

where $T = 2\pi/\omega_n$ is the natural period of the responding system.

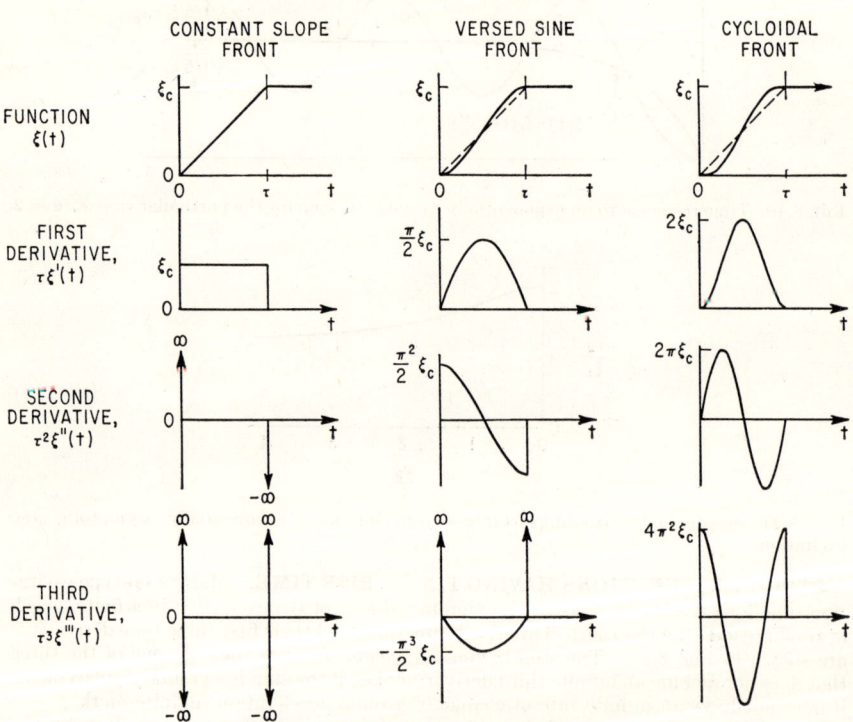

Fig. 8.12. Three step-type excitation functions and their first three time-derivatives. (*Jacobsen and Ayre.*[22])

In the case of step-type excitations, the maximax response occurs after the excitation has reached its constant maximum value ξ_c and is related to the residual response amplitude by

$$\nu_M = \nu_R + \xi_c \quad (8.27)$$

Figure 8.13 shows the spectra of maximax response versus step rise time τ expressed relative to the natural period T of the responding system. In Fig. 8.13A the comparison is based on *equal rise times*, and in Fig. 8.13B it relates to *equal maximum slopes of the step fronts*. The residual response amplitude has values of zero ($\nu_M/\xi_c = 1$) in all three cases; for example, the step excitation having a constant-slope front results in zero residual amplitude at $\tau/T = 1, 2, 3, \ldots$.

Figure 8.12 and the other illustrations crediting Ref. 22 are reprinted by permission from "Engineering Vibrations," by L. S. Jacobsen and R. S. Ayre. Copyright, 1958. McGraw-Hill Book Company, Inc., New York.

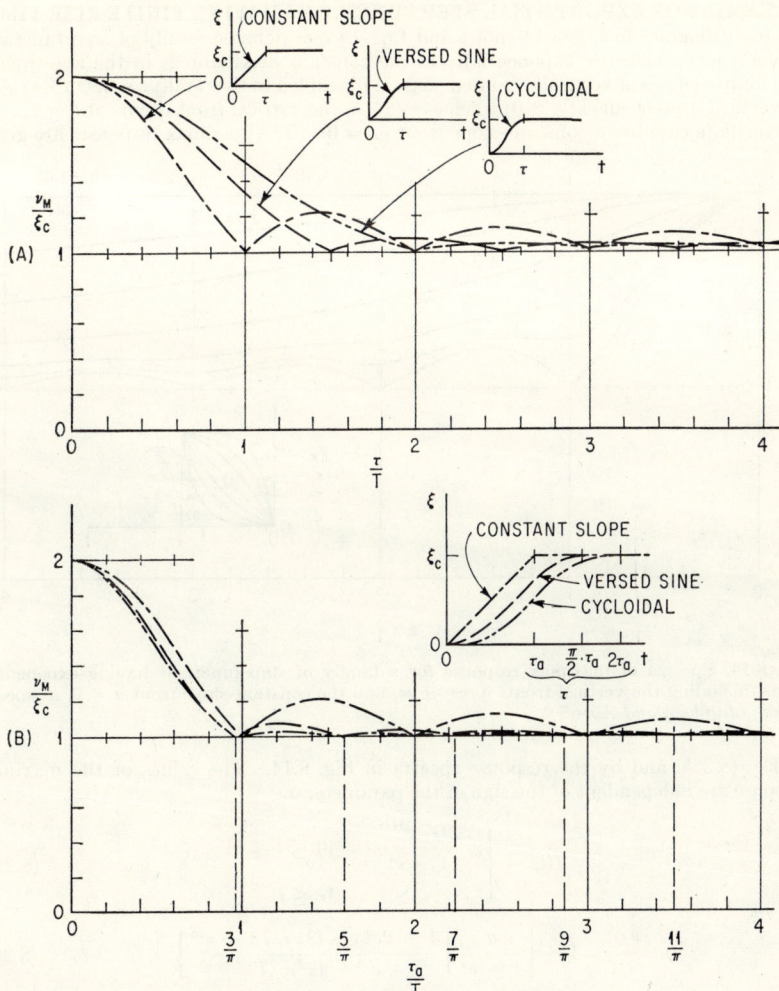

FIG. 8.13. Spectra of maximax response resulting from the step excitation functions of Fig. 8.12. (A) For step functions having equal rise time τ. (B) For step functions having equal maximum slope ξ_c/τ_a. (*Jacobsen and Ayre.*[22])

A FAMILY OF EXPONENTIAL STEP FUNCTIONS HAVING FINITE RISE TIME.

The inset diagram in Fig. 8.14 shows and Eqs. (8.28a) define a family of step functions having fronts which rise exponentially to the constant maximum ξ_c in the rise time τ. Two limiting cases of vertically fronted steps are included in the family: When $a \to -\infty$, the vertical front occurs at $t = 0$; when $a \to +\infty$, the vertical front occurs at $t = \tau$. An intermediate case has a constant-slope front ($a = 0$). The maximax responses are given

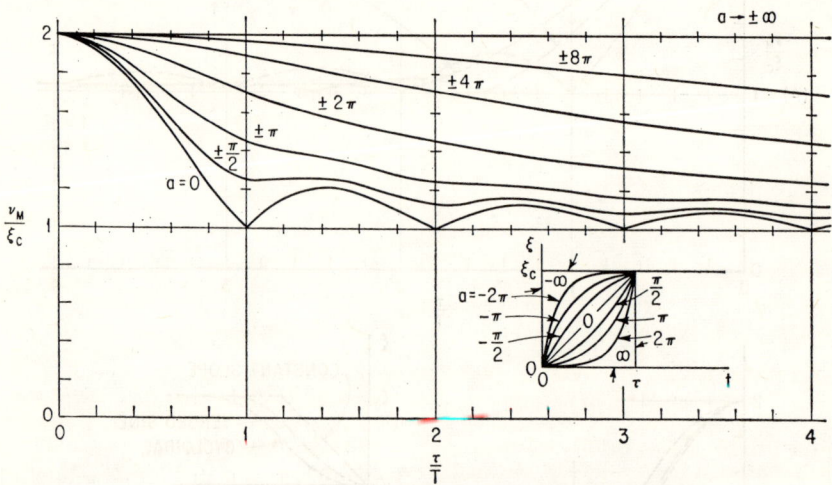

Fig. 8.14. Spectra of maximax response for a family of step functions having exponential fronts, including the vertical fronts $a \to \pm \infty$, and the constant-slope front $a = 0$, as special cases. (Jacobsen and Ayre.[22])

by Eq. (8.28b) and by the response spectra in Fig. 8.14. The values of the maximax response are independent of the sign of the parameter a.

$$\xi(t) = \begin{cases} \xi_c \dfrac{1 - e^{at/\tau}}{1 - e^a} & [0 \leq t \leq \tau] \\ \xi_c & [\tau \leq t] \end{cases} \quad (8.28a)$$

$$\frac{\nu_M}{\xi_c} = 1 + \left| \frac{a}{1 - e^a} \left[\frac{1 - 2e^a \cos(2\pi\tau/T) + e^{2a}}{a^2 + 4\pi^2\tau^2/T^2} \right]^{1/2} \right| \quad (8.28b)$$

where T is the natural period of the responding system.

There are zeroes of residual response amplitude ($\nu_M/\xi_c = 1$) at finite values of τ/T only for the constant-slope front ($a = 0$). Each of the step functions represented in Fig. 8.13 results in zeroes of residual response amplitude, and each function has antisymmetry with respect to the half-rise time $\tau/2$. This is of interest in the selection of cam and control-function shapes, where one of the criteria of choice may be *minimum residual amplitude of vibration* of the driven system.

PULSE-TYPE EXCITATION FUNCTIONS

THE SIMPLE IMPULSE. If the duration τ of the pulse is short relative to the natural period T of the system, the response of the system may be determined by equating the impulse J, i.e., the force-time integral, to the momentum $m\dot{x}_J$:

$$J = \int_0^\tau F(t)\, dt = m\dot{x}_J \quad (8.29a)$$

RESPONSE FUNCTIONS AND RESPONSE SPECTRA 8–21

Thus, it is found that the impulsive velocity $\dot{x}_J$ is equal to J/m. Consequently, the velocity-time response is given by $\dot{x} = \dot{x}_J \cos \omega_n t = (J/m) \cos \omega_n t$. The displacement-time response is obtained by integration, assuming a start from rest,

$$x = x_J \sin \omega_n t$$

where

$$x_J = \frac{J}{m\omega_n} = \omega_n \int_0^\tau \frac{F(t)\, dt}{k} \tag{8.29b}$$

The impulse concept, used for determining the response to a short-duration force pulse, may be generalized in terms of ν and ξ by referring to Table 8.1. The *generalized impulsive response* is

$$\nu = \nu_J \sin \omega_n t \tag{8.30a}$$

where the amplitude is

$$\nu_J = \omega_n \int_0^\tau \xi(t)\, dt \tag{8.30b}$$

The impulsive response amplitude ν_J and the generalized impulse $k \int_0^\tau \xi(t)\, dt$ are used in comparing the effects of various pulse shapes when the pulse durations are short.

SYMMETRICAL PULSES. In the following discussion a comparison is made of the responses caused by single symmetrical pulses of rectangular, half-cycle sine, versed-sine, and triangular shapes. The excitation functions and the time-response equations are given by Eqs. (8.31) to (8.34). Note that the residual response amplitude factors are set in brackets and are identified by the time interval $\tau \leq t$.

Rectangular:

$$\left. \begin{array}{l} \xi(t) = \xi_p \\ \nu = \xi_p(1 - \cos \omega_n t) \end{array} \right\} \qquad [0 \leq t \leq \tau] \tag{8.31a}$$

$$\left. \begin{array}{l} \xi(t) = 0 \\ \nu = \xi_p \left[2 \sin \dfrac{\pi \tau}{T} \right] \sin \omega_n \left(t - \dfrac{\tau}{2} \right) \end{array} \right\} \qquad [\tau \leq t] \tag{8.31b}$$

Half-cycle sine:

$$\left. \begin{array}{l} \xi(t) = \xi_p \sin \dfrac{\pi t}{\tau} \\ \nu = \dfrac{\xi_p}{1 - T^2/4\tau^2} \left(\sin \dfrac{\pi t}{\tau} - \dfrac{T}{2\tau} \sin \omega_n t \right) \end{array} \right\} \qquad [0 \leq t \leq \tau] \tag{8.32a}$$

$$\left. \begin{array}{l} \xi(t) = 0 \\ \nu = \xi_p \left[\dfrac{(T/\tau) \cos (\pi \tau/T)}{(T^2/4\tau^2) - 1} \right] \sin \omega_n \left(t - \dfrac{\tau}{2} \right) \end{array} \right\} \qquad [\tau \leq t] \tag{8.32b}$$

Versed-sine:

$$\left. \begin{array}{l} \xi(t) = \dfrac{\xi_p}{2} \left(1 - \cos \dfrac{2\pi t}{\tau} \right) \\ \nu = \dfrac{\xi_p/2}{1 - \tau^2/T^2} \left(1 - \dfrac{\tau^2}{T^2} + \dfrac{\tau^2}{T^2} \cos \dfrac{2\pi t}{\tau} - \cos \omega_n t \right) \end{array} \right\} \qquad [0 \leq t \leq \tau] \tag{8.33a}$$

$$\left. \begin{array}{l} \xi(t) = 0 \\ \nu = \xi_p \left[\dfrac{\sin \pi \tau/T}{1 - \tau^2/T^2} \right] \sin \omega_n \left(t - \dfrac{\tau}{2} \right) \end{array} \right\} \qquad [\tau \leq t] \tag{8.33b}$$

Triangular:

$$\left.\begin{aligned}\xi(t) &= 2\xi_p \frac{t}{\tau} \\ \nu &= 2\xi_p \left(\frac{t}{\tau} - \frac{T}{\tau}\frac{\sin \omega_n t}{2\pi}\right)\end{aligned}\right\} \qquad \left[0 \leq t \leq \frac{\tau}{2}\right] \qquad (8.34a)$$

$$\left.\begin{aligned}\xi(t) &= 2\xi_p \left(1 - \frac{t}{\tau}\right) \\ \nu &= 2\xi_p \left(1 - \frac{t}{\tau} - \frac{T}{\tau}\frac{\sin \omega_n t}{2\pi} + \frac{T}{\tau}\frac{\sin \omega_n(t - \tau/2)}{\pi}\right)\end{aligned}\right\} \left[\frac{\tau}{2} \leq t \leq \tau\right] \quad (8.34b)$$

$$\left.\begin{aligned}\xi(t) &= 0 \\ \nu &= \xi_p \left[2\frac{\sin^2(\pi\tau/2T)}{\pi\tau/2T}\right] \sin \omega_n(t - \tau/2)\end{aligned}\right\} \qquad [\tau \leq t] \qquad (8.34c)$$

where T is the natural period of the responding system.

Equal Maximum Height of Pulse as Basis of Comparison. Examples of time response, for six different values of τ/T, are shown separately for the rectangular, half-cycle sine, and versed-sine pulses in Fig. 8.15, and for the triangular pulse in Fig. 8.22B. The basis of comparison is equal maximum height of excitation pulse ξ_p.

Residual Response Amplitude and Maximax Response. The spectra of maximax response ν_M and residual response amplitude ν_R are given in Fig. 8.16 by (A) for the rectangular pulse, by (B) for the sine pulse, and by (C) for the versed-sine pulse. The maximax response may occur either within the duration of the pulse or after the pulse function has dropped to zero. In the latter case the maximax response is equal to the residual response amplitude. In general, the maximax response is given by the residual response amplitude only in the case of short-duration pulses; for example, see the case $\tau/T = \frac{1}{4}$ in Fig. 8.15 where T is the natural period of the responding system. The response spectra for the triangular pulse appear in Fig. 8.24.

Maximax Relative Displacement When the Excitation Is Ground Displacement. When the excitation $\xi(t)$ is given as *ground displacement* $u(t)$, the response ν is the absolute displacement x of the mass (Table 8.1). It is of practical importance in the investigation of the maximax *distortion* or *stress* in the elastic element to know the maximax value of the relative displacement. In this case the relative displacement is a *derived quantity* obtained by taking the difference between the response and the excitation, that is, $x - u$ or, in terms of the general notation, $\nu - \xi$.

If the excitation is given as ground acceleration, the response is determined directly as relative displacement and is designated δ_x (Table 8.1). To avoid confusion, relative displacement determined as a *derived quantity*, as described in the first case above, is designated by $x - u$; relative displacement determined directly as the *response variable* (second case above) is designated by δ_x. The distinction is made readily in the general notation by use of the symbols $\nu - \xi$ and ν, respectively, for relative response and for response. The maximax values are designated $(\nu - \xi)_M$ and ν_M, respectively.

The maximax relative response may occur either within the duration of the pulse or during the residual vibration era ($\tau \leq t$). In the latter case the maximax relative response is equal to the residual response amplitude. This explains the discontinuities which occur in the spectra of maximax relative response shown in Fig. 8.16 and elsewhere.

The meaning of the relative response $\nu - \xi$ may be clarified further by a study of the time-response and time-excitation curves shown in Fig. 8.15.

Equal Area of Pulse as Basis of Comparison. In the preceding section on the comparison of responses resulting from pulse excitation, the pulses are assumed of equal maximum height. Under some conditions, particularly if the pulse duration is short relative to the natural period of the system, it may be more useful to make the comparison on the basis of equal pulse area; i.e., equal impulse (equal time integral).

The *areas* for the pulses of maximum height ξ_p and duration τ are as follows: rectangle, $\xi_p\tau$; half-cycle sine, $(2/\pi)\xi_p\tau$; versed-sine $(\frac{1}{2})\xi_p\tau$; triangle, $(\frac{1}{2})\xi_p\tau$. Using the area of the

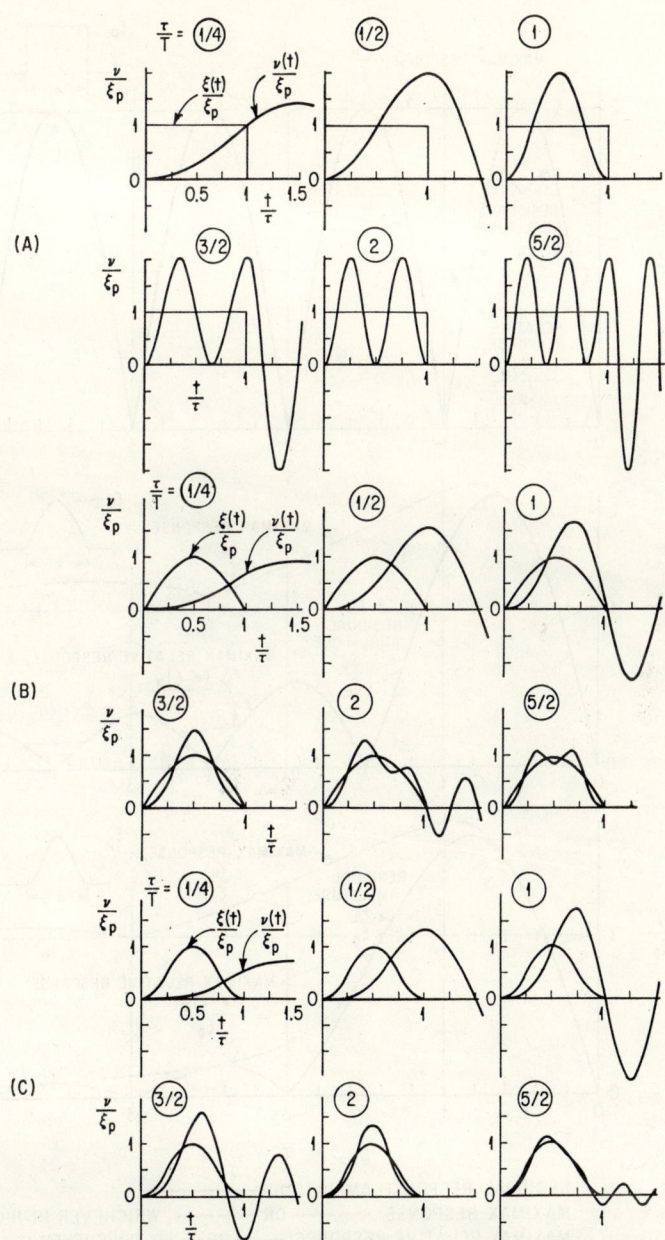

Fig. 8.15. Time response curves resulting from single pulses of (A) rectangular, (B) half-cycle sine, and (C) versed-sine shapes.[19]

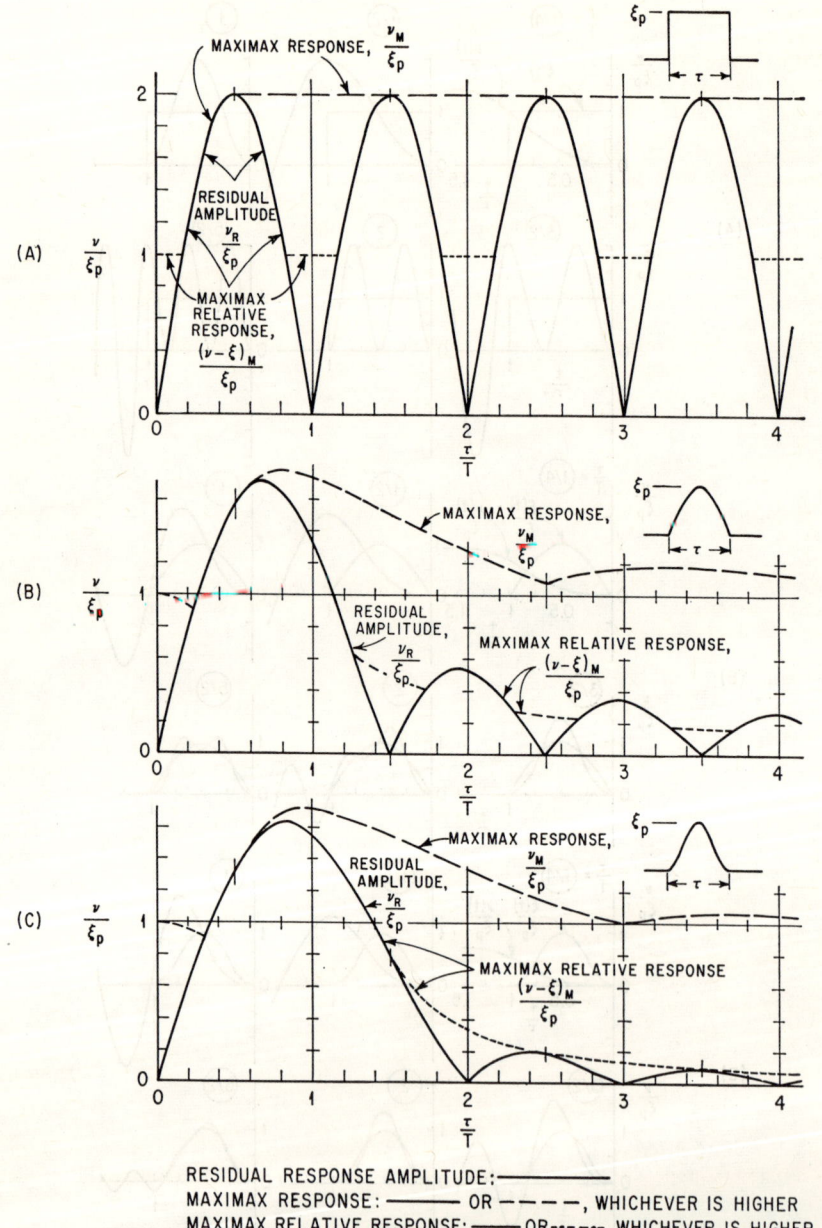

Fig. 8.16. Spectra of maximax response, residual response amplitude, and maximax relative response resulting from single pulses of (A) rectangular, (B) half-cycle sine, and (C) versed-sine shapes.[19] The spectra are shown on another basis in Fig. 8.18.

triangular pulse as the basis of comparison, and requiring that the areas of the other pulses be equal to it, it is found that the pulse *heights*, in terms of the height ξ_{po} of the *reference triangular pulse*, must be as follows: rectangle, $(\frac{1}{2})\xi_{po}$; half-cycle sine, $(\pi/4)\xi_{po}$; versed-sine, ξ_{po}.

Figure 8.17 shows the time responses, for four values of τ/T, redrawn on the basis of *equal pulse area* as the criterion for comparison. Note that the response reference is the constant ξ_{po}, which is the height of the triangular pulse. To show a direct comparison, the response curves for the various pulses are superimposed on each other. For the shortest duration shown, $\tau/T = \frac{1}{4}$, the response curves are nearly alike. Note that the responses to two different rectangular pulses are shown, one of duration τ and height $\xi_{po}/2$, the other of duration $\tau/2$ and height ξ_{po}, both of area $\xi_{po}\tau/2$.

The response spectra, plotted on the basis of equal pulse area, appear in Fig. 8.18. The residual response spectra are shown altogether in (A), the maximax response spectra in (B), and the spectra of maximax relative response in (C).

Since the pulse area is $\xi_{po}\tau/2$, the generalized impulse is $k\xi_{po}\tau/2$, and the amplitude of vibration of the system computed on the basis of the generalized impulse theory, Eq. (8.30b), is given by

$$\nu_J = \omega_n \xi_{po} \frac{\tau}{2} = \pi \frac{\tau}{T} \xi_{po} \quad (8.35)$$

A comparison of this straight-line function with the response spectra in Fig. 8.18B shows that *for values of τ/T less than one-fourth the shape of the symmetrical pulse is of little concern.*

Family of Exponential, Symmetrical Pulses. A continuous variation in shape of pulse may be investigated by means of the family of pulses represented by Eqs. (8.36a) and shown in the inset diagram in Fig. 8.19A:

$$\xi(t) = \begin{cases} \xi_p \dfrac{1 - e^{2at/\tau}}{1 - e^a} & \left[0 \leq t \leq \dfrac{\tau}{2}\right] \\ \xi_p \dfrac{1 - e^{2a(1-t/\tau)}}{1 - e^a} & \left[\dfrac{\tau}{2} \leq t \leq \tau\right] \\ 0 & [\tau \leq t] \end{cases} \quad (8.36a)$$

The family includes the following special cases:

$a \to -\infty$: rectangle of height ξ_p and duration τ;
$a = 0$: triangle of height ξ_p and duration τ;
$a \to +\infty$: spike of height ξ_p and having zero area.

The residual response amplitude of vibration of the system is

$$\frac{\nu_R}{\xi_p} = \frac{2aT}{\pi\tau}\left(\frac{e^a - \cos(\pi\tau/T) - (aT/\pi\tau)\sin(\pi\tau/T)}{(1 - e^a)(1 + a^2T^2/\pi^2\tau^2)}\right) \quad (8.36b)$$

where T is the natural period of the responding system. Figure 8.19A shows the spectra for residual response amplitude for seven values of the parameter a, compared on the basis of *equal pulse height*. The zero-area spike ($a \to +\infty$) results in zero response.

The area of the general pulse of height ξ_p is

$$A_p = \xi_p \frac{\tau}{a}\left(\frac{1 - e^a + a}{1 - e^a}\right) \quad (8.36c)$$

If a comparison is to be drawn on the basis of *equal pulse area* using the area $\xi_{po}\tau/2$ of the

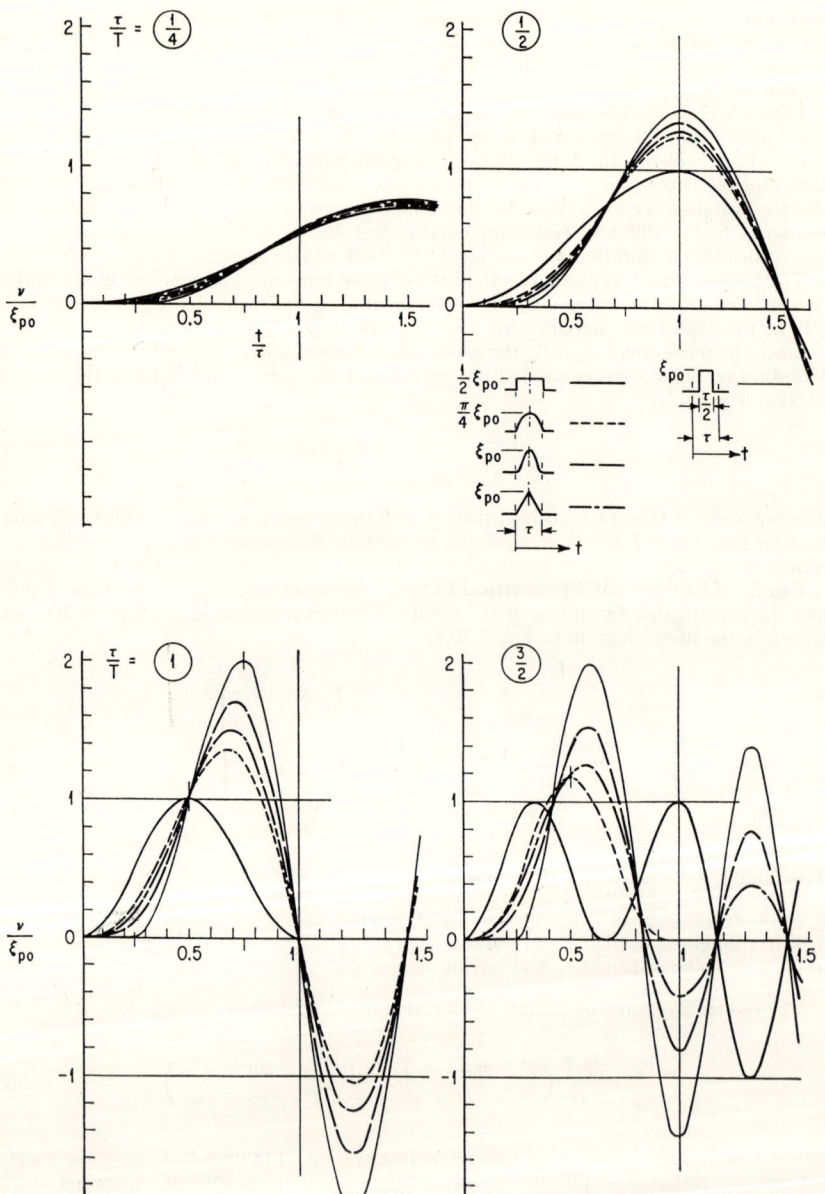

Fig. 8.17. Time response to various symmetrical pulses having equal pulse area, for four different values of τ/T. (Jacobsen and Ayre.[22])

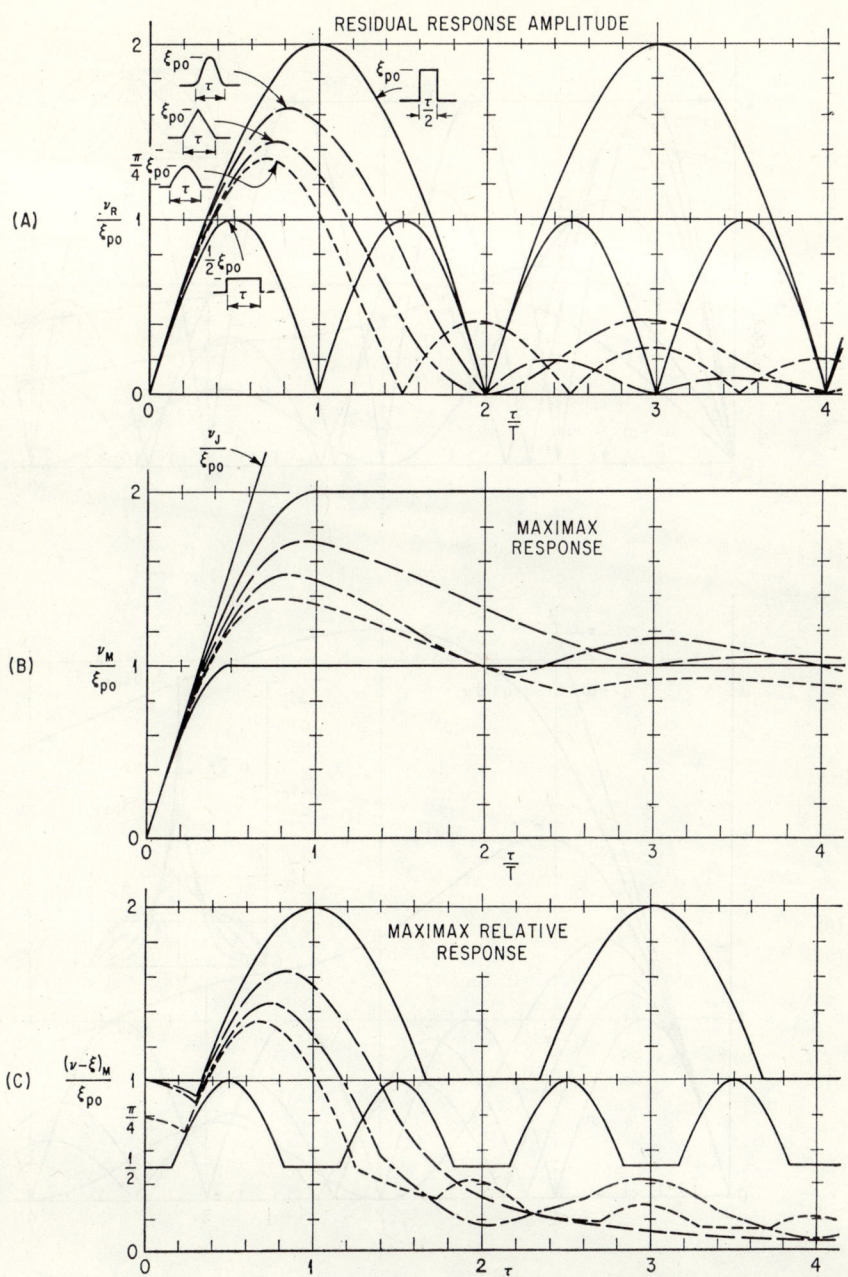

Fig. 8.18. Response spectra for various symmetrical pulses having equal pulse area: (*A*) residual response amplitude, (*B*) maximax response, and (*C*) maximax relative response. (*Jacobsen and Ayre.*[22])

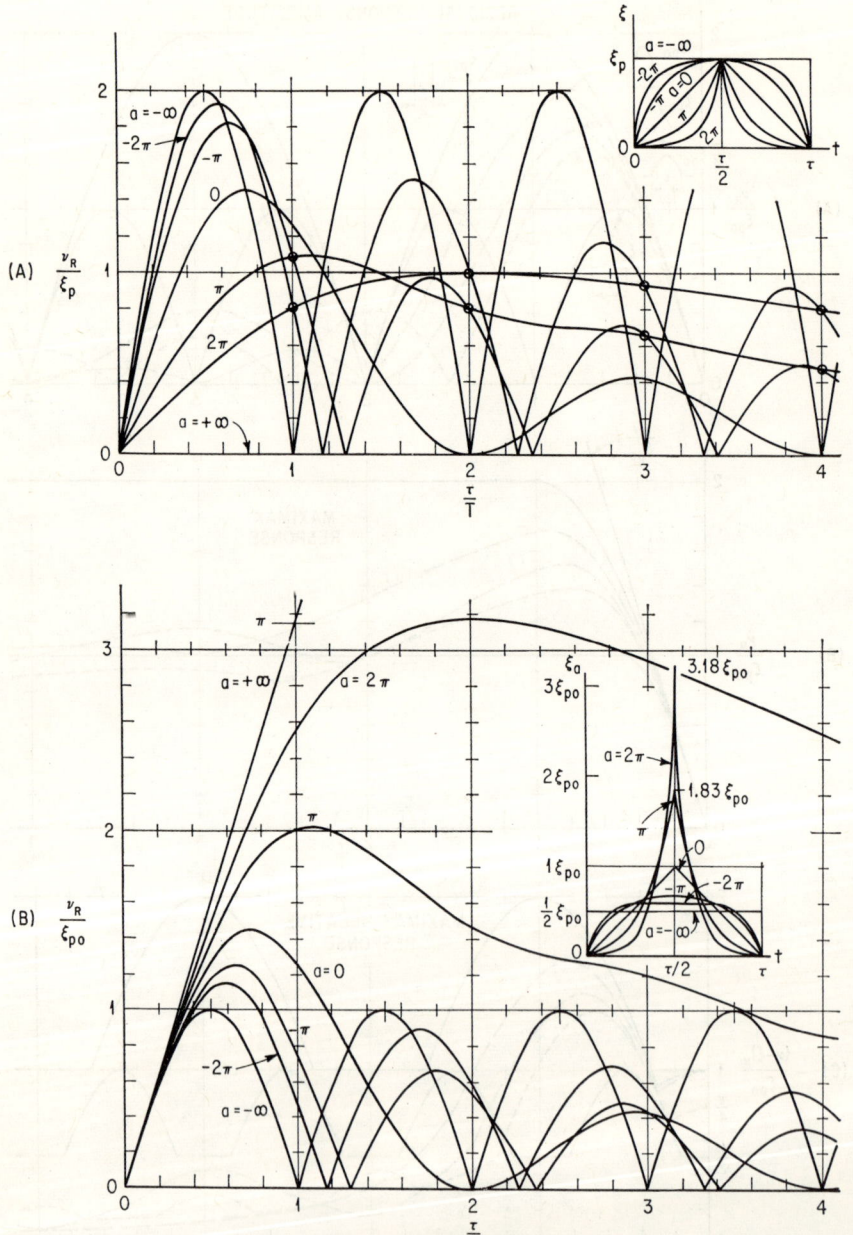

Fig. 8.19. Spectra for residual response amplitude for a family of exponential, symmetrical pulses: (A) pulses having equal height; (B) pulses having equal area. (*Jacobsen and Ayre.*[22])

triangular pulse as the reference, the height ξ_{pa} of the general pulse is

$$\xi_{pa} = \xi_{po} \frac{a}{2} \left(\frac{1 - e^a}{1 - e^a + a} \right) \tag{8.36d}$$

The residual response amplitude spectra, based on the equal-pulse-area criterion, are shown in Fig. 8.19B. The case $a \to +\infty$ is equivalent to a generalized impulse of value $k\xi_{po}\tau/2$ and results in the straight-line spectrum given by Eq. (8.35).

Symmetrical Pulses Having a Rest Period of Constant Height. In the inset diagrams of Fig. 8.20 each pulse consists of a rise, a central rest period or "dwell" having constant height, and a decay. The expressions for the pulse *rise* functions may be obtained from Eqs. (8.24a), (8.25a), and (8.26a) by substituting $\tau/2$ for τ. The pulse *decay* functions are available from symmetry.

If the *rest* period is long enough for the maximax displacement of the system to be reached during the duration τ_r of the pulse rest, the maximax may be obtained from Eqs. (8.24b), (8.25b), and (8.26b) and, consequently, from Fig. 8.13. The substitution of $\tau/2$ for τ is necessary.

Equations (8.37) to (8.39) give the residual response amplitudes. The spectra computed from these equations are shown in Fig. 8.20.

Constant-slope rise and decay:

$$\frac{\nu_R}{\xi_p} = \frac{2T}{\pi\tau} \left[1 - \cos\frac{\pi\tau}{T} + \frac{1}{2}\cos\frac{2\pi\tau_r}{T} - \cos\frac{\pi(\tau + 2\tau_r)}{T} + \frac{1}{2}\cos\frac{2\pi(\tau + \tau_r)}{T} \right]^{1/2} \tag{8.37}$$

Versed-sine rise and decay:

$$\frac{\nu_R}{\xi_p} = \frac{1}{1 - \tau^2/T^2} \left[1 + \cos\frac{\pi\tau}{T} - \frac{1}{2}\cos\frac{2\pi\tau_r}{T} - \cos\frac{\pi(\tau + 2\tau_r)}{T} - \frac{1}{2}\cos\frac{2\pi(\tau + \tau_r)}{T} \right]^{1/2} \tag{8.38}$$

Cycloidal rise and decay:

$$\frac{\nu_R}{\xi_p} = \frac{2T/\pi\tau}{1 - \tau^2/4T^2} \left[\cos\frac{\pi\tau_r}{T} - \cos\frac{\pi(\tau + \tau_r)}{T} \right] \tag{8.39}$$

Note that τ in the abscissa is the sum of the rise time and the decay time and is *not* the total duration of the pulse. Attached to each spectrum is a set of values of τ_r/T where T is the natural period of the responding system.

When $\tau_r/T = 1, 2, 3, \ldots$, the residual response amplitude is equal to that for the case $\tau_r = 0$, and the spectrum starts at the origin. If $\tau_r/T = \frac{1}{2}, \frac{3}{2}, \frac{5}{2}, \ldots$, the spectrum has the maximum value 2.00 at $\tau/T = 0$. The *envelopes* of the spectra are of the same forms as the *residual-response-amplitude* spectra for the related *step functions*; see the spectra for $[(\nu_M/\xi_c) - 1]$ in Fig. 8.13A. In certain cases, for example, at $\tau/T = 2, 4, 6, \ldots$, in Fig. 8.20A, $\nu_R/\xi_p = 0$ for all values of τ_r/T.

UNSYMMETRICAL PULSES. Pulses having only slight asymmetry may often be represented adequately by symmetrical forms. However, if there is considerable asymmetry, resulting in appreciable steepening of either the rise or the decay, it is necessary to introduce a parameter which defines the *skewing* of the pulse.

The ratio of the rise time to the pulse period is called the *skewing constant*, $\sigma = t_1/\tau$. There are three special cases:

$\sigma = 0$: the pulse has an *instantaneous* (vertical) *rise*, followed by a decay having the duration τ. This case may be used as an elementary representation of a *blast pulse*.

$\sigma = \frac{1}{2}$: the pulse may be *symmetrical*.

$\sigma = 1$: the pulse has an *instantaneous decay*, preceded by a rise having the duration τ.

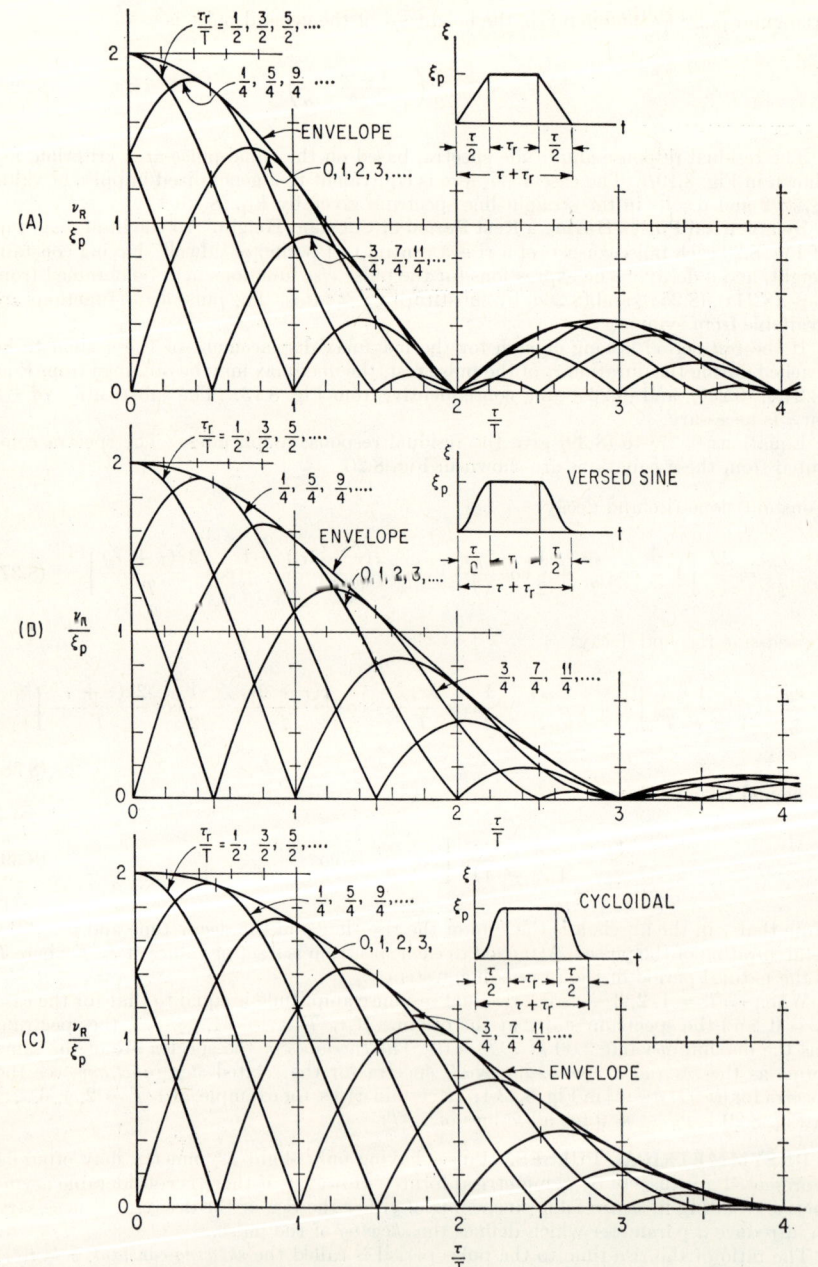

FIG. 8.20. Residual response amplitude spectra for three families of symmetrical pulses having a central rest period of constant height and of duration τ_r. Note that the abscissa is τ/T, where τ is the sum of the rise time and the decay time. (A) Constant-slope rise and decay. (B) Versed-sine rise and decay. (C) Cycloidal rise and decay. (*Jacobsen and Ayre.*[22])

Triangular Pulse Family. The effect of asymmetry in pulse shape is shown readily by means of the family of triangular pulses (Fig. 8.21). Equations (8.40) give the excitation and the time response.

Rise era: $0 \leq t \leq t_1$

$$\xi(t) = \xi_p \frac{t}{t_1}$$
$$\nu = \xi_p \left(\frac{t}{t_1} - \frac{T}{2\pi t_1} \sin \omega_n t \right)$$
(8.40a)

Decay era: $0 \leq t' \leq t_2$, where $t' = t - t_1$

$$\xi(t) = \xi_p \left(1 - \frac{t'}{t_2} \right)$$
$$\nu = \xi_p \left[1 - \frac{t'}{t_2} + \frac{T}{2\pi t_2} \left(1 + 4 \frac{t_2}{t_1} \frac{\tau}{t_1} \sin^2 \frac{\pi t_1}{T} \right)^{\frac{1}{2}} \sin(\omega_n t' + \theta') \right]$$
(8.40b)

where

$$\tan \theta' = \frac{\sin(2\pi t_1/T)}{\cos(2\pi t_1/T) - \tau/t_2}$$

Residual-vibration era: $0 \leq t''$, where $t'' = t - \tau = t - t_1 - t_2$

$$\xi(t) = 0$$

$$\nu = \xi_p \frac{1}{\pi} \left[\frac{T}{t_1} \frac{T}{t_2} \left(\frac{\tau}{t_1} \sin^2 \frac{\pi t_1}{T} + \frac{\tau}{t_2} \sin^2 \frac{\pi t_2}{T} - \sin^2 \frac{\pi \tau}{T} \right) \right]^{\frac{1}{2}} \sin(\omega_n t'' + \theta_R)$$
(8.40c)

where

$$\tan \theta_R = \frac{(\tau/t_2) \sin(2\pi t_2/T) - \sin(2\pi \tau/T)}{(\tau/t_2) \cos(2\pi t_2/T) - \cos(2\pi \tau/T) - t_1/t_2}$$

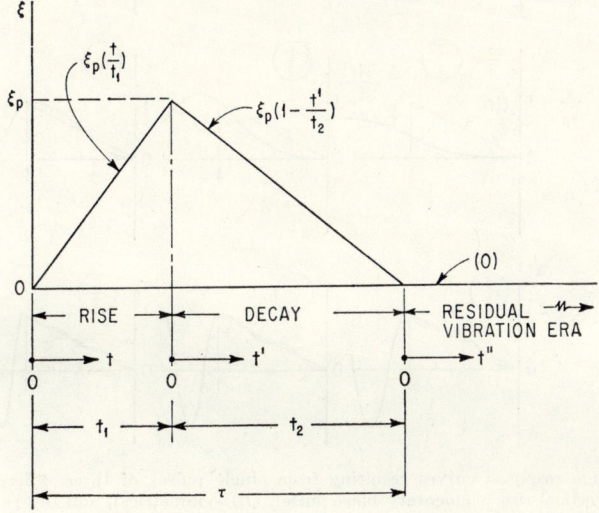

Fig. 8.21. General triangular pulse.

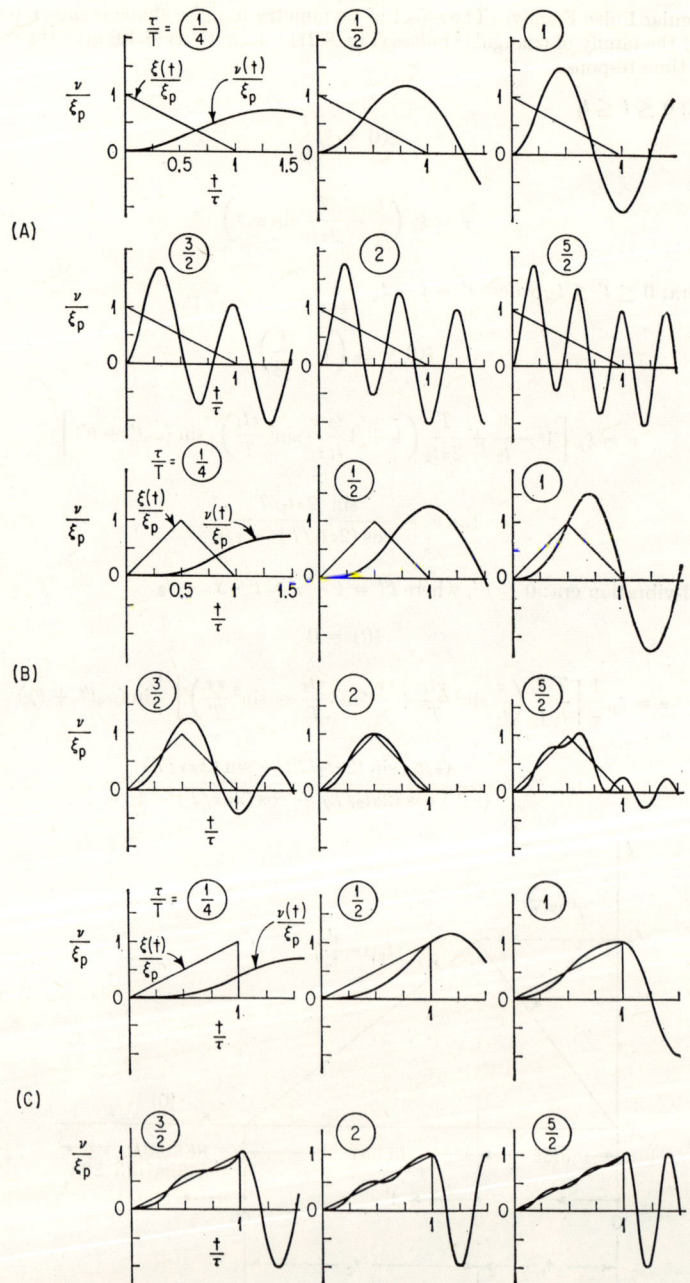

Fig. 8.22. Time response curves resulting from single pulses of three different triangular shapes: (A) vertical rise (elementary blast pulse), (B) symmetrical, and (C) vertical decay.[19]

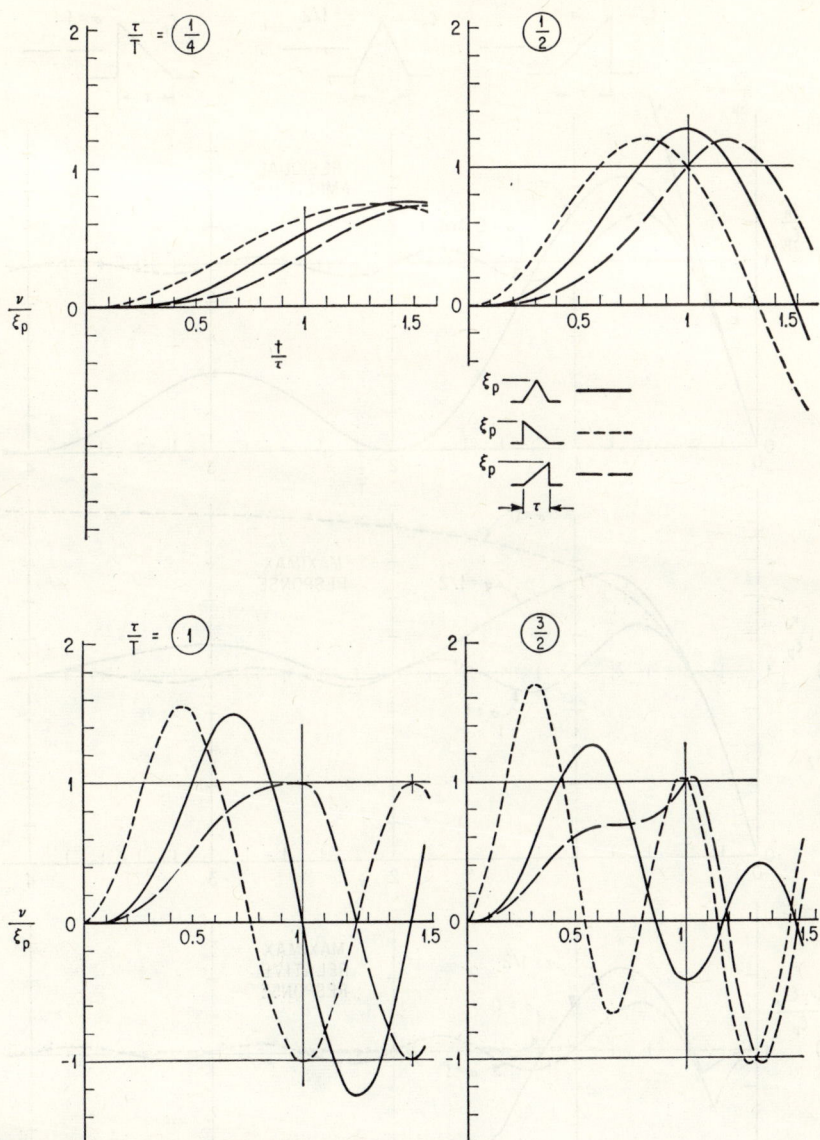

FIG. 8.23. Time response curves of Fig. 8.22 superposed, for four values of τ/T. (*Jacobsen and Ayre.*[22])

For the special cases $\sigma = 0$, ½ and 1, the time responses for six values of τ/T are shown in Fig. 8.22, where T is the natural period of the responding system. Some of the curves are superposed in Fig. 8.23 for easier comparison. The response spectra appear in Fig. 8.24. The straight-line spectrum ν_J/ξ_p for the amplitude of response based on the impulse theory also is shown in Fig. 8.24A. In the two cases of extreme skewing, $\sigma = 0$ and

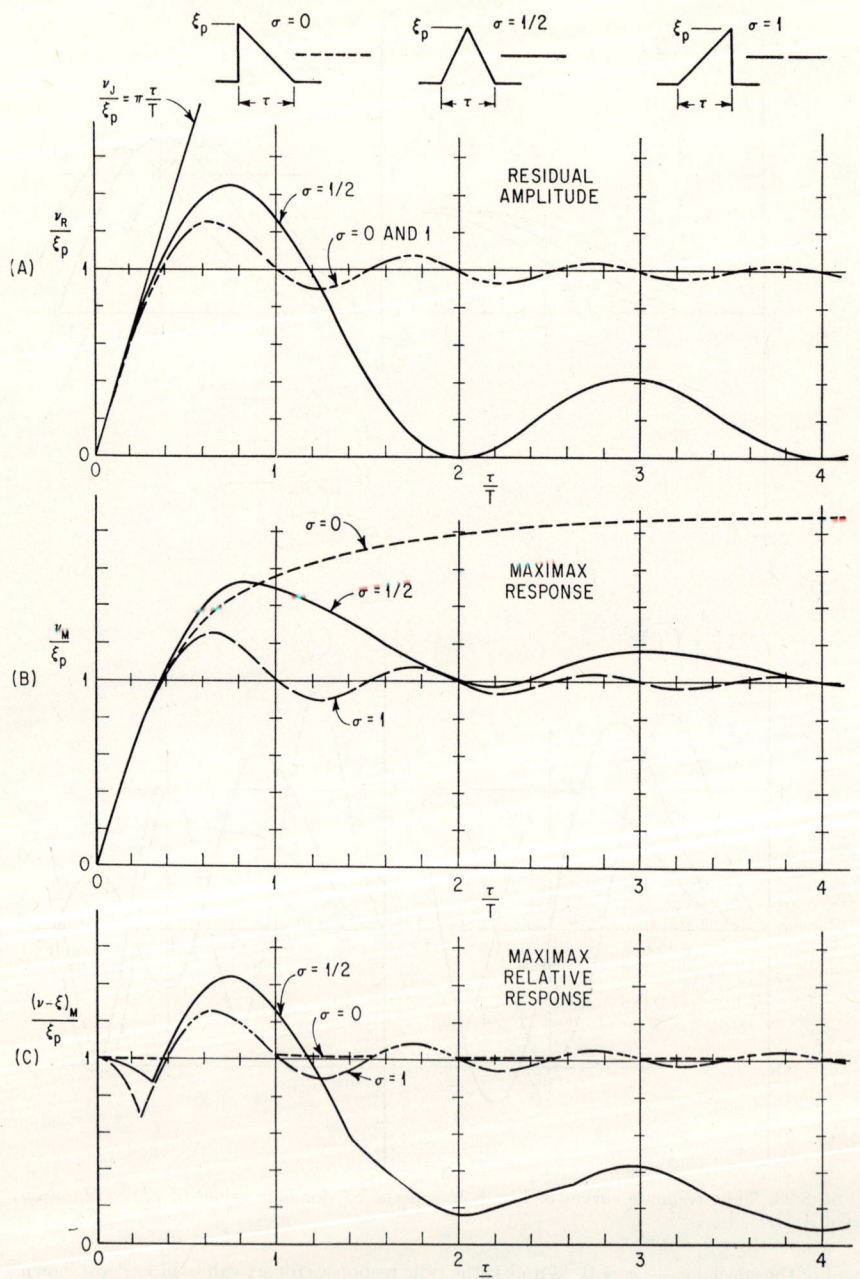

FIG. 8.24. Response spectra for three types of triangular pulse: (A) Residual response amplitude. (B) Maximax response. (C) Maximax relative response. (*Jacobsen and Ayre.*[22])

$\sigma = 1$, the residual amplitudes are *equal* and are given by Eq. (8.41a). For the symmetrical case, $\sigma = \frac{1}{2}$, ν_R is given by Eq. (8.41b).

$$\sigma = 0 \text{ and } 1: \quad \frac{\nu_R}{\xi_p} = \left[1 - \frac{T}{\pi\tau} \sin \frac{2\pi\tau}{T} + \left(\frac{T}{\pi\tau}\right)^2 \sin^2 \frac{\pi\tau}{T} \right]^{\frac{1}{2}} \quad (8.41a)$$

$$\sigma = \frac{1}{2}: \quad \frac{\nu_R}{\xi_p} = 2 \frac{\sin^2 (\pi\tau/2T)}{\pi\tau/2T} \quad (8.41b)$$

The residual response amplitudes for other cases of skewness may be determined from the amplitude term in Eqs. (8.40c); they are shown by the response spectra in Fig. 8.25. The residual response amplitudes resulting from single pulses that are mirror images of

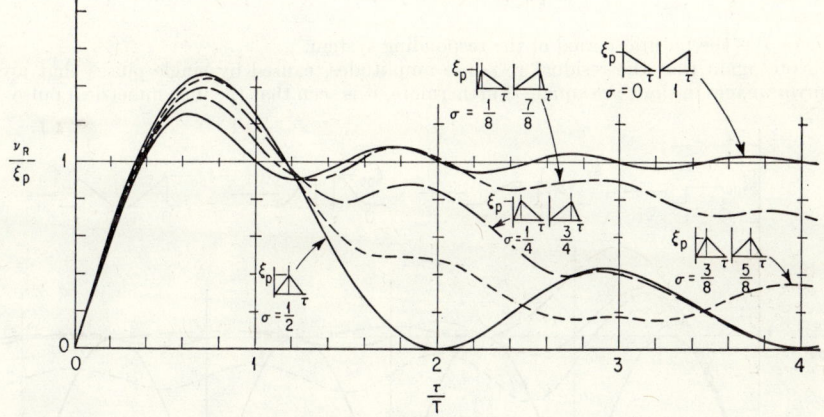

Fig. 8.25. Spectra for residual response amplitude for a family of triangular pulses of varying skewness. (Jacobsen and Ayre.[22])

each other in time are equal. In general, the phase angles for the residual vibrations are unequal.

Note that in the cases $\sigma = 0$ and $\sigma = 1$ for *vertical rise* and *vertical decay*, respectively, there are *no zeroes of residual amplitude*, except for the trivial case, $\tau/T = 0$.

The family of triangular pulses is particularly advantageous for investigating the effect of varying the skewness, because both criteria of comparison, equal pulse height and equal pulse area, are satisfied simultaneously.

Various Pulses Having Vertical Rise or Vertical Decay. Figure 8.26 shows the spectra of residual response amplitude plotted on the basis of *equal pulse area*. The rectangular pulse is included for comparison. The expressions for residual response amplitude for the rectangular and the triangular pulses are given by Eqs. (8.31b) and (8.41a), and for the quarter-cycle sine and the half-cycle versed-sine pulses by Eqs. (8.42) and (8.43).

Quarter-cycle "sine":

$$\xi(t) = \xi_p \begin{cases} \sin \dfrac{\pi t}{2\tau} & \text{for vertical decay} \\ \text{or} \\ \cos \dfrac{\pi t}{2\tau} & \text{for vertical rise} \end{cases} \quad [0 \leq t \leq \tau]$$

$$\xi(t) = 0 \quad [\tau \leq t]$$

$$\frac{\nu_R}{\xi_p} = \frac{4\tau/T}{(16\tau^2/T^2) - 1} \left(1 + \frac{16\tau^2}{T^2} - \frac{8\tau}{T} \sin \frac{2\pi\tau}{T} \right)^{\frac{1}{2}} \quad (8.42)$$

8-36   TRANSIENT RESPONSE TO STEP AND PULSE FUNCTIONS

Half-cycle "versed-sine":

$$\xi(t) = \xi_p \begin{cases} \frac{1}{2}\left(1 - \cos\frac{\pi t}{\tau}\right) & \text{for vertical decay} \\ \text{or} \\ \frac{1}{2}\left(1 + \cos\frac{\pi t}{\tau}\right) & \text{for vertical rise} \end{cases} \quad [0 \leq t \leq \tau]$$

$$\xi(t) = 0 \quad [\tau \leq t]$$

$$\frac{\nu_R}{\xi_p} = \frac{1/2}{(4\tau^2/T^2) - 1}\left[1 + \left(1 - \frac{8\tau^2}{T^2}\right)^2 - 2\left(1 - \frac{8\tau^2}{T^2}\right)\cdot\cos\frac{2\pi\tau}{T}\right]^{1/2} \quad (8.43)$$

where T is the natural period of the responding system.

Note again that the residual response amplitudes, caused by single pulses that are mirror images in time, are equal. Furthermore, it is seen that the unsymmetrical pulses,

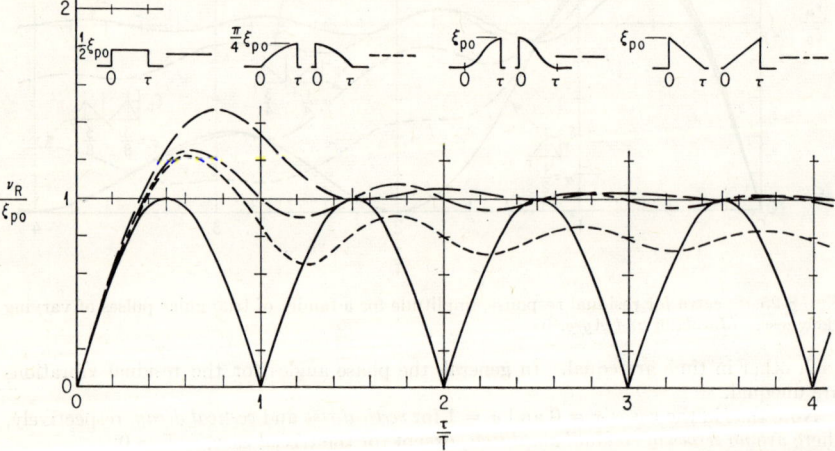

FIG. 8.26. Spectra for residual response amplitude for various unsymmetrical pulses having either vertical rise or vertical decay. Comparison on the basis of equal pulse area.[19]

having either vertical rise or vertical decay, result in no zeroes of residual response amplitude, except in the trivial case $\tau/T = 0$.

Exponential Pulses of Finite Duration, Having Vertical Rise or Vertical Decay. Families of exponential pulses having either a vertical rise or a vertical decay, as shown in the inset diagrams in Fig. 8.27, can be formed by Eqs. (8.44a) and (8.44b).

Vertical rise with exponential decay:

$$\xi(t) = \begin{cases} \xi_p\left(\frac{1 - e^{a(1 - t/\tau)}}{1 - e^a}\right) & [0 \leq t \leq \tau] \\ 0 & [\tau \leq t] \end{cases} \quad (8.44a)$$

Exponential rise with vertical decay:

$$\xi(t) = \begin{cases} \xi_p\left(\frac{1 - e^{at/\tau}}{1 - e^a}\right) & [0 \leq t \leq \tau] \\ 0 & [\tau \leq t] \end{cases} \quad (8.44b)$$

RESPONSE FUNCTIONS AND RESPONSE SPECTRA 8-37

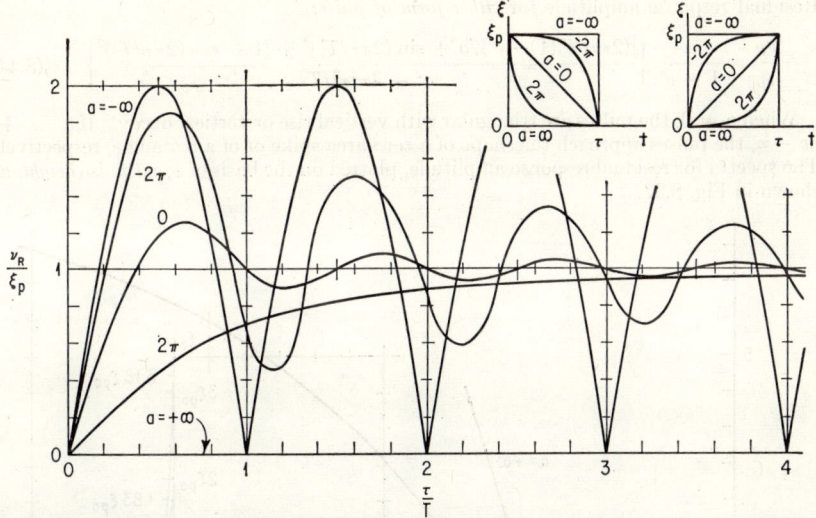

Fig. 8.27. Spectra for residual response amplitude for unsymmetrical exponential pulses having either vertical rise or vertical decay. Comparison on the basis of equal pulse height.[19]

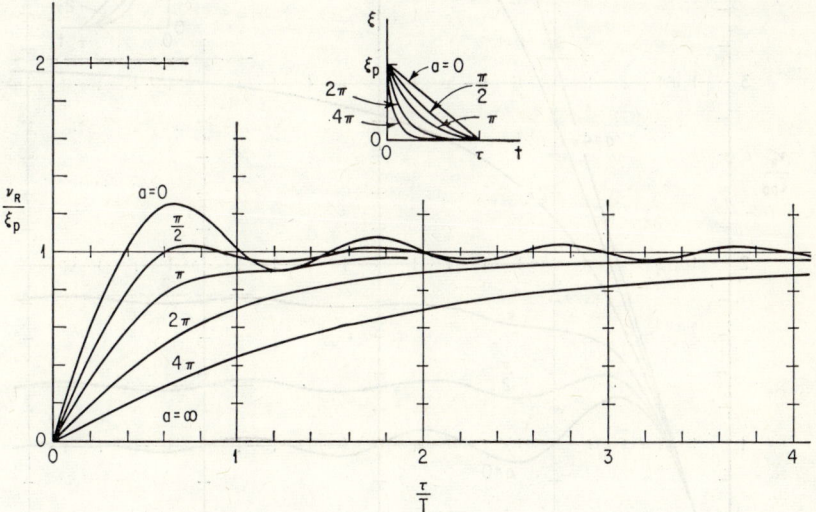

Fig. 8.28. Spectra for residual response amplitude for a family of simple blast pulses, the same family shown in Fig. 8.27 but limited to positive values of the exponential decay parameter a. Comparison on the basis of equal pulse height. (These spectra also apply to mirror-image pulses having vertical decay.) (*Jacobsen and Ayre.*[22])

8-38 TRANSIENT RESPONSE TO STEP AND PULSE FUNCTIONS

Residual response amplitude for *either form of pulse*:

$$\frac{\nu_R}{\xi_p} = \frac{a}{1 - e^a} \left\{ \frac{[(2\pi\tau/T)(1 - e^a)/a + \sin(2\pi\tau/T)]^2 + [1 - \cos(2\pi\tau/T)]^2}{a^2 + 4\pi^2\tau^2/T^2} \right\}^{1/2} \quad (8.44c)$$

When $a = 0$, the pulses are triangular with vertical rise or vertical decay. If $a \to +\infty$ or $-\infty$, the pulses approach the shape of a zero-area spike or of a rectangle, respectively. The spectra for residual response amplitude, plotted on the basis of *equal pulse height*, are shown in Fig. 8.27.

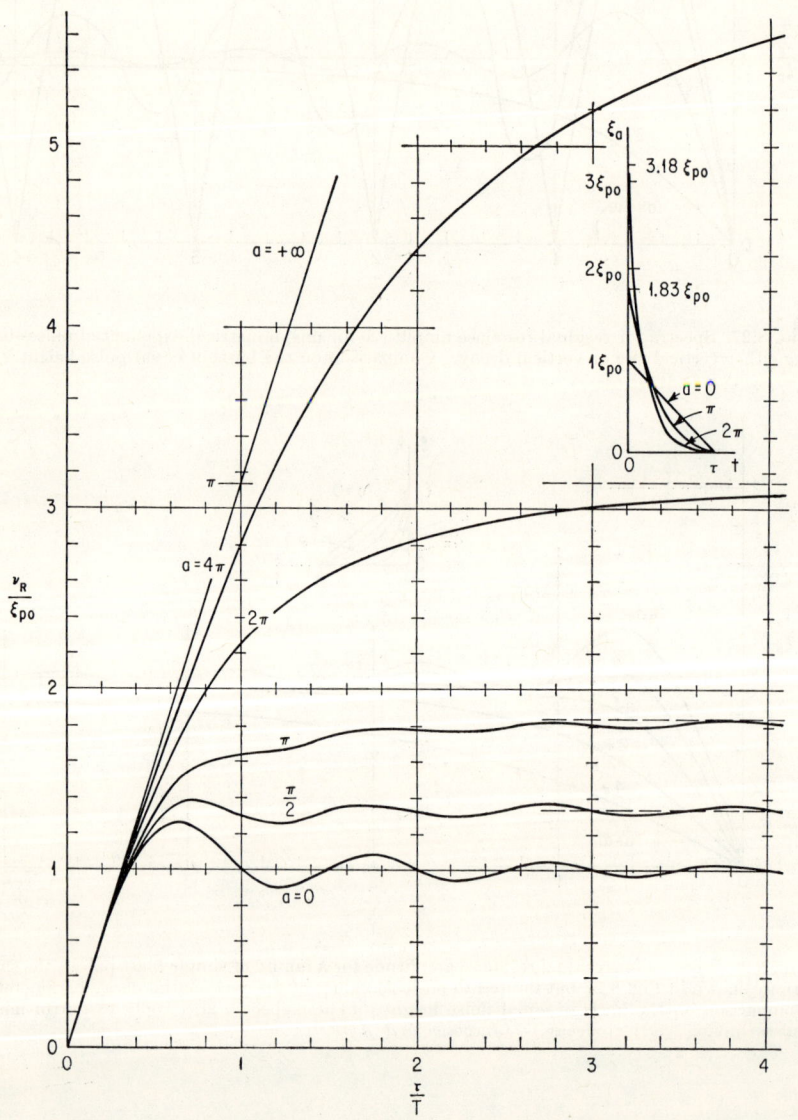

FIG. 8.29. Spectra for residual response amplitude for the family of simple blast pulses shown in Fig. 8.28, compared on the basis of equal pulse area. (*Jacobsen and Ayre*.[22])

Figure 8.28 shows the spectra of residual response amplitude in greater detail for the range in which the parameter a is limited to positive values. This group of pulses is of interest in studying the effects of a simple form of blast pulse, in which the peak height and the duration are constant but the rate of decay is varied.

The areas of the pulses of equal height ξ_p, and the heights of the pulses of equal area $\xi_{po}\tau/2$ are the same as for the symmetrical exponential pulses [see Eqs. (8.36c) and (8.36d)]. If the spectra in Fig. 8.28 are redrawn, using *equal pulse area* as the criterion for comparison, they appear as in Fig. 8.29. The limiting pulse case $a \to +\infty$ repre-

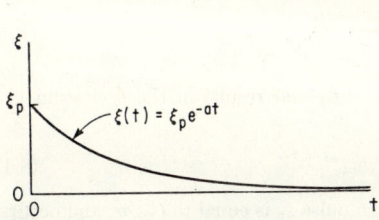

FIG. 8.30. Pulse consisting of vertical rise followed by exponential decay of infinite duration.

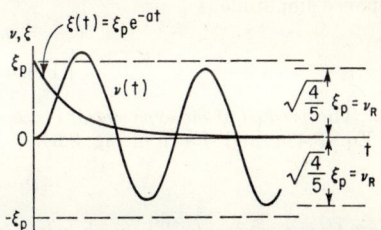

FIG. 8.31. Time response to the pulse having a vertical rise and an exponential decay of infinite duration (Fig. 8.30), for the particular case $\omega_n/a = 2$.

sents a generalized impulse of value $k\xi_{po}\tau/2$. The asymptotic values of the spectra are equal to the peak heights of the equal area pulses and are given by

$$\frac{\nu_R}{\xi_{po}} \to \frac{a(1 - e^a)}{2(1 - e^a + a)} \quad \text{as} \quad \frac{\tau}{T} \to \infty \tag{8.44d}$$

Exponential Pulses of Infinite Duration. Five different cases are included as follows:
1. The excitation function, consisting of a vertical rise followed by an exponential decay, is

$$\xi(t) = \xi_p e^{-at} \quad [0 \le t] \tag{8.45a}$$

It is shown in Fig. 8.30. The response time equation for the system is

$$\nu = \xi_p \frac{(a/\omega_n) \sin \omega_n t - \cos \omega_n t + e^{-at}}{1 + a^2/\omega_n^2} \tag{8.45b}$$

and the asymptotic value of the residual amplitude is given by

$$\frac{\nu_R}{\xi_p} \to \frac{1}{\sqrt{1 + a^2/\omega_n^2}} \tag{8.45c}$$

The maximax response is the first maximum of ν. The time response, for the particular case $\omega_n/a = 2$, and the response spectra are shown in Figs. 8.31 and 8.32, respectively.

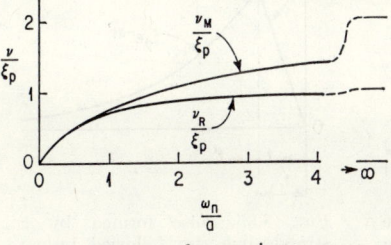

FIG. 8.32. Spectra for maximax response and for asymptotic residual response amplitude, for the pulse shown in Fig. 8.30.

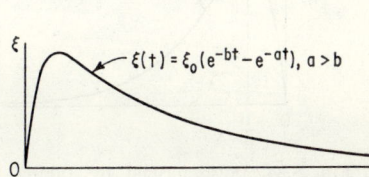

FIG. 8.33. Pulse formed by taking the difference of two exponentially decaying functions.

2. The *difference of two exponential functions*, of the type of Eq. (8.45a), results in the pulse given by Eq. (8.46a):

$$\xi(t) = \xi_0(e^{-bt} - e^{-at}) \qquad (8.46a)$$

$$a > b \qquad [0 \leq t]$$

The shape of the pulse is shown in Fig. 8.33. Note that ξ_0 is the ordinate of each of the exponential functions at $t = 0$; it is *not* the pulse maximum. The asymptotic residual response amplitude is

$$\nu_R \to \xi_0 \frac{(b/\omega_n) - (a/\omega_n)}{[(1 + a^2/\omega_n^2)(1 + b^2/\omega_n^2)]^{1/2}} \qquad (8.46b)$$

3. The *product of the exponential function* e^{-at} *by time* results in the excitation given by Eq. (8.47a) and shown in Fig. 8.34.

$$\xi(t) = C_0 t e^{-at} \qquad (8.47a)$$

where C_0 is a constant. The peak height of the pulse ξ_p is equal to C_0/ae, and occurs at the time $t_1 = 1/a$. Equations (8.47b) and (8.47c) give the time response and the asymptotic residual response amplitude:

$$\nu = \xi_p \frac{ae/\omega_n}{(1 + a^2/\omega_n^2)^2} \left\{ \left[\frac{2a}{\omega_n} + \left(1 + \frac{a^2}{\omega_n^2}\right) \omega_n t \right] e^{-at} - \frac{2a}{\omega_n} \cos \omega_n t - \left(1 - \frac{a^2}{\omega_n^2}\right) \sin \omega_n t \right\}$$

(8.47b)

$$\frac{\nu_R}{\xi_p} \to \frac{e}{(a/\omega_n) + (\omega_n/a)} \qquad (8.47c)$$

The *maximum* value of ν_R occurs in the case $a/\omega_n = 1$, and is given by

$$(\nu_R)_{\max} = \xi_p e/2 = 1.36 \xi_p$$

Both of the excitation functions described by Eqs. (8.46a) and (8.47a) include finite times of rise to the pulse peak. These rise times are dependent on the exponential decay constants.

4. The rise time may be made independent of the decay by inserting a separate rise function before the decay function, as in Fig. 8.35, where a *straight-line rise precedes the exponential decay*. The response-time equations are as follows:

Pulse rise era:

$$\nu = \xi_p \frac{\omega_n t - \sin \omega_n t}{\omega_n t_1} \qquad [0 \leq t \leq t_1] \qquad (8.48a)$$

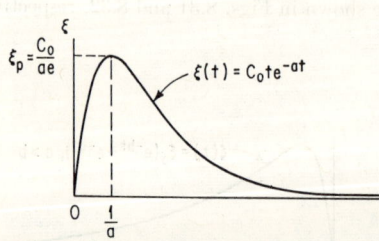

FIG. 8.34. Pulse formed by taking the product of an exponentially decaying function by time.

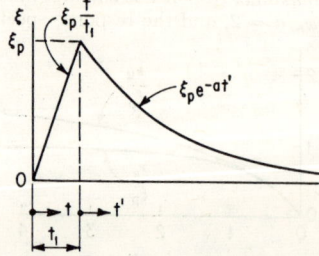

FIG. 8.35. Pulse formed by a straight-line rise followed by an exponential decay asymptotic to the time axis.

Pulse decay era:

$$\nu = \xi_p \left[\frac{e^{-at'}}{1 + a^2/\omega_n^2} + \left(\frac{a^2/\omega_n^2}{1 + a^2/\omega_n^2} - \frac{\sin \omega_n t_1}{\omega_n t_1} \right) \cos \omega_n t' \right.$$
$$\left. + \left(\frac{a/\omega_n}{1 + a^2/\omega_n^2} + \frac{1 - \cos \omega_n t_1}{\omega_n t_1} \right) \sin \omega_n t' \right] \quad (8.48b)$$

where $t' = t - t_1$ and $0 \leq t'$.

5. Another form of pulse, which is a more complete representation of a blast pulse since it includes the possibility of a negative phase of pressure,[14] is shown in Fig. 8.36. It consists of a straight-line rise, followed by an exponential decay through the positive phase, into the negative phase, finally becoming asymptotic to the time axis. The rise time is t_1 and the duration of the positive phase is $t_1 + t_2$.

Unsymmetrical Exponential Pulses with Central Peak. An interesting family of unsymmetrical pulses may be formed by using Eqs. (8.36a) and changing the sign of the exponent of e in both the numerator and the denominator of the second of the equations. The resulting family consists of pulses whose maxima occur at the mid-period time and which satisfy simultaneously both criteria for comparison (equal pulse height and equal pulse area).

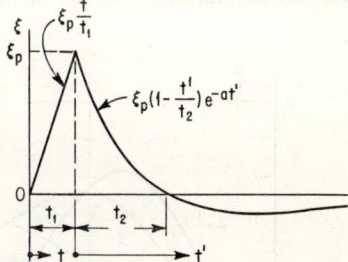

Fig. 8.36. Pulse formed by a straight-line rise followed by a continuous exponential decay through positive and negative phases. (*Frankland*.[14])

Figure 8.37 shows the spectra of residual response amplitude and, in the inset diagrams, the pulse shapes. The limiting cases are the symmetrical triangle of duration τ and height ξ_p, and the rectangles of duration $\tau/2$ and height ξ_p. All pulses in the family have the area $\xi_p \tau/2$. Zeroes of residual response amplitude occur for all values of a, at

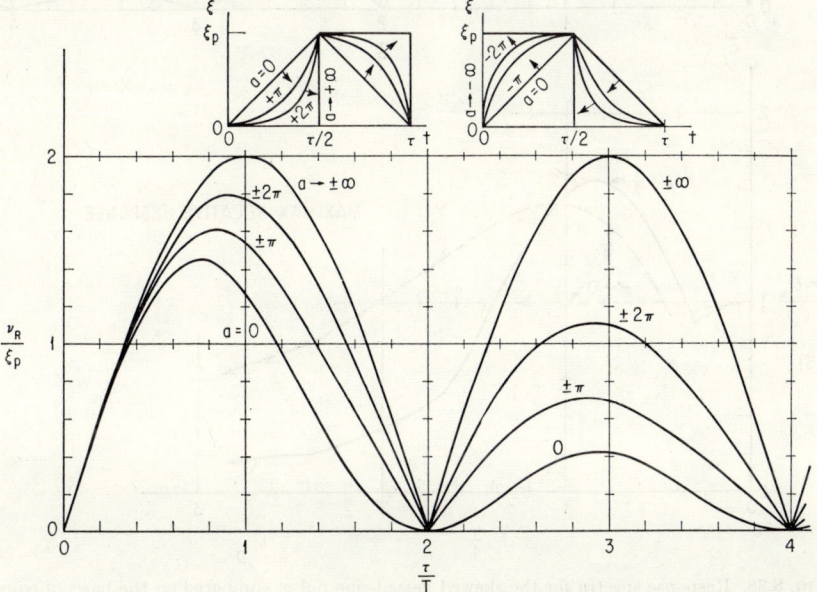

Fig. 8.37. Spectra of residual response amplitude for a family of unsymmetrical exponential pulses of equal area and equal maximum height, having the pulse peak at the mid-period time. (*Jacobsen and Ayre*.[22])

8–42 TRANSIENT RESPONSE TO STEP AND PULSE FUNCTIONS

even integer values of τ/T. The residual response amplitude is

$$\frac{v_R}{\xi_p} = \frac{aT/\pi\tau}{1 - \cosh a} \cdot$$

$$\left[\frac{\cosh 2a - \cosh a - (1 - \cosh a)\cos(2\pi\tau/T) + (1 - \cosh 2a)\cos(\pi\tau/T)}{1 + a^2T^2/\pi^2\tau^2}\right]^{1/2} \quad (8.49)$$

Pulses which are mirror images of each other in time result in equal residual amplitudes.

Skewed Versed-sine Pulse. By taking the product of a decaying exponential and the versed-sine function, a family of pulses with varying skewness is obtained.[13,22] The

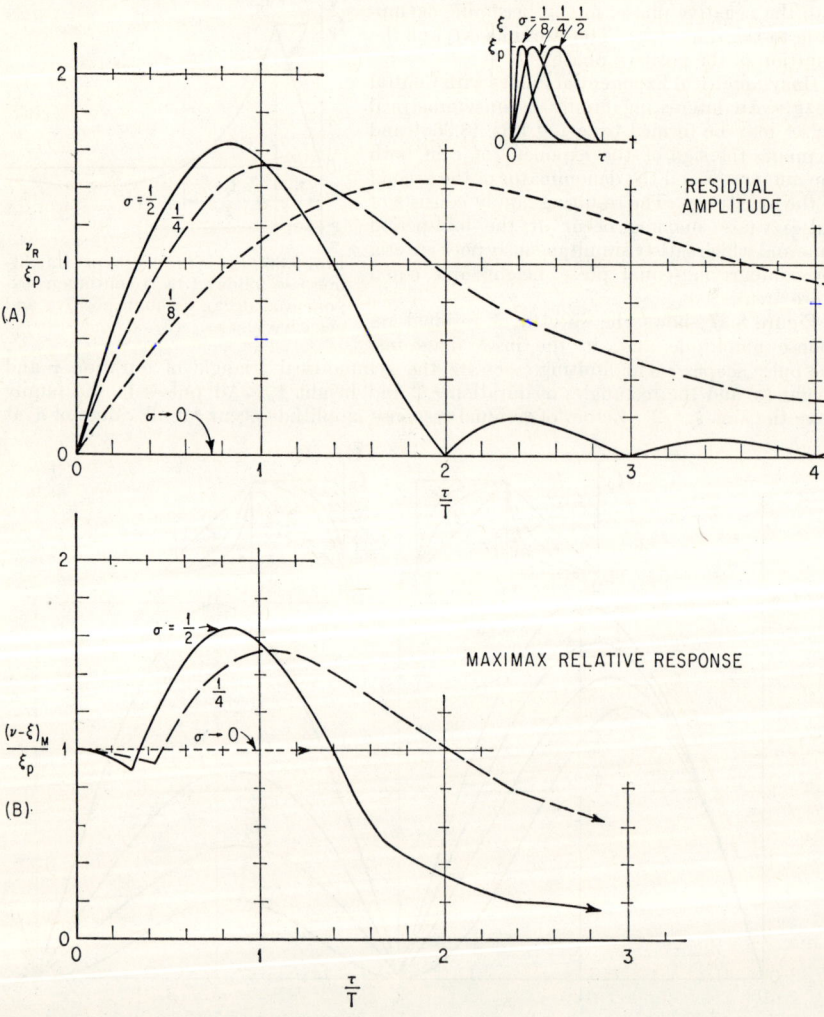

FIG. 8.38. Response spectra for the skewed versed-sine pulse, compared on the basis of equal pulse height: (A) Residual response amplitude. (B) Maximax relative response. (*Jacobsen and Ayre.*[22])

family is described by the following equation:

$$\xi(t) = \begin{cases} \xi_p \dfrac{e^{2\pi(\sigma - t/\tau)\cot \pi\sigma}}{1 - \cos 2\pi\sigma}(1 - \cos 2\pi t/\tau) & [0 \leq t \leq \tau] \\ 0 & [\tau \leq t] \end{cases} \qquad (8.50)$$

These pulses are of particular interest when the excitation is a ground displacement function because they have continuity in both velocity and displacement; thus, they do not involve theoretically infinite accelerations of the ground. When the skewing constant σ equals one-half, the pulse is the symmetrical versed sine. When $\sigma \to 0$, the front of the pulse approaches a straight line with infinite slope, and the pulse area approaches zero.

The spectra of residual response amplitude and of maximax relative response, plotted on the basis of equal pulse height, are shown in Fig. 8.38 for several values of σ. The residual response amplitude spectra are reasonably good approximations to the spectra of maximax relative response except at the lower values of τ/T.

Figure 8.39 compares the residual response amplitude spectra on the basis of equal pulse area. The required pulse heights, for a constant pulse area of $\xi_{po}\tau/2$, are shown in the inset diagram. On this basis, the pulse for $\sigma \to 0$ represents a generalized impulse of value $k\xi_{po}\tau/2$.

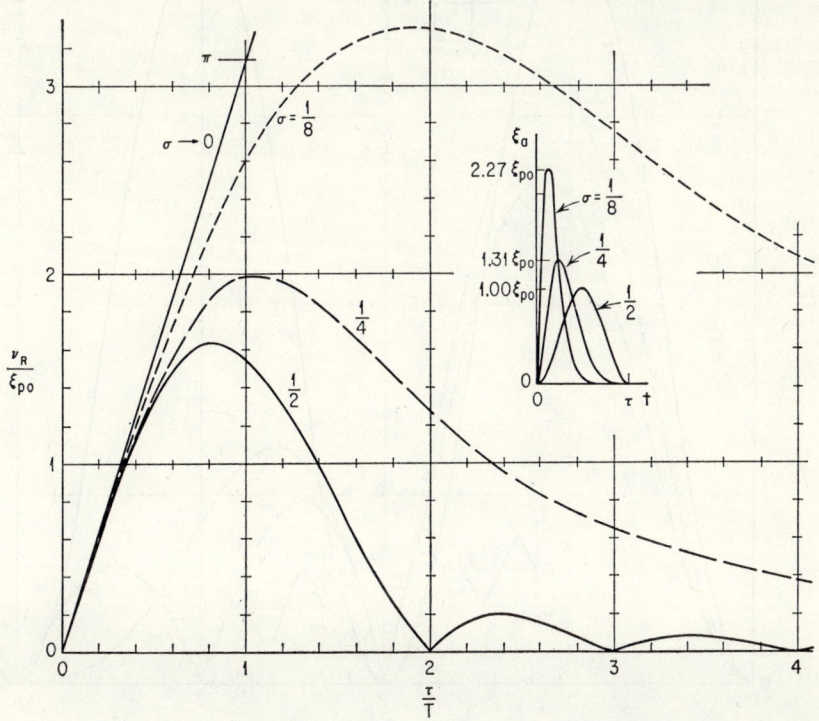

Fig. 8.39. Spectra of residual response amplitude for the skewed versed-sine pulse, compared on the basis of equal pulse area.[19]

FULL-CYCLE PULSES (FORCE-TIME INTEGRAL = 0). The residual response amplitude spectra for three groups of *full-cycle pulses* are shown as follows: in Fig. 8.40 for the rectangular, the sinusoidal, and the symmetrical triangular pulses; in Fig. 8.41 for three types of pulse involving sine and cosine functions; and in Fig. 8.42 for three forms of triangular pulse. The pulse shapes are shown in the inset diagrams. Expressions for the residual response amplitudes are given in Eqs. (8.51) to (8.53).

Full-cycle rectangular pulse:

$$\frac{\nu_R}{\xi_p} = 2 \sin \frac{\pi \tau}{T} \left[2 \sin \frac{\pi \tau}{T} \right] \qquad (8.51)$$

Full-cycle "sinusoidal" pulses:
  Symmetrical half cycles

$$\frac{\nu_R}{\xi_p} = 2 \sin \frac{\pi \tau}{T} \left[\frac{T/\tau}{(T^2/4\tau^2) - 1} \cos \frac{\pi \tau}{T} \right] \qquad (8.52a)$$

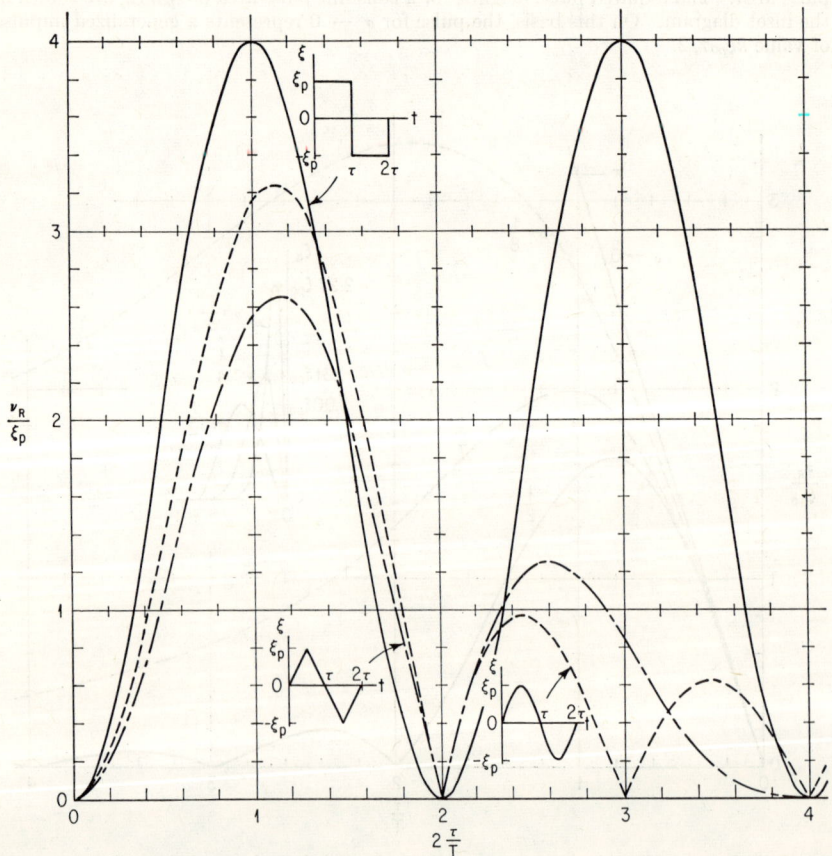

FIG. 8.40. Spectra of residual response amplitude for three types of full-cycle pulses. Each half cycle is symmetrical.[19]

Vertical front and vertical ending

$$\frac{\nu_R}{\xi_p} = \frac{2}{1 - T^2/16\tau^2} \cos \frac{2\pi\tau}{T} \qquad (8.52b)$$

Vertical jump at mid-cycle

$$\frac{\nu_R}{\xi_p} = \frac{2}{1 - T^2/16\tau^2} \left(1 - \frac{T}{4\tau} \sin \frac{2\pi\tau}{T}\right) \qquad (8.52c)$$

Full-cycle triangular pulses:
Symmetrical half cycles

$$\frac{\nu_R}{\xi_p} = 2 \sin \frac{\pi\tau}{T} \left[\frac{4T}{\pi\tau} \sin^2 \frac{\pi\tau}{2T}\right] \qquad (8.53a)$$

Vertical front and vertical ending

$$\frac{\nu_R}{\xi_p} = 2 \left(\frac{T}{2\pi\tau} \sin \frac{2\pi\tau}{T} - \cos \frac{2\pi\tau}{T}\right) \qquad (8.53b)$$

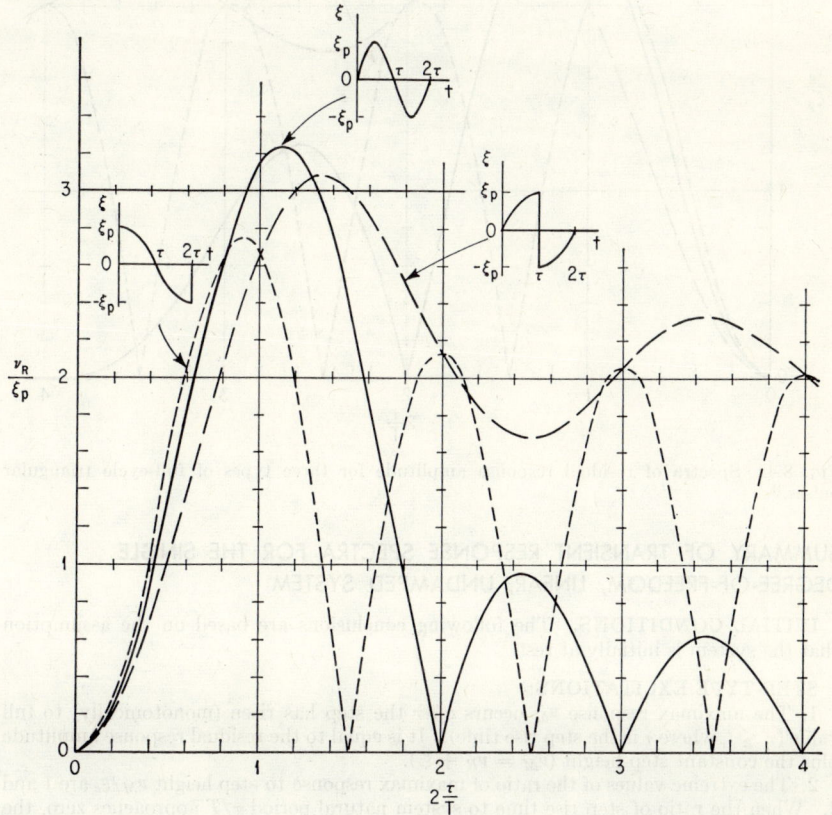

FIG. 8.41. Spectra of residual response amplitude for three types of full-cycle "sinusoidal" pulses.[19]

Vertical jump at mid-cycle

$$\frac{\nu_R}{\xi_p} = 2\left(1 - \frac{T}{2\pi\tau}\sin\frac{2\pi\tau}{T}\right) \tag{8.53c}$$

In the case of full-cycle pulses having symmetrical half cycles, note that the residual response amplitude equals the residual response amplitude of the symmetrical one-half-cycle pulse of the same shape, multiplied by the dimensionless residual response amplitude function $2\sin(\pi\tau/T)$ for the single rectangular pulse. Compare the bracketed functions in Eqs. (8.51), (8.52a), and (8.53a) with the bracketed functions in Eqs. (8.31b), (8.32b), and (8.34c), respectively.

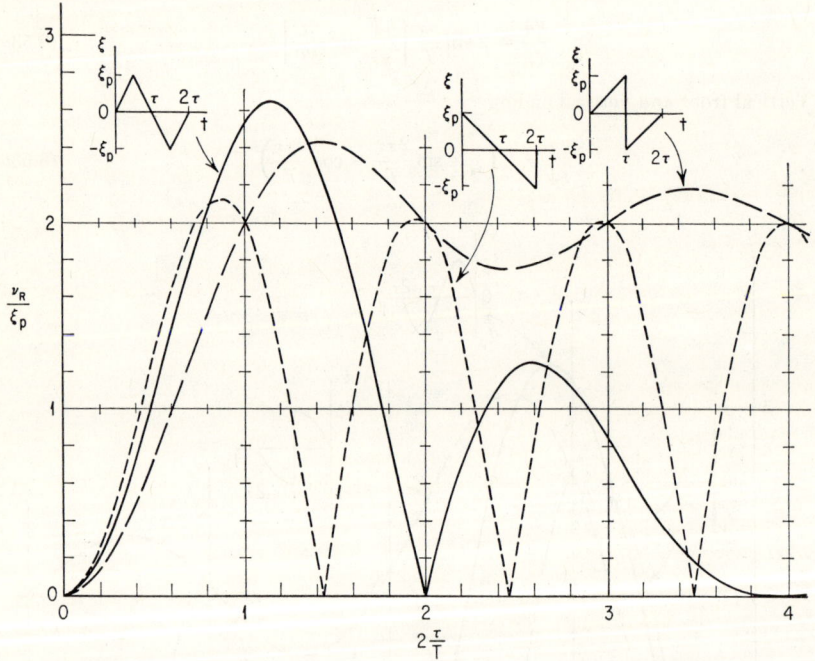

FIG. 8.42. Spectra of residual response amplitude for three types of full-cycle triangular pulses.[19]

SUMMARY OF TRANSIENT RESPONSE SPECTRA FOR THE SINGLE DEGREE-OF-FREEDOM, LINEAR, UNDAMPED SYSTEM

INITIAL CONDITIONS. The following conclusions are based on the assumption that the system is initially at rest.

STEP-TYPE EXCITATIONS:

1. The maximax response ν_M occurs *after* the step has risen (monotonically) to full value ($\tau \leq t$, where τ is the step rise time). It is equal to the residual response amplitude plus the constant step height ($\nu_M = \nu_R + \xi_c$).

2. The extreme values of the ratio of maximax response to step height ν_M/ξ_c are 1 and 2. When the ratio of step rise time to system natural period τ/T approaches zero, the step approaches the simple rectangular step in shape and ν_M/ξ_c approaches the upper extreme of 2. If τ/T approaches infinity, the step loses the character of a dynamic ex-

citation; consequently, the inertia forces of the system approach zero and ν_M/ξ_c approaches the lower extreme of 1.

3. For some particular shapes of step rise, ν_M/ξ_c is equal to 1 at certain finite values of τ/T. For example, for the step having a constant-slope rise, $\nu_M/\xi_c = 1$ when $\tau/T = 1, 2, 3, \ldots$. The lowest values of $\tau/T = (\tau/T)_{\min}$, for which $\nu_M/\xi_c = 1$, are, for three shapes of step rise: constant-slope, 1.0; versed-sine, 1.5; cycloidal, 2.0. The lowest possible value of $(\tau/T)_{\min}$ is 1.

4. In the case of step-type excitations, when $\nu_M/\xi_c = 1$ the residual response amplitude ν_R is zero. Sometimes it is of practical importance in the design of cams and dynamic control functions to achieve the smallest possible residual response.

SINGLE-PULSE EXCITATIONS:

1. When the ratio τ/T of pulse duration to system natural period is less than $\frac{1}{2}$, the time shapes of certain types of equal area pulses are of secondary significance in determining the maxima of system response [maximax response ν_M, maximax relative response $(\nu - \xi)_M$, and residual response amplitude ν_R]. If τ/T is less than $\frac{1}{4}$, the pulse shape is of little consequence in almost all cases and the system response can be determined to a fair approximation by use of the simple impulse theory. If τ/T is larger than $\frac{1}{2}$, the pulse shape may be of great significance.

2. The maximum value of maximax response for a given shape of pulse, $(\nu_M)_{\max}$, usually occurs at a value of the period ratio τ/T between $\frac{1}{2}$ and 1. The maximum value of the ratio of maximax response to the reference excitation, $(\nu_M)_{\max}/\xi_p$, is usually between 1.5 and 1.8.

3. If the pulse has a *vertical rise*, ν_M is the first maximum occurring, and $(\nu_M)_{\max}$ is an asymptotic value approaching $2\xi_p$ as τ/T approaches infinity. In the special case of the rectangular pulse, $(\nu_M)_{\max}$ is equal to $2\xi_p$ and occurs at values of τ/T equal to or greater than $\frac{1}{2}$.

4. If the pulse has a *vertical decay*, $(\nu_M)_{\max}$ is equal to the maximum value $(\nu_R)_{\max}$ of the residual response amplitude.

5. The maximum value $(\nu_R)_{\max}$ of the residual response amplitude, for a given shape of pulse, often is a reasonably good approximation to $(\nu_M)_{\max}$, except if the pulse has a steep rise followed by a decay. A few examples are shown in Table 8.3. Furthermore, if $(\nu_M)_{\max}$ and $(\nu_R)_{\max}$ for a given pulse shape are approximately equal in magnitude, they occur at values of τ/T not greatly different from each other.

6. Pulse shapes that are mirror images of each other in time result in equal values of residual response amplitude.

7. The residual response amplitude ν_R generally has zero values for certain finite values of τ/T. However, if the pulse has either a vertical rise or a vertical decay, but not both, there are no zero values except the trivial one at $\tau/T = 0$. In the case of the rectangular

Table 8.3. Comparison of Greatest Values of Maximax Response and Residual Response Amplitude

Pulse shape	$(\nu_M)_{\max}/(\nu_R)_{\max}$
Symmetrical:	
Rectangular	1.00
Sine	1.04
Versed sine	1.05
Triangular	1.06
Vertical-decay pulses	1.00
Vertical-rise pulses:	
Rectangular	1.00
Triangular	1.60
Asymptotic exponential decay	2.00

pulse, $\nu_R = 0$ when $\tau/T = 1, 2, 3, \ldots$. For several shapes of pulse the values of $(\tau/T)_{\min}$ (lowest values of τ/T for which $\nu_R = 0$) are as follows: rectangular, 1.0; sine, 1.5; versed-sine, 2.0; symmetrical triangle, 2.0. The lowest possible value of $(\tau/T)_{\min}$ is 1.

8. In the formulation of pulse as well as of step-type excitations, it may be of practical consequence for the residual response to be as small as possible; hence, attention is devoted to the case, $\nu_R = 0$.

SINGLE DEGREE-OF-FREEDOM LINEAR SYSTEM WITH DAMPING

The calculation of the effects of damping on transient response may be laborious. If the investigation is an extensive one, use should be made of an analog computer.

DAMPING FORCES PROPORTIONAL TO VELOCITY (VISCOUS DAMPING)

In the case of steady forced vibration, even very small values of the viscous damping coefficient have great effect in limiting the system response at or near resonance. If the excitation is of the single step- or pulse-type, however, the effect of damping on the maximax response may be of relatively less importance, unless the system is highly damped.

For example, in a system under steady sinusoidal excitation at resonance, a tenfold increase in the fraction of critical damping c/c_c from 0.01 to 0.1 results in a theoretical tenfold decrease in the magnification factor from 50 to 5. In the case of the same system, initially at rest and acted upon by a half-cycle sine pulse of "resonant duration" $\tau = T/2$, the same increase in the damping coefficient results in a decrease in the maximax response of only about 9 per cent.

HALF-CYCLE SINE PULSE EXCITATION. Figure 8.43 shows the spectra of maximax response for a viscously damped system excited by a half-cycle sine pulse.[12, 29] The

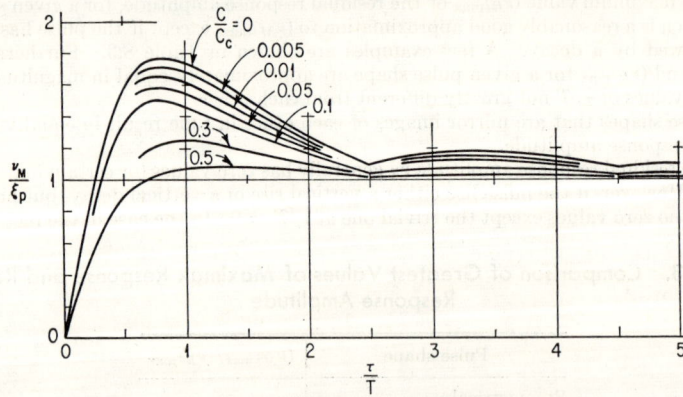

FIG. 8.43. Spectra of maximax response for a viscously damped single degree-of-freedom system acted upon by a half-cycle sine pulse. (R. D. Mindlin, F. W. Stubner, and H. L. Cooper.[29])

system is initially at rest. The results apply to the cases indicated by the following differential equations of motion:

$$\frac{m\ddot{x}}{k} + \frac{c\dot{x}}{k} + x = \frac{F_p}{k}\sin\frac{\pi t}{\tau} \qquad (8.54a)$$

$$\frac{m\ddot{x}}{k} + \frac{c\dot{x}}{k} + x = u_p \sin\frac{\pi t}{\tau} \qquad (8.54b)$$

SINGLE DEGREE-OF-FREEDOM LINEAR SYSTEM WITH DAMPING

$$\frac{m\ddot{\delta}_x}{k} + \frac{c\dot{\delta}_x}{k} + \delta_x = \frac{-m\ddot{u}_p}{k} \sin \frac{\pi t}{\tau} \tag{8.54c}$$

and in general

$$\frac{m\ddot{\nu}}{k} + \frac{c\dot{\nu}}{k} + \nu = \xi_p \sin \frac{\pi t}{\tau} \tag{8.54d}$$

where $0 \leq t \leq \tau$.

For values of t greater than τ, the excitation is zero. The distinctions among these cases may be determined by referring to Table 8.1. The fraction of critical damping c/c_c in Fig. 8.43 is the ratio of the damping coefficient c to the critical damping coefficient $c_c = 2\sqrt{mk}$. The damping coefficient must be defined in terms of the velocity $(\dot{x}, \dot{\delta}_x, \dot{\nu})$ appropriate to each case. For $c/c_c = 0$, the response spectrum is the same as the spectrum for maximax response shown for the undamped system in Fig. 8.16B.

OTHER FORMS OF EXCITATION; METHODS. Qualitative estimates of the effects of viscous damping in the case of other forms of step or pulse excitation may be made by the use of Fig. 8.43 and of the appropriate spectrum for the undamped response to the excitation in question.

Quantitative calculations may be effected by extending the methods described for the undamped system. If the excitation is of general form, given either numerically or graphically, the *phase-plane-delta* [21,22] method described in a later section of this chapter may be used to advantage. Of the analytical methods, the *Laplace transformation* is probably the most useful. A brief discussion of its application to the viscously damped system follows.

LAPLACE TRANSFORMATION. The differential equation to be solved is

$$\frac{m\ddot{\nu}}{k} + \frac{c\dot{\nu}}{k} + \nu = \xi(t) \tag{8.55a}$$

Rewriting Eq. (8.55a),

$$\frac{\ddot{\nu}}{\omega_n^2} + \frac{2\zeta\dot{\nu}}{\omega_n} + \nu = \xi(t) \tag{8.55b}$$

where $\zeta = c/c_c$ and $\omega_n^2 = k/m$.

Applying the operation transforms of Table 8.2 to Eq. (8.55b), the following algebraic equation is obtained:

$$\frac{1}{\omega_n^2}[s^2 F_r(s) - sf(0) - f'(0)] + \frac{2\zeta}{\omega_n}[sF_r(s) - f(0)] + F_r(s) = F_e(s) \tag{8.56a}$$

The *subsidiary* equation is

$$F_r(s) = \frac{(s + 2\zeta\omega_n)f(0) + f'(0) + \omega_n^2 F_e(s)}{s^2 + 2\zeta\omega_n s + \omega_n^2} \tag{8.56b}$$

where the initial conditions $f(0)$ and $f'(0)$ are to be expressed as ν_0 and $\dot{\nu}_0$, respectively.

By performing an *inverse* transformation of Eq. (8.56b), the response is determined in the following operational form:

$$\nu(t) = \mathcal{L}^{-1}[F_r(s)]$$

$$= \mathcal{L}^{-1}\left[\frac{(s + 2\zeta\omega_n)\nu_0 + \dot{\nu}_0 + \omega_n^2 F_e(s)}{s^2 + 2\zeta\omega_n s + \omega_n^2}\right] \tag{8.57}$$

Example 8.9. Rectangular Step Excitation. Assume that the damping is less than critical ($\zeta < 1$), that the system starts from rest ($v_0 = \dot{v}_0 = 0$), and that the system is acted upon by the rectangular step excitation: $\xi(t) = \xi_c$ for $0 \leq t$. The transform of the excitation is given by

$$F_e(s) = \mathcal{L}[\xi(t)] = \mathcal{L}[\xi_c] = \xi_c \frac{1}{s}$$

Substituting for v_0, $\dot{v}_0$ and $F_e(s)$ in Eq. (8.57), the following equation is obtained:

$$v(t) = \mathcal{L}^{-1}[F_r(s)] = \xi_c \omega_n^2 \mathcal{L}^{-1}\left[\frac{1}{s(s^2 + 2\zeta\omega_n s + \omega_n^2)}\right] \quad (8.58a)$$

Rewriting,

$$v(t) = \xi_c \omega_n^2 \mathcal{L}^{-1}\left[\frac{1}{s[s + \omega_n(\zeta - j\sqrt{1-\zeta^2})][s + \omega_n(\zeta + j\sqrt{1-\zeta^2})]}\right] \quad (8.58b)$$

where $j = \sqrt{-1}$.

To determine the inverse transform $\mathcal{L}^{-1}[F_r(s)]$, it may be necessary to expand $F_r(s)$ in partial fractions as explained previously. However, in this particular example the transform pair is available in Table 8.2 (see item 16). Thus, it is found readily that $v(t)$ is given by the following:

$$v(t) = \xi_c \omega_n^2 \left[\frac{1}{ab} + \frac{be^{-at} - ae^{-bt}}{ab(a-b)}\right] \quad (8.59a)$$

where $a = \omega_n(\zeta - j\sqrt{1-\zeta^2})$ and $b = \omega_n(\zeta + j\sqrt{1-\zeta^2})$. By using the relations, $\cos z = (\frac{1}{2})(e^{jz} + e^{-jz})$ and $\sin z = -(\frac{1}{2})j(e^{jz} - e^{-jz})$, Eq. (8.59a) may be expressed in terms of *cosine* and *sine* functions:

$$v(t) = \xi_c \left[1 - e^{-\zeta\omega_n t}\left(\cos \omega_d t + \frac{\zeta}{\sqrt{1-\zeta^2}}\sin \omega_d t\right)\right] \quad [\zeta < 1] \quad (8.59b)$$

where the *damped* natural frequency $\omega_d = \omega_n\sqrt{1-\zeta^2}$.

If the damping is negligible, $\zeta \to 0$ and Eq. (8.59b) reduces to the form of Eq. (8.22) previously derived for the case of zero damping:

$$v(t) = \xi_c(1 - \cos \omega_n t) \quad [\zeta = 0] \quad (8.22)$$

CONSTANT (COULOMB) DAMPING FORCES; PHASE-PLANE METHOD

The phase-plane method is particularly well suited to the solving of transient response problems involving Coulomb damping forces.[21,22] The problem is truly a stepwise linear one, provided the usual assumptions regarding Coulomb friction are valid. For example, the differential equation of motion for the case of ground displacement excitation is

$$m\ddot{x} \pm F_f + kx = ku(t) \quad (8.60a)$$

where F_f is the Coulomb friction force. In Eq. (8.60b) the friction force has been moved to the right side of the equation and the equation has been divided by the spring constant k:

$$\frac{m\ddot{x}}{k} + x = u(t) \mp \frac{F_f}{k} \quad (8.60b)$$

The effect of friction can be taken into account readily in the construction of the phase trajectory by modifying the ordinates of the stepwise excitation by amounts equal to $\mp F_f/k$. The quantity F_f/k is the Coulomb friction "displacement," and is equal to one-fourth the decay in amplitude in each cycle of a *free* vibration under the influence of Coulomb friction. The algebraic sign of the friction term changes when the velocity changes sign. When the friction term is placed on the right-hand side of the differential equation, it must have a *negative* sign when the velocity is positive.

Example 8.10. Free Vibration. Figure 8.44 shows an example of free vibration with the initial conditions $x = x_0$ and $\dot{x} = 0$. The locations of the arc centers of the phase trajectory alternate each half cycle from $+F_f/k$ to $-F_f/k$.

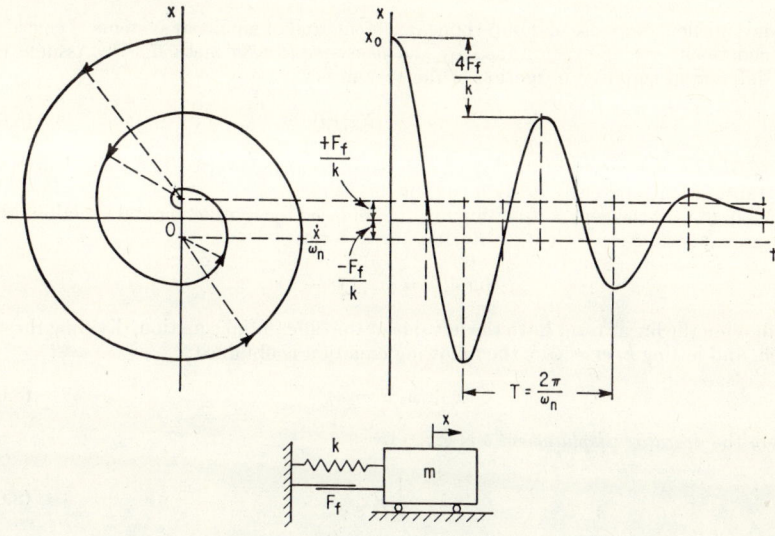

FIG. 8.44. Example of phase-plane solution of free vibration with Coulomb friction;[2] the natural frequency is $\omega_n = \sqrt{k/m}$.

Example 8.11. General Transient Excitation. A general stepwise excitation $u(t)$ and the response x of a system under the influence of a friction force F_f are shown in Fig. 8.45. The case of zero friction is also shown. The initial conditions are $x = 0$, $\dot{x} = 0$. The arc centers are located at ordinates of $u(t) \mp F_f/k$. During the third step in the excitation, the velocity of the system changes sign from positive to negative (at $t = t_2'$); consequently, the friction displacement must also change sign, but from negative to positive.

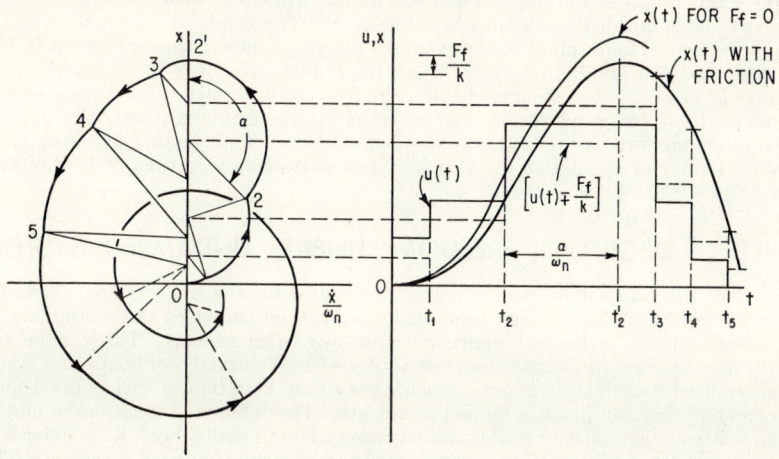

FIG. 8.45. Example of phase-plane solution for a general transient excitation with Coulomb friction in the system.[2]

SINGLE DEGREE-OF-FREEDOM NONLINEAR SYSTEMS

PHASE-PLANE-DELTA METHOD

The transient response of damped linear systems and of nonlinear systems of considerable complexity can be determined by the *phase-plane-delta* method.[21, 22] Assume that the differential equation of motion of the system is

$$m\ddot{x} = G(x,\dot{x},t) \tag{8.61a}$$

where $G(x,\dot{x},t)$ is a general function of x, $\dot{x}$, and t to any powers. The coefficient of $\ddot{x}$ is constant, either inherently or by a suitable division.

In Eq. (8.61a) the general function may be replaced by another general function minus a linear, constant-coefficient, restoring force term:

$$G(x,\dot{x},t) = g(x,\dot{x},t) - kx$$

By moving the linear term kx to the left side of the differential equation, dividing through by m, and letting $k/m = \omega_n^2$, the following equation is obtained:

$$\ddot{x} + \omega_n^2 x = \omega_n^2 \delta \tag{8.61b}$$

where the *operative displacement* δ is given by

$$\delta = \frac{1}{k} g(x,\dot{x},t) \tag{8.61c}$$

The separation of the kx term from the G function does not require that the kx term exist physically. Such a term can be separated by first adding to the G function the *fictitious* terms, $+kx - kx$.

With the differential equation of motion in the δ form, Eq. (8.61b), the response problem can now be solved readily by stepwise linearization. The left side of the equation represents a simple, undamped, linear oscillator. Implicit in the δ function on the right side of the equation are the nonlinear restoration terms, the linear or nonlinear dissipation terms, and the excitation function.

If the δ function is held constant at a value δ for an interval of time Δt, the response of the linear oscillator in the phase-plane is an arc of a circle, with its center on the X axis at δ and subtended by an angle equal to $\omega_n \Delta t$. The graphical construction may be similar in general appearance to the examples already shown for linear systems in Figs. 8.5, 8.44, and 8.45. Since in the general case the δ function involves the dependent variables, it is necessary to estimate, before constructing each step, appropriate average values of the system displacement and/or velocity to be expected during the step. In some cases, more than one trial may be required before suitable accuracy is obtained.

Many examples of solution for various types of systems are available in the literature.[3, 5, 6, 8, 13, 15, 20, 21, 22, 25]

MULTIPLE DEGREE-OF-FREEDOM, LINEAR, UNDAMPED SYSTEMS

Some of the transient response analyses, presented for the single degree-of-freedom system, are in complete enough form that they can be employed in determining the responses of linear, undamped, multiple degree-of-freedom systems. This can be done by the use of *normal (principal) coordinates*. A system of normal coordinates is a system of generalized coordinates chosen in such a way that vibration in each normal mode involves only one coordinate, a normal coordinate. The differential equations of motion, when written in normal coordinates, are all independent of each other. Each differential equation is related to a particular normal mode and involves only one coordinate. The differential equations are of the same general form as the differential equation of motion for the single degree-of-freedom system. The response of the system in terms of the

MULTIPLE DEGREE-OF-FREEDOM, LINEAR, UNDAMPED SYSTEMS

physical coordinates, for example, displacement or stress at various locations in the system, is determined by superposition of the normal coordinate responses. Normal coordinates are discussed in Chaps. 2 and 7, and in Refs. 37 and 38.

Example 8.12. Sine Force Pulse Acting on a Simple Beam. Consider the flexural vibration of a prismatic bar with simply supported ends, Fig. 8.46. A sine-pulse concentrated force $F_p \sin(\pi t/\tau)$ is applied to the beam at a distance c from the left end (origin of coordinates). Assume that the beam is initially at rest. The displacement response of the beam, during the time of action of the pulse, is given by the following series:

$$y = \frac{2F_p l^3}{\pi^4 EI} \sum_{i=1}^{i=\infty} \frac{1}{i^4} \sin\frac{i\pi c}{l} \sin\frac{i\pi x}{l} \left[\frac{1}{1 - T_i^2/4\tau^2}\left(\sin\frac{\pi t}{\tau} - \frac{T_i}{2\tau}\sin\omega_i t\right)\right] \quad [0 \leq t \leq \tau] \quad (8.62a)$$

where $\quad i = 1, 2, 3, \ldots; T_i = \dfrac{2\pi}{\omega_i} = \dfrac{2l^2}{i^2\pi}\sqrt{\dfrac{A\gamma}{EIg}} = \dfrac{T_1}{i^2}$, sec

A comparison of Eqs. (8.62a) and (8.32a) shows that the time function $[\sin(\pi t/\tau) - (T_i/2\tau) \sin \omega_i t]$ for the ith term in the beam-response series is of exactly the same form as the time function $[\sin(\pi t/\tau) - (T/2\tau) \sin \omega_n t]$ in the response of the single degree-of-freedom system. Furthermore, the magnification factors $1/(1 - T_i^2/4\tau^2)$ and $1/(1 - T^2/4\tau^2)$ in the two equations have identical forms.

Following the end of the pulse, beginning at $t = \tau$, the vibration of the beam is expressed by

$$y = \frac{2F_p l^3}{\pi^4 EI} \sum_{i=1}^{i=\infty} \frac{1}{i^4} \sin\frac{i\pi c}{l} \sin\frac{i\pi x}{l} \left[\frac{(T_i/\tau)\cos(\pi\tau/T_i)}{(T_i^2/4\tau^2) - 1}\sin\omega_i\left(t - \frac{\tau}{2}\right)\right] \quad [\tau \leq t] \quad (8.62b)$$

A comparison of Eqs. (8.62b) and (8.32b) leads to the same conclusion as found above for the time era $0 \leq t \leq \tau$.

Excitation and Displacement at Mid-span. As a specific case, consider the displacement at mid-span when the excitation is applied at mid-span ($c = x = l/2$). The even-numbered terms of the series now are all zero and the series take the following forms:

$$y_{l/2} = \frac{2F_p l^3}{\pi^4 EI} \sum_{i=1,3,5,\ldots}^{\infty} \frac{1}{i^4}\left[\frac{1}{1 - T_i^2/4\tau^2}\left(\sin\frac{\pi t}{\tau} - \frac{T_i}{2\tau}\sin\omega_i t\right)\right] \quad [0 \leq\, \leq \tau] \quad (8.63a)$$

$$y_{l/2} = \frac{2F_p l^3}{\pi^4 EI} \sum_{i=1,3,5,\ldots}^{\infty} \frac{1}{i^4}\left[\frac{(T_i/\tau)\cos(\pi\tau/T_i)}{(T_i^2/4\tau^2) - 1}\sin\omega_i\left(t - \frac{\tau}{2}\right)\right] \quad [\tau \leq t] \quad (8.63b)$$

Assume, for example, that the pulse period τ equals two-tenths of the fundamental natural period of the beam ($\tau/T_1 = 0.2$). It is found from Fig. 8.16B, by using an abscissa value of 0.2, that the maximax response in the *fundamental* mode ($i = 1$) occurs in the residual vibration era ($\tau \leq t$). The value of the corresponding ordinate is 0.75. Consequently, the maximax response for $i = 1$ is 0.75 $(2F_p l^3/\pi^4 EI)$.

In order to determine the maximax for the *third* mode ($i = 3$), an abscissa value of $\tau/T_i = i^2\tau/T_1 = 3^2 \times 0.2 = 1.8$, is used. It is found that the maximax is greater than the residual amplitude and consequently that it occurs during the time era $0 \leq t \leq \tau$. The value of the corresponding ordinate is 1.36; however, this must be multiplied by $1/3^4$, as indicated by the series. The maximax for $i = 3$ is thus 0.017 $(2F_p l^3/\pi^4 EI)$.

The maximax for $i = 5$ also occurs in the time era $0 \leq t \leq \tau$ and the ordinate may be estimated to be about 1.1. Multiplying by $1/5^4$, it is found that the maximax for $i = 5$ is approximately 0.002 $(2F_p l^3/\pi^4 EI)$, a negligible quantity when compared with the maximax value for $i = 1$.

To find the maximax total response to a reasonable approximation, it is necessary to sum on a time basis several terms of the series. In the particular example above, the maximax total response occurs in the residual vibration era and a reasonably accurate value can be obtained by considering only the first term ($i = 1$) in the series, Eq. (8.63b).

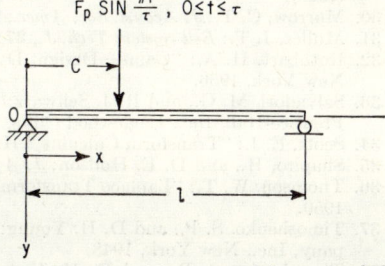

FIG. 8.46. Simply supported beam loaded by a concentrated force sine pulse of half-cycle duration.

GENERAL INVESTIGATION OF TRANSIENTS

An extensive (and efficient) investigation of transient response in multiple degree-of-freedom systems requires the use of an automatic computer.[26] In some of the simpler cases, however, it is feasible to employ numerical or graphical methods. For example, the phase-plane method may be applied to multiple degree-of-freedom linear systems [1,2] through the use of normal coordinates. This involves independent phase-planes having the coordinates q_i and $\dot{q}_i/\omega_i$, where q_i is the ith normal coordinate.

REFERENCES

1. Ayre, R. S.: *J. Franklin Inst.*, **253**:153 (1952).
2. Ayre, R. S.: *Proc. World Conf. Earthquake Eng.*, 1956, p. 13-1.
3. Ayre, R. S., and J. I. Abrams: *Proc. ASCE*, EM 2, Paper 1580, 1958.
4. Biot, M. A.: *Trans. ASCE*, **108**:365 (1943).
5. Bishop, R. E. D.: *Proc. Inst. Mech. Engrs. (London)*, **168**:299 (1954).
6. Braun, E.: *Ing.-Arch.*, **8**:198 (1937).
7. Bronwell, A.: "Advanced Mathematics in Physics and Engineering," McGraw-Hill Book Company, Inc., New York, 1953.
8. Bruce, V. G.: *Bull. Seismol. Soc. Amer.*, **41**:101 (1951).
9. Cherry, C.: "Pulses and Transients in Communication Circuits," Dover Publications, New York, 1950.
10. Crede, C. E.: "Vibration and Shock Isolation," John Wiley & Sons, Inc., New York, 1951.
11. Crede, C. E.: *Trans. ASME*, **77**:957 (1955).
12. Criner, H. E., G. D. McCann, and C. E. Warren: *J. Appl. Mechanics*, **12**:135 (1945).
13. Evaldson, R. L., R. S. Ayre, and L. S. Jacobsen: *J. Franklin Inst.*, **248**:473 (1949).
14. Frankland, J. M.: *Proc. Soc. Exptl. Stress Anal.*, **6**:2, 7 (1948).
15. Fuchs, H. O.: *Product Eng.*, August, 1936, p. 294.
16. Gardner, M. F., and J. L. Barnes: "Transients in Linear Systems," vol. I, John Wiley & Sons, Inc., New York, 1942.
17. Hartman, J. B.: "Dynamics of Machinery," McGraw-Hill Book Company, Inc., New York, 1956.
18. Hudson, G. E.: *Proc. Soc. Exptl. Stress Anal.*, **6**:2, 28 (1948).
19. Jacobsen, L. S., and R. S. Ayre: "A Comparative Study of Pulse and Step-type Loads on a Simple Vibratory System," *Tech. Rept.* N16, under contract N6-ori-154, T. O. 1, U.S. Navy, Stanford University, 1952.
20. Jacobsen, L. S.: *Proc. Symposium on Earthquake and Blast Effects on Structures*, 1952, p. 94.
21. Jacobsen, L. S.: *J. Appl. Mechanics*, **19**:543 (1952).
22. Jacobsen, L. S., and R. S. Ayre: "Engineering Vibrations," McGraw-Hill Book Company, Inc., New York, 1958.
23. Kelvin, Lord: *Phil. Mag.*, **34**:443 (1892).
24. Kornhauser, M.: *J. Appl. Mechanics*, **21**:371 (1954).
25. Lamoen, J.: *Rev. universelle mines*, ser. 8, **11**:7, 3 (1935).
26. McCann, G. D., and J. M. Kopper: *J. Appl. Mechanics*, **14**:A127 (1947).
27. Meissner, E.: *Schweiz. Bauzt.*, **99**:27, 41 (1932).
28. Mindlin, R. D.: *Bell System Tech. J.*, **24**:353 (1945).
29. Mindlin, R. D., F. W. Stubner, and H. L. Cooper: *Proc. Soc. Exptl. Stress Anal.*, **5**:2, 69 (1948).
30. Morrow, C. T.: *J. Acoust. Soc. Amer.*, **29**:596 (1957).
31. Muller, J. T.: *Bell System Tech. J.*, **27**:657 (1948).
32. Rothbart, H. A.: "Cams—Design, Dynamics and Accuracy," John Wiley & Sons, Inc., New York, 1956.
33. Salvadori, M. G., and R. J. Schwarz: "Differential Equations in Engineering Problems," Prentice-Hall, Inc., Englewood Cliffs, N.J., 1954.
34. Scott, E. J.: "Transform Calculus," Harper & Brothers, New York, 1955.
35. Shapiro, H., and D. E. Hudson: *J. Appl. Mechanics*, **20**:422 (1953).
36. Thomson, W. T.: "Laplace Transformation," Prentice-Hall, Inc., Englewood Cliffs, N.J., 1950.
37. Timoshenko, S. P., and D. H. Young: "Advanced Dynamics," McGraw-Hill Book Company, Inc., New York, 1948.
38. Timoshenko, S. P., and D. H. Young: "Vibration Problems in Engineering," 3d ed., D. Van Nostrand Company, Inc., Princeton, N.J., 1955.
39. Walsh, J. P., and R. E. Blake: *Proc. Soc. Exptl. Stress Anal.*, **6**:2, 151 (1948).
40. Williams, H. A.: *Trans. ASCE*, **102**:838 (1937).

9

EFFECTS OF IMPACT ON STRUCTURES

W. H. Hoppmann II
Rensselaer Polytechnic Institute

INTRODUCTION

This chapter discusses a particular phenomenon in the general field of shock and vibration usually referred to as impact.[1] An impact occurs when two or more bodies collide. An important characteristic of an impact is the generation of relatively large forces at points of contact for relatively short periods of time. Such forces sometimes are referred to as *impulse-type* forces.

Three general classes of impact are considered in this chapter: (1) impact between spheres or other rigid bodies, where a body is considered to be rigid if its dimensions are large relative to the wavelengths of the elastic stress waves in the body; (2) impact of a rigid body against a beam or plate that remains substantially elastic during the impact; and (3) impact involving yielding of structures.

DIRECT CENTRAL IMPACT OF TWO SPHERES

The elementary analysis of the central impact of two bodies is based upon an experimental observation of Newton.[2] According to that observation, the relative velocity of two bodies after impact is in constant ratio to their relative velocity before impact and is in the opposite direction. This constant ratio is the *coefficient of restitution;* usually it is designated by e.[3]

Let $\dot{u}$ and $\dot{x}$ be the components of velocity along a common line of motion of the two bodies before impact, and $\dot{u}'$ and $\dot{x}'$ the component velocities of the bodies in the same direction after impact. Then, by the observation of Newton,

$$\dot{u}' - \dot{x}' = -e(\dot{u} - \dot{x}) \qquad (9.1)$$

Now suppose that a smooth sphere of mass m_u and velocity $\dot{u}$ collides with another smooth sphere having the mass m_x and velocity $\dot{x}$ moving in the same direction. Let the coefficient of restitution be e, and let $\dot{u}'$ and $\dot{x}'$ be the velocities of the two spheres, respectively, after impact. Figure 9.1 shows the condition of the two spheres just before collision. The only force acting on the spheres during impact is the force at the point of contact, acting along the line through the centers of the spheres.

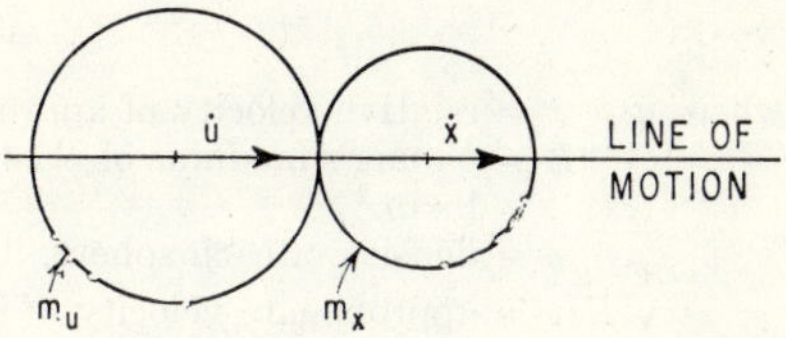

Fig. 9.1. Positions of two solid spheres at instant of central impact.

According to the law of conservation of linear momentum:

$$m_u \dot{u}' + m_x \dot{x}' = m_u \dot{u} + m_x \dot{x} \qquad (9.2)$$

Solving Eqs. (9.1) and (9.2) for the two unknowns, the velocities $\dot{u}'$ and $\dot{x}'$ after impact,

$$\dot{u}' = \frac{(m_u\dot{u} + m_x\dot{x}) - em_x(\dot{u} - \dot{x})}{m_u + m_x}$$

$$\dot{x}' = \frac{(m_u\dot{u} + m_x\dot{x}) + em_u(\dot{u} - \dot{x})}{m_u + m_x}$$

(9.3)

This analysis yields the resultant velocities for the two spheres on the basis of an experimental law and the principle of the conservation of momentum, without any specific reference to the force of contact F. A similar result is obtained for a ballistic pendulum used to measure the muzzle velocity of a bullet. A bullet of mass m_u and velocity $\dot{u}$ is fired into a block of wood of mass m_x which is at rest initially and finally assumes a velocity $\dot{x}'$ after the impact. Using only the principle of the conservation of momentum,

$$\dot{u} = \frac{(m_u + m_x)\dot{x}'}{m_u}$$

(9.4)

No knowledge of the complicated pattern of force acting on the bullet and the pendulum during the embedding process is required.

These simple facts are introductory to the more complicated problem involving the vibration of at least one of the colliding bodies, as discussed in a later section.

HERTZ THEORY OF IMPACT OF TWO SOLID SPHERES

The theory of two solid elastic spheres which collide with one another is based upon the results of an investigation of two elastic bodies pressed against one another under purely statical conditions.[4] For these static conditions, the relations between the sum of the displacements at the point of contact in the direction of the common line of motion and the resultant total pressure have been derived. The sum of these displacements is equal to the relative approach of the centers of the spheres, assuming that the spheres act as rigid bodies except for elastic compression at the point of contact. The relative approach varies as the two-thirds power of the total pressure; a formula is given for the time of duration of the contact.[4] The theory is valid only if the duration of contact is long in comparison with the period of the fundamental mode of vibration of either sphere.

The range of validity of the Hertz theory is related to the possibility of exciting vibration in the spheres.[5] The dimensionless ratio of the maximum kinetic energy of vibration to the sum of the kinetic energies of the two spheres just before collision is approximately

$$R = \frac{1}{50} \frac{\dot{u} - \dot{x}}{\sqrt{E/\rho}}$$

(9.5)

where $\dot{u} - \dot{x}$ = relative velocity of approach, in./sec
E = Young's modulus of elasticity, assumed to be the same for each sphere, lb/in.^{2}
ρ = density of each sphere, lb-sec^2/in.^{4}
$\sqrt{E/\rho}$ = approximate velocity of propagation of dilatational waves, in./sec

The ratio R usually is a very small quantity; thus, the theory of impact set forth by Eq. (9.5) has wide application because vibration is not generated in the spheres to an appreciable degree under ordinary conditions. The energy of the colliding spheres remains translational, and the velocities after impact are deducible from the principles of energy and of momentum. The important point of plastic deformation at the point of contact is discussed in a later section.

Formulas for force between the spheres, the radius of the circular area of contact, and the relative approach of the centers of the spheres, all as functions of time, can be determined for any two given spheres.[6]

IMPACT OF A SOLID SPHERE ON AN ELASTIC PLATE

An extension of the Hertz theory of impact to include the effect of vibration of one of the colliding bodies involves a study of the transverse impact of a solid sphere upon an infinitely extended plate.[7] The plate has the role of the vibrating body. The coefficient of restitution is an important element in any analysis of the motion ensuing after the collision of two bodies.

The analysis is based on the assumption that the principal elastic waves of importance are flexural waves of half-period equal to the duration of impact. Let $2h$ and $2D$ be the thickness of plate and diameter of sphere, respectively, ρ_1, ρ_2 their densities, E_1, E_2 their Young's moduli, ν_1, ν_2 their values of Poisson's ratio, and τ_H the duration of impact. The velocity c of long flexural waves of wavelength λ in the plate is given by

$$c^2 = \frac{4\pi^2}{3} \frac{h^2}{\lambda^2} \frac{E_1}{\rho_1(1-\nu_1^2)} \tag{9.6}$$

The radius a of the circle on the plate over which the disturbance has spread at the termination of impact is given by

$$a = c\tau_H = \frac{\lambda}{2} \tag{9.7}$$

Combining Eqs. (9.6) and (9.7),

$$a^2 = \pi\tau_H h \sqrt{\frac{E_1}{3\rho_1(1-\nu_1^2)}} \tag{9.8}$$

The next step is to find the kinetic and potential energies of the wave motion of the plate. The kinetic energy may be determined from the transverse velocity of the plate at each point over the circle of radius a covered by the wave. Figure 9.2 shows an approximate distribution of velocity over the circle of radius a at the end of impact.[8] The direction of the impact also is shown. The kinetic energy in the wave at the end of impact is

$$T = \int_0^a \tfrac{1}{2} \cdot 2h \cdot \rho_1 \cdot 2\pi R \cdot \dot{w}^2 \, dR \tag{9.9}$$

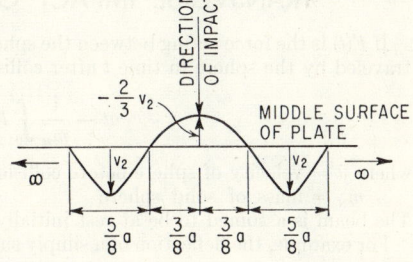

Fig. 9.2. Distribution of transverse velocities in plate as a result of impact by a moving body. (*After Lamb*.)

where $\dot{w}$ is the transverse velocity at distance R from the origin. As an approximation it is assumed that the sum of the potential energy and the kinetic energy in the wave is $2T$. With considerable effort these energies can be calculated in terms of the motion of the plate, although the calculation may be laborious.

The impulse in the plate produced by the colliding body is

$$J = \int_0^a \tfrac{1}{2} \cdot 2h \cdot \rho_1 \cdot 2\pi R \cdot \dot{w} \, dr \tag{9.10}$$

The integration should be carried out with due regard to the sign of velocity. If m_u is the mass of colliding body, $\dot{u}$ its velocity before impact, and e the coefficient of restitution, the following relations are obtained on the assumption that the energy is conserved:

$$\tfrac{1}{2} m_u \dot{u}^2 (1 - e^2) = 2T \tag{9.11}$$

$$m_u \dot{u} (1 + e) = J \tag{9.12}$$

Equation (9.11) represents the energy lost to the moving sphere as a result of impact and Eq. (9.12) represents the change in momentum of the sphere.

The coefficient of restitution e is determined by evaluating the integrals for T and J and substituting their values in Eq. (9.12). The necessary integrations can be performed by taking the function for transverse velocity in Fig. 9.2 as arcs of sine curves. The resultant expression for e is

$$e = \frac{h\rho_1 a^2 - 0.56 m_u}{h\rho_1 a^2 + 0.56 m_u} \quad (9.13)$$

where a, the radius of the deformed region, is given by Eq. (9.8) and τ_H, the time of contact between sphere and plate, is given by Hertz's theory of impact to a first approximation.[4] The mass of the sphere is m_u; the mass of the plate is assumed to be infinite. Large discrepancies between theory and experiment occur when the diameter of the sphere is large compared with the thickness of the plate. The duration of impact τ_H is

$$\tau_H = 2.94 \frac{\alpha}{\dot{u}} \quad (9.14)$$

where

$$\alpha = \left[\frac{15}{16} \nu_1^2 \left(\frac{1 - \nu_1^2}{E_1} + \frac{1 - \nu_2^2}{E_2} \right) m_u \right]^{2/5} R_s^{-1/5}$$

The radius of the striking sphere is R_s and its velocity before impact is $\dot{u}$. Subscripts 1 and 2 represent the properties of the sphere and plate, respectively. The value of τ_H may be substituted in Eq. (9.8) above.

Experimental results verify the theory when the limitations of the theory are not violated. The velocity of impact must be sufficiently small to avoid plastic deformation. When the collision involves steel on steel, the velocity usually must be less than 1 ft/sec. However, useful engineering results can be obtained with this approach even though plastic deformation does occur locally.[9, 10]

TRANSVERSE IMPACT OF A MASS ON A BEAM

If $F(t)$ is the force acting between the sphere and the beam during contact, the distance traveled by the sphere in time t after collision is [11]

$$\dot{u}t - \frac{1}{m_u} \int_0^t F(t_v)(t - t_v)\, dt_v \quad (9.15)$$

where $\dot{u}$ = velocity of sphere before collision (beam assumed to be at rest initially)
m_u = mass of solid sphere

The beam is assumed to be at rest initially.

For example, the deflection of a simply supported beam under force $F(t_v)$ at its center is

$$\sum_{1,3,5\ldots}^{\infty} \frac{1}{m_b} \int_0^t F(t_v) \frac{\sin \omega_n(t - t_v)}{\omega_n}\, dt_v \quad (9.16)$$

where m_b = one-half of mass of beam
ω_n = angular frequency of the nth mode of vibration

Equation (9.16) represents the transverse vibration of a beam. While the present case is only for direct central impact, the cases for noncentral impact depend only on the corresponding solution for transverse vibration. Oblique impact also is treated readily.

The expression for the relative approach of the sphere and beam; i.e., penetration of beam by sphere, is [11]

$$\alpha = \kappa_1 F(t)^{2/3} \quad (9.17)$$

where κ_1 is a constant depending on the elastic and geometrical properties of the sphere and the beam at the point of contact, and α is given by Eq. (9.14). Consequently, the equation that defines the problem is

$$\alpha = K_1 F^{2/3} = \dot{u}t - \frac{1}{m_u} \int_0^t F(t_v)(t - t_v)\, dt_v - \sum_{1,3,5}^{\infty} \frac{1}{m_b} \int_0^t F(t_v) \frac{\sin \omega_n(t - t_v)}{\omega_n}\, dt_v \quad (9.18)$$

Equation (9.18) has been solved numerically for two specific problems by subdividing the time interval 0 to t into small elements and calculating, step by step, the displacements of the sphere.[11] The results are not general but rather apply only to the cases of beam and sphere.

For the impact of a mass on a beam, the sum of the kinetic and the potential energies may be expressed in terms of the unknown contact force.[12] Also, the impulse integral J in terms of the contact force may be expressed as

$$J = \int_0^t F(t)\,dt = m_u \dot{u}(1 + e) \tag{9.19}$$

A satisfactory approximation to $F(t)$ is defined in terms of a normalized force $\bar{F}$:

$$F(t) = m_u \dot{u}(1 + e)\bar{F}(t) \tag{9.20}$$

Thus, from Eqs. (9.19) and (9.20),

$$\int_0^t \bar{F}\,dt = 1 \tag{9.21}$$

The value of this integral is independent of the shape of $F(t)$. The normalized force is defined such that its maximum value equals the maximum value of the corresponding normalized Hertz force.[12] To perform the necessary integrations, a suitable function for defining $F(t)$ is chosen as follows:

$$\begin{aligned}\bar{F}(t) &= \frac{\pi}{2\tau_L}\sin\frac{\pi}{\tau_L}t && [0 < t < \tau_L] \\ \bar{F}(t) &= 0 && [|t| > \tau_L]\end{aligned} \tag{9.22}$$

Results for particular problems solved in this manner agree well with those obtained for the same problems by the numerical solution of the exact integral equation.[12]

To apply these results to a specific beam impact problem, it is necessary to express the deflection equation for the beam in terms of known quantities. One of these quantities is the coefficient of restitution; a formula must be provided for its determination in terms of known functions. This is given by Eq. (9.31).

IMPACT OF A RIGID BODY ON A DAMPED ELASTICALLY SUPPORTED BEAM

For the more general case of impact of a rigid body on a damped, elastically supported beam, it is assumed that there is external damping, damping determined by the Stokes' law of stress-strain, and an elastic support attached to the beam along its length in such a manner that resistance is proportional to deflection.[13] The differential equation for the deflection of the beam is

$$EI\frac{\partial^4 w}{\partial x^4} + c_1 I\frac{\partial^5 w}{\partial x^4 \partial t} + c_2\frac{\partial w}{\partial t} + kw + \rho S\frac{\partial^2 w}{\partial t^2} = F(x,t) \tag{9.23}$$

where w = deflection, in.
 E = Young's modulus, lb/in.^{2}
 I = moment of inertia for cross section (constant), in.^{4}
 c_1 = internal damping coefficient, lb/in.^{2}-sec (Stokes' law)
 c_2 = external damping coefficient, lb/in.^{2}-sec
 k = foundation modulus, lb/in.^{2}
 ρ = density, lb-sec^2/in.^{4}
 S = area of cross section (constant), in.^{2}
 $\dfrac{\partial^2 w}{\partial t^2}$ = acceleration, in./sec^2
 t = time, sec
 $F(x,t)$ = driving force per unit length of beam, lb/in.

For example, to illustrate the application of specific boundary conditions, consider a simply supported beam of length l. The moments and deflections must vanish at the ends. The beam is assumed undeflected and at rest just before impact, and central impact is assumed although with some additional computation this restriction may be dropped. The solution may be written as follows:

$$w(x,t) = \sum_{}^{\infty} \sin \frac{n\pi x}{l} \sin \frac{n\pi}{2} \frac{1}{m} \frac{1}{\sqrt{\omega_n^2 - \delta_n^2}}$$

$$\times \int_0^t e^{-\delta_n(t-\tau)} \sin\left[\sqrt{\omega_n^2 - \delta_n^2} \cdot (t-\tau)\right] F_1(\tau) \, d\tau \quad (9.24)$$

where e = base of natural logarithms

δ_n = damping numbers = $\dfrac{1}{2}\left(r_i \dfrac{n^4 \pi^4}{l^4} + r_e\right)$

$r_i = \dfrac{c_1 I}{\rho S}$

$r_e = \dfrac{c_2}{\rho S}$

ω_n = angular frequencies

$m = \tfrac{1}{2}\rho A l$

A satisfactory analytical expression for the contact force $F_1(t)$, a particularization of $F(x,t)$ in Eq. (9.23), must be developed. Although $F_1(t)$ is assumed to act at the center of the beam, the methods apply with only minor alterations if the impact occurs at any other point of the beam.

One of the conditions which the contact force must satisfy is that its time integral for the duration of impact equal the change in momentum of the striking body. The change of momentum is

$$m\dot{z} - m\dot{z}' = m\dot{z}\left(1 - \frac{\dot{z}'}{\dot{z}}\right) \quad (9.25)$$

where m = mass of rigid body, lb-sec^2/in.

$\dot{z}$ = velocity of rigid body just before collision, in./sec

$\dot{z}'$ = velocity of rigid body just after collision, in./sec

When the velocity of the beam is zero, Eq. (9.1) may be written

$$e = -\frac{\dot{z}'}{\dot{z}} \quad (9.26)$$

Equation (9.26) may be written

$$m\dot{z}\left(1 - \frac{\dot{z}'}{\dot{z}}\right) = m\dot{z}(1 + e) \quad (9.27)$$

From the equivalence of impulse and momentum:

$$\int_0^{\tau_0} F_1(t) \, dt = m\dot{z}(1 + e) \quad (9.28)$$

where τ_0 is the time of contact.

It can then be shown [13] that the impact force may be written

$$F_1(t) = m\dot{z}(1+e) \frac{\pi}{2\tau_L} \sin \frac{n\pi t}{\tau_L} \quad [0 < t < \tau_L]$$
$$F_1 = 0 \quad [t > \tau_L] \quad (9.29)$$

IMPACT OF RIGID BODY ON DAMPED ELASTICALLY SUPPORTED BEAM

It can be shown further [13] that

$$\tau_L = 3.28 \left[\frac{m^2}{\ddot{z}R} \cdot \frac{(1-\nu^2)}{E^2} \right]^{1/5} \quad (9.30)$$

where R = radius of sphere, in.
ν = Poisson's ratio

The time interval τ_L is a special value of the time of contact T_0. It agrees well with experimental results.

The coefficient of restitution e is [13]

$$e = \frac{1 - \dfrac{m}{m_b} \sum_1^\infty \Phi_n - \dfrac{m}{m_b} \sum_1^\infty \Psi_n}{1 + \dfrac{m}{m_b} \sum_1^\infty \Phi_n + \dfrac{m}{m_b} \sum_1^\infty \Psi_n} \quad (9.31)$$

where m = mass of sphere
m_b = half mass of beam

The functions Φ_n and Ψ_n are given in the form of curves in Figs. 9.3 and 9.4; the symbol $\beta_n = \delta_n/\omega_n$ represents fractional damping and $Q_n = \omega_n \tau_L/2\pi$ is a dimensionless frequency where ω_n = angular frequency of nth mode of vibration of undamped vibration

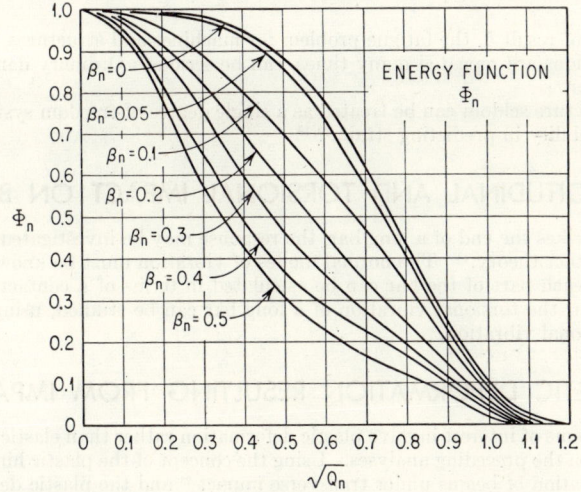

FIG. 9.3. Energy functions ϕ_n used with Eq. (9.31) to determine the coefficient of restitution from the impact of a rigid body on a damped elastically supported beam.

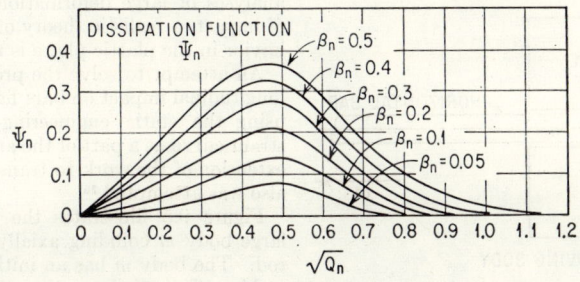

FIG. 9.4. Dissipative (damping) functions ψ_n used with Eq. (9.31) to determine the coefficient of restitution from the impact of a rigid body on a damped elastically supported beam.

of beam, rad/sec, and τ_L = length of time the sinusoidal pulse is assumed to act on beam [see Eq. (9.30)]. If damping is neglected, the functions Ψ_n vanish from Eq. (9.31).

The above theory may be generalized to apply to the response of plates to impact. The deflection equation of a plate subjected to a force applied at a point is required. The various energy distributions at the end of impact are arrived at in a manner analogous to that for the beam.

The theory has been applied to columns and continuous beams [14,15] and also could be applied to transverse impact on a ring. Measurement of the force of impact illustrates the large number of modes of vibration that can be excited by an impact.[16,17,22]

Principal qualitative results of the foregoing analysis are:

1. Impacts by bodies of relatively small mass moving with low velocities develop significant bending strains in beams.

2. External damping of the type assumed above has a rapidly decreasing effect on reducing deflection and strain as the number of the mode increases.

3. Internal damping of the viscous type here assumed reduces deflection and strain appreciably in the higher modes. For a sufficiently high mode number, the vibration becomes aperiodic.

4. Increasing the modulus for an elastic foundation reduces the energy absorbed by the structure from the colliding body.

5. Impacts from collision produce sharp initial rises in strain which are little influenced by damping.

6. Because of result 5, the fatigue problem for machines and structures, in which the impact conditions are repeated many times, can be serious. Ordinary damping affords little protection.

7. The structure seldom can be treated as a single degree-of-freedom system with any degree of reliability in predicting strain.[13,19]

LONGITUDINAL AND TORSIONAL IMPACT ON BARS

If a mass strikes the end of a long bar, the response may be investigated by means of the Hertz contact theory.[11] The normal modes of vibration must be known so the displacement at each part of the bar can be calculated in terms of a contact force. In a similar manner, the torsional vibration of a long bar can be studied, using the normal modes of torsional vibration.

PLASTIC DEFORMATION RESULTING FROM IMPACT

Many problems of interest involve plastic deformation rather than elastic deformation as considered in the preceding analyses. Using the concept of the plastic hinge, the large plastic deformation of beams under transverse impact [23] and the plastic deformation of free rings under concentrated dynamic loads [24] have been studied. In such analyses, the elastic portion of the vibration usually is neglected. To make further progress in analyses of large deformations as a result of impact, a realistic theory of material behavior in the plastic phase is required.

An attempt to solve the problem for the longitudinal impact on bars has been made using the static engineering-type stress-strain curve as a part of the analysis.[25] An extension of the work to transverse impact also was attempted.[26]

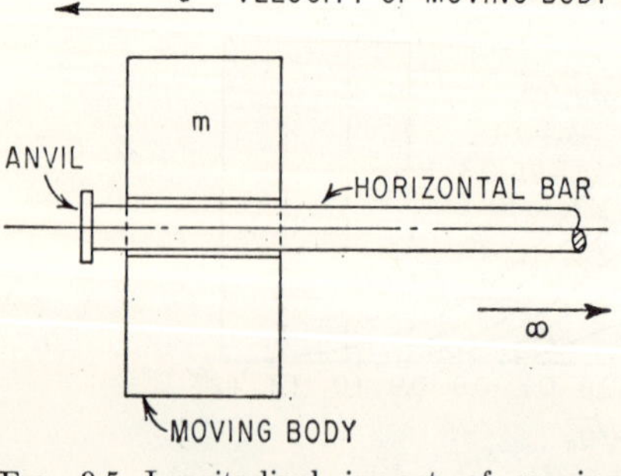

Figure 9.5 illustrates the impact of a large body m colliding axially with a long rod. The body m has an initial velocity $\dot{u}$ and is sufficiently large that the end of the rod may be assumed to move with constant

FIG. 9.5. Longitudinal impact of moving body on end of rod.

velocity $\dot{u}$. At any time t a stress wave will have moved into the bar a definite distance; by the condition of continuity (no break in the material), the struck end of the bar will have moved a distance equal to the total elongation of the end portion of the bar:

$$\dot{u}t = \epsilon \cdot l \qquad (9.32)$$

The velocity c of a stress wave is $c = l/t$, and Eq. (9.32) becomes

$$\epsilon = \frac{\dot{u}}{c} \qquad (9.33)$$

The stress and strain in an elastic material are related by Young's modulus. Substituting for strain from Eq. (9.33),

$$\sigma = \epsilon \cdot E = E\frac{\dot{u}}{c} \qquad (9.34)$$

where $\dot{u}$ = velocity of end of rod, in./sec
l = distance stress wave travels in time t, in.
t = time, sec
σ = stress, lb/in.^{2}
ϵ = strain (uniform), in./in.
E = Young's modulus, lb/in.^{2}
c = velocity of stress wave (dilatational), in./sec

When the yield point of the material is exceeded, Eq. (9.34) is inapplicable. Extensions of the analysis, however, lead to some results in the case of plastic deformation.[25] The differential equation for the elastic case is

$$E\frac{\partial^2 u}{\partial x^2} = \rho \frac{\partial^2 u}{\partial t^2} \qquad (9.35)$$

where u = displacement, in.
x = coordinate along rod, in.
t = time, sec
E = Young's modulus, lb/in.^{2}
ρ = mass density, lb-sec^2/in.^{4}

The velocity of the elastic dilatational wave obtained from Eq. (9.35) is

$$c = \sqrt{\frac{E}{\rho}}$$

The modulus E is the slope of the stress-strain curve in the initial linear elastic region. Replacing E by $\partial\sigma/\partial\epsilon$ for the case in which plastic deformation occurs, the slope of the static stress-stress curve can be determined at any value of the strain ϵ.[25] Equation (9.35) then becomes

$$\frac{\partial\sigma}{\partial\epsilon}\frac{\partial^2 u}{\partial x^2} = \rho \frac{\partial^2 u}{\partial t^2} \qquad (9.36)$$

Equation (9.36) is nonlinear; its general solution never has been obtained. For the simple type of loading discussed above and an infinitely long bar, the theory predicts a so-called critical velocity of impact because the velocities of the plastic waves are much smaller than those for the elastic waves and approach zero as the strain is indefinitely increased.[25] Since the impact velocity $\dot{u}$ is an independent quantity, it can be made larger and larger while the wave velocities are less than the velocity for elastic waves. Hence a point must be reached at which the continuity of the material is violated. Experimental data illustrate this point.[27]

ENERGY METHOD

Many problems in the design of machines and structures require knowledge of the deformation of material in the plastic condition. In statical problems the method of limit design [28] may be used. In dynamics, the most useful corresponding concept is less theoretical and may be termed the energy method; it is based upon the impact test used for the investigation of brittleness in metals. Originally, the only purpose of this test was to break a standard specimen as an index of brittleness or ductility. The general method, using a tension specimen, may be used in studying the dynamic resistance of materials.[27] An axial force is applied along the length of the specimen and causes the material to rupture ultimately. The energy of absorption is the total amount of energy taken out of the loading system and transferred to the specimen to cause the plastic deformation. The elastic energy and the specific mode of build-up of stress to the final plastic state are ignored. Such an approach has value only to the extent that the material has ductility. For example, in a long tension-type specimen of medium steel, the energy absorbed before neck-down and rupture is of the order of 500 ft-lb per cubic inch of material. Thus, if the moving body in Fig. 9.5 weighs 200 lb and has an initial velocity of 80 ft/sec, it represents 20,000 ft-lb of kinetic energy. If the tension bar subjected to the impact is 10 in. long and 0.5 in. in diameter, it will absorb approximately 1,000 ft-lb of energy. Under these circumstances it will rupture. On the other hand, if the moving body m weighs only 50 lb and has an initial velocity of 30 ft/sec, its kinetic energy is approximately 700 ft-lb and the bar will not rupture.

If the tension specimen were severely notched at some point along its length, it would no longer absorb 500 ft-lb per cubic inch to rupture. The material in the immediate neighborhood of the notch would deform plastically; a break would occur at the notch with the bulk of the material in the specimen stressed below the yield stress for the material. A practical structural situation related to this problem occurs when a butt weld is located at some point along an unnotched specimen. If the weld is of good quality, the full energy absorption of the entire bar develops before rupture; with a poor weld, the rupture occurs at the weld and practically no energy is absorbed by the remainder of the material. This is an important consideration in applying the energy method to design problems.

REFERENCES

1. Love, A. E. H.: "The Mathematical Theory of Elasticity," p. 25, Cambridge University Press, New York, 1934.
2. Timoshenko, S., and D. H. Young: "Engineering Mechanics," p. 334, McGraw-Hill Book Company, Inc., New York, 1940.
3. Loney, S. L.: "A Treatise on Elementary Dynamics," Cambridge University Press, p. 199, New York, 1900.
4. Hertz, H.: *J. Math. (Crelle)*, pp. 92, 155, 1881.
5. Rayleigh, Lord: *Phil. Mag.* (ser. 6), **11**:283 (1906).
6. Timoshenko, S.: "Theory of Elasticity," p. 350, McGraw-Hill Book Company, Inc., New York, 1934.
7. Raman, C. V.: *Phys. Rev.*, **15**, 277 (1920).
8. Lamb, H.: *Proc. London Math. Soc.*, **35**, 141 (1902).
9. Hoppmann, II, W. H.: *Proc. SESA*, **9**:2, 21 (1952).
10. Hoppmann, II, W. H.: *Proc. SESA*, **10**:1, 157 (1952).
11. Timoshenko, S.: "Vibration Problems in Engineering," 3d ed., p. 413, D. Van Nostrand Company, Inc., Princeton, N.J., 1955.
12. Zener, C., and H. Feshbach: *Trans. ASME*, **61**:a-67 (1939).
13. Hoppmann, II, W. H.: *J. Appl. Mechanics*, **15**:125 (1948).
14. Hoppmann, II, W. H.: *J. Appl. Mechanics*, **16**:370 (1949).
15. Hoppmann, II, W. H.: *J. Appl. Mechanics*, **17**:409 (1950).
16. Goldsmith, W., and D. M. Cunningham: *Proc. SESA*, **14**:1, 179 (1956).
17. Barnhart, Jr., K. E., and Werner Goldsmith: *J. Appl. Mechanics*, **24**:440 (1957).
18. Emschermann, H. H., and K. Ruhl: *VDI-Forschungsheft* 443, Ausgabe B, Band 20, 1954.
19. Hoppmann, II, W. H.: *J. Appl. Mechanics*, **19** (1952).
20. Wenk, E., Jr.: Dissertation, The Johns Hopkins University, 1950, and *David W. Taylor Model Basin Rept.* 704, July, 1950.

REFERENCES

21. Compendium, "Underwater Explosion," O. N. R., Department of the Navy, 1950.
22. Prager, W.: James Clayton Lecture, *The Institution of Mechanical Engineers, London*, 1955.
23. Lee, E. H., and P. S. Symonds: *J. Appl. Mechanics*, **19**:308 (1952).
24. Owens, R. H., and P. S. Symonds: *J. Appl. Mechanics*, **22** (1955).
25. Von Kármán, T.: *NDRC Rept.* A-29, 1943.
26. Duwez, P. E., D. S. Clark, and H. F. Bohnenblust: *J. Appl. Mechanics*, **17**, 27 (1950).
27. Hoppmann, II, W. H.: *Proc. ASTM*, **47**:533 (1947).
28. Symposium on the Plastic Theory of Structures, Cambridge University, September, 1956, *British Welding J.*, **3**(8) (1956); **4**(1) (1957).

10

MECHANICAL IMPEDANCE AND MOBILITY

Elmer L. Hixson
The University of Texas

INTRODUCTION

The differential equations of motion of a linear mechanical system can be expressed in terms of the driving force and the acceleration, velocity, or displacement of the elements of the system. With a sinusoidal force driving a linear system, the steady-state response exhibits sinusoidal acceleration, velocity, and displacement at the driving frequency and with a fixed phase relation to the driving force. The relationship between driving force and the motion resulting at various points of the system can be expressed by algebraic equations involving complex numbers. The force and motion characteristics of the system are defined as described in the following paragraphs.

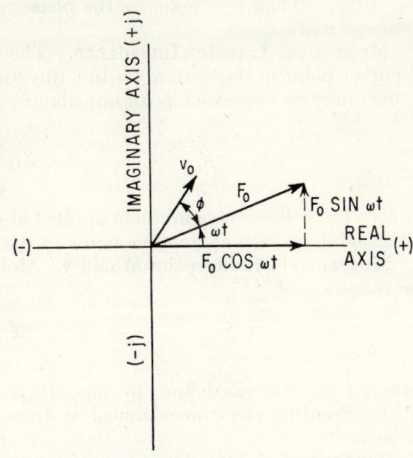

FIG. 10.1. Complex-plane representation for sinusoidal force and velocity.

Force. A sinusoidal force can be represented as a rotating phasor (to differentiate from a vector having magnitude and direction) on the complex plane, as shown in Fig. 10.1. The phasor has a length or magnitude F_0; it rotates counterclockwise with an angular velocity ω, and at some time t subtends an angle ωt with the $(+)$ real axis. The instantaneous force is the projection of the phasor on the real axis, $F_0 \cos \omega t$. The complex representation of the force in terms of real and imaginary components is

$$F = F_0 \cos \omega t + jF_0 \sin \omega t \tag{10.1}$$

The force also can be expressed in terms of magnitude and phase angle:

$$F = F_0 e^{j\omega t} \tag{10.2}$$

Velocity. A sinusoidal velocity also can be represented as a phasor, as shown in Fig. 10.1. When v_0 is rotating at the same angular velocity ω as F_0, it can be represented as

$$v = v_0[\cos(\omega t + \phi) + j \sin(\omega t + \phi)] = v_0 e^{j(\omega t + \phi)} \tag{10.3}$$

where ϕ is the phase angle between F and v. The phase reference in Eq. (10.3) is the $(+)$

real axis. When F is taken as the phase reference, v may be written

$$v = v_0 e^{j\phi} \tag{10.4}$$

Displacement. Displacement in sinusoidal motion is readily determined from velocity by expressing velocity in the form of Eq. (10.2), $v = v_0 e^{j\omega t}$, and noting that $x = \int v\, dt$:

$$x = \frac{v}{j\omega} \tag{10.5}$$

Acceleration. Acceleration is derived from velocity by using the relation $\ddot{x} = dv/dt$:

$$\ddot{x} = j\omega v \tag{10.6}$$

Mechanical Driving-point Impedance. The ratio of the driving force acting on a system to the resulting velocity of the system is *mechanical impedance.** If the velocity is measured at the point of application of the driving force, the ratio of force to velocity is *mechanical driving-point impedance* Z defined as follows:[1,2,3]

$$Z = \frac{F}{v} \tag{10.7}$$

where F is the applied force and v is the resultant velocity in the direction of the force. The quantities F and v are represented by complex numbers, as indicated by Eqs. (10.1) to (10.3). When F is taken as the phase reference, $\phi = \theta$ where θ is defined as the *impedance angle*.

Mechanical Transfer Impedance. The ratio of the driving force to the velocity at another point in the system (or in a direction different from the direction of the applied force) may be expressed as an impedance:

$$Z_{12} = \frac{F_1}{v_2} \tag{10.8}$$

where F_1 is the sinusoidal force applied at one point in a system and v_2 is the resulting sinusoidal velocity at another point in the system, F_1 and v_2 being complex quantities.

Mechanical Driving-point Mobility. Mobility, the inverse of impedance, is determined as follows:[4,5,6]

$$\mathfrak{M} = \frac{v}{F} \tag{10.9}$$

where F and v are as defined by Eqs. (10.1) and (10.3). The concept of mobility is useful in representing electromechanical systems. Mobility sometimes is called *mechanical admittance*.

Dynamic Modulus. Dynamic modulus is defined as follows:[7,8]

$$D = \frac{F}{x} \tag{10.10}$$

where F is the applied sinusoidal force and x is the resulting displacement in the direction of the force, both expressed as complex quantities.

Receptance. Receptance is the reciprocal of dynamic modulus; it is defined as follows:[9,10]

$$R = \frac{x}{F} \tag{10.11}$$

where F and x are defined as above.

---

* The ratio of driving force to resulting displacement sometimes is termed "mechanical impedance." This is not recommended usage.

Table 10.1. Units of Parameters Used in Analysis by Mechanical Impedance and Mobility Methods

Quantity	Units		
	cgs	mks	English gravitational
Velocity, v	cm/sec	meters/sec	in./sec
Force, F	dynes	newtons	lb
Mass, m	grams	kilograms	lb-sec^2/in.
Spring stiffness, k	dynes/cm	newtons/meter	lb/in.
Damping coefficient, c	dynes/cm/sec	newtons/meter/sec	lb/in./sec
Impedance, Z	dyne-sec/cm grams/sec	newton-sec/meter kilograms/sec	lb-sec/in.
Mobility, $\mathfrak{M}$	cm/dyne/sec sec/gram	meters/newton/sec sec/kilogram	in./lb/sec
Power, P	dyne-cm/sec ergs/sec	newton-meters/sec joules/sec	in./lb-sec

In the ensuing discussion, only impedance and mobility are used; the discussion is limited to linear systems in which the motion is translational in the direction of the driving force. Other systems that are analogous to translational systems may be analyzed by the same methods using analogous parameters. The several quantities used in impedance methods may be expressed in any consistent system of units. Three such systems are given in Table 10.1.

Impedance methods allow a complete analysis of the motions and forces acting on each part of a physical system; however, the differential equations of Newtonian mechanics become algebraic equations. In addition, a "black box" concept is introduced in which the forces and motions at one or two points of major interest in a system are determined without a complete analysis of the entire system. These methods are adaptable to the use of results of physical measurements on systems that are so complicated that a complete analysis becomes impractical. Typical curves illustrating the quantitative aspect of impedance as measured on a wide range of structure types are presented.

IMPEDANCE OF MECHANICAL ELEMENTS

Three idealized mechanical system elements with lumped constants are conventionally assembled to form linear physical systems. These elements are resistance (or damping), spring, and mass.

MECHANICAL RESISTANCE (DAMPER). A mechanical resistance is a device in which the relative velocity between its end points is proportional to the force applied to the end points. Such a device can be represented by the dashpot of Fig. 10.2 in which the force resisting the extension (or compression) of the dashpot is the result of viscous friction. An ideal resistance is assumed to be made of massless, infinitely rigid elements.

In Fig. 10.2 the velocity of point A with respect to point B is

$$v = (v_1 - v_2) = \frac{F_a}{c} \qquad (10.12)$$

where c is the constant of proportionality called the *mechanical resistance* or *damping constant*. For there to be a relative velocity v as a result of force at A, there must be an equal reaction force at B. Thus, it may be considered that a transmitted force F_b is

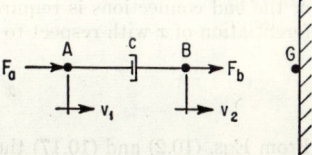

Fig. 10.2. Schematic representation of an ideal mechanical resistance element.

equal to F_a. The velocities v_1 and v_2 are measured with respect to the stationary reference G; their difference is the relative velocity between the end points of the resistance.

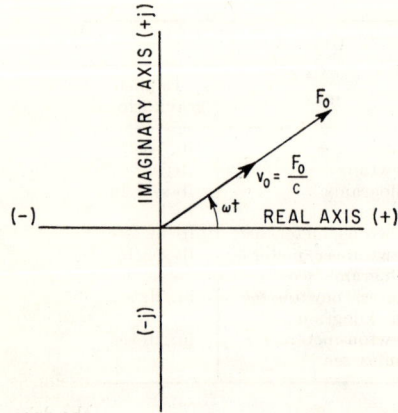

FIG. 10.3. Force and velocity relationships at the connections of an ideal resistance element.

With the sinusoidal force of Eq. (10.2) applied to point A of Fig. 10.2 and point B attached to a fixed (immovable) point, the velocity v_1 is obtained from Eq. (10.12):

$$v_1 = \frac{F_0 e^{j\omega t}}{c} = v_0 e^{j\omega t} \quad (10.13)$$

The force and velocity phasors thus are rotating at the same angular frequency and are said to be "in phase," as shown in Fig. 10.3.

The mechanical impedance of the resistance is obtained by substituting from Eqs. (10.2) and (10.13) in Eq. (10.7):

$$Z_c = \frac{F}{v} = \frac{F_0 e^{j\omega t}}{\frac{F_0}{c} e^{j\omega t}} = c \quad (10.14)$$

The mechanical impedance of a resistance is the value of its damping constant c; i.e., when a resistance is placed in a mechanical system, the ratio of the transmitted force to the relative velocity of its end points is its impedance c.

SPRING. A linear spring is a device for which the relative displacement between its end points is proportional to the force applied. It is illustrated in Fig. 10.4 and can be represented mathematically as follows:

$$\delta x = x_1 - x_2 = \frac{F_a}{k} \quad (10.15)$$

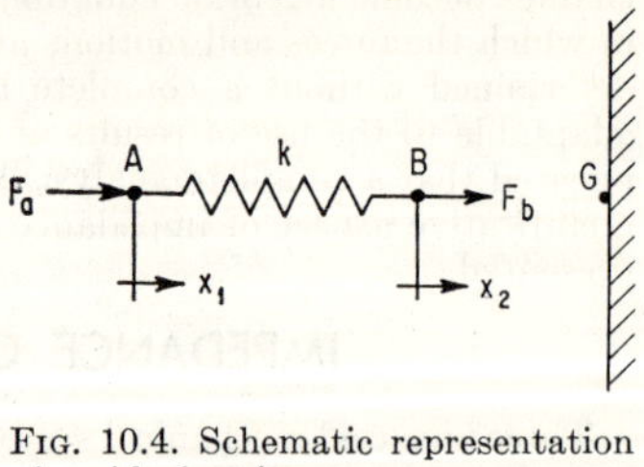

where k is the *spring stiffness* and x_1, x_2 are displacements relative to the reference point G. The stiffness k can be expressed alternately in terms of a *compliance* $C = 1/k$. The spring transmits the applied force so that $F_b = F_a$.

FIG. 10.4. Schematic representation of an ideal spring.

With the force of Eq. (10.2) applied to point A and with point B fixed, the displacement of point A is given by Eq. (10.15):

$$x_1 = \frac{F_0 e^{j\omega t}}{k} = x_0 e^{j\omega t} \quad (10.16)$$

The displacement is thus sinusoidal and in phase with the force. The relative velocity of the end connections is required for impedance calculations and is given by the differentiation of x with respect to time:

$$\dot{x} = v = \frac{j\omega F_0 e^{j\omega t}}{k} = jv_0 e^{j\omega t} \quad (10.17)$$

From Eqs. (10.2) and (10.17) the impedance of the spring is

$$Z_k = \frac{F_0 e^{j\omega t}}{j\omega F_0 e^{j\omega t}/k} = \frac{k}{j\omega} = -\frac{jk}{\omega} \quad (10.18)$$

IMPEDANCE OF MECHANICAL ELEMENTS

The ratio of force to velocity for a spring thus depends on the stiffness and frequency, and is an imaginary number.

To find the phase relation between v and F for a spring, Eq. (10.17) may be written in the form of Eq. (10.1):

$$v = \frac{F_0}{k}(-\omega \sin \omega t + j\omega \cos \omega t) = \frac{\omega}{k} F_0 e^{j(\omega t + 90°)}$$

The velocity is always 90° ahead of, or leading, the applied force, as shown in Fig. 10.5. This is indicated by the factor j in Eq. (10.17).

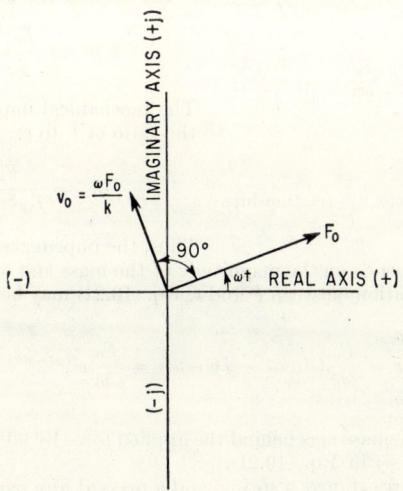

Fig. 10.5. Force and velocity relationships at the connections of an ideal spring.

MASS. In the ideal mass illustrated in Fig. 10.6A, the acceleration $\ddot{x}$ of the rigid body is proportional to the applied force F:

$$\ddot{x}_1 = \frac{F_a}{m} \tag{10.19}$$

where m is the mass of the body and $\ddot{x}_1 = \ddot{x}_2$ since the element is rigid. By Eq. (10.19), the force F_a is required to give the mass the acceleration $\ddot{x}_1$ and the transmitted force F_b is zero. The representation of Fig. 10.6B is used commonly to indicate that one end of a

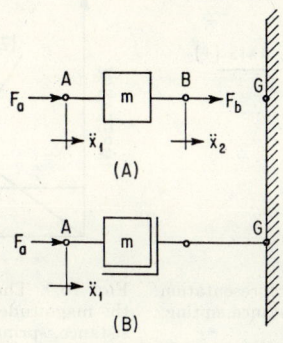

Fig. 10.6. Two schematic representations of an ideal mass.

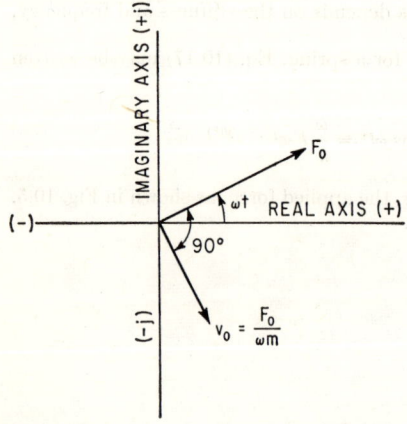

FIG. 10.7. Force and velocity relationships for an ideal mass.

mass is connected rigidly to an inertial reference point, leaving it with only one end which is free to move. When a sinusoidal force is applied, Eq. (10.19) becomes

$$\ddot{x}_1 = \frac{F_0 e^{j\omega t}}{m} \quad (10.20)$$

The acceleration is sinusoidal and in phase with the applied force.

Integrating Eq. (10.20) to find velocity

$$\dot{x} = v = \frac{F_0 e^{j\omega t}}{j\omega m} = -jv_0 e^{j\omega t} \quad (10.21)$$

The mechanical impedance of the mass is the ratio of F to v:

$$Z_m = \frac{F_0 e^{j\omega t}}{F_0 e^{j\omega t}/j\omega m} = j\omega m \quad (10.22)$$

Thus, the impedance of a mass is an imaginary quantity that depends on the magnitude of the mass and on the frequency.

To find the phase relation between F and v, Eq. (10.21) may be written:

$$v = \frac{F_0}{\omega m}(\sin \omega t - j \cos \omega t) = \frac{F_0}{\omega m} e^{j(\omega t - 90°)}$$

Thus the velocity of the mass lags behind the applied force by 90° as shown in Fig. 10.7, and indicated by the $-j$ in Eq. (10.21).

The impedances of a resistance, a spring, and a mass at any particular frequency may be represented on the complex plane as shown in Fig. 10.8. Their variation with frequency is indicated in Fig. 10.9. The impedances of elements and several combinations of elements are given in Table 10.2.

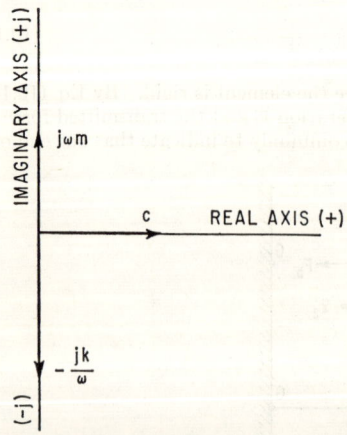

FIG. 10.8. A complex-plane representation for the impedance of ideal resistance, spring, and mass at a fixed frequency.

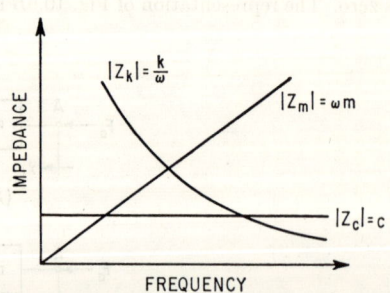

FIG. 10.9. The variation with frequency of the magnitude of the impedance of ideal resistance, spring, and mass.

Table 10.2. Driving-point Impedance and Mobility of Ideal Mechanical Elements and Lumped Parameter Systems

When the system of elements shown includes a mass, the impedance or mobility given is the relationship between force and velocity at the one available connection, the other connection being attached to the inertial reference plane. When no mass is included, the impedance or mobility describes the relation between the force applied to the two system connections and the resulting relative velocity between these connections. Graphs of magnitude of impedance vs. ω and magnitude of mobility vs. ω are plotted on a log-log scale.

DIAGRAM OF SYSTEM	MATHEMATIC FORMULAS: IMPEDANCE Z – EQ.(10.7) MOBILITY $\mathcal{M}$ – EQ.(10.9)	IMPEDANCE IN THE COMPLEX PLANE	MOBILITY IN THE COMPLEX PLANE	MAGNITUDE OF IMPEDANCE	MAGNITUDE OF MOBILITY	IMPEDANCE ANGLE θ FIG.10.34
1. (dashpot c)	$Z = c$ $\mathcal{M} = 1/c$					
2. (spring k)	$Z = \dfrac{k}{j\omega}$ $\mathcal{M} = \dfrac{j\omega}{k}$			SLOPE=-1	SLOPE=+1	+90°, -90°
3. (mass m)	$Z = j\omega m$ $\mathcal{M} = \dfrac{1}{j\omega m}$			SLOPE=+1	SLOPE=-1	+90°, -90°

Table 10.2. (Continued)

DIAGRAM OF SYSTEM	MATHEMATIC FORMULAS: IMPEDANCE Z - EQ.(10.7) MOBILITY $\mathcal{M}$ - EQ.(10.9)	IMPEDANCE IN THE COMPLEX PLANE	MOBILITY IN THE COMPLEX PLANE	MAGNITUDE OF IMPEDANCE	MAGNITUDE OF MOBILITY	IMPEDANCE ANGLE θ FIG. 10.34
4. 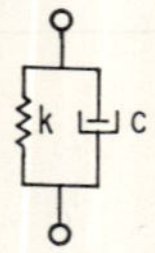 (c and k in parallel)	$Z = c + \dfrac{k}{j\omega}$ $\mathcal{M} = \dfrac{c - \dfrac{k}{j\omega}}{c^2 + (k/\omega)^2}$ $\omega_1 = \dfrac{k}{c}$	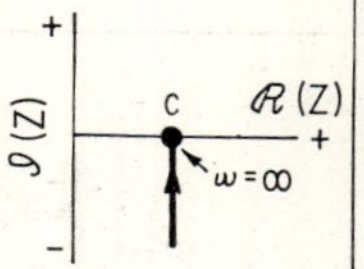	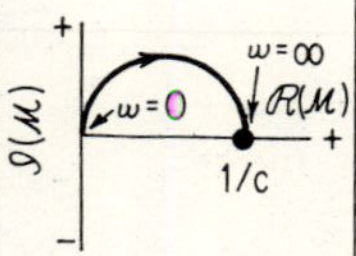	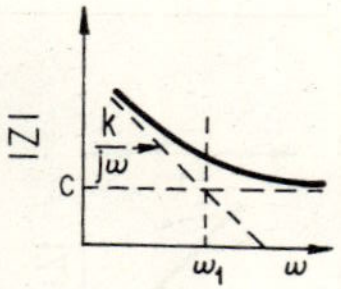	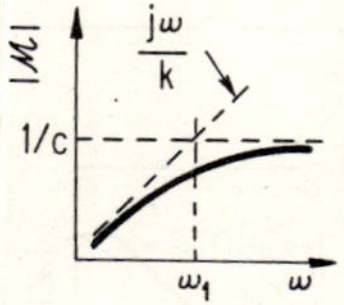	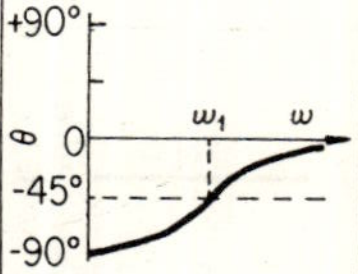
5. 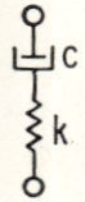 (c and k in series)	$Z = \dfrac{1/c - j\omega/k}{(1/c)^2 + (\omega/k)^2}$ $\mathcal{M} = \dfrac{1}{c} + \dfrac{j\omega}{k}$ $\omega_1 = \dfrac{k}{c}$	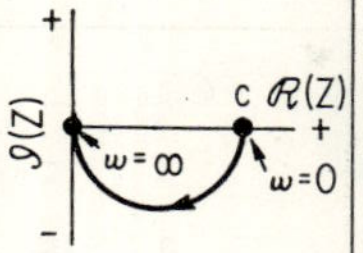	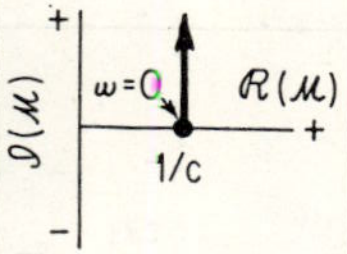	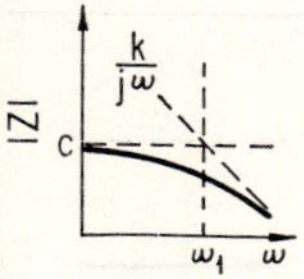	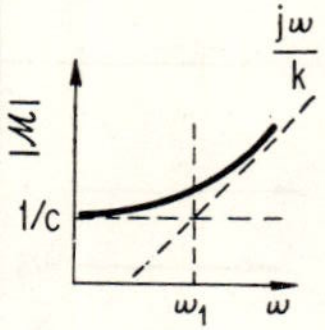	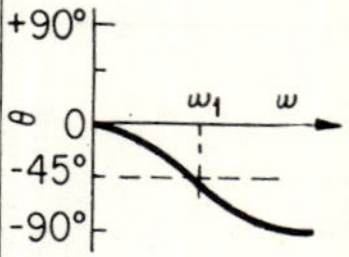
6. 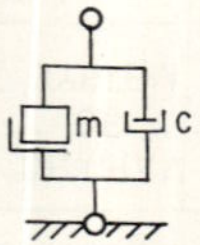 (c, m in parallel with spring to ground)	$Z = c + j\omega m$ $\mathcal{M} = \dfrac{c - j\omega m}{c^2 + \omega^2 m^2}$ $\omega_1 = \dfrac{c}{m}$	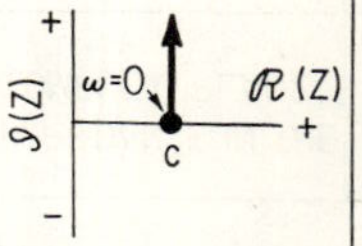	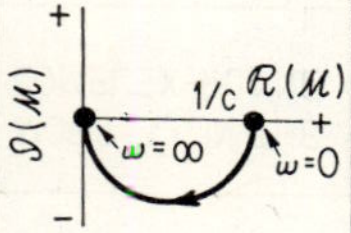	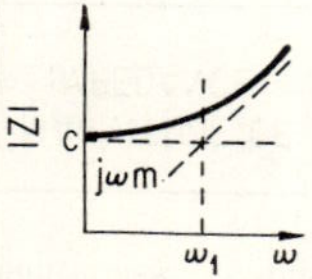	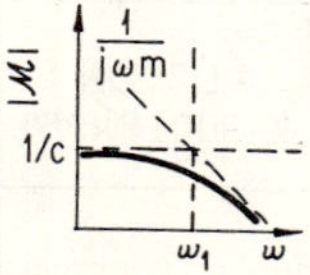	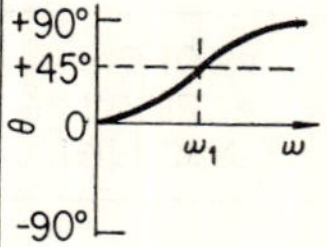

10-8

DIAGRAM OF SYSTEM	MATHEMATIC FORMULAS: IMPEDANCE Z - EQ.(10.7) MOBILITY $\mathcal{M}$ - EQ.(10.9)	IMPEDANCE IN THE COMPLEX PLANE	MOBILITY IN THE COMPLEX PLANE	MAGNITUDE OF IMPEDANCE	MAGNITUDE OF MOBILITY	IMPEDANCE ANGLE θ FIG.10.34
7.	$Z = \dfrac{1/c + j/\omega m}{(1/c)^2 + (1/\omega m)^2}$ $\mathcal{M} = 1/c + 1/j\omega m$ $\omega_1 = \dfrac{c}{m}$					
8.	$Z = j\omega m + \dfrac{k}{j\omega}$ $\mathcal{M} = \dfrac{-j}{\omega m - k/\omega}$ $\omega_0 = \sqrt{\dfrac{k}{m}}$					
9.	$Z = \dfrac{-j}{\omega/k - 1/\omega m}$ $\mathcal{M} = j\omega/k + 1/j\omega m$ $\omega_0 = \sqrt{\dfrac{k}{m}}$					

Table 10.2. (Continued)

DIAGRAM OF SYSTEM	MATHEMATIC FORMULAS: IMPEDANCE Z - EQ.(10.7) MOBILITY $\mathcal{M}$ - EQ.(10.9)	IMPEDANCE IN THE COMPLEX PLANE	MOBILITY IN THE COMPLEX PLANE	MAGNITUDE OF IMPEDANCE	MAGNITUDE OF MOBILITY	IMPEDANCE ANGLE θ FIG.10.34
10.	$Z = c + j(\omega m - k/\omega)$ $\mathcal{M} = \dfrac{c - j(\omega m - k/m)}{c^2 + (\omega m - k/m)^2}$ $\omega_0 = \sqrt{\dfrac{k}{m}}$					
11.	$Z = 1/c - j(\omega/k - 1/\omega m)$ $\mathcal{M} = 1/c + j(\omega/k - 1/\omega m)$ 					
11.	$Z = \dfrac{1/c - j(\omega/k - 1/\omega m)}{(1/c)^2 + (\omega/k - 1/\omega m)^2}$ $\mathcal{M} = 1/c + j(\omega/k - 1/\omega m)$ $\omega_0 = \sqrt{\dfrac{k}{m}}$					
12.	$Z = \dfrac{\dfrac{c_1 + c_2}{c_1^2} + \dfrac{c_2}{\omega^2 m^2} + \dfrac{j}{\omega m}}{(1/c_1)^2 + (1/\omega m)^2}$ $\mathcal{M} = \dfrac{c_1 + c_2}{c_1^2} + \dfrac{c_2}{\omega^2 m^2} - \dfrac{j}{\omega m} \Big/ \left(\dfrac{c_1 + c_2}{c_1}\right)^2 + \left(\dfrac{c_2}{\omega m}\right)^2$					

DIAGRAM OF SYSTEM	MATHEMATIC FORMULAS: IMPEDANCE Z - EQ.(10.7) MOBILITY $\mathcal{M}$ - EQ.(10.9)	IMPEDANCE IN THE COMPLEX PLANE	MOBILITY IN THE COMPLEX PLANE	MAGNITUDE OF IMPEDANCE	MAGNITUDE OF MOBILITY	IMPEDANCE ANGLE θ FIG.10.34
13.	$Z = \dfrac{c_1\left(\dfrac{c_1+c_2}{c_2}\right) + \dfrac{\omega^2 m^2}{c_2} + j\omega m}{\left(\dfrac{c_1+c_2}{c_2}\right)^2 + \left(\dfrac{\omega m}{c_2}\right)^2}$ $\mathcal{M} = \dfrac{c_1\left(\dfrac{c_1+c_2}{c_2}\right) + \dfrac{\omega^2 m^2}{c_2} - j\omega m}{c_1^2 + \omega^2 m^2}$					
14.	$Z = \dfrac{\dfrac{1}{c}+j\left[\dfrac{\omega m_2}{c^2} + \dfrac{1}{\omega m_1}\left(\dfrac{m_1+m_2}{m_1}\right)^2\right]}{(1/c)^2 + (1/\omega m_1)^2}$ $\mathcal{M} = \dfrac{\dfrac{1}{c}-j\left[\dfrac{\omega m_2}{c^2} + \dfrac{1}{\omega m_1}\left(\dfrac{m_1+m_2}{m_1}\right)^2\right]}{\left(\dfrac{m_1+m_2}{m_1}\right)^2 + \left(\dfrac{c}{\omega m_1}\right)^2}$ $\omega_1 = \dfrac{c}{m_2}\qquad \omega_2 = \dfrac{c}{m_1}\left(\dfrac{m_1+m_2}{m_1}\right)$					
15.	$Z = \dfrac{\dfrac{1}{c} + \dfrac{j}{\omega}\left[\dfrac{1}{m}-k\left(\dfrac{1}{c^2}+\dfrac{1}{\omega^2 m^2}\right)\right]}{(1/c)^2 + (1/\omega m)^2}$ $\mathcal{M} = \dfrac{\omega^2 m^2 \dfrac{c}{c} + j\dfrac{\omega^2 m^2 k}{c^2} + \dfrac{k}{\omega}-\omega m}{(mk/c)^2 + (\omega m - k/\omega)^2}$ $\omega_0 = \sqrt{\dfrac{k}{m - \dfrac{km^2}{c^2}}}$					

Table 10.2. (Continued)

DIAGRAM OF SYSTEM	MATHEMATIC FORMULAS: IMPEDANCE Z - EQ. (10.7) MOBILITY $\mathcal{M}$ - EQ. (10.9)	IMPEDANCE IN THE COMPLEX PLANE	MOBILITY IN THE COMPLEX PLANE	MAGNITUDE OF IMPEDANCE	MAGNITUDE OF MOBILITY	IMPEDANCE ANGLE θ FIG. 10.34
16. 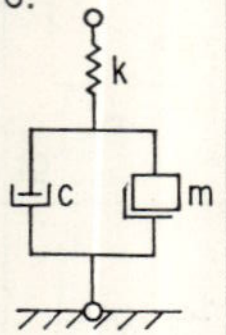	$Z = \dfrac{ck^2}{\omega^2} - jkm\dfrac{\left[(\omega m - k/\omega) + \dfrac{c^2 k}{\omega m}\right]}{c^2 + (\omega m - k/\omega)^2}$ $\mathcal{M} = \dfrac{c + j\omega\left(\dfrac{c^2}{k} + \omega^2\dfrac{m^2}{k} - m\right)}{c^2 + \omega^2 m^2}$ $\omega_0 = \sqrt{\dfrac{k}{m} - \dfrac{c^2}{m^2}}$	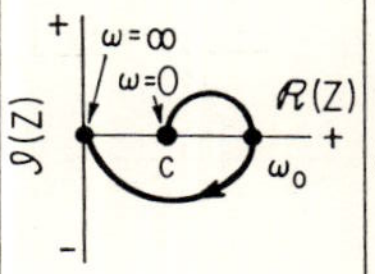	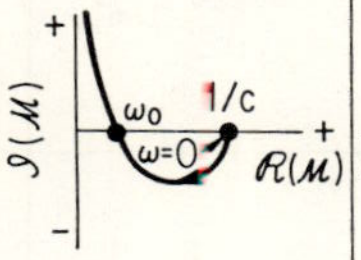	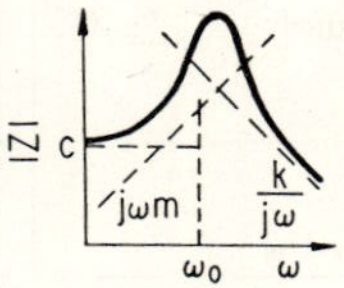	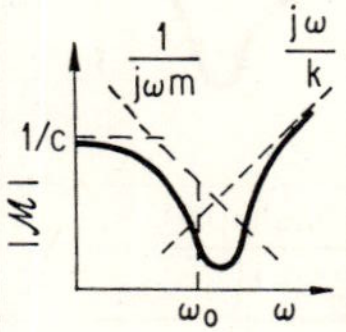	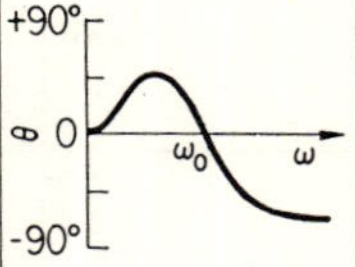
17. 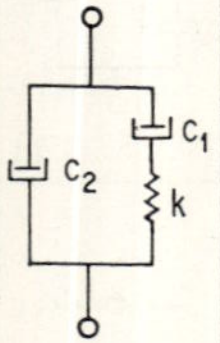	$Z = \dfrac{\dfrac{c_1 + c_2}{c_1^2} + \dfrac{c_2\omega^2}{k^2} - j\dfrac{\omega}{k}}{(1/c_1)^2 + (\omega/k)^2}$ $\mathcal{M} = \dfrac{\dfrac{c_1 + c_2}{c_1^2} + \dfrac{c_2\omega^2}{k^2} + \dfrac{j\omega}{k}}{\left(\dfrac{c_1 + c_2}{c_1}\right)^2 + \left(\dfrac{c_2\omega}{k}\right)^2}$	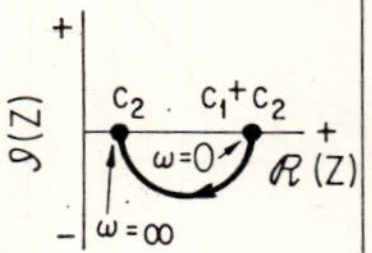	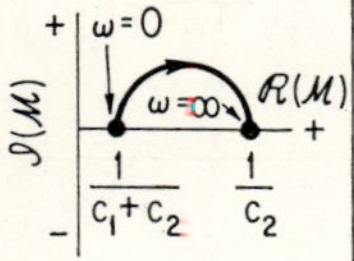	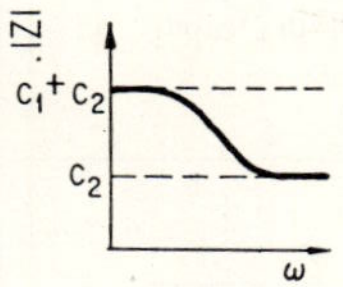	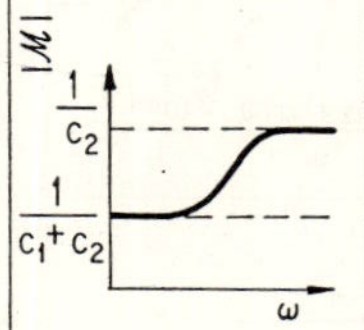	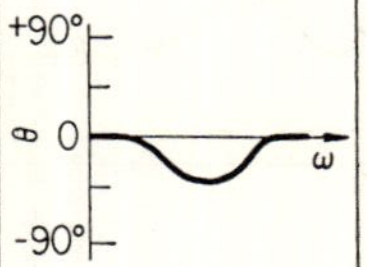
18. 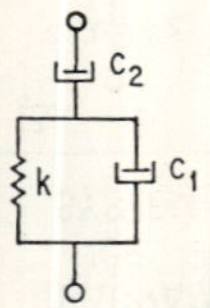	$Z = \dfrac{c_1\left(\dfrac{c_1 + c_2}{c_2}\right) + \dfrac{k^2}{c_2\omega^2} - j\dfrac{k}{\omega}}{\left(\dfrac{c_1 + c_2}{c_2}\right)^2 + \left(\dfrac{k}{\omega c_2}\right)^2}$ $\mathcal{M} = \dfrac{c_1\left(\dfrac{c_1 + c_2}{c_2}\right) + \dfrac{k^2}{c_2\omega^2} + \dfrac{jk}{\omega}}{c_1^2 + (k/\omega)^2}$	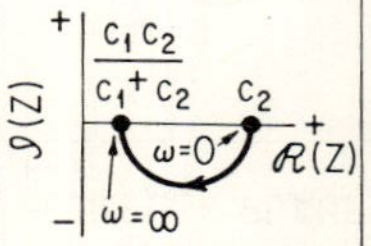	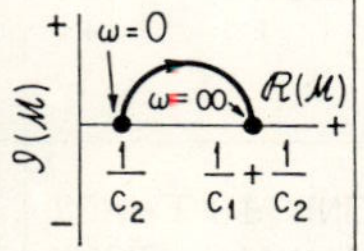	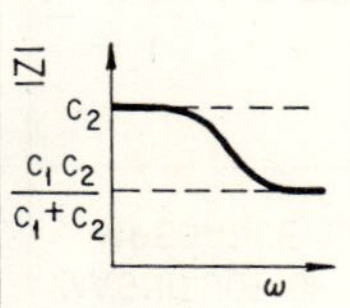	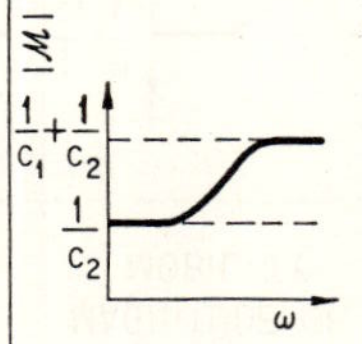	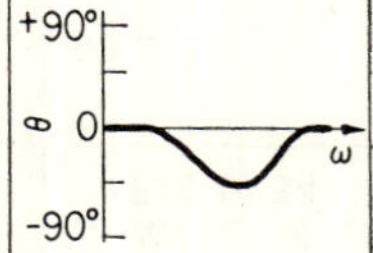

10-12

DIAGRAM OF SYSTEM	MATHEMATIC FORMULAS: IMPEDANCE Z – EQ. (10.7) MOBILITY $\mathcal{M}$ – EQ. (10.9)	IMPEDANCE IN THE COMPLEX PLANE	MOBILITY IN THE COMPLEX PLANE	MAGNITUDE OF IMPEDANCE	MAGNITUDE OF MOBILITY	IMPEDANCE ANGLE θ FIG. 10.34
19. (c, k_1 parallel, in series with k_2)	$Z = \dfrac{\dfrac{1}{c} - j\left[\dfrac{k_2}{\omega c^2} + \dfrac{\omega}{k_1}\left(\dfrac{k_1+k_2}{k_1}\right)\right]}{(1/c)^2 + (\omega/k_1)^2}$ $\mathcal{M} = \dfrac{\dfrac{1}{c} + j\left[\dfrac{k_2}{\omega c^2} + \dfrac{\omega}{k_1}\left(\dfrac{k_1+k_2}{k_1}\right)\right]}{\left(\dfrac{k_1+k_2}{k_1}\right)^2 + \left(\dfrac{k_2}{\omega c}\right)^2}$					
20. (k_1 in series with parallel k_2, c)	$Z = \dfrac{c - j\left[\dfrac{\omega c^2}{k_2} + \dfrac{k_1}{\omega}\left(\dfrac{k_1+k_2}{k_2}\right)\right]}{\left(\dfrac{k_1+k_2}{k_2}\right)^2 + \left(\dfrac{\omega c}{k_2}\right)^2}$ $\mathcal{M} = \dfrac{c + j\left[\dfrac{\omega c^2}{k_2^2} + \dfrac{k_1}{\omega}\left(\dfrac{k_1+k_2}{k_2}\right)\right]}{c^2 + (k_1/\omega)^2}$					
21. (k, c, m)	$Z = \dfrac{\dfrac{1}{c} + j\omega\left[m\left(\dfrac{1}{c^2} + \dfrac{\omega^2}{k^2}\right) - \dfrac{1}{k}\right]}{(1/c)^2 + (\omega/k)^2}$ $\mathcal{M} = \dfrac{\dfrac{1}{c} - j\omega\left[m\left(\dfrac{1}{c^2} + \dfrac{\omega^2}{k^2}\right) - \dfrac{1}{k}\right]}{\left(\dfrac{\omega m}{c}\right)^2 + \left(1 - \dfrac{\omega^2 m}{k}\right)^2}$ $\omega_0 = \sqrt{\dfrac{k}{m} - \dfrac{k^2}{c^2}}$					

10–13

Table 10.2. (Continued)

DIAGRAM OF SYSTEM	MATHEMATIC FORMULAS: IMPEDANCE Z - EQ.(10.7) MOBILITY $\mathcal{M}$ - EQ.(10.9)	IMPEDANCE IN THE COMPLEX PLANE	MOBILITY IN THE COMPLEX PLANE	MAGNITUDE OF IMPEDANCE	MAGNITUDE OF MOBILITY	IMPEDANCE ANGLE θ FIG.10.34
22. 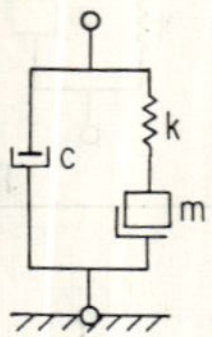	$Z = c + \dfrac{j\omega mk}{k - \omega^2 m}$ $\mathcal{M} = \dfrac{\dfrac{j\omega mk}{k - \omega^2 m}}{c^2 + \left(\dfrac{\omega mk}{k - \omega^2 m}\right)^2}$ $\omega_0 = \sqrt{\dfrac{k}{m}}$	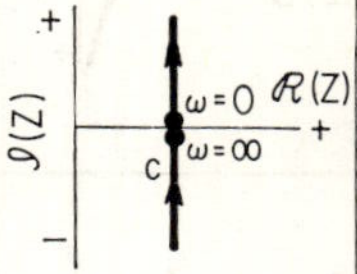	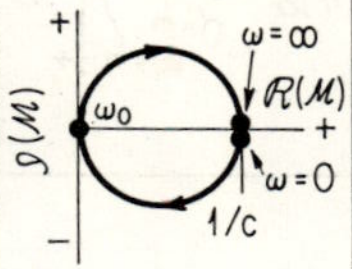	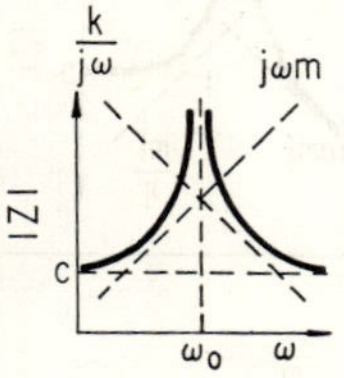	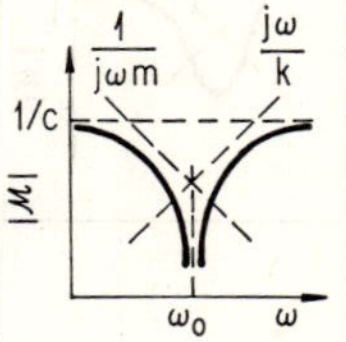	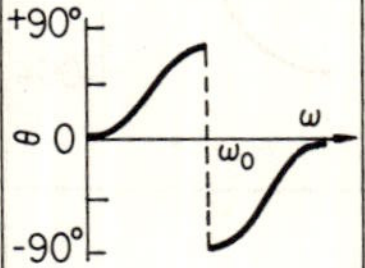
23. 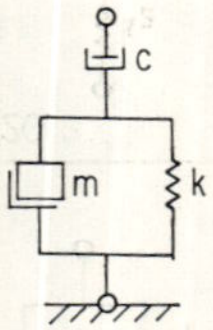	$\dfrac{1}{Z} = \dfrac{j\omega}{c} - \dfrac{j\omega}{k - \omega^2 m}$ $Z = \dfrac{1}{\left(\dfrac{1}{c}\right)^2 + \left(\dfrac{\omega}{k - \omega^2 m}\right)^2}$ $\mathcal{M} = \dfrac{1}{c} + \dfrac{j\omega}{k - \omega^2 m}$ $\omega_0 = \sqrt{\dfrac{k}{m}}$	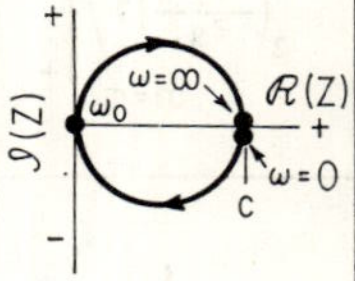	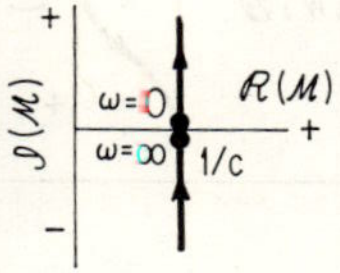	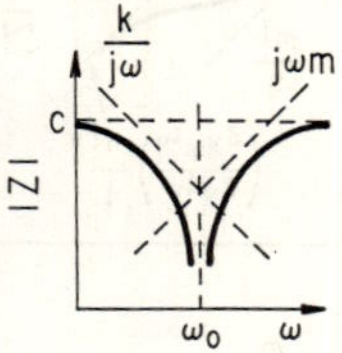	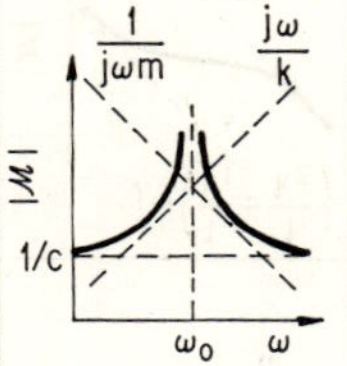	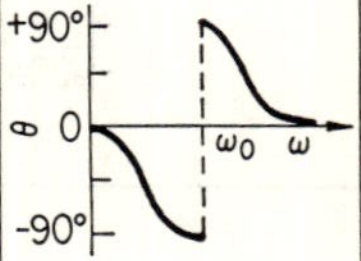
24. 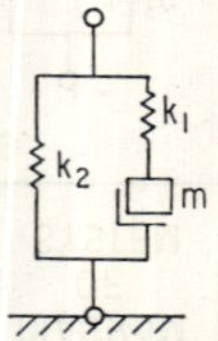	$Z = \dfrac{-j\left(\dfrac{k_1 + k_2}{k_1} - \dfrac{k_2}{\omega^2 m}\right)}{\omega/k_1 - 1/(\omega m)}$ $\mathcal{M} = \dfrac{+j(\omega/k_1 - 1/\omega m)}{\left(\dfrac{k_1 + k_2}{k_1}\right) - \dfrac{k_2}{\omega^2 m}}$ $\omega_1 = \sqrt{\dfrac{1}{m}\left(\dfrac{k_1 k_2}{k_1 + k_2}\right)}$ $\omega_2 = \sqrt{\dfrac{k_1}{m}}$	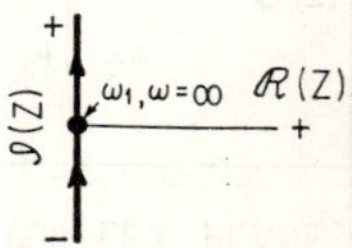	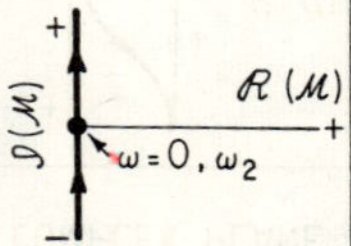	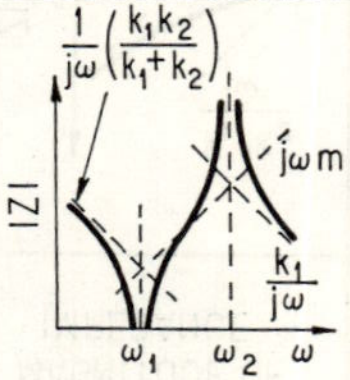	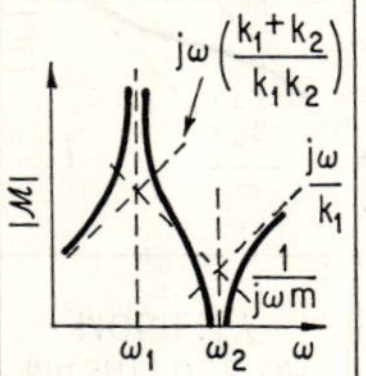	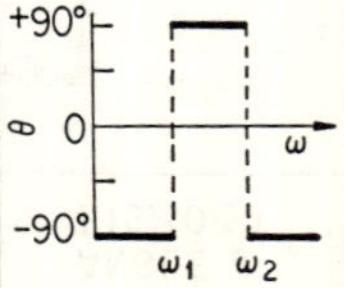

10-14

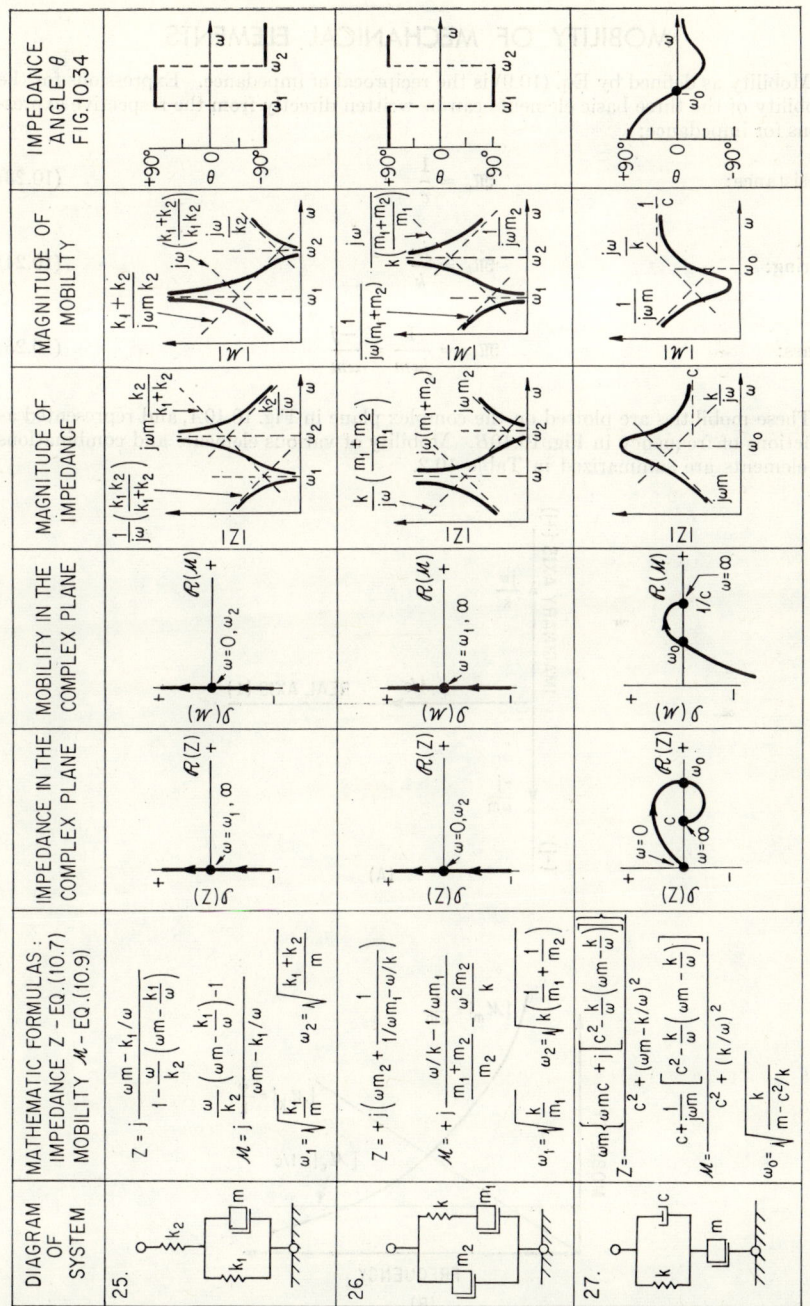

MOBILITY OF MECHANICAL ELEMENTS

Mobility as defined by Eq. (10.9) is the reciprocal of impedance. Expressions for the mobility of the three basic elements can be written directly from the respective expressions for impedance:

Resistance: $$\mathfrak{M}_c = \frac{1}{c} \tag{10.23}$$

Spring: $$\mathfrak{M}_k = \frac{j\omega}{k} \tag{10.24}$$

Mass: $$\mathfrak{M}_m = \frac{1}{j\omega m} = \frac{-j}{\omega m} \tag{10.25}$$

These mobilities are plotted on the complex plane in Fig. 10.10A, and represented as functions of frequency in Fig. 10.10B. Mobility of various elements and combinations of elements are summarized in Table 10.2.

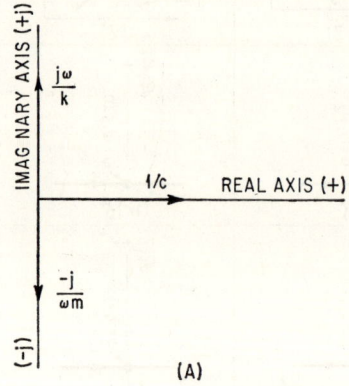

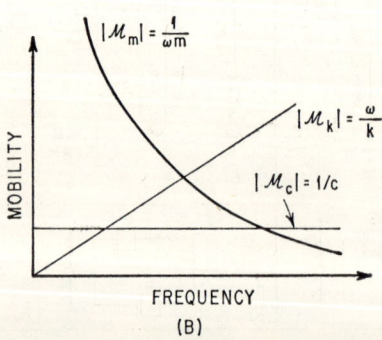

Fig. 10.10. The complex-plane representation of the mobility of an ideal resistance, spring, and mass at a fixed frequency is shown in (A); the variation of the mobility magnitude with frequency is shown in (B).

MECHANICAL SYSTEMS

The properties of a mechanical system that are used in analyses involving mechanical impedance and mobility are points of force application, paths for transmitting forces, and points of common velocities. The velocity at each connection point and the forces exerted on each element may be calculated by the proper application of Eq. (10.7) or Eq. (10.9) if the impedance of each element in the system and the forces or velocities imposed on points in the system are known. In many applications, only the motion at a few points is of primary interest. Then it is advantageous to combine groups of elements into single impedances. Methods for calculating the impedance of such combined elements are given in this section.

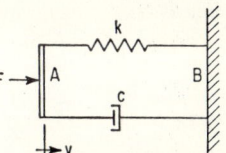

FIG. 10.11. Schematic representation of a parallel spring-resistance combination.

PARALLEL ELEMENTS

To illustrate the concept of parallel elements only ideal elements, multiply connected across two points, will be considered. Two ideal elements connected in parallel are shown in Fig. 10.11. Both are constrained to have the same relative velocities between their connections. The force required to give the resistance the velocity v is found from Eqs. (10.7) and (10.14):

$$F_c = vZ_c = vc$$

The force required to give the spring this same velocity is, from Eq. (10.18),

$$F_k = vZ_k = \frac{vk}{j\omega}$$

The total force F is

$$F = F_c + F_k = v\left(c + \frac{k}{j\omega}\right)$$

Since $Z = F/v$,

$$Z = c + \frac{k}{j\omega}$$

Thus, the over-all impedance is the sum of the impedances of the two elements.

By extending this concept to any number of parallel elements, the driving force F equals the sum of the resisting forces:

$$F = \sum_{i=1}^{n} vZ_i = v\sum_{i=1}^{n} Z_i \quad \text{and} \quad Z_p = \sum_{i=1}^{n} Z_i \qquad (10.26)$$

where Z_p is the total mechanical impedance of the parallel combination of the individual elements Z_i.

When the properties of the parallel elements are expressed as mobilities, the total mobility of the combination follows from Eqs. (10.9) and (10.26):

$$\frac{1}{\mathfrak{M}_p} = \sum_{i=1}^{n} \frac{1}{\mathfrak{M}_i} \qquad (10.27)$$

SERIES ELEMENTS

In determining the impedance presented by the end of a number of series-connected elements, consider the arrangement shown in Fig. 10.12. Elements Z_1 and Z_2 must have no mass since a mass always has one end connected to a stationary inertial reference. However, the impedance Z_3 may be a mass. The relative velocities between the end connections of each element are indicated by v_a, v_b, and v_c; the velocities of the connections with respect to the stationary reference point G are indicated by v_1, v_2, and v_3:

$$v_3 = v_c \qquad v_2 = v_3 + (v_2 - v_3) = v_c + v_b$$
$$v_1 = v_2 + (v_1 - v_2) = v_a + v_b + v_c$$

The impedance at point 1 is F/v_1, and the force F is transmitted to all three elements. The relative velocities are

$$v_a = \frac{F}{Z_1} \qquad v_b = \frac{F}{Z_2} \qquad v_c = \frac{F}{Z_3}$$

Thus, the total impedance is defined by

$$\frac{1}{Z} = \frac{F/Z_1 + F/Z_2 + F/Z_3}{F} = \frac{1}{Z_1} + \frac{1}{Z_2} + \frac{1}{Z_3}$$

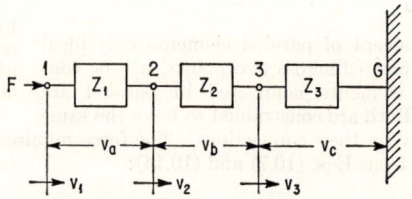

Fig. 10.12. Generalized three-element system of series-connected mechanical impedances.

Extending this principle to any number of massless series elements,

$$\frac{1}{Z_s} = \sum_{i=1}^{n} \frac{1}{Z_i} \tag{10.28}$$

where Z_s is the total mechanical impedance of the elements Z_i connected in series.

Using Eq. (10.9), the total mobility of series connected elements (expressed as mobilities) is

$$\mathfrak{M}_s = \sum_{i=1}^{n} \mathfrak{M}_i \tag{10.29}$$

SYSTEM SIMPLIFICATION

When one point of attachment of a multiple element system is of interest, the above methods for calculating the impedance of combined elements may be used to give one expression for the impedance at that point. When a force is applied to the point of attachment, the resulting velocity can be calculated from Eq. (10.7). In general, such an expression involves real and imaginary terms. When an *impedance* expression is obtained, the two terms can be considered to represent a purely resistive element connected in *parallel* with a purely imaginary or reactive element. A *mobility* expression may be considered to represent a resistive element in *series* with a reactive element. Either of these two representations has the characteristics of the original multiple element system and is called an *equivalent system*.

The calculation of the impedance of a multielement system using Eqs. (10.26) through (10.29) requires considerable manipulation of complex quantities.

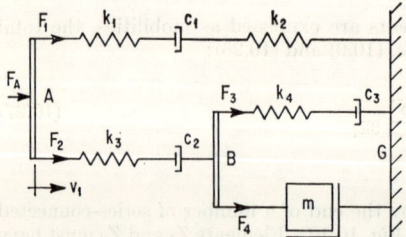

Fig. 10.13. System of several ideal elements analyzed in Example 10.1.

Example 10.1. For the system of Fig. 10.13, calculate the impedance at point A so that the velocity v_1 as a result of the application of force F_A may be determined. The procedure to be followed is basically as follows:

1. When series elements such as k_4 and c_3 are paralleled by other elements (such as m), add the mobilities of the series elements and convert the resulting expression to an impedance which may be added directly to the impedance of the parallel elements.

2. When parallel connected elements such as k_4, c_3, and m are connected in series with other elements (such as k_3 and c_2), convert the impedance of the parallel elements to a mobility and add to the mobility of the series elements.

In accordance with (1), add the mobilities of k_4 and c_3 to obtain the mobility of the branch at B:

$$\mathfrak{M}_{B1} = \frac{1}{c_3} + \frac{j\omega}{k_4}$$

The impedance of this branch is then

$$Z_{B1} = \frac{1}{\mathfrak{M}_{B1}} = \frac{1/c_3 - j\omega/k_4}{(1/c_3)^2 + (\omega/k_4)^2}$$

$$= \frac{1/c_3}{(1/c_3)^2 + (\omega/k_4)^2} - \frac{j\omega/k_4}{(1/c_3)^2 + (\omega/k_4)^2}$$

The system at B is then the parallel combination shown in Fig. 10.14A where $\mathfrak{R}(Z_{B1})$ represents the real part of Z_{B1}, and $\mathfrak{I}(Z_{B1})$ is the imaginary part of Z_{B1}. The parallel impedance at B including the impedance of mass m is

$$Z_B = \frac{1/c_3}{(1/c_3)^2 + (\omega/k_4)^2} + j\left[\omega m - \frac{\omega/k_4}{(1/c_3)^2 + (\omega/k_4)^2}\right]$$

The corresponding mobility is

$$\mathfrak{M}_B = \frac{\mathfrak{R}(Z_B)}{|Z_B|^2} - \frac{j\mathfrak{I}(Z_B)}{|Z_B|^2}$$

where $|Z_B|$ is the magnitude of Z_B which is represented by Fig. 10.14B. Adding the mobilities of c_2 and k_3, the mobility presented at A by this branch is

$$\mathfrak{M}_{A2} = \mathfrak{R}(\mathfrak{M}_B) + \frac{1}{c_2} + j\left(\mathfrak{I}(\mathfrak{M}_B) + \frac{\omega}{k_3}\right)$$

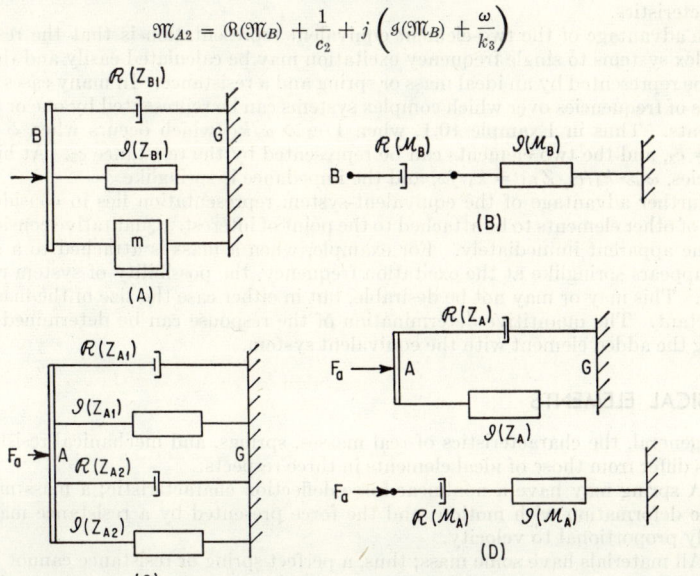

Fig. 10.14. Equivalent representations of parts of the system of Fig. 10.13 used in reducing that system to a representation equivalent to a single impedance or mobility.

To add the effect of this branch to that of the parallel branch consisting of k_1, k_2, and c_1, the impedances of both branches must be found:

$$Z_{A2} = \frac{1}{\mathfrak{M}_{A2}} = \frac{\mathfrak{R}(\mathfrak{M}_{A2})}{|\mathfrak{M}_{A2}|^2} - j\frac{\mathfrak{g}(\mathfrak{M}_{A2})}{|\mathfrak{M}_{A2}|^2}$$

$$\mathfrak{M}_{A1} = \frac{1}{c_1} + \frac{k_1 + k_2}{j\omega}$$

$$Z_{A1} = \frac{1/c_1}{(1/c_1)^2 + [(k_1 + k_2)/\omega]^2} + j\frac{(k_1 + k_2)/\omega}{(1/c_1)^2 + [(k_1 + k_2)/\omega]^2}$$

The impedances Z_{A1} and Z_{A2} then can be represented by the four parallel elements of Fig. 10.14C. The total impedance at A is

$$Z_A = Z_{A1} + Z_{A2} = \mathfrak{R}(Z_{A1}) + \mathfrak{R}(Z_{A2}) + j[\mathfrak{g}(Z_{A1}) + \mathfrak{g}(Z_{A2})]$$

and can be represented by the two parallel elements of Fig. 10.14D. The mobility expression at A is $1/Z_A$; it can be represented by the series elements of Fig. 10.14D.

As an aid to system simplification problems, the impedances and mobilities of several combinations of elements are given in Table 10.2.

EQUIVALENT SYSTEMS

The ratio of an applied force to a resulting velocity at the point A of the eight-element system in Fig. 10.13 is finally expressed analytically by an impedance having real and imaginary terms. The impedance characteristics at A are represented by the two-element systems of Fig. 10.14D. The two elements are not necessarily the ideal resistance, spring, or mass. The resistance term may vary with frequency and the reactance term may appear springlike, masslike, or be zero—depending on frequency. Even the impedance Z_{B1} of the two series elements c_3 and k_4 in Example 10.1 exhibits some of these characteristics.

The advantage of the two-element equivalent representation is that the response of complex systems to single frequency excitation may be calculated easily and the system may be represented by an ideal mass or spring and a resistance. In many cases there are ranges of frequencies over which complex systems can be represented by one or two ideal elements. Thus in Example 10.1, when $1/c_3 \gg \omega/k_4$ (which occurs when $\omega \ll k_4/c_3$), $Z_{B1} = c_3$, and the two elements can be represented by the resistance c_3. At higher frequencies, $\omega \gg k_4/c_3$, $Z_{B1} = k_4/j\omega$, and the impedance is springlike.

A further advantage of the equivalent-system representation lies in considering the effect of other elements to be attached to the point of interest. Qualitative considerations become apparent immediately. For example, when a mass is attached to a structure that appears springlike at the excitation frequency, the possibility of system resonance exists. This may or may not be desirable, but in either case the size of the mass is very important. The quantitive determination of the response can be determined by combining the added element with the equivalent system.

PHYSICAL ELEMENTS

In general, the characteristics of real masses, springs, and mechanical resistance elements differ from those of ideal elements in three respects:

1. A spring may have a nonlinear force-deflection characteristic; a mass may suffer plastic deformation with motion; and the force presented by a resistance may not be exactly proportional to velocity.

2. All materials have some mass; thus, a perfect spring or resistance cannot be made. Some compliance or spring effect is inherent in all elements. Energy can be dissipated in a system in several ways: friction, acoustic radiation, hysteresis, etc. Such a loss can be represented as a resistive component of the element impedance.

MECHANICAL SYSTEMS

3. Elements can differ from the ideal elements at high frequencies or when long connecting elements are used because the element length becomes comparable to a wavelength of stress waves in the material. Then a lumped system analysis does not apply, and analysis on the basis of wave motion in a transmitting medium must be made.

LUMPED ELEMENTS

A piece of high-density metal of spherical or cubical shape provides a nearly ideal mass. Wave motion effects do not occur until very high frequencies are reached, and a rigid attachment is feasible. However, supporting means for the mass may involve friction, and the attaching elements may have significant compliance. A parallel representation of the impedance of a physical mass at low frequencies is

$$Z_m = c + j\omega m + \frac{k}{j\omega}$$

If the mass is freely suspended by fairly rigid supports, the $j\omega m$ term predominates over a wide frequency range and Eqs. (10.22) and (10.25) may be used.

A helical wound spring can be made to have a nearly linear force-displacement characteristic, thereby attaining a constant value of k. But, under dynamic conditions the mass of the spring must be considered. When one end of a spring is rigidly fixed and the other end attached to a vibrating point, one-third of the mass of the spring may be considered to be lumped at the vibrating end of the spring. When both ends are vibrating, from one-third to one-half the mass of the spring may be considered lumped at each end. At low frequencies the mass reactance may be small compared to the compliant effect so that either approximation for the mass of the spring may be adequate.

Hysteresis effects in the spring metal can cause energy loss resulting in a resistive component in the spring impedance. The impedance of a spring can be represented by a parallel connection of the three ideal elements, but in most cases the resistive term is negligible. At low frequencies, the impedance of a spring when one end is fixed is approximately

$$Z_k = \frac{k}{j\omega} + j\omega \frac{m_k}{3} \qquad (10.30)$$

In many cases a mechanical resistance represents damping which is inherent in physical systems. In the case of sliding friction the reaction force may not be proportional to velocity but may be a constant, independent of velocity. Such a characteristic can be represented by a constant-force generator, 180° out-of-phase with the relative velocity across the sliding elements.

Resistive effects also may arise from the losses in rubber, felt, cork, or other vibration isolation materials. With such materials the damping coefficient may vary with the amplitude of motion.

One type of resistive element can be constructed by an arrangement of parts so arranged that the applied force pushes a body through a liquid or subjects it to other viscous forces. Because of the mass and compliance of the parts, such a device will not constitute an ideal resistance. Elements can be made sufficiently rigid so that the spring term is negligible; however, the effects of the mass of the elements may not be negligible.

Although ideal elements do not actually exist, it is common in the use of impedance methods to represent physical members by one or more ideal elements. Figure 10.15A shows a system of five physical elements as represented, for analysis at low frequency, by ideal elements. Many of the masses are essentially in parallel and can be added together as follows to produce the system of Fig. 10.15B:

$$m' = \frac{m_{k1}}{2} + \frac{m_{c1}}{2} \qquad m'' = m + \frac{m_{c2}}{2} + \frac{m_{k2}}{3} + \frac{m_{k1}}{2} + \frac{m_{c1}}{2}$$

Since one end of k_2 is rigidly fixed, one-third of its mass is lumped on the moving end. The masses on the fixed ends of c_2 and k_2 are not included in Fig. 10.15B because they do not enter into the dynamic problem.

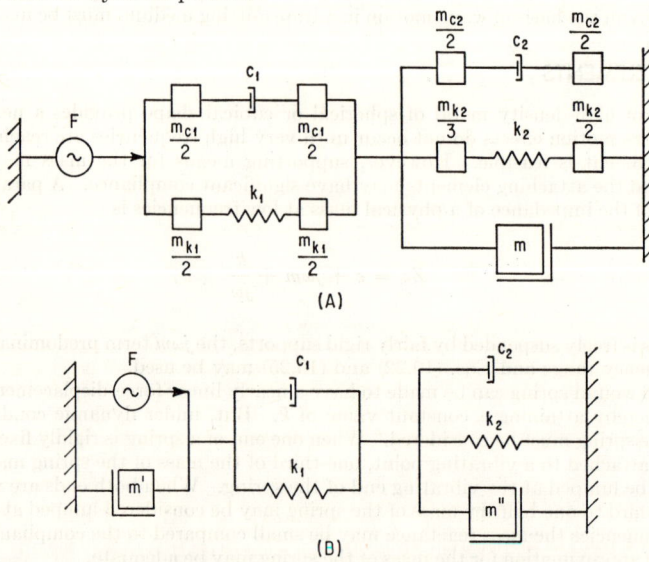

FIG. 10.15. The schematic representation applicable at low frequencies of a system of physical elements is shown in (A); the redrawn system suitable for impedance method analysis is shown in (B).

ELEMENTS WITH DISTRIBUTED MASS

All mechanical elements have distributed mass since mass cannot be concentrated at a point. If the vibration frequency is low enough that the time of travel of stress waves in an element of length l is very small compared to a period (i.e., if $l/\lambda \ll 1/2\pi$), the mass can be considered as a lumped element.[11]

The velocity of compressional stress waves in an elastic medium is $c_s = \sqrt{E/\rho}$, where E is Young's modulus and ρ is mass density. In long thin rods the velocity is c_s if the diameter is small compared to a wavelength, but when the diameter is as great as 2λ, the velocity drops to about $0.6c_s$.[12]

The force transmitted by and the velocities at the ends of a distributed mass element may be determined by transmission line methods.[13,14,15] For example, consider a uniform bar, as shown in Fig. 10.16A. (Other element shapes may be approximated by sections of uniform bars.) The bar has a length l, a cross-sectional area S, a density ρ, and a stress-wave velocity c_s. The force and velocity at the driven end (vibrational source attachment) are represented by F_i and v_i; F_r and v_r are the force and velocity at the load end. The material

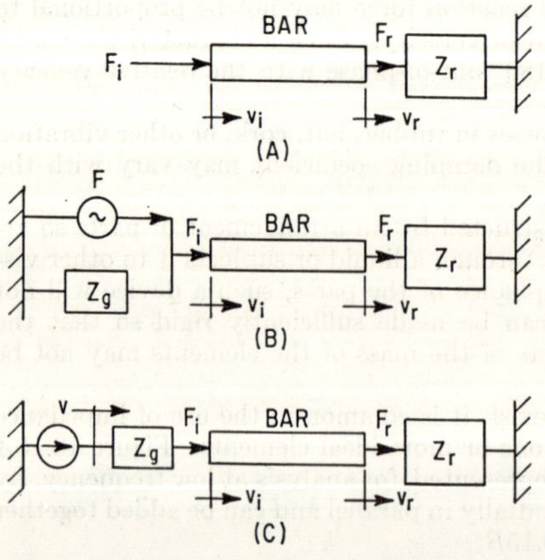

FIG. 10.16. The schematic representation of an element having distributed mass and its representation when it is connected to a source of vibration energy.

is assumed to be lossless, and the bar is characterized by a *propagation constant* $\gamma = j\beta = j\omega/c_s$ (where $\beta = \omega/c_s$) and a *characteristic impedance* (the impedance seen at the driven end for an infinitely long bar) $Z_0 = \rho c_s S$. Analysis by transmission line methods with the load Z_r attached gives the following relationships:

Load velocity in terms of the driven end velocity:

$$v_r = \frac{v_i}{\cos \beta \ell + j(Z_r/Z_0) \sin \beta \ell} \qquad (10.31)$$

Load force in terms of the driving end force:

$$F_r = \frac{F_i}{\cos \beta \ell + j(Z_0/Z_r) \sin \beta \ell} \qquad (10.32)$$

Driving end impedance:

$$Z_i = Z_0 \frac{Z_r \cos \beta \ell + jZ_0 \sin \beta \ell}{Z_0 \cos \beta \ell + jZ_r \sin \beta \ell} \qquad (10.33)$$

When the bar is driven by a force generator with a paralleled impedance Z_g the schematic representation is that of Fig. 10.16B. The load element force and velocity, and the driven end force and velocity, may be determined in terms of the known force F by the following expressions:

$$F_i = \frac{F}{1 + Z_i/Z_g} \qquad V_i = \frac{F}{Z_i(1 + Z_g/Z_i)}$$

$$F_r = \frac{F}{(1 + Z_g/Z_i)[\cos \beta \ell + j(Z_0/Z_r) \sin \beta \ell]} \qquad (10.34)$$

$$v_r = \frac{F}{(Z_g + Z_i)[\cos \beta \ell + j(Z_r/Z_0) \sin \beta \ell]}$$

where Z_g is the impedance of the generator.

When the bar is driven from a velocity generator with a series impedance $\mathfrak{M}_g$, as in Fig. 10.16C, the quantities of Eq. (10.34) are given in terms of the known velocity v and the mobility:

$$\mathfrak{M}_0 = \frac{1}{Z_0} \qquad \mathfrak{M}_r = \frac{1}{Z_r} \qquad \mathfrak{M}_g = \frac{1}{Z_g} \quad \text{and} \quad \mathfrak{M}_i = \frac{1}{Z_i}$$

$$v_i = \frac{v}{1 + \mathfrak{M}_g/\mathfrak{M}_i} \qquad F_i = \frac{v}{\mathfrak{M}_g + \mathfrak{M}_i}$$

$$v_r = \frac{v}{(1 + \mathfrak{M}_g/\mathfrak{M}_i)[\cos \beta \ell + j(\mathfrak{M}_0/\mathfrak{M}_r) \sin \beta \ell]} \qquad (10.35)$$

$$F_r = \frac{v}{(\mathfrak{M}_g + \mathfrak{M}_i)[\cos \beta \ell + j(\mathfrak{M}_r/\mathfrak{M}_0) \sin \beta \ell]}$$

Equation (10.33) can be applied to a mass at frequencies for which the lumped approximation is no longer valid. In this case $Z_r = 0$ and

$$Z_m = jZ_0 \tan \beta \ell \qquad (10.36)$$

where $Z_0 = \rho c_s S$. This expression may be compared with Eq. (10.22), the impedance of an ideal mass, $Z_m = j\omega m$.

THEOREMS

The following theorems are statements of basic principles (or combination of them) which apply to elements of a mechanical system; they are in a form convenient to analysis by the impedance method. In all but Kirchhoff's laws, these theorems apply only to systems made up of linear, bilateral elements. Linearity implies that the system elements can be represented by combinations of ideal elements in which c, k, and m are constants regardless of motion amplitude. A bilateral element is one in which forces are transmitted equally well in either direction across its connections.

Kirchhoff's Laws. 1. *The sum of all the forces acting at a point (common connection of several elements) is zero:*

$$\sum_i^n F_i = 0 \quad \text{(at a point)} \qquad (10.37)$$

This follows directly from the consideration leading to Eq. (10.26).

2. *The sum of the relative velocities across the connections of series mechanical elements taken around a closed loop is zero:*

$$\sum_i^n v_i = 0 \quad \text{(around a closed loop)} \qquad (10.38)$$

This follows from the considerations leading to Eq. (10.28).

Kirchhoff's laws apply to any system even when the elements are not linear or bilateral.

Superposition Theorem. *If a mechanical system of linear bilateral elements includes more than one vibration source, the force or velocity response at a point in the system can be determined by adding the response to each source, taken one at a time (the other sources supplying no energy but replaced by their internal impedances).*

The internal impedance of a vibrational generator is that impedance presented at its connection point when the generator is supplying no energy. This theorem finds useful application in systems having several sources. A very important application arises when the applied force is nonsinusoidal but can be represented by a Fourier series. Each term in the series can be considered a separate sinusoidal generator. The response at any point in the system can be calculated for each generator by using the impedance values at that frequency. Each response term becomes a term in the Fourier series representation of the total response function. The over-all response as a function of time then can be synthesized from the series.

Reciprocity Theorem. *If a force generator operating at a particular frequency at some point (1) in a system of linear bilateral elements produces a velocity at another point (2), the generator can be removed from (1) and placed at (2); then the former velocity at (2) will exist at (1), provided the impedances at all points in the system are unchanged.* This theorem also can be stated in terms of a vibration generator that produces a certain velocity at its point of attachment (1), regardless of force required, and the force resulting on some element at (2).

Reciprocity is an important characteristic of linear bilateral elements. It indicates that a system of such elements can transmit energy equally well in both directions. It further simplifies calculation on two-way energy transmission systems since the characteristics need be calculated for only one direction.

Thévenin's Equivalent System. *If a mechanical system of linear bilateral elements contains vibration sources and produces an output to a load at some point at any particular frequency, the whole system can be represented at that frequency by a single constant-force generator F_c in parallel with a single impedance Z_i connected to the load.* Thévenin's equivalent-system representation for a physical system may be determined by the following experimental procedure: Denote by F_c the force which is transmitted by the attachment point of the system to an infinitely rigid fixed point; this is called the *clamped force.* When the load connection is disconnected and perfectly free to move, a free velocity v_f is measured. Then the parallel impedance Z_i is F_c/v_f. The impedance Z_i also can be determined by measuring the internal impedance of the system when no source is supplying motional energy.

If the values of all the system elements in terms of ideal elements are known, F_c and Z_i may be determined analytically. A great advantage is derived from this representation in that attention is focused on the characteristics of a system at its output point and not on the details of the elements of the system. This allows an easy prediction of the response when different loads are attached to the output connection. After a final load condition has been determined, the system may be analyzed in detail for strength considerations.

Norton's Equivalent System. *A mechanical system of linear bilateral elements having vibration sources and an output connection may be represented at any particular frequency by a single constant-velocity generator v_f in series with an internal impedance Z_i.*

This is the series system counterpart of Thévenin's equivalent system where v_f is the free velocity and Z_i is the impedance as defined above. The same advantages in analysis exist as with Thévenin's parallel representation. The most advantageous one to use depends upon the type of structure to be analyzed. In the experimental determination of an equivalent system, it is usually easier to measure the free velocity than the clamped force on large heavy structures, while the converse is true for light structures. In any case, one representation is easily derived from the other. When v_f and Z_i are determined, $F_c = v_f Z_i$.

T and π Equivalent System. *Any mechanical system having an input and output connection and which is composed of linear bilateral elements may be represented by a three-element system of the T or π configuration.* The required equivalent system impedances

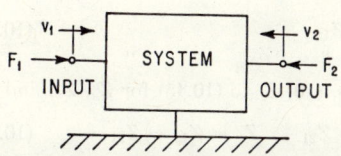

Fig. 10.17. Generalized two-connection mechanical system.

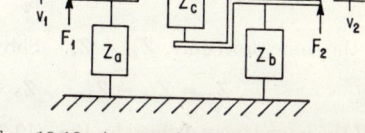

Fig. 10.18. A π equivalent system representation for a passive two-connection system.

determined below may not be obtainable physically, but the mathematical model represents the original system accurately.

π System. A generalized system is shown in Fig. 10.17. If a force F_1 is applied to the input connection, a velocity v_1 results. The ratio of F_1 to v_1 is the *input impedance*. If a force F_2 is applied to the output connection, a velocity v_2 results; the ratio of F_2 to v_2 is the *output impedance*. When the force F_1 is applied to the input, a velocity v_2 at the output results; the ratio of F_1 to v_2 is the *reverse transfer impedance*. If the force F_2 is applied to the output and a velocity v_1 results, the ratio of F_2 to v_1 is the *forward transfer impedance*.

These definitions can be used to find the impedances of an equivalent π system, as shown in Fig. 10.18, that are required to represent any general system, such as Fig. 10.17. First, find the input, output, and transfer impedances of the π system under the conditions specified below:

$$Z_{11} = \left.\frac{F_1}{v_1}\right|_{v_2=0} \tag{10.39}$$

where Z_{11} is the input impedance with the output connected to a rigid point so that

$$v_2 = 0$$

i.e., output clamped.

$$Z_{22} = \left.\frac{F_2}{v_2}\right|_{v_1=0} \tag{10.40}$$

where Z_{22} is the output impedance with the input clamped.

$$Z_{12} = \left.\frac{F_1}{v_2}\right|_{v_1=0} \tag{10.41}$$

where Z_{12} is the reverse transfer impedance with the input clamped and F_1 is the force required to maintain the input velocity $v_1 = 0$.

$$Z_{21} = \left.\frac{F_2}{v_1}\right|_{v_2=0} \tag{10.42}$$

where Z_{21} is the forward transfer impedance with the output clamped and F_2 is the force required to clamp the output.

Applying Eq. (10.39) to the system of Fig. 10.18, the input impedance when the output is clamped consists of Z_a and Z_c in parallel:

$$Z_{11} = Z_a + Z_c \tag{10.43}$$

Applying Eq. (10.40),

$$Z_{22} = Z_b + Z_c \tag{10.44}$$

When the input is clamped and a force F_2 is applied to the output, the velocity of the output v_2 results. The force F_1 then is transmitted by Z_c to the clamping point, and v_2 is the relative velocity of the connections of Z_c. By Eq. (10.7),

$$\left.\frac{F_1}{v_2}\right. = Z_c = Z_{12} \tag{10.45}$$

By the same procedure, $Z_{21} = Z_c$. Solving Eqs. (10.43) to (10.45) for Z_a, Z_b, and Z_c,

$$Z_a = Z_{11} - Z_{12} \qquad Z_b = Z_{22} - Z_{21} \qquad Z_c = Z_{12} = Z_{21} \tag{10.46}$$

If the impedances defined by Eqs. (10.39) to (10.42) can be determined for a system by either analytical or experimental means, the three impedances of an equivalent π system can be determined by Eqs. (10.46).

T System. An equivalent T system is shown in Fig. 10.19. The system values are most conveniently found in terms of mobilities. The necessary mobilities for a generalized system such as Fig. 10.17 are as follows:

The input mobility with the output "free" (no restraining force):

$$\mathfrak{M}_{11} = \left.\frac{v_1}{F_1}\right|_{F_2=0} \tag{10.47}$$

The output mobility with the input "free":

$$\mathfrak{M}_{22} = \left.\frac{v_2}{F_2}\right|_{F_1=0} \tag{10.48}$$

The reverse transfer mobility with the input "free":

$$\mathfrak{M}_{12} = \left.\frac{v_1}{F_2}\right|_{F_1=0} \tag{10.49}$$

where v_1 is the velocity of the "free" input point when the force F_2 is applied to the output.
The forward transfer mobility with the output "free":

$$\mathfrak{M}_{21} = \left.\frac{v_2}{F_1}\right|_{F_2=0} \tag{10.50}$$

Finding these quantities for the T system of Fig. 10.19,

$$\mathfrak{M}_{11} = \mathfrak{M}_x + \mathfrak{M}_z \qquad \mathfrak{M}_{22} = \mathfrak{M}_y + \mathfrak{M}_z \qquad \mathfrak{M}_{12} = \mathfrak{M}_{21} = \mathfrak{M}_z \tag{10.51}$$

Solving Eqs. (10.51) for $\mathfrak{M}_x$, $\mathfrak{M}_y$, and $\mathfrak{M}_z$,

$$\mathfrak{M}_x = \mathfrak{M}_{11} - \mathfrak{M}_{12} \qquad \mathfrak{M}_y = \mathfrak{M}_{22} - \mathfrak{M}_{21} \qquad \mathfrak{M}_z = \mathfrak{M}_{12} = \mathfrak{M}_{21} \tag{10.52}$$

ANALYSIS METHODS

If the mobilities required by Eqs. (10.47) to (10.50) can be determined analytically or experimentally for a system, the three mobilities of an equivalent T system can be obtained from Eqs. (10.52).

The forward and reverse transfer impedances and mobilities are equal for both π and T systems. This is a characteristic of even very complicated systems when the system elements are linear and bilateral; it is not true if the system contains sources of vibration energy. However, systems with vibration sources can be represented by π and T equivalent systems of a slightly more complex form.

A system of mechanical elements with one connection point can be represented by a single impedance. Thévenin's or Norton's equivalent systems represent systems having vibration sources and an output connection by a source with a constant output and a single impedance element. The π or T equivalent systems can represent connecting elements between a vibration source and some load system. Combining the three representations gives a system that is simple to analyze; it focuses attention on the force and motion at the load connection and at the output connection of the source. The effects of changes in the source, load, or connecting elements then can be readily calculated.

Another advantage of the π and T representations is that the equivalent system can be obtained from direct measurements on complicated systems that are almost impossible to represent by a collection of idealized mechanical elements. An analytical expression for the system thus is obtained that allows an analysis by the methods discussed in the next section. After the values of the equivalent elements are determined at the operating frequencies, it may become evident that the original system may be replaced by a much simpler system of physical elements, based on the equivalent representation.

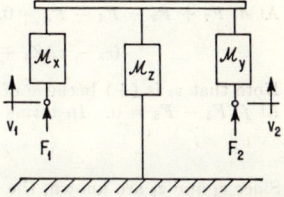

Fig. 10.19. A T equivalent system representation for a passive two-connection system.

Despite the many advantages of equivalent system representations, the complete schematic representation of a system often is desired. This representation is necessary for a complete analysis to determine the stresses on all of the parts of the system. To draw such a representation, a detailed study is required that can lead to a better understanding of the operation of the system.

In all but Kirchhoff's laws, the above theorems require linear, bilateral elements. This does not prevent their use in nonlinear, nonbilateral systems. Equivalent system representations with linear bilateral elements can be determined in many cases that represent accurately the nonlinear system over a limited range of operation.

ANALYSIS METHODS

The analysis of a system by impedance methods leads to the determination of forces acting on elements and the velocities of their connection points. From these quantities stresses may be determined, and accelerations and displacements may be calculated from Eqs. (10.5) and (10.6). When the motion is nonsinusoidal, the forces and velocities can be determined with the impedance approach, using the methods of analysis presented in this section. These methods are not the only ones available, but they are basic and adequate for many problems. Other methods may be found in the literature on electric circuit theory.

COMPLETE ANALYSIS

Kirchhoff's laws provide a sufficient number of simultaneous equations to determine all forces and velocities in a mechanical system. Equation (10.37) allows a force equation to be written for each independent connection point (force node). These equations usually are written in terms of impedances and velocities since a connection point ensures that one end of each element connected there has the same velocity when referred to a stationary reference. Equation (10.38) allows a velocity equation to be written for each independent loop, usually in terms of mobilities and forces. Velocity equations, best suited to series systems, do not find much usage in physical systems. Because the masses associated with physical elements have one connection effectively attached to a stationary reference (inertial reference), such a system seldom includes ideal series connected elements.

Example 10.2. Find the velocity of all the connection points and the forces acting on the elements of the system shown in Fig. 10.20. The system contains two velocity generators v_1 and v_6. Their magnitudes are known, their frequencies are the same, and they are 180° out-of-phase.

A. Using Eq. (10.37), write a force equation for each connection point except a and e.
At point b: $F_1 - F_2 - F_3 = 0$. In terms of velocities and impedances:

$$(v_1 - v_2)Z_1 - (v_2 - v_3)Z_2 - (v_2 - v_4)Z_4 = 0 \tag{a}$$

At c, the two series elements have the same force acting: $F_2 - F_2 = 0$. In terms of velocities and impedances:

$$(v_2 - v_3)Z_2 - (v_3 - v_4)Z_3 = 0 \tag{b}$$

At d: $F_2 + F_3 - F_4 - F_5 = 0$. In terms of velocities and impedances:

$$(v_3 - v_4)Z_3 + (v_2 - v_4)Z_4 - (v_4 + v_6)Z_5 - (v_4 - v_5)Z_6 = 0 \tag{c}$$

Note that v_6 is $(+)$ because of the 180° phase relation to v_1.
At f: $F_5 - F_5 = 0$. In terms of velocities and impedances:

$$(v_4 - v_5)Z_6 - v_5 Z_7 = 0 \tag{d}$$

Since v_1 and v_6 are known, the four unknown velocities v_2, v_3, v_4, and v_5 may be determined by solving the four simultaneous equations above. After the velocities are obtained, the forces may be determined from the following:

$$F_1 = (v_1 - v_2)Z_1 \qquad F_2 = (v_2 - v_3)Z_2 = (v_3 - v_4)Z_3$$
$$F_3 = (v_2 - v_4)Z_4 \qquad F_4 = (v_4 + v_6)Z_5$$
$$F_5 = (v_4 - v_5)Z_6 = v_5 Z_7$$

B. The method of *node forces*. Equations (a) through (d) above can be rewritten as follows:

$$v_1 Z_1 = (Z_1 + Z_2 + Z_3)v_2 - Z_2 v_3 - Z_4 v_4 \tag{a'}$$

$$0 = -Z_2 v_2 + (Z_2 + Z_3)v_3 - Z_3 v_4 \tag{b'}$$

$$0 = -Z_4 v_2 - Z_3 v_3 + (Z_3 + Z_4 + Z_5 + Z_6)v_4 - Z_6 v_5 \tag{c'}$$

$$-v_6 Z_5 = -Z_6 v_4 + (Z_6 + Z_7)v_5 \tag{d'}$$

These equations can be written by inspection of the schematic diagram by the following rule: *At each point with a common velocity (force node), equate the force generators to the sum of the impedances attached to the node multiplied by the velocity of the node, minus the impedances multiplied by the velocities of their other connection points.*

When the equations are written so that the unknown velocities form columns, the equations are in the proper form for a determinant solution for any of the unknowns. Note that the

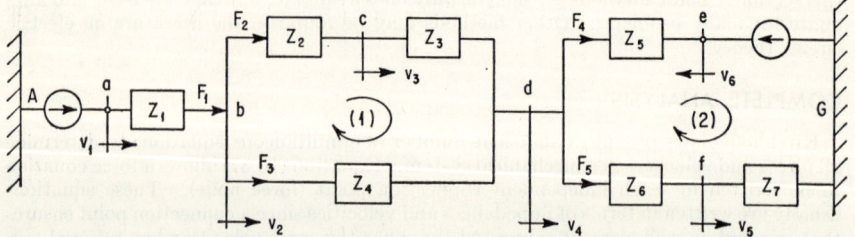

FIG. 10.20. System of mechanical elements and vibration sources analyzed in Example 10.2 to find the velocity of each connection and the force acting on each element.

determinant of the Z's is symmetrical about the main diagonal. This condition always exists and provides a check for the correctness of the equations.

C. Using Eq. (10.38), write a velocity equation in terms of force and mobility around enough closed loops to include each element at least once. In Fig. 10.20, note that

$$F_3 = F_1 - F_2 \quad \text{and} \quad F_5 = F_1 - F_4$$

Around loop (1):

$$F_2(\mathfrak{M}_2 + \mathfrak{M}_3) - (F_1 - F_2)\mathfrak{M}_4 = 0 \tag{e}$$

The minus sign preceding the second term results from going across the element 4 in a direction opposite to the assumed force acting on it.

Around loop (2):

$$F_4 \mathfrak{M}_5 - v_6 - (F_1 - F_4)(\mathfrak{M}_6 + \mathfrak{M}_7) = 0 \tag{f}$$

A summation of velocities from A to G along the upper path forms the following closed loop:

$$v_1 + F_1 \mathfrak{M}_1 + F_2(\mathfrak{M}_2 + \mathfrak{M}_3) + F_4 \mathfrak{M}_5 - v_6 = 0 \tag{g}$$

Equations (e), (f), and (g) then may be solved for the unknown forces F_1, F_2, and F_4. The other forces are $F_3 = F_1 - F_2$ and $F_5 = F_1 - F_4$. The velocities are:

$$v_2 = v_1 - F_1 \mathfrak{M}_1 \qquad v_3 = v_2 - F_2 \mathfrak{M}_2 \qquad v_4 = v_2 - F_3 \mathfrak{M}_4 \qquad v_5 = F_5 \mathfrak{M}_7$$

When a system includes more than one source of vibration energy, a Kirchhoff's law analysis with impedance methods can be made only if all the sources are operating at the same frequency. This is the case because sinusoidal forces and velocities can add as phasors only when their frequencies are identical. However, they may differ in magnitude and phase. Kirchhoff's laws still hold for instantaneous values and can be used to write the differential equations of motion for any system.

SUPERPOSITION

The superposition method of system analysis is advantageous when there are several vibration sources in a system. This method is illustrated by an example involving two force generators.

Example 10.3. In the system of Fig. 10.21A, the force generators F_1 and F_2 operate at the same frequency and their magnitudes are known. Determine the velocity of the common connection point c by the superposition theorem. First, determine the velocity v_c by Kirchhoff's laws for comparison with the results obtained by superposition. Writing the force-node equations at a, b, and c,

At a: $F_1 = Z_1 v_a - Z_1 v_c$

At c: $0 = -Z_1 v_a + (Z_1 + Z_2 + Z_3)v_c - Z_2 v_b$

At b: $F_2 = -Z_2 v_c + Z_3 v_b$

The impedance Z_1 is c_1 and k_1 in parallel, Z_2 is c_2 and k_2 in parallel, and Z_3 is c_3, k_3, and m_3 in parallel. Solving the equations for forces at a, b, and c to obtain v_c,

$$v_c = \frac{F_1 + F_2}{Z_3}$$

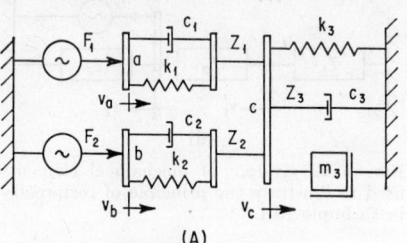

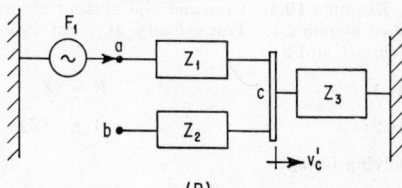

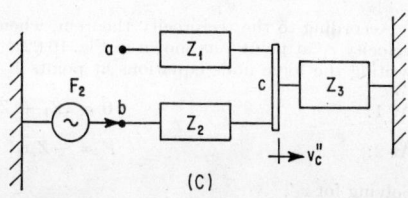

Fig. 10.21. System of mechanical elements including two force generators used to illustrate the principle of superposition in Example 10.3.

Using the superposition theorem, let the force F_2 be zero, and redraw the system as in Fig. 10.21B. Then let the velocity at point c be v_c' when only F_1 is applied. Since the force F_1 is transmitted by Z_1 to Z_3, v_c' becomes

$$v_c' = \frac{F_1}{Z_3}$$

Now letting $F_1 = 0$, applying F_2 as in Fig. 10.21C, and denoting the velocity of point c by v_c'',

$$v_c'' = \frac{F_2}{Z_3}$$

By the superposition theorem,

$$v_c = v_c' + v_c'' = \frac{F_1 + F_2}{Z_3}$$

If the force generators in Example 10.3 operate at different frequencies, the velocities are $v_c' = F_1/Z_3'$ and $v_c'' = F_2/Z_3''$, where Z_3' is the value of Z_3 at the frequency of force generator 1 and Z_3'' is the value of Z_3 at the frequency of generator 2. The instantaneous value of v_c, when phase angles are neglected, is

$$v_c = |v_c'|\sin \omega_1 t + |v_c''|\sin \omega_2 t$$

The mean square value is

$$v_c^2 = |v_c'|^2 + |v_c''|^2$$

Since the solution for a velocity or force by the superposition method involves only one source at a time, Kirchhoff's laws and impedance methods can be used in finding the responses to each source.

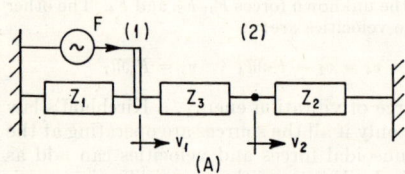

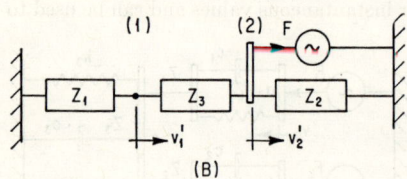

Fig. 10.22. System of mechanical elements used to illustrate the principle of reciprocity in Example 10.4.

RECIPROCITY

Since reciprocity is an inherent property of linear bilateral systems the statement of the theorem can be verified by analyzing such a system.

Example 10.4. Consider the system shown in Fig. 10.22A with the force generator F applied at point 1. The velocity at point 2 is found by solving the force node equations at points 1 and 2:

At 1: $\qquad F = (Z_1 + Z_3)v_1 - Z_3 v_2$

At 2: $\qquad 0 = -Z_3 v_1 + (Z_2 + Z_3)v_2$

Solving for v_2,

$$v_2 = \frac{Z_3 F}{Z_1 Z_2 + Z_2 Z_3 + Z_1 Z_3}$$

According to the reciprocity theorem, when the force generator is attached at point 2, the velocity v_1' at point 1 as shown in Fig. 10.22B is the same as v_2 above. This can be verified by writing the force node equations at points 1 and 2:

At 1: $\qquad 0 = (Z_1 + Z_3)v_1' - Z_3 v_2'$

At 2: $\qquad F = -Z_3 v_1' + (Z_2 + Z_3)v_2'$

Solving for v_1',

$$v_1' = \frac{Z_3 F}{Z_1 Z_2 + Z_2 Z_3 + Z_1 Z_3}$$

The velocity v_1' is identical to v_2 above.

ANALYSIS METHODS

The conditions of Eqs. (10.46) and (10.51), $Z_{12} = Z_{21}$ and $\mathfrak{M}_{12} = \mathfrak{M}_{21}$, often are used as a criterion for reciprocity. When these transfer functions as measured on a physical system are equal, the system obeys the law of reciprocity.

SYSTEMS WITH ONE CONNECTION

In one-connection systems considered here, one connection is attached to a rigid immovable point or inertial reference and another connection is available for external connection. These systems differ from multiple-connection systems in that they do not transmit a force or velocity from one system to another. One-connection systems can be divided into two types:

1. Passive systems, in which no sources of vibration energy are included. These systems can be represented by a single impedance or mobility function as shown in the section on *system simplification*. Such systems often are connected to, and considered as loads on, active systems as defined below.

2. Active systems, in which one or more sources of vibration energy is included. A complete schematic representation of such a system and its external connections may be drawn and analyzed by Kirchhoff's laws and superposition. The simplified representations of Thévenin's and Norton's equivalent systems are considered here.

THÉVENIN'S EQUIVALENT SYSTEM. Example 10.5. The system of Fig. 10.23A has one force generator, a number of elements, and a connection point b. Find Thévenin's equivalent system for this system. First combine the parallel elements c_1, k_1, and m_1 into one impedance Z_1, combine c_2 and k_2 into Z_2, and express m_3 in terms of Z_3. The schematic representation is shown in Fig. 10.23B. Determine the value of the force transmitted to a rigid immovable structure when point b is clamped to it. Point b is shown clamped in Fig. 10.23C, and Z_3 is omitted because it has no motion. Impedances Z_1 and Z_2 then are in parallel, and the force F divides between them to give the force transmitted to Z_2:

$$F_c = \frac{FZ_2}{Z_1 + Z_2} \qquad (a)$$

The impedance required for the Thévenin equivalent system can be determined from the system as rearranged in Fig. 10.23D. The force generator is removed and the impedance at b

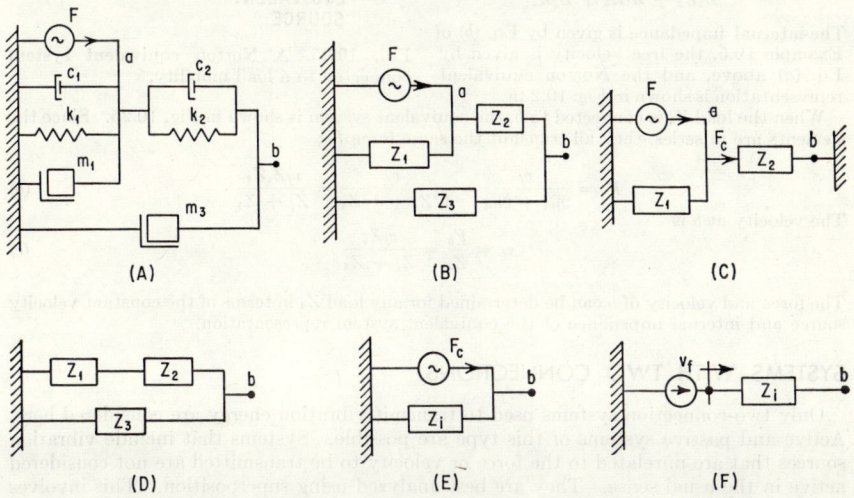

Fig. 10.23. System of mechanical elements including a force generator for which a Thévenin equivalent system is determined in Example 10.5 and a Norton equivalent system is determined in Example 10.6.

is calculated:
$$Z_i = Z_3 + \frac{1}{1/Z_1 + 1/Z_2} = \frac{Z_1Z_2 + Z_1Z_3 + Z_2Z_3}{Z_1 + Z_2} \quad (b)$$

The clamped force is given by Eq. (a) above, and the internal impedance is given by Eq. (b). The Thévenin equivalent system is shown in Fig. 10.23E.

When a load Z_4 is connected to point b, the equivalent representation is shown in Fig. 10.24. Under these conditions the velocity v_b of point b and the force applied to Z_4 can be calculated. Writing the force-node equation at b,

$$F_c = (Z_i + Z_4)v_b$$

Solving for v_b,
$$v_b = \frac{F_c}{Z_i + Z_4} \quad (c)$$

The force at b is then $F_b = v_b Z_4$, or
$$F_b = \frac{F_c Z_4}{Z_i + Z_4} \quad (d)$$

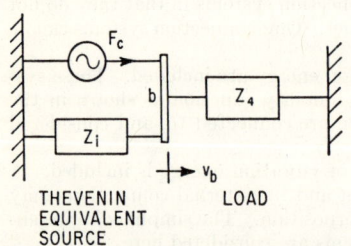

FIG. 10.24. A Thévenin equivalent system connected to a load impedance.

For any load impedance Z_4, the force and velocity at b can be calculated from Eqs. (c) and (d).

NORTON'S EQUIVALENT SYSTEM. Example 10.6. Find Norton's equivalent representation for the system of Fig. 10.23A, and find the force and velocity of point b when a load is attached to that point.

For this representation the free velocity at point b is required. This velocity v_f can be calculated from the system as drawn in Fig. 10.23B. Write the force-node equations at a and b, and let v_b be called v_f:

$$F = (Z_1 + Z_2)v_a - Z_2 v_f$$
$$0 = -Z_2 v_a + (Z_2 + Z_3)v_f$$

Solving for v_f,
$$v_f = \frac{FZ_2}{Z_1 Z_2 + Z_1 Z_3 + Z_2 Z_3} \quad (a)$$

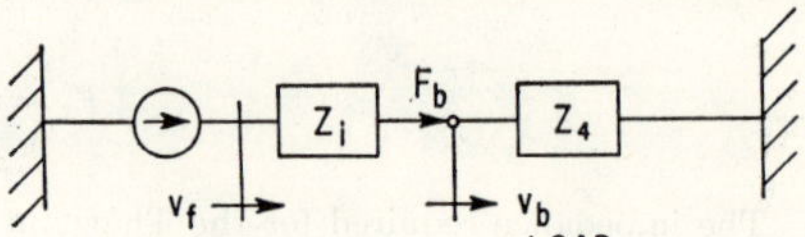

The internal impedance is given by Eq. (b) of Example 10.5, the free velocity is given by Eq. (a) above, and the Norton equivalent representation is shown in Fig. 10.23F.

FIG. 10.25. A Norton equivalent system connected to a load mobility.

When the load Z_4 is connected to b, the equivalent system is shown in Fig. 10.25. Since the elements are in series, they all transmit the same force F_b:

$$F_b = \frac{v_f}{\mathfrak{M}_i + \mathfrak{M}_4} = \frac{v_f}{1/Z_i + 1/Z_4} = \frac{v_f Z_i Z_4}{Z_i + Z_4} \quad (b)$$

The velocity at b is
$$v_b = \frac{F_b}{Z_4} = \frac{v_f Z_i}{Z_i + Z_4} \quad (c)$$

The force and velocity of b can be determined for any load Z_4 in terms of the constant velocity source and internal impedance of the equivalent system representation.

SYSTEMS WITH TWO CONNECTIONS

Only two-connection systems used to transmit vibration energy are considered here. Active and passive systems of this type are possible. Systems that include vibration sources that are unrelated to the force or velocity to be transmitted are not considered active in the usual sense. They are best analyzed using superposition. This involves determining the responses at the two connections due to the internal sources by Kirchhoff's laws, finding the transmitted motion and force by the methods that follow, and adding the results.

An active two-connection system by the usual definition is one in which an input function applied to one connection controls an energy source so as to produce an output function of the same shape as the input, but with an increased amplitude. Such a device is called an *amplifier*. The hydraulically powered vibration generator that has its control valve driven by a small electrodynamic vibration exciter is an example of this type of system. Many amplifying systems produce mechanical output functions, but they involve other forms of energy and dynamic media. In such cases, analysis is possible by impedance methods when compatible dynamical analogies and suitable coupling factors between dynamic media are used.

EQUIVALENT π SYSTEMS. Example 10.7. When a passive two-connection system is represented by an equivalent π system as shown in Fig. 10.18, its operation as a device to transmit vibration energy may be analyzed by adding the vibration source and the load impedance to the schematic representation. With the equivalent π system, it is convenient to use a Thévenin equivalent system for the source and a single impedance Z_c to represent the load. The complete system then is represented as shown in Fig. 10.26A. Since Z_i and Z_a as well as Z_b and Z_L are in parallel, they may be combined as in Fig. 10.26B where

$$Z_a' = Z_i + Z_a \quad \text{and} \quad Z_b' = Z_L + Z_b$$

Writing force node equations in terms of v_1 and v_L,

$$F_c = (Z_a' + Z_c)v_1 - Z_c v_L$$
$$0 = -Z_c v_1 + (Z_b' + Z_c)v_L$$

Solving for v_1 and v_L,

$$v_1 = \frac{F_c(Z_b' + Z_c)}{Z_a' Z_b' + Z_a' Z_c + Z_b' Z_c}$$

$$v_L = \frac{F_c Z_c}{Z_a' Z_b' + Z_a' Z_c + Z_b' Z_c}$$

The force applied to the load is $F_L = v_L Z_L$.

EQUIVALENT T SYSTEM. Example 10.8. When the T equivalent system representation is used, it is convenient to use the Norton equivalent representation for the vibration source and a mobility representation for the load. The complete representation is shown in Fig. 10.27A. Mobilities $\mathfrak{M}_i$ and $\mathfrak{M}_x$ as well as $\mathfrak{M}_2$ and $\mathfrak{M}_y$ are in series and can be combined as in Fig. 10.27B in which $\mathfrak{M}_x' = \mathfrak{M}_x + \mathfrak{M}_i$ and $\mathfrak{M}_y' = \mathfrak{M}_y + \mathfrak{M}_L$. A determination of the velocity v of the common connection allows a simple determination of F_1 and F_L. Since $\mathfrak{M}_y$ and $\mathfrak{M}_L$ are in series, the force F_L is the force applied to the load. The force F_1 is supplied by the vibration source. Proceeding to find v,

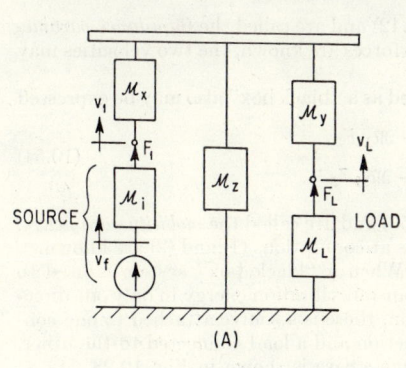

Fig. 10.26. A π equivalent system representation for a two-connection element used to transmit vibration energy, as analyzed in Example 10.7.

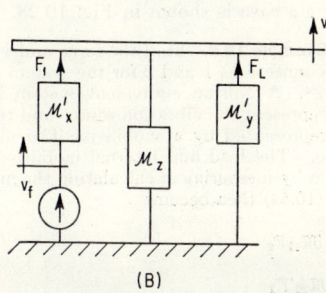

Fig. 10.27. A T equivalent system representation for the two-connection system analyzed in Example 10.8.

write a force equation at the common connection by Eq (10.37):

$$\frac{v_f - v}{\mathfrak{M}_x'} = \frac{v}{\mathfrak{M}_y'} + \frac{v}{\mathfrak{M}_z} \qquad v = \frac{v_f}{1 + \mathfrak{M}_x'/\mathfrak{M}_y' + \mathfrak{M}_x'/\mathfrak{M}_z}$$

The force acting on the load is then $F_L = v/\mathfrak{M}_L'$, and the load velocity is

$$v_L = F_L \mathfrak{M}_L = v_L \mathfrak{M}_L/\mathfrak{M}_y'$$

The force produced by the source which acts at the input connection is

$$F_1 = (v_f - v)/\mathfrak{M}_x'$$

and the velocity of the input connection v_1 is

$$v_1 = v + F_1 \mathfrak{M}_x = \frac{v\mathfrak{M}_i + v_f \mathfrak{M}_x}{\mathfrak{M}_x'}$$

IMPEDANCE AND MOBILITY PARAMETERS. In Examples 10.7 and 10.8 the actual system is represented by an equivalent system made up of fictitious impedance or mobility elements. Another approach is to represent a two-connection passive system by a "black box," attached to an inertial reference, that has input and output connections. Such a "black box" is illustrated in Fig. 10.17. When the elements in the system are linear and bilateral, and force generators are attached to the connections, the relationship between the velocities and forces may be expressed:

$$F_1 = Z_{11}v_1 + Z_{12}v_2$$
$$F_2 = Z_{21}v_1 + Z_{22}v_2 \qquad (10.53)$$

where the Z's are defined by Eqs. (10.39) to (10.42) and are called the *impedance parameters*. When the impedance parameters and two forces are known, the two velocities may be obtained by solving Eqs. (10.53).

The forces and velocities of a system considered as a "black box" also may be expressed as

$$v_1 = \mathfrak{M}_{11}F_1 + \mathfrak{M}_{12}F_2$$
$$v_2 = \mathfrak{M}_{21}F_1 + \mathfrak{M}_{22}F_2 \qquad (10.54)$$

Here the $\mathfrak{M}$'s are defined by Eqs. (10.47) to (10.50) and are called the *mobility parameters*. This representation is useful when the velocities at connections (1) and (2) are known.

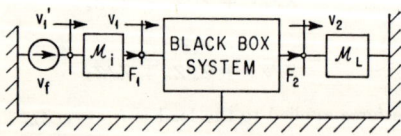

Fig. 10.28. A "black box" representation of a two-connection system analyzed by the mobility parameter method in Example 10.9.

When a "black box" system is used to transmit vibration energy in only one direction, there is a source attached to one connection and a load connected to the other. Such a case is shown in Fig. 10.28.

Example 10.9. Find the force and velocity at connections 1 and 2 for the system of Fig. 10.28. A Norton equivalent system is used to represent the vibration source and the load is represented by a mobility. The mobility parameters are thus the most advantageous to use. The load and internal mobility of the source may be included with the "black box" system by measuring or calculating the mobility parameters with $\mathfrak{M}_L$ and $\mathfrak{M}_i$ in place. Equations (10.54) then become

$$v_f = \mathfrak{M}_{11}'F_1 + \mathfrak{M}_{12}'F_2$$
$$0 = \mathfrak{M}_{21}'F_1 + \mathfrak{M}_{22}'F_2$$

Forces and velocities are considered at points 1' and 2. The velocity v_1 becomes v_f, and v_2 is zero since no external force is applied. The mobility $\mathfrak{M}_{11}'$ is $\mathfrak{M}_{11}$ of Eq. (10.47) determined at point 1', and $\mathfrak{M}_{22}'$ is $\mathfrak{M}_{22}$ of Eq. (10.48) with $\mathfrak{M}_L$ in place. Mobilities $\mathfrak{M}_{12}$ and $\mathfrak{M}_{21}$ are not dependent on the external connections. Solving for F_1 and F_2:

$$F_1 = \frac{v_f \mathfrak{M}_{22}'}{\mathfrak{M}_{11}' \mathfrak{M}_{22}' - \mathfrak{M}_{12} \mathfrak{M}_{21}}$$

$$F_2 = \frac{-v_f \mathfrak{M}_{21}}{\mathfrak{M}_{11}' \mathfrak{M}_{22}' - \mathfrak{M}_{12} \mathfrak{M}_{21}}$$

The force applied to the load is thus $F_L = F_2$, and the load velocity is $v_L = F_2 \mathfrak{M}_L$. The force at connection 1 is F_1, since $\mathfrak{M}_i$ transmits this force and the input connection velocity is

$$v_1 = v_f - F_1 \mathfrak{M}_i$$

FOUR-POLE PARAMETERS. The relationship between input and output forces and velocities can be written in still another form:

$$F_1 = \alpha_{11} F_2 + \alpha_{12} v_2$$
$$v_1 = \alpha_{21} F_2 + \alpha_{22} v_2 \tag{10.55}$$

The α's are called *four-pole parameters* and are defined as follows:

$$\alpha_{11} = \frac{F_1}{F_2}\bigg|_{v_2=0} \qquad \alpha_{12} = \frac{F_1}{v_2}\bigg|_{F_2=0}$$

$$\alpha_{21} = \frac{v_1}{F_2}\bigg|_{v_2=0} \qquad \alpha_{22} = \frac{v_1}{v_2}\bigg|_{F_2=0}$$

The notation $v_2 = 0$ indicates that the output connection (2) is *clamped* and $F_2 = 0$ indicates the output is *free*. The quantities α_{11} and α_{22} are the force and velocity transfer functions, while α_{12} is an impedance and α_{21} is a mobility. "Black box" systems may be analyzed by the use of these parameters and Eqs. (10.55). In addition, the system may be analyzed by the method that uses the four-pole parameters of the ideal elements and rules for combining their parameters when they are series or parallel connected.[16]
To establish the four-pole parameters for the ideal elements, the relationships between applied and transmitted forces and the velocities of the connections are written in the form of Eqs. (10.55); then the α's are noted as the coefficients of the F_2 and v_2 terms.

Example 10.10. Determine the four-pole parameters for a mass, spring, and resistance.
Mass. Since a mass may be considered as a rigid body, the velocities of its connections are equal. The force required to give the mass a velocity v_1 is $j\omega m v_1$ (or $j\omega m v_2$). If a force F_2 is transmitted, this must be added to $j\omega m v_2$ to determine F_1. Writing these relationships in the form of Eqs. (10.55),

$$F_1 = F_2 + j\omega m v_2$$
$$v_1 = v_2$$

The α's are $\alpha_{11} = 1$, $\alpha_{12} = j\omega m = Z_m$, $\alpha_{21} = 0$, and $\alpha_{22} = 1$.
Spring. The ideal spring transmits an applied force so that $F_1 = F_2$. The relative velocity is $(v_1 - v_2) = j\omega F_2/k$. Expressing these relationships in the form of Eqs. (10.55),

$$F_1 = F_2$$
$$v_1 = \frac{j\omega}{k} F_2 + v_2$$

The α's are $\alpha_{11} = 1$, $\alpha_{12} = 0$, $\alpha_{21} = j\omega/k = \mathfrak{M}_k$, and $\alpha_{22} = 1$.
Resistance. For the resistance, $F_1 = F_2$ and the relative velocity of the connection is $(v_1 - v_2) = F_2/c$.
In the form of Eqs. (10.55):

$$F_1 = F_2$$
$$v_1 = \frac{F_2}{c} + v_2$$

The α's are $\alpha_{11} = 1$, $\alpha_{12} = 0$, $\alpha_{21} = 1/c = \mathfrak{M}_c$, and $\alpha_{22} = 1$.

Note that the determinant of the α's in each case in Example 10.10 is equal to unity. This is a characteristic of all systems made up of linear bilateral elements and is another indication of reciprocity. The unity value of the α determinant greatly simplifies the solution of Eqs. (10.55) by the determinant or matrix method.

Matrix methods are convenient in system analysis by the four-pole parameter methods. The parameters for a series-connected system, expressed in terms of the individual element parameters, are determined as follows:

SERIES-CONNECTED SYSTEMS. Two such elements are shown in Fig. 10.29A. The four-pole parameter equations for both elements in matrix form are

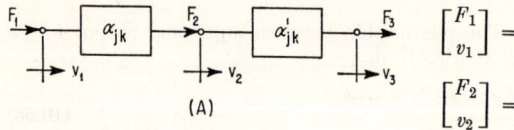

$$\begin{bmatrix} F_1 \\ v_1 \end{bmatrix} = \begin{bmatrix} \alpha_{11} & \alpha_{12} \\ \alpha_{21} & \alpha_{22} \end{bmatrix} \begin{bmatrix} F_2 \\ v_2 \end{bmatrix}$$

$$\begin{bmatrix} F_2 \\ v_2 \end{bmatrix} = \begin{bmatrix} \alpha_{11}' & \alpha_{12}' \\ \alpha_{21}' & \alpha_{22}' \end{bmatrix} \begin{bmatrix} F_3 \\ v_3 \end{bmatrix}$$

Combining these to find the output at (3) in terms of the input at (1):

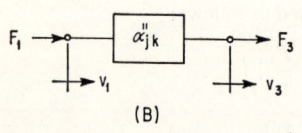

$$\begin{bmatrix} F_1 \\ v_1 \end{bmatrix} = \begin{bmatrix} \alpha_{11} & \alpha_{12} \\ \alpha_{21} & \alpha_{22} \end{bmatrix} \begin{bmatrix} \alpha_{11}' & \alpha_{12}' \\ \alpha_{21}' & \alpha_{22}' \end{bmatrix} \begin{bmatrix} F_3 \\ v_3 \end{bmatrix}$$

$$\begin{bmatrix} F_1 \\ v_1 \end{bmatrix} = \begin{bmatrix} \alpha_{11}'' & \alpha_{12}'' \\ \alpha_{21}'' & \alpha_{22}'' \end{bmatrix} \begin{bmatrix} F_3 \\ v_3 \end{bmatrix}$$

FIG. 10.29. Series-connected mechanical elements considered in the four-pole parameter method.

The single system of Fig. 10.29B is represented by the α''''s and replaces the two series elements. The final four-pole parameters in terms of the original element parameters are

$$\alpha_{11}'' = \alpha_{11}\alpha_{11}' + \alpha_{12}\alpha_{21}' \qquad \alpha_{12}'' = \alpha_{11}\alpha_{12}' + \alpha_{12}\alpha_{22}'$$
$$\alpha_{21}'' = \alpha_{21}\alpha_{11}' + \alpha_{22}\alpha_{21}' \qquad \alpha_{22}'' = \alpha_{22}\alpha_{22}' + \alpha_{21}\alpha_{12}' \qquad (10.56)$$

This process can be continued to combine any number of elements in a series string. Note that this method allows for a series mass and any number of elements to be considered series-connected.

PARALLEL-CONNECTED ELEMENTS. The canonical equations for parallel-connected elements, as shown in Fig. 10.30A, can be written in the form of Eqs. (10.55)

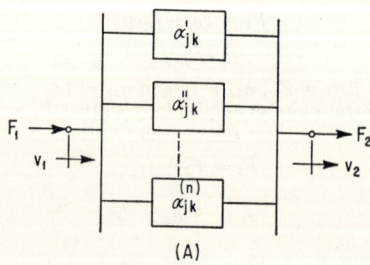

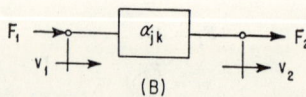

FIG. 10.30. Parallel-connected mechanical elements considered in the four-pole parameter method.

where the α's for the single resultant element are defined in terms of the individual elements in the system:

$$\alpha_{11} = \frac{A}{B} \qquad \alpha_{12} = \frac{AC}{B} - B \qquad \alpha_{21} = \frac{1}{B} \qquad \alpha_{22} = \frac{C}{B} \quad (10.57)$$

where
$$A = \sum_{i=1}^{n} \frac{\alpha_{11}{}^{(i)}}{\alpha_{21}{}^{(i)}} \qquad B = \sum_{i=1}^{n} \frac{1}{\alpha_{21}{}^{(i)}} \qquad C = \sum_{i=1}^{n} \frac{\alpha_{22}{}^{(i)}}{\alpha_{21}{}^{(i)}}$$

The factors A, B, and C depend on the parameters of the individual elements as indicated by $i = 1, 2, \ldots, n$. The parallel combination then is represented by a single element as in Fig. 10.30B.

VIBRATION SOURCES. Thévenin's and Norton's equivalent system representations are convenient in this method. Consider a Thévenin representation such as Fig. 10.24. The force F_1 applied to a load and the velocity of the load input v_1 are related to the clamped force F_c and internal impedance Z_i as follows:

$$F_1 = F_c - Z_i v_1 \quad (10.58)$$

For the Norton representation in Fig. 10.25, the force F_1 is related to the v_1 as follows:

$$F_1 = Z_i v_c - Z_i v_1 \quad (10.59)$$

Example 10.11. By the four-pole parameter method, determine the velocity of a mass driven by a vibration source. The mass is shown driven by a Thévenin equivalent source in Fig. 10.31. When there is no force transmitted by the mass, the four-pole parameter equations for the mass are

$$\begin{bmatrix} F_1 \\ v_1 \end{bmatrix} = \begin{bmatrix} 1 & j\omega m \\ 0 & 1 \end{bmatrix} \begin{bmatrix} 0 \\ v_2 \end{bmatrix}$$

In conventional form, $F_1 = j\omega m v$ and $v_1 = v_2$. The source equation $F_1 = F_c - Z_i v_1$ is combined with the mass equation to give

$$F_c - Z_i v_1 = j\omega m v_2 = j\omega m v_1$$

Solving for v_1,

$$v_1 = \frac{F_c}{Z_i + j\omega m}$$

When the mass is driven by a Norton equivalent system with free velocity v_f, the combination of equations yields

$$Z_i v_f - Z_i v_1 = j\omega m v_1$$

Solving for v_1,

$$v_1 = \frac{Z_i v_f}{Z_i + j\omega m}$$

The "black box" concept is evident in each of the above methods of analysis for two-connection systems. This approach focuses attention on two points in a system, usually the more important points. In the case of vibration isolators, the performance is measured in terms of the velocities or forces at the two connections when the devices are used to connect two systems. The above methods allow a calculation of the response of such a device in any application when the impedances of the systems to be connected are known, and one of the sets of the above parameters is known. All the parameters lend themselves to direct measurement. Vibration response characteristics of very complex systems under many different conditions may be determined with a minimum of analysis time when one of the sets of parameters is known. All the sets of parameters are defined in terms of sinusoidal forces and velocities, and they must be measured with this type of driving function.

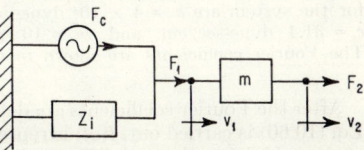

Fig. 10.31. A mass driven by a Thévenin equivalent source, as analyzed by the four-pole parameter method in Example 10.11.

NONSINUSOIDAL FUNCTIONS

The forces generated by the sources of vibration energy encountered in the normal operation of many mechanical systems are not sinusoidal in form. The responses of systems to nonsinusoidal driving functions can be determined by the impedance method. The Fourier analysis methods and the principle of superposition make this possible.

Excitations of three types frequently encountered are:

1. Continuous, periodic but nonsinusoidal force or velocity functions. Such functions can be represented by a Fourier series and the response to each term found by the impedance method.

2. Nonrepeated transient excitation, usually called shock excitation. The Fourier integral transform of the function yields a continuous frequency spectrum. When impedance expressions are considered as frequency functions, the frequency spectrum of the response may be determined. By the inverse Fourier integral transform the response as an instantaneous time function may be obtained.

3. Continuous, nonperiodic random excitation. The Fourier integral transform applies, and the methods are similar to those for (2) above. Random vibration is discussed in Chap. 11.

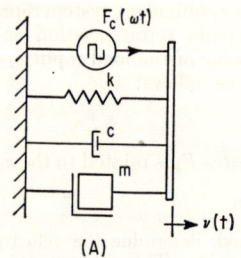

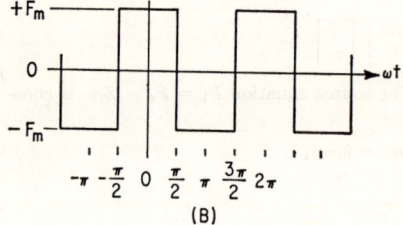

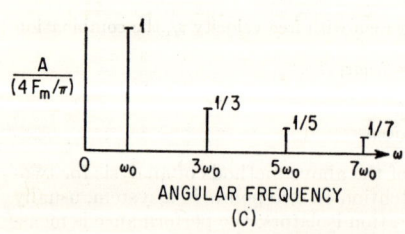

Fig. 10.32. A mechanical system (A) driven by a square-wave force (B) as analyzed in Example 10.12. The physical constants for the system are $k = 4 \times 10^4$ dynes/cm, $c = 31.4$ dyne-sec/cm, and $m = 10$ gm. The Fourier coefficients are shown in (C).

PERIODIC FUNCTIONS. The Fourier series representation for a periodic function * is

$$f(\omega t) = a_0 + \sum_{n=1}^{\infty} (a_n \cos n\omega_0 t + b_n \sin n\omega_0 t)$$

(10.60)

where $\omega_0 = 2\pi/\tau$; τ is the period of the function and the coefficients are

$$a_0 = \frac{1}{2\pi} \int_0^{2\pi} f(\omega t) \, d\omega t$$

$$a_n = \frac{1}{\pi} \int_0^{2\pi} f(\omega t) \cos n\omega t \, d\omega t \quad (10.61)$$

$$b_n = \frac{1}{\pi} \int_0^{2\pi} f(\omega t) \sin n\omega t \, d\omega t$$

In these expressions $f(\omega t)$ is the analytical expression for the waveform in the interval $0 \leq \omega t \leq 2\pi$. Any other interval, such as $\theta_1 \leq \omega t \leq \theta_1 + 2\pi$, is acceptable, provided one full period is included. If the function is discontinuous, the integration can be carried out piecewise over each continuous region.

After the Fourier coefficients are determined from Eqs. (10.61), and the summation of Eq. (10.60) is carried out, $f(\omega t)$ is represented as a constant term a_0 and an infinite series of sinusoidal functions of different frequency and amplitude. The frequencies are harmonically related in that they are integral multiples of the fundamental angular frequency ω_0. The response of a system to each term can be calculated and the results

* Also see Chap. 22.

added, as allowed by the superposition theorem, to produce a series expression for the complete response.

Example 10.12. Consider the force generator of Fig. 10.32A that produces the clamped force shown in Fig. 10.32B. Determine the velocity $v(\omega t)$ when this square wave of force is applied to the mechanical system. By Eq. (10.60), the force can be represented by a series of sinusoidal functions:

$$F_c(\omega t) = a_0 + \sum_{n=1}^{\infty} (a_n \cos n\omega_0 t + b_n \sin n\omega_0 t)$$

Using Eqs. (10.61) to find the coefficients, choose the zero point as shown and consider the interval $-\pi < \omega t < +\pi$:

$$a_0 = \frac{1}{2\pi} \int_{-\pi}^{\pi} F(\omega t) \, d\omega t$$

Using piecewise integration,

$$a_0 = \frac{1}{2\pi} \left[\int_{-\pi}^{-\pi/2} (-F_m) \, d\omega t + \int_{-\pi/2}^{+\pi/2} F_m \, d\omega t + \int_{+\pi/2}^{+\pi} (-F_m) \, d\omega t \right] = 0$$

$$a_n = \frac{1}{\pi} \left[\int_{-\pi}^{-\pi/2} (-F_m) \cos n\omega_0 t \, d\omega t + \int_{-\pi/2}^{+\pi/2} F_m \cos n\omega_0 t \, d\omega t + \int_{+\pi/2}^{+\pi} (-F_m) \cos n\omega_0 t \, d\omega t \right]$$

$$= \frac{4F_m}{n\pi} \sin \frac{n\pi}{2}$$

The evaluation of b_n yields $b_n = 0$. Then the series representation for $F_c(\omega t)$ becomes

$$F_c(\omega t) = \frac{4F_m}{\pi} (\cos \omega_0 t - \tfrac{1}{3} \cos 3\omega_0 t + \tfrac{1}{5} \cos 5\omega_0 t - \cdots)$$

The magnitudes of these terms can be plotted versus ω to give the frequency function shown in Fig. 10.32C. This type of function usually is referred to as a *line amplitude spectrum*.

The superposition theorem allows each term in the series to be considered as a separate generator of angular frequency $n\omega_0$, and the velocity resulting from each to be calculated. The resulting velocities as time functions then can be added to produce the complete velocity response. Since each generator produces s sinusoidal force, the response is calculated readily by the impedance method. From the expression for $\mathfrak{M}(\omega)$, $v(\omega t)$ can be calculated:

$$v(n\omega_0) = F(n\omega_0) \mathfrak{M}(n\omega_0)$$

Values of $\mathfrak{M}(\omega)$ at $\omega = \omega_0, 3\omega_0, 5\omega_0, \ldots$ are required. Equation (10.60) indicates that an infinite number of terms are required to represent completely the force wave form. However, for most physically obtainable functions, the series converges rapidly, as indicated by the amplitude spectrum of Fig. 10.32C. The number of terms required depends on the accuracy desired.

The mobility of the elements k, c, and m in parallel in Fig. 10.32A is

$$\mathfrak{M}(\omega) = \frac{1}{c + j(\omega m - k/\omega)} = \frac{1}{c + j\sqrt{km}\left(\dfrac{\omega}{\omega_1} - \dfrac{\omega_1}{\omega}\right)}$$

where $\omega_1 = (k/m)^{1/2} = 63$ rad/sec, or $f_1 = 10$ cps.

Consider the case when the period of the applied square wave of force is $\tau = 0.1$ sec. Then $f_0 = 1/\tau = 10$ cps, or $\omega_0 = \omega_1$. Then $|\mathfrak{M}(\omega)|$ becomes

$$|\mathfrak{M}(\omega)| = \frac{1}{\sqrt{c^2 + km\left(\dfrac{\omega}{\omega_0} - \dfrac{\omega_0}{\omega}\right)^2}}$$

The mobility angle is

$$\theta = \tan^{-1} \frac{\sqrt{km}}{c} \left(\frac{\omega_0}{\omega} - \frac{\omega}{\omega_0}\right)$$

Substituting the values for the system and evaluating $\mathfrak{M}(\omega)$:

$$\mathfrak{M}(\omega_0) = 3.18 \times 10^{-2} e^{j0°}$$
$$\mathfrak{M}(3\omega_0) = 5.90 \times 10^{-4} e^{-j89°}$$
$$\mathfrak{M}(5\omega_0) = 3.29 \times 10^{-4} e^{-j90°}$$
$$\mathfrak{M}(7\omega_0) = 2.31 \times 10^{-4} e^{-j90°}$$

Then the velocities are as follows:

$$v(\omega_0) = F(\omega_0)\mathfrak{M}(\omega_0) = \frac{4F_m}{\pi} 3.18 \times 10^{-2} e^{-j90°}$$

$$v(3\omega_0) = \frac{4F_m \times 5.90 \times 10^{-4}}{3\pi} e^{j1°}$$

$$v(5\omega_0) = \frac{4F_m \times 3.29 \times 10^{-4}}{5\pi} e^{j0°}$$

$$v(5\omega_0) = \frac{4F_m \times 3.29 \times 10^{-4}}{5\pi} e^{j0°}$$

$$v(7\omega_0) = \frac{4F_m \times 2.31 \times 10^{-4}}{7\pi} e^{j0°}$$

The Fourier series representation for the velocity then becomes

$$v(\omega t) = \left(\frac{4F_m}{\pi}\right)(3.18 \times 10^{-2})[\cos \omega_0 t - 6.2 \times 10^{-3} \sin(3\omega_0 t + 1°)$$
$$+ 2.07 \times 10^{-3} \sin 5\omega_0 t - 1.04 \times 10^{-3} \sin 7\omega_0 t \dots]$$

The shape of this velocity waveform can be produced by plotting all significant terms and adding them point by point to give the complete function. In this example, the velocity of the $\omega_0 t$ term is over 100 times as great as the largest of the remaining terms; thus, the velocity is essentially the cosine wave shown in Fig. 10.33 because the system is resonant with the fundamental frequency of the driving force.

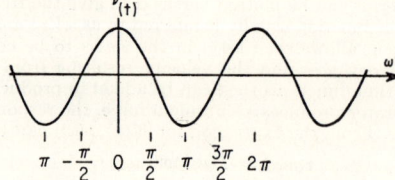

Fig. 10.33. Velocity response of system with excitation shown in Fig. 10.32.

Shock Excitation. A shock excitation method very similar to the analysis procedure for periodic functions can be used for calculating shock response. This method uses the Fourier integral transform pair. The first of these is analogous to Eq. (10.60):

$$F(t) = \int_{-\infty}^{+\infty} G(\omega)e^{j\omega t} \, d\omega \qquad (10.62)$$

This expression combines the sines and cosines into the complex form, and $G(\omega)$ replaces the constants a_n, b_n. The second expression [Eq. (10.63)] gives amplitude and phase as a function of frequency in terms of the driving function $f(t)$:

$$G(\omega) = \frac{1}{2\pi} \int_{-\infty}^{+\infty} f(t)e^{-j\omega t} \, dt \qquad (10.63)$$

where $G(\omega)$ is in general a continuous function of frequency. A plot of $G(\omega)$ versus ω provides a *continuous amplitude spectrum* instead of the line spectrum of Fig. 10.32C. Such a plot usually is referred to as the *frequency spectrum* of the original time function.

This frequency spectrum allows impedance methods to be used in calculating the shock response of mechanical systems. Consider a system as shown in Fig. 10.32A, but with an applied force that is a nonrepetitive transient. If the shape of the force waveform is known, the frequency spectrum $G(\omega)$ can be calculated by Eq. (10.63). The resulting velocity as a function of frequency is then

$$\mathcal{V}(\omega) = G(\omega) \mathfrak{M}(\omega)$$

After the velocity spectrum has been calculated, the velocity as a function of time can be calculated from Eq. (10.62):

$$v(t) = \int_{-\infty}^{\infty} v(\omega) e^{j\omega t}\, d\omega$$

Note that the Fourier method transforms a function in the time domain to one in the frequency domain. This frequency function can be multiplied by an impedance, mobility, or transfer function to obtain the frequency function at some point in a physical system. Finally, the inverse transformation allows the output function in the time domain to be determined. The application of the Fourier integral is considered in more detail in Chap. 23.

PRESENTATION AND EVALUATION OF IMPEDANCE DATA

At a particular frequency the mechanical impedance (or mobility) of a system is a complex number which can be expressed as the sum of a real and an imaginary component or as a magnitude and impedance or mobility angle.

$$Z = \mathcal{R}(Z) + j\mathcal{I}(Z) = |Z|e^{j\theta}$$

The magnitude $|Z|$ and angle θ are

$$|Z| = \sqrt{[\mathcal{R}(Z)]^2 + [\mathcal{I}(Z)]^2} \qquad \theta = \tan^{-1}\frac{\mathcal{I}(Z)}{\mathcal{R}(Z)}$$

These relations also may be expressed

$$\mathcal{R}(Z) = |Z|\cos\theta \qquad \mathcal{I}(Z) = |Z|\sin\theta$$

Impedance at a particular frequency may be represented as a directed line on the complex plane, as shown in Fig. 10.34. Thus, two quantities are required to specify completely an impedance or mobility.

The impedances of the three ideal elements (mass, spring, and resistance) are functions of frequency as indicated by Eqs. (10.14), (10.18), and (10.22). It is convenient to plot impedance data on graph paper with logarithmic impedance and frequency scales (usually referred to as log-log paper). The impedance curves of Fig. 10.9 are replotted on log-log paper in Fig. 10.35. Curves for impedance of masses have a slope of $(+1)$;

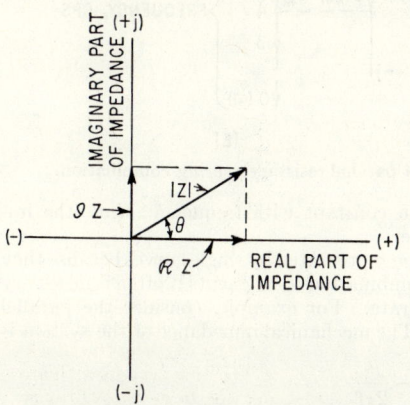

FIG. 10.34. The complex-plane representation of magnitude and angle of impedance, by real and imaginary parts.

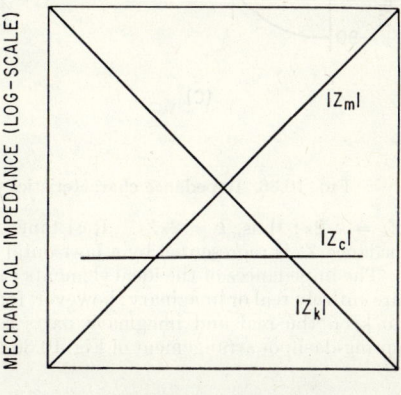

FIG. 10.35. The magnitude of the impedance as a function of frequency of ideal resistance, spring, and mass plotted to a log-log scale.

they cross the line $f = 1$ cps at $Z_m = 2\pi m$. Thus $m = Z_m/2\pi$. The value of m for an unknown mass thus can be obtained from the impedance curve. The curves for impedance of ideal springs have slopes of (-1) and they intercept the line $f = 1$ cps at

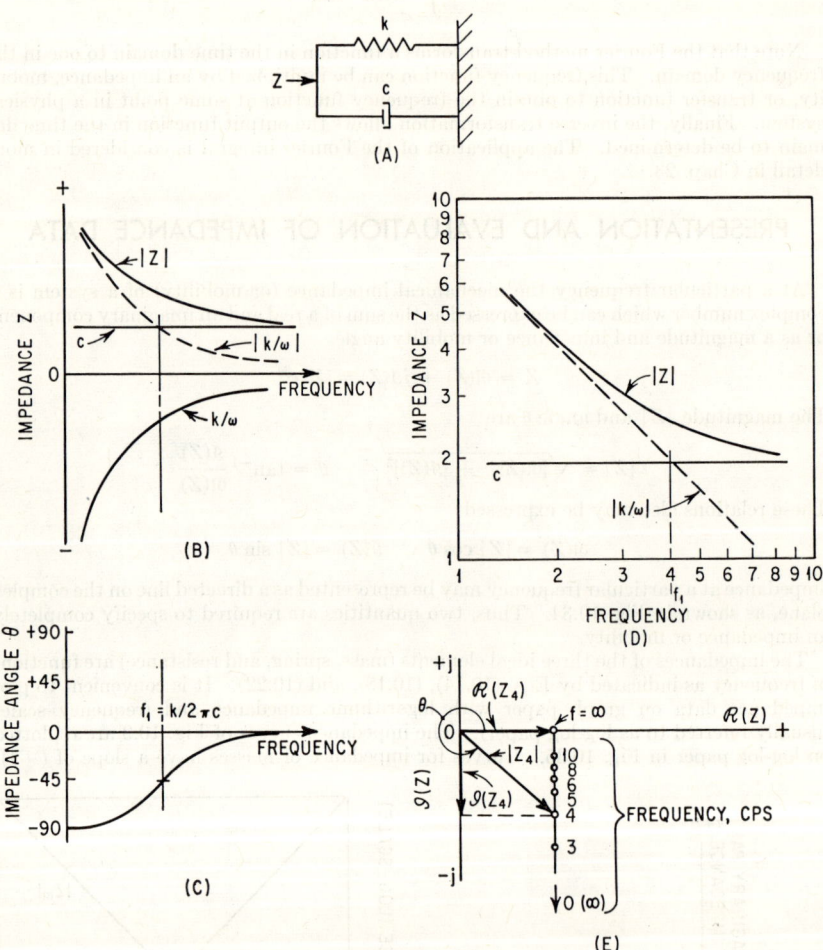

FIG. 10.36. Impedance characteristics of a parallel resistance-spring combination.

$Z_k = k/2\pi$; thus, $k = 2\pi Z_k$. Resistance is a constant with frequency; thus, the impedance Z_c is represented by a horizontal line.

The impedances of the ideal elements can be represented by single curves because they are entirely real or imaginary; however, for combinations of elements it often is necessary to keep the real and imaginary parts separate. For example, consider the parallel spring-dashpot arrangement of Fig. 10.36A. The mechanical impedance of the system is

$$Z = c - \frac{jk}{2\pi f}$$

Expressed as magnitude and angle:

$$|Z| = \sqrt{c^2 + \left(\frac{k}{2\pi f}\right)^2} \qquad \theta = \tan^{-1}\left(-\frac{k}{2\pi f c}\right)$$

where the real part $\mathcal{R}(Z) = c$ and the imaginary part $\mathcal{I}(Z) = -k/2\pi f$ are plotted in Fig. 10.36B to linear coordinates. The magnitude $|Z|$ is plotted on the same set of coordinates. The imaginary part $\mathcal{I}(Z)$ is shown dotted in the positive impedance region to indicate that $|Z|$ approaches the imaginary term at low frequencies and the real term at high frequencies. The angle of the impedance is plotted in Fig. 10.36C. The curves shown in Fig. 10.36B to linear coordinates are shown in Fig. 10.36D to logarithmic coordinates.

An alternate method for presenting impedance data is shown in Fig. 10.36E. All four factors required to specify impedance are presented on one curve. At each frequency the real and imaginary parts of impedance are plotted as a point on the complex plane, corresponding to Fig. 10.34. These points are connected to form the impedance curve. As shown in Fig. 10.36E for a frequency of 4 cps, the coordinates of the point are the real and imaginary parts of Z. The length of the radial line from the origin to the point is then $|Z|$, and the angle that this line makes with the (+) real axis is the angle of impedance. Frequency appears as a parameter in this presentation.

As shown in Fig. 10.36E, the impedance becomes real at high frequencies since $|Z|$

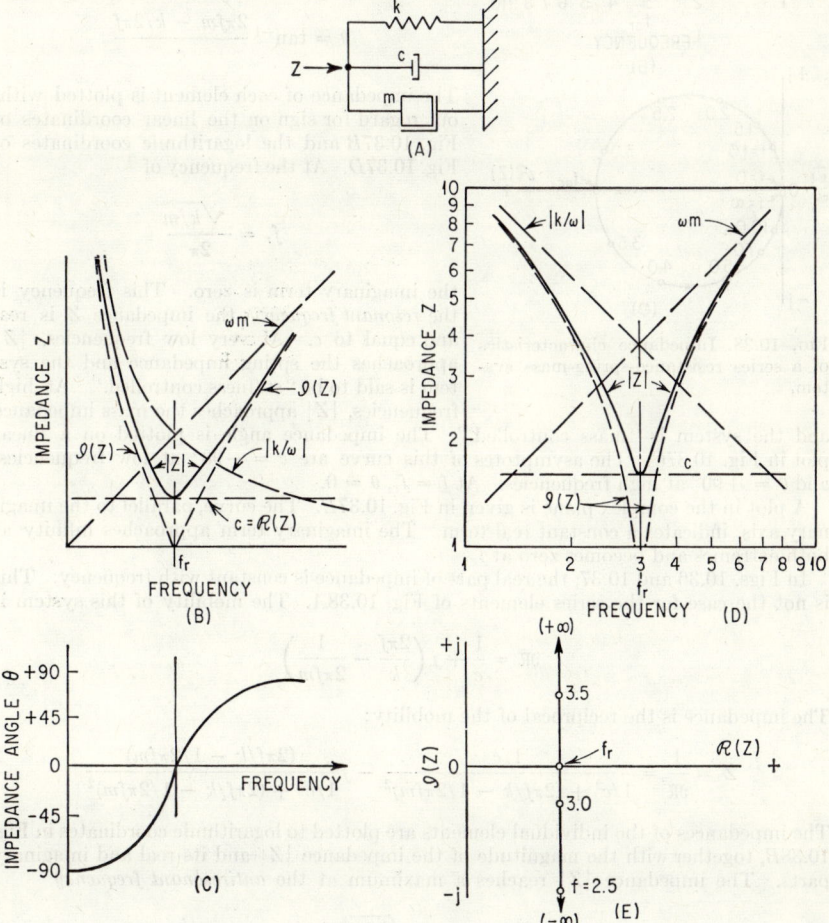

FIG. 10.37. Impedance characteristics of a parallel resistance-spring-mass system.

approaches c and θ approaches 360° (0°). At low frequencies $|Z|$ becomes large, θ approaches 270° (−90°), and the negative imaginary impedance indicates springlike action. The asymptotes of $|Z|$ indicate the characteristics of the system at frequency extremes.

The addition of a mass to the parallel system of Fig. 10.36A gives the system of Fig. 10.37A. The impedance at the driving point is

$$Z = c + j\left(2\pi f m - \frac{k}{2\pi f}\right)$$

The magnitude and angle are

$$|Z| = \sqrt{c^2 + \left(2\pi f m - \frac{k}{2\pi f}\right)^2}$$

$$\theta = \tan^{-1}\frac{2\pi f m - k/2\pi f}{c}$$

The impedance of each element is plotted without regard for sign on the linear coordinates of Fig. 10.37B and the logarithmic coordinates of Fig. 10.37D. At the frequency of

$$f_r = \frac{\sqrt{k/m}}{2\pi}$$

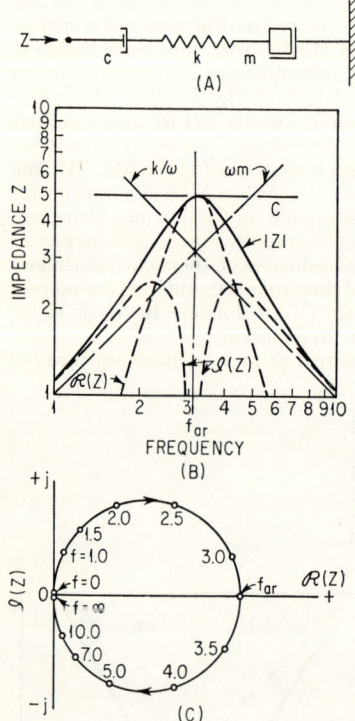

Fig. 10.38. Impedance characteristics of a series resistance-spring-mass system.

the imaginary term is zero. This frequency is the *resonant frequency*; the impedance Z is real and equal to c. At very low frequencies, $|Z|$ approaches the spring impedance and the system is said to be "stiffness controlled." At high frequencies, $|Z|$ approaches the mass impedance and the system is "mass controlled." The impedance angle is plotted on a linear plot in Fig. 10.37C. The asymptotes of this curve are $\theta = -90°$ at low frequencies, and $\theta = +90°$ at high frequencies. At $f = f_r$, $\theta = 0$.

A plot in the complex plane is given in Fig. 10.37E. The curve, parallel to the imaginary axis, indicates a constant real term. The imaginary term approaches infinity at both extremes and becomes zero at f_r.

In Figs. 10.36 and 10.37, the real part of impedance is constant with frequency. This is not the case for the series elements of Fig. 10.38A. The mobility of this system is

$$\mathfrak{M} = \frac{1}{c} + j\left(\frac{2\pi f}{k} - \frac{1}{2\pi f m}\right)$$

The impedance is the reciprocal of the mobility:

$$Z = \frac{1}{\mathfrak{M}} = \frac{1/c}{1/c^2 + (2\pi f/k - 1/2\pi f m)^2} - j\frac{(2\pi f/k - 1/2\pi f m)}{1/c^2 + (2\pi f/k - 1/2\pi f m)^2}$$

The impedances of the individual elements are plotted to logarithmic coordinates in Fig. 10.38B, together with the magnitude of the impedance $|Z|$ and its real and imaginary parts. The impedance $|Z|$ reaches a maximum at the *antiresonant frequency*

$$f_{ar} = \frac{\sqrt{k/m}}{2\pi}$$

When the system is driven by a sinusoidal excitation of constant-force amplitude, the velocity amplitude at the driving point is at a minimum at the antiresonant frequency.

The complex plane representation is shown in Fig. 10.38C. At low frequencies the impedance is small, the imaginary term is positive, and the angle of the impedance is nearly $+90°$. Masslike action is indicated. At high frequencies, the impedance is small but the imaginary term is negative and the angle is $-90°$. The system is then "stiffness controlled." At intermediate frequencies the locus of the impedance graph is a circle having a diameter c with its center on the $+$ real axis. At f_{ar} the impedance is real and equal to c.

Note that the mobility of the series-connected elements of Fig. 10.38A is given by an expression of the same form as that for the impedance of the parallel-connected elements of Fig. 10.37A. Thus, when mobility is plotted in the same way as impedance, the shapes of the resulting curves for the series elements (Fig. 10.39) are the same as for the impedance of the parallel elements shown in Fig. 10.37.

The mobility of the parallel elements of Fig. 10.37A takes the same form as the impedance of the series elements of Fig. 10.38A. Conversely, the mobility curves for the

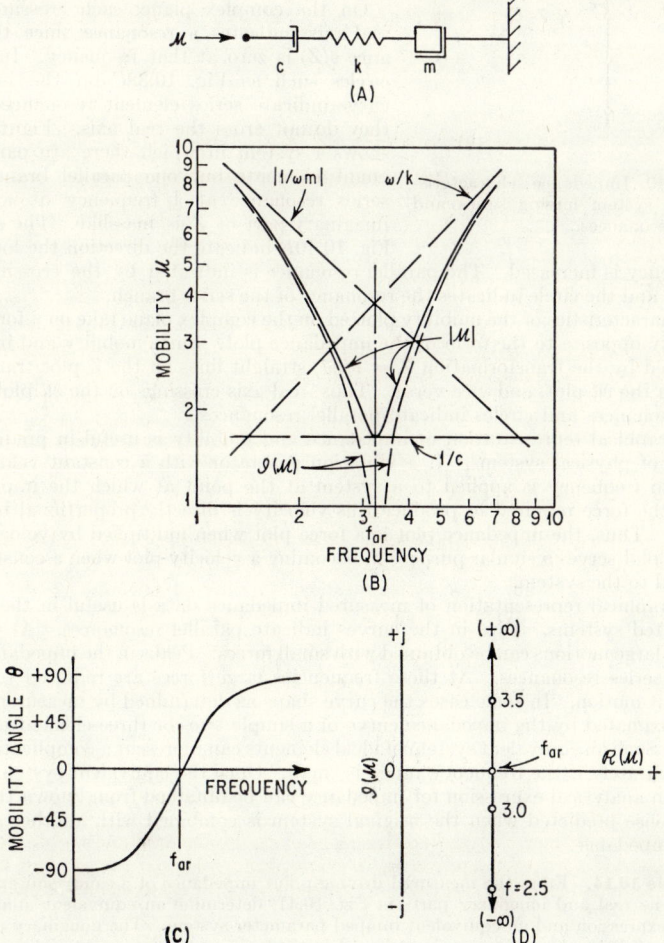

FIG. 10.39. Mobility characteristics of a series resistance-spring-mass system.

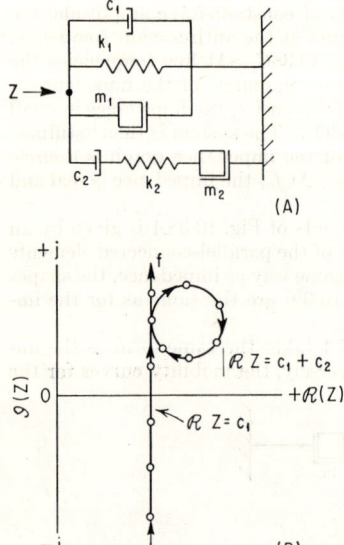

Fig. 10.40. Impedance characteristics of a system having series and parallel resonances.

parallel-connected elements have the same form as the impedance curves for the series-connected elements of Fig. 10.38.

Complicated mechanical structures may have many resonances at different frequencies. If the impedance is measured and plotted on log-log paper, results similar to those presented in Figs. 10.37 and 10.39 are useful in analyzing the system qualitatively. Positive peaks on the impedance curve indicate the resonance of series elements, and negative peaks indicate the resonance of parallel elements. Frequency regions where the impedance curve has a constant positive slope of $+1$ indicate masslike action with a positive impedance angle. Similarly, regions of the curve with a slope of -1 indicate springlike action with a negative impedance angle.

On the complex plane, each crossing of the real axis indicates a resonance since the reactance $\mathcal{J}(Z)$ is zero at that frequency. In general, circles such as Fig. 10.38C on the impedance curve indicate series element resonances even if they do not cross the real axis. Figure 10.40A shows a system in which there are parallel resonant elements but one parallel branch has a series resonance at a frequency at which the imaginary part of Z is masslike. The arrows in Fig. 10.40B indicate the direction the locus takes as frequency is increased. The parallel resonance is indicated by the crossing of the real axis, and the circle indicates the resonance of the series branch.

The characteristics of the mobility plotted on the complex plane take on a form that is essentially opposite to the form of the impedance plot. Since mobility and impedance are related by the transformation $Z = 1/\mathfrak{M}$, straight lines on the Z plot transform to circles on the $\mathfrak{M}$ plot, and vice versa. Thus, real axis crossings on the $\mathfrak{M}$ plot indicate series resonances and circles indicate parallel resonances.

The graphical representation of impedance and mobility is useful in predicting the response of physical systems. If a vibration generator with a constant velocity with respect to frequency is applied to a system at the point at which the impedance is known, the force required to produce this velocity is directly proportional to the impedance. Thus, the impedance plot is a force plot when multiplied by velocity. The mobility plot serves a similar purpose in becoming a velocity plot when a constant force is applied to the system.

The graphical representation of measured impedance data is useful in the study of complicated systems. Dips in the curves indicate parallel resonances. At these frequencies large motions can be obtained with small forces. Peaks in the impedance curves indicate series resonances. At these frequencies large forces are required to produce significant motion. In some cases the curve shape as determined by measurement may be approximated by the impedance curve of a simple two- or three-element equivalent system. Such an equivalent system of ideal elements can represent a complicated system quite accurately if the frequency range is small. Once the equivalent system is determined, an analytical expression for impedance can be obtained from known theory and the response predicted when the original system is combined with another system of known impedance.

Example 10.14. From the measured driving-point impedance of a sandy soil expressed in terms of its real and imaginary parts in Fig. 10.41, determine an equivalent analytical impedance expression and an equivalent lumped parameter system. The imaginary part of the impedance Z has a slope of approximately -1 in the low-frequency range; the dip in the curve indicates a parallel resonance at $f_r = 700$ cps. A finite resistance indicates the parallel system of a mass, spring, and resistance shown in Fig. 10.41. A line with slope -1 that best fits the experimental points in the low-frequency region represents the impedance of the equivalent spring. Noting $Z_k = 3.2 \times 10^6$ dyne-sec/cm at $f = 100$ cps and calculating k from Eq. (10.18):

$$k = 2\pi \times 100 \times 3.2 \times 10^6 = 2 \times 10^9 \text{ dynes/cm}$$

The measured data do not extend into a frequency region in which the slope is $+1$. However, at the resonant frequency f_r, the spring impedance equals the equivalent mass impedance; then

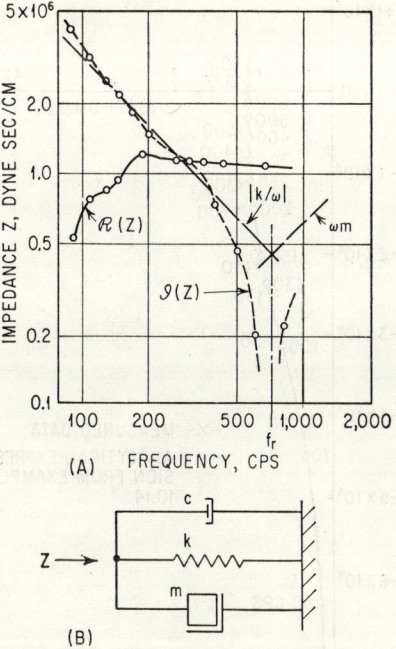

FIG. 10.41. The measured driving-point impedance characteristics of a sandy soil are shown in (A); the equivalent system representation as determined in Example 10.14 is shown in (B).

the mass can be found as follows:

$$m = \frac{k}{4\pi^2 f_r^2} = 100 \text{ gm}$$

The resistive term can be determined from the curve for $\mathcal{R}(Z)$. In this case the value of $\mathcal{R}(Z)$ is not constant so that an ideal resistive element does not represent this component. Note that $\mathcal{R}(Z)$ is approximately constant and equal to $c = 1.1 \times 10^6$ gm/sec from 200 to 800 cps. An equivalent system with a constant resistance term then represents the actual system from 200 to 800 cps.

The imaginary term for an analytical impedance expression is the sum of the mass and spring impedances. The curve for $\mathcal{R}(Z)$ can be approximated by an expression of the form

$$c = \frac{c_0}{1 + (f_0/f)^n}$$

where c_0 is the constant-resistance value given above, $c_0 = 1.1 \times 10^6$ dyne-sec/cm. The frequency f_0 is the frequency at which $c = c_0/2$; thus, $f_0 = 90$ cps. The exponent n is a measure of the rate that c increases; it is obtained from the physical slope of the curve for $\mathcal{R}(Z)$ on the log-log plot in the frequency range less than 200 cps. Here $n = 3$. The final expression for mechanical impedance is

$$Z = \frac{1.1 \times 10^6}{1 + (90/f)^3} + j(100\omega - 2 \times 10^9/\omega)$$

This expression is plotted on the complex plane in Fig. 10.42 with the original experimental data to indicate the accuracy of representation by the equivalent system.

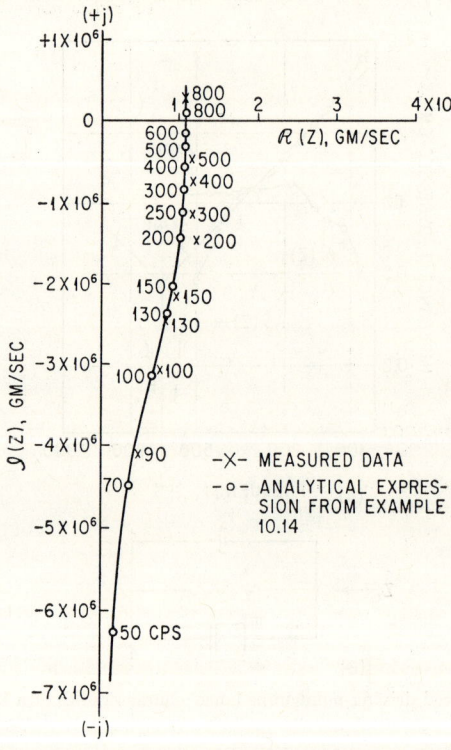

Fig. 10.42. Comparison of the analytical impedance representation of a sandy soil with the measured impedance.

Figure 10.43A shows a plot in the complex plane of the driving-point impedance of clay soil. This soil can be represented by the equivalent system shown in Fig. 10.43B. The resistance c_1 has the same form as that shown in Fig. 10.42 and the two series branches resonate, one at a frequency of approximately 130 cps and the other at a frequency of approximately 400 cps. Since the two series elements are springlike above resonance, they combine with k_1 to resonate with m_1 at about 1,000 cps.

The asymptotic lines that are approached by the log impedance and log mobility vs. log-frequency curves of Figs. 10.36 to 10.39 are useful in synthesizing impedance and mobility curves for physical systems.[17] In Example 10.14 these asymptotic lines are used to determine an equivalent system. When the normal mode concept is used, the total mobility is the sum of the mobilities presented by each mode. In a distributed system, such as a free-free beam, these modes consist of the several possible antiresonances of the beam. At each mode, the frequency of antiresonance and an effective mass may be determined[18] from which a mobility curve can be synthesized. The complete curve then can be obtained by adding all the curves for each mode. In most cases, at frequencies far removed from the resonance, the mobility is small. Thus, in the region of a particular resonance only one or two adjacent resonances need be considered. The concept of normal modes is equally applicable to lumped parameter systems, such as the equivalent system of Fig. 10.43. The two series branches can be considered as normal modes of a composite system, and the above method used to form the approximate mobility curve.

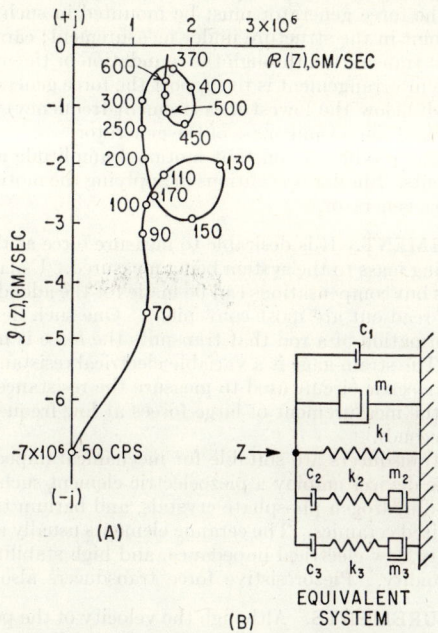

Fig. 10.43. Measured driving-point impedance of a clay soil is shown in (A); the equivalent system representation is shown in (B).

THE MEASUREMENT OF IMPEDANCE AND MOBILITY

Possibly the greatest advantage to the impedance method of solving dynamic problems is that systems may be analyzed from the measured impedance at a point; a complete analysis including every part in the system is not required. In the above section on analysis, most examples involve equivalent systems determined from driving point impedance or mobility and transfer impedance, or from force and velocity ratios. These factors may be calculated if a system is specified completely, but in many cases systems are so complicated that a measurement of the factors is obtained more easily.

Impedance measurement methods may be divided into two types, direct and indirect. In the direct method, a sinusoidal force is applied to the point at which the impedance is desired, and the magnitude of the force and resulting velocity is measured. The phase angle between force and velocity also is required; then impedance is the ratio of force to velocity, when they are considered as phasors. Mobility is the ratio of velocity to force. In the indirect method a sinusoidal force is applied through some intermediate device. The characteristics of the intermediate device are measured without and with the system attached, and impedance is calculated from the change in the characteristic of the device.

DIRECT METHODS

FORCE GENERATION. Many devices are available for the generation of vibration forces (see Chap. 25). The basic requirement is that the force must be sinusoidal since impedance is defined only for sinusoidal forces and velocities. The excitation frequency should be variable readily over the range of interest. The magnitude of force need be only large enough to produce a measurable motion; even for large structures such as ship hulls,[19] a peak force of 10 lb usually is sufficient. The force magnitude should be variable, and the element that applies the force should be fixed rigidly to the structure to be measured. This is particularly important at frequencies above a few hundred cycles per second. The force generator must be mounted in such a way that force is applied only at one point in the structure under measurement; care should be taken to ensure that force is not transmitted through the foundation or the mounting of the force generator. A convenient arrangement is to support the force generator on weak springs (natural frequency well below the lowest measurement frequency) and apply the force between the structure and the seismic mass of the generator.

Vibration sources that produce a constant motional amplitude are equally useful in impedance measurements. Similar precautions in applying the motion must be observed as in the constant-force generator.

FORCE MEASUREMENT. It is desirable to measure force at the point of force application without adding mass to the system being measured. Usual methods of attachment do not allow this but compensations can be made for the added mass. Force transducers with electrical read-out are most convenient. One such device is a load cell in which the elastic deformation of a rod that transmits the force is measured by a strain gage (see Chap. 17). The strain gage is a variable electrical resistance element that produces a voltage in an electric circuit used to measure the resistance. Load cells of this type are suitable for the measurement of large forces at low frequencies (below several thousand cycles per second).

Piezoelectric force transducers are suitable for mechanical impedance measurements (see Chap. 16). These devices employ a piezoelectric element such as quartz, Rochelle salts and ammonium dihydrogen phosphate crystals, and barium titanate and lead zirconium titanate polarized ceramics. The ceramic elements usually are preferred because of their high sensitivity, low electrical impedance, and high stability during changes in temperature and humidity. Piezoresistive force transducers also may be employed.

VELOCITY MEASUREMENTS. Although the velocity of the point of force application is required for impedance calculations, acceleration or displacement may be measured and the velocity determined from the relationships of Eqs. (10.5) and (10.6). Many motion transducers with electrical output voltages proportional to displacement, velocity, and acceleration are available (see Chap. 12). The desirable characteristics of such a device are that it add no mass or spurious resonances to the system being measured and that it measure the motion at the point of force application. Allowance can be made for added mass but undesired resonances are less predictable. For measurements on rigid structures at low frequencies, the motion may be measured at some distance from the point of force application; at high frequencies (above several hundred cycles per second), very few structures can be considered rigid and the motion must be measured as close to the point of force application as possible.

ELECTRICAL MEASUREMENTS. The magnitudes of electrical voltages from the force and motion transducers, and the phase angles between the voltages, must be determined. From these quantities and the calibration factors of the transducers, impedance may be calculated. The frequency range encountered may be from a few cycles per second to about 10,000 cps. Voltages may range from a few microvolts to several volts, and the phase angle between force and velocity voltages may vary from $-90°$ to $+90°$ for the driving-point impedance measurements of passive systems. Phase angles for transfer impedances, mobilities, or force and velocity ratio measurements may lie in any quadrant.

In many cases the electrical impedance of the transducer may be so high that cable capacitance and stray pickup may prevent accurate measurements. Then impedance transformation and amplification may be required near the transducer.

Electronic voltmeters are most convenient for measurement of the voltages from the transducers used for impedance measurements. Voltages may be measured with errors less than 2 per cent with such devices; the magnitudes are given in terms of the effective value of a sine wave. Sine waves are employed in impedance measurements, and peak voltages ($\sqrt{2}$ times the effective value) should be used. However, since ratios of voltages are used to calculate impedance, any consistent measuring system is satisfactory.

Several commercially available electronic phase meters and a specially adapted method [20] are available with which accuracies of 1° can be attained over the required frequency range. In most cases angles from 0 to 360° are read directly from a meter. Phase-angle measurements may be made with less accuracy by the use of a laboratory oscilloscope. If the force and motion voltages are applied to the vertical and horizontal deflection channels, an elliptical pattern such as that of Fig. 10.44 is obtained. If the ellipse is centered, the phase angle ϕ between the voltages is given by $\sin \phi = a/b$.

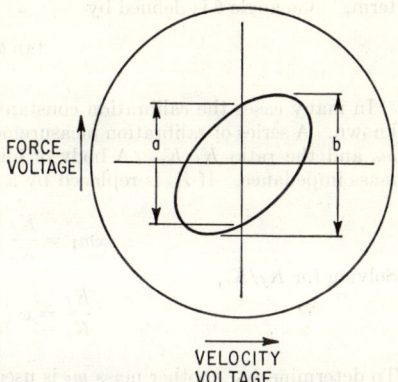

Fig. 10.44. Oscilloscope method for determining the phase angle between force and velocity voltages.

When displacement or acceleration is measured, the angle between the voltages is not the angle between force and velocity. For displacement measurements with force as the phase reference, velocity is

$$v = j\omega x = \omega x_0 e^{j\psi + 90°}$$

where ψ is the angle between the force and displacement voltages. The angle θ between force and velocity then is $\theta = \psi + 90°$.

For acceleration measurements, velocity is

$$v = \frac{\ddot{x}}{j\omega} = \frac{\ddot{x} e^{j\phi - 90°}}{\omega}$$

where the angle θ between force and velocity is

$$\theta = \phi - 90°$$

CALIBRATION. An impedance-measuring device is shown schematically in Fig. 10.45. When the calibration constants for the transducers and the mass added by the transducers are known, the unknown impedance Z_x may be determined as follows:

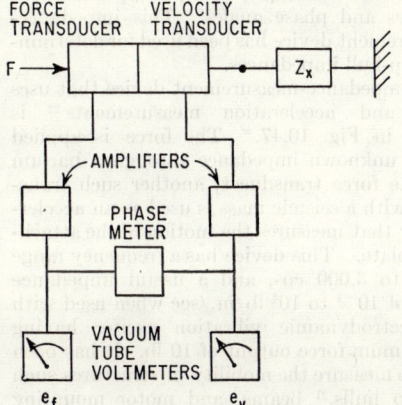

Fig. 10.45. Arrangement of equipment for the measurement of mechanical impedance or mobility.

$$Z_x = \frac{K_f |e_f|}{K_v |e_v|} e^{j\beta} - j\omega m_o$$

$$= \frac{K_f |e_f|}{K_v |e_v|} \cos \beta + j\left(\frac{K_f |e_f|}{K_v |e_v|} \sin \beta - \omega m_o\right)$$

(10.64)

where K_f is in lb/volt and K_v is in in./sec/volt in the English system of units. The voltages e_f and e_v are those produced by the force and velocity transducers, β is the phase angle by which e_v lags e_f, and m_o is the added mass. Impedance then is given in lb/in./sec. Any other consistent system of units, such as those in Table 10.1, may be used. The angle β between force and velocity voltages is not the impedance angle θ because of the $j\omega m_o$

term. The angle θ is defined by

$$\tan\theta = \frac{\mathcal{I}(Z_x)}{\mathcal{R}(Z_x)}$$

In many cases the calibration constants K_f, K_v, and the added mass m_0 may not be known. A series of calibration measurements on known impedances may be used to find m_0 and the ratio K_f/K_v. A body of high-density material represents nearly an ideal mass impedance. If Z_x is replaced by a mass m_1, Eq. (10.64) becomes

$$j\omega m_1 = \frac{K_f}{K_v}\frac{|e_{f1}|}{|e_{v1}|} e^{j90°} - j\omega m_0$$

Solving for K_f/K_v,

$$\frac{K_f}{K_v} = \omega \frac{|e_{v1}|}{|e_{f1}|}(m_1 + m_0)$$

To determine m_0, another mass m_2 is used and the voltages determined again. Then

$$\frac{K_f}{K_v} = \omega \frac{|e_{v2}|}{|e_{f2}|}(m_2 + m_0)$$

The added mass m_0 is determined from the last two expressions for K_f/K_v:

$$m_0 = \frac{\dfrac{e_{v2}}{e_{f2}}m_2 - \dfrac{e_{v1}}{e_{f1}}m_1}{\dfrac{e_{v1}}{e_{f1}} - \dfrac{e_{v2}}{e_{f2}}}$$

where the voltages e_v, e_f are the magnitudes read from a voltmeter. Substituting the calculated value of m_0 above, the ratio K_f/K_v may be determined.

IMPEDANCE-MEASURING DEVICES. A system that applies the force F through a thin-walled aluminum tube having a strain gage affixed thereto, and measures velocity adjacent to the point of force application [21] is shown schematically in Fig. 10.46. The force is produced by an electrodynamic vibration exciter. Because of the difficulty of obtaining good sinusoidal waveforms at low frequencies, the frequency range of this device is about 7.5 to 2,500 cps; the impedance is calculated from Eq. (10.64), using measured voltages and phase angles. This impedance-measurement device has been used for determining ship-hull impedances.

An impedance-measurement device that uses force and acceleration measurements [22] is shown in Fig. 10.47.* The force is applied to the unknown impedance through a barium titanate force transducer; another such transducer with a seismic mass is used as an accelerometer that measures the motion of the attachment plate. This device has a frequency range of 10 to 3,000 cps, and a useful impedance range of 10^{-1} to 10^4 lb/in./sec when used with an electrodynamic vibration exciter having a maximum force output of 10 lb. It has been used to measure the mobility of structures such as ship hulls,[21] beams, and motor mounting plates.[22]

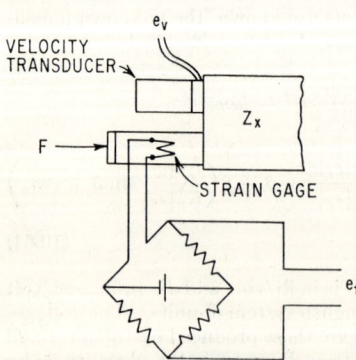

Fig. 10.46. Mechanical impedance-measurement equipment in which force and velocity are measured directly. (Coleman.[21])

* Manufacturers of devices of this type include Wilcoxon Research, Bethesda, Md., and Evdevco Corp., Pasadena, Calif.

When acceleration is measured, the expression for calculating impedance becomes

$$Z_x = \frac{K_f|e_f|}{K_a|e_a|/j\omega} e^{j\beta_a} - j\omega m_0$$

$$= -\omega \frac{K_f|e_f|}{K_a|e_a|} \sin \beta_a + j\omega \left(\frac{K_f|e_f|}{K_a|e_a|} \cos \beta_a - m_0 \right) \quad (10.65)$$

where e_a is the acceleration voltage, K_a is the accelerometer calibration constant in in./sec^2/volt (or any other system of units), and β_a is the phase angle by which e_a lags e_f.

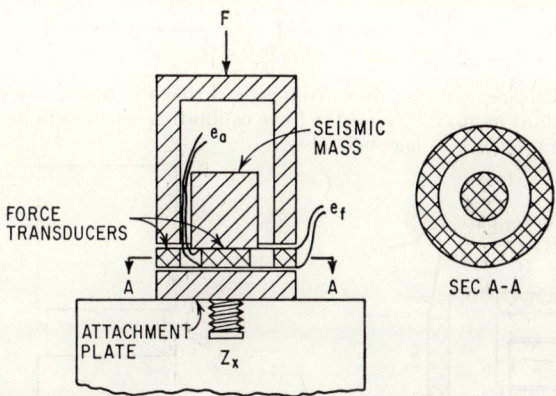

Fig. 10.47. Device for the measurement of mechanical impedance in which force and acceleration are measured. (Plunkett.[22])

A modification of the above method is shown in Fig. 10.48. This device was designed to measure the driving-point impedance of soils.[23,24] A small force is produced by passing a sinusoidal current through the geophone; this force is coupled to the soil through a ceramic force transducer and a flat plate which is 2.9 in. in diameter. In this case m_0 is the mass of the flat plate. Motion is measured by an accelerometer mounted on top of the force generator. For the motion of the soil surface to be measured accurately, the force transducer and geophone case must be quite rigid and the acceleration produced must be less than $1.0g$ since the device is not clamped to the soil. The useful frequency range is approximately 50 to 900 cps (although resonances in the force generator cause spurious responses at about 400 cps). A relatively simple analog computer for use with this device has been designed to solve Eq. (10.65).[25] A complex plane impedance plot is produced as frequency is varied.

An impedance device using displacement measurements for measuring the driving-point impedance of the human forehead and mastoid in hearing-aid calibrations is shown in Fig. 10.49.[26] The force is produced by a voltage applied to a ceramic cylinder force transducer. Force is measured by a ceramic plate force transducer and the motion of the force coupling element is measured by a capacity-type displacement gage. This gage operates on the same principle as the con-

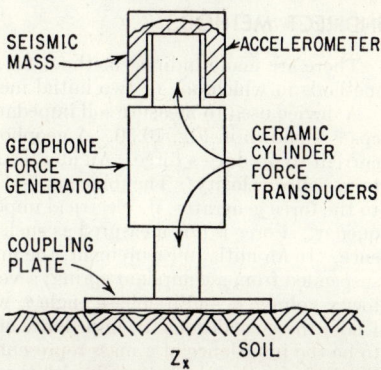

Fig. 10.48. Device for measuring the mechanical impedance of soils in which force and acceleration are measured. (Runyon and Anderson;[23] Hixson and Wittenborn.[24])

denser microphone. The motion causes the distance between the electrodes to change. This causes a change in capacity which produces a change in an applied polarizing voltage; the voltage change is proportional to displacement. For accurate motion measurements the force transducer and coupling element must be rigid. This impedance device is used to obtain measurements of impedance from 40 to 10,000 cps.

Impedance may be calculated from displacement measurements as follows:

$$Z_x = \frac{K_f|e_f|}{j\omega K_x|e_x|} e^{j\beta_d} - j\omega m_o$$

$$= \frac{K_f|e_f|}{\omega K_x|e_x|} \sin\beta_d - j\left(\frac{K_f|e_f|}{\omega K_x|e_x|}\cos\beta_d + \omega m_o\right) \qquad (10.66)$$

where e_x is the displacement transducer voltage, e_f is the force transducer voltage, K_x the calibration constant in in./volt, K_f is the force calibration constant in lb/volt, and β_d is the phase angle by which e_x lags behind e_f.

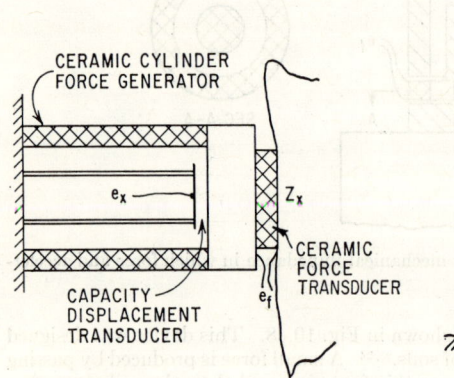

FIG. 10.49. Device for measuring the mechanical impedance of the human forehead and mastoid in which force and displacement are measured. (*Corliss and Koidan*.[26])

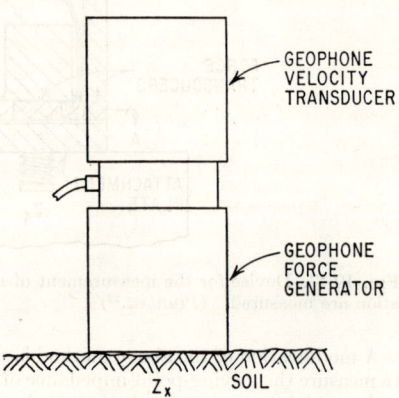

FIG. 10.50. Device using an indirect method for the measurement of the impedance of soils. (*Washburn and Wiley*.[27])

INDIRECT METHODS

There are many indirect methods for measuring impedance. All employ substitution methods in which one or two initial measurements are made on a known impedance.

A device used to measure soil impedance over a frequency range from about 50 to 200 cps [27] is shown in Fig. 10.50. A geophone velocity transducer is driven by a sinusoidal current to produce a force. An identical geophone coupled to the force generator is used to measure velocity. The force applied to the soil is proportional to the voltage applied to the force generator, its electrical impedance, the total mass of the device, and the frequency. Force is not measured as such but the voltage applied is used as a phase reference. In an initial measurement, the impedance device (force and velocity geophone) is suspended from a compliant spring; a voltage is applied to the force generator, and a velocity voltage e_c and its phase angle θ_c with respect to the driving voltage are measured. Under these conditions the impedance of the device itself is measured. This is presumed to be the impedance of a mass representing the weight of the impedance device. When the device is placed on the soil, with the same driving voltage and frequency applied, the soil impedance is

$$Z_x = j\omega m \left(\frac{|e_c|}{|e_x|} e^{j(\theta_c - \theta_x)} - 1\right) \qquad (10.67)$$

where m is the mass of the impedance device and e_x and θ_x are the velocity voltage and phase angle read when the device is placed on the soil. This measurement method employs a simple, easy-to-construct mechanical procedure but the impedance calculation is quite tedious.

This type of substitution method has been used to measure the impedance of floors, walls, and ceilings for room acoustic studies.[28] An electromagnetic force generator was used and acceleration was measured.

A device with limited usefulness is shown in Fig. 10.51.[29] The force necessary for impedance measurements is produced by rotating eccentric masses driven by an electric motor. The masses are counterrotating and are phased to produce a force acting along a vertical axis.

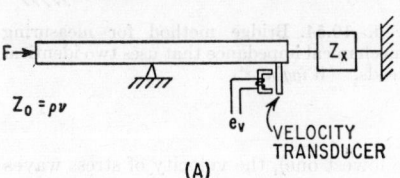

FIG. 10.51. An eccentric-mass force generator for use in determining mechanical impedance. (*Korn and Kirschner*.[29])

The phase angle between force and velocity is determined by using the velocity voltage to control a stroboscopic light. This device is suspended from a compliant spring, and a velocity voltage and phase position are noted. When applied to a structure, the impedance is given by Eq. (10.67).

A variation of the above method has been used to measure the impedance of the human mastoid.[30,31] An inertia-reaction drive unit as used in bone-conduction hearing-aid receivers is used as a force generator. Motion is measured by an accelerometer. The device is suspended from a compliant spring, a voltage e_0 is applied to the force generator, and the accelerometer voltage and its phase angle with respect to e_0 are noted. The free motion thus is measured. Under these same conditions a known impedance Z_k is added, and a voltage Δe_k is added to e_0. The voltage Δe_k is adjusted in magnitude and phase until the original accelerometer voltage and phase angle are obtained. Then Z_k is removed, and the device is applied to an unknown impedance Z_x. Again a voltage Δe_x is added to e_0 in the proper magnitude and phase to produce the original free motion. For a mass m_k as the known impedance, the unknown impedance Z_x is

$$Z_x = j\omega m_k \frac{|\Delta e_x|}{|\Delta e_k|} e^{j(\theta_x - \theta_k)}$$

Since the original free motion of the impedance device must be imparted to the unknown impedance, the maximum value of impedance measured is determined by the minimum measurable free motion and the maximum available force. The accuracy depends on the force generator being truly linear, and on the accuracy with which voltages Δe_x, Δe_k and phase angles θ_x, θ_k can be measured. With elaborate voltage and phase measuring equipment, an error of less than 1 per cent from 400 to 4,000 cps can be obtained.

A substitution method using a resonant rod has been used to measure the mechanical impedance of the skin of the human arm.[32] As shown in Fig. 10.52A, a vibratory force

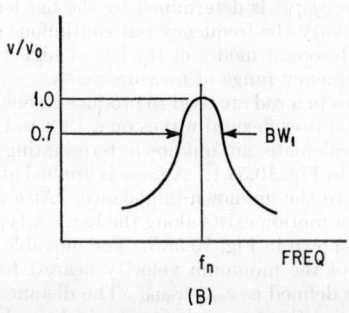

FIG. 10.52. Method for determining impedance that uses the effect of the unknown impedance Z_x on the resonance characteristics of a metal bar. (*von Gierke*.[32])

of constant amplitude is applied to one end of the bar, and the motion is measured with a variable reluctance velocity transducer. With the bar freely suspended, a resonance curve such as that shown in Fig. 10.52B is determined. The resonant frequency f_n is determined, and the frequency bandwidth between the points where the velocity is 0.7 of the maximum is noted as bw_1. When an unknown impedance is connected to the free end, a new resonance curve is measured. As a result of the load, the frequency of resonance shifts by an amount Δf_0 determined by $\mathcal{I}(Z_x)$; the bandwidth is bw_2, a function of $\mathcal{R}(Z_x)$. The impedance Z_x then is

$$Z_x = \frac{Z_0 \pi l}{v}[(bw_2 - bw_1) + j2\,\Delta f_0]$$

where Z_0 is the characteristic impedance and l is the length of the bar. From a measurement with a known mass m_0 on the free end, $Z_0 = 2f_0 m_0/\Delta f_0$. From the initial free

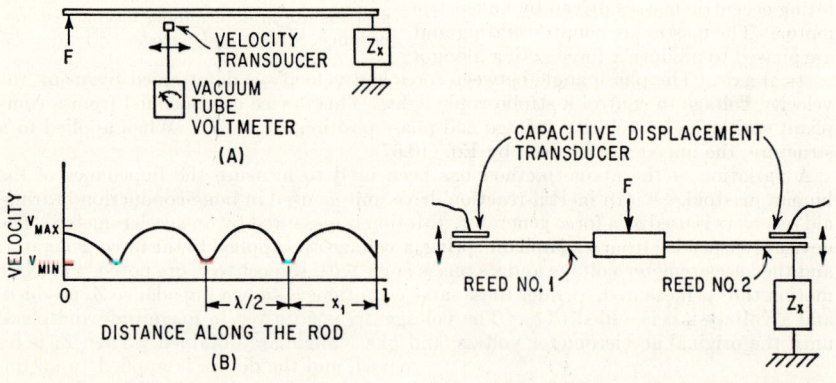

Fig. 10.53. Method for determining impedance that uses the effect of the unknown impedance Z_x on the standing wave of velocity in a bar excited in a bending mode. (*von Vogel*.[33])

Fig. 10.54. Bridge method for measuring mechanical impedance that uses two identical reeds. (*Wiggins*.[34])

measurement (if the resonance measured is the lowest one), the velocity of stress waves in the bar is $v = 4f_0 l$.

This method is useful at high frequencies. It has been used from 1,400 to 18,000 cps. The impedance-measurement frequency is $f_0 + \Delta f_0$; it is determined by the bar length and $\mathcal{I}(Z)$. Several bar lengths may be used to vary the frequency but continuous data over a frequency range cannot be obtained. Resonant modes of the bar at higher frequencies may be used to extend the upper frequency range of measurements.

In the above method, longitudinal stress waves in a rod are used to produce resonances for impedance measurements. A similar method uses flexural waves on a thin rod and more general transmission line methods to calculate an unknown terminating impedance.[33] The physical arrangement is shown in Fig. 10.53A. A force is applied at one end of the bar and the other end is connected to the unknown impedance. At a fixed frequency of force excitation, a standing wave of motion exists along the bar. A typical plot of the magnitude of velocity of the bar is shown in Fig. 10.53B. The movable motion detector is used to measure the position of the minimum velocity nearest to the unknown impedance; the *standing-wave ratio* is defined as $v_{\max}/v_{\min}$. The distance between adjacent minima also is needed to determine the wave velocity on the bar. From these quantities and the characteristic impedance of the bar, the value of Z_x may be calculated from transmission-line theory.

Measurements on sand and its effectiveness in damping vibration have been made over the frequency range of 100 to 2,000 cps with this method. Rotational motion is intro-

THE MEASUREMENT OF IMPEDANCE AND MOBILITY   10-57

duced, thereby requiring compensation to measure true impedance. The exact positions of the minima are difficult to determine, particularly when Z_x has about the same magnitude as Z_0 for the bar.

A final substitution method, usually known as a mechanical impedance bridge,[34] contains two identical reeds which are driven by a sinusoidal force as shown in Fig. 10.54. Their motions are measured by capacity displacement transducers. The motion is proportional to the force applied to each reed; when the unknown is not attached, both transducer voltages are equal. When the two voltages are subtracted, a zero or null is obtained. When the unknown is attached to reed 2, the motion of this reed becomes different and the difference voltage is no longer zero. The value of Z_x is then

$$Z_x = \frac{e_2 - e_1}{e_1}\left(Z_m - \frac{Z_m^2}{Z_k}\right)$$

where $e_2 - e_1$ is the difference voltage, e_1 is the voltage from reed 1, Z_m is the impedance of the effective mass of the reed, and Z_k is the impedance of the effective stiffness of the reed.

Accurate measurements can be made up to a frequency of about 0.7 times the resonant frequency of the reed. At frequencies well below resonance, the motion becomes very small so that accuracy is limited. To cover a wide frequency range, several reeds may be required. This method has been used to measure the mechanical impedance of phonograph pickups from 30 to 10,000 cps.

A single impedance device will not satisfy all measurement needs. The types of structures and the ranges in impedance magnitude are so great that it usually is necessary to construct a device for the particular application.

TYPICAL MEASURED IMPEDANCE AND MOBILITY CURVES

Typical soil impedance curves are presented in Figs. 10.41 and 10.43. Measured impedance and mobility curves for several structures including the human head are shown in Figs. 10.55 to 10.59.

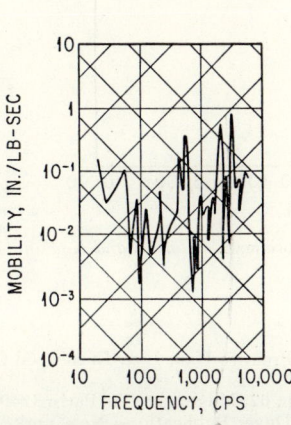

Fig. 10.55. Mobility measured near one end of a rectangular cast-iron motor base of dimensions 16¾ by 29½ by ½ in. with webbing reinforcing. (Plunkett.[22])

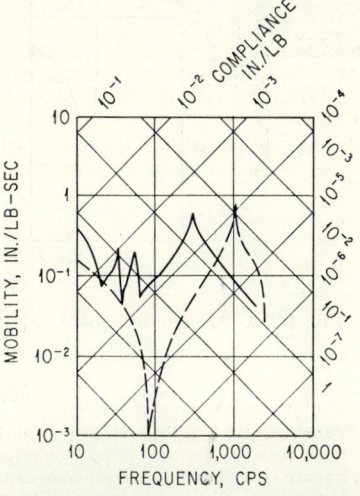

Fig. 10.56. Mobility of a bench and bench grinder. The mobility of the bench is shown in solid lines, and the mobility of the grinder in dotted lines. (Plunkett.[36])

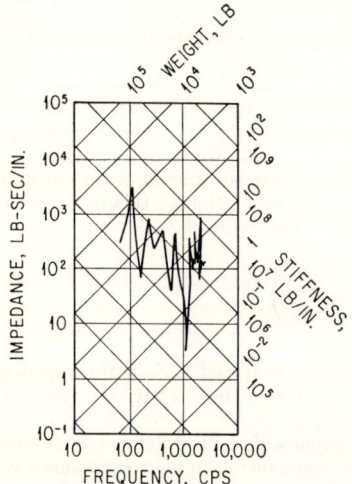
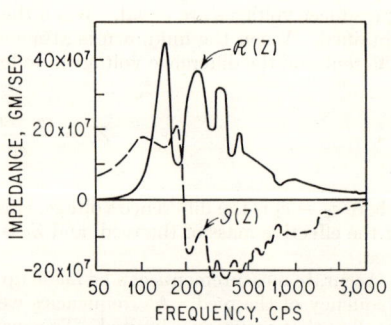

Fig. 10.57. Impedance measured on the keel of a submarine. (*Coleman*.[35])

Fig. 10.58. Mechanical impedance of a floor with linoleum covering. (*von Elling*.[28])

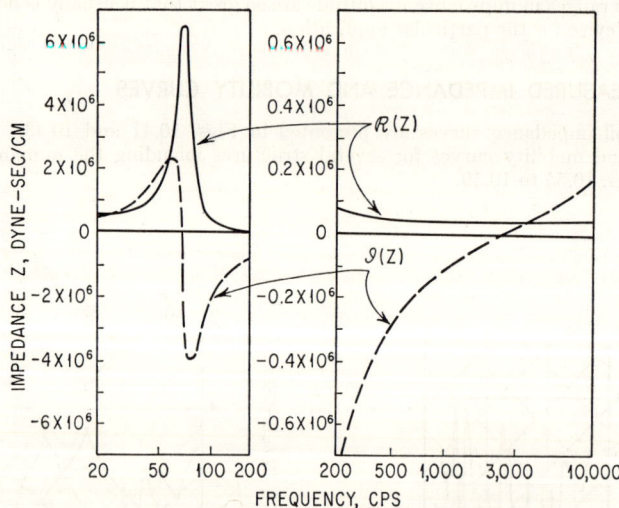

Fig. 10.59. Mechanical impedance of the human forehead. (*Corliss and Koidan*.[26])

REFERENCES

1. Thomson, W. T.: "Mechanical Vibrations," p. 70, Prentice-Hall, Inc., Englewood Cliffs, N.J., 1948.
2. Rocard, Y.: "Dynamique générale des vibrations," p. 62, Masson et Cie, Paris, 1949.
3. McLachlan, N. W.: "Theory of Vibrations," p. 27, Dover Publications, New York, 1941.
4. Firestone, F. A.: *J. Appl. Phys.*, **9**:373 (1938).
5. Freeberg, C. R., and E. N. Kemler: "Elements of Mechanical Vibrations," p. 179, John Wiley & Sons, Inc., New York, 1949.
6. Firestone, F. A.: *J. Acoust. Soc. Amer.*, **28**:1117 (1956).

REFERENCES

7. Biot, M. A.: *J. Aeronaut. Sci.*, **7**:376 (1940).
8. Von Kármán, T., and M. A. Biot: "Mathematical Methods in Engineering," McGraw-Hill Book Company, Inc., New York, 1940.
9. Duncan, W. J.: *British Aeronautical Research Committee R & M* 2000, 1947.
10. Bishop, R. E. D.: *J. Roy. Aeronaut. Soc.*, **58**:703(1954).
11. Harrison, M., A. O. Sykes, and P. G. Marcotti: *J. Acoust. Soc. Amer.*, **24**:384(1952).
12. Bancroft, D.: *Phys. Revs.*, **59**:588(1941).
13. Sykes, A. O.: "The Effects of Machine and Foundation Resilience and of Wave Propagation on the Isolation Provided by Vibration Mounts," SAE National Aeronautic Meeting, Los Angeles, Calif., Oct. 2, 1957.
14. Wright, D. V.: *Colloq. Mechanical Impedance Methods Mechanical Vibrations, ASME*, Dec. 2, 1958, pp. 19–42.
15. Chenea, P. F.: *J. Appl. Mechanics*, **20**:233 (1953).
16. Molloy, C. T.: *J. Acoust. Soc. Amer.*, **29**:842 (1957).
17. Plunkett, R.: *Proc. 2nd U.S. Natl. Congr. Appl. Mechanics*, 1954, p. 121.
18. Young, D., and R. P. Felgar: "Tables of Characteristic Function of a Beam," *Univ. Texas, Bur. Eng. Research*, no. 44, July, 1949.
19. Muster, D. F.: *Colloq. Mechanical Impedance Methods Mechanical Vibrations, ASME*, Dec. 2, 1958, p. 87.
20. Redfern, J. T.: "An Electrical Phase Shifter and Meter," U.S. Navy Electronics Laboratory, San Diego, Calif., *Rept.* 591, March 9, 1955.
21. Coleman, G. M.: *Colloq. Mechanical Impedance Methods Mechanical Vibration, ASME*, Dec. 2, 1958, pp. 69–75.
22. Plunkett, R.: ASME Paper 53-A-45, *J. Appl. Mechanics*, **21**:250 (1954).
23. Runyon, W. R., and R. E. Anderson: *J. Acoust. Soc. Amer.*, **28**:73 (1956).
24. Hixson, E. L., and A. F. Wittenborn: *Paper* 202, SAE National Aeronautical Meeting, Sept. 30–Oct. 5, 1957.
25. Hixson, E. L.: *Proc. 10th Southwest IRE Conf.*, April, 1958, p. 14.
26. Corliss, E. L. R., and W. Koidan: *J. Acoust. Soc. Amer.*, **27**:1164 (1955).
27. Washburn, H., and H. Wiley: *Geophysics*, **6**:116 (1941).
28. von Elling, W.: *Acustica*, **4**:396 (1954).
29. Korn, T. S., and F. Kirschner: *Acustica*, **4**:671 (1954).
30. Morton, J. Y.: *Acustica*, **4**:117 (1954).
31. Ayers, E. W., E. Aspinall, and J. Y. Morton: *Acustica*, **6**:11 (1956).
32. von Gierke, H. E.: "Measurement of the Acoustic Impedance and the Acoustic Absorption Coefficient of the Surface of the Human Body," *A.F. Tech. Rept.* 6010, March, 1950, U.S. Air Force, Air Material Command, Wright-Patterson AFB, Dayton, Ohio.
33. von Vogel, S.: *Acustica*, **6**:511 (1956).
34. Wiggins, A. M.: *J. Acoust. Soc. Amer.*, **15**:50 (1943).
35. Coleman, G. M.: "Evaluation of Hull Damping Effectiveness by the Impedance Method U.S.S. MACKEREL (SST-1)," *Confidential Rpt.*, U.S. Navy Electronics Laboratory, San Diego, Calif., TM-247, June 12, 1957.
36. Plunkett, R.: *Noise Control*, **4**:18 (1958).

11

STATISTICAL CONCEPTS IN VIBRATION

John W. Miles
University of California, Los Angeles

W. T. Thomson
University of California, Los Angeles

INTRODUCTION

A random vibration is one whose instantaneous value is not predictable. Such vibration is generated, for example, by the rocket engines of guided missiles and by aerodynamic turbulence and gusts. Figure 11.1A is an acceleration-time record of a vibration measured in the vicinity of a rocket engine. Clearly there are no well-defined periodicities, and the value of the acceleration at any instant of time is unrelated to that at any other instant and cannot be predicted.

Although the time variation of a random vibration cannot be predicted, the probability of a given value being within a certain range is predictable on a statistical basis. For example, consider a short sample of the above record. The acceleration $\ddot{x}$, at any instant of time t, varies in a random manner about a mean value $\bar{\ddot{x}}$. Suppose the vertical scale is divided into small divisions $\Delta\ddot{x}$. Now determine the statistical probability that the value will be within $\Delta\ddot{x}$. This can be done by measuring the time Δt_n during which the signal has amplitude values between $\ddot{x}$ and $\ddot{x} + \Delta\ddot{x}$ relative to the total length of time T over which the phenomenon is studied. Thus *amplitude probability* P in this example is given by

$$P(\ddot{x}, \ddot{x} + \Delta\ddot{x}) = \frac{\sum_i \Delta t_n}{T}$$

and the *amplitude density* p is given by

$$p(\ddot{x}) = \lim_{\Delta\ddot{x} \to 0} \frac{P(\ddot{x}, \ddot{x} + \Delta\ddot{x})}{\Delta\ddot{x}}$$

The probability of the instantaneous value being infinite is zero; the probability of its peak value being zero or infinite is zero; and the probability of any other value may be distributed as indicated in Fig. 11.1B.

This chapter deals with the problem of a linear, vibrating system subjected to a random excitation, including (1) a statistical description of the random input (*excitation*), (2) a description of the linear system in terms that permit the calculation of its response over the frequency spectrum of the input, and (3) a statistical description of the random output (*response*). Mathematical concepts are discussed first, followed by illustrative applications to mechanical systems characterized by one or more resonances and applica-

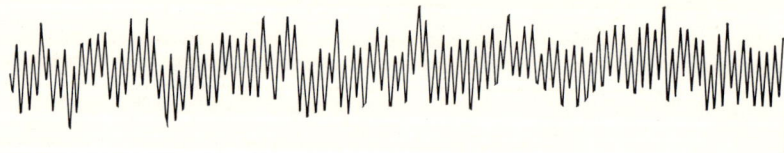

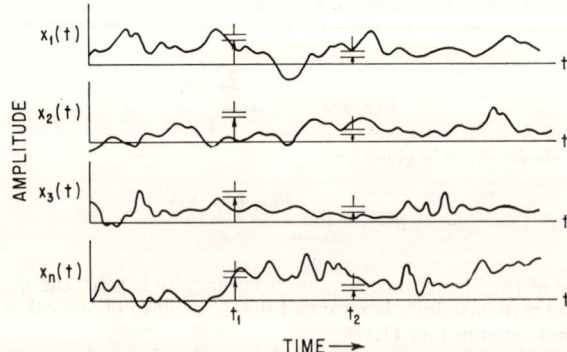

Fig. 11.1. Examples of random vibration. (A) Acceleration vs. time record of vibration measured in the vicinity of a rocket engine. (B) Example of narrow-band random vibration; the curves along the vertical axis give the probability distribution of the instantaneous (dashed curve) and peak (solid curve) values.

Fig. 11.2. Assembly of random time functions representing a random phenomenon $x(t)$.

tions to the problem of fatigue. A study of the statistical properties of the spectra of a vibrating system often is useful in leading to an understanding of the physical mechanism producing the vibration. Furthermore, it provides information which is useful in the simulation of vibration and useful for design purposes.

STATISTICAL CONCEPTS

FIRST PROBABILITY DISTRIBUTION

The statistical properties of a random time function, such as the acceleration-time record shown in Fig. 11.1A, are related to its so-called "probability distribution." For the determination of the statistical properties of a random function, it is generally necessary to consider a large assembly of random time records, as shown in Fig. 11.2. In this illustration, x is a function that may represent displacement, velocity, or acceleration. Let n represent the number of the members of this family of functions (called an *assembly* or *ensemble*). This chapter is restricted to a consideration of "stationary random processes."

A random process (an ensemble of time functions) is said to be *stationary* if any translation of the time origin leaves its statistical properties unaffected. The *ergodic hypothesis* states that assembly averaging in stationary random processes can be replaced by time averaging over a single record of very long duration. Time averaging of random processes greatly simplifies the task of data analysis.

At any time t_1, the value of $x(t_1)$ will differ in the various records; however, it is possible to count the fraction ν/n of the total number of records for which $x(t_1)$ will be less than some specified value x and plot it as shown in Fig. 11.3A. Such a curve is known as the *first probability distribution*, $P_1(x,t_1)$, at time t_1, and it will tend to a stationary value as $n \to \infty$. The first probability distribution $P_1(x,t_1)$ is a nondecreasing function of x with the following bounds (see Fig. 11.3):

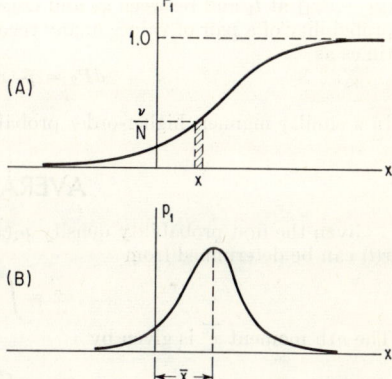

FIG. 11.3. (*A*) First probability distribution of the assembly of time functions shown in Fig. 11.2 at time t. (*B*) The probability density function corresponding to (*A*).

The fraction having a value less than ∞ at time t_1 is 1, i.e., the probability distribution is given by

$$P_1(\infty,t_1) = 1$$

The fraction having a value less than $-\infty$ at time t_1 is 0:

$$P_1(-\infty,t_1) = 0$$

The fraction having a value less than some intermediate value x at time t_1 is given by

$$0 \leq P_1(x,t_1) \leq 1$$

If the derivative of $P_1(x,t_1)$ exists, the density of the *first probability distribution* $p_1(x,t_1)$ is given by

$$p_1(x,t_1) = \frac{d}{dx} P_1(x,t_1) \qquad (11.1)$$

as shown in Fig. 11.3B.

The probability of $x(t_1)$ having a value between x_1 and $x_1 + dx$, as shown in Fig. 11.3, is

$$dP_1 = P_1(x_1 + dx, t_1) - P_1(x_1,t_1) = p_1(x_1,t_1)\, dx \qquad (11.2)$$

It follows that

$$P_1(x,t_1) = \int_{-\infty}^{x} p_1(x,t_1)\,dx \qquad (11.3)$$

$$P_1(\infty,t_1) = \int_{-\infty}^{\infty} p_1(\xi,t_1)\,d\xi = 1 \qquad (11.4)$$

SECOND PROBABILITY DISTRIBUTION

The *second probability distribution* P_2 determines the likelihood of pairs of values occurring a specified time interval apart. Suppose this time interval to be τ sec, so that $t_2 = t_1 + \tau$. Then a count of the number of records in which $x(t)$ lies between x_1 and $(x_1 + dx_1)$ at t_1 and between x_2 and $(x_2 + dx_2)$ at t_2, as shown in Fig. 11.2, defines the probability of a pair of values in any record being in the specified ranges at the specified times as

$$dP_2 = p_2(x_1,t_1,\,x_2,t_2)\,dx_1\,dx_2 \qquad (11.5)$$

In a similar manner, higher-order probability functions can be determined.

AVERAGE VALUE

Given the first probability density $p_1(x)$, the average value $\bar{x}$ of the random function $x(t)$ can be determined from

$$\bar{x} = \int_{-\infty}^{\infty} x p_1(x,t_1)\,dx \qquad (11.6)$$

The nth moment $\overline{x^n}$ is given by

$$\overline{x^n} = \int_{-\infty}^{\infty} x^n p_1(x,t_1)\,dx \qquad (11.7)$$

For stationary random processes, Eqs. (11.6) and (11.7) reduce to the following time averages by letting $n = 2$ in Eq. (11.7):

Average value:

$$\bar{x} = \lim_{T \to \infty} \frac{1}{2T} \int_{-T}^{T} x(t)\,dt \qquad (11.8)$$

Mean-square value:

$$\overline{x^2} = \lim_{T \to \infty} \frac{1}{2T} \int_{-T}^{T} x^2(t)\,dt \qquad (11.9)$$

GAUSSIAN (NORMAL) DISTRIBUTION

The probability distribution that is most frequently encountered (at least as an acceptable approximation) in random vibration problems and the only such distribution that is amenable to extensive analysis is the *Gaussian or normal distribution*. Suppose the amplitude-density curve is normalized:

$$\int_{-\infty}^{\infty} p(x)\,dx = 1$$

where $p(x)$ is the amplitude density. Then for a Gaussian distribution the amplitude density at time t_1 is given by

$$p_1(x,t_1) = \frac{1}{\sigma\sqrt{2\pi}} e^{-\frac{(x-\bar{x})^2}{2\sigma^2}} \qquad (11.10)$$

POWER SPECTRAL DENSITY

where $\bar{x}$ is the mean value of x and σ, the *standard deviation* (i.e., the rms value), is defined by

$$\sigma^2 = \int_{-\infty}^{\infty} (x - \bar{x})^2 p_1(x,t_1)\, dx = \overline{x^2} - (\bar{x})^2 \tag{11.11}$$

The quantity σ^2 is a measure of the dispersion of the random function and is called the *variance*.

The Gaussian distribution defined by Eq. (11.10) is plotted in Fig. 11.4. It is the only distribution (for instantaneous values) considered in this chapter. The assumption that a distribution is Gaussian automatically excludes such properties as skewness (asymmetry of the distribution). Conversely, the existence of some degree of skewness in the plot of probability density vs. $(x - \bar{x})$ is a priori evidence of departure from a Gaussian distribution (see Fig. 11.5).

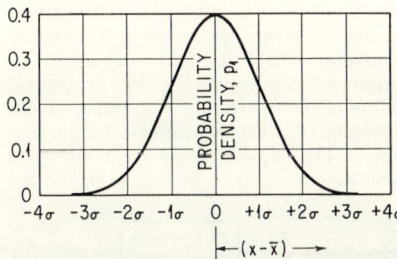

Fig. 11.4. Gaussian (also called normal) distribution. This represents the probability that the magnitude of the instantaneous value of x exceeds its rms value. For example, the probability of x exceeding 2σ is 4.6 per cent.

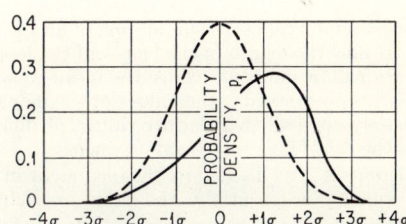

Fig. 11.5. Example of skewed distribution compared with normal distribution.

POWER SPECTRAL DENSITY

Power, which is the rate of doing work, is proportional to the square of the amplitude of a harmonic vibration. If two frequencies are present in a vibration, the power is proportional to the sum of the squares of the individual amplitudes associated with the two frequencies.

A random vibration can be considered to be a sum of a large number (tending to infinity) of harmonic vibrations of appropriate amplitudes and phase. The total power is again the sum of the power of the component harmonic vibrations, and it is of interest to know how this power is distributed as a function of frequency. Therefore *power spectral density* is defined as the power per unit frequency interval. A plot of this quantity indicates the frequency distribution of power.

DETERMINATION OF POWER SPECTRAL DENSITY

The analysis of random vibration always involves a record of finite length. Assume that such a record has been recorded on magnetic tape and is to be analyzed by an electronic analyzer as a stationary process; the tape can be formed into a loop of arbitrary length and made to repeat indefinitely on the analyzer. By such a procedure, an arbitrary fundamental period corresponding to the length of the loop is established, and the contents of the record may be defined in terms of the Fourier coefficients of integral multiple harmonics of the loop frequency ω_0.

The function $x(t)$ then can be represented by the real part of the Fourier series,

$$x(t) = \Re \sum_{n=-\infty}^{\infty} C_n e^{jn\omega_0 t} = C_0 + 2 \sum_{n=1}^{\infty} |C_n| \cos(n\omega_0 t - \alpha_n) \tag{11.12}$$

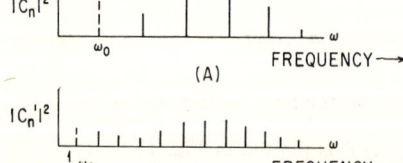

FIG. 11.6. Spectrum analysis of random vibration, as determined from a magnetic-tape loop. In (A) the period is T. In (B) the period is $2T$. Note that doubling loop period doubles number of spectral lines and reduces $|C_n|^2$ to one-half the former value.

where the complex amplitude C_n is defined by the equation

$$C_n = \frac{1}{2T}\int_{-T}^{T} x(\xi)e^{-jn\omega_0 \xi}\,d\xi \quad (11.13)$$

and $2T$ is the loop period.

The mean-square value $\overline{x^2}$ is a stationary property of $x(t)$ that is determined by Eq. (11.9) as

$$\overline{x^2} = \frac{1}{2T}\int_{-T}^{T} x^2(t)\,dt = C_0^2 + 2\sum_{n=1}^{\infty}|C_n|^2 \quad (11.14)$$

This equation indicates that the contribution to the mean-square value in any frequency interval is the summation of all the components within that frequency band.

If now the length of the loop and the loop period are doubled, the number of spectral lines will be doubled. Since the mean-square value of the entire spectrum must remain essentially constant, the values of the new coefficients $|C_n'|^2$ must decrease to approximately one-half their former values, as illustrated in Fig. 11.6. The sum of the lengths of the $|C_n|^2$ lines, up to any frequency, is the same in each diagram of Fig. 11.6, and the increment in $\overline{x^2}$ divided by the increment in frequency is essentially the same in each case,

$$\frac{\Delta \overline{x^2}}{\Delta \omega} = \frac{2|C_n|^2}{\omega_0} = \frac{2|C_n'|^2}{\tfrac{1}{2}\omega_0} = W(\omega_n) \quad (11.15)$$

The quantity $W(\omega_n)$ is the *power spectral density* at $\omega = \omega_n$. The dimensions are the square of the parameter represented by $\Delta \overline{x^2}$ in Eq. (11.15) per unit of frequency.

As the length of the loop is increased, more and more spectral lines are introduced; the magnitude of $|C_n|^2$ decreases correspondingly, but $W(\omega_n)$ remains finite. In the limiting case, as $T\to\infty$ or $\Delta\omega\to 0$, the spectrum becomes continuous, and the discrete values of $W(\omega_n)$ approach a smooth function $W(\omega)$:

$$W(\omega) = \lim_{\Delta\omega\to 0} \frac{\Delta \overline{x^2}}{\Delta\omega} = \frac{d\overline{x^2}}{d\omega} \quad (11.16)$$

The mean-square value can be expressed by the integral

$$\overline{x^2} = \int_0^{\infty} W(\omega)\,d\omega \quad (11.17)$$

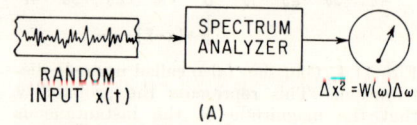

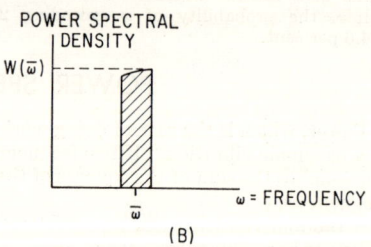

FIG. 11.7. The experimental determination of the power spectral density $W(\omega)$ of a random function $x(t)$. The function $x(t)$ is recorded on a magnetic tape and fed into a spectrum analyzer that transmits only those frequencies within the passband $\bar{\omega}\pm\tfrac{1}{2}\Delta\omega$. The output in the passband, namely, $\Delta\overline{x^2}$, is recorded on a mean-square meter, and the mean power spectral density over the passband is determined by dividing the meter reading by $\Delta\omega$. Thus, as shown in (B), the meter reading, $\Delta\overline{x^2}$, appears in the power spectral density vs. frequency plane as an increment of area of width $\Delta\omega$ and mean ordinate $f(\bar{\omega})$.

The power spectral density of a random time function may be determined by (see Fig. 11.7) feeding the function $x(t)$ into a spectrum analyzer that transmits only those frequency components within the passband of the analyzer, $\bar{\omega} \pm \tfrac{1}{2}\Delta\omega$. The output in the passband, namely, $\Delta\overline{x^2}$ is then indicated by a "mean-square" meter. The mean power spectral density $W(\bar{\omega})$ over the passband, at $\bar{\omega}$, is determined by dividing the meter reading by $\Delta\omega$. This division may be incorporated directly in the calibration of the meter.

INPUT-OUTPUT RELATIONSHIP

The complete distribution of $W(\omega)$ then is determined by changing the frequency $\bar{\omega}$ in increments of $\Delta\omega$.

TYPICAL POWER SPECTRAL DENSITY

Typical power spectral densities may be characterized *broadly* by two parameters: the mean-square value, given by Eq. (11.17), and some characteristic frequency—say ω_1. For example, the *white noise* * distribution with a cutoff frequency ω_1 is described by the equation

$$W(\omega) = \begin{cases} \overline{x^2}/\omega_1 & [\omega < \omega_1] \\ 0 & [\omega > \omega_1] \end{cases} \quad (11.18)$$

If $W(\omega)$ is (at least approximately) a monotonically decreasing function of ω, it is convenient to choose $\omega_1 = \omega_{1/2}$, where $\omega_{1/2}$ defines (at least roughly) the *half-power point* according to

$$\int_0^{\omega_{1/2}} W(\omega)\,d\omega \doteq \int_{\omega_{1/2}}^{\infty} W(\omega)\,d\omega \quad (11.19)$$

If $W(\omega)$ exhibits a definite peak at some frequency appreciably greater than zero, say ω_m, it is convenient to choose $\omega_1 = \omega_m$.

Examples of these two possible types of power spectral density are given by

Monotonic:
$$W_A(\omega) = \left(\frac{2\overline{x^2}}{\pi^{1/2}\omega_1}\right) e^{-(\omega/\omega_1)^2} \quad (11.20)$$

Peaked:
$$W_B(\omega) = \left(\frac{4\overline{x^2}}{\pi^{1/2}\omega_1}\right)\left(\frac{\omega}{\omega_1}\right)^2 e^{-(\omega/\omega_1)^2} \quad (11.21)$$

and are plotted in Fig. 11.8. The actual half-power point for $W_A(\omega)$ occurs at $\omega = 0.48\omega_1$, while the peak of $W_B(\omega)$ occurs at $\omega = \omega_1$. It should be emphasized that these examples are hypothetical; although they are representative of the more important types, most spectra are so complicated that they can be defined only graphically. Whether or not either of the (closely related) methods discussed in this section is the most expedient for the actual determination of the power spectral density from a given random function depends on the equipment available, the accuracy, and the ultimate application of the data. Another procedure for obtaining $W(\omega)$ is based on the autocorrelation of $x(t)$, making use of Eq. (11.60).

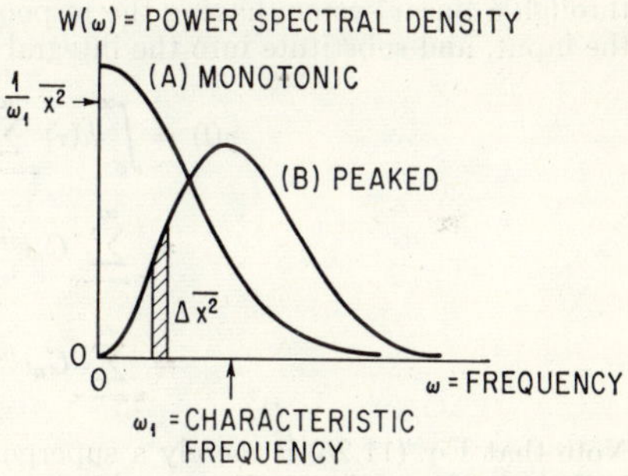

Fig. 11.8. Typical power spectral densities: (A) as given by Eq. (11.20) and (B) by Eq. (11.21). Note that $\overline{x^2} = \int_0^{\infty} W(\omega)\,d\omega$.

INPUT-OUTPUT RELATIONSHIP

The relationship between the input and output of a linear system for any time function $x(t)$, whether random or not, may be expressed in terms of the integral

$$y(t) = \int_{-\infty}^{t} x(\tau) h(t-\tau)\,d\tau = \int_0^{\infty} x(t-\xi) h(\xi)\,d\xi \quad (11.22)$$

where $h(t)$ is the response of the system to a unit impulse $\delta(t)$ and $x(t)$ is the input, which may have existed from $t = -\infty$. In the usual form of the above integral, the input is

---
* The power spectral density of true white noise is a constant for all ω.

assumed to be zero for $t < 0$, in which case the range of integration becomes 0 to t in both integrals.

The spectral aspects of the system are defined by a response function which is directly significant only for harmonic excitation. Letting $x(t) = e^{j\omega t}$ represent such a harmonic excitation, and noting that there is no provision for starting or stopping such a function, its substitution into Eq. (11.22) results in the following:

$$y(t) = \int_0^\infty e^{j\omega(t-\tau)}h(\tau)\,d\tau = e^{j\omega t}\int_0^\infty e^{-j\omega\tau}h(\tau)\,d\tau$$

$$= A(j\omega)e^{j\omega t} \tag{11.23}$$

where
$$A(j\omega) = \int_0^\infty e^{-j\omega\tau}h(\tau)\,d\tau \tag{11.24}$$

is defined as the response function of the system. Thus the response to a harmonic excitation is the excitation multiplied by the response function $A(j\omega)$ which then serves to define the spectral characteristics of the system. However, a sinusoidal input is given by either the real or the imaginary part of $e^{j\omega t}$, and either the real or the imaginary part of the response $y(t)$ represents the solution to such an excitation. Since $h(\tau)$ is zero for $\tau < 0$, the lower limit of Eq. (11.24) can be extended to $-\infty$ without altering the integral. Thus $A(j\omega)$ is the Fourier transform of the impulsive response function $h(\tau)$.

Suppose that a random time function (over a finite time) is expressible in terms of a Fourier series. Determine the spectral form of the output after the function has passed through a linear system having the response function $A(j\omega)$. Let Eq. (11.12) represent the input, and substitute into the integral of Eq. (11.22) for the response:

$$y(t) = \int_0^\infty h(\tau) \sum_{n=-\infty}^\infty C_n e^{jn\omega_0(t-\tau)}\,d\tau$$

$$= \sum_{n=-\infty}^\infty C_n e^{jn\omega_0 t} \int_0^\infty h(\tau) e^{-jn\omega_0\tau}\,d\tau$$

$$= \sum_{n=-\infty}^\infty C_n e^{jn\omega_0 t} A(jn\omega_0) \tag{11.25}$$

Note that Eq. (11.25) is merely a superposition of harmonic components, one of which is given by Eq. (11.23). For a real input, the output must be real; then Eq. (11.25) becomes

$$y(t) = C_0 A(0) + 2\sum_{n=1}^\infty |C_n||A(jn\omega_0)|\cos(n\omega_0 t + \beta_n) \tag{11.26}$$

From Eq. (11.9), the mean-square value of the response is

$$\overline{y^2} = \frac{1}{2T}\int_{-T}^T y^2(t)\,dt = C_0^2 A^2(0) + 2\sum_{n=1}^\infty |C_n|^2|A(jn\omega_0)|^2 \tag{11.27}$$

The increment of the mean-square value of the response at $\omega = n\omega_0$ is

$$\Delta\overline{y^2} = 2|C_n|^2|A(jn\omega_0)|^2 \tag{11.28}$$

and the power spectral density of the output for the discrete spectrum, with the frequency increment of $\omega_0 = \pi/T$, is

$$W_r(\omega_n) = \frac{2|C_n|^2|A(jn\omega_0)|^2}{\omega_0} \tag{11.29}$$

Since the power spectral density of the input $W_e(\omega_n)$ is given by Eq. (11.15), the input and output spectra are related by the equation

$$W_r(\omega_n) = W_e(\omega_n)|A(jn\omega_0)|^2 \tag{11.30}$$

INPUT-OUTPUT RELATIONSHIP   11-9

If the period $2T$ is increased without limit, the discrete spectrum becomes continuous; then $n\omega_0 = \omega_n = \omega$, so that Eq. (11.30) may be written as

$$W_r(\omega) = W_e(\omega)|A(j\omega)|^2 \tag{11.31}$$

The mean-square output is given by the integral

$$\overline{y^2} = \int_0^\infty W_r(\omega)\,d\omega = \int_0^\infty W_e(\omega)|A(j\omega)|^2\,d\omega \tag{11.32}$$

Thus the mean-square output (which is a statistical property of the output) is determined by the integral, over frequency, of the power spectral density of the input as modified by the response function of the system. When the time variation of the input is random and unknown, the time variation of the output is also random and unknown, and the relationship between the output and input is the statistical one given by Eqs. (11.31) and (11.32).

RESPONSE OF A SINGLE DEGREE-OF-FREEDOM SYSTEM

According to Eq. (2.27), the differential equation for a mechanical system of one degree-of-freedom (referred to as a "simple oscillator") is

$$m\ddot{y} + c\dot{y} + ky = F_0 \sin \omega t \tag{11.33}$$

Let

$\omega_n = \sqrt{\dfrac{k}{m}}$ = natural angular frequency of the oscillator

$c_c = 2\sqrt{km}$ = critical damping

$\zeta = \dfrac{c}{c_c}$ = fraction of critical damping

If the excitation is due to the motion $u(t)$ of the support instead of the force $F_0 \sin \omega t$, according to Eq. (2.47), the equation of motion of the mass has the same general form:

$$m\ddot{z} + c\dot{z} + kz = -m\ddot{u} \tag{11.34}$$

where $z = (y - u)$ is the relative displacement.

The response function of the simple oscillator can be determined from its Fourier transform, or more simply from the relation

$$z(t) = \ddot{u}_0 e^{j\omega t} A_1(j\omega) \tag{11.35}$$

which expresses its steady-state response to the harmonic input $\ddot{u}_0 e^{j\omega t}$. The square of the absolute value of the response function for Eqs. (11.33) and (11.34) is

$$|A_1(j\omega)|^2 = \frac{1}{\omega_n^4\left\{\left[1 - \left(\dfrac{\omega}{\omega_n}\right)^2\right]^2 + \left(2\zeta\dfrac{\omega}{\omega_n}\right)^2\right\}} \tag{11.36}$$

The power spectral density of the response then may be obtained from Eq. (11.32). In general this integration can be performed graphically, but it sometimes may be possible to effect analytical approximations, as in the following paragraph.

RESPONSE OF A NARROW-BAND OSCILLATOR

An important case, illustrated in Fig. 11.9, arises when the bandwidth of the oscillator, $\zeta\omega_n$, is small compared with the frequency interval over which the power spectral density is substantial and when $W_e(\omega_n)$, the power spectral density of the input at the resonant frequency of the oscillator, is of the same order of magnitude as $W_e(\omega_1)$. The response

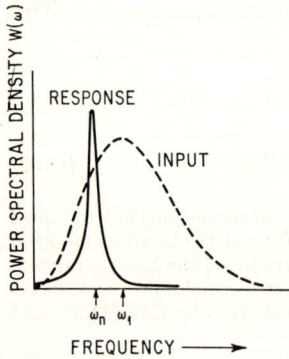

Fig. 11.9. Power spectral density of the output response of a simple oscillator for an input having the power spectral density shown by the dashed curve [see Eq. (11.37)].

of the oscillator then is essentially resonant, and it receives energy only in the neighborhood of its resonant frequency. Then its power spectral density of relative displacement may be approximated by

$$W_r(\omega) = W_e(\omega_n)|A_1(j\omega)|^2 \quad (11.37)$$

where $W_r(\omega)$ is the mean-square displacement density of relative motion z; $W_e(\omega)$ is the mean-square acceleration density of the input u; and $A_1(j\omega)$ is the response function defined by Eq. (11.36). The corresponding mean-square response then is given by Eq. (11.32) as

$$\overline{z^2} = W_e(\omega_n) \int_0^\infty |A_1(j\omega)|^2 \, d\omega = \frac{\pi}{4\zeta} \frac{W_e(\omega_n)}{\omega_n^3} \quad (11.38)$$

where the integral has been evaluated by substituting $|A(j\omega)|^2$ from Eq. (11.36) and assuming $\zeta \ll 1$ *after* carrying out the integration (which is given in integral tables). It may be assumed, without loss of generality, that the mean response $\bar{z}$ is zero (since, if $\bar{z} \neq 0$, it is necessary only to measure z from $\bar{z}$, i.e., replace z in the following analysis by $z - \bar{z}$); it then follows that $\sigma^2 = \overline{z^2}$.

The resonant response of Eq. (11.38) is simple in that, whatever the probability distribution of the input $x(t)$, the probability distribution of the response $z(t)$ tends to the Gaussian distribution of Eq. (11.10) as $\zeta\omega_n/\omega_1$ tends to zero, i.e., as the bandwidth of the oscillator becomes small compared with that of the input.* (Note that this last qualification excludes a periodic input.) The statistical properties of this essentially resonant response then are *completely* characterized by its resonant frequency ω_n and its mean-square value (compare the *broad* characterization of the input by its characteristic frequency ω_1 and mean-square value $\overline{x^2}$).

Note that if a random input is applied to a lightly damped oscillator, it acts as a narrow bandpass filter—passing only those frequencies in the neighborhood $\omega_n \pm \zeta\omega_n$ [cf. Eq. (11.37) and Fig. 11.9]. Because of the phenomenon of *beats* (wherein the addition of two harmonic waves of approximately the same frequency gives rise to a wave oscillating with the mean of the two frequencies and having an amplitude envelope that fluctuates at a rate equal to the difference frequency), the addition of a large number (tending to infinity) of harmonic waves having frequencies in the approximate band $\omega_n \pm \zeta\omega_n$ and randomly distributed phases will give rise to a wave oscillating with frequency ω_n and having an amplitude envelope that exhibits a random fluctuation, the rapidity of which must be of the order of $2\zeta\omega_n$. Such a response (illustrated in Fig. 11.1B) is called a *random sine wave*.

The probability distribution of this envelope (as opposed to the distribution of the instantaneous response) may be determined by synthesizing it from the randomly oriented vectors representing the individual harmonic waves.†

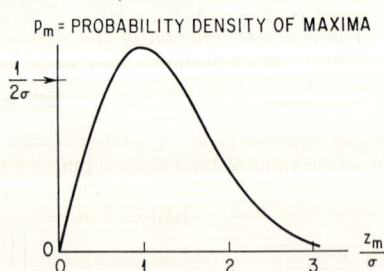

Fig. 11.10. Probability density distribution of an idealized random amplitude vibration of narrow bandwidth (nearly single frequency). This curve, known as a "Rayleigh distribution," gives the probability density of the maxima for essentially resonant loading as defined by Eq. (11.39).

* It is generally argued that this result follows directly from the "central-limit theorem."
† This problem, known as the two-dimensional *random walk* problem, was first solved by Lord Rayleigh.

RAYLEIGH DISTRIBUTION

The probability that the envelope $z_m(t)$ lies between z and $z + dz$ is

$$P\{z < z_m(t) < z + dz\} = p_m\, dz = \sigma^{-2} e^{-\frac{1}{2}(z/\sigma)^2} z\, dz \tag{11.39}$$

where σ, the standard deviation or rms value, is defined by Eq. (11.11). The quantity p_m, known as a *Rayleigh distribution* density, is plotted in Fig. 11.10.

The Rayleigh distribution of Eq. (11.39) describes the variation in amplitudes of the envelope of the random sine wave; the probability distribution of the instantaneous value is given by the Gaussian distribution of Eq. (11.10). Thus, the probability that $|z|$ will exceed a specified value z_1 is determined by

$$P\{|z| > z_1\} = \frac{2}{\sqrt{2\pi}} \int_{z_1}^{\infty} e^{-\frac{1}{2}(z/\sigma)^2} d\left(\frac{z}{\sigma}\right) = \mathrm{erfc}\, \frac{z_1}{\sigma\sqrt{2}} \tag{11.40a}$$

$$P\{|z| > z_1\} \simeq \sqrt{\frac{2}{\pi}} \left(\frac{\sigma}{z_1}\right) e^{-\frac{1}{2}(z_1/\sigma)^2} \left[1 - \left(\frac{\sigma}{z_1}\right)^2 + \cdots \right] \tag{11.40b}$$

where erfc denotes the complementary error function and Eq. (11.40b) gives its asymptotic approximation for $z_1 \gg \sigma$. Equations (11.40a) and (11.40b) may be used to determine the probability of exceeding the ultimate stress, whereas the Rayleigh distribution P_m is employed in studies of fatigue failure.

RESPONSE OF MULTIPLE DEGREE-OF-FREEDOM AND CONTINUOUS SYSTEMS

Every elastic system has associated with it certain normal-mode properties. Because of the orthogonal character of normal modes, such a system can be treated in terms of simple oscillators.[2]

The displacement response of a linear structure can be expressed in terms of its normal modes $\varphi_n(x,y,z)$: *

$$u(x,y,z,t) = \sum_n q_n(t)\varphi_n(x,y,z) \tag{11.41}$$

Assuming structural damping (damping in phase with the velocity and proportional to the generalized displacement of each mode), the Lagrange equations of the system have the form

$$\ddot{q}_n + (1 + j\gamma_n)\omega_n^2 q_n = \frac{Q_n(t)}{M_n} \tag{11.42}$$

where γ_n = coefficient of structural damping

ω_n = nth normal angular frequency

$$M_n = \int \varphi_n^2(x,y,z)\, dm = \text{generalized mass} \tag{11.43}$$

$$Q_n(t) = \int f(x,y,z,t)\varphi_n(x,y,z)\, dv = \text{generalized force} \tag{11.44}$$

$dv = dx\, dy\, dz$

The generalized mass and generalized force are obtained by integration over the structure.

---

* The concept of normal modes is discussed in detail under *Normal Modes of Vibration* in Chaps. 2 and 7. In particular, see Eq. (2.52) for a discussion of structural damping and Eq. (2.71) for a discussion of generalized mass and generalized force.

To establish the response function, a harmonic force can be applied. In this case the generalized force becomes

$$Q_n(t) = e^{j\omega t} \int F(x,y,z)\varphi_n(x,y,z)\, dv = e^{j\omega t} w_0 w_n \quad (11.45)$$

where
$$w_0 = \int F(x,y,z)\, dv = \text{amplitude of the total applied force} \quad (11.46)$$

and
$$w_n = \frac{1}{w_0}\int F(x,y,z)\varphi_n(x,y,z)\, dv = \text{mode participation factor of load} \quad (11.47)$$

The steady-state solution for q_n is

$$q_n = \frac{w_0 w_n e^{j(\omega t + \theta_n)}}{M_n \omega_n{}^2 \left\{ \left[1 - \left(\frac{\omega}{\omega_n}\right)^2\right]^2 + \gamma_n{}^2 \right\}^{1/2}} \quad (11.48)$$

which results in the response function

$$A(j\omega) = \sum_n \frac{w_n \varphi_n(x,y,z) e^{j\theta_n}}{M_n \omega_n{}^2 \left\{ \left[1 - \left(\frac{\omega}{\omega_n}\right)^2\right]^2 + \gamma_n{}^2 \right\}} \quad (11.49)$$

The square of the response function, required for the evaluation of the mean-square response, is then

$$|A(j\omega)|^2 = A(j\omega)A(-j\omega)$$

$$= \sum_n \frac{w_n{}^2 \varphi_n{}^2(x,y,z)}{M_n{}^2 \omega_n{}^4 \left\{ \left[1 - \left(\frac{\omega}{\omega_n}\right)^2\right]^2 + \gamma_n{}^2 \right\}} \quad (11.50)$$

Utilizing again the approximations allowed for small damping, the mean-square response for the structure becomes

$$\overline{u^2} = \frac{\pi}{2} \sum_n \frac{w_n{}^2 \varphi_n{}^2(x,y,z) W(\omega_n)}{\gamma_n M_n{}^2 \omega_n{}^3} \quad (11.51)$$

Equation (11.51) is expressed in terms of the normal modes of the structure and the power spectral density of the excitation, which allows a simple calculation for any linear structure.

CORRELATION FUNCTIONS

In statistical studies, the correlation function indicates the correlation between two sets of numbers. The autocorrelation, $R(t_1,\tau)$, of a random function $x(t)$ is a measure of the correlation between two values of $x(t)$ spaced τ apart. Crosscorrelation indicates the correlation between sets of numbers in two quantities $x(t)$ and $y(t)$.

AUTOCORRELATION FUNCTIONS

Autocorrelation is found by averaging the product of the two values $x(t_1)$ and $x(t_1 + \tau)$ of each record over the assembly of records. The autocorrelation function $R(t,\tau)$ is related to the second probability density as follows:

$$R(t_1,\tau) = \overline{x_i(t_1)x_i(t_1 + \tau)}$$

$$= \int_{-\infty}^{\infty}\int_{-\infty}^{\infty} x_1 x_2 p_2(x_1,t_1,x_2,t_2)\, dx_1\, dx_2 \quad (11.52)$$

CORRELATION FUNCTIONS

For a stationary function, the autocorrelation function $R(\tau)$ is determined by time averaging.

$$R(\tau) = \lim_{T \to \infty} \frac{1}{2T} \int_{-T}^{T} x(t) x(t + \tau) \, dt \tag{11.53}$$

From Eq. (11.12) the two terms to be multiplied and averaged are

$$x(t) = C_0 + 2 \sum_{n=1}^{\infty} |C_n| \cos(n\omega_0 t - \alpha_n) \tag{11.54}$$

$$x(t + \tau) = C_0 + 2 \sum_{n=1}^{\infty} |C_n| \{\cos(n\omega_0 t - \alpha_n) \cos n\omega_0 \tau - \sin(n\omega_0 t - \alpha_n) \sin n\omega_0 \tau\} \tag{11.55}$$

and the correlation function $R(\tau_n)$ for a finite period τ_n becomes

$$R(\tau_n) = \frac{1}{2T} \int_{-T}^{T} x(t) x(t + \tau) \, dt = C_0^2 + 2 \sum_{n=1}^{\infty} |C_n|^2 \cos n\omega_0 \tau \tag{11.56}$$

The increment in $R(\tau_n)$ divided by the increment in frequency is then

$$\frac{\Delta R(\tau_n)}{\Delta \omega} = \frac{2|C_n|^2}{\omega_0} \cos n\omega_0 \tau = W(\omega_n) \cos n\omega_0 \tau \tag{11.57}$$

As $T \to \infty$, $n\omega_0 = \omega$, $W(\omega_n) \to W(\omega)$, $R(\tau_n) \to R(\tau)$, and

$$\frac{dR(\tau)}{d\omega} = W(\omega) \cos \omega \tau \tag{11.58}$$

Thus, by integration, the relationship between the power spectral density $W(\omega)$ and the autocorrelation function $R(\tau)$ is

$$R(\tau) = \int_0^{\infty} W(\omega) \cos \omega \tau \, d\omega \tag{11.59}$$

It is evident from this expression that $R(\tau)$ is an even function of τ. In any practical mechanical system $W(\omega)$ approaches zero as $\omega \to \infty$, and $R(\infty) \to 0$. Also if $\tau \to 0$, the value of the function at each time increment is multiplied by itself; then Eq. (11.59) and Eq. (11.17) are identical, indicating that $R(0)$ is equal to the mean-square value $\overline{x^2}$. For $0 < \tau < \infty$, $R(\tau)$ may experience a change in sign and the curve for $R(\tau)$ may appear as in Fig. 11.11.

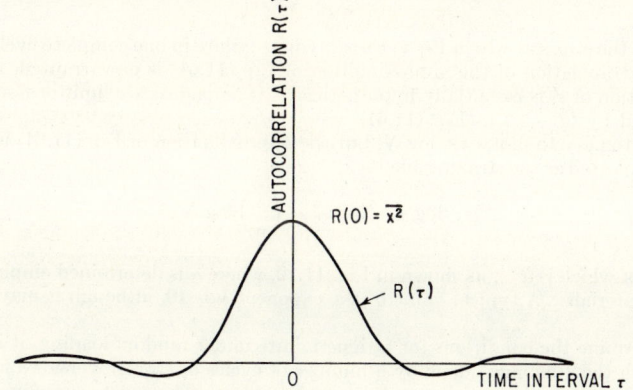

Fig. 11.11. Typical autocorrelation, an even function of τ [see Eq. (11.59)].

Equation (11.59) may be identified as a Fourier cosine transform, whence the power spectral density may be determined from the autocorrelation function by taking the inverse transform to obtain

$$W(\omega) = \frac{2}{\pi} \int_0^\infty R(\tau) \cos \omega\tau \, d\tau \qquad (11.60)$$

It is often desirable to obtain $W(\omega)$ for a given record by first obtaining $R(\tau_n)$ according to Eq. (11.56), with T finite, and then evaluating $W(\omega)$ from Eq. (11.60). These operations are mathematically equivalent to those discussed under *power spectral density*, and the ultimate choice depends on the equipment available, the required accuracy, and the intended application of the results as indicated in Chap. 22.

STRUCTURAL FATIGUE UNDER RANDOM LOADING

This section is concerned with the calculation of the probable fatigue life and equivalent fatigue stress for a lightly damped, single degree-of-freedom structure subjected to random loading.[3] It is based on the concept of cumulative damage.[4,5,6] (See *Fatigue Properties of Metals* and *Concept of Cumulative Fatigue Damage* in Chap. 24.) The concept of cumulative damage is likely to be more adequate for random loading, where the stress amplitudes are distributed continuously over a wide range, than for simple harmonic loading at a single amplitude.[3]

The only information commonly available for the (phenomenological) description of fatigue is the s,N curve, which gives the number of cycles N of complete stress reversal (i.e., $-s$ to $+s$) of fixed amplitude s required to produce failure for a given structure (see Fig. 24.5). For this type of loading, there sometimes exists a minimum stress, designated as the endurance limit s_e, below which fatigue failure never occurs (i.e., $N = \infty$ if $s < s_e$). However, the existence of an endurance limit depends on changes in the material produced by alternating stress at a low level ($s < s_e$), and intermediate excursions to higher stress levels may counteract these changes; moreover, only a few materials (notably steel) seem to possess well-defined endurance limits. Thus, it is reasonable to ignore the possibility of an endurance limit when, as in the problem at hand, stress amplitudes are distributed over a wide range. The introduction of an endurance limit has, in any event, no important effect on the results presented here, provided that the root-mean-square (rms) stress is large compared with this limit.

Experimental fatigue data usually exhibit a wide scatter on an s,N plot, but within the limits of this scatter an adequate approximation is furnished by [5]

$$N(s) = \left(\frac{s_1}{s}\right)^\alpha \qquad (11.61)$$

where s_1 is the stress at which Eq. (11.61) predicts failure in one complete cycle [although such an extrapolation of the approximation of Eq. (11.61) is unwarranted, so that this interpretation of s_1 is essentially hypothetical]. If an endurance limit is assumed, s can be replaced by $(s - s_e)$ in Eq. (11.61).

It is customary to plot s vs. log N, but the approximation of Eq. (11.61) is more conveniently plotted as the straight line

$$\log s = \log s_1 - \left(\frac{1}{\alpha}\right) \log N \qquad (11.62)$$

the slope of which is α^{-1}, as shown in Fig. 11.12, where α is determined empirically from tests of materials. A typical value of the exponent α is 10, although it may be as high as 25.[3]

To determine the conditions for fatigue failure under random loading, it is necessary to establish the "damage" done by a number of cycles of stress reversals n less than the number N that produces failure at a given level s. The most common assumption (*Miner's rule*) is that the damage accumulates linearly, so that if a structure is subjected to n_i stress reversals at a level s_i the partial damage is $n_i/N(s_i)$, and the cumulative damage D is given by [4]

$$D = \sum_i \left(\frac{n_i}{N_i}\right) \qquad N_i = N(s_i) \tag{11.63}$$

with fatigue failure occurring at $D = 1$. Assuming the s,N law of Eq. (11.61), then Eq. (11.63) becomes

$$D = \sum_i n_i \left(\frac{s_i}{s_1}\right)^\alpha \tag{11.64}$$

The cumulative damage law of Eq. (11.63) ignores the temporal sequence of loadings. Various alternative hypotheses have been offered, but it may be argued that presently

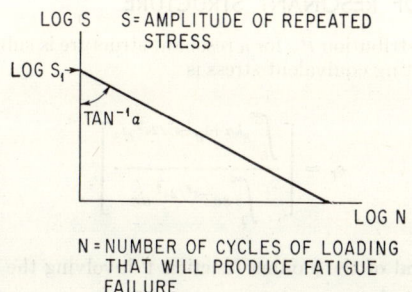

Fig. 11.12. Logarithmic plot of typical s,N curve, as given by Eq. (11.62).

available fatigue data scarcely warrant the additional complications in so far as the prediction of the fatigue life of a *single sample* is concerned. Moreover, the introduction of cumulative-damage hypotheses more closely in accord with appropriately averaged experimental results has no appreciable effect when, as in the present application, the stress amplitudes are continuously distributed over a wide range.[3]

In formulating design criteria, it is convenient to introduce an equivalent stress (often called *reduced stress*) s_r that would produce the same fatigue damage as the spectrum $(n_1, s_1), (n_2, s_2), \ldots$ after the same (total) number of cycles of loading. On Miner's hypothesis,[4] it follows from Eq. (11.64) that

$$s_{rM} = \left(\frac{\sum_i n_i s_i^\alpha}{\sum_i n_i}\right)^{1/\alpha} \tag{11.65}$$

where the exponent $\dfrac{1}{\alpha}$ can be determined either empirically or on the basis of some physical model (with or without direct reference to the slope of the s,N curve). An alternative form is

$$s_r = \left(\frac{\sum_i n_i s_i^{k\alpha}}{\sum_i n_i}\right)^{1/k\alpha} \tag{11.66}$$

where k is a constant, with 2 as a (probably) conservative upper limit, and α is defined as in Eq. (11.62).

In the calculation of probable fatigue life of a lightly damped structure under random loading,* the probable number of cycles of loading having an amplitude in the range

---

* The implied assumption of completely reversed loading is consistent with the result that the rate of fluctuation of the envelope is of the order of $2\zeta\omega_n$, so that successive maxima and minima differ only by a fraction of order ζ.

$(s, s + ds)$ is obtained by multiplying the total number of cycles (at frequency $\omega_n/2\pi$) by $P_m(s)\,ds$. [See Eq. (11.39).] It then follows from Eq. (11.66) that the probable equivalent stress is given by

$$s_r = \left(\frac{\int_0^\infty s^{k\alpha} P_m(s)\,ds}{\int_0^\infty P_m(s)\,ds} \right)^{1/k\alpha} \tag{11.67}$$

The (angular) frequency at which the equivalent stress s_r may be supposed to oscillate is ω_n.

FATIGUE FAILURE OF RESONANT STRUCTURE

If the probability distribution P_m for a resonant structure is substituted in Eq. (11.67) from (11.39), the resulting equivalent stress is

$$s_r = \left[\frac{\int_0^\infty s^{k\alpha+1} e^{-s^2/2\overline{s^2}}\,ds}{\int_0^\infty s e^{-s^2/2\overline{s^2}}\,ds} \right]^{1/k\alpha} \tag{11.68}$$

Introducing the integral of the gamma function Γ involving the arbitrary variable z,

$$\Gamma(z+1) = 2^{-z} \int_0^\infty x^{2z+1} e^{-x^2/2}\,dx \tag{11.69}$$

Eq. (11.68) becomes

$$s_r = [\Gamma(\tfrac{1}{2}k\alpha + 1)]^{1/k\alpha} (2\overline{s^2})^{1/2} \tag{11.70}$$

If (since α is always large) the gamma function is approximated by Stirling's formula, namely,

$$\Gamma(z+1) \simeq e^{-z} z^{z+1/2} \sqrt{2\pi} \qquad [z \gg 1] \tag{11.71}$$

the equivalent stress becomes

$$s_r = (\pi k\alpha)^{1/2k\alpha} e^{-1/2} (k\alpha \overline{s^2})^{1/2} \simeq e^{-1/2} (k\alpha \overline{s^2})^{1/2} \qquad [\alpha \gg 1] \tag{11.72}$$

The determination of the equivalent stress under random loading depends essentially on the determination of the rms stress produced by the same loading. If s_0 denotes the static stress that would be produced by an input having the mean-square value $\overline{x^2}$, then the mean-square stress produced by the random loading $x(t)$, having the power spectral density $W(\omega)$, may be obtained from Eq. (11.40) according to

$$\frac{\overline{s^2}}{s_0^2} = \left(\frac{\pi}{4\zeta} \right) \left[\frac{\omega_n W(\omega_n)}{\int_0^\infty W(\omega)\,d\omega} \right] \tag{11.73}$$

where ζ is the fraction of critical damping and the integral in the denominator is simply $\overline{x^2}$. Both sides of Eq. (11.73) are dimensionless, and $x(t)$ need not have the dimensions of displacement but may be any input to which the load is linearly related, e.g., force.

To form some estimate of the magnitudes of s_r given by Eq. (11.72), the equivalent stresses based on the representative power spectral density distributions of Eqs. (11.20) and (11.21) will be evaluated. Substituting these distributions in Eqs. (11.72) and (11.73) yields

Monotonic:
$$\left(\frac{s_r}{s_0}\right)_a = \frac{\pi^{1/4}}{(2e)^{1/2}}\left(\frac{k\alpha}{\zeta}\right)^{1/2}\left(\frac{\omega_n}{\omega_1}\right)^{1/2} e^{-1/2(\omega_n/\omega_1)^2} \tag{11.74}$$

Peaked:
$$\left(\frac{s_r}{s_0}\right)_b = \frac{\pi^{1/4}}{e^{1/2}}\left(\frac{k\alpha}{\zeta}\right)^{1/2}\left(\frac{\omega_n}{\omega_1}\right)^{3/2} e^{-1/2(\omega_n/\omega_1)^2} \tag{11.75}$$

These results are plotted in Fig. 11.13.

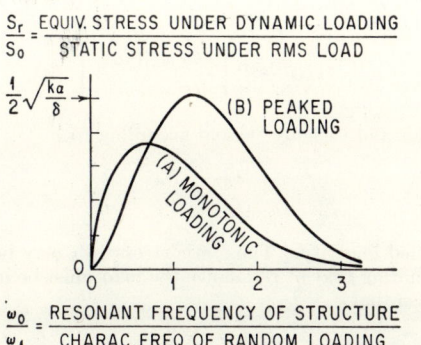

FIG. 11.13. The stress ratios given by Eqs. (11.74) and (11.75).

It is probable that Eq. (11.75) is representative of random load distributions of the type encountered in practice. The corresponding upper bound for the equivalent stress is given by

$$\left(\frac{s_r}{s_0}\right)_{max} \simeq 0.51\left(\frac{k\alpha}{\zeta}\right)^{1/2} \tag{11.76}$$

For $\alpha = 10$, $k = 2$, and $\zeta = 0.02$ (a typical value for structural damping of a panel) this last result yields 16, while for $\alpha = 25$, $k = 2$, and $\zeta = 0.01$ (probably the worst possible case that could be obtained, even under laboratory conditions) the ratio is 36.

RELATION BETWEEN FATIGUE LIVES UNDER SINUSOIDAL AND RANDOM LOADING

The following problem sometimes arises in fatigue testing: A sinusoidal load $\sqrt{2}\,F_h \sin(\omega_n t)$ is applied to a structure having the resonant frequency ω_n, and the fatigue life T_h (h implies *harmonic*) is measured; it then is required to estimate the probable fatigue life T_r (r implies *random*) of an identical structure under a random load having the rms value F_r and a relatively flat power spectral density $W(\omega)$ in the neighborhood of ω_n.

Let the equivalent fatigue stress s_r now be defined as the amplitude of a sinusoidal stress of frequency ω_n that would yield the same fatigue life as the random stress produced by the random loading of rms value F_r. (Note that the amplitude of the sinusoidal load that would be required to produce the hypothetical stress would not be $\sqrt{2}\,F_r$; conversely s_r is not the sinusoidal stress amplitude that would be produced by a sinusoidal load of amplitude $\sqrt{2}\,F_r$.) The fatigue life implied by s_r is

$$T_r = \left(\frac{2\pi}{\omega_n}\right) N(s_r) \tag{11.77}$$

where $(2\pi/\omega_n)$ is the period of the oscillator, and $N(s_r)$ is the number of cycles given by the s,N curve at $s = s_r$. In addition, let the sinusoidal stress amplitude s_h be defined as the value inferred from the measured fatigue life T_h according to

$$T_h = \left(\frac{2\pi}{\omega_n}\right) N(s_h) \tag{11.78}$$

and the *static stresses* $(s_0)_h$ and $(s_0)_r$ be defined as the stresses that would be produced by the static application of the rms loads F_h and F_r, respectively.

The equivalent fatigue stress for a structure approximated by a single degree-of-freedom linear oscillator of resonant frequency ω_n and fraction of critical damping ζ subjected to a random load having the power spectral density $W(\omega)$ is given approximately by Eqs. (11.72) and (11.73) as

$$s_r = (\pi k\alpha)^{(\frac{1}{2}k\alpha)} \left[\frac{\pi k\alpha}{4e\zeta} \frac{\omega_n W(\omega_n)}{F_r^2}\right]^{\frac{1}{2}} (s_0)_r \tag{11.79}$$

The static stresses $(s_0)_r$ and $(s_0)_h$ are related according to

$$\frac{(s_0)_r}{(s_0)_h} = \frac{F_r}{F_h} \tag{11.80}$$

in virtue of the assumed linearity. The static stress $(s_0)_h$ may be related to s_h directly through Eq. (11.33), noting that at resonance the load must be in equilibrium with the damping force alone, so that

$$(s_0)_h = \sqrt{2} \, \zeta s_h \tag{11.81}$$

Comparing Eqs. (11.79) through (11.81), it follows that

$$\frac{s_r}{s_h} = (\pi k\alpha)^{(\frac{1}{2}k\alpha)} \left[\frac{\pi k\alpha \zeta}{2e} \frac{\omega_n W(\omega_n)}{F_h^2}\right]^{\frac{1}{2}} \tag{11.82}$$

The corresponding ratio of fatigue lives, on the approximation that N is proportional to $s^{-\alpha}$ of Fig. 11.12 (now assumed to be valid at $s = s_h$ as well as $s = s_r$), is given by

$$\frac{T_r}{T_h} = \left(\frac{s_r}{s_h}\right)^{-\alpha} = (\pi k\alpha)^{-\frac{1}{2}k} \left[\frac{2e}{\pi k\alpha \zeta} \frac{F_h^2}{\omega_n W(\omega_n)}\right]^{\alpha/2} \tag{11.83}$$

If the conservative value $k = 2$ is assumed,

$$\frac{T_r}{T_h} = 0.63\alpha^{-\frac{1}{4}} \left[\frac{0.87}{\alpha \zeta} \frac{F_h^2}{\omega_n W(\omega_n)}\right]^{\alpha/2} \tag{11.84}$$

If $F_h = F_r$, then T_r probably would be much larger than T_h for a lightly damped (i.e., $\zeta \ll 1$) structure.

Nonlinear effects might render ζ a function of amplitude. It is possible to compensate *partially* for these effects by allowing ζ to have the different values ζ_r and ζ_h in Eqs. (11.79) and (11.80), respectively, which has the end effect of replacing ζ by

$$\zeta = \frac{\zeta_h^2}{\zeta_r} \tag{11.85}$$

in Eqs. (11.83) and (11.84). Alternatively, nonlinear effects can be taken into account more precisely by adjusting the amplitude of the sinusoidal forcing function such that s_h would be equal to s_r, as given by Eq. (11.79). It then would follow that $T_r = T_h$, but this probably would increase the testing time appreciably.

If a straight line does not furnish an adequate approximation to the log s vs. log N curve, despite the assumption of this s,N approximation in deriving Eq. (11.79), it is likely that a better estimate of T_r than that provided by Eqs. (11.83) and (11.84) may be obtained by the procedure implied by Eqs. (11.77), (11.78), and (11.82), namely:

1. Determine s_h from the measured life T_h, using Eq. (11.78) in conjunction with the actual s,N curve.
2. Use this value of s_h to determine s_r from Eq. (11.82); the value of α to be used in this determination should be the local slope of the log s vs. log N curve at $s = s_r$ (this requires an iterative calculation of s_r, starting with an initial estimate of α).
3. Determine T_r from Eq. (11.77), using s_r and the actual s,N curve.

The principal shortcomings of the foregoing procedure—as opposed to a direct measurement of the fatigue life under random loading—are (1) the assumption of linearity (but note the penultimate paragraph), (2) the difficulty of determining the actual power spectral density in the neighborhood of the structure, and (3) the approximations inherent in the result of Eq. (11.79), particularly the neglect of random phasing of the spatial loading. This last effect will be negligible if the wavelength of the spatial loading distribution is large compared with the dimensions of the structure (a condition that often will be met if the loading is that resulting from sound pressure).

REFERENCES

1. Rice, S. O.: *Bell System Tech. J.*, **23**:282 (1944) and **24**:44 (1945); reprinted in "Noise and Stochastic Processes," N. Wax (ed.), Dover Publications, New York, 1954.
2. Thomson, W. T., and M. V. Barton: *J. Appl. Mechanics*, **24**:248 (1957).
3. Miles, J. W.: *J. Aeronaut. Sci.*, **21**:753 (1945).
4. Miner, M. A.: *J. Appl. Mechanics*, **12**:159 (1945).
5. Shanley, F. R.: *Rand Corporation, Rept. P-350*, Santa Monica, Calif., 1952.
6. Freudenthal, A. M., and R. A. Heller: *J. Aeronaut. Sci.*, **26**:431 (1959).
7. Bendat, J. S.: "Principles and Applications of Random Noise Theory," John Wiley & Sons, Inc., New York, 1958.

12
INTRODUCTION TO SHOCK AND VIBRATION MEASUREMENTS

Wilson Bradley, Jr.
Endevco Corporation

Eldon E. Eller
Endevco Corporation

INTRODUCTION

This chapter is the first in a group of nine chapters on the measurement of shock and vibration. The following chapters describe in detail various types of instruments and measuring systems, and set forth primary considerations in the calibration and field use of such instruments and systems. This chapter defines the terms and describes the general principles of shock- and vibration-measuring instruments; it also sets forth the mathematical basis for the measurement of shock and vibration, and includes a brief description of important instruments and instrumentation systems.

DEFINITION OF TERMS

A *transducer* (or *"pickup"*) is a device which converts shock or vibratory motion into an optical, a mechanical, or, most commonly, an electrical signal that is proportional to a parameter of the experienced motion.

A *transducing element* is the part of the transducer that accomplishes the conversion of motion into the signal.

A *measuring instrument or system* converts shock and vibratory motion into an *observable* form that is directly proportional to a parameter of the experienced motion. It may consist of a transducer with transducing element, signal-conditioning equipment, and device for displaying the signal. An *instrument* contains all of these elements in one package (see Chap. 13), while a *system* utilizes separate packages.

An *accelerometer* is a transducer whose output is proportional to the acceleration input.

A *velocity pickup* is a transducer whose output is proportional to the velocity input.

A *displacement pickup* is a transducer whose output is proportional to the displacement input.

CLASSIFICATION OF INSTRUMENTS

In principle, shock and vibration are measured with reference to a point fixed in space.* This may be accomplished by either of two fundamentally different types of instruments:

---
* For special purposes, the motion of one point may be measured relative to another point. Then one terminal of the instrument is attached to each of the points.

12-2 INTRODUCTION TO SHOCK AND VIBRATION MEASUREMENTS

1. In a *fixed reference instrument*, one terminal of the instrument is attached to a point that is fixed in space and the other terminal is attached (e.g., mechanically, electrically, optically, etc.) to the point whose motion is to be measured.

2. In a *mass-spring instrument* (*seismic instrument*), the only terminal is the base of a mass-spring system; this base is attached at the point where the shock or vibration is to be measured. The motion at the point is inferred from the motion of the mass relative to the base.

FIXED REFERENCE INSTRUMENTS

Figure 12.1 illustrates schematically the principle of the fixed reference instrument for measuring the motion of the "vibrating part" relative to the "fixed reference." By attaching a scale to the fixed reference and a pointer to the vibrating part, as shown in

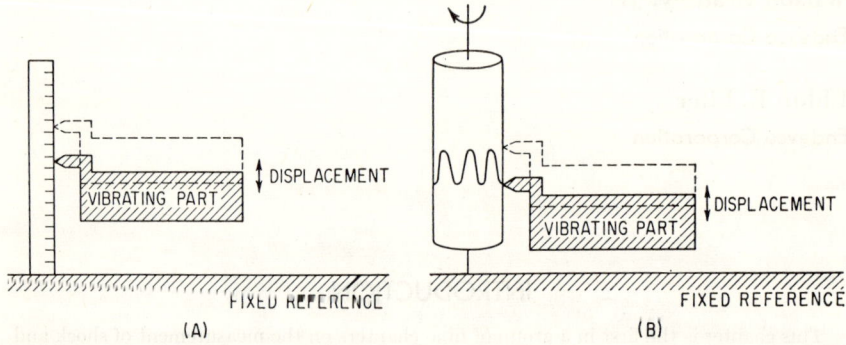

FIG. 12.1. Measurement of vibration with a fixed reference system: (*A*) Displacement of the vibrating part is indicated for direct observation, with or without optical enlargement techniques. (*B*) Displacements are recorded on a rotating drum, describing the waveform as a function of time.

Fig. 12.1*A*, the limits of motion of the vibrating part can be determined visually. If the pointer inscribes a mark upon the scale, the peak-to-peak displacement can be determined by noting the length of the inscribed line. Additional information can be obtained by substituting a rotating drum for the scale, as illustrated in Fig. 12.1*B*. Then the inscribed trace represents the time-history of the displacement of the vibrating part relative to the fixed reference.

Various means are used to perform the functions indicated schematically in Fig. 12.1. Usually, the displacement associated with vibratory motion is relatively small; thus, magnifying means must be used to obtain useful sensitivity and accuracy of measurement. This may be done mechanically by incorporating a lever with the pointer or scriber to amplify the motion to a more readily observed size (see *Vibrographs* in Chap. 13). Because of mechanical limitations, a lever is useful only for the measurement of shock and vibration having a relatively large displacement and a relatively low frequency. Electrical means may be employed to indicate the motion of the vibrating part relative to the fixed reference; such means are described in this chapter under *Characteristics of the Transducing Elements*. Optical or visual methods also are used wherein the observer occupies the position of the scale or drum in Fig. 12.1 and is provided with means to determine the displacement of the vibrating part.

VISUAL DISPLACEMENT INDICATORS. The accuracy of an optical system at a high frequency of vibration is limited by the resolution and rigidity of the fixed reference used in the system. With good technique, it is possible to achieve accurate measurements at frequencies as high as 2,000 cps. Under highly specialized conditions, good results may be obtained at frequencies as high as 20,000 cps. Inasmuch as displacements at high frequency usually are measured in microinches or microns, it is essential that the optical system be secured firmly to the fixed reference.

Fixed reference instruments of the visual observation type are described in Chap. 13. Such an instrument employs an appropriate "target" which is attached to the surface of the vibrating part in position for visual observation. A number of different types of targets are utilized, several of which are shown in Fig. 12.2. (Also see Figs. 18.11 and 13.40, which illustrate a method of vibration displacement measurement that uses a vibrating wedge.)

Figure 12.2A shows a "target" made of a material containing reflecting spots b as it appears through a microscope equipped with a reticle scale c. Typical reflective materials are Scotchlite, fine garnet paper, or a blackened mirror with a pinhole through the blacking. Each of these materials reflects a fine pinpoint of light when illuminated by a light source. When such a reflector is vibrated, persistence of vision causes each reflecting point to appear as a straight line d in Fig. 12.2A. The length of the line may be determined by the use of the microscope reticle, and indicates the peak-to-peak displacement of the vibrating part.

When it is desired to set a vibratory displacement to a known and fixed value, a target of the type shown in Fig. 12.2B is useful. This target is a circle with its diameter d equal to the desired peak-to-peak displacement. During vibration two circles appear because the velocity of the moving target is zero at times of maximum displacement. These circles move further apart as the displacement is increased. The desired value of displacement is attained when the circles appear to just touch one another, as shown in the

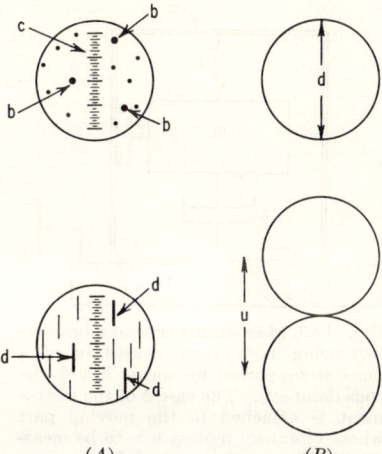

FIG. 12.2. Visual and optical targets for measurement of vibratory displacement. The appearance of the targets when stationary is shown in the upper views; the corresponding appearance when vibrating is shown in the lower views.

(A) Illuminated target containing small reflecting spots observed through microscope equipped with reticle. Under vibration the spots become lines. The observer selects the most prominent line and measures its length by the scale in the reticle. This is a measure of the peak-to-peak displacement u multiplied by the fixed power of the optical system.

(B) A circular target of diameter d equal to a desired peak-to-peak displacement u will appear as two touching circles when u equals d.

lower view of Fig. 12.2B. Because of persistence of vision, the target appears motionless at these points. The required size of target circles may be too small to be drawn accurately to size; then the circle may be drawn on a larger scale and reduced photographically.

STROBOSCOPE. A stroboscope is a light source that can be adjusted to flash at a desired rate. It may be used to illuminate the vibrating surface in a fixed reference system (see Chap. 13). If the flashing frequency is the same as the vibration frequency or some submultiple thereof, the vibrating surface is in the same position each time it is illuminated. Therefore, the vibrating surface appears motionless due to persistence of vision. If the flashing frequency is slightly different from the vibration frequency, the vibratory displacement appears in slow motion. The stroboscope may be calibrated and used to measure the frequency of vibration.

MASS-SPRING TRANSDUCERS (SEISMIC TRANSDUCERS)

In many applications, such as moving vehicles or missiles, it is impossible to establish a fixed reference for shock and vibration measurements. Therefore, many transducers use the response of a mass-spring system to measure shock and vibration. A mass-spring transducer is shown schematically in Fig. 12.3; it consists of a mass m suspended from the transducer case a by a spring of stiffness k. The motion of the mass within the case may be damped by a viscous fluid or electric current symbolized by a dashpot with damp-

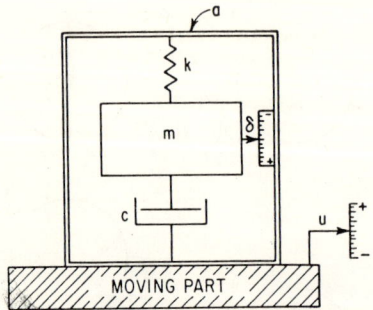

Fig. 12.3. Mass-spring type of vibration-measuring instrument consisting of a mass m supported by spring k and viscous damper c. The case a of the instrument is attached to the moving part whose vibratory motion u is to be measured. The motion u is inferred from the relative motion δ between the mass m and the case a.

ing coefficient c. It is desired to measure the motion of the moving part whose displacement with respect to fixed space is indicated by u. When the transducer case is attached to the moving part, the transducer may be used to measure displacement, velocity, or acceleration, depending on the portion of the frequency range which is utilized and whether the relative displacement δ or relative velocity $d\delta/dt$ is sensed by the transducing element. The typical response of the mass-spring system is analyzed in the following paragraphs and applied to the interpretation of transducer output.

Consider an instrument whose case experiences a motion u, and let the relative displacement between the mass and the case be δ. Then the motion of the mass with respect to a reference fixed in space is $\delta + u$, and the force causing its acceleration is $m[d^2(\delta + u)/dt^2]$. Thus, the force applied by the mass to the spring and dashpot assembly is $-m[d^2(\delta + u)/dt^2]$. The force applied by the spring is $-k\delta$, and the force applied by the damper is $-c(d\delta/dt)$. Adding all force terms and equating the sum to zero,

$$-m\frac{d^2(\delta + u)}{dt^2} - c\frac{d\delta}{dt} - k\delta = 0 \qquad (12.1)$$

Equation (12.1) may be rearranged:

$$m\frac{d^2\delta}{dt^2} + c\frac{d\delta}{dt} + k\delta = -m\frac{d^2u}{dt^2} \qquad (12.2)$$

Assume that the motion u is sinusoidal, $u = u_0 \cos \omega t$, where $\omega = 2\pi f$ is the angular frequency in radians per second and f is expressed in cycles per second. Neglecting transient terms, the response of the instrument is defined by $\delta = \delta_0 \cos(\omega t - \theta)$; then the solution of Eq. (12.2) is

$$\frac{\delta_0}{u_0} = \frac{\omega^2}{\sqrt{\left(\frac{k}{m} - \omega^2\right)^2 + \left(\omega \frac{c}{m}\right)^2}} \qquad (12.3)$$

$$\theta = \tan^{-1} \frac{\omega \frac{c}{m}}{\frac{k}{m} - \omega^2} \qquad (12.4)$$

The undamped natural frequency f_n of the instrument is the frequency at which

$$\frac{\delta_0}{u_0} = \infty$$

when the damping is zero ($c = 0$), or the frequency at which $\theta = 90°$. From Eqs. (12.3) and (12.4), this occurs when the denominators are zero:

$$\omega_n = 2\pi f_n = \sqrt{\frac{k}{m}} \quad \text{rad/sec} \qquad (12.5)$$

Thus, a stiff spring and/or light mass produces an instrument with high natural frequency. A heavy mass and/or compliant spring produces an instrument with a low natural frequency.

The damping in a transducer is specified as a *fraction of critical damping*. Critical damping c_c is the minimum level of damping that prevents a mass-spring transducer from oscillating when excited by a step function or other transient. It is defined by

$$c_c = 2\sqrt{km} \tag{12.6}$$

Thus, the fraction of critical damping ζ is

$$\zeta = \frac{c}{c_c} = \frac{c}{2\sqrt{km}} \tag{12.7}$$

It is convenient to define the excitation frequency ω for a transducer in terms of the undamped natural frequency ω_n by using the dimensionless frequency ratio ω/ω_n. Substituting this ratio and the relation defined by Eq. (12.7), Eqs. (12.3) and (12.4) may be written:

$$\frac{\delta_o}{u_o} = \frac{\left(\dfrac{\omega}{\omega_n}\right)^2}{\sqrt{\left[1 - \left(\dfrac{\omega}{\omega_n}\right)^2\right]^2 + \left(2\zeta\dfrac{\omega}{\omega_n}\right)^2}} \tag{12.8}$$

$$\theta = \tan^{-1}\frac{2\zeta\dfrac{\omega}{\omega_n}}{1 - \left(\dfrac{\omega}{\omega_n}\right)^2} \tag{12.9}$$

The response of the mass-spring transducer given by Eq. (12.8) may be expressed in terms of the acceleration $\ddot{u}$ of the moving part by substituting $\ddot{u}_0 = -u_0\omega^2$. Then the ratio of the relative displacement amplitude δ_0 between the mass m and transducer case a to the impressed acceleration amplitude $\ddot{u}_0$ is

$$\frac{\delta_o}{\ddot{u}_o} = -\frac{1}{\omega_n^2}\left[\frac{1}{\sqrt{\left[1 - \left(\dfrac{\omega}{\omega_n}\right)^2\right]^2 + \left[2\zeta\dfrac{\omega}{\omega_n}\right]^2}}\right] \tag{12.10}$$

The relation between δ_0/u_0 and the frequency ratio ω/ω_n is shown graphically in Fig. 12.4 for several values of the fraction of critical damping ζ. Corresponding curves for $\delta_0/\ddot{u}_o$ are shown in Fig. 12.5. The phase angle θ defined by Eq. (12.9) is shown graphically in Fig. 12.6, using the scale at the left side of the figure. Corresponding phase angles between the relative displacement δ, and the velocity $\dot{u}$ and acceleration $\ddot{u}$, are indicated by the scales at the right side of the figure.

APPLICATION OF MASS-SPRING TRANSDUCERS

The mass-spring transducer may be adapted for the measurement of displacement, velocity, or acceleration, depending upon the natural frequency of the transducer and the type of transducing element used. Considerable versatility frequently is available in instrumentation systems for transforming one parameter to another. For example, velocity may be measured with an accelerometer by integrating the signal from the transducer. Often, this is done electrically and the recorded result indicates velocity directly.

12-6 INTRODUCTION TO SHOCK AND VIBRATION MEASUREMENTS

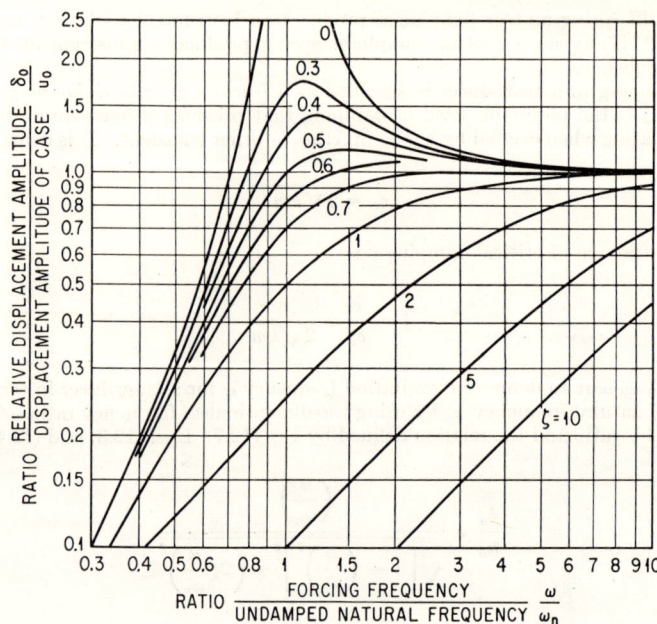

Fig. 12.4. Displacement response δ_o/u_o of a mass-spring system subjected to a sinusoidal displacement $u = u_o \sin \omega t$. The fraction of critical damping ζ is indicated for each curve.

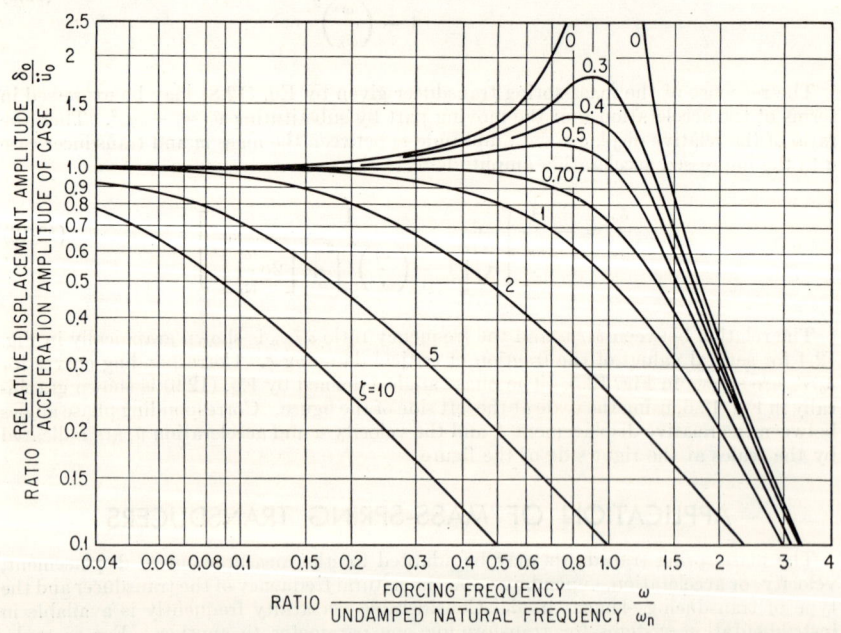

Fig. 12.5. Relationship between the relative displacement amplitude δ_o of a mass-spring system and the acceleration amplitude $\ddot{u}_o$ of the case. The fraction of critical damping ζ is indicated for each response curve.

DISPLACEMENT-MEASURING TRANSDUCERS

When ω/ω_n is substantially greater than 1.0, the ratio δ_o/u_o approaches the constant value of 1.0, approximately independent of frequency. If the signal obtained from the transducing element is directly proportional to δ_o, the transducer measures the displacement amplitude u_o to the approximation and over the frequency range indicated by Fig. 12.4. Therefore, the instrument is a displacement-measuring instrument. The addition of substantial damping makes the transducer useful for indicating displacement amplitude over a wider range of frequencies, but introduces limitations in the form of wave distortion (see *Phase Shift* in this chapter).

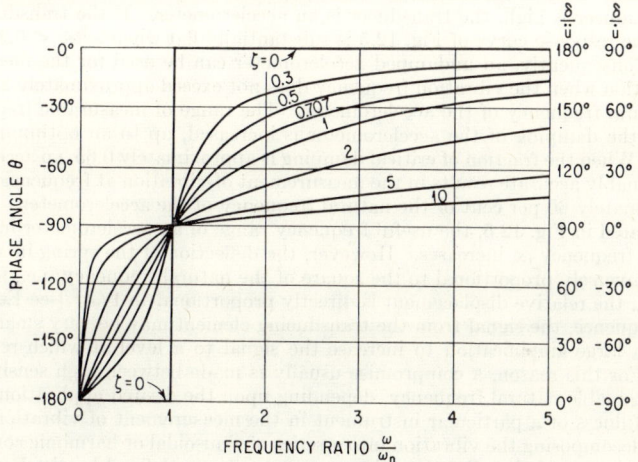

Fig. 12.6. Phase angle of a mass-spring transducer when used to measure sinusoidal vibration. The phase angle θ on the left-hand scale relates the relative displacement δ to the impressed displacement, as defined by Eq. (12.9). The right-hand scales relate the relative displacement δ to the impressed velocity and acceleration.

DISPLACEMENT LIMITS. Mass-spring transducers operating above their natural frequencies have several general characteristics in common, independent of the type of transducing element employed. The size of the transducer is determined by the requirement that the mass remain nearly stationary in space while the transducer case is moved about it. Thus, the internal clearance in both plus and minus directions of the sensitive axis must be larger than the largest displacement amplitude to be measured. Most transducers of this type contain "stops" to limit the relative displacement between case and mass, thereby defining the displacement limits of the transducer.

HIGH-FREQUENCY RESPONSE. The curves of Fig. 12.4 indicate that the response δ_o/u_o of the transducer is constant to infinitely high frequencies. This assumes that the mass and transducer case are ideal rigid structures and that the spring and dashpot are massless. In practical transducer designs, however, an upper limit of frequency is imposed by secondary resonances in the spring. A lower limit of displacement is imposed by noise or lack of resolution in the measuring system; this limit becomes significant at high frequency because displacements tend to become small as the frequency increases.

VELOCITY-MEASURING TRANSDUCERS *

A mass-spring transducer for measuring velocity usually is a transducer with a low natural frequency, utilizing the frequency range above its natural frequency. The rel-

---
* Several velocity-measuring transducers are described in Chap. 15.

ative velocity $d\delta/dt$ between the mass and case is measured with a velocity-sensing transducing element, such as a coil moving in a magnetic field. Since the mass-spring system of a velocity transducer operates in the same frequency range as that of a displacement transducer, it has the same limitation of size, input displacement, frequency range, and phase shift as a displacement-measuring instrument.

ACCELERATION-MEASURING TRANSDUCERS

As indicated in Fig. 12.5, the relative displacement amplitude δ_o is directly proportional to the acceleration amplitude $\ddot{u}_o = -u_o\omega^2$ of the sinusoidal vibration being measured, at small values of the frequency ratio ω/ω_n. Thus, when the natural frequency ω_n of the transducer is high, the transducer is an accelerometer. If the transducer is undamped, the response curve of Fig. 12.5 is substantially flat when $\omega/\omega_n < 0.2$, approximately. Consequently, an undamped accelerometer can be used for the measurement of acceleration when the vibration frequency does not exceed approximately 20 per cent of the natural frequency of the accelerometer. The range of measurable frequency increases as the damping of the accelerometer is increased, up to an optimum value of damping. When the fraction of critical damping is approximately 0.65, an accelerometer gives reasonably accurate results in the measurement of vibration at frequencies as great as approximately 60 per cent of the natural frequency of the accelerometer.

As indicated in Fig. 12.5, the useful frequency range of an accelerometer increases as its natural frequency ω_n increases. However, the deflection of the spring in an accelerometer is inversely proportional to the square of the natural frequency; i.e., for a given value of $\ddot{u}_o$, the relative displacement is directly proportional to $1/\omega_n^2$ [see Eq. (12.10)]. As a consequence, the signal from the transducing element may be very small, thereby requiring a large amplification to increase the signal to a level at which recording is feasible. For this reason, a compromise usually is made between high sensitivity and highest attainable natural frequency, depending upon the desired application.

The usefulness of a particular instrument in the measurement of vibration is determined by decomposing the vibration into a series of sinusoidal or harmonic components. For steady-state periodic vibration, these components are defined by the Fourier coefficients which give the amplitudes at the discrete frequencies of the components (see Chap. 22). Then it becomes possible to determine which of the components can be measured effectively by a particular instrument. A nonperiodic motion can be defined by its Fourier spectrum, which, in general, is a continuous function of frequency; i.e., all frequencies within a given range are present (see Chap. 23). A comparison of the Fourier spectrum with the frequency-response characteristics of the instrument, such as Figs. 12.4 and 12.5, indicates the extent to which the instrument can measure the nonperiodic motion effectively. If the spectrum includes frequency components beyond the range of capability of the instrument, these are deleted or modified as indicated by the instrument-response characteristics. Thus, Figs. 12.4 and 12.5 indicate in general the capability of an instrument in measuring both periodic and nonperiodic vibration.

RESPONSE TO SHOCK. The capability of an accelerometer in the measurement of shock may be evaluated by noting the response of the accelerometer to acceleration pulses. Ideally, the response of the accelerometer (i.e., the output of the transducing element) should correspond identically with the pulse. In general, this result may be approached but not attained exactly. Three typical pulses and the corresponding responses [1] of accelerometers are shown in Figs. 12.7 to 12.9. The pulses are shown in dotted lines, a sinusoidal pulse in Fig. 12.7, a triangular pulse in Fig. 12.8, and a rectangular pulse in Fig. 12.9. Curves of the response of the accelerometer are shown in solid lines. For each of the three pulse shapes, the response is given for ratios τ_n/τ of 1.014, 0.338, and 0.203, where τ is the pulse duration and $\tau_n = 1/f_n$ is the natural period of the accelerometer. These response curves, computed for the fraction of critical damping $\zeta = 0, 0.4, 0.7$, and 1.0, indicate the following general relationships:

1. The response of the accelerometer follows the pulse most faithfully when the natural period of the accelerometer is smallest relative to the period of the pulse. For example, the responses at A in Figs. 12.7 to 12.9 show considerable deviation between the pulse

APPLICATION OF MASS-SPRING TRANSDUCERS

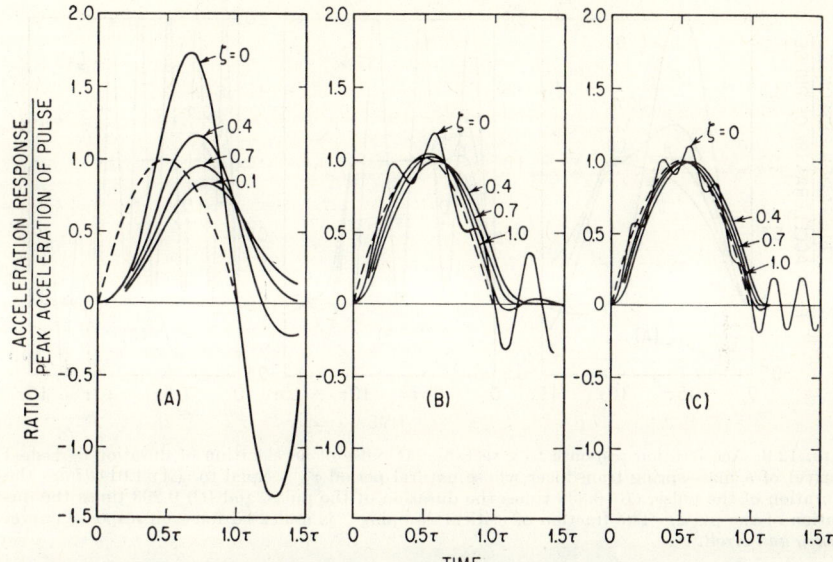

FIG. 12.7. Acceleration response to a half-sine pulse of acceleration of duration τ (dashed curve) of a mass-spring transducer whose natural period τ_n is equal to: (A) 1.014 times the duration of the pulse, (B) 0.338 times the duration of the pulse, and (C) 0.203 times the duration of the pulse. The fraction of critical damping ζ is indicated for each response curve. (Levy and Kroll.[1])

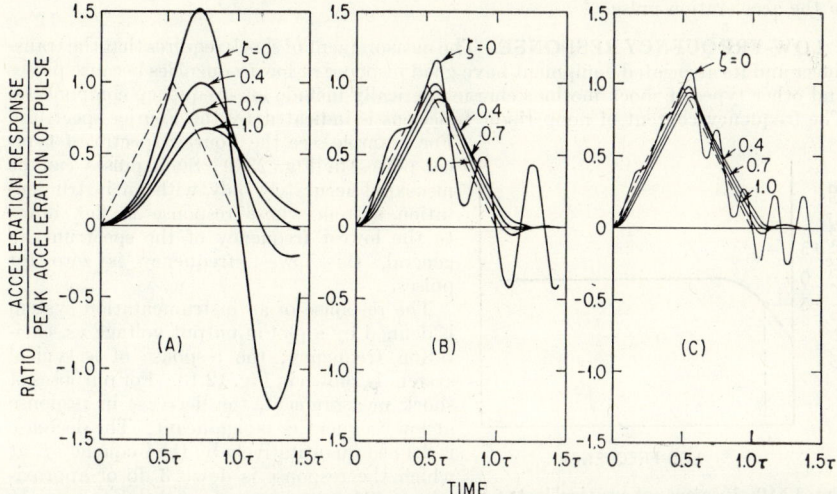

FIG. 12.8. Acceleration response to a triangular pulse of acceleration of duration τ (dashed curve) of a mass-spring transducer whose natural period τ_n is equal to: (A) 1.014 times the duration of the pulse, (B) 0.338 times the duration of the pulse, and (C) 0.203 times the duration of the pulse. The fraction of critical damping ζ is indicated for each response curve. (Levy and Kroll [1])

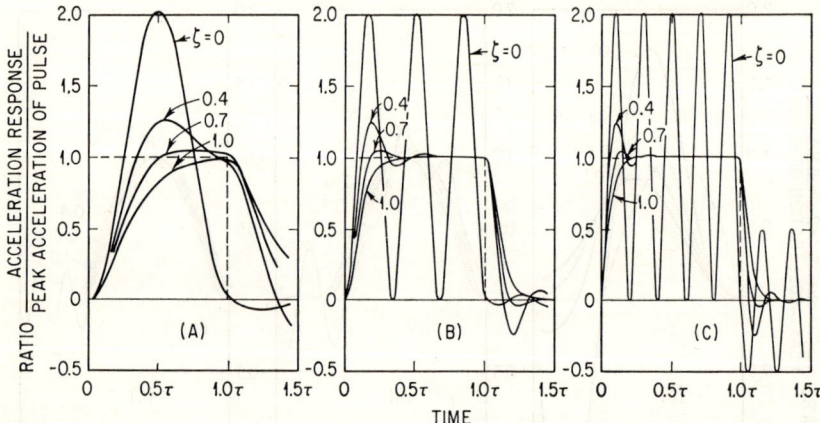

Fig. 12.9. Acceleration response to a rectangular pulse of acceleration of duration τ (dashed curve) of a mass-spring transducer whose natural period τ_n is equal to: (A) 1.014 times the duration of the pulse, (B) 0.338 times the duration of the pulse, and (C) 0.203 times the duration of the pulse. The fraction of critical damping ζ is indicated for each response curve. (*Levy and Kroll.*[1])

and the response; this occurs when τ_n is approximately equal to τ. However, when τ_n is small relative to τ (Figs. 12.7C to 12.9C), the deviation between the pulse and the response is much smaller.

2. Damping in the transducer reduces the response of the transducer at its own natural frequency; i.e., it reduces the transient vibration superimposed upon the pulse. Damping also reduces the maximum value of the response to a value lower than the actual pulse in the case of large damping. For example, in some cases a fraction of critical damping $\zeta = 0.7$ provides an instrument response that does not reach the peak value of the acceleration pulse.

LOW-FREQUENCY RESPONSE. The measurement of shock requires that the transducer and its associated equipment have good response at low frequencies because pulses and other types of shock motions characteristically include low-frequency components. The frequency content of nonperiodic functions is indicated by the Fourier spectrum. For example, see the Fourier spectra of typical pulses in Fig. 23.3. Such pulses can be measured accurately only with an instrumentation system whose response is flat down to the lowest frequency of the spectrum; in general, this lowest frequency is zero for pulses.

The response of an instrumentation system is defined by a plot of output voltage vs. excitation frequency; the response of a typical system is shown in Fig. 12.10. For purposes of shock measurement, the decrease in response at low frequencies is significant. The decrease is defined quantitatively by the frequency f_c at which the response is down 3 db or approximately 30 per cent below the flat response which exists at the higher frequencies. The distortion which occurs in the measurement of a pulse is related to the frequency f_c in a manner illustrated with respect to the example [2] of Fig. 12.11.

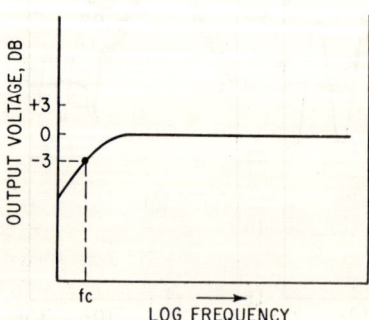

Fig. 12.10. Response of a typical instrument for measuring shock, illustrating drop-off at low frequencies. The low-frequency cutoff point f_c is the frequency at which the response is down 3 db or 30 per cent below the flat portion of the response curve.

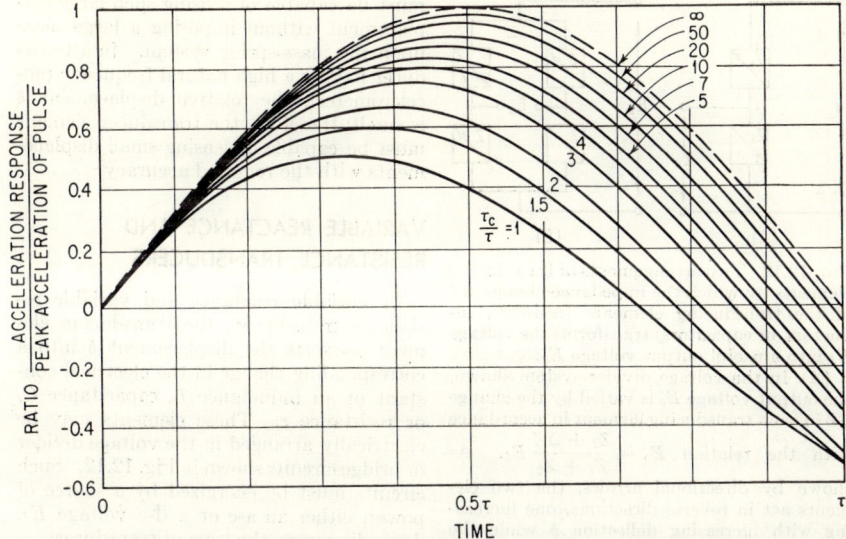

FIG. 12.11. Signal from a mass-spring accelerometer (solid line) in response to a half-sine pulse of acceleration (dotted line). The parameter of the family of curves is τ_c/τ, where τ is the duration of the acceleration pulse and $\tau_c = 1/f_c$ where f_c is the frequency defined in Fig. 12.10. The signal is given relative to the signal obtained when $f_c = 0$; i.e., when τ_c/τ is infinite. (*Lawrence*.[2])

Figure 12.11 shows by the dotted line a half-sinusoidal pulse of acceleration having a duration τ, and in solid lines the signal from a transducer having various values of the frequency f_c defined in Fig. 12.10. The frequency f_c is expressed in terms of its period $\tau_c = 1/f_c$; the parameter for the family of signal curves is τ_c/τ, where τ is the duration of the pulse shown by the dotted line. When $f_c = 0$, τ_c is infinitely great and the signal from the instrumentation system corresponds to the pulse. For values of f_c significantly greater than zero, the signal deviates from the pulse in the following respects: (1) The maximum value of the signal is lower than the maximum value of the pulse and (2) the signal exhibits a negative value near the termination of the pulse. As indicated in Fig. 12.11, this distortion becomes severe for a half-sinusoidal pulse when the period τ_c is of the same order of magnitude as the period τ of the pulse. Values of τ_c in excess of fifty times τ are desirable.

CHARACTERISTICS OF THE TRANSDUCING ELEMENTS

Most transducers for measurement of shock and vibration use a mechanical-to-electrical transducing element to transform the displacement δ between the transducer case and the spring-supported mass * into an electrical signal. Such a transducing element is used to convert motion into its electrical analog because (1) the electrical signal may be transmitted over considerable distances and (2) it may be used as the input to amplifiers, filters, analyzers, and recorders for data record and reduction purposes (see Chap. 19). The time-history of the electrical signal may be used to provide information concerning the frequency and waveform of the vibration as well as its magnitude.

In a transducer having a low resonant frequency, used for measurement of displacement and velocity, the relative displacement δ is large; then the transducer element

---

* This discussion relates explicitly to mass-spring transducers. However, it is equally applicable to the use of a fixed reference transducer by considering the transducing element to be interposed between the moving part and the fixed reference.

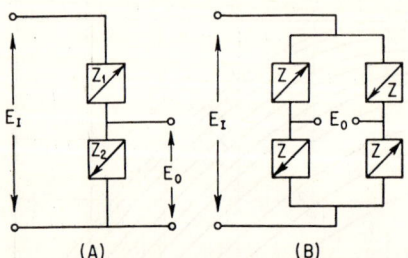

FIG. 12.12. Two arrangements of transducing elements by which the impedance change ΔZ of the transducing elements (resistors, inductors, or capacitors) transforms the voltage E_I into a useful output voltage E_o:

(A) In the voltage divider system shown, the output voltage E_o is varied by the change ΔZ in each transducing element in accordance with the relation $E_o = \dfrac{Z_2 + \Delta Z}{Z_1 + Z_2} E_I$. As shown by directional arrows, the two elements act in reverse directions, one increasing with increasing deflection δ while the other decreases with increasing deflection.

(B) In the full bridge shown, all elements are of the same value; thus, there is no static voltage when the transducer is at equilibrium. The bridge may contain (1) only one active transducing element, the other three remaining constant in value regardless of deflection δ, or (2) as many as four active transducing elements. The output voltage is $E_o = n \dfrac{\Delta Z}{4Z} E_I$ where the number of active elements is symbolized by n. Where more than one element is active, each must act in reverse direction from both adjacent elements, as shown by the directional arrows.

must be capable of sensing such large displacement without imposing a large force upon the mass-spring system. In a transducer having a high natural frequency (accelerometer), the relative displacement δ is small; therefore, the transducer element must be capable of sensing small displacements with the required accuracy.

VARIABLE REACTANCE AND RESISTANCE TRANSDUCERS

In variable reactance and variable resistance transducers, the transducing element converts the displacement δ into a corresponding change in the electrical constant of an inductance L, capacitance C, or resistance r. These elements may be electrically arranged in the voltage divider or bridge circuits shown in Fig. 12.12. Such circuits must be energized by a source of power, either an a-c or a d-c voltage E_I, depending upon the type of transducer:

1. A variable reactance transducer is operative only when the voltage E_I is an a-c voltage (carrier voltage) because the inductance or capacitance used as the transducing element responds only to an alternating voltage.

2. A variable resistance transducer is operative when the voltage E_I is either a-c or d-c.

In the circuit diagrams shown in Fig. 12.12, Z indicates a circuit element of known impedance. At least one of these elements is the transducing element of the transducer; its characteristics change in response to the relative deflection δ and a consequent change in voltage E_o results. A transducer may embody several similar transducing elements arranged in the circuit to increase the output of the circuit, with corresponding increase in sensitivity of the instrumentation system. For example, for a given deflection δ, the impedance Z_1 in Fig. 12.12A may increase and the impedance Z_2 may decrease correspondingly.

If the voltage E_I in Fig. 12.12A is d-c, a d-c voltage E_o results when the transducer is inactive. When the transducer is subjected to a motion, the impedance of the transducing element varies in a manner dictated by the motion and the voltage E_o changes accordingly. If the voltage E_I is a-c, the voltage E_o is a-c at the same frequency. Upon activation of the transducer, the impedance of the transducing element varies in a manner dictated by the motion and modulates the voltage E_I accordingly. For example, Fig. 12.13A shows the carrier voltage E_I, and Fig. 12.13B shows the change in impedance of the transducing element as a function of time. The modulated voltage E_o is shown in Fig. 12.13C. A voltage similar to the impedance variation of Fig. 12.13B is obtained from E_o by removing the carrier frequency (demodulation) by filters called discriminators or detectors. Demodulation cannot be carried out effectively unless the frequency of the carrier voltage is much greater;* e.g., ten times as great as the highest frequency component in the vibration being measured.

* There often is a practical upper limit to carrier frequency because the complexity of the supporting equipment increases as the carrier frequency increases.

A principal advantage of using the a-c carrier voltage E_I is that the instrumentation system is usable to zero frequency without the use of d-c amplifiers. Such amplifiers tend to experience zero drift over relatively long periods of time. An a-c amplifier is relatively stable; with the use of the a-c carrier voltage, it is possible to use a-c amplifiers throughout and to demodulate the signal immediately prior to recording.

A variable reactance transducing element also can be used as a part of a tuned circuit; the circuit oscillates at a fixed frequency until the transducing element (an inductance or capacitance) changes its value due to the relative displacement δ within the transducer.

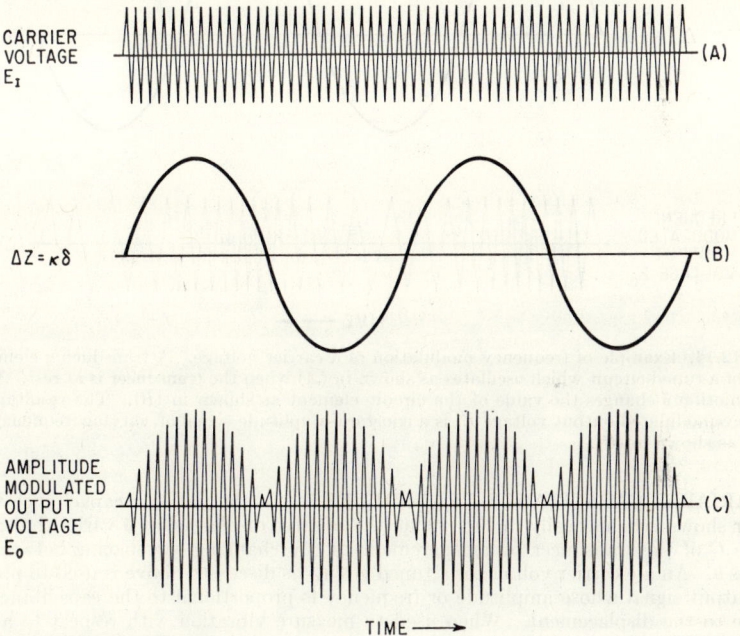

FIG. 12.13. Example of amplitude modulation of a carrier voltage. When a carrier voltage (A) is applied to a voltage bridge containing a transducing element experiencing a vibratory motion δ as shown in (B), the resultant amplitude-modulated signal E_o appears as shown in (C).

The oscillating frequency of the tuned circuit is shifted (frequency-modulated) by the change in the value of reactance. Figure 12.14 shows a typical frequency-modulated waveform obtained from such a system. The upper frequency limit of such a system is comparable to that of the amplitude modulation system discussed above.

VARIABLE INDUCTANCE TRANSDUCERS.* In the variable inductance transducer shown schematically in Fig. 12.15, a displacement of the mass results in a variation in the reluctance of the magnetic circuit; thus, the inductance L of the coil is varied by the increment ΔL. The auxiliary equipment used with a transducer of this type produces an output signal that is proportional in either amplitude or frequency to the inductance of the coil, and hence to the displacement. The resolution of a variable inductance transducing element usually is not adequate for very small displacements. Therefore, it is used most widely in displacement-measuring transducers or in accelerometers with relatively low natural frequencies, i.e., transducers with usable frequencies below approximately 100 cps.

* Several variable inductance transducers are described in Chap. 15.

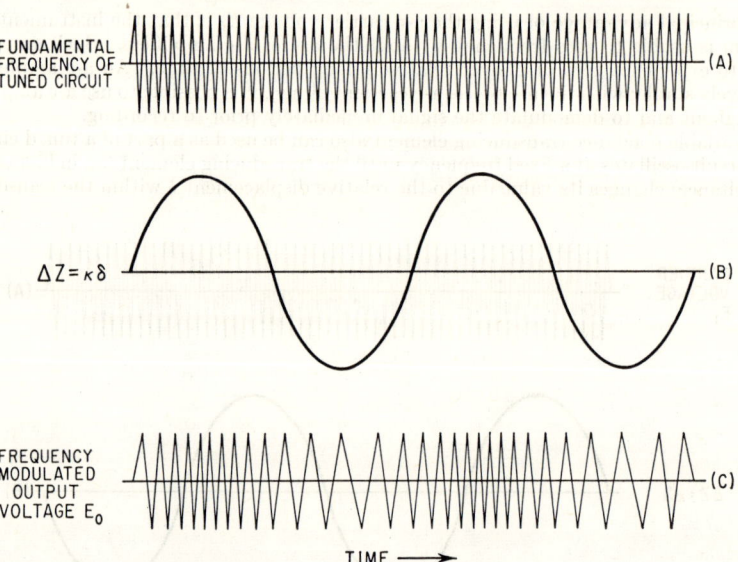

Fig. 12.14. Example of frequency modulation of a carrier voltage. A transducing element is part of a tuned circuit which oscillates as shown in (A) when the transducer is at rest. Vibratory motion δ changes the value of the circuit element as shown in (B). The resultant frequency-modulated output voltage E_o is a constant-amplitude signal of varying frequency content, as shown in (C).

VARIABLE CAPACITANCE TRANSDUCERS. In the variable capacitance transducer shown schematically in Fig. 12.16, the relative displacement δ varies the capacitance C of the transducer by an increment ΔC by changing the spacing between the plates b. An a-c carrier voltage or a tuned circuit as discussed above is used to produce an output signal whose amplitude or frequency is proportional to the capacitance and hence to the displacement. When used to measure vibration with respect to a fixed reference, one plate may be attached to the reference and the other to the moving part whose vibration is to be measured, with the resultant advantage that the vibrating part

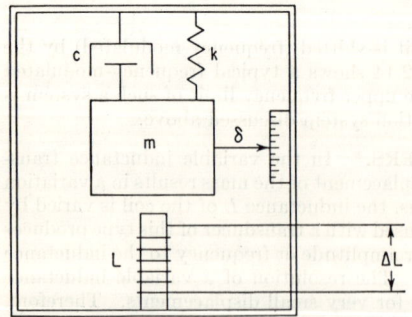

Fig. 12.15. Variable inductance transducer in which the relative motion δ of the internal mass changes the air gap of an inductive coil, causing a change in the inductance of the coil.

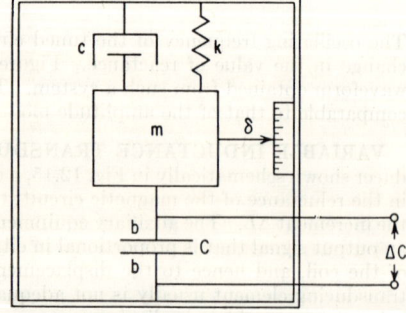

Fig. 12.16. Variable capacitance transducer in which the relative motion δ of the internal mass causes a change in the capacitance between fixed and moving plates.

is loaded only by the relatively small mass of the capacitor plate. A further advantage results from the high sensitivity of this transducing element to small displacements. Although a capacitance-type vibration pickup itself is relatively simple, the required auxiliary equipment is relatively complex, particularly because high-frequency carriers must be used to cover the normal vibration frequencies of interest. See Table 14.1 and Figs. 14.1 to 14.9 for additional information on variable capacitance transducers.

VARIABLE POTENTIOMETERS. One type of variable resistance transducer consists of a potentiometer with a resistance winding attached to the case and a "wiper" at-

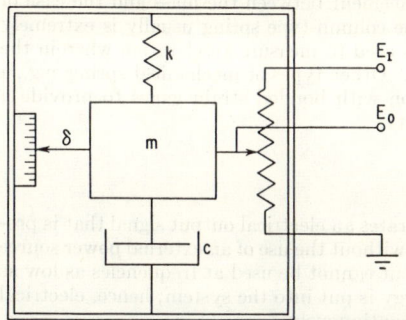

Fig. 12.17. Potentiometer transducer which experiences a change in resistance proportional to the relative displacement δ of the internal mass.

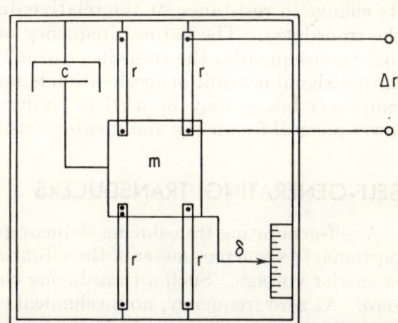

Fig. 12.18. Unbonded strain-gage transducer embodying four resistance wires r arranged to support the mass m to form a mass-spring system. The resistance of each wire experiences a change proportional to the relative displacement δ of the mass.

tached to the mass of a mass-spring system. The relative displacement δ between the case and the mass results in a proportionate change in resistance and hence a change in electrical output voltage ΔE_o when a voltage E_I is applied as shown in Fig. 12.17. A potentiometer-type transducer usually provides a relatively large output signal, and is usable at a frequency as low as zero. Because the resolution is limited by the diameter of the resistance wire in the potentiometer, a relatively large displacement δ is required to produce a usable signal. Hence, this type of transducer is used mainly in displacement-measuring instruments or in very low-frequency accelerometers.

STRAIN-GAGE TRANSDUCERS.* In a strain-gage transducer, the transducing element is one or more fine wires arranged to be elongated or relaxed in response to the relative displacement δ. The change in length of the wire causes a change in its resistance. The wires are arranged to form the elements Z of a bridge, as illustrated in Fig. 12.12B; such a bridge is operative if only one element responds to the relative displacement δ but has maximum sensitivity if all four elements are made responsive. A transducer may embody either unbonded or bonded wires.

Figure 12.18 shows schematically a type of variable resistance transducer which uses an *unbonded* strain gage. The mass m is supported by strain-sensitive wires r. These wires act both

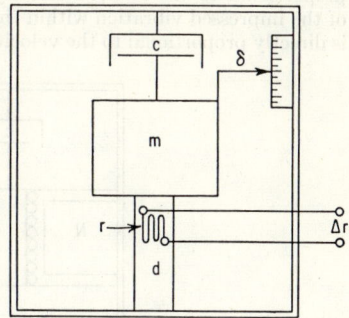

Fig. 12.19. Bonded strain-gage transducer embodying resistance strain gage r cemented to the support column (spring) for mass m. The resistance of the wire experiences a change proportional to the relative displacement δ of the mass.

---

*Strain-gage transducers are discussed in detail in Chap. 17.

as the spring in the mass-spring system and as the mechanical-to-electrical transducing element. As the wires are stretched and relaxed by relative displacements δ between the mass and case, they increase and decrease in resistance by an increment Δr. Such wires have a wide range of stiffnesses, giving possible natural frequencies of the mass-spring system from 100 to 2,000 cps. Transducers using unbonded strain gages usually are acceleration-measuring instruments.

Figure 12.19 shows one type of *bonded* strain-gage transducer in which the strain-sensitive wire r is bonded (e.g., cemented) to a column d which supports the mass. The column acts as the spring in the mass-spring system, and the strain gage measures by its change in resistance Δr the relative displacement between the mass and the case of the transducer. The natural frequency of the column-type spring usually is extremely high; consequently, the transducer usually is used to measure acceleration wherein the output signal per unit of acceleration is small. Other types of mechanical spring, e.g., a cantilever beam, may be used in conjunction with bonded strain gages to provide a lower natural frequency and greater sensitivity.

SELF-GENERATING TRANSDUCERS

A self-generating transducing element generates an electrical output signal that is proportional to some parameter of the vibration, without the use of an external power source or carrier voltage. Such a transducing element cannot be used at frequencies as low as zero. At zero frequency, no mechanical energy is put into the system; hence, electrical energy cannot be removed from the system continuously.

VELOCITY TRANSDUCERS.* The self-generating transducer shown schematically in Fig. 12.20 contains (1) a coil of wire mounted on relatively compliant springs to produce a low natural frequency and (2) a magnet to create a magnetic field in which the coil moves. Such a transducer is used to measure vibration having a frequency greater than its natural frequency so that the relative displacement δ is substantially equal to the impressed vibratory displacement u, according to Fig. 12.4. A voltage E_o is generated across the terminals of the coil according to the following equation:

$$E_o = -\frac{d\Phi}{dt}$$

where Φ is the magnetic flux and $d\Phi/dt$ is the rate of change of flux. The term $d\Phi/dt$ is proportional to the relative velocity $\dot{\delta}$, which is substantially equal to the velocity $\dot{u}$ of the impressed vibration within the useful frequency range; thus the electrical output is directly proportional to the velocity of vibration being measured.

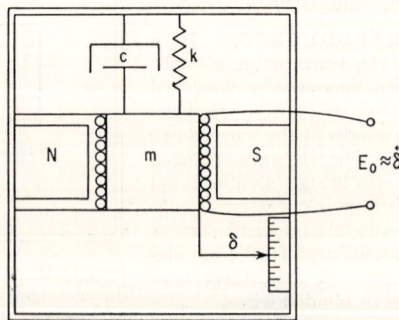

FIG. 12.20. Transducer for sensing velocity. It includes a magnet and a coil attached to the mass of the mass-spring system, the coil moving in the field of the magnet. The voltage generated in the coil is proportional to the relative velocity $\dot{\delta}$ of the mass.

*Transducers for measurement of velocity are discussed in Chap. 15.

This type of transducer produces large signals that can be integrated or differentiated to obtain displacement or acceleration. It is limited in low-frequency response by its natural frequency, often approximately 10 cps. Its high-frequency limit of 500 to 2,000 cps is determined by the usually small velocity values at high frequencies and by secondary resonances in the mass-spring system. A velocity pickup tends to be relatively large in size, both because a permanent magnet is required and because the coil must have adequate clearance to remain stationary in space as the case moves about it.

PIEZOELECTRIC TRANSDUCERS.* Figure 12.21 shows schematically an accelerometer in which the transducing element is a small disc of piezoelectric material K— a dielectric that generates an electric charge when it is compressed or extended. The

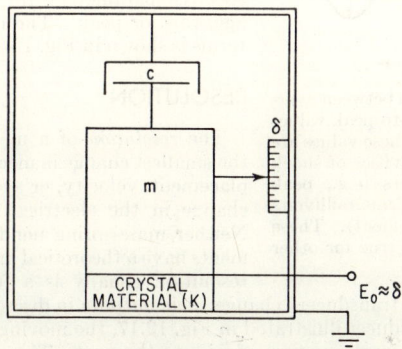

FIG. 12.21. Piezoelectric transducer in which the voltage generated is proportional to the relative motion δ of the mass; i.e., to the deflection of the crystal *between* its ends.

piezoelectric element may be either a natural or synthetic crystal or a ceramic material, such as barium titanate. Materials of this type are very stiff, and usually act as a part of the spring in the mass-spring system. A transducer which uses the piezoelectric element in compression may have its fundamental natural frequency as high as 100,000 cps but the natural frequencies of commercially available transducers usually are between 25,000 and 75,000 cps.† It is sensitive to small displacements, and provides large output signals. Furthermore, it can be made extremely small in size. The stiffness of the piezoelectric material makes such transducers most usable as accelerometers. The output impedance of a piezoelectric transducer usually is in the range of 100 to 10,000 mmfd of capacity. This capacitance and the input resistance of the circuit to which it is connected establish a time constant which limits the low-frequency response of the transducer. Therefore, piezoelectric transducers usually require cathode followers or other high impedance matching circuits for the measurement of vibration of low frequency.

ELECTROCHEMICAL TRANSDUCERS. An electrolytic cell can be built so that a flow of electrolytic solution is induced or modulated by either shock or vibration. This phenomenon is employed in several special transducers. See Chap. 14.

IMPORTANT CHARACTERISTICS OF SHOCK- AND VIBRATION-MEASURING INSTRUMENTS

SENSITIVITY

The *sensitivity* of a shock- and vibration-measuring instrument is the ratio of its electrical output to its mechanical input. The output usually is expressed in terms of voltage

---

* Piezoelectric transducers are discussed in detail in Chap. 16.
† The piezoelectric element also may be used in bending; then the natural frequency is substantially lower (see Chap. 16).

12–18 INTRODUCTION TO SHOCK AND VIBRATION MEASUREMENTS

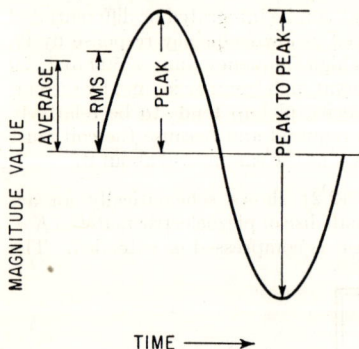

FIG. 12.22. Relationships between average, rms, peak, and peak-to-peak values for a simple sine wave. These values are used in specifying sensitivities of shock and vibration transducers (e.g., peak millivolts per peak g, or rms millivolts per peak-to-peak displacement). These relationships do *not* hold true for other than simple sine waves.

per unit of displacement, velocity, or acceleration. This specification of sensitivity is sufficient for instruments which generate their own voltage independent of an external power source. However, the sensitivity of an instrument requiring an external voltage usually is specified in terms of output voltage per unit of voltage supplied to the instrument per unit of displacement, velocity, or acceleration, e.g., millivolts per volt per g of acceleration. It is important to note the terms in which the respective parameters are expressed; e.g., average, rms, or peak. The relation between these terms is shown in Fig. 12.22. Also see Table 1.3.

RESOLUTION

The *resolution* of a measuring instrument is the smallest change in mechanical input, i.e., displacement, velocity, or acceleration, for which a change in the electrical output is discernible. Neither mass-spring nor fixed reference instruments have a theoretical limit of resolution. The resolution usually is a function of the transducing element. Some transducers change their outputs in discrete increments; e.g., in the potentiometer transducer illustrated in Fig. 12.17, the moving element in the transducer transfers contact from one turn of wire to the next. Therefore, the resolution is fixed by the resistance value of a single turn of the resistance wire.

Recording equipment, indicating equipment, and other auxiliary equipment used with the vibration-measuring instruments often establish the resolution of the over-all measurement system. If the electrical output of an instrument is indicated by a meter, the resolution may be established by the smallest increment that can be read from the meter. Resolution can be limited by noise levels in the instrument or in the system. In general, any signal change smaller than the noise level will be obscured by the noise, thus determining the resolution of the system.

TRANSVERSE SENSITIVITY *

If a transducer is subjected to vibration of unit amplitude along its axis of maximum sensitivity, the amplitude of the voltage output e_{max} is the sensitivity. The sensitivity e_θ along the X axis, inclined at an angle θ to the axis of e_{max}, is $e_\theta = e_{max} \cos \theta$, as illustrated in Fig. 12.23. Similarly, the sensitivity along the Y axis is $e_t = e_{max} \sin \theta$. In general, the sensitive axis of a transducer is designated. Ideally, the X axis would be designated the sensitive axis, and the angle θ would be zero. Practically, θ can be made

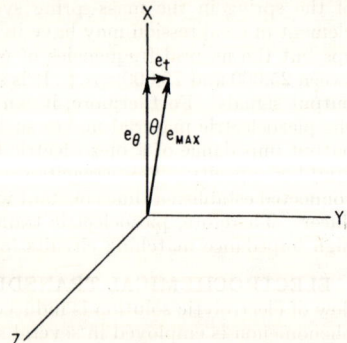

FIG. 12.23. The designated sensitivity e_θ and cross-axis sensitivity e_t that result when the axis of maximum sensitivity e_{max} is not aligned with the axis of e_θ.

only to approach zero because of manufacturing tolerances and/or unpredictable variations in the characteristics of the transducing element. Then the transverse sensitivity is expressed as the tangent of the angle θ; i.e., the ratio of e_t to e_θ:

$$\frac{e_t}{e_\theta} = \tan \theta$$

* Also called *lateral* or *cross-axis sensitivity*.

In practice, tan θ is between 0.01 and 0.10, and is expressed as a percentage. For example, if tan θ = 0.05, the transducer is said to have a transverse sensitivity of 5 per cent.

Figure 12.24 shows typical equipment used to determine transverse sensitivities. The table of the vibration machine on which the transducer is mounted may be rotated 360°

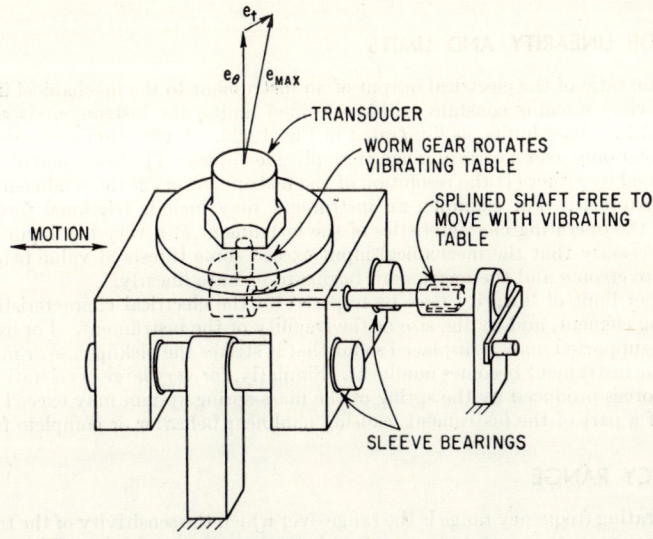

FIG. 12.24. Apparatus for investigating transverse sensitivity. The transducer being investigated is mounted on a surface vibrating at a known magnitude on an axis *normal* to the designated sensitive axis e_θ of the pickup. The mount is so constructed that while the transducer is in motion it may be rotated about its designated sensitive axis, thereby indicating transverse sensitivity along *all* axes normal to e_θ.

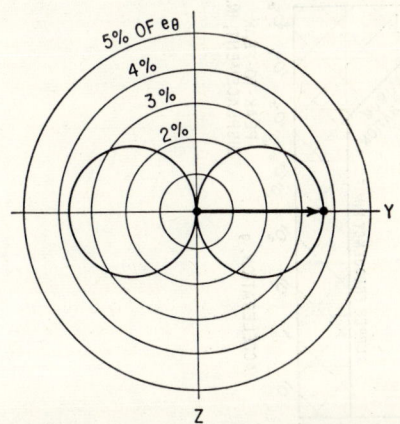

FIG. 12.25. Plot of transducer sensitivity in all axes normal to the designated axis e_θ, plotted according to axes shown in Fig. 12.23. Cross-axis sensitivity reaches a maximum e_t along the Y axis and a minimum value along the Z axis.

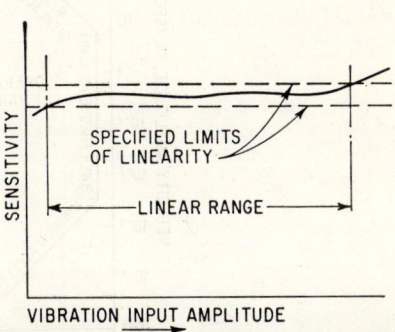

FIG. 12.26. Typical plot of sensitivity as a function of amplitude for a shock and vibration transducer. The *linear range* is established by the intersection of the sensitivity curve and the specified limits (dashed lines).

while the transducer is being vibrated normal to its axis of designated sensitivity e_θ. A typical result is the polar plot shown in Fig. 12.25. If the transverse sensitivity is due to a misalignment angle θ, the transverse sensitivity has both a major and a minor axis. The major axis sensitivity reaches a maximum value of e_t when the vibration is in the direction of the Y axis, and reaches a theoretical value of zero in the direction of the Z axis.

AMPLITUDE LINEARITY AND LIMITS

When the ratio of the electrical output of an instrument to the mechanical input (i.e., the sensitivity) remains constant within specified limits, the instrument is said to be "linear" within those limits, as illustrated in Fig. 12.26. A vibration-measuring instrument is linear only over a certain range of amplitude values. The lower end of this range is determined by either (1) the resolution of the instrument or (2) the nonlinear behavior of the instrument. For example, an instrument may include frictional forces which determine the operating characteristics of the instrument at a very low input level. It may be necessary that the mechanical input exceed some threshold value before static friction is overcome and the instrument begins to operate linearly.

The upper limit of linearity may be imposed by the electrical characteristics of the transducing element, and by the size or the fragility of the instrument. For example, if the spring-supported mass is displaced so far that it strikes the pickup case or mechanical "stops," the instrument becomes nonlinear. Similarly, for very large acceleration values, the large forces produced by the spring of the mass-spring system may exceed the yield strength of a part of the instrument, causing nonlinear behavior or complete failure.

FREQUENCY RANGE

The operating frequency range is the range over which the sensitivity of the transducer does not vary more than a stated percentage from the rated sensitivity. This range may be limited by the electrical or mechanical characteristics of the pickup or by its associated auxiliary equipment. These limits can be added to amplitude linearity limits to define completely the operating ranges of the instrument, as illustrated in Fig. 12.27.

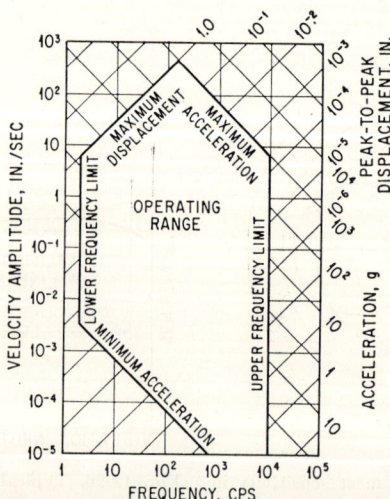

Fig. 12.27. Linear operating range of a transducer. Amplitude linearity limits are shown as a combination of displacement and acceleration values. The lower amplitude limits usually are expressed in acceleration values as shown.

SHOCK- AND VIBRATION-MEASURING INSTRUMENTS

LOW-FREQUENCY LIMIT. The *mechanical* response of a mass-spring transducer does not impose a low-frequency limit for an acceleration transducer because the transducer responds to vibration with frequencies less than the natural frequency of the transducer; however, there is a low-frequency limit for a displacement (or velocity) transducer because it responds significantly only to vibration whose frequency is greater than the natural frequency of the transducer.

In evaluating the low-frequency limit, it is necessary to consider the electrical characteristics of both the transducer and the associated equipment. In general, a transducing element that utilizes external power or a carrier voltage does not have a lower frequency limit, whereas a self-generating transducing element is not operative at zero frequency. The frequency response of amplifiers and other circuit components may limit the lowest usable frequency of an instrumentation system; frequently, an instrumentation system may be used to very low frequencies with a-c amplifiers through the use of a carrier voltage together with a discriminator placed after the amplifiers in the circuit.

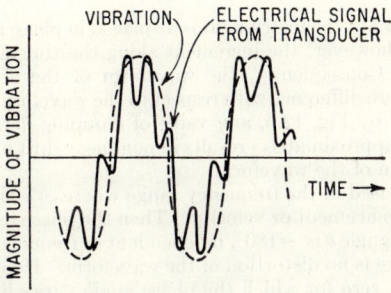

FIG. 12.28. Distorted response (solid line) of a lightly damped ($\zeta < 0.1$) mass-spring accelerometer to vibration (dashed line) containing a small harmonic content of the same frequency as the natural frequency of the accelerometer.

HIGH-FREQUENCY LIMIT. In principle, the mechanical response of a mass-spring transducer does not impose a high-frequency limit for a displacement (or velocity) transducer because the transducer responds to all vibration with frequencies greater than the natural frequency of the transducer. In practice, however, the transducer is usable only at frequencies below the natural frequencies of mechanical components of the transducer.

An acceleration transducer (accelerometer) has an upper usable frequency limit because it responds to vibration whose frequency is less than the natural frequency of the transducer. The limit is a function of (1) the natural frequency and (2) the damping of the transducer, as discussed with reference to Fig. 12.5. An attempt to use such a transducer beyond this frequency limit may result in distortion of the signal, as illustrated in Fig. 12.28.

The upper frequency limit for slightly damped vibration-measuring instruments is important because these instruments exaggerate the small amounts of harmonic content that may be contained in the motion, even when the operating frequency is well within the operating range of the instrument. The result of exciting an undamped instrument at its natural frequency may either damage the instrument or obscure the desired measurement. Figure 12.28 shows how a small amount of harmonic distortion in the vibratory motion may be exaggerated by an undamped transducer.

The upper frequency limit of transducers may be established by the electrical characteristics of the transducing element. For example, in some transducers a fixed frequency carrier voltage is required to excite the transducing element. The signal from the transducing element can be separated from the carrier signal only if the frequency of the vibration is a small percentage of the carrier frequency. This establishes an upper limit for the operating frequency range.

PHASE SHIFT

Phase shift is the time delay between the mechanical input and the electrical output signal of the instrumentation system. Unless the phase-shift characteristics of an instrumentation system meet certain requirements, a distortion may be introduced that consists of the superposition of vibration at several different frequencies. Consider first an accelerometer, for which the phase angle θ_1 is given by Fig. 12.6. If the accelerometer is undamped, $\theta_1 = 0$ for values of ω/ω_n less than 1.0; thus, the phase of the relative displacement δ is equal to that of the acceleration being measured, for all values of frequency within the useful range of the accelerometer. Therefore, an undamped accelerometer measures acceleration without distortion of phase. If the fraction of critical damping ζ for the accelerometer is 0.65, the phase angle θ_1 increases approximately linearly with the frequency ratio ω/ω_n, within the useful frequency range of the accelerometer. Then the expression for the relative displacement may be written

$$\delta = \delta_0 \cos(\omega t - \theta) = \delta_0 \cos(\omega t - a\omega) = \delta_0 \cos \omega(t - a)$$

Thus, the relative motion δ of the instrument is displaced in phase relative to the acceleration $\ddot{u}$ being measured; however, the increment along the time axis is a constant independent of frequency. Consequently, the waveform of the accelerometer output is undistorted but has a phase difference with respect to the waveform of the vibration being measured. As indicated by Fig. 12.6, any value of damping in an accelerometer other than $\zeta = 0$ or $\zeta = 0.65$ (approximately) results in nonlinear shift of phase with frequency and consequent distortion of the waveform.

When a transducer is used in the frequency range where ω/ω_n is substantially greater than 1.0, it measures displacement or velocity. Then the phase relations are less favorable. If $\zeta = 0$, the phase angle θ is $-180°$, independent of frequency; this represents only a change of sign and there is no distortion of the waveform. However, there is no value of damping greater than zero for which the phase angle varies linearly with frequency throughout the range of frequency for which the transducer is useful unless the damping is very large. Thus, a displacement or velocity-measuring instrument with a moderate degree of damping always introduces distortion of complex or superimposed waveforms, depending on the degree of damping in the instrument.

CALIBRATION REQUIREMENTS

The calibration of shock- and vibration-measuring instruments consists in determining the relationship of the output (e.g., electrical or mechanical) of the transducer to the input (displacement, velocity, or acceleration), i.e., determining its sensitivity. Calibration methods are described in detail in Chap. 18. The type and amount of calibration information required depend both on the type of instrument and on its intended application; in general, the following information is required:

1. *The sensitivity over the frequency range of interest.* It is necessary to know the ratio of the output to the input at all frequencies where measurements are to be made. Usually this ratio is measured at a frequency near the center of the range; then a plot is made of the deviation of this value as the frequency is varied.

2. *The sensitivity over an environmental range of interest.* If accurate results are to be obtained over a wide range of environmental conditions, it is necessary to determine the effects upon the characteristics of the transducer of temperature, supply voltage variation, radiation, acoustic noise, electromagnetic field, altitude, and humidity.

3. *The sensitivity over an amplitude range of interest.* If the vibration (or shock) level to be measured is expected to have a large range of magnitude, the characteristics of the transducer should be measured at both high and low vibration (or shock) levels to determine the effects of noise, resolution, and nonlinearity.

4. *Stability of calibration with time.* The sensitivity of most vibration- and shock-measuring instruments is relatively stable with time. However, instruments should be recalibrated on a regular basis to ensure continued accuracy and reliability.

ENVIRONMENTAL EFFECTS

TEMPERATURE. The sensitivity, natural frequency, and damping of a transducer may be affected by temperature. The specific effects produced depend on the type of transducer and the details of its design. For example, Fig. 12.29 shows the variation of damping with temperature for several different damping media. Either of two methods may be employed to compensate for temperature effects: (1) the temperature of the pickup may be held constant by local heating or cooling or (2) the pickup characteristics may be measured as a function of temperature; if necessary, the appropriate corrections can then be applied to the measured data.

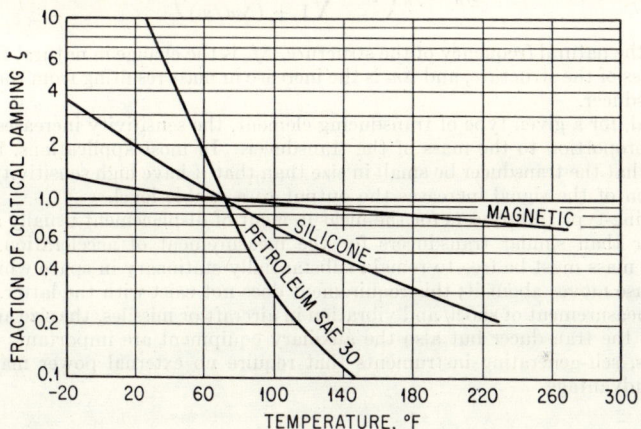

Fig. 12.29. Variation of damping with temperature for different damping means. The ordinate indicates the fraction of critical damping ζ at various temperatures assuming $\zeta = 1$ at 70°F.

HUMIDITY. Humidity may affect the characteristics of certain types of vibration instruments. In general, a transducer which operates at a high electrical impedance is affected by humidity more than a transducer which operates at a low electrical impedance. It usually is impractical to correct the measured data for humidity effects. However, instruments that might otherwise be adversely affected by humidity often are sealed hermetically to protect them from the effects of moisture.

ACOUSTIC NOISE. Acoustic energy of high intensity often accompanies vibration. If structures of the transducer or auxiliary equipment can be excited acoustically to vibrate, there may be large error signals. Sometimes the vibration to be measured is the source of the acoustic energy; in other cases, the acoustic energy induces the vibration to be measured. Only in the latter case is the ratio of the acoustic energy to the vibration energy high enough to produce serious errors. In general, an accelerometer subjected to acoustic noise does not produce an electrical output equivalent to the voltage produced by an acceleration of $1g$ until the sound pressure level of the acoustic noise exceeds 150 db. With such a high sound pressure level it is likely that the vibration level is large and the error introduced by the acoustic noise is not apt to be important. Figure 16.21 shows the output of several types of piezoelectric accelerometers as a function of the sound pressure level of the sound field to which they are exposed. Where it is necessary to measure vibration of small magnitude in a high-intensity acoustic field, the response of the instrument to the acoustic field alone should be measured to determine the effect of the noise on the measurement of vibration. In performing such a test, the transducer is mounted so that its mounting will not be set into vibration by the acoustic field.

PHYSICAL PROPERTIES

Size and weight of the transducer are very important considerations in many vibration and shock measurements. A large instrument may require a mounting structure that will change the local vibration characteristics of the structure whose vibration is being measured. Similarly, the added mass of the transducer may produce substantial changes in the vibratory response of such structure. Generally, the natural frequency of a structure is lowered by the addition of mass; specifically, for a simple spring-mass structure:

$$\Delta f_n = f_n \left(1 - \sqrt{\frac{1}{1 + (\Delta m/m)}}\right)$$

where f_n is the natural frequency of the structure, Δf_n is the change in natural frequency, m is the mass of the structure, and Δm is the increase in mass resulting from the addition of the transducer.

In general, for a given type of transducing element, the sensitivity increases approximately in proportion to the mass of the transducer. In most applications, it is more important that the transducer be small in size than that it have high sensitivity because amplification of the signal increases the output to a usable level.

Mass-spring-type transducers for the measurement of displacement usually are larger and heavier than similar transducers for the measurement of acceleration. In the former, the mass must be free to remain substantially stationary in space while the instrument case moves about it; this requirement does not exist with the latter.

For the measurement of shock and vibration in aircraft or missiles, the size and weight of not only the transducer but also the auxiliary equipment are important. In these applications, self-generating instruments that require no external power may have a significant advantage.

AUXILIARY EQUIPMENT

CABLE EFFECTS. Proper functioning of a vibration-measuring system is dependent on adequate electrical and mechanical characteristics of the cable connecting the transducer to the associated auxiliary equipment. A cable that is either stiff or heavy may alter the vibration response of the structure to which the instrument is mounted. The electrical characteristics of cables are particularly important with instruments having high electrical impedance, e.g., piezoelectric accelerometers. The mechanical distortion of the cable may generate an electrical signal or introduce extraneous electrical noise into the circuit, as explained in Chap. 19. Special cables are available to minimize these effects. The resistance, capacitance, and electrical impedance of the cable are also factors affecting the frequency response of many different types of transducing elements.

AUXILIARY ELECTRONIC EQUIPMENT.* External power frequently is required for energizing the transducer as well as for amplifying and other signal-conditioning functions. The auxiliary equipment is an important factor in determining the over-all accuracy of the system. Important factors are total size, weight, reliability, maintenance, and cost.

REFERENCES

1. Levy, S., and W. D. Kroll: *Research Paper* 2138, *J. Research Natl. Bur. Standards,* **45**:4 (1950).
2. Lawrence, A. F.: "Crystal Accelerometer Response to Mechanical Shock Impulses," *Rept.* 168-56-16, Sandia Corp., Albuquerque, N.M.

* Auxiliary electronic equipment is discussed in detail in Chap. 19.

13

SELF-CONTAINED VIBRATION-MEASURING INSTRUMENTS

Fred Mintz
Lockheed Aircraft Corporation

John J. Dreher
Lockheed Aircraft Corporation

INTRODUCTION

This chapter describes some well-known and commercially available instruments for the measurement of vibration. In general, only so-called "self-contained" instruments are included, i.e., those instruments whose components are confined to a single, easily portable case (with possibly the addition of a transducer which is connected to the case by lead wires). This type of instrument combines the advantages of portability and ease of operation with the disadvantages inherent in most relatively simple instrument packages: the read-out capability does not contain either the degree of accuracy or the precision obtainable by using relatively more delicate and complex laboratory-type equipment. A number of rugged and efficient instruments are described which are suitable for use, both in the field and laboratory, by either the engineer or technician.

CLASSIFICATION OF INSTRUMENTS

The instruments described in this chapter are classified in the following way:

MASS-SPRING VIBRATION INSTRUMENTS
Vibrographs
Vibrometers
Frequency-indicating vibrometers
Mechanical recording accelerometers
Seismographs
Electronic vibration meters and recorders
Shock recorders
Impact recorders
"Step gages" for shock measurement
Torsional-vibration recorders

13-1

FIXED REFERENCE INSTRUMENTS

Wedge-type visual displacement indicator
Visual techniques—measuring microscope
Visual techniques—stroboscopes
Framing cameras
Streak photography
Mechanical strain recorders

The principal design and operational features of these classes of instruments are described in succeeding sections, followed by specific detailed data on commercially available instruments.

MASS-SPRING TYPE VIBRATION INSTRUMENTS

PRINCIPLES OF OPERATION

The ideal mass-spring type pickup which is shown in Fig. 13.1 (sometimes called a "seismic instrument") consists of a mass supported on a spring and dashpot in a case. The pickup case is attached to the vibrating surface by suitable means and its motion is essentially the same as that of the surface. The relative motion between the case and the spring-supported mass is the response of the instrument. The natural frequency of the mass-spring system determines the response characteristics of the instrument.

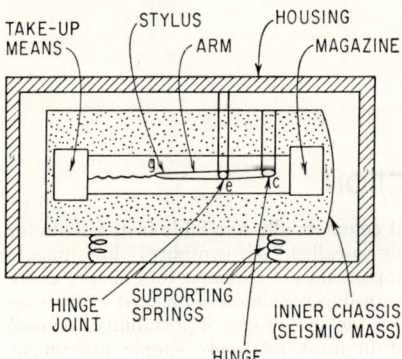

FIG. 13.1. Principle of operation of the vibrograph.

DISPLACEMENT MEASUREMENT. As shown in Chap. 12, for a sinusoidal applied motion, if the frequency of applied motion is much higher than the natural frequency of the mass-spring system, then the response displacement amplitude is approximately equal to the displacement amplitude of applied motion. This is the characteristic of a displacement pickup of this type. If the frequency of applied vibration is greater than three times the natural frequency of the instrument, the error in determining displacement amplitude is less than 10 per cent, becoming smaller as the frequency ratio increases. Physically, this means that for a high ratio of frequency of applied motion to natural frequency of the mass-spring system, the mass remains essentially motionless. This provides a virtually stationary frame-of-reference against which to measure the applied motion. Since the natural frequency of the mass-spring system must be considerably lower than the frequency of the vibration to be measured, the vibration pickup usually has a "soft" spring suspension to give it a suitably low natural frequency (usually 10 cps or lower).

Many of the self-contained commercially available vibration pickups are designed, for the sake of convenience, to be hand-held rather than to be fastened to the vibrating surface. In a device of this type, the instrument has a prod which is held against the surface whose vibration is to be measured (see Fig. 13.2). The prod is forced against the surface by a spring fastened to the case of the instrument. The spring, which pushes against the case also, acts as the spring element of the mass-spring system—the mass of which is provided by the remainder of the instrument. In principle, the hands of the operator support the case in such a way that the case acts as the mass. In practice, the operator's hands introduce an indeterminate and variable amount of mass, stiffness, and damping. Consequently, the hand vibration pickup is not a true mass-spring type of instrument. However, when used within its frequency limitations, the hand-held vibration pickup often provides a satisfactory approximation to a true mass-spring type of displacement pickup.

ACCELERATION MEASUREMENT. The acceleration pickup has essentially the same features of design, construction, and behavior as the mass-spring type of displacement pickup. In contrast to the latter, however, the acceleration pickup has an element having a relatively stiff spring to give it a relatively high natural frequency. Physically, the behavior of the accelerometer pickup may be described as follows: Consider a mass-spring system which is less than critically damped. Suppose that the system is subjected to a sinusoidal vibration and has a natural frequency several times greater than the frequency of applied vibration. Then, the spring experiences an alternating force of amplitude equal to the mass multiplied by the acceleration amplitude; consequently, the spring deflection (i.e., the response of the system) is proportional to the applied acceleration.

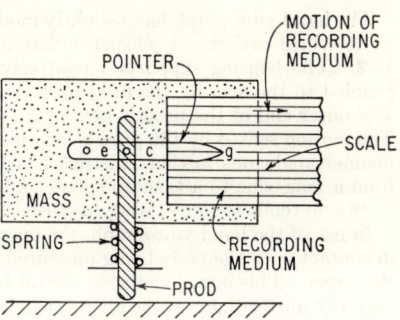

FIG. 13.2. Schematic diagram of the hand vibrograph. The pointer arm is supported by the housing (mass) and is coupled to a prod. The end of the prod is held against the vibrating surface. Motion of the prod is magnified by the ratio $(e-g)/(e-c)$ and inscribed on a moving recording medium. The recording medium and the associated drive mechanism are a part of the mass.

In the acceleration pickup, adequate damping (i.e., 0.5 to 0.7 of critical damping) is more important than in the displacement pickup: (1) to minimize free vibration of the seismic mass when the instrument is subjected to a transient motion; if this vibration were not damped rapidly, the free oscillation would obscure the record of the applied motion; (2) to permit the use of the instrument at frequencies up to 0.6 of its natural frequency with an acceleration-amplitude error not exceeding 5 per cent, and up to its natural frequency with an acceleration amplitude error not exceeding 20 per cent.

VIBROGRAPHS

A vibrograph is a mass-spring type of displacement pickup with self-contained means for recording the relative motion between the mass and the case consisting of a recording medium contained on a supply spool, a transport mechanism for the recording medium, and a take-up spool. The principle of operation of the vibrograph is shown by the schematic diagram of Fig. 13.1. A housing supports an inner chassis (which acts as the mass) on supporting springs. A relatively light arm is supported from the chassis by the hinge attached to one end of the arm. A hinge joint attaches the arm to the housing as shown. The relative motion between the housing and the chassis is inscribed on a recording medium by the stylus, being magnified by the lever ratio $(e-g)/(e-c)$. The ratio between the displacement of the stylus and the relative motion between the recorder and inertia mass is called the *magnification ratio*.

The recording medium is in strip form feeding from the magazine to the takeup means. In general, the paper * speed may be varied between given limits and means are provided in the various commercially available instruments for determining the paper speed accurately. (For simplicity, the paper-drive mechanism is not shown; in some vibrographs, it is part of the chassis; in others, the recording mechanism is supported on the housing.)

A *pallograph* is a low-frequency vibrograph of the type designed and developed at the David Taylor Model Basin of the U.S. Navy, based on the original pallographs described in Ref. 1. These instruments are especially useful for recording vibratory motions of ship hulls, structure, and machinery.

---

* The word "paper" is used here in a generic sense to indicate a material on which the stylus writes. The various commercially available instruments use different materials, as described later with reference to the specific instruments.

The hand vibrograph has a slightly modified design, to allow for the action of the prod, as described earlier. A schematic diagram of this type of vibrograph is shown in Fig. 13.2. The housing supports a relatively light arm upon a fixed pivot; a prod also is coupled to the arm by the prod pivot and extends through an opening in the housing. The outer end of the prod is held against the body whose vibration is to be measured. The motion sensed by the prod is inscribed on paper by the stylus, being magnified in a manner analogous to that illustrated in Fig. 13.1. The paper is in strip form, feeding from a magazine to a take-up spool. A spring interposed between the housing and the lever arm tends to force the stylus toward the prod side of the instrument.

In use of the hand vibrograph, the operator holds the housing and maintains the prod in contact with the body being measured, applying enough force to center the stylus on the paper. This introduces a significant limitation in the use of the instrument, namely, that any unsteadiness of the operator in holding the instrument is recorded on the paper. It sometimes is difficult to distinguish this unsteadiness from low-frequency vibration, particularly if the measurements are made in a moving vehicle or under other conditions where steadiness is difficult to maintain. A steady, vibration-free support for the instrument is advantageous but often unavailable. Some commercially available instruments * may be used either as conventional mass-spring type of instruments, or with especially constructed supports.

If the vibratory accelerations are high, because of large amplitudes or high frequencies of motion, the prod may not be able to follow the motion; instead it may chatter. In some instruments, the spring force is adjustable and may be increased to keep the prod in contact with the vibrating surface at relatively high accelerations.

The principal advantages of the vibrograph are (1) a permanent record of vibratory displacement as a function of time is available immediately, thus giving an indication of the waveshape and frequency content of both steady-state and transient vibration; and (2) the instrument is portable, convenient, and ready for immediate use without assembling auxiliary equipment.

There is a pronounced disadvantage in the form of the record—it is not well adapted to detailed data reduction except by laborious graphical methods, although estimates of frequency and amplitude may be obtained by inspection. The instrument is limited to the measurement of vibration whose frequency is relatively low and whose amplitude is relatively large, although the amplitude can be extended to a lower value by the use of a microscope to analyze the record. The improved resolution obtained by use of a microscope generally is limited somewhat by the "fuzziness" of the trace at moderate magnification. Certain recording media permit greater useful magnification than others; for example, a trace engraved on cellulose-acetate film is smoother than one inscribed on wax-coated paper.

Another significant disadvantage of the vibrograph is the relatively large loading imposed on the vibrating surface which usually makes it impractical for use in recording vibration of relatively light structures. Most vibrographs weigh several pounds, although some are many times heavier. The loading imposed on the structure being measured is that due to the "unsprung" weight of the vibrograph, plus that due to the reaction forces in the springs supporting the mass of the mass-spring system. For hand vibrographs, only the spring-reaction loading is imposed via the prod; however, even this can modify the vibration of a light structure, thereby introducing error into the measurement.

The upper limit of the useful frequency range frequently is determined by the dynamics of the mechanical linkage and the recorder stylus.

ASKANIA HAND VIBROGRAPH.† This vibrograph is a lightweight instrument for the recording of mechanical vibration. It is comprised of two principal parts: a housing with feeler tube and stylus assembly, and a chassis which includes a spring-motor drive, battery housing, and time markers. The assembly is held together by two lateral holding screws; the release of these screws allows the two major parts to separate. The feeler tube contains a lightweight rod with a prod on the end, which moves in the longitudinal

---

* The Westinghouse LE, Davey, and Cambridge vibrographs.
† Manufactured by Askania-Werke AG, Berlin; distributed by EPIC, Inc., New York.

direction. It is pressed against the object under test by means of a spring whose tension is adjustable. For light structures, a weak contact pressure usually is used; when measuring vibration with high acceleration, stronger contact pressure is required which imposes greater loading on the surface being measured. The feeler rod transmits the movement to the stylus, which is pivoted at the feeler tube. The stylus has a sapphire point which records on waxed paper tape.

Windows on both the front and rear of the instrument allow observation of the recording during measurement. The rear window, when open, allows removal of the used tape, which is not taken up on a spool but allowed to accumulate in a small compartment. A hinged mirror over the front window permits observation of the stylus when the unit is being used for vertical vibration measurement.

The Askania Hand Vibrograph has both the advantages and disadvantages of the typical hand vibrograph described above. The fact that the record paper is not taken up on a spool, but is allowed to accumulate in a little chamber, offers a slight advantage in that the record may be inspected readily without unrolling a previously accumulated record or disassembling the instrument; however, because only a limited length of record can be accumulated in the take-up chamber, separate records must be torn off and rolled up at intervals, increasing the possibility of loss or damage to the record.

Special accessories:

1. Feeler tubes which may be selected to provide the following magnification factors: 1:1, 5:1, 20:1, 50:1 (a 5:1 ratio usually is provided).
2. An attachment which provides motion reductions of 5:1 or $2\frac{1}{2}:1$; it can be used with any of the feeler tubes.
3. A clamping device which provides a firm base for supporting the vibrograph when it is used for relative motion measurements.
4. A rod providing an extension of 100 mm; several rods may be used in cascade.
5. A "joint connection" which provides direct connection of the feeler point with a vibrating surface.
6. A roller fork, which permits making measurements on a rotating shaft.
7. A pressure-measuring attachment, consisting of a fitting that includes a calibrated diaphragm whose displacement is recorded.
8. A torsional-vibration-measuring attachment (see *torsional-vibration pickups*).
9. An electromagnetic remote control for actuating the paper drive.
10. An evaluation plate and magnifier which facilitate the reading of displacement and frequency on a record. This glass plate has a coordinate-lined section with millimeter line markings. The magnifier is a three-power unit which has a millimeter coordinate scale on the lower side.

Approximate amplitude limits available with the various magnifications are shown in Table 13.1. When the unit is used in the conventional manner as a hand-held vibrograph, the lower limit of the useful frequency range (indicated below) is determined by the nominal natural frequency of 5 cps. The upper frequency limit of about 250 cps (this

Table 13.1. Approximate Limits of Useful Range of Askania Hand Vibrograph

Feeler tube ratio	Displacement amplitude		Acceleration limit *
	Min., mm	Max., mm	
1:1	±0.1	±12	±50g (Model A)
			±100g (Model B)
5:1	±0.02	±2.4	±20g (Model A)
			±50g (Model B)
20:1	±0.005	±0.6	±20g (Model A)
			±50g (Model B)
50:1	±0.003	±0.2	

* At highest spring tension.

Note: Models A and B represent the same instrument with different tension springs; Model B has a heavier spring

SELF-CONTAINED VIBRATION-MEASURING INSTRUMENTS

upper limit varies with amplitude, as shown in Fig. 13.3) is set by two factors: (1) the extremely small displacements associated with high-frequency vibration that are within the acceleration limits of the instrument; (2) resonance of the mechanical elements of the instrument linkages, including the stylus.

Technical Data

Natural frequency:	5 cps (approximately)
Frequency range:	10 to 250 cps (approximately)
Magnification ratio:	1:1; 5:1; 20:1; 50:1 (with different feeler tubes; motion reduction attachment providing ratios of 1:2.5 and 1:5 may be used with any feeler)
Recording medium:	Waxed tape 25 mm wide, 9 meters (30 ft) long in roll
Paper drive:	Key-wound spring with governor
Paper speed:	About 40 mm/sec
Time-and-event markers:	(1) Internally energized
	(2) Externally energized magnetic marker
Dimensions:	25 by 14 by 8 cm (10 by 5½ by 3¼ in.)
Weight:	1.7 kg (3¾ lb)
Power supply:	Self-contained flashlight cell

ASKANIA UNIVERSAL VIBROGRAPH. The Askania Universal Vibrograph, shown in Fig. 13.4, is a self-powered portable vibration recorder of the mass-spring type. The

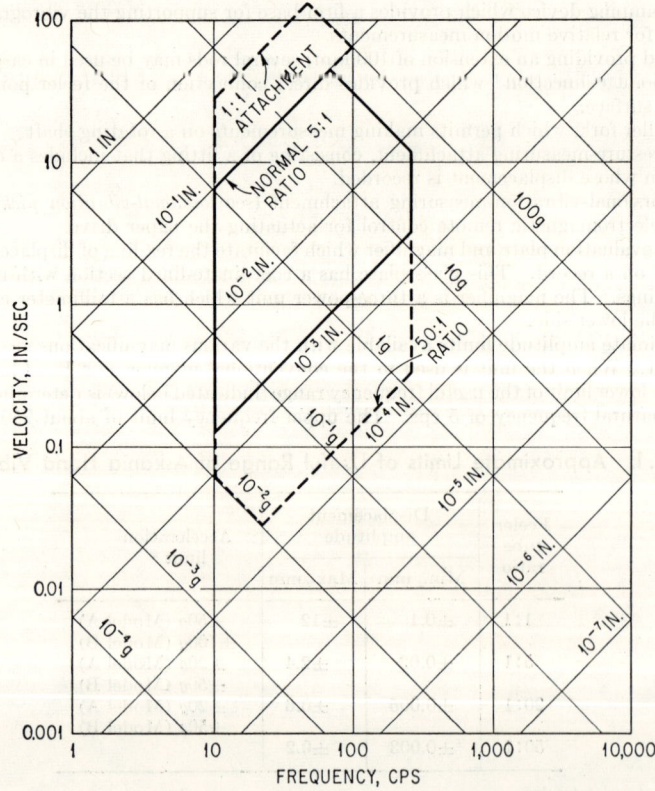

FIG. 13.3. Useful range of amplitude, Askania Hand Vibrograph.

MASS-SPRING TYPE VIBRATION INSTRUMENTS 13-7

mass is supported and guided on four leaf springs; an additional helical spring provides adjustment of the zero position so that the vibrograph may be used for measuring either horizontal or vertical vibrations. A mechanical linkage drives a recording arm in straight-line motion. The arm has a sapphire-tipped stylus which records on wax-coated tape. Three amplification ratios are available. The paper may be driven either by a spring motor or an electric-motor drive. Air damping is provided by the use of bellows and an air nozzle which yield damping of about half the critical value. This feature allows the vibrograph to be used for the recording of displacements down to 5 cps (its natural frequency). (A seismic instrument with damping ratio of 0.5 records displacement with an error of less than 20 per cent down to its natural frequency). This damping also allows the instrument to be used as an accelerometer for vibrations below 5 cps. (Further details are given in the section on *Accelerometers*.)

This compact vibrograph imposes considerable loading on the object being measured because of its moderately heavy weight. Therefore its use is limited to relatively large and heavy machines, vehicles, and structures. Since it is powered by a self-contained flashlight battery, it can be used conveniently anywhere in the field.

Technical Data. The useful frequency range is shown in Fig. 13.4A.

Natural frequency:	5 cps
Damping:	Air bellows; 0.5 critical
Magnification ratio:	4:1; 10:1; 20:1
Minimum displacement amplitude:	±0.002 in.; ±0.0004 in.
Maximum displacement amplitude:	±0.2 in.; ±0.4 in.
Minimum acceleration amplitude:	±0.01g; ±0.004g; ±0.002g
Maximum acceleration amplitude:	±1.0g; ±0.4g; ±0.2g

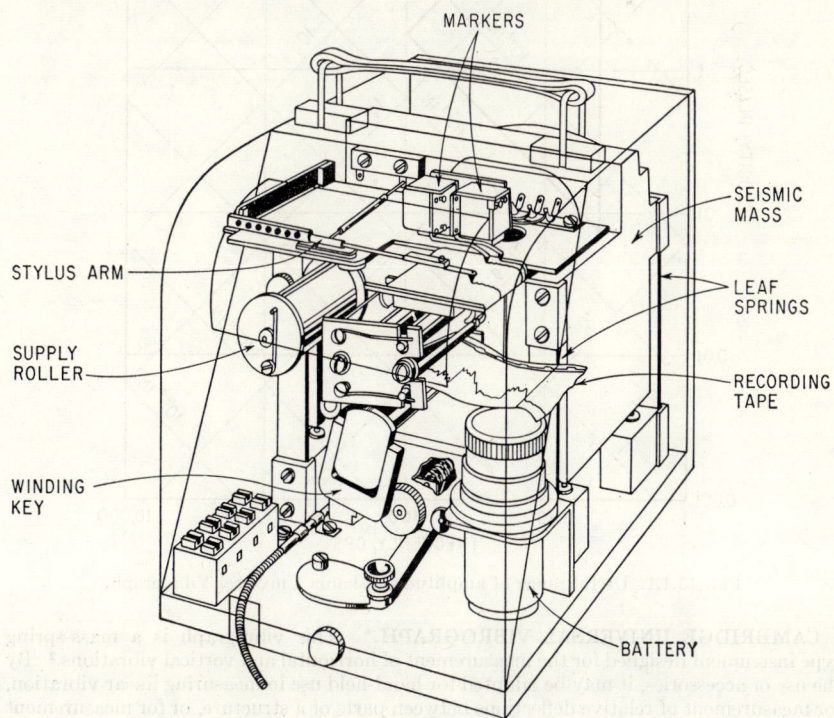

FIG. 13.4. Schematic diagram of Askania Universal Vibrograph.

13-8 SELF-CONTAINED VIBRATION-MEASURING INSTRUMENTS

Recording medium:	Wax-coated paper tape, 50 mm wide by 30 ft long (40 mm of tape width available for recording motion; 10 mm available for event marker)
Paper drive:	Spring motor with governor or 6-volt electric motor
Tape speed:	10 mm/sec or 30 mm/sec with spring-motor drive; 20 mm/sec with electric-motor drive
Time-and-event markers:	Operated by flashlight cell; one marker for 1-sec intervals; the other marker for external events
Dimensions:	15 by 17 by 23 cm (6 by 7 by 9 in.)
Weight:	Mass of mass-spring system: 6 kg (13 lb); total weight, 12 kg (26 lb)
Power:	Self-contained flashlight cell for event markers; external 6-volt battery for optional electric-motor drive

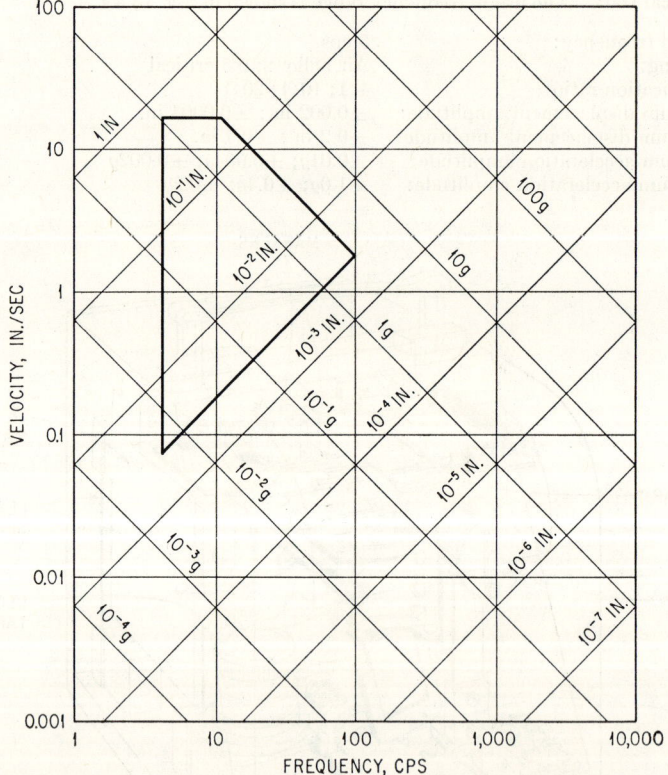

Fig. 13.4A. Useful range of amplitude, Askania Universal Vibrograph.

CAMBRIDGE UNIVERSAL VIBROGRAPH.* This vibrograph is a mass-spring type instrument designed for the measurement of horizontal and vertical vibrations.[2] By the use of accessories, it may be adapted for hand-held use in measuring linear vibration, for measurement of relative deflections between parts of a structure, or for measurement of torsional vibration.

* Manufactured by Cambridge Instrument Co., Ltd., London.

MASS-SPRING TYPE VIBRATION INSTRUMENTS 13-9

The basic unit is the recorder assembly which contains a clockwork drive for the film, a recording stylus with linkage system, a time-marking stylus with a clockwork system, and a signal-marking stylus. Controls are provided for film speed, for stylus pressure, and for zero centering of the recording stylus.

A stylus having a spherical tip of about 0.02 mm lightly pressed into contact with cellulose-acetate film is the recording member. This method of recording, in which the stylus pressure causes the plastic to flow, yields a smooth line, capable of large magnification; yet the frictional drag is relatively small. With suitable optical magnification, readings accurate to 0.001 mm (0.00004 in.) are possible, and measurements to 0.01 mm (0.0004 in.) can be made with ease.

For the recording of vertical or horizontal vibration, a mass (mounted on a spring stirrup) is attached to the recording unit. The complete assembly then is mounted on a base in one of two positions (one for horizontal vibration, the other for vertical), the mass being mounted so that there is freedom to vibrate in the desired direction. The instrument set up for measurement of vertical vibration is shown in Fig. 13.5.

When used as a hand vibrograph, the recording unit has a handle fitted to one end; a feeler toe, attached to the control spring, replaces the mass. The feeler toe is pressed lightly on the vibrating surface. With a minor modification, this arrangement is used for measuring relative motions between parts of a structure—the recorder unit being mounted on the base, which is attached to the fixed reference point. See Fig. 13.5A.

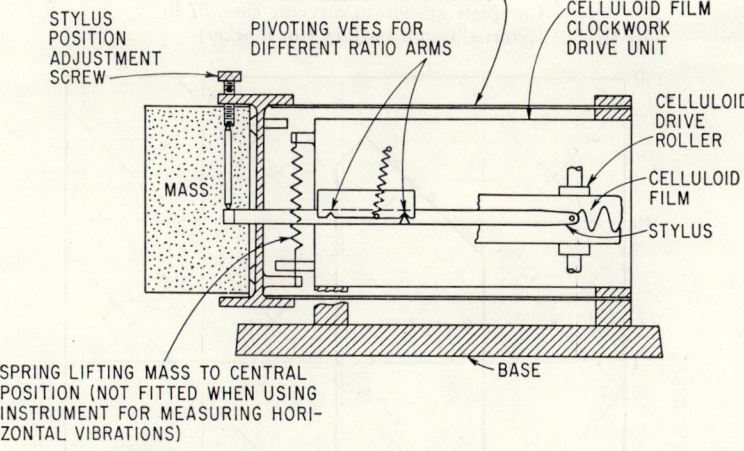

Fig. 13.5. Cambridge Universal Vibrograph set up for recording vertical vibration.

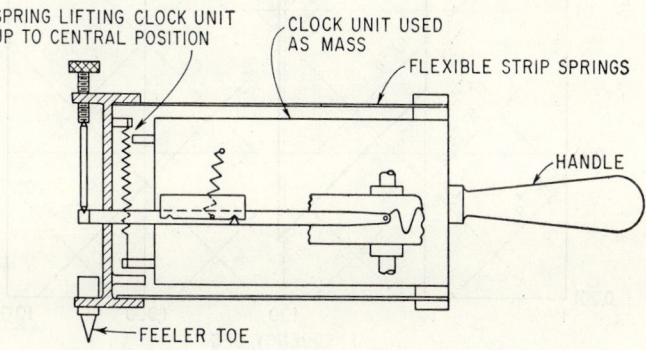

Fig. 13.5A. Cambridge Universal Vibrograph as a hand-held unit.

13-10 SELF-CONTAINED VIBRATION-MEASURING INSTRUMENTS

The vibrograph may be modified for use in measuring torsional vibrations by mounting a flywheel unit on the base and attaching it to the recording unit (see *Torsional-vibration Recorders*).

The entire system is fitted into a portable wood case, together with a 6-volt battery. The case also contains an optical system with a ground-glass screen for enlarging and viewing the records. Illumination is provided by a 6-volt lamp. An optical magnification of 10:1 is provided by this arrangement.

Technical Data. The useful frequency range is shown in Fig. 13.6.

Natural frequency:	4 cps
Frequency range:	Approximately 10 to 110 cps
Magnification ratio:	1:1 or 5:1 (selected by changing stylus arm)
Disp. amp. range:	5:1 linkage, $\pm 2 \times 10^{-5}$ to 0.06 in.; 1:1 linkage, $\pm 10^{-4}$ to 0.3 in.
Recording medium:	Cellulose-acetate tape, 20 mm wide, 10 ft long
Tape drive:	Spring clockwork drive
Tape speed:	3 to 21 mm/sec; continuously adjustable
Markers:	Time marker provides 0.1- and 1-sec marks; external-event marker
Dimensions:	11 by 5¼ by 6¾ in. (when used as a hand vibrograph, handle and contact toe add 6 in. to length)
Weight:	14 lb (as seismic vibrograph); 8 lb (as hand-held vibrograph) Complete system in carrying case, 37 lb
Power required:	External 6-volt battery (for marker)

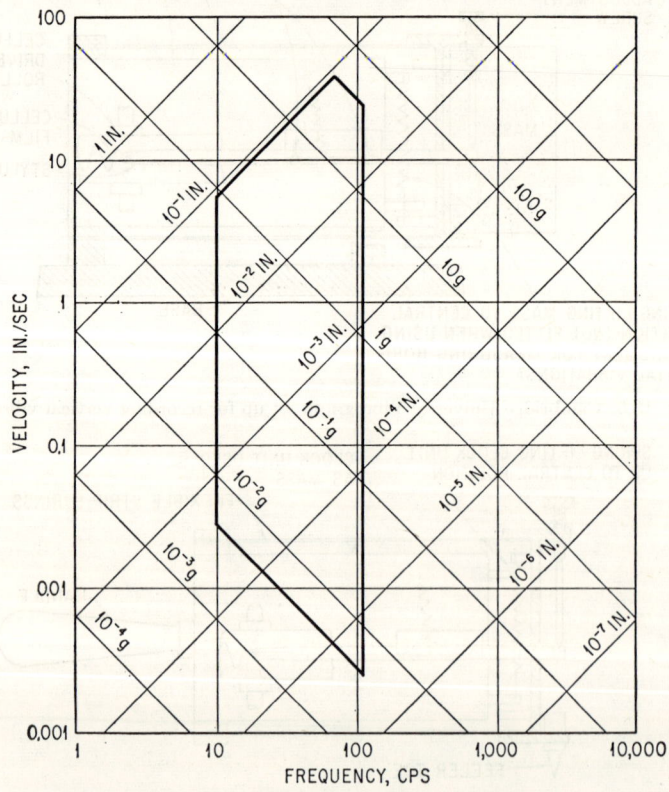

Fig. 13.6. Useful range of amplitude, Cambridge Universal Vibrograph.

MASS-SPRING TYPE VIBRATION INSTRUMENTS

DAVEY HAND VIBROGRAPH.* This vibrograph is a compact, readily portable instrument which requires no external power. It is a typical hand-held vibrograph in which the case and mechanism serves as the mass of the mass-spring system (additional inertia is provided by the operator's hands). A spring-loaded feeler prod projects from the side of the instrument. Motion is transmitted from the pickup pin, through a variable-amplification linkage, to a sapphire-pointed stylus 2¼ in. long, which records the vibrational waveform on a moving waxed-paper chart.

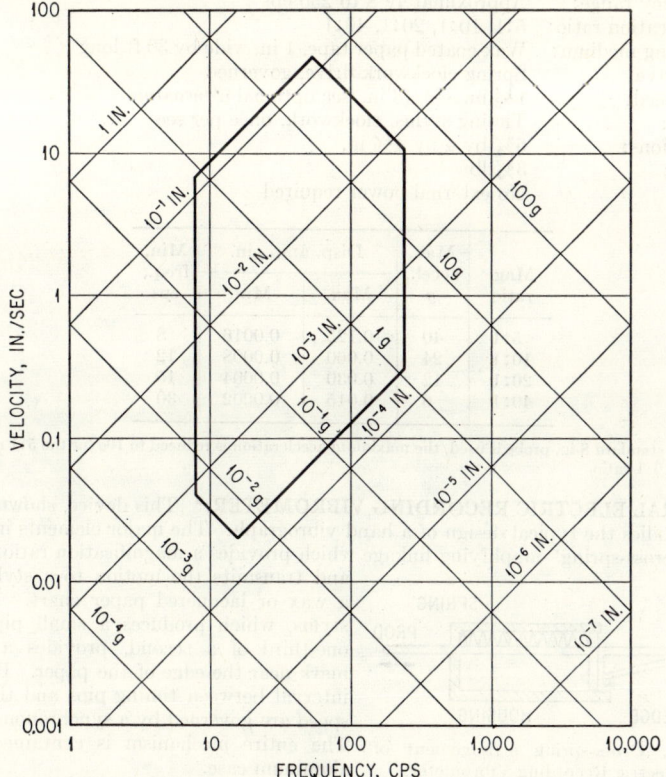

Fig. 13.7. Useful range of amplitude, Davey Hand Vibrograph.

A timing stylus, actuated by the clockwork motor, makes a notch on the edge of the chart once per second.

A small aperture with a hinged metal cover permits viewing of the stylus tip and a portion of the chart. A sliding transparent window protects the mechanism but allows access to the chart for adding record identification marks. A similar aperture at the rear of the case permits removal of the paper. Pressure of the stylus on the paper is regulated by a knob in the top of the case.

A simple adjustment allows selection of the magnification ratio. A pickup pin is mounted in a member which slides along a slot in the front of the instrument. The ratio values are etched in a plate on the slider which is kept in place at the desired location by a ball detent.

The vibrograph is supplied with a three-power magnifier which has an etched scale plate on its underside; amplitude scales are marked in thousandths of an inch (mils) for

* Manufactured by Vibroscope Co., Glenford, N.Y.

13-12 SELF-CONTAINED VIBRATION-MEASURING INSTRUMENTS

three of the four mechanical ratios provided (40:1, 20:1 and 10:1 or 5:1). In addition, there is a scale marked in time intervals of 1/10 and 1/20 sec to simplify determination of the frequency of the recorded vibration.

A detachable probe tube is available as an optional accessory. Standard length of the probe is 8 in., but other lengths can be used.

Technical Data. The useful frequency range is shown in Fig. 13.7.

Natural frequency:	Approximately 5 cps
Frequency range:	Approximately 8 to 250 cps
Magnification ratio:	5:1, 10:1, 20:1, 40:1
Recording medium:	Wax-coated paper tape, 1 in. wide by 30 ft long
Tape drive:	Spring clockwork drive, governed
Tape speed:	1½ in./sec, (3 in./sec optional alternate)
Marker:	Timing stylus, clockwork, once per sec
Dimensions:	6¾ by 4 by 4½ in.
Weight:	3¾ lb
Power:	No external power required

Mag. ratio	Max. accel., g	Disp. amp., in.		Min. freq., cps
		Max.	Min.	
5:1	40	0.125	0.0016	8
10:1	24	0.060	0.0008	12
20:1	12	0.030	0.0004	15
40:1	6	0.015	0.0002	30

When the standard 8-in. probe is used, the maximum acceleration is reduced to $10g$ for the 5:1 ratio, and $15g$ for the 10:1 ratio.

GENERAL ELECTRIC RECORDING VIBROMETER.* This device, shown in Fig. 13.8, embodies the typical design of a hand vibrograph. The major elements include a prod, a "cross-spring" amplifying linkage which provides a magnification ratio of 10:1 and transmits the motion to a stylus, and a wax or lacquered paper chart. Another stylus, which produces a small pip every one-third of a second, provides a timing mark near the edge of the paper. Both the interval between timing pips and the chart speed are governed by a synchronous motor. The entire mechanism is contained in an aluminum case.

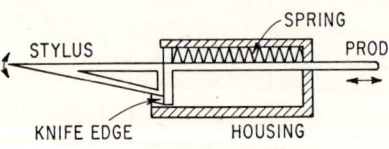

Fig. 13.8. Cross-spring arrangement of General Electric Recording Vibrometer.

The prod, which extends from one end of the vibrometer, is held against the vibrating body. Motion of the stylus on the chart can be observed through the window provided in the top of the aluminum case. One side of the case can be removed for rapid replacement of the wax paper roll.

Technical Data. The useful frequency range is shown in Fig. 13.9.

Natural frequency:	5 cps (approximate)
Magnification ratio:	10:1
Recording medium:	Wax- or lacquer-coated paper tape about 1½ in. wide by 50 ft long
Tape drive:	Synchronous motor
Tape speed:	1 and 3 in./sec (push-button-selected)
Markers:	One stylus, marking ⅓-sec intervals
Dimensions:	8 by 4½ by 4⅛ in.; prod extends 1⅝ in.
Weight:	7 lb
Power:	115 volts, 60 cps

* Manufactured by General Electric Company, West Lynn, Mass.

MASS-SPRING TYPE VIBRATION INSTRUMENTS 13-13

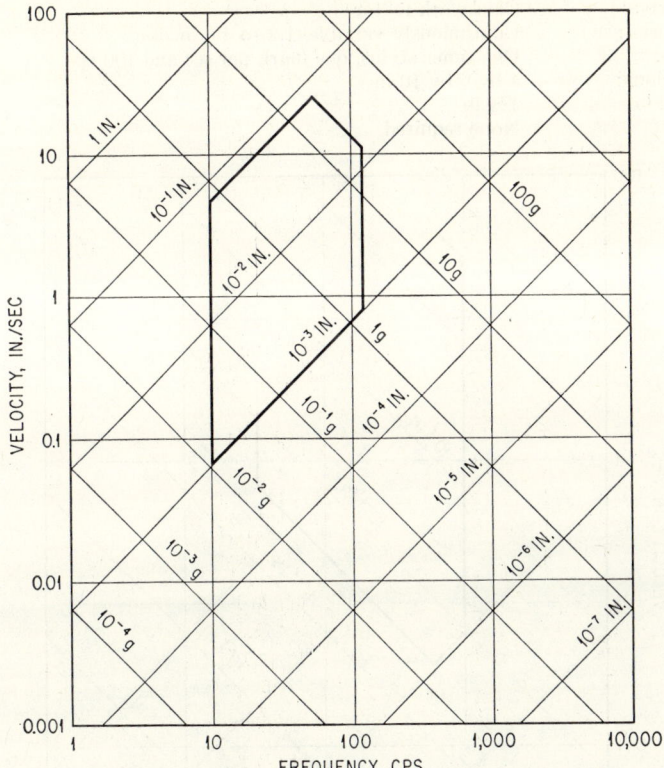

Fig. 13.9. Useful range of amplitude, General Electric Recording Vibrometer.

KORFUND HAND VIBROGRAPH.* This vibrograph embodies the typical features of a hand-held vibration-recording instrument; i.e., a spring-loaded probe, a motion amplifying and transmitting linkage and a motor-driven recording medium. The entire mechanism except for the probe arm is contained in the case.

In this instrument the recording medium may be either a wax-coated paper tape on which a stylus traces the record or a plain paper tape; in the latter case, a pen is substituted for the stylus. The record is stored on a take-up spool.

Standard accessories include: an extension rod and a roller contact for use in measuring lateral vibration of shafts. Optional accessories include: a reduction lever for recording of large amplitudes; a laboratory measuring stand to provide a fixed base for relative-displacement measurements; slow-motion clockwork, allowing continuous recording over a long period of time; a remote magnetic control for starting and stopping the motor and for rewinding; an extra event marker; and a pressure gage attachment.

Technical Data. The useful frequency range is shown in Fig. 13.10. It may be extended to 0 cps if a fixed base is used.

Natural frequency: 5 cps
Magnification ratio: 2:1, 4:1, 5:1, 10:1, 20:1
Disp. amp. range: 0.0004 in. (minimum) to 0.8 in.
Recording medium: Wax-coated paper tape (for stylus recording) or plain paper tape (for pen recording) 1 in. wide by 350 in. long

* Distributed by Korfund Co., Inc., Long Island City, N.Y.

Tape drive: Clockwork motor
Tape speed: Continuously variable, 1.2 to 4.6 in./sec
Marker: One timer stylus, one mark per sec and 100 cps
Dimensions: 4 by 6 by 10 in.
Weight: 4¾ lb
Power: None required

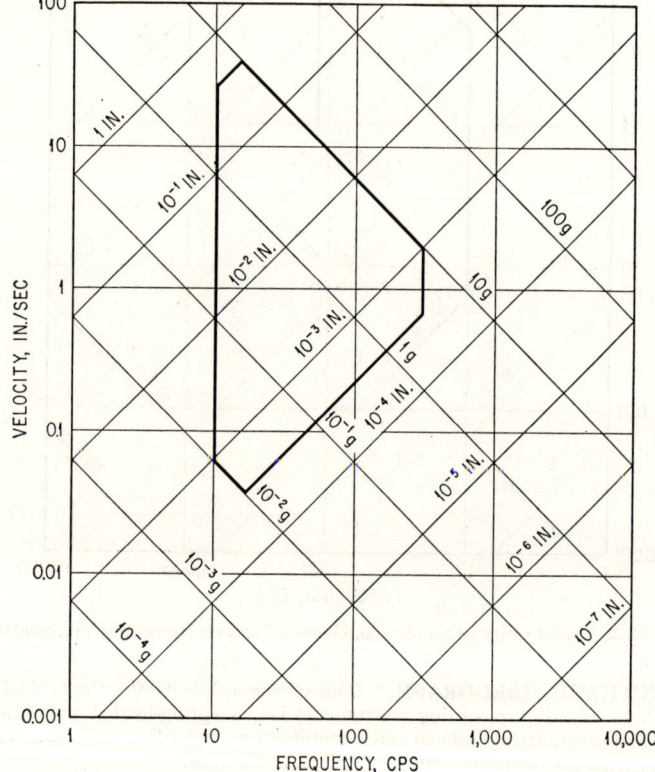

Fig. 13.10. Useful range of amplitude, Korfund Hand Vibrograph.

TYPE-B PALLOGRAPH.* This low-frequency vibrograph is shown in Fig. 13.11. This instrument incorporates, in simple form, all the basic features of the typical vibrograph. It is intended primarily for recording vertical vibration, but may be adapted for measuring horizontal motions. In the standard mode of operation, for recording vertical vibration, the mass is suspended as a horizontal pendulum by an arm which is supported by flexure springs attached to the frame. In order to achieve the desired low natural frequency (about ¼ cps), the principle of instability is employed in the design of the suspension. The restoring spring is attached in such a way that its lever arm decreases as the spring is extended, and vice versa. Thus, the product of change of lever arm and change in restoring force (which determines the restoring torque as the angular position of the pendulum changes) can be made very small, resulting in an extremely low natural frequency. An air dashpot provides adjustable damping.

Motion of the pendulum is transmitted and amplified, by a simple mechanical linkage, to the stylus arm. The stylus records by scribing on waxed paper which is transported

* Developed by the David Taylor Model Basin, U.S. Navy.

from the supply spool across a platen to the take-up spool. With this type of drive, the paper speed varies with the amount of paper on the take-up spool. When the pallograph is adapted for measurement of horizontal vibration, the instrument is mounted on a bracket so that the pendulum is vertical. Then the spring tension is adjusted to allow the pendulum—and stylus—to center at the proper position. In this mode of operation, the "instability principle" is not used; the natural frequency of 1 cps is that of a pendulum with compliant springs.

Because of its low natural frequency, this pallograph is capable of recording vertical vibration at frequencies as low as 0.8 cps. The inertia associated with the relatively large components of the stylus linkage limits its upper usable frequency to about 17 cps.

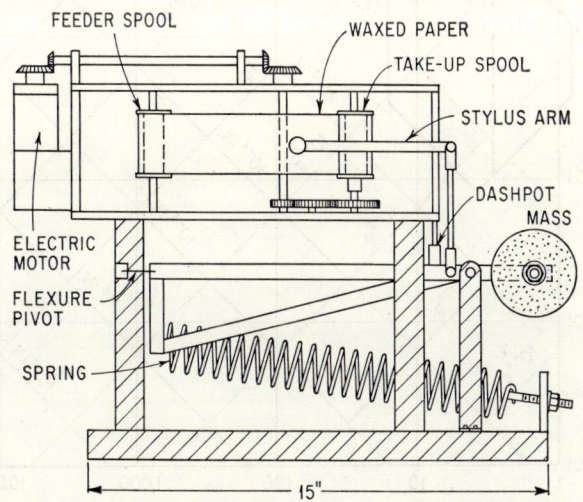

Fig. 13.11. The TMB Type-B Pallograph. This instrument has a dashpot with adjustable air damping which is particularly useful in shipboard work when the ship is pitching or rolling.

The lower limit is determined mainly by the smallest oscillations that can be read accurately on the record. Using high optical magnification does not improve resolution because the record line traced on the waxed paper has a rather rough edge when observed under magnification.

Technical Data. The useful frequency range is shown in Fig. 13.12.

Natural frequency:	Vertical model, 15 cpm (0.25 cps); horizontal model, 60 cpm (1.0 cps)
Frequency range:	Vertical, 50 to 1,000 cpm (0.8 to 17 cps); horizontal, 170 to 1,000 cpm (3 to 17 cps)
Magnification ratio:	2:1
Disp. amp. range:	±0.003 to ±0.225 in.
Recording medium:	Wax-coated paper tape, 2 in. wide
Tape drive:	Geared electric motor
Tape speed:	Varies·along roll; about 2 in./sec
Markers:	One time marker (1-sec marks provided by use of accessory synchronous timer)
Dimensions:	20 by 8 by 14 in.
Weight:	40 lb
Power:	115 volts, 60 cps

13-16 SELF-CONTAINED VIBRATION-MEASURING INSTRUMENTS

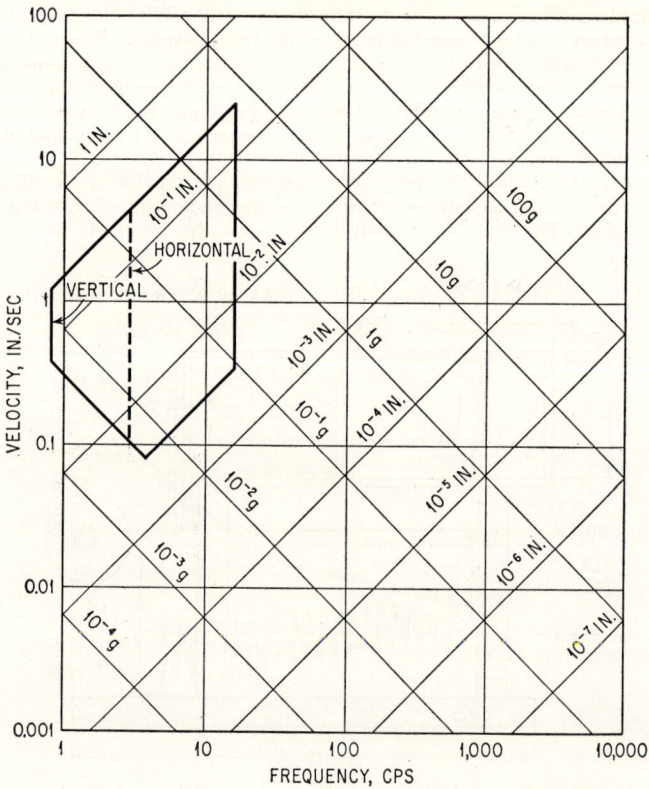

Fig. 13.12. Useful range of amplitude, TMB Type-B Pallograph.

TYPE-C PALLOGRAPH.* This low-frequency vibrograph, shown schematically in Fig. 13.13, is a somewhat smaller instrument than the Type-B, designed specifically for recording the horizontal vibration of ship superstructures. It is similar to the Type-B, except that the mass is an inverted pendulum having a natural frequency of 1 cps. Two

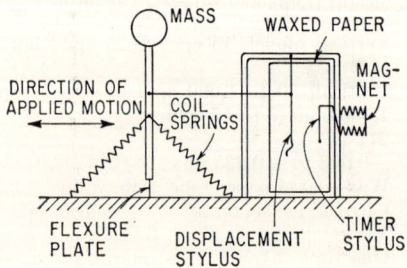

Fig. 13.13. The TMB Type-C Pallograph. This instrument was developed specifically to record vibration of ship superstructures. It is a horizontal vibrograph, weighing 25 lb. It has the same type of magnification, recording, and timing as the Type-B, but its natural frequency is slightly higher. The mass of the mass-spring system is supported as an inverted pendulum with two springs, one on each side, used to restore it to its vertical position.

* Developed by the David Taylor Model Basin, U.S. Navy.

restoring springs are used, one on each side of the pendulum. It employs the same wax-coated recording paper and chart drive as the Type-B unit.

The stylus linkage provides a magnification ratio of 3:1, making it possible to record vibration displacements between 0.002 and 0.150 in., at frequencies from 2.5 cps (150 cpm) to about 17 cps (1,000 cpm).

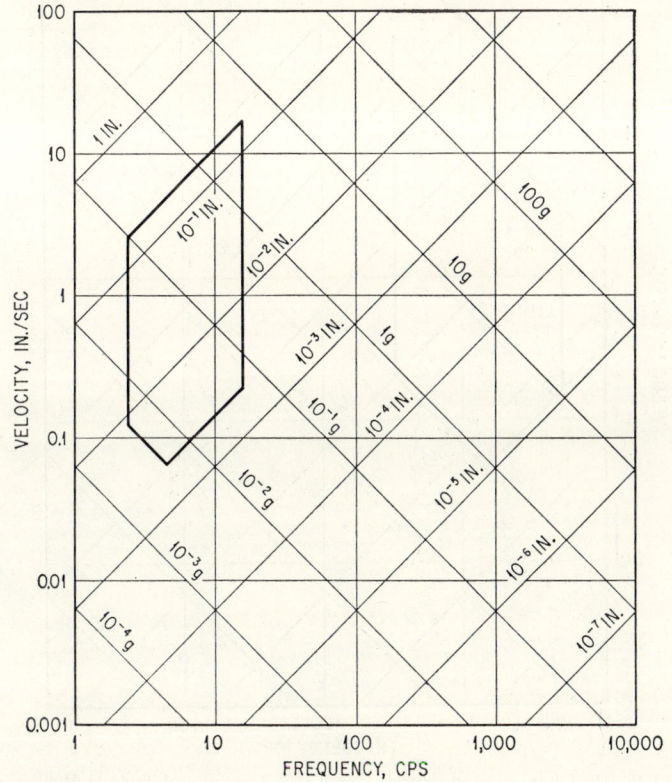

Fig. 13.14. Useful range of amplitude, the TMB Type-C Pallograph.

Technical Data. The useful frequency range is shown in Fig. 13.14.

Natural frequency: 60 cpm (1 cps)
Frequency range: 150 to 1,000 cpm (2.5 to 17 cps)
Magnification ratio: 3:1
Disp. amp. range: ±0.002 to ±0.150 in.
Recording medium: Wax-coated paper tape, 2 in. wide
Tape drive: Geared electric motor
Tape speed: Varies along roll; about 2 in./sec
Markers: One time marker (1-sec marks provided by use of accessory synchronous timer)
Dimensions: 16 by 8 by 11 in.
Weight: 25 lb
Power: 115 volts, 60 cps

13-18 SELF-CONTAINED VIBRATION-MEASURING INSTRUMENTS

TYPE-V AND TYPE-H PALLOGRAPHS.* Modern versions of the Type-B and Type-C Pallographs are designated as Type-V and Type-H, respectively. Almost identical in basic design, these latter types of low-frequency vibrographs have greater amplitude range than their prototypes. Adjustable linkage ratios of 6:1, 4:1, and 2:1 are available for selection.

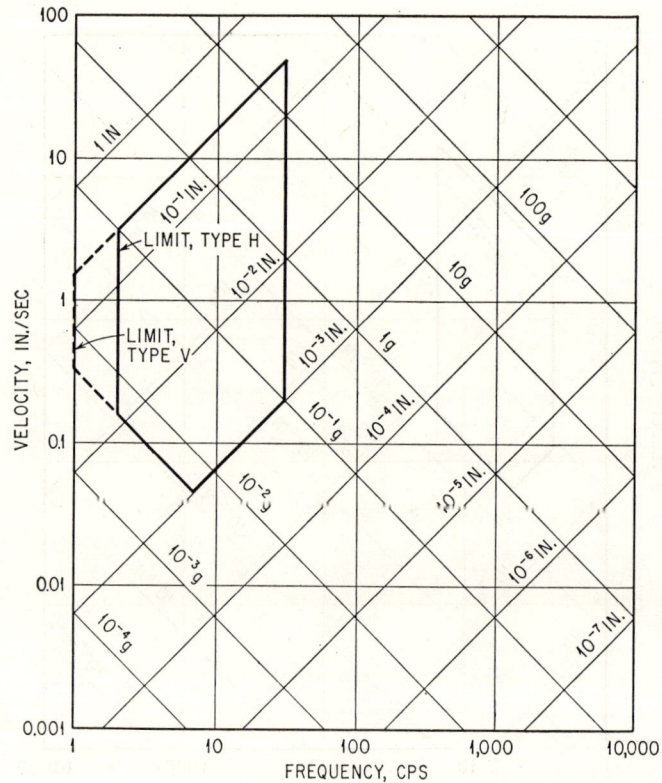

FIG. 13.15. Useful range of amplitude, the TMB Type-V and Type-H Pallographs.

Technical Data. The useful frequency ranges of these instruments are shown in Fig. 13.15.

	Type-H	*Type-V*
Natural frequency:	40 cpm (2/3 cps)	20 cpm (1/3 cps)
Frequency range:	120 to 1,800 cpm (2 to 30 cps)	60 to 1,800 cpm (1 to 30 cps)
Magnification ratio:	Adjustable 2:1, 4:1, or 6:1	Same
Disp. amp. range:	±0.001 to ±0.250 in.	Same
Recording medium:	Wax-coated paper tape, 2 in. wide	Same
Tape drive:	Geared electric motor	Same
Tape speed:	1 or 2 in./sec, adjustable	Same
Markers:	Two: one time marker (1-sec marks with accessory synchronous timer); one event marker	
Dimensions:	18 by 9 by 12 in.	19 by 10 by 15 in.
Weight:	25 lb	60 lb
Power:	115 volts ac, 60 cps	Same

* Developed by the David Taylor Model Basin, U.S. Navy.

MASS-SPRING TYPE VIBRATION INSTRUMENTS

TMB TWO-COMPONENT PALLOGRAPH.* This low-frequency vibrograph records vertical and horizontal vibrations simultaneously. Because the mass-spring system has a very low natural frequency, it is useful for recording vibrations as low as 2 cps. Although basically similar in design to earlier models, this recorder incorporates a number of significant refinements. The recording medium is a heat-sensitive paper on which traces are made by a heated-wire stylus. The paper, 3 in. wide, is available in rolls 80 ft

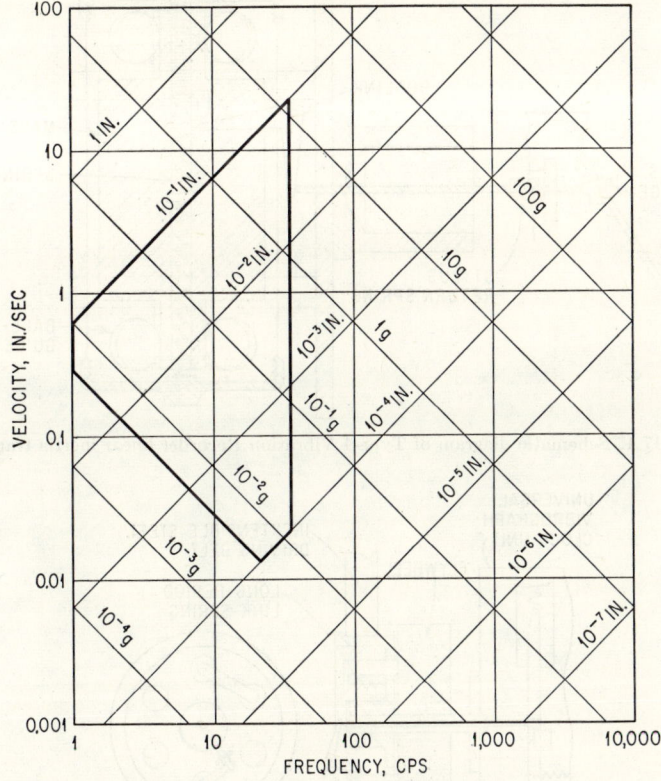

FIG. 13.16. Useful range of amplitude, the TMB 2-Component Pallograph.

long. Paper speed depends on the drive motor installed, but usually is from 0.2 to 3 in./sec.

Two timing markers are installed; one provides time synchronization among several different instruments; the other supplies 1-sec markings from an internal synchronous source.

Technical Data. The useful frequency range is shown in Fig. 13.16.

Natural frequency: $\frac{2}{3}$ cps (both components)
Magnification ratio: 5:1
Disp. amp. range: ± 0.0001 to ± 0.100 in.

TYPE-4 VIBRATION RECORDER.† This device (a modern version of the Geiger Vibrograph) is a portable instrument for recording torsional or linear vibration on a wax-coated strip chart. The vibration is mechanically detected, amplified, and re-

---
* Developed by the David Taylor Model Basin, U.S. Navy.
† Manufactured by Industrial Manufacturing and Tool Co., Hazel Park, Mich.

corded. Event markers supply timing intervals and revolution or dead-center indications.

One of the major components of the device is the electric-motor paper drive. This unit consists of a small synchronous motor, gears, clutches, and controls for driving the

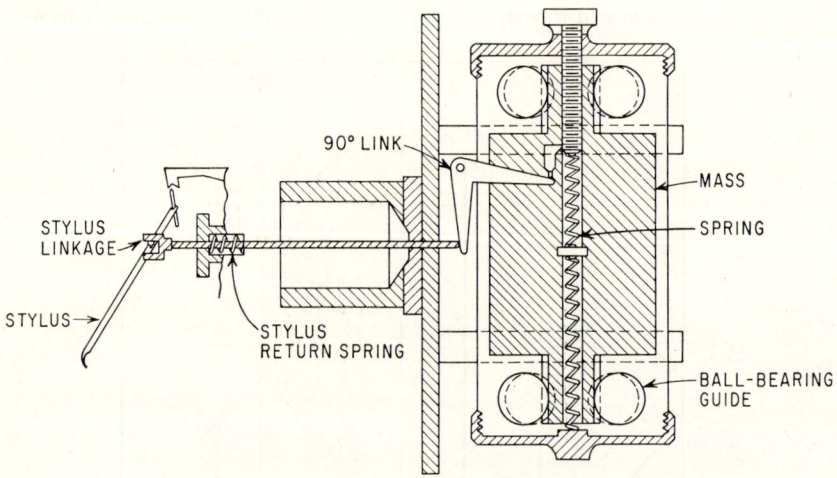

FIG. 13.17A. Schematic diagram of Type-4 Vibration Recorder linear inertia translator.

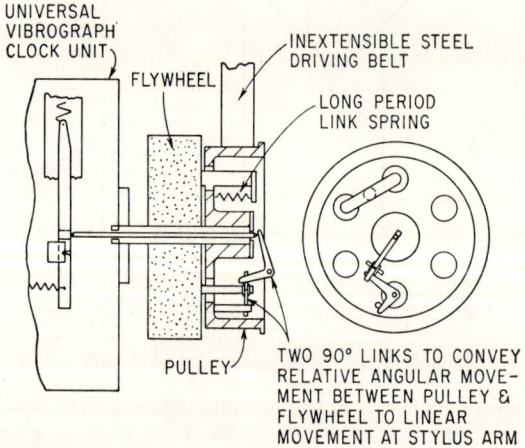

FIG. 13.17B. Schematic diagram of the Type-4 Vibration Recorder with torsional translator attached.

chart feed mechanism at any one of several paper speeds. The recorder chassis houses the recorder pen, linkage and instrument controls on a rigid framework with a heavy base which is suitable for bolting to a flat surface. The vibration recording stylus may be set for one of several magnification ratios.

Four types of mechanical translators are designed for use with the recorder: a linear inertia translator (Fig. 13.17A), a torsional translator (Fig. 13.17B), a linear contact translator, and a linear action translator.

For use as a vibrograph, the "Type-4 Linear Inertia Translator" is used. The prin-

MASS-SPRING TYPE VIBRATION INSTRUMENTS 13-21

cipal component of this unit is a guided cylindrical mass, of the mass-spring system, with antifriction bearings at both ends. The bearings are mounted in a surrounding shell, attached to an angularly graduated disc and quill. The quill fits into the opening of the recorder housing. When the instrument is mounted on a vibrating surface, the relative motion between the recorder and the inertia mass is transmitted through a bell crank to the recording stylus. The translator may be rotated to respond to vibration in any selected direction at right angles to the quill axis.

Technical Data. The useful frequency range of the linear translator illustrated in Fig. 13.17A is shown in Fig. 13.18.

Natural frequency:	2 cps
Frequency range:	4 to 160 cps
Magnification ratio:	3:1; 6:1; 12:1; 24:1
Disp. amp. range:	With 3:1 magnification, ± 0.002 to ± 0.25 in.; with 24:1 magnification, ± 0.0003 to ± 0.03 in.
Recording medium:	Wax-coated paper tape, $1\tfrac{31}{32}$ in. wide
Tape drive:	Geared synchronous electric motor
Tape speeds:	0.25, 0.50, 1.0, 2.5, 5.0, 10.0 in./sec
Markers:	One timing marker, $\tfrac{1}{5}$ or 1 sec; one revolution indicator
Dimensions:	Recorder chassis: 12 by $8\tfrac{1}{2}$ by $9\tfrac{1}{2}$ in.; linear inertia translator: 6 in. in diameter by $5\tfrac{3}{4}$ in. long
Weight:	Recorder chassis: 40 lb; linear inertia translator: 17 lb
Power required:	115 volts, 60 cps for chart drive motor; 6 volts d-c for event markers

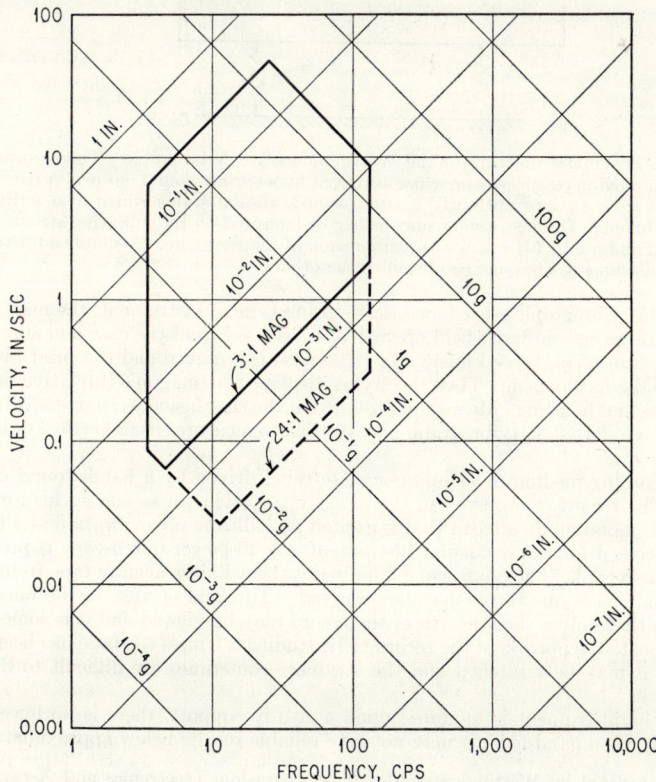

FIG. 13.18. Useful range of amplitude, Type-4 Vibration Recorder illustrated in Fig. 13.17A.

This instrument is rugged, making it suitable for a variety of vibration measurements on machinery and large structures; its relatively large weight makes it somewhat unsuitable for use on light structures or vehicles, or for exploratory measurements where the recorder must be moved from place to place during a short time interval. The need for external 115-volt, a-c power also restricts its mobility to some extent (as compared to a hand vibrograph with a clockwork paper drive, for example).

WESTINGHOUSE LE VIBROGRAPH.* This device is a compact, portable instrument designed for hand-held use—the entire case and mechanism (plus the operator's hands) acting as the mass in the mass-spring system—or as a self-contained mass-spring instrument. When used as a hand-held instrument, one end of the probe is held against the vibrating structure by a spring which acts between the case and the probe. The other end of the probe drives the displacement stylus, which is pivoted near the probe.

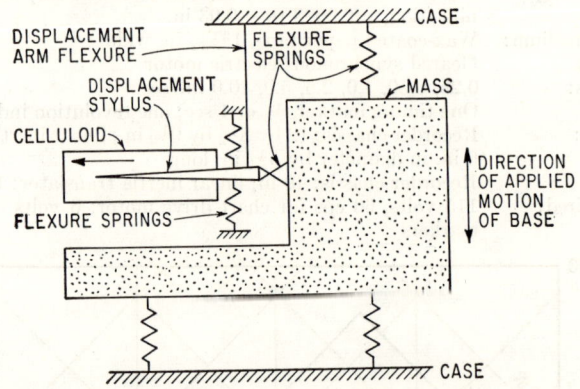

FIG. 13.19. Schematic diagram of the Westinghouse LE Vibrograph. The recording stylus scribes on a moving cellulose tape which is driven by a spring-wound motor. A record of time is obtained from an intermittently excited, tuned, timing stylus which also scribes on the moving celluloid. The instrument may either be mounted on the vibrating structure or held in the hand and used with a probe for exploration of the structure. A simple adjustment converts the instrument from one condition to the other.

When this vibrograph is used as a mass-spring type of instrument, the mass, which is clamped to the case in hand-held operation, is released from the case and supported by a set of flexure springs (see Fig. 13.19). The prod is removed and the prod pivot point is fixed to the mechanism. Thus the stylus displays the (magnified) relative motion between mass and housing. Means are provided in the suspension to compensate for gravitational force so that the vibrograph may be used to measure either vertical or horizontal vibration.

The recording medium is cellulose-acetate tape, driven by a hand-wound clockwork motor. The record is engraved in the tape by the stylus; a second stylus for marking purposes is mounted on an arm that is excited periodically into vibration at a frequency that is specified for the particular instrument. A 40-power microscope is provided for reading the record. The microscope illuminates the cellulose-acetate tape from the back side so that the stylus traces may be observed. The field of view of the microscope is limited so that only a short length of the record may be viewed and it is sometimes difficult to locate the portion of the record to be studied. Unless the tape has been handled carefully, it may be scratched and the scratches sometimes are difficult to distinguish from traces.

When the instrument is mounted upon a steady support, there is no lower limit of frequency; when hand-held, it may not give reliable results below approximately 10 cps

* Manufactured by Westinghouse Electric Corporation, Electronics and X-ray Division, Baltimore, Md.

MASS-SPRING TYPE VIBRATION INSTRUMENTS

because of motion of the operator's hands. When used as a mass-spring type of instrument, the lower usable frequency is approximately 15 cps; the upper frequency limit decreases as the displacement increases.

Technical Data. The useful range of frequency is given in Fig. 13.20.

Natural frequency:	Hand-held: 5 cps; mass-spring: 7 cps
Frequency range:	10 to 250 cps
Magnification ratio:	8½:1
Disp. amp. range:	±0.0001 to ±0.025 in.
Acceleration amplitude limit:	$6.25g$
Recording medium:	Cellulose-acetate tape, 1 in. wide
Tape drive:	Hand-wound clockwork motor
Tape speed:	Approximately 1 in./sec
Markers:	Time stylus, intermittently excited in free vibration at a specified frequency
Dimensions:	7½ by 5 by 4¾ in.
Weight:	10 lb
Power required:	None

VIBROMETERS

A *vibrometer* is a displacement pickup of the mass-spring type, with self-contained means of indicating relative motion between the mass and the case. It is essentially the

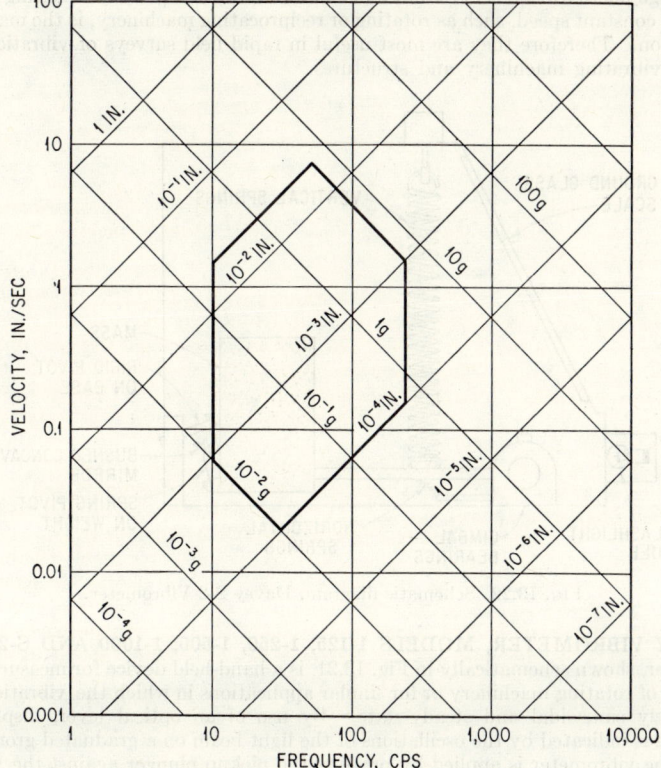

FIG. 13.20. Useful range of amplitude, Westinghouse LE Vibrograph.

same as the vibrograph, described earlier, except that the response of the instrument is displayed as a nonpermanent reading rather than as a tape or chart tracing. This display may take the form of an indication on a dial or the deflection of a pointer against a graduated scale. Vibrometers may be either hand-held or mounted on the vibrating surface. As with other amplitude-indicating devices, they usually are most useful when

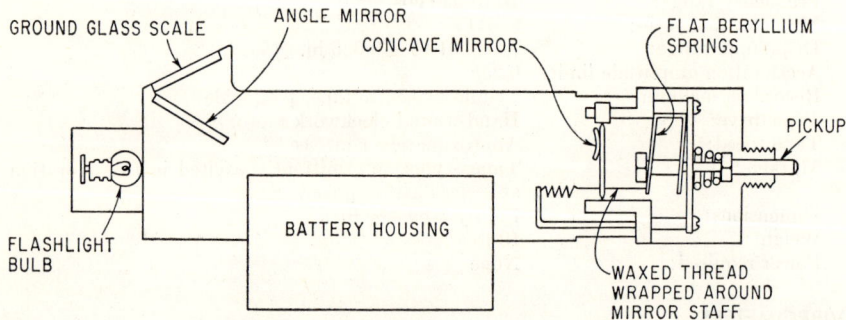

Fig. 13.21. Schematic diagram, Davey Vibrometer.

the frequency of the vibration is known and when the vibration itself is fairly steady and free of large fluctuations. This situation is found wherever a powerful exciting source of relatively constant speed, such as rotating or reciprocating machinery, is the major cause of vibration. Therefore they are most useful in rapid field surveys of vibration amplitudes in vibrating machinery and structures.

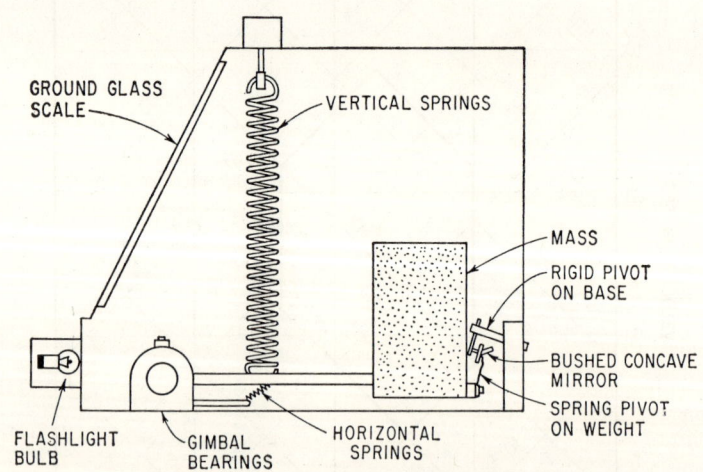

Fig. 13.22. Schematic diagram, Davey S-2 Vibrometer.

DAVEY VIBROMETER, MODELS 1-125, 1-250, 1-500, 1-1000 AND S-2.* This vibrometer, shown schematically in Fig. 13.21, is a hand-held device for measurement of vibration of rotating machinery or for similar applications in which the vibration is approximately sinusoidal and steady-state. By use of an optical lever, displacement amplitude is indicated by the oscillations of the light beam on a graduated ground-glass scale. The vibrometer is applied by pressing the pickup plunger against the vibrating

* Manufactured by The Vibroscope Co., Glenford, N.Y.

surface, either vertically or horizontally. Several models are available—the model number indicating the magnification ratio, i.e., 125:1 for Model 1-125, 1,000:1 for Model 1-1000, etc. A two-to-one reduction ratio changer is available for all units except Model 1-125. A two-component unit, Model S-2, also is available, for measuring vertical and horizontal vibration simultaneously. This instrument is shown schematically in Fig. 13.22.

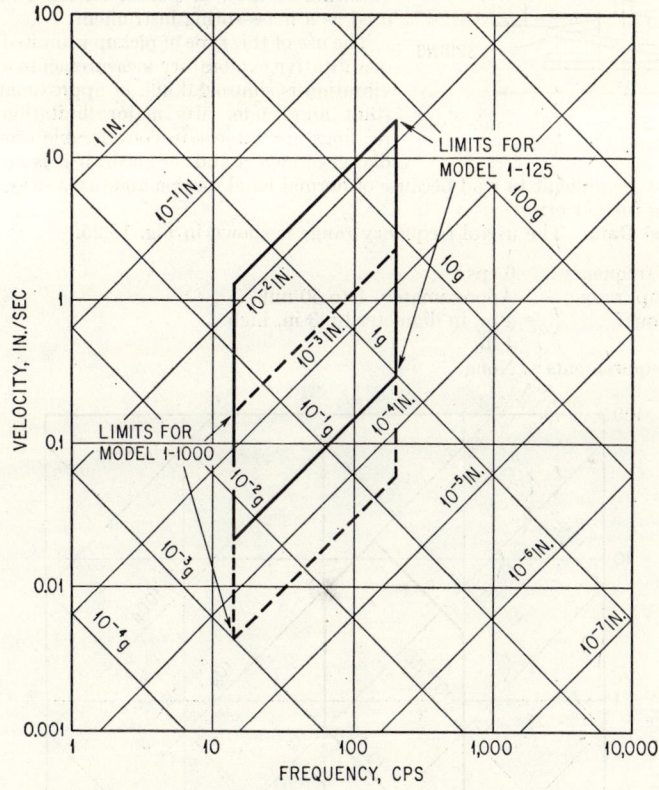

Fig. 13.23. Useful range of amplitude, Davey Hand Vibrometer.

Technical Data. The useful frequency range is shown in Fig. 13.23.

Natural frequency: 10 cps
Frequency range: 15 to 200 cps (0 cps lower limit if rigidly supported)
Amplitude range: ±0.05 mil to ±2 mils (1,000:1 magnification) (Model 1-1000); ±0.25 mil to ±16 mils (125:1 magnification) (Model 1-125)
Dimensions: 5 in. wide by 6 in. long by 2 in. high
Weight: 3¾ lb
Power requirements: Three flashlight batteries in accessory housing; also operable from 115 volts a-c with accessory resistance unit

WESTINGHOUSE HAND VIBROMETER.* This device is typical of a simple vibrometer employing a dial gage as the motion-amplifying and indicating device; see Fig.

* Manufactured by Westinghouse Electric Corporation, Electronics and X-ray Division, Baltimore, Md.

13-26 SELF-CONTAINED VIBRATION-MEASURING INSTRUMENTS

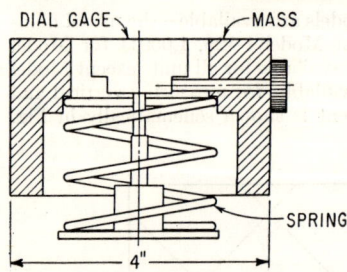

FIG. 13.24. Schematic diagram, Westinghouse Hand Vibrometer.

13.24. It is comprised of a cylindrical mass resting on a coil spring and a dial gage firmly attached to the mass. The gage contacts a base which rests on the vibrating surface; displacement amplitude is indicated by the band swept by the dial-gage pointer. It may be held by hand against the vibrating surface, or placed on the surface and used as a mass-spring instrument.

The use of this type of pickup is limited to semi-quantitative exploratory measurements where the vibration is sinusoidal and of approximately constant amplitude. Its major limitation is that readings are subject to considerable error due to dial-gage inertia and backlash; if it is used hand-held, it may be difficult to read because of normal hand motion and body sway, thereby introducing further error.

Technical Data. The useful frequency range is shown in Fig. 13.25.

Natural frequency: 6 cps
Disp. amp. range: Approximately 1 to 50 mils
Dimensions: 4 in. in diameter by 3 in. high
Weight: 4 lb
Power requirements: None

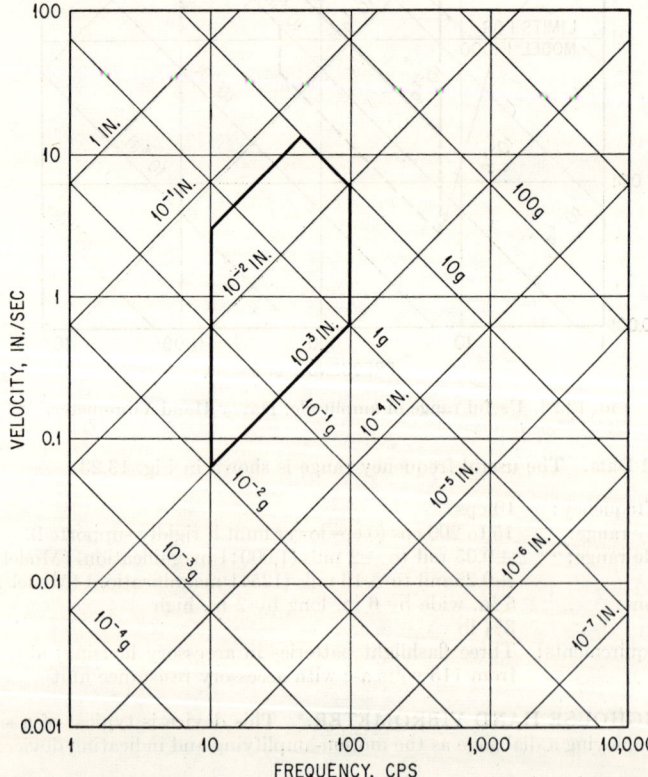

FIG. 13.25. Useful range of amplitude, Westinghouse Hand Vibrometer.

REED-TYPE FREQUENCY-INDICATING VIBROMETERS

In measurement problems involving relatively steady vibration, one of the quantities of major interest is the dominant frequency. Determination of this frequency often provides the information needed to identify an unknown source of vibration. For example, the source of vibration in a building can be traced to an offending compressor. Measurement of vibration frequency also can be used conveniently to determine the rotational speed of an engine or motor.

Reed-type frequency-indicating vibrometers are useful devices for such measurements. Such instruments depend on the resonant vibration of a cantilever reed with low damping to indicate the frequency of the exciting vibration; low damping is achieved by careful design and assembly to minimize losses at the clamping end. These devices are of two types: (1) multiple-reed gages which employ a number of reeds of different natural frequencies that are closely spaced and (2) single-reed, adjustable-length gages.

The fundamental natural frequency f_n of a cantilever reed of the type used in these instruments is given by [3]

$$f_n = \frac{3.52}{2\pi} \sqrt{\frac{EI}{\mu l^4}} \quad \text{cps} \tag{13.1}$$

where E is modulus of elasticity, I the section moment of inertia, μ the mass per unit length of the reed, and l the length of the reed. In the multiple-reed type of frequency indicator, each reed generally has a small mass on the end. For a cantilever with a mass on the end, the fundamental natural frequency is given by the expression [4]

$$f_n = \frac{1}{2\pi} \sqrt{3EI/l^3(m_2 + 0.23m_1)} \quad \text{cps} \tag{13.2}$$

where m_1 is the mass of the reed and m_2 is the value of the mass added to the end.

In order to determine the frequency of vibration, the reed vibrometer is held or placed with its base in contact with the vibrating surface. A multiple-reed instrument contains a number of reeds of different natural frequency, usually progressing in uniform increments of frequency between adjacent reeds from the lowest to the highest. If the vibration being observed is steady and has a major frequency component within the range covered by the reeds, several of the reeds closest in natural frequency to this component will vibrate with relatively large excursions, the one which is closest in frequency having the largest excursion. The instrument has a frequency scale indicating the natural frequency of each reed, so that the vibration frequency may be determined visually.

In vibrometers having a single, adjustable reed, the clamped end of the reed is constrained between two spring-loaded rollers; its length (and therefore its natural frequency) is adjusted by turning a knob fastened to one of the rollers. In use, with the instrument held against the vibrating surface, the length of the reed is varied slowly until the reed vibrates steadily with its greatest excursion. The nonvibrating portion of the reed has an index mark on its very end, which moves along a scale calibrated in frequency as the free length of the reed is changed. The natural frequency of the reed is read on this scale.

Although this type of instrument is convenient to use because it is extremely light and compact and can be used anywhere the observer can reach, it possesses a serious limitation. Adjustment of the reed length, in the process of tuning for resonance, as well as inadvertent motions of the operator's hand, may excite the reed into vibration at its natural frequency; since the reed is very lightly damped, such spurious oscillations continue for some time. This often makes tuning to the frequency of applied vibration time-consuming—particularly when the vibration is varying in amplitude. However, for relatively pure sinusoidal vibration, there is a marked indication at resonance, and the frequency can be determined readily.

FRAHM'S MULTIPLE-REED GAGES. A typical multiple-reed gage is shown schematically in Fig. 13.26. The reeds all have different natural frequencies, designed to give continuous frequency coverage throughout a given range of frequencies. This indication is given by resonant vibration of the reed at, or closest to, the frequency of the vibrating body. The reeds have white marks on their ends to aid in determining visually which reed is in resonance. If a multiple-reed gage of this type is subject to vibration of discrete frequency, the reed corresponding most nearly in natural frequency to the exciting frequency will show the greatest displacement amplitude; if the exciting frequency is midway between the natural frequencies of two adjacent reeds, they will show equal displacement amplitudes.

These instruments have the advantage of being small, portable, reasonably accurate and reliable; they do not require a source of power. They are particularly useful for indicating speeds of rotating machinery. A major limitation is that they are effective only for rather steady vibration. In addition, a relatively limited frequency range is covered by individual units.

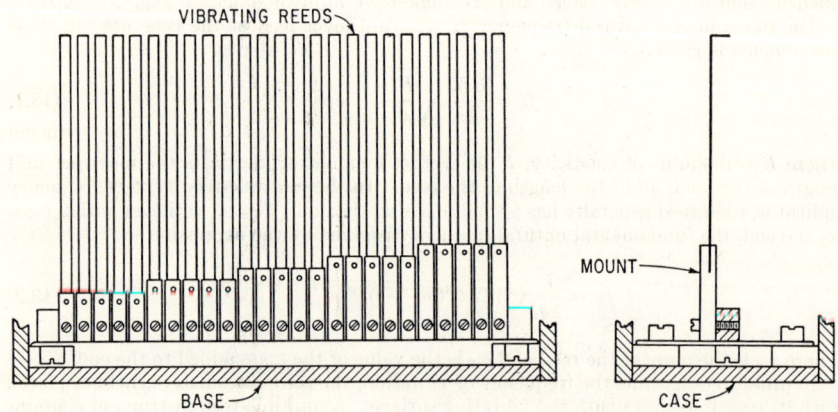

Fig. 13.26. Schematic diagram, Frahm's reeds.

The minimum displacement amplitude which gives an observable response is about 0.1 mil. The maximum allowable (applied directly) is about 2 to 3 mils, but the instrument can be located on a portion of structure that is not vibrating excessively. To avoid excessive fatigue, reeds should not be allowed to vibrate continuously with peak-to-peak displacement greater than ¾ in. for frequencies up to 60 cps, and correspondingly less for higher frequencies. Where vibration is excessive, the instrument should be cushioned with the hand rather than applied directly to the structure being tested.

One manufacturer * has reed gages covering the following frequency ranges: 23.4 to 33.4 cps (1,400 to 2,000 cpm), in single-row units; and 13.3 to 23.4 cps (800 to 1,400 cpm), 25 to 50 cps (1,500 to 3,000 cpm), 41.6 to 83.4 cps (2,500 to 5,000 cpm), and 75 to 150 cps (4,500 to 9,000 cpm) in double-row units. Special units are constructed having an upper frequency limit of approximately 500 cps.

Another manufacturer † produces gages covering ranges as low as 13 to 17.7 cps (780 to 1,070 cpm) and as high as 1,300 to 1,775 cps (78,000 to 107,000 cpm).

DAVEY REED VIBROMETER.‡ This vibrometer is a single-reed, adjustable-length, frequency indicator in which the nonvibrating portion of the reed is stored in a circular coil, much like a pocket measuring tape. The free length of the reed is adjusted by turning a thumb nut at the top of the case. Frequency markings, in cycles per minute (cpm), are graduated on the outer edge of a disc on the front face of the instrument which turns as the reed length changes.

* James C. Biddle Co., Philadelphia, Pa.
† Hartmann and Braun, Frankfurt, Germany; distributed by EPIC, Inc., New York.
‡ Manufactured by The Vibroscope Co., Glenford, N.Y.

MASS-SPRING TYPE VIBRATION INSTRUMENTS   13-29

The base of the instrument is held against the vibrating surface. When the reed length has been adjusted to give maximum excursion (i.e., to resonance), the frequency is indicated on the graduated scale by an index mark along the periphery. The frequency range covered is 7.5 to 830 cps (450 to 50,000 cpm).

WESTINGHOUSE REED VIBROMETER.* This vibrometer is a single-reed, adjustable-length frequency indicator. Shown schematically in Fig. 13.27, this device consists of a thin steel cantilever reed, clamped between rollers. Rotation of one of the rollers, by means of a thumb nut, changes the length of the free reed, resulting in a corresponding change in natural frequency. The frequency is indicated by the position of a pointer on the fixed end of the reed along a calibrated frequency scale on the frame. To measure vibration frequency, the head of the instrument is held against the vibrating surface and the reed length is slowly extended until the reed vibrates at maximum amplitude. The motion of the end of the reed at resonance provides an approximate indication of the amplitude of vibration applied, as the resonant magnification is approximately constant from about 300:1 to 500:1.

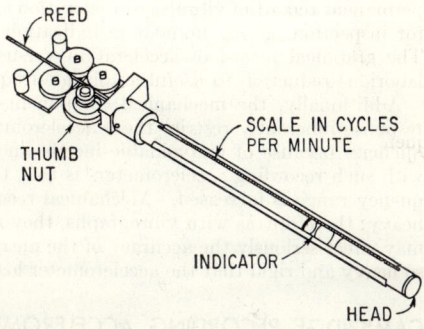

Fig. 13.27. Schematic diagram, Westinghouse Reed Vibrometer.

The frequency range covered is 8.3 to 333 cps (500 to 20,000 cpm). The minimum detectable displacement is about 0.1 mil. The vibrometer should not be used directly on an object vibrating with a displacement greater than ±0.010 in. since the reed vibration, when tuned to resonance, will be excessive; however, there is no practical maximum limit since the pickup can be placed on a portion of structure that is not vibrating excessively. The dimensions of this device are about 9 by 1¾ by 1¼ in.; its weight is about ½ lb.

KORFUND REED VIBROMETER.† This vibrometer is a single-reed, adjustable-length frequency indicator. It has a reed in the form of a steel strip whose length can be varied by turning a knurled thumb nut on the side of the instrument, thus varying its natural frequency. The design is generally similar to that shown in Fig. 13.27. The vibrometer has a probe attached which is held against the vibrating structure. When the reed is tuned to the proper frequency and resonates, the value of the frequency is indicated by a marker (attached to the enclosed end of the reed) on a frequency scale engraved on the tube surrounding the enclosed end which serves as a handle.

This vibrometer has a built-in electromagnetic pickup which provides an electrical signal that can be amplified and recorded or viewed on an oscilloscope. In addition, an amplitude comparator scale attached to the vibrometer base allows the operator to read visually the excursion of the end of the reed; this permits determination of variations in amplitude at different frequencies of vibration or at different locations.

The vibrometer covers the frequency range from 2 to 250 cps (120 to 15,000 cpm). The low-frequency range, 120 to 600 cpm (2 to 10 cps), is attained by use of a weight on the end of the reed. The free (unweighted) reed covers the range from 600 to 15,000 cpm (10 to 250 cps). In the low range, this vibrometer can be used only with the reed vertical, as the static deflection of the end of the reed causes it to sag off-scale if held horizontally. This instrument is 9¼ in. by 1⅝ in. by 1 in. in over-all dimensions, and it weighs about 1 lb.

* Manufactured by Westinghouse Electric Corporation, Electronics and X-ray Division, Baltimore, Md.
† The Korfund Co., Inc., Long Island City, N.Y.

MECHANICAL RECORDING ACCELEROMETERS

The self-contained mechanical recording accelerometer has basically the same design and construction features as the vibrograph; however, the mass-spring system usually has a fairly high natural frequency (relative to the frequencies of vibration being measured) and a damping coefficient fairly close to 0.7 of the critical value. The useful frequency range, for dynamic error less than 20 per cent, extends from zero frequency (steady acceleration) to about the natural frequency if damping is from 0.5 to 0.7 of the critical value.

The advantages of this type of instrument are similar to those of the mechanical vibrograph. These recording accelerometers are relatively portable and rugged, simple to set up and operate, and require little or no accessory equipment and power supplies. A permanent record of vibratory acceleration as a function of time is available immediately for inspection, giving immediate indication of waveshape and frequency components. The graphical record of acceleration versus time is subject to the same problems of laborious reduction to useful amplitude-frequency data as the vibrograph records.

Additionally, the mechanical components used in transmitting and amplifying the response inherently restrict these accelerometers to measurement of relatively low frequencies because of unavoidable inertia and flexibility limitations. Another difficulty with such recording accelerometers is that their sensitivities decrease as the useful frequency range is increased. Mechanical recording accelerometers usually are relatively heavy; therefore, as with vibrographs, they may load the structure being measured and may affect seriously the accuracy of the measurement unless the structure in question is so heavy and rigid that the accelerometer load may be neglected.

CAMBRIDGE RECORDING ACCELEROMETER *

This two-component mechanical recording accelerometer shown in Fig. 13.28 is capable of recording accelerations in two different planes simultaneously. Two masses are mounted on flat steel springs and oriented to move at right angles to each other. These elements are critically damped by means of an electromagnet. The movement of each mass, relative to the instrument base, is transmitted by a mechanical linkage to a stylus point. The two stylus points, situated side by side, mark on the cellulose-acetate film; a third stylus, recording on the other side of the film, is an event marker, which can provide time or other reference marks. The record is formed by plastic flow of the cellulose acetate under the light pressure of the spherical stylus point. Film speed is adjustable between 3 and 20 mm/sec.

Mass-spring systems of various natural frequencies (from 20 to 70 cps) and maximum acceleration ranges (from 1 to 20 g respectively) are available.[5] Since the elements are critically damped, the useful frequency range extends almost up to the natural frequency of the element.

The instrument is mounted in a rugged metal case with controls and event marker terminals on the front cover and equipped with a carrying handle. Various accessories are available, including a microscope and a portable viewer-enlarger for reading the records. The viewer enlarges the record 10 times, making it convenient to read records to about 0.1 mm. With the microscope, which has a calibrated reticle, films can be read accurately to 0.01 mm, or $\frac{1}{200}$ the

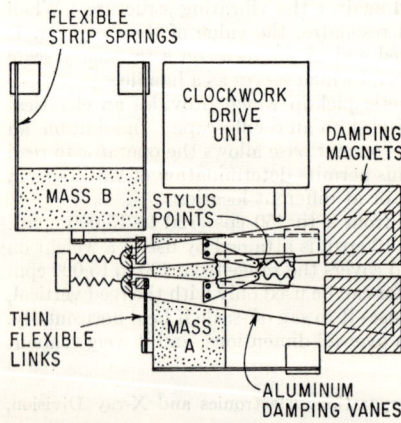

FIG. 13.28. Schematic diagram, Cambridge Recording Accelerometer.

* Manufactured by Cambridge Instrument Co., Ltd., London.

nominal maximum acceleration. If needed, greater optical magnification is possible because of the characteristic of the record, which is produced by cold flow of the celluloid under the stylus pressure. A smooth line results, thereby permitting the use of high optical magnification so that small deflections of the stylus can be resolved. This feature provides a wide useful amplitude range in spite of the fact that the nominal maximum excursion on the film is only 2 mm.

Because the natural frequencies of the seismic elements are relatively low, use of the accelerometer is restricted to measurements of acceleration associated with low-frequency vibration or gross motions such as vehicle acceleration and ride motions. For example, the 1g element, which has a natural frequency of 20 cps and is near critically damped, is useful up to about 15 cps. Because of its weight, this instrument is not suitable for measurement of the motions of relatively light and flexible structures but is limited to use on relatively heavy machinery, structures, and vehicles.

Technical Data. The useful frequency range is given in Fig. 13.29.

Useful frequency range:	0 to 15 cps, for $f_n = 20$ cps unit (upper limit proportionately higher for other units)
Recording medium:	Cellulose-acetate film, 20 mm wide (supplied in spools of 10 ft)
Tape drive:	Spring motor, governed
Dimensions:	10 by 6 by 10¾ in. (25 by 14 by 27 cm)
Weight:	25.5 lb (11.6 kg)
Power required:	None

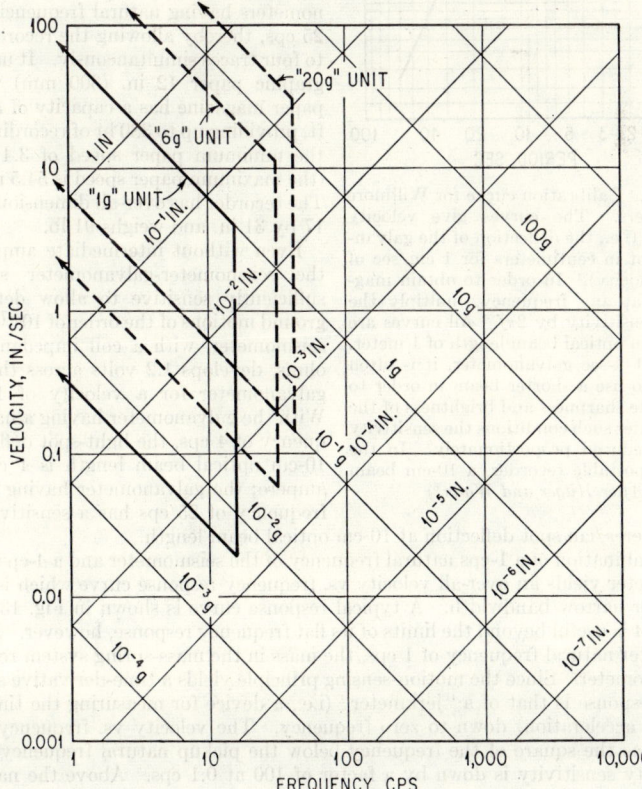

FIG. 13.29. Useful range of amplitude, Cambridge Recording Accelerometer.

SEISMOGRAPHS

"Self-contained" multichannel recording systems can be assembled from a selection of commercially available vibration pickups plus a commercially available oscillograph with self-contained battery power supply.

WILLMORE SEISMOGRAPH.* This self-contained instrument is designed for seismographic measurements in the field.[5] The system consists of one or more velocity-sensing seismometer pickups, a photographic-recording galvanometer oscillograph, and a required power supply. This seismometer is a velocity-sensing pickup employing a large spring-supported magnet as the mass of a mass-spring system, the coil being fixed to the case. It has a natural frequency of only 1 cps; it is 5½ in. in diameter, 11½ in. long, and weighs 21 lb. This seismometer may be used for the measurement of either horizontal or vertical vibration. The damping is adjustable, but usually is set at about 0.7 of the critical value.

The recorder accommodates four galvanometers having natural frequencies of 4 or 25 cps, thereby allowing the recording of up to four traces simultaneously. It uses photographic paper 12 in. (300 mm) wide; the paper magazine has a capacity of about 300 ft, providing up to 380 hr of recording time at the minimum paper speed of 3.4 mm/min (the maximum paper speed is 54.5 mm/min). The recorder has over-all dimensions of 10 by 17 by 31 in. and weighs 91 lb.

Even without intermediate amplification, the seismometer-galvanometer system is sufficiently sensitive to allow detection of ground motions of the order of 10^{-7} cm. The seismometer, with a coil impedance of 500 ohms, develops 1.2 volts across the 70-ohm galvanometer for a velocity of 1 cm/sec. With the galvanometer having a natural frequency of 4 cps, the light-spot deflection for 10-cm optical beam length is 1 cm/microampere; the galvanometer having a natural frequency of 25 cps has a sensitivity of 2.5

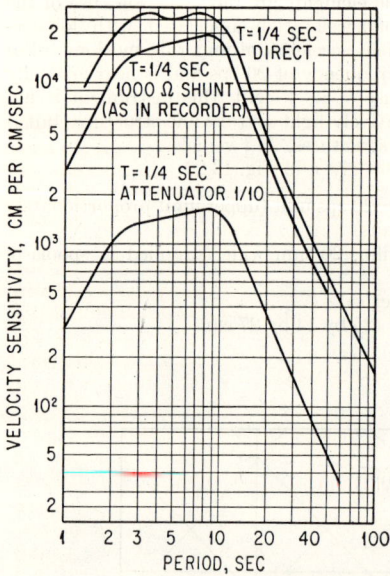

FIG. 13.30. Calibration curve for Willmore seismometers. The curves give velocity sensitivity (i.e., the deflection of the galvanometer spot in centimeters for 1 cm/sec of ground velocity). In order to obtain magnification at any frequency, multiply the velocity sensitivity by $2\pi f$. All curves are given for an optical beam length of 1 meter. With the 1/4-sec galvanometer, it is often desirable to use a shorter beam in order to improve the sharpness and brightness of the trace. Under such conditions the sensitivity must be reduced proportionately. In the Willmore portable recorder, a 10-cm beam is used. (*After Hilger and Watts.*[6])

microamperes/cm spot deflection at 10-cm optical beam length.

The combination of a 1-cps natural frequency of the seismometer and a 4-cps or 25-cps galvanometer yields an over-all velocity vs. frequency response curve which is flat only in a rather narrow bandwidth. A typical response curve is shown in Fig. 13.30. The instrument is useful beyond the limits of its flat frequency response, however. Below the seismometer natural frequency of 1 cps, the mass in the mass-spring system responds as an accelerometer. Since the motion-sensing principle yields a time-derivative signal, the over-all response is that of a "jerkmeter" (i.e., a device for measuring the time rate of change of acceleration) down to zero frequency. The velocity vs. frequency response decreases as the square of the frequency below the pickup natural frequency. Hence, the velocity sensitivity is down by a factor of 100 at 0.1 cps. Above the natural fre-

---

* Manufactured by Hilger and Watts, Ltd., London.

quency of the galvanometer, the response of the galvanometer decreases as the square of the frequency; consequently, over-all sensitivity is down by a factor of 100 at 10 times the galvanometer natural frequency. The damping provided in both the seismometer pickup and the galvanometer circuit helps flatten the response in the regions of resonance.

Because of its extreme sensitivity and long-time recording capabilities, this instrument is a useful device for field seismometry. Its bulk and weight, however (it requires a 6-volt storage battery plus a vibrator pack for field use), make it unsuitable for general use in the measurement of vehicle or machinery vibrations.

ELECTRONIC VIBRATION METERS AND RECORDERS

This section describes "self-contained" electronic vibration meters and recorders which employ electrical vibration pickups as sensing elements. Such an instrument may be a relatively simple vibration meter or a more complex electronic system that records acceleration, velocity, or displacement. The term *vibration meter* is applied to a self-contained measurement system comprised of a vibration pickup, adjustable attenuator, self-contained amplifier, and direct-reading meter.

These types of vibration measurement systems offer substantial advantages over the vibrograph. Their frequency and amplitude ranges are larger, while the use of a separate pickup, which can be small and light (e.g., 1 or 2 in. or less in maximum dimension, and a few ounces in weight), permits measurement of vibration of objects and structures the motion of which would be affected seriously by the load of a vibrograph. Since the output signal can be transmitted conveniently to a remote recorder, it is practical to record vibration at relatively inaccessible, inconvenient, or hazardous locations and at many points simultaneously with a multichannel recorder. Some physical limitations of mechanical recording devices are eliminated, providing much greater dynamic amplitude and frequency range of measurement than is possible with a vibrograph. Of particular importance is the great ease and practicality of frequency analysis of the motion by the use of magnetic recording, automatic frequency analysis, and data-reduction equipment. Also useful is the capability of electrically filtering out unwanted signal components and of integrating and differentiating the signal to provide, respectively, displacement and acceleration data. Furthermore, by the use of meter damping, rapid variations in amplitude can be averaged out, allowing a more reliable visual estimate of a short-time average of the meter reading.

In devices of this type, connectors usually are provided for oscilloscope presentation, for headphone listening, and for connection to a vibration analyzer which can measure the amplitude and frequency of the various components of the vibration.* If only the frequency of a steady-state vibration is required, the frequency may be determined from an oscilloscope presentation (see Fig. 19.52) by the use of an electronic frequency counter, or by use of special frequency analyzers.†

In using vibration-measurement systems of the electronic type in very high noise levels, spurious readings may be obtained as a result of microphonics (in the vacuum tubes) which are sonically induced. Under such field conditions, this effect should always be evaluated.

Several manufacturers ‡ of sound-level meters produce a "control box" which can be used to convert a sound-level meter into a "vibration meter"; an accelerometer pickup is connected to the input of the control box and the output is connected to the sound-level meter. By means of integrating networks in the control box, voltages can be delivered to the sound-level meter which are proportional to acceleration, velocity, or displacement—the desired response being selected by a switch.

GENERAL RADIO TYPE-761A VIBRATION METER AND PICKUP.§ This device consists of a Rochelle salt crystal pickup (which delivers a voltage proportional to the

---

* For example, General Radio Company's Type 762-B or 1554-A vibration analyzers.

† One such special frequency analyzer is the Stroboconn (manufactured by G. C. Conn, Ltd., Elkhart, Ind.). This device can indicate the frequency of an electrical signal (supplied by a pickup) in the range from 33 to 4,186 cps to an accuracy of 0.05 per cent.

‡ For example, General Radio Co., West Concord, Mass.; H. H. Scott, Inc., Maynard, Mass.

§ Manufactured by General Radio Company, West Concord, Mass.

13-34 SELF-CONTAINED VIBRATION-MEASURING INSTRUMENTS

acceleration of the vibratory motion), an adjustable attenuator, an amplifier, and a direct-reading indicating meter. An integrating network can be switched to convert the output of the vibration pickup to a voltage proportional to either displacement or velocity.

The meter is calibrated to indicate readings in rms values of microinches, microinches per second, or microinches per second per second. The instrument requires no external power, and results, read on the meter, are immediately available. The piezoelectric accelerometer is held against, screw-mounted to, or attached by means of a special permanent-magnet clamp to the vibrating structure.

The meter itself should not be exposed to intense noise or vibration levels when measuring small amplitudes since an error may result from microphonics. Because of the large R-C components (time constants) in this instrument which enable it to operate at low frequencies, transients due to switching require 5 to 10 sec to stabilize before readings are taken. The piezoelectric pickup should not be subjected to rms acceleration values greater than $10g$ or to temperatures exceeding 120°F.

Technical Data. Frequency ranges are shown in Fig. 13.31.

Pickup natural frequency: 1,500 cps
Disp. amp. range: 0.05 to 30,000 mils
Indication method: Meter reading of amplitude
Power requirement: Self-contained batteries
Dimensions: Pickup: 2½ by 2½ by 2 in.; case: 13 by 9 by 13 in.
Weight: Pickup, 0.8 lb; case, 23 lb

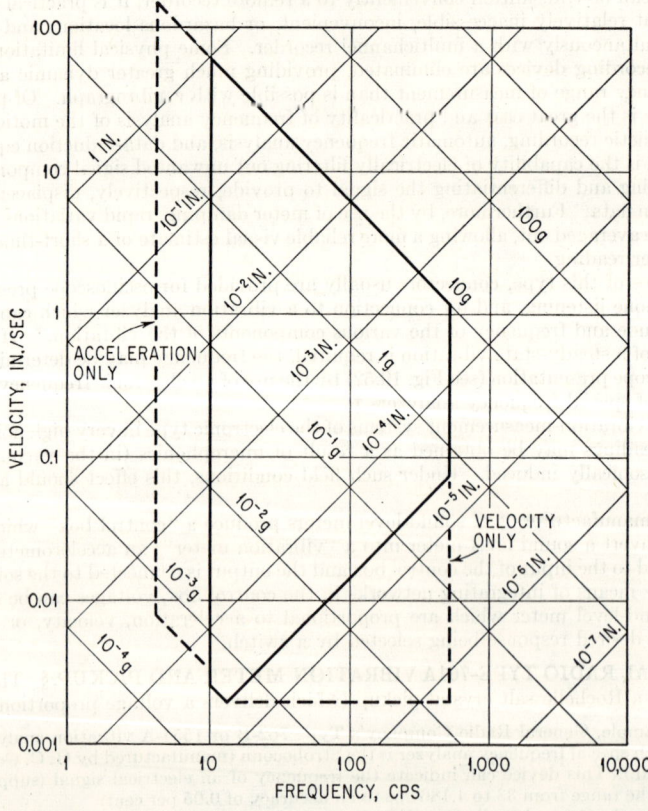

Fig. 13.31. Useful range of amplitude; rms values are shown. General Radio Type 761-A Vibration Meter.

MB VIBRATION METER, MODEL M1A.* This portable battery-operated vibration meter is used in conjunction with a velocity pickup (see Chap. 15) so that it indicates velocity amplitudes directly. Integrating and differentiating circuits, either of which may be switch-selected, provide displacement or acceleration amplitude values, respectively. The amplitude values are read on a calibrated meter scale. An output jack permits the feeding of the output signal to an external oscilloscope for monitoring or record-

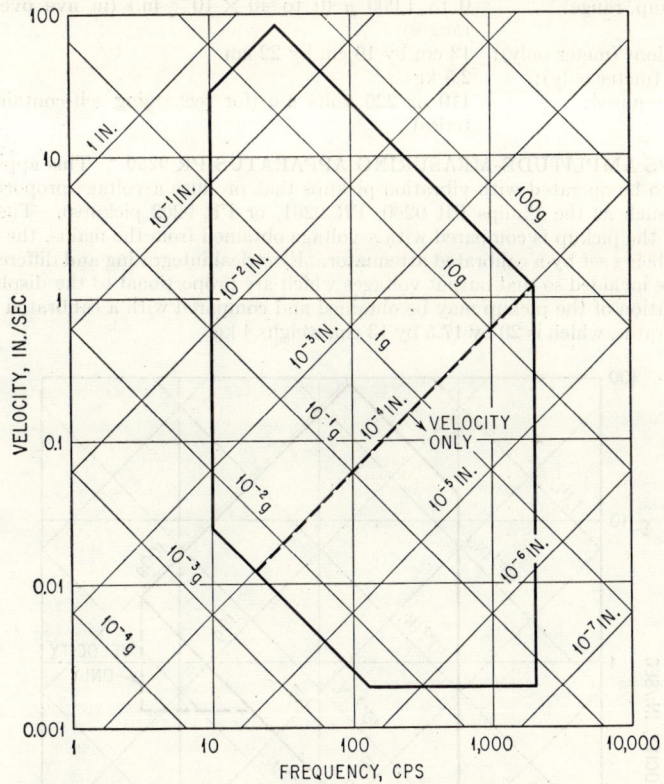

FIG. 13.32. Useful range of amplitude, MB Model M1A Vibration Meter.

ing. The pickup is held against the vibrating surface by hand, by screws, or by some means of clamping.

Technical Data. The useful frequency range is shown in Fig. 13.32.

Pickup natural frequency: Approximately 8 cps
Disp. amp. range: Approximately 0.0001 to 0.5 in. (smaller displacements measurable as velocity reading; see Fig. 13.26)
Dimensions: Pickup: 1 9/32 by 2 13/32 in.; meter: 13 1/2 by 7 by 9 in.
Weight: Pickup: 9.7 oz; meter: 12 lb
Power required: Self-contained batteries

PHILIPS VIBRATION METER PR 9252.† A battery-operated, transistorized vibration meter, in combination with either Philips electrodynamic vibration pickup PR 9260 or 9261, forms a self-contained measurement system. A moving-coil meter provides

* Manufactured by MB Manufacturing Company, New Haven, Conn.
† Manufactured by Philips, Eindhoven, Holland.

13-36 SELF-CONTAINED VIBRATION-MEASURING INSTRUMENTS

direct readings in terms of microns peak value. An output terminal is provided for connecting an oscilloscope or high-speed level recorder to the vibration meter. The apparatus is fed from five self-contained chargeable nickel-cadmium cells. The instrument also contains a charging circuit for connection to a-c mains.

Technical Data

Frequency range:	10 to 1,000 cps
Disp. amp. range:	0 to 1,000 μ (0 to 40×10^{-3} in.) (in five overlapping ranges)
Dimensions (meter only):	13 cm by 10 cm by 22 cm
Weight (meter only):	2.5 kg
Power required:	110 or 220 volts a-c (for recharging self-contained batteries)

PHILIPS AMPLITUDE-MEASURING APPARATUS PR 9250.* This apparatus is designed to be operated with vibration pickups that produce a voltage proportional to velocity (such as the Philips PR 9260, PR 9261, or PR 9262 pickups). The output voltage of the pickup is compared with a voltage obtained from the mains, the value of the latter being set by a calibrated attenuator. Electrical integrating and differentiating circuits are included so that output voltages which are proportional to the displacement or acceleration of the pickup may be obtained and compared with a calibrated voltage. This apparatus, which is 20 by 17.5 by 13 cm, weighs 4 kg.

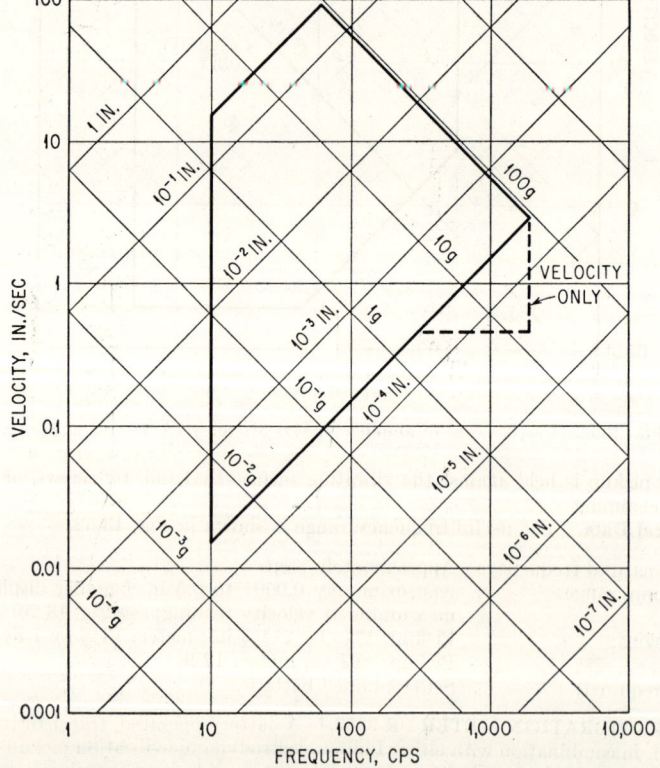

FIG. 13.33. Useful range of amplitude, CEC Model 1-110B Vibration Meter.

* Manufactured by Philips, Eindhoven, Holland.

MASS-SPRING TYPE VIBRATION INSTRUMENTS 13-37

CEC VIBRATION METER (TYPE 1-110B).* This portable a-c powered unit has a built-in amplifier which provides optional direct or integrated output; average velocity or peak-to-peak displacement amplitude is indicated on the calibrated meter. It employs a small, light, magnetic pickup which yields a signal proportional to vibration velocity. Full-scale setting on the meter is selected by a rotary switch attenuator, covering ranges from 5 to 500 mils peak-to-peak (± 0.0025 to ± 0.25 in.).

This vibration meter has a wide useful range of frequency and amplitude, and may be used for measurement of (approximate) rms values of complex steady motions. Output jacks are provided, thus permitting the signal to be fed into an oscilloscope for monitoring purposes or directly into a low resistance galvanometer for oscillographic recording of steady-state or transient signals.

Technical Data. The useful frequency range is shown in Fig. 13.33.

Natural frequency:	9 cps
Disp. amp. range:	± 0.00025 to 0.25 in.
Dimensions:	Pickup: 2 in. in diameter by 2¾ in.; meter: 9 by 10½ by 8¼ in.
Weight:	Pickup: 0.6 lb; meter: 8 lb
Power requirements:	115 volts a-c

GLENNITE MODEL AR-1 ACCELEROMETER RECORDER.[7]† This recording accelerometer employs a seismic transducer incorporating a spring suspension which is

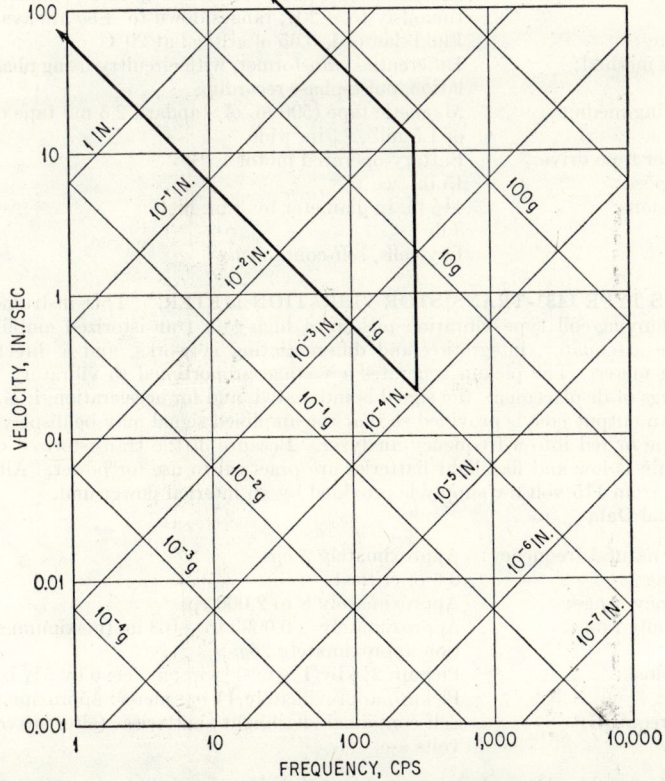

FIG. 13.34. Useful range of amplitude, Glennite Model AR-1 Accelerometer

* Manufactured by Consolidated Electrodynamics Corp., Pasadena, Calif.
† Manufactured by Gulton Industries, Inc., Metuchen, N.J.

combined with a differential transformer sensing element so as to reduce the sensitivity to acceleration in directions other than that being measured. Normally the unit is supplied with the transducer mounted on the inside of the top plate housing so that the axis of sensitivity is parallel to the axis of the cylinder. However, the unit can be obtained with the transducer mounted separate from the recorder; then the axis of sensitivity is perpendicular to the mounting flange of the transducer.

The second part of this system is a lightweight recorder whose circuitry is completely transistorized to minimize current drain, size and weight, and to provide instantaneous starting and ruggedness under high accelerations. Pulse-position modulation as a function of acceleration is employed, resulting in a phase-modulated signal.

The magnetic-tape drive mechanism is powered by dry cells. Thirty seconds of recording time are provided by using standard ¼ in. acetate-base tape. For reproducing the signals that have been recorded on tape, a standard magnetic-tape playback drive mechanism is employed. A special demodulator circuit is used which is compatible with the recorder. Its output is directly proportional to the mechanical signal input to the recorder.

A similar three-channel recorder playback system Model AR-3 also is available.

Technical Data. The useful frequency range is shown in Fig. 13.34.

Natural frequency:	450 cps
Over-all frequency range:	0 to 300 cps
Acceleration range:	$\pm 60g$ full-scale maximum (see Fig. 13.34); adjustable continuously to $\pm 20g$; ranges down to $\pm 5g$ are available
Damping:	Fluid-damped, 0.65 of critical at 20°C
Sensing method:	Differential transformer, with circuitry using phase modulation, pulse-phase recording
Recording medium:	Magnetic tape (500 in. of standard 2.5 mil tape or 800 in. of 1.5 mil), ¼ in. wide
Recorder tape drive:	Battery-operated motor
Tape speed:	15 in./sec
Dimensions:	4½ in. in diameter by 3 in. high
Weight:	3 lb
Power:	Dry cells, self-contained

DAWES TYPE 1431-TRANSISTOR VIBRATION METER.* This instrument comprises a moving-coil type vibration pickup, a high-gain transistorized amplifier with adjustable attenuator, integrating and differentiating networks, and a direct-reading indicating meter. The pickup generates a voltage proportional to vibratory velocity; for readings of displacement, the signal is integrated, and for acceleration, it is differentiated. An output jack is provided so that the amplified signal may be displayed on an oscilloscope or fed into a frequency analyzer. Because of the transistorized circuitry, power drain is low and flashlight batteries are practical to use for power. Alternative operation from 115 volt a-c supply is provided by an internal power unit.

Technical Data

Pickup natural frequency:	Approximately 5 cps
Damping:	0.6 of critical
Frequency range:	Approximately 8 to 2,000 cps
Disp. amp. range:	Approximately ± 0.0003 to ± 0.3 in. (maximum acceleration approximately $26g$)
Dimensions:	Pickup: 2½ by 1⅝ by 1½ in.; meter: 6 by 5¾ by 8¾ in.
Weight:	Pickup: approximately 11 oz; meter: approximately 6 lb
Power required:	Self-contained flashlight batteries (alternatively, 115 volts a-c)

* Manufactured by Dawes Instruments Limited, Ealing, London.

MASS-SPRING TYPE VIBRATION INSTRUMENTS 13-39

VIBROTEST.* This instrument employs a piezoelectric acceleration pickup together with a compact, self-powered amplifier-meter combination to give readings of acceleration amplitude or, by use of integration circuit, velocity amplitude of the vibration being measured. The meter scale is calibrated directly in g units of acceleration or in meters per second (m/sec) units of velocity. An output jack is provided.

Technical Data

Pickup natural frequency:	Approximately 18,000 cps (or higher)
Frequency range:	30 to 12,000 cps (maximum error ±20 per cent)
Amplitude range:	Acceleration, 0.1 to 300g (in six overlapping ranges). Velocity, 0.005 to 10 m/sec (in six overlapping ranges) (approximately 0.2 to 390 in./sec)
Dimensions:	Pickup: approximately 2 by 2 by 1 cm (approximately 1 by 1 by ½ in.); meter: 30 by 8.5 by 8.5 cm (approximately 12 by 3½ by 3½ in.)
Weight:	Pickup: 28 gm (approximately 1 oz); meter: 2.2 kg (approximately 5 lb)
Power required:	Self-contained batteries

SHOCK RECORDERS

The term "shock recorder" refers to any self-contained instrument having a mass-spring system which is intended to provide a true time-history record of shock motions. This category includes "ruggedized" vibrograph-type devices for recording shock displacements, ruggedized vibration meters of the acceleration-recorder type for recording single shocks, and especially designed reed gages. (This class does not include the so-called "impact recorders" or "ride recorders," discussed in the next section, which provide a long-duration record of the occurrence and magnitude of individual shocks or impacts.)

LEACH TRI-AXIAL ACCELEROMETER.† This device is a miniature three-component mechanical shock recorder designed for use in missile testing and related applications. It incorporates three cantilever reeds, oriented to respond to motion in three mutually perpendicular directions.

A stainless-steel stylus on the end of each reed inscribes the reed motion on the "gold-flashed" surface of a stainless-steel ball-bearing race which serves as the recording member. The reeds are arranged so that the vertical-responding reed (Z axis) records on the outer peripheral surface, while the horizontal-responding reeds (X and Y axis) record on the flat cylindrical portion of the race. A battery-driven electric motor turns the race at 10 rps, giving a total recording time for 1 revolution of 100 milliseconds (msec).

This instrument is provided with either of two sets of reeds for different acceleration ranges. The natural frequencies and associated acceleration ranges are listed below (since the reeds are relatively undamped, upper useful frequency is about one-third the natural frequency):

Range no.	Transverse reeds		Long'tudinal reeds	
	Nat. freq., cps	Accel. range, g	Nat. freq., cps	Accel. range, g
-1	1,500	150–7,500	2,000	250–10,000
-2	250	4– 200	500	16– 800

* Manufactured by Rohde & Schwarz, Munich.
† Manufactured by Leach Corp., Inet Division, Compton, Calif.

Technical Data

Dimensions:	$2\frac{1}{16}$ by $1\frac{11}{16}$ by $1\frac{7}{8}$ in.
Weight:	0.4 lb
Power:	5 to 7 volts d-c at 0.5 amp
Operating temperature range:	-50 to $160°F$
Operating relative humidity:	Up to 100 per cent

Because of its small size and low weight, this accelerometer is well suited for measurement applications where weight and space are at a premium. In some such applications, the fact that it can be used for only one event recording is not a severe handicap. The instrument is relatively insensitive due to lack of mechanical amplification. However, high sensitivity is not essential in many shock measurements. Optical magnification cannot completely eliminate this lack of sensitivity as lack of definition of the line edges generally limits the amount of magnification that can be used successfully. Detailed evaluation and analysis of the records is somewhat inconvenient, not only because of the graphic form of the record but also because of the need for a microscope or large-magnification optical comparator to observe the record.

REED GAGE. Of the shock-recording gages that do not provide a time-history of motion, perhaps the best known and most widely used is the so-called "reed gage,"* also known as the "multifrequency reed gage" and "multifrequency impact gage."[8,9] This device consists of a rigid frame supporting a series of mass-spring elements in the form of mass-loaded cantilever reeds of different natural frequencies. Natural frequencies of these reeds range about from 25 to 2,000 cps. The response of each reed is recorded by a stylus on the end of the reed scribing on waxed paper.

In current practice, the peak deflection of each reed is measured and generally is converted into an equivalent acceleration value by multiplication by $(2\pi f_0)^2$, where f_0 is the natural frequency of the reed in units of cycles per second.[10] The equivalent peak accelerations are then plotted against reed natural frequency, the resulting curve being known as a "shock spectrum"[10] (see Chap. 23 for detailed discussion). The utility of this method of reducing and displaying the data is that, although it does not describe what the applied shock motion is, it does provide an "operational" definition of the motion by describing what the motion does, i.e. by describing the effect of the motion on simple mass-spring systems of various natural frequencies.

A significant limitation is that resonant responses may give misleading indications, as damping of reeds is not consistent. Also, since there is no mechanical amplification, deflections of the higher-frequency reeds are extremely small, even for rather high accelerations, while accuracy and resolution are rather poor. For example, a 1,000-cps reed will deflect only 0.01 in. for a $1,000g$ acceleration. This deflection can barely be resolved satisfactorily on the record because the trace of the (ruggedized) stylus on the waxed paper has a breadth of similar magnitude.

A Type II gage has also been designed and built at the David Taylor Model Basin. It differs from the Type I gage described above mainly in that provision is incorporated for moving the recording surface during a shock; consequently, the occurrence of multiple shocks of large magnitude can be detected.

Technical Data

Frequency range:	Natural frequencies of elements: 25 to 2,000 cps
Amplitude range:	5 to $20,000g$
Dimensions:	$9\frac{1}{2}$ by $3\frac{1}{2}$ by 9 in.
Weight:	$19\frac{1}{2}$ lb
Power requirements:	None

LEACH MULTIPLE-RECORDING ACCELEROMETER MRA-440.† This is a miniaturized version of the reed gage, much smaller and lighter than the instrument dis-

---

* Not commercially available; designed and built at the David Taylor Model Basin and U.S. Naval Research Laboratory. A version of this instrument also has been manufactured privately for the Naval Ordnance Test Station.

† Manufactured by Leach Corp., Inet Division, Compton, Calif.

cussed on the preceding page. It is comprised of an array of eight reeds mounted radially on a plate 2½ in. in diameter with a ¾-in.-diameter hub. The reeds are supported on the outer edge of the plate; the response of each reed is recorded by means of a stainless-spring-steel stylus on the end of the reed which inscribes on the recording hub. The hub, which is stainless steel, has a specially treated surface finished in soft gold; the record is obtained as a fine, bright scratch on the gold finish.

Two units are available, the MRA-440-1 and the MRA-440-2, differing only in the frequency ranges covered. The -1 model has reeds ranging in natural frequency from 950 to 2,000 cps in approximately 150-cps increments. The reeds of the -2 model have natural frequencies from 350 to 870 cps in approximately 70-cps increments. The two units may be mounted together at the same measuring point by ganging.

Practical measuring range of the -1 model is considered to be from 250 to 10,000g, while the -2 model covers from 60 to 5,000g. Accuracy is ±50 per cent at the low accelerations, improving to ±5 per cent for the high acceleration values. Records are read with a toolmaker's microscope or an optical comparator at 50× magnification. The width of the recorded trace is about 0.001 in. and the resolution obtainable is considered to be about half the line width, or about 0.0005 in.

The instrument is 2½ in. in diameter and about 7/16 in. thick. It weighs 3 oz (about 85 gm).

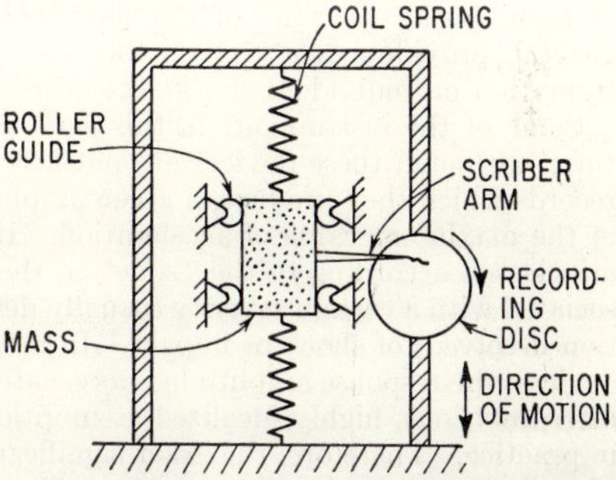

TMB SHOCK-DISPLACEMENT GAGE.* This device is an extremely rugged version of a vibrograph, designed for the measurement of relatively large displacements associated with shocks in naval vessels due to underwater explosions. It is simple in construction, as shown schematically in Fig. 13.35, the housing being a heavy steel framework. The mass of the spring-mass system, suspended from the case by helical coil springs at either end, is guided by rollers to move along the axis of the instrument. A rigid steel scriber attached directly to the mass records, without mechanical amplification, on a rotating chrome-plated steel disc which is driven by a heavy-duty geared electric motor.

Fig. 13.35. The TMB shock displacement gage employs as the inertia element a spring-mounted seismic weight restrained by rollers to single degree-of-freedom. It is similar to the pallographs in principle but was designed primarily for the purpose of obtaining time-displacement records of large-amplitude shock displacements. The instrument records on a rotating chrome-plated disc by a steel scriber attached directly to the seismic element.

Technical Data

Natural frequency:	8½ cps (520 cpm)
Practical frequency range:	800 to 10,000 cpm (13 to 160 cps)
Magnification:	1:1
Disp. amp. range:	0.01 to 1.0 in.
Recording medium:	Chrome-plated steel disc
Recording-medium drive:	Geared electric motor
Markers:	None
Dimensions:	25 by 12 by 15 in.
Weight:	89 lb
Power required:	115 volts, 60 cps

The major advantage of this instrument is its simplicity and ruggedness, which make it suitable for use under severe shock conditions. It gives a time-history of shock motions directly (within its amplitude and frequency limitations). Major limitations are its low sensitivity due to the lack of mechanical magnification in the record and its

* Developed by the David Taylor Model Basin, U.S. Navy.

weight, which loads the structure substantially. These factors limit its application to the measurement of gross shock motions of large vehicles and structures.

IMPACT RECORDERS

"Impact recorders"* are used widely in the transportation and package-handling fields for recording mechanical shock.[11,12] The typical impact recorder has a paper speed (a few inches per hour) which is so slow that a time-history of motion is not obtained. Instead, an impact produces a single excursion from the neutral position of the recording stylus. A record obtained over a period of time therefore shows the occurrence of a number of shocks over this time span, and it provides some indication of the severity of each shock. Because of the slow tape speed, this type of instrument does not, in general, provide a true time-history record which yields waveshape and frequency information on individual shocks (as do *Shock Recorders* described above).

Some of the instruments in this class are labeled "accelerometers" by the manufacturer, although these devices are not accelerometers in the usual sense; in interpreting records which they produce, a given amplitude of response is considered to be indicative of the maximum value of acceleration. In other instruments, the response amplitudes are classed according to the "zone" of the record into which they fall; each zone is associated with a certain severity (usually defined in terms of number of g units of acceleration involved) of shock or impact. In interpreting the records obtained with still other devices, the response amplitude is associated with a given velocity of impact. In all such interpretations, highly idealized assumptions are made which are not generally realized in practice. Therefore, the exact significance of the response as displayed on the record of an impact recorder is uncertain. The major utility of this type of instrument is in recording the occurrence of impact or shock events over a long period of time. If such records are correlated with information on the history of a shipment, the sources of real or potential damage to shipments can be identified. Further, if all of the shocks are similar in waveshape, then they can be compared directly in severity.

SAVAGE IMPACT REGISTER.† This small, sturdy instrument, when secured longitudinally in a freight car, measures and records the severity of end impacts received by the car, and records the times at which they occur.[13] It also may be used in the laboratory to test packaging, crating, and merchandise on the Conbur (see Chap. 26) or other testing machines.

The portion of the instrument which is affected by longitudinal impacts consists of a weight mounted on two rods riding on rollers, the movement of which is restrained by springs on either side. The rate of each spring is 6.38 lb/in. The maximum movement on either side of the zero point is 2.13 in. The natural frequency for this system is about 6.5 cps.

The direct-recording stylus is mounted in the center of the moving weight; the total weight is 1½ lb. Longitudinal impacts are recorded from both sides of the center line in five zones of impact; the width of each zone after the first is 0.47 in. The instrument has a static calibration of $2g$ per zone.

The wax-coated charts are 5 in. wide and 61 ft long. The standard chart speed is 3 ft/day. The time is printed on the chart. The chart is driven by a double-spring eight-day movement, with special features which enable the movement to withstand rough treatment. Also available are 16-day and 30-day clock movements which make possible the use of the register for long shipments.

The register case is supplied with a hasp which may be secured with an ordinary car seal. The instrument may be secured to the floor or side wall of the car. The dimensions of this device are 8 by 15 by 8 in. and its weight is 15 lb.

Impact registers of this type are also built for use in railroad passenger cars. They are similar to the freight-type, with the following exceptions:

1. The passenger-car impact-measuring device has a heavier weight and lighter springs, which make it more than ten times as sensitive as the freight-car type registers.

---

* Sometimes termed "ride recorders."
† Manufactured by Impact Register Co., Champaign, Ill.

2. The chart speed is 6 in./hr, so as to give more accurate determination of time of impact and better resolution of impacts which occur in rapid succession.

3. It is not necessary to secure this register to the floor.

A two-component version of the above instrument, the R-S Two-way Ride Recorder, simultaneously records both longitudinal and vertical impacts on one chart. The longitudinal indications are used to define end impacts of freight cars, and in testing packages, crates and merchandise on Conbur or other testing machines in the laboratory. The vertical indications are used to compare riding qualities of freight cars and to check roadbeds. In the laboratory, the vertical records show the acceleration to which crating and merchandise are subjected when tested on vibration tables. The range of both longitudinal and vertical impacts to be recorded can be adjusted by changing springs.

On the outside of one end of the frame is a 26¾-oz weight which is suspended from a spring and held upward against a stop. This weight is separated from its stop when the recorder receives upward accelerations and/or downward decelerations in excess of the initial tension; the distance it moves varies with the severity of the vertical accelerations.

The following arrangement is used for recording vertical accelerations. A spring, holding the weight, is adjusted with sufficient initial tension so that an added load equal to 10 per cent of the weight ($0.1g$) may be applied without moving the weight; the spring is of such scale that an additional static or constant load of $0.2g$ will produce a record 0.1 in. long and each additional similar load of $0.2g$ will produce an additional 0.1 in. record, up to 1.0 in. which is produced by a load of $2.1g$.

This device has a magnification ratio of 10:1. The theoretical natural frequency of this spring and weight system is approximately 14 cps, if it were not preloaded to one side. This preloading prevents the weight from vibrating on both sides of its zero position and consequently reduces the possibility of resonant magnification of the accelerations producing the records.

MODEL RM-A RECORDER.* This railroad-car impact recorder is intended to provide data on the number of impacts occurring in a railway freight car. It is similar in principle and applicability to the Savage Impact Register—except that it does not provide information on the time of occurrence of impacts. Instead, an impact-actuated ratchet mechanism advances the recording chart 0.06 in. on each impact involving a velocity of 3 mph or higher.

The mass-spring system of this device is a steel bar pivoted at one end, with a stainless-steel stylus at the other end. Restoring forces are provided by helical steel springs bearing against the bar about 1 in. from the pivot point. The stylus records on a wax-coated paper chart 5 in. wide, marked off in zones ($2g$ per zone 0.47 in. wide).

This device is 6¼ by 2½ by 10 in. in dimensions and 4.2 lb in weight.

MODEL 1-M ACCELEROMETER.* Although this device is labeled an "accelerometer," it is actually a modified form of impact recorder, intended for use in laboratory tests in conjunction with other instruments. It employs, as a spring-mass system, a cylindrical mass and helical steel spring. (Various mass-spring systems are available to cover a wide range of accelerations and frequencies.) A record of the excursion of the mass is made by a spring-loaded stainless-steel stylus on a 1-in. wide paper chart which remains stationary during the impact. After each test, the section of paper chart containing the record is removed and a new section of paper moved into place. The readings obtained with this type of instrument indicate maximum acceleration only if the duration of the impact is long compared with the natural period of the mass-spring system. This compact, rugged device weighs 22 oz and is 2½ by 2 by 6 in. in over-all dimensions.

MODEL 3AR ACCELEROMETER.* This is a maximum response instrument for recording impacts in three mutually perpendicular directions. Although the instrument is called an "accelerometer," it is not an acceleration recorder in the usual sense, since it provides only an indication of the maximum response of a single mass-spring system. Its basic utility and application is similar to other impact recorders described here.

* Manufactured by Impact Register Co., Champaign, Ill.

Three separate mass-spring systems are used, one for each direction. A stylus, directly connected to each mass, records on a separate 1½-in. wide wax-coated chart, supplied in 8-ft lengths. Displacements of the stylus $\pm \frac{1}{2}$ in. from neutral position can be recorded. Depending on the mass-spring system used, this stylus motion corresponds to a maximum acceleration amplitude as low as $\pm 5g$ or as high as $\pm 100g$. The chart is driven by impact-actuated adjustable weights which advance the tape about 0.6 in. on each impact by means of a ratchet. The weights may be adjusted to advance the chart according to a selected predetermined value of acceleration. This permits the recording of a large number of separate shocks but it does not record information regarding time of occurrence.

Dimensions of this device are 5½ by 5½ by 6½ in.; its weight is 8 lb, 4 oz.

IMPACT-O-GRAPH.* This type of device is intended primarily for use in recording of shock and impact events in package-handling operations and in vehicles used for transport of packages and other lading. The instrument contains a low-frequency mass-spring system. In all models but one, a record is made of acceleration in three mutually perpendicular directions. Three sensing elements are mounted on a "stylus bracket" which also supports each stylus assembly, including a 10:1 amplifying linkage. The stylus motion is recorded on a wax- or plastic-coated tape.

Natural frequencies of mass-spring systems are available from 60 cpm to 1,800 cpm. Since the record obtained represents the response to an impulsive motion, "frequency range" is not a meaningful quantity. The amplitude range depends on the natural frequency of its mass-spring system. Ranges available are ± 0 to $1g$, 0.8 to $6.1g$, 2.7 to $15g$, 4 to $30g$, 3 to $70g$, 12 to $100g$, and 20 to $300g$.

Some models employ a continuously moving tape, transported by a clock-spring or an electric-motor drive. With the clock drive, record durations of 28 days may be obtained—providing a continuous history of impact events. At this extremely slow tape speed, each event is recorded as a single peak response.

Other models employ a ratchet mechanism which advances the tape each time the recorder is subjected to an impact. This allows the recording of a large number of separate shocks but does not provide information as to time of occurrence.

Models H and HS:	Designed for use in trailers and trucks; 28-day operation; weight 11 lb; dimensions about 11 by 9 by 4½ in.
Model R:	Designed for use in freight cars; other features similar to Models H and HS
Model RR:	One-directional recorder (otherwise same as Model R)
Models SS and L:	Sensitive one-directional recorders
Model YS:	Ratchet-type recorder; weighs 1½ lb; dimensions about 6 by 3½ by 3 in.
Model YC:	Peak-indicating type (nonrecording) instrument which can be equipped with counter to show number of impacts over a specified level

IMPACT REGISTER,† MODEL 1A RECORDING ACCELEROMETER. This device is a direct-recording mechanical accelerometer intended for use in determining riding qualities, and acceleration and deceleration characteristics of automobiles, railroad cars, and other vehicles. The mass-spring system incorporates a ball-bearing-guided mass and interchangeable springs to provide a selection of various natural frequencies. Air damping, adjustable by use of various sizes of orifices, also is provided. A ball-bearing-mounted actuating rod with a stylus on the end transmits the motion to the 5-in. wide recording chart paper, wax-coated, which is supplied in 91-ft rolls.

Chart drives can be provided either with clock-spring motors giving chart speeds ranging from ½ in./hr to 6 in./min or with electric motors. Electric-motor drives are available for 28 volts d-c, 115 volts at 400 cps or 115 volts at 60 cps. With electric-motor drive, chart speeds range from ½ in./hr to 60 in./min (1 in./sec).

* Manufactured by The Impact-O-Graph Corp., Cleveland, Ohio.
† Manufactured by Impact Register Company, Champaign, Ill.

The accelerometer is contained in a rugged but light case of ¼-in. thick magnesium. It is 5½ by 10 by 6½ in. in dimensions and weighs 11 lb.

A three-component recording accelerometer of the same general design and characteristics designated the Model 3A Accelerometer also is available. This instrument, which records accelerations in three mutually perpendicular directions, is 8 by 18 by 8 in. in dimensions; it weighs 18 lb.

STEP GAGES

A "step gage" is a type of instrument which indicates that motion of a specified severity has occurred.[9] Generally, the severity of the motion is defined in terms of a value of maximum acceleration which has been reached or exceeded. By using a number of such gages, each set to give an indication at a different level of acceleration, the peak value of acceleration can be bracketed between the highest value which is indicated to have occurred and the lowest which does not indicate occurrence. Reasonably good data on the relative severity of similar shocks can be obtained with step gages.

This type of instrument sometimes is used in measurement of shock motions because it is in general compact, simple to install, rugged in use, and easy to read. However, it is rarely used where precise or reliable measurements are required. A major source of error is introduced because the mass of the mass-spring system does not follow the maximum acceleration which actually occurs. In general, the mass-spring systems of step gages have natural frequencies from a few tens to a few thousands of cycles per second—no higher (and often lower) than the vibration frequencies associated with high-acceleration

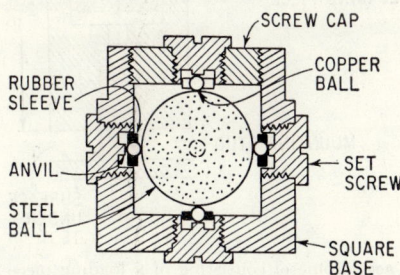

Fig. 13.36. Copper-ball accelerometer is designed to indicate accelerations in three directions. It consists of a steel ball held in a case by point contact with six annealed copper balls, two in each direction of motion, which are held in place by setscrews in the instrument case. When the pickup is accelerated, the steel ball deforms the copper balls by an amount which depends on the magnitude and duration of the acceleration.

shocks. Consequently, large dynamic errors may occur, and resonance effects may be introduced.

The major utility of step gages, in consequence of the foregoing, is in comparison of the severity of similar shocks; in this case, reasonably good data on relative severity can be provided by step gages.

COPPER-BALL ACCELEROMETER. This instrument [14] is designed to indicate magnitude of peak acceleration by the deformation of small copper balls acted on by a large steel ball. Various models are available. One configuration, a three-component model, is shown in Fig. 13.36. It consists of a steel ball held in a case by point contact with six annealed copper balls, two in each direction of motion, which are held in place by setscrews in the instrument case. When the pickup is accelerated, the copper balls are deformed by the steel ball. The deformation of the copper balls depends upon the magnitude and duration of the acceleration.

The instrument is not recommended for the measurement of motions involving successive accelerations since, after the initial acceleration, a space is created between the steel and copper balls and a hammering effect is produced by successive accelerations. A later design of the device employs a spring-loaded wedge between the copper balls and the instrument case to keep the steel and copper balls in contact throughout the motion being measured. The amount of flattening of the copper balls is assumed to be proportional to the magnitude of peak acceleration. The copper-ball accelerometer shown in cross section in Fig. 13.36 has the following characteristics:

Technical Data

Single amplitude range:	10 to $3,200g$
Damping method:	None
Method of indication:	Flattening of small copper ball by large steel ball
Power requirement:	None
Dimensions:	2½ by 2½ by 2½ in.
Weight:	1.5 lb

MASS-PLUG ACCELEROMETER. This instrument [15] consists of a "loading piece" of known mass which screws onto one end of a Bakelite plug which has a reduced section

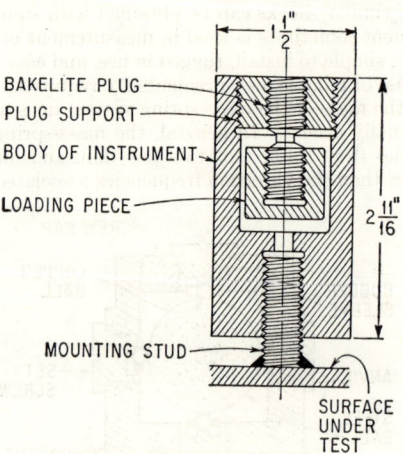

Fig. 13.37. A mass-plug accelerometer consisting of a loading piece of known weight which screws onto one end of a bakelite plug which has a reduced section at mid-length. The other end of the plug screws into the plug support which screws into the body of the instrument. The accelerometer may be attached to the test body by screwing it onto a mounting stud which is welded to the latter.

at mid-length, as shown in Fig. 13.37. The other end of the plug screws into the plug support in the body of the instrument. The accelerometer may be attached to the test body by screwing it onto a mounting stud which is welded to the latter. It is assumed that the tensile force at the reduced section of the plug is the same as the equivalent force acting on the loading piece when the case is accelerated. It is further assumed that when the plug breaks, the instrument has been subjected to an acceleration equal in value to the static force required to break the plug divided by the mass of the loading piece. In general, this assumption is not valid; dynamic forces due to the oscillatory motion of the system can cause fracture of the plug at a value of acceleration either lower or higher than the nominal value.

Figure 13.37 shows one such device which has the following characteristics:

Acceleration range:	$10g$ to $4,500g$
Damping method:	None
Natural frequency:	500 to 2,600 cps
Method of indication:	Fracture of Bakelite plug
Power requirement:	None
Dimensions:	1½ by 2¾ in.
Weight:	1¼ lb

PUTTY GAGE. This instrument,[16] which provides an indication of peak acceleration, consists of a number of spring-loaded masses whose motion results in the indentation of a puttylike material. Each mass has a cylindrical rod with a conical tip, which runs axially through the spring. When the gage is subjected to acceleration, the motion of the rod against the spring force produces an indentation in a plasticene cylinder, the depth of which is an indication of peak acceleration. Plasticene is superior to putty because it retains its properties over a longer period of time. The chief source of error in this device

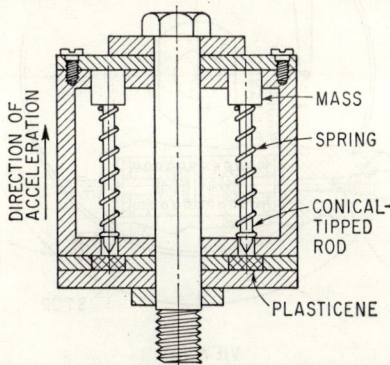

Fig. 13.38. A putty gage for measuring peak values of acceleration. This gage consists of eight spring-loaded masses of different frequencies ranging from 35 to 200 cps. Each mass has a cylindrical rod with a conical tip, which runs axially through the spring. When the instrument is subjected to an acceleration, the motion of the rod against the spring force produces an indentation in a plasticene cylinder which is an indication of the peak acceleration.

results from variations in the properties of this plastic material. Figure 13.38 shows a cross section of one such putty gage having the following characteristics:

Acceleration range:	$25g$ to $2,000g$
Damping method:	Inherent friction plus putty
Natural frequency:	35 to 200 cps
Indication method:	Amount of putty indentation
Mounting method:	Bolted to structure
Power requirement:	None
Dimensions:	3 by 6 in.
Weight:	6 lb

TORSIONAL-VIBRATION RECORDERS

HAND VIBROGRAPHS WITH TORSIONAL-VIBRATION ATTACHMENT. The Askania Hand Vibrograph may be provided with an attachment for torsional-vibration measurement. The attachment clamps to the feeler tube of the vibrograph and has a contact wheel which is pressed against the rotating shaft and assumes the same peripheral speed. Changes in speed due to torsional vibration cause relative motions which are transmitted through a linkage to the tip of the vibrograph pickup arm. Three different sizes of contact wheels and two different linear magnifications are available. The useful frequency range is 5 to 100 cps; the amplitude range is ±0.02 to 10°. Other characteristics are those of the Askania Hand Vibrograph, described earlier.

Similar modifying attachments are available for the Cambridge Universal Vibrograph and the Korfund Hand Vibrograph.

GENERAL MOTORS MECHANICAL TORSIOGRAPH. This instrument is a rotational-displacement type of pickup consisting of an inertia flywheel driven through an elastic connection with the shaft under test. An indicating stylus finger is connected to

13-48 SELF-CONTAINED VIBRATION-MEASURING INSTRUMENTS

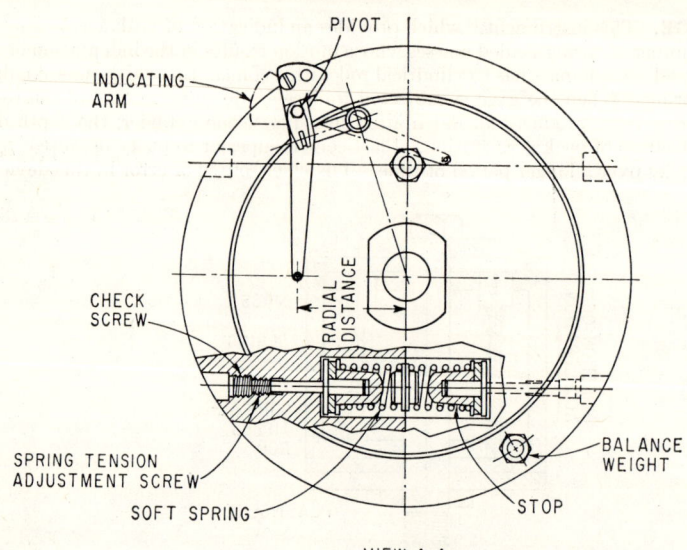

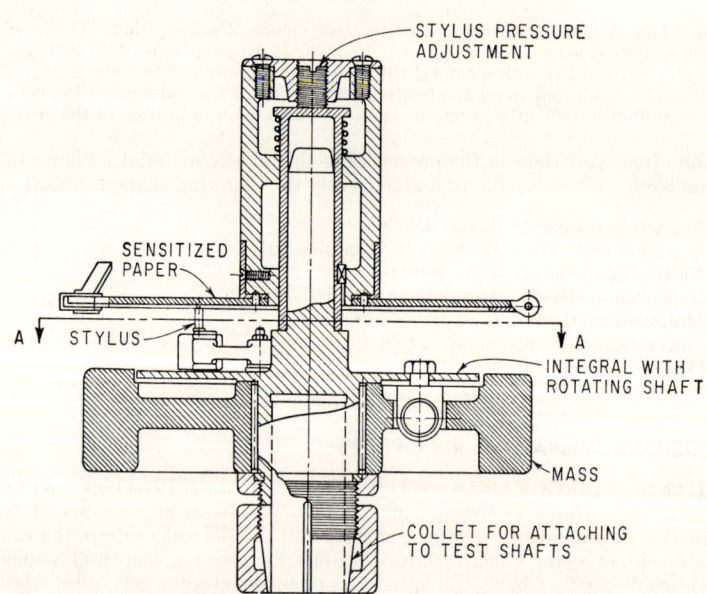

Fig. 13.39. General Motors Mechanical Torsiograph.

FIXED REFERENCE INSTRUMENTS

the seismic mass and to the shaft through a mechanical-linkage system. Any angular displacement between the seismic mass and the rotating shaft actuates the stylus finger, which records the motion on chemically sensitized paper attached to a hand-held member which is momentarily held against the stylus. Thus a polar diagram of the torsional vibration is obtained, the amplitude being read on a transparent scale. The polar diagram permits direct determination of the "order" of the torsional vibration, i.e., the number of cycles of oscillation per shaft revolution, but it does not provide frequency information. A separate measurement of shaft speed is required. Figure 13.39 is a cross-sectional diagram of the pickup with recording member in place.

Technical Data

Frequency range:	16.7 to 334 cps (1,000 to 20,000 cpm)
Amplitude range:	±0.05 to 2.50°
Recording method:	Stylus on sensitized paper
Natural frequency:	8 cps
Power requirement:	None
Attachment method:	Fastened to end of shaft with 1-in. collet
Dimensions:	6¼ by 10 in.
Weight:	9½ lb

TYPE-4 VIBRATION RECORDER WITH TORSIONAL-VIBRATION TRANSLATOR. By replacing the linear inertia translator with a torsional-vibration translator, the Type-4 Vibration Recorder, described earlier, may be converted into an instrument for recording amplitudes and frequencies of torsional vibration. The principal component of the torsional-vibration translator is a light aluminum pulley 100 mm in diameter, which is driven from the test shaft by a short, stiff belt. Within the pulley is a flywheel which is elastically connected to the pulley; this flywheel tends to revolve at a uniform speed due to its inertia. The fluctuations of rotation cause relative motions between the pulley and the flywheel which are transmitted by a rocker arm and connecting rod through a hollow spindle to the recording stylus. The stylus records on waxed paper which may be driven at six different speeds by a synchronous motor or may be driven by the pulley. Magnification of 3:1, 6:1, 12:1, or 24:1 may be obtained by adjustment of the stylus and support arm.

Technical Data

Frequency range:	10 to 83.4 cps (600 to 5,000 cpm)
Recording method:	Stylus on waxed paper
Natural frequency:	5 cps
Power requirement:	115 volts a-c and 6 volts d-c
Attachment method:	Belt and pulley
Dimensions:	12 by 9¼ by 9½ in.
Weight:	28½ lb

FIXED REFERENCE INSTRUMENTS

WEDGE-TYPE VISUAL VIBRATION-DISPLACEMENT INDICATOR

A quasi-optical magnification is provided in the vibration-displacement indicator shown in Fig. 13.40. A gage of this type can be made up in almost any amplitude range by drawing a wedge of suitable dimensions on a piece of paper. The drawing or a photographic reproduction is attached by glue or by pressure-sensitive adhesive tape to the structure whose vibration is to be measured. The indicator relies on persistence of vision to produce the visual impression of a dark triangle with gray edges, the position of the apex of the triangle indicating the value of the vibrational amplitude. The effective magnification obtained is the ratio of the altitude of the triangle to its base length; a ratio of about 20 is the maximum that is practical to use. (See the discussion associated with Fig 18.11.)

Technical Data

Frequency range:	The low-frequency limit about 15 cps (limited by persistence of vision). High-frequency limit: none in principle; smallest displacement amplitude practical to measure: about 0.005 in., corresponding to $40g$ acceleration amplitude at 200 cps
Disp. amp. range:	Minimum practical displacement about 5 mils; no inherent maximum except dimensions of object being observed
Size and weight:	Can be made to any practical size; base length usually about 2 or 3 in.; can be extremely light (a few milligrams)
Power requirements:	None
Advantages:	Simple and inexpensive to prepare, apply, and use
Limitations:	Useful for sinusoidal motions only; errors in reading can be relatively high

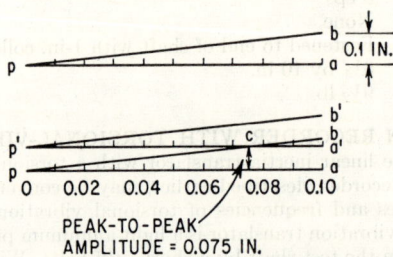

FIG. 13.40. Method of displacement measurement of relatively large amplitudes by the vibrating-wedge technique. A card is attached to the body whose motion is to be measured. The card is oriented so that at rest the line p-a is perpendicular to the expected motion as illustrated in the upper figure. When the body vibrates, two angles are seen. Apparent intercept c permits direct reading of peak-to-peak amplitude pp'. (ASA Standard S-2.2, 1959.[4])

VISUAL TECHNIQUES—MEASURING MICROSCOPES

Under some conditions, the excursion of a given point on a vibrating surface may be directly scaled with a measuring microscope.

The normal unaided eye sees an object with greatest clarity at a distance of approximately 25 cm; any decrease of this distance seems to make the object both larger in size and less distinct in form, due to the eye's inability to accommodate. The microscope's function is to bring this image into sharp focus while preserving the enlarged size. This is accomplished by a system of lenses, whose complexity varies with the number and type of elements, as shown in Fig. 13.41.

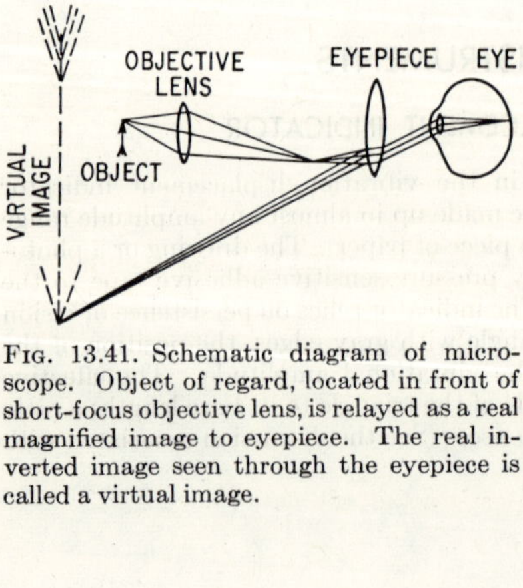

FIG. 13.41. Schematic diagram of microscope. Object of regard, located in front of short-focus objective lens, is relayed as a real magnified image to eyepiece. The real inverted image seen through the eyepiece is called a virtual image.

Used in conjunction with the microscope are the reticle eyepiece, the micrometer eyepiece, and the micrometer stage. The reticle eyepiece, a simple magnifier with an interposed measuring grid, is easy and quick to use over an area being measured, possesses a minimum of parallax when used as the eyepiece of a low-power microscope, but has low primary magnification and introduces significant parallax when used alone. The necessity to interpolate impairs its accuracy when used with the microscope. The micrometer eyepiece, while limited in range, has the advantage of greater accuracy of measurement. The micrometer stage is the most accurate of the three instruments; it can be used with high magnification and is not so limited in range as the other two. Its chief disadvantages are its high cost and somewhat greater difficulty in use.

Advantages of this method of measurement which are not possessed by other techniques include: (1) Displacement amplitudes as large as 1 in. and as small as 0.0001 in. can be measured. (2) The only power requirement is that of the light source. (3) This technique of vibration-amplitude measurement is inherently not frequency limited. Additional advantages include simplicity, precision of calibration, and the absence of loading on the vibrating surface to be measured. Disadvantages include the requirement for an extremely rigid support in close proximity to the vibrating surface, the need for steady vibrating motion, and the fact that the method is often cumbersome and slow.

VISUAL TECHNIQUES—STROBOSCOPIC SYSTEMS

Observation of the displacement amplitude by a microscope is greatly facilitated by the use of an auxiliary stroboscopic light whereby the body is viewed under the illumination of cyclically pulsed light. A common technique is to affix a "whisker" (i.e., a "target") or a bit of glass normal to the surface of vibration, as shown in Fig. 13.42. The microscope is placed at right angles to the whisker, with the tip of the whisker in the microscope field. The stroboscope which illuminates the whisker against a dark field is then adjusted so that its frequency is near to that of the observed vibration rate such that the excursion of the whisker may be seen through the microscope.* Since the excursion of the measuring whisker is observed against a reticle in the eyepiece of the microscope, a firm support independent of the vibrating system must be provided.

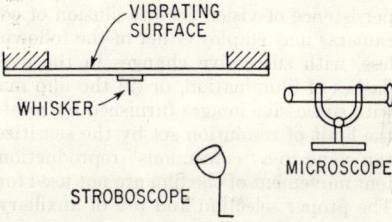

FIG. 13.42. Schematic diagram showing placement of microscope, vibrating plate, measuring whisker, and stroboscope.

In essence all stroboscopic methods permit a determination of the frequency of the observed motion, the frequency being the same as that indicated for the stroboscope's intermittent illumination; in this determination, no load is placed on the vibrating system. An important advantage of this method of observation is that it is possible to see vibration or rotation at slow speed. This is done by adjusting the frequency of the electronically controlled stroboscope until it differs from that of the vibration being observed by some small amount. The mechanism then will apparently move at the difference rate. By this means it often is possible to detect or observe chatter, vibration at critical speeds, torsional and transverse whipping and vibration, and commutator eccentricity, as well as vibration of various other components of a machine.

PHOTOGRAPHIC TECHNIQUES

Under some conditions, an actual picture of the motion of a vibrating surface which may be examined is more satisfactory than the visual techniques described above. In general, photographic techniques employ bulky and complex equipment, are relatively costly, and require special lighting and associated accessories, extra labor and time in processing, and fragile lenses and mounts. These limitations apply most accurately to direct high-speed photography of vibration phenomena. Often a very satisfactory and considerably more reasonable photographic approach is possible with the use of stroboscopic lighting techniques. This is accomplished by the combined use of a motion-picture camera of the "framing" type (discussed more fully below) and a system of strobo-

---

* For example, the following stroboscopes are manufactured by the General Radio Company, West Concord, Mass.: Strobotac, Strobolux, and Strobolume. This manufacturer published a booklet, "Eyes for Industry," describing various applications for these devices. Other stroboscopes are manufactured by Philips, Eindhoven, Holland. High-speed stroboscopes of special design are manufactured by Edgerton, Germeshausen and Grier, Inc., Boston, Mass.

scopic illumination by which the apparent movement of the specimen may be made to vary in its cycle at the control of the operator. Such a system optimally employs a camera with its film transport speed synchronized to the flash rate of the illumination, although satisfactory results often may be obtained with a conventional motion-picture camera if some compromise can be made regarding evenness of illumination and continuity of apparent motion. Both the speed (usually from about 64 to 2,500,000 pictures per second) and the type of camera operation appropriate to a given job must be determined by the requirements of the particular motion to be inspected.

This section describes slow-speed synchronized photographic technique, streak photography, and the high-speed framing camera. Several camera principles are employed. Basically, the image of the target is reproduced, usually by some type of lens system, upon some sensitized film.

FRAMING CAMERAS

The framing camera differs in its action from the continuous-streak camera in that it takes a series of discrete pictures of the phenomenon photographed, and it depends upon persistence of vision for the illusion of continuity when the film is projected. Framing cameras may employ either of the following techniques: (1) the film may remain motionless, with successive changes in the event recorded in superposition by intermittent flashes of illumination, or (2) the film may move either intermittently or continuously, with successive images furnished by rotating shutters or prisms. In both these methods the limit of resolution set by the sensitized material determines how nearly the display can come to a "continuous" reproduction of the actual event. Cameras with intermittent movement of the film are not used for rates exceeding about 200 pictures per second. The proper selection and use of auxiliary equipment (e.g., markers, indicators, stroboscopic and flood lights, exposure meters, etc.) is in most cases of extreme importance in producing satisfactory results.

SLOW-TO-MODERATE-SPEED CAMERAS. This class of camera usually is made with a variety of speeds and magazine capacities.

Stroboscopic Techniques. Often, with the use of a high-speed stroboscopic light, a moderately priced motion-picture camera can supply data almost equivalent to a very complex instrument, depending upon the range of movement involved. Other advantages to be gained by the use of synchronized stroboscopic technique include: (1) Because of the relatively slow film speed in the synchronized system, the shooting time may be increased by several orders of magnitude. Therefore the chance of photographing a random failure may be increased. (2) The camera can be started or stopped any time during a test. (3) The slow-motion effect can be seen during filming, whereas with high-speed photography the slow-motion inspection must await processing. (4) The synchronized system's slow-motion projection rate is constant for any frequency of vibration, while the slow-motion rate with high-speed photography is a function of the ratio of film speed to vibration frequency.

Stroboscopic Ciné System. One commercially available camera system * employing stroboscopic light consists of a combination of special camera, "slave" stroboscopic light, and synchronized unit. With this system it is possible to produce slow-motion movies of vibration tests using stroboscopic illumination and a pulse camera, both synchronized with the vibration. It utilizes a synchronized camera and stroboscopic light units for the automatic and continuous visual observation of vibration phenomena in the following way.

The sine-wave signal driving a vibration shaker is fed into a commercial device called a Slip-Sync which generates pulses. These pulses drive the Strobex stroboscopic lights at a flash rate differing from the vibration frequency by 1 cps (adjustable from $\frac{1}{3}$ to 3 cps). Illumination for this source produces an apparent slow motion at 1 cps. The pulses also serve to control the drive of a special camera † so that the picture frames move in synchronism with the stroboscopic lights, taking one frame with each flash of the strobo-

---

* Slip-Sync System, Model 360, Chadwick-Helmuth Co., Monrovia, Calif.
† For example, Pulse Camera, Model 360, Chadwick-Helmuth Co.

scopic lights. The result is a slow-motion film which duplicates visual stroboscopic observation under these conditions. This system operates at frequencies of from 5 cps to 10,000 cps.

Conventional Ciné Systems. This group includes cameras that (1) have a maximum framing rate of 300 or more frames per second, and (2) produce a ciné image, i.e., an image that can be projected to give the illusion of motion. In general, such cameras are of the rotating-prism type. The film is driven at high speeds over a rotating cylinder. Between the lens and shutter is a glass prism which rotates in synchronism with the film. The rotating prism bends the light to follow along with the rapidly moving film so that the image is stationary on the film. Then the light is cut off while the glass spins into position for the next frame. This operation is shown schematically in Fig. 13.43. Characteristics of many cameras of this type are listed in Table 13.2.

ULTRAHIGH-SPEED CAMERAS. Ultrahigh-speed cameras, having speeds exceeding 1,000,000 pictures per second, are useful in exploring extremely rapid transient phenomena, such as explosions. For example, see the extensive bibliography on high-speed techniques given in Ref. 17. Their use for general vibration work is normally prohibited by their high cost, the need for highly trained operating personnel, and limited availability. Dependent upon the resolution and speed desired, a wide variety of methods of image relay and formation are used. One such method,[18] employed in the Russian FP-22 camera, is shown schematically in Fig. 13.44.

STREAK PHOTOGRAPHY

Photographing an object with a moving-film shutterless camera with ordinary illumination yields streaks which indicate the positions of the more contrasting areas of the image. It is possible to obtain time-displacement curves of this type which can include amplitudes up to several inches with accuracies within a few thousandths of an inch and a few hundred thousandths of a second. The image recorded on film by the streak camera is a continuous, wavy line against a contrasting background. It can be described as a simple cartesian plot of displacement against time. A bright spot may be obtained from the reflection of light from a dot of white paint, a small steel ball, fine wire, or scratch

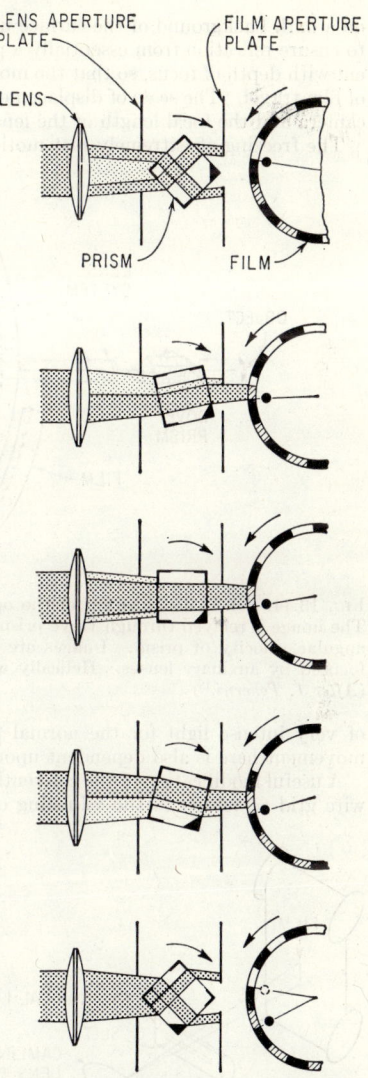

□ UNEXPOSED FRAME
▒ INEFFECTIVE PORTION OF LIGHT RAYS
▓ LIGHT TRANSMITTED THRU PRISM
▨ EXPOSED FRAME
■ INTERVAL BETWEEN FRAMES
▶ CYCLE REFERENCE ON PRISM
● CYCLE REFERENCE ON FILM

FIG. 13.43. Operation of rotating prism. Film aperture prevents stray light rays from striking edges of frame. Image is moved by prism in relation to film transport such that there is no relative movement between image and film.

on a dark background of the surface under study. These spots, which ideally are small to ensure reflection from essentially a point source, are placed in the field of view consistent with depth of focus, so that the motion to be studied is at right angles to the motion of film travel. The scale of displacement depends upon the distance from the spot to the camera and the focal length of the lens.

The freezing of extremely fast motion may be effected by substituting a short pulse

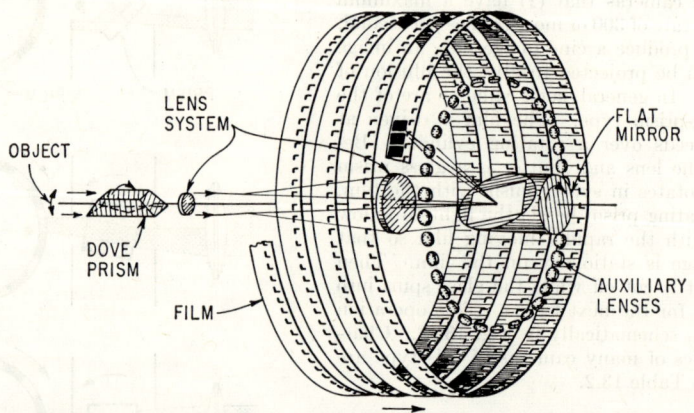

FIG. 13.44. Schematic diagram of the optical system in the FP-22 ultrahigh-speed camera. The image is relayed through Dove prism and lens system to flat mirror rotating at twice the angular velocity of prism. Images are formed on film frames as a result of the reflection focused by auxiliary lenses. Helically wound film is transported to maintain synchronism. (After J. Tcherni.[18])

of very intense light for the normal photoflood illumination. The scale of vibratory movement here is also dependent upon the rate of film travel.

A useful modification of the conventional streak photograph [19] employs a fine, polished wire grid cemented to the vibrating object or white-filled sharply scribed lines on the vibrating surface from which light is reflected. The use of such a grid or lines provides a much sharper image than can be obtained from the reflection of light from steel balls. The presence of a number of lines constituting the grid makes possible determination of displacements as great as 4 to 5 in. to an accuracy of several thousandths of an inch.

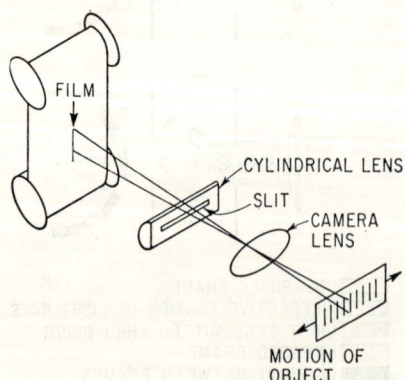

FIG. 13.45. Schematic layout of component parts in modified method of streak photography. Distance from slit to film is small compared with lens-film distance. (After I. Vigness and R. C. Nowak.[19])

By means of slit near the film plane the images of the wire change position in relation to the vibration, with one line moving out of the field as another line moves in, as shown in Fig. 13.45.

A number of continuous-writing drum cameras are commercially available.* A typical 35-mm camera of this type may produce a streak image 10 to 50 in. long, have a time resolution of 10^{-6} to 10^{-8} sec, and have a sweep rate of from 100 to 300 meters/sec.

A summary of available high-speed cameras is given in Table 13.2.

* For example, Models 194, 224, and 307 manufactured by Beckman & Whitley, San Carlos, Calif.

Table 13.2. Summary of High-speed Ciné Camera Systems *

Camera	Frame	Max. film load, ft	Speeds, pps †	Weight, lb
Beckman & Whitley Co., San Carlos, Calif.				
Magnifax Model 333 and 334	1,200, 3,200 pps	100	1,200, 3,200	32
Benson-Lehner, Los Angeles, Calif.				
HS-16B	16-mm full or double	400	100–400 with high-speed motor; 16–64 with low	8½
Vinten HS-300	35-mm full frame	400	100–300 (24–300 with accessories)	75
Ercona Corp., New York, N.Y.				
Zeiss Ikon ZLI	18 × 22 mm or 9 × 22 mm on 35-mm film	150	250, 2,000 (18 × 22) or 500/4,000 (9 × 22)	770 (complete system)
Fairchild Camera & Instrument Corp., Yonkers, N.Y.				
Model HS 100 (airborne model)	16-mm full	100	16–2,200 (depending on motor)	11
Model HS 101 (industrial model)	16-mm full	100	16–8,000 (depending on motor)	10
Model HS 108	8-mm full or 16-mm half	100	32–16,000	10
Model HS 116	Quarter frame 16 mm	100	64–32,000	16
Model HS 401	16-mm full	400	25– 6,000	24
Model HS 408	8-mm full or 16-mm half	400	50–16,000	24
Model HS 416	Quarter frame 16-mm	400	100–32,000	30
D. B. Milliken Co., Arcadia, Calif.				
DBM 3	16-mm full	100	4–400	7¼
DBM 4	16-mm full	200 (will accept 400-ft magazines)	4–400	8
DBM 5	16-mm full	400	4–400	11
Photo Sonics, Inc., Burbank, Calif.				
16MM-1A	16-mm full	100	300	20
16MM-1B	16-mm full	1,200	12–1,000	9
16MM-1-C	16-mm full	400	To 4,000	32
35MM-4B	35-mm full	1,000	250–2,800	80
70MM-1A	0.218 × 2.25 in.	80	100–400	30
70MM-1B	0.218 × 2.218 in.	1,000	100–400	85

* After data given in *Photo Methods for Industry*, **2**:8 (1959).
† The camera speeds refer to the number of pictures per second (abbreviated pps).

Table 13.2 (*Continued*)

Camera	Frame	Max. film load, ft	Speeds, pps †	Weight, lb
Wollensak Optical Co., Rochester, N.Y.				
Fastair FA-16	16-mm full	50, 100, or 200 (depending on model)	12–680 (depending on motor)	8
Fastax WF1	8-mm single or double width on 16-mm film with 8-mm perforations	100	300–16,000	25
Fastax WF2	As above	400	700–18,000	38
Fastax WF3	16-mm full	100	150– 8,000	25
Fastax WF3T	16-mm full	100	150– 6,000	25
Fastax WF4	16-mm full	400	350– 9,000	38
Fastax WF4T	16-mm	400	350– 6,000	38
Fastax WF4S	16-mm full	400	350– 9,000	45
Fastax WF4ST	16-mm	400	350– 6,000	45
Fastax WF5	35-mm half	100	100– 6,000	29
Fastax WF8	35-mm full	500	200– 2,500	62
Fastax WF8A	35-mm full	500	20– 2,000	105
Fastax WF9	35-mm half (picture and oscillogram ½ frame each)	100	100– 6,000	29
Fastax WF14	16-mm full	400	650– 8,000	38
Fastax WF14T	16-mm full	400	350– 6,000	38
Fastax WF15	8-mm single or double width on 16-mm film with 8-mm perforations	400	1,300–16,000	38
Fastax WF17	16-mm full	100	150– 8,000	25
Fastax WF17T	16-mm full	100	150– 6,000	25
Fastax WF21	Same as WF-15	100	300–16,000	25
Fastax WF22	16-mm full	400	350– 9,000	45
Fastax WF 23	16-mm	400	350– 9,000	38

MECHANICAL STRAIN RECORDERS

Mechanical strain recorders (also known as *stress recorders*) measure strain by recording the relative motion between two points on the surface of a structure. By recording on a suitable moving medium, a time-history of the strain (or stress) is obtained. Although this type of device is not necessarily a fixed reference instrument, it is included here for convenience. The advantage of this type of instrument is that it provides a direct measure of the stresses in, and the forces acting on, a structure. Since it does not depend on the response of a mass-spring system, it can be used to measure vibration down to zero frequency. The upper frequency limit is determined by the inertia of the moving elements and linkages. (Also see *Strain Gages*, Chap. 17.)

THE DE FOREST SCRATCH GAGE.* This is a self-contained recording strain gage, weighing less than 2 gm, which records deformations below the elastic limit and ranging from 0.0001 to 0.050 in. over a 2-in. gage length. It also may be used to measure vibration well into the sonic range. The presence of a torsional strain and its frequency in terms of longitudinal vibration are indicated but are not measured quantitatively. The record is presented in the form of a scratch which indicates the exact value of the defor-

* Manufactured by Baldwin-Lima-Hamilton Corp., Eddystone Division, Philadelphia, Pa.

mation. It can be measured or photographed at suitable magnification under a metallographic microscope. The record does not, however, give an indication of time or frequency.

The gage may be fastened to a member in various ways; screws, solder, spot welding or clamps are used. Because of its small size and light weight, it may be attached to such members as airplane propellers or rapidly moving parts of high-speed machinery.

The scratch gage consists of only two parts: the *scratch arm* and the *target*. The target and the fulcrum end of the scratch arm are attached rigidly to the member which is to be tested, in such a way that the scratch arm is parallel to the direction of strain and lies closely on the center line of the target (see Fig. 13.46).

The tip of the scratch arm carries a special abrasive, and is in contact with the polished,

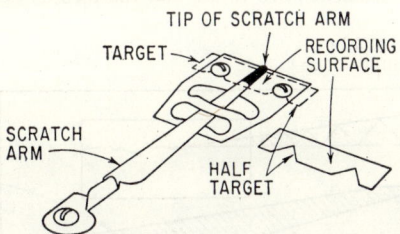

Fig. 13.46. The deForest Scratch Gage consists of a *scratch arm* and a *target*. The target and the fulcrum end of the scratch arm are rigidly attached to the member to be tested in such a way that the scratch arm is parallel to the direction of the strain, and lies closely on the center line of the target. The tip of the scratch arm carries abrasive and maintains contact with the polished chrome-plated recording surface of the target. Deformation of the member causes the scratch arm and target to move longitudinally in relation to each other; this movement is recorded as a scratch on the recording surface.

chrome-plated recording surface of the target. Deformation of the member causes the scratch arm and target to move longitudinally in relation to each other. This movement is recorded as a scratch on the recording surface.

The unique feature of the gage is the method by which the scratch arm is caused to progress across the target at a right angle to the direction of strain, so as to produce a continuous record. The portion of the scratch arm near the fulcrum is designed so that it constitutes a spring. When the gage is made ready for the test, the tip of the scratch arm is pushed to the left or right. This sets up a restoring force in the fulcrum spring and marks a zero line on the recording surface. One part of the target functions as a friction bar, pressing down on the scratch arm with enough force to hold it in place when the test surface is static. When strain occurs, however, this friction is insufficient, and the force exerted by the fulcrum spring causes the scratch arm to move slightly toward the center of the target at the same time that the deformation of the member causes movement in a longitudinal direction. The force of the fulcrum spring may be reduced by reducing its thickness; friction may be varied by bending the friction bar. By varying these, the record may be closely packed or spread out, as desired.

CAMBRIDGE STRESS RECORDER.* This light, rugged instrument is used for measuring and recording directly the stresses in bridges and other structures subject to moving loads. The recording is produced by a moving stylus on cellulose film.

The recorder stands on three sharply pointed toes, two at the rear, rigidly fixed to the case of the instrument, and one at the front, which is free to move longitudinally. By a system of levers the movements of this front toe are caused to operate a stylus moving over the surface of a traveling cellulose film driven electrically at a steady speed which can be set to 5, 20, or 100 mm/sec. The electric motor is energized from a 12-volt battery. A specially designed gearbox, incorporated in the instrument, enables the speeds to be changed without switching off the motor. Both these controls can be operated

* Manufactured by Cambridge Instrument Co., Ltd., London.

from a distance by means of a switchboard supplied with each instrument. The recorder is clamped in the desired position on a girder, flange, or buttress by a clamp applied to the projecting part of a spring plunger; the toes, which are of hardened steel alloy, bite firmly into the girder; and the spring plunger ensures uniformity of pressure by the clamp.

When in use, the instrument operates as an extensometer measuring the extension or contraction between the pointed toes at each end. Figure 13.47 is a schematic representation of a mechanical extensometer of this type. Two alternative sizes of the stress recorder are made, having base lengths (between the moving and fixed toes) of 15 in. and 10 in., respectively. With the 15-in. base length, measurements of the records to 0.0001 in. represent approximately 0.1 ton/in.^{2} for a steel structure. The mechanical magnification of the instrument is 10 times, and the records can be magnified further

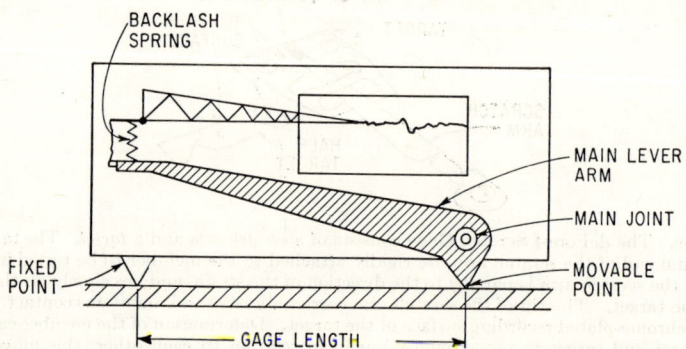

Fig. 13.47. Schematic of mechanical stress recorder. Strain at the movable point is translated into irregular line on recording medium by coupled lever system. A variation of this principle is used in the Cambridge Stress Recorder. (*After Bernhard.*[20])

and viewed through a portable viewer which has a magnification of 10 times, or through a portable microscope.

Often in practice a number of recorders clamped to various parts of a structure are employed, and their records are synchronized. Each instrument, in addition to the recording stylus, is fitted with two extra styli, recording on the underside of the film, operated by an electromagnet. One stylus may be used in connection with a suitable electrical circuit to mark a synchronizing point on the respective records, while the other may be connected to a time-marking mechanism. A 0.1-sec contact clock can be supplied.

REFERENCES

1. Steuding, H.: "Messung mechanischer Schwingungen," Springer-Verlag, Berlin, 1928.
2. Cambridge Instrument Co., Ltd., London, Folder 62.
3. Den Hartog, J. P.: "Mechanical Vibrations," 4th ed., McGraw-Hill Book Company, Inc., New York, 1956.
4. Ref. 3, p. 188.
5. Brochure on Recording Accelerometer, Cambridge Instrument Co., Ltd., London.
6. "The Willmore Seismograph Handbook," Hilger and Watts, Ltd., London.
7. Ericksen, H. W., and D. J. Ettelman: *IRE Trans. on Instrumentation*, September, 1957, p. 178.
8. Davidson, S., and E. Adams: "Theoretical Study of the Multifrequency Reed Gage for Measuring Shock Motion," *David Taylor Model Basin Rept.* 613, 1948.
9. Vigness, I., E. W. Kammer, and S. G. Holt: "Shock and Vibration Instrumentation and Measurements," *U.S. Naval Research Lab. Rept.* O-2645, 1945.
10. Walsh, J. P., and R. E. Blake: "The Equivalent Static Acceleration of Shock Motions," *U.S. Naval Research Lab. Rept.* F-3303, 1948.
11. The Model 3A 3-Component Recording Accelerometer, and other brochures, Impact Register Co., Champaign, Ill.

REFERENCES

12. Impact-O-Graph brochure, Impact-O-Graph Co., Cleveland, Ohio.
13. Savage Impact Register brochure, Impact Register Co., Champaign, Ill.
14. *U.S. Dept. of Commerce Rept.* PB 122 062, Office of Technical Services, Washington, D.C.
15. Smith, D. M.: "Experimental and Theoretical Investigation of the Mass-plug Accelerometer," *David Taylor Model Basin Rept.* R-267, 1946.
16. Allnutt, R. B., and F. Mintz: "Instruments at the David Taylor Model Basin for Measuring Vibration and Shock on Ship Structures and Machinery," *David Taylor Model Basin Rept.* 563, 1948.
17. Garvin, E. L.: "Bibliography on High-speed Photography," Eastman Kodak Co., Rochester, N.Y.
18. Tcherni, J.: *Proc. 3rd Intern. Congr. on High-speed Photog.*, R. B. Collins (ed.), Butterworths Scientific Publications, London, 1957.
19. Vigness, I., and R. C. Nowak: *J. Appl. Phys.*, **21**:445 (1950).
20. Bernhard, R. K.: "Mechanical Vibrations, Theory & Application," Pitman Publishing Corporation, New York, 1943.

14

SPECIAL-PURPOSE AND MISCELLANEOUS SHOCK AND VIBRATION TRANSDUCERS

Charles S. Duckwald
General Electric Company

Burton S. Angell
General Electric Company

INTRODUCTION

This chapter considers special-purpose and miscellaneous vibration transducers and measurement systems. In all of these systems, the vibratory motion produces an electrical indication which is proportional to displacement, velocity, or acceleration amplitude. These devices are of the following types:
 Capacitance variation between stationary and moving plates
 Potentiometer
 Electrochemical
 Mechanoelectronic
 Light-beam electronic
 Optical interferometer
 Vibrating wire
 Tuning fork
 Interference of waves reflected from moving surface
 Gas-discharge transducer
 Servo accelerometer
Of the types listed above, capacitance-type and potentiometer-type transducers are employed most often.

CAPACITANCE-TYPE TRANSDUCER

The capacitance-type transducer is basically a displacement-sensitive device. Its output is proportional to the change in capacitance between two plates caused by the change of relative displacement between them as a result of the motion to be measured. Appropriate electronic equipment is used to generate a voltage corresponding to the change in capacitance.

Transducers based upon the variation of capacitance are primarily special-purpose devices. This is principally due to the fact that auxiliary electronic equipment required for their operation is inherently more complicated than that required for the more popular types of transducers. The capacitance-type transducer's main advantages are (1) its simplicity in installation, (2) its negligible effect on the operation of the vibrating system since it is a proximity-type pickup which adds no mass or restraints, (3) its extreme sensitivity, (4) its wide displacement range, due to its low background noise, and (5) its wide frequency range, which is limited only by the electric circuit used.

The capacitance-type transducer often is applied to a conducting surface of a vibrating system by using this surface as the ground plate of the capacitor. In this arrangement,

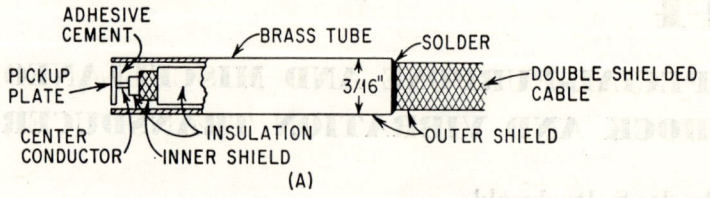

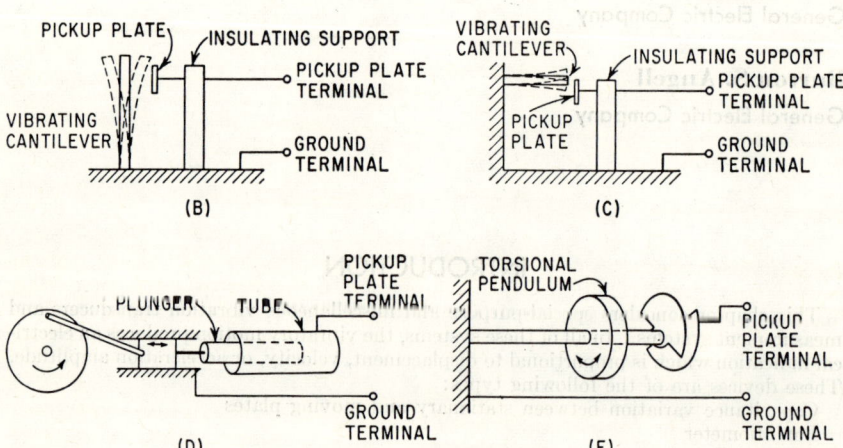

FIG. 14.1. Capacitance-type transducers and their application: (A) construction of typical assembly, (B) gap length or spacing sensitive pickup for transverse vibration, (C) area sensitive pickup for transverse vibration, (D) area sensitive pickup for axial vibration, and (E) area sensitive pickup for torsional vibration.

the insulated plate of the capacitor should be supported on a rigid structure close to the vibrating system. Figure 14.1A shows the construction of a typical capacitance pickup; Figs. 14.1B, C, D, and E show a number of possible methods of applying this type of transducer. In each of these, the metallic vibrating system is the ground plate of the capacitor. Where the vibrating system at the point of instrumentation is an electrical insulator, the surface can be made slightly conducting and grounded by using a metallic paint or by rubbing the surface with graphite.

The maximum operating temperature of the transducer is limited by the insulation breakdown of the plate supports and leads. Bushings made of alumina No. 38900 are commercially available and provide adequate insulation at temperatures as high as 2000°F (1093°C).

THE CAPACITOR

DISPLACEMENT-TO-CAPACITANCE SENSITIVITY. The capacitance C between two parallel conducting plates that are insulated from each other is directly proportional to the effective area S of the plates and is inversely proportional to the distance d between the plates, i.e.,

$$C = 0.225 \frac{S \text{ (in.}^2)}{d \text{ (in.)}} = 0.0885 \frac{S \text{ (cm}^2)}{d \text{ (cm)}} \quad \text{mmfd} \tag{14.1}$$

CAPACITANCE-TYPE TRANSDUCER

Table 14.1. Formulas for Calculating Capacitance and Change in Capacitance for Various Geometries of Air-dielectric-type Capacitors

	Capacitance in air, mmfd		Change in capacitance due to change in position ΔC, mmfd
	Dimensions, in.	Dimensions, cm	
	$C = 0.225 \dfrac{ab}{d}$	$C = 0.0885 \dfrac{ab}{d}$	$\Delta C \simeq C \dfrac{\Delta d}{d}$
	$C = 0.225 \dfrac{ab}{d}$	$C = 0.0885 \dfrac{ab}{d}$	$\Delta C \simeq C \dfrac{\Delta a}{a}$
	$C = 0.225 \dfrac{\pi R^2}{d}$	$C = 0.0885 \dfrac{\pi R^2}{d}$	$\Delta C \simeq C \dfrac{\Delta d}{d}$
	$C = 0.353 \dfrac{R_1{}^2 - R_2{}^2}{d}\left(\dfrac{\theta}{\pi}\right)$	$C = 0.139 \dfrac{R_1{}^2 - R_2{}^2}{d}\left(\dfrac{\theta}{\pi}\right)$	$\Delta C \simeq C \dfrac{\Delta \theta}{\theta}$
	$C = \dfrac{0.613 l}{\log \dfrac{R_1}{R_2}}$	$C = \dfrac{0.242 l}{\log \dfrac{R_1}{R_2}}$	$\Delta C \simeq C \dfrac{\Delta l}{l}$

for an air-dielectric capacitor at standard conditions. Equation (14.1) shows that the capacitance variation can be obtained either by changing the effective area S or by changing the spacing d. Table 14.1 shows this equation applied to various geometric shapes of capacitors; it also gives the relationship for the change in capacitance ΔC due to a small change in plate position. (Fringing effects are neglected.)

CAPACITANCE VERSUS DISPLACEMENT NONLINEARITY. Variable-capacitance transducers have three principal sources of nonlinearity in the relationship of output voltage to mechanical displacement which should be considered in any application of the transducer: (1) the effect of fringing in the electric field between the capacitor plates, (2) the fact that the capacitance is inversely, and not directly, proportional to the spacing, and (3) the electric circuit associated with the transducer. The effect of fringing depends to such an extent upon the transducer plate design and its shielding that it is difficult to determine analytically; fringing effects should be checked experimentally in each installation.

The capacitance variation, which is obtained when the spacing between the plates is changed, is not a linear function of the spacing. This can lead to serious distortion unless the percentage change in the spacing is limited. For example, assume a sinusoidal variation in spacing d:

$$d = d_0 + d_s \sin(2\pi f t) \quad (14.2)$$

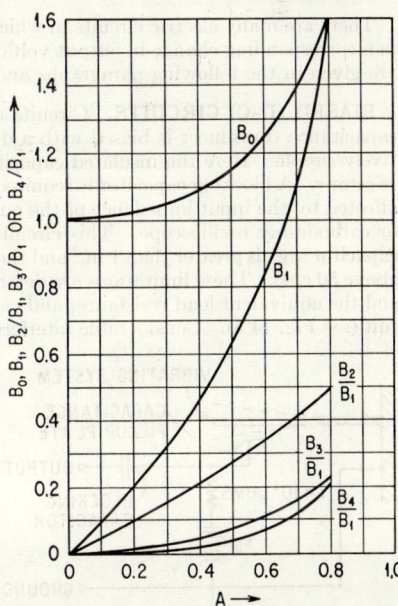

Fig. 14.2. Harmonic coefficients in Eq. (14.5) as a function of the variation in spacing of the condenser plates.

where d_0 = average plate spacing
d_s = amplitude of spacing variation
f = vibration frequency, cps
t = time, sec

Equation (14.2) can be rewritten as

$$d = d_0(1 + A \sin \omega t) \qquad (14.3)$$

where $A = d_s/d_0$ and $\omega = 2\pi f$. Substituting Eq. (14.3) into Eq. (14.1), the capacitance can be expressed as

$$C = \frac{0.225 S}{d_0(1 + A \sin \omega t)} \quad \text{mmfd} \qquad (14.4)$$

Expanding Eq. (14.4) in a Fourier series,

$$C = 0.225 \frac{S}{d_0}\left[B_0 + B_1\left(\sin \omega t + \frac{B_2}{B_1}\cos 2\omega t\right.\right.$$
$$\left.\left. + \frac{B_3}{B_1}\sin 3\omega t + \frac{B_4}{B_1}\cos 4\omega t + \cdots\right)\right] \quad \text{mmfd} \quad (14.5)$$

where the values of $B_0, B_1, B_2, B_3, B_4 \ldots$ are functions of the variation of the spacing ratio A. The coefficients for the first four harmonics as a function of A are given in Fig. 14.2. The deviation of B_0 from the value 1.0 indicates the amount of zero shift obtained as the value of A increases. The conditions for the capacitance change to be proportional to the vibration amplitude are $B_1 = A$, and $B_2 = B_3 = B_4 = 0$. The harmonic distortion is indicated by the curves for B_2/B_1, B_3/B_1, and B_4/B_1 for the second, third, and fourth harmonics, respectively.

ELECTRICAL CIRCUITS

There are many electric circuits in which a small change in the capacitance will result in a corresponding change in output voltage. Descriptions of a number of these circuits are given in the following paragraphs and in Refs. 1 and 2.

BIASED (D-C) CIRCUITS. Circuits of the type shown in Fig. 14.3, in which the capacitance transducer is biased with a d-c voltage through a high resistance, are relatively simple. Here the insulated capacitor plate is biased to a voltage e_b through the resistor r. A blocking capacitor is required so that the biasing voltage is not significantly affected by the input impedance of the output device, such as a vacuum-tube voltmeter or cathode-ray oscilloscope. This circuit is effective for many applications where the capacitor area is greater than 1 in.[2] and the vibration frequency is relatively high (usually above 50 cps). These limitations are determined by the value of the average capacitance and the equivalent load resistance, and can be evaluated by means of an equivalent circuit (see Fig. 14.5). Considerable interference may result from electrical pickup and the motion of nearby objects since neither the capacitor plate nor the lead to the output device are shielded from stray electric fields.

The circuits shown in Figs. 14.4A and 14.4B are superior to that in Fig. 14.3 in many respects. The capacitor plate and its lead are fully shielded, which reduces pickup from stray electric fields and nearby moving objects. The inner shield is approximately at the potential of the signal voltage; this reduces the shunting effect of the lead capacitance and practically eliminates the microphonic or vibration sensi-

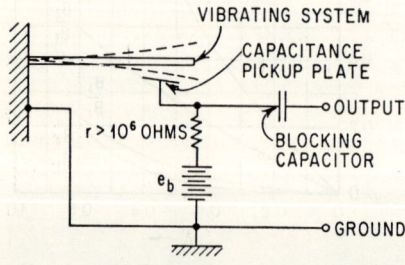

Fig. 14.3. Biasing circuit used with capacitance-type transducer.

tivity of the lead. For the values of r_c, R_b, r_g and the type of vacuum tube shown in Fig. 14.4A, the input impedance of the circuit is approximately 300 megohms. The circuit shown in Fig. 14.4B has an input impedance of approximately 1,000 megohms. In these circuits it is necessary to select the vacuum tube carefully to obtain a low thermal noise and a high input impedance. In practice it has been found that one tube in five ordinarily is satisfactory. With these circuits it is possible to obtain output voltages greater than 10 millivolts for vibration displacements of 10 microinches above 50 cps.

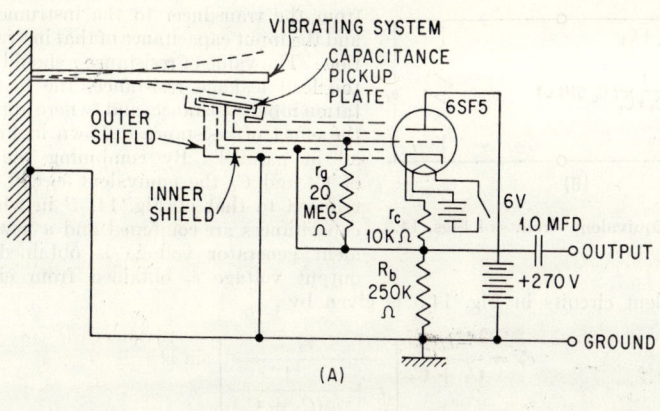

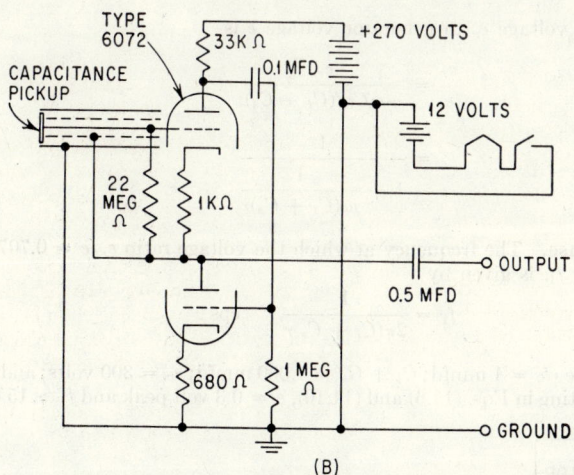

Fig. 14.4. Biasing circuits used with capacitance-type transducer which are superior to that shown in Fig. 14.3. Input impedance in (A) is 300 megohms; in (B) it is 1,000 megohms.

Equivalent-circuit Analyses. The sensitivity and frequency response of the d-c biased capacitance-type transducer shown in Figs. 14.3 and 14.4 can be calculated in terms of the equivalent circuits of the system, as shown in Fig. 14.5. The equivalent generator voltage e in Fig. 14.5A is given by

$$e = -\left(\frac{e_b}{C_t}\right)(C_0 \sin \omega t) \tag{14.6}$$

14-6 SPECIAL-PURPOSE SHOCK AND VIBRATION TRANSDUCERS

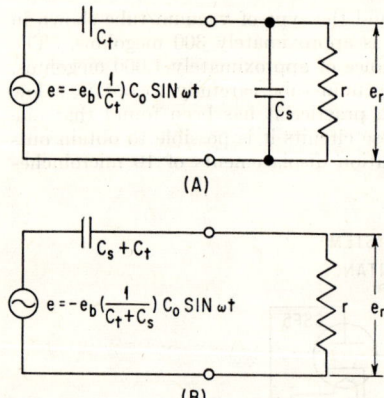

FIG. 14.5. Equivalent circuits of Figs. 14.3 and 14.4.

where C_t = total capacitance of transducer only
C_0 = peak value of sinusoidal change in capacitance
e_b = average or d-c voltage across capacitance C_t
e = instantaneous voltage across C_t

The value of capacitance C_s in Fig. 14.5A should include the capacitance of the leads from the transducer to the instrumentation and the input capacitance of that instrumentation. The value of resistance r should include the lead leakage resistance, the instrumentation input resistance, and (where applicable) the effect of resistance r shown in Fig. 14.3, all in parallel. By combining the effects of C_s and C_t the equivalent circuit can be reduced to that of Fig. 14.5B in which the capacitances are combined and a new equivalent generator voltage is obtained. The output voltage e_r obtained from either of the equivalent circuits in Fig. 14.5 is given by

$$e_r = \frac{-e_b C_0}{C_t + C_s} \left[\frac{r}{r + \frac{1}{j\omega(C_t + C_s)}} \right] \sin \omega t \qquad (14.7)$$

The ratio of output voltage e_r to generator voltage e is

$$\frac{e_r}{e} = \frac{e_r}{-e_b C_0/(C_t + C_s)} \qquad (14.8)$$

$$= \frac{1}{1 + \frac{1}{j\omega(C_t + C_s)r}} \qquad (14.9)$$

Frequency Response. The frequency at which the voltage ratio $e_r/e = 0.707$, called the cutoff frequency f_1, is given by

$$f_1 = \frac{1}{2\pi(C_t + C_s)r} \quad \text{cps} \qquad (14.10)$$

For example, suppose $C_0 = 1$ mmfd; $C_t + C_s = 1,000$ mmfd; $e_b = 300$ volts; and $r = 10$ megohms. Substituting in Eqs. (14.6) and (14.10), $e = 0.3$ volt peak and $f_1 \simeq 15$ cps.

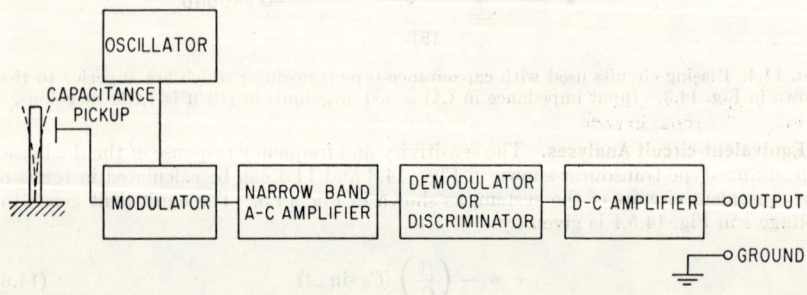

FIG. 14.6. Schematic diagram of a-c carrier system for use with a capacitance-type transducer.

CAPACITANCE-TYPE TRANSDUCER

CIRCUITS EMPLOYING AN A-C CARRIER VOLTAGE. In a-c carrier-voltage circuits [1,2,4] used with capacitance-type transducers the vibration signal either amplitude or frequency modulates the carrier. The modulated carrier then is amplified by a narrow-band a-c amplifier which is terminated in a demodulating stage. A system of this type is illustrated by the block diagram in Fig. 14.6. The output of the demodulating stage is a function of the change in capacitance. In contrast to circuits which employ d-c bias, these circuits can be used for very low frequencies or static measurements as well as for higher frequencies. The carrier frequency must be at least five times the highest vibration frequency to be measured.

Bridge Circuits. [1,2,5–8] The bridge circuit shown in Fig. 14.7 is typical of a circuit in which variations in capacitance of a transducer are used to amplitude modulate the carrier.

FIG. 14.7. Bridge-type amplitude-modulating circuit employing an a-c carrier system for use with a capacitance-type transducer. (*Courtesy of Fielden Instrument Division, Robertshaw-Fulton Controls Company.*)

Advantages of this type of circuit are (1) the vibrating system can be one of the plates of the capacitor, (2) it is relatively insensitive to electric noise pickup, and (3) there is provision for a double-shielded lead which reduces the microphonics and vibration sensitivity of the lead and increases the transducer's sensitivity.

The value of the balancing resistor r in Fig. 14.7 should be one-quarter the capacitive reactance X_C of the capacitor C. This reactance can be determined from the relation

$$X_C = \frac{1}{2\pi f_c C} \qquad \text{ohms} \qquad (14.11)$$

where f_c equals the carrier frequency in cycles per second. The coarse balance adjustment on the bridge should cover the range of capacitance expected in the sum of the transducer capacitance plus the fine balance capacitor.

The carrier frequency should be at least five times the highest vibration frequency; otherwise distortion may become excessive. If a narrow-band a-c amplifier is employed, as shown in Fig. 14.6, its bandwidth should be equal to twice the value of the highest vibration frequency to be measured.

Feedback Amplifier Circuit. A feedback amplifier circuit is shown in block diagram form in Fig. 14.8. The capacitance C_t represents a capacitance-type transducer which is connected between the output and input terminals of a high-gain amplifier. The carrier voltage e_c from an isolating transformer is applied to the input terminals of the high-gain amplifier through a reference capacitor C_r. Initially the values of C_t and C_r are adjusted so that no current flows through r, in which case the output voltage e_{out} is zero. When

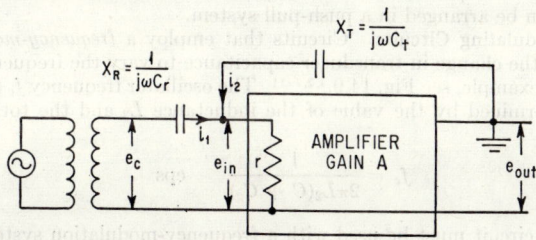

FIG. 14.8. Schematic diagram of feedback amplifier circuit employing an a-c carrier system for use with a capacitance-type transducer.

C_t is varied, a corresponding unbalance current flows in r and, due to the high gain of the amplifiers, a large output voltage is generated that is proportional to the unbalance. The output voltage is directly proportional to the spacing d between the capacitance transducer plates, i.e., $e_{\text{out}} = Bd$.

From Fig. 14.8,

$$i_1 = \frac{e_c - e_{\text{in}}}{X_{C_r}} = \frac{e_c - (i_1 + i_2)r}{X_{C_r}}$$

$$i_2 = \frac{e_{\text{out}} - e_{\text{in}}}{X_{C_t}} = \frac{e_{\text{out}} - (i_1 + i_2)r}{X_{C_t}}$$

where X_{C_t} and X_{C_r} are the reactances of the capacitors C_t and C_r, respectively. Solving these two expressions for i_1 and i_2 and substituting into the following equation:

$$e_{\text{out}} = (i_1 + i_2)rA$$

the following relation is obtained for the output voltage:

$$e_{\text{out}} = \frac{e_c X_{C_t} rA}{X_{C_t} X_{C_r} + r(X_{C_t} + X_{C_r}) - rX_{C_r}A} \quad (14.12)$$

where A represents the amplifier gain. Circuit constants are chosen so that X_{C_r} is comparable to X_{C_t}, and so that $X_{C_t}X_{C_r} + r(X_{C_t} + X_{C_r}) \ll rX_{C_r}A$. The latter condition holds since $A \gg 1$. With these constants, the relation for e_{out} reduces to

$$e_{\text{out}} = -e_c \frac{X_{C_t}}{X_{C_r}} = -e_c \frac{C_r}{C_t} \quad (14.13)$$

In the case of a parallel-plate capacitor (see Table 14.1) having dimensions $a \times b$ and separated by a distance d,

$$C_t = \left(\frac{0.225ab}{d}\right)$$

Equation (14.13) becomes

$$e_{\text{out}} \simeq -e_c C_r \left(\frac{d}{0.225ab}\right) = -Bd \quad (14.14)$$

where B is a constant of proportionality.

Other Amplitude-modulating Circuits. Other amplitude-modulated circuits include:
1. The detuning of an L-C circuit by placing the capacitance transducer in parallel with the C of the L-C circuit.[1,15,16]
2. The detuning of the grid circuit of a tuned-grid, tuned-plate oscillator.[12-14] This system also can be used as the demodulator, since a change in tuning also affects the plate current in the oscillator tube.
3. A capacitance divider. This circuit requires a large percentage change in capacitance to obtain a useful per cent modulation; it is best used with capacitance transducer elements that can be arranged in a push-pull system.

Frequency-modulating Circuits. Circuits that employ a *frequency-modulated carrier* customarily use the change in transducer capacitance to vary the frequency in an oscillator circuit; for example, see Fig. 14.9.[1,2,7-11] The oscillator frequency f_c (approximately 1,000 kc) is determined by the value of the inductance L_2 and the total capacitance, $C + C_g$:

$$f_c = \frac{1}{2\pi L_2(C + C_g)} \quad \text{cps} \quad (14.15)$$

A demodulation circuit must be used with a frequency-modulation system. This type of circuit is usually more complex than that used with an amplitude-modulated system. One demodulation scheme which is commonly employed with frequency-modulated systems is to amplify the FM carrier, to amplitude-limit the resultant signal, and then to use this signal to drive a sharply tuned L-C circuit slightly to one side of resonance. This then transforms the frequency-modulated carrier to an amplitude-modulated carrier that can be demodulated with an AM detector.

The heterodyne principle [17-19] can be used with a frequency-modulation system to increase the sensitivity of the over-all system. In the heterodyne circuit the voltage from a very stable oscillator is used to beat with that from an FM oscillator; the frequency of the former is a few cps above or below the maximum frequency swing of the latter. The resultant voltage contains a component at the difference frequency. Thus the percentage swing in the difference frequency will be much greater than the corresponding percentage frequency swing in the original frequency-modulated voltage. The difference frequency can be measured on a standard frequency meter. With the same

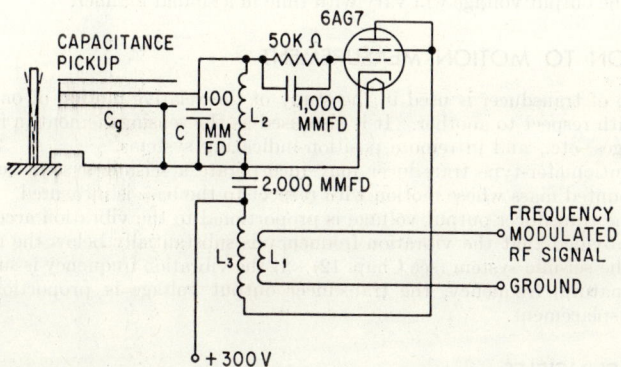

Fig. 14.9. Frequency-modulation circuit employing an a-c carrier system for use with a capacitance-type transducer. (*After E. V. Potter.*[11])

meter, an output voltage can be obtained which is proportional to the frequency, and thus proportional to the original displacement of the capacitance transducer. Similarly, a frequency-sensitive circuit can be used to obtain a voltage proportional to the frequency.[20]

COMMERCIALLY AVAILABLE CAPACITANCE-TYPE TRANSDUCERS

Commercially available vibration-measuring equipment that uses the capacitance-variation principle includes:
1. The Wayne Kerr Corporation—Vibration Meter Model B-731A.
   Circuit: Feedback amplifier with 50-kc carrier.
   Power supply: 110 volts a-c, 40 to 60 cps, 50 watts.
   Ranges (full-scale): 0.001 in., 0.01 in., 0.1 in., 0.5 in.
   Accuracy: 2 per cent of full-scale deflection at 1,000 cps. Better than 5 per cent at 5 cps and 10,000 cps (both figures apply to flat-surface probes parallel to test surface).
   Dimensions: 17 by 7½ by 11½ in.
   Weight: 26 lb.
2. Photocon Research Products—Proximity Transducer Model PT3-1 with a set of 11 probes up to 1 in. in diameter. Electrical equipment: Dynagage Models DG-400, DG-500, DG-600.
   Circuit: Transducer capacitance change affects the tuning of an IF transformer. Approximately 900-kc carrier.
   Power supply: 110 volts a-c, 60 cps, 50 watts.
   Displacement amplitude range: One microinch (2.54×10^{-6} cm) to 2 in. (5 cm).
   Frequency range: 0 to 10,000 cps.
   Accuracy: Approximately ±5 per cent, 0 to 10,000 cps.
   Dimensions: 8½ by 11½ by 7 in.
   Weight: 25 lb.

3. Fielden Instrument Division, Robertshaw-Fulton Controls Company—Proximity Meter (probe plate supplied by customer according to recommended specifications).
Circuit: Bridge, 500-kc carrier.
Power supply: 115 volts a-c, 60 cps, 60 watts.
Sensitivity: 0.02×10^{-12} farad change in capacity produces full-scale deflection of 1 volt or 1 ma into 1,000-ohm load.
Frequency range: 0 to 15,000 cps (flat within ± 4 db over this range).
Dimensions: 15 by 9 by 9 in.

POTENTIOMETER-TYPE TRANSDUCER

The potentiometer-type transducer is a variable-voltage divider in which the output voltage is a function (usually linear) of a displacement. If the displacement varies with time, then the output voltage will vary with time in a similar manner.

APPLICATION TO MOTION MEASUREMENT

This type of transducer is used in the study of the relative motion of one part of a structure with respect to another. It is also used as the sensing element in force gages, pressure gages, etc., and in remote position-indicating systems.

The potentiometer-type transducer may incorporate a seismic system consisting of a spring-mounted mass whose motion with respect to the base is measured. In this application, the transducer output voltage is proportional to the vibration acceleration of the base, provided that the vibration frequency is substantially below the natural frequency of the seismic system (see Chap. 12). If the vibration frequency is substantially above the natural frequency, the transducer output voltage is proportional to the vibration displacement.

GENERAL PRINCIPLES

The potentiometer-type transducer circuit is shown schematically in Fig. 14.10A and its equivalent circuit in Fig. 14.10B. The potentiometer is energized with the d-c voltage e. Its total circuit resistance is r_p which includes the resistance of the d-c source, the potentiometer, and the line between the source and potentiometer. The potentiometer senses a displacement by moving a variable contact through a distance which is proportional to the displacement; this movement causes a corresponding change in resistance r_2, or in the resistance ratio $\alpha = r_2/r_p$ in Fig. 14.10A.

When there is no electrical circuit load on the output terminal of the potentiometer, the output voltage $e_{out} = e\alpha$ (the equivalent generator voltage of Fig. 14.10B). If an indicating instrument or load having a resistance r_L is connected across the potentiometer output terminals, the loading effect [21-23] introduces an error or nonlinearity in the transducer action as shown by Eq. (14.16). Then the output voltage e_{out} is not a linear function of α, but is given by

$$e_{out} = \frac{\dfrac{r_L}{r_p}\dfrac{r_2}{r_p}e}{\dfrac{r_L}{r_p} + \dfrac{r_2}{r_p} - \dfrac{(r_2)^2}{(r_p)^2}} = \left[\frac{\beta}{\beta + \alpha - \alpha^2}\right]e\alpha$$

(14.16)

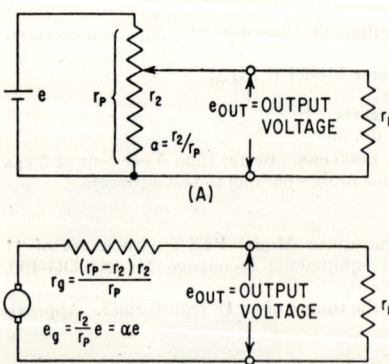

Fig. 14.10. Potentiometer-type transducer circuits: (A) typical circuit and (B) equivalent circuit.

This nonlinearity is illustrated by the family of curves in Fig. 14.11A. The per cent by which the output voltage is below the correct

POTENTIOMETER-TYPE TRANSDUCER   14–11

value $(e\alpha)$ is given in Fig. 14.11B as a function of the relative slider position for a number of load resistance ratios $\beta = r_L/r_p$. The maximum per cent error occurs when the movable contact is at the center ($\alpha = 0.5$). This curve, plotted in Fig. 14.11C, shows that the error is less than 2.5 per cent if the load resistance r_L is greater than 10 times the potentiometer circuit resistance r_p.

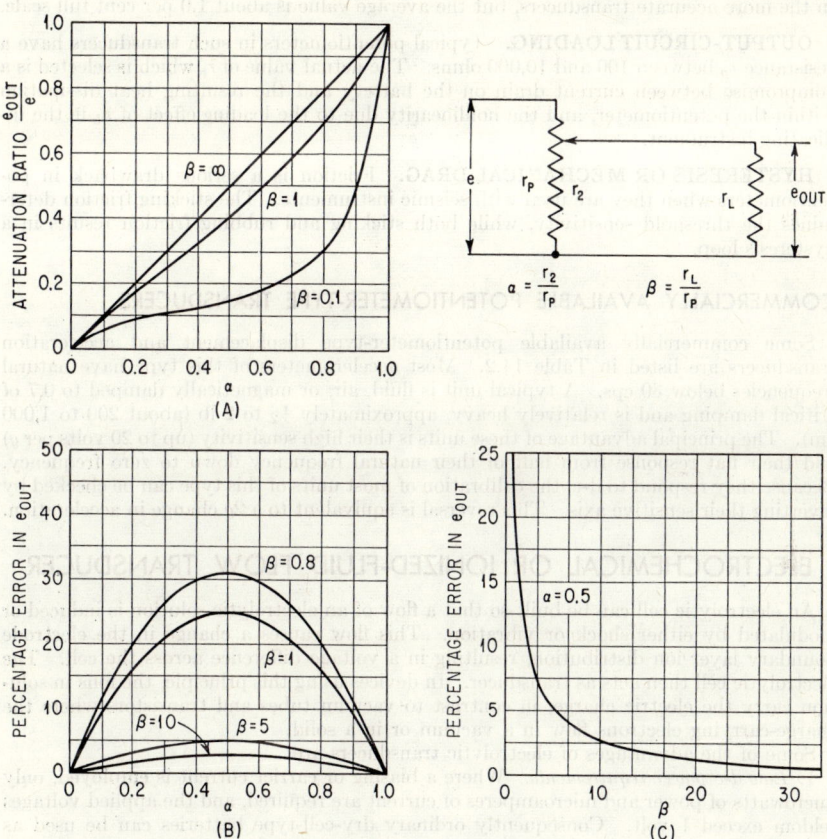

FIG. 14.11. Characteristics of potentiometer-type transducer circuit as a function of the resistance ratio α: (A) nonlinearity for various loading conditions, (B) error for various loading conditions, and (C) maximum error for various loading conditions.

FACTORS AFFECTING USE

SENSITIVITY. The full-scale travel of the movable contact in the potentiometer depends upon the particular transducer. In some accelerometers, a full-scale travel of less than 0.1 in. is used. If the transducer is used in the study of the motion of controls, control surfaces, actuators, etc., a travel up to 10 in. may be required.

The sensitivity in terms of output voltage per unit travel of the movable contact is seldom listed by the manufacturer. The general practice is to give the operating limitations, and to let the user calculate the sensitivity according to his specific operating conditions. Manufacturers recommend either the maximum energizing voltage, the maximum energizing current, or the maximum power input to the potentiometer. Knowing these and the full-scale travel, it is possible to determine the sensitivity.

RESOLUTION. The resolution of potentiometers used in transducers is usually between 0.1 and 0.5 per cent of the full range. It represents the step change in voltage obtained when the movable contact moves from one turn to the next in the potentiometer resistance winding.

LINEARITY. The potentiometer can be linear to within 0.1 per cent full scale in the more accurate transducers, but the average value is about 1.0 per cent full scale.

OUTPUT-CIRCUIT LOADING. Typical potentiometers in such transducers have a resistance r_p between 100 and 10,000 ohms. The actual value of r_p which is selected is a compromise between current drain on the battery and the resulting heat dissipation within the potentiometer, and the nonlinearity due to the loading effect of r_L in the indicating instrument.

HYSTERESIS OR MECHANICAL DRAG. Friction is a serious drawback in potentiometers when they are used with seismic instruments. The sticking friction determines the threshold sensitivity, while both sticking and rubbing friction results in a hysteresis loop.

COMMERCIALLY AVAILABLE POTENTIOMETER-TYPE TRANSDUCERS

Some commercially available potentiometer-type displacement and acceleration transducers are listed in Table 14.2. Most accelerometers of this type have natural frequencies below 30 cps. A typical unit is fluid, air, or magnetically damped to 0.7 of critical damping and is relatively heavy, approximately ½ to 2 lb (about 200 to 1,000 gm). The principal advantage of these units is their high sensitivity (up to 20 volts per g) and their flat response from half of their natural frequency down to zero frequency. Because they respond to d-c, the calibration of most units of this type can be checked by inverting their sensitive axis. This reversal is equivalent to a $2g$ change in acceleration.

ELECTROCHEMICAL OR IONIZED-FLUID FLOW TRANSDUCER

An electrolytic cell can be built so that a flow of an electrolytic solution is induced or modulated by either shock or vibration. This flow causes a change in the electrode boundary layer ion distribution, resulting in a voltage difference across the cell. The electrolytic cell then acts as transducer. In devices using this principle, the ions in solution carry the electric charge, in contrast to vacuum tubes and transistors where the charge-carrying electrons flow in a vacuum or in a solid.

Some of the advantages of electrolytic transducers are:

1. *Low d-c power requirements.* Where a biasing or carrier current is employed, only microwatts of power and microamperes of current are required, and the applied voltages seldom exceed 1 volt. Consequently ordinary dry-cell-type batteries can be used as power sources.

2. *No moving mechanical parts.* The signals in the electrolytic transducer are obtained from the minute motion of the electrolytic fluid.

3. *Low sensitivity to bias or carrier voltage changes.* Current-carrying capacity of the electrolyte is governed primarily by diffusion effects, rather than by the applied voltage.

Some of the disadvantages of electrolytic transducers are:

1. Certain electrolytic transducers are *suitable only for low-frequency applications.*
2. Some of the electrolytic transducers are *very temperature-sensitive.*
3. *Extreme chemical purity and freedom from included gases must be maintained* in the electrolytic solutions.

The transducer systems described in the following sections represent three different approaches to the use of electrochemical ion transfers.

SOLION TRANSDUCER [24-27]

In this type of transducer, an external voltage source establishes an ion concentration around the electrodes of an electrolytic cell. Shock- or vibration-induced flow of the

ELECTROCHEMICAL OR IONIZED-FLUID FLOW TRANSDUCER

electrolyte disturbs this ion concentration, resulting in an increase of current flow in the external circuit. A transducer utilizing this principle is known as a "solion" (i.e., solution of ions) transducer.

The solion transducer is quite sensitive, requires extremely low power for its operation, has no moving mechanical parts, has a long shelf and operational life, occupies little space, may be coupled directly to recording circuits without requiring electronic equipment, and is not affected by an external electromagnetic field. However, this transducer is inherently a low-frequency device, usually operating best in a range from approximately 0 to 20 cps, with 200 cps the maximum upper limit. Because of its temperature dependence (operational range 28 to 90°F, −2 to 32°C), the solion must include careful temperature compensation, such as negative temperature coefficient resistors or nonlinear resistive elements.

The transducer shown in cross section in Fig. 14.12 is typical of those employing solion action.[26] This transducer contains two anodes, usually made of platinum gauze, connected together electrically, and a cathode with a small orifice in the center, usually a narrow slit. These electrodes are mounted in a plastic insulating housing that forms two isolated cavities which are filled with the electrolytic solution, in this case iodine–potassium iodide. The fluid flows between the two cavities through the orifice in the cathode. Energy is supplied to the system by a small, low-voltage battery in series with a resistive load. When the electrodes are energized, a concentration of ions is built up around the cathode orifice area (see Fig. 14.12). This concentration (polarization of the cathode) limits the effective current flow to the transducer. Thus, in the static state, the iodine solution in the vicinity of the orifice is reduced—acquiring electrons from the cathode metal; iodide

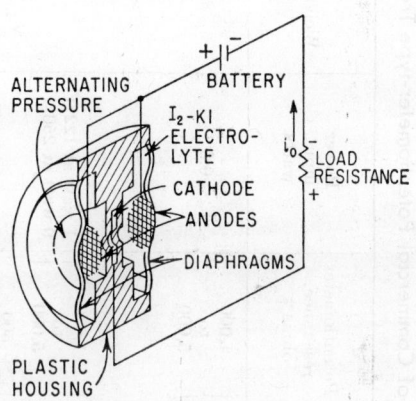

Fig. 14.12. Schematic cross section of typical solion vibration transducer. (*Courtesy of National Carbon Co.*[26])

in the vicinity of the anode is oxidized—giving up electrons to the anode metal. This equal and opposite reaction at anode and cathode, occurring so that there is no net change in concentration as current flows through the cell, is commonly called a *redox system*. The reaction can produce a voltage difference of as much as 50 to 100 millivolts between the two electrodes, depending upon the concentration difference. This action establishes a small static current (called the *diffusion current*) through the external circuit. This current is limited by the electrode area and by the diffusion coefficient of the iodine ions.

If an alternating pressure is imposed upon the diaphragms of Fig. 14.12, the electrolyte is pumped into the orifice, thus bringing new ions into the vicinity of the cathode and causing a correspondingly large increase in the current through the external circuit. In the case of vibration acceleration normal to the diaphragms, the inertia of the electrolyte provides the agitation required to produce the flow and consequently the electrical output.

The frequency response of a solion device is limited because of the high ratio of ion-to-electron mass. Since the same forces act on both the ions and the electrons, the ions will move much more slowly because of their much greater mass. For iodine, this ratio is approximately 233,000:1. This limits the upper useful frequency to about 20 cps.

POROUS DISC TRANSDUCER

The operation of this transducer utilizes an electrokinetic phenomenon known as "streaming potential"—a phenomenon which occurs when a polar liquid (such as methanol, water, or acetonitrile) is forced through a porous disc.[28-34] When the liquid flows

Table 14.2. Typical Characteristics * of Commercial Potentiometer-type Transducers

Manufacturer	Model	Function	Range Acceleration, full scale	Range Natural frequency, cps	Potentiometer resistance, ohms †	Power input, watts	Weight, oz	Ratio of critical damping for temperatures indicated
Bourns Lab.	602A	Acceleration	±1g to ±10g	6 18	1,000 to 5,000	0.5	8	0.7 to 0.6 (−65 to 200°F)
	604	Acceleration	±5g to ±30g	40 100				
Edcliff Instruments	605	Acceleration	±0.5g to ±5.0g	5 15	1,000 to 5,000	1 watt at 122°F 0 watt at 250°F	11	0.6 ± 0.1 (−67 to 248°F)
	7-31	Acceleration	±0.5g to ±100g	.....	500 to 20,000	0.5 watt at 68°F	8	0.65 ± 0.05 (at 68°F)
Genisco, Inc.	GLH	Acceleration	±1g to ±8g	6 17	2,000 to 10,000	1	32	0.7 (at 75°F) 0.9 to 0.6 (−65 to 180°F)
	DDL	Acceleration	±1g to ±8g	5 13	2,000 to 10,000	2	40	0.7 (at 75°F) 0.9 to 0.5 (−65 to 275°F)
	GMO	Acceleration	±2g to ±30g	12 52	5,000 to 10,000	2	8	0.7 (at 75°F) 2.0 to 0.3 (−10 to 180°F)
Giannini Controls Corp.	24117	Acceleration	±2.5g to ±20g	31 89	2,000 to 5,000	0.5	10	0.6 (at 77°F) 1.0 to 0.4 (−65 to 160°F)
	24142	Acceleration	±1g to ±40g	8 54	10,000	2.25	16	1.5 to 0.2 (−65 to 160°F)

14–14

Model	Measures	Range	(col 4)	(col 5)	(col 6)	(col 7)	(col 8)
24144	Acceleration	±2.5g to ±30g	2,000	11 / 38	0.5	8	1.0 to 0.2 (−65 to 160°F)
24145	Acceleration	±2g to ±30g	2,000	6 / 35	0.5	17	1.5 to 0.2 (−65 to 160°F)
Humphrey, Inc.							
LA-03-0102-1	Acceleration	±10g	5,000	35	0.5	5	0.7 to 0.5 (−65 to 165°F)
LA-07-0103-1	Acceleration	−1g to +3g	20,000	10	0.5	18	0.7 ± 0.1 (−65 to 185°F)
LA-12-0101-1	Acceleration	±5g	1,000	8	0.5	5.5	0.6 ± 0.1 (−65 to 185°F)
RP04-0101-1[2]	Displacement	4.13 in.	30,000	...	2.0	2.5	
RP01-0310-1[2]	Displacement	6.00 in.	9,000 (dual pot.)	...	3.0 (each)	7.5	
Minneapolis-Honeywell (Boston Div.)							
LA-108	Acceleration	±5g	2,000	11	0.5	16	0.6 ± 0.2 (at 71°F)
-111	Acceleration	±20g		15.8			0.65 ± 0.2 (at 77°F)
-113	Acceleration	±3g		8.5			0.6 min (at 77°F)
-114	Acceleration	±10g		11.5			0.6 min (at 77°F)
-301	Acceleration	−3g to +7g		13			1.5 to 0.5 (−65 to 160°F)
LA-500	Acceleration	±1g to ±60g	1,000 to 14,000	5 / 60	0.5	16	0.75 ± 0.3 (−65 to 175°F)
Pacific Scientific Co.							
4205[1]	Acceleration	+35g, −10g	5,000	45	...	...	
Research, Inc.							
4040[1]	Displacement	10 ft					
4046[1]	Displacement	0.5 in. to 3.5 in.	5,000		1.25		0.6 ± 0.1 (at 68°F)

* Taken from manufacturer's literature.

† In general, resistance value increases with improved resolution required.

Note 1: These transducers are made up of cable and spring-loaded (tension) reel assemblies; the cable turns the reel which rotates the potentiometer.

Note 2: Designed to measure relative displacement with solid mechanical connections.

through the pores, a voltage drop is generated across the disc which is in phase with, and directly proportional to, the differential pressure across the faces of the disc. When the direction of flow is reversed, the polarity of the electrical signal also is reversed. The cell employing this principle sometimes is called an *electroosmotic cell*.

The frequency response of vibration transducers employing this principle is flat within ±3 db from 3 cps to 60,000 cps.[39] The sensitivity depends upon the density of the cell liquid and column length; it generally is a few millivolts per g, but some accelerometers have been constructed with a sensitivity of 500 millivolts/g from 0.5 cps to over 1,000 cps. The operating temperature range is determined largely by the boiling and freezing points of the liquid which is used and by the hydrostatic pressures encountered. For acetonitride, the sea-level range is limited to between $-40°F$ ($-40°C$) and $+176°F$ ($+80°C$). For high impedance loads, no temperature compensation is necessary; when the load impedance is comparable to the cell impedance, the effect of temperature on sensitivity must be compensated by means of an external circuit. The high mechanical impedance and resistive electrical impedance make possible the use of relatively simple electrical and mechanical impedance matching methods. The operating characteristics of this type of transducer can be made to vary over extremely wide ranges by varying the density of the working fluid.

The primary elements of an electrokinetic cell are shown schematically in Fig. 14.13. The cell consists of a microporous disc glazed into the center of an impermeable ring. Diaphragms tightly sealed on each side of the ring retain the polar liquid which fills the intervening space. A wire mesh electrode is mounted on each side of the porous disc and is attached to external leads. When an external force is applied to one of the diaphragms, a pressure difference is developed across the porous disc. If the whole assembly shown in Fig. 14.13 is given an acceleration normal to the face of the diaphragm or discs, a pressure difference is built up across the porous disc, caused by the accelerating forces on the electrokinetic fluid. This pressure difference results in a flow through the disc and a "streaming potential" which is proportional to the applied pressure. The potential is independent of the area and length of the plug and, except for very fine material, the size of the pores. In this case, the pressure difference is directly proportional to the acceleration. The unit may be made more sensitive as an accelerometer by mounting weights on the diaphragms, which essentially increases the pressure difference across the disc. The same effect can be produced by using longer liquid columns or a more dense liquid.

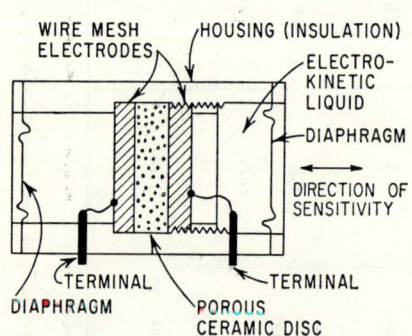

FIG. 14.13. Schematic cross section of an electrokinetic cell used as an acceleration transducer. (*After E. V. Hardway, Jr.*[30,32-34])

This instrument is analogous to an electric circuit containing resistance, inductance, and capacitance in which the resistive component predominates. In such a circuit, the output across the resistance does not vary with frequency except at very high and very low frequencies. Therefore, variations in frequency within the limits of the frequency band have little effect on the output of this type of transducer.

MERCURY-ELECTROLYTE TRANSDUCER

This type of transducer is an application of an electrokinetic phenomenon whereby mechanical energy is transformed into electrical energy. This transformation occurs when relative motion takes place between a mercury electrolyte system and the capillary tube containing it (see Figs. 14.14 and 14.15). The generated voltage is, within limits, proportional to the relative displacement of the electrolyte in the tube.

Advantages of mercury-electrolyte transducers include: (1) These devices have high sensitivity combined with high power output. (2) Units are either self-generating or require only a very small battery supply due to extremely low current drain (under 100 microamperes) when no input stimulus is present. (3) The tube can be constructed as a fairly low internal impedance voltage generator (in contrast to relatively high-impedance streaming-potential type of transducers). (4) Seismometers with low natural frequencies can be constructed which produce voltage outputs proportional to displacement in the frequency range from 20 cps to over 200 cps. (5) Units can be made quite small compared to connecting leads, so that their attachment to light structures should not appreciably affect the vibration characteristics of the structure to be measured.

A disadvantage of this type of transducer is its sensitivity to mechanical shock. A drop of the transducer from only a few centimeters to a hard surface can cause the mercury column to break.[35]

Although the output stability of these sensors appears satisfactory at room temperature, they are unsatisfactory at 122°F (50°C). The sensitivity of a unit under test decreased after the unit was subjected to vibration at this temperature for 30 min.*

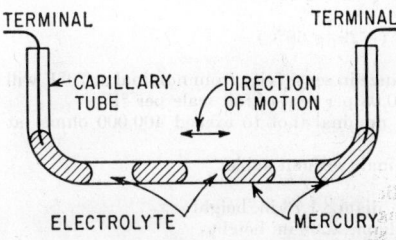 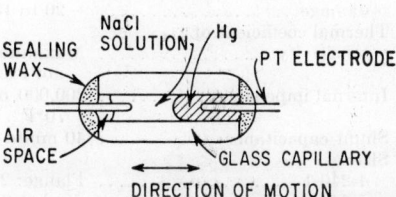

FIG. 14.14. Schematic cross section of a mercury-electrolyte vibration transducer.

FIG. 14.15. Diagram of a mercury-electrolyte transducer; typical construction of tube with one interface. (*After W. W. Fain, S. L. Brown, and A. E. Lockenvitz.*[35])

A typical mercury-electrolyte transducer is shown in Fig. 14.14. It is made of alternate slugs of mercury and electrolyte solution, such as potassium iodide or sodium chloride placed in a glass capillary tube with electrodes contacting the mercury slugs at the ends. A voltage is generated between the electrodes when the tube is shaken. If the tube ends are sealed so that relative motion cannot take place between the capillary tube and the mercury-electrolyte system, a small output voltage is produced. This voltage may be attributed to periodic changes in the interfacial area associated with the mercury in contact with an electrolytic solution as a result of the acceleration forces.[36,37] When the tube ends are arranged so that the mercury-electrolyte system is free to move relative to the glass capillary walls, a considerably larger voltage results; this may be attributed to the relative motion between the capillary and the mercury-electrolyte system (called the U-effect II [38] or the Latour effect [35]).

Experimental results on such systems show that for a constant frequency of vibration, an output voltage is obtained which is approximately proportional to the relative displacement between the mercury-slug-electrolyte system and the capillary tube walls. The output voltage is a function of the number and length of the slugs, vibration amplitude, and frequency. If the vibration amplitude is held constant and the vibration frequency varied in the range from 10 to 200 cps, the output voltage increases with increasing frequency until an apparent resonant frequency of the system is reached; with further increase in frequency, the output voltage remains constant.

A single interface of mercury, a sodium chloride electrolyte, and platinum electrodes have been used as indicated in Fig. 14.15, which shows the diagram of a tube with one interface. For small displacement amplitudes of vibration (i.e., up to 0.00022 in. peak), the output voltage is linear with amplitude for fixed frequencies and there is no noticeable distortion of waveform in the frequency range from 300 to 1,100 cps.[35] For amplitudes above about 0.00022 in., the output waveform is distorted and the reproducibility is quite poor. The amplitude at which waveform distortion first occurs is a function of the

---

* Unpublished tests by A. J. Yerman, General Electric Company.

Table 14.3. Specifications for Commercial Electrokinetic-type Acceleration Transducers *

Type 4-240-1 and Type 4-240-2:	
Acceleration range............	0.01g to 1,000g
Maximum acceleration without damage....................	5,000g
Nominal sensitivity...........	Greater than 3 mv/g at 70°F into a resistive load not less than 2 megohms
Frequency response...........	Flat from 5 cps to 40 kc ±12 per cent at 70°F; flat from 10 cps to 30 kc ±6 per cent. Deviation from flat response will not exceed 50 per cent at 3 cps and 60 kc (see Fig. 14.16B)
Linearity....................	For a constant applied frequency with a sensitive axis parallel to the direction of acceleration, the pickup sensitivity will not vary more than 0.1 per cent of full scale within its acceleration limits as specified
Operating temperature range....	−10 to 140°F (−29 to 60°C)
Temperature range without damage....................	−20 to 150°F (−29 to 66°C)
Thermal coefficient of sensitivity.................	Maximum change in sensitivity from nominal at 70°F will not exceed 0.08 per cent of full scale per °F
Internal impedance...........	200,000 ohms nominal (not to exceed 400,000 ohms) at 70°F
Shunt capacitance............	40 mmfd nominal (without cable)
Size:	
4-240-1...................	Flange: 2.7 in. diam.; 1.73 in. height
4-240-2...................	Stud: 1.9 in. diam.; 2.68 in. height
Weight......................	11 oz (including cable)
Electrical connection.........	18 in. of twin shielded Teflon-wrapped lead; low-capacitance cable is used to minimize reduction in high-frequency response
Type 4-242-1 and Type 4-242-2:	
Acceleration range............	0.01g to 1,000g
Maximum acceleration without damage....................	5,000g
Nominal sensitivity...........	Greater than 1.3 mv/g at 70°F into a resistive load not less than 2 megohms
Frequency response...........	Flat from 50 cps to 60 kc at 70°F; deviation from flat response will not exceed 50 per cent at 25 cps and 80 kc (see frequency-response illustration Fig. 14.16A)
Linearity....................	For a constant applied frequency with the sensitive axis parallel to the direction of acceleration, the pickup sensitivity will not vary more than 0.5 per cent of full scale within its acceleration limits as specified.
Operating temperature range...	−10 to 140°F (−24 to 60°C)
Temperature range without damage....................	−20 to 150°F (−29 to 66°C)
Thermal coefficient of sensitivity.................	Maximum change in sensitivity from nominal at 70°F will not exceed 0.08 per cent of full scale per °F
Internal impedance...........	200,000 ohms nominal (not to exceed 400,000 ohms) at 70°F
Shunt capacitance............	12 mmfd nominal (without cable)
Size:	
4-242-1...................	Flange: ¾ in. diam.; ⅝ in. height
4-242-2...................	Stud: ⅝ in. diam.; 2 in. height including 0.415 in. stud bolt and connector
Weight:	
4-242-1...................	8 gm (including cable)
4-242-2...................	20 gm (not including cable)
Electrical connection.........	18 in. of twin shielded lead is supplied with pickup; low-capacitance cable is used to minimize reduction in high-frequency response

* Consolidated Electrodynamics Corporation.[39]

frequency, becoming a minimum at the resonant frequency. The shape of the output vs. frequency curve is similar at both high and low amplitudes of vibration between 20 and 1,400 cps, but reproducibility is poor at the higher amplitudes. The output voltage for fixed amplitudes of vibration increases from zero at zero frequency to a definite peak at the resonant frequency. Above this frequency, the output drops off rapidly to a value much lower than that of the peak but essentially remains constant throughout the rest of the frequency range. The resonant frequency is closely associated with the mechanical structure of the tube.[35] The character of the output-impedance curves during vibration for constant unloaded tube output is the same for tubes of similar construction, even though their resonant frequencies are widely separated.

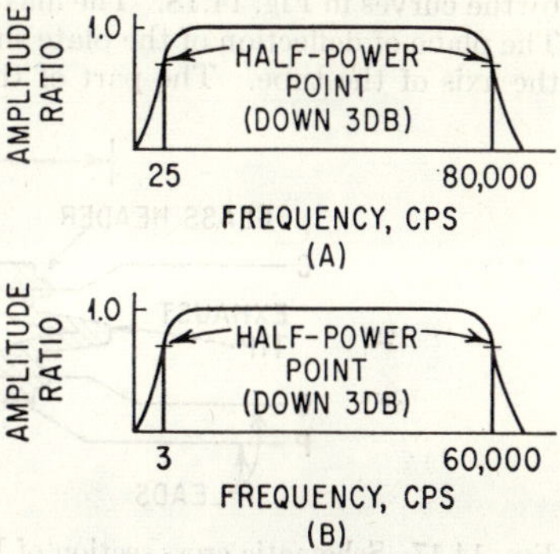

COMMERCIALLY AVAILABLE ELECTRO-CHEMICAL TRANSDUCERS

Commercial devices utilizing the solion principle have been explored.[24, 25, 27] A complete vibration meter * has been developed which uses the "solion" principle.[26] The instrument is in the shape of a cylinder 3¾ in. long and 2 in. in diameter. It is not considered a precision instrument, but does give a reliable and reproducible measurement up to approximately 200 cps.

Specifications for the three available types of porous disc electrokinetic transducers † are given in Table 14.3.[39] Representative response curves are shown in Fig. 14.16 A and B.

FIG. 14.16. Typical frequency-response curves of porous disc electrokinetic transducers: (A) accelerometer, Type 4-242; (B) accelerometer, Type 4-240. (*Courtesy Consolidated Electrodynamics Corp.*[39])

MECHANOELECTRONIC TRANSDUCER

The mechanoelectronic transducer is an electron tube which develops a change in voltage when the relative spacing between the elements (plate, grids, and cathode) is changed. Both acceleration [45–47] and displacement [41, 43] types of mechanoelectronic transducers have been developed.

The mechanoelectronic transducer possesses the following fundamental advantages: (1) the vibrating element in the transducer can be made very small with a resultant combination of light weight and high sensitivity, (2) less voltage gain and fewer components are needed in the auxiliary equipment because of the inherent gain in the tube itself, and (3) for complex transducers, there are innumerable possibilities in control, mixing, etc. The principal disadvantages of this transducer are (1) an unpredictable "zero drift" and (2) a "zero shift" following impacts of short duration. Furthermore, it is difficult to manufacture a hermetically sealed shell through which controlled vibration can be transferred. Units that have been developed are fragile and require special handling.

DISPLACEMENT TRANSDUCER (RCA TYPE 5734)

In order to produce a relative motion between two elements inside a vacuum tube, it is necessary to couple at least one of the elements to the external motion to be observed. This is usually accomplished by a flexible diaphragm that forms part of the tube shell.[40–42]

In the RCA Type 5734 mechanoelectronic transducer,[43] the plate support shaft extends through a thin diaphragm in the end of the tube. A tilting motion of this support shaft around the diaphragm as a pivot changes the distance between the fixed grid and the

* Manufactured by the National Carbon Company.
† Manufactured by Consolidated Electrodynamics Corporation.

plate, resulting in a change in the plate current. Figure 14.17 shows a sectional view of this transducer indicating the relative positions of the cathode, grids, and the movable plate shaft (anode rod). This tube can be mounted in any position. General application data for the tube are given in Table 14.4 and the average tube characteristics are given by the curves in Fig. 14.18. The maximum angular deflection of the plate shaft is ±0.5°. The plane of deflection of the plate shaft coincides with the plane through terminal 5 and the axis of the tube. The part of the plate shaft within the tube has a minimum free

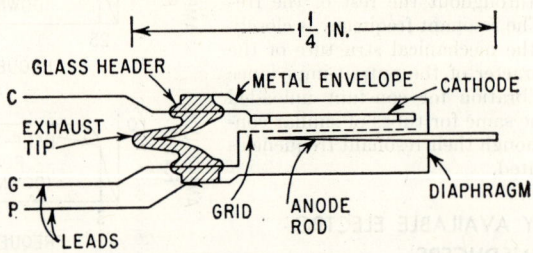

Fig. 14.17. Schematic cross section of RCA Type 5734 mechanoelectronic transducer. (*After H. F. Olson.*[41])

cantilever resonant frequency of 12,000 cps; with suitable mechanical coupling to the external end of the shaft, this permits measurement of vibration having a frequency as great as 12,000 cps.[43]

The transducer may be mounted by means of a supporting clamp which should firmly grip the metal shell of the tube. It is essential, however, that the pressure exerted on the shell by the clamp be held to a minimum to prevent possible fracture of the seals. The plate shaft should not be displaced from its normal position by more than 0.5°, as a

Table 14.4. Application Data, RCA Type 5734 Mechanoelectronic Transducer

Heater voltage	6.3 a-c or d-c volts
Heater current	0.15 amp
Maximum ratings, *design-center values:*	
D-C plate-supply voltage	300 volts
D-C plate current	5 ma
Plate dissipation	0.4 watt
Peak heater-cathode voltage:	
Heater negative with respect to cathode	90 volts
Heater positive with respect to cathode	90 volts
Typical operation:	
D-C plate-supply voltage	300 volts
D-C grid voltage	0 volt
Amplification factor *	20
Plate resistance *	72,000 ohms
Transconductance *	275 microhms
D-C plate current *	1.5 ma
Load resistance	75,000 ohms
Deflection sensitivity †	40 volts/degree 2,300 volts/radian
Moment of inertia of plate ‡	3.4 mg-cm²
Rotation compliance of diaphragm ‡	0.0013 × 10⁻³ radian/dyne-cm 0.075°/gm-cm

* For plate shaft in undeflected position.
† Average change in voltage across 75,000-ohm plate load resistor when the plate shaft is deflected from −0.5° to +0.5°. The plane of deflection of the plate shaft must coincide with the plane through terminal 5 and the axis of the tube.
‡ Based on external plate-shaft length of ⅛ in. (0.32 cm) and the center of the diaphragm as pivot.
Source: Radio Corporation of America.[43]

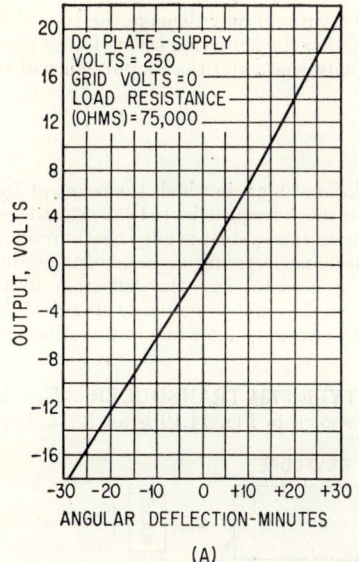
(A)

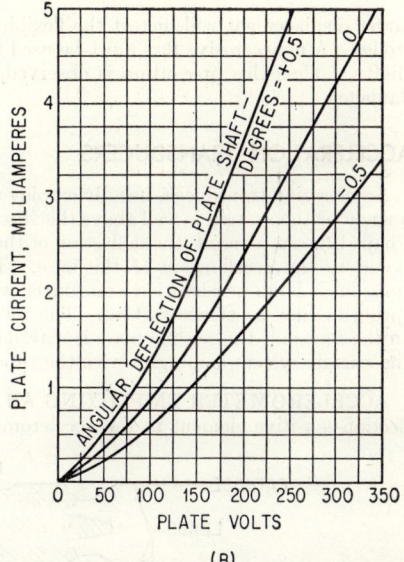
(B)

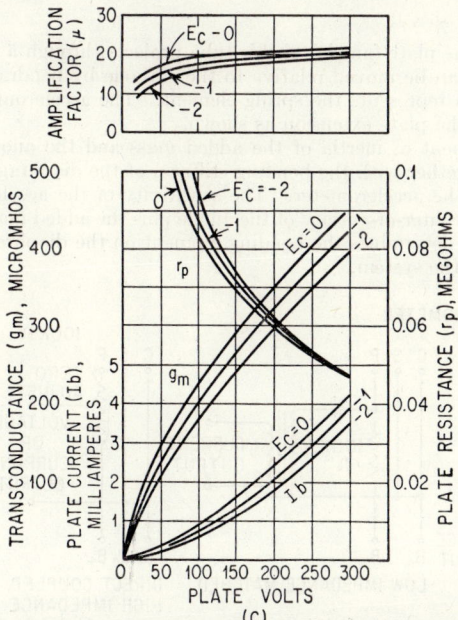
(C)

FIG. 14.18. Average tube characteristics of RCA Type 5734 mechanoelectronic transducer: (A) Output voltage vs. angular deflection of the plate shaft from its undeflected position with respect to the diaphragm as a fulcrum; (B) plate current vs. plate voltage for various angular deflections of the plate shaft; (C) transconductance, plate resistance, and amplification factor vs. plate voltage with the plate shaft in the undeflected position. (*Courtesy of Radio Corporation of America.*[43])

14-22 SPECIAL-PURPOSE SHOCK AND VIBRATION TRANSDUCERS

larger displacement will distort the flexible diaphragm and may damage the tube electrodes. A noncorrosive flux must be used in soldering the actuating stylus to the plate shaft. Unless this precaution is observed, the plate shaft and the diaphragm will be damaged.

ACCELERATION TRANSDUCERS

Acceleration transducers usually employ some form of mass in which the resonant frequency of the system is well above the vibration or shock frequencies to be measured (see Chap. 12). As a result, the deflection of the spring in this system is very nearly proportional to the acceleration of the base. The mechanoelectronic displacement-sensing transducer is an acceleration transducer when it is used to sense the deflection of the spring in such a seismic system. The following accelerometers employ this principle. In the first unit the seismic mass is external to the electron tube while in the other two the seismic system is integral with the tube.

ACCELEROMETER EMPLOYING AN RCA TYPE 5734 TRANSDUCER. The deflection-sensitive element in the accelerometer [44] shown in Fig. 14.19 is an RCA Type

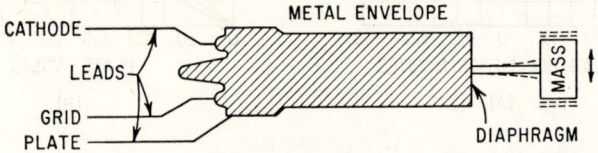

Fig. 14.19. Sketch of an accelerometer employing an RCA Type 5734 mechanoelectronic transducer.

5734 transducer.[43] The plate (anode) of this tube projects through a diaphragm in the end of the tube and can be moved relative to the cathode by bending the diaphragm. Hence, the diaphragm represents the spring element. The accelerometer is completed by adding a mass to the plate extension as shown.

The combined moment of inertia of the added mass and the anode rod about the diaphragm center, together with the bending stiffness of the diaphragm, determine the natural frequency of the accelerometer. The sensitivity of the accelerometer depends on the position of the center-of-gravity of the anode plus the added mass with respect to the diaphragm. This determines the bending moment on the diaphragm due to transverse accelerations of the system.

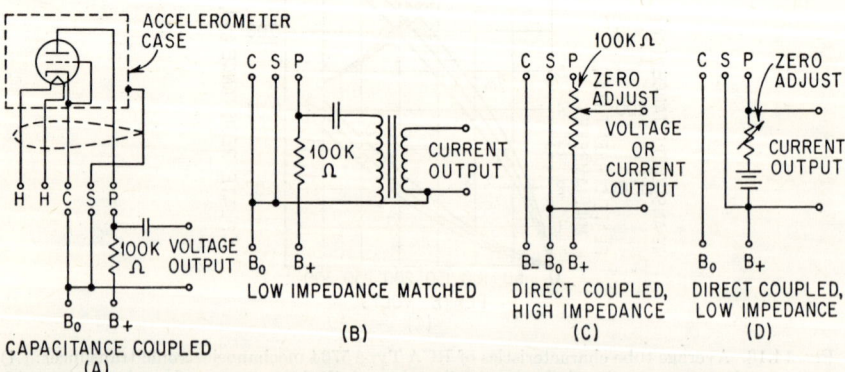

Fig. 14.20. Typical output circuits for use with an RCA Type 5734 mechanoelectronic transducer as an accelerometer: (A) capacitance coupled; (B) low impedance; (C) direct coupled, high impedance; (D) direct coupled, low impedance. (After R. C. Lewis.[44])

Output circuits that can be used with this transducer are represented schematically in Fig. 14.20. The circuits in Figs. 14.20C and 14.20D have response down to 0 cps (d-c); thus they are sensitive to zero drift of the transducer. Where low-frequency response is not required, the circuits shown in Figs. 14.20A and 14.20B should be used.

RAMBERG VACUUM-TUBE ACCELEROMETER. The Ramberg electron-tube accelerometer [45-47] contains two elastically mounted parallel plates with a rigidly fixed cathode between them forming a dual diode (see Fig. 14.21). Acceleration in a direction normal to the plane of the plates causes a displacement of the plates relative to the cathode. This displacement increases the current between one plate and the cathode and decreases the current be-

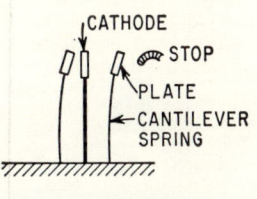

FIG. 14.21. Schematic of Ramberg vacuum-tube accelerometer. (*After S. Levy.*[48])

tween the other plate and the cathode. The use of symmetrically placed twin plates makes the output of the tube relatively insensitive to fluctuations in the total electron current from the cathode, particularly when the accelerometer is connected in an electrical bridge circuit. The changes in the currents are proportional to the component of acceleration normal to the plane of the plates. Therefore the sensitivity of the device is relatively independent of vibration frequency for frequencies that are low compared to the natural frequencies of the plates on their supports.

The Ramberg [45-47] electron-tube accelerometer (Sylvania Type SD759) has a fundamental natural frequency of approximately 750 cps, a flat response for acceleration up to 200 cps, and a linear range of $\pm 150g$. The output at an acceleration of $10g$ is sufficient to drive a high-frequency recording galvanometer directly, without an amplifier. The principal disadvantages of the tube are an unpredictable "zero drift" of a voltage equivalent to $\pm 1g$ in 15 min and a zero shift (and sometimes a change in calibration factor) following impacts of very short duration, for example, impacts which occur while tapping the envelope of the tube or from dropping it on a hard surface from a height of 1 in.

RAMBERG-LEVY VACUUM-TUBE ACCELEROMETER.[48] By employing stops and improved manufacturing techniques, this accelerometer overcomes the above disadvantages and is approximately twenty-five times more sensitive. The increase in sensitivity is obtained at the sacrifice of frequency range; its natural frequency is about 160 cps. Figure 14.22 gives the output vs. acceleration characteristics of a typical Ramberg-Levy vacuum-tube accelerometer, showing its linearity. The characteristics of the particular unit tested are (see Fig. 14.23A):

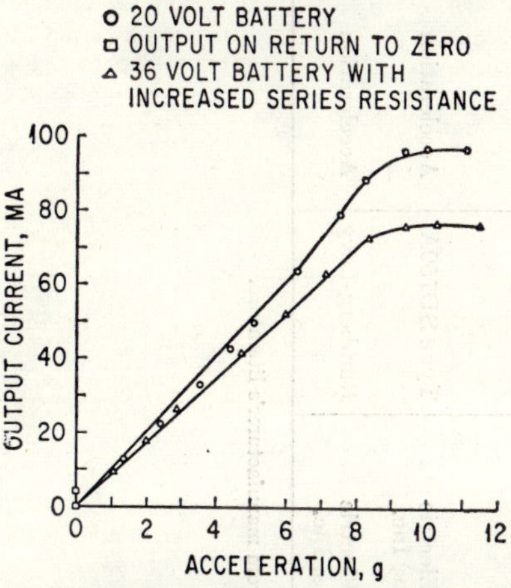

Plate resistances: $r_{p1} = 194$ ohms, $r_{p2} = 118$ ohms.
Heater voltage: 7 volts.
Plate voltage: 10 volts.
Range: $+9.4g$ to $-11.1g$.
Max. drift in 4-hr period: $0.032g$ (after 20 min. warmup).
Calibration factor: 0.116.
Sensitivity: 1.0 volts/g (corresponds to 7.3 ma/g with 11.5 ohms in meter circuit).

FIG. 14.22. Output vs. acceleration characteristics of a typical Ramberg-Levy vacuum-tube accelerometer. (*After S. Levy.*[48])

The circuits in Fig. 14.23A and B can be used with either the Ramberg or the Ramberg-Levy accelerometers. The circuit in Fig. 14.23B is an adaptation of the Kelvin double bridge and is intended to minimize the effect of contact resistance both in the tube socket

Table 14.5. Typical Characteristics* of Commercial Mechanoelectronic Transducers

Manufacturer	Type or Model	Function measured	Rating range available				Average weight	Remarks
			Amplitude	Frequency, cps	Sensitivity			
Radio Corp. of America	Type 5734	Displacement	±0.5°	0–12,000	40 volts/degree			Small triode with externally movable anodes
The Ling-Calidyne Co.	Model 18	Acceleration	$1g$–$70g$	80–450	15 volts/g 0.21 volt/g		2 oz (56.8 gm)	D-C operated will deliver 400 microamperes to 1,000 ohms or 20 volts to 10^6 ohms
Sylvania Electric Products, Inc.	Type SD759A	Acceleration	±150g	0–200				Dual diode in glass envelope mounting on loctol base
Sylvania Electric Products, Inc.	Ramberg-Levy	Acceleration	±10g	0–20	7.3 ma/g 1 volt/g			11.5 ohms in meter circuit

* Taken from manufacturer's literature.

and in the adjustable rheostat. The blocking filter preceding the galvanometer is intended to eliminate possible output at the natural frequency of the accelerometer. The frequency response of the over-all system (including the Ramberg-Levy accelerometer) shown in Fig. 14.23B is flat within 5 per cent up to 20 cps and then drops off to nearly zero output at 160 cps.

A representative list of mechanoelectronic transducers commercially available is given in Table 14.5.

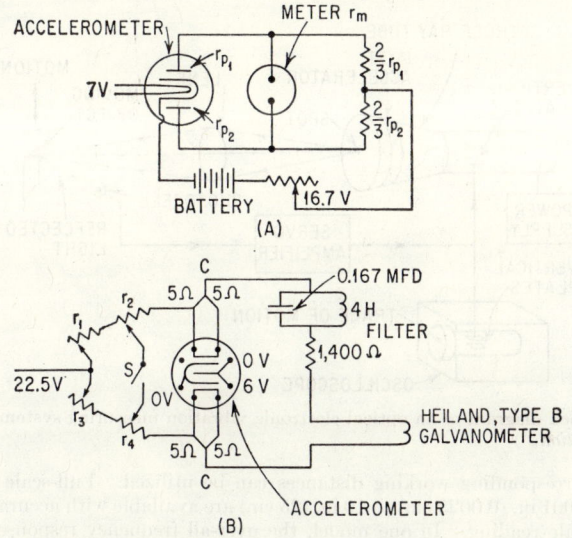

Fig. 14.23. Circuit diagrams for Ramberg or Ramberg-Levy accelerometers.

OPTICAL-ELECTRONIC TRANSDUCER

The Optron [49] is a commercially available noncontacting transducer for the measurement of displacement, which incorporates an optical system in an electronic circuit. The exact waveshape of the motion showing all transients and other characteristics of the displacement are displayed on an oscilloscope. This device can be used to measure amplitude and frequency of shake tables and accelerometers, runout of rotating equipment, or the response of cam followers. Its main advantage is that it does not require contact with the work; working distance may be as great as 21 in. For example, it can measure the vibration of individual blades of an impeller while running; as each blade comes past, the device can be "locked in" on the motion of the blade, following it through a segment of its revolution.

The principle of operation is illustrated in block diagram form in Fig. 14.24. The output signal is proportional to the voltage required to force a spot of illumination to follow the edge or boundary of a moving object. The intense spot from a special cathode-ray tube is projected by a lens onto the edge of the moving object. Reflected light is picked up by a 45° mirror behind the lens and directed to a photocell. The vertical deflection plates of the cathode-ray tube are biased so that with no light falling on the photocell, the imaged spot is driven down. However, when light is directed to the photocell, the amplified output causes the spot to be driven up. Hence an object placed in front of the lens reflects light to the photocell and drives the spot up. When it reaches the edge, it sees less (or no) light, and hence is driven down to "fix" on the edge. In this way the projected spot will fix on the edge of any reflecting material. If there is no convenient edge which serves the purpose, then a target of black and white can be painted on the part. When the edge of the object is moved or vibrated in the direction of the sensitive axis of the instrument, the illuminating spot travels with it. The electrical signal driving the spot is proportional to the displacement and can be observed or measured by standard instruments such as a cathode-ray oscilloscope.

As normally used, the scanning device is set on a tripod or stand and pointed at the moving object so that the intense spot from the special cathode-ray tube is imaged on the moving object and follows its motion. The device can be calibrated dynamically by scanning a known displacement. By changing the optical system, different displacement

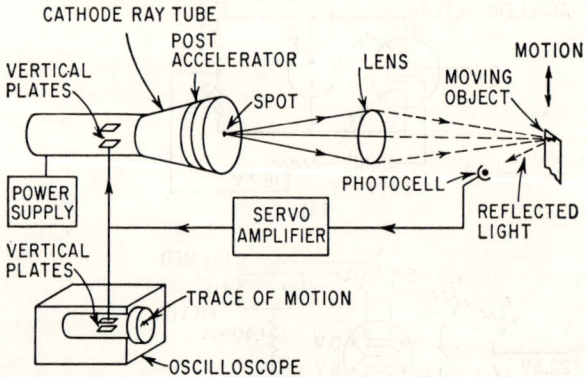

FIG. 14.24. Block diagram of an optical-electronic vibration measuring system. (*Courtesy of Optron Corporation.*[49])

ranges and corresponding working distances can be utilized. Full-scale displacement ranges from 0.001 in. (0.0025 cm) to 10 in. (25 cm) are available with accuracies of 0.1 per cent of full-scale reading. In one model, the over-all frequency response is from 0 to 100,000 cps; the output voltage for full-scale reading is approximately 250 volts.

A daylight model is also available which can be operated in an ambient illumination of up to 40 foot-candles of either daylight or incandescent light (usual illumination in a machine shop or on an office desk). Full-scale displacement ranges from 0.050 in. (0.13 cm) to 0.50 in. (1.3 cm) are available for this model with the same degree of accuracy. The over-all frequency range is from 0 to 5,000 cps. The output voltage for full-scale reading is approximately 8 volts.

OPTICAL INTERFEROMETER

An interferometer divides a beam of light into two parts which travel different paths and recombine to form interference fringes. The form of these fringes is determined by the difference in the optical path traveled by the two beams. This optical method is adaptable as a means of absolute calibration above vibration frequencies of 1,000 cps where the vibration displacements encountered are low (Chap. 18).

The optical interferometer method of vibration measurement [50-53] is best suited for measuring sinusoidal displacements at amplitudes below ±30 microinches (about 80×10^{-6} cm) in the frequency range from 30 to over 10,000 cps. When used with monochromatic light of 5,461 angstroms (from mercury-vapor lamp with filter), the smallest displacement that can be detected is about 4 microinches (10.4×10^{-6} cm).

GENERAL PRINCIPLE OF OPERATION

In the interferometers shown schematically in Fig. 14.25A and B, a beam of light from a monochromatic source is reflected in part from one surface and the balance from another; the resulting beam of light is recombined and viewed. If the effective length of the light path from the two mirrors is equal, then the resulting light beam has the same intensity as the original or incident beam. If the light is split into equal parts and totally reflected from the two mirrors, and if the effective light path length for one mirror is a half wavelength longer than that for the other, then when the two light beams are again recombined, cancellation will cause a dark field. If one of the two mirrors is not perfectly perpendicular to its incident beam of light, then the combined beam of light will have dark and light bands, as shown in Fig. 14.26A.

The interferometer when used for vibration measurements is arranged so that mirror D (Fig. 14.25A and B) is mounted on the vibrating surface whose motion is essentially rectilinear. When the vibration amplitude of this mirror is slowly increased, the fringes will disappear very sharply at set amplitudes. As the amplitude is further increased, the interference bands will again appear, and as they disappear the second time, a second known displacement amplitude is determined. The order of the disappearance and the resultant displacement amplitude are given in Table 18.1 for monochromatic light of 5,461 angstroms.

ANALYSIS OF OPERATING PRINCIPLES

The intensity of the fringe pattern I produced by an interferometer, if the mirrors are stationary, is given by

$$I = K\left(1 + \cos 2\pi \frac{x}{h}\right) \quad (14.17)$$

where K = constant
x = distance measured perpendicularly across the fringes
h = fringe width

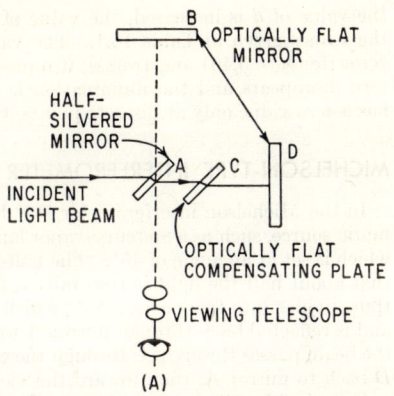

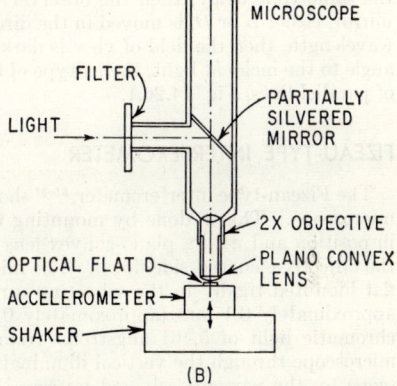

FIG. 14.25. Schematic diagram of optical paths of interferometers used for vibration-amplitude measurement. (A) Michelson type. (After E. I. Feder and A. M. Gillen.[62]) (B) Fizeau type. (After S. Edelman, E. Jones, and E. R. Smith.[56])

If the mirror D in Fig. 14.25 is vibrated sinusoidally in a direction perpendicular to its plane, the fringe pattern is set into simple harmonic motion. If the frequency is greater than 30 cps, or faster than the eye can follow, then according to Eq. (18.29) the time average of the light intensity is [51]

$$I = K\left[1 + J_0\left(\frac{2\pi d}{\lambda}\right)\cos\frac{2\pi x}{h}\right] \quad (14.18)$$

where $J_0(2\pi d/\lambda)$ is a Bessel function of order zero and where
λ = wavelength of light used
d = peak-to-peak amplitude of the vibrating surface

The value of $J_0(2\pi d/\lambda)$ is dependent only upon the amplitude of vibration and the wavelength λ of the light source. As

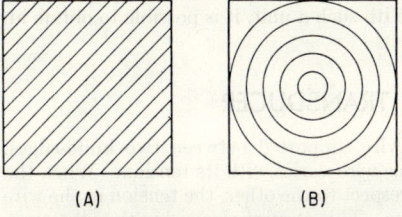

FIG. 14.26. Optical interferometer interference patterns: (A) Michelson type; (B) Fizeau type.

the value of d is increased, the value of $J_0(2\pi d/\lambda)$ will go through zero successively at the values given in Table 18.1. The value of d is tabulated for each of the successive zeros (for $\lambda = 5{,}461$ angstroms); it represents those values of d for which the fringe pattern disappears and the illumination is uniform. Since the Bessel function $J_0(2\pi d/\lambda)$ has a zero value only at discrete points, this fringe disappearance is very sharp.

MICHELSON-TYPE INTERFEROMETER

In the Michelson interferometer,[50-53] Fig. 14.25A, a beam of light from a monochromatic source, such as a mercury-vapor lamp, is directed toward the half-silvered mirror A which is set at an angle of 45°. The half-silvered mirror A has a coating which is so thin that about half the light is transmitted through it and the rest is reflected. The beam thus divides into two parts, one of which is reflected to the completely silvered mirror B and is reflected back through mirror A toward the viewing telescope. The other part of the beam passes through A, through the compensating plate C, and is reflected by mirror D back to mirror A, then toward the viewing telescope.

If the incident and reflected beams from both B and D are parallel and involve exactly the same time delay, then the observer will see a uniformly illuminated field. If one mirror, either B or D, is moved in the direction of the light beam a distance of a quarter wavelength, then the field of view is dark. If either of the two mirrors is set at a slight angle to the incident light, then a type of fringe pattern is viewed which contains a series of parallel lines, Fig. 14.26A.

FIZEAU-TYPE INTERFEROMETER

The Fizeau-type interferometer,[54-57] shown in Fig. 14.25B, can be made by modifying a microscope. This is done by mounting the microscope vertically with a 2X objective in position and with a plano-convex lens (focal length about 1,000 mm) at the focus of the objective, curved side down. The microscope should be mounted so that an optical flat mounted rigidly to the vibrating surface and normal to the direction of motion is approximately 0.1 mm (approximately 0.004 in.) from the plano-convex lens. Monochromatic light of 5,461 angstroms from a mercury-vapor lamp and filter enters the microscope through the vertical illuminator and is deflected down the axis of the microscope by the partially silvered mirror. It is then partially reflected from the curved surface of the plano-convex lens and from the optical flat D mounted on the vibrating surface. As a result of interference between these two reflected beams (similar to that described for the Michelson interferometer), a set of Newton's rings or fringes, Fig. 14.26B, is formed which can be viewed through the eyepiece of the microscope.

COMMERCIALLY AVAILABLE INTERFEROMETERS FOR VIBRATION MEASUREMENT *

Many interferometers have been assembled for special-purpose vibration tests.[52-54, 58-62] Special packaged interferometers designed for high-frequency vibration calibration of instruments are commercially available.[57, 62] With such a unit, it is possible to obtain an absolute calibration from 800 to 20,000 cps.

VIBRATING-WIRE TRANSDUCER

The vibrating-wire transducer has a taut wire supported between two knife-edges. For a fixed length, the natural frequency of this wire varies with its tension. When the support for one end of the wire is moved with respect to the other, the tension in the wire changes causing the natural frequency to change. This property is used in the vibrating-

---

* The following companies manufacture interferometers for the high-frequency calibration of vibration pickups: Gulton Industries, Inc., Metuchen, N.J.; Gaertner Scientific Corporation, Chicago, Ill.

wire transducer. It has been utilized in the measurement of the deformation of a structure,[63,64] in "the acoustic strain gage,"[65,66] and in a "dual-range strain accelerometer."[67]

Figure 14.27A shows the vibrating-wire transducer in its simplest form. A wire is driven by a small electromagnet which is powered by an oscillator and power amplifier. The detector is a similar electromagnet which is connected to a vacuum-tube voltmeter or an oscilloscope. The oscillator frequency is varied until a maximum amplitude is indicated by the vacuum-tube voltmeter. This system may be used at very low fre-

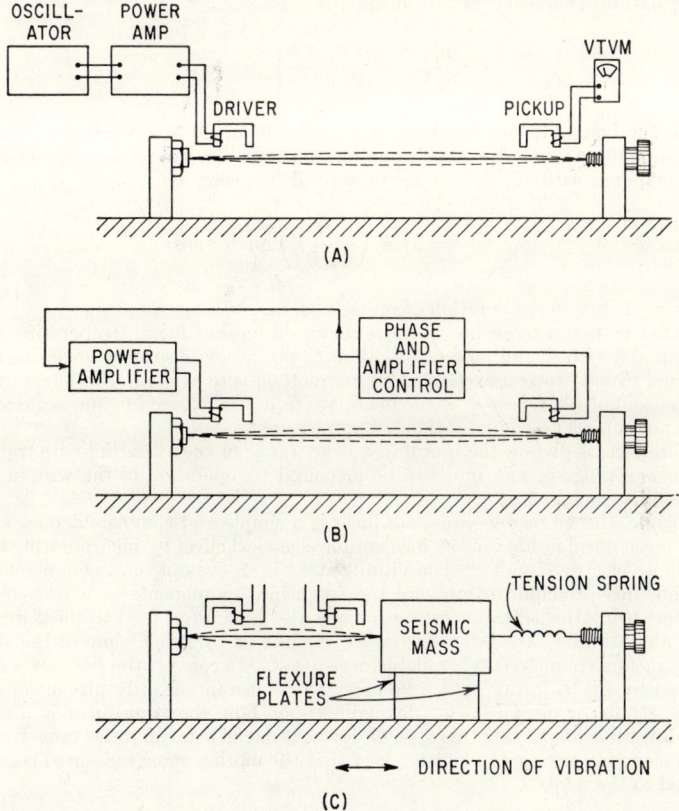

Fig. 14.27. Block diagrams of vibrating-wire transducer circuit: (A) force-driven system; (B) feedback system; (C) system incorporating an accelerometer.

quencies. When vibration measurements are made, it is necessary to use a system that supplies energy to the vibrating wire at its resonant frequency, even though the resonant frequency varies rapidly as a result of the vibration. An arrangement for accomplishing this is shown in Fig. 14.27B. Here the vibration of the wire is detected by the pickup coil whose output voltage is amplified and shifted in phase by the control circuit. The output of the control circuit signal is further amplified and is used to drive the electromagnet which in turn drives the wire. This circuit is a self-oscillating feedback-type system that will always oscillate, when properly adjusted, at the resonant frequency of the wire.

The vibrating-wire transducer can be employed in an accelerometer [67] essentially as shown in Fig. 14.27C. When the system is accelerated in a direction parallel to the axis of the wire, the accelerating force on the mass W/g will change the tension in the wire, thereby changing its resonant frequency. With this arrangement the change in vibrating frequency will be proportional to the acceleration as derived below. The lowest natural frequency for a vibrating wire of weight W' and length l is given by

$$f = \frac{1}{2}\left[\frac{T}{lW'/g}\right]^{1/2} \quad \text{cps} \tag{14.19}$$

where T is the tension in the wire. The tension T at any instant can be expressed as the sum of the initial tension of the wire T_0 plus the acceleration force of the mass W/g. Then the natural frequency at any instant is

$$f = \frac{1}{2}\left[\frac{T_0 + (W/g)a}{lW'/g}\right]^{1/2} \quad \text{cps} \tag{14.20}$$

where a is the instantaneous acceleration of mass W/g. If f_0 is the natural frequency corresponding to an initial tension T_0, and if the change Wa/g in tension during operation is small compared with T_0, the change in natural frequency is

$$\Delta f = \left(\frac{W}{8f_0 lW'}\right)\Delta a \quad \text{cps} \tag{14.21}$$

where Δa = change in acceleration of mass W/g.

Care must be taken to ensure that the change in tension due to temperature effects is small compared with the initial tension. This can be accomplished by enclosing the wire element in a sealed protective case. Another method is to evaluate this effect by means of a reference unit placed close to the operative unit and exposed to the same temperature, but not subject to the vibration being measured.

Space limitations dictate the useful frequency range of the vibrating-wire transducer. The frequency range of any unit can be increased by operation of the wire in a mode higher than its fundamental.

In principle, the vibrating-wire transducer is a simple and convenient means of converting a mechanical motion into a measurable electrical effect by incorporating it as one element of an electric circuit.[68] The Vibrotron [69] * is an example of a commercial transducer using this principle to measure the frequency components of a pressure pulse. Transducers using the vibrating-wire principle also have been used to measure torque, pressure, and strain.[70] By using tuned L-C circuits or by using some of the standard frequency meters commercially available, it is possible to convert the frequency modulation caused by the vibration into a voltage that is instantaneously proportional to acceleration, strain, or displacement, depending upon how the transducer is used. The primary difficulty in applying this type of device is one of attaching the taut wire to the structure being instrumented in such a way that the motion being measured is faithfully transferred to the wire.

TUNING FORK RATE-OF-TURN TRANSDUCER

Commercially available tuning forks can be used to measure rate of turn (i.e., angular rate). If a fork is gripped firmly by the handle, and if it is excited and rotated about the axis of symmetry, a vibration is felt in the hand. This vibration, which is torsional and at the same frequency as that of the prongs, has an amplitude which is proportional to the rate of rotation, i.e., to the angular velocity; it reverses its phase when the direction of turn is reversed.

The angular moment of inertia of the tuning fork is greatest when the prongs are farthest apart, and least when the prongs are nearest to each other. Thus to conserve angular momentum, when freely rotating in free space, the tuning fork will rotate faster during the latter condition. This rotation acts as a variable means of coupling between the resonant system of the tuning fork and the resonant system of the torsional structure.

* Manufactured by the B. J. Corporation Division of the Borg-Warner Corporation.

An example of a device utilizing this principle is the Sperry Rate Gyrotron * or vibratory gyroscope,[71-73] shown schematically in Fig. 14.28A. The steady-state vibration of the tuning fork prongs is maintained by use of the fork pickup coil whose signal is amplified and used to energize the tuning fork drive coil. This creates a self-excited feedback loop that always drives the fork at its resonant frequency. The tuning fork is mounted on a heavy base through a damped torsional spring between the handle of the tuning fork and the base. The resonant frequency of the torsional system consisting of the tuning fork, torsional spring, and base system is adjusted to be the same as that of the tuning-fork prongs. When the steadily vibrating tuning fork is rotated around its axis

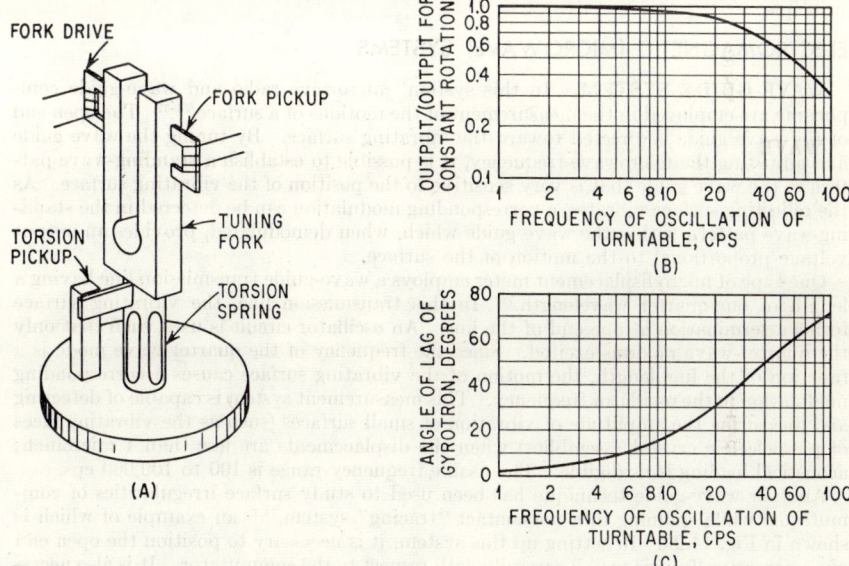

Fig. 14.28. Gyrotron tuning fork torsional vibration transducer: (A) schematic diagram, (B) frequency-response curve, and (C) phase shift curve. (After C. T. Morrow.[72,73])

of symmetry, a torsional vibration will be experienced by the torsional spring. The twist in the torsional spring is detected by the torsional pickup shown in Fig. 14.28A. The output of the torsion pickup is an a-c voltage at the fork frequency and the amplitude is proportional to the rate of turn. This signal may be considered as a carrier whose modulation is an indication of the measured rate of turn. It is a suppressed carrier, since the amplitude ideally goes to zero when the rate of turn is zero. The frequency response and phase shift for a representative Gyrotron having a Q equal to 30 and a natural frequency of 2,000 cps is shown in Fig. 14.28B and C. These curves represent the relative amplitude and phase, respectively, of the carrier envelope compared to a sinusoidal variation in the rate of turn of the base. The advantage of a tuning fork for this application, compared to many other possible configurations, is that the torsion pickup can be placed at a portion of a structure which is essentially free of vibration in the absence of an impressed rate of turn. This makes the instrument less critical to the balance of the pickup and extends the range of useful operation down to lower rates of turn. In addition, it is inherently more free of difficulty from spurious sensitivities about other than the nominal axis as compared to a rate gyroscope.

* Registered U.S. Patent Office.

REFLECTED-WAVE TRANSDUCER SYSTEMS

The motion of some vibrating systems can be detected by reflections from the vibrating surface of electromagnetic or ultrasonic waves which impinge on the surface. In each of the systems described below, there is no physical contact between the vibrating surface and the source of wave motion. Therefore, an instrument using the reflected-wave principle offers several advantages: (1) the instrument will not load or disturb the vibrating surface under observation, (2) a large number of regions can be examined in rapid succession, and (3) an extremely wide bandwidth can be achieved, since there are no mechanical resonances.

ELECTROMAGNETIC (MICROWAVE) SYSTEMS

WAVE-GUIDE SYSTEM. In this system, microwave radio and wave-guide components are employed in the measurement of the motions of a surface.[74-77] The open end of the wave guide is directed toward the vibrating surface. By tuning the wave guide and adjusting the microwave frequency, it is possible to establish a standing-wave pattern in the wave guide that is very sensitive to the position of the vibrating surface. As the reflecting surface vibrates, a corresponding modulation can be detected in the standing-wave pattern within the wave guide which, when demodulated, provides an output voltage proportional to the motion of the surface.

One type of microdisplacement meter employs a wave-guide transmission line having a length of one-quarter wavelength.[74] In this transmission line, the vibrating surface forms a termination at one end of the line. An oscillator circuit is used such that only the quarter-wave mode is excited. Since the frequency of the quarter-wave mode is a function of the line length, the motion of the vibrating surface causes a corresponding modulation in the oscillator frequency. This measurement system is capable of detecting and measuring the amplitude of vibration of small surfaces (such as the vibrating faces of piezoelectric crystal assemblies) when the displacements are less than 1 microinch; acoustical loading is negligible. The usable frequency range is 100 to 100,000 cps.

Another wave-guide technique has been used to study surface irregularities of commutators while spinning in a noncontact "tracing" system,[75,76] an example of which is shown in Fig. 14.29. In setting up this system, it is necessary to position the open end of a microwave "magic tee"[77] carefully with respect to the commutator. It is also necessary to adjust the shorting plug and the klystron oscillator frequency properly in order to obtain the proper sensitivity in the standing-wave pattern at the detector section of the tee. For the system described, the maximum displacement that can be measured is in the order of one-eighth wavelength of the microwave carrier used, or 0.050 in. (0.13 cm), while displacements of 100 microinches (25×10^{-6} cm) give readable deflections (0.1 in.) in a high-gain cathode-ray oscilloscope. The frequency range over which measurement can be made is determined primarily by the crystal detector and its associated amplifiers. The technique may be used from 0 cps to approximately one-hundredth of the carrier frequency.

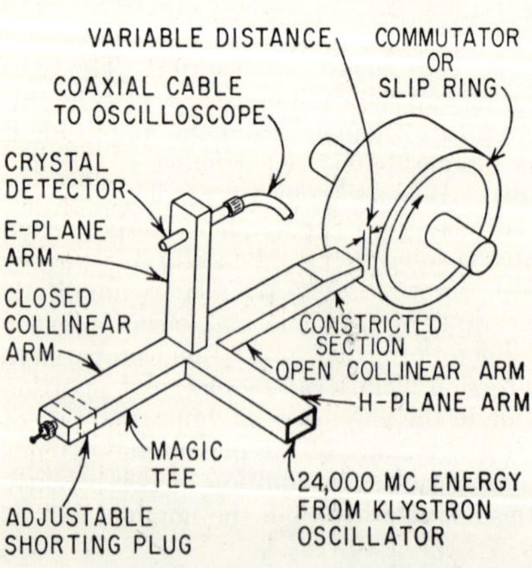

Fig. 14.29. Schematic of wave-guide vibration transducer system. (*After A. H. Ryan and S. D. Summers.*[75])

Another measurement system,[78] similar in principle to that described above, is shown in Fig. 14.30A. A reflex-type klystron oscillator supplies approximately 30 mw of power at a wavelength of 3.2 cm to an electromagnetic horn-type wave guide. The open end of the horn contains a dielectric lens that focuses the microwave energy on the vibrating object. The vibrating surface reflects some of the energy back into the wave guide, modulating the standing-wave amplitude in the wave guide and thus the output of the klystron oscillator. This change in the efficiency of the oscillator causes a corresponding change in the klystron plate current, as shown in Fig. 14.30B. The output of the transformer in the plate circuit is then a voltage which is proportional

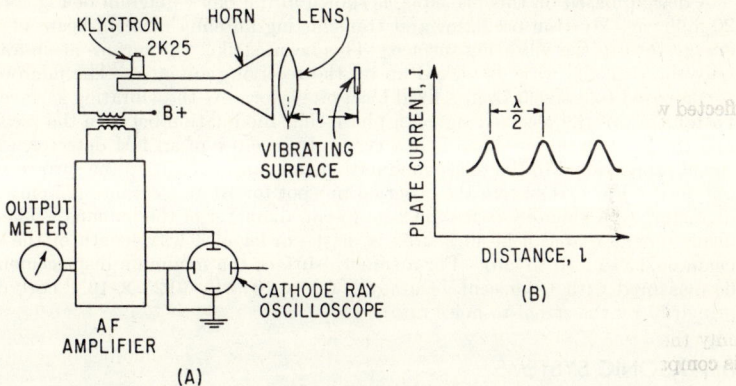

Fig. 14.30. Horn-type wave-guide vibration transducer: (A) circuit diagram; (B) change in plate current with vibration displacement. (*After S. Satomura et al.*[78])

to the vibration of the object. With a spacing (l in Fig. 14.30A) of 1.26 in. (3.2 cm), it is possible to measure vibration displacment amplitudes as large as 0.15 in. (0.38 cm) to approximately 40×10^{-6} in. (1×10^{-4} cm) at vibration frequencies from 5 to over 100,000 cps, depending primarily upon the type of transformer used.

MICROWAVE-BEAM REFLECTING SYSTEM. If a narrow beam of microwave electromagnetic energy is transmitted toward a vibrating surface, the phase of the re-

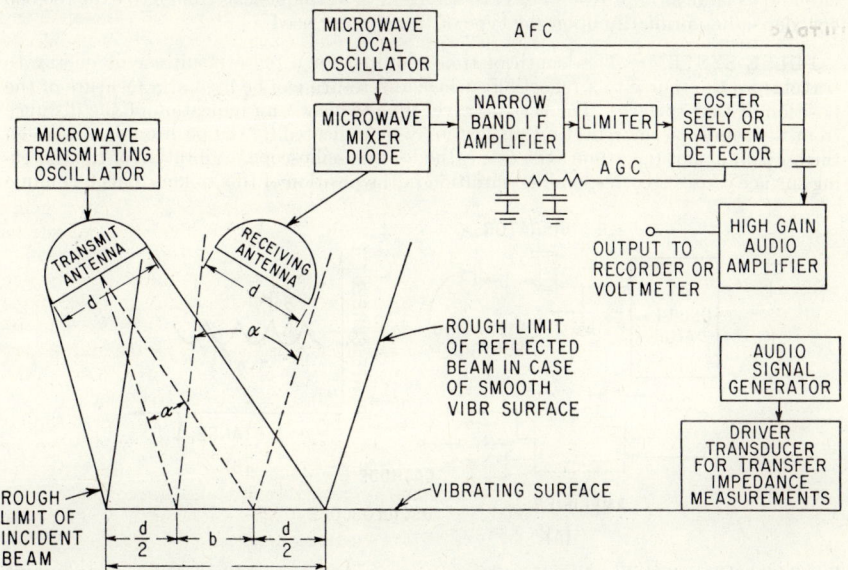

Fig. 14.31. Block diagram of microwave reflected-beam system. (*After C. Stewart.*[79])

flected wave will be modulated in proportion to the vibration amplitude and at the vibration frequency. By detecting this phase modulation, an electrical voltage is obtained which is proportional to the vibration velocity of the vibrating surface. This measurement is free from effects of mechanical loading and surface reflectivity variation, without requiring close proximity to the surface.

A design, based on this principle, is shown in the block diagram of Fig. 14.31.[79] The 20-milliwatt klystron oscillator and transmitting antenna direct a beam of microwave energy toward the vibrating surface. This beam strikes the surface at an angle so that only the reflected wave is picked up by the receiving antenna. The microwave signal is compared to a signal from a fixed local oscillator. As the vibrating surface varies the path length of the received signal, a phase-shift modulation between the received signal and the local oscillator signal is observed. By the use of an FM detector, an electrical signal proportional to the phase modulation and consequently to the surface vibration is obtained. For this system the observation spot for 10 in. clearance (25 cm) will have a diameter approximately twice the 2-in. (5-cm) diameter of the antenna. The maximum displacement that can be measured is of the order of a wavelength of the microwave beam or 0.34 in. (0.85 cm). For a smooth surface the minimum displacement that can be measured with 1 per cent accuracy is 0.1×10^{-6} in. (0.25×10^{-6} cm), determined primarily by the signal-to-noise ratio.

ULTRASONIC SYSTEMS

REFLECTED ENERGY SYSTEM. The total power radiated by an ultrasonic transducer is affected by both the amount and phase of the ultrasonic energy reflected back upon the transducer. When the reflecting surface is vibrating, there is a corresponding modulation in the radiated energy. A sensitive vibration-measuring system using this principle is shown in the block diagram of Fig. 14.32A.[77] The plate current of the RF oscillator is modulated by the reflected energy seen by the transducer. The change in plate current with relative position of the reflecting surface is illustrated by the curve in Fig. 14.32B. The sensitivity and selectivity of this technique can be increased if the transducer is concave so as to concentrate or focus the ultrasonic energy on a small area; it is possible to measure vibration displacements as small as 10×10^{-6} in. (0.25×10^{-4} cm) and as large as approximately 1/16 wavelength at frequencies from 5 to over 100,000 cps, depending primarily upon the type of transformer used.

PULSE SYSTEM. The length of time required for a pulse of ultrasonic energy to travel from the source to a reflecting surface and return can be used as a measure of the position of that surface. The actual travel time (which is an indication of the distance from the source to the reflecting surface) may be indicated by the position of a spike on the horizontal axis (i.e., time axis) of a cathode-ray oscilloscope. Vibration of the reflecting surface causes a corresponding variation in the position of this spike. This technique

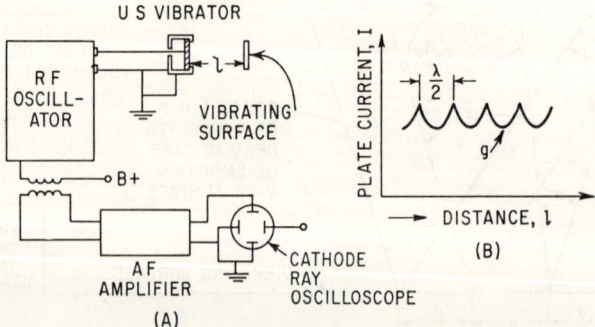

FIG. 14.32. Ultrasonic vibration transducer: (A) circuit diagram; (B) change in plate current of RF oscillator with vibration displacement. (*After S. Satomura et al.*[78])

can be used in vibration measurements having peak displacements as small as ±0.01 in. (0.025 cm).

ULTRASONIC DOPPLER-SHIFT SYSTEM. In this system, the velocity of a vibrating surface is measured by determining the Doppler-frequency shift in an ultrasonic carrier-frequency beam reflected from a vibrating surface.[80, 81] It is a frequency-modulation system, so that its calibration does not depend upon the distance between the ultrasonic pickup transducer and the surface being measured; its output is proportional to the velocity of motion of the surface.

For example, in the design of one system, the ultrasonic source and receiving transducers are barium titanate crystals operating at 110 kc.[80] The ultrasonic beam width is approximately 15°. The unit can be operated at distances from about 2 in. to 6 ft (i.e., from about 5 to 180 cm). The frequency response of the system is flat between 0 and 700 cps. The minimum velocity that can be measured is approximately 0.01 in./sec (0.025 cm/sec).

The reflected-energy system described in Ref. 77 also can be used as a Doppler-shift system since there is the same shift in the frequency of the reflected wave proportional to the normal velocity of the reflecting surface.

GAS-DISCHARGE TRANSDUCER

When a gas-discharge tube containing two electrodes is placed in a radio-frequency electric field, a d-c voltage may be developed between the electrodes. Figure 14.33 shows two possible configurations for gas-discharge tubes. If the electrodes A and B are not symmetrically located with respect to the radio-frequency electric field, then a d-c voltage appears between them. Transducers of this type have been built with displacement sensitivities as high as 0.05 volt/microinch, and in some units d-c output voltages up to 60 volts can be obtained.[82–85]

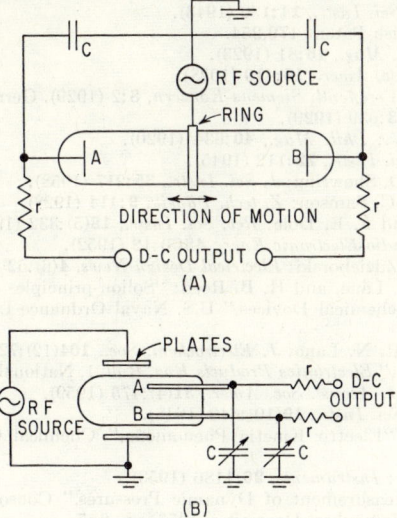

FIG. 14.33. Gas-discharge transducer and basic circuit. (*Courtesy of Decker Aviation Corp.*[88])

SERVO ACCELEROMETER

The displacement or relative motion of a seismic mass, which can be sensed by almost any standard transducer, represents the error in a servo accelerometer. This error signal establishes a restoring force which is applied to the seismic system such that it tends to reduce to zero the initial error signal. In this type of accelerometer, the restoring force is proportional to the acceleration and is used to produce the output signal.

The servo accelerometer does not represent a new transducer principle, but rather employs a standard transducer in a unique circuit. The resulting over-all system can be made very sensitive, accurate, and stable.[86-88] Linearity of better than 0.02 per cent full scale and threshold sensitivities of $10^{-5}g$ have been obtained.[83] However, this type of accelerometer is limited to low frequencies (generally below 100 cps) because of its relatively slow response.

REFERENCES

1. Roberts, H. C.: "Mechanical Measurements by Electrical Methods," The Instruments Publishing Co., Inc., Pittsburgh, Pa., 1951.
2. Attree, V. H.: *Electronic Engr.*, **27**:308 (1955).
3. Shafer, S. N., and R. Plunkett: *Proc. Soc. Exptl. Stress Anal.*, **13** (1):123 (1955).
4. Olken, H.: *Instruments*, **5**:33 (1932).
5. Fielden Proximity Meter, *Bull.* 951 and *Eng. Spec.* 951-1, Fielden Instrument Div., Robertshaw-Fulton Controls Co., Philadelphia, Pa.
6. Cook, G. W.: *Electronics*, **26**:105 (1953).
7. Thompson, N. J., and E. W. Counsin: *Electronics*, **20**:90 (1947).
8. Koidan, W.: *J. Acoust. Soc. Amer.*, **26**:428 (1954).
   Shattuck, R. D.: *J. Acoust. Soc. Amer.*, **31**:1297 (1959).
   Rule, E., F. J. Suellentrop, and T. A. Perls: *J. Acoust. Soc. Amer.*, **33**:33 (1961).
9. Eoley, G. M.: *Trans. ASME*, **67**(1):553 (1945).
10. Arndt, J. P., Jr.: *J. Acoust. Soc. Amer.*, **21**:385 (1949).
11. Potter, E. V.: *Rev. Sci. Instr.*, **14**:130 (1943).
12. Dowling, J. J.: British Patent 179,254.
13. Dowling, J. J.: *Phil. Mag.*, **46**:81 (1923).
14. Obata, J.: *J. Opt. Soc. Amer.*, **16**:419 (1928).
15. Gerdien, H.: *Wiss. Veröffentl. Siemens-Konzern*, **8**:2 (1929), German Patent 436,157.
16. Thoma, H.: *VDI*, **73**:639 (1929).
17. Whiddington, R. W.: *Phil. Mag.*, **40**:634 (1920).
18. Bradshaw, E.: *J. Sci. Instr.*, **22**:112 (1945).
19. Nordbury, J., and D. Shewring: *J. Sci. Instr.*, **35**:217 (1958).
20. Leobe, W. W., and C. Samson: *Z. tech. Physik*, **9**:414 (1928).
21. Nettleton, L. A., and F. E. Dole: *Rev. Sci. Instr.*, **18**(5):332 (1947).
22. Kaufman, A. B.: *Radio-Electronic Engr.*, **48**(3):12 (1952).
23. Post, R. E., and J. Zdzieborski: *Electrical Design News*, **4**(6):52 (1959).
24. Hurd, R. M., R. N. Lane, and H. B. Reed: "Solion-principles of Electro-chemistry and Low-power Electro-chemical Devices," U.S. Naval Ordnance Laboratory, Silver Spring, Md., 1957.
25. Hurd, R. M., and R. N. Lane: *J. Electrochem. Soc.*, **104**(12):727 (1957).
26. "Solion for Industry," *Electronics Products Eng. Bull.* 1, National Carbon Co., Nov., 1957.
27. Wittenborn, A. F.: *J. Acoust. Soc. Amer.*, **31**(4):475 (1959).
28. Williams, M.: *Rev. Sci. Instr.*, **19**(10):640 (1948).
29. Abramson, H. A.: "Electro Kinetic Phenomena," Chemical Catalog Company, Inc., New York, 1934.
30. Hardway, E. V., Jr.: *Instruments*, **26**:1186 (1953).
31. "Electro Kinetic Measurement of Dynamic Pressures," Consolidated Electrodynamics Corp. Recordings, November, December, 1956, pp. 6, 7.
32. Hardway, E. V., Jr.: "Electro-kinetic Measuring Instruments," U.S. Patent 2,661,430, Dec. 1, 1953.
33. Hardway, E. V., Jr.: "Electro-kinetic Transducer," U.S. Patent 2,769,929, Nov. 6, 1956.
34. Hardway, E. V., Jr.: "Electro-kinetic Transducing Devices," U.S. Patent 2,782,394, Feb. 19, 1957.
35. Fain, W. W., S. L. Brown, and A. E. Lockenvitz: *J. Acoust. Soc. Amer.*, **29**:902 (1957).
36. Ueda, Wantanabe, Tsuji, and Nishizaur: *J. Electrochem. Soc. Japan*, **20**:605 (1952); **21**:267 (1953).

37. Yeager, E., and F. Hovorka: *J. Acoust. Soc. Amer.*, **25**:447 (1953).
38. Podalsky, B., G. Kuschevics, and J. L. Revers: *J. Appl. Phys.*, **28**:357 (1957).
39. *Bulls.* 1600 and 1602, Consolidated Electrodynamics Corp.
40. McArthur, E. D.: *Electronics*, **10**:16 (1937).
41. Olson, H. F.: *J. Acoust. Soc. Amer.*, **19**(2):307 (1947).
42. Gunn, R.: *J. Appl. Mechanics*, **62**:A49 (1940); bound with *ASME Transactions*, vol. **62**, 1940.
43. *Tube Manual*, Radio Corp. of America.
44. Lewis, R. C.: *J. Acoust. Soc. Amer.*, **22**:357 (1950).
45. Ramberg, W.: *Elec. Eng.*, **66**(4):402 (1947).
46. Ramberg, W.: *Elec. Eng.*, **66**(6):555 (1947).
47. Ramberg, W.: *J. Research Natl. Bur. Standards*, **33**(6):391 (1946), Research Paper RP 1754.
48. Levy, S.: *Proc. Soc. Exptl. Stress Anal.*, **9**(1):151 (1951).
49. "Optron—Displacement Follower," Optron Corp., Santa Barbara, Calif.
50. Thomas, H. A., and G. W. Warren: *Phil. Mag.* **5**:1125 (1928).
51. Ostenberg, H.: *J. Opt. Soc. Amer.*, **22**:19 (1932).
52. Kennedy, W. J.: *J. Opt. Soc. Amer.*, **31**:99 (1941).
53. Ziegler, C. A.: *J. Acoust. Soc. Amer.*, **25**:135 (1953).
54. Smith, D. H.: *Proc. Phil. Soc. (London)*, **57**:534 (1945).
55. Hunton, R. D., A. Weis, and W. Smith: *J. Opt. Soc. Amer.*, **44**:264 (1954).
56. Edelman, S., E. Jones, and E. R. Smith: *J. Acoust. Soc. Amer.*, **27**(4):732 (1955).
57. Orlacchio, A. W.: *Elect. Mfg.*, **59**(1):78 (1957).
58. Smith, E. R., S. Edelman, V. A. Schmidt, and E. Jones: *J. Acoust. Soc. Amer.*, **30**:867 (1948).
59. Davis, S. M.: *Nature*, **178**:161 (1956).
60. Barber, G. J.: *J. Sci. Instr.*, **32**:7 (1955); *Phys. Ber.*, 9577, December, 1955.
61. Huntoon, R. D., A. Weis, and W. Smith: *J. Opt. Soc. Amer.*, **44**:264 (1954).
62. Feder, E. I., and A. M. Gillen: *IRE Trans. on Instrumentation*, **I-6**(2):98 (1957).
63. Davidenkoff, D. N.: *Proc. ASTM*, **34**:857 (1945, Part 2).
64. Pabst, W., and W. Kroll: U.S. Patent 2,306,137.
65. Jerrett, R. S.: *J. Sci. Instr.*, **22**:29 (1945).
66. *Electronics*, **18**:160 (1945).
67. Allan, W. H.: U.S. Patent 2,725,492; reviewed *J. Acoust. Soc. Amer.*, **28**:744 (1956).
68. Dickson, A. W., and W. P. Murden: *Electronics*, **26**(9):164 (1953).
69. *B. J. Electronics—Tech. Bull.* 58-130, June, 1958.
70. Chapman, J. C.: *The Engineer*, **206**:640 (1958).
71. Morrow, C. T.: *J. Acoust. Soc. Amer.*, **27**:56 (1955).
72. Morrow, C. T.: *J. Acoust. Soc. Amer.*, **27**:62 (1955).
73. Morrow, C. T.: *J. Acoust. Soc. Amer.*, **27**:581 (1955).
74. Post, R. F., and R. L. Howard: *J. Acoust. Soc. Amer.*, **19**(1):283 (1947).
75. Ryan, A. H., and S. D. Summers: *Elect. Engr.*, **73**:251 (1945).
76. Cohen, G. I., and B. Ebstein: *Proc. Natl. Electronics Conf.*, **12**:982 (1956).
77. Pound, R. V., and E. Durand: "Microwave Mixers," MIT Radiation Laboratory Series, vol. 16, p. 259, McGraw-Hill Book Company, Inc., New York, 1948. Also see: Gik, L. D., L. Ya. Mizuk, and K. B. Karandeyev, *Byulleten izobreteniy (USSR)*, **11**:76 (1958).
78. Satomura, S., S. Matsubara, and M. Yoshioka: *Mem. Inst. Sci. Ind. Research, Osaka Univ.*, **13**:125, 1956.
79. Stewart, C.: *J. Acoust. Soc. Amer.*, **30**:644 (1958).
80. Hardy, H. C., H. H. Hall, D. B. Callaway, and D. S. Schorer: *J. Acoust. Soc. Amer.*, **27**:1004 (1955).
81. Hardy, H. C.: U.S. Patent 2,733,597, February, 1956, assigned to Armour Research Foundation.
82. Lion, K. S.: *Rev. Sci. Instr.*, **27**:222 (1956).
83. *Tele-Tech & Electronic Ind.*, **13**(11):67 (1954).
84. *Industrial Laboratories*, **5**(9): (1954).
85. Decker Aviation Corporation, Industrial Data Sheets—103-1, 303-1, 901-1, *Tech. Bull.* 01.
86. Pope, K. E.: *Control Eng.*, **5**(11):97 (1958).
87. Donner Scientific Company Data File No. 410, November, 1957.
88. Kearfott Company, Inc., Technical Data D1754-01 and D1756-01, March, 1959.

15

INDUCTIVE-TYPE PICKUPS

R. R. Bouche
National Bureau of Standards,* Washington, D.C.

INTRODUCTION

An inductive-type shock and vibration pickup is an electromechanical device in which the magnetic characteristics of its electric circuit change in response to the motion of an object. Inductive-type pickups have several distinct advantages. Inherently, they are extremely stable and consequently can be employed in very accurate measurements. They have low electrical impedance so that they usually can be used with very long cables without significantly affecting their output voltages.

The *self-generating type* of inductive pickup generates an electromotive force in the pickup in response to the motion of an object. There are four different self-generating inductive types: electrodynamic, electromagnetic, eddy-current, and magnetostrictive. These pickups operate on the principle that a voltage is generated when a conductor cuts through a magnetic field or when the magnetic field is varied about a stationary conductor. The electrodynamic velocity types have the advantage of producing very large output voltages, so that they can be used without amplifiers. It is usually not practical, however, to use them to measure vibration above several thousand cycles per second. The output voltages of electrodynamic-type pickups are proportional to velocity; their outputs, corresponding to given accelerations, therefore decrease with increasing frequency. For accelerations several times the acceleration of gravity, the outputs usually are very small at frequencies above several thousand cycles per second.

In the *passive type* of inductive pickup, a change in the inductance of the coils in the pickup is caused by the motion of an object. There are three basically different passive inductive types: mutual inductance, differential transformer, and variable reluctance. Their principle of operation is that the inductance of the coils is changed by varying the configuration of the magnetic flux path.

TYPES OF PICKUPS USING INDUCTIVE PRINCIPLES

An inductive-type pickup may be referred to as a displacement, velocity, acceleration, or jerk pickup, depending upon whether its electrical output throughout a specified range of frequencies is proportional to displacement, velocity, acceleration, or jerk. Inductive-type pickups also may be classified as proximity pickups, movable-core pickups, or seismic pickups, as indicated in Table 15.1.

PROXIMITY PICKUPS

No part of a proximity pickup touches the object whose motion it is measuring. The pickup is usually attached to a fixture so that there is a small gap between the pickup

* Now at Endevco Corp.

15-1

Table 15.1. Types of Shock and Vibration Pickups Using Inductive Principles

			Seismic	
	Proximity (see Table 15.2)	Movable-core (see Table 15.3)	Used above natural frequency (see Table 15.4)	Used below natural frequency (see Table 15.5)
Displacement	Mutual inductance	Differential transformer Variable reluctance		
Velocity	Electromagnetic	Electrodynamic	Electrodynamic	
Acceleration			Eddy-current	Differential transformer Variable reluctance
Jerk				Magnetostrictive

and the object. The proximity pickup has the advantage that it does not mechanically load the vibrating object. For this reason, it can be used to measure the vibration of very thin diaphragms, for example. The mutual inductance proximity pickup measures the displacement of the vibrating object. The electromagnetic proximity pickup measures the velocity of the vibrating object. These pickups are described in the sections entitled *Mutual-inductance Pickups* and *Electromagnetic Pickups*.

MOVABLE-CORE PICKUPS

In a movable-core pickup, only the core is mechanically attached to the vibrating object. The movable-core pickup can be used without modification to measure very-low-frequency vibratory motions; it also can be constructed into a seismic pickup (see section on *Seismic Pickups*, below). Movable-core pickups are described in sections on the differential transformer, variable reluctance, and electrodynamic type of pickups, according to the inductive principles which they employ.

SEISMIC PICKUPS

Ideally a seismic pickup consists of a mass element connected to the housing (or case) of the pickup by a massless spring and a massless damping device. Equation (12.8) expresses the displacement amplitude of the mass element relative to the housing of the pickup. As indicated in Fig. 12.4 at high frequencies, the mass element remains fixed in space as the housing of the pickup vibrates. If the output of the pickup is proportional to the displacement amplitude of the mass δ_0, the displacement sensitivity of the pickup is constant over a considerable range of frequencies above the natural frequency of the pickup; the pickup is then a displacement pickup. If the output is proportional to $\delta_0\omega$, it is a velocity pickup because the sensitivity of the pickup is the output divided by the applied velocity $u_0\omega$. Table 15.1 indicates that the electrodynamic principle is employed in the seismic velocity pickup normally used above its natural frequency. Pickups whose outputs are proportional to $\delta_0\omega^2$ and $\delta_0\omega^3$ are acceleration and jerk pickups, respectively; such pickups operate at frequencies *above* their natural frequencies. For example, the eddy-current pickup listed in Table 15.1 is an acceleration pickup of this type.

Many inductive-type pickups are designed to be used at frequencies well *below* their natural frequencies. Thus, below its natural frequency a pickup whose transducing element output is proportional to δ_0 is called an "acceleration pickup" because its acceleration sensitivity (its output divided by the applied acceleration $u_0\omega^2$) is constant over a considerable range of frequencies. Differential-transformer and variable-reluctance principles are used in this type of acceleration pickup.

The so-called "jerk pickup" has an output which is proportional to $\delta_0\omega$; its sensitivity (output divided by $u_0\omega^3$) is constant in the frequency range below its natural frequency. The magnetostrictive pickup, Table 15.1, is a jerk pickup. It is interesting to note that if the electrodynamic velocity seismic pickup (normally used *above* its natural frequency) is used *below* its natural frequency, it acts as a jerk pickup. This is because its sensitivity, output divided by $u_0\omega^3$, is constant at frequencies below about one-third of the natural frequency.

SELF-GENERATING-TYPE PICKUPS

Self-generating inductive pickups make use of either a steady magnetic field or a variable magnetic field. That is, a voltage can be induced in a conductor (1) by changing the position of the conductor in a steady magnetic field or (2) by using a fixed conductor about an electromagnet of variable field strength. The electrodynamic and eddy-current pickups are based on the steady magnetic field principle; the voltage induced in the conductor is proportional to the velocity of the conductor relative to the magnet. Electromagnetic and magnetostrictive pickups generate a voltage proportional to the time rate of change of the magnetic field about the conductor.

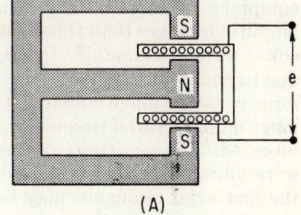

ELECTRODYNAMIC PICKUPS

The output voltage of the electrodynamic pickup is proportional to the relative velocity between the coil and the magnetic flux lines being cut by the coil. For this reason it is commonly called a velocity pickup. The principle of operation of the device is illustrated in Fig. 15.1A. A magnet has an annular gap in which a coil wound on a hollow cylinder of nonmagnetic material moves. Usually a permanent magnet is used, although an electromagnet may be used. The pickup also can be designed with the coil stationary and the magnet movable. The open-circuit voltage e generated in the coil is [1,2]

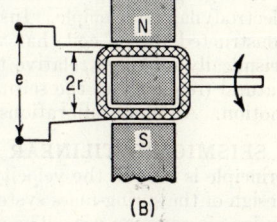

$$e = -Blv(10^{-8}) \quad \text{volts} \quad (15.1)$$

Fig. 15.1. Electrodynamic principle: (A) rectilinear; (B) rotational. The voltage e generated in the coil is proportional to the velocity of the coil relative to the magnet.

where B is the flux density in gausses; l is the total length in centimeters of the conductor in the magnetic field; and v is the relative velocity in centimeters per second between the coil and magnetic field. The magnetic field decreases sharply outside the space between the pole pieces; therefore, the length of coil wire outside the gap generates only a very small portion of the total voltage.

The basis of the rotational electrodynamic principle is illustrated in Fig. 15.1B. A soft-iron core is placed within the coil so that the magnetic field in the gaps remains sufficiently intense. The output voltage e is

$$e = -Blr \frac{d\phi}{dt}(10^{-8}) \quad \text{volts} \tag{15.2}$$

where r is half the distance in centimeters between the lengths of wire in the two gaps, and ϕ is the angle of the rotation in radians of the rectangular-shaped coil about the geometric axis of the iron core.

MOVABLE-CORE ELECTRODYNAMIC PICKUPS. The electrodynamic pickup is the only velocity-type movable-core transducer in use. The pickup consists of a permanent bar magnet that moves in the core of a long coil wound on a form of nonmagnetic material. The self-generated voltage is proportional to the relative velocity between the magnet core and coil. For a constant velocity, a d-c voltage is generated. For vibratory motions, a-c voltage is generated which is an accurate analog of the mechanical motion. The performance [3] characteristics of the pickup are indicated in Table 15.3. It has good stability, and it has a large output which is a linear function of velocity if the magnet is kept well within the coil. Considerable off-centering of the longitudinal axis of the magnet relative to the longitudinal axis of the coil does not significantly affect the sensitivity or linearity nor does the presence of metallic objects outside the coil have any appreciable effect.[3] In fact, it is sometimes desirable to enclose the coil in a soft-iron shield to reduce the stray magnetic field from the magnet that may affect other instruments. The output resistance and number of turns are kept reasonably small so that the output is not excessively reduced by the electrical loading of the input impedance of the voltage-measuring equipment used with the pickup. Pickups which are designed for shorter displacement amplitude ranges than those listed in Table 15.3 have smaller output impedances. If the coil is wound on metallic forms, the sensitivity is reduced somewhat at high frequencies due to eddy-current flow. Some improvement in this respect can be achieved if the coil form contains longitudinal slots. The pickup normally is used below the elastic body longitudinal natural frequency of the bar or coil-form assembly, depending upon which is vibrated. The sensitivity of the pickup would be affected if the bar or coil-form assembly were vibrated at their longitudinal resonances. This is usually not a problem because the first axial resonance may be as high as 1,000 cps or 10,000 cps, depending upon the size of the pickup. Normally transverse resonance will not affect the performance if the transverse motion is small compared to the gap between the magnet and coil form.

A large percentage of all vibration pickup calibrators incorporate the movable-core electrodynamic principle. Instead of using a movable magnet, the calibrator usually is constructed with a coil that vibrates within the annular gap of a permanent magnet, seismically mounted relative to the frame of the calibrator. At frequencies above the natural frequency of the seismic magnet, the coil measures the absolute velocity of the motion. Absolute calibrations [4] of these coils can be made up to 5,000 cps.

SEISMIC RECTILINEAR ELECTRODYNAMIC PICKUPS. The electrodynamic principle is used in the velocity-type seismic pickup. There are many variations in the design of the spring-mass system used in the pickup. Figure 15.2 illustrates three of the designs in common use. The mass element of the pickup in Fig. 15.2A consists of an extremely light coil attached to a support arm which is pivoted on a shaft with bearings. A spiral spring attached to the arm and the housing of the pickup completes the seismic system. The Alnico permanent magnet rigidly attached to the housing of the pickup provides the steady magnetic flux field about the coil. When the housing of the pickup is vibrated above its natural frequency, the support arm rotates about a point near its center-of-gravity. Depending upon the center of rotation of the support arm, the displacement of the coil may be more or less than the displacement applied to the housing. In either case, the output voltage of the coil is proportional to the rectilinear velocity of the case. The permanent magnet of the pickup shown in Fig. 15.2B also is attached to the housing. The coil is suspended by springs from the housing. Above its natural frequency, the coil stands still in space when the housing is vibrated. In another common design, the coil is attached rigidly to the housing, as shown in Fig. 15.2C, and the permanent magnet is mounted seismically through a spring to the housing of the pickup.

Some of the performance characteristics of such pickups are listed in Table 15.4. Usually the pickup is used only at frequencies above its natural frequency. It is not very useful at frequencies above several thousand cycles per second. The sensitivity of most pickups of this type is quite large, particularly at low frequencies where their output voltage is larger than that of many other types of pickups. The coil impedance is quite small even at relatively high frequencies so that the output voltage can be measured directly with a high-impedance voltmeter.

This type of pickup is designed to measure quite large displacement amplitudes. Its use in some applications is limited by the effect of its relatively large weight on the structural response. For a pickup in which the coil is seismically mounted, the mechanical loading on the structure corresponds closely to the total weight of the pickup. When the permanent magnet is the seismic mass element, the effective mass attached to the vibrating structure is less than the total weight of the pickup. The actual effect on the structure is dependent on the mechanical impedance of the pickup. The pickup must be attached quite rigidly to the structure, particularly for measurements at very high frequencies. The contact resonance, i.e., the resonance between the pickup and the structure, should be considerably higher than the highest frequency of motion to be measured. The contact resonance of the Model PR 9260 pickup listed in Table 15.4, for example, is about 3,000 cps when attached to the structure by a screw of approximately 0.4 in. diameter.[5]

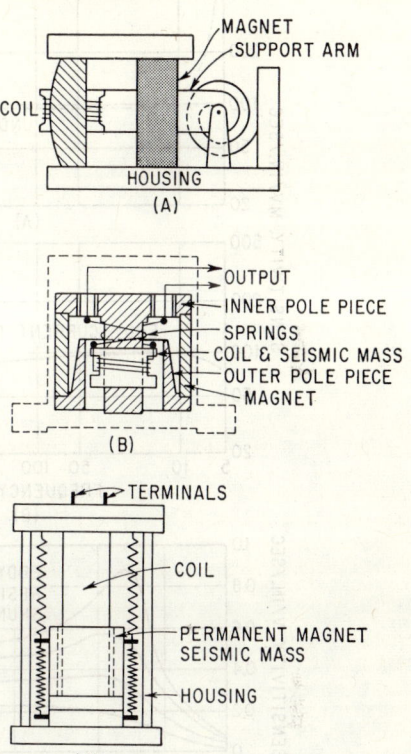

Fig. 15.2. Rectilinear velocity pickups based on the electrodynamic principle. The pickup is used at frequencies above its natural frequency. [(A) *Courtesy of MB Electronics*. (B, C) *Courtesy of Consolidated Electrodynamic Corp.*]

Most of the pickups can be used to measure vibration in either the vertical or horizontal direction. In some cases it is necessary to adjust the springs in the pickup to compensate for the deflection of the mass element due to its own weight when the position of the pickup is changed. Regardless of position, the transverse sensitivity of the pickup is usually near zero. However, for a pickup in which frictional contact can exist on the moving parts inside the pickup, the output may not indicate very accurately the motion along the sensitive axis when the pickup is vibrated in another direction.

The damping in a pickup is a factor in determining the shape of its frequency-response curve. Figure 15.3 shows typical response curves for an undamped pickup and for other pickups having various types of damping. The sensitivity of the undamped pickup, Fig. 15.3A, is constant for frequencies above several times its natural frequency. However, the sensitivity increases to a very large value near its natural frequency. Figure 15.3B illustrates eddy-current damping. For this pickup, the damping is achieved by eddy-current flow in the coil form (see section on *Eddy-current Pickups* below); the amount of damping increases at the higher frequencies, and the sensitivity drops off significantly. Eddy-current damping also can be achieved by placing a relatively small resistive shunt on the coil so that appreciable current flows in the coil. The effects of shunts of different resistance are shown in Fig. 15.3C. If the resistance of the shunt is reduced, the damping is increased; also, the sensitivity drops off at high frequencies

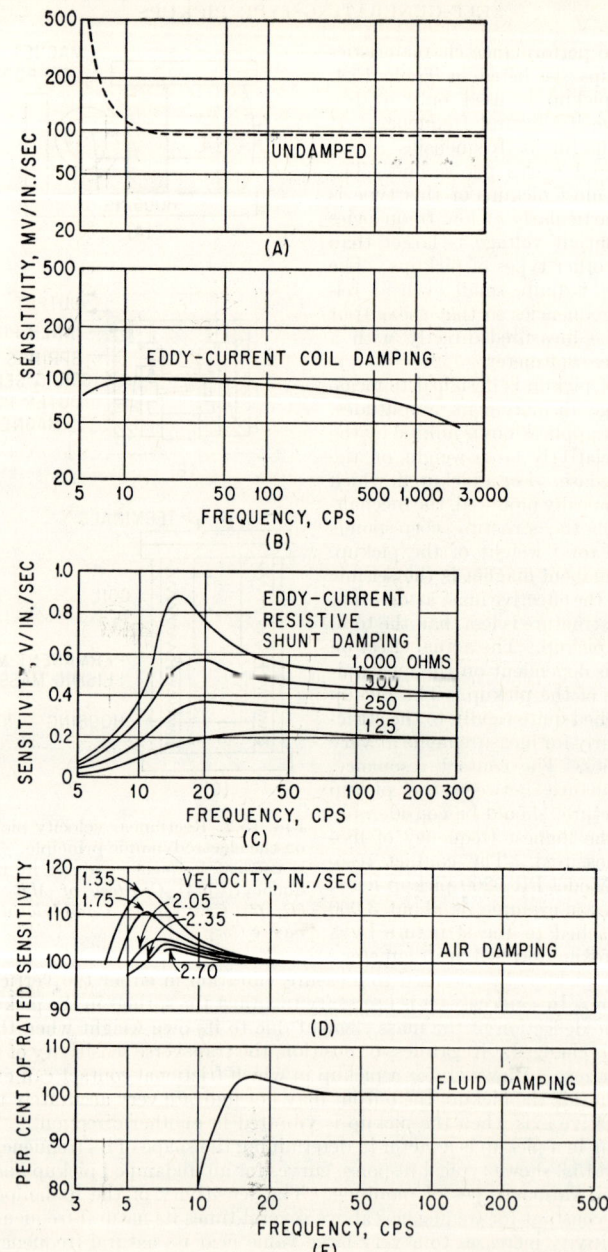

Fig. 15.3. Typical response curves of rectilinear velocity pickups with various types of damping. (A) MB Model 125; (B) MB Model 124. (*Courtesy of MB Electronics.*) (C) Electro-Tech. Model EVS-2. (*Courtesy of Electro-Technical Labs.*) (D) Sperry Model 1777291. (*Courtesy of Sperry Gyroscope Co.*) (E) I. R. D. Model 545. (*Courtesy of International Research and Development Corp.*) The sensitivity of the undamped pickup is constant at frequencies above the natural frequency. The undamped pickup is used in applications where the motion does not excite the natural frequency of the pickup. The damped pickups are used in rugged laboratory tests.

SELF-GENERATING-TYPE PICKUPS

since the impedance of the coil is increased and the effect of the resistive shunt is greater.[6] The response of a pickup with air damping is shown in Fig. 15.3D. The damping is not proportional to the velocity. Over an appreciable range of velocities in the operating range of the pickup, the sensitivity of the pickup is nearly constant. A typical response characteristic of a fluid-damped pickup is shown in Fig. 15.3E. The sensitivity does not change greatly at high frequencies. However, fluid damping has the disadvantage

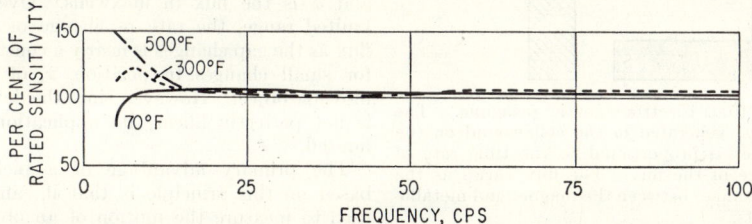

Fig. 15.4. Effect of temperature on response of a rectilinear velocity pickup. (*Courtesy of MB Electronics.*)

that it changes considerably with temperature.[7] The effect of temperature on a pickup having eddy-current damping is shown in Fig. 15.4. There is little change in sensitivity within the operating frequency range for temperatures up to 500°F (260°C). Another eddy-current damped pickup [5] experiences a 1 per cent change in the sensitivity for a temperature change of approximately 18°F (10°C).

The variation of phase lag of the output voltage relative to the applied velocity is shown in Fig. 15.5 for the two pickups whose sensitivities are shown in Fig. 15.3A and Fig. 15.3B. The phase lag is constant for the undamped pickup and varies linearly with frequency for the damped pickup at frequencies well above the natural frequency of the pickup.

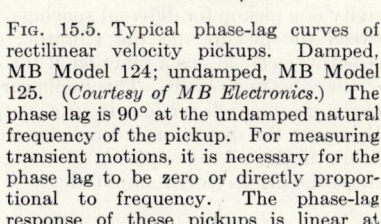

Fig. 15.5. Typical phase-lag curves of rectilinear velocity pickups. Damped, MB Model 124; undamped, MB Model 125. (*Courtesy of MB Electronics.*) The phase lag is 90° at the undamped natural frequency of the pickup. For measuring transient motions, it is necessary for the phase lag to be zero or directly proportional to frequency. The phase-lag response of these pickups is linear at frequencies significantly above the natural frequency.

SEISMIC ANGULAR ELECTRODYNAMIC PICKUPS. The construction of the electrodynamic angular vibration pickup which is listed in Table 15.4 is basically that shown in Fig. 15.1B. A permanent magnet is seismically mounted to the housing and the coil is rigidly attached to the housing. The pickup is used in torsional-vibration applications, e.g., on the arbor of a milling machine. At frequencies above its natural frequency, the voltage output of the pickup is proportional to the angular velocity of torsional oscillations superimposed on the rotational speed of the shaft to which the pickup is attached. The weight of the pickup, 7 lb (3.2 kg), limits its use to vibration testing of large machinery.

ELECTROMAGNETIC PICKUPS

The operation of electromagnetic pickups is based on the principle that a voltage is induced in a conductor when the magnetic flux about it is varied. This is illustrated in Fig. 15.6. When a metallic object is moved relative to the permanent magnet, the magnetic flux density changes and a voltage is induced in the coil. The voltage induced e is proportional to the time rate of change of the flux, and is proportional to the

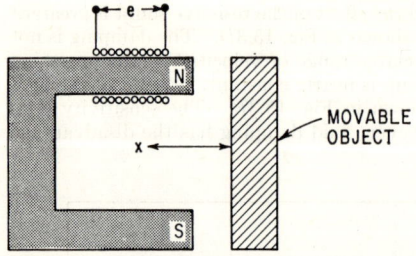

Fig. 15.6. Electromagnetic principle. The voltage generated in the coil wound on the magnet is proportional to the time rate of change of the flux. The flux varies as the gap changes between the magnet and metallic objects.

velocity $\dot{x}$ of the object, in centimeters per second, i.e.,

$$e = -n \frac{d\phi}{dx} \dot{x}(10^{-8}) \qquad \text{volts} \qquad (15.3)$$

where n is the number of turns in the coil and ϕ is the flux in maxwells. Over a limited range, the rate of change of the flux as the gap changes is nearly a constant for small changes in position, x, of the movable object. However, since the device is not perfectly linear, its application is limited.

The primary advantage of a pickup based on this principle is that it can be used to measure the motion of an object without touching it. By placing the electromagnet in the proximity of the object, a voltage is induced when the gap is changed. If the object is fabricated of magnetic material, the reluctance of the flux path is increased when the gap is decreased; the flux increases and thus generates a voltage in the coil. If the object is of nonmagnetic material, currents are set up in the object which oppose the magnetic field, thereby reducing the flux and inducing a voltage in the coil. Actually both effects may be present; if so, they tend to cancel each other. For best results, the material should either have a high permeability or be very nearly nonmagnetic.

The last two pickups listed in Table 15.2 are typical proximity pickups of the electromagnetic type. They are used for a variety of applications; for example, in speed measurements they are placed about 0.005 in. from a rotating object such as a rotating toothed gear. The output voltage increases as the speed of the gear increases. In vibration measurements, the voltage generated by the pickup is nearly proportional to the velocity of a metallic object in its proximity. Figure 15.7 shows the variation in sensitivity of a pickup for different spacings of the gap between the pickup and the vibrating object. The sensitivity changes as a function of the displacement of the vibrating object. This nonlinear behavior introduces some distortion in the output voltage. However, the distortion is small for small displacements if the average spacing of the gap is several times the displacement amplitude.

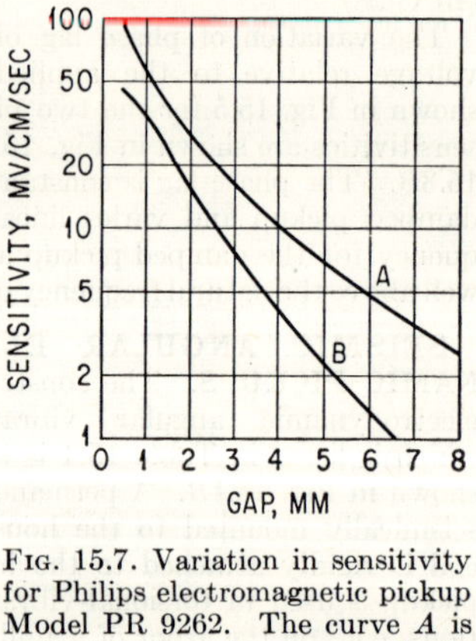

Fig. 15.7. Variation in sensitivity for Philips electromagnetic pickup Model PR 9262. The curve A is plotted for large ferromagnetic objects; curve B for nonmagnetic objects to which silicon-iron discs have been attached. (*After Anon.*[5])

EDDY-CURRENT PICKUPS

Consider a nonferrous plate of low resistivity which is moved in a direction perpendicular to the flux lines of a magnet, as indicated in Fig. 15.8. A current is generated in the plate which is proportional to the velocity of the plate; these eddy currents set up a magnetic field in a direction opposing the magnetic field that creates them. Thus, the resulting magnetic field varies as the generated current changes. The output voltage e of a coil of wire wound on the magnet is proportional to the time rate of change of the eddy currents, and therefore to the acceleration of the motion of the plate relative to the flux field. The above principle forms the basis of the eddy-current pickup. Such a device has an advantage over an electromagnetic pickup in that the voltage is linearly related to the motion since the gap between the plate and electromagnet remains constant.

SELF-GENERATING-TYPE PICKUPS

Table 15.2. Typical Characteristics * of Proximity Pickups

Manufacturer	Model	Principle	Frequency range, cps	Amplitude limit, in.	Temperature range, °F	Output coil impedance, ohms
Tel-Instrument Electronics Corp.	501	Mutual inductance	10–20,000	0.020		
Electro Products Laboratories	3055	Electromagnetic	. . . .	. . . . .	−400 to 500	90
Philips (Holland)	PR 9262	Electromagnetic	0–2,000	0.3	. . . .	1,260

* Taken from manufacturers' literature which gives detailed specifications and information on additional models available.

A significant feature of a pickup based on the eddy-current principle is that the force produced by the flux field from the eddy currents opposes the motion of the plate and is proportional to its velocity. Thus mechanical damping is achieved. Such a pickup has inherent damping in the same plate that is used to cause a voltage to be generated in the coil.

In the pickup which is based on the eddy-current principle listed in Table 15.4, a permanent magnet and coil are rigidly attached to the housing of the pickup. A rectangular plate is seismically mounted on the housing. The eddy currents generated in the plate induce a voltage in the coil and at the same time provide damping in the pickup. Above the natural frequency of the pickup, the generated voltage is proportional to the acceleration applied to the pickup. The frequency response of the pickup is given in Fig. 15.9. This type

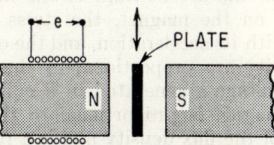

Fig. 15.8. Eddy-current principle. Current generated in the plate is proportional to the velocity of the plate as it cuts the flux field in the gap of the magnet. The current reduces the flux field, and the voltage e generated in the coil is proportional to the acceleration of the plate.

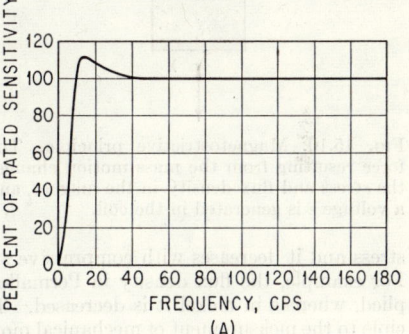

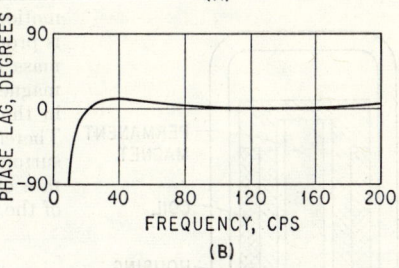

Fig. 15.9. Response curves of a prototype model of an acceleration pickup based on the eddy-current principle; (A) sensitivity and (B) phase angle. The pickup is used above its natural frequency. The undamped natural frequency of the pickup is 11 cps. (*Courtesy of The Calidyne Company.*)

of response is nearly ideal for measuring transient motion. The phase lag between the output voltage and acceleration applied to the pickup is close to 0° at frequencies above its natural frequency, and the percentage increase in the sensitivity is very small near the natural frequency of the pickup. These characteristics reduce the distortion in the output voltage of the pickup.

MAGNETOSTRICTIVE PICKUPS

The magnetic flux density in a ferromagnetic material changes when its length is altered due to the application of stress. This characteristic is used in magnetostrictive pickups. The flux density in some materials increases with the application of tensile

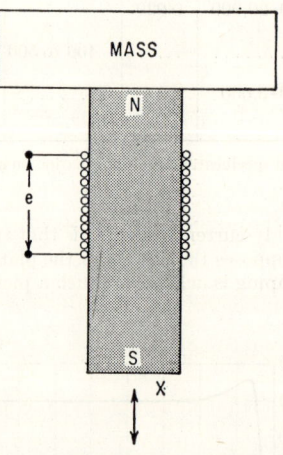

FIG. 15.10. Magnetostrictive principle. The force resulting from the mass motion changes the stress and flux density in the magnet, and a voltage e is generated in the coil.

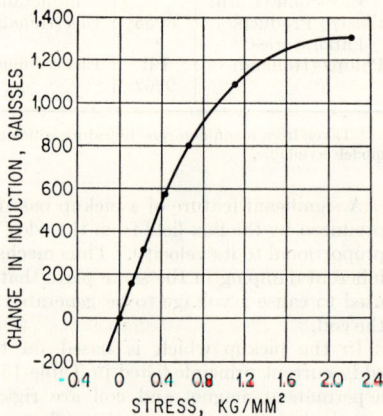

FIG. 15.11. Magnetostrictive effect in 45 Permalloy. The change in inductive flux density is linear only over a limited stress range. (*After R. M. Bozarth.*[8])

stress and it decreases with compressive stress; the opposite is true in other materials. For example, the flux density in Permalloy [8] is increased when tensile stresses are applied, whereas in nickel it is decreased. The application of the magnetostrictive principle to the measurement of mechanical motion is illustrated in Fig. 15.10. If mechanical motion is applied to the bar magnet, a force which is proportional to the acceleration of the attached mass is exerted on the magnet, the stress in the magnet varies with its acceleration, and the changes in the flux density are proportional to the stress. Therefore the voltage e generated in a coil which surrounds the magnet is proportional to the time rate of change of the flux density and is a function of the time rate of change of the acceleration:

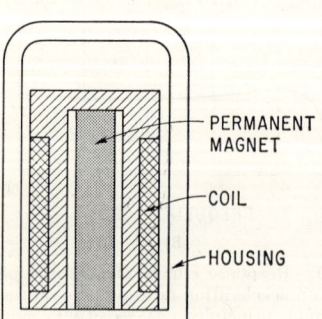

FIG. 15.12. Jerk pickup based on magnetostrictive principle. The voltage output of the coil is proportional to the time rate of change of the acceleration applied to the housing of the pickup. (*Courtesy of Sperry Gyroscope Company.*)

$$e = CG(\sigma)\frac{d^3x}{dt^3} \quad \text{volts} \qquad (15.4)$$

where C is a constant and $G(\sigma)$ is the slope of the flux density vs. applied stress. The slope $G(\sigma)$ is constant only over a limited range of stress, as indicated in Fig. 15.11. In the range of constant slope, the generated voltage is proportional to the time rate of change of the acceleration. Therefore magnetostrictive pickups are linear over only a small range of stresses in the magnet.

A cross-sectional view of a magnetostrictive jerk pickup is shown in Fig. 15.12. These pickups are used primarily for measuring the vibration of internal-combustion-engine cylinder heads in evaluating engine knock, valve timing, ignition, and mechanical malfunction, etc. The sensitivity of one of these pickups is approximately 51×10^{-9} mv/in./sec^3 (see Table 15.5). A half-sine-wave shock pulse of $100g$ with a time duration of 1 millisecond produces an output voltage of about 6 millivolts. Another magnetostrictive jerk pickup [9] is designed for the measurement of acceleration pulses of 1 millisecond duration up to more than $1,000g$. However, because of its low sensitivity, frequency components at low accelerations and frequencies are difficult to measure.

PASSIVE INDUCTIVE-TYPE PICKUPS

There are three types of passive inductive pickups: mutual-inductance, differential-transformer, and variable-reluctance. In these devices the inductance changes as a component of the pickup changes its position. This results in a corresponding change in output voltage. Pickups of this type can be constructed which have a usable response down to zero frequency.

MUTUAL-INDUCTANCE PICKUPS

The operation of a pickup based on the mutual-inductance principle is illustrated in Fig. 15.13. Two coils are wound on a form which is fabricated of nonmetallic material. The primary coil is energized by an alternating current i_p. The magnetic field produced by the primary coil sets up a field which is opposed by that produced by eddy-current flow in the metal surface. The resulting magnetic field induces a voltage in the secondary coil. As the metal surface moves closer to the secondary coil, the opposing eddy-current field increases, and the output of the secondary coil is reduced. The voltage e in the secondary coil is

$$e = M\omega i_p \qquad (15.5)$$

where M is the mutual inductance in henries, ω the angular frequency in radians per second, and i_p the alternating current in the primary coil in amperes. Figure 15.14 shows the variation in mutual inductance for a nonmagnetic metal surface. The mutual inductance can be made nearly linear over a surface displacement range up to approximately 5 per cent of the coil diameter. For a metal surface of ferromagnetic material, the intercept of the curve is greater and its slope is less because the high flux in the surface creates a field that opposes the field due to the eddy currents. Therefore, a nonmagnetic vibrating surface induces a larger output in the secondary coil per unit of displacement of the surface than does a surface of magnetic material.

Under static conditions, the output of the secondary coil is a voltage having the frequency of the primary current. If the surface vibrates, this voltage is frequency-modulated at a frequency corresponding to the motion of the metal surface, and the output voltage varies in proportion to the surface-to-coil spacing. The distinct advantage of a pickup of this type is that it can measure motions from d-c to very high frequencies without touching the vibrating object.

The properties of a typical mutual-inductance

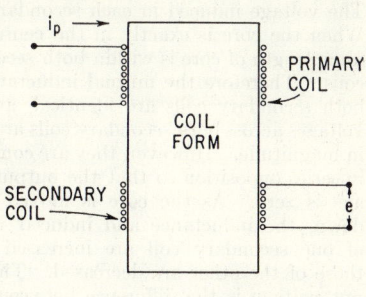

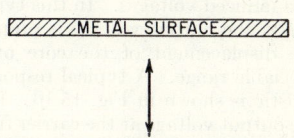

Fig. 15.13. Mutual-inductance principle. The mutual inductance of the coils changes as the gap between the coil form and metal surface is varied. The output voltage e of the secondary coil is proportional to the mutual inductance when the input current i_p is maintained constant. (*After M. L. Greenough.*[10])

INDUCTIVE-TYPE PICKUPS

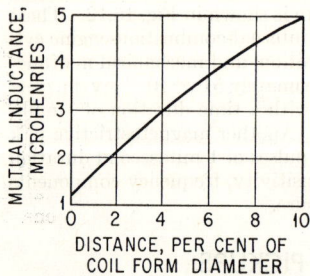

Fig. 15.14. Typical variation of mutual inductance for the displacement-measuring device. The variation is nearly linear for changes in the distance between the coil form and metal surface up to 5 per cent of the diameter of the coil form.

pickup are listed in Table 15.2. This pickup is used to measure the vibration of nonmagnetic metallic objects by bringing the two-coil probe to within 0.03 in. of the object; for other materials it is recommended that a disc at least 1 in. in diameter and 0.01 in. thick be attached to the vibrating object.[11,12] The output of the pickup is sufficiently linear so that it is practicable to measure vibration up to 0.020 in. amplitude. This limits its practical usefulness at very low frequencies where large displacements may occur. Electronic instrumentation is available for the pickup which includes a regulated 2.5-megacycle oscillator for the primary coil and a voltmeter to read the output. The variation in mutual inductance can also be used to vary the frequency of an oscillator.[13]

This type of pickup can be used to measure the extraneous motion of linear vibration exciters.[14] It also can be used for many other vibration-measuring applications, particularly when it is undesirable to load mechanically the vibrating object with a seismic-type vibration pickup. Since it is a zero-frequency displacement-type pickup, it may be calibrated statically with a dial micrometer or other accurate static displacement device, although it is desirable to calibrate it throughout the frequency range of intended use.

DIFFERENTIAL-TRANSFORMER PICKUPS

The operation of a pickup of the differential-transformer type is similar to one of the mutual-inductance type in that its output depends upon the mutual inductance between a primary and secondary coil. The basic components are shown in Fig. 15.15. The pickup consists of a core of magnetic material, a primary coil, and two secondary coils. The voltage induced in each secondary coil is equivalent to that given by Eq. (15.5). When the core is exactly in the center, the same length of core is within both secondary coils. Therefore the mutual inductances of both secondary coils are identical, and the voltages across both secondary coils are equal in magnitude. However, they are connected in series opposition so that the output voltage is zero. As the core is moved up or down, the inductance and induced voltage of one secondary coil are increased while those of the other are decreased. The output voltage is the difference between these two induced voltages. In this type of transducer, the output voltage is proportional to the displacement of the core over an appreciable range. A typical response characteristic is shown in Fig. 15.16. In practice, the output voltage at the carrier frequency of the primary current is not exactly zero when the core is centered, and the output near the center position is not exactly linear. When the core is vibrated, the output voltage is a carrier wave, modulated at a frequency and amplitude corresponding to the motion of the core relative to the coils.

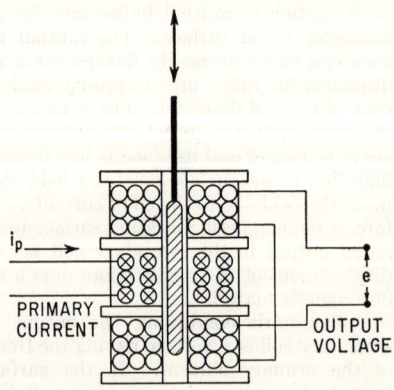

Fig. 15.15. Differential-transformer principle. The inductance of the coils changes as the core is moved. For constant input current i_p to the primary coil, the output voltage e is the difference of the voltages in the two secondary coils which are wound in series opposition. (*Courtesy of Automatic Timing and Controls, Inc.*)

PASSIVE INDUCTIVE-TYPE PICKUPS

There are a considerable number of ways [15, 16, 17] that these cores and coils can be arranged. For example, the core may be attached to one object and the coils to another, i.e., as a movable-core pickup. It also is common to use the core as the mass element of a seismic acceleration pickup operated below its natural frequency.

MOVABLE-CORE DIFFERENTIAL-TRANSFORMER PICKUPS. Table 15.3 lists typical movable-core differential-transformer pickups and some of their important characteristics. These pickups are used for very-low-frequency measurements. The sensitivities listed in Table 15.3 give an approximate indication of the output voltages obtainable. Actually the sensitivity

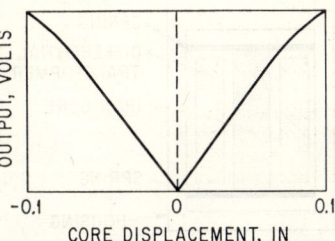

Fig. 15.16. Variation in output voltage with core displacement for a differential-transformer pickup. (*After H. Schaevitz.*[17])

varies with the carrier frequency of the current in the primary coil. A typical plot of the sensitivity vs. carrier frequency is given in Fig. 15.17. The carrier frequency should be at least ten times the highest frequency of the mechanical motion to be measured in order that the output voltage, when demodulated, be an accurate representation of the mechanical motion. Therefore the carrier frequency usually is above 500 cps. The last column in Table 15.3 gives the d-c resistance of the secondary coils. At high carrier frequencies, the reactive component of the coil impedances becomes large, and the output impedance of the coil is significantly greater than its d-c resistance. The load impedance of the voltage-measuring instrument should be very large compared to the output impedance of the differential transformer in order to avoid excessive reduction in the output voltage.

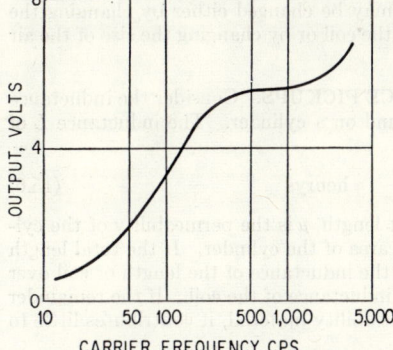

Fig. 15.17. Variation in output voltage with carrier frequency of the input current for a differential-transformer pickup. Usually a carrier frequency of several hundred cycles per second is used. (*After H. Schaevitz.*[17])

Note that if the primary coil is energized with direct current, the differential-transformer type of pickup becomes an electrodynamic pickup. However, few pickups are used this way since the sensitivity of the resulting electrodynamic pickup usually would be less than that of a permanent-magnet electrodynamic pickup.

RECTILINEAR SEISMIC DIFFERENTIAL-TRANSFORMER PICKUPS. A typical design of a differential-transformer seismic pickup is shown in Fig. 15.18. The flat diaphragm spring permits the mass-element iron core to move axially in the differential transformer. A pickup of this type is used in the frequency range from 0 cps up to about two-thirds of its natural frequency. Table 15.5 lists typical acceleration ranges. Pickups with the higher acceleration ranges have larger usable frequency ranges; however, their sensitivities are correspondingly smaller.

Eddy-current damping is used frequently in pickups with small frequency ranges. The effect of temperature on these pickups is small. Fluid damping is used frequently in pickups with large frequency ranges, but is significantly affected by temperature. If the pickup were used only for sinusoidal motion measurements at frequencies up to one-third of its undamped natural frequency, it would be preferable to use no damping. However, in general-purpose testing and in measuring transient motions when significant components of the motion are present at frequencies near the natural frequency, damping is necessary in order to avoid excessive voltage outputs and to avoid contact between the mass element and housing of the pickup as a result of excessive displacements. For ac-

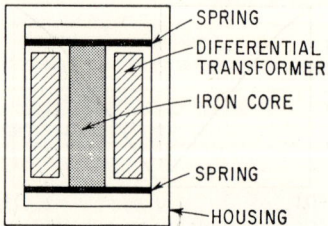

Fig. 15.18. Typical differential-transformer acceleration pickup used below its natural frequency. The output voltage of the differential transformer is proportional to the acceleration of the housing in the operating frequency range of the pickup. (*Courtesy of Gulton Industries, Inc.*)

curate measurements of transient motion, it is also necessary for the phase angle of the output voltage and input motion to be zero or to be directly proportional to frequency. According to the curves of Fig. 12.26 the phase response of such a device should be nearly proportional to frequency within its operating range.

ANGULAR SEISMIC DIFFERENTIAL-TRANSFORMER PICKUPS. A differential-transformer seismic angular-acceleration pickup (used below its natural frequency) is listed in Table 15.5. Other models with natural frequencies as high as 100 cps are available. The performance characteristics are similar to those of the rectilinear pickup. A differential-transformer seismic angular-displacement pickup (used above its natural frequency) has been built [18] which has a natural frequency of 2.5 cps and constant sensitivity to 100 cps.

VARIABLE-RELUCTANCE PICKUPS

The operation of a variable-reluctance type of pickup is based on the variation of inductance that is produced in a coil when the reluctance of the magnetic flux path about the coil is changed. The inductance of a coil may be changed either by changing the length of a magnetic core which is inserted into the coil or by changing the size of the air gap of an electromagnet.

MOVABLE-CORE VARIABLE-RELUCTANCE PICKUPS. Consider the inductance of a simple coil of small diameter which is wound on a cylinder. The inductance L of such a coil is, approximately,

$$L = 4\pi n^2 \mu S l (10^{-9}) \quad \text{henrys} \quad (15.6)$$

where n is the number of turns of wire per unit length, μ is the permeability of the cylinder, l is the length of the cylinder, and S is the area of the cylinder. If the total length of the coil is nearly filled with a magnetic core, the inductance of the length of coil over the core contributes the major part of the total inductance of the coil. If the remainder of the coil is filled with air or a core of low-permeability material, it contributes little to

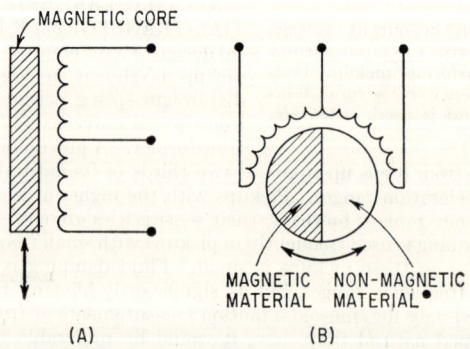

Fig. 15.19. Variable-reluctance principle, solenoid type. The inductance of the coils changes as the reluctance of the flux path is changed due to motion of the core. The core is made of magnetic material. The coils are usually arranged as part of a Wheatstone bridge circuit. Design (*A*) is for measuring rectilinear motion and (*B*) for rotational motion. In the latter, only half of the cylinder is made of magnetic material.

PASSIVE INDUCTIVE-TYPE PICKUPS

the total inductance of the coil. Therefore, for small displacements of the core, the inductance of the coil is nearly proportional to the length of core within the coil. The coil usually is center-tapped as shown in Fig. 15.19A. When the core is moved up or down, the inductance of half the coil is increased while that of the other half is decreased. The two halves of the coil are arranged as two arms of a Wheatstone bridge and the output is linear [19] over an appreciable range. Because the use of this type of pickup is limited to the very low-frequency range, input carrier frequencies as low as 60 cps are used, although the recommended carrier frequency for most models is between 600 and 50,000 cps. The effects of the carrier frequency on the performance of the pickup are similar to those given for differential transformers. This type of pickup has a very low output impedance.

Table 15.3 lists some commercial pickups of this type. Other models are available for the measurement of both rectilinear and angular motion. In such commercial units, the core which moves within the coil usually is made of high-permeability stainless-steel rod. A nonmagnetic extension is attached to the rod so that the core can be mechanically attached to an object. The output from the bridge is proportional to the relative displacement between the core and coil. One of the designs [19] of this type of pickup (in which solid ceramic potting is used) may be operated at temperatures as high as 1300°F (704°C).

Angular displacement can be measured by using a cylindrical core with only half of its cross section made of magnetic material as indicated by the shaded area in Fig. 15.19B. The core is normal to the axis of the coil. Its principle of operation is similar to the rectilinear coil in that the inductances of the two halves of the coil depend upon the amount of magnetic material in their flux paths.

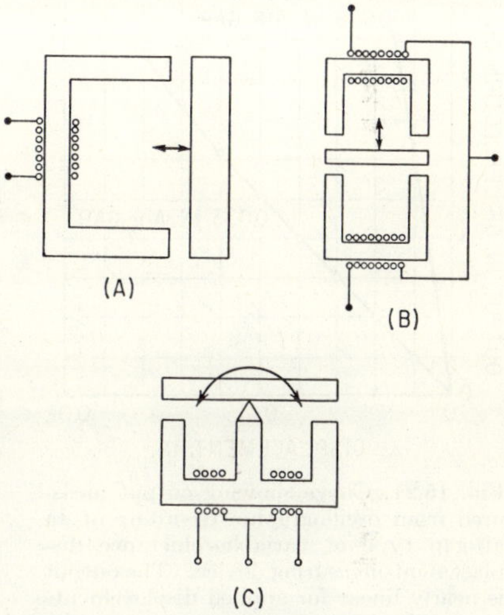

SEISMIC VARIABLE-RELUCTANCE PICKUPS.

Fig. 15.20. Variable-reluctance principle, air-gap type. The inductance of the coils is changed by varying the air gap between the movable and stationary parts of the magnets. Designs including (A) a single magnet, (B) a double magnet, and (C) an E-core magnet are used. The coils are usually arranged as part of a Wheatstone bridge circuit.

The variable-reluctance principle also can be applied in pickups by varying the thickness of a small air gap in the magnetic flux path of an electromagnet. In the pickups illustrated in Fig. 15.20A, B, and C, the air gap is very small compared to the total length of magnetic core material through which the intense flux field passes. For such an arrangement the inductance of the coil L is approximately

$$L = \frac{4\pi n^2 S(10^{-9})}{l} \quad \text{henrys} \quad (15.7)$$

where S is the area of the air gap perpendicular to the direction of flux flow, l is its thickness, and n is the total number of coil turns. For small air gaps, the inductance of the coil is proportional to the change in air-gap thickness over an appreciable range. Usually one movable armature is used between two electromagnets, each of which has coils wound on it as shown in Fig. 15.20B, or an armature is pivoted on an E-core magnet as shown in Fig. 15.20C. These two coils form two arms of a Wheatstone bridge. Figure 15.21 illustrates the linear range of such an arrangement of coils.

The armature of the air-gap variable-reluctance pickup is usually made the mass element of a seismic system. Variable-reluctance pickups of the rectilinear type are in most general use, although angular-acceleration pickups have been built.[21] The characteristics of several acceleration pickups of the rectilinear type are listed in Table 15.5. A diagram of a typical design is shown in Fig. 15.22. The mass element and the magnetic armature lever rotate and change the gap dimensions when the pickup is subjected to vibration in the direction of the arrows. Single-coil variable-reluctance pickups with the magnetic armature suspended by diaphragm springs, e.g., the Model 20J22 pickups in Table 15.5, and other designs are used.[22] This type of variable-reluctance pickup is used at fre-

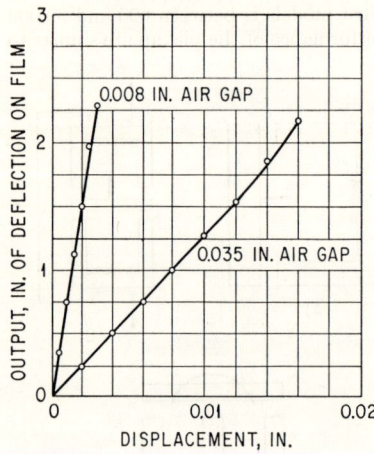

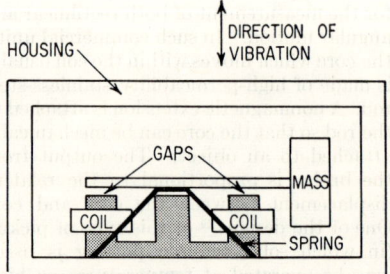

FIG. 15.21. Curve showing output measured from oscillographic recording of an air-gap type of variable-reluctance displacement-measuring device. The output is nearly linear for applied displacements up to about one-fourth of the air gap. (*After B. F. Langer.*[20])

FIG. 15.22. Typical variable-reluctance acceleration pickup used below its natural frequency. The inductance of the coils changes as the gaps vary when the pickup is vibrated. (*Courtesy of Wiancko Engineering Company.*)

quencies up to about two-thirds of its natural frequency. In addition to those listed in Table 15.5, pickups with acceleration ranges as small as ±0.5g and as large as ±1,000g are available. A carrier frequency of several thousand cycles per second is fed into the

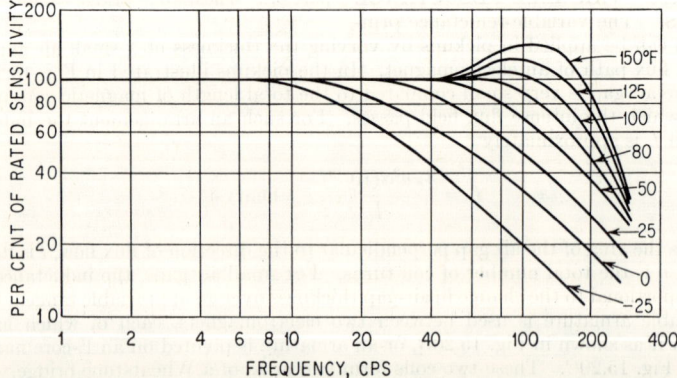

FIG. 15.23. Effect of temperature on response of CEC Model 4-260 variable-reluctance acceleration pickup. The operating frequency range within which the sensitivity is constant is reduced at low temperatures by an increase in the damping. (*Courtesy of Consolidated Electrodynamics Corp.*)

input when the coils are arranged in a Wheatstone bridge. The output voltage is amplitude-modulated at the frequency of the vibratory motion.

Both eddy-current and fluid damping are used. The effect of temperature on a pickup with fluid damping is indicated in Fig. 15.23. Near room temperature, the damping is about 0.6 of critical damping and the frequency range of flat response is large compared to that which would exist were no damping present. At very high and low temperatures, the damping is changed so that the usable frequency range is much less than that obtained at room temperature. The considerations regarding damping and phase angle discussed in the section *Rectilinear Seismic Differential-transformer Pickups* also apply to this type of pickup.

REFERENCES

1. Mason, W. P.: "Electromechanical Transducers and Wave Filters," p. 187, D. Van Nostrand Company, Inc., Princeton, N.J., 1948.
2. Nelson, H. M.: "Moving Coil Vibration Generators and Pickups," *Instruments Note* 54, Aeronautical Research Laboratories, Australia, 1954.
3. Perls, T. A., and E. Buchmann: *Rev. Sci. Instr.*, **22**:475 (1951).
4. Levy, S., and R. R. Bouche: *J. Research Natl. Bur. Standards*, **57**:227 (1956).
5. Anon.: "Philips Vibration Measuring and Exciting Instruments," Philips, Eindhoven, Holland.
6. Leslie, C. B., J. M. Kendall, and J. L. Jones: *J. Acoust. Soc. Amer.*, **28**:711 (1956).
7. White, G. E.: *ASME Symposium on Shock and Vibration Instr.*, vol. 10, 1952.
8. Bozarth, R. M., and H. J. Williams: *Revs. Modern Phys.*, **17**(1):72 (1945).
9. Wilde, H., and E. Eisele: *Z. angew. Phys.*, **1**(8):359 (1949).
10. Greenough, M. L.: *Trans. Am. Inst. Elect. Engineers*, **67**(I):589 (1948).
11. Yates, W., and M. Davidson: *Electronics*, **26**(9):183 (1953).
12. Schacher, D. L.: *Instr. and Automation*, **30**:470 (1957).
13. Clark, H. F.: *Trans. AIEE*, **74**(I): 186 (1955).
14. Elliott, W. R.: *Proc. Instr. Soc. Amer.*, **10**(55-21-1) (1955).
15. Boggis, A. G.: *Proc. Soc. Exptl. Stress Analysis*, **9**(2):171 (1952).
16. MacGeorge, W. D.: *Instruments*, **23**(6):610 (1950).
17. Schaevitz, H.: *Proc. Soc. Exptl. Stress Analysis*, **4**(2):79 (1947).
18. DeMichele, D. J.: *Proc. Instr. Soc. Amer.*, **10**(55-12-1) (1955).
19. Sawyer, E. V.: *Proc. Instr. Soc. Amer.*, **10**(55-7-2) (1955).
20. Langer, B. F.: *Rev. Sci. Instr.*, **2**:336 (1931).
21. Wiancko, T. H.: U.S. Patent 2,759,157, 1956.
22. Dranetz, A. I.: *Machine Design*, **30**(1):120 (1958).

Table 15.3. Typical Characteristics * of Movable-core Pickups

Manufacturer	Model	Principle	Amplitude limit, in.	Sensitivity, mv/volt/in.	Temperature range, °F	Output coil impedance, ohms
Automatic Timing Controls, Inc.	6205	Differential transformer	2.5	5	500	
	6207	Differential transformer	0.01	55	500	
Crescent Engineering Research Co.	MA-4A-0.3	Differential transformer	0.15	6,000 †		6,000
	MB-4C-0.3	Differential transformer	0.15	2,000 †	−160 to 300	600
International Resistance Co.	70-3105	Differential transformer	1	1,000	−65 to 225	
	70-3151	Differential transformer	0.005	700	−65 to 225	
Minneapolis Honeywell	MOD2	Differential transformer	0.1	1,740	−65 to 250	
	1C-020C ‡	Differential transformer	70 ‡	3 ‡	−65 to 250	
Philips (Holland)	PR 9310	Differential transformer	0.039	3,180	158	
Sanborn Company	576DT-1000	Differential transformer	1	700 †	−50 to 205	490
	576DT-050	Differential transformer	0.050	700 †	−50 to 205	2,130
Schaevitz Engineering	1000S-L	Differential transformer	1	1,530 †		2,100
	005M-L	Differential transformer	0.005	4,530 †		85
Crescent Engineering Research Co.	HHC-16.0	Reluctance	4	50 §§	−160 to 1,300	400 §§
	LC-2-0.1	Reluctance	0.015	1,500 §§	−160 to 300	60 §§
	VCH-16 ¶	Electrodynamic	6	50 ¶	−60 to 300	3,000
Sanborn Company	7LV9 ¶	Electrodynamic	4.5	350 † ¶	−50 to 200	17,000

* Taken from manufacturers' literature which gives detailed specifications and information on additional models available.
† Open circuit.
‡ Angular-displacement pickup; amplitude limit in degrees; sensitivity in mv/volt/degree.
§ Inductance-bridge operation at 3,000 cps carrier frequency.
¶ Velocity pickup; sensitivity in mv/in./sec.

Table 15.4. Typical Characteristics of Seismic Pickups Used above Natural Frequency

Manufacturer	Model	Principle	Sensitivity, mv/in./sec	Natural frequency, cps	Frequency range, cps	Amplitude limit, in.	Temperature range, °F	Damping Fraction of critical	Damping Type	Weight, oz	Coil resistance, ohms
Consolidated Electrodynamics Corp.	4-102A	Electrodynamic	110	9	8 to 700	1	0 to 150	...	...	10	925
	4-120	Electrodynamic	105	...	40 to 2,000	0.05	−65 to 500	0.7	Eddy-current	5.15	575
Electro Mechanisms (England)	9/B1	Electrodynamic	88	8	Above 10	0.125	...	...	...	4.75	400
Electro-Technical Labs...	EVS-4	Electrodynamic	710	7.5	...	...	...	0.7	Eddy-current	9.5	215
Hilger & Watts (England)...	......	Electrodynamic	3,000	1	...	...	...	...	...	336	500
International Research and Development Corp.	545	Electrodynamic	1,250	15	20 to 500	...	−60 to 125	0.6	Oil	19	7,000
Keystone Instruments...	......	Electrodynamic	470	24	Above 30	...	...	0.50	Eddy-current	4.5	500
Micro Balancing, Inc...	11620	Electrodynamic	260	4	8 to 2,000	0.125	−60 to 250	0–1	Oil	10	800
M B Electronics Co...	125	Electrodynamic	96.4	4.75	10 to 2,000	0.2	−50 to 250	0	...	9.7	650
	122	Electrodynamic	96.4	4.75	7 to 2,000	0.2	−50 to 500	0.65	Eddy-current	11.2	650
Philips (Holland)...	PR 9260	Electrodynamic	765	12	Above 5	0.16	...	0.5	Eddy-current	20	2,250
Southwestern Industrial Electronics	S-16	Electrodynamic	......	18	To 5,000	...	...	0.3	Eddy-current	15	...
Sperry Gyroscope Co...	1777291-1	Electrodynamic	10	4.5	15 to 2,000	0.25	−65 to 500	0.65	Air	7.5	30
Texas Instruments...	S-36	Electrodynamic	5,000	2	Above 1	...	...	0.5†	Eddy-current	272	4,000
The Calidyne Co...	Prototype ‡	Eddy-current	41 ‡	11	10 to 200	0.15	...	0.6	Eddy-current	12	7,340
Consolidated Electrodynamics Corp.	9-102 §	Electrodynamic	9 §	3	10 to 1,000	2 §	0 to 150	0.64	Oil	112	700

* Taken from manufacturers' literature which gives detailed specifications and information on additional models available.
† With 10,000 ohms external resistor.
‡ Acceleration pickup; sensitivity in mv/g.
§ Angular-velocity pickup; sensitivity in mv/degree/sec; amplitude limit in degrees.

15–19

Table 15.5. Typical Characteristics * of Seismic Pickups Used below Natural Frequency

Manufacturer	Model	Principle	Sensitivity, mv/volt/g	Natural frequency, cps	Accel. amp. limit, g	Temperature range, °F	Damping Fraction of critical	Damping Type	Weight, oz	Output coil impedance, ohms
Edcliff Instruments............	....	Differential transformer	500	......	1	....	....	....	....	....
Gulton Industries, Inc.........	ADT-910-5	Differential transformer	26	30	5	−65 to 250	0.65	Eddy-current	28	5,000
Schaevitz Engineering.........	VG-700	Differential transformer	0.1	620	700	....	0.6 to 0.7	Oil	4	....
	F-200 †	Differential transformer	1.2 †	25	220 †	....	0.6 to 0.7	Oil	....	....
Consolidated Electrodynamics Corp.........	4-260	Reluctance	6.4	200	12	−25 to 150	....	....	2	100
Electro Mechanisms Ltd. (England).........	IT1-22F-31	Reluctance	....	300	90	−4 to 158	0.6	Oil	2.75	750
Genisco, Inc.................	GAL	Reluctance	36	18	10	−65 to 185	0.7	Eddy-current	43	....
Gulton Industries, Inc.........	AVR-250-30	Reluctance	1	250	30	0 to 165	0.7	....	0.09	400
Office National d'Etudes et de Recherches Aeronautiques (France).........	20J22	Reluctance	....	500	100	−50 to 200	0.6 to 0.75	Oil	1.6	....
Wiancko Engineering Co......	A1001	Reluctance	0.8	400	100	−94 to 500	0	....	7	500
Electro Products Labs........	3020-A ‡	Magnetostrictive	....	14,000	500	....	....	....	2.25	....
Sperry Gyroscope Co.........	610956 ‡	Magnetostrictive	51×10^{-9} ‡	12,000	......	−50 to 300	0	....	4	110

* Taken from manufacturers' literature which gives detailed specifications and information on additional models available.
† Angular acceleration pickup; sensitivity in mv/volt/radian/sec^2; acceleration amplitude limit in radians/sec^2.
‡ Jerk pickup; sensitivity in mv/in./sec^3.

16

PIEZOELECTRIC AND PIEZORESISTIVE PICKUPS

PART I: PICKUP CHARACTERISTICS

Abraham I. Dranetz
Gulton Industries, Inc.

Anthony W. Orlacchio
Gulton Industries, Inc.

INTRODUCTION

Certain solid-state materials are electrically responsive to mechanical force; they often are used as the mechanical-to-electrical transduction devices in shock and vibration pickups. Generally exhibiting high elastic stiffness, these materials can be divided into two categories: *the self-generating type*, in which electric charge is generated as a direct result of applied force, and the *passive-circuit type*, in which applied force causes a change in the electrical characteristics of the material. *Piezoelectric* materials are of the self-generating type. In the linear elastic range, a piezoelectric material produces an electric charge proportional to stress. Passive-circuit types of materials include *magnetostrictive* and *piezoresistive* materials. A magnetostrictive material has a magnetic permeability which depends upon applied force; a piezoresistive material has an electrical resistance which depends upon applied force.

A solid-state material whose electrical resistance depends upon the applied magnetic field is said to be *magnetoresistive*. Their use is restricted to vibration pickups having highly compliant spring-mass combinations, and hence low resonant frequency.

The first half of this chapter discusses the theory, design, and characteristics of solid-state pickups, with particular emphasis on piezoelectric devices, since the latter are used most widely. Basic design concepts of piezoresistive pickups also are discussed. The second half of the chapter describes the characteristics of piezoelectric and piezoresistive materials.

PIEZOELECTRIC ACCELEROMETERS

PRINCIPLE OF OPERATION OF SIMPLE SEISMIC SYSTEM

An accelerometer of the type shown in Fig. 16.1A is a linear seismic transducer (i.e., pickup) utilizing a piezoelectric element in such a way that an electric charge is produced which is proportional to the applied acceleration. This "ideal" seismic piezoelectric pickup can be represented (over most of its frequency range) by the elements shown in Fig. 16.1B. A mass is supported on a linear spring which is fastened to the frame of the instrument. The piezoelectric crystal which produces the charge acts as the spring.

16–2 PIEZOELECTRIC AND PIEZORESISTIVE PICKUPS

Viscous damping between the mass and the frame is represented by dashpot c. In Fig. 16.1C the frame is given an acceleration upward to a displacement of u, thereby producing a compression in the spring equal to δ. The displacement of the mass relative to the frame is dependent upon the applied acceleration of the frame, the spring stiffness, the mass and the viscous damping between the mass and the frame, as indicated in Eq. (12.10) and illustrated in Fig. 12.5.

For frequencies far below the resonant frequency of the mass and spring, this displacement is directly proportional to the acceleration of the frame and is independent of frequency. At low frequencies, the phase angle of the relative displacement x, with respect

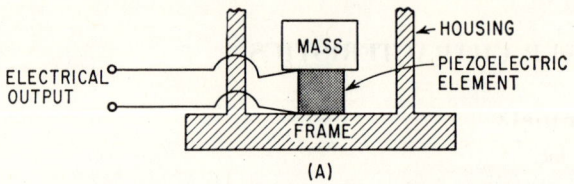

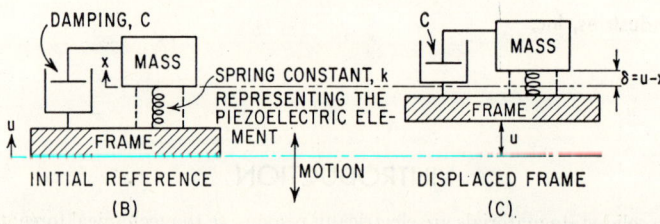

Fig. 16.1.(A) Schematic diagram of a linear seismic piezoelectric accelerometer. (B) A simplified representation of the accelerometer shown in (A) which applies over most of the useful frequency range. A mass m rests on the piezoelectric element which acts as a spring having a spring constant k. The damping in the system, represented by the dashpot, has a damping coefficient c. (C) The frame is accelerated upward, producing a displacement u of the frame, moving the mass from its initial position by an amount x, and compressing the spring by an amount δ.

to the applied acceleration, is proportional to frequency. As indicated in Fig. 12.6, for low fractions of critical damping which is characteristic of many piezoelectric pickups, the phase angle is proportional to frequency at frequencies below 30 per cent of the resonant frequency.

In Fig. 16.1, inertial force of the mass causes a mechanical strain in the piezoelectric element which produces an electric charge proportional to the stress * and, hence, proportional to strain and acceleration.[1] If the dielectric constant of the piezoelectric material does not change with electric charge, the voltage generated also is proportional to acceleration. Metallic electrodes are applied to the piezoelectric element and electrical leads are connected to the electrodes for measurement of the electrical output of the piezoelectric element.

In the ideal seismic system shown in Fig. 16.1, the mass has infinite stiffness, the spring has zero mass, viscous damping exists only between mass and frame, and the frame has infinite stiffness. In practical piezoelectric pickups, these assumptions cannot be fulfilled. For example, the mass may have as much compliance as the piezoelectric element. In some seismic elements (particularly those which produce voltage from bending of the piezoelectric element) the mass and spring are inherently a single structure. Furthermore, in many practical designs where the frame is used to hold the mass and

---

* The piezoelectric element contained in an accelerometer normally is used within its linear elastic range; hence, one may consider that the charge is a result of the stress applied or of the strain which results from the stress.

piezoelectric element, distortion of the frame may produce mechanical forces upon the seismic element.[2] All these factors may change the performance of the seismic system from those calculated using equations based on an ideal system. In particular, the resonant frequency of the piezoelectric combination may be substantially lower than that indicated by theory. Nevertheless, the equations for an ideal system are useful both in design and application of piezoelectric accelerometers.

ELECTRICAL ANALOG OF PIEZOELECTRIC ACCELEROMETER. The electrical analog of a mechanical system often offers a useful representation, particularly to those familiar with electric circuit analysis. Such analogs are described in Chap. 29. An *electrical analog* is particularly useful in describing an electromechanical device, such as a

Table 16.1. Analogous Parameters for an Electric Circuit Which Is the Direct Analogy of a Mechanical System

Electrical parameters		Mechanical parameters	
Symbol	Description	Symbol	Description
q	Charge	x	Displacement
$\dot{q}, i$	Current	$\dot{x}$	Velocity
$\ddot{q}, \dfrac{di}{dt}$	Rate of change of current	$\ddot{x}$	Acceleration
e	Voltage	F	Force
L	Inductance	m	Mass
R	Resistance	c	Damping coefficient
C	Capacitance	$1/k$	Compliance

piezoelectric pickup, since it provides a description for the entire device of a consistent set of electrical parameters.

Employing Eq. (12.1), a simple substitution of mechanical parameters by analogous electrical parameters in the equation of motion for the simple seismic system shown in Fig. 16.1 yields

$$L(\ddot{q}_2 - \ddot{q}_1) + R\dot{q}_2 + \frac{1}{C}q_2 = 0 \quad (16.1)$$

where inductance L is analogous to mass m, resistance R is analogous to damping coefficient c, capacitance C is analogous to the reciprocal of spring constant k, charge q_1 is analogous to input frame displacement u, and charge q_2 is analogous to spring compression δ. The analogous quantities are summarized in Table 16.1. This type of analog is called a "direct analog" or "force-voltage" analog. It can be shown that Eq. (16.1) is satisfied by the electric circuit of Fig. 16.2A, in which the frame acceleration is represented by $\ddot{q}_1$ and the spring distortion is represented by q_2.

In a simple seismic accelerometer, the deflection of the spring is a measure of the acceleration. However, in the analysis of a seismic accelerometer sometimes it is easier to relate the piezoelectric constants to the applied force rather than to the deflection. Assuming linear elastic characteristics of the piezoelectric element, the force F and strain δ are related by

$$\delta = \frac{F}{k} \quad \text{meters} \quad (16.2)$$

where k = stiffness of the piezoelectric element in newtons/meter and F = force in newtons.

As a result of mechanical stress in the piezoelectric element, an electric charge q is generated and built up across the internal electrical capacitance C_E of the piezoelectric element. The electric charge is given by

$$q = A\delta = \frac{AF}{k} \quad \text{coulombs} \quad (16.3)$$

where A is a constant (expressed in coulombs per meter) determined by the size, shape and material properties of the piezoelectric element and by the position of the electrodes. For certain simple configurations, the constant A is given by

$$A = d_{ij}k \qquad \text{coulombs/meter} \qquad (16.4)$$

where d_{ij} is the piezoelectric constant of the material relating applied stress to generated electric charge. Subscripts i and j refer to orientation of electrodes and direction of ap-

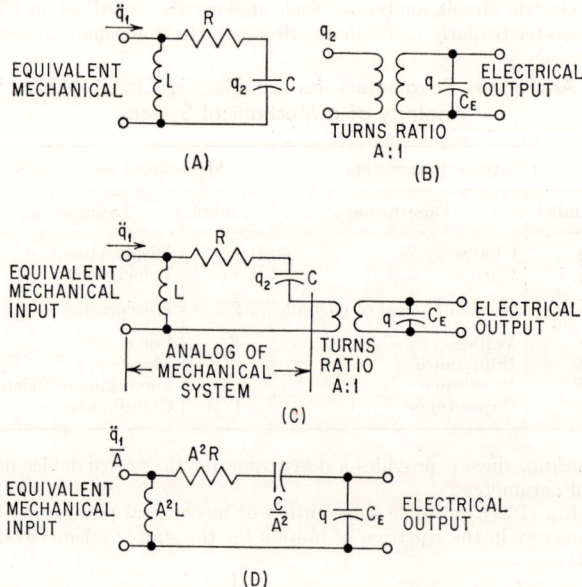

FIG. 16.2. Direct analog and equivalent circuit of a piezoelectric accelerometer employing the seismic system of Fig. 16.1A. (A) Direct analog of simple seismic system of Fig. 16.1A. (B) Electrical equivalent of a piezoelectric element. (C) Electrical equivalent circuit of a piezoelectric accelerometer. (D) Electrical equivalent circuit of a piezoelectric accelerometer which is an alternate form of the circuit shown in (C).

plied stress [see Eq. (16.8)]. (Values of d_{ij} for several typical materials are given in Tables 16.3 and 16.4.)

The equivalent circuit representing the generation of electric charge across the piezoelectric element is shown in Fig. 16.2B. The transformer is a perfect transformer having no loss between zero frequency and the highest frequency for which the piezoelectric element is operative; it has a turns ratio of $A:1$.

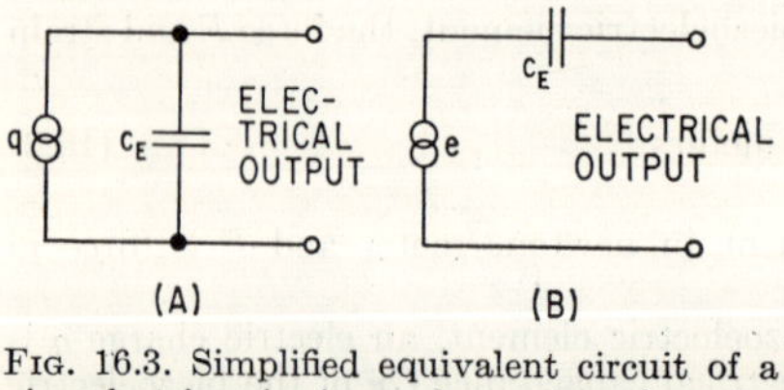

The electric equivalent circuit of a piezoelectric accelerometer (assuming a simple seismic system of the type indicated in Fig. 16.1B) is shown in Fig. 16.2C which represents the combination of Figs. 16.2A and 16.2B. Another form of this electric equivalent circuit is shown in Fig. 16.2D. Often the viscous damping coefficient is negligible in piezoelectric accelerometers so that the circuit element, R in Fig. 16.2C and D, is small.

At frequencies well below the resonant

FIG. 16.3. Simplified equivalent circuit of a piezoelectric seismic accelerometer when the frequency of applied acceleration is far below the resonant frequency. (A) Charge generating equivalent circuit. (B) Voltage generating equivalent circuit in which $e = q/C_E$.

frequency of the pickup, the electric charge generated is directly proportional to acceleration, as given, for example, by Eq. (16.10). Therefore, at these low frequencies, the equivalent circuit of the pickup, as seen at the electrical output terminals, is that given in Fig. 16.3. A *charge equivalent circuit* is shown in Fig. 16.3*A*, where the electric charge generated by acceleration is represented by q and the electrical capacitance of the piezoelectric element is represented by C_E. The open-circuit output voltage generated by the pickup is given by

$$e = \frac{q}{C_E} \quad \text{volts} \quad (16.5)$$

The *charge equivalent circuit* of Fig. 16.3*A* is equivalent to the *voltage equivalent circuit* of Fig. 16.3*B*. These equivalent circuits are particularly useful in determining the effect of coupling the output of a pickup to other electronic equipment.

CALCULATION OF PERFORMANCE CHARACTERISTICS OF AN ACCELEROMETER. The approximate performance characteristics of a practical piezoelectric accelerometer can be calculated, if it is assumed that the behavior is that of an ideal simple seismic system. For example, consider the pickup shown in Fig. 16.4. This piezoelectric accelerometer consists of a frame to which is fastened piezoelectric disc 1. This disc is provided with metallic electrodes on its upper and lower faces. A metallic disc (for electrical contact) is fastened to the upper face of piezoelectric disc 1; also bonded to the metallic disc is piezoelectric disc 2 (electroded on upper and lower faces);

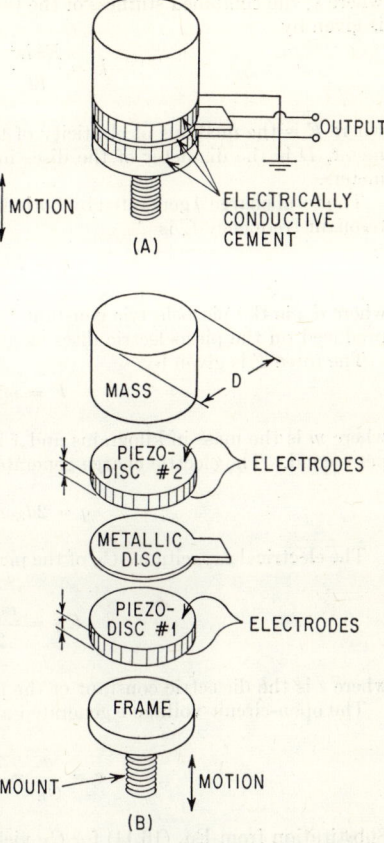

Fig. 16.4. Compression-type piezoelectric accelerometer without housing: (*A*) assembled and (*B*) "exploded" view.

a mass is bonded to piezoelectric disc 2.* Acceleration in the direction indicated causes compressive or tensile forces on the piezoelectric elements. For this reason, such a pickup is termed a *compression-type accelerometer*. The piezoelectric elements are electrically connected so as to add the electric charges which are created by each element when compressed. Assuming negligible damping, by methods outlined below, one can calculate (1) the resonant frequency f_n, (2) *charge* and *voltage sensitivities* † at frequencies far below the resonant frequency (e.g., below 0.3 f_n), and (3) the electromechanical conversion efficiency. From such calculations one may determine reasonable design compromises among sensitivity, frequency response, and weight.

Approximate Formulas for Transducer Characteristics. The resonant frequency of the pickup shown in Fig. 16.4 is given by

$$f_n = \frac{1}{2\pi} \left(\frac{k}{m}\right)^{1/2} \quad \text{cps} \quad (16.6)$$

---

* Often only one piezoelectric disc is used. The use of two piezoelectric discs, however, makes it possible to connect the mass to the housing electrically, without the use of electrical insulators necessary in a single disc design.

† *Charge sensitivity* is defined as the electric charge generated per unit of acceleration (given by $S_q = q/\ddot{x}$). *Voltage sensitivity* is defined as the open-circuit voltage generated per unit of acceleration (given by $S_v = e/\ddot{x}$).

where k, the combined stiffness of the two piezoelectric discs (one-half that of each disc), is given by

$$k = \frac{E\pi D^2}{8t} \quad \text{newtons/meter} \tag{16.7}$$

where E is the modulus of elasticity of the piezoelectric material in newtons per square meter, D is the diameter of the discs in meters, and t is the thickness of each disc in meters.

The total charge q generated by both discs due to vibration at frequencies far below the resonant frequency f_n is

$$q = 2d_{33}F \quad \text{coulombs} \tag{16.8}$$

where d_{33} is the piezoelectric constant * of the piezoelectric material and F is the force produced on the piezoelectric discs as a result of acceleration.

The force F is given by

$$F = m\ddot{x} \quad \text{newtons} \tag{16.9}$$

where m is the mass in kilograms and $\ddot{x}$ is the applied acceleration in meters per second per second. The electric charge generated thus is

$$q = 2d_{33}m\ddot{x} \quad \text{coulombs} \tag{16.10}$$

The electrical capacitance C_E of the pickup (due to the two discs in parallel) is given by

$$C_E = \frac{\epsilon \pi D^2}{2t} \quad \text{farads} \tag{16.11}$$

where ϵ is the dielectric constant of the piezoelectric material in farads per meter.

The open-circuit voltage e generated at frequencies well below f_n is

$$e = \frac{q}{C_E} = \frac{2d_{33}m\ddot{x}}{C_E} \quad \text{volts} \tag{16.12}$$

Substitution from Eq. (16.11) for C_E yields

$$e = \frac{4d_{33}m\ddot{x}}{\epsilon \pi D^2} = \frac{4g_{33}m\ddot{x}}{\pi D^2} \quad \text{volts} \tag{16.13}$$

where $g_{33} = \dfrac{d_{33}}{\epsilon}$ is a piezoelectric constant relating open-circuit voltage to applied stress. Values of g_{ij} for various materials are given in Tables 16.3 and 16.4.

The electrical energy generated by the pickup is given by

$$E = \frac{q^2}{2C_E} = \frac{(2d_{33}m\ddot{x})^2 t}{\epsilon \pi D^2} \quad \text{joules} \tag{16.14}$$

Example of Calculations for an Accelerometer. Consider a seismic accelerometer, similar in design to that shown in Fig. 16.4, which utilizes a polycrystalline ceramic composed of 96 per cent barium titanate, and 4 per cent lead titanate. However, instead of

---

* The piezoelectric constant d_{ij} relates electric charge to the applied stress. Subscripts 33 refer to the simple case of a piezoelectric ceramic disc undergoing a compressive (or tensile) force in the direction of the created electrical field (i.e., perpendicular to the metallic electrodes). Other piezoelectric configurations also are used. For example, if the stress is applied parallel to (rather than perpendicular to) the electrodes, the d_{31} constant is used. In some cases, a mechanical shear stress may be applied, in which case the d_{15} constant is applicable (see Part II of this chapter).

mechanical bonding, a rugged housing is utilized to clamp the entire seismic assembly under a heavy preloading force (up to 400 lb/in.^{2}). Assume the following constants:

$D = 0.175$ in. $= 4.45 \times 10^{-3}$ meter
$t = 0.020$ in. $= 5.09 \times 10^{-4}$ meter
$E = 1.13 \times 10^{12}$ dynes/cm^2 $= 1.13 \times 10^{11}$ newtons/meter^2
$m = 1.1$ gm $= 1.1 \times 10^{-3}$ kg
$\epsilon = 900 \, \epsilon_0 = 7.95 \times 10^{-9}$ farad/meter
$d_{33} = 280 \times 10^{-8}$ statcoulomb/dyne $= 9.32 \times 10^{-11}$ coulomb/newton
$g = 9.8$ meter/sec^2

From Eqs. (16.6), (16.10), (16.11), and (16.13) the following characteristics of the pickup may be calculated: natural frequency, 198,000 cps; capacitance, 490 mmfd; charge sensitivity, 2.0×10^{-12} coulomb/g; and voltage sensitivity, 3.1×10^{-3} volt/g. In practical designs, the calculated values may differ considerably from the measured values because the calculations neglect effects such as clamping by the housing and because of resonances introduced by the distributed parameter system.

TYPICAL ACCELEROMETER CONSTRUCTIONS

Piezoelectric accelerometers utilize a variety of seismic element configurations. Most pickups are constructed of polycrystalline ceramic piezoelectric materials because of their ease of manufacture, high piezoelectric sensitivity, and excellent time and temperature stability. These seismic devices may be classified either as *compression-* or *bender-type* accelerometers.

COMPRESSION-TYPE ACCELEROMETER. The compression-type seismic accelerometer, in its simplest form, consists of a piezoelectric disc and a mass placed on a frame as shown in Fig. 16.4. Motion in the direction indicated causes compressive (or tensile) forces to act on the piezoelectric element, producing an electrical output proportional to acceleration. In this example, the mass is cemented with a conductive material to the piezoelectric element which, in turn, is cemented to the frame. The components must be cemented firmly so as to avoid being separated from each other by the applied acceleration.

In the unit shown in Fig. 16.5 the mass is held in place by means of a stud extending from the frame through the ceramic. This accelerometer (NRL Type C-4), employing a ceramic element, has the following characteristics: [3]

Diameter = 0.80 in. = 2.03 cm
Thickness = 0.30 in. = 0.76 cm
$d_{33} = 2.5 \times 10^{-6}$ coulombs/newton
$\epsilon/\epsilon_0 = 1,000$ (where ϵ_0 is dielectric constant of free space)

Sensitivity = 100 millivolts/g (open circuit)
Acceleration range = 1 to 5,000g
Resonant frequency = 14,000 cps
Useful frequency range = 10 to 4,000 cps
Temperature range = $-40°$ to $+75°$C

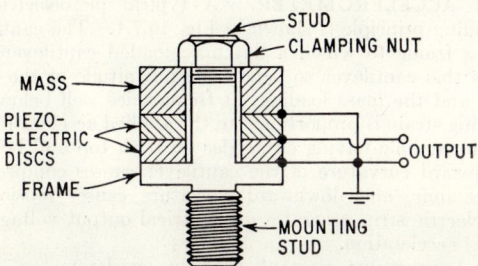

Fig. 16.5. NRL-type piezoelectric accelerometer. This pickup has a sensitivity of 100 millivolts/g, a resonant frequency of 14,000 cps, and a useful acceleration up to 5,000g. (*Courtesy of the U.S. Naval Research Laboratory.*)

A typical commercial accelerometer is shown in Fig. 16.6. This unit consists of a housing, one piezoelectric ceramic element, a seismic mass, and a spring and cap for clamping the mass against the ceramic element. The compression cap applies the necessary preloading through the case isolation spring. The resonant frequency of the pickup is always lower than that of the mass-ceramic combination alone. This type of accelerometer must be attached to the structure with care in order to minimize distortion of the housing which may affect the sensitivity of the accelerometer. For this reason a recommended value of mounting torque in attaching the unit and mounting surface often is specified.[4]

In the accelerometer shown in Fig. 16.6, an insulating material is placed between the housing base and the lower electrode of the piezoelectric element in order to provide an insulated output lead. The upper electrode of the ceramic is connected electrically to the housing through the mass-spring combination. The preloading of the spring is adjusted so as to prevent "chattering" of the mass for values of acceleration up to the rated acceleration range of the instrument. If a clamp or cementing technique is not used for maintaining the mass in contact with the ceramic element in the above type of construction, the output will be proportional to acceleration only if the forces due to the applied acceleration do not exceed the bias force. An unclamped seismic system is sometimes used to calibrate vibrating systems, by observing the occurrence of chatter,[4] which takes place whenever the downward acceleration exceeds $1g$ (see Chap. 18).

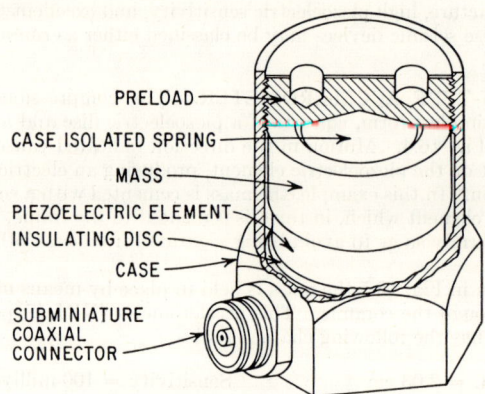

Fig. 16.6. Commercial compression-type accelerometer. The spring provides the necessary preload for the mass and piezoelectric element, so as to maintain intimate contact throughout the rated range of acceleration. (*Courtesy of Endevco Corp.*)

BENDER-TYPE ACCELEROMETER. A typical piezoelectric seismic element operating on a bending principle is shown in Fig. 16.7A. The cantilever construction shown consists of a frame to which a flat mass-loaded cantilever strip is attached. Acceleration causes this cantilever to bend—the magnitude of the deflection depending on its stiffness and the mass loading; at frequencies well below the resonant frequency f_n the bending strain is proportional to the applied acceleration of the frame. A strip of piezoelectric ceramic having electrodes on both top and bottom is bonded to the cantilever. Upward curvature of the cantilever causes compression in length of the piezoelectric ceramic, and downward curvature causes tension. The resulting strain in the piezoelectric strip generates an electrical output voltage which is proportional to the applied acceleration.

Another typical bender-type element used in accelerometers is shown in Fig. 16.7B; this type is supported at its center and has a mass at each end with a piezoelectric strip bonded to each side. A means is provided for mounting the element to the frame. In this configuration the metallic support for the element is connected to the output ground; the two outer piezoelectric surfaces are silvered and are connected together to the "high" side of the output. When acceleration is applied to the frame, tensile stresses on one ceramic element and compressive stresses on the other are produced simultaneously. The two piezoelectric ceramic strips are oriented so as to generate electric charge of the same polarity when one is compressed and the other is extended so that the generated charges are additive. Typical characteristics of such a device are: length, 1 in.; mass loading, 3 gm; sensitivity, 150 mv/g; capacitance, 1,000 mmfd; and resonant frequency, 3,600 cps. It also is possible to connect the output of the ceramic strips in series rather than in parallel; this doubles the sensitivity and reduces the capacitance by a factor of 4.

In Fig. 16.7C a circular bender-type accelerometer is shown. This unit has an element which consists of a circular metal disc which is supported at its center by a mounting stud; a circular ceramic disc is cemented to the flat circular portion of the metal. The sensitivity and resonant frequency of this seismic element may be changed by mass loading in the form of a peripherical ring. Acceleration along the axial direction, as shown, results in a bending of the disc which strains the ceramic in a radial direction. The ceramic produces an output voltage proportional to stress. The base of the unit is comprised of two molded ceramic insulating rings, an outer metal and inner metal ring, and the mounting stud. The piezoelectric element is clamped in the inner metal ring and electrical connection is made from the inner metal ring to the protective housing, which acts as the negative electrical lead. The positive electrical lead is attached to the top of the ceramic and is connected to the insulated pin of the coaxial connector. The purpose of electrically insulating the seismic element from the mounting stud is to isolate *ground-loop* voltages from the electrical output terminals of the accelerometer (see *Ground Loops*, Chap. 19).

UNIAXIAL AND TRIAXIAL ACCELEROMETERS

The majority of piezoelectric accelerometers manufactured are of the uniaxial type, i.e., for measuring acceleration

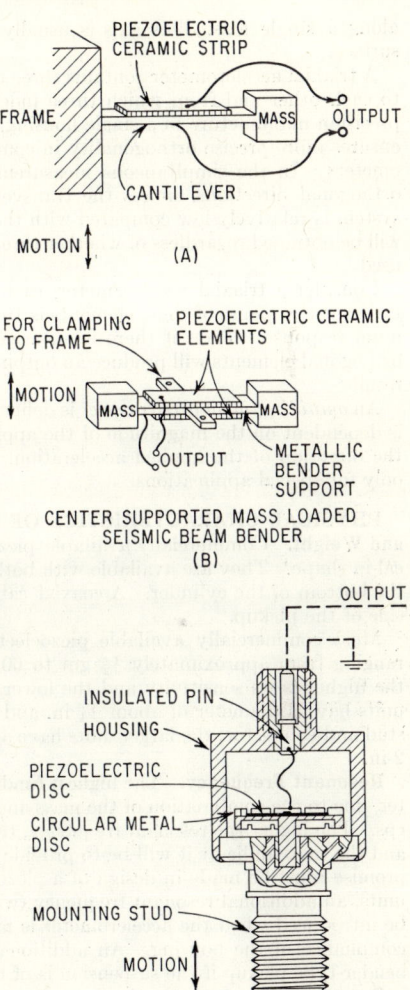

FIG. 16.7. (A) A typical piezoelectric seismic element employing the bending principle, showing the operation of a seismic system utilizing a mass-loaded cantilever. Below the natural frequency f_n of the cantilever the amplitude of bending strain is proportional to the applied acceleration. The strain in the piezoelectric strip generates an electrical output voltage proportional to the applied acceleration. (B) Two ceramic strips are connected in parallel so that they create a charge of the same polarity when strained. (C) A commercial bender-type seismic accelerometer. Acceleration in the axial direction produces a strain due to bending of the metal disc, resulting in an output proportional to acceleration. (*Gulton Industries, Inc.*)

along a single axis. This axis is usually perpendicular to the plane of the mounting surface.

A triaxial accelerometer contains three uniaxial seismic elements which are orthogonal to each other and from which three independent electrical outputs are obtained. By precision manufacture in a single housing, a triaxial accelerometer is more compact and ensures more precise orthogonality in comparison with three individual uniaxial accelerometers In the simultaneous measurement of acceleration components along three orthogonal directions, unless the transverse sensitivity * of each piezoelectric seismic system is relatively low compared with the maximum sensitivity,[6] erroneous indications will be obtained regardless of whether three uniaxial pickups or a single triaxial pickup is used.

Consider a triaxial accelerometer, each element of which is identical. Suppose the transverse response of these elements is 10 per cent of the value in the direction of maximum response. Then if there is an acceleration of $10g$ in a transverse direction, the orthogonal elements will produce an output equivalent to $1g$; an error of 10 per cent may result.

An *omnidirectional accelerometer* is defined as one which produces a voltage output that is dependent on the magnitude of the applied acceleration, but which is independent of the direction of the applied acceleration. Usually, such accelerometers are fabricated only for special applications.

PHYSICAL CHARACTERISTICS OF PIEZOELECTRIC PICKUPS. Shape, Size, and Weight. Commercially available piezoelectric accelerometers usually are cylindrical in shape. They are available with both attached and detachable mounting studs at the bottom of the cylinder. A coaxial cable connector is provided at either the top or side of the pickup.

Most commercially available piezoelectric pickups † are relatively light in weight, ranging from approximately ½ gm to 60 gm. Usually, the larger the accelerometer, the higher is its sensitivity and the lower is its resonant frequency. The smallest-size units hav a diameter of about ¼ in. and a height of ¼ in., exclusive of the mounting stud and connector; the larger units have a diameter of about 2 in. and a height of about 2 in.

Resonant Frequency. The highest fundamental resonant frequency of an accelerometer, due to the combination of the mass and piezoelectric element, may be above 100,000 cps. The higher the resonant frequency, the lower will be the capacitance or sensitivity, and the more difficult it will be to provide mechanical damping. Therefore, some compromise must be made in design of a piezoelectric accelerometer. In compression-type units, an additional resonant frequency (which is lower in frequency and of low Q) may be introduced when the accelerometer is mounted because of the seismic mass and the compliance of the housing. An additional resonant frequency is not introduced in a bender-type pickup if the suspension is of the type shown in Fig. 16.7C.

Damping. The *amplification ratio* of an accelerometer is defined as the voltage sensitivity at its resonant frequency to the voltage sensitivity in the frequency band in which sensitivity is independent of frequency. This ratio depends on the amount of damping in the seismic system; it decreases with increasing damping. Most piezoelectric accelerometers are essentially undamped, having magnification ratios between 5 and 50. Damping is employed in some pickups having resonant frequencies below 20,000 cps by the use of silicone oil which has a kinematic visco ity ‡ of from 1.0 to 200,000 centistokes.

---

* Also called *lateral sensitivity* or *cross-axis sensitivity*.

† Manufacturers of commercially available piezoelectric accelerometers include: Clevite Brush Development Co., Cleveland, Ohio; Columbia Research Laboratories, Woodlyn, Pa.; Endevco Corp., Pasadena, Calif.; Gulton Industries, Inc., Metuchen, N.J.; Kintel, San Diego, Calif.; Kistler Instrument Co., North Tonawanda, N.Y.; Bruel and Kjaer, Naerum, Denmark; Philips, N. V., Eindhoven, Netherlands.

‡ The viscosity of the oil is defined as the ratio of the shearing stress (dyne/cm^2) to the rate of shear (sec^{-1}). The unit of viscosity is dyne-sec/cm^2 which is equal to 1 *poise*. The *kinematic viscosity* is defined as the viscosity of the oil divided by its density. If the density is expressed in gm/cm^3, the unit of kinematic viscosity is the *stoke* (1 centistoke = 0.01 stoke). See Chap. 32 for further details.

The temperature dependence of the kinematic viscosity of a typical silicone oil having a relatively low viscosity-temperature dependence (Dow-Corning 200) is shown in Fig. 16.8. As an example of the use of the data given in this figure, consider a seismic system having a resonant frequency of 4,000 cps with a fluid damping equal to 0.25 of critical damping at 25°C; at −30°C, the damping is 0.8 of critical; at +80°C, the damping is 0.1 of critical.

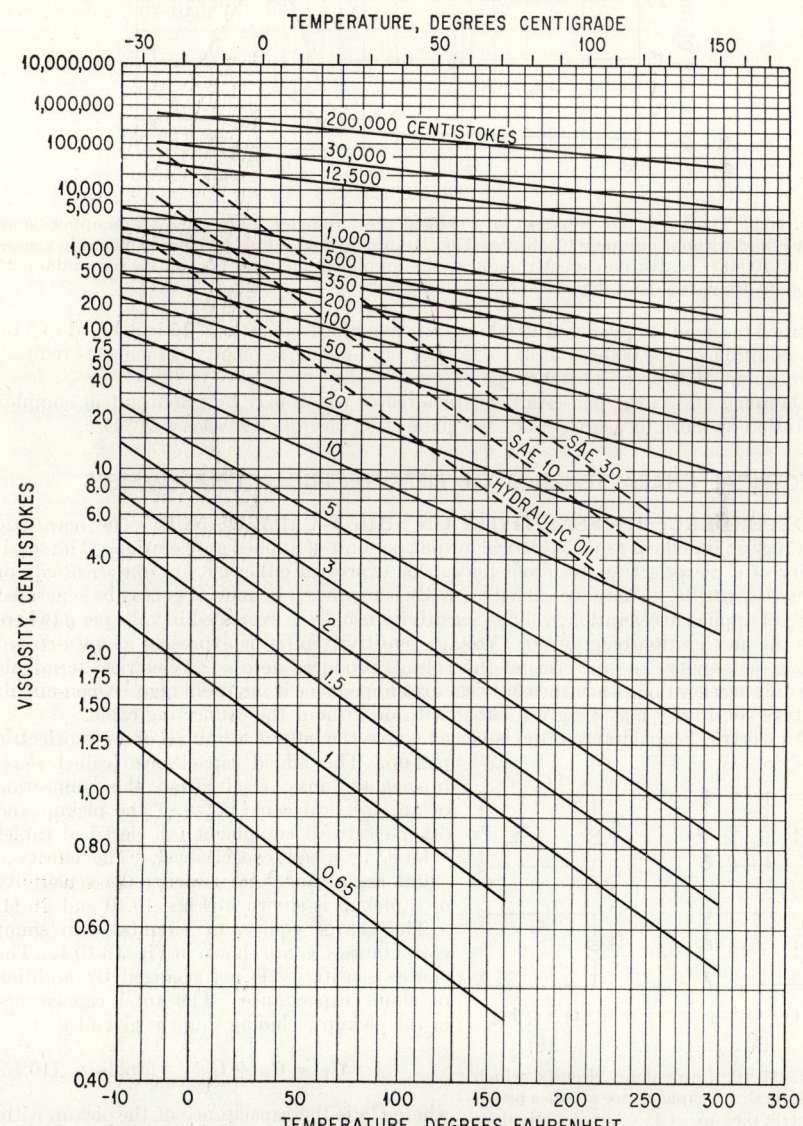

FIG. 16.8. Viscosity-temperature relationship for Dow Corning Type No. 200 Silicone Fluid. Similar data are shown (as dashed lines) for hydraulic oil, and for SAE 10 and 30 petroleum oils. (*Courtesy of Dow Corning Corp.*)

16–12 PIEZOELECTRIC AND PIEZORESISTIVE PICKUPS

The frequency response of a typical piezoelectric accelerometer using a bender-type center-supported cantilever is shown in Fig. 16.9. The dashed curve shows the frequency-response characteristics for the accelerometer without damping. The solid curve shows the characteristics when damping is added. The piezoelectric element has an un-

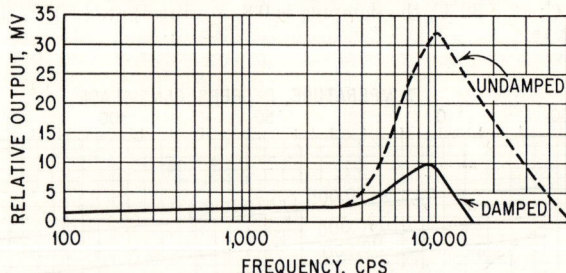

Fig. 16.9. The frequency response of a typical accelerometer is shown with damping (solid curve) and without damping (dashed curve). Damping with a fluid having a kinematic viscosity of 100,000 centistokes greatly reduces the amplification ratio and shifts the undamped resonant frequency to a slightly lower value.

damped resonant frequency of 10,000 cps; its magnification ratio of 15 is reduced to 5 by the addition of 100,000-centistoke damping oil, and its resonant frequency is reduced from about 10,000 cps to 9,000 cps.

Damping affects the phase-shift characteristics which may be important in complex vibration or shock measurements. See the section on *Output Response to Shock*.

ELECTRICAL CHARACTERISTICS OF PIEZOELECTRIC ACCELEROMETERS

DEPENDENCE OF SENSITIVITY ON SHUNT CAPACITANCE.

The *sensitivity* of a pickup is defined as the electrical output per unit of applied acceleration. The sensitivity of a piezoelectric accelerometer can be expressed either as a *charge sensitivity* or *voltage sensitivity*. Charge sensitivity usually is expressed in units of coulombs generated per g of applied acceleration; voltage sensitivity usually is expressed in volts per g (where g is the acceleration of gravity). Voltage sensitivity often is expressed as *open-circuit voltage* sensitivity, i.e., in terms of the voltage produced across the electrical terminals per unit acceleration, when the electrical load impedance is infinitely high. Open-circuit voltage sensitivity may be given either with or without the connecting cable.

An electrical capacitance often is placed across the output terminals of a piezoelectric pickup. This added capacitance (called *shunt capacitance*) may result from the connection of an electrical cable between the pickup and other electrical equipment (all electrical cables exhibit interlead capacitance). The effect of shunt capacitance* in reducing the sensitivity of a pickup is shown in Figs. 16.10 and 16.11.

The charge equivalent circuits, with shunt capacitance C_S, are shown in Fig. 16.10A. The charge sensitivity is not changed by addition of shunt capacitance. The total capacitance of the pickup including shunt is given by

$$C_T = C_E + C_S \quad \text{farads} \quad (16.15)$$

where C_E is the capacitance of the pickup without shunt capacitance.

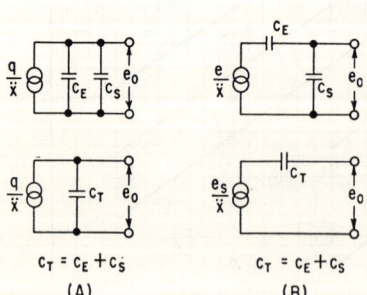

Fig. 16.10. Equivalent circuits which include shunt capacitance across a piezoelectric pickup. (A) Charge equivalent circuit. (B) Voltage equivalent circuit.

---

* In order to eliminate the effect of cable shunt capacitance, accelerometers are available with the coupling electronic circuits built into the same housing as the seismic element. These devices employ either vacuum-tube or transistor amplification circuits.

PIEZOELECTRIC ACCELEROMETERS

The voltage equivalent circuits are shown in Fig. 16.10B. With the shunt capacitance $C\hat{S}$, the total capacitance is given by Eq. (16.15) and the open-circuit voltage sensitivity is given by

$$\frac{e_s}{\ddot{x}} = \left(\frac{q}{\ddot{x}}\right)\left(\frac{1}{C_E + C_S}\right) \quad \frac{\text{volt-sec}^2}{\text{meter}} \tag{16.16}$$

where $q/\ddot{x}$ is the charge sensitivity. The voltage sensitivity without shunt capacitance is given by

$$\frac{e}{\ddot{x}} = \left(\frac{q}{\ddot{x}}\right)\left(\frac{1}{C_E}\right) \quad \frac{\text{volt-sec}^2}{\text{meter}} \tag{16.17}$$

Therefore the effect of the shunt capacitance is to reduce the voltage sensitivity by a factor:

$$\frac{e_s/\ddot{x}}{e/\ddot{x}} = \frac{C_E}{C_E + C_S} \tag{16.18}$$

Figure 16.11 shows a plot of the factor $C_E/(C_E + C_S)$ vs. C_S/C_E.

Because the addition of a shunt capacitance reduces voltage sensitivity of a piezoelectric pickup, a shunt capacitance sometimes is added to adjust the sensitivity of a pickup to a value which is set by the dynamic range of the auxiliary equipment associated with the pickup.

OUTPUT RESPONSE TO VIBRATION. The sensitivity of a piezoelectric accelerometer is dependent upon the frequency of the applied vibration as shown in Fig. 12.5. Over a given frequency region, the sensitivity is independent of the frequency (this is called the *mid-band range*); however, at frequencies below and above this band the sensitivity is dependent upon the frequency of the applied vibration.

Low-frequency Response. At low frequencies (i.e., below the mid-band range), the frequency response is dependent upon the electrical capacitance C_T of the accelerometer (including its cable capacitance and other shunt capacitance) and the input resistance R of the coupling amplifier (or cathode follower) to which the pickup output is connected. The magnitude of the sensitivity at low frequencies is given by

$$\frac{e_f}{\ddot{x}} = \frac{e_s}{\ddot{x}} \frac{1}{\left[1 + \left(\frac{1}{2\pi f R C_T}\right)^2\right]^{1/2}} \quad \frac{\text{volt-sec}^2}{\text{meter}} \tag{16.19}$$

where f is the frequency and $e_s/\ddot{x}$ is the open-circuit voltage sensitivity.

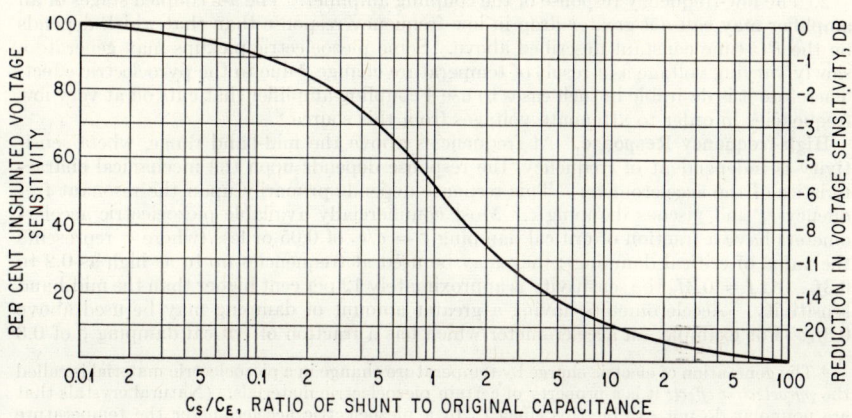

Fig. 16.11. Curve showing the reduction of open-circuit voltage sensitivity due to shunt capacitance across a pickup; the charge output from the transducer remains constant, regardless of the capacitive loading.

Equation (16.19) shows that the frequency dependence of sensitivity is a function of the product RC. This product is expressed in units of seconds and is called the *RC time constant*. Thus the combination $R = 10$ megohms and $C = 10,000$ mmfd provides the same low frequency response as the combination $R = 100$ megohms and $C = 1,000$ mmfd. Plots of e_f/e_s (i.e., open-circuit voltage sensitivity) vs. frequency are given in Fig. 16.12 for three values of RC. Because an increase in the RC time constant improves low-frequency response, a common technique used to obtain better low-frequency re-

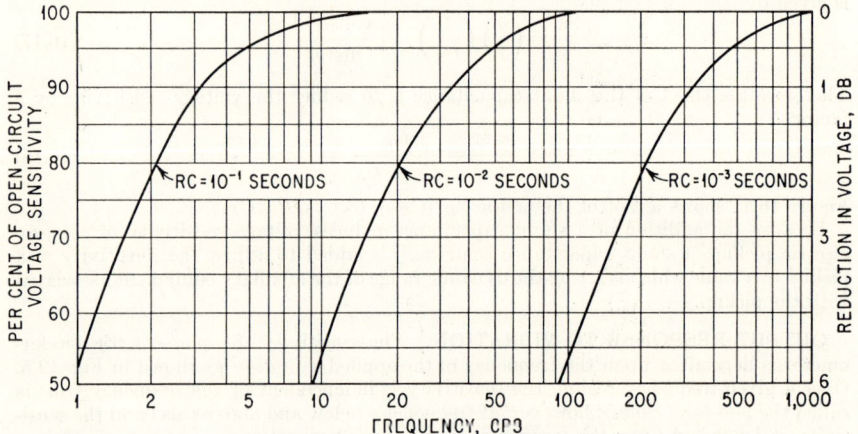

FIG. 16.12. Ratio of voltage output to open-circuit voltage vs. frequency for a piezoelectric pickup having RC time constants of 0.1, 0.01, and 0.001 sec.

sponse is to shunt the accelerometer with a large capacitor. This has the disadvantage of reducing voltage sensitivity throughout the mid-band frequency range.

The low-frequency response may be limited by the following factors in addition to the load resistance R of the coupling amplifier:

1. The internal shunt resistance of the piezoelectric or insulating materials within the accelerometer. Normally this shunt resistance is extremely high (100,000 megohms or more), and its effect can be neglected; however, at high temperatures it may be less than 100 megohms.

2. The low-frequency response of the coupling amplifier. The a-c coupled stages of an amplifier may cause a greater drop in low-frequency response than that which depends on the RC time constant described above. Some piezoelectric pickups may generate a slowly varying voltage as a result of temperature change * due to the pyroelectric effect. Therefore it is desirable in such cases to use a coupling amplifier that cuts off at very low frequencies in order to attenuate voltages from this source.[8]

High-frequency Response. At frequencies above the mid-band range, where sensitivity is independent of frequency, the response depends upon the mechanical characteristics of the accelerometer. This response depends primarily upon the resonant frequency f_n and viscous damping c. Most commercially available piezoelectric accelerometers have a fraction of critical damping $\zeta = c/c_c$ of 0.05 or less (where c_c represents the value of critical damping); they may be used at frequencies up to as high as 0.2 to $0.3f_n$. At $f = 0.3f_n$ the sensitivity is approximately 12 per cent higher than the mid-band sensitivity. Accelerometers having a greater amount of damping may be used above $0.3f_n$. For example, an accelerometer which has a fraction of critical damping ζ of 0.3

---

* The generation of electric charge by temperature change in a piezoelectric material is called the *pyroelectric effect;* it is a property of certain piezoelectric materials. (Natural crystals that are nonpolar do not exhibit this effect.) In a piezoelectric accelerometer the temperature changes within the piezoelectric material often occur at considerably slower rate than 0.01°F/sec, which allows much of the output to be attenuated in an a-c amplifier.

may be used up to approximately $0.4f_n$; and an accelerometer which has a fraction of critical damping ζ of 0.7 may be used to $0.7f_n$.

OUTPUT RESPONSE TO SHOCK. Faithful reproduction of a shock pulse may be attained with a piezoelectric accelerometer if the duration of the shock pulse is sufficiently short and the rise time of the shock pulse is sufficiently long.

Dependence of Response on Pulse Duration. When an amplifier is connected to a piezoelectric accelerometer, an electrical resistance path is placed across the pickup through which the electric charge may leak off. This leaking off of charge results in a continuous decay of the output voltage with time. For example, if an acceleration is applied suddenly to the pickup and is maintained at this fixed level, the output voltage e of the pickup is given as a function of time by

$$e = e_0 e^{-t/RC_T} \tag{16.20}$$

where e_0 is the voltage generated immediately after application of the acceleration, C_T is the combined electrical capacitance of the accelerometer and connecting cable, R is the input resistance of the coupling amplifier, and t is the time after application of the acceleration. The voltage decay is dependent upon the ratio of time t to the RC time constant, as indicated by Eq. (16.20). Over extended durations of time, the decay can be a large percentage of the voltage e_0. For example, when t exceeds $0.1RC$, the voltage e is less than $0.9e_0$.

While the allowable decay and overshoot depend upon the specific requirement of measurement accuracy, it is recommended that for most applications the RC time constant be at least ten times the pulse length in seconds.

As examples of the dependence of the response on pulse duration, consider the re-

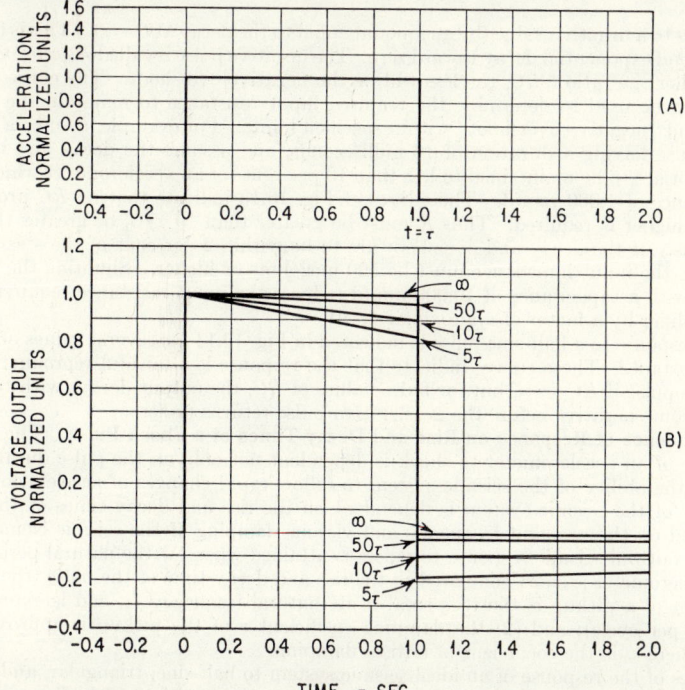

Fig. 16.13. Response of a piezoelectric accelerometer to a square acceleration pulse for RC time constants of 5τ, 10τ, 50τ, and ∞. (A) Input acceleration vs. time and (B) output voltage vs. time.

sponses of an accelerometer to square-wave and half-sine-wave pulses. A square-wave acceleration pulse is one which exhibits the following characteristic: Prior to time $t = 0$, the acceleration is zero; at time $t = 0$, the acceleration suddenly is increased to a fixed level and is maintained at this level for a time $t = \tau$, at which time the acceleration instantaneously returns to zero and is maintained at zero thereafter. The response of a piezoelectric pickup to a square-wave pulse is shown in Fig. 16.13 for four values of RC. During the acceleration pulse, the output exhibits an exponential decay (except for the curve which corresponds to $RC = \infty$); when the acceleration returns to zero, the output

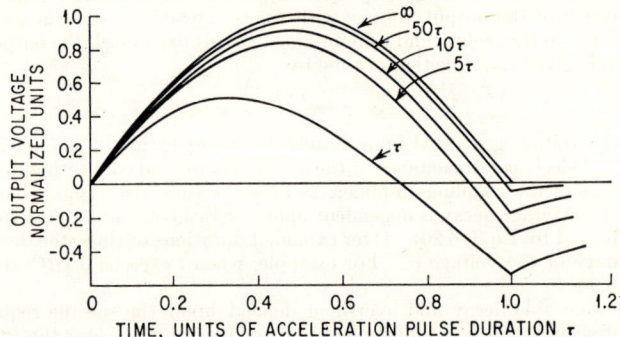

FIG. 16.14. Response of a piezoelectric accelerometer to a half-sine acceleration pulse for RC time constants equal to τ, 5τ, 10τ, 50τ, and ∞, where τ is equal to the duration of the half-sine pulse.[10]

decreases to a negative value (by an amount equal to the decay at $t = \tau$). This is followed by a second exponential decay toward zero. The negative pulse is called *negative overshoot*; the smaller the ratio t/RC, the less will be the negative overshoot. The curves of Fig. 16.13 can be used to determine the required input resistance to maintain the voltage decay and "negative overshoot" within specified limits. For example, assume a square-wave pulse having a duration of 10 milliseconds, and assume the decay and negative overshoot are to be maintained to less than 10 per cent for an accelerometer which has a capacitance of 1,000 mmfd. The curves of Fig. 16.13 indicate that an RC product of 10τ or higher is required. Thus R must be greater than $10\tau/C$, or greater than 100 megohms. If the decay and overshoot are to be within 2 per cent of the steady-state voltages, the input impedance must be 500 megohms or higher. Shunting the accelerometer with a capacitance of 10,000 mmfd reduces the input resistance requirement of the amplifier by a factor of approximately 10.

The response to a half-sine pulse is indicated in Fig. 16.14 for several values of the RC time constant.[9] These curves indicate that the response is a faithful reproduction of a half-sine pulse if $RC = \infty$, but for finite values of RC, the output decays with time and may become negative before the acceleration pulse returns to zero.

Dependence of Response on Rise and Decay Times of a Shock Pulse The faithful response of an accelerometer to shock is dependent not only on the pulse duration but also on the ability of the seismic system to follow rapid changes of acceleration. The response of the seismic system is dependent on the rise and decay times of the shock pulse and on the resonant frequency and viscous damping of the seismic element. In general, faithful seismic response to pulses is attained when (1) the natural period τ_n of the accelerometer is short compared to the rise and decay time of the pulse (the natural period τ_n of a pickup is the reciprocal of its natural frequency f_n, and is expressed in seconds per cycle); and (2) the damping coefficient c of the pickup is approximately $0.7c_c$, where c_c is the coefficient of critical damping.

Curves of the response of an ideal seismic system to half-sine, triangular, and square-wave pulses of acceleration are shown in Figs. 16.15, 16.16, and 16.17.* For each of the three pulse shapes, the response is given for ratios τ_n/τ of 1.014, 0.338, and 0.203 where

* Also see the discussion of Figs. 12.7 to 12.9.

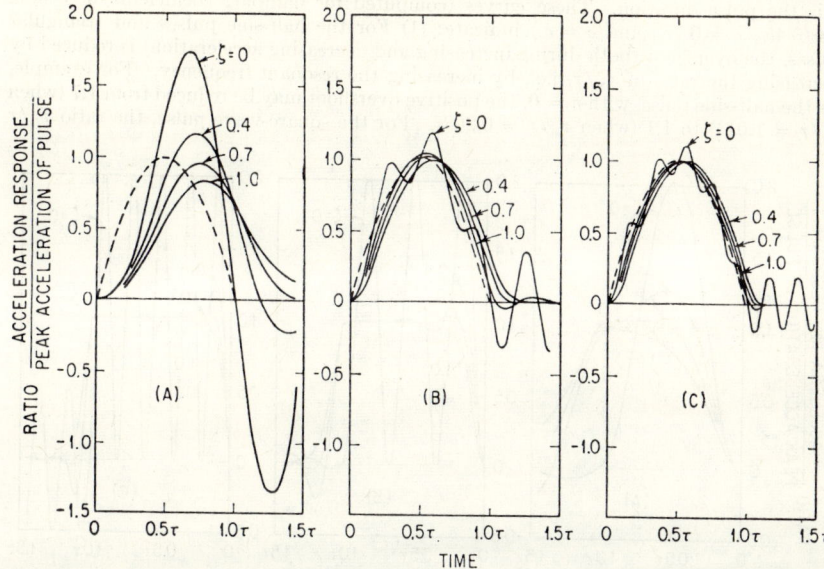

Fig. 16.15. Acceleration response to a half-sine pulse of acceleration of duration τ (dashed curve) of an ideal seismic transducer whose natural period τ_n is equal to: (A) 1.014 times the duration of the pulse, (B) 0.338 times the duration of the pulse, and (C) 0.203 times the duration of the pulse. The fraction of critical damping, $\zeta = c/c_c$, is indicated for each response curve.[10]

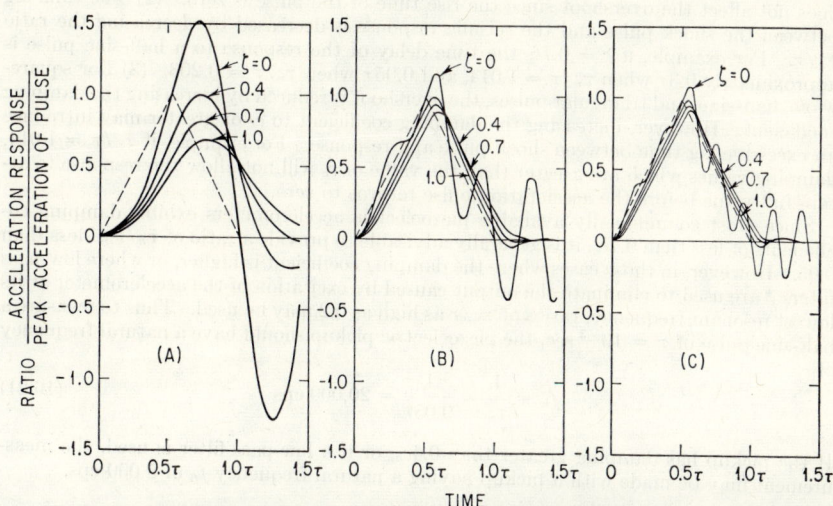

Fig. 16.16. Acceleration response to a triangular pulse of acceleration of duration τ (dashed curve) of an ideal seismic transducer whose natural period τ_n is equal to: (A) 1.014 times the duration of the pulse, (B) 0.338 times the duration of the pulse, and (C) 0.203 times the duration of the pulse. The fraction of critical damping ζ is indicated for each response curve.[10]

τ is the pulse duration. These curves (computed for damping coefficients of $c = 0$, $c = 0.4c_c$, $c = 0.7c_c$, and $c = c_c$) indicate: (1) For the half-sine pulses and triangular pulses, the overshoot (both during increasing and decreasing acceleration) is reduced by decreasing the ratio of τ_n/τ, i.e., by increasing the resonant frequency. For example, for the half-sine pulse, with $c = 0$, the positive overshoot may be reduced from 1.7 (when $\tau_n/\tau = 1.014$) to 1.1 (when $\tau_n/\tau = 0.203$). For the square-wave pulse, the ratio τ_n/τ

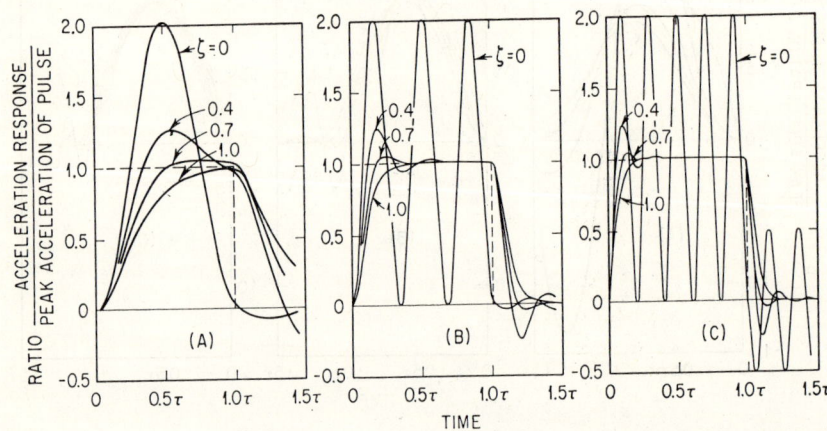

FIG. 16.17. Acceleration response to a rectangular pulse of acceleration of duration τ (dashed curve) of a mass-spring transducer whose natural period τ_n is equal to: (A) 1.014 times the duration of the pulse, (B) 0.338 times the duration of the pulse, and (C) 0.203 times the duration of the pulse. The fraction of critical damping ζ is indicated for each response curve.[10]

does not affect the overshoot since the rise time of the pulse is zero. (2) The time lag between the shock pulse and the seismic response is decreased by decreasing the ratio τ_n/τ. For example, if $c = 0.7c_c$ the time delay of the response to a half-sine pulse is approximately 0.3τ when $\tau_n/\tau = 1.014$, and 0.05τ when $\tau_n/\tau = 0.203$. (3) For square-wave, half-sine, and triangular pulses, the overshoot is reduced by increasing the damping coefficient. However, increasing the damping coefficient to c_c or greater may introduce an excessive lag time between shock pulse and response. For example, if $\tau_n/\tau = 1.014$, damping values which are greater than the value of c_c will not allow the response to attain full value before the acceleration pulse returns to zero.

Since most commercially available piezoelectric accelerometers exhibit damping coefficients of less than $0.1c_c$, it is generally advisable to provide a ratio of τ_n/τ of less than 0.05. However, in those cases where the damping coefficient is higher, or where low-pass filters * are used to eliminate the output caused by excitation of the accelerometer at its lowest resonant frequency, ratios of τ_n/τ as high as 0.2 may be used. Thus to measure a half-sine pulse of $\tau = 10^{-3}$ sec, the piezoelectric pickup should have a natural frequency of

$$f_n = \frac{1}{\tau_n} = \frac{1}{0.05\tau} = 20{,}000 \text{ cps} \qquad (16.21)$$

If the pickup has damping greater than $0.4c_c$, or if a low-pass filter is used, the measurement may be made with a pickup having a natural frequency f_n of 5,000 cps.

* A low-pass filter may be inserted following a high-impedance coupling device (e.g., a cathode follower) if the resonant frequency of the pickup is low enough so that significant output results from shock excitation. In the passband, and for the range of amplitudes employed, the filter must have the following characteristics: the input-output characteristic must be linear, the amplitude response must be linear, and the phase response must be linear. It is important that the transient characteristics of the filter be evaluated.

PIEZOELECTRIC ACCELEROMETERS

ACCELERATION-AMPLITUDE CHARACTERISTICS. Piezoelectric accelerometers are generally useful for the measurement of acceleration of magnitudes of from $10^{-4}g$ to more than $10^4 g$. The lowest value of acceleration which can be measured is approximately that which will produce an output voltage equivalent to the electrical input noise of the coupling amplifier connected to the accelerometer when the pickup is at rest. Over its useful operating range, the output of a piezoelectric pickup is directly and continuously proportional to the input acceleration. A single accelerometer often can be used to provide measurements over a dynamic amplitude range of 10,000 to 1 (80 db), which is substantially greater than the dynamic range of most of the associated transmission, recording, and analysis equipment.

Commercial accelerometers generally exhibit excellent linearity of output voltage vs. input acceleration and virtually no mechanical hysteresis * under normal usage. A linearity of better than 2 per cent of actual reading up to $4,000g$ and better than 5 per cent of actual reading up to $20,000g$ is possible.[11]

At very high values of acceleration (depending upon the design characteristics of the particular pickup), nonlinearity, hysteresis, or damage may occur. For example, large dynamic forces may produce voltage outputs sufficient to reduce permanently the sensitivity of the piezoelectric material. Further, if the dynamic forces exceed the biasing or clamping forces, the seismic element may "chatter" or fracture, although such a fracture might not be observed in subsequent low-level acceleration calibrations. High dynamic accelerations also may cause a slight physical shift in position of the piezoelectric element in the accelerometer—sometimes sufficient to cause a change in sensitivity. The upper limit of acceleration measurements depends upon the specific design and construction details of the pickup, and may vary considerably from one accelerometer to another, even though the design is the same. In certain designs the upper limit is dependent upon the machining accuracy of the parts. It is not always possible to calculate the upper acceleration limit of a pickup. Therefore one cannot assume linearity at acceleration levels for which calibration data cannot be obtained.

DIRECTIONAL SENSITIVITY—TRANSVERSE SENSITIVITY. The output of an ideal piezoelectric pickup depends upon the direction of applied acceleration and is given by

$$e_\theta = e_{\max} \cos \theta \tag{16.22}$$

where $e_{\max}$ is the output in the direction of maximum sensitivity and θ is the angle between the direction of maximum sensitivity and the direction of acceleration. The directional dependence of sensitivity of an ideal pickup is shown in Fig. 16.18. If the maximum sensitivity is in the Y direction, the sensitivity in the XY or YZ planes is described by a *figure-eight* locus. The sensitivity at $\theta = 90°$ is called the *transverse sensitivity*.† In an ideal pickup the transverse sensitivity is zero.

It is not practical to construct an accelerometer having a transverse sensitivity of zero. Because of practical fabrication tolerances, the piezoelectric element in a pickup may not seat tightly against the frame, the mass may have an axial tilt, or the mounting stud or flat may vary slightly from the ideal direction. These factors create an effective angular tilt of the axis of maximum sensitivity.

The effect of tilting the axis of maximum sensitivity by an angle θ_1 in the XY plane is shown in Fig. 16.19. As indicated in Fig. 16.19A, in the XY plane this results in a voltage output of

$$e_\theta = e_{\max} \cos(\theta - \theta_1) \tag{16.23}$$

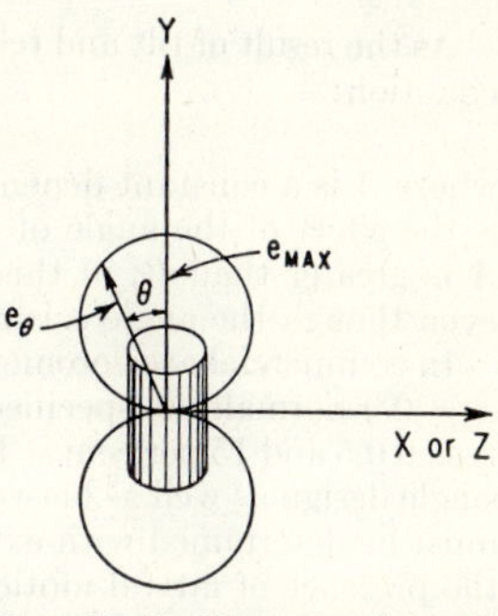

Fig. 16.18. Directional sensitivity of an ideal seismic piezoelectric pickup.

---

* Mechanical hysteresis in this context is the change in sensitivity of a pickup at one level of acceleration after being subjected to another level of acceleration within the acceleration-measuring range of the instrument.

† The terms cross-axis sensitivity and lateral sensitivity are used as synonyms for transverse sensitivity.

Along the Y axis, the following voltage is produced:

$$e_y = e_{max} \cos \theta_1 \qquad (16.24)$$

Along the X axis, the following voltage results:

$$e_x = e_{max} \sin \theta_1 \qquad (16.25)$$

In the YZ plane a typical figure-eight locus is obtained (see Fig. 16.19B) which has a maximum sensitivity (represented by the diameter of each locus circle) given by Eq. (16.24). The output in the transverse plane (XZ) is shown in Fig. 16.19C. In the transverse plane, the sensitivity at an angle ϕ is given by

$$e_\phi = e_{max} \sin \theta_1 \cos \phi \qquad (16.26)$$

In an ideal pickup, $e_\phi = 0$. Output also can be produced by transverse acceleration which creates small shear forces in the piezoelectric material; these forces are independent of the direction of the transverse acceleration in the XZ plane, i.e., independent of ϕ.

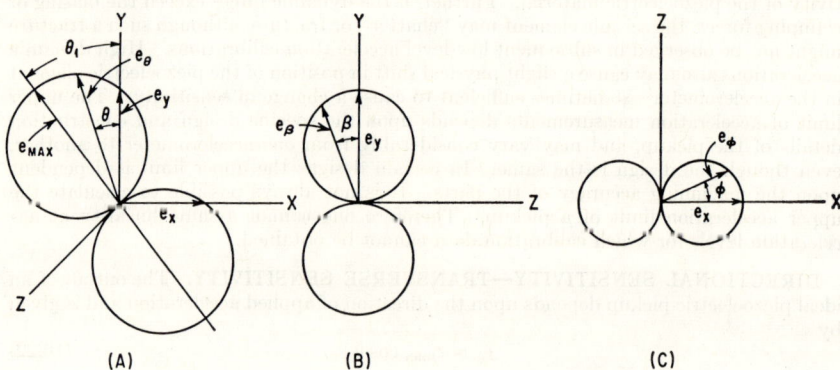

FIG. 16.19. Directional sensitivity of a pickup tilted from the Y axis by an angle θ_1. (A) Output in the XY plane, (B) output in the YZ plane, and (C) output in the (transverse) XZ plane.

As the result of tilt and related effects, the transverse sensitivity usually is given by the equation:

$$e_\phi = e_{max} (A + B \cos \phi) \qquad (16.27)$$

where A is a constant depending upon effects such as the shear effect described above, B is the effect of the angle of tilt, and ϕ is the angle in the transverse plane. Sometimes A is greater than B; in this case the transverse sensitivity may not pass through zero even though the angle ϕ is rotated through 360°.

In commercial accelerometers, the maximum transverse sensitivity (which occurs when $\phi = 0°$) normally is specified in per cent of the maximum sensitivity and is usually between 0.5 and 15 per cent. The maximum transverse sensitivity varies between units of a single design as well as between different designs. The directional sensitivity of a pickup must be determined with extreme care. If a vibration machine is used for this purpose, the presence of lateral motion in the vibration table may produce significant errors (see Chap. 18).

EFFECTS OF MOUNTING ON PICKUP CHARACTERISTICS. The torque used to mount an accelerometer in place should be controlled to prevent damage to the pickup. Excessive torque can strip or break mounting threads and studs. The application of insufficient torque will result in a mounting that is not secure, thereby causing large errors in high-frequency vibration measurements. Mounting torques of the order of 6 to 18 in.-lb usually are recommended by manufacturers.

Because of design limitations, some compression-type accelerometers have sensitivities that are a function of mounting torque.[5] This characteristic is called *torque sensitivity*. Since *torque sensitivity* generally is attributed to bending stresses induced in the housing of the accelerometer, the sensitivity may also depend upon the location of intimate contact between the pickup and the mounting surface. If an accelerometer exhibits *torque sensitivity*, one should use the same type of mounting and the same mounting torque in each application in order to obtain reproducible and accurate data.

The condition of the mounting surface to which the accelerometer is attached is important (see Fig. 20.2). If the surface is rough, a tight joint cannot be made between the accelerometer and the structure. As a result, the accelerometer will be decoupled from the vibratory motion of the structure. Any mounting plate, jig, or fixture between the accelerometer and the structure should be examined to ensure that a tight joint is obtained. The mounting jig itself is effectively a spring which becomes softer with increasing thickness and stiffer with increasing diameter. Many mounting fixtures have their first resonance below 2,000 cps and therefore can severely affect the accuracy of the measurement being made with the accelerometer. Consideration also should be given to the characteristic of any mounting fixture subject to transverse acceleration. Transverse motion may excite resonances which usually occur at lower frequencies than resonances in the axial direction.

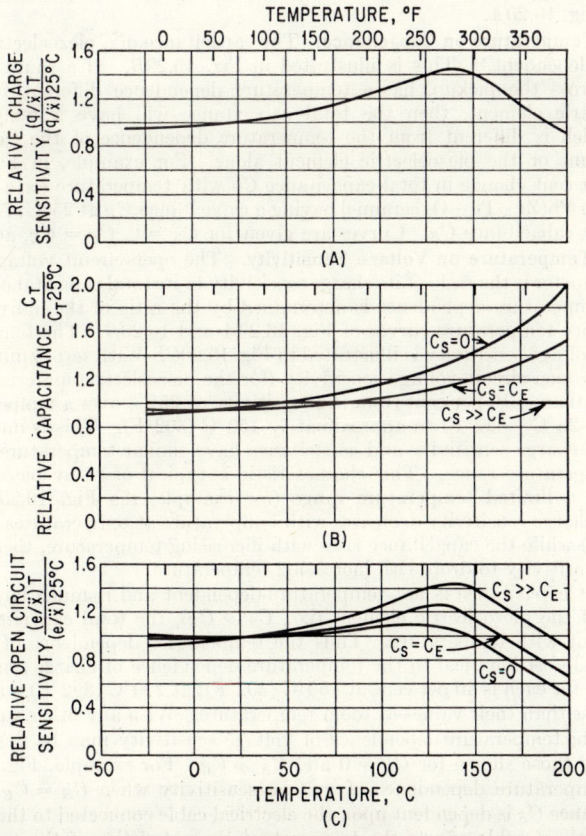

FIG. 16.20. Effect of temperature on the characteristics of a piezoelectric pickup employing a $Pb(Zr_{.53}Ti_{.47})O_3$ ceramic. (A) Temperature dependence of charge sensitivity. The charge sensitivity vs. temperature is shown relative to the value at 25°C. (B) Temperature dependence of capacity. (C) Temperature dependence of open-circuit voltage sensitivity.

Mounting. Misalignment in the mounting angle of an accelerometer will produce an effect equivalent to a component of transverse sensitivity, as indicated by Eq. (16.23) and Fig. 16.19. For example, a misalignment of 4° from the true perpendicular will produce an effect equivalent to a transverse sensitivity of 7 per cent along one axis.

EFFECTS OF TEMPERATURE ON A PICKUP. Piezoelectric pickups are available which may be used in the temperature range from −300°F (−185°C) to above 525°F (274°C) without the aid of external cooling. The voltage sensitivity, charge sensitivity, capacitance, and frequency response of a pickup depend upon the ambient temperature of the pickup. This temperature dependence is due primarily to variations in the characteristics of the piezoelectric material, of which several types are described in the second half of this chapter, but it also may be due to variations in the insulation resistance of cables and connectors—especially at high temperatures.[12] Consider the following example of the temperature dependence of a pickup based upon the use of a typical piezoelectric material—$Pb(Zr_{.53}Ti_{.47})O_3$.

Effects of Temperature on Charge Sensitivity. The charge sensitivity of a piezoelectric pickup is directly proportional to the d_{ij} piezoelectric constant of the material used in the piezoelectric element. As indicated in Fig. 16.20, the d_{ij} constants of most piezoelectric materials vary with temperature. The change in charge sensitivity with temperature for one type of accelerometer using $Pb(Zr_{.53}Ti_{.47})O_3$ as the piezoelectric ceramic is shown in Fig. 16.20A.

Effects of Temperature on Capacitance. The capacitance of a piezoelectric element is temperature-dependent.[13] This is illustrated in Fig. 16.20B. If a shunt capacitance, connected across the pickup, has a temperature dependence different from that of the piezoelectric element, then the total capacitance will have a temperature dependence which is different from the temperature dependence of the capacitance of either the shunt or the piezoelectric element alone. For example, in Fig. 16.20B is shown the per cent change in total capacitance C_T with temperature for a piezoelectric element [using $Pb(Zr_{.53}Ti_{.47})O_3$ ceramic] having a capacitance C_E at 25°C (77°F) shunted by a constant capacitance C_S. Curves are given for $C_S = 0$, $C_S = C_E$, and $C_S \gg C_E$.

Effects of Temperature on Voltage Sensitivity. The open-circuit voltage sensitivity of an accelerometer is the ratio of its charge sensitivity to its total capacitance ($C_S + C_E$). Hence the temperature dependence is determined by the ratio of the individual charge and capacitance temperature curves of Fig. 16.20A and 16.20B. The temperature dependence of voltage sensitivity is illustrated in Fig. 16.20C. With zero shunt capacitance ($C_S = 0$), the open-circuit voltage sensitivity (for the piezoelectric material illustrated) varies by less than +10 per cent from the sensitivity at 25°C, over a temperature range from about −25°C (−13°F) to approximately 150°C (302°F). This is due to the fact that both the charge sensitivity and capacitance have similar temperature dependence over this temperature range. This characteristic is typical of many piezoelectric materials within a limited temperature range (for example, see Fig. 16.20). For this material the charge sensitivity decreases with temperature at temperatures greater than 150°C (302°F) while the capacitance rises with increasing temperature, thereby causing the voltage sensitivity to drop with increasing temperature.

If the shunt capacitance is not temperature-dependent and is much larger than the capacitance of the piezoelectric element (i.e., $C_S \gg C_E$), the total capacitance is practically constant with temperature. Thus the temperature dependence of the voltage sensitivity is almost identical to the temperature dependence of charge sensitivity; the maximum rise for each is 40 per cent at 150°C (302°F); at 200°C (392°F) both are about 12 per cent less than their values at room temperature. With any other value of shunt capacitance, the temperature dependence of voltage sensitivity may be described by a curve between those shown for $C_S = 0$ and $C_S \gg C_E$. For example, Fig. 16.20C also shows the temperature dependence of voltage sensitivity when $C_S = C_E$. Since the shunt capacitance C_S is dependent upon the electrical cable connected to the pickup, the type and length of cable affects the temperature characteristics of the accelerometer-cable combination.

The above temperature dependence of characteristics is typical of certain commercial accelerometers but is not indicative of the state of the art. Materials are available for use over broader temperature ranges with less temperature dependence than those illustrated. Accelerometer manufacturers should be consulted for data regarding individual accelerometers.

Effects of Temperature on Low-frequency Response. As indicated in the section on *Low-frequency Response*, see Fig. 16.12, the voltage output at low frequencies depends on the shunt resistance across the pickup, both external and internal. Since the insulation resistance of all insulating materials and piezoelectric materials decreases exponentially with increasing temperature, the low-frequency response may be affected at very high temperatures.

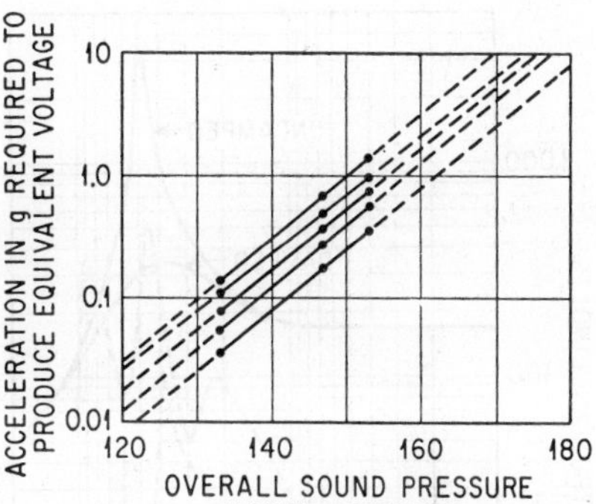

Effects of Temperature on High-frequency Response. To some extent, the temperature of a pickup affects its damping coefficient and (to a much lesser extent) its resonant frequency. The resonant frequency is affected because the elastic coefficients of the materials of which it is constructed vary with temperature. Damping is typically dependent upon temperature, as illustrated in Fig. 16.8 for silicone oil. Because most piezoelectric pickups are used at frequencies far below their lowest natural frequency (e.g., below $0.3f_n$), changes in resonant frequency and damping coefficient due to temperature variations have little effect upon their usable high-frequency response.

Fig. 16.21. The electrical outputs of several representative types of piezoelectric accelerometers as a result of acoustic excitation. The output voltage is expressed in terms of the acceleration in g required to produce an equivalent voltage. The sound source is random noise having uniform spectral density in the frequency range from 150 to 9,600 cps. The over-all sound-pressure level of this noise is expressed in decibels referred to 0.0002 dyne/cm². (*After W. Bradley, Jr.*[15])

Effects of Temperature on Shock Pulse Response. The response of a pickup to a shock pulse may be temperature dependent because of variation in capacitance and insulation resistance with temperature. These variations affect the electrical RC time constant, and hence may affect the ability of the pickup to reproduce long shock pulses faithfully. The response also may be temperature-dependent because of the variation in damping coefficient with temperature, which may affect the short risetime response. Results of these variations are described in previous sections of this chapter.

Effects of Transient Temperature. The voltage sensitivity of a piezoelectric pickup is not affected by the *rate of change* of the temperature of the instrument. However, electrical voltages (as large as 10 volts) due to a temperature fluctuation may be produced at the output terminals of the pickup in the following ways: (1) Differential expansion or compression of the mechanical and piezoelectric elements of the pickup may produce slowly varying forces on the piezoelectric element, and hence produce an output. (2) Many piezoelectric materials exhibit a *pyroelectric effect*, which is the generation of electric charge when the temperature of the piezoelectric material is changed. In general, the charge produced is proportional to the temperature change.

The voltage produced by temperature fluctuations occurs at a very slow rate and follows the temperature changes. Thermal insulation of the seismic element often is used to increase the time lag. Such pyroelectrically generated voltages may be filtered out electrically by means of a high-pass filter. Such a filter (having a cutoff frequency of the order of 5 cps) may be incorporated in the coupling amplifier. However, if the input voltage which results from this source is sufficiently high, the operation of the amplifier may be affected even though such a filter is used.

EFFECTS OF HIGH-INTENSITY SOUND. As indicated in Chap. 48, high-intensity sound can be an important source of vibratory excitation. Thus when vibration measurements are made on a structure that is exposed to such sound, the structure may vibrate with a considerable acceleration amplitude as a result of this source of ex-

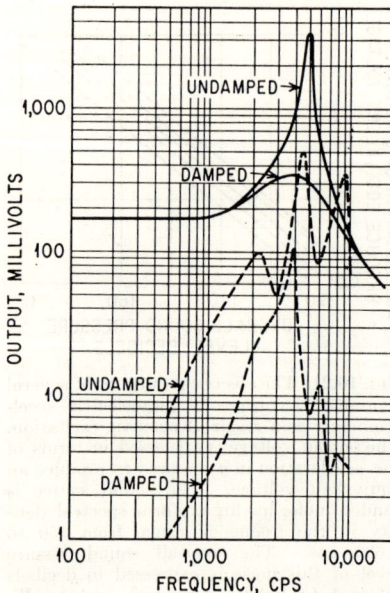

FIG. 16.22. Frequency-response curves for a piezoelectric accelerometer, illustrating the effects of damping on acoustic sensitivity. The solid lines show the usual sensitivity vs. frequency-response curves for 1g acceleration for the undamped and damped accelerometer; damping reduces the amplification ratio. The dashed curves show the voltage output when the accelerometers are exposed to a sound field of 150 db (referred to 0.0002 dyne/cm^2), for sine-wave excitation.

citation. The acoustic excitation also may affect the operation of an accelerometer used in the measurements by mechanical excitation of the piezoelectric element used in the accelerometer.

An illustration of the effect of high-intensity sound in producing an electrical output is shown in Fig. 16.21. In this example, a group of pickups are exposed to a random noise sound source having a uniform pressure spectral density in the frequency range from 150 to 9,600 cps. The output of the accelerometer resulting from the acoustic excitation is expressed in terms of the acceleration required to produce an equivalent voltage. These data are given for various models of a piezoelectric pickup.[15] For example, if one of these pickups is exposed to a sound pressure level of 170 db, the acoustical excitation produces an electrical signal equivalent to that produced by an acceleration of 10g; at 150 db, the acoustic excitation produces an electrical signal equivalent to an acceleration of 1g; and at 130 db, the corresponding value is only 0.1g.

When a piezoelectric accelerometer is exposed to noise, the electrical output which is produced will be greatest at the resonant frequency of the seismic system. This is illustrated by the dashed curves in Fig. 16.22 for an undamped seismic system which is exposed to (essentially) a 150-db sine-wave acoustic excitation. For acoustic excitation at 3,000 cps at a sound pressure level of 150 db, a voltage is generated which is greater than that which would be produced by a vibratory acceleration of approximately 1g.

Figure 16.22 shows that damping in the accelerometer reduces the effect of acoustic excitation, which effect can also be obtained by substantially increasing the resonant frequency of the accelerometer.

PIEZORESISTIVE ACCELEROMETERS

Piezoresistive solid-state materials can be employed as strain elements in accelerometers so as to provide a much higher sensitivity (e.g., by a factor of 50 to 100) than conventional wire strain-gage accelerometers.* A major advantage of the piezoresistive-type accelerometers is that they have good frequency response down to d-c (0 cps) along with a relatively good high-frequency response. The physical characteristics of piezoresistive materials are described in Part II of this chapter. Here they are applied to the design of a practical accelerometer.

DESIGN PARAMETERS

Many different configurations are possible for an accelerometer of this type (for example, see Ref. 16). For purposes of illustration, the design parameters are considered

---

* Manufacturers include: Gulton Industries, Inc., Metuchen, N.J.; Fairchild Controls Corp., Hicksville, L.I., N.Y.; Statham Instruments, Inc., Los Angeles, Calif.; Electro-Optical Systems, Inc., Pasadena, Calif.; Consolidated Electrodynamics Corp., Pasadena, Calif.

for a piezoresistive accelerometer which has a cantilever arrangement as shown in Fig. 16.23A. This uniformly stressed cantilever beam is loaded at its end with mass m. In this arrangement, four identical piezoresistive elements are used—two cemented to each side of the beam whose length is L in. These elements, whose resistance is R, form the

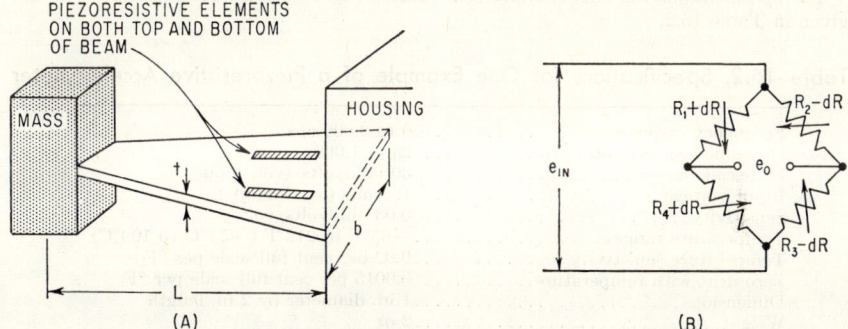

Fig. 16.23. (A) Schematic drawing of a piezoresistive accelerometer of the cantilever beam type. (B) Four piezoresistive elements are used—two cemented to each side of the uniformly stressed beam. These elements are connected in the above bridge circuit.

active arms of the balanced bridge shown in Fig. 16.23B. A change of length ΔL of the beam produces a change in resistance ΔR in each element. The gage factor K for each of the elements [defined by Eq. (17.1)] is

$$K = \frac{\Delta R/R}{\Delta L/L} = \frac{\Delta R/R}{\epsilon}$$

where ϵ is the strain induced in the beam, expressed in inches/inch, at the surface where the elements are cemented. If the resistances in the four arms of the bridge are equal, then the ratio of the output voltage of the bridge circuit to the input is

$$\frac{e_o}{e_{in}} = \frac{dr}{R} \simeq \frac{\Delta R}{R} = K\epsilon$$

Values of the gage factor K can be obtained from Eq. (16.29) and Table 16.5.

For the accelerometer shown in Fig. 16.23, suppose that
- b = width of beam at the support, in.
- t = thickness of beam, in.
- E = Young's modulus of beam, lb/in.^{2}
- m = mass of load at end of beam, lb-sec^2/ft

Then the following characteristics of the *uniformly stressed* beam, which is approximated by the configuration shown in Fig. 16.23A, may be derived which affect the response of this piezoresistive accelerometer:

Natural frequency:

$$f_n \simeq \frac{1}{2\pi}\sqrt{\frac{bEt^3}{6L^3m}} \quad \text{cps}$$

Deflection x of end of beam for mass m:

$$x \simeq \frac{mgL^3}{2EI} = \frac{6mg}{Eb}\left(\frac{L^3}{t^3}\right) = \epsilon\left(\frac{L^2}{t}\right) \quad \text{in.}$$

Strain at the surface to which the elements are bonded:

$$\epsilon \simeq \frac{6mgL}{Ebt^2}$$

The specifications for one example of a piezoresistive accelerometer of this type are given in Table 16.2.

Table 16.2. Specifications for One Example of a Piezoresistive Accelerometer

Frequency response	0 to 2,000 cps
Dynamic acceleration range	$2g$ to $1,000g$
Full-scale output	30 millivolts/(volt input)
Input voltage	10 volts (maximum)
Sensitivity	0.03 millivolts/volt/g
Temperature range	−65°F to 212°F (−54°C to 100°C)
Temperature sensitivity	0.02 per cent full scale per °F
Zero drift with temperature	0.0015 per cent full scale per °F
Dimensions	1 in. diameter by 2 in. length
Weight	2 oz
Output impedance	250 to 350 ohms
Transverse response	Less than 2 per cent of the maximum sensitivity
Linearity (including hysteresis)	±1 per cent

REFERENCES FOR PART I

1. Guttwein, G. K., and A. I. Dranetz: *Electronics*, **24**:120 (1951).
2. Orlacchio, A. W., and G. Hieber: *Electronic Inds.*, **16**:75 (1957).
3. Peters, R. J.: "NRL Type C-4 Barium Titanate Accelerometers," Proceedings of Symposium on Barium Titanate Accelerometers, *NBS Rept.* 2654, 1953, p. 57.
4. Perls, T. A.: "Determination of Sinusoidal Acceleration at Peak Levels Near That of Gravity by the Chatter Method," *NBS Rept.* 3399, September, 1954.
5. Perls, T. A., and C. W. Kissinger: "A Barium Titanate Accelerometer with Wide Frequency and Acceleration Ranges," *NBS Rept.* 2390, April, 1953.
6. Kissinger, C. W.: "Transverse Response," Proceedings of Symposium on Barium Titanate Accelerometers, *NBS Rept.* 2654, 1953, p. 119.
7. Dranetz, A. I.: "Natural Frequency and Frequency Response," Proceedings of Symposium on Barium Titanate Accelerometers, *NBS Rept.* 2654, 1953, p. 95.
8. Fleming, S. T.: "Cathode Follower Design," Proceedings of Symposium on Barium Titanate Accelerometers, *NBS Rept.* 2654, 1953, p. 175.
9. Lawrence, A. F.: "Crystal Accelerometer Response to Mechanical Shock Impulses," *Shock and Vibration Bull.* 24, Office of Secretary of Defense, February, 1957, p. 298.
10. Levy, S., W. D. Knoll, and D. Wilhelmia: *J. Research Natl. Bur. Standards*, **45**: 303/RP2138 (1950).
11. Perls, T. A.: "Tests of Accelerometer Linearity at High Accelerations," Proceedings of Symposium on Barium Titanate Accelerometers, *NBS Rept.* 2654, 1953, p. 133.
12. Orlacchio, A. W.: *Elec. Mfg.*, **59**:78 (1957).
13. Rudnick, N.: "Variation of the Dielectric Properties of $BaTiO_3$ Ceramic with Temperature," *NBS Rept.* 2654, 1953, p. 37.
14. Connelly, D. B.: "Characteristics of a Lead Metaniobate Accelerometer," *Proc. Instr. Soc. Amer.*, vol. 11, Paper 56-10-1, September, 1956.
15. Bradley, W., Jr.: "Effects of High Intensity Acoustic Fields on Crystal Vibration Pickups," *Proc. 26th Shock and Vibration Symposium*, Office of Secretary of Defense, 1958.
16. Padgett, E. D., and W. V. Wright: "Silicon Piezoresistive Devices," *Proc. Instr. Soc. Amer.*, vol. 15, Paper 42-NY60, 1960. Also, Xavier, M. A., and C. O. Vogt: "Characteristics and Applications of a Semiconductor Strain Gage," *Proc. Instr. Soc. Amer.*, vol. 15, Paper 16-NY60, 1960.

16

PART II: PROPERTIES OF PIEZOELECTRIC AND PIEZORESISTIVE MATERIALS

W. P. Mason
Bell Telephone Laboratories

This part of Chap. 16 describes the properties of solid-state materials that are useful in connection with the design and application of electromechanical transducers which are employed in shock and vibration work. These include piezoelectric and piezoresistive materials.

PIEZOELECTRIC MATERIALS

A piezoelectric material is one that generates an electric charge when it is subjected to a physical stress. Such materials, when employed in vibration-measuring devices, are capable of generating relatively large voltages. For example, in a typical piezoelectric material, a stress resulting in a strain of 10^{-2} microinch/in. may produce as much as 0.2 volt. Conversely, a piezoelectric crystal undergoes a change in dimensions when a voltage is applied to it. In general, this change is proportional to the applied electrical potential. This effect is made use of in the piezoelectric vibration machines (frequently employed as calibration "shake tables") described in Chap. 25.

Piezoelectric materials can be divided into (1) *natural crystals* (such as quartz) and *synthetic crystalline materials* (such as ammonium dihydrogen phosphate or lithium sulfate) and (2) *polarized ferroelectric ceramics* such as the barium titanates. In natural and synthetic crystalline materials, the relation between applied stress and the generated charge depends on the symmetry of the crystal, the direction of the applied stress, and the location of the electrodes. In the polarized ferroelectric ceramics, the relationship between the applied stress and the generated charge depends on the direction and magnitude of the induced polarization, the direction of the applied stress, and the location of the electrodes.

The most widely used ceramic ferroelectric material is barium titanate ($BaTiO_3$). Ceramics of this material are made by sintering together powders of barium oxide (BaO) and titanium dioxide (TiO_2). The grain size of the ceramic depends on the temperature and length of baking. In general, smaller grain sizes appear to result in higher coercive forces and more stable materials.

To obtain the piezoelectric effect in a ceramic ferroelectric material, one must polarize the material with a constant voltage, or apply a voltage and remove it, leaving a remnant polarization. The latter process sometimes is carried out by raising the temperature of the ceramic above its Curie temperature (the temperature above which the material loses its ferroelectric properties), or to a phase transition temperature for which the coercive field is low, applying a biasing voltage, and then cooling the ceramic slowly under the influence of the bias. For PZT, bias voltages usually are between 30,000 volts/in. and 100,000 volts/in. thickness.*

---

* For properties of PZT, see Ref. 20

Because of their higher dielectric constant and greater sensitivity, polarized ferroelectric ceramics usually can be used to measure smaller stresses extending over a longer time than can piezoelectric crystals such as quartz. This factor together with the ease of fabrication and the lower cost makes such materials particularly applicable for use in accelerometers and stress-measuring devices. These materials suffer from the disadvantage that their sensitivities change appreciably with time as a result of aging,[27] and from the fact that very large stresses and high temperatures may cause them to become depolarized. Hence, devices using such ceramics cannot be classed as primary standards giving reproducible results under all circumstances. For such standards, quartz crystals and piezoresistive crystals are superior.

Tables 16.3 and 16.4 give representative values for several fundamental constants for piezoelectric materials. The values obtained for a given specimen of ferroceramic material may differ somewhat from the values given in Table 16.4 since they depend on the degree of polarization, on the method of fabrication, and on the time subsequent to polarization (because of possible aging effects which cause the dielectric constant and the piezoelectric constant to decrease with time, and cause the elastic stiffness and the Q factors to increase with time).

CRYSTAL CUTS AND THEIR PROPERTIES

Piezoelectric elements are cut from crystals such as those shown in Fig. 16.24. They are designated by the axis perpendicular to the largest face of the cut; they are further identified by the angular orientation with respect to the other two axes. Thus a 45° X-cut represents a crystal with its thickness direction along the X axis, and its length

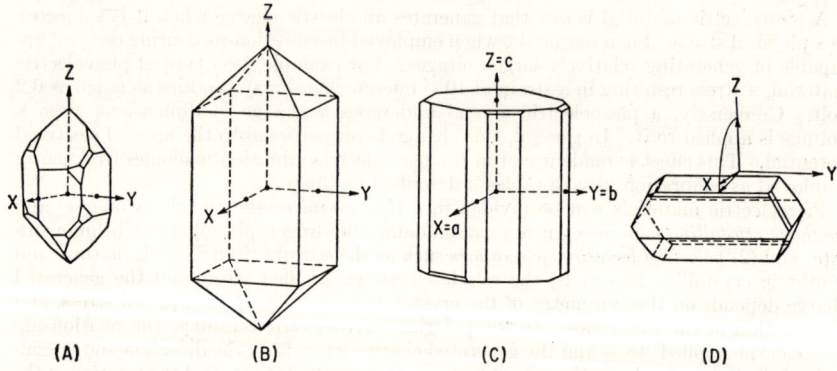

Fig. 16.24. Crystal forms, showing the major crystal axes: (A) quartz, (B) ammonium dihydrogen phosphate (a tetragonal crystal), (C) Rochelle salt (an orthorhombic crystal), and (D) lithium sulfate monohydrate (a monoclinic crystal).[32]

axis 45° between the other two; the output voltage is measured along the thickness direction. Table 16.3 gives some fundamental constants for a number of well-known crystal cuts.

The quartz crystal cuts which have been most widely used for stress and acceleration measurements are the first three listed in Table 16.3. The first quartz cut is used to measure stresses at relatively low frequencies. The second cut is widely used for measuring very rapidly varying stresses and accelerations. The third cut responds to shear stresses and is used to detect shear waves in solids. Other crystals used in measuring stresses are lithium sulfate monohydrate (LH) and tourmaline.

For a given crystal material and for a given cut, the voltage generated depends on the mode of strain: i.e., whether it is thickness shear, thickness expansion, face shear, or transverse expansion. These basic types of deformation are illustrated in Fig. 16.25.

Table 16.3. Properties of Piezoelectric Crystals

(After W. P. Mason.[23])

Crystal;* cut	Mode†	Elastic constant,‡ m^2/newton	Piezoelectric constant, coulombs/newton	Dielectric constant ϵ, farads/m	Open-circuit voltage, $g = d/\epsilon$, $\left(\dfrac{\text{volts}}{\text{m}}\right)\bigg/\left(\dfrac{\text{newton}}{\text{m}^2}\right)$
Quartz, X-cut, length Y	L.L.	1.27×10^{-11}	$d_{21} = 2.25 \times 10^{-12}$	$\epsilon_1 = 4.06 \times 10^{-11}$	0.055
Quartz, X-cut, thickness X	T.L.	1.16 §	$d_{11}\S = -2.04$	$\epsilon_1 = 4.06$	0.050
Quartz, Y-cut, thickness Y	T.S.	2.57 §	$2d_{11}\S = +4.4$	$\epsilon_1 = \epsilon_2 = 4.06$	0.108
Rochelle salt, 45° X-cut	L.L.	6.7	$\dfrac{d_{14}}{2} = 435$	$\epsilon_1 = 444.0$	0.098
Rochelle salt, 45° Y-cut	L.L.	9.89	$\dfrac{d_{25}}{2} = -28.4$	$\epsilon_2 = 9.85$	0.200
ADP; 45° Z-cut	L.L.	5.3	$\dfrac{d_{36}}{2} = 24.6$	$\epsilon_3 = 13.8$	0.178
KDP; 45° Z-cut	L.L.	4.85	$\dfrac{d_{36}}{2} = 10.7$	$\epsilon_3 = 19.6$	0.058
EDT; Y-cut, length X	L.L.	3.88	$d_{21} = 11.3$	$\epsilon_2 = 7.40$	0.152
DKT; 45° Z-cut	L.L.	4.25	$d_{31} = -12.2$	$\epsilon_3 = 5.80$	0.210
LH; Y-cut	T.L.	2 §	$d_{31} = 15$	$\epsilon_2 = 9.15$	0.165
LH (hydrostatic)	H.		$d_{21} + d_{22} + d_{23} = 13$	$\epsilon_2 = 9.15$	0.143
Tourmaline, Z-cut	T.L.	0.61 §	$d_{33}\S = -1.84$	$\epsilon_3 = 6.65$	0.0275
Tourmaline (hydrostatic)	H.		$d_{31} + d_{33} = -2.16$	$\epsilon_3 = 6.65$	0.032

* ADP = ammonium dihydrogen phosphate; KDP = potassium dihydrogen phosphate; EDT = ethylene diamine tartrate; LH = lithium sulfate monohydrate.
† Abbreviations: L. L. = length longitudinal; T. L. = thickness longitudinal; T. S. = thickness shear; H = hydrostatic.
‡ Measured in a constant electric field.
§ Indicates effective values of compliance and piezoelectric constant for thickness mode.

Table 16.4. Properties of Ferroelectric Ceramics *

Material	Young's modulus, newtons/m²	Piezoelectric constant d_{31}, coulombs/newton	Piezoelectric constant d_{33}, coulombs/newton	Dielectric capacitivity ϵ, farads/m	Open-circuit voltage $g_{33} = d_{33}/\epsilon$, $\left(\dfrac{\text{volts}}{\text{m}}\right) \Big/ \left(\dfrac{\text{newtons}}{\text{m}^2}\right)$
Commercial BaTiO₃ ceramics	1.18×10^{11}	-5.6×10^{-11}	$13\text{-}16 \times 10^{-11}$	$1{,}250 \times 10^{-11}$	0.0106
97% BaTiO₃, 3% CaTiO₃	1.22	-5.3	13.5	1,230	0.0111
96% BaTiO₃, 4% PbTiO₃	1.14	-3.8	10.5	880	0.012
90% BaTiO₃, 4% PbTiO₃, 6% CaTiO₃	1.24	-4.0	11.5	710	0.016
84% BaTiO₃, 8% PbTiO₃, 8% CaTiO₄	1.31	-2.7	8.0	530	0.015
80% BaTiO₃, 12% PbTiO₃, 8% CaTiO₃	1.28	-2.0	6.0	400	0.015
PZT 4 †	0.815	-9.7	23.5	875	0.0268
PZT 5 †	0.675	-14.0	32.0	1,200	0.0266
PZT 6 †	0.865	-7.8	19.1	860	0.022
NbO₃ (K₁ 50%; Na 50%)	1.02	-3.2	8.0	235	0.034
Pb(NbO₃)₂	0.35	-1.1	8.0	200	0.040

* These data represent values at 25°C.
† PZT is a copyrighted name for a ceramic composition of the Clevite Brush Co.; data are from Clevite Brush Co.

PIEZOELECTRIC MATERIALS

RELATIONSHIP BETWEEN APPLIED FORCE AND GENERATED CHARGE

Equation (16.8) relates applied force to the generated charge for a widely used configuration of piezoelectric material. For other arrangements, the constant d_{ij} in this equation is different. In the most general case, the three longitudinal strains and the three shearing strains require twenty-one elastic constants to relate them to the six

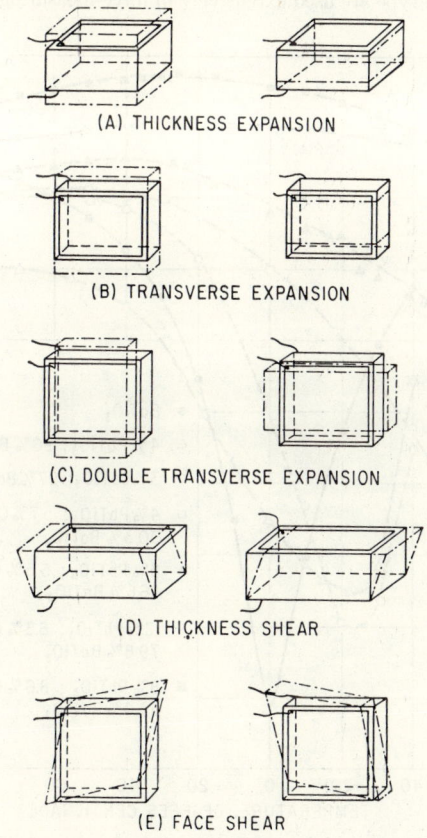

(A) THICKNESS EXPANSION

(B) TRANSVERSE EXPANSION

(C) DOUBLE TRANSVERSE EXPANSION

(D) THICKNESS SHEAR

(E) FACE SHEAR

FIG. 16.25. Basic deformations of a crystal.

stresses that can be applied.* The effect of symmetry is to reduce the number of elastic and piezoelectric constants required.

Constants for a number of piezoelectric materials are given in Tables 16.3 and 16.4. Also included with these data is the open-circuit voltage per unit of applied force; the *open-circuit voltage* for a unit cube of piezoelectric material acted on by a force of 1 newton is the quantity $g = d_{33}/\epsilon$ given in these tables.

EFFECTS OF TEMPERATURE AND HUMIDITY ON PROPERTIES OF PIEZOELECTRIC MATERIALS

The properties of ordinary barium titanate become very temperature dependent just below room temperature. This can be seen from the data of Figs. 16.26 and 16.27 which

* For definitions of stresses, strains, and elastic, piezoelectric, and dielectric constants in crystals, see Ref. 17.

show its elastic, piezoelectric, and dielectric constants as a function of temperature. By adding certain amounts of lead titanate ($PbTiO_3$) and calcium titanate ($CaTiO_3$), this effect is reduced considerably and the Curie temperature is raised. As can be seen from Figs. 16.26 and 16.27, all three properties are improved materially with temperature, although both the piezoelectric constant and the dielectric constant are lowered. Their ratio, which determines the open-circuit voltage, is not reduced significantly. Hence ceramics of this general type are used extensively in force-measuring devices.

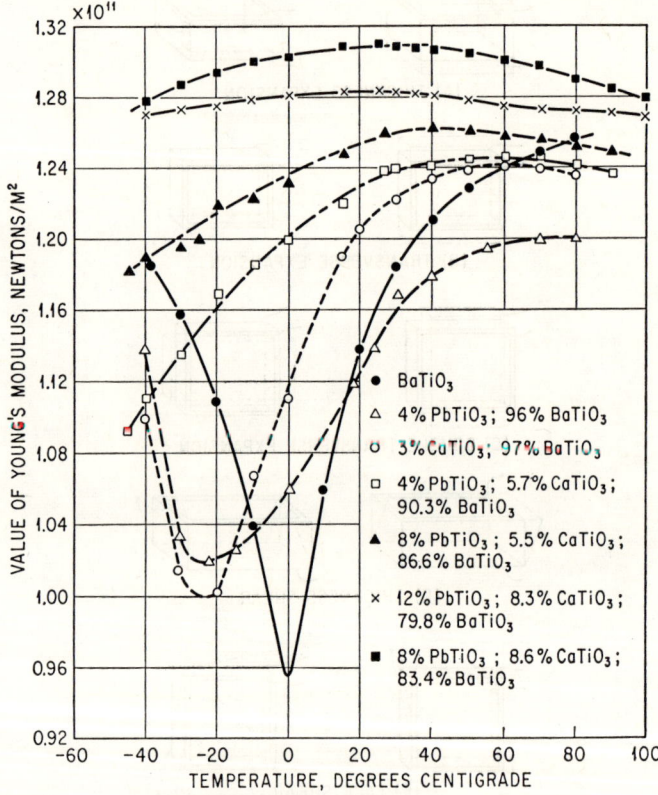

FIG. 16.26. Value of Young's modulus vs. temperature for several compositions of barium titanate ($BaTiO_3$), lead titanate ($PbTiO_3$), and calcium titanate ($CaTiO_3$). (*After W. P. Mason.*[27])

Barium titanate has a Curie temperature of about 130°C, but its piezoelectric properties do not remain stable much above 90°C. Other ceramics are available which have considerably higher Curie and transition temperatures, for example: lead titanate-zirconate mixture [20] $[x(PbTiO_3) + (1 - x)PbZrO_3]$, and lead metaniobate [22] $[Pb(NbO_3)_2]$, where x is the mole fraction of $PbTiO_3$ and $(1 - x)$ is the mole fraction of $PbZrO_3$.

FREQUENCY LIMITATIONS OF PIEZOELECTRIC MATERIALS

In general, piezoelectric crystals and ferroelectric ceramics are useful for measuring stresses which vary rapidly with time. For static stresses or very slowly varying stresses, the leakage resistance present for all dielectric materials causes the charge developed to leak off to $1/eth$ of its original value in a time t given by

$$t = RC$$

where R is the leakage resistance of the crystal and associated amplifier tube and C is the total capacitance of the crystal and amplifier tube. For example, in a typical accelerometer-amplifier combination, the capacitance C of the accelerometer and cable may be between 0.1×10^{-9} and 5×10^{-9} farad, and the resistance may be between 10^8 and 10^9 ohms, providing a time constant between 0.01 and 5 sec. Higher time constants may be

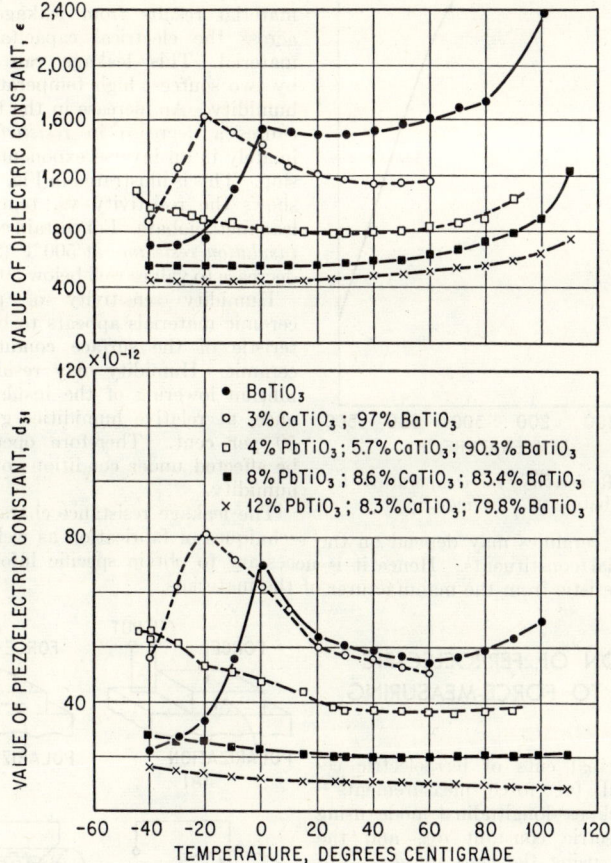

Fig. 16.27. Values of dielectric constant and effective piezoelectric constants vs. temperature for polarized compositions of $BaTiO_3$, $PbTiO_3$, and $CaTiO_3$. (*After W. P. Mason.*[27])

obtained (up to several hours) by the use of quartz piezoelectrics and electrometer amplifiers having a combined leakage resistance of approximately 10^{14} ohms. As indicated in the next section, the presence of high humidity or a relatively high temperature, which causes the leakage resistance to decrease exponentially with temperature, may lower this time still further.

The high-frequency limit for pickups which are fabricated of any of the materials which are described in this chapter is set by the resonant frequencies for the modes of motion employed. For thin sections of material which are mounted on backing plates, stresses having a duration as short as 0.1 microseconds are measured easily. Hence by the use of solid-state materials, faithful reproductions of stress, strain, or acceleration functions can be obtained from static values up to about 10 megacycles per second. By using quartz as an element in a wave guide, measurements have been made at frequencies up to 10^3 megacycles per sec.

LEAKAGE RESISTANCE OF PIEZOELECTRIC MATERIALS

As indicated above, a limitation in the low-frequency response of a piezoelectric material results from leakage resistance across the electrical capacitance of the material. This leakage can be caused by two sources: high temperature or high humidity. An increase in the temperature causes a decrease in *resistivity*—approximately in an inverse exponential relationship. This is illustrated in Fig. 16.28 which shows the resistivity vs. temperature of lead metaniobate. For certain ceramics, the *insulation resistance* at 500°F (260°C) may decrease to values well below 500 megohms.

Humidity sensitivity of piezoelectric ceramic materials appears to be a characteristic of the surface condition of the ceramic. Humidity may result in a significant lowering of the insulation resistance, at relative humidities greater than 50 per cent. Therefore operation may be affected under conditions of very high humidity.

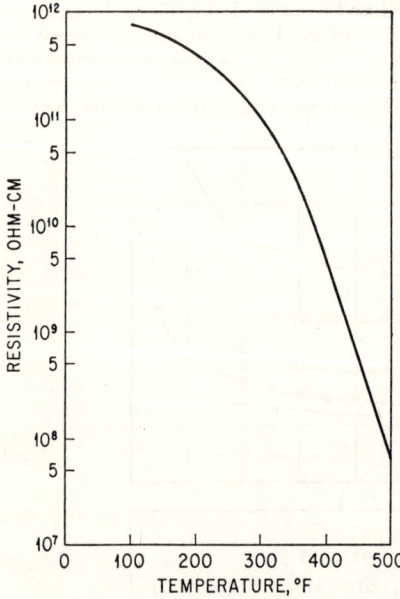

FIG. 16.28 Resistivity vs. temperature for lead metaniobate. (*G. Goodman.*[20])

The leakage resistance characteristics of piezoelectric ceramics may depend on the techniques of fabrication as well as on the basic material constituents. Hence it is necessary to obtain specific information on this characteristic from the manufacturer of the materials.

APPLICATION OF FERROELECTRIC CERAMICS TO FORCE-MEASURING DEVICES

The principal cuts of ferroelectric ceramics used for force measurements [25] use the thickness-longitudinal mode using the piezoelectric constant d_{33}, and the shear mode using the constant d_{15}. For the first cut, shown in Fig. 16.29A, the ceramic is polarized along the same direction that the force is applied and the voltage is measured in the same direction. The second cut, shown in Fig. 16.29B, can be used for measuring tangential forces. The ceramic is polarized in the same direction as that in which the force is applied, but the voltage is measured at right angles to the direction of polarization. Other arrangements used for measuring torques are the radial disc of Fig. 16.29C and the cylinder polarized as shown by Fig. 16.29D.

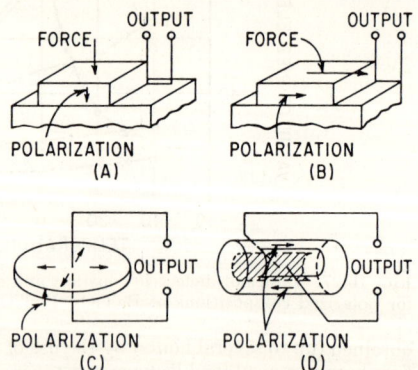

FIG. 16.29. Principal cuts for ferroelectric ceramics used in force measurements. (*A*) Ceramic is polarized along same direction as force is applied; voltage is measured in the same direction. (*B*) Ceramic is polarized along same direction as force is applied; voltage is measured in a direction at right angles to the direction of polarization. (*C*) Radial disc for measuring torque. (*D*) Cylinder for torque measurements.

PIEZORESISTIVE MATERIALS

The piezoresistance effect is the change of resistivity of a semiconductor or other material as a function of the applied stress. The stress may be a tension, a hydrostatic pressure, a shear, or a torque. Materials such as metals produce this effect but the percentage change is at least two orders of magnitude higher for certain semiconductors such as silicon, germanium, indium antimonide, and gallium arsenide than it is for metals. Since the resulting strain is proportional to the applied stress, piezoresistive semiconductors can be very useful for measuring small strains. These semiconductors occur in the form of stable, single crystals. Therefore, the sensitivities of transducers which make use of the piezoresistive effect are not affected by time, nor are their sensitivities affected by relatively high static pressures. Such materials also have the advantage that they can be calibrated by static weights since they are responsive to static forces. Therefore, they may be used in devices having a frequency response from d-c to a very high frequency. An added advantage is that they make use of the circuitry (described in Chap. 17) which has been developed for strain gages.

LONGITUDINAL PIEZORESISTANCE COEFFICIENT

Suppose a bar of material which exhibits the piezoresistive effect is subject to simple tension T along the bar. If T is expressed in dynes per square centimeter, then the fractional change in resistivity due to a tension is

$$\frac{\Delta \rho}{\rho} = \pi_l T \qquad (16.28)$$

where π_l is the longitudinal piezoresistive coefficient, expressed in square centimeters per dyne, which depends on the direction of the crystal structure in the bar. The actual resistance change in a tension member will be the result of the combined effects of this resistivity change and dimension changes. A tension always results in an increase in length and a decrease in cross-sectional area, and thus tends to increase the resistance because of dimension changes. The effect of the resistivity change adds to the effect of dimension changes if π_l is positive, and subtracts if π_l is negative. In many cases, the small effect of dimensional changes can be neglected in comparison with the larger effect due to the change in resistivity.

If the direction of the length of the sample has the direction cosines l, m, and n with respect to the X, Y, and Z crystallographic axes, then the piezoresistive coefficient π_l along the length of the sample is related to three fundamental constants, whose values are given in Table 16.5, by the equation [30a]

$$\pi_l = \pi_{11} + 2(\pi_{44} + \pi_{12} - \pi_{11})(l^2 m^2 + l^2 n^2 + m^2 n^2)$$

This relationship applies for cubic crystals, such as germanium, silicon, and indium antimonide. Since the factor $(l^2m^2 + l^2n^2 + m^2n^2)$ is a maximum in the direction making equal angles with the crystal axes, i.e., in the [111] crystallographic directions, it follows that π_l is a maximum or minimum along the [111] directions provided $(\pi_{44} + \pi_{12} - \pi_{11})$ has the same sign as π_{11}, or if $\frac{2}{3}|\pi_{44} + \pi_{12} - \pi_{11}| > |\pi_{11}|$. Otherwise the maximum effect occurs along the [100] axis. For P- and N-type germanium and for P-type silicon, the maximum occurs along the [111] axes. For N-type silicon the maximum occurs along the [100] axes.

USE AS A STRAIN GAGE

A thin rod or bar of any material exhibiting a large piezoresistance effect can be used in the same way as the wire strain gage described in Chap. 17. The gage factor K [defined by Eq. (17.1)] for a piezoresistive strain gage is

$$K = 1 + 2\sigma + E\pi_l \qquad (16.29)$$

Table 16.5. Adiabatic Piezoresistance Coefficients at Room Temperature
(*After W. P. Mason and R. N. Thurston.*[30a])

Materials	ρ, ohm-cm	π_{11}, 10^{-12} dyne/cm^2	π_{12}, 10^{-12} dyne/cm^2	π_{44}, 10^{-12} dyne/cm^2	π_1, 10^{-12} dyne/cm^2	E, 10^{12} dynes/cm^2	$(\pi_1)E$ (dimensionless)
Germanium:							
N-type....	1.5	−2.3	−3.2	−138.1	−94.9	1.55 *	−147
	5.7	−2.7	−3.9	−136.8	−94.7	1.55 *	−147
	9.9	−4.7	−5.0	−137.9	−96.9	1.55 *	−150
	16.6	−5.2	−5.5	−138.7	−101.2	1.55 *	−157
P-type....	1.1	−3.7	+3.2	+96.7	+65.4		+101.5
	15.0	−10.6	+5.0	+46.5	+31.4		+48.7
Silicon:							
P-type....	7.8	+6.6	−1.1	+138.1	93.6	1.87 *	+175
N-type....	11.7	−102.2	+53.4	−13.6	−102	1.30 †	−133
Indium-antimonide	....	−81.6	−114.2	+33.0	−81.3	0.74 †	−60.5

* Young's modulus in the [111] direction.
† Young's modulus in the [100] direction.

where E is Young's modulus for the material and σ denotes Poisson's ratio—the ratio of the magnitude of transverse strain to longitudinal strain resulting from the simple tension. The first term on the right expresses the resistance change due to the change in length; the second term is due to the change in cross-sectional area, and the third is due to the resistivity change. Since $0 < \sigma < \frac{1}{2}$ for all materials, the contribution of the first two terms is always between 1 and 2.

As indicated in Chap. 17, most commonly used metal-wire strain gages have gage factors between 2 and 4. Somewhat higher magnitudes are obtained with platinum, and values for nickel are in the range from 12 to 20. In contrast, germanium and silicon often have a gage factor of over 150. Table 16.5 includes values of the product $E\pi_1$ in the most sensitive direction, which is the principal part of the gage factor for these materials; the maximum value of this product occurs along the [111] direction for all the materials listed in Table 16.5 except N-type silicon, in which product the maximum is along a crystal axis.

Measurements on the effect of doping (i.e., the addition of small amounts of impurities) indicate that considerable improvement in the temperature stability of the gage factor can be obtained by employing higher doping values than those used in the materials whose characteristics are described in this section. For the most sensitive material, P-type silicon (cut along a [111] direction), the gage factor remains approximately constant within the range 60 to 65 for a range of temperature from −100°C to +150°C. The characteristics of a highly doped material (e.g., one having 1×10^{20} boron atoms/cm^3) are linear for strains as high as 2,500 microinches/inch. To achieve strains of this magnitude consistently requires that one of the dimensions of the gage be less than 0.002 in., which can be obtained by a grinding and etching process. Such highly doped materials have a resistivity of about 0.001 ohm-cm. This value does not vary by more than ± 10 per cent over a temperature range from −100°C to +200°C. These materials can be employed in a variety of transducers for the measurement of pressure, force, acceleration, and extension.

FORCE OR DISPLACEMENT MEASUREMENT

The longitudinal piezoresistance effect may be used in measuring forces or very small displacements by an arrangement in which a bar of silicon or other suitable material is placed under simple tension or compression. The resulting resistance change then can be related to the force or displacement being measured. The associated electric circuitry required is essentially the same as for resistance strain gages. This method is particularly advantageous in cases where it is desired to measure a steady or slowly varying force while permitting only negligible movement. For example, one arrangement that is very sensitive is shown in Fig. 16.30. It consists of two thin strips of piezoresistive materials joined together with a thin layer of nonconducting glue, cement, or other adhesive. For highest sensitivity, if germanium is used, the length of the strips should be along the [111] crystallographic direction, since this direction has the largest value of π_1. When a force F is applied, the device deflects as a cantilever beam, with tension in

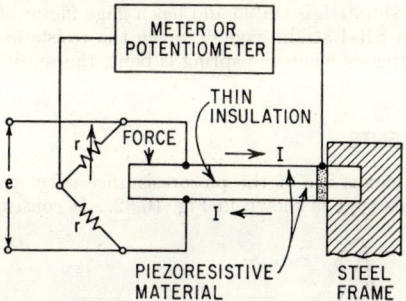

Fig. 16.30. Bridge connections for measurement of force or displacement. The sensitive element consists of two strips of piezoresistive materials which are cemented together. (*W. P. Mason and R. N. Thurston.*[30a])

the upper half and compression in the lower. This causes resistance changes of equal magnitude and opposite sign in the two halves, whose magnitude is given by Eq. (16.28).

A sensitive displacement gage of this type is shown in Fig. 16.31. Two strips of germanium 0.25 in. long by 0.006 in. on a side are cemented with an epoxy resin cement to opposite sides of a cantilever beam having the dimensions shown. Since the resistance of one side decreases while that of the other side increases when the cantilever is bent, the sensitivity is twice that of a single element. The resistivity of the germanium is 0.033 ohm-cm, giving a total resistance of 55 ohms for each of the germanium strips. Small

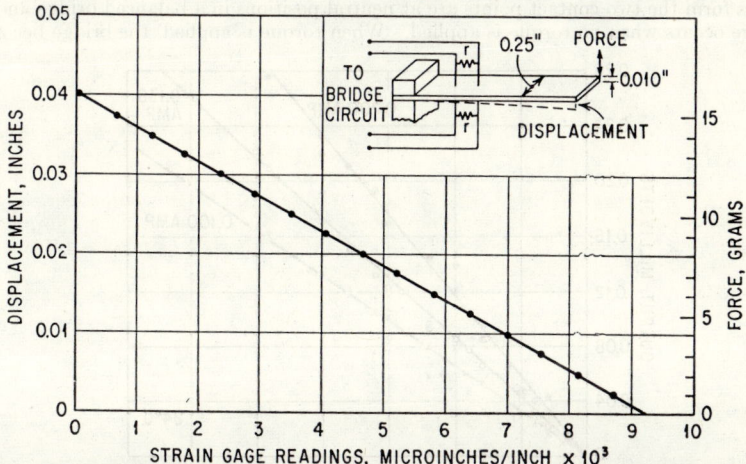

Fig. 16.31. Characteristics of a strain gage consisting of two germanium piezoresistive strips which are mounted on a thin steel cantilever beam showing displacement and force vs. meter reading in strain-gage circuit. (*J. J. Forst and F. T. Geyling.*[31])

gold-lead wires, containing 0.1 per cent antimony, are used in order to obtain a low-resistance contact with germanium. The wires are connected to the germanium by the use of pressure and heat at a temperature slightly above the gold-germanium alloying temperature (about 350°C). The germanium strips are connected to the two arms of a bridge circuit as indicated in Fig. 16.30. The device is so sensitive that ordinary building vibration produces a significant output. This extreme sensitivity may be reduced by shunting each element with a resistance. For example, Fig. 16.31 shows the reading of the meter, with a 12-ohm shunt, in the strain-gage circuit calibrated in microinches per inch, as a function of the applied force or the displacement of the end of the cantilever spring. The device, so shunted, is stable and has a gage factor of 30, i.e., it is about 14 times as sensitive as an SR-4 strain gage.[31] Since the resistance of one side decreases while the other side increases when the spring is bent, the sensitivity is twice that of a single element.

TORQUE MEASUREMENTS

A transducer which makes use of the piezoresistance effect can be used to measure torque. A device of this type is shown in Fig. 16.32. It consists of a cylinder of ger-

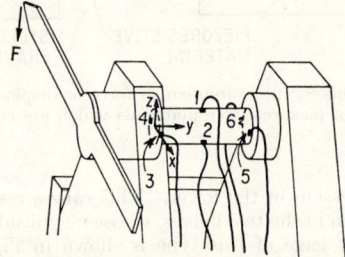

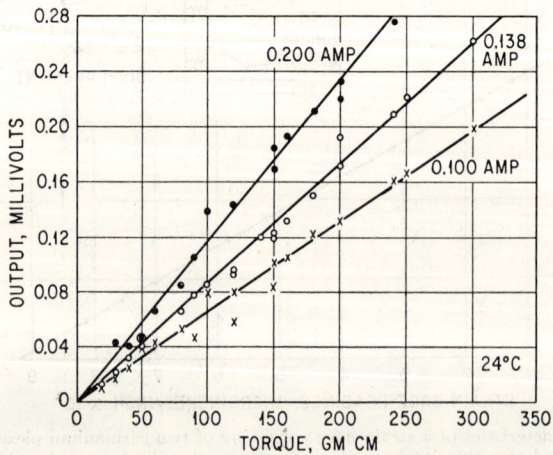

FIG. 16.32. A germanium piezoresistive transducer for use as a torque measurement device. The output voltage appears across terminals 1 and 2. (*After W. P. Mason and R. N. Thurston.*[30a])

manium (whose length is along a [100] direction) and six electrodes arranged as shown. In this form the two contact points are at neutral positions in a balanced bridge and no voltage occurs when no torque is applied. When torque is applied, the bridge becomes

FIG. 16.33. Voltage vs. torque relations for the germanium torque meter shown in Fig. 16.32.[30a]

unbalanced in proportion to the applied torque and a linear voltage-torque curve results. The output voltage appears across terminals 1 and 2. A distribution of current density is required which can be obtained by making electrodes 3 and 6 of one polarity, and 4 and 5 of the opposite polarity. Such a device can indicate both the magnitude and sign of an applied torque. For example, in the transducer of Fig. 16.32, the cylindrical portion has a diameter of 0.13 in. and a length of 0.5 in. Both ends are a little larger than the active cylindrical part. One end is cemented to a base with an epoxy resin cement: a torque arm is cemented to the front surface. To calibrate the device, known torques are applied by hanging weights on the torque arm. Figure 16.33 shows typical calibration curves so obtained. For a given biasing current, the output voltage increases in proportion to the torque. The current values shown include the unavoidable current from electrodes 3 to 4 and 6 to 5 as well as the useful currents from 3 to 5 and 6 to 4. Such a device must be shielded from strong light since germanium is photosensitive. Advantages of this type of torque meter are its high sensitivity, its response to steady torques as well as alternating torques, and negligible angular displacement of the transducer itself.

REFERENCES FOR PART II

17. Mason, W. P.: "Piezoelectric Crystals and Their Applications to Ultrasonics," Appendix, D. Van Nostrand Company, Inc., Princeton, N.J., 1950. Cady, W. G.: "Piezoelectricity," McGraw-Hill Book Company, Inc., New York, 1946.
18. Mason, W. P.: "Electromechanical Transducers and Wave Filters," chap. VI and Sec. 6.32, D. Van Nostrand Company, Inc., Princeton, N.J., 1948.
19. King, J. C.: Fifth Interim Report, Contract DA 36-039SC-64586, Signal Corps Engineering Laboratories, May, 1956.
20. Jaffee, B., R. S. Roth, and S. Marzullo: $J. Appl. Phys.$, **25**:809 (1954). $Natl. Bur. Standards Bull.$, **39**:149 (1944). Shirane, G., K. Suzuki, and Tokeda: $Phys. Soc. Japan$, **7**:12 (1956). Shirane, G., and K. Suzuki: $J. Phys. Soc. Japan$, **7**:333 (1952). Bulletins of the Clevite Brush Co.
21. Egerton, L.: $J. Am. Ceram. Soc.$, **42**:438 (1959).
22. Goodman, G.: $J. Am. Ceram. Soc.$, **36**:368 (1953).
23. Mason, W. P.: "American Institute of Physics Handbook," Sec. 3g, McGraw-Hill Book Company, Inc., New York, 1957.
24. Mason, W. P., and S. D. White: $Bell System Tech. J.$, **31**:469 (1952).
25. Mason, W. P.: Ref. 17, p. 494.
26. Thurston, R. N., and P. Andreatch: $IRE Convention Record$, Part 9, March, 1955.
27. Mason, W. P.: $J. Acoust. Soc. Amer.$, **27**:73 (1955).
28. Smith, C. S.: $Phys. Rev.$, **94**:42 (1954).
29. Burns, F. P., and A. A. Fleischer: $Phys. Rev.$, **107**(5):81 (1957).
30a. Mason, W. P., and R. N. Thurston: $J. Acoust. Soc. Amer.$, **29**:1096 (1957).
30b. Mason, W. P., J. J. Forst, and L. M. Tornillo: "Recent Developments in Semiconductor Strain Transducers," $Instr. Soc. Amer.$, Preprint 15 NY 60, Sept. 26, 1960.
31. Forst, J. J., and F. T. Geyling: $Bell System Tech. J.$, **39**:705 (1960).
32. Anon.: Standards on Piezoelectric Crystal. $Proc. IRE$, **37**:1.378 (1949).

17

STRAIN-GAGE INSTRUMENTATION

C. C. Perry
Wayne State University

Herbert R. Lissner
Wayne State University

INTRODUCTION

The resistance strain gage may be employed in shock or vibration instrumentation in either of two ways. The strain gage may be the active element in a commercial or special-purpose transducer or pickup, or it may be bonded directly to a critical area on a vibrating member. Both of these applications are considered in this chapter, together with a discussion of strain-gage types and characteristics, cements and bonding techniques, circuitry for signal enhancement and temperature compensation, and related aspects of strain-gage technology.

The electrical resistance strain gage discussed in this chapter is basically a piece of very fine wire or thin foil which exhibits a change in resistance proportional to the mechanical strain imposed on it. In order to handle such a delicate filament, it is either mounted on or bonded to some type of carrier material or else is wound on a jig or fixture. The former is known as the *bonded strain gage*, and the latter as the *unbonded strain gage*. The properties of semiconductor (piezoresistive) strain gages are discussed in Chap. 16.

The strain gage is used universally by stress analysts in the experimental determination of stresses. Since strain always accompanies vibration, the strain gage or the principle by which it works is broadly applicable in the field of shock and vibration measurement. Here it serves to determine not only the magnitude of the strains produced by the shock or vibration, but also the entire time-history of the event, no matter how great the frequency of the phenomenon.

BASIC STRAIN-GAGE THEORY AND CONSTRUCTION

The relationship between resistance change and strain in the wire or foil used in strain-gage construction can be expressed as

$$\frac{\Delta L}{L} = \frac{1}{K} \frac{\Delta R}{R}$$

or

$$K = \frac{\Delta R/R}{\Delta L/L}$$

(17.1)

where K is defined as the *gage factor* of the wire, ΔR is the resistance change due to strain, R is the initial resistance, ΔL is the change in length, L is the original length of the wire or foil, and $\Delta L/L$ is the unit strain to which the wire or foil is subjected.

All materials do not exhibit this strain-sensitivity effect and different materials have different gage factors. Filament materials in common use in strain gages are Constantan (Ni 0.45, Cu 0.55), which has a gage factor of approximately $+2.0$, and Iso-elastic (Ni 0.36, Cu 0.08, Fe 0.52, and Mo 0.005), which has a gage factor of about $+3.5$.*

STRAIN-GAGE CONSTRUCTION

Since the wire or foil used in a strain gage must be very fine or thin to have a relatively high electrical resistance, it is difficult to handle. For example, the wire used in typical gages is 1 mil (0.001 in.) in diameter, which is about one-third the diameter of a human hair. The foil used in gages is about 0.1 mil in thickness.

In order to handle this wire or foil, it must be provided with a "carrier medium," usually a piece of paper or plastic to which the filament is cemented. A piece of paper,

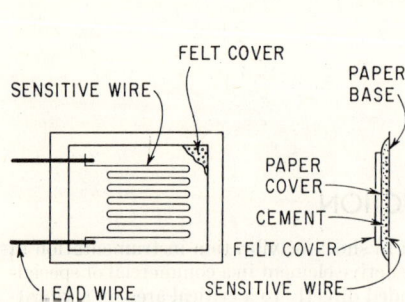

Fig. 17.1. Typical bonded wire strain-gage construction, with thickness greatly exaggerated to show details. Felt cover is not used on all gages.

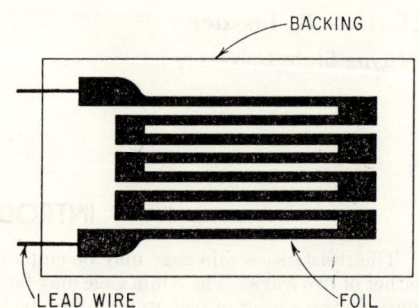

Fig. 17.2. Foil strain gage with construction similar to wire gage in Fig. 17.1. Ends of foil grid elements are made wide to minimize transverse sensitivity. Foil gage also has better heat-dissipation characteristics and allows larger gage currents.

plastic, or felt is usually cemented over the wire (Fig. 17.1), but the top of the foil is generally not covered. The fine gage wires are soldered or welded to larger-diameter lead wires which are also cemented in the sandwich for a considerable length to provide strong elements to which electrical connections can be made. Lead wires are generally not provided on foil gages. The cement and paper or plastic sandwich perform another very important function in addition to providing ease of handling and simplicity of application. The cement provides so much lateral resistance to the wire that it can be shortened by as much as 3 per cent without buckling; then compressive as well as tensile strains can be measured.

A length of about 5 in. of wire is required to give adequate resistance (about 120 ohms) to the gage. Since a gage 5 in. long has very limited use, the wire must be doubled back and forth to provide a shorter gage length. For very short gage lengths the wire is wound in a coil around a paper tube (like a soda straw). The tube is then flattened and cemented between two more pieces of paper to provide the necessary stiffness and insulation. Foil gages are formed in a grid as shown in Fig. 17.2.

TRANSVERSE SENSITIVITY

Because of its construction, a portion of the wire in each gage lies in the transverse direction and will respond to transverse strain. Therefore the gage factor K of a gage †

---

* Semiconductor strain gages have much higher gage factors, e.g., 100. See piezoresistive devices and materials, Chap. 16.

† In determining the *gage factor* of the gage it is assumed that the gage is mounted on a material having a Poisson's ratio of 0.285 and subjected to uniaxial stress in the direction of the gage axis.

is always slightly smaller than the gage factor of the wire of which it is fabricated. Since only about 4 per cent of the wire is in the transverse direction and subject to transverse strain, usually only a very small error is introduced into the strain reading due to this cause. When extreme accuracy is required, it is possible to make mathematical corrections to eliminate this error since the transverse sensitivity of each gage type is known.[1]

One of the desirable features of foil-type gages is their low transverse sensitivity. In this case, the gage consists of a flat foil grid; a sufficiently large amount of the foil is left at the ends of each strand to reduce the transverse sensitivity of the gage to one-half the value for wire gages for some types and to essentially zero for others.

STRAIN-GAGE CLASSIFICATIONS

Strain gages are classified in several ways. One classification cites the purpose for which the gage is to be used, that is, for static or dynamic strain measurement. Static gages are made up with Constantan wire, which has a minimum change of resistance with temperature. Dynamic strain gages are made up with Iso-elastic wire, which provides a greater gage factor than Constantan. The dynamic gages, while having a much greater resistance change for a given strain than the static gages, also are much more sensitive to changes in temperature. They are used only where the phenomenon to be measured is so short in time duration that no temperature change of any consequence can occur during the time of measurement. Gages also are available for the measurement of very large strains (up to 10 per cent) occurring in the plastic region of the material as distinguished from the more common gages which are used to measure elastic strains (up to 1 per cent). Table 17.1 lists commercially available bonded strain gages and gage manufacturers.

BONDED STRAIN GAGES

DUCO OR PAPER GAGES. Bonded strain gages may be classified according to the cement used in their manufacture. Thus, paper-backed gages are manufactured using a nitrocellulose cement, similar to Duco Household Cement,* which sets and hardens with evaporation of its solvent. Gages of this type are referred to as "Duco" or "paper" gages. They can be applied with the same kind of cement, although other adhesives can be used for this application. The cement tends to be hygroscopic; therefore the gage should not be used in moist or humid environments unless adequately protected from moisture. The gages cannot be used in applications where the temperature exceeds 180°F (80°C).

BAKELITE GAGES. In the Bakelite gage, the strain-sensitive wire is embedded in a thin sheet of phenolic thermosetting resin. This type of gage is usually applied with Bakelite Cement † and the application must be heated to polymerize the cement and make the gage adhere. The gages are waterproof and can be used in some applications at temperatures up to 500°F (260°C). When a gage must remain operative for a long period of time, this type is usually used, although under certain conditions Duco gages also can be used in long-time installations.

UNBONDED STRAIN GAGES

Unbonded strain gages operate on the same principle as the bonded gage.[2] In unbonded gages the strain-sensitive wire is wound on a fixture. Motion of one end of the fixture with respect to the other end changes the length of the wire, thus changing its resistance. For example, in Fig. 17.3A and B the mass is constrained by a pair of cantilever leaf springs. When the mass is moved in one direction with respect to the case, two sets of wires are subject to an increase in tension and the other two undergo a decrease in tension; stops prevent excessive strains. Figure 17.3C shows the Wheatstone bridge arrangement to which the four sets of wires are connected. This bridge requires an ex-

* Manufactured by E. I. du Pont de Nemours & Company.
† Manufactured by Union Carbide Plastics Co. Division of Union Carbide Corp.

Table 17.1. Bonded Strain-gage Manufacturers

Manufacturer	Trade name or gage designation	Gages produced
United States		
Baldwin-Lima-Hamilton Corp., Waltham, Mass.	SR-4	Wire, foil; static, dynamic; room-temp., high-temp., very high-temp.; paper, Bakelite, ceramic
High Temperature Instrument Eng. Co., Bala-Cynwyd, Pa.	HT	Wire; static, dynamic; high-temp.; very high-temp.; ceramic
Budd Instruments Division, Phoenixville, Pa.	Metalfilm	Foil; static, dynamic; room temp., high-temp., very high-temp.; ceramic
Transonics, Inc., Bedford, Mass...	Surface transferable resistors	Wire; ceramic; dynamic; very high-temp.
England		
Saunders-Roe, Ltd., East Cowes, Isle of Wight	....	Foil; static, dynamic; room-temp., high-temp., very high-temp.; ceramic
Teddington Industrial Equipment, Ltd., Sudbury on Thames, Middlesex	....	Wire; static, dynamic; room-temp.; paper
H. Tinsley & Co., Ltd., London..	....	Wire; static, dynamic; room-temp., high-temp.; paper, Bakelite
Rotol, Ltd., Gloucester..........	....	Wire; static, dynamic; room-temp., high-temp.; paper, thermosetting resin
Japan		
Kyowamusen Kenkyujo Co., Ltd., Tokyo	....	Wire, foil; static, dynamic; room-temp., high-temp.; paper, Bakelite, Polyester
Toko Sokki Kenkyujo Co., Ltd., Tokyo	TML	Wire; static, dynamic; room-temp.; Polyester
Shinkoh Tsushinkogyo Co., Ltd., Kanagawa-ken	Shinkoh	Wire, foil; static, dynamic; room-temp., high-temp.; paper, Bakelite
Switzerland		
Huggenberger-Zurich, Zurich....	Tepic	Wire; static, dynamic; room-temp., high-temp.; paper, thermosetting resin
Germany		
Hottinger Messtechnik GmbH, Darmstadt	Impa (also distributor for SR-4 gages)	Wire; static, dynamic; room-temp.; acrylic resin
Netherlands		
Philips, N.V., Eindhoven	PR	Wire; static, dynamic; room-temp., high-temp.; paper, thermosetting resin
Sweden		
Herbert Lembcke, Stockholm K..	Gustafsson	Wire; static, dynamic; room-temp., high-temp., thermosetting resin
Norway		
Electrometer, Trondheim........	....	Wire, foil; static, dynamic; room-temp., high-temp.; paper, Bakelite

Note: Room-temperature gages can be used up to 100 or 150°F (40 or 65°C). High-temperature gages can be used up to 350 or 500°F (175 or 260°C). Very-high-temperature gages can be used up to 1500 or 1800°F (820 or 980°C).

BONDED-GAGE CHARACTERISTICS 17–5

ternal power supply of approximately 10 volts. This device is insensitive to temperature changes, since all arms of the bridge are affected in the same manner. In general, the unbonded wire strain gage must be fastened to the structure being investigated. Advantages of this type of gage are that it can be removed and used repeatedly on different structures, and its transverse sensitivity is zero. However, its mass is very great in comparison to the mass of the bonded gage—a factor which may affect the vibration characteristics of the part to which it is attached.

Unbonded strain gages are commonly used in accelerometers (for example, see Fig. 17.15) and other transducers.

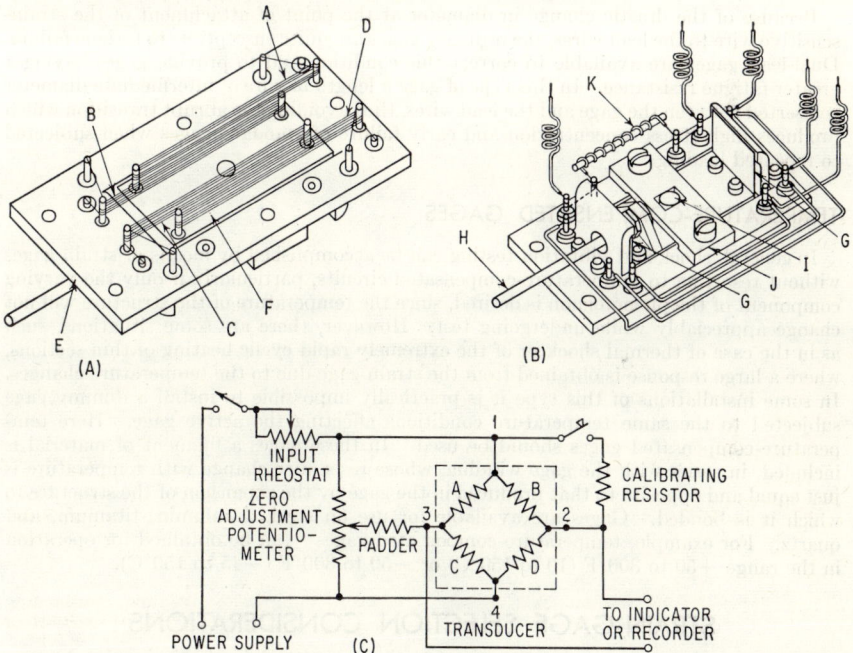

FIG. 17.3. Unbonded wire strain gage. (A) Wire elements A, B, C, D attached at one end to movable armature F and at the other end to the pickup body E. (B) View of the underside of (A) showing armature cantilever springs G; motion-limiting stop I; trimmer resistance K. Probe H is attached to the armature to apply motion externally. (C) A bridge arrangement for using the transducer. (*Courtesy of Statham Instruments, Inc.*)

BONDED-GAGE CHARACTERISTICS

Bonded wire strain gages are cemented to a structure in the region where the strain is to be determined. They can be used only in the location where they are originally applied and cannot be removed successfully for subsequent use. These gages are characterized by their extremely small mass in comparison to the mass of most structures to which they are applied. This characteristic is valuable in vibration measurements since the use of the gage will not affect the vibration pattern of the structure to which it is bonded. The gage is rugged, has high sensitivity, is very easily handled, and is applied easily in most cases. Since the strain to which the gage is subjected produces very small changes in electrical resistance, a very sensitive measuring device is required to detect the strain magnitude accurately.

In physical dimensions, bonded gages vary in length from $\frac{1}{32}$ to 6 in. and in widths from a single strand of 1/1,000-in.-diameter wire to $\frac{3}{4}$ in. Generally the **gages of shorter** length are also the narrower ones. In electrical characteristics, the gage resistances vary from 40 to 2,000 ohms. The most commonly used gages have a resistance of 120 ohms. Gage factors range from 1.5 to 3.5, with dynamic gages having the higher value and most static gages having a factor of about 2.0.

Paper gages are available which are fabricated with extra thin paper so that drying of the gage is hastened. This special paper permits the more rapid evaporation of the solvent in Duco cement and thus allows earlier use of the gage after its application.

FATIGUE-RESISTANT GAGES

Because of the drastic change in diameter at the point of attachment of the strain-sensitive wire to the lead wires, the ordinary gages are quite susceptible to fatigue failure. Dual-lead gages are available to correct this condition and to provide gages having a greater fatigue resistance. In this type of gage a length of wire of intermediate diameter is inserted between the gage and the lead wires, thus avoiding the abrupt transition which produces high stress concentration and early failure of standard gages when subjected to repeated loading.

TEMPERATURE-COMPENSATED GAGES

In general, shock and vibration testing can be accomplished by means of strain gages without resorting to temperature-compensated circuits, particularly if only the varying component of the phenomenon is desired, since the temperature of the structure will not change appreciably while undergoing test. However, there are some situations, such as in the case of thermal shock or of the extremely rapid cyclic heating of thin sections, where a large response is obtained from the strain gage due to the temperature changes. In some installations of this type it is practically impossible to install a dummy gage subjected to the same temperature conditions affecting the active gage. Here temperature-compensated gages should be used. In these gages a filament of material is included, in series with the gage winding, whose resistance change with temperature is just equal and opposite to that produced in the gage by the expansion of the structure to which it is bonded. Gages are available for use on steel, duralumin, titanium, and quartz. For example, temperature-compensated gages * can be obtained for operation in the range +50 to 300°F (10 to 150°C) or −50 to 300°F (−45 to 150°C).

STRAIN-GAGE SELECTION CONSIDERATIONS

SHOCK MEASUREMENTS

In the case of shock measurements, a transient may be applied to the structure that is under investigation only once or it may be repetitive. Shock is of very short time duration, and the problem of temperature compensation is nonexistent because in most cases the temperature does not have time to change during the impact. For this reason a dynamic-type gage usually can be employed for the measurement of shock. This type of gage has the advantage of a higher gage factor than the static gage so that it will provide the greatest possible electrical signal for a given strain. If the gage is to remain on the structure for a long period of time, or if it is to be subjected to temperatures between 180 and 500°F (80 and 260°C), a Bakelite gage should be selected. If the gage is to be serviceable for a period during which the structure is subjected to a large number of shocks, a fatigue-resistant dual-lead type of gage should be employed.

VIBRATION MEASUREMENTS

For vibration measurement the type of gage selected is dependent on the kind of information desired. If only the frequency of vibration and the magnitude of the cyclic stresses are desired, dynamic-type gages can be used since temperature changes will not

* Manufactured by the Baldwin-Lima-Hamilton Corporation (designated by the letter *E*).

affect the results obtained unless the temperature fluctuates at the same rate as the stress. If, however, a measurement of the static or slowly varying component of the stress is also to be determined (i.e., if the absolute values of the stresses are desired), a static-type gage must be employed. Since changes in temperature will affect the gage reading, temperature compensation must be incorporated to obtain true values of stress.

STRAIN GRADIENT

Gage selection is dependent on the space limitation and steepness of strain gradient in any region. The strain gage indicates the average strain over the length of the gage; in a region of steep strain gradient, this indicated value may be much less than the maximum strain. The shorter the gage used in such a region, the closer is the gage indication to the maximum strain (Fig. 17.4). However, two possible objectives must be considered quite carefully in selecting a gage for a particular installation: (1) the determination of the frequency of vibration, or comparison of relative amplitudes and frequencies with different conditions of excitation, and (2) the determination of the maximum stress pattern resulting from the vibration set up.

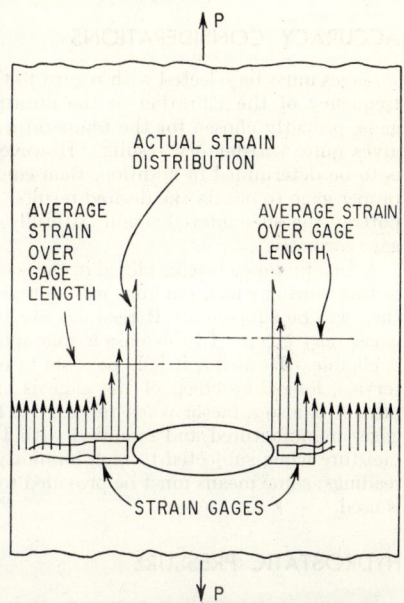

Fig. 17.4. Effect of gage length on indicated strain in the presence of a severe strain gradient. The shorter gage on the right indicates a higher strain. An infinitesimal gage length would be necessary to indicate the peak strain.

In the first case there is considerable freedom with regard to the location of the gage on the structure, and therefore with the selection of the gage itself. In the second case severe restrictions exist in regard to the region of application of the gage and its possible dimensions. In general, very short gage-length gages are more difficult to apply properly. Therefore it is desirable to employ gage lengths of ½ in. or longer whenever possible. When the actual magnitude of the maximum stress resulting from shock or vibration is to be determined, a much more complicated system of gages must be employed. A single gage can be used in only the very limited case where a stress exists in one direction only, and that direction must be known. If stresses exist in several directions, or if the direction of a singly existing stress is unknown, a strain-gage rosette consisting of three or more gages must be employed.[3]

PHYSICAL ENVIRONMENT

The physical environment of the applied gage is an important factor which must be considered in gage selection and protective treatment. Temperature, pressure, humidity, oil, corrosive acid, abrasive action, and possible electromagnetic, neutron, and radiation fields are conditions which affect the choice of gage and its required protection. If high temperatures (up to 500°F or 260°C) are to be encountered, a Bakelite or other high-temperature-type gage must be selected. If even higher temperatures must be withstood, a ceramic-type gage should be employed. Gages of this sort are used at temperatures as high as 2000°F (1100°C). If the temperature never exceeds 180°F (80°C), however, any type of gage can be used. Most gages operate satisfactorily at very low temperatures.

ACCURACY CONSIDERATIONS

Gages must be selected with regard to the desired precision of the results. If only the frequency of the vibration or the duration of a shock wave is required, almost any gage, properly chosen for the temperature and humidity conditions to be encountered, gives quite satisfactory results. However, if the magnitude of the stresses produced is to be determined in addition, then considerable care must be exercised to select the proper gage to obtain the desired results. Not only must the gage be the proper one to portray the encountered strain faithfully, but precautions must be taken to install the gage correctly.

A final factor to be considered in the selection of gages for a particular job is the length of time during which readings will be taken and the number of cycles of strain to which they will be subjected. If readings are to be obtained over a long period of time, the gages may creep. In this case a gage must be selected whose creep characteristics are negligible. However, if it is possible to obtain no-load or zero readings at periodic intervals, long-time creep of the gage is of no consequence. The amount of moisture present is also a factor when the gage is to be used over a long period of time. Paper gages manufactured and mounted with Duco cement are hygroscopic and will absorb moisture when subjected to high humidity. This condition leads to inaccuracy in gage readings; some means must be provided to keep moisture out of the gage if a paper gage is used.

HYDROSTATIC PRESSURE

In some instances it is necessary to install strain gages which are exposed to compressive forces normal to the surface of the gage. For example, in shock testing of pressure vessels, gages might be installed in locations where high pressures are exerted on the surface of the gage. In such cases great care must be taken to provide a smooth surface to which the gage is bonded to prevent the gage filaments being forced into holes, pits, or depressions by the pressure, thus giving rise to incorrect readings. Wrap-around-type gages cannot be used in installations when the gages are subject to high normal or hydrostatic pressures. Pressures as high as 25,000 lb/in.^{2} do not affect the output of wire strain gages to any appreciable degree [4] (2 to 5 microinches/in. of indicated compressive strain per 1,000 lb/in.^{2}). In general, this indicates that transverse pressure effects on wire strain gages can be neglected. One precaution to be observed is in the mounting and placement of a dummy gage in connection with such an installation. If the dummy gage is mounted on a block of material which is subjected to the same pressure to which the active gage is subjected, an entirely different strain pattern generally will exist in the dummy gage block than exists in the material to which the active gage is attached.[5] An appreciable error will be introduced into the readings obtained unless a correction for this effect is made. If adequate temperature compensation can be obtained, the best practice is not to subject the dummy gage to the hydrostatic pressure.

REPEATED LOADING

As indicated above under *Fatigue-resistant Gages*, when a gage is to function during a large number of strain cycles it is subject to fatigue failure. This failure generally occurs at the point where the lead wire is attached to the gage wire, and is due to the large difference in the stiffness of the wires. Dual-lead gages are available which have a transition between the lead wire and the gage wire to increase the resistance of the gage to fatigue. Their endurance limit or life at any given strain level is much greater than that of ordinary gages.

It is imperative to select a strain gage specifically recommended by the gage manufacturer for application to alternating stress service, since strain gages are subject to fatigue failures like many other components. The operating life of the strain gage can be extended by limiting the maximum alternating strain range to 1,000 microinches/in. or less if practicable.

MAGNETOSTRICTIVE EFFECTS

One possible source of error when using dynamic-type gages of Iso-elastic or nickel wire or foil is due to the magnetostrictive effect.[6] Voltages are generated in these materials by straining the gage material itself. This voltage is a function of the rate of strain. Tests have shown that these self-generated voltages may approach one millivolt for Iso-elastic gages and exceed several millivolts for nickel gages. Static-type gages of Constantan do not exhibit this characteristic.

A further result of the magnetostrictive effect is to change the gage factor of the Iso-elastic and nickel gages with strain. The gage factor for Iso-elastic gages may change as much as 3 per cent for elastic strains, and for nickel the gage factor may change by many times this amount. Where very precise stress determinations are required, the magnetostrictive effect must be taken into account.

NUCLEAR RADIATION

Strain gages have been subjected to high-intensity nuclear radiation fields to determine the effect of neutron and gamma-ray bombardment on their behavior.[7] For severe neutron bombardment, paper gages are unsatisfactory. Constantan foil-type high-temperature gages (900°F or 480°C), applied with Allen P-1 cement,* are quite stable when subjected to severe neutron and gamma-ray bombardment at temperatures below 150°F (66°C). For Nichrome V gages, the resistance of the gage increases with increasing neutron radiation while the resistance of the Constantan gage remains essentially constant. However, for both gages the insulation resistance to ground decreases steadily with increasing exposure time until it approaches a value of 10,000 ohms after cumulative exposure of 3×10^8 thermal and 1.5×10^{17} fast neutrons per square centimeter. After this exposure the resistance to ground may become erratic. Soldered connections to lead wires are unsuitable for use in operating nuclear reactors; lead wires should be spot-welded to gage leads for such use. Proper lead wire insulation is also important. Wires insulated with Delta Beston † insulation have stood up under prolonged neutron and gamma radiation.

BONDING TECHNIQUES

The proper functioning of a strain gage is completely dependent on the bond which holds it to the structure undergoing test. If the bond does not faithfully transmit the strain from the test piece to the wire or foil of the gage, the results obtained cannot be accurate. The methods of strain-gage application, described in Appendix 17.1, closely follow those recommended by the manufacturers of the gages. Some of the instructions given are very detailed and meticulous because the results obtained from the gage are entirely dependent on its bonding so that it acts as one with the piece undergoing test. Failure to bond over even a minute area of the gage will result in incorrect strain indications. The greatest weakness in the entire technique of strain measurement by means of wire or foil gages is in the bonding of the gage to the test piece.

Many cements are available for bonding strain gages. Table 17.2 lists the characteristics of a number of common cements as well as the manufacturers or suppliers.

* Manufactured by Robert G. Allen Co., Mechanicville, N.Y.
† Manufactured by General Electric Co.

Table 17.2. Strain-gage Cements

Name	Characteristics	Manufacturer or supplier
Duco	Nitrocellulose cement for room-temperature use-cure by evaporation of solvent (24 hr usually required)	Can be obtained from any department store as Duco Household Cement
SR-4	Nitrocellulose cement for room-temperature use-cure by evaporation of solvent (24 hr usually required)	Baldwin-Lima-Hamilton Corp. Waltham, Mass.
Bakelite	Thermosetting phenolic resin Cure 1 hr at 140°F (60°C) 2 hr at 175°F (80°C) 2 hr at 250°F (120°C) For use to 500°F (260°C)	Union Carbide Plastics Co., Division of Union Carbon & Carbide Co., Bound Brook, N.J.
Araldite	Epoxy resin Cure 1 hr at 160°F (70°C); for use to 200°F (95°C)	Ciba Co., Fair Lawn, N.J.
Armstrong A-1 A-2 C-2 A-6	Epoxy resin Cure 1 hr at 160°F (70°C); for use to 200°F (95°C)	Armstrong Products Co., Warsaw, Ind.
F-88 Dental Cement	Cures in minutes on mixing cement with activator; for room-temperature use under conditions of extreme moisture	Industrial Division, American Consolidated Mfg. Co., Inc., Philadelphia, Pa.
Cyano-acrylate (Eastman 910)	Cures practically instantaneously upon contact with activator; for room-temperature use	Budd Instruments Division, Phoenixville, Pa.
Allen P-1	Ceramic cement for use to 1800°F (980°C) Cure 1 hr at room temperature 1 hr at 200°F (95°C) 1 hr at 600°F (315°C)	Robert G. Allen Co. Mechanville, N.Y.
BLH RX-1	For use to 600°F (315°C) Cure 1 hr at 100°F (40°C) 4 to 6 hr at 150 to 180°F (65 to 80°C)	Baldwin-Lima-Hamilton Corp., Waltham, Mass.
Quigley 1925	Ceramic cement for use to 1500°F (815°C) Cure 2 hr at 150°F (65°C) 1 hr at 1700°F (930°C)	Quigley Co., San Francisco, Calif.
de Khotinsky	Thermoplastic cement melts at 300°F (150°C) for room-temperature use	Central Scientific Co., Chicago, Ill.

Note: Most cements can be obtained from gage manufacturers.

TRANSDUCING ELEMENTS WITH BONDED RESISTANCE STRAIN GAGES

BASIC PROPERTIES OF STRAIN GAGES AND CIRCUITRY

The resistance strain gage, because of its inherent linearity, very small mass, wide frequency response (from zero to more than 50,000 cps), general versatility, and ease of installation in a variety of applications, is an ideal sensitive component for electrical transducers for use in shock and vibration instrumentation.[8] The Wheatstone bridge circuit, described in a subsequent section, can be used to extend the versatility of the strain gage to still broader applications by performing mathematical operations on the strain-gage output signals. The combination of these two devices can be used effectively for the measurement of acceleration, displacement, force, torque, pressure, and similar mechanical variables. Other useful attributes include the capacity for separation of forces and moments, vector resolution of forces and accelerations, and cancellation of undesired vector components.

The usual technique for employing a strain gage as a transducing element is to attach the gage to some form of mechanical member which is loaded or deformed in such a manner as to produce a signal in the strain gage proportional to the variable being measured. The mechanical member can be utilized in tension, compression, bending, torsion, or any combination of these. All strain-gage-actuated transducers can be considered as either force- or torque-measuring instruments. Any mechanical variable which can be predictably manifested as a force or a couple can be instrumented with strain gages.

There are a number of precautions which should be observed in the design and construction of custom-made strain-gage transducers.[9] First, the elastic member on which the strain gage is to be mounted should be characterized by very low mechanical hysteresis and should have a high ratio of proportional limit to modulus of elasticity (i.e., as large an elastic strain as possible). Although aluminum, bronze, and other metals are often employed for this purpose, steel is the most common material. An alloy steel such as SAE 4340, heat-treated to a hardness of RC 30–40, will ordinarily function very satisfactorily.

The physical form of the elastic member, and the location of the strain gages thereon, are not subject to specific recommendation, but vary with the special requirements of each individual instrumentation task. When no such requirements exist, a standard commercial transducer ordinarily should be used. In general, the shape of the member should be such as to (1) allow adequate space for mounting strain gages (preferably in regions of zero or near-zero strain gradient), (2) provide the desired natural frequency, (3) produce a strain in the gages which is great enough at low values of the measured variable to result in an output signal readily subject to accurate indication or recording, and not so great as to cause nonlinearities or abbreviated gage life at peak load values, (4) provide temperature compensation and/or signal augmentation (as described in a subsequent section) whenever feasible, and (5) allow for simplicity of machining, ease of gage attachment and wiring, and, if necessary, protection of the gages.

The strain gages should be cemented to the elastic member with the usual care and cleanliness necessary to all strain-gage applications, special attention being given to minimizing the bulk of the installation if the added mass is significant to the frequency response of the instrument. Other considerations vital to successful strain-gage-application technique are described elsewhere in this chapter.

DISPLACEMENT MEASUREMENT

Measurement of displacement with strain gages can be accomplished by exploiting the fact that the deflection of a beam or other loaded mechanical member is ordinarily proportional to the strain at every point in the member as long as all strains are within the elastic limit.

For small displacements at low frequencies, a cantilever beam arranged as shown in Fig. 17.5 can be employed. The beam should be mounted with sufficient preload on

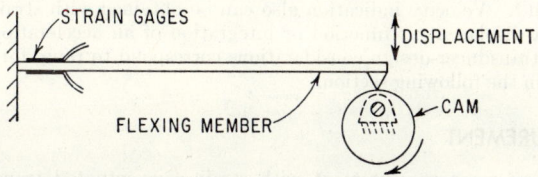

FIG. 17.5. Strain gages mounted on a cantilever beam for displacement measurement produce electrical signal proportional to cam motion.

the moving surface that continuous contact at the maximum operating frequency is assured. In the case of higher frequency applications the beam can be held in contact with the moving surface magnetically or by a fork or yoke arrangement, as illustrated in Fig. 17.6. It is necessary to make certain that the measuring beam will not affect the

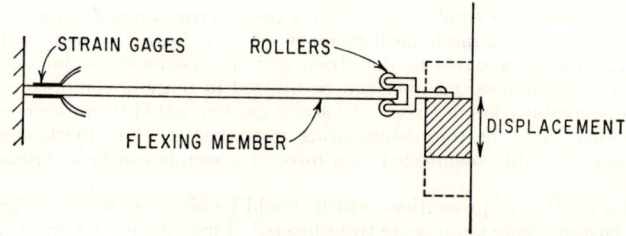

Fig. 17.6. Displacement transducer designed for continuous, positive contact with moving object.

displacement to be instrumented, and that no natural mode of vibration of the beam itself will be excited.[10]

The measurable displacement magnitude can be increased above that for the cantilever beam by employing other schemes such as the "clip gage" shown in Fig. 17.7. This gage is constructed by bonding strain gages to the upper and lower sides of a piece of channel-shaped spring steel, as shown in Fig. 17.7. The assembly is then clipped or otherwise mounted on the test specimen so that the legs deflect as the specimen is strained, thus straining the backbone of the clip gage to a greater or lesser extent. Any desired reduction in strain magnitude can be obtained in this manner by merely altering the proportions of the clip gage. Unfortunately, the maximum allowable frequency generally decreases as the displacement amplitude increases, since stiffness and natural frequency tend to change together. Displacement also can be measured through the use of the relative motion of a seismically mounted mass of much lower natural frequency than the applied frequency.

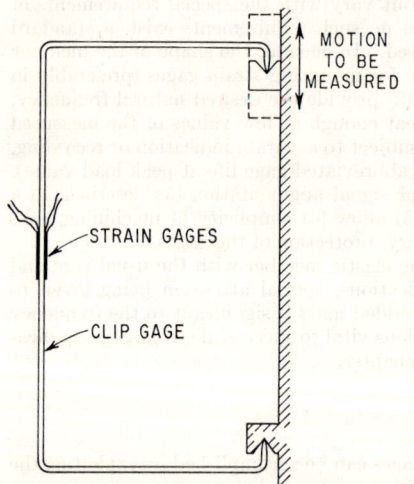

Fig. 17.7. Clip gage for instrumenting large displacements. Proportions of clip gage are designed to keep strain well within the proportional limit of the material.

VELOCITY

Velocities can be measured directly with strain-gage transducers only by producing a force such as viscous damping or hydro- or aerodynamic drag force which is uniquely related to velocity. Velocity indication also can be obtained with strain gages by differentiation of a displacement function or integration of an acceleration function. In either case, the transducer-design considerations correspond to those for force measurement described in the following section.

FORCE MEASUREMENT

The principle of force measurement with strain-gage-actuated transducers is very similar to that for displacement.[11-14] The procedure consists of placing a strain-gage-instrumented elastic member in series with the force to be measured. The strain in the transducer, and thus the output signal, is proportional to the force if all stresses are kept within the elastic limit. The proportionality constant between strain and force must be obtained by calibration if precise results are desired. Otherwise, tolerances on the gage factor of the strain gage, and uncertainty as to the elastic properties of the instrumented member, can produce errors of 10 per cent or greater—even for transducer configurations with readily calculable strain distributions.

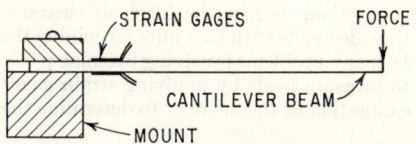

FIG. 17.8. Cantilever force-measuring transducer, consisting of beam with load applied at free end. Gage strain is a linear function of the force if the proportional limit is not exceeded.

Figure 17.8 illustrates a common form of force transducer, the cantilever beam. Strain gages are mounted on the top and bottom of the beam, producing double sensitivity (output) and virtually complete temperature compensation. While this type of transducer is probably best suited to static or quasi-static measurements such as reaction forces, it also can be used very successfully for many shock and vibration problems as long as the natural frequency of the beam is higher than the frequency of the force being

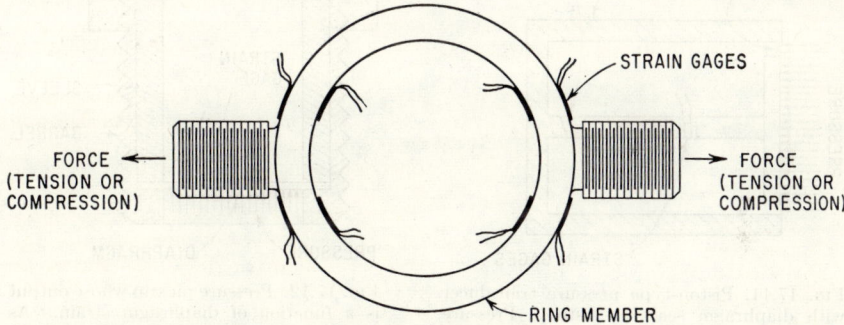

FIG. 17.9. Ring gage for force measurement. This type of gage provides sensitive axial load measurement without undue loss of rigidity or ruggedness.

measured. The ring gage (Fig. 17.9) can be categorized with the cantilever beam, and is equally applicable to static or dynamic force measurement within the limitations imposed by its comparatively low natural frequency.

For most dynamic force-instrumentation problems a small compression or tension member (Fig. 17.10) is ordinarily employed. If the load is characterized by alternation between compression and tension, the transducer must be designed for a rigid, integral

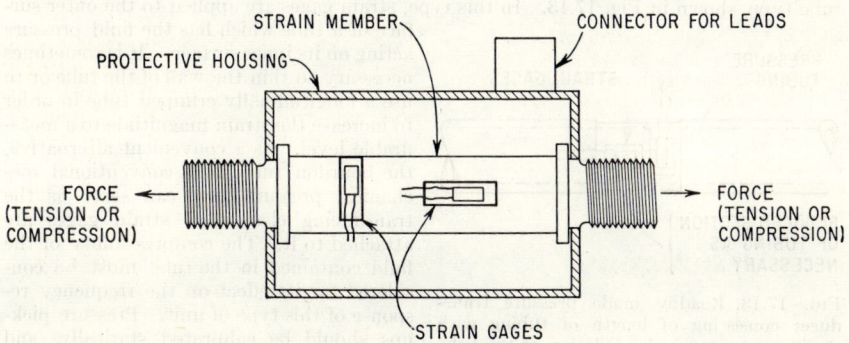

FIG. 17.10. Widely used commercial form of axial force transducer for large loads.

connection, with no backlash or clearance. This can be accomplished by employing threaded ends with lock nuts for joining the transducer to the remainder of the assembly. In many problems involving machine parts or other mechanical components it is possible to measure loads by applying strain gages to the machine member itself, necessitating calibration of the member to determine the relationship between force and strain.

PRESSURE

In hydraulic and aerodynamic devices, pressure fluctuations are often associated with vibration phenomena—either as cause or effect. Strain-gage transducers are widely used in such situations.[15-18]

Pressure pickups based on strain gages are commonly one of three principal types: piston, diaphragm, or tube. In the piston type the pressure acts against a freely mov-

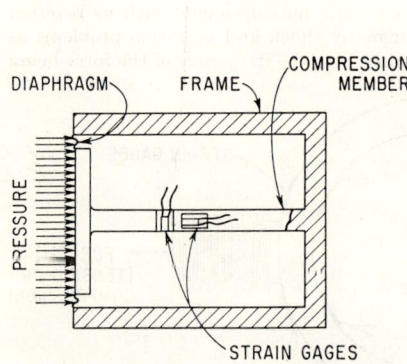

Fig. 17.11. Piston-type pressure transducer with diaphragm seal for piston. Pressure load on piston head is sensed by strain gages on supporting column.

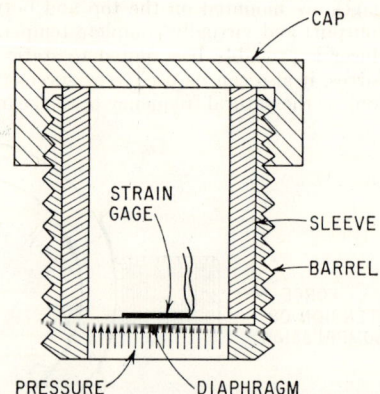

Fig. 17.12. Pressure pickup whose output is a function of diaphragm strain. As diaphragm deforms under pressure, strain is transmitted to gage to produce electrical signal.

able flat surface (which may be either a piston or diaphragm), the motion of which is inhibited by an elastic member instrumented with strain gages to measure the force (Fig. 17.11).

Diaphragm-type pressure transducers, shown in Fig. 17.12, have the strain gages applied directly to the back surface of the diaphragm so that diaphragm strain is a measure of pressure.[19] The simplest form of pressure transducer to construct is the tube type, shown in Fig. 17.13. In this type, strain gages are applied to the outer surface of a tube which has the fluid pressure acting on its inner surface. It is sometimes necessary to thin the wall of the tube or to use a longitudinally crimped tube in order to increase the strain magnitude to a measurable level. As a convenient alternative, the bourdon tube in a conventional mechanical pressure gage can serve as the transducing element if strain gages are attached to it. The compressibility of the fluid contained in the tube must be considered for its effect on the frequency response of this type of unit. Pressure pickups should be calibrated statically, and preferably dynamically, prior to use.

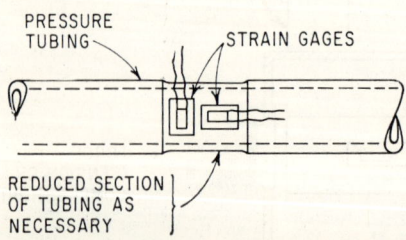

Fig. 17.13. Readily made pressure transducer consisting of length of tubing with strain gages attached. Dilation of the tube with pressure creates strain in gages.

ACCELERATION

Strain-gage accelerometers are similar in design to force-sensing units. Ordinarily they consist of a mass mounted so as to deform an elastic member in an amount proportional to the inertia force.[20-22] Strain-gage accelerometers, in contrast to the more commonly used piezoelectric types, are suitable for use at very low frequencies. However,

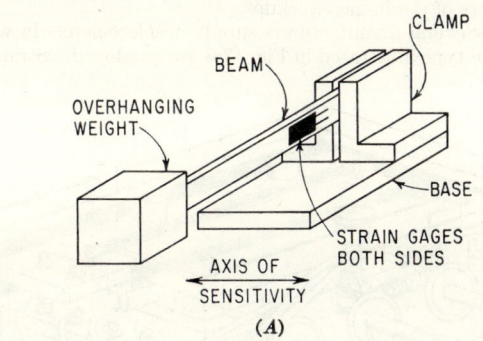

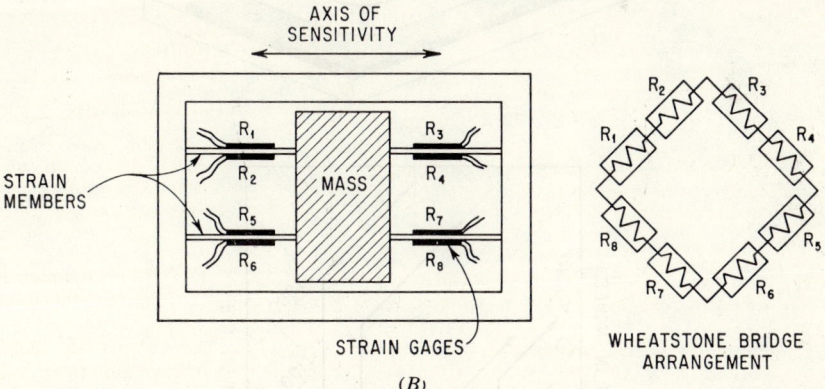

Fig. 17.14. Common forms of strain-gage accelerometer. (A) Weight on free end of cantilever beam. Inertia forces from acceleration cause beam to bend and strain the gages proportionally. (B) High-g accelerometer with seismic mass producing axial strain in instrumented members.

the output signal from any strain-gage device is considerably smaller than that from a piezoelectric transducer.

Two types of strain-gage accelerometers are shown in Fig. 17.14 in greatly simplified form. In addition to the requirement for a high natural frequency, the other criteria for successful operation include selection of fatigue-resistant strain gages, placement of gages whenever possible for augmented output and temperature compensation, limiting maximum strain so as to enhance linearity and minimize hysteresis, and calibration of the completed transducer. Damping of the accelerometer can be accomplished by enclosing the unit in a case filled with a dielectric damping fluid. Except for details associated with the strain gages themselves, the over-all design considerations for strain-gage accelerometers are the same as those for other seismic accelerometers (see Chap. 12).

COMMERCIAL STRAIN-GAGE PICKUPS

Although it is sometimes necessary to design and construct custom-made transducers with resistance strain gages to perform specific jobs, commercially available instruments should ordinarily be given first consideration. In general, the commercial transducers are characterized by better linearity, greater accuracy, and superior reliability than can be achieved by the ordinary individual with a custom-built unit. Table 17.3 lists representative manufacturers of strain-gage pickups.

Unbonded Gages. Several manufacturers supply accelerometers in which unbonded wire strain gages of the type illustrated in Fig. 17.3 are employed. Strain-sensitive wire

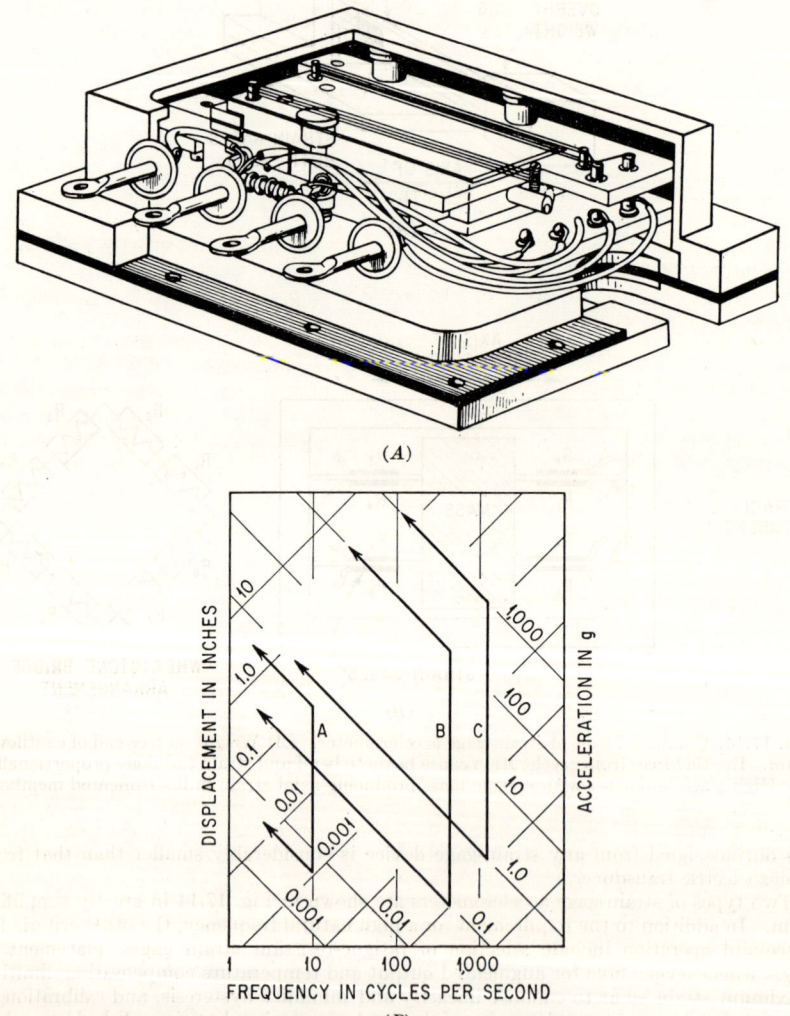

Fig. 17.15. (A) Unbonded strain-gage accelerometer. (B) Frequency-amplitude range limits for three models of unbonded accelerometers. (*Courtesy of Statham Instruments, Inc.*)

Table 17.3. Representative Manufacturers of Strain-gage Transducers for Acceleration, Force, Torque, Pressure, etc.

United States:
  Baldwin-Lima-Hamilton Corp., Waltham, Mass.
  Consolidated Electrodynamics Corp., Pasadena, Calif.
  Control Engineering Corp., Canton, Mass.
  Detroit Controls Division of American Standard Corp., Norwood, Mass.
  Statham Instruments, Inc., Los Angeles, Calif.
England:
  Boulton Paul Aircraft, Ltd., London
  Southern Instruments, Ltd., Camberly, Surrey
France:
  Sexta, S.A., Bagneux, Seine
Japan:
  Shinkoh Tsushinkogyo Co., Ltd., Kanagawa-ken
  Toyo Measuring Instruments Co., Ltd., Tokyo
Netherlands:
  Philips, N.V., Eindhoven

is wound around a series of pins or on an armature with initial tension in such a manner that relative motion accompanying acceleration causes the filament to be strained to a greater or lesser amount. Figure 17.15A illustrates an unbonded strain-gage accelerometer which operates in this manner.

Unbonded strain-gage accelerometers provide a relatively low signal output level. Their maximum full-scale output is generally about 50 millivolts open circuit. However, their internal impedance is low, so that they may be used to drive sensitive oscillographic galvanometers directly. Other advantages are the linearity of output with input, constancy of calibration, and the feasibility of using much of the large amount of auxiliary equipment that has been developed for the bonded-wire gage for amplifying and recording the output of these accelerometers.

The natural frequencies of unbonded accelerometers extend to about 5,000 cps. A damping fluid usually is used which is effective for units having natural frequencies lower than a few hundred cycles per second, but which is not easily maintained at the required value for high-frequency units. The frequency-amplitude ranges for three models of commercial unbonded strain-gage accelerometers are shown in Fig. 17.15B. The arrows indicate that the low-frequency response extends to 0 cps.

STRAIN-GAGE CIRCUITRY AND INSTRUMENTATION

The output of a resistance strain gage usually is from 10 to 1,000 microvolts for practicable strain ranges. In order to study the detailed cyclic nature of vibration problems or the transient phenomena commonly associated with mechanical shock, it is necessary to obtain some form of oscillographic record of the events. Oscillographic recording equipment commonly requires an input signal several orders of magnitude greater than the strain-gage output. Thus, some form of electrical amplification is necessary between these two components. The strain gage also needs a stable source of electric current, or excitation, to produce an output voltage proportional to resistance change. These two factors are of primary importance in determining the nature of the electrical instrumentation system which can be used satisfactorily with the resistance strain gage.

THE POTENTIOMETER CIRCUIT

The simplest circuit arrangement for supplying a strain gage with excitation current, and obtaining a signal corresponding to deformation of the gage, is known as the *potenti-*

ometer circuit (Fig. 17.16). It is very well suited to the instrumentation of dynamic or fluctuating strains, but is totally unsuited for the measurement of static strains or the static component of a combined static and dynamic strain.

In Fig. 17.16, R_B represents a ballast resistor, the principal function of which is to maintain the current flow in the circuit relatively constant and independent of small resistance changes in the strain gage R_G.

Under steady-state, zero-strain conditions, the output voltage e_0 across the strain gage is

$$e_0 = \left(\frac{R_G}{R_B + R_G}\right)e \qquad (17.2)$$

where e is the supply voltage. A typical strain gage may have a resistance of about 120 ohms and an allowable current rating of 30 milliamperes. Thus, there is a 3.6-volt

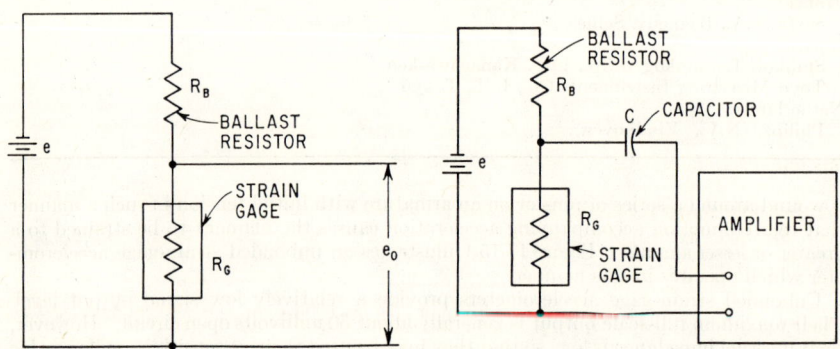

FIG. 17.16. Potentiometer circuit for dynamic strain signals. Nearly constant current through the circuit, combined with varying gage resistance, produces output signal.

FIG. 17.17. Over-all arrangement of circuits for instrumenting dynamic strain. Signal from gage is taken to a-c amplifier through isolating capacitor.

drop across the strain gage under zero-strain conditions. Assume that the gage is subjected to an axial strain of 1,000 microinches/in. (corresponding to approximately 30,000 lb/in.² stress in steel) and that the gage factor of the strain gage is 2.0; then the change in resistance of the gage due to the above strain is 0.24 ohm, and the change in output voltage (e_0) is 0.0072 volt if the ratio R_B/R_G is large. In this case it is only the change in voltage which is significant as a measure of the external variable which caused it; such a small percentage change in the total voltage across the strain gage is difficult to measure accurately. For this reason it is the usual practice in dynamic strain applications to block the steady-state (zero-strain) portion of the output voltage so that only the fluctuating component is measured. This is done by inserting a capacitor between the potentiometer circuit output and the input of the following amplifying or indicating equipment which may be employed in conjunction with the potentiometer circuit. The over-all arrangement is illustrated in Fig. 17.17. An alternating signal, representing the alternations in the strain to which the gage is subjected, is transmitted through the capacitor to the input terminals of an audio-frequency amplifier.

Any influences in addition to strain which may act to modify the resistance of the strain gage (for example, temperature changes) also produce output voltages in this circuit. Since the capacitor coupling to the amplifier is essentially a high-pass filter, temperature-induced output voltage changes are attenuated severely unless the frequency of such changes is high enough to be of the same order of magnitude as that of the alternating strain. Fortunately, most temperature changes which may affect strain gages occur too slowly to be carried through this instrumentation system. Exceptions arise in strain-gage installations where the cyclic frequency of temperature change is about the same as that of the strain.

POTENTIOMETER-CIRCUIT OUTPUT. The general equation for the electrical output of the potentiometer circuit in terms of the circuit parameters, the strain-gage characteristics, and the dynamic strain can be obtained as follows:

The instantaneous voltage across the strain gage is

$$e_G = \left(\frac{R_G}{R_B + R_G}\right) e \qquad (17.3)$$

By differentiation,

$$de_G = \frac{R_B R_G}{(R_B + R_G)^2} e \frac{dR_G}{R_G}$$

But, from the fundamental equation relating strain to resistance change,

$$\frac{dR_G}{R_G} = K \frac{dL}{L} = K\epsilon$$

so that

$$de_G = \frac{R_B R_G}{(R_B + R_G)^2} K\epsilon e = \frac{R_B/R_G}{(R_B/R_G + 1)^2} K\epsilon e \qquad (17.4)$$

This equation states that the strain-gage output voltage is proportional to the circuit voltage, the strain, the gage factor of the strain gage, and to a function of the ratio R_B/R_G, which function might be termed the "circuit factor." For the special case in which R_B/R_G is unity, the circuit factor reduces to the value ¼, and the output voltage becomes

$$de_G = \frac{K\epsilon e}{4} \qquad (17.5)$$

CIRCUIT CONSTANTS (RATIO R_B/R_G). The selection of the ratio R_B/R_G is based upon two somewhat conflicting considerations. As can be seen from Eq. (17.4), the output voltage of the potentiometer circuit is not a linear function of the resistance change in the strain gage unless the resistance of the gage is small with respect to the ballast resistor. From a practical viewpoint, however, the nonlinearity is negligibly small if R_B/R_G is unity or greater. In addition, a high value of R_B/R_G has the advantage of increasing the output signal (de_G) across the strain gage for a fixed nominal current through the gage. The practical limits to the selection of a very large R_B/R_G ratio are (1) that the supply voltage requirements become excessive in order to maintain a given nominal current through the system and (2) that the relative gains in linearity and output signal fall off rapidly with large ratios and virtually disappear at $R_B/R_G = 10$. In contrast to these considerations, the output signal across the strain gage is a maximum for a fixed supply voltage when the ratio R_B/R_G is unity. This can be demonstrated by expanding Eq. (17.4) and differentiating it with respect to R_B/R_G for a maximum. Figure 17.18 illustrates the variation of relative output voltage with the ratio R_B/R_G for a fixed supply voltage; and Figs. 17.19 and 17.20 show, respectively, the output voltage and the required supply voltage as functions of R_B/R_G for a fixed value of current through the strain gage. In practice the ratio R_B/R_G is usually selected in the range from 1.0 to 3.0.[23,24]

Other factors remaining constant, the output signal across the strain gage varies linearly with the nominal current through the gage; thus, the signal magnitude can be increased by increasing this current. The upper limit is set by the overheating of the strain gage. The upper limit varies with the size, shape, specific heat, and thermal conductivity of the member upon which the gage is mounted, and somewhat with the

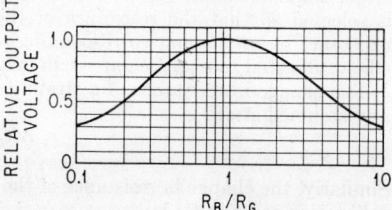

FIG. 17.18. Variation of relative output voltage with ratio R_B/R_G for a fixed supply voltage.

ambient conditions in which the installation operates. Conventional wire strain gages have been operated at currents of 50 milliamperes and greater, but for dependable functioning over a wide range of conditions, the manufacturer ordinarily recommends a maximum of 30 milliamperes through the gage. Foil-type strain gages, which are characterized by a much higher ratio of conductor surface area to cross-sectional area, can carry correspondingly larger currents.

OUTPUT FROM BALLAST RESISTOR. The output from the potentiometer circuit can be taken from across the ballast resistor instead of the strain gage if desired. In this

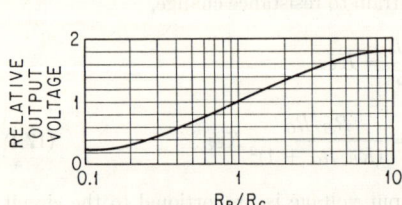

Fig. 17.19. Relative output voltage as a function of the ratio R_B/R_G for a fixed current through the strain gage.

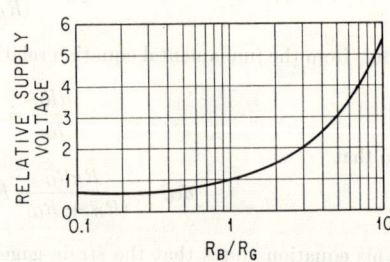

Fig. 17.20. Required supply voltage as a function of the ratio R_B/R_G for a fixed current through the strain gage.

case the output signal is identical in magnitude and opposite in sign to that appearing across the strain gage:

$$de_B = -\frac{R_B/R_G}{(R_B/R_G + 1)^2} K\epsilon e \tag{17.6}$$

If the supply voltage remains constant, irrespective of small changes in the resistance of the strain gage, any increase in voltage across the gage must be accompanied by a corresponding simultaneous decrease in voltage across the ballast resistor in order that the sum of the voltages around the complete circuit equal zero. The principal factor in selecting the strain gage or the ballast resistor as the output element is one of impedance considerations if R_B/R_G is other than unity. The effects of R_B/R_G on the relative output signal magnitude described in the preceding paragraph also apply when the signal is taken from the ballast resistor.

CALIBRATION WITH THE POTENTIOMETER CIRCUIT. A representative instrumentation system which can be employed with the potentiometer circuit for indicating dynamic strains is shown in Fig. 17.21. In order to interpret the signal amplitudes on the screen of the oscilloscope in terms of strains, it is necessary to generate some form of standard calibrating signal in the potentiometer circuit. The common technique for accomplishing this employs a large resistor which can be switched in parallel with either the ballast resistor or the strain gage. The size of the calibration resistance R_C is selected so that the resistance of the parallel combination is less than the initial resistance by an amount corresponding to the resistance change in the strain gage itself when subjected to a particular strain magnitude.

The change in resistance of a strain gage (with known resistance and gage factor) for any assumed strain is

$$\Delta R_g = K\epsilon R_g \tag{17.7}$$

Similarly, the change in resistance of the parallel combination of the strain gage and the calibration resistor R_C is

$$\Delta R = R_G - \frac{R_G R_C}{R_G + R_C} \tag{17.8}$$

Equating Eqs. (17.7) and (17.8) and solving for R_C yields

$$R_C = \frac{R_G(1 - K\epsilon)}{K\epsilon} \qquad (17.9)$$

Since the factor $K\epsilon$ is commonly smaller than 0.005, Eq. (17.9) reduces to the following close approximation:

$$R_C \simeq \frac{R_G}{K\epsilon} \qquad (17.10)$$

As a numerical example, assume that it is wished to display a calibration pulse on an oscilloscope representative of 1,000 microinches/in. strain in a 120-ohm strain gage

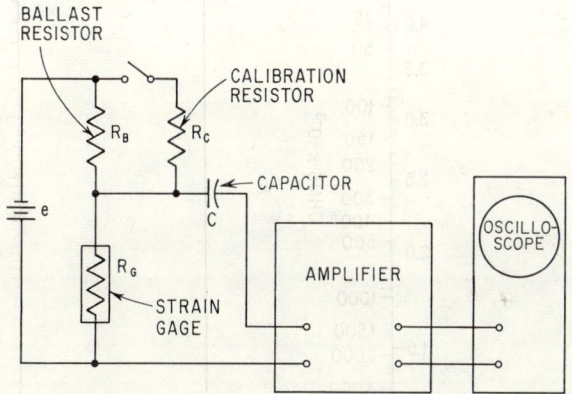

FIG. 17.21. Instrumentation for use with the potentiometer circuit. Calibration resistor is repetitively switched in parallel with the ballast resistor to produce a dynamic calibration signal on the oscilloscope.

having a gage factor of 2.0. From Eq. (17.10) the calibration resistor has a value

$$R_C = \frac{120}{2.0 \times 0.001} = 60,000 \text{ ohms}$$

According to Eq. (17.9) the exact value is 59,880 ohms, and the difference in amplitude of the calibration pulses for the two cases is 0.2 per cent.

Calibration can also be accomplished by shunting the ballast resistor instead of the strain gage if this is more convenient. In this case,

$$R_C = \frac{R_B(R_B - K\epsilon R_G)}{K\epsilon R_G} \qquad (17.11)$$

And if $K\epsilon$ is small compared to unity,

$$R_C \simeq \frac{R_B^2}{K\epsilon R_G} = \left(\frac{R_B}{R_G}\right)^2 \frac{R_G}{K\epsilon} \qquad (17.12)$$

Equations (17.10) and (17.12) for calculating calibration resistor sizes are equally applicable whether the output voltage signal is taken from across the strain gage or the ballast resistor.

The nomograph in Fig. 17.22 provides a method for rapid determination of calibration resistor sizes for shunting the strain gage. If the ballast resistor is shunted instead of the strain gage, the result from the nomograph should be multiplied by $(R_B/R_G)^2$, as indicated by a comparison of Eqs. (17.10) and (17.12).

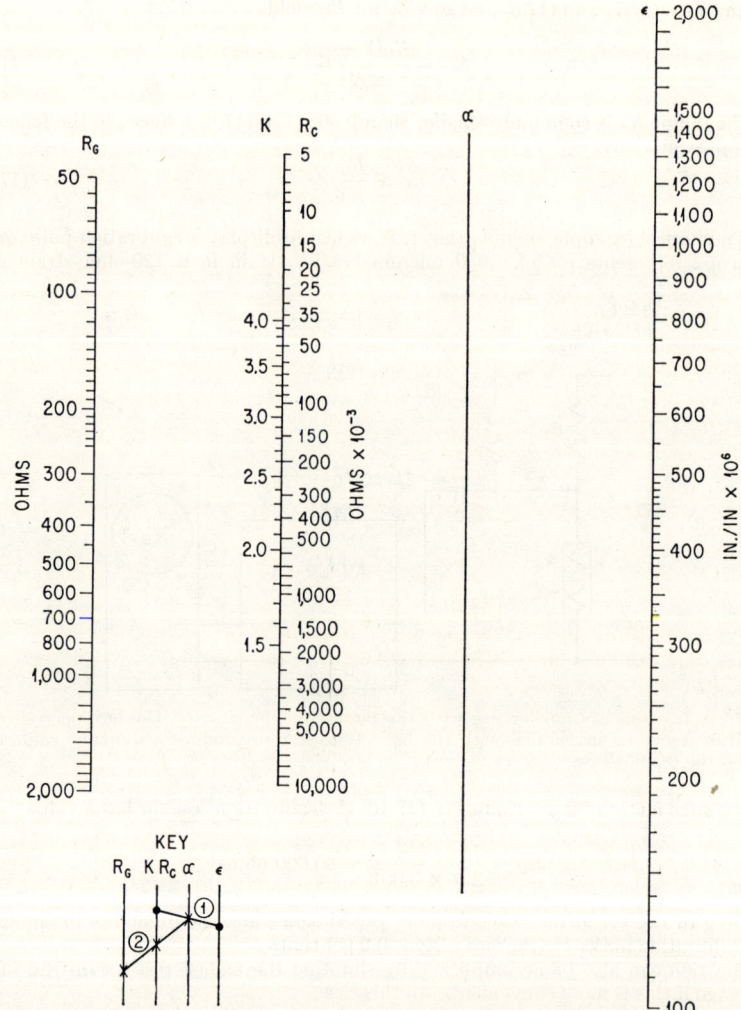

Fig. 17.22. Nomograph for determining calibration resistor sizes or for the synthetic strain signal resulting from a particular calibration resistor. To find calibration resistor size, construct a line joining the desired synthetic strain signal on the ϵ scale and the gage factor on the K scale. Construct a second line from the strain-gage resistance on the R_G scale to the intersection of the first line with the index line α. The second line crosses the R_C scale at the proper value of the calibration resistance. To find the synthetic strain signal from a particular calibration resistance, construct a line from the gage resistance on the R_G scale to the resistance of the calibration resistor on the R_C scale. Project this line to the index line α. A second line, constructed from the gage factor on the K scale through the intersection of the first line with the index line α, to the ϵ scale, yields the magnitude of the synthetic strain signal.

THE WHEATSTONE BRIDGE

In the potentiometer circuit it is necessary to block the d-c component of the output voltage with a capacitor before feeding the signal to the input of an amplifier. The same effect can be achieved by suppressing the d-c component of the signal by connecting two potentiometer circuits in parallel and taking the output signal from corresponding points in the two branches of the resulting network as shown in Fig. 17.23. This circuit arrangement is generally referred to as a *Wheatstone bridge*, and represents one of the most precise methods known for measuring (or comparing) resistances. Advantages of the Wheatstone bridge over the potentiometer circuit are (1) much greater flexibility in circuit arrangements for signal augmentation, temperature compensation, and cancellation or separation of variables, (2) capacity for accurately indicating combined static and dynamic strains, and (3) virtually complete freedom from error due to resistive changes in the conductors connecting the supply voltage to the network. As an example of the significance of the last point, consider the effect of the contact resistance variations which might occur in a set of slip rings being used in conjunction with a test of torsional vibration in a rotating shaft.

CIRCUIT OUTPUT. In Fig. 17.23 the output voltage e_0 of the Wheatstone bridge is the difference between the voltages at points A and C:

$$e_0 = e_A - e_C = \left(\frac{R_1}{R_1 + R_4}\right)e - \left(\frac{R_2}{R_2 + R_3}\right)e$$

$$= e\left[\frac{R_1}{R_1 + R_4} - \frac{R_2}{R_2 + R_3}\right] \qquad (17.13)$$

If one of the resistors constituting the bridge network is a strain gage (say R_1) then the bridge output as a function of resistance change in R_1 can be obtained by differentiating Eq. (17.13) with respect to R_1:

$$\frac{de_0}{dR_1} = \frac{R_4 e}{(R_1 + R_4)^2}$$

or

$$de_0 = \frac{R_1 R_4}{(R_1 + R_4)^2} K \epsilon e \qquad (17.14)$$

The Wheatstone bridge can be used in either of two modes of operation: *unbalanced*, for dynamic, or combined static and dynamic strains, and *null balance*, largely for static strains. Equation (17.14) corresponds to unbalanced operation, and demonstrates that under these conditions the output signal from the bridge circuit varies with the supply voltage. For static strain measurements which might take place over a time interval long enough to involve variations in the supply voltage, the resulting errors ordinarily will be intolerable unless a regulated power supply is employed. The same problem exists when attempting to record both components of a combined static and dynamic strain.

UNBALANCED OPERATION. Dynamic strain-measuring instruments almost universally employ the potentiometer or unbalanced Wheatstone bridge circuits. Errors produced by drift of the power-supply voltage are ordinarily negligible since

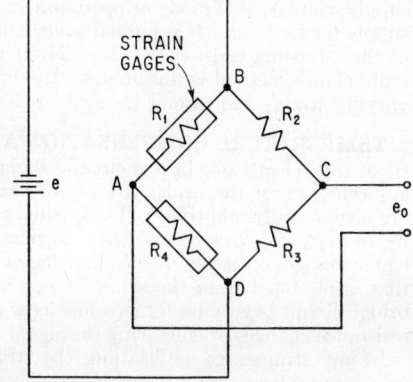

FIG. 17.23. Wheatstone-bridge circuit for static and dynamic strain measurement.

such variations occur at too low a frequency to be sensed by the amplifying and indicating portions of the instrument.

The Wheatstone-bridge circuit in the unbalanced mode of operation is characterized by a slight nonlinearity due to changes in current which accompany resistance changes. An approximate expression for the nonlinearity error as a fraction of the output signal is

$$\frac{e_e}{de_0} \simeq \frac{R_1}{R_1 + R_4} K\epsilon \qquad (17.15)$$

where e_e is the output error in volts. This nonlinearity error is always opposite in sign to the output signal. Thus de_0 is always too small by the magnitude e_e. Equation (17.15) shows, however, that inaccuracies from this source are, for elastic strains in metals, only a fraction of 1 per cent of the output signal.

The nonlinearity increases with the extent of the Wheatstone-bridge unbalance. Therefore, it is common practice to operate the bridge in an unbalanced mode around a point of initial balance for zero strain. This procedure has the advantage of retaining the proper zero base line for interpreting combined static and dynamic strains. In most commercial strain-measuring instruments such balancing is accomplished by shunting one or more of the bridge-circuit resistors with a variable balancing resistance. Because this technique invariably produces a reduction in sensitivity of the strain gage, calibration is required or sizable inaccuracies may result.

The output equations given above assume an infinite impedance in the input circuit of the amplifier or indicating instrument which accompanies the Wheatstone bridge. For a finite input impedance Z, the bridge output will be attenuated by the ratio

$$\frac{Z}{Z + R}$$

where R is the resistance of one leg of a Wheatstone bridge having four equal legs. The effect of this factor can be accounted for by calibration.

NULL-BALANCE OPERATION. If a Wheatstone bridge is "in balance" (i.e., whenever the voltage difference from A to C in Fig. 17.23 is zero), then

$$\frac{R_1}{R_4} = \frac{R_2}{R_3} \qquad (17.16)$$

If R_1 is a strain gage, its resistance change due to strain can be accurately determined by adjusting a calibrated balancing resistor (R_4, for example) to bring the circuit back to a state of balance. Since the determining relationship [Eq. (17.16)] does not involve the supply voltage, this mode of operation can give precise results, independent of power-supply fluctuations. It is limited principally by the calibration precision and constancy of the adjusting resistor R_4. Because of the time required to balance the bridge, the null-balance method is unsuited for the measurement of anything but static or slowly varying strains and cannot be used in shock or vibration instrumentation.

TEMPERATURE COMPENSATION AND SIGNAL ENHANCEMENT. An analysis of the Wheatstone-bridge circuit characteristics indicates that resistance changes in adjacent legs of the bridge are algebraically subtractive, while those in opposite legs are algebraically additive.[25] Thus, equal resistance changes of the same sign in R_1 and R_2, or in R_1 and R_4, etc., will cause no change in the output voltage. On the other hand, equal changes of like sign in R_1 and R_3, or R_2 and R_4, result in twice the output voltage that is obtained from the same change in any single resistor. This attribute of the bridge circuit is very useful as a means of either compensating for temperature-induced resistance changes or enhancing the signal output from the bridge, or both.

In any strain-gage application, the strain gage measures strain *primarily along the gage axis*. Therefore, any circuit arrangements intended to cancel undesired thermal or elastic effects, or to augment signal levels, should be selected with this consideration in mind.[26,27]

Temperature compensation can be achieved by using strain gages for both R_1 and R_2 (the gages should preferably be taken from the same package, so as to ensure optimum similarity of characteristics). One of the gages is employed as an "active" gage to sense the varying strains in the part being studied. The second gage (referred to as a "dummy") is mounted on an unstrained block of the same material as the test part, being kept at the same temperature as the part. Any resistance changes due to temperature will be the same in both gages and thus will be canceled in the bridge circuit, leaving the net effect of the strain changes to appear as an output signal. When cir-

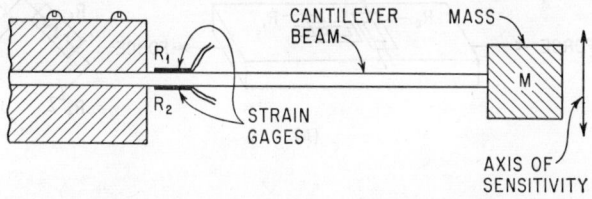

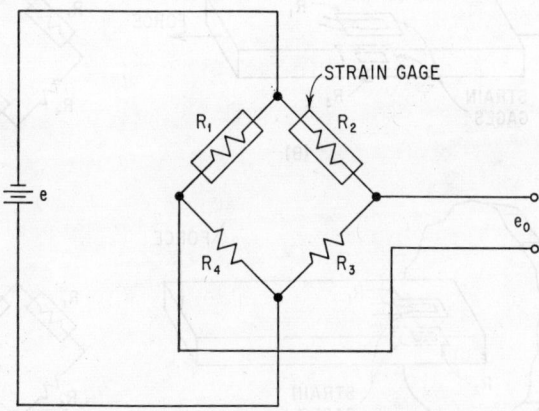

Fig. 17.24. Seismic transducer with strain gages arranged physically and electrically to produce complete compensation of temperature effects and double the output signal of a single gage.

cumstances permit, the strain gage being used as a dummy can serve to augment the output as well as to compensate for temperature changes. For example, in the seismic transducer shown in Fig. 17.24, the temperature effects in gages R_1 and R_2 are always the same in magnitude and sign, while the mechanical strains are equal and opposite. The over-all result is that temperature changes are completely nullified, while the bridge output signal for a particular strain in the beam due to bending is doubled. This presumes that no significant temperature gradients exist in the beam. Use of the above technique is restricted to members in which there are two strains always equal in magnitude and opposite in sign—the common examples being torsion and bending.

The output signal also can be augmented by using pairs of gages in opposite legs such as R_1 and R_3, or R_2 and R_4. In this case, the mechanical strain in both gages must be equal and of the same sign, and temperature compensation can be achieved only by using dummy strain gages, mounted on an unstrained piece of the same material as the test part, for the remaining two legs in a complete Wheatstone-bridge circuit. If the dummy gages are not used, the temperature errors in this type of system will be doubled in comparison to a single gage without compensation. The need for this type of circuit arrangement may occur, for example, when instrumenting a simple tension or compression member, because the axial surface strain is everywhere the same in both sign and magnitude. An alternative arrangement for treating simple tension or compression problems involves mounting gages at right angles to one another, aligned with the two principal axes, so that one gage senses the primary axial strain and the other gage is subjected to a smaller strain of the opposite sign due to Poisson's ratio. The two strain gages must be connected in adjacent legs of the Wheatstone bridge to obtain temperature compensation, and the system must be calibrated mechanically for load vs. strain, since

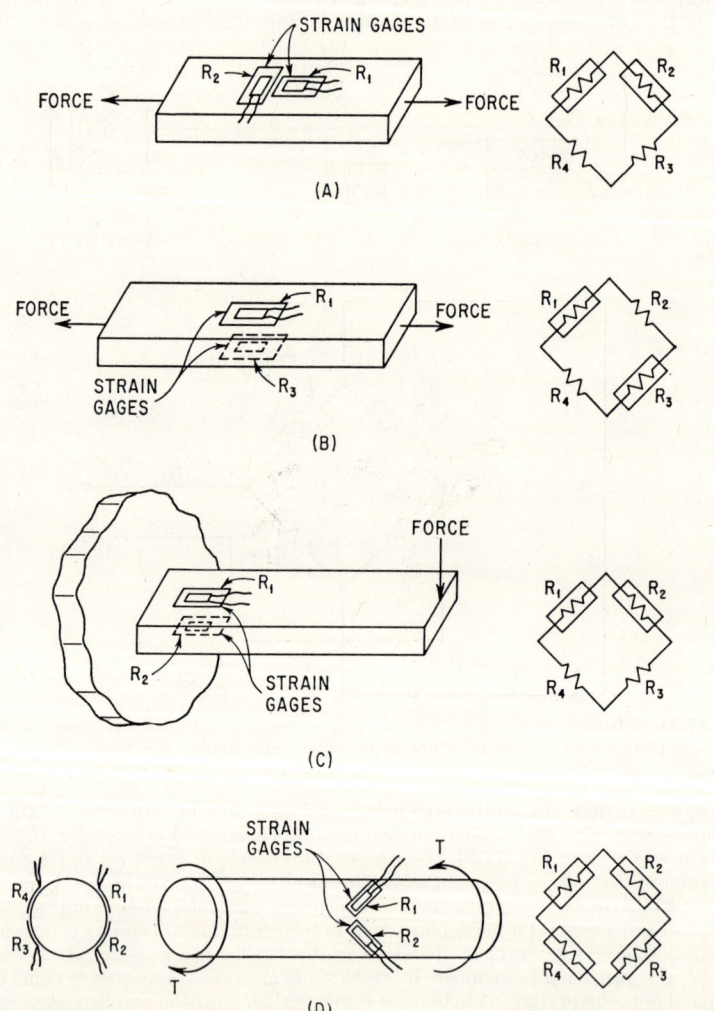

Fig. 17.25. Various techniques for temperature compensation, signal augmentation, and separation of variables, with characteristics as follows: (A) *Axial load (tension or compression)*: temperature effects are nullified; signal augmentation factor is approximately $1.3\,(1+\mu)$; must be calibrated for Poisson's ratio, or the ratio must be assumed. (B) *Axial load (tension or compression)*: temperature effects are doubled unless dummy strain gages are used for R_2 and R_4; signal augmentation factor is 2.0; bending strains are canceled. (C) *Bending load:* temperature effects are nullified; signal augmentation factor is 2.0; strains due to axial loading are canceled. (D) *Torsional load:* temperature effects are nullified; signal augmentation factor is 4.0; strains due to bending and axial loading are canceled.

there is a fixed (but not precisely known) relationship between the axial and Poisson strains. In Fig. 17.25 are given several techniques, with the physical and electrical arrangements of the strain gages, for accomplishing temperature compensation and/or signal enhancement. In the measurement of dynamic strains (with no static component of interest) any of the signal-augmentation techniques can usually be applied without regard to temperature compensation if the temperature variations occur at a much lower frequency than the strain variations.

SELECTION OF INSTRUMENTS FOR STRAIN MEASUREMENT

As shown by Eqs. (17.5) and (17.14), the output voltage from a strain-gage potentiometer circuit or Wheatstone bridge is, for elastic strain magnitudes in metals, very small. Electrical amplification is required to bring the signal to a level where it can be used conveniently for indication or recording. To assure satisfactory performance and precision, the entire instrument system, from power supply to recording instrument, should be considered as a unit. Figure 17.26 illustrates in block form the basic elements of a strain-gage instrumentation system. The criteria for selecting the individual components of such a system are fixed by the nature of the strain being studied, the type of information required from the system, and the mutual compatibility of the various system components. Consideration should be given to the required frequency response, the input and output impedances of the units in the system, the signal amplitudes being dealt with, and the accuracy of measurement desired. In general, it is safe to assume that the strain gage will respond to frequencies considerably higher than any mechanical device to which it may be attached. In the case of small members vibrating at high frequencies the limitation is more apt to arise from the change in mass due to the presence of the gage and its lead wires.

Commerical instruments for use with strain gages usually combine into a single unit several if not all of the components of Fig. 17.26. The limitations of such devices should be investigated prior to purchase. For example, the instrument may include an alternating-frequency source of power for the Wheatstone bridge. This can lead to difficulties in the measurement of high-frequency strains. The frequency of the strain being measured (which will modulate the power supply in the bridge circuit) is limited to approximately 10 to 20 per cent of the carrier frequency. If the carrier frequency is high enough to overcome this objection, the capacitive unbalance and pickup in the strain-gage leads is apt to be excessive.

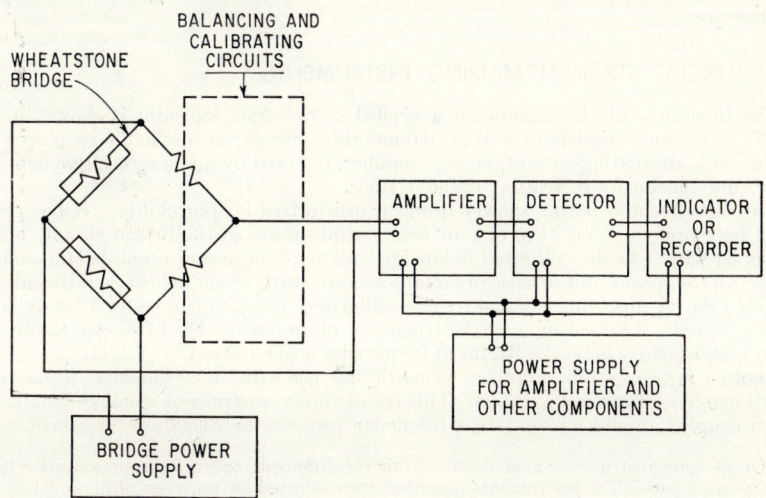

FIG. 17.26. Block diagram of basic elements of strain-gage instrumentation system.

Alternating-current bridge supply sources usually range from 500 to 5,000 cps, and impose corresponding limitations on the strain frequency which can be studied with precision. Commercial instruments usually include convenient, built-in circuits for balancing and calibration. These features should be investigated and tested for the distortion and gage desensitization effects which they may introduce into the system. The carrier systems have several advantages: a high degree of stability, capacity for handling low-frequency and static components of strain as well as higher-frequency signals, and (in conjunction with a phase-sensitive demodulator) the ability to preserve the sign (tensile or compressive) of the strain. For studying mechanical vibration at frequencies above 1,000 cps it is generally advisable to employ a battery or regulated

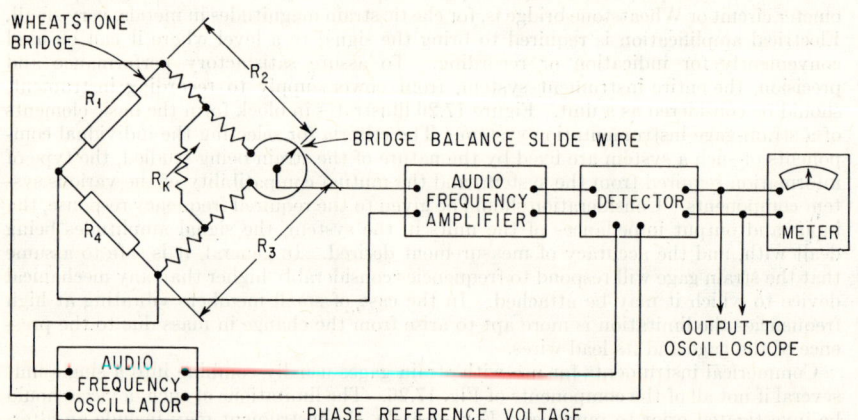

Fig. 17.27. Block diagram of commercial static strain indicator. It employs a-c bridge powered by oscillator at approximately 1,000 cps. Bridge output is amplified and then rectified in a phase-sensitive detector to give a d-c signal proportional to strain, and of a corresponding sign. This indicator can be used with an oscilloscope for dynamic strains at frequencies up to approximately 100 cps.

d-c source to supply the bridge circuit, a direct-coupled amplifier, and a cathode-ray oscilloscope.

COMMERCIAL STRAIN-MEASURING INSTRUMENTS *

The functional block diagram for a typical static strain indicator is shown in Fig. 17.27. This unit consists of a Wheatstone-bridge circuit, an oscillator to power the bridge with alternating current, an a-c amplifier, followed by a phase-sensitive demodulator, and a meter for indicating bridge balance.

The instrument is battery-powered and transistorized for portability. It has provisions for employing one, two, or four active strain gages in the bridge circuit, and is equipped with a finely calibrated balancing slide wire for accurate null-balance operation. Although this unit is basically a null-balance static strain indicator, it includes an output jack for supplying a signal to an oscilloscope, and thus can be used for dynamic strains as well. Limited by a carrier frequency of approximately 1,000 cps, the instrument is satisfactory for cyclic strains at frequencies below 100 cps.

Another form of strain amplifier, primarily for use with an oscilloscope, is shown in functional block form in Fig. 17.28. This transistorized instrument employs a stabilized direct-coupled amplifier to obtain a frequency response to 20,000 cps or above. The

* Of the many instruments available for strain measurement, only a few representative types are discussed here. The instruments described were selected for purposes of illustration, and are not necessarily those recommended by the authors.

battery-supplied d-c Wheatstone bridge is arranged to accept one, two, or four active strain gages.

Figure 17.29 is the block diagram for a popular type of carrier-frequency instrument for use with a direct-writing (pen-and-ink or electric stylus) recorder. This instrument

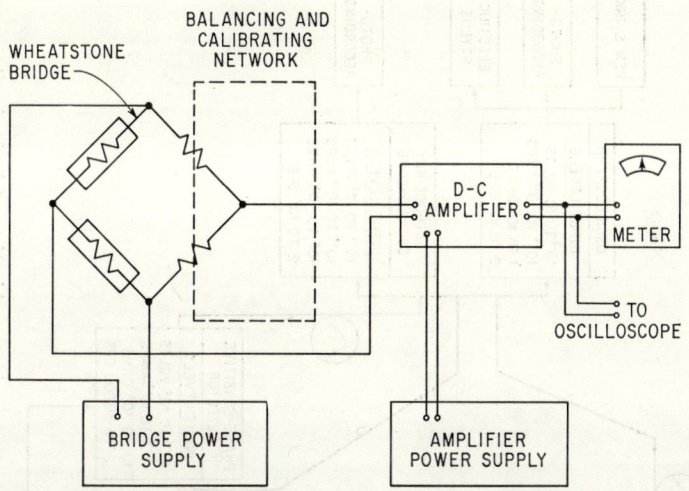

Fig. 17.28. Block diagram of strain amplifier with direct-connected circuitry, intended primarily for use with an oscilloscope.

employs alternating current (approximately 2,500 cps) to supply the Wheatstone-bridge circuit. The bridge output is first augmented in an a-c amplifier and subsequently demodulated and amplified again in a direct-coupled amplifier to a level where the signal is capable of driving the galvanometer in a direct-writing oscillograph. Instruments of this type generally include provisions for balancing the bridge and for obtaining a con-

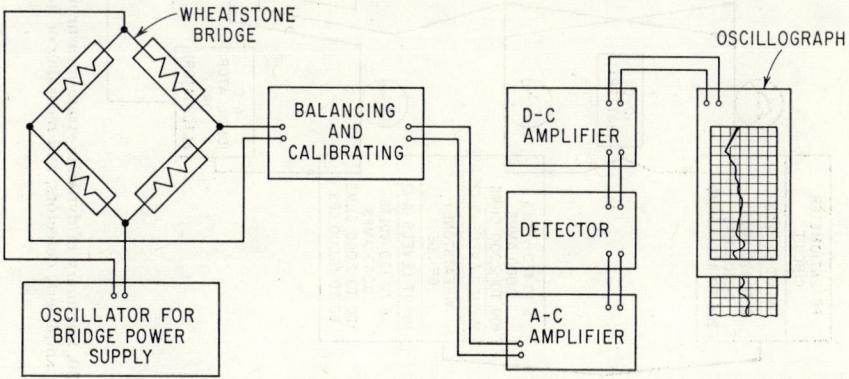

Fig. 17.29. Block diagram of carrier-frequency instrument for use with direct-writing recorder.

venient calibration signal. Limited largely by the frequency response of the stylus-driving galvanometer, these units are applicable to vibration problems with frequencies of less than 100 cps.

Higher-frequency strain-recording instruments (up to 5,000 cps) employ mirror galvanometers and light-beam recording on photosensitive paper. These instruments gen-

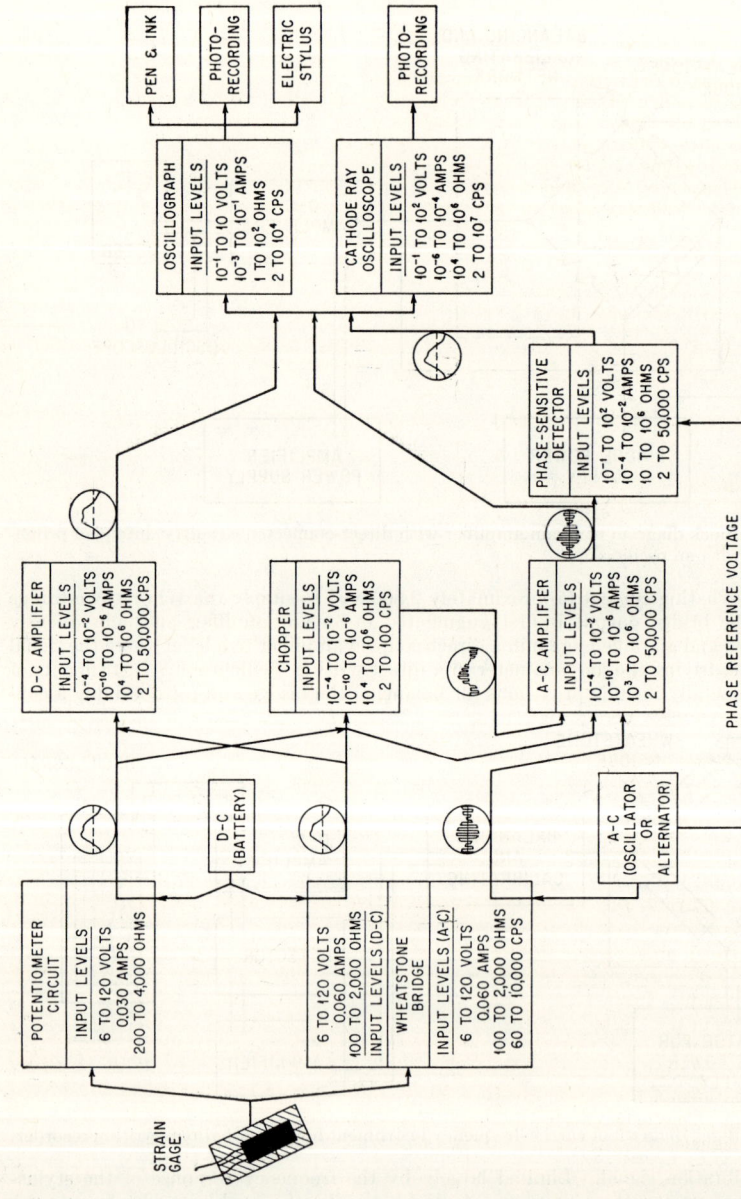

Fig. 17.30. Summary of dynamic strain-measuring instrumentation, illustrating various combinations of basic circuits, power supplies, amplifiers, detectors, and recording or indicating equipment.

Table 17.4. Representative Manufacturers of Strain-gage Instruments
(Amplifiers, Wheatstone Bridges, Strain Indicators, etc.)

United States:
  Allegany Instrument Co., Cumberland, Md.
  Arthur R. Anderson Co., Tacoma, Wash.
  Baldwin-Lima-Hamilton Corp., Waltham, Mass.
  Brush Instrument Division of Clevite Corp., Cleveland, Ohio
  Consolidated Electrodynamics Corp., Pasadena, Calif.
  Ellis Associates, Pelham, N.Y.
  Hathaway Instrument Division of Hamilton Watch Co., Denver, Colo.
  Heiland Division of Minneapolis-Honeywell Corp., Denver, Colo.
  Sanborn Co., Waltham, Mass.
Denmark:
  Brüel and Kjaer, Naerum
England
  H. Tinsley and Co., Ltd., London
  Kelvin and Hughes, Ltd., London
  McMichael Radio, Ltd., Slough, Bucks.
  Savage and Parsons, Ltd., Watford, Herts.
  Southern Instruments, Ltd., Camberly, Surrey
France:
  Ribet-Desjardins, Montrouge, Seine
  Sexta, S. A.: Bagneux, Seine
Germany:
  Elektro Spezial GmbH, Hamburg
  Hottinger Mess-Technik GmbH, Darmstadt
Japan:
  Kyowamusen Kenkyujo Co., Ltd., Tokyo
  Shinkoh Tsushinkogyo Co., Ltd., Kanagawa-ken
  Toko Sokki Kenkyujo Co., Ltd., Tokyo
  Toyo Measuring Instruments Co., Ltd., Tokyo
Netherlands:
  Peekel, Rotterdam
  Philips, N. V., Eindhoven
Norway:
  Elektrometer A9S, Trondheim
Sweden:
  Elema-Schönander, Stockholm

erally use the techniques already described for powering the Wheatstone bridge and for accomplishing balancing and calibration.

The circuitry described in the preceding paragraphs for amplifying and recording strain-gage outputs can be readily arranged for multichannel operation to record a number of related events simultaneously. Most galvanometer oscillographs are available in multichannel units having from 2 to 24 or more channels.

Figure 17.30 summarizes the common techniques which can be used in instrumenting dynamic strain. Table 17.4 lists a number of the manufacturers of strain-measuring instruments in the United States and other countries. Although the design and operational details may vary from instrument to instrument, strain-measuring equipment is generally very similar in principle in all countries, and most instruments are variations or combinations of the types described here. Permanent records of dynamic and transient strain phenomena can be obtained by the standard techniques employed for other electrical signals.

APPENDIX 17.1
STRAIN-GAGE APPLICATION TECHNIQUES

In applying and using bonded strain gages considerable attention must be given to apparently minute and insignificant detail if the maximum possible accuracy of the gage is to be obtained.

BONDING TECHNIQUES

SURFACE PREPARATION. Before cementing any type of strain gage to a structure being tested, certain surface requirements must be met in order to ensure a complete bond between the gage and the surface.

Surface Roughness. The surface to which the gage is to be bonded should be smooth in a gross sense, but not too highly polished since this condition does not promote adhesion. All scale, rust, paint, etc., should first be removed from the surface; if the surface is pitted or rough, as is apt to be the case for forgings or castings, it may be necessary to grind lightly a spot a little larger than the gage. If the surface is highly polished to start with, better bonding will be obtained by roughening it slightly with a fine grade of emery paper.

Surface Cleanliness. It is necessary to clean the surface thoroughly since the bonding cements used will adhere well only to a clean surface. Some volatile solvent such as carbon tetrachloride or acetone will serve for this purpose. It is imperative that cleanliness be maintained until the gage is firmly in place. A clean solvent and clean cloths should be used, and fingers should be kept off the prepared surface and the back of the gage. Special care must be taken not to contaminate the solvent that remains in the container for future use. Gauze sponges, such as are used by the medical profession, or cotton swabs (Q-Tips) serve admirably for the purpose of cleaning an area with carbon tetrachloride or acetone. Duco-type strain gages should not be saturated with liquid acetone since the acetone will dissolve the cement used in gage construction. However, it is good practice to wipe the back of any wire gage with gauze or cotton slightly dampened with acetone. It is extremely important, after the surface of the metal and the back of the gage have been cleaned, that neither is touched by the fingers or any object which may contaminate them, thereby preventing proper adherence of the gage. One additional step which frequently is desirable is to coat the metal surface with a solution which promotes adhesion of the cement to it. The sources for such a solution and other items of value in strain-gage application are listed in Table 17.5.

PAPER GAGES—APPLICATION. After the surface preparations have been completed, the gage is cemented in place. For paper gages the manufacturer supplies a cellulose adhesive called SR-4 cement; Duco household cement also is satisfactory for this purpose. In the cementing process a fairly thick layer of cement is spread on both the gage and the prepared surface and the gage is set in place at once. In the first few seconds after the gage has been applied to the surface, it is possible to slide it around slightly in order to orient it in the desired direction. It is easier to locate the gage properly if center lines have previously been lightly scribed on the surface to which the gage is to be bonded. After the gage is in the proper position, it is pressed or rolled with the finger to squeeze out the excess cement. A fair amount of pressure with the finger is desirable since this prevents the gage from slipping when clamping pressure

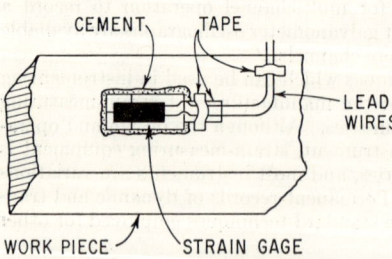

FIG. 17.31. A properly applied strain gage, with excess cement around entire periphery of gage, lead wires securely fastened to workpiece, and soldered connections insulated from workpiece.

is applied to it. A properly applied gage looks like Fig. 17.31, which shows a gage cemented in place having a small bead of excess cement around the edge. Scraping off this excess cement helps shorten the drying time. While the gage is drying, it should be held in place with a clamping force of about 1 lb. Too great a pressure at this time may result in grounding of the gage or lead wires to the underlying material. If conditions permit, a 1-lb weight placed on the gage makes a very suitable clamp. For gages which are mounted so that it is impossible to use a weight, some form of spring clamp should be used. In any case, the clamping force should be applied to the gage through a resilient material, such as a Neoprene pad, to ensure an even distribution of the pressure over the entire gage surface.

Table 17.5. Auxiliary Materials Useful in Mounting and Moistureproofing Strain Gages

Name	Supplier	Use
Bakelite Varnish B-51	Union Carbide Plastics Co., Division of Union Carbide Corp., Bound Brook, N.J.	Simple protective coating for gages not exposed to severe moisture or erosion
DC-4 Silicone Grease	Dow Corning Co., Midland, Mich.	First coating of a multilayer built-up protection
Glyptal Varnish	General Electric Co., Schenectady, N.Y.	Simple protective coating for gages not exposed to severe moisture or erosion
Petrosene-A Wax	General Petroleum Co., San Francisco, Calif.	First coating of a multilayer built-up protection
Ozite-B Bitumastic Compound	G and W Electric Specialty Co., Chicago, Ill.	First coating of a multilayer built-up protection
Di Jell 171 Wax	L. Sonneborn Sons, Inc., Building Products Division, New York, N.Y.	First coating of a multilayer built-up protection
Zophar C-276 Wax	Zophar Mills, Inc., Brooklyn, N.Y.	First coating of a multilayer built-up protection
Rubber to Metal Cement	General Cement Co., Rockford, Ill.	Used to cement a protective rubber shield over gage installation
EC-864 Synthetic / EC-801 Rubber	Minnesota Mining and Mfg. Co., Detroit, Mich.	For coating over Di Jell, Petrosene-A Wax, etc.
EC-807 Accelerator	Minnesota Mining and Mfg. Co., Detroit, Mich.	To be mixed with synthetic rubber
EC-853 Metal Primer	Minnesota Mining and Mfg. Co., Detroit, Mich.	Used to prepare metal for adhesion of synthetic rubber
Herecrol RD-9 Primer	Heresite and Chemical Co., Manitowoc, Wis.	For coating and sealing synthetic-rubber installation
Adhesive Cellophane and Electrical Tape	Minnesota Mining and Mfg. Co., Detroit, Mich.	For insulation and mounting gage leads
Thiokol EC 755	Minnesota Mining and Mfg. Co., Detroit, Mich.	Used as a protective waterproof coating
Ten X Sealing Compound	Electro Cote Co., St. Paul, Minn.	Used as a protective waterproof coating
SR-4 Cement Precoat	Baldwin-Lima-Hamilton Corp., Waltham, Mass.	Applied as undercoat before cementing paper gage in place
Metal Prep No. 10	Nielson Chemical Co., Detroit, Mich.	Used as a metal surface conditioner before cementing gage in place
NBS No. L-6AC Ceramic Precoat	O. Hommel Co., Pittsburgh, Pa.	Used to precoat surfaces for Bakelite-gage application to temperatures of 500°F (260°C); must be fired at 1750°F (950°C)

Gage Curing. The drying time allowed for paper strain gages which are bonded with Duco-type cement depends on the degree of stability required of the gage, the type of gage construction, and the temperature and humidity of the surrounding atmosphere. Strain gages are available made with a special thin paper which allows extra fast evaporation of the cement solvent and thus permits the gage to dry in a shorter time period. With standard flat grid gages and Duco-type cement, the clamp can be left on the gage for an hour or so and then removed and the gage left to air-dry for 18 to 24 hr at room temperature.

The drying process can be accelerated by heating the strain gage. Temperatures for the thermoplastic gages cannot be increased too rapidly because bubbles may form in the cement. It is advisable to heat the metal adjacent to the gage rather than to heat the gage itself. The heat then travels to the cement through the metal and drives the solvent up through the gage. After the gage has been left to air-dry for about an hour, the temperature can be gradually increased to a value as high as 120°F (50°C). At this temperature the strain gage will be dry in a few hours.

If during drying the electrical resistance is measured between the strain-gage lead and the metal surface to which the gage is being bonded, it will be found that as the cement dries, resistance increases. When the resistance no longer increases, the gage is dry and ready for use. It is also possible to accelerate the drying process somewhat by thinning the Duco cement slightly with acetone before applying it, or by using the SR-4 cement which is comparatively fast drying. Specific recommendations for drying procedures for the various Duco gages are given in the manufacturer's bulletins. In general, however, it is convenient merely to let the gages dry for a day; this results in essentially complete evaporation of the solvent. Two major exceptions to this are cases in which the gages are located where the humidity is very high or the temperature very low. In the latter case, particularly, heat drying is an absolute necessity in order to drive the solvent out of the cement. Care must be exercised in the drying of the wrap-around-type gages of very short lengths such as the SR-4, A-19 gage. Gages of this type must be dried for a relatively long period of time before use if any consistent operation is to be obtained—a drying time as long as 7 days being desirable before they are put into service.

BAKELITE GAGES—APPLICATION. Bakelite strain gages are considerably more difficult to attach than paper gages. Therefore, as a rule they are used only under one or more of the following circumstances: (1) where temperatures are between 180 and 500°F (80 and 260°C), (2) where gage stability is required despite high humidity conditions, and (3) where long-time gage stability is necessary.

Preparation of the surface to which the strain gage is bonded is the same for all gages whether paper, Bakelite, or other types using the quick-drying cements. The back of the Bakelite gage should be thoroughly cleaned with acetone or carbon tetrachloride before attempting to cement it to the prepared surface. A treatment of the surface with a solution designed to promote adhesion of the cement (see Table 17.5) also is valuable but may be dispensed with if a good clean area is obtained. It is very important that neither the gage nor the surface to which it is to be cemented be touched by the fingers or otherwise contaminated after cleaning, since any oil or dirt will prevent proper bonding. After the surfaces are properly prepared, a coat of Bakelite cement is applied to both the metal surface and the underside of the gage, eliminating any air bubbles that may form. The gage is placed in the proper position while the cement is wet by means of guide lines previously scribed in the surface. Next, the gage can be covered by a strip of cellophane (or masking tape) followed by a Neoprene pad on the cellophane for distributing the clamping pressure uniformly. A metal plate ¼ in. thick or thicker should be placed over the pad and clamping pressure applied to it. The clamping force must be exerted through some resilient member, such as a compression spring or a thick piece of Neoprene in order that the pressure be uniformly distributed over the gage and remain essentially constant while the cement is being cured. For best results a clamping pressure of 100 to 200 lb/in.^{2} (7 to 14 kg/cm^2) should be employed, although satisfactory bonding can be accomplished with pressures as low as 30 lb/in.^{2} (2 kg/cm^2).

Gage Curing. The clamping pressure should be applied during the following baking cycle: (1) 1 hr at 140°F (60°C), (2) 2 hr at 175°F (80°C), and (3) 2 hr at 250°F (120°C).

Gage stability will be improved noticeably if the gage is given an additional hour of heating at 275°F (135°C) after removing the clamp. The best practice is to apply the heat to the metal surface immediately adjacent to the gage rather than to the gage itself, thus polymerizing the cement from the metal out to the gage. If the strain gage is to be used at temperatures above 275°F (135°C), it should be cycled to the maximum expected temperature once or twice before being put in use in order to obtain optimum stability. After the gage has been applied and heat-treated according to the foregoing recommendations, it is ready for lead-wire attachment and checking.

OTHER CEMENTS USED FOR APPLYING STRAIN GAGES. Epoxy Resin Cements. A popular type of cement for bonding strain gages is an epoxy resin such as Armstrong A-1 Cement * or Araldite.† This type of adhesive can be used with either Duco, Bakelite, or similar strain gages, will withstand temperatures up to 200°F (95°C), is essentially nonhygroscopic and unaffected by oils and most common solvents, and can be cured rapidly enough to permit the gage to be used within an hour after cementing it in place. This type of adhesive is marketed as two components, a resin and an activator, packaged separately. The user mixes a small amount of the resin and activator in the prescribed proportions and applies the mixture to the test surface and to the back of the strain gage in the same manner in which Duco cement is applied. The adhesive is self-polymerizing at room temperature and begins to cure immediately upon mixing. To obtain a good bond it is important to observe the specified proportions of resin and activator; the quantities used should be measured carefully. Other major requirements for a good bond are clean surfaces and freedom from disturbance during the curing process. While the epoxy resin cement needs little or no clamping pressure during curing, it has extremely low cohesive strength when wet; thus if a strain gage is applied to a curved surface, some form of clamping or binding is required to keep the gage in intimate contact with the surface to which it is being bonded. This lack of cohesive strength is very noticeable after using Duco cement in which the gage adheres very strongly to the surface while the cement is still wet. In the case of an epoxy resin cement, if the piece were inverted, the gage in many cases would simply drop off. This cement develops half its strength in 16 hr at room temperature and full strength after about a week. Heating to 160°F (70°C) drastically shortens the curing time to about an hour. A method sometimes used in curing an epoxy resin cement is to pass from 30 to 50 milliamperes of current through the strain gage (120 ohms). The heat generated accelerates the curing process at least as effectively as an external heat source. The actual amount of current used depends on the type of metal and the size and shape of the piece to which the gage is being bonded. All other application procedures such as surface cleanliness are identical with those described previously.

de Khotinsky Cement. A strain-gage adhesive having unique properties is de Khotinsky ‡ cement, a thermoplastic material which acts as a fusible adhesive. The cement comes in stick form and is applied by heating the test surface until it melts the cement from the stick. The actual procedure used in bonding a strain gage with this cement is first to clean and prepare the surface in the usual manner and then to heat the surface to approximately 285°F (140°C). The end of the cement stick is next applied to the hot area and a thin layer deposited over a space slightly larger than the strain gage. The gage is immediately set in place and held there under light pressure for several minutes until the cement has cooled and hardened. This bonding procedure is recommended for use only with Bakelite gages because of the high temperature to which the gage is subjected when first installed. Gages applied with this cement can be removed later by reheating the cement to its melting point. They are suitable for reuse, if handled with reasonable care.

Dental Cement. Dental cement has very fast drying characteristics. For example, Type F-88 Dental Cement § consists of a white powder to which an activator is added just before the cement is to be used. The activator and powder are thoroughly mixed and applied to the prepared surface and back of the gage; the gage is put into place

---
* Armstrong Products Company, Warsaw, Ind.
† Ciba Products Co., Fair Lawn, N.J.
‡ Central Scientific Co., Chicago, Ill.
§ Industrial Division, American Consolidated Mfg. Co., Inc., Philadelphia, Pa.

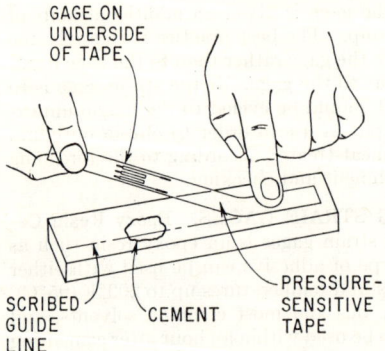

Fig. 17.32. Applying a foil gage with acrylic or other contact cement. Gage must be correctly aligned when pressed into contact with the cement on the work surface.

at once, being pressed down firmly to squeeze out excess cement. The cement begins to set as soon as the activator is added, hardening in a few minutes. Therefore, it is important that the entire operation of mixing and applying the cement, and properly placing the gage, be performed as rapidly as possible. In addition to its rapid setting, this cement sets under conditions of severe moisture, even when the surface to which it is applied is wet or under water. In any gage installation where it is impossible to provide a dry surface this cement is invaluable.

Cyano-acrylate Cement. Cyano-acrylate Eastman 910 * cement is useful for the rapid bonding of foil-type gages. The surface is first prepared in the manner previously described. After cleaning with carbon tetrachloride it is coated with a solution † designed to promote adhesion of the cement. A neutralizer,* a strong base, is applied to the area, which is then wiped dry. The gage itself is removed from its backing material by pressing a strip of masking tape or self-adhesive transparent tape over the top of the gage. The gage then adheres to the tape and the backing material is pulled away. A drop of plasticizer * is spread over the back of the gage, and cyano-acrylate cement is applied to the test surface. With one end of the masking tape stuck down to the part to which the gage is to be applied and the other end of the tape being held free so that the tape makes an angle of about 30° with the surface, the finger is run along the tape, pulling it down to bring it into contact with the test structure, as shown in Fig. 17.32. The tape is positioned in such a way that as it is applied to the surface, the gage is brought into contact with the cemented area and rolled into place as the finger passes over the tape. The curing of this cement is practically instantaneous and as soon as it has been put down the tape can be pulled free and the gage will be properly cemented in place. Since the foil in this gage is not covered by insulation, it may be necessary to provide a protective insulating coating to the surface after the gage has been applied.

TESTING THE GAGE APPLICATION

After the cement appears to be dry there are several quick tests which can be made to ascertain if the gage will function properly. First, the gage resistance is checked; if it shows an open circuit or if the measured resistance is noticeably different in value from the manufacturer's specification, this indicates damage in the handling or bonding of the gage. Next, the resistance between the strain-gage filament and the metal surface to which the gage is bonded should be measured. A thoroughly dried or cured gage has a resistance of 1,000 megohms or higher, although considerably lower resistance (a minimum of 50 megohms) can be tolerated without serious consequences. Gages having a resistance to ground of less than 10 megohms should not be used except for dynamic strain indications. Such a low resistance indicates that electrical leakage to ground is occurring; this results in gage instability and zero drift. In measuring the high resistance which exists between the gage filament and ground, a vacuum-tube high-range ohmmeter should be used. Measuring devices which place a high voltage across the resistance should not be used since there is danger of damaging the gage by high-voltage breakdown of the insulation, particularly if the gage is not thoroughly dry when the voltage is applied.

* Supplied by Budd Instruments Division, Phoenixville, Pa.
† For example, Metal Prep.

A final check on the gage, especially with regard to the integrity of the bond between the gage and the surface to which it is cemented, can be made by the following process. Insert the gage in a Wheatstone-bridge circuit such as the SR-4 strain indicator and balance the bridge, bringing the meter reading to zero. Then press lightly on the strain gage with the eraser end of a lead pencil. When this pressure is applied, the meter should show a very slight deflection corresponding to a few microinches per inch of strain. When the pressure is released, the meter needle should return to zero. If the meter fails to return to zero or becomes unsteady, or if large strain indications occur when the gage is pressed on by the eraser, then the gage is probably either imperfectly bonded or actually damaged.

LEAD-WIRE INSTALLATION

The probability of damaging a gage during wiring will be reduced materially if the instrument leads are firmly tied or cemented to the structure at a point near the gage before making the soldered connection to the gage leads. To prevent the occurrence of any strain on the gage leads once the installation is completed, it is recommended that a small loop be placed in each of the leads, as shown in Fig. 17.31. The usual procedure is to place a small piece of tape on the surface immediately adjacent to the gage and tape the soldered joints down at this point. This provides a shock-absorbing effect and may prevent an accidentally shorted gage circuit as well as eliminate external strains on the gage. Careful wiring of strain gages, including thoroughly soldered joints, results in increased stability and reliability of gage readings. All lead wires from the measuring instrument to the strain gage should be securely taped or cemented to the structure bearing the gages so that no relative motion occurs.

CONNECTIONS TO ROTATING OR OSCILLATING BODIES. Strain-gage connections to rotating bodies can be made through slip rings and brushes. Special circuits [28] have been developed to reduce the noise introduced by the slip rings to a minimum value. Slip rings of silver and brushes of silver-graphite are commonly used. Monel metal slip rings using silver-graphite brushes give excellent results when fairly high brush pressures are used, namely, 95 lb/in.^{2} (7 kg/cm^2). For operation in oil the best results are obtained with shim brass slip rings and silver-graphite brushes operating at a pressure of 175 lb/in.^{2} (12 kg/cm^2). Slip rings and brushes can be used in high-speed operations.[29] In this case, it is important to ensure that a very low eccentricity exists, the maximum permissible being in the order of magnitude of 0.0003 in. (0.0008 cm). In all cases several brushes should be used with each slip ring and it is good practice to use different spring rates for each brush since this provides different resonant frequencies and minimizes the possibility of all brushes leaving the ring at the same time, in the event of operation at a critical speed.

To instrument oscillating motion, sliding contacts may be used in place of slip rings. However, it is desirable and frequently possible to mount the lead wires in such a way that sliding contacts are unnecessary. This can be done through the use of a number of articulating arms which follow the strain gage in its motion and carry the lead wires. By forming the wires into a coil of one or more loops at each joint, fatigue failure of the leads under continuous flexing can be prevented.[30]

HIGH-TEMPERATURE GAGE APPLICATION

Two sources of difficulty are encountered when gages are operated at high temperatures. The first source of difficulty is that the properties of the wire change at high temperatures and it does not give consistent results with respect to the strain to which it is subjected. The second difficulty results from the breakdown of the electrical resistance of the carrier and bonding material used with high-temperature gages. If a breakdown of electrical resistance occurs, the resistance to ground for the gage will become too low for its satisfactory operation. The common strain-sensitive wires are generally unsuitable for high-temperature operation and the conventional gage bonding agents are incapable of withstanding such temperatures.

Table 17.6. Properties of Strain-gage Filament Materials

Material	Composition	Gage factor	Temp. coef. of res.*	Max. temp.
Iso-elastic............	0.36 Ni, 0.52 Fe, 0.08 Cr, 0.005 Mo	3.5	+470	500°F (260°C)
Constantan or Advance..........	0.55 Cu, 0.45 Ni	2	+2	900°F (480°C)
Karma..............	0.74 Ni, 0.20 Cr, 0.03 Fe, 0.03 Cu	2	+2	1500°F (815°C)
Nichrome V.........	0.80 Ni, 0.20 Cr	2	+400	2000°F (1100°C)
Platinum-iridium.....	0.80–0.90 platinum 0.20–0.10 iridium	6	+800	2000°F (1100°C)

* Micro-ohms per ohm per degree centigrade.
These values may vary appreciably with slight changes in composition and at elevated temperatures.

WIRE MATERIALS. The wire materials that are employed in strain gages designed for use in the range 1000 to 2000°F (540 to 1090°C) include Constantan, Nichrome V, and several platinum-iridium alloys. Table 17.6 lists the approximate composition and electrical properties of these materials. The platinum-iridium alloys have a strain-sensitivity factor roughly three times that of the ordinary gage materials and they are characterized by inherent chemical stability even at high temperatures. The preceding advantages are partly discounted, however, by a low resistance and high thermal coefficient of resistance. This alloy is sometimes used for high-temperature measurement of dynamic strains.

Up to approximately 900°F (480°C), Karma alloy displays many of the characteristics of an ideal high-temperature strain-gage material. Although its strain-sensitivity factor is only about one-third that of the platinum-iridium alloys, it has a high resistance and a low thermal coefficient of resistance, is stable in the presence of ceramic cements, and is well suited to the measurement of static strains. Above 900°F (480°C) Karma alloy undergoes metallurgical changes and develops a varying thermal coefficient of resistance.

Many conventional strain gages are unsuitable for high-temperature use because they are subject to galvanic corrosion under these conditions. Thus it is evident that precise static strain measurements above 900°F (480°C) are difficult to obtain. Current practice is to employ Karma alloy or Nichrome V up to their maximum usable temperature. Above 900°F (480°C), platinum-iridium alloys usually are employed. For example, they have been used in dynamic strain measurements on turbine blades at 1500°F (815°C).

CERAMIC CEMENTS. The following ceramic cements are used for bonding high-temperature strain gages. Sauereisen cement No. 31 has been successfully employed by a number of laboratories and Quigley AAA high-temperature paints are generally satisfactory as strain-gage precoats and covercoats. Paints AAA Nos. 1900, 1920, and 1925 are most commonly used. The NACA has employed these cements in constructing strain gages for use on gas-turbine blades.[31] Most commercial manufacturers of high-temperature strain gages recommend Allen P-1 ceramic cement * for the installation of their gages.

COMMERCIAL HIGH-TEMPERATURE STRAIN GAGES. Ordinary Bakelite gages can be used up to 500°F (260°C) if a ceramic precoat is first applied to the metal as an undercoat for the gage. Commercial gages for operation at 900°F (480°C), designated as Type 64-1 surface transferable resistors, are manufactured by Transonics, Inc.† These consist of a fine Constantan wire grid embedded in a ceramic matrix, and are applied to the test surface with a ceramic cement which is cured at 600°F (315°C). These

* Robert G. Allen Co., Mechanicville, N.Y.
† Transonics, Inc., Bedford, Mass.

gages can be used for either static or dynamic measurements. Type 64-2 resistors using Nichrome wire are applied in the same manner and can be used for the dynamic measurement of strain at temperatures as high as 2000°F (1095°C), depending to some extent upon the particular metal to which the resistor is bonded.

The foil gage is more widely used for high-temperature work than the wire gage. It is of the so-called "strippable" type, in which the carrier is stripped from the foil when the gage is cemented in place. The gages are bonded with Allen P-1 ceramic cement which must be cured at 600°F (315°C).

Table 17.7 is a summary of commercially available high-temperature strain gages, indicating the maximum temperatures to which they can be used.

WATERPROOFING AND PROTECTIVE COATINGS

Occasionally gages must operate under adverse conditions of humidity and moisture, submerged in oil or water, or subjected to abrasive action or to a combination of the

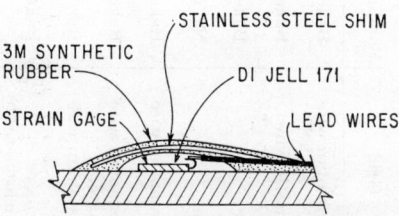

FIG. 17.33. Cross section of strain-gage waterproofing by the Dean shim-cap method. Gage is protected from moisture by successive layers of dielectric jelly, rubber, stainless steel, and more rubber.

foregoing. Under such circumstances, electrical and mechanical protection for the gage and for the lead wires must be provided. Various techniques have been devised to obtain such protection. The following method has been developed for gages subjected to severe abrasive conditions in underwater service on ship hulls and propeller shaft supports. It is generally applicable and may be modified for less severe conditions.[32]

For gages mounted on carbon steel, stainless steel, or S-T types of aluminum, carefully clean the gaging area to insure that all the surfaces to be waterproofed are free from grease, oil and finger prints. Using absorbent cotton dampened with acetone, swab around the gage and up and down the lead wires until a fresh piece of cotton shows no discoloration. For paper base gages, apply a wax buffer precoat of Zophar C-276 or Di Jell 171 * over each gage. This prevents direct contact of the 3-M † compounds with gages. Bakelite gages ordinarily do not require this precoat. If 3-M, EC-864 synthetic rubber is to be used as the principal coating, a single thin coating of a 3-M metal primer should be applied next. No primer is needed for the number EC-801 synthetic rubber as it contains added bonding resins. EC-853 primer thinned 50 per cent with methyl isobutyl ketone may be used for all steels, including stainless, and for S-T types of aluminum. Brush the primer over the area of freshly cleaned metal around the gage. The primer may not be applied over the wax buffer coating nor over Bakelite gages, but should cover all adjacent bare metal surfaces thoroughly. Allow the primer to dry for at least one hour at room temperature or longer if humidity is high. Mild heat up to 130°F (55°C) will speed the drying. No adverse effects have been noted if the primer is allowed to dry for an extended time period prior to the application of 3-M compound, provided the installation has been kept free of oils, finger prints, dust, etc. Clean the plastic insulation on connecting wires with acetone and coat thinly with EC-1217 thinned 50 per cent with methyl isobutyl ketone. Clean rubber insulation with acetone and naphtha solvent and then coat thinly with EC-853. Primer application brushes should be washed out with acetone. A partial coating of either EC-864 or EC-801 mixed with EC-807 accelerator is applied next. Mix ten parts by weight of the base to one part of the accelerator in absolutely clean mixing vessels and do all mixing thoroughly. If the accelerator has settled out in storage, stir or shake jar vigorously until any

* L. Sonneborn & Sons, Inc., Building Products Division, New York.
† Minnesota Mining & Manufacturing Co., Detroit, Mich.

Table 17.7. High-temperature Strain Gages

Gage	Manufacturer	Designation	Grid	Bonding cement	Max. temp.	Static or dynamic
SR-4*	Baldwin-Lima-Hamilton Corp., Waltham, Mass.	AB	Constantan wire	Bakelite	350°F (175°C)	S or D
SR-4*	Baldwin-Lima-Hamilton Corp., Waltham, Mass.	CB	Isoelastic wire	Bakelite	350°F (175°C)	D only
Electrometer*	Electrometer, Trondheim, Norway	B	Wire	Bakelite	350°F (175°C)	S or D
Tepic*	Huggenberger-Zurich, Zurich, Switzerland	BL	Wire	NL Cement	350°F (175°C)	S or D
Kyowamusen*	Kyowamusen Kenkyujo Co., Ltd., Tokyo, Japan	KB	Constantan wire	Bakelite	350°F (175°C)	S or D
Shinkoh*	Shinkoh Tsushinkogyo Co., Ltd., Kanagawa-ken, Japan	B	Constantan wire	Bakelite	350°F (175°C)	S or D
Rotol*	Rotol, Ltd., Gloucester, England	J	Constantan wire	Bakelite	390°F (200°C)	S or D
Rotol*	Rotol, Ltd., Gloucester, England	J	Karma or Nichrome wire	Bakelite	390°F (200°C)	D only
Philips*	Philips, N.V., Eindhoven, Netherlands	PR-B	Wire	Araldite XV	390°F (200°C)	D only
Gustafsson	Herbert Lembcke, Stockholm K, Sweden	A-V	Wire	AR Cement	570°F (300°C)	S or D
Gustafsson	Herbert Lembcke, Stockholm K, Sweden	B-V	Wire	AR Cement	570°F (300°C)	S or D
Tinsley	H. Tinsley and Co., Ltd., London, England	2D	Constantan wire	Bakelite	570°F (300°C)	S or D
Tinsley	H. Tinsley and Co., Ltd., London, England	2C	Karma wire	Bakelite	570°F (300°C)	D only

Saunders-Roe	Saunders-Roe, Ltd., East Cowes, Isle of Wight, England	1AS	Ferry foil	Bakelite	570°F (300°C)	S or D
Metalfilm	Budd Instruments Division, Phoenixville, Pa.	100	Unmounted foil	Allen P-1	600°F (320°C)	S or D
SR-4	Baldwin-Lima-Hamilton Corp., Waltham, Mass.	HFA	Constantan foil	Allen P-1	600°F (320°C)	S or D
HT	High Temperature Instrument Eng. Co., Bala-Cynwyd, Pa.	600	Wire	Allen P-1	600°F (320°C)	S or D
Surface Transferable Resistor	Transonics, Inc., Bedford, Mass.	64-1	Constantan wire	Ceramic Cement	900°F (480°C)	D only
HT	High Temperature Instrument Eng. Co., Bala-Cynwyd, Pa.	100	Karma wire	Allen P-1	900°F (480°C)	D only
Metalfilm	Budd Instruments Division, Phoenixville, Pa.	500	Unmounted foil	Allen P-1	950°F (510°C)	D only
Saunders-Roe	Saunders-Roe, Ltd., East Cowes, Isle of Wight, England	......	Unmounted Nichrome foil	Ceramic Cement	1500°F (800°C)	D only
Metalfilm	Tatnall Measuring Systems, Phoenixville, Pa.	400	Unmounted foil	Allen P-1	1800°F (980°C)	D only
SR-4	Baldwin-Lima-Hamilton Corp., Waltham, Mass.	HFN	Nichrome V unmounted foil	Allen P-1	1800°F (980°C)	D only
Surface Transferable Resistor	Transonics, Inc., Bedford, Mass.	64-2	Nichrome V wire	Ceramic Cement	2000°F (1090°C)	D only

Note: All gages with asterisk can be used up to 500°F (260°C) if cemented to a ceramic precoated surface to minimize impedance reduction at this temperature. Precoat may be NBS L-6AC supplied by O. Hommel Co., Pittsburgh, Pa., designated No. 3E2334. This precoat must be fired at 1750°F (950°C).

Bonding cements can be procured from the gage manufacturer. In each case the manufacturer's instructions for applying the gage with the recommended cement should be followed in detail.

top fluid is completely blended. EC-864 may be mixed in clean cans or bowls. EC-801 requires more thorough mixing on a flat surface such as a slab of safety glass. Stir and fold in the accelerator with a stiff spatula. Do not permit the accelerator to dry out around the edges and flake into the fresh mix. Do not mix more material than can be used in the next 30 minutes. EC-864 and EC-801 are available in one pint cans and the proper bonding accelerator is furnished in separate glass jars. Apply the mixed compound over the gage area with a putty knife or spatula to the desired thickness. A cap of stainless steel shim stock 0.002 in. thick is rolled on and pressed down into the fresh 3-M waterproofing compound and then the coating of synthetic rubber is applied over the entire placement. When using EC-864, the stainless steel shim cap requires a primer such as EC-853. With the use of EC-801, however, no primer coat is required for the stainless steel shim stock. The total thickness of the build-up over the gage will be approximately $\frac{1}{8}$ in. After the 3-M compound has cured, apply several coats of Herecrol RC-9 primer as a surface sealer. This is quick drying and may be applied with a brush. A cross section of a completed waterproofing installation is shown in Fig. 17.33. This type of protection has worked satisfactorily at the David Taylor Model Basin for gages subjected to extremely severe conditions of submersion, erosion, and long-time applications.

Table 17.5 lists additional waterproofing and protective materials suitable for use under less severe conditions.

EFFECT OF LEAD WIRES ON STRAIN INDICATION

In the selection of lead wires to be run from the strain gage to the measuring instrument certain basic factors must be taken into account. Since the measuring instruments respond to the ratio of change in resistance to total resistance, the resistance of the lead wires is of significance because it adds to the resistance of the strain gage. If the lead-wire resistance is large, $\Delta R/R$ is smaller (R is the resistance of the strain gage and lead wire) and a decrease in sensitivity results. Therefore, lead wires must be selected to keep their resistance to a minimum value.

A second factor which must be considered when using measuring instruments employing carrier waves is the possible reactance effect of the lead wires. Since such instruments respond to inductance and capacitance as well as resistance changes, care must be taken not to introduce reactance effects into the wires; this distorts the signal and decreases the sensitivity of the gage. In general, for an instrument of the SR-4 strain indicator type, leads shorter than 50 ft in length will not affect significantly the signal obtained; leads longer than 50 ft may result in loss in sensitivity in amounts varying with the length and configuration of the leads.

The insulation on the lead wires must be as good or better than the insulation of the gage itself. It is apparent that if the resistance between the wires and ground is less than the minimum value prescribed for the gage, difficulties will ensue. Insulation must also protect the conductors from moisture penetration in regions of high humidity. A stranded-wire conductor having a good plastic insulation provides a satisfactory lead wire.* Wire size No. 20 is recommended, although any number from 20 through 28 can be used.

In certain applications, electrical shielding of the lead wires is necessary, as, for example, when 60-cps pickup from an adjacent piece of equipment occurs. The shield should have an insulating cover and only one end of the shield should be grounded (see discussion on *Ground Loops*, Chap. 19). The lead wires should be fastened to the work at intervals either by cementing them in place or by securing them by means of some type of adhesive tape. It is very important that they be fastened securely so that no force applied to the lead wires can be transmitted to the strain gage itself.

* For example, Belden Manufacturing Co. No. 8014 wire, which has chrome vinyl plastic insulation, is satisfactory.

REFERENCES

1. Perry, C. C., and H. R. Lissner: "The Strain Gage Primer," p. 141, McGraw-Hill Book Company, Inc., New York, 1955.
2. Statham, L.: "Characteristics and Applications of Resistance Strain Gages," *Natl. Bur. Standards Circ.* 528, p. 31, 1954.
3. Ref. 1, p. 117.
4. Guerard, J. P., and G. F. Weissman: *Proc. Soc. Exptl. Stress Anal.*, **16**(1):151 (1958).
5. Shanley, F. R.: "Strength of Materials," p. 140, McGraw-Hill Book Company, Inc., New York, 1957.
6. Vigness, I.: *Proc. Soc. Exptl. Stress Anal.*, **14**(2):139 (1957).
7. Smith, R. C., and N. J. Rendler: *Proc. Soc. Exptl. Stress Anal.*, **17**(2):73 (1959).
8. Fyffe, R. J., and A. Arobone: *Prod. Eng.*, **23**:121 (1952).
9. Dranetz, A. I.:*Machine Design*, **30**:120 (1958).
10. Cleveland, A. E.: *J. Soc. Auto. Engrs.*, **59**:34 (July, 1951).
11. Conover, R. E., *Instruments*, **23**:445 (1950).
12. Kaufman, A. B.: *Radio and Television News*, vol. 43 [*Radio-Electronic Eng.*, **14**:7 (1950)].
13. Kaufman, A. B.: *Radio and Television News*, vol. 44 [*Radio-Electronic Eng.*, **15**:99 (1950)].
14. Ref. 7, p. 193.
15. Jasper, N. H.: *Proc. Soc. Exptl. Stress Anal.*, **8**(2):83 (1951).
16. Wenk, E., Jr.: *Proc. Soc. Exptl. Stress Anal.*, **8**(2):90 (1951).
17. Warshaw, H. D.: *Rev. Sci. Instr.*, **23**:493 (1952).
18. Kinslow, R.: *Design News*, **13**:140 (1958).
19. Werner, F. D.: *Proc. Soc. Exptl. Stress Anal.*, **11**(1):137 (1953).
20. Kammer, E. W., and S. Holt, Jr.: *Proc. Soc. Exptl. Stress Anal.*, **6**(2):53 (1948).
21. Boggis, A. G.: *J. Sci. Instr.*, **27**:212 (1950).
22. Weiss, D. E.: *Proc. Soc. Exptl. Stress Anal.*, **4**(2):89 (1947).
23. Geldmacher, R. C.: *Proc. Soc. Exptl. Stress Anal.*, **12**(1):27 (1954).
24. Meier, J. H.: *Proc. Soc. Exptl. Stress Anal.*, **12**(1):33 (1954).
25. Murray, W. M., and P. K. Stein: "Strain Gage Techniques," p. 111, Society for Experimental Stress Analysis, Westport, Conn., 1956.
26. Cunningham, D. M., and G. W. Brown: *Proc. Soc. Exptl. Stress Anal.*, **9**(2):75 (1952).
27. Loewen, E. G., and M. C. Shaw: *Proc. Soc. Exptl. Stress Anal.*, **8**(2):1 (1951).
28. Warshawsky, I.: "A Multiple Bridge for Elimination of Contact Resistance Errors in Strain Gage Measurements," *NACA Tech. Note* 1031, 1946.
29. Rebeske, J. J., Jr.: "Investigation of a NACA High Speed Strain Gage Torquemeter," *NACA Tech. Note* 2003, 1950.
30. Wallace, W. A., and W. A. Casler: *Proc. Soc. Exptl. Stress Anal.*, **4**(2):52 (1947).
31. Kemp, R. H., W. C. Morgan, and S. S. Manson: "The Application of High-temperature Strain Gages to the Measurement of Vibratory Stresses in Gas-turbine Buckets," *NACA Tech. Note* 1174, 1947.
32. Dean, M., III: "Strain Gage Water-proofing Methods and Installation of Gages on Propeller Strut of USS Saratoga," *David W. Taylor Model Basin Rept.* 1146, 1958.

18

CALIBRATION OF PICKUPS

Samuel Levy
General Electric Co.

Russell H. Bickford
General Electric Co.

INTRODUCTION

This chapter describes various methods of calibrating shock and vibration pickups. Each method is subject to inherent mechanical and electrical limitations such as the frequency and amplitude of vibration, electrical noise, and the weight and size of the pickup. For these reasons, it is often necessary to apply several calibration methods in order to determine the sensitivity of a single vibration pickup. A calibration method should be selected which provides the required information in the range of frequencies and amplitude levels in which the pickup will operate.

An appreciation of the difficulty of accurate calibration of vibration pickups is obtained when one considers that vibration at an acceleration amplitude of 5g at 1,000 cps corresponds to a peak-to-peak displacement of only 0.0001 in. If this calibration is to be done by measuring the peak-to-peak displacement and if an accuracy of greater than 1 per cent is to be attained, the accuracy of measurement of the peak-to-peak displacement must be to somewhat better than 0.000,001 in. Correspondingly, when considering a shock pulse with a peak value of acceleration of 5,000g and a duration of 100 microseconds, extreme accuracy of time measurement is required, which may present a problem.

CALIBRATION FACTOR

The calibration factor of a linear pickup is defined as the ratio of electrical output to mechanical input over a useful range of amplitude and frequency. The electrical output is usually a voltage but can be a current or the charge on a condenser. The mechanical input can be a displacement, velocity, or acceleration depending on the function of the pickup being tested. The calibration factor may provide both amplitude and phase information if it is

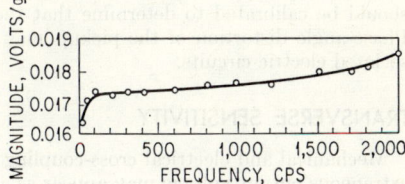

FIG. 18.1. Calibration factor for a vibration pickup (piezoelectric accelerometer) calibrated by the National Bureau of Standards using a calibrator which had been calibrated by the reciprocity method of calibration as described in the text. The magnitude is plotted as a function of frequency. (*After S. Levy and R. R. Bouche.*[16])

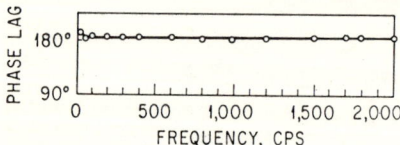

Fig. 18.2. Calibration factor; phase lag vs. frequency for the pickup of Fig. 18.1. (*After S. Levy and R. R. Bouche.*[16])

expressed as the complex ratio of output to input. The calibration factor describes the pickup performance with a stated accuracy only over a limited frequency and amplitude range.

The calibration factor of a pickup varies with frequency; for example, see Figs. 18.1 and 18.2, which show amplitude and phase measurements for a typical pickup. In the recommended usable range of the pickup, the variation must be acceptably small. There are many reasons why the calibration factor may vary with frequency. Most important among these is the effect of resonance of the seismic element, resulting in a rising response as the vibration period approaches the natural period of the pickup. Other factors which cause the calibration factor to vary with frequency are resonances in the pickup case and external electrical loading resulting from the limitations of auxiliary circuits. In some instances, damping fluid may develop springlike properties at high frequencies or the fluid may experience a change in temperature altering the damping rate. Joints which behave nearly elastically at low frequencies can develop appreciable damping at higher frequencies. Generally these effects upon frequency response occur simultaneously, adding to the variation of the calibration factor.

AMPLITUDE RANGE

In the recommended frequency range of operation of a pickup the calibration factor usually is nearly constant for a wide range of amplitudes (see Fig. 18.1). At amplitudes above this range, the calibration factor may vary because of nonlinear performance of the mechanical or electrical elements, amplitude limitations set by mechanical stops, or for other reasons. At amplitudes below this range the calibration factor may vary because of the limited resolving power of the sensing element, electrical noise, or sticking where rubbing friction is present.

PHASE-ANGLE DISTORTION

When a mechanical vibration containing several frequency components is applied to a vibration-measuring instrument and it is found that the waveform of the electrical output is undistorted from the vibration input waveform, phase distortion is said not to exist. A distortion-free condition exists if the phase shift is 0° or 180°, or if it is proportional to the frequency. As indicated in Chap. 12, accelerometers with damping about 0.65 of critical have a phase shift which is almost proportional to frequency over their usable range. Therefore, pickups that are intended for use in measuring either transient disturbances or vibrations containing several frequency components simultaneously should be calibrated to determine that the phase-angle distortion is acceptably small. Phase-angle distortion of the pickup may be either increased or compensated by associated electric circuits.

TRANSVERSE SENSITIVITY

Mechanical and electrical cross-coupling may exist in vibration pickups such that an extraneous output voltage may appear as a result of input vibration which is in a direction not parallel to the axis of maximum sensitivity of the pickup. The transverse sensitivity of a vibration pickup usually is expressed as a percentage of the maximum sensitivity. [For example, see Fig. 12.23 and Eqs. (16.23) to (16.27).] The methods of measurement of transverse sensitivity are described later in this chapter.

INTRODUCTION

ENVIRONMENTAL FACTORS

It is unusual in measuring vibration to have measurement conditions which are ideal or even approximately constant. In general, a vibration pickup is required to function over a wide range of environmental conditions. The behavior of the calibration factor of the pickup under combined environmental conditions is difficult to ascertain, often requiring special research techniques and an accurate knowledge of pickup design. Such environmental changes affect the sensitivity of the pickup. Thus, a calibration performed at ambient room temperature may have little or no meaning if the pickup is sub-

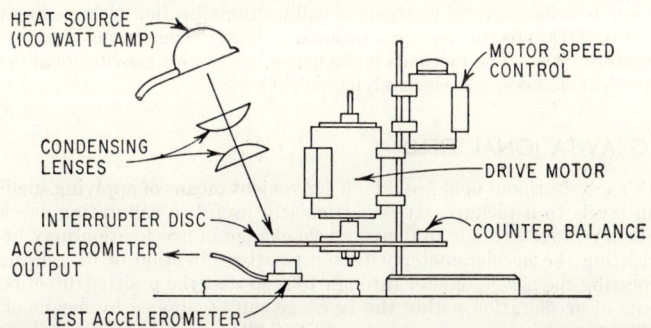

FIG. 18.3. Laboratory arrangement for studying dynamic temperature changes on piezoelectric accelerometer. Heat source shines on pickup through interrupter disc. (*After T. A. Perls.*[1])

jected to a temperature of 600°F (315°C). Ambient temperature changes, rapidly fluctuating temperatures, magnetic fields, pressure variations, intense sound fields, resonances in the mounting structure, mounting techniques, etc., all may have a marked effect upon the sensitivity of the pickup. It has become increasingly important to determine the effects of these environmental factors, singly or in combination, upon the performance of a vibration pickup as they are reflected as variations in the calibration factors.

The output of a piezoelectric crystal pickup is affected by dynamic temperature changes to which the pickup is subjected; it is proportional to the heat absorbed by the crystal. An example of an experimental arrangement [1] for studying such temperature effects is shown in Fig. 18.3. A typical accelerometer response to transient heating is shown in Fig. 18.4. As a result of thermal lag, this is a low-frequency phenomenon.

Effects of mounting techniques and mounting-bracket design may be explored dynamically, using a suitable shaker. As the frequency is varied, it is possible to excite case- and mounting-structure resonances. Changes in calibration factor resulting from various installation conditions can be determined by the calibration procedures. Examples [2,3] of different mounting techniques of piezoelectric accelerometers and the corresponding frequency characteristics are shown in Fig. 20.2. It is observed that substantial differences are present in the upper-frequency range. Therefore when calibrating a pickup, the mounting condition should be simulated as nearly as possible.

Moisture generally has little effect upon well-designed vibration pickups which are properly serviced. However, moisture can seriously affect the performance of piezo-

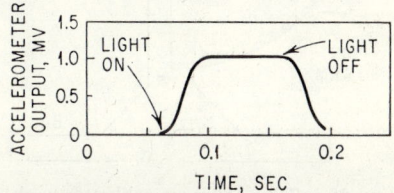

FIG. 18.4. Accelerometer response to dynamic heating, obtained from the experimental arrangement in Fig. 18.3. Accelerometer output increases to full value in about 0.04 sec. (*After T. A. Perls.*[1])

electric accelerometers by reducing the insulation resistance and thereby influencing the lower limiting frequency response.

Generally the effects of salt spray, sand, pressure variations, vacuum, etc., and their relationship upon the calibration factor of the pickup can be examined in the laboratory employing commercially available environment simulators.

CALIBRATION METHODS SUITABLE FOR LOW FREQUENCY AND LOW ACCELERATION

This section describes several methods of calibrating vibration pickups in a frequency range from 0 to 2,000 cps, in the acceleration amplitude range from a fraction of $1g$ up to approximately $100g$. The methods make use of the earth's gravitational field, centrifuges, mechanical shakers, and electrodynamic shakers.

EARTH'S GRAVITATIONAL FIELD

The earth's gravitational field provides a convenient means of applying small constant acceleration levels to a pickup. It is particularly useful in calibrating accelerometers whose frequency range extends to 0 cps. A $2g$ change in acceleration may be obtained by first orienting the accelerometer with the positive direction of its sensing axis up, and then rotating the accelerometer through 180° so that the positive direction is down.

Increments of acceleration within the $1g$ range can be applied by means of a tilting-support calibrator with an accuracy better than $0.004g$ near the horizontal and $0.0003g$ near the vertical.[4] The pickup to be calibrated is fastened as shown in Fig. 18.5 to a platform at the end of an arm. The arm may be set at any angle ϕ between 0 and 180° relative to the vertical. It is furnished with a pointer to indicate the angle ϕ. Care should be taken to level the base when the arm is set at $\phi = 0°$. Positioning of the arm angle ϕ to $\pm 0.2°$ or less is possible with an accurately divided circle. The component of acceleration along the arm is given by

$$a = g \cos \phi \qquad (18.1)$$

The pickup is subjected to a component of acceleration at right angles to its sensitive direction equal to

$$a_t = g \sin \phi \quad \text{(transverse)} \qquad (18.2)$$

Because of the transverse component, this method is not recommended for calibration of pickups for which the transverse sensitivity is significant.

The calibration factor is obtained by plotting the response of the accelerometer as a function of the acceleration a, given by Eq. (18.1), for successive values of ϕ and determining the slope of the straight line fitted through the points.

PHYSICAL PENDULUM CALIBRATOR

The physical pendulum calibrator is a simple device for imparting to a pickup a transient pulse of acceleration of as great as $10g$ with a duration of the order of a second.[4,5] It imparts a transverse and an angular acceleration as well as the accelera-

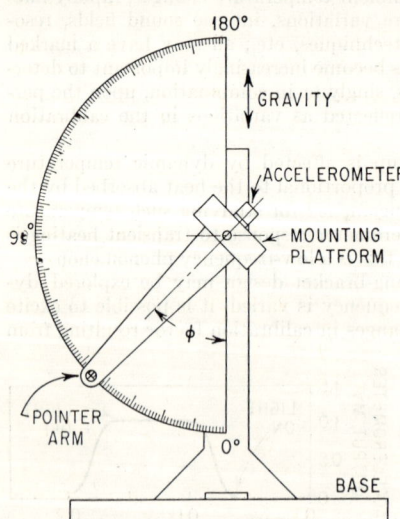

FIG. 18.5. Tilting-support calibrator used to supply incremental static accelerations of $1g$ or less to accelerometers which have zero frequency response such as strain-gage accelerometers. (*After ASA Standard S 2.2-1960.*[4])

tion inward from its center of rotation. Therefore it should be used with caution in calibrating pickups sensitive to these extraneous motions.

The physical pendulum calibrator in Fig. 18.6 consists of a beam and platform with its center-of-gravity pivoted so as to swing about a horizontal axis. The pickup to be calibrated is attached to the platform, and is carefully aligned so that the pickup axis of sensitivity is along a radius of the beam. Electrical connections can be brought from the pickup along the beam bridging to the support frame. A well-constructed pendulum has negligible damping. If the pendulum is released to swing after being raised to an angle ϕ above the vertical, the increment a in acceleration along the radius vector, from the time of release to the bottom of the swing, is

$$a = \left(g + \frac{8\pi^2 l}{\tau^2}\right)(1 - \cos \phi) \quad (18.3)$$

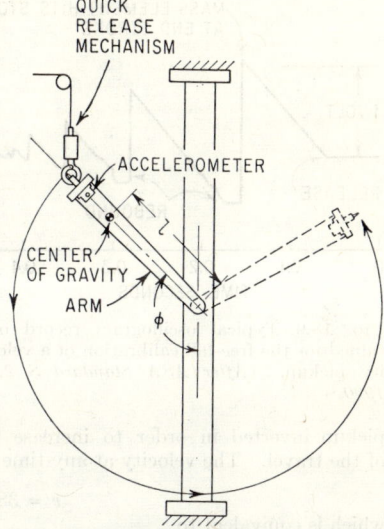

Fig. 18.6. Physical-pendulum calibrator for determining the calibration factor by measuring an acceleration pulse of relatively long duration, approximately 1 sec. Pickup to be calibrated is mounted on table at end of arm. (*After H. Levy.*[5])

where g is the acceleration of gravity, l is the distance from the center-of-gravity of the mass element to the pivot, τ is the period of the pendulum for small amplitudes, and ϕ is equal to the average of the release angle and the angle to which the pendulum rises after passing the bottom of its swing.

The calibration factor normally is determined by plotting the peak response

$$R = \frac{1}{2}(e_1 + e_2)$$

of the pickup as a function of the acceleration a as given by Eq. (18.3) for successive values of ϕ, and determining the slope of the straight line fitted through the data. The error is ordinarily below 1 per cent, where care is exercised. A typical oscillogram record of an accelerometer response as a function of time, obtained from a calibration using the physical pendulum, is shown in Fig. 18.7; a block diagram of an instrumentation recording system used in this method is shown in Fig. 18.8.

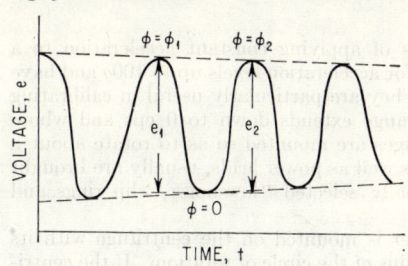

Fig. 18.7 Typical oscillogram of an accelerometer response to the physical-pendulum calibrator in Fig. 18.6. The release angle is ϕ_1, and the angle to which the pendulum rises after passing the bottom of its swing is ϕ_2. The average angle $(\phi_1 + \phi_2)/2$ corresponds to the average response $(e_1 + e_2)/2$. (*After H. Levy.*[5])

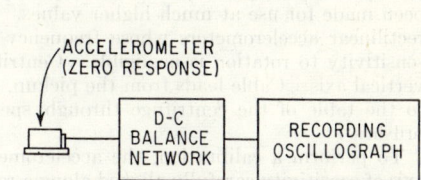

Fig. 18.8. Block diagram of the instrumentation recording system used to obtain the oscillogram for Fig. 18.7. (*After H. Levy.*[5])

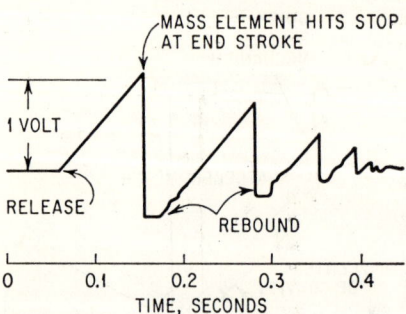

Fig. 18.9. Typical oscillogram record obtained for the free-fall calibration of a velocity pickup. (*After ASA Standard S 2.2-1960.*[4])

FREE-FALL CALIBRATION OF VELOCITY PICKUP

The amplitude sensitivity of an undamped velocity pickup may be determined by removing the support springs of the mass element, placing the base of the pickup on a horizontal surface, raising the mass element to the top of its stroke, and then releasing the mass to fall freely while recording the pickup output as a function of time on an oscillograph.[4] A typical record so obtained is shown in Fig. 18.9. The record shows, in addition to the usable output for the initial drop, the extraneous outputs corresponding to the drops after successive rebounds from the bottom. The test usually is repeated with the pickup inverted in order to increase the accuracy of the measurement at the ends of the travel. The velocity at any time after release is

$$v = 386t \quad \text{in./sec}$$

which is equivalent to

$$v = 980t \quad \text{cm/sec} \tag{18.4}$$

where t = time in seconds.

The voltage output e is measured in volts from the oscillograph record. A second trace of a calibrated time base is included on the record. The calibration factor is given by the ratio of e to v. Since it may be desired to determine the calibration factor as a function of position of the mass along the travel, the distance dropped can be computed from

$$d = 193t^2 \quad \text{in.}$$

which is equivalent to

$$d = 490t^2 \quad \text{cm} \tag{18.5}$$

A check on the calibration can be obtained by comparing the calibration factors determined before and after inverting the pickup. Another check may be obtained by comparing the maximum travel with the value of d computed from Eq. (18.5) with t taken as the measured time from release to rebound.

CENTRIFUGE

A centrifuge provides a convenient means of applying constant acceleration to a pickup. Centrifuges can be obtained readily for acceleration levels up to $100g$ and have been made for use at much higher values. They are particularly useful in calibrating rectilinear accelerometers whose frequency range extends down to 0 cps and whose sensitivity to rotation is negligible. Centrifuges are mounted so as to rotate about a vertical axis. Cable leads from the pickup, as well as power leads, usually are brought to the table of the centrifuge through specially selected "low-noise" slip rings and brushes.

To perform a calibration, the accelerometer is mounted on the centrifuge with its axis of sensitivity carefully aligned along a radius of the circle of rotation. If the centrifuge rotates with an angular velocity of ω radians/sec, the acceleration acting on the pickup is

$$a = \omega^2 r \tag{18.6}$$

where r is the distance from the center-of-gravity of the mass element of the pickup to the axis of rotation. Where the exact location of the center-of-gravity of the mass in the pickup is not known, mount the pickup with its positive sensing axis first outward and then inward; then compare the average response with the average acceleration acting on the pickup computed from Eq. (18.6), where r is taken as the mean of the radii to a given point on the pickup base. Care should be taken to mount the pickup at a distance from the axis of rotation such that the deflection of the mass element is negligible in the determination of r.

The calibration factor usually is determined by plotting the output e of the pickup as a function of the acceleration a given by Eq. (18.6) for successive values of ω and determining the slope of the straight line fitted through the data. Error results primarily from difficulty in measuring the angular velocity accurately and in holding the angular velocity constant during the time required to take a reading.

ROTATING-TABLE METHOD (CENTRIFUGAL FIELD METHOD)

In the rotating-table method of calibration, the accelerometer is rotated at a uniform angular rate about a horizontal axis.[6,7] The experimental arrangement is shown in Fig. 18.10. The pickup is mounted so as to rotate about a horizontal axis; the sensitive axis of the accelerometer rotates in a vertical plane. Thus it is possible to apply very low-frequency sinusoidal motion in the region where sinusoidal linear translation of large amplitude would otherwise be required to give useful accelerations. As a result, a sinusoidal acceleration, having a peak amplitude of $1g$ at the rotation frequency, is superposed on the centrifugal acceleration. By this method it is possible to obtain the response of the accelerometer under both static and dynamic conditions in the same test setup. The rotating-table method not only provides excitation at very low frequencies but also provides an accurate constant level of acceleration. The basic consideration in applying this method is that the vibration pickup be sufficiently sensitive to have adequate output in the range of acceleration up to $1g$. The threshold sensitivity and the repeat-

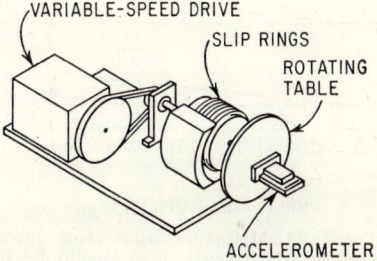

Fig. 18.10. Rotating-table calibrator which is used for both static and dynamic calibration with a peak acceleration amplitude of $1g$. Table rotates in a vertical plane. Pickup mounted on table is alternately subjected to plus and minus $1g$ as table rotates. (*After W. A. Wildback and R. O. Smith.*[7])

ability of the pickup under test must be such that a periodic excitation of $\pm 1g$ or less produces a response that is representative of the instrument's behavior. Excitation at less than $\pm 1g$ can be provided by inclining the axis of rotation. The useful frequency range over which this calibration technique applies is from about 0 cps to 10 cps; this is the frequency range in which it is difficult to obtain a calibration by means of an electrodynamic exciter.

The chief limitations of the rotating-table method of dynamic calibration are (1) the rotation speed of a table is limited; (2) the acceleration level is limited to $1g$; (3) the response to the centrifugal acceleration tends to become infinite at the natural frequency of the pickup.

Variations in the angular velocity, such as those produced by dynamic unbalance of the rotating system, may introduce extraneous accelerations with consequent error in the calibration. Electrical "noise" generated between brushes and slip rings may decrease the precision of measurement. If the pickup under test does not have an electrical output (as is the case for many devices described in Chap. 13), special optical arrangements are necessary to permit the reading or photographing of the indicated response.

MECHANICAL SHAKERS

Rectilinear motion can be produced by mechanical shaker systems of the type described under *Direct-drive Mechanical Vibration Machine* in Chap. 25. At low frequencies, a mechanical drive consisting of an electric motor which turns a cam or drives a linkage can be used to produce nearly sinusoidal motion. Rotating eccentric weights also can be used to produce sinusoidal motion of a spring-supported table.[8] This equipment is described under *Rotating Unbalance Vibration Machine* in Chap. 25. In both of these machines, the amplitude of the table is nearly independent of frequency. The usable frequency range is from a few cycles per second to about 100 cps. Although the displacement is essentially sinusoidal, appreciable distortion of acceleration from a pure sinusoid usually is present as a result of bearing tolerances and the geometry of linkage systems.

Some commercially available shakers provide a means of indicating the displacement amplitude of the table as part of their standard equipment. One method of measuring relatively large displacement amplitudes is shown in Fig. 18.11.[4] A card is attached

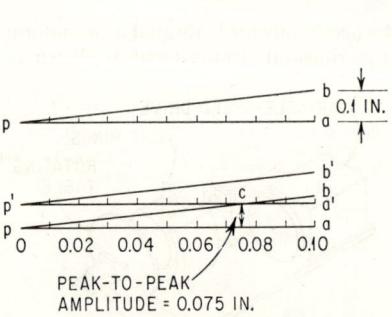

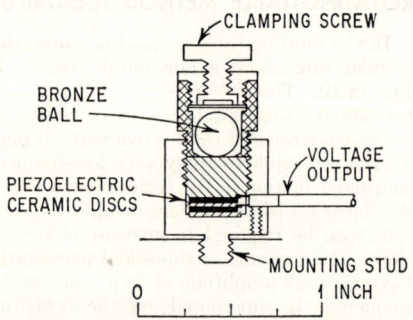

FIG. 18.11. Method of displacement measurement of relatively large amplitudes by the vibrating-wedge technique. A card is attached to the body whose motion is to be measured. The card is oriented so that at rest the line p–a is perpendicular to the expected motion as illustrated in the upper figure. When the body vibrates, two angles are seen. Apparent intercept c permits direct reading of peak-to-peak amplitude pp'. (*ASA Standard S 2.2-1960*.[4])

FIG. 18.12. Cross-sectional drawing of an experimental model of a piezoelectric chatter accelerometer used to measure $1g$ acceleration amplitude. Bouncing of ball indicates when acceleration on down stroke exceeds $1g$, resulting in separation from ball. (*After C. W. Kissinger.*[9])

to the body whose vibratory motion is to be measured. This card contains two lines that intersect to form an angle as shown in the upper portion of the illustration. With the body at rest, the card is oriented so that a straight line p–a is perpendicular to the expected motion. Line p–b intersects p–a and forms a small angle with it. A scale is marked along p–a at each tenth of its length and the length of a–b is set at 0.1 in. When vibrating, two angles apb and $a'p'b'$ are seen, corresponding to the extremes of motion, and an apparent intercept c appears. In the example the intercept c falls 0.75 of the distance from p to a so that the peak-to-peak displacement amplitude p–p' is read as 0.075 in. (The use of this method with the triangular area apb filled in to form a wedge results in a penumbra error, depending on the observer, and would require calibration.)

The pickup response e is plotted against peak-to-peak amplitude $2x_0$, and the slope of a line fitted through the points at a given frequency f is taken as the calibration factor of the vibration pickup at that frequency.

"CHATTER" METHOD OF CALIBRATING ACCELEROMETERS

The chatter method utilizes the earth's gravitational field as a means for calibrating accelerometers at constant low acceleration levels. In this technique, especially designed accelerometers such as the one shown in Fig. 18.12, or especially designed electromechanical shakers such as the one shown in Fig. 18.13, present an indication when an acceleration of $1g$ has been exceeded.[9,10] If a mass is placed at the center of an electromechanical shaker table, and if the mass rests loosely on it, separation will occur whenever the downward acceleration exceeds 1g. Contact occurs during the deceleration portion of the cycle and may be detected by an electrical or audible indication.[9]

An electrical indication that separation has occurred is provided by the electrical output of the chatter accelerometer in Fig. 18.12. The accelerometer output voltage represents the acceleration of the shaker table to which it is attached. When separation of the bronze ball occurs, the accelerometer sensitivity is reduced in the ratio of the mass of the ball to the total mass which loads the crystal. Since the ball makes contact as deceleration of the table occurs, the force exerted at each bounce provides a sharp peak voltage which is readily observable on the screen of a cathode-ray oscillograph. This particular accelerometer design, when used as a chatter accelerometer, has an upper-frequency limit of 150 cps and provides an accurate indication of the ±1g point with

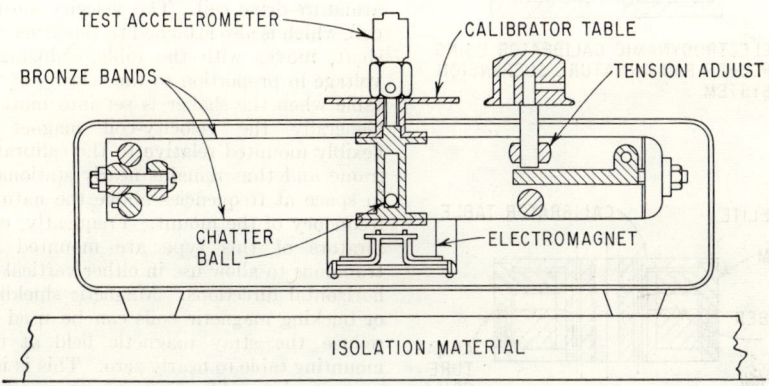

Fig. 18.13. Cross-sectional drawing of a commercial portable shake table with a tuned resonant frequency of 60 cps. The chatter ball is used to indicate a 1g acceleration amplitude of the calibrator table. (*After Bruel and Kjaer*.[10])

approximately 1 per cent error. When used as a self-calibrating secondary standard with the ball clamped, the frequency response has been determined flat to 1,200 cps within ±3.6 per cent.

The portable shaker table shown in Fig. 18.13 contains a built-in chatter ball within a hollow armature.[10] The test accelerometer is mounted on the table and the amplitude of vibration is adjusted until audible chatter occurs. A more precise indication of chatter contact can be obtained by isolating the shaker and providing a sensitive microphone and observing its output with a cathode-ray oscilloscope. The system shown provides a one-point calibration of 1g at 60 cps within an accuracy of ±5 per cent.

ELECTRODYNAMIC CALIBRATOR

The rectilinear electrodynamic shaker (see Chap. 25) is widely used in calibration equipment. It provides a means for applying sinusoidal motion over a range of displacement amplitudes from about 1 in. at low frequencies to about the amplitude which is equivalent to an acceleration amplitude of approximately 50g at high frequencies. Over a frequency range from about 5 to 10,000 cps electrodynamic calibrators are available which provide relatively undistorted sinusoidal waveform.[4] Ordinarily, several shakers are needed to cover this broad frequency range since it is necessary to avoid resonant frequencies of the shaker when using it for calibration of pickups. Freedom from harmonics in the motion is largely dependent on the pure sinusoidal waveform of the driving current. In some cases ripple in the direct current used to energize the field coil of the shaker introduces small amplitude spurious motions by transformer coupling with the armature. (The latter is evident only when the armature circuit is closed.)

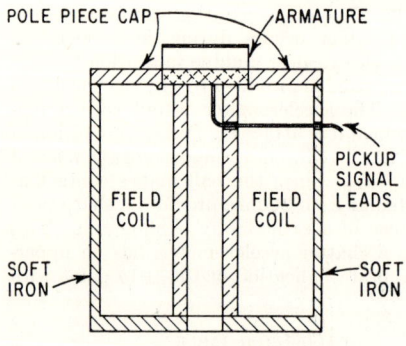

(A) ELECTRODYNAMIC CALIBRATOR USING RUBBER PAD ARMATURE SUSPENSION SYSTEM

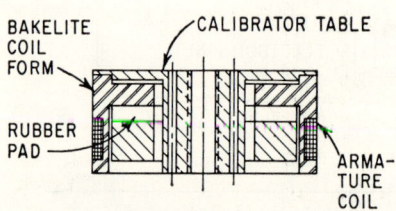

(B) ARMATURE OF SPECIAL ELECTRODYNAMIC CALIBRATOR

FIG. 18.14. Cross-sectional drawing of an experimental electrodynamic calibrator developed at the National Bureau of Standards. A rubber pad is used as the armature suspension system. The upper frequency limit of this design is 500 cps. (*After S. Edelman, E. Jones, and E. R. Smith.*[11])

The construction of typical electrodynamic vibration pickup calibrators is illustrated schematically in Figs. 18.14 and 18.15.[4, 11, 12] * In Fig. 18.15 the pickup to be calibrated is attached to the calibrator table. The armature coil and the table are rigidly connected to the armature shaft. Flexures support the armature shaft from extensions on the field-magnet housing. The flexures are designed to constrain the motion of the moving parts to an axial direction. The table is set in motion by passing the driving current through the armature-drive coil. The velocity-sensing coil, which is also attached to the armature shaft, moves with the table, inducing a voltage in proportion to the velocity of the table when the shaker is set into motion. Generally the velocity-coil magnet is flexibly mounted relative to the calibrator frame and thus remains nearly stationary in space at frequencies above the natural frequency of the mount. Frequently, calibrators of this type are mounted on trunnions to allow use in either vertical or horizontal directions. Magnetic shielding or bucking magnetic coils can be used to reduce the stray magnetic field at the mounting table to nearly zero. This is important when calibrating pickups sensitive to magnetic fields, e.g., electrodynamic velocity-type pickups. To achieve isolation from building vibration, a calibrator may be mounted on a heavy rigid mass which in turn is supported by soft springs.

Calibrators purchased commercially ordinarily come with a calibration factor for the velocity-sensing coil. Where a calibration factor is not available, it can be determined conveniently by the reciprocity method, since two electrodynamic transducer coils are part of the calibrator. A secondary standard also can be used to calibrate the velocity coil. In some cases, a displacement method is applicable.

CALIBRATION OF PICKUPS BY DIRECT MEASUREMENT

A general arrangement of equipment for the calibration of vibration pickups is shown in Fig. 18.16. The pickup to be calibrated is attached rigidly to the calibrator table with its center-of-gravity centered on the table. Pickup lead wires are secured tightly to avoid whipping. Particular care must be taken in servicing cables from a pickup that is sensitive to distortion of its case. The pickup output voltage then is fed into suitable auxiliary instrumentation. The oscillator frequency is adjusted to desired values throughout the frequency range of the pickup, while the amplifier gain is set to give desired amplitudes. Waveform normally is monitored with an oscilloscope or measured by a wave analyzer to avoid distortion resulting from nonsinusoidal motion of the armature.

* A small battery-operated unit is the General Radio Co. Vibration Calibrator Type 1557-A. It produces an acceleration of $1g$ rms at 100 cps. The entire device weighs 3¼ lb.

ACCELERATION PICKUPS

In the case of the acceleration pickup, at a given frequency f of excitation, the response e of the pickup and the velocity v indicated by the velocity coil are read for a series of values of the driving current in the armature coil. The values of e are then plotted against $2\pi f v$ and the slope of the line through the points is taken as the pickup calibration factor for frequency f. The calibration factor in g units is obtained by multiplying by the acceleration of gravity in units consistent with those used in measuring v. The calibration can be repeated at other frequencies in the operating range of the pickup. The phase shift between the pickup output and the table motion can be obtained by measuring the phase angle between the response e and the velocity-coil output with a

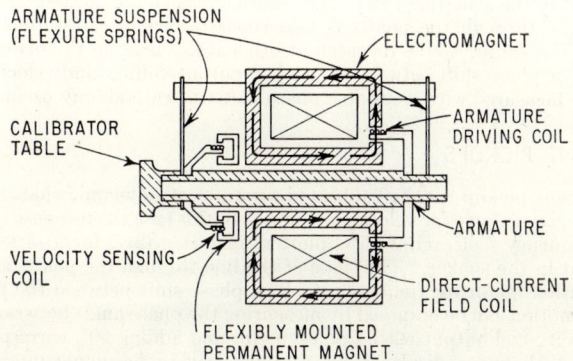

FIG. 18.15. Cross-sectional drawing of a typical rectilinear electrodynamic vibration pickup calibrator. The armature shaft is supported by flexure springs. (*After K. Unholtz.*[12])

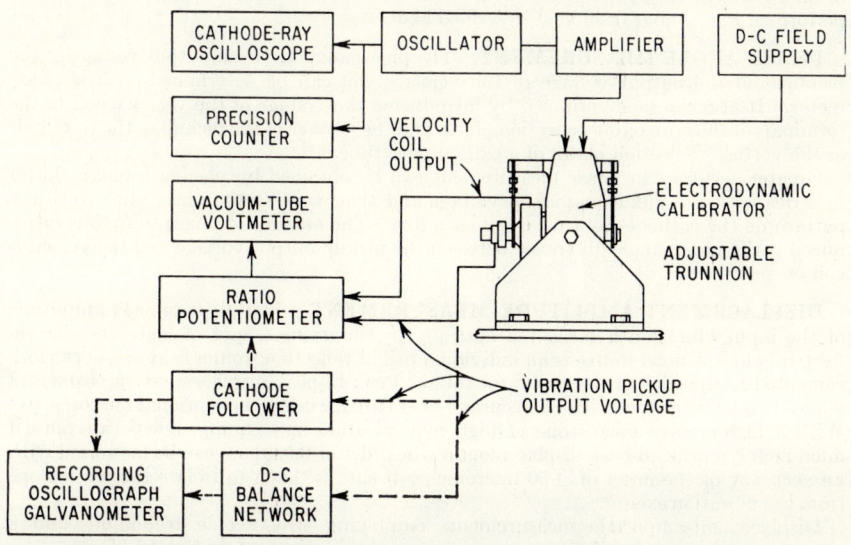

FIG. 18.16. General arrangement of an electrodynamic calibrator and the instrumentation measuring system used in the calibration of displacement, velocity, and acceleration pickups. Calibrations are performed by direct measurement and/or comparison using the ratio potentiometer. (*After K. Unholtz.*[12])

phase-shift meter and subtracting 90°, corresponding to the lag which exists between velocity and acceleration in harmonic motion. (At frequencies near the frequency for axial resonance between the shake table and the velocity coil, calibration may show that a significant phase shift is present between the table motion and the velocity-coil output. If circumstances require use of a shaker in this frequency range, the phase shift between the table motion and velocity-coil output also should be subtracted from the phase meter reading to obtain the phase shift of the pickup itself.)

VELOCITY PICKUPS

In calibrating a velocity pickup at a frequency f the response e of the pickup and the table velocity v, indicated by the velocity coil, are determined for a series of values of driving current in the armature coil. The values of e then are plotted against v and the slope of the line through the points is taken as the pickup calibration factor at the frequency f. The calibration is repeated at other frequencies in the operating range of the pickup. The phase shift between the pickup output voltage and velocity-coil output voltage can be measured with a suitable phase meter or cathode-ray oscilloscope.

DISPLACEMENT PICKUPS

A displacement pickup can be calibrated on an electrodynamic shaker in a manner similar to that given above for velocity pickups. In this case the response e of the pickup at a given frequency f of excitation is plotted against $v/(2\pi f)$ for a series of values of driving current in the shaker. The slope of the line through the points is taken as the pickup calibration factor for frequency f. The phase shift between the pickup output and the table motion can be obtained by measuring the phase angle between the response e and the velocity-coil output with a phase meter and adding 90°, corresponding to the lag which exists between displacement and velocity in harmonic motion. The applicability of shakers to the calibration of displacement pickups is limited at high frequencies by the small available displacement amplitude.

In calibrating a probe-type pickup, the pickup should be mounted as it is to be used or on a massive block; the probe should bear on the center of the table. Calibration is performed in a similar fashion as described above.

PHASE-ANGLE MEASUREMENT. The phase angle between output voltage of the pickup and the output voltage of the velocity coil can be determined with a phase meter. It also can be determined by introducing the voltage of the velocity coil to the terminals of the horizontal-deflection plates and the voltage of the pickup to the terminals of the vertical-deflection plates of a cathode-ray tube.[12, 13]

Greater accuracy in phase measurement can be obtained by placing a phase shifter in series with the vibration-pickup voltage and then varying the phase shift until the pattern on the cathode-ray tube becomes a line. The amount of phase shift thus introduced is the phase-angle difference between the pickup output voltage and the velocity-coil output voltage.

DISPLACEMENT AMPLITUDE MEASUREMENT. The displacement amplitude of the input vibration is measured optically by observing a spot of high-intensity reflected light. A good source is an individual bright reflection from a field of emery cloth cemented to the edge of the calibrator table. Peak-to-peak displacement amplitudes of 0.5 in. can be measured with an accuracy of ± 0.01 in. using conventional microscopes.[4] With a high-quality microscope of high magnification and equipped with an optical micrometer, a peak-to-peak displacement of the order of 0.0001 in. can be measured with an accuracy of the order of ± 50 microinches if care is taken to isolate the microscope from the vibration exciter.

Displacement-amplitude measurements employing stroboscopic techniques and a vibrating wedge are described in other sections of this chapter (see Fig. 18.11).

CALIBRATION BY COMPARISON—POTENTIOMETER METHOD. The voltage amplitudes can be measured directly by a vacuum-tube voltmeter. In many cases greater accuracy is attained by a comparison of the voltage outputs of the built-in velocity coil and the vibration pickup in conjunction with a specially calibrated potentiometer.[12] The calibration point is determined by adjusting the velocity-coil voltage by means of a potentiometer to equal the output voltage of the vibration pickup. The experimental arrangement is shown in Fig. 18.17. All wiring must be thoroughly shielded. The voltage outputs of the pickup and the velocity coil are connected to the terminals of the ratio network containing the potentiometer and switching panel. The ratio network includes a 20,000-ohm calibrated potentiometer placed across the velocity-coil terminals and a switch which will allow alternate switching of the reduced voltage of the velocity coil and the voltage of the pickup to the measuring circuit.

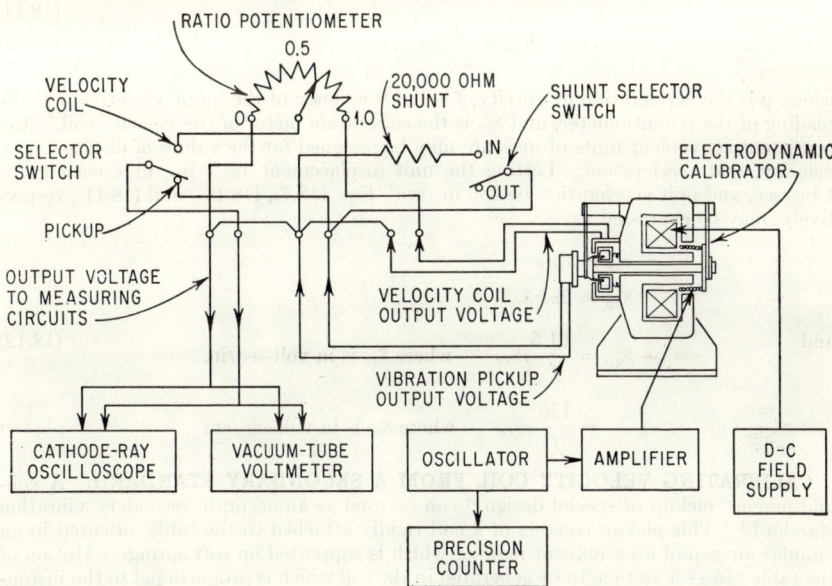

Fig. 18.17. Schematic diagram of an electrodynamic calibrator and the instrumentation measuring system used in calibration by the comparison method (ratio-potentiometer method). (After K. Unholtz.[12])

In addition, a 20,000-ohm (or other) shunt can be provided for shunting the pickup output terminals where the pickup would normally drive a meter with a 20,000-ohm (or other) resistance. The vibration-pickup output voltage and the potentiometer voltage are alternately switched to the voltmeter while making adjustment to the potentiometer such that the same scale deflection is observed. If the voltage output of the vibration pickup is larger than that of the velocity coil, the pickup voltage may be attenuated by placing across its terminals a relatively high resistance potentiometer set so as to reduce the input by a known percentage and bring it into the range of the velocity-coil voltage.

The calibration factor S_{vp} for a velocity pickup is given by

$$S_{vp} = rS_{vc} \tag{18.7}$$

where the calibration factor S_{vc} of the velocity coil is the ratio of velocity-coil voltage to table velocity and r is the reading on the potentiometer. The calibration factor S_{xp} for amplitude x_0 of the displacement pickup is expressed as

$$S_{xp} = \frac{v}{x_0} rS_{vc} \tag{18.8}$$

Making use of the relationship
$$v = 2\pi f x_0 \qquad (18.9)$$
Eq. (18.8) may be written as
$$S_{xp} = 2\pi f r S_{vc} \qquad (18.10)$$

where f is equal to the frequency of the input vibration, r is the reading on the potentiometer, and S_{vc} is the calibration factor of the velocity coil.

The calibration factor S_{ap} for the acceleration pickup is expressed as

$$S_{ap} = \frac{g}{a} v r S_{vc} = \frac{g}{2\pi f v} v r S_{vc}$$
$$S_{ap} = \frac{g}{2\pi f} r S_{vc} \qquad (18.11)$$

where g is the acceleration of gravity, f is the frequency of the input vibration, r is the reading of the potentiometer, and S_{vc} is the calibration factor of the velocity coil. Any consistent convenient units of measure may be assigned for the values of displacement, velocity, and acceleration. Letting the unit displacement be 1 in., unit velocity be 1 in./sec, and unit acceleration be 386 in./sec^2, Eqs. (18.7), (18.10), and (18.11), respectively, may be expressed as

$$S_{vp} = r S_{vc}$$
$$S_{xp} = 2\pi f r S_{vc}$$

and
$$S_{ap} = \frac{61.5}{f} r S_{vc} \quad \text{where } S_{vc} \text{ is in volt-sec/in.} \qquad (18.12)$$
$$= \frac{156}{f} r S_{vc} \quad \text{where } S_{vc} \text{ is in volt-sec/cm}$$

CALIBRATING VELOCITY COIL FROM A SECONDARY STANDARD. A coil-and-magnet pickup of special design [12] can be used as an accurate secondary vibration standard.[14] This pickup consists of a coil rigidly attached to the table, oriented in an annular air gap of a permanent magnet which is supported on soft springs. Motion of the table causes a voltage to be generated in the coil which is proportional to the instantaneous value of table velocity referred to a fixed coordinate system. Figure 18.18 is a schematic drawing of the setup. In this example the coil is about three times as long as the magnet pole length and weighs about 0.03 lb. The voltage from the coil mounted on the table is compared with the voltage from the velocity coil, through the frequency range of calibration. Since the calibration factor for the coil-and-magnet assembly has been determined previously, the calibration factor of the velocity coil is determined by simple ratios. The calibration factor of a coil-and-magnet assembly may be obtained through a service established by the National Bureau of Standards for calibrating vibration pickups. In a typical NBS calibration of such a coil-and-magnet assembly,[14] the calibration factor is constant within 1 per cent up to 2,000 cps for accelerations up to 10g.

DISPLACEMENT METHOD FOR CALIBRATING THE VELOCITY COIL OF AN ELECTRODYNAMIC SHAKER. The velocity-coil calibration factor S_{vc} can be obtained by direct measurement of frequency, displacement amplitude, and velocity-coil output voltage. The input frequency of the calibrator is usually set between 40 and 50 cps, where the amplitude of vibration is large. The frequency f is accurately determined by measurement with a precision counter calibrated against a frequency standard or may be conveniently determined by Lissajous figures presented on a cathode-ray oscilloscope. Peak-to-peak displacement $2x_0$ is measured by a microscope with a calibrated eyepiece. The voltage e generated by the velocity coil is obtained from a vacuum-tube voltmeter or by the resistor insert method,[12] which employs a vacuum thermocouple, a microammeter, a precision potentiometer, and a standard cell. The velocity-coil calibration factor is

$$S_{vc} = \frac{e}{2\pi f x_0} \tag{18.13}$$

RECIPROCITY CALIBRATION OF VELOCITY COIL. The velocity coil of an electrodynamic shaker may be calibrated by the reciprocity technique.[15, 16] For this technique to be applicable, it is necessary that the shaker system be linear, i.e., that

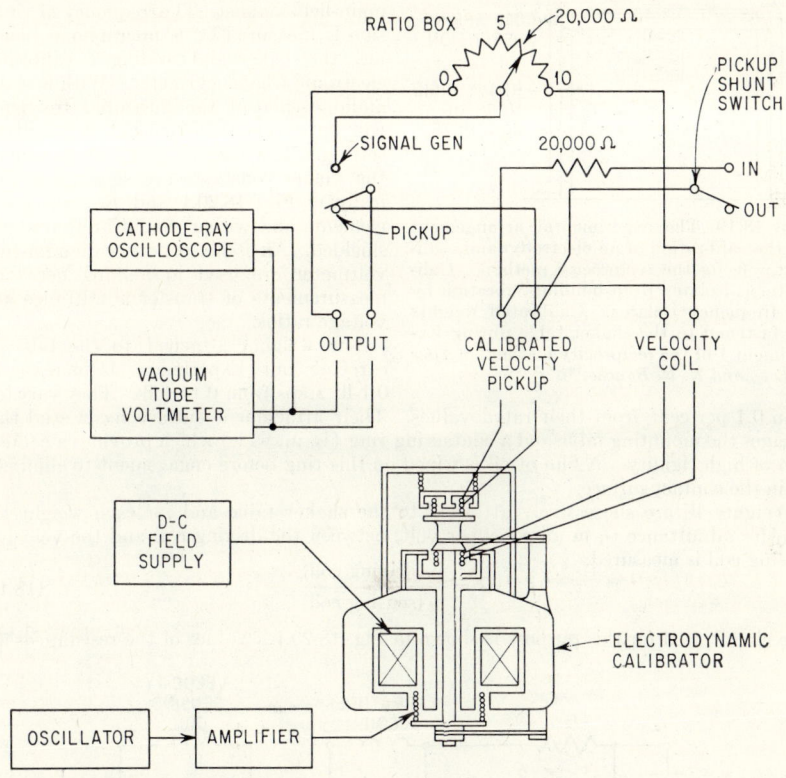

FIG. 18.18. Schematic drawing of the experimental setup for calibrating the velocity coil, using a velocity pickup as a secondary vibration standard. (*After R. R. Bouche*.[14])

displacement, velocity, and current increase linearly with force and voltage. An experimental arrangement for the calibration of an electrodynamic shaker is shown in Fig. 18.19. Neither the internal structure, the magnet structure, the shaker, nor mounting need be considered rigid. However, all joints should be tight. The reciprocity method is not recommended for calibration of the velocity coil near shaker resonant frequencies. At these frequencies the inertia and elastic forces are nearly in balance. As a result, small changes in temperature with their corresponding effect on elasticity may affect the accuracy of calibration significantly.

The positive terminals of the driving and velocity coils are designated as those having positive voltages when the mounting table is moved inward. Conversely, positive current in the coils produces outward velocity at the mounting table. Calibration of the shaker (i.e., measurement of its vibration constants) is achieved by performing the following two experiments:

Experiment 1. The equipment used in this experiment is similar to that shown in Fig. 18.16. The electrodynamic shaker, as shown in Fig. 18.19, is mounted on a con-

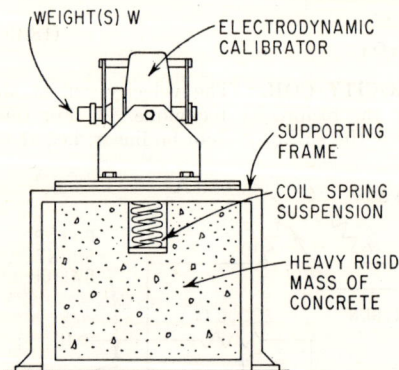

crete and steel mass supported by springs from a framework resting on the floor. A mount having a natural frequency of about 1 cps provides adequate isolation from building vibration. The shaker is driven by an oscillator and appropriate amplifiers; a d-c power supply may be required for the main-field magnet. The frequency of vibration is measured by a microphone placed near the shaker and feeding a calibrated electronic frequency meter. With a stable audio oscillator, the measured frequency will be within 0.2 per cent of oscillator setting. The circuits used in measuring the shaker constants are shown schematically in Fig. 18.20A and B. All circuit elements and wiring should be thoroughly shielded. An oscilloscope and vacuum-tube voltmeter are used in making necessary measurements of transfer admittance and voltage ratios.

The weights attached to the table in carrying out Experiment 1 increase in 0.1-lb steps from 0 to 1 lb. They vary less

FIG. 18.19. The experimental arrangement for the calibration of an electrodynamic calibrator using the reciprocity method. Calibrator is isolated from building vibration by low-frequency isolator. Calibrated weights are fastened to the shaker table during Experiment 1 of the reciprocity method. (*After S. Levy and R. R. Bouche.*[16])

than 0.1 per cent from their rated values. Their attachment surface has a stud that engages the mounting table and a contacting ring (⅛ in. wide) which provides a connection of high rigidity. A film of oil is wiped on this ring before engagement to eliminate air in the contact surface.

Weights W are successively attached to the shaker table and for each weight the transfer admittance G, in amperes per volt, between the driving coil and the velocity-sensing coil is measured.

$$G = \frac{i_1 \text{ (driving coil)}}{e_2 \text{ (sensing coil)}} \tag{18.14}$$

The circuit used for this purpose is shown in Fig. 18.20A. Values of the resistances are

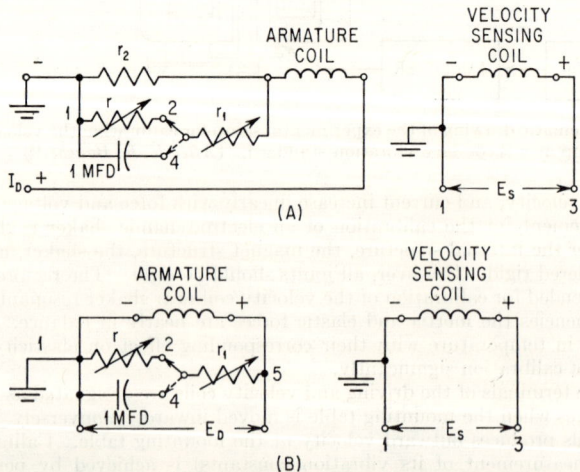

FIG. 18.20. Circuits used in the measurement of constants in the reciprocity method of calibrating an electrodynamic calibrator: (A) measurements of transfer admittance and (B) measurements of voltage ratio. (*After S. Levy and R. R. Bouche.*[16])

chosen to load the amplifier suitably. Typical values used are $r_2 = 10$ ohms and $r + r_1 = 10,000$ ohms (roughly, 1,000 times r_2). With the switch in the "up" position at terminal 2, the values of r and r_1 are adjusted until the voltage drops $|e_{12}|$ (across terminals 1 and 2) and $|e_{13}|$ (across the velocity-sensing coil, terminals 1 and 3) are equal as measured on a high-impedance voltmeter. The magnitude of G is given by

$$|G| = \frac{r + r_1 + r_2}{rr_2} \qquad (18.15)$$

To determine the phase angle of G, two additional measurements are made. In the first measurement the voltage $|e_{23}|$ is read from the voltmeter so that the voltage polygon

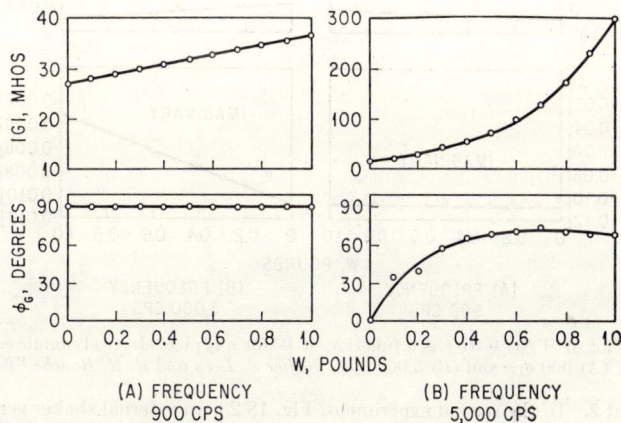

(A) FREQUENCY 900 CPS (B) FREQUENCY 5,000 CPS

Fig. 18.21. Variations of transfer admittance G with weight W on the calibrator mounting table for a typical electrodynamic calibrator at frequencies of (A) 900 cps and (B) 5,000 cps. (*After S. Levy and R. R. Bouche.*[16])

$|e_{12}|$, $|e_{23}|$, $|e_{13}|$ can be constructed. From this the phase angle φ_G of G is given by

$$\varphi_G = \cos^{-1}\left(\frac{1 - |e_{23}^2|}{2|e_{13}^2|}\right) \qquad (18.16)$$

Note that r and r_1 were adjusted to make $|e_{12}| = |e_{13}|$. In the second measurement, the switch, Fig. 18.20A, is set in the "down" position to terminal 4 and the value of r_1 is adjusted until $|e_{14}|$ equals $|e_{13}|$. A check is made to determine if the product $(r_1\omega \times 10^{-6})$ is greater than 100. If this is not the case, r_2 is increased and the first step repeated. Then $|e_{13}|$ and $|e_{34}|$ are read on the voltmeter. An approximate value of phase angle for G is given by

$$\varphi_G \text{ (approx.)} = 90° - \cos^{-1}\left(\frac{1 - |e_{34}^2|}{2|e_{13}^2|}\right) \qquad (18.17)$$

Equation (18.17) should be used together with Eq. (18.16) to determine the quadrant of φ_G; Eq. (18.16) gives the most accurate measure of the angle.

The variation of transfer admittance G with weight W on the mounting table for a typical vibration pickup calibrator at frequencies of 900 and 5,000 cps, respectively, is shown in Fig. 18.21A and B. In Fig. 18.22A and B are shown (for the same cases) plots of the real and imaginary parts of $W/(G - G_0)$ against W where G_0 is the value of G when W is zero. (G_0 is best determined by plotting G against W and taking G_0 as the intercept at $W = 0$.) The ordinate intercept J and the slope Q of the function, when plotted against the weight W attached to the mounting table, are needed subsequently in de-

CALIBRATION OF PICKUPS

termining the calibration factor. For example, for the curves shown in Fig. 18.22A and B at 900 cps, the magnitudes and phase angle of the complex quantities are given by

$$J = 0.1089 / -94.2° \text{ lb-ohm}$$

and

$$Q = 0.00767 / -118° \text{ ohm}$$

while at 5,000 cps

$$J = 0.01094 / -77.4° \text{ lb-ohm}$$

and

$$Q = 0.00765 / 100.4° \text{ ohm}$$

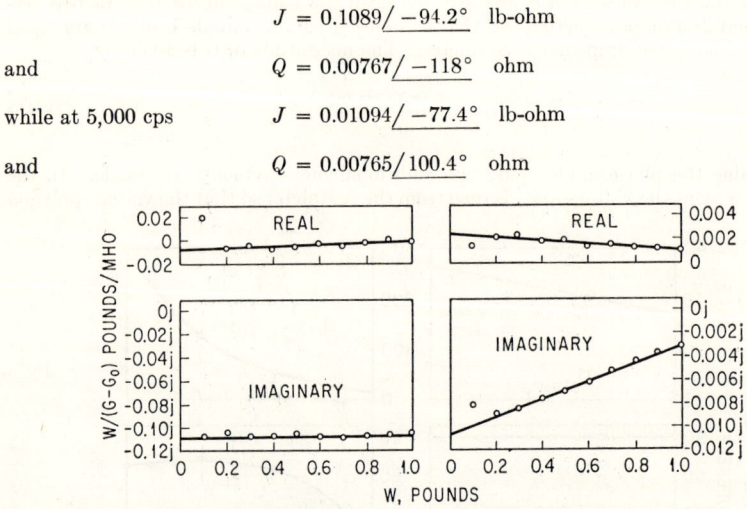

FIG. 18.22. Plot of $W/(G - G_0)$ as a function of W for a typical electrodynamic calibrator at frequencies of (A) 900 cps and (B) 5,000 cps. (After S. Levy and R. R. Bouche.[16])

Experiment 2. In the second experiment, Fig. 18.23, an external shaker is required to provide sinusoidal motion to the shaker being calibrated. The purpose of this experiment is to measure the ratio R of the open-circuited voltages generated in the velocity-sensing and driving coils of the shaker when it is being driven by an external shaker. It is important that the field strength and temperature equilibrium of the shaker being calibrated be the same in Experiment 2 as it was in Experiment 1.

The circuit shown in Fig. 18.20B is used to measure the ratio R of the open-circuited voltages in the velocity-sensing and driving coils of the calibrator. With the switch in the "up" position at terminal 2 and with r_1 set at 10,000 ohms, resistor r is adjusted until the voltages across terminals 1 and 2 and across terminals 1 and 3, as measured by a high-impedance voltmeter, are equal. The magnitude of the voltage ratio R is then given by

$$|R| = \frac{r}{10,000 + r} \quad (18.18)$$

In the case of a calibrator for which 10,000 ohms is a significant electrical load for the armature coil, r_1 should be set to a higher value and the constant in Eq. (18.18) appropriately increased to the value of r_1.

To determine the phase angle of R, i.e., φ_R, the voltages $|e_{12}| = |e_{13}|$ as well as $|e_{23}|$ are measured with the switch in the "up" position. The phase angle φ_R is obtained

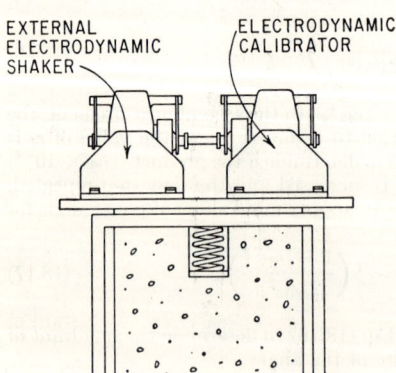

FIG. 18.23. Measurement of constants of an electrodynamic calibrator. The external electrodynamic shaker is required to provide sinusoidal motion to the calibrator during Experiment 2 of the reciprocity method. (After S. Levy and R. R. Bouche.[16])

CALIBRATION OF PICKUPS BY DIRECT MEASUREMENT

from the relation

$$\varphi_R = \cos^{-1}\left(\frac{1-|e_{23}{}^2|}{2|e_{13}{}^2|}\right) \tag{18.19}$$

Unfortunately Eq. (18.19) gives large errors in φ_R for small errors in $|e_{23}|$ when φ_R is near either 0 or 180°, the usual angles for a shaker with negligible mechanical damping. For this reason Eq. (18.19) is used only to determine the quadrant of φ_R, while the magnitude is determined by a second measurement in which the switch in Fig. 18.20B is set in the "down" position to terminal 4. Resistor r_1 is adjusted until $|e_{14}|$ equals $|e_{13}|$, and $|e_{43}|$ and $|e_{13}|$ are measured by a high-impedance voltmeter. Again using a polygon of voltages,

$$\varphi_R = \cos^{-1}\left(\frac{1-|e_{43}{}^2|}{2|e_{13}{}^2|}\right) - \cos^{-1}|R| \tag{18.20}$$

This equation is used to determine the magnitude of φ_R.

Computational Procedure. With J, Q, and R measured by the method described in the preceding experiments the calibration factor S_{vc} is given by [16]

$$S_{vc} = 0.0711(j\omega JR)^{\frac{1}{2}} + 6.601Q\left(\frac{R}{j\omega J}\right)^{\frac{1}{2}} Z_P \quad \text{volt-sec/in.}$$

$$= 0.00674(j\omega JR)^{\frac{1}{2}} + 2.600Q\left(\frac{R}{j\omega J}\right)^{\frac{1}{2}} Z_P \quad \text{volt-sec/cm} \tag{18.21}$$

where the numerical constants are conversion factors to conventional units, ω is the angular frequency in radians per second, j is the unit imaginary vector, and Z_P is the mechanical impedance of the pickup attached to the table. Note that the second term in Eq. (18.21) is negligible except at frequencies so high that there is significant stretch between the shaker table and the velocity-sensing coil.

The mechanical impedance Z_P is equal to $j\omega m$, where m is the pickup mass, for frequencies below the pickup resonant frequency. The impedance Z_P can generally be determined readily with the calibrated shaker itself if G, Eqs. (18.15) to (18.17), is measured with and without the pickup added. Let G_P be the value of G with the pickup attached and G_0 the value without a pickup. Then

$$Z_P = j\left(\frac{\omega}{386}\right)\left[\frac{J-(G_P-G_0)}{1-Q(G_P-G_0)}\right] \tag{18.22}$$

The magnitude and phase angle of the calibration factor for the velocity coil of a typical electrodynamic shaker are shown in Fig. 18.24 as a function of frequency. The

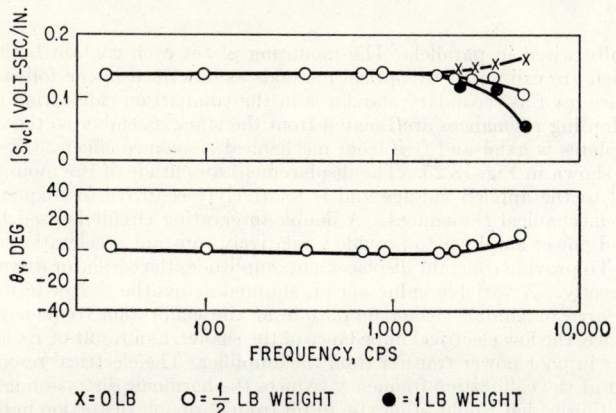

FIG. 18.24. Variations of the calibration factor of the velocity-sensing coil with frequency and pickup weight for a typical electrodynamic calibrator. (*After S. Levy and R. R. Bouche.*[16])

mechanical impedances for the equivalent pickups correspond to weights of 0, 0.5, and 1.0 lb ($Z_P = j\omega W/386$).

PIEZOELECTRIC CALIBRATOR. The piezoelectric calibration system of the type shown in Fig. 18.25 (also see Fig. 25.47) is used to obtain calibration factors for vibration pickups whose response range extends from approximately 1,000 to 9,000 cps at acceleration amplitudes from less than $1g$ to $3g$.[17, 18, 19] Generally the mass of the pickup to be calibrated is limited by the allowable mass loading of the crystal stack. The piezoelectric shaker shown in Fig. 18.25 consists of a stack of polarized piezoelectric ceramic discs bonded together with titanium mounting plates at either end of the stack. The discs

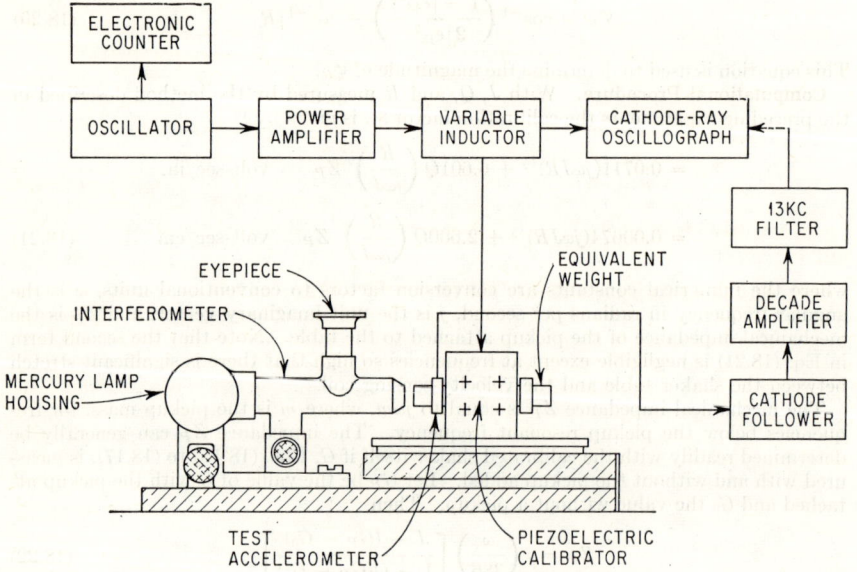

FIG. 18.25. General equipment arrangement for high-frequency calibration of accelerometers using a piezoelectric calibrator and interferometer for direct amplitude measurements. Calibrations can be performed in the frequency range from 1,000 to 9,000 cps. (*After Gulton Industries.*[17])

are electrically wired in parallel. The mounting plates each contain built-in accelerometers which are calibrated by optical methods as described in the following section, and which are used as secondary standards in the comparison calibration of vibration pickups. Housing resonances are isolated from the stack assembly so that the motion of the end plates is axial and free from mechanical resonance effects. The calibrator is driven as shown in Fig. 18.25. The displacement amplitude of the mounting plate is proportional to the applied voltage and is relatively insensitive to frequency changes except near mechanical resonances. A double-integrating circuit is used between the oscillator and power amplifier to provide a relatively constant acceleration output with frequency. To provide constant displacement amplitude, the oscillator drives the power amplifier directly. A variable inductance is shunted across the shaker terminals and is adjusted to give maximum shaker amplitude at the calibration frequency. Without this inductance, the low electrical impedance of the shaker, as a result of its large capacitance, results in poor power transfer from the amplifier. The electrical resonance of the tuned circuit at the calibration frequency reduces the harmonic distortion in the mounting-plate motion which would otherwise result from harmonic distortion in the amplifier output. This effect is important since harmonic distortion in displacement motion is magnified when considering acceleration.

CALIBRATION OF PICKUPS BY DIRECT MEASUREMENT 18-21

The accelerometer to be calibrated is attached as rigidly as possible to one mounting plate as shown in Fig. 18.25 and a "dummy" mass of equivalent weight is attached to the opposite mounting plate. The frequency of the oscillator signal is determined. The pickup acceleration amplitude a can be determined by using the built-in accelerometers in the end plates as secondary standards. The acceleration can also be determined to a lesser accuracy from the relationship

$$a = 2\pi^2 f^2 Be \qquad (18.23)$$

where e is the voltage applied to the shaker terminals and B is a shaker constant (ratio of mounting-plate peak-to-peak displacement to shaker voltage) determined by the fringe disappearance technique which is described in the following section. This method is applicable only when the pickup weight is less than 2 oz for the shaker in Fig. 18.25 or in general when it is small enough to have no effect on B. A third method is to use the fringe disappearance technique directly to measure the peak-to-peak displacement amplitude $2x_0$ and compute the acceleration from

$$a = 4\pi^2 f^2 x_0 \qquad (18.24)$$

By utilizing the fringe disappearance technique for displacement measurement it is possible to obtain calibration factors with an accuracy of about ±5 per cent, corresponding to displacements up to about 4 microinches and frequencies from 1,000 to 9,000 cps.[17]

OPTICAL INTERFEROMETRIC METHOD OF CALIBRATION. The phenomenon of interference band disappearance in an optical interferometer can be used as a precision means of determining amplitude of motion. Figure 18.26 shows the principle of operation of the optical interferometer employed in this technique.[18, 19] One of the mirrors D, Fig. 18.26A, is attached to the mounting plate of the calibrator. It is possible to obtain interference fringe patterns which are composed of straight lines, or bands, as shown in Fig. 18.26B. The intensity I of light for 50 per cent transmission through the silvered surface when K is the illuminating intensity is given by

$$I = K\left(1 + \cos\frac{2\pi x}{h}\right) \qquad (18.25)$$

where h is the separation of bands and x is the lateral displacement from a point midway between two bands.

A movement of one of the mirrors D in the direction of the light path by an amount r causes a change in the length of the light path of the amount $2r$ and a shift in the posi-

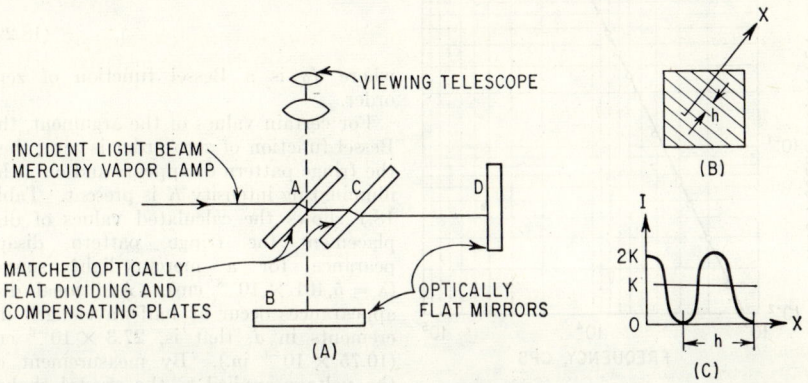

FIG. 18.26. The principles of operation of an interferometer: (A) optical system, (B) the observed interference pattern, and (C) the light intensity along the X axis. The system is used in the fringe disappearance technique for measuring displacement.

tion of the pattern by the amount x'. When $2r$ is equal to the wavelength λ of the light used, the shift x' equals h. When $2r$ has other values, the shift is given by

$$x' = \frac{2rh}{\lambda} \qquad (18.26)$$

When the mirror D vibrates sinusoidally with a frequency f and a peak-to-peak displacement amplitude of $2x_0$, its position is given by

$$r = x_0 \sin \omega t \qquad (18.27)$$

and therefore

$$x' = \frac{2x_0 h}{\lambda} \sin \omega t \qquad (18.28)$$

where $\omega = 2\pi f$.

Table 18.1. Optical Interferometric Method of Calibration

Values of peak-to-peak displacement amplitude d for which fringe pattern vanishes for a mercury-light source ($\lambda = 5{,}461 \times 10^{-8}$ cm).

Root number	Displacement, d	
1	8.18×10^{-6} in.	20.8×10^{-6} cm
2	18.90	47.8
3	29.60	75.0

The time average of the light intensity at position x when t is zero, as a result of the varying band shift x', is given by

$$I = \frac{K\omega}{2\pi} \int_0^{2\pi/\omega} \left[1 + \cos\frac{2\pi(x - x')}{h}\right] dt$$

Substituting for x' its value as given by Eq. (18.28) and evaluating the integral,

$$I = K\left[1 + J_0\left(\frac{2\pi d}{\lambda}\right)\cos\left(\frac{2\pi x}{h}\right)\right] \qquad (18.29)$$

where J_0 is a Bessel function of zero order.

For certain values of the argument, the Bessel function of zero order is zero; then the fringe pattern disappears and only the illuminating intensity K is present. Table 18.1 shows the calculated values of displacement for fringe pattern disappearance for a mercury-light source ($\lambda = 5{,}461 \times 10^{-8}$ cm). Additional disappearances occur at half-wavelength increments in d, that is, 27.3×10^{-6} cm (10.75×10^{-6} in.). By measurement of the voltage applied to the crystal shaker at selected frequencies, while observing the fringe pattern disappearance, the cali-

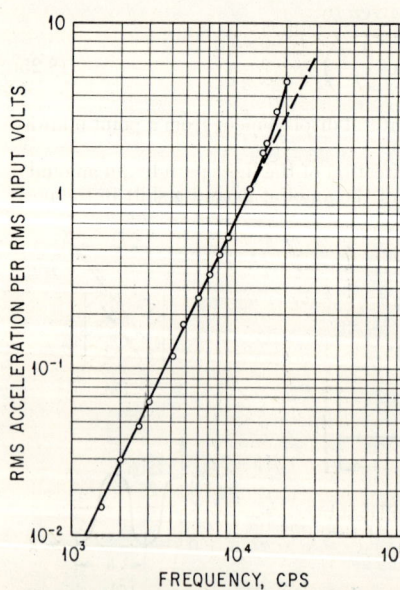

FIG. 18.27. Calibration factor of a typical piezoelectric calibrator vs. frequency. (*After Gulton Industries.*[17])

bration factor at those frequencies is obtained. A plot of calibration as a function of frequency is given in Fig. 18.27. The rise in acceleration response at the higher frequencies indicates that the frequency of mechanical resonance of the crystal stack is being approached. This places an upper frequency limit on the use of the applied voltage as a basis for calibration.

The procedure described above also can be used to calibrate the auxiliary accelerometer in the mounting plate for subsequent use as a secondary standard.

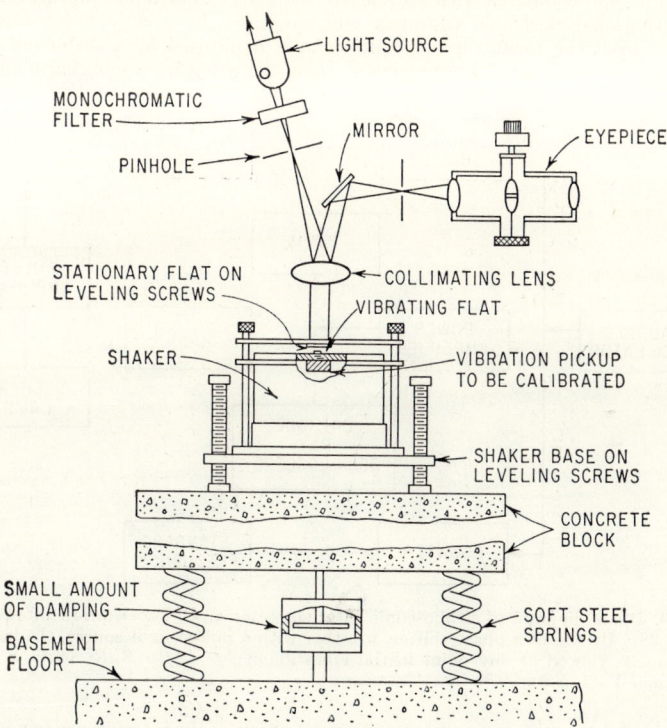

FIG. 18.28. Piezoelectric calibrator and a stroboscopic interferometer used in the calibration of vibration pickups in the frequency range from 100 to 20,000 cps. (*After E. R. Smith, S. Edelman, E. Jones, and V. A. Schmidt.*[20])

A Stroboscopic Interferometer Technique for Displacement-amplitude Measurement. The piezoelectric system shown in Fig. 18.28 employs a stroboscopic interferometer of the Fizeau type to measure displacement.[20] The position of the fringe in relation to a reference mark in the field of view, at any instant, is determined by the distance between the stationary surface and the vibrating surface at that instant. The fringes move laterally in unison with the vibratory motion. The stroboscopic effect is obtained by driving the lamp by voltage pulses which are synchronized with the vibration. A flash of light occurs at each cycle of vibration. If the phase of the flash is constant relative to the phase of the motion, the interference pattern appears to be motionless. If the phase of the light flash is varied slowly, the pattern appears to move in the direction perpendicular to the fringes. The amplitude of the motion of the fringes is proportional to the amplitude of vibration. The amplitude of vibration is determined by measuring the excursion x of the fringe and the distance h between fringes on the same scale. Their ratio is equal to the ratio of twice the displacement amplitude $2x_0$ to the half-wavelength of the light used. Thus

$$x_0 = \frac{\lambda x}{4h} \tag{18.30}$$

The block diagram of the electronic circuit for doing this is shown schematically in Fig. 18.29. The light flashes are synchronized with the motion by using the same oscillator to drive the lamp and the shaker. The frequency is determined by means of the cycle counter and compared with a standard frequency. The input voltage is measured by a conventional electronic voltmeter and monitored by a cathode-ray oscilloscope. The signal from the pickup under calibration is monitored by a distortion meter at intervals to ensure sinusoidal excitation. Slight distortion from pure harmonic motion

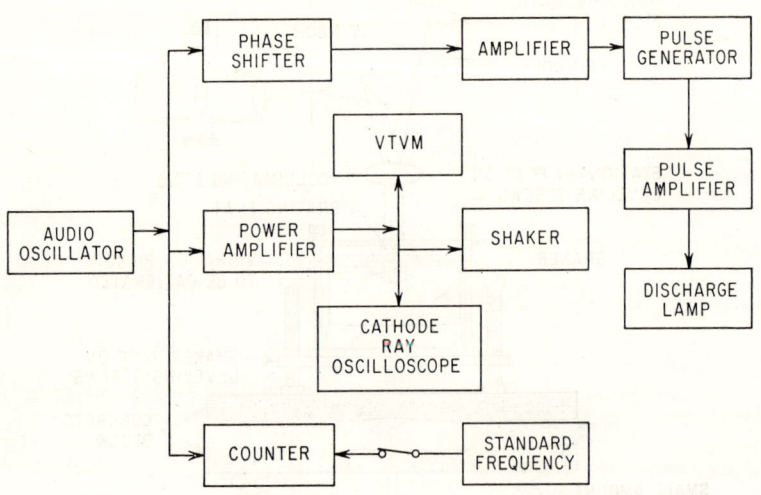

FIG. 18.29. Block diagram of stroboscopic interferometer and piezoelectric calibrator shown in Fig. 18.28. By use of a phase shifter, together with a pulsed light source, the interference pattern can be viewed at any point in the vibration cycle. (*After E. R. Smith, S. Edelman, E. Jones, and V. A. Schmidt.*[20])

is detectable by nonuniform, or nonrectilinear, motion of the interference fringe pattern. The scatter of repeated measurements using this method can be kept below ± 6 per cent. The variations are attributed to irregularities in the pulse triggering voltage, lack of precision in the measurement of the voltage, and lack of flatness in the reflecting surfaces.

RECIPROCITY METHOD OF CALIBRATION OF PICKUPS. Linear bilateral pickups can be calibrated by the reciprocity method.[15,21] Pickups are bilateral if velocity at the mechanical terminal results in the generation of an electrical output voltage and, conversely, current through the electrical terminals results in force at the mechanical terminal. Pickups which employ purely electromagnetic, electrostatic, or piezoelectric coupling satisfy the bilateral requirement. Pickups, such as those employing variable reluctance or resistance as a sensing means, do not satisfy the bilateral requirement and are described as unilateral.

Calibration by reciprocity methods ordinarily uses two similar bilateral pickups (although it is possible to use the method when the pickups are different and even when one of the pair is unilateral).

Complex notation is used to represent the harmonically varying current, voltage, force, sensitivity, etc., as is done in alternating-current theory. The positive terminal of the pickup is designated as that having a positive voltage when the mechanical terminal is moving inward. Conversely, positive current produces outward velocity of the mechanical terminal. The pickup to be calibrated is designated by subscript 1 and the second pickup by subscript 2. To perform the calibration for a given frequency f two experiments are performed.

Experiment 1. Rigidly mount the two pickups as shown schematically in Fig. 18.30A on the opposite ends of a right circular cylinder whose mass has been determined from the measured value of its weight. Apply a driving voltage of frequency f to pickup 1 and measure the ratio

$$G = \frac{i_1}{e_2} \tag{18.31}$$

where i_1 is equal to the current in the first pickup and e_2 is equal to the voltage generated

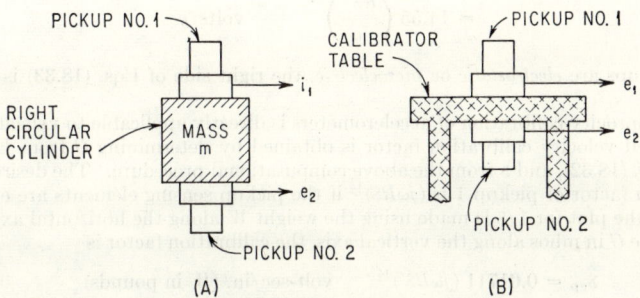

Fig. 18.30. Vibration pickup arrangement for reciprocity calibration at high frequencies: (A) Experiment 1 and (B) Experiment 2. (*After H. M. Trent.*[15])

in the open-circuited second pickup. Repeat for a series of cylinders of increasing mass m.

Experiment 2. Mount the pickups on a sinusoidal shaker capable of driving the pickups simultaneously as shown in Fig. 18.30B with velocities of equal magnitude at frequency f and measure the ratio

$$R = \frac{e_1}{e_2} \tag{18.32}$$

where e_1 is equal to the open-circuit voltage generated in pickup 1 and e_2 is equal to the open-circuit voltage generated in pickup 2.

Computational Procedure. Plot m along the horizontal axis against G along the vertical axis and determine the slope $(1/S)$ of the resulting line. [Since G is a complex number, it may be necessary to plot its real and imaginary portions separately, the corresponding slopes being the real and imaginary parts of $1/S$. When $1/S = a + jb$, $S = (a - jb)/(a^2 + b^2)$.] The desired acceleration calibration factor of pickup 1 is $(RS/j\omega)^{1/2}$ if pickup 1 is electromagnetic, and $(jRS/\omega)^{1/2}$ if it is electrostatic or piezoelectric. The slope S may be determined from the data analytically, using the method of least squares if desired.

It should be noted that the plot of m against G should be a straight line and that the ratio R should be independent of the velocity at which it is measured. If such is not the case, it usually indicates that the linear range of the pickup has been exceeded and the calibration is not applicable.

The presentation given is for a consistent set of units. If the plot to obtain the slope $1/S'$ is constructed using the weight W in pounds along the horizontal axis in place of the mass m and if the transfer admittance G is measured in mhos, the acceleration calibration factor S_{ap} of pickup 1 (if the pickups are *electromagnetic*) is

$$S_{ap} = 0.01711 \left(\frac{RS'}{j\omega}\right)^{\frac{1}{2}} \quad \text{volt-sec}^2/\text{in.}$$

$$= 6.60 \left(\frac{RS'}{j\omega}\right)^{\frac{1}{2}} \quad \text{volts}/g$$

(18.33a)

where g is the acceleration of gravity. If the plot to obtain the slope $1/S''$ is constructed using the mass m along the horizontal axis in kilograms, the acceleration calibration factor S_{ap} of pickup 1 (if the pickups are *electromagnetic*) is

$$S_{ap} = 0.01482 \left(\frac{RS''}{j\omega}\right)^{\frac{1}{2}} \quad \text{volt-sec}^2/\text{cm}$$

$$= 14.55 \left(\frac{RS''}{j\omega}\right)^{\frac{1}{2}} \quad \text{volts}/g$$

(18.33b)

If the pickups are *electrostatic* or *piezoelectric*, the right side of Eqs. (18.33) is multiplied by j.

The reciprocity calibration of accelerometers is directly applicable to velocity pickups. The desired velocity calibration factor is obtained by determining G from Eq. (18.31), R from Eq. (18.32), and S from the above computational procedure. The desired velocity calibration factor of pickup 1 is $(j\omega RS)^{\frac{1}{2}}$ if the pickup sensing elements are electromagnetic. If the plot for $1/S$ is made using the weight W along the horizontal axis with the admittance G in mhos along the vertical axis, the calibration factor is

$$S_{vp} = 0.01711 \ (j\omega RS')^{\frac{1}{2}} \quad \text{volt-sec/in.} \ (W \text{ in pounds})$$

$$= 0.01482 \ (j\omega RS'')^{\frac{1}{2}} \quad \text{volt-sec/cm} \ (W \text{ in kilograms})$$

(18.34)

RESONANT-BAR CALIBRATOR

The use of a resonant bar, Figs. 18.31 and 18.32, to apply sinusoidal accelerations for calibration purposes has several advantages:[24, 25, 26] (1) The frequency is inherently constant and (2) very large amplitudes of acceleration (as high as 4,000g) with very little distortion in the waveform can be attained with moderate driving force. A disadvantage of this type of calibrator is that only the resonant frequencies are available for use.

In a typical mounting, Fig. 18.31, the bar is supported at its nodal points and the pickup to be calibrated is mounted at the mid-length of the bar.[24, 25] The bar is energized by an electromagnet. The magnet current is obtained by superposing on a polarizing direct current the output of a power amplifier energized by an oscillator. Self-excitation is possible if the oscillator is replaced by the output of a second pickup mounted on the beam, in conjunction with a phase shifter to form a closed servo loop. Beams of about 2-in. thickness vary in natural frequency from about 400 cps for a 31-in.-length beam to about 2,000 cps for a 14-in.-length beam. Acceleration amplitudes of several thousand g are obtained. By employing smaller beams, higher frequencies and acceleration levels can be excited. For example, calibrations can be performed at acceleration levels of 12,000g employing bars in axial resonance.

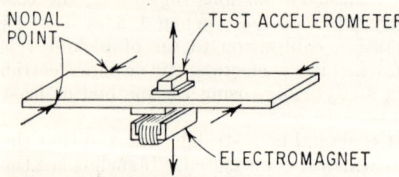

Fig. 18.31. Resonant-bar calibrator supported at the nodal points for free-free bending vibration. The bar at its natural frequency is driven by an electromagnet. The pickup is mounted at the mid-length of the bar. (*After ASA Standard S 2.2-1960.*[4])

The displacement at the point of attachment of the pickup is measured optically since displacements encountered are adequately large. The response R and the peak-to-peak displacement $2x_0$ are read for a series of values of driving force. The values of R are then plotted against $4\pi^2 f^2 x_0$ and the slope of the line through the points is taken as the pickup calibration factor for the frequency f of resonance of the bar. To calibrate a pickup at other frequencies, the bar must ordinarily be changed and the procedure repeated.

The resonant-bar calibrator shown in Fig. 18.32 is primarily limited in amplitude by the fatigue resistance of the bar.[26] Using aluminum bars, levels up to $500g$ have been attained without special design criteria. For levels up to $4,000g$ a bar material such as tempered vanadium steel machined to have a mounting boss at its mid-length is more suitable. The resonant bar is mounted at its mid-length on a conventional electrodynamic shaker. The accelerometer being calibrated is mounted at one end of the bar and an equivalent balance weight is mounted at the opposite end in the same relative position. The calibration procedure uses an optical means for measuring displacement amplitude from which the applied acceleration amplitude is computed as $4\pi^2 f^2 x_0$, where f is the frequency and $2x_0$ is the peak-to-peak displacement. (It should be noted that a pickup mounted at the end of a resonant bar is subjected to a rocking motion in addition to the desired translation. High-frequency pickups generally are unaffected by this extraneous rocking motion.)

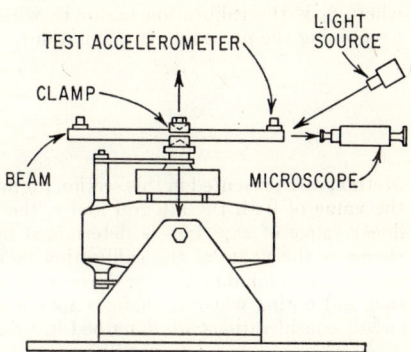

Fig. 18.32. Resonant-bar calibrator driven by an electrodynamic shaker. The pickup is mounted at one end and a counterbalancing weight is at the other. (*After E. I. Feder and A. M. Gillen.*[26])

HIGH-FREQUENCY, HIGH-ACCELERATION CALIBRATION METHODS (IMPACT METHOD)

There are several methods by which sudden velocity change may be applied to pickups designed for high-frequency acceleration measurement. Any method which generates a reproducible velocity change and time duration can be used to obtain the calibration factor.[4] Impactive techniques can be employed to obtain calibrations over an amplitude range from a few to over $30,000g$. An accurate determination of shock performance of an accelerometer depends not only upon frequency response, and mechanical and electrical characteristics designed into the transducer, but also upon the characteristics of the instrumentation and recording equipment. It is often best to perform system calibrations to determine the linearity of the vibration pickup as well as the linearity of the recording instrumentation in the approximate range under observation. Each of the following methods makes use of the fact that the velocity change during a transient pulse is equal to the time integral of acceleration:

$$v = \int_{t_1}^{t_2} a \, dt \tag{18.35}$$

where the initial or final velocity is taken as reference zero and the integration is performed to or from the time at which the velocity is constant.[23]

In this section, several methods are presented for applying known velocity changes v to a pickup. The voltage output e of a pickup of calibration factor S_1 for an acceleration a is

$$e = \frac{S_1 a}{g} \tag{18.36}$$

where S_1 is the calibration factor in volts per g (g being the acceleration of gravity). Combining the preceding two equations,

$$S_1 = \frac{\int_{t_1}^{t_2} eg\, dt}{v} \tag{18.37}$$

Methods are presented in this section for measuring the integral in this equation. Having the value of both the integral and v, the calibration factor S_1 can be computed. The linear range of a pickup is determined by noting at what magnitude of the velocity change v the value of the calibration factor S_1 deviates appreciably from its previous values. The minimum pulse duration is similarly found by shortening the pulse duration and noting when S_1 changes appreciably from previous values. Typical auxiliary circuit considerations are described in this section and typical calibrations are given.

BALLISTIC PENDULUM CALIBRATOR

The ballistic pendulum calibrator provides a means for applying a sudden velocity change to a vibration pickup. The calibrator consists of two masses which are suspended by wires or metal ribbons. These ribbons restrict the motion of the masses to a common vertical plane.[22, 23, 24] This arrangement, shown in Fig. 18.33, maintains horizontal alignment of the principal axes of the masses in the direction parallel to the direction of motion at impact. The velocity attained by the anvil mass as the result of the sudden impact is determined.

The accelerometer to be calibrated is mounted to an adapter which attaches to the forward face of the anvil. The hammer is raised to a predetermined height and held in the release position by a solenoid-actuated clamp. Since the anvil is at rest prior to impact, recording and measurement of the change in velocity of the anvil and the transient waveform on a calibrated time base are required. One method of measurement of the velocity change is performed by focusing a light beam through a grating attached to the anvil, as shown in Fig. 18.34. The slots modulate the light beam intensity, thus varying the phototube output which is recorded with the pickup output. Since the distance between grating lines is known, the velocity of the anvil is calculated directly, assuming that the velocity is essentially constant over the distance between successive grating lines. Other schemes may be used to measure velocity change, such as a sliding contactor or any suitable linear displacement device which can be timed as it crosses a sensitive region

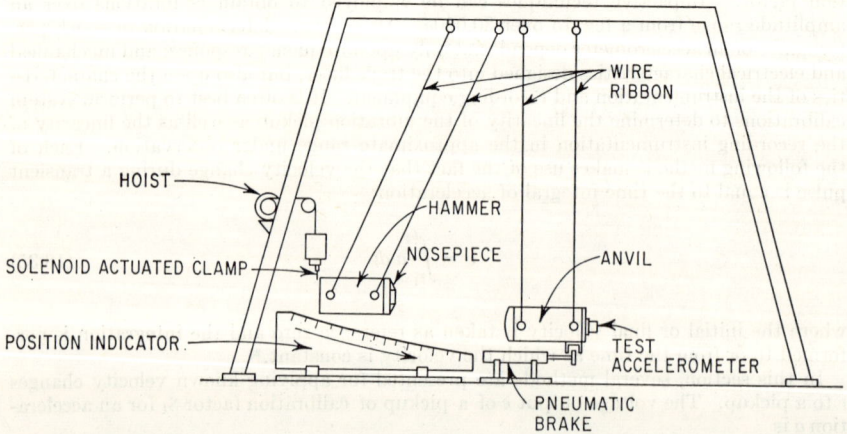

Fig. 18.33. Schematic of a typical ballistic pendulum calibrator system used to apply a sudden velocity change to a vibration pickup. (*After E. I. Feder and A. M. Gillen.*[26])

HIGH-FREQUENCY, HIGH-ACCELERATION CALIBRATION METHODS   18-29

whose dimensions are known. The velocity of the anvil in each case is determined directly; the time relation between initiation of the velocity and the pulse at the output of

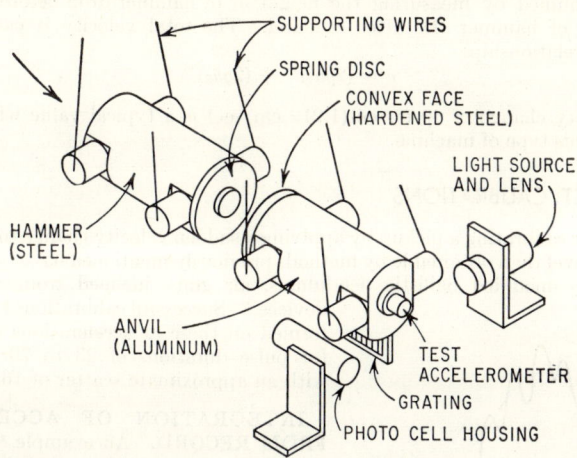

FIG. 18.34. Schematic arrangement of the ballistic pendulum with phototube and light grating to determine the anvil-velocity change during impact. (*After R. W. Conrad and I. Vigness.*[23])

the pickup is obtained by recording both signals on the same time base. The most frequently used method infers the anvil velocity from its vertical rise by measuring the maximum horizontal displacement and making use of the geometry of the pendulum system. This method has been proved quite satisfactory and provides results that correlate with the more precise techniques described above to within a few per cent.

The duration of the pulse, which is the time during which the hammer and anvil are in contact, can be varied within close limits.[23] In Fig. 18.34 the hammer nosepiece is a disc with a raised spherical surface. It develops a contact time of 0.55 millisecond. For larger periods, ranging up to 1 millisecond, the stiffness of the nosepiece is decreased by bolting a hollow ring between it and the hammer. Pulses longer than 1 millisecond may be obtained by placing various compliant materials, such as lead, between the contacting surfaces.

DROP-TEST CALIBRATOR

Another frequently used impulsive device is the drop tester shown in Fig. 18.35. The pickup is attached to the hammer using a suitable adapter plate. An impact is produced as the guided hammer falls under the influence of gravity and strikes the fixed anvil. To determine the velocity change, measurement is made of the time required for a contactor to pass over a known region just prior

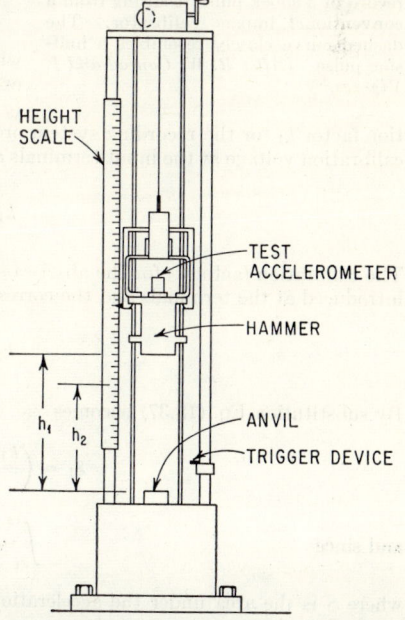

FIG. 18.35. Schematic of a conventional drop tester used to apply a sudden velocity change to a vibration pickup. (*After R. W. Conrad and I. Vigness.*[23])

to and after impact. The pickup output and the contactor indication are recorded simultaneously in conjunction with a calibrated time base. The velocity change also may be determined by measuring the height h_1 of hammer drop before rebound and the height h_2 of hammer rise after rebound. The total velocity is calculated from the following relationship:

$$v = (2gh_1)^{1/2} + (2gh_2)^{1/2} \tag{18.38}$$

A total velocity change of 40 ft/sec (1,219 cm/sec) is a typical value which has been achieved by this type of machine.

HIGH-IMPACT CALIBRATIONS

Methods for calibrating a pickup by applying a sudden velocity change to it at a higher acceleration level than obtainable by methods previously mentioned have been developed using specially modified ballistic pendulums, air guns, inclined troughs, and other devices.[22] Successful calibrations have been performed on these at accelerations up to 40,000g for pulse durations of 23 to 70 microseconds with an approximate scatter of 16 per cent.

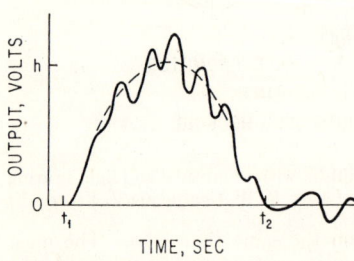

FIG. 18.36. Typical acceleration-time record of a shock pulse resulting from a conventional impact calibrator. The dashed curve closely resembles a half-sine pulse. (*After R. W. Conrad and I. Vigness.*[23])

INTEGRATION OF ACCELERATION FROM RECORD. An example [23] of a typical acceleration vs. time record describing the characteristic response of a vibration pickup to a pulse input is shown in Fig. 18.36. The following linear relationship holds between acceleration a and voltage output e:

$$S_1 = \frac{e}{a/g} \tag{18.39}$$

where S_1 is the accelerometer calibration factor expressed in output volts per g acceleration (g is the acceleration of gravity). The calibration factor k_1 for the recording system ordinate scale is obtained by applying a known calibration voltage at the input terminals and noting the resulting deflection y:

$$k_1 = \frac{e}{y} \tag{18.40}$$

The calibration factor k_2 for the abscissa scale is obtained from a precision timing trace introduced at the terminals and the corresponding deflection x:

$$k_2 = \frac{t}{x} \tag{18.41}$$

By substitution Eq. (18.37) becomes

$$S_1 = \left(\frac{k_1 k_2 g}{v}\right) \int_{x_1}^{x_2} y \, dx \tag{18.42}$$

and since

$$\int_{x_1}^{x_2} y \, dx = S \tag{18.43}$$

where S is the area under the acceleration vs. time curve, the calibration factor for the test accelerometer expressed in terms of volts per unit gravity is

$$S_1 = \frac{k_1 k_2 S g}{v} \tag{18.44}$$

For piezoelectric accelerometers the calibration factor S_2 is often expressed in terms of charge $Q = eC'$ per unit of gravity:

$$S_2 = \frac{Q}{a/g} = \frac{k_1 k_2 S C' g}{v} \tag{18.45}$$

where C' is the capacity in farads of the pickup plus all the parallel capacities charged by the voltage e and Q is the charge in coulombs developed by the pickup.

The area S can be determined by use of a planimeter or by numerical integration. If the output closely resembles a half-sine pulse, the area is equal to $2hx/\pi$, where h is the height of the pulse and x its width.

ELECTRICAL INTEGRATION OF PICKUP RESPONSE. The integration of the area response S may be performed by electrical integration. It is necessary to select constants for the integrating network that will not appreciably load the voltage e. If an RC network consisting of a series resistance R followed by a parallel capacitance C, as in Fig. 18.37, is used to integrate the accelerometer output voltage, the maximum integrated output voltage is

$$e_1 = \frac{1}{RC} \int_{t_1}^{2} e\, dt \tag{18.46}$$

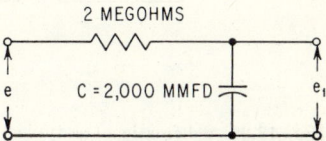

It follows from Eq. (18.37) that

$$S_1 = \frac{gRCe_1}{v} \tag{18.47}$$

Fig. 18.37. Single-stage integration network. The resistance $R = 2$ megohms; the parallel capacitance $C = 2,000$ mmfd. (*After T. A. Perls and C. W. Kissinger.*[22])

Several restrictions are imposed by this method of calibration and generally are related to the low-frequency response of the electrical integrator and the natural period of the accelerometer. In the case of piezoelectric pickups, the distortion in response at low frequency is limited by the time constant derived from the total capacitance of the accelerometer and cables and the input impedance of the conversion system. The response at high frequency is limited by the natural period of the accelerometer and the nonlinearity of the integration network.

The theoretical limit of calibration accuracy [23] for accelerometers with little or no damping is 2 per cent for ratios of pulse duration to natural period greater than 6. In practice, calibration errors may vary from as little as 5 per cent to as much as 20 per cent, depending upon the region of frequency response that is explored. Experimental transient studies [23] conducted for two accelerometers over a range of pulse duration greater and less than the natural period are shown in Fig. 18.38. The variations in relative response become greater for those values of the ratio of pulse duration to natural period greater than 6.

AUXILIARY CIRCUIT EFFECTS ON VELOCITY-CHANGE CALIBRATION METHOD. Cathode followers and amplifying and recording equipment ordinarily are used with high-frequency, high-acceleration pickups. Distortion in these auxiliary circuits can be more important in high-level pickups than in low-level pickups. Therefore, the accurate determination of the performance of a pickup by the velocity-change method depends on the lack of frequency distortion and nonlinearity in the auxiliary circuits for the range of frequencies and amplitudes excited by the shock motion.

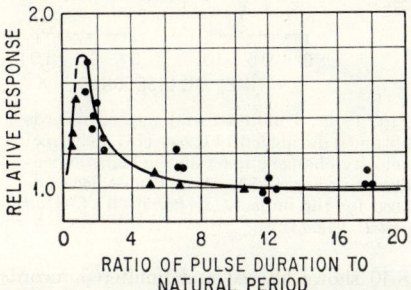

Fig. 18.38. Experimental data on the variation of relative response with relative pulse duration for two commercial piezoelectric accelerometers. (*After T. A. Perls and C. W. Kissinger.*[22])

18-32   CALIBRATION OF PICKUPS

A low-pass filter sometimes can be used, as shown in Fig. 18.39, to eliminate unwanted high-frequency high-amplitude portions of the response. Where this is done, it is important that the linear range not be exceeded in portions of the circuit preceding the filter and that the upper cutoff frequency of the filter be well above the frequency range over which the calibration is desired.

Wherever possible it is desirable to determine that the response of circuit elements is unaffected by frequency distortion in the auxiliary circuits, and that the expected volt-

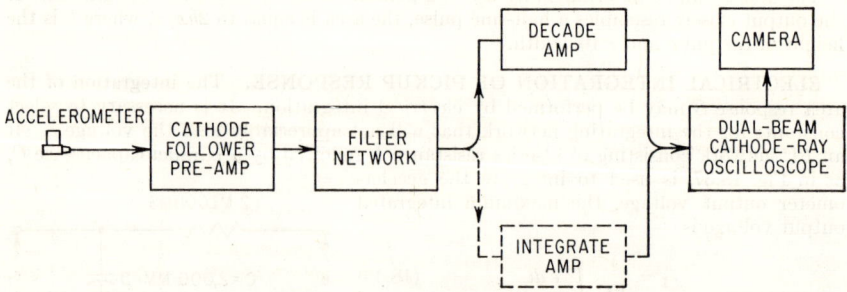

Fig. 18.39. Schematic of a typical circuit arrangement used to calibrate piezoelectric accelerometers by the velocity-change method. Dual-beam oscilloscope is used to record velocity before and after impact on one channel and pickup output on the other.

ages will not exceed the linear range of any element. A more detailed and general discussion of calibration of auxiliary circuits is given later in this chapter.

TYPICAL EXAMPLES OF IMPACT CALIBRATION

Typical records of accelerometer outputs are shown in Figs. 18.40 to 18.42 displaying representative waveforms produced by devices in Figs. 18.34 and 18.35 as well as illustrat-

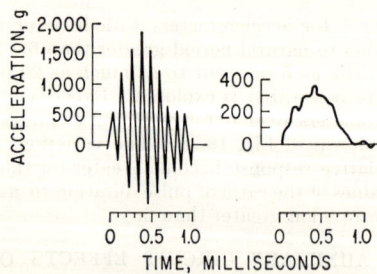

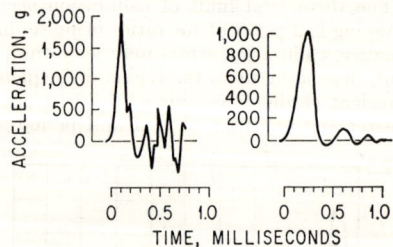

Fig. 18.40. Unfiltered and filtered records obtained during a 4.92 ft/sec (150 cm/sec) velocity-change impact on the ballistic pendulum, Fig. 18.34. A spring-disc nosepiece was used for the impact. (*After R. W. Conrad and I. Vigness.*[23])

Fig. 18.41. Unfiltered and filtered records obtained during a 6.14 ft/sec (187.1 cm/sec) velocity-change impact on the ballistic pendulum in Fig. 18.34. A solid nosepiece was used for the impact. (*After R. W. Conrad and I. Vigness.*[23])

ing accelerometer characteristics. Figure 18.40 shows filtered and unfiltered records taken during a 4.92 ft/sec (150 cm/sec) velocity change achieved with the spring-disc nosepiece in Fig. 18.34. The 9,000-cps frequency present on both the filtered and unfiltered records is the fundamental longitudinal mode of the anvil and not the natural frequency of the accelerometer which is known to be 14,000 cps. The area under the two curves is nearly the same, thus giving the same value for the calibration factor. A velocity change of 6.14 ft/sec (187.1 cm/sec) can be obtained as shown in Fig. 18.41 by using the solid nosepiece. Again the area under the two curves is nearly the same, giving the same value for the calibration factor. A principal difference between the filtered and unfiltered records is a decrease in the amplitude and an increase in the apparent period. In addition, it is easier to evaluate the area for the filtered record. The error induced by an incorrectly selected filter is usually small if some of the high-frequency response comes through.

A typical record taken on a drop tester similar to that in Fig. 18.35 is shown in Fig. 18.42. A multi-channel recording oscillograph is required to record impact and rebound velocity. The unfiltered waveform has only a small proportion of high-frequency components as a result of placing a small lead-ball slug between the impacting surfaces. Very often this is necessary to avoid distortions resulting from ringing of the hammer in this type of calibration system.[23]

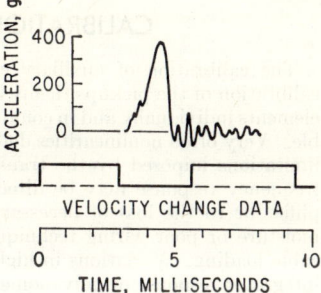

FIG. 18.42. Unfiltered record obtained during an 11.37 ft/sec (346.6 cm/sec) velocity-change impact on drop tester similar to that shown in Fig. 18.35. A lead slug between the surfaces eliminated high-frequency components and increased the pulse duration. (*After R. W. Conrad and I. Vigness.*[23])

CALIBRATION BY THE COMPARISON METHOD

A rapid and convenient method of obtaining the calibration factor of a vibration pickup is by direct comparison of the pickup voltage to that of a second pickup used as a secondary standard and calibrated for frequency and phase response by one of the methods described in this chapter.[4]

Pickups are mounted back-to-back as shown in Fig. 18.43. The most important consideration is to ensure that each pickup experiences the same motion. If both pickups are rectilinear and are placed on the calibrator table, the angular rotation of the table should be small to avoid any difference in excitation between the two pickup locations. The error due to rotation may be reduced by carefully locating the pickups firmly on opposite faces with the center-of-gravity of the pickups located at the center of the table. Relative differences in pickup excitation may be observed by reversing the pickup locations and observing if the voltage ratio is the same in both positions.

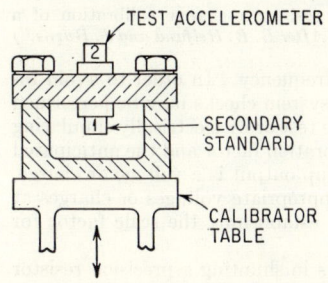

FIG. 18.43. Comparison method of calibration. Pickup 2 is calibrated against pickup 1 as the secondary standard. The two pickups are excited by any of the means described in this chapter. (*After ASA Standard S 2.2-1960.*[4])

Calibration by the comparison method is limited to the range of frequencies and amplitudes for which the secondary pickup has been calibrated. If both pickups are linear, the amplitude sensitivity S_2 of a vibration pickup can be determined by

$$S_2 = \frac{e_2}{e_1} S_1 \qquad (18.48)$$

where S_1 is the sensitivity of the calibrated pickup and e_2 and e_1 are the corresponding measured voltages. The phase relationship is obtained by phase measurement between output voltages and phase calibration of the secondary-standard pickup.

CALIBRATION OF AUXILIARY CIRCUITS

The calibration of auxiliary circuits for vibration pickups is as important as the calibration of the pickups themselves. Calculation of the performance of various circuit elements individually and in combination in a system is difficult and in most cases impossible. Very often nonlinearities detected in the calibration of pickups can be attributed to limitations imposed by the transformation and recording system. Variations in low-frequency response may be directly related to an incorrectly selected matching amplifier or to the loss of necessary high insulation resistance as the result of excessive moisture or poor wiring techniques. Loss in sensitivity frequently occurs because of cable loading. Variations in high-frequency response may be caused by limitations of integrator response or galvanometer response or, in the case of magnetic tape, by head

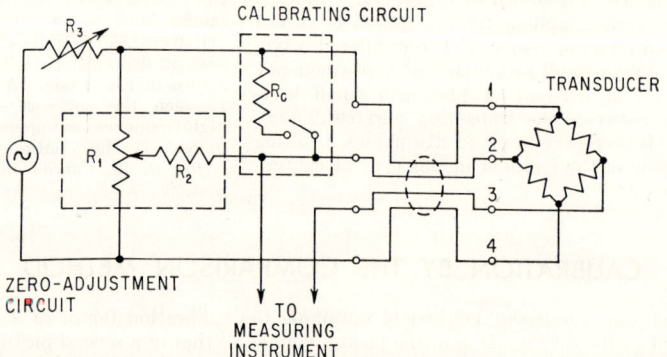

FIG. 18.44. Schematic diagram of circuit details in the resistance-change calibration of a Wheatstone-bridge unbonded-strain-gage accelerometer. (*After B. B. Helfand and J. Burns.*[27])

gap, tape speed, or the allowable deviation of center frequency. In order to avoid inadvertent errors in circuit selection and construction, system checks may be performed by calibrating through the system to be used during the test or by electrically simulating physical quantities based upon the predetermined calibration factor and the anticipated range of operation. In those instances where the pickup output is a voltage or charge, the auxiliary circuits can be calibrated by applying appropriate voltages or charges at selected frequencies and recording the response, thus establishing the scale factor for the instrumentation system.

A simple illustration of a system calibration consists in shunting a precision resistor across one arm of a resistance-bridge-type vibration pickup.[27] The pickup output is essentially in the form of a resistance change. The effect simulates the combined resistance changes of the active bridge arms due to acceleration. Electrical calibration in this manner does not remove the necessity for a precise calibration of the system. A calculated value of resistance is introduced during the physical calibration of such magnitude that the resulting indication is equal to that resulting from direct excitation. Since the resistance change appears as a step function, it is important that the auxiliary circuits have a flat response down to 0 cps. The circuit schematic of a typical system is shown in Fig. 18.44.

The calculation of the calibrating resistor value corresponding to an equivalent acceleration is given as

$$R_c = \frac{10^6 r}{2NS} \tag{18.49}$$

where R_c is the calibrating resistor value in ohms, r is the resistance of each of the four bridge arms in ohms, S is the calibration factor in microvolts open-circuit output per volt input applied across the bridge per unit acceleration determined by physical calibration, and N is the mechanical acceleration input simulated by the calibrating resistor. The accuracy of the calibrating resistor method depends upon the stability of the calibrating resistor R_c, the bridge-arm resistance r, and the calibration factor S.

A complete system checkout is important in order to provide assurance that the vibration-measuring system will function properly under test conditions. It is desirable to supplement the laboratory calibration of a vibration pickup and measuring circuit with at least a one-point calibration of the field installation. Small portable shakers, Fig. 18.45, may be used to perform this function. By such means an over-all transducer-to-read-out calibration can be made of conventional or telemetered channels assigned to vibration measurement.[28]

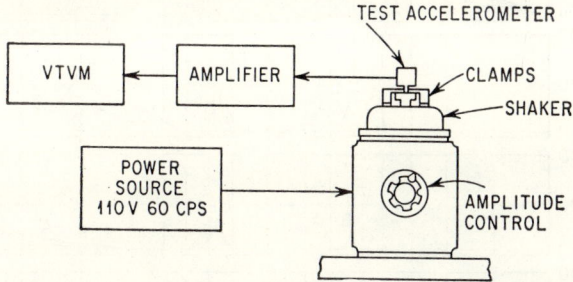

Fig. 18.45. Portable electrodynamic calibrator for system field calibration of small vibration pickups. The calibrator is driven at power frequency and the amplitude is adjusted by two reference circles similar to the vibrating-wedge technique described in the text. (*After E. Rule, T. A. Perls, and F. J. Sullentrap.*[28])

MEASUREMENT OF TRANSVERSE (CROSS-AXIS) SENSITIVITY

The characteristics of a vibration pickup may be such that an extraneous output voltage is generated as a result of vibration which is in a direction at right angles to the axis of maximum sensitivity of the pickup [see Fig. 12.19 and Eq. (16.26)]. In order to measure the transverse (cross-axis) sensitivity of a pickup, it is first necessary to define the absolute motion of the calibrator table since the percentage of transverse motion in the shaker establishes the lowest determinable value of transverse sensitivity present in the pickup. Several different modes of vibration of the calibrator may occur throughout its range. It may translate along or rotate about each of the three axes, sometimes independently but frequently simultaneously. In one method used to determine the absolute motion of the calibrator table, the motion is measured by two noncontacting mutual-inductance probes.[29] Summary curves of a typical investigation are shown in Fig. 18.46. Transverse calibrations of vibration pickups are best performed within the range of the curves where errors resulting from extraneous motions are small. Typical transverse calibrations of piezoelectric accelerometers are shown in Fig. 18.47[30] and Fig. 18.48. The curve shown in Fig. 18.47 presents the sensitivity of the pickup to vibration in a plane including the sensitive axis. The curve shown in Fig. 18.48 presents the pickup sensitivity to vibration in the plane normal to the sensitive axis. Where a pickup is intended to be used in a general vibration field, it is desirable to determine that the transverse sensitivity is sufficiently small, for it is not uncommon to encounter a transverse output which is as much as 10 per cent of the output for excitation in the sensitive direction.

Another method[14] for determining the transverse motion in a plane near and parallel to the surfaces of the mounting table, used by the National Bureau of Standards in evaluating vibration standards, is illustrated in Fig. 18.49. Three piezoelectric accelerometers with nearly equal sensitivities are mounted on the standard with their principal sensitive axes mutually perpendicular to each other. The response of these accelerometers in planes transverse to the principal axis is approximately 5 per cent of their response along the principal axis. One accelerometer is mounted with its sensitive

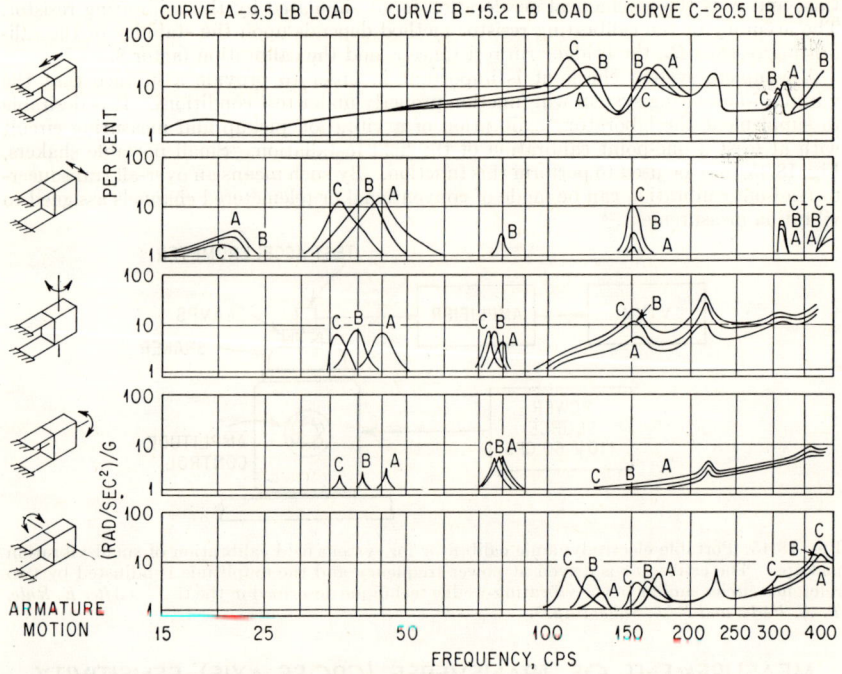

Fig. 18.46. Summary response curves of extraneous motions of the armature of a typical electrodynamic shaker. (*After W. R. Elliot*.[29])

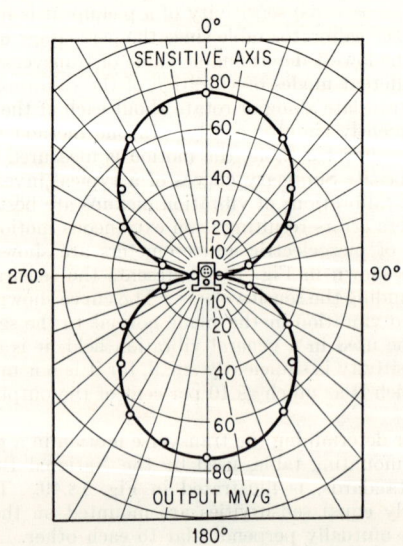

Fig. 18.47. Transverse sensitivity of a piezoelectric accelerometer at a frequency of 200 cps vibrating in a plane including the axis of principal sensitivity. (*After Gulton Industries*.[30])

axis parallel to the axial motion of the standard. In Fig. 18.49, the axial motion is in a horizontal plane alternately to the right and left across the page. The second accelerometer has its sensitive axis parallel to the flexure-plate springs. It measures the transverse motion in a direction alternately up and down in the plane of the page. The third accelerometer measures the motion perpendicular to the page in a horizontal plane.

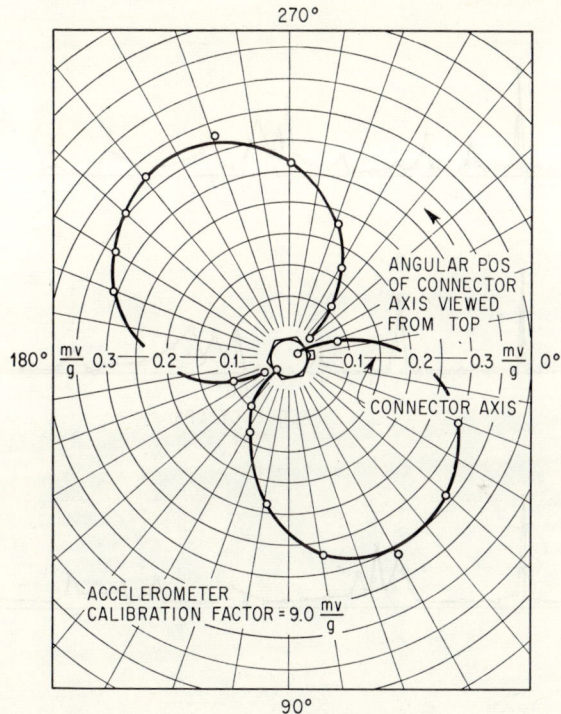

FIG. 18.48. Transverse sensitivity of a piezoelectric accelerometer to vibration in the plane normal to the sensitive axis. (*After Gulton Industries.*[30])

This motion is referred to as the transverse motion in the horizontal direction or horizontal acceleration. Two dummy accelerometers (one of which is hidden from view in Fig. 18.49) are attached to balance the load on the mounting table.

Using the above method, the transverse motion in several electrodynamic vibration standards equipped with flexure-plate springs is measured. The results on two standards

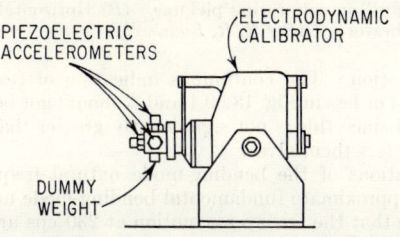

FIG. 18.49. Test setup for measuring transverse motion of a typical electrodynamic calibrator used as a "standard" in calibrating pickups. The three piezoelectric accelerometers attached to the mounting table indicate the acceleration in three mutually perpendicular directions. (*After R. R. Bouche.*[14])

are shown in Fig. 18.50A and B. In one standard, Fig. 18.50A, the transverse motion at 230 cps in the vertical direction is 1.5 times the axial motion. Significant transverse motion in the vertical direction also is noted at 580, 640, 690, 1,050, 1,260, 1,350, and 1,500 cps. In the horizontal direction, significant transverse motion is noted at 150, 220, 230, 570, 990, and 1,500 cps. In the second standard, Fig. 18.50B, there is also con-

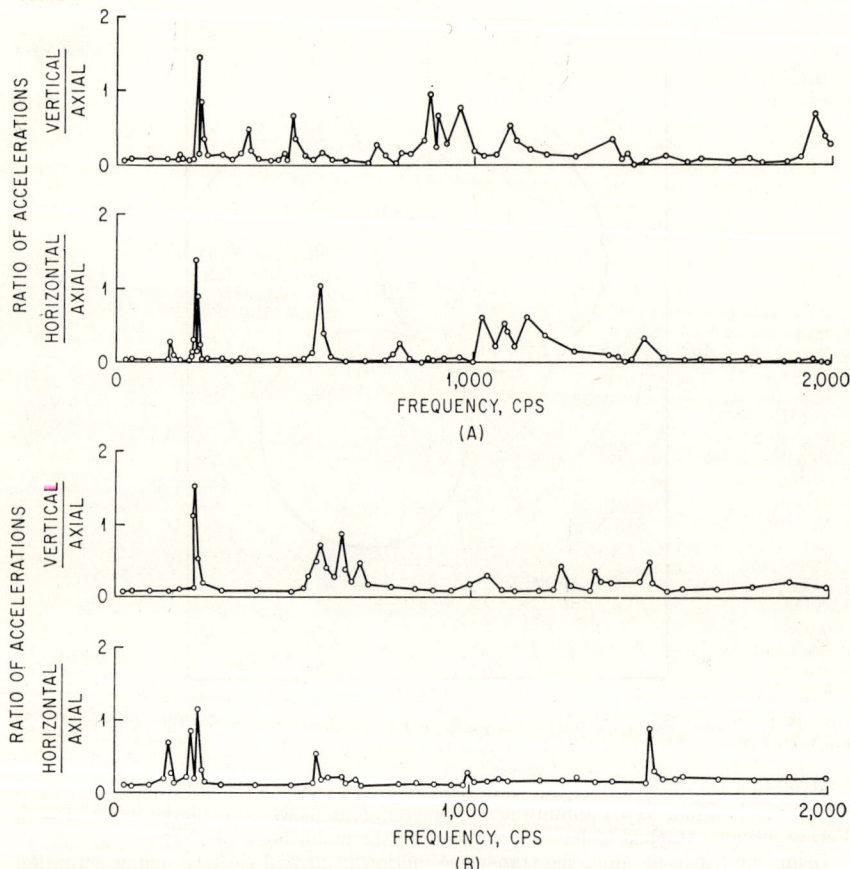

Fig. 18.50. (A) Horizontal and vertical transverse motions of a typical electrodynamic calibrator used as a "standard" in calibrating pickups. (B) Horizontal and vertical transverse motions of a similar calibrator. (After R. R. Bouche.[14])

siderable transverse motion. The continuous indication of transverse motion of the order of 0.1 (10 per cent) or less in Fig. 18.50A and B should not be ascribed to transverse motion in the standard since this is not significantly greater than the transverse sensitivity of the accelerometers themselves.

Theoretical computations of the bending-mode natural frequencies of the flexure-plate springs and the approximate fundamental bending-mode natural frequency of the shaft and coils indicate that the transverse motion at 230 cps and near 600 cps is due to resonant response in the flexure-plate springs and shaft, respectively. The transverse motions at a few other frequencies correspond closely to higher natural modes of vibration in the flexure-plate springs. A substantial reduction in transverse motion can be achieved by replacing the flexure-plate springs by tri-wire supports.[14]

MEASUREMENT OF TRANSVERSE (CROSS-AXIS) SENSITIVITY 18-39

A low-frequency shaker,[31] having provision to rotate the accelerometer under test while the calibrator is in motion, may be used to give a direct polar plot of transverse sensitivity. At frequencies well below any resonance of the pickup this method of determining transverse sensitivity gives good results.

REFERENCES

1. Perls, T. A.: "Temperature and Other Extraneous Effects of Piezoelectric Vibration Pickups," *Proc. 26th Shock and Vibration Symposium*, May, 1953.
2. Bruel and Kjaer: *Bulls.* 4308 and 4309, Naerum, Denmark, January, 1958.
3. Fig. 16.19.
4. American Standard Methods for the Calibration of Shock and Vibration Pickups, S 2.2-1960, American Standards Association.
5. Levy, H.: "Description of Pendulum Accelerometer Calibrator," *Naval Air Mat. Cen. (Philadelphia) Rept.* AML. NAM 2425, part V, Aug. 25, 1945.
6. White, G. E., and S. Kerstner: "Basic Methods for Accelerometer Calibration," *Statham Lab. Instr. Notes* Nos. 17 and 18, October, 1950.
7. Wildback, W. A., and R. O. Smith: *Proc. Instr. Soc. Amer.*, vol. 9, part 5, Paper 54-40-3, 1954.
8. Levy, S., A. E. McPherson, and E. V. Hobbs: Calibration of Accelerometers, *J. Research Natl. Bur. Standards*, **41**(5):359 (1948).
9. Kissinger, C. W.: *Proc. Instr. Soc. Amer.*, vol. 10, Paper 54-40-1, 1955.
10. Bruel and Kjaer: Instruction and Application No. 1606, January, 1958.
11. Edelman, S., E. Jones, and E. R. Smith: *J. Acoust. Soc. Amer.*, **27**(4):728 (1955).
12. Unholtz, K.: "The Calibration of Vibration Pickups to 2,000 Cps," *MB Mfg. Co. Bull.* 112M6, 1956.
13. Fig. 19.53.
14. Bouche, R. R.: "Improved Standard for the Calibration of Vibration Pickups," *Proc. SESA Paper* 570, May, 1959.
15. Trent, H. M.: *J. Appl. Mechanics*, **15**:49 (1948).
16. Levy, S., and R. R. Bouche: *J. Research Natl. Bur. Standards*, **57**:227 (1956).
17. Instruction Manual: Glennite Crystal Vibrator, Gulton Industries, Inc., December, 1957.
18. Instruction Manual: Glennite Interferometer, Gulton Industries, Inc., November, 1957.
19. Orlacchio, A. W.: *Elec. Manufacturing*, January, 1957, p. 78.
20. Smith, E. R., S. Edelman, E. Jones, and V. A. Schmidt: *J. Acoust. Soc. Amer.*, **30**:867 (1958).
21. Harrison, M., A. O. Sykes, and P. G. Marcotte: "The Reciprocity Calibration of Piezoelectric Accelerometers," *David W. Taylor Model Basin Rept.* 811, March, 1952.
22. Perls, T. A., and C. W. Kissinger: *Proc. Instr. Soc. Amer.*, vol. 9, Paper 54-40-2, 1954.
23. Conrad, R. W., and I. Vigness: *Proc. Instr. Soc. Amer.*, vol. 8, Paper 53-11-3, 1953.
24. Perls, T. A., C. W. Kissinger, and D. R. Paquette: *Bull. Am. Phys. Soc.*, **30**(3):34 (1955).
25. Tyzzer, F. G., and H. C. Hardy: *J. Acoust. Soc. Amer.*, **22**:454 (1950).
26. Feder, E. I., and A. M. Gillen: *IRE Trans. on Instr.*, 1–6 (2), June, 1957.
27. Helfand, B. B., and J. Burns: "Calibration of Resistance-bridge Transducer Circuits under Temperature Extremes," *Statham Lab. Instr. Notes* No. 14.
28. Rule, E., T. A. Perls, and F. J. Sullentrap: "Pickup to Read-out Calibration of Vibration Channels," Lockheed Aircraft Corp., LSMD, August, 1958.
29. Elliot, W. R.: *Proc. Instr. Soc. Amer.*, vol. 10, part 2, Paper 55-21-1, 1955.
30. *Gulton Industries, Inc., Bull.* A500, 1955.
31. Endevco Series 2200 Accelerometers Manual, 1959, p. 12.

19

SHOCK AND VIBRATION INSTRUMENTATION

Robert W. Conrad
National Aeronautics and Space Administration

INTRODUCTION

This chapter describes the equipment, instruments, and procedures employed in shock and vibration tests, from the transducer output up to the finished record of the test.

The material divides logically into sections equivalent to the component elements of the over-all instrumentation system. Signal transmission cables are considered first. Included in this section are the effects of cable parameters on system response for different transducers; sources of noise and interference; and remedial techniques for eliminating their consequences. The next section describes operational electronic circuits which prepare the signal for recording. A very elementary summary of vacuum-tube and transistor electronics is included as a convenience for the nonelectronic reader who may be forced by necessity to employ electronic circuits in his work. Brief descriptions of the circuits normally encountered in shock and vibration instrumentation setups also are presented here. A section on filter characteristics follows. In the next section the construction, arrangements, characteristics, and performance of recorders are given. Recorders considered are pen recorders, magnetic oscillographs, cathode-ray oscillographs, and magnetic-tape recorders. The final section provides an elementary introduction to telemetry with particular emphasis on those methods which have been standardized.

SIGNAL TRANSMISSION CABLES

The electrical cable connecting a transducer to the electronic components with which it is associated is a very important element of the over-all measurement system. Cable types and applications range from well-shielded coaxial cables employed over the long distances encountered in some types of field tests to short lengths of unshielded twisted-pair wires which are widely used in strain-gage measurements in the laboratory. Choice of a suitable cable depends on the particular application, transducer, cable length, and environmental conditions.

Cables normally are considered passive elements. However, under certain conditions of use and environment, spurious signals may be induced in, or generated by, the connecting cables. The signals at the receiving end of the line then contain components that were not initially present in the transducer output. The most common cause of interference is electrical "pickup" from strong electromagnetic fields associated with nearby power lines, electrical machinery, switching transients, or radar and commercial radio transmitters. "Ground loops," i.e., circulating ground currents, also can produce spurious signals very similar in appearance to electrical pickup, but by an entirely different mechanism. In addition to these external sources of interference, untreated cables operated at high-impedance levels are prone to exhibit a self-generated type of "cable noise" when mechanically agitated or distorted.

These departures from the normal passive nature of signal transmission cables play an important part in selecting a suitable cable for a particular application, and affect the quality and reliability of the recorded data. The following sections discuss the circumstances under which these cable effects arise, and describe practical measures and techniques which negate their consequences.

TRANSMISSION-LINE CHARACTERISTICS

CABLE PARAMETERS. Characteristics of cables and transmission lines are determined by their fundamental physical properties.* For the idealized transmission line of Fig. 19.1 having distributed parameters, the series impedance Z and shunt admittance Y per unit length of line are given by

$$Z = R + j\omega L \quad \text{ohms/unit length} \quad (19.1a)$$

$$Y = G + j\omega C \quad \text{mhos/unit length} \quad (19.1b)$$

where R = series resistance, ohms/unit length
L = series inductance, henrys/unit length
G = shunt conductance, mhos/unit length
C = shunt capacitance, farads/unit length

As a circuit element, the electrical properties of a cable are more conveniently expressed in terms of its *characteristic* or *surge* impedance Z_0 and its *propagation constant* γ. These, derived from Eq. (19.1), are defined as

$$Z_0 = \left(\frac{Z}{Y}\right)^{1/2} \quad \text{ohms} \quad (19.2)$$

$$\gamma = (ZY)^{1/2} = \alpha + j\beta \quad (19.3)$$

where α = attenuation constant, nepers
β = phase constant, radians

In the general case, both Z_0 and γ are complex quantities. However, on high-quality low-loss cables or at radio frequencies where the reactive components dominate the expression of Eqs. (19.1), the characteristic impedance becomes resistive and is independent of frequency, i.e.,

$$Z_0 = (L/C)^{1/2} \quad \text{ohms} \quad (19.4)$$

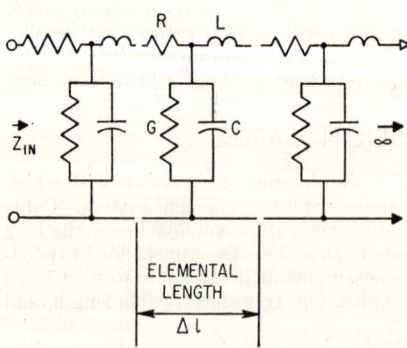

FIG. 19.1. Equivalent circuit of an infinitely long transmission line with distributed constants.

This is the impedance seen looking into the input terminals of an infinitely long cable. For finite lengths the input impedance depends on the value of terminating impedance relative to the characteristic impedance. When a cable is terminated by a resistance equal to its characteristic impedance, its input impedance also is equal to Z_0, and all the energy in the incident wave will be absorbed in the terminating resistance. For all other values of terminating impedance, the receiving end represents a discontinuity which is respon-

* The material presented here is restricted to elementary fundamentals of transmission-line theory. It is developed only to the extent required to cover most shock and vibration applications. The reader desiring more comprehensive treatment of transmission-line theory is directed to one of the numerous texts on the subject.[1-3]

sible for propagating a reflected wave back to the sending end. Standing-wave patterns then appear on the line, the amplitude of which depend on the amount of mismatch and line losses.

The propagation constant γ contains a real part α, the *attenuation constant*, and an imaginary part β, the *phase constant*. These constants represent the rate of loss in amplitude and the rate of phase variation, respectively, as the wave is propagated down the line.

Nominal values of characteristic impedance range from about 35 to 150 ohms for coaxial cables, and considerably higher for open or twisted pair. The attenuation and phase constants are frequency-dependent, and they become quite small for frequencies below 1 megacycle.

Cable parameters and characteristics, outlined above, provide the basis of transmission-line theory, which is usually concerned with circumstances where the cable is many wavelengths long. Voice frequencies over long-distance telephone lines, and television or other UHF frequencies over short cables are typical applications. In shock and vibration work, cable lengths seldom exceed a small fraction of a wavelength for the highest frequencies of interest, even with the long cable lengths required with some field tests. Except for a few special cases, general transmission-line theory is not applicable. The following sections deal with the physical properties of cables as they affect the performance of transducers and circuits normally encountered in shock and vibration applications.

LOW-FREQUENCY CHARACTERISTICS OF SHORT CABLES

CABLE CONSTANTS. When operated in the audio-frequency range the series inductance L and the shunt leakage G of good-quality, short cables are negligibly small in comparison with the other parameters and may be neglected. Figure 19.2A shows the equivalent low-frequency representation of a cable with distributed constants. For

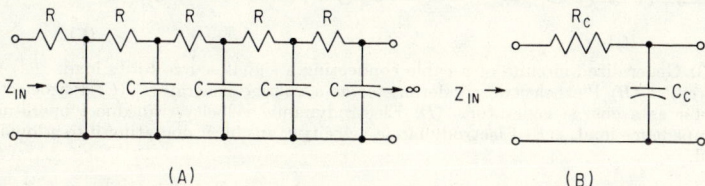

FIG. 19.2. Successive approximations in the representation of a short high-quality transmission line at audio frequencies. (A) Distributed constant configuration, neglecting series inductance and shunt leakage. (B) Lumped-constant configuration.

most purposes in shock and vibration applications, the simpler lumped-constant arrangement of Fig. 19.2B is sufficiently accurate. The quantities R_c and C_c are now the total ohmic resistance of the conductors and total capacity between them, respectively. Values for a typical coaxial cable, such as RG-58A/U,* are $R_c = 0.01$ ohm/ft, $C_c = 29$ mmfd/ft. The nominal characteristic impedance of 50 ohms for this cable has no significance in these short low-frequency applications. The open-circuit input impedance of the cable is almost exclusively capacitive. When terminated, it takes on the impedance of the load, modified by the series and shunt parameters.

When used as a circuit element to connect a transducer to its associated electronic equipment, or to couple two pieces of equipment together, the capacity of the cable has, by far, the most detrimental effects on system performance. The extent of this shunt loading depends on the impedance levels involved in relation to the magnitude of the cable capacity, and the maximum frequencies of interest. The presence of the cable

---

* Coaxial cables of this type are manufactured by numerous companies, among which are Amphenol Cable and Wire Div., Tensolite Insulated Wire Co., Inc.

series resistance can be neglected except when it becomes comparable to either the source or the terminating impedance.

TRANSDUCER CABLES. A generalized circuit consisting of a cable connecting a signal source to its load is given by Fig. 19.3A. The generator may be a transducer or some preceding electronic stage having an open-circuit output voltage e_1. Its internal impedance Z_1 may be resistive or reactive. The load R_L is usually resistive, i.e., the input of an electronic circuit or a galvanometer. Typical applications are given below.

Piezoelectric Accelerometer Cables. As indicated in Chap. 16, piezoelectric accelerometers are charge generators with a very high (capacitive) internal impedance. Typical values of capacitance for barium-titanate accelerometers range around 500

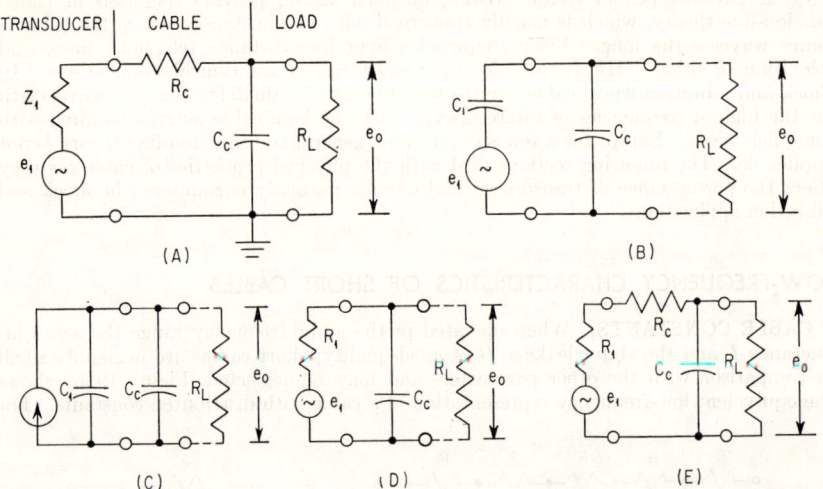

FIG. 19.3. Generalized circuits of a cable connecting a signal source to its load. (A) Generalized circuit. (B) Piezoelectric accelerometer as a voltage generator. (C) Piezoelectric accelerometer as a charge generator. (D) Electrodynamic velocity transducer operating into a high-impedance load. (E) Electrodynamic velocity transducer operating into a low-impedance load.

mmfd. Such pickups generally operate into a cathode follower whose input impedance is so high that the loading effect of the input impedance may be neglected. The shunt cable capacity C_c then forms a capacitive voltage divider with the internal accelerometer impedance C_1 (Fig. 19.3B) to yield an output voltage

$$e_0 = \frac{C_1}{C_1 + C_c} e_1$$

Since the open-circuit voltage depends on the accelerometer's charge sensitivity according to

$$\frac{e_1}{g} = \frac{Q/g}{C_1} \tag{19.5}$$

a substitution can be made yielding

$$\frac{e_0}{g} = \frac{Q/g}{C_1 + C_c} \tag{19.6}$$

for which the revised circuit diagram of Fig. 19.3C applies.

Equation (19.6) shows that the *voltage sensitivity* of a piezoelectric accelerometer is inversely proportional to the total shunt capacitance across its terminals. For this reason the accelerometer's sensitivity must be recalculated each time its cable length is changed. This method of adding shunt capacitors is sometimes used at the amplifier input to trim the voltage sensitivity to a particular value.

Cable loading effects on piezoelectric accelerometers are quite large. For example, the voltage sensitivity of a typical small barium-titanate accelerometer having an internal impedance of 500 mmfd, including a short length of cable, is reduced to two-thirds of its original value by the addition of only 8 ft of RG-58A/U cable. Because of these extreme variations, voltage-sensitivity figures for piezoelectric accelerometers are only approximate unless exact loading conditions are specified.

Velocity Transducer Cables. Small velocity pickups usually are of the electrodynamic type and have a resistive internal impedance R_1 of about 600 to 1,000 ohms. Usually they are connected to a cathode-ray oscillograph or other recorder whose input impedance R_L is several orders of magnitude greater than their own internal impedance. Output loading, if any, then will be due principally to the cable shunt capacitance C_c.

Figure 19.3D shows a circuit representing an electrodynamic velocity pickup operating into a high-impedance load. Neglecting R_L and the series cable resistance R_c, the ratio of output voltage to open-circuit voltage is

$$\frac{e_0}{e_1} = \frac{Z_c}{Z_1 + Z_c} = \frac{1}{1 + \omega R_1 C_c} \tag{19.7}$$

which indicates that the error increases with capacitive loading and frequency. For example, if typical conditions are assumed:

$$R_1 = 1{,}000 \text{ ohms}$$
$$f_{max} = 2{,}000 \text{ cps}$$
$$\frac{e_0}{e_1} = 0.95 \text{ (5 per cent error)}$$

the capacitance C_c can be as great as 4,200 mmfd, equivalent to approximately 137 ft of RG-58A/U cable. The relatively long cable length permitted by these transducers, due to their low internal impedance, is the principal reason for their popularity in field measurements. No corrections need be made for normal laboratory lengths of connecting cable.

Occasionally a very low-impedance velocity pickup is operated directly into a magnetic oscillograph galvanometer or is terminated by a low-valued resistor to provide proper damping. In these instances, Fig. 19.3E, the cable series resistance R_c and connector contact resistances become more important than the shunt capacitance C_c. For example, if $R_1 = 0.5$ ohm and $R_L = 25$ ohms, the output voltage

$$\frac{e_0}{e_1} = \frac{R_L}{R_1 + R_c + R_L} \tag{19.8}$$

is 5 per cent down when the cable resistance R_c equals 0.8 ohm, amounting to approximately 80 ft of RG-58A/U cable.

Cables for Other Transducers. Cable effects for strain gages, potentiometers, and most other types of transducers, having relatively low-resistive internal impedances, are generally intermediate to the two cases cited above for velocity transducers. Transducer impedance, load impedance, and cable characteristics will determine which parameters are important and which can be neglected. In multichannel arrangements, phase shift and time delays caused by the reactive components of the input network and cable should be considered in establishing a common time base for different transducers.

INTERFERENCE AND NOISE IN TRANSMISSION CABLES

In an electrical sense, any unintentional or undesirable component of an electrical signal is described as "noise." Noise may be produced by many causes, but the types considered here are restricted to those which arise out of the use of connecting cables.

ELECTRICAL INTERFERENCE. The most common and troublesome sources of noise in instrumentation systems result from electrical interference or "pickup." These are noise components superimposed on the desired signal due to the proximity of the connecting cable to the electromagnetic field of an electrical disturbance. Hum from power lines, pickup from radio-frequency transmitters, and "static-type" disturbances from switching transients are the worst offenders. In general, precautionary procedures against electrical interference take the form of short cables, shielding, grounding, and, where permissible, the lowering of the line impedance.

Shielding. Shielding is accomplished by completely surrounding a susceptible circuit by a conductive surface which keeps the enclosed space free of external fields. Requirements for magnetic and electrostatic shielding are somewhat different. *Magnetic shields* are effective partially due to short circuiting of magnetic flux lines by low-reluctance paths and partially from self-annihilation due to opposing fields set up by eddy currents. Accordingly, they are made from high-permeability materials such as Permalloy, and are as thick as possible and contain a minimum of joints, holes, etc. A good magnetic shield is also a good electrostatic shield, but the converse is not true. *Electrostatic shields* provide a conducting surface for the termination of electrostatic flux lines, but need not be magnetic material. Stranded braid, mesh, and screens of good electrical conductors such as aluminum or copper are good electrostatic shields. Most shielded cables use copper braid as the outer conductor and shield.

Grounds. A circuit is *grounded* when one terminal is connected to the "earth" by a low-impedance path. Grounding removes the potential difference between that side of the circuit and earth, and eliminates the variable stray capacitances which tend to induce voltages in "floating" (i.e., ungrounded) systems. A good ground has the ability to absorb electrons without change in potential. In the laboratory, water and steam pipes make good ground connections because of their intimate contact with the earth. Metal building frames and power cable armor may be less satisfactory because of poorer bonding at junctions. A metal pipe driven several feet into moist soil makes a good ground in the field. At sea, the ship's metal hull and piping make good grounds.

Line Impedance. The magnitude of interfering components is proportional to cable impedance. High-impedance lines are more susceptible to undesirable pickup than low-impedance lines. Cable impedances therefore should be kept as low as time constants and loading effects will permit. If the interfering signals are considerably higher in frequency than the desired components, as is usually the case with radio-frequency transmitter interference, a shunt capacitor across the line will serve to short out the high-frequency pickup. The capacitor should be chosen small enough not to attenuate the highest frequencies desired, but large enough to be an effective short at the interfering frequency. Shunt capacitors are also useful for removing sharp spikes associated with interference from switching transients.

INTERFERENCE-SUPPRESSION TECHNIQUES. Double Shielding. Single-shielded cable is sufficient under ordinary operating conditions to keep electrical pickup levels to an acceptable minimum. For those installations where the cable lengths are long, where the impedance levels are high, or the interference is overly objectionable, a double-shielded cable is available. In this type cable, a second shielding braid is woven over the cable jacket, electrically insulated from the inner shield. The inner braid furnishes additional shielding against electrostatic fields which penetrate the first shield. For best results the shields should be connected together and grounded at only one point, preferable at the input to the associated electronic equipment.

Magnetic Fields. Magnetic fields associated with current-carrying power lines, electronic equipment power transformers, and line regulators are the most troublesome sources of magnetic interference in instrumentation setups. Interference is chiefly audio "hum" at the power-line frequency and its harmonics. Fields from power transformers produce a readily recognizable waveform containing a large third-harmonic component. At power-line frequencies, cable shielding provides rather ineffective protection. Since these fields attenuate rapidly away from the offending source, the most practical solution for this type of interference is to keep signal cables as far removed as possible. Occasionally, movable equipment can be reoriented to produce a null in induced voltage in the affected cable.

Ground Loops. Ground loops are formed when a common connection in a system is grounded at more than one point, Fig. 19.4A. If circulating ground currents cause a potential difference e_{gnd} to develop between these grounding points, the normal input signal will be modulated by hum and pickup from such a potential difference. The voltage seen by the input amplifier is then the sum of the normal input signal e_1 in series with the hum voltage e_{gnd} as given by the equivalent circuit, Fig. 19.4B. It is not uncommon for ground potential differences of several volts to exist between grounded electrical items only a few feet apart.

The principal source of ground loops in instrumentation setups results from the established practice of connecting the "low" side of transducers and electronic signal circuits to the instrument case. When several such items in a system are remotely located, Fig. 19.5A, and each housing is connected to the common signal lead and to a

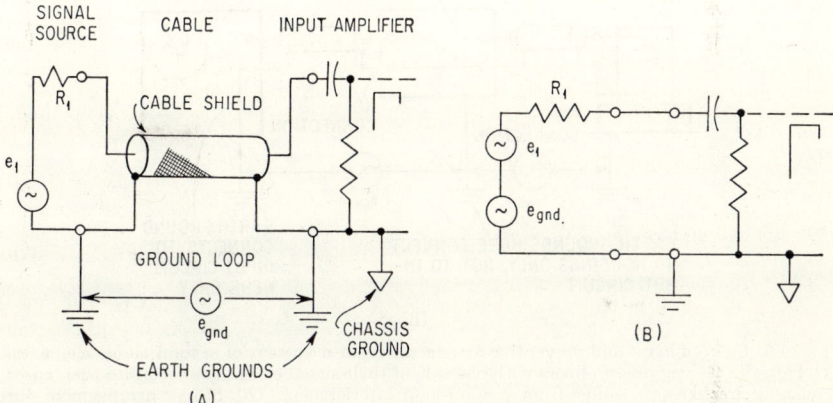

Fig. 19.4. Ground loops. (A) Multiple grounds in a system form ground loops which can cause interference if a ground potential difference exists. (B) Equivalent circuit of a system with a ground loop.

local earth ground, circulating ground currents can flow in the signal circuit if ground potential differences exist. The situation is further complicated if a number of independent data channels culminate in a single, multichannel recorder. Common lead grounding generally does not cause trouble if the separate equipments are physically close together or otherwise situated so that ground potential differences do not occur.

Ground-loop interference can be eliminated only by avoiding multiple grounds on the signal circuit; *the circuit should be grounded at only one point.* When it is possible to disconnect the low-signal side from the various housings, the arrangement of Fig. 19.5B is satisfactory. Although the case of each unit is connected to its own local ground, the signal circuit is grounded only at the recorder input and the previous ground loops are open-circuited. When single-conductor shielded cable is used with this arrangement, care must be taken to ensure that cable and connector shields do not inadvertently come into contact with grounded objects.

Some equipment, for example small piezoelectric accelerometers, Fig. 19.6A, are constructed so that the case is irrevocably part of the signal circuit. Ground loops can be broken with this type equipment by electrically insulating the housing from ground, as in Fig. 19.6B. Since potential differences are seldom over several volts, electrical breakdown is not a problem, and virtually any nonconductor will suffice. For transducers, the insulating material should be quite stiff so as not to filter out the higher-frequency components of the motion being measured. Dental cement is quite useful for this purpose; not only is it a rigid insulator but it sets up quickly and eliminates drilling holes in the test structure. Preamplifiers can be wrapped in electrical tape or merely placed on a pad of paper or cardboard, if they need not be tied down. Sponge-rubber pads also reduce microphonics from structure-borne vibration.

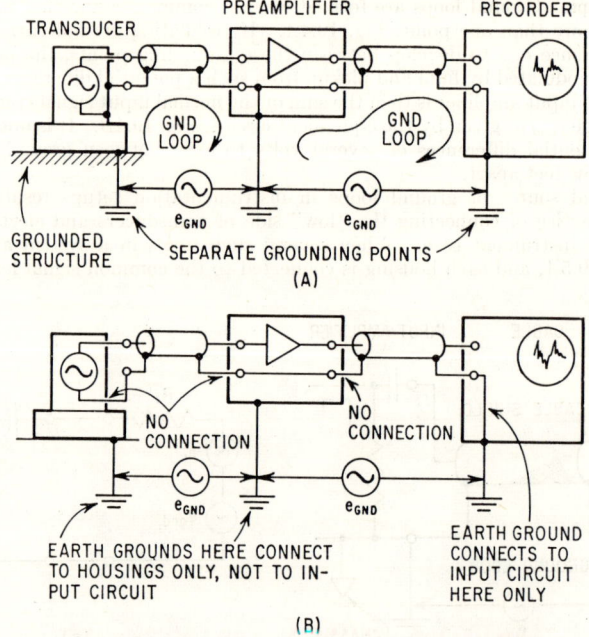

Fig. 19.5. Ground loops and preventive arrangements in a system of several series elements. (A) Transducers and preamplifiers with one side of their signal circuit connected to their cases or chassis are likely to suffer from ground-loop interference. (B) Proper arrangement for isolating the low side from individual chassis to open-circuit ground-loop potentials.

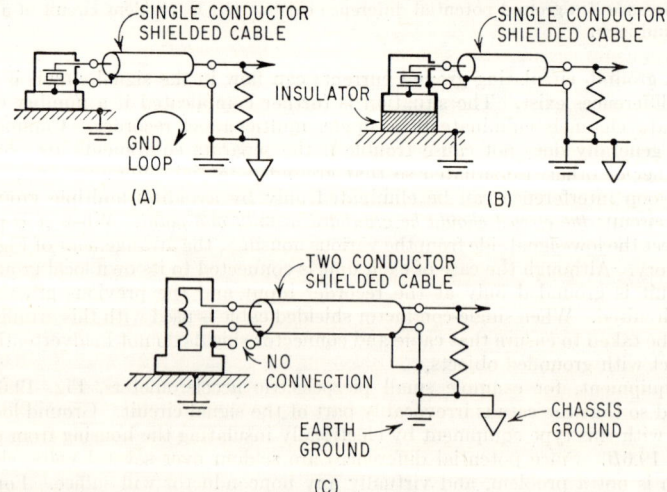

Fig. 19.6. Transducer cable connections for eliminating ground loops. (A) Many small piezoelectric accelerometers use the transducer case as the ground return and form ground loops when mounted on grounded structures. (B) Ground loops can be broken on grounded transducers by insulating the case of the transducer from ground. (C) Proper connections for transducers with floating terminals. Shield does not carry signal currents and is grounded at equipment input only.

SIGNAL TRANSMISSION CABLES   19–9

Two-conductor shielded cables should preferably be used when the signal circuit can be completely isolated from the transducer case, Fig. 19.6C. This relieves the shield of the responsibility for one side of the signal circuit and permits both sides of the line to be shielded. Normally the shield is connected to ground at one end only. However, little interference would result if it were accidentally grounded at a second place since it is not part of the signal circuit.

In all these circuits, care should be taken to maintain complete electrostatic shielding of the cable wires against radiated interference. With multiconductor cables this can be done, while preventing ground-loop currents through the shield, by connecting the shield to the connector shell at one end only. At the other end the shield is carried into the connector, but electrically insulated from it.

The most satisfactory place to make the system ground is at the recorder input. This is the only location in multichannel circuits that will not cause ground loops between

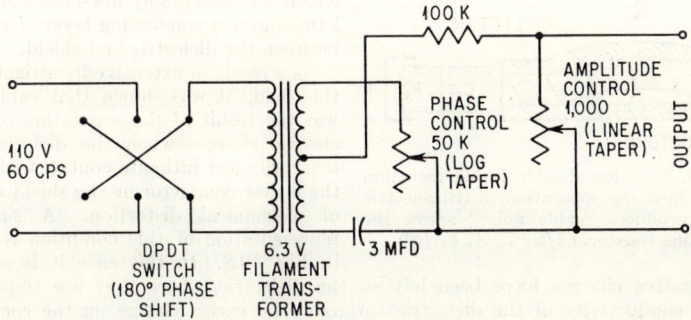

FIG. 19.7. Hum-compensating circuit with amplitude and phase controls.

channels. Grounding the recorder, which is frequently a-c operated, shorts out the large stray capacity to the power-line ground and greatly reduces system noise and hum. It also reduces the danger of shock to operating personnel who must make adjustments to the equipment. Preamplifiers ahead of the recorder are often battery-operated and remote, and may be left "floating" without introducing appreciable capacitive hum pickup or becoming a shock hazard.

In some special cases, the system ground must be made at a point other than at the recorder input—for example, when measuring the line current across a shunt with an unbalanced input circuit. Operator safety dictates that the shunt should be in the line with the lowest potential to ground. The shunt then becomes the ground point of the entire system, and may make the recorder "hot" (i.e., at a high potential) with respect to other grounded objects. Consequently, it should be ungrounded and insulated. If it is part of a multichannel system it is a good precaution to fuse the line between shunt and recorder low side in the event an accidental earth ground develops on any other part of the system. With the capacity of the generator available, short-circuit current can be very damaging to electronic circuit parts and wiring.

Hum Compensators. Ungrounded transducers, or those not readily adapted to shielding, pickup, and leakage within a-c operated equipment, are common sources of hum interference which are difficult or inconvenient to eliminate. When the residual noise in a system cannot be reduced to an acceptable level by careful application of interference-suppression techniques, and consists mostly of hum at the power-line frequency, considerable improvement in signal-to-noise ratio can be obtained with *hum compensators* or *hum buckers*. These networks introduce into the circuit an additional signal derived from the power line, whose amplitude and phase can be adjusted to exactly cancel the residual hum interference. A typical circuit is shown in Fig. 19.7. The circuit permits independent adjustment of amplitude and phase. A filament transformer supplies low-voltage power to the phase-shift network and isolates it from the power-line ground. The hum compensator can be connected either in series or parallel with the signal, depending on the components selected, local d-c levels, and impedance at the point of injection.

TRIBOELECTRIC CABLE NOISE. One of the early difficulties encountered with accelerometer systems in shock work was a spurious "cable noise" generated by the cable whenever it was suddenly squeezed, bent, struck, or mechanically distorted. Peak noise voltages from this source were frequently as large as the actual acceleration signals being recorded. Little was then known of the mechanisms which produced cable noise, except that it was found experimentally that cables of certain type and manufacture were better than others and that good cables deteriorated rather rapidly in shock test use. One homemade design, consisting of a center conductor encased in $\frac{1}{4}$-in. copper tubing and then filled with beeswax, proved far superior to any of the then available commercial microphone or coaxial cables. Its principal disadvantage was its rigidity. The same objection pertained to a thick underwater cable used somewhat later, but which had acceptably low-noise characteristics due to a conducting layer of graphite between the dielectric and shield.

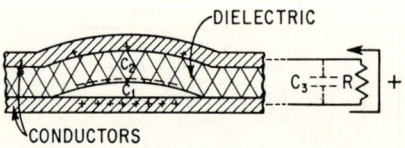

Fig. 19.8. A section of cable during distortion showing how the separation of triboelectric charges produces "cable noise" across the terminating resistor. (*After T. A. Perls.*[4])

As a result of extensive investigations in this field,[4] it was shown that cable noise was the result of the separation of triboelectric charges when the dielectric momentarily lost intimate contact with either the center conductor or the shield because of mechanical distortion. A simplified representation of this condition is shown in Fig. 19.8. In this case it is assumed that negative charges have been left on the dielectric, where they are trapped by the low conductivity of the dielectric surface. The excess charge on the conductor, being free to move, is neutralized by a transfer of electrons between conductors through the terminating impedance at a rate determined by the time constant. The transfer produces a voltage pulse on the input of the following electronic equipment. On reestablishment of dielectric-conductor contact, the excess negative charge on the conductor-dielectric interface is again neutralized by charge redistribution, resulting in a pulse of the opposite polarity. Peak amplitudes of the voltages developed depend on the affected area, charge separation, total cable capacity, and the ratio of the two local capacities formed by separation.

From the above analysis, it is apparent that flexible low-noise cables can be produced if the dielectric surfaces are covered by a conducting coating. The conductivity of the coating need not be high, just as long as it provides a leakage path on the dielectric surface for charges which would otherwise be immobilized there during mechanical separation. The charges then are redistributed through a local path rather than through the terminating impedance where they are indistinguishable from accelerometer signals.

Miniaturized, flexible, low-noise cables using this coating technique are available from several manufacturers.* These treated cables exhibit very low cable-noise characteristics and are generally capable of withstanding considerable abuse before the coatings develop enough holes to become noisy. In this respect it is important in assembling these cables, as in splicing or fitting them with connectors, that none of the conducting material be allowed to form a leakage path between conductors. Carbon tetrachloride and xylene are satisfactory solvents and cleaning agents.

VACUUM-TUBE AND TRANSISTOR ELECTRONICS

The study of vacuum tubes or transistors, and their behavior as circuit elements, is properly the subject of a separate text. However, their use is so universal in all fields of scientific endeavor that an understanding of basic electronic principles is a fundamental requirement in any area where measurements are to be made.

* Manufacturers include Microdot Div., Type 50-3804; Gulton Mfg. Co., Type Glennite C-5.

VACUUM-TUBE AND TRANSISTOR ELECTRONICS

Material in this section is restricted to the dynamic properties of the system, i.e., its performance with a small steady-state sinusoidal signal applied. Static properties, which set the operating point and no-signal conditions, generally are omitted except where they are vital to an understanding of the circuit. A brief review of vacuum-tube and transistor parameters and their equivalent circuits is given as an aid to circuit operations which follow. Circuits selected are representative of the types commonly encountered in instrumenting shock and vibration tests.

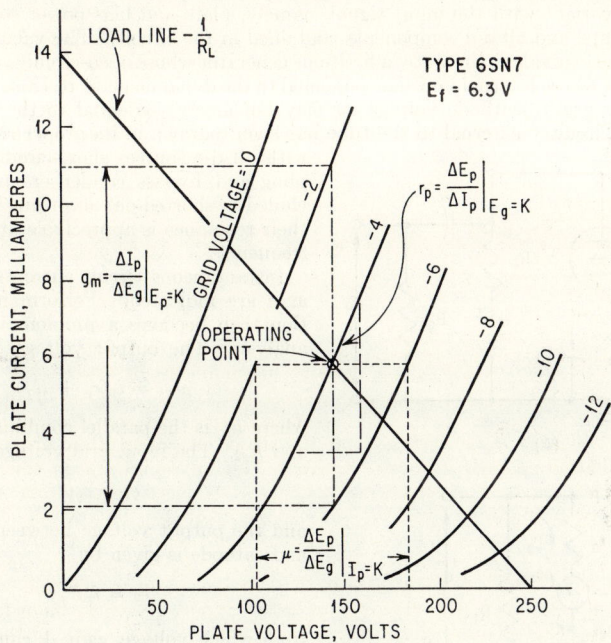

FIG. 19.9. Family of plate characteristics for a typical triode. Tube coefficients are evaluated at the operating point. Examples: $\mu = 80$ volts/4 volts $= 20$, $r_p = 25$ volts/4 milliamperes $= 6{,}250$ ohms, $g_m = 9$ milliamperes/4 volts $= 2{,}250$ μmhos. The plate current I_p is given as a function of plate voltage E_p for various values of grid voltage E_g.

VACUUM-TUBE CIRCUITS

VACUUM-TUBE COEFFICIENTS. The dynamic properties of a vacuum tube are determined by three vacuum-tube coefficients: They are represented graphically in Fig. 19.9.

Amplification factor:

$$\mu = -\frac{\Delta E_p}{\Delta E_g} \qquad I_p = \text{constant} \qquad (19.9)$$

Plate resistance:

$$r_p = \frac{\Delta E_p}{\Delta I_p} \qquad E_g = \text{constant} \qquad (19.10)$$

Mutual transconductance:

$$g_m = \frac{\Delta I_p}{\Delta E_g} \qquad E_p = \text{constant} \qquad (19.11)$$

At a particular operating point they are related by

$$\mu = g_m r_p \qquad (19.12)$$

Tube coefficients are not constants, but vary widely with plate current because of curvature in the plate-circuit characteristics. Under small-signal conditions, however, they may be considered constant in the region around the operating point, at the particular values determined by the static plate current.

EQUIVALENT CIRCUITS. So far as its performance as a circuit element is concerned, the behavior of a vacuum tube, for small input signals, may be determined from its equivalent circuit, Fig. 19.10. This type of presentation ignores all circuit features which are invariant with the input signal, namely, plate and bias power supplies, bypassed elements, and all d-c components contained in the circuit. The vacuum tube is replaced in the equivalent circuit by a fictitious generator whose open-circuit output voltage is μe_{gk} and whose internal impedance is equal to the dynamic plate resistance r_p. The instantaneous grid-to-cathode voltage e_{gk} may not always be equal to the input grid voltage e_1. Circuitry external to the tube has been redrawn in its proper relationship, with all d-c sources short-circuited. Coupling and bypass condensers may be included or shorted-out according to whether their reactance is appreciable at the signal frequency.

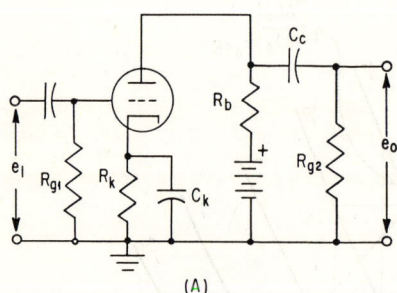

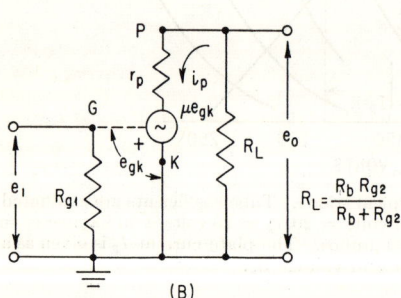

FIG. 19.10. R-C coupled triode amplifier stage and its equivalent circuit. (A) Complete circuit diagram. (B) Equivalent circuit. The total plate load is the parallel combination of R_b and R_{g2}. C_c and C_k are assumed to have negligible reactance at the operating frequency.

Instantaneous circuit currents and voltages are indicated. Performance evaluation then becomes a problem in network analysis. The output voltage is

$$\mu e_{gk} = i_p(r_p + R_L)$$

where R_L is the parallel combination of R_b and R_{g2}. The input grid voltage is

$$e_1 = e_{gk}$$

and the output voltage between the plate and cathode is given by

$$e_0 = i_p R_L$$

from which voltage gain A can be determined:

$$A = \frac{e_0}{e_1} = \frac{\mu R_L}{r_p + R_L} \quad (19.13)$$

Two cases are of special interest. When the plate resistance r_p is considerably smaller than the load resistance R_L, Fig. 19.11A, the output voltage becomes essentially

$$e_0 \simeq -\mu e_{gk} \quad (19.14)$$

and is relatively insensitive to changes in load resistance. The tube then appears to its load as a constant-voltage generator. Triodes, because of their low plate resistance, customarily are represented as constant-voltage generators in this manner.

On the other hand, when the load resistance is small compared to the plate resistance, plate current is dominated primarily by r_p and is virtually independent of R_L:

$$i_p \simeq \frac{\mu e_{gk}}{r_p} \quad [R_L \ll r_p] \quad (19.15)$$

This is the constant-current generator equivalent circuit exemplified by pentodes because of their extremely high plate resistance. The output voltage is

$$e_0 = i_p R_L \cong \frac{\mu e_{gk} R_L}{r_p} \cong g_m e_{gk} R_L \quad (19.16)$$

and is most conveniently represented by the constant-current equivalent circuit of Fig. 19.11B.

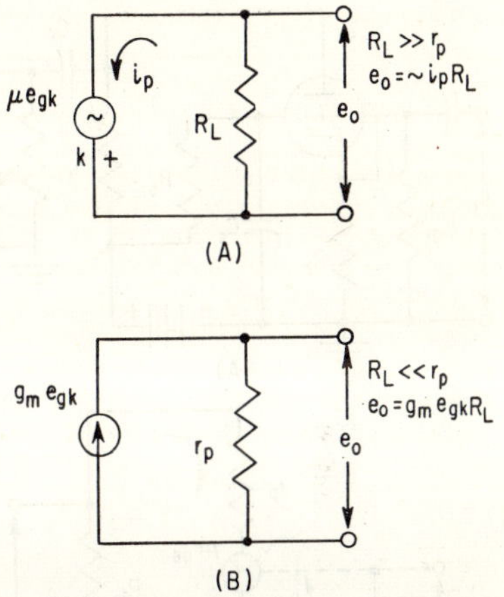

AMPLIFIERS

Amplifiers are employed to raise the voltage or power level, or both, available from a signal source. In most applications they are expected to perform this function without distortion, i.e., the output waveform should be a precise replica of the input in all its ramifications, but to a different amplitude scale.

Amplifiers are classified according to their principal function, type of coupling, special circuitry or characteristics, and type of amplifying element. Circuits are usually designed primarily as voltage or power amplifiers, although both objectives are frequently obtained simultaneously. Current amplifiers per se are rare; current-operated devices such as galvanometers and meters are generally driven from a power amplifier whose internal impedance is low enough to provide the required current.

FIG. 19.11. Simple vacuum-tube equivalent circuits. (A) Constant-voltage generator circuit for R_L greater than r_p. (B) Constant-current generator circuit for R_L smaller than r_p.

With respect to coupling, amplifiers are classified as a-c coupled if they pass only the alternating components of the signal and block the steady-state components, or as d-c coupled if their low-frequency response extends to d-c (zero frequency). Resistance-, impedance-, and transformer-coupled circuits are the usual types of a-c coupling. D-C amplifiers require careful design, maximum compensation, or special circuitry to be sufficiently free of drift and other instabilities. This is because any voltage change in the first stages caused by plate- or filament-voltage variations, warmup, or aging is amplified by following stages as if it were a valid input signal. Special circuits, such as carrier amplifiers and chopper-stabilized amplifiers, utilize various schemes to provide a d-c response with the high gain and stability inherent in a-c amplifier circuits. Another widely used circuit is the *cathode follower*, which is an amplifier having a high input impedance and a low output impedance. All these circuits are available in both vacuum-tube and transistor versions.

RESISTANCE-CAPACITANCE COUPLED AMPLIFIERS. This type of voltage amplifier, often abbreviated as *R-C amplifier*, is the most popular in use because of its simplicity and economy of size, weight, and cost of components used. When properly designed, it is capable of good stage gain, wide frequency response, and excellent stability.

R-C amplifiers, Figs. 19.10 and 19.12, are characterized by a resistive plate load R_b and an output coupling condenser C_c which passes the a-c component of the signal, but blocks the d-c component of plate voltage. The output voltage is 180° out of phase with the input. The value of C_c is chosen such that the time constant ($t = RC$) formed by C_c and the next stage load R_{g2} is large enough to pass the lowest frequency of interest without appreciable attenuation. The resistance R_b is selected as a compromise between high-stage gain (large R_b) and adequate high-frequency response (low R_b). Tables of component values for various performance criteria are given by tube manufacturers' handbooks.[5]

In the triode circuit of Fig. 19.10, grid bias is obtained by the positive voltage developed across the cathode resistor R_k due to plate current flowing through it. The cathode bypass condenser C_k is made large enough so that signal components of plate current do not appreciably alter the bias voltage for the lowest frequency to be passed. For

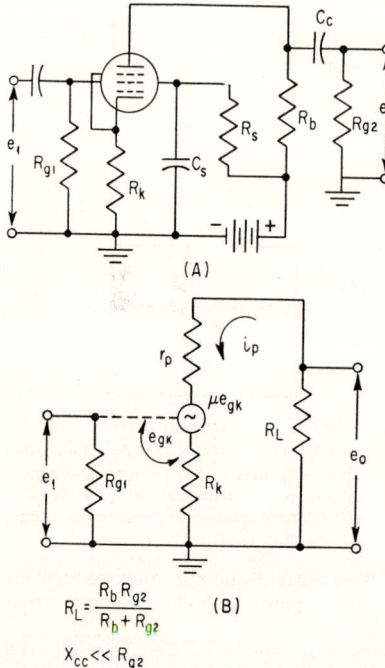

FIG. 19.12. *R-C* coupled pentode amplifier stage with cathode degeneration. (*A*) Complete circuit diagram. (*B*) Equivalent circuit. Note that the effective grid-to-cathode voltage is not the same as the input voltage e_1, because of the $i_p R_k$ voltage drop.

mid-range frequencies the equivalent circuit is then as given by Fig. 19.10*B*, and the stage gain can be calculated from Eq. (19.13):

$$A = \frac{e_0}{e_1} = \frac{\mu R_L}{r_p + R_L}$$

where R_L is the parallel combination of R_b and R_{g2}.

In the pentode circuit of Fig. 19.12, the screen is bypassed to prevent screen-current variations from altering operating conditions. For this illustrative circuit, the cathode resistor is purposely left unbypassed to stabilize the stage through cathode degeneration. This is one form of negative feedback. Performance is as follows: A positive signal on the control grid permits more plate current to flow, increasing the voltage drop across R_k, which raises the cathode voltage and thereby reduces the grid-to-cathode voltage that would have accrued if the cathode voltage had not changed. The plate-current change, and therefore the stage gain, are not as great as in the bypassed case, and represent a penalty which must be paid for the extra stability obtained. From the equivalent circuit, Fig. 19.12*B*,

$$\mu e_{gk} = i_p(r_p + R_L + R_k) \quad (19.17)$$

$$e_{gk} = e_1 - i_p R_k, \quad e_0 = i_p R_L, \quad \text{and}$$

$$A = \frac{e_0}{e_1} = \frac{\mu R_L}{r_p + R_L + (\mu + 1) R_k} \quad (19.18)$$

The gain A is considerably less than that given for the circuit without degeneration shown in Eq. (19.13), but the circuit is less dependent on variations of tube coefficients, as can be seen by dividing Eq. (19.18) through by μ:

$$A \simeq \frac{R_L}{1/g_m + R_L/\mu + R_k} \quad (19.19)$$

The effect of tube parameter variations is less pronounced due to the addition of R_k in the denominator.

Coupling-network Response. The resistance-capacitance interstage coupling network used in *R-C* amplifiers is largely responsible for the performance of the entire amplifier. Component values usually are chosen such that the capacitive reactance is negligible compared to the grid-leak resistance above a minimum frequency for which the equipment is designed. Over a wide range of frequencies, *R-C* networks exhibit uniform response with negligible attenuation or phase shift. Response falls off at the high end when tube input, output, and stray wiring capacities, in shunt across the stage output, become comparable to the load resistance and tend to lower the stage gain. With usual plate loads, gain begins to decrease in the neighborhood of 15,000 to 20,000 cps. This point is somewhat beyond the normal range of interest for shock and vibration applications and usually can be neglected in equipment of this type.

However, the low-frequency range of *R-C* networks is extremely important in these applications. As the frequency becomes lower, the coupling capacitor reactance increases, forming a voltage divider with the grid-leak resistance of the next stage. Low-frequency attenuation and phase-shift characteristics of $R\text{-}C$ networks are shown in Fig. 19.13.

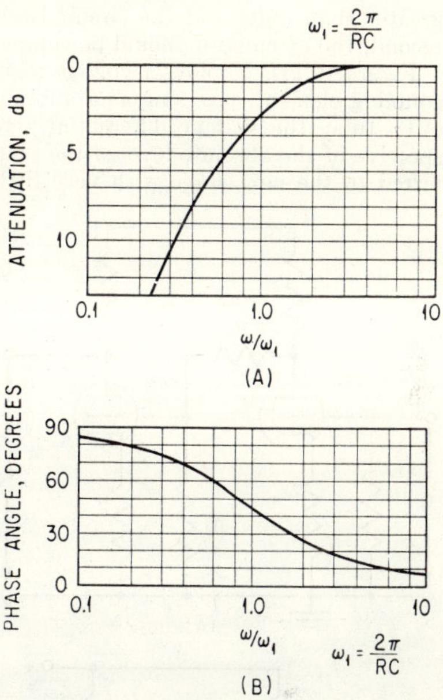

It is standard practice to specify the flat frequency-response range as the bandwidth between the 3-db down points on the amplifier-response curve. These are the lower and upper frequencies where the capacitive reactance and resistance of the network are equal (45° phase shift). This practice is satisfactory for performance evaluation if it is realized that -3 db (voltage) represents a 30 per cent loss of signal. When system errors are limited to 5 per cent (0.4 db) or less, the usable low-frequency limit of the equipment is approximately three times as high as that given by the -3-db point. For example, equipment rated as down 3 db at 10 cps can be used only to about 30 cps if errors of less than 5 per cent are required.

An alternative method of evaluating the low-frequency response limit of an $R\text{-}C$ network is from its time constant, given as the product of the resistance and capacitance. This figure is a mathematical convenience which gives the time (in seconds) required for the network output voltage to drop to $1/e$ of the initial value of a suddenly applied step change. With sinusoidal excitation the lowest frequency which can be passed with no more than 5 per cent amplitude error has a period

Fig. 19.13. Low-frequency attenuation and phase-shift characteristics of $R\text{-}C$ coupling networks. (A) Low-frequency attenuation characteristics. (B) Low-frequency phase-shift characteristics.

roughly twice that given by the time constant. For example, a coupling network comprising a 1-mfd capacitor and a 1-megohm resistor has a time constant of 1 sec, and can pass a 2-sec period (0.5 cps) sine wave with about 5 per cent amplitude error. With amplifiers of more than one stage, coupling errors are compounded and the minimum frequency correspondingly increased.

D-C AMPLIFIERS. D-C amplifiers have a low-frequency limit which extends down to zero frequency (d-c). This characteristic is essential for measuring potentials that change very slowly with time, or those that may contain both static and dynamic components where both must be preserved. D-C amplifiers are classified according to their circuit processes and may be of the following types: direct-coupled, carrier, or chopper stabilized.

Direct-coupled Amplifiers. Direct-coupled amplifiers employ normal amplifier stages similar to those discussed under $R\text{-}C$ amplifiers, except that the coupling between stages contains only resistive elements; reactive elements are employed only where they do not impair low-frequency performance. For this reason coupling and bypass capacitors cannot be allowed; peaking circuits which influence only the high-frequency response are permissible, however. Because the plate is at a higher d-c voltage than its grid, cascading several direct-coupled stages necessitates an extremely high plate-supply voltage to provide sufficient voltage for later stages. The high d-c level of the output is also disadvantageous. Several methods of direct interstage coupling that alleviate the power-supply problem are shown in Fig. 19.14.

In Fig. 19.14A, resistors R_1 and R_2 form a voltage divider to reduce the static plate voltage to a lower d-c level for the following grid. Unfortunately it also reduces signal voltages in the same proportion, so that the stage gain is reduced considerably. Additional stages must be used to obtain the required over-all gain. For this circuit to be useful, the product of the stage gain A and attenuation ratio $R_2/(R_1 + R_2)$ must be greater than unity. If the circuit is expected to operate above several kilocycles per second, the attenuator should be compensated (see *Cathode-ray Oscillographs*).

Figure 19.14B employs a voltage regulator tube of the gaseous-discharge type as the coupling element. As long as a minimum current flows through the voltage regulator (VR) tube, the voltage differential across it will remain substantially constant, irrespective of the absolute level of the plate voltage. Therefore signal voltages are transferred to the next stage with very little loss in level, whereas the d-c levels between stages are changed appreciably. This circuit has several disadvantages which limit its usefulness. Regulator tubes have a minimum current which must be supplied if constant voltage is to be maintained. This current is appreciable (generally 5 milliamperes or greater) and limits the size of R_b and R_{g2} to small values which result in poor stage gain. It also imposes extra drain on the power supply. In addition, gas tubes are inherently noisy, tend to change ionization potential spuriously, and otherwise give erratic performance.

Direct-coupled amplifiers usually are unstable and drift continuously because a variation of any sort which results in a change in plate voltage, especially in low-level stages, is amplified by succeeding stages as if it were part of the input signal. Design techniques for stabilizing direct-coupled amplifiers to the point of usability include:

1. Large amounts of negative feedback around the entire circuit to render insignificant any internally initiated variations.
2. Regulated plate and heater power supplies, the latter usually only on input stages.
3. Complementary push-pull design which tends to cancel out common-mode signals occurring in both halves simultaneously.
4. Special circuits which compensate automatically for tube coefficient variations with temperature, voltage, age, etc.

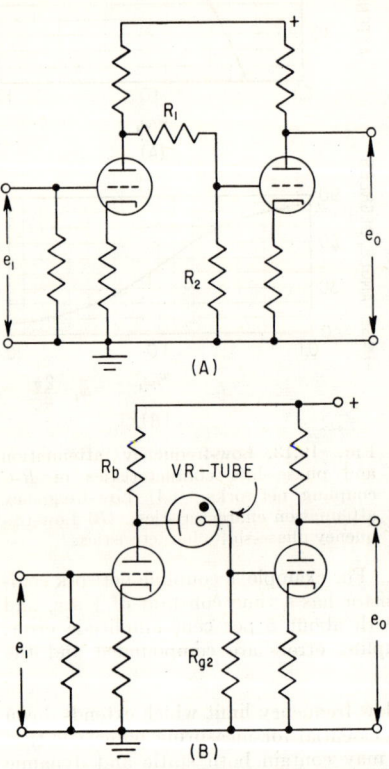

Fig. 19.14. Direct-coupled amplifier circuits. (A) D-C amplifier with a resistive voltage-divider coupling network. (B) D-C amplifier with gaseous-discharge tube for interstage coupling.

Carrier Amplifiers. A-C coupled amplifiers are relatively free of drift because the coupling capacitors between stages do not support slowly varying components. Carrier amplifiers, Fig. 19.15, make use of this property by having the d-c input signal linearly modulate a locally generated audio-frequency carrier. From this point on, the input signal rides on the carrier as a modulation component. Since the signal is constant in frequency and varies only in amplitude, a-c coupled amplifiers can be used to provide the necessary gain. A demodulator recovers the signal which is then thoroughly filtered to remove any trace of the carrier.

Direct modulation of the local carrier is practical only for signals having a voltage above about 100 millivolts because of the inevitable minor variations of conversion gain at the modulator which cannot be compensated by feedback. Compensation by feedback processes is not practical because of the difference in signal structure. Excellent low-level response is possible if the input transducer can be operated as part of a bridge circuit, Fig. 19.16. Differential transformers, strain gages, and strain-gage-type accelerometers are well suited to this technique. Excitation for the bridge is obtained from the local oscillator. Transducer variations alter the bridge output proportionally, without the need for a separate modulator. This procedure permits lower levels through improved stability. Since the bridge is usually balanced initially, transducer signals of either polarity will unbalance the bridge. Polarity of the input modulation can be resolved by a phase-sensitive demodulator, referenced to the local oscillator.

Because of sideband problems with modulated carriers, and the extensive smoothing required of the low-pass filter to remove carrier-frequency components, carrier-amplifier bandwidth cannot be much greater than about one-tenth of the carrier frequency. Carrier frequencies between 3,000 and 5,000 cps are common, and represent a compro-

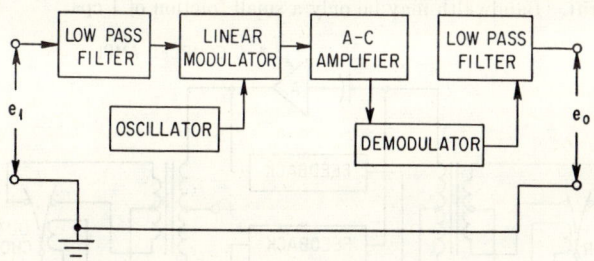

Fig. 19.15. Block diagram of a d-c carrier-amplifier system.

mise between a maximum bandwidth and freedom from problems associated with high frequencies in amplifier and bridge balancing circuits.

Chopper Amplifiers. "Chopper amplifiers" resort to a-c coupled amplifiers to circumvent drift and instability problems of direct-coupled d-c amplifiers. In this method, the d-c input signal is chopped into a succession of square waves whose amplitude is proportional to the input voltage. Conversion takes place in the input, and may be accomplished by a vibrating reed (chopper) excited by an electromagnet; contacts carried by the vibrating reed alternately open and close as the reed vibrates, periodically interrupting the input voltage.

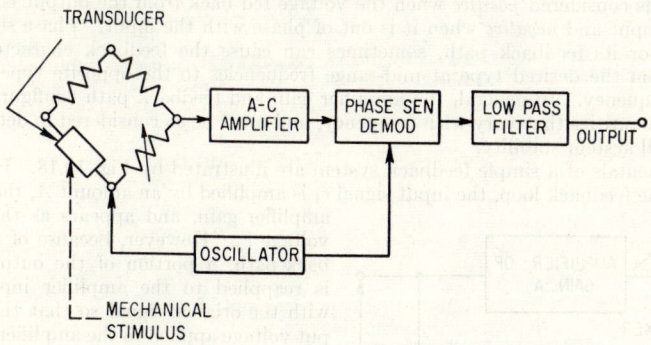

Fig. 19.16. Carrier amplifier with signal transducer as part of the input bridge.

Choppers may be arranged in several ways. The most popular single-pole double-throw type is arranged to reverse alternately the direction of current flow in the input transformer, Fig. 19.17, or to connect the amplifier alternately between the input signal and ground. The latter method has only half the sensitivity of the current-reversing method but does not require a transformer. This technique provides a ground reference marker which otherwise would be lost in transmission through the a-c coupled amplifier stages. Phase-sensitive demodulators are required with these arrangements to distinguish input voltage polarity. The demodulator may be a second chopper operated in synchronism with the first or an extra set of contacts on the input chopper. Reeds are usually tuned to, and driven by, the line frequency.

Choppers, which are essentially electromechanical switches, are subject to wear and deterioration that degrade their performance after extended periods of time. Their operating life normally exceeds the 1,000-hr guarantee. Because the allowable bandwidth is only a few per cent of the reed frequency, noise spikes from switching transients are relatively unimportant in this application. As a general rule, the bandwidth becomes progressively less as the gain of the amplifier is increased. Because of their extreme stability, chopper amplifiers can be designed to operate in the 1-microvolt region without appreciable drift. Bandwidth may be only a small fraction of 1 cps.

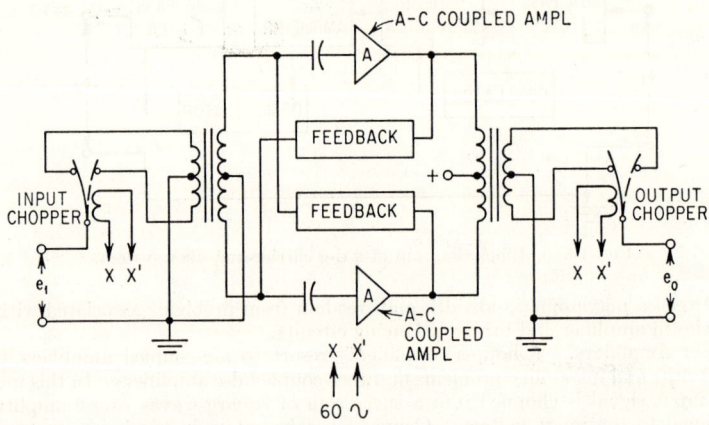

Fig. 19.17. D-C chopper amplifier with synchronous chopper demodulator.

FEEDBACK IN AMPLIFIERS. Amplifier characteristics can be radically modified when a portion of the output signal is fed back and combined with the input signal. Feedback is considered *positive* when the voltage fed back from the output is in phase with the input and *negative* when it is out of phase with the input. Phase shift in an amplifier, or its feedback path, sometimes can cause the feedback characteristic to change from the desired type at mid-range frequencies to the opposite type at some remote frequency. In general, the amplifier gain and feedback path configuration are complex quantities that vary with frequency, and must be so considered in determining the over-all system stability.

Fundamentals of a simple feedback system are illustrated by Fig. 19.18. In the absence of the feedback loop, the input signal e_1 is amplified by an amount A, the over-all amplifier gain, and appears as the output voltage e_0. However, because of the feedback path, a portion of the output signal is reapplied to the amplifier input along with the original signal so that the net input voltage applied to the amplifier becomes the vector sum of these two sources.

An amplifier with feedback has an effective gain of

$$A_f = \frac{A}{1 - A\beta} \quad (19.20)$$

where A_f = gain with feedback
A = gain without feedback
β = fraction of output voltage returned to the input

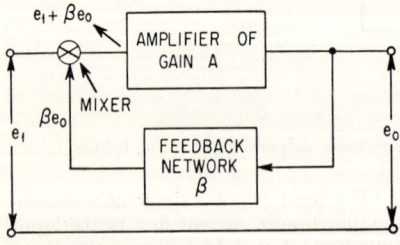

$(e_1 + \beta e_0)A = e_0$
$e_0/e_1 = A_f = \dfrac{A}{1-A\beta}$

Fig. 19.18. Basic feedback circuit showing relationships of the various parameters.

The sign of the feedback factor $A\beta$ is positive for positive feedback, and is negative for negative feedback.

Positive feedback produces instability in amplifiers because it inaugurates a cyclic build-up of the net amplifier input voltage, limited only by increased losses. When $A\beta$ equals unity, the gain theoretically becomes infinite and the amplifier becomes unstable. Oscillators and multivibrators employ positive feedback to sustain their operations. Positive feedback is generally undesirable in amplifiers, since it adversely affects input and output impedance and frequency range as well as instability.

Since $A\beta$ is negative with negative feedback, the amplifier is unconditionally stable. However, the phase opposition between input signal and feedback voltage reduces the effective input signal to the amplifier and consequently results in a decrease in over-all system gain. Loss in gain, which can easily be recovered by additional stages, is more than compensated for by better performance.

When the amplifier gain is very large, such that $|A\beta| \gg 1$, the gain with negative feedback is approximately

$$A_f \simeq -\frac{1}{\beta} \tag{19.21}$$

and is virtually independent of the absolute value of A. This means that the gain characteristics of the amplifier depend only on the passive elements comprising the feedback loop, and are relatively insensitive to changes in tube coefficients with age, heat, power-supply variations, or tube substitutions. In designing amplifiers with negative feedback, care must be taken to ensure that the frequency and phase characteristics of the amplifier and feedback loop yield negative feedback over the entire passband.

Negative feedback is instrumental in improving the range of uniform frequency response and in reducing noise and distortion generated within the amplifier. Internal noise, hum, and distortion are reduced in a manner similar to gain:

$$N_f = \frac{NA}{1 - A\beta} \tag{19.22}$$

and

$$D_f = \frac{DA_f}{A} \tag{19.23}$$

where N_f and D_f are the noise and distortion present in an amplifier with feedback compared to the noise N and the distortion D of the same amplifier without feedback. The improved characteristics obtained with negative feedback should not be an excuse for poor initial design, but rather should be considered as a means for further improving a good design.

The amount of feedback present in an amplifier is customarily expressed in decibels and represents the loss in gain due to feedback.

$$\text{Amount of feedback} = 20 \log \frac{A}{A_f} \quad \text{db}$$

Negative feedback may be obtained from either the output voltage or current, or both. Its effect on amplifier output impedances, when derived from *voltage* feedback, is given by

$$R_{0f} = \frac{R_0}{1 - A\beta} \tag{19.24}$$

and

$$R_{0f} = R_0 - A\alpha R_1 \tag{19.25}$$

when derived from *current* feedback; where α is defined as the ratio of feedback voltage applied to the input to the voltage developed across R_1 in series with the load; R_0 is the amplifier output impedance without feedback, and R_{0f} is the amplifier output impedance with feedback; other notations are defined above.

SPECIAL CIRCUITS

CATHODE FOLLOWERS. A cathode-follower circuit has the entire load in the cathode circuit. Typical configurations are shown by Fig. 19.19. It is essentially an amplifier circuit with 100 per cent negative feedback; it is characterized by a gain less than unity and a low output impedance. Cathode followers are used extensively in shock and vibration work as an impedance transformer to present a high-impedance load to a crystal accelerometer and a low-impedance source for feeding long cable leads.

The circuit of Fig. 19.19A is the basic cathode-follower configuration. From its equivalent circuit, Fig. 19.19B,

$$\mu e_{gk} = (r_p + R_k) i_p \qquad (19.26)$$

where $\qquad e_{gk} = e_1 - i_p R_k \qquad$ and $\qquad e_0 = i_p R_k$

from which
$$A = \frac{e_0}{e_1} = \frac{\mu R_k}{r_p + (\mu + 1) R_k} \simeq \frac{g_m R_k}{1 + g_m R_k} \qquad (19.27)$$

The gain of a cathode follower can never exceed unity; typical values range from about 0.8 to 0.9, increasing with R_k. Neglecting stray capacities and transit-time effects, the

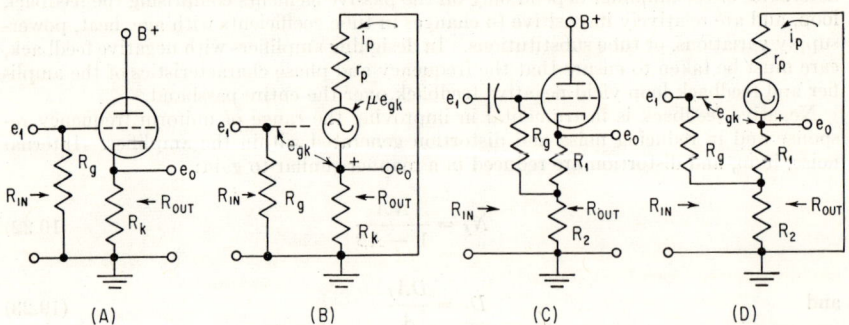

FIG. 19.19. Cathode followers and their equivalent circuits. (A) Basic cathode-follower circuit. (B) Cathode-follower equivalent circuit. (C) Modified cathode-follower circuit with grid leak returned to a tap on the cathode resistor. (D) Equivalent circuit of modified cathode follower.

input resistance R_{in} is equal to the grid-leak resistance R_g, and is limited by grid-current considerations to a maximum of several megohms. The output resistance R_o is the parallel resistance of r_p and R_k:

$$R_o = \frac{r_p R_k}{r_p + R_k} \qquad (19.28)$$

and is generally low, in the order of several hundred ohms. Input and output voltages of a cathode follower are in phase.

In the modified circuit of Fig. 19.19C, the grid leak is returned to a tap on the cathode resistor. This connection allows the grid bias voltage to be set at normal values, while permitting the large cathode resistor (and large voltage drop) required to give voltage gains closer to unity. The cathode tap has an extremely important additional advantage, in that it makes the input circuit resistance appear to be much larger than its ohmic value. As far as the external circuit is concerned, the input resistance it sees (given by the ratio of input voltage to current) is

$$R_{in} = \frac{e_1}{i_1} \qquad (19.29)$$

But since $e_1 = i_1 R_g + e_2$, where $e_2 = e_0 R_2/(R_1 + R_2)$,

then
$$i_1 = \frac{1}{R_g}\left(e_1 - \frac{e_0 R_2}{R_1 + R_2}\right) \quad (19.30)$$

from which
$$R_{in} = \frac{R_g}{1 - A\left[\dfrac{R_2}{(R_1 + R_2)}\right]} \quad (19.31)$$

where $A = e_0/e_1$. Since A and $R_2/(R_1 + R_2)$ are individually only slightly less than unity, the denominator becomes extremely small, and the apparent input resistance cor-

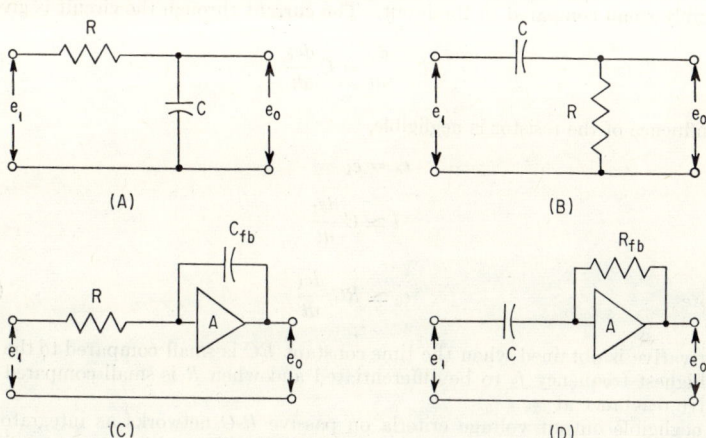

FIG. 19.20. Integrator and differentiator circuits. (A) Passive integrator. (B) Passive differentiator. (C) Operational integrator. (D) Operational differentiator.

respondingly larger than R_g. An input resistance up to several hundred megohms may be obtained.

This connection also permits handling abnormally large signals because the actual grid-to-cathode voltage is only $\{1 - A[R_2/(R_1 + R_2)]\}$ of the actual input voltage. The permissible peak-to-peak input voltage is approximately equal to the plate-supply voltage. In d-c circuits the d-c potential on the grid may be a possible disadvantage.

Cathode followers can distort high-frequency signals which swing in the negative direction if the cathode-circuit time constant is greater than the negative rise time of the pulses. This is because the cathode voltage cannot drop at a faster rate than allowed by its time constant, even if the plate current is suddenly cut off completely. Positive pulses do not experience this distortion since the cathode-circuit capacitances can be charged quickly from the low-impedance power supply.

INTEGRATORS AND DIFFERENTIATORS. Integrator and differentiator circuits perform the equivalent of these mathematical operations by producing an output voltage which is the time integral $\int e_1\, dt$ or time derivative de_1/dt of an applied input voltage e_1.

The voltage e_c across the capacitance of a simple R-C circuit, Fig. 19.20A, is the integral of the input voltage when the output voltage is negligibly small compared to the input. When the reactance of condenser C is negligible compared with R,

$$i \simeq \frac{e_1}{R}$$

From simple transient circuit theory

$$e_0 = \frac{1}{C} \int i \, dt \tag{19.32}$$

which reduces to

$$e_0 \simeq \frac{1}{RC} \int e_1 \, dt \tag{19.33}$$

when $e_0 \ll e_1$. This condition is fulfilled when the time constant $(t = RC)$ is large compared to the period of the lowest frequency f_1 to be integrated.

The voltage across the resistance of an R-C circuit, Fig. 19.20B, is proportional to the derivative of the input voltage under the identical assumption that the output voltage is negligibly small compared to the input. The current through the circuit is given by

$$i = \frac{dq}{dt} = C \frac{de_c}{dt} \tag{19.34}$$

If the influence of the resistor is negligible,

$$e_c = e_1$$

and

$$i \simeq C \frac{de_1}{dt}$$

Therefore

$$e_0 \simeq RC \frac{de_1}{dt} \tag{19.35}$$

The derivative is obtained when the time constant RC is small compared to the period of the highest frequency f_2 to be differentiated and when R is small compared to the capacitive reactance at f_2.

The negligible output voltage criteria on passive R-C networks as integrators and differentiators seriously limit their usefulness. Voltage amplifiers are required to produce adequate signal levels. Special advantages are obtained by the operational amplifier circuits of Fig. 19.20C and D, where the shunt element of the R-C network is connected in a feedback loop around the amplifier. When the phase difference between input and output is 180°, the feedback element appears to be shunted across the input in its proper place, but modified in value by the amplifier gain. A further consequence of this circuit is that when A is very large, circuit performance becomes independent of the absolute value of A and depends only on circuit elements.

The shunt capacity C of the operational integrator, Fig. 19.20C, neglecting grid-plate capacitance, has an effective value greater than the feedback condenser C_{fb} given by

$$C = C_{fb}(1 + A) \tag{19.36}$$

When this value is substituted in Eq. (19.33), and allowance made for the amplifier gain A, the output voltage becomes

$$e_0' = Ae_0 \simeq \frac{A}{RC_{fb}(1+A)} \int e_1 \, dt$$

and since A usually is large compared with 1,

$$e_0' \simeq \frac{1}{RC_{fb}} \int e_1 \, dt \tag{19.37}$$

In a similar manner the resistance R of the operational differentiator, Fig. 19.20D, appears to have an effective value less than the feedback resistor R_{fb}:

$$R = \frac{R_{fb}}{1 + A} \tag{19.38}$$

and the output voltage, from Eq. (19.35), becomes

$$e_0' = Ae_0 \simeq \frac{R_{fb}CA}{(1+A)}\frac{de_1}{dt} \simeq R_{fb}C\frac{de_1}{dt} \qquad (19.39)$$

TRANSISTOR CIRCUITS

Transistors are semiconductor devices which possess amplifying properties. In many electronic circuits they can be made to perform the same functions as vacuum tubes. Their principal advantages are derived from their small size and weight, low power requirements, absence of a heater, instant operation, and an almost unlimited life when operated within their ratings. Disadvantages include a rather large temperature dependence, an abnormally high noise level, limited power-handling capabilities, and a somewhat restricted high-frequency limit.

TRANSISTOR PROPERTIES. The physics of semiconductor materials is somewhat complex.[6-8] Only the more important of the transistor mechanisms, needed to understand their operation, are presented here. The main emphasis is on their external characteristics as a circuit element.

Characteristics of Semiconductors. In their pure, or intrinsic, state, both germanium and silicon, the materials commonly used for transistors, have a tetragonal crystalline structure such that each of their four valence electrons forms a covalent bond with a similar electron in a neighboring atom. The valence electrons are consequently tightly bound in the lattice structure, and are not readily available as carriers of electricity. Pure germanium (or silicon *) therefore exhibits a very low conductivity and is virtually an insulator. On occasion an electron will acquire enough energy, for example from thermal agitation, to break its bonds and become free to wander about in the crystal. The vacancy left by the electron is termed a "hole." Electron-hole pairs can exist in the neutral lattice until they disappear through recombination.

Under an applied electric field, electrons will drift to the positive terminal, some will combine with holes they encounter, and some will enter the external circuit. The holes, on the other hand, are fixed in the lattice and are immobile. However, they appear to drift toward the negative terminal since recombinations with drifting electrons leave an uncovered hole displaced toward the negative terminal. The over-all appearance is the same as if the hole were a positive entity with a charge equal in magnitude, but opposite in sign, to that of an electron. By this interpretation, both electrons and holes become carriers of electricity.

Semiconductor materials used for transistors are carefully "doped" with impurities to increase greatly their conductivity over that of the intrinsic material. Impurities with five-valence electrons (antimony, phosphorus, or arsenic) enter the tetravalent crystal, leaving one electron unbound to the crystal and available as a free current carrier. This type of semiconductor material has an excess of negative current carriers and is termed an "N-type" material. Trivalent impurities (gallium or indium) do not have enough valence electrons to complete all the lattice bonds. The incomplete bond constitutes a hole. Current transport in this material is by holes, or positive carriers, and it is consequently termed a "P-type" material.

JUNCTION TRANSISTORS. The junction transistor is roughly analogous to a triode vacuum tube. As shown in Fig. 19.21, it consists of two regions of one type, separated by a region of the opposite type. When a small forward bias is applied to the first junction, and a large reverse bias to the second, the system behaves much like a vacuum-tube triode. Most germanium transistors are PNP, while most silicon transistors are NPN.

The center section is termed the *base*. It is generally made quite thin, in the order of several mils or less, to keep the transit time across the base short enough to preserve the high-frequency response. The outer sections are termed the *emitter* and the *collector*.

---

* Except for absolute levels, comments concerning germanium can generally be applied to silicon as well.

Although of the same semiconductor type, the emitter and collector are not interchangeable since the concentration of majority carriers in the emitter is many orders of magnitude greater than that in the collector.

Bias Polarities. Figure 19.21 indicates the proper direction of bias for PNP and NPN transistors. With respect to the base, the emitter is forward-biased, i.e., in the direction to reduce the potential barrier for the majority carriers at the emitter-base junction. The collector is reverse-biased, i.e., it increases the potential barrier for majority carriers at the collector-base junction. For a PNP transistor, the emitter bias potential is positive, while the collector bias potential is negative, both with respect to the base. Bias potentials are reversed for an NPN transistor. Typical magnitudes are between 0.05 and 0.2 volt for the emitter, and between 3 and 30 volts for the collector.

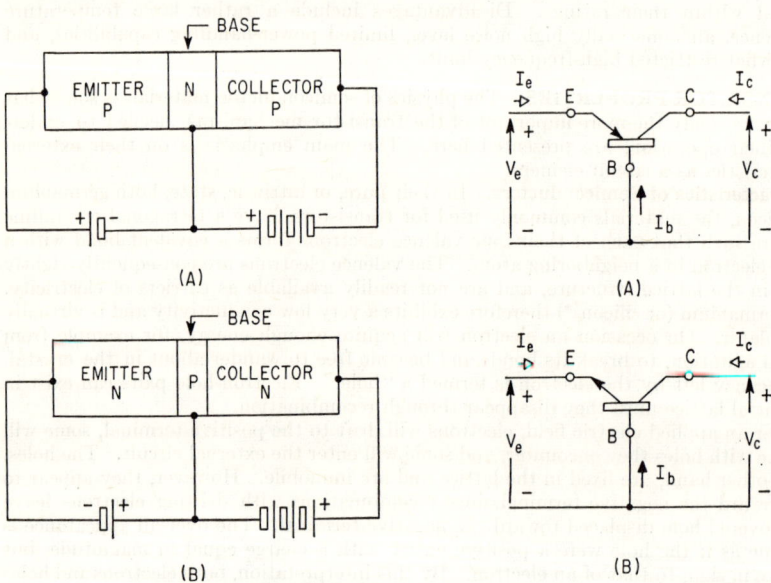

Fig. 19.21. Bias potentials for PNP and NPN transistors. (A) A PNP transistor. (B) An NPN transistor.

Fig. 19.22. Standard symbols, bias polarities, and directions of standard current flow for PNP and NPN transistors. (A) A PNP transistor. (B) An NPN transistor.

Standard Notations. The symbols and conventions which have been adopted for transistor electronics are indicated in Fig. 19.22. The symbol is a very close approximation to the actual configuration of the original point-contact transistor and conveys a picture of transistor action. It serves equally well for junction transistors. The emitter E is distinguished from the collector C by an arrowhead pointed in the direction of conventional current flow (opposite to electron flow) in the emitter circuit. Irrespective of the type of transistor represented, normal electrode polarities are always considered positive with respect to the base, and the positive direction of current flow is always toward the junction. As a result of these conventions some characteristics are given in terms of negative quantities. The symbol V is used to denote voltage in lieu of E to avoid confusion with the E representing the emitter.

TRANSISTOR-CIRCUIT CONFIGURATIONS. A transistor has three possible circuit arrangements, Fig. 19.23. These are termed common- (or grounded-) base, common- (or grounded-) emitter, and common- (or grounded-) collector. Their counterparts in vacuum-tube circuitry are the grounded grid, grounded cathode, and cathode follower (grounded plate). The connotation of "grounded" in this terminology does not necessarily mean that a particular element is shorted to ground, but rather that the other electrodes are referenced to it. The "common-" designation is somewhat more descriptive since it indicates that a particular electrode is common to both the input and output circuits.

Transistor Equivalent Circuits. Equivalent circuits for the three transistor configurations are also given by Fig. 19.23. Each electrode has associated with it an internal resistance. These meet in a common point at the junction. An equivalent voltage

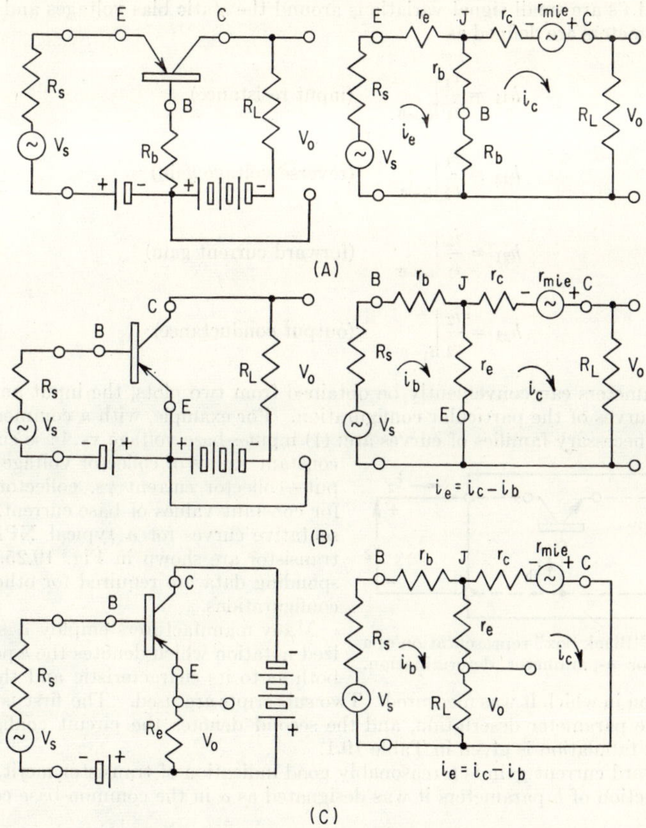

FIG. 19.23. The three basic transistor configurations and their equivalent circuits. (A) Common- or grounded-base configuration. (B) Common- or grounded-emitter configuration. (C) Common- or grounded-collector configuration.

generator $r_m i_e$ is included in the collector circuit to account for the active nature of the network. The symbol r_m is a mutual resistance; i_e is the dynamic emitter current and is the algebraic sum of the loop currents flowing in the emitter.

The dynamic resistances represented by r_e, r_b, r_c, and r_m can be obtained from plots of static characteristics, but take four separate plots to yield all four parameters. Two of these plots define useful characteristics of the transistor; the other two do not. As with vacuum-tube parameters, the internal electrode resistances are dynamic resistances, obtained for small signal variations at a particular set of bias conditions. Their values will change with the operating point.

Because the above notation is inconvenient to apply, an alternate method using hybrid (h) parameters has become more popular. This technique considers the transistor as a linear, four-pole "black box," and evaluates its characteristics from external measurements without regard to what might be inside. For the four-pole network of Fig. 19.24 the following relations are obtained:

$$v_1 = h_{11}i_1 + h_{12}v_2 \tag{19.40}$$

$$i_2 = h_{21}i_1 + h_{22}v_2 \tag{19.41}$$

The v's and i's are small signal variations around the static bias voltages and currents. The h parameters are defined as

$$h_{11} = \left.\frac{v_1}{i_1}\right|_{v_2=0} \quad \text{(input resistance)} \tag{19.42}$$

$$h_{12} = \left.\frac{v_1}{v_2}\right|_{i_1=0} \quad \text{(reverse voltage gain)} \tag{19.43}$$

$$h_{21} = \left.\frac{i_2}{i_1}\right|_{v_2=0} \quad \text{(forward current gain)} \tag{19.44}$$

$$h_{22} = \left.\frac{i_2}{v_2}\right|_{i_1=0} \quad \text{(output conductance)} \tag{19.45}$$

The h parameters can conveniently be obtained from two plots, the input and output family of curves of the particular configuration. For example, with a common-emitter circuit the necessary families of curves are: (1) input—base voltage vs. base current for constant values of collector voltage; (2) output—collector current vs. collector voltage for constant values of base current. Representative curves for a typical NPN silicon transistor are shown in Fig. 19.25. Corresponding data are required for other circuit configurations.

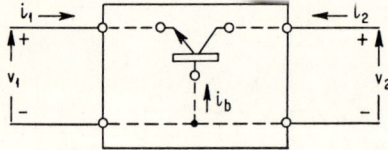

Fig. 19.24. "Black box" representation of a transistor for h-parameter determination.

Many manufacturers employ a standardized notation which denotes the h parameter both as to its characteristic and the circuit configuration in which it was measured. Two subscripts are used. The first is the first letter of the parameter description, and the second denotes the circuit configuration. A complete tabulation is given in Table 19.1.

The forward current gain is a reasonably good indication of transistor merit. Before the introduction of h parameters it was designated as α in the common-base configura-

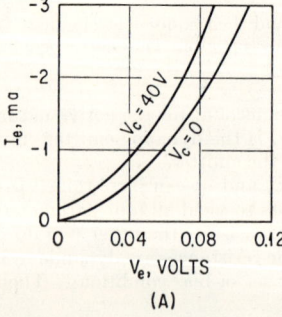
(A)

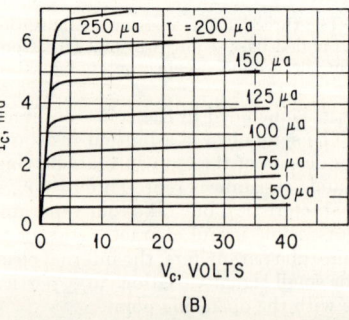
(B)

Fig. 19.25. Families of characteristic curves for a typical NPN silicon transistor. (A) Common-emitter input characteristics, I_e vs. V_e. (B) Common-emitter output characteristics, I_c vs. V_c.

Table 19.1. Transistor *h*-parameter Nomenclature

Characteristic	Input-output	Common base	Common emitter	Common collector
Input resistance	h_{11}	h_{ib}	h_{ie}	h_{ic}
Reverse voltage gain	h_{12}	h_{rb}	h_{re}	h_{rc}
Forward current gain	h_{21}	h_{fb}	h_{fe}	h_{fc}
Output admittance	h_{22}	h_{ob}	h_{oe}	h_{oc}

tion and β in the common-emitter and common-collector configurations. Thus

$$-h_{fb} = \alpha \quad \text{and} \quad h_{fe} = h_{fc} = \beta \tag{19.46}$$

with the additional relation that

$$\beta = \frac{\alpha}{1 - \alpha} \tag{19.47}$$

The quantity α is very nearly unity, usually between 0.95 and 0.99$^+$. Therefore β becomes very large as α approaches unity. Typical values for β range from 20 to 200.

Circuit Characteristics. The input and output impedance and the voltage, current, and power gains of transistor circuits can be evaluated from their equivalent circuits. Formulas for these parameters are not particularly difficult except that they contain many terms and are therefore difficult to commit to memory. The usual transistor circuit has the following characteristics:

1. Input impedance of transistors is low: approximately 50 ohms for common-base configurations, an order of magnitude higher with common-emitter, and highest—possibly to 0.5 megohm—for common-collector stages. It varies with output load over a rather wide impedance range.

2. Output impedance of transistor circuits tends to be high. Typical values are: 400,000 ohms for common-base, 50,000 ohms for common-emitter, and 25 ohms for common-collector stages.

3. Voltage gains are highest (500) with common-emitter stages, somewhat less (150) with common-base, and always less than unity (0.99) with common-collector stages.

4. Current gains for common-emitter and common-collector transistor configurations are approximately equal and range around 30. Common-base current gain is always less than unity (0.96).

5. Power gain is greatest (6,000) with common-emitter circuits, moderate (500) with common-base stages, and negligible with the common-collector configuration.

Magnitudes given above are representative of average transistors in usual circuit arrangements and can be expected to vary widely.

Amplifier Circuits. Figure 19.26 shows typical, simple transistor amplifier circuits.

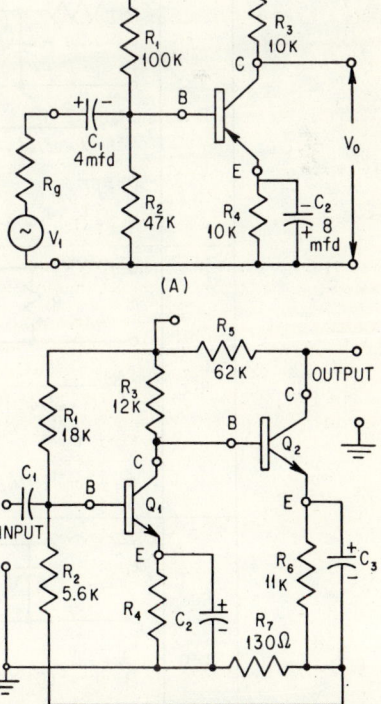

FIG. 19.26. Typical transistor amplifier circuits. (*A*) Single-stage PNP transistor audio amplifier. (*B*) Two-stage, direct-coupled transistor amplifier with feedback. (*After R. B. Hurley.*[12])

A single-stage common-emitter circuit is shown in Fig. 19.26A. Input is to the base through capacitor C_1, which must be large enough in combination with the transistor input resistance and R_2 to maintain the required low-frequency response. Base bias is provided by the voltage divider R_1 and R_2 across the power supply. The emitter resistance R_4 and bypass capacitor C_2, in conjunction with the base bias, set the operating point of the transistor. Output voltage is taken from the collector.

The circuit shown in Fig. 19.26B is a two-stage amplifier, with feedback from the second emitter to the first base. Both stages are common-emitter stages to take advantage of the high gain of this configuration. The circuit has low distortion, good frequency-response characteristics, and excellent stability.

TRANSISTOR DEFICIENCIES. Transistors have several characteristics which must be circumvented or nullified before they can be used with the effectiveness and reliability of vacuum tubes. Transistors are temperature-sensitive; germanium-type transistors are much more temperature-sensitive than transistors of the silicon type. The principal effect of elevated temperatures is to increase drastically the leakage current I_{co}. This increases the collector current, which produces more heating, making the cycle cumulative, and may lead to "thermal runaway" and eventual destruction of the tran-

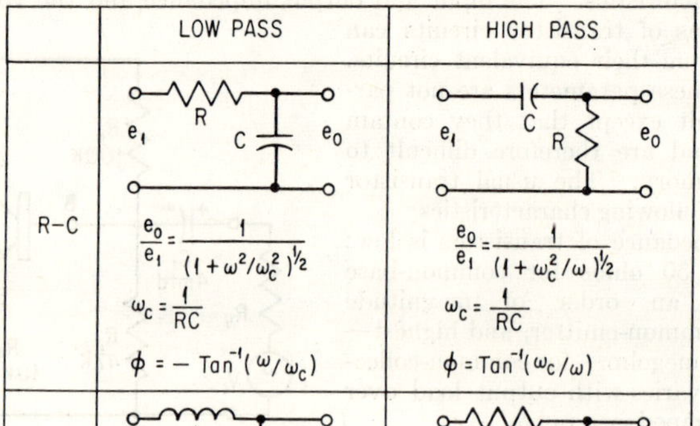

	LOW PASS	HIGH PASS
R-C	$\dfrac{e_0}{e_1} = \dfrac{1}{(1+\omega^2/\omega_c^2)^{1/2}}$ $\omega_c = \dfrac{1}{RC}$ $\phi = -\mathrm{Tan}^{-1}(\omega/\omega_c)$	$\dfrac{e_0}{e_1} = \dfrac{1}{(1+\omega_c^2/\omega^2)^{1/2}}$ $\omega_c = \dfrac{1}{RC}$ $\phi = \mathrm{Tan}^{-1}(\omega_c/\omega)$
R-L	$\dfrac{e_0}{e_1} = \dfrac{1}{(1+\omega^2/\omega_c^2)^{1/2}}$ $\omega_c = R/L$ $\phi = -\mathrm{Tan}^{-1}(\omega/\omega_c)$	$\dfrac{e_0}{e_1} = \dfrac{1}{(1+\omega_c^2/\omega^2)^{1/2}}$ $\omega_c = R/L$ $\phi = \mathrm{Tan}^{-1}(\omega_c/\omega)$
L-C	$\dfrac{e_0}{e_1} = \dfrac{1}{1-\omega^2 LC}$ $\omega_c = \dfrac{1}{(LC)^{1/2}}$ $\phi = 0, f < f_0$ $\quad = \pi, f > f_0$	$\dfrac{e_0}{e_1} = \dfrac{1}{1-\dfrac{1}{\omega^2 LC}}$ $\omega_c = \dfrac{1}{(LC)^{1/2}}$ $\phi = \pi, f < f_0$ $\quad = 0, f > f_0$

Fig. 19.27. Simple *R-C*, *R-L*, and *L-C* filter circuits and equations.

sistor unless limited by circuit components. Various bias schemes, feedback circuits, and temperature-sensitive elements are employed to neutralize temperature effects.

Noise is appreciably higher in transistors than in vacuum tubes because of the randomness of motion of the electrons and holes. The frequency distribution of noise components very closely approximates a characteristic which varies inversely with frequency, and is somewhat dependent on bias conditions. It is several orders of magnitude higher than what would be expected from thermal noise alone. At very low frequencies, near d-c, the instability and flutter components of the inherent noise preclude the use of transistors for amplifying signals in the range below about 100 microvolts.

FILTERS

An electrical network which discriminates against particular frequencies, while passing components in a different portion of the frequency spectrum, is termed an *electrical filter*. Filters are classified according to the relative range of frequencies they pass or reject, i.e., low-pass, high-pass, bandpass, or band-rejection. Different types of filters may be cascaded to produce any combination of pass and reject characteristics. Classification also can be carried further to include design parameters, configuration, and construction features. Common designations include: R-C, R-L, constant-k, balanced-T, crystal, etc. Filters may be either passive or active, depending on the nature of their network elements.

The material presented here on filter design is restricted to basic considerations of the design approach and fundamental properties of filters in general. Many treatises are available.[9-11]

ELEMENTARY R-C, R-L, AND L-C NETWORKS

Simple R-C, R-L, and L-C networks, Fig. 19.27, exhibit broad frequency-discriminating characteristics useful where a high attenuation rate and sharp cutoff are not required. R-C and R-L networks of the same type and design cutoff frequency f_c have identical characteristics. Attenuation rates for these networks increase gradually through the transition region, and reach a maximum of 6 db/octave. Sections may be cascaded to yield an additional 6 db/octave/section, provided the input impedance of following sections is high compared to the shunt impedance of preceding sections.

Ohmic resistance of the inductance used in R-L and L-C networks alters the performance somewhat from calculated response. Idealized L-C equations indicate zero attenuation in the passband and infinite attenuation outside, with an abrupt 180° phase reversal at the cutoff frequency. The effect of inductor resistance in this circuit is to moderate these discontinuities in response, the rate of change from one state to the other becoming more gradual with increasing resistance.

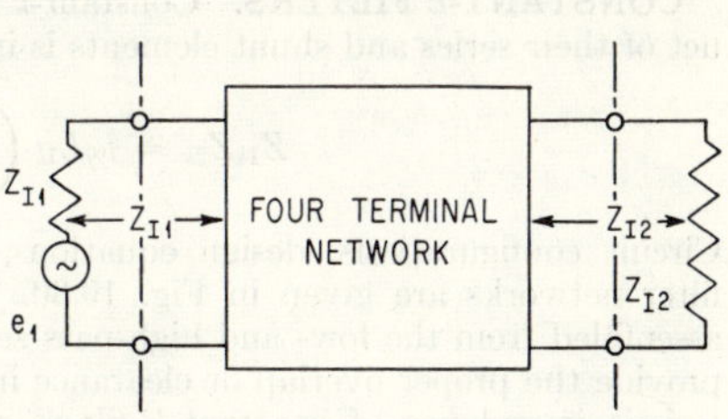

L-C FILTERS

FILTER DESIGN METHODS. For purposes of design analysis, a filter is considered a four-terminal network composed of linear, resistive, and reactive elements. It is terminated on an *image impedance* basis if the following conditions apply simultaneously: (1) the impedance looking into the input terminals is Z_{I_1} when the output is shunted by Z_{I_2} and (2) the impedance looking into the output terminals is Z_{I_2} when the input is being fed by a source of internal impedance Z_{I_1}. As shown in Fig. 19.28, the impedance looking in either direction at the input and output terminals is then Z_{I_1} and Z_{I_2}, respectively. The special case when Z_{I_1} equals Z_{I_2} is important in image impedance filter design, and defines the *characteristic impedance* Z_0 of the network.

FIG. 19.28. A four-terminal network terminated on an image impedance basis.

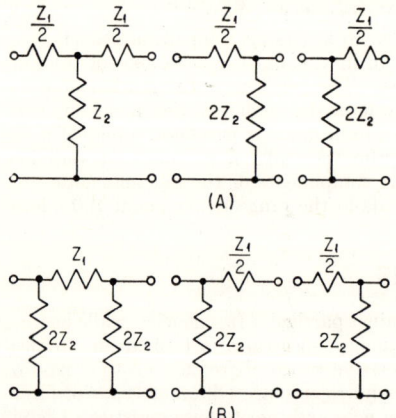

FIG. 19.29. T and π networks. (A) Full- and half-section T networks. (B) Full- and half-section π networks.

Equations for filters designed on the image impedance basis consider the network elements as idealized components, i.e., without losses due to ohmic resistance of windings, leakage, or stray capacitances between elements and ground. Fortunately, most filter applications are sufficiently liberal in their requirements to the extent that the approximate results obtained by image impedance methods can be tolerated.

For more precise results, *modern network theory* is employed. In this method the network mesh equations are solved in terms of the filter parameters, and the actual values adjusted to produce the desired response. While more accurate and versatile, modern network design theories are more time-consuming since each network becomes a special case.

Symmetrical T and π sections are the basic units of filter networks. The series and shunt elements are labeled as in Fig. 19.29 to make design equations for both types compatible. Values calculated from filter design equations then are multiplied or divided by 2, as required, to obtain the size of the proper component in a particular configuration.

When a network is terminated in its characteristic impedance, there are no reflections. The ratio of input current I_1 to output current I_2 then is

$$\frac{I_1}{I_2} = e^\gamma \tag{19.48}$$

which defines the *propagation constant* γ. In general γ is a complex number:

$$\gamma = \alpha + j\beta \tag{19.49}$$

where α is the *attenuation constant* in nepers (1 neper = 8.69 db) and β is the *phase constant* in radians.

CONSTANT-k FILTERS. Constant-k filters possess the characteristic that the product of their series and shunt elements is independent of frequency. Thus

$$Z_{1k}Z_{2k} = j\omega L_{1k}\left(\frac{1}{j\omega C_{2k}}\right) = \frac{L_{1k}}{C_{2k}} = R_k^2 \tag{19.50}$$

Circuit configurations, design equations, and response characteristics for constant-k filter networks are given in Fig. 19.30. Bandpass and band-rejection filters can be assembled from the low- and high-pass sections by adjusting the cutoff frequencies to provide the proper overlap or clearance in the passband region. Although the characteristic impedance of constant-k filters remains resistive in the passband region, its magnitude varies widely with frequency. The design value of Z_0 is obtained only at zero frequency for low-pass filters and at infinity for high-pass filters. Single-section constant-k filters have rather low attenuation rates near cutoff. Sharper cutoff can be obtained by cascading additional sections.

m-DERIVED FILTER SECTIONS. Characteristics of constant-k filters can be improved substantially by the addition of m-derived sections, Fig. 19.31. The parameter m is a number having any value between zero and one.

These m-derived sections have two important applications. When they act as terminating half sections for constant-k filters (with $m = 0.6$), they maintain the characteristic input impedance uniform within 5 per cent of its design value over 85 per cent

	LOW PASS	HIGH PASS
T-SECTION CONFIGURATION	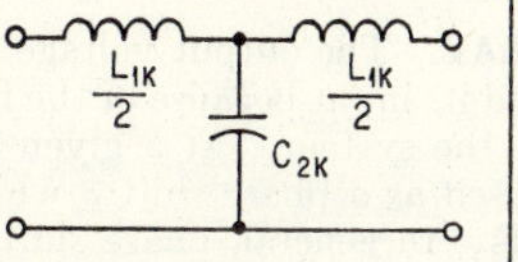 Series $L_{1K}/2$, $L_{1K}/2$; shunt C_{2K}	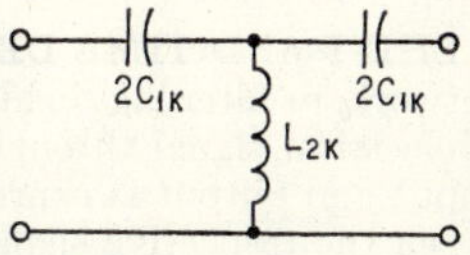 Series $2C_{1K}$, $2C_{1K}$; shunt L_{2K}
π-SECTION CONFIGURATION	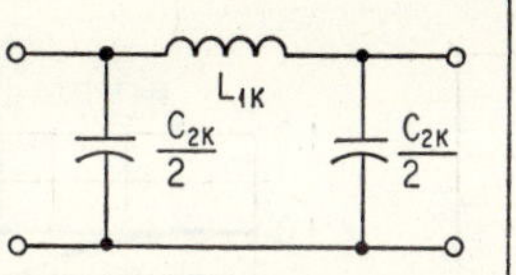 Shunt $C_{2K}/2$, $C_{2K}/2$; series L_{1K}	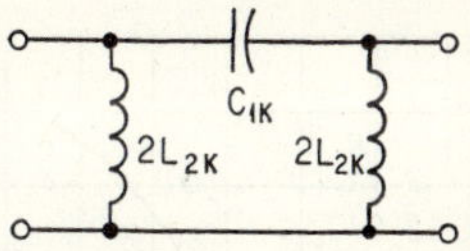 Shunt $2L_{2K}$, $2L_{2K}$; series C_{1K}
CHARACTERISTIC IMPEDANCE Z_0	FOR $\omega = 0$ $Z_0 = R_K = (L_{1K}/C_{2K})^{1/2}$ FOR $Z_0(\omega)$ $Z_0(T-) = R_K(1 - \omega^2/\omega_c^2)^{1/2}$ $Z_0(\pi-) = R_K/(1 - \omega^2/\omega_c^2)^{1/2}$	FOR $\omega = \infty$ $Z_0 = R_K = (L_{2K}/C_{1K})^{1/2}$ FOR $Z_0(\omega)$ $Z_0(T-) = R_K(1 - \omega_c^2/\omega^2)^{1/2}$ $Z_0(\pi-) = R_K/(1 - \omega_c^2/\omega^2)^{1/2}$
CUT OFF FREQUENCY ω_c	$\omega_c = \dfrac{2}{(L_{1K}C_{2K})^{1/2}}$	$\omega_c = \dfrac{1}{2(C_{1K}L_{2K})^{1/2}}$
ATTENUATION α	$\omega < \omega_c$ $\alpha = 0$ $\omega > \omega_c$ $\alpha = \cosh^{-1}\left(\dfrac{\omega^2 L_{1K} C_{2K}}{2} - 1\right)$	$\omega < \omega_c$ $\alpha = \cosh^{-1}\left(\dfrac{1}{2\omega^2 C_{1K} L_{2K}} - 1\right)$ $\omega > \omega_c$ $\alpha = 0$
PHASE SHIFT β	$\omega < \omega_c$ $\beta = \cos^{-1}\left(1 - \dfrac{\omega^2 L_{1K} C_{2K}}{2}\right)$ $\omega > \omega_c$ $\beta = \pi$	$\omega < \omega_c$ $\beta = -\pi$ $\omega > \omega_c$ $\beta = \cos^{-1}\left(1 - \dfrac{1}{2\omega^2 C_{1K} L_{2K}}\right)$
CIRCUIT VALUES	$L_{1K} = \dfrac{2R_K}{\omega_c}$ $C_{2K} = \dfrac{2}{R_K \omega_c}$	$C_{1K} = \dfrac{1}{2R_K \omega_c}$ $L_{2K} = \dfrac{R_K}{2\omega_c}$

FIG. 19.30. Constant-k filter parameters.

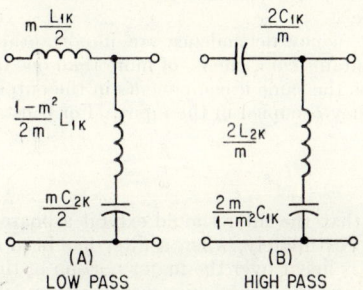

FIG. 19.31. Terminating filter half sections of the m-derived type.

of the passband. When they act as full sections (with $m = 0.2$ to 0.3) they can produce extremely high attenuation rates near cutoff or at any other critical frequency in the stopband.

PHASE SHIFT AND TIME DELAY. The output voltage from a transmission line, reactive network, or filter lags behind its input because of the finite velocity of propagation of an electrical signal through the system. At a given frequency, the difference between input and output is expressed as a phase shift β which denotes the electrical angle between the respective signals. In general, phase shift is frequency-dependent, and varies from one frequency to another according to the arctangent of a complex ratio

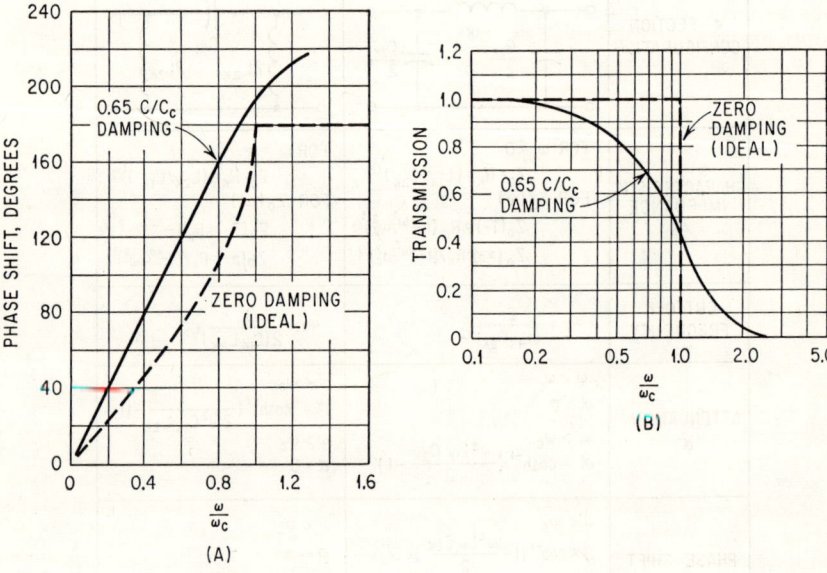

FIG. 19.32. Characteristics of ideal and practical filters. (A) Phase-shift characteristics. (B) Transmission characteristics.

of reactances and resistances. Phase shift of an ideal low-pass constant-k filter is shown by Fig. 19.32A.

At a particular frequency, phase shift corresponds to a retardation of the output signal in time with respect to the input, as given by

$$\beta = \omega \tau \tag{19.51}$$

where β = phase shift, radians
ω = angular frequency ($2\pi f$), radians/sec
τ = delay time, sec

Delay time and its frequency dependence are important characteristics of filters. When the input signal contains components of more than one frequency, the delay time for all frequencies must be the same if components in the output are to retain the same relative phase positions they occupied in the input. For constant delay, it follows that

$$\tau = \frac{\beta}{\omega} \tag{19.52}$$

This condition indicates that the filter should exhibit a phase-shift characteristic proportional to frequency. Fortunately, when a filter has been optimumly damped, the phase-shift characteristic is linear over the major portion of the passband, as indicated in Fig. 19.32A. The slope of the phase-shift characteristic, which is delay time, is related to the amount of damping.

PRACTICAL FILTER DESIGN CONSIDERATIONS

Equations for passive filters designed on the image impedance basis neglect the resistive components of inductor windings and other losses. The errors in steady-state response which are introduced by neglecting these factors are generally negligible when the Q of the coil is greater than about 10.

In their idealized form these filters are unsuitable for complex vibratory or shock signals. Because of negligible damping, the L-C components tend to "ring" when excited by transients, and introduce spurious components in the passed signal. The network oscillates at the cutoff frequency. Transmitted signals are also subject to delay errors due to the nonuniform phase-shift characteristic in the passband. Characteristics of an idealized constant-k low-pass filter are indicated in Figs. 19.32B and 19.33.

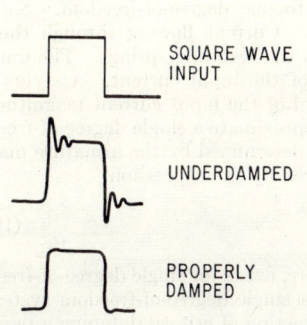

FIG. 19.33. Square-wave response of ideal and practical filters.

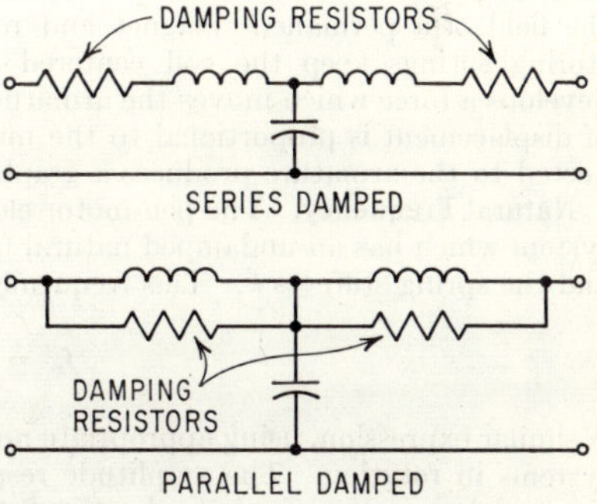

FIG. 19.34. Series and parallel damping circuits for ideal and practical filters.

Both of these undesirable conditions can be alleviated by the addition of damping (resistance) into the network. Series and shunt methods are shown in Fig. 19.34. Both give approximately the same relative frequency-response characteristic when the damping resistances are properly adjusted. Optimum performance is obtained when the network response to an impressed square wave one decade lower than the design cutoff frequency produces the steepest rise and sharpest turnover possible without appreciable overshoot or ringing. Damping resistances are best adjusted experimentally. Their effect on frequency response is to modify the abrupt discontinuity at the cutoff frequency to yield a more gradual transition from the pass to reject band.

The principal difference between series and shunt damping methods lies in the absolute magnitude of their response in the pass and reject regions. Series damping resistances, although small (about 0.2 to 0.3 of the characteristic impedance Z_0), form a voltage divider with the terminating resistor which may reduce the output level in the passband to 50 per cent of the input level. They do not adversely affect the high attenuation level in the reject region. Damping resistances for the shunt method are of the same order of magnitude as Z_0, but since they are in parallel with the low ohmic resistances of the inductors, they have no appreciable effect in the passband. However, at frequencies above cutoff, when the inductive reactance becomes high, the damping resistances form shunt paths around the inductances which set the maximum attenuation obtainable from the filter network.

RECORDERS

PEN RECORDERS

Pen recorders are direct-writing oscillographs which produce permanent, immediately available records of electrical phenomena in the low audio-frequency range. The recording method is either an ink-fed or an electric stylus writing on a suitable paper chart. The basic components of a pen recorder are a pen motor for converting electrical signal currents into corresponding mechanical motions, a stylus for transcribing the motions of the pen motor, the chart paper, and a chart-drive mechanism for moving the chart paper past the stylus tip at a uniform speed. A pen motor is an electromechanical transducer which transforms an electrical input signal into a related mechanical output motion. Pen motors are current-operated devices and require more power than is usually available directly from the input transducer. Amplifiers with suitable voltage and power gains are therefore necessary to drive the pen motor. In addition to gain, these amplifiers are designed to complement the pen-motor characteristics to yield a wider, more uniform frequency-response range. When operated with their companion amplifiers, pen recorders have a flat (± 10 per cent) frequency response which extends from d-c to about 100 cps.

PEN-MOTOR CHARACTERISTICS. D'Arsonval and electrodynamic-type constructions are most common. The pen motor consists of an armature positioned in the field of a permanent magnet and restrained to one degree-of-freedom. Soft restoring springs keep the coil centered statically. Current flowing through the coil develops a force which moves the armature against the restoring springs. The amount of displacement is proportional to the magnitude of the input current. A stylus connected to the armature produces a graphical record of the input current magnitude.

Natural Frequency. The pen-motor elements approximate a single degree-of-freedom system which has an undamped natural frequency determined by the armature mass m and the spring stiffness k. This frequency f_n is given by the expression

$$f_n = \frac{1}{2\pi} \left(\frac{k}{m} \right)^{1/2} \tag{19.53}$$

A similar expression, using appropriate nomenclature, exists for single degree-of-freedom systems in rotation. The amplitude response of a single degree-of-freedom system to sinusoidal excitations is shown by Fig. 2.13, with fraction of critical damping as a parameter. For moderate amounts of damping (0.40 to 0.70 of critical) the frequency response remains substantially constant from d-c to frequencies near resonance; beyond resonance it falls off rapidly. The pen-motor natural frequency is consequently made as high as possible, consistent with other design factors, to provide a maximum upper frequency limit. Equation (19.53) indicates that this is accomplished by making the spring stiffness large and the mass small. Second-order effects (lower current sensitivity, increased torque, and stylus bending and whipping) preclude pen-motor natural frequencies much higher than the usual values of 35 to 60 cps.

Damping. Vibrating systems which are damped dissipate energy. Damping is termed viscous when the damping force is proportional to velocity. The damping coefficient c is important only in comparison to other system parameters, and is therefore usually expressed in terms of the *critical damping coefficient* c_c, which is the smallest damping for return to zero without overshoot. The addition of viscous damping to a simple system depresses the resonant amplitude peak and lowers the resonant frequency. At about 0.65 of critical damping, the system has the most uniform response over the widest frequency range. This is generally referred to as the *optimum damping ratio*. Pen motors usually are designed to this value.

Viscous damping in pen motors can be produced by three methods: air damping, magnetic damping from a shorted-turn metallic coil form, and magnetic damping resulting from back-generated currents flowing in the driving coil. The first is fixed by mechanical design. The second may be omitted to keep the coil weight low. Taken together, these two are not generally sufficient in D'Arsonval movements to provide the optimum damping, but can be employed effectively in electrodynamic types. Therefore the principal damping for D'Arsonval types is derived from the driver output impedance and is adjusted, in conjunction with the built-in damping, to raise the total damping to 0.65 of critical. Manufacturer's literature lists the correct drive impedance.

Phase Shift. The output of a damped single degree-of-freedom system lags its input by an amount which varies with frequency, as indicated by the phase-shift curves of Fig. 2.14. In order that components of different frequency in the output bear the same phase relationships to each other as do these components in the input, the delay time must be the same for all frequencies. Delay time for mechanical systems is defined in the same manner as for electrical filters. A damping which is 0.65 of critical produces the best combination of phase shift and delay.

Design Considerations. In addition to the conditions discussed above, the design of a pen motor represents a compromise of many other conflicting, interrelated requirements. To keep the armature mass small, the coil is made small dimensionally, and is often wound without a mandrel. Stylus mass constitutes a large percentage of the total armature mass, and is also kept to a minimum. These parts cannot be drastically re-

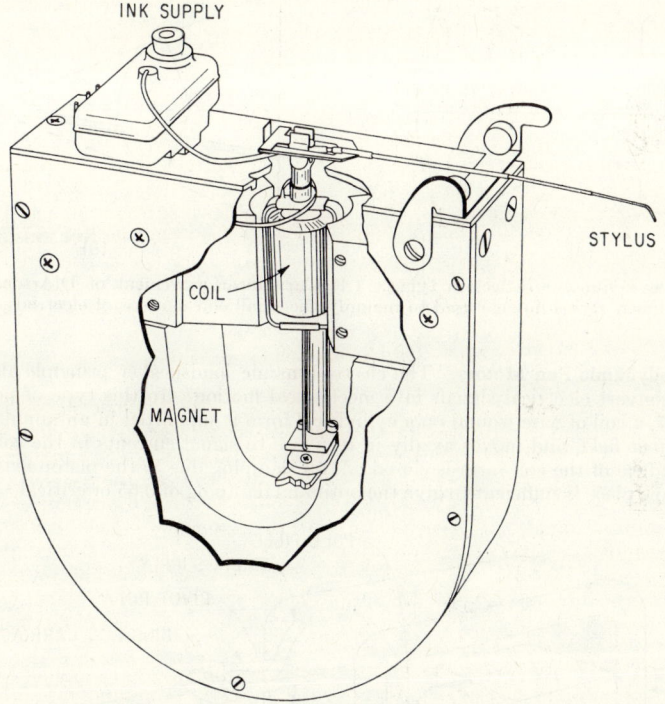

FIG. 19.35. D'Arsonval-type pen motor. (*Courtesy of Brush Instrument Division.*)

duced in size without increasing the demands on the other elements of the system. For example, the pen-motor torque, which must be sufficient to overcome stylus drag and produce a good speed of response, is proportional to coil size, coil turns, flux density, and drive current. Reducing the coil size and/or turns to reduce its mass requires that either or both the flux density or drive current be increased to maintain the same torque. Both are equally undesirable. Similarly, amplitude linearity requires that the stylus be made sturdy enough to preclude flexing during high amplitude vibration at the maximum operating frequency. Pen-motor spring stiffness is governed by two opposing requirements: the stiffness should be high for maximum natural frequency, but low for maximum deflection sensitivity.

D'Arsonval Pen Motors. The majority of pen-motor designs are essentially D'Arsonval-type movements, similar to that used in indicating meters, except that they are designed to develop the higher torque necessitated by the recording stylus. Figure 19.35 shows their principal features. An armature, wound of many turns of fine wire, is positioned between soft-iron pole pieces of a permanent magnet. It is supported axially on jeweled thrust bearings, but is free to rotate against the restoring springs. These springs, which may be torsion rods, also provide the zero centering of the mechanical system and may be used to connect the armature winding to the external circuit. Electromechanical coupling, between signal currents in the armature winding and the permanent-magnet field, produces a torque on the armature directly proportional to the signal current. The armature then rotates through an angle proportional to the magnitude of the exciting current, carrying the stylus with it. Stylus deflection is a section of an arc, producing records in curvilinear coordinates, as in Fig. 19.36A.

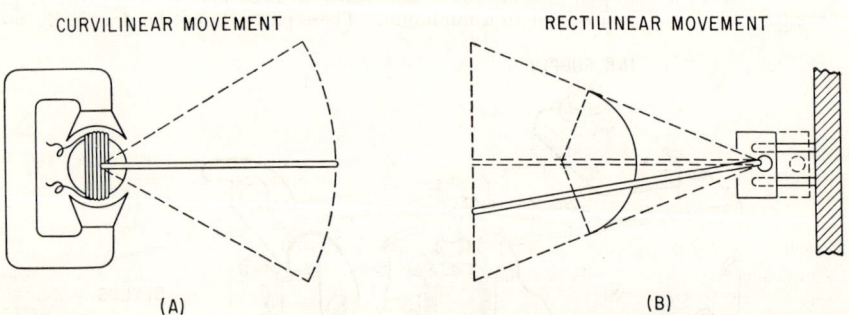

Fig. 19.36. Pen-motor deflection types. (A) Curvilinear movement of D'Arsonval types. (B) Rectilinear stylus linkage used to magnify the small coil motions of electrodynamic pen motors.

Electrodynamic Pen Motors. The electrodynamic loudspeaker principle also can be used to convert electrical signals into mechanical motion. In this type of pen motor, Fig. 19.37, a coil of wire wound on a cylindrical form is supported in an annular permanent-magnet field, and moves axially in response to signal currents in the coil. When the outer face of the coil form is closed off, air damping due to the piston action of the central pole piece is sufficient to give the optimum damping of 0.65 of critical. Since the

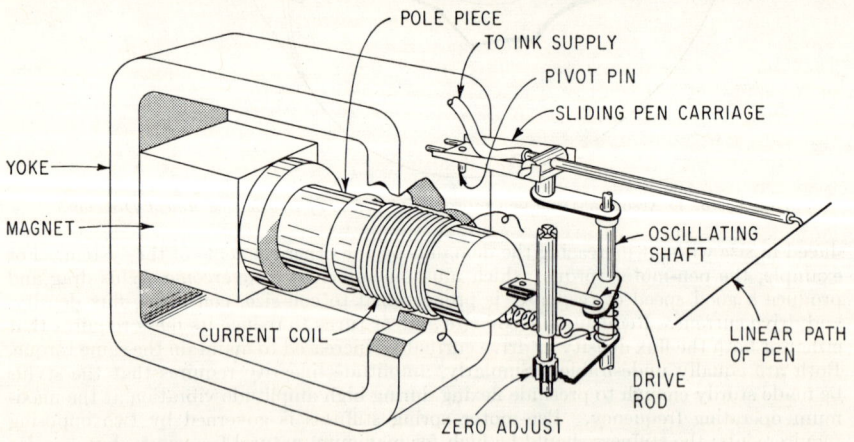

Fig. 19.37. Electrodynamic pen motor. (*Courtesy of Massa Laboratories.*)

damping is self-contained in the pen motor, the driver-amplifier output impedance can be any convenient value, provided it is not so low as to cause appreciable magnetic damping.

The maximum excursions produced by electrodynamic pen motors are limited by mechanical, electrical, and magnetic design factors, and in practical configurations are too small for direct recording. These motions can be magnified to a suitable size by a system of cranks and lever arms. In its simplest form a pivoted stylus is used, with the drive coupled close to the pivot point. Unfortunately this destroys the inherent rectilinear displacement of this movement and yields curvilinear records. The more complicated linkage system of Fig. 19.36B reconstructs the recorded signal in rectilinear coordinates by employing a sliding, rather than a fixed, pivot point. Stylus systems with extra linkages and bearings are more prone to damage from rough handling and to errors resulting from bearing backlash.

Friction. Friction between the stylus tip and chart paper has a very pronounced effect on the over-all performance of pen recorders. It tends to make the recorded amplitudes somewhat smaller than they should be, and is most noticeable for signals that have discrete steps or periodically return to zero. Because of friction, the stylus appears overdamped and approaches its steady-state position in a gradual manner. This sluggishness sometimes can be compensated by adjusting the amplifier damping to permit a small amount of overshoot. Friction also destroys the fine detail in the signal structure since these components usually produce torques which are smaller than the frictional threshold level.

The motion of the chart paper past the stylus tip aids in reducing the steady-state positional error caused by friction. The relative motion produces a drag force which is essentially in space phase quadrature with the stylus torque, rather than in phase opposition as occurs when the chart is stationary. The resultant of these forces is in a direction to reduce the position error and is most effective near zero where the restoring torque is small.

Typical stylus position errors due to friction in a well-adjusted pen motor are less than 0.25 mm, or approximately half the normal ink trace width.

PEN-MOTOR AMPLIFIERS. Sensitivity. Pen-motor power requirements are too high, and their internal impedance too low, to be driven directly from the signal source. Hence electronic amplifiers generally are supplied as an integral part of the recording system. In addition to their basic function of increasing the over-all sensitivity of the system, special characteristics are built into these amplifiers which (1) increase the flat frequency-response range of the entire system, (2) provide positioning controls for the stylus, and (3) protect the pen motor against inadvertent overload voltages. Direct-coupled amplifiers, which take advantage of the intrinsic d-c response of pen motors, are available with sensitivities of 50 millivolts/mm at good stability. Chopper stabilization of the amplifier can increase this sensitivity to 100 microvolts/mm. Carrier systems with built-in demodulators sensitive to 1 microvolt/mm also are available for those transducers which require external excitation.

Compensation. The upper limit of uniform pen-motor frequency response can be extended by compensation in the amplifiers. High-frequency boost circuits in the driver amplifier are adjusted to increase the amplifier gain with frequency at the same rate as the pen-motor rolloff, and substantially increase the range of uniform frequency response. Compensation by this method is feasible to about twice the pen-motor natural frequency. Beyond this frequency, the amount of compensation is too great to be handled by practical amplifier circuits. In this region the response degrades very rapidly, and may give rise to "ringing" on transients if the discontinuity in the frequency-response characteristic is too sharp. The frequency response of a typical compensated pen motor, Fig. 19.38, can be made flat up to about 100 cps or higher. Because the compensating networks are custom-fitted to pen motors of

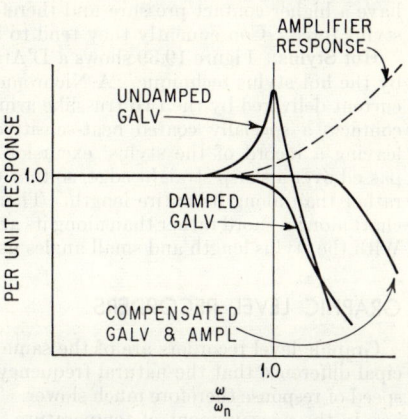

Fig. 19.38. Amplitude response of a compensated amplifier and pen motor.

a particular design, it is not generally feasible to interconnect pen motors and amplifiers of different manufacture without first checking the frequency response to see that it is satisfactory.

PEN-MOTOR STYLI. Ink Pens. The most common method of recording employs ink from a small pen at the stylus tip, writing on a paper chart pulled past the tip at a uniform speed. Capillary action draws the ink down the stylus tube from a stationary inkwell to which it is attached by a small piece of flexible tubing. This method is simple, economical, and where normal care is taken of the ink feed system, convenient. Clogging is the biggest trouble encountered with ink systems. It results from ink contamination from dust-laden air, and from sedimentation and drying as the result of long standby periods or too infrequent cleaning.

Ink splatter, due to the large centrifugal forces developed at the tip, sometimes is encountered with high-frequency, high-amplitude signals. It can best be eliminated by reducing the excursion of the stylus. Ink is normally suitable for recording over the temperature range from 25 to 100°F. With special inks this range can be extended to −45 to 135°F. Standard pen styli are approximately 3 in. long and produce a trace width of 0.015 in. Longer pens can be used to give greater deflection when signal frequencies are very low.

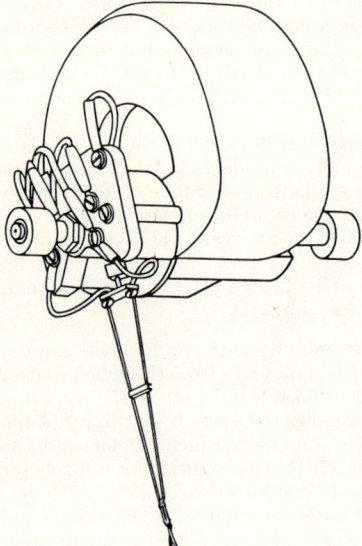

Fig. 19.39. D'Arsonval galvanometer equipped with a hot wire stylus. (*Courtesy of Sanborn Co.*)

Electric Styli. Electric styli give more reliable operation under adverse conditions of long unattended standby periods and low temperature. In this system the stylus is tipped with a small wire probe, and is connected to a source of high voltage (500 volts). Current flowing from the stylus tip through a specially prepared chart paper to the backing platen exposes a dark carbon layer under the chart surface, thus recording the passage of the stylus over the chart. Trace width from a 0.006-in.-diameter tip is approximately 0.003 in. Because of the smaller tip diameter, electric styli have a higher contact pressure and therefore more friction than ink styli for the same stylus force. Consequently they tend to be less accurate.

Hot Stylus. Figure 19.39 shows a D'Arsonval-type pen motor equipped for recording by the hot stylus technique. A Nichrome ribbon at the end of the stylus is heated by current delivered by the two trusslike arms of the structure. Where this heated ribbon contacts a specially coated heat-sensitive chart paper, it removes the surface coat, leaving a record of the stylus' excursion. In the unit shown the recording paper is passed over a sharp straight edge, so that the ribbon contacts the chart paper at a point rather than along its entire length. Then as the stylus deflects, the ribbon wipes the chart along a chord rather than along its arc, producing records in rectilinear coordinates. With the stylus length and small angles used, the amplitude error is negligible.

GRAPHIC LEVEL RECORDERS

Graphic level recorders are of the same generic type as pen recorders, with the principal difference that the natural frequency of their movements is much lower and their speed of response therefore much slower. They are most useful in the quasi-static range, i.e., in the measurement of temperature, current, pressure, etc., where the levels vary slowly with time. In this sense, they are level recorders which plot the value of a function averaged over a long period of time, rather than an oscillograph which plots the instantaneous magnitude with respect to time. Frequency range and speed of response are sacrificed for increased sensitivity.

TECHNIQUES. Level recorders can be divided into groups according to the methods employed to drive the recording stylus. Galvanometric types are simple meter movements which produce enough torque from the signal current to operate the stylus directly. Electronic amplification of the signal ahead of the galvanometer is possible, but requires highly stabilized circuits to eliminate amplifier instabilities over long periods of time at near d-c frequencies. Drift is eliminated when the circuit is arranged as a servo follow-up system. This technique employs an amplified error signal to drive

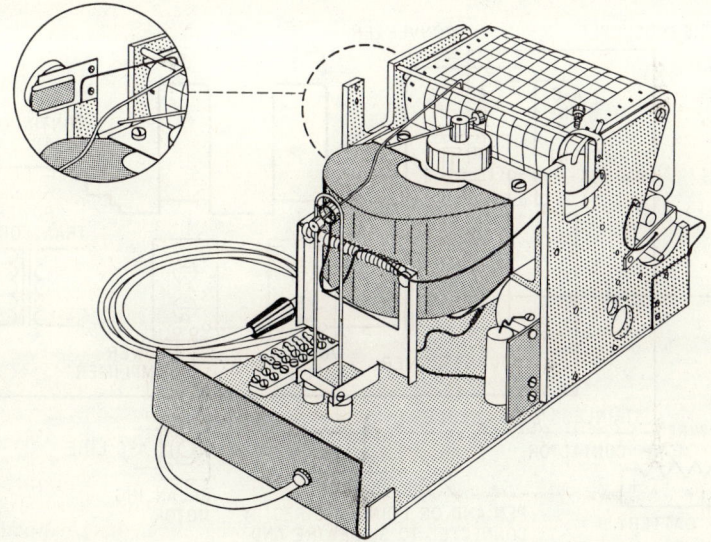

Fig. 19.40. Graphic level recorder with rectilinear stylus linkage. (*Courtesy of Texas Instrument Co.*)

the stylus in the proper direction to restore balance. Self-balancing systems most often employ a potentiometer or photoelectric element in the error-determining network, although other error-sensing elements may be used. Recording is usually by ink on a rolled paper chart contained within the instrument, and may be curvilinear or rectilinear depending on the configuration of the stylus system. The paper is pulled past the stylus at a uniform speed by a suitable motor and speed selector mechanism. When the follow-up system develops considerable torque, wax-coated paper charts, ball-point pen styli, and electric recording also can be employed. So-called X-Y recorders are useful where it is desired to plot a dependent variable as a function of an independent variable other than time. Both the X and Y axes are servo-controlled, and are mechanically arranged to operate over their full-scale ranges without cross coupling.

Galvanometric Types. In this type instrument, Fig. 19.40, the torque developed by the galvanometer is sufficient to operate the stylus directly. Meter movements may be either an electrodynamometer type for measuring a-c components at power-line frequencies or the D'Arsonval type for d-c measurements. Undamped natural frequencies are usually less than 2 cps. The compliant restoring springs and large coil sizes permitted by these low natural frequencies allow sensitivities down to 1 milliampere for full-scale deflection. For more sensitive units the torque must be increased by auxiliary means. Damping is mostly magnetic, obtained from an aluminum vane or a nonmagnetic coil mandrel, moving between the poles of a permanent magnet. Stylus flexibility is no problem at these low frequencies. The stylus can be made quite long to give a large tip deflection for a small deflection angle, with good amplitude linearity. Charts are customarily 6 in. wide with an active width of 4½ in., and are ruled in curvilinear coordinates to match the stylus arc. Level recorders are also available which employ a linkage system to convert the customary curvilinear motions to rectilinear motions.

Self-balancing Potentiometers. When it is desired to record signals smaller than can be handled conveniently on galvanometric-type recorders, auxiliary means must be provided to increase the signal levels to usable values. The self-balancing potentiometer system of Fig. 19.41 is simple, stable, and capable of great sensitivity and accuracy. This system operates on an error signal, using a closed-loop servo follow-up system to adjust continuously the voltage across a motor-operated potentiometer to be equal to the input-

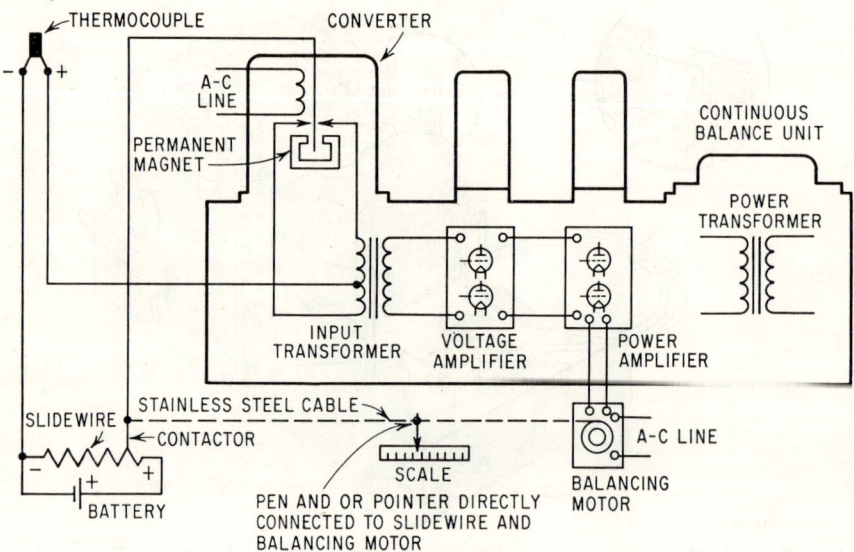

Fig. 19.41. Block diagram of a self-balancing potentiometer system. (*Courtesy of Brown Instrument Division.*)

signal voltage. The recorder reference voltage and input voltage form two legs of a bridge. They are compared at the amplifier input by a synchronous chopper running at line frequency. The circuit is so arranged that no error voltage is induced at balance, but when a change in input signal occurs, an error current flows in the transformer primary whose phase depends on the relative magnitudes of the input and reference voltages. The error signal is amplified by a conventional high-gain a-c amplifier and used to excite one winding of a two-phase induction motor. The other winding is connected to the a-c line. The phase of the amplified error signal relative to the line is always such that the motor rotates in the direction which will reduce the error signal to zero, i.e., restore balance between input and reference signals. A stylus connected to the potentiometer's movable arm records the corrections necessary to restore equilibrium and is therefore a plot of the input voltage.

The accuracy of this system depends only on the precision of the potentiometer and is entirely independent of the amplifier gain. Furthermore, a-c amplifier principles can be used on the chopped input signal to eliminate drift problems and permit the high gains necessary to minimize backlash and balancing errors. Controllable velocity-type damping is usually included as part of the electronic circuitry to minimize overshoot, jitter, and hunting.

Graphic level recorders of the self-balancing potentiometer type often are used as level recorders in acoustic measurements. Reference potentiometers with logarithmic tapers are used in this work, producing records in decibels above a reference datum. Full-scale response time is approximately 1 sec.

Photoelectric Galvanometer Recorder. In this type of recorder the measuring galvanometer is relieved of the stylus burden and is required only to rotate a small mirror to control a closed-loop servo follow-up system. The stylus is attached to a second, more powerful galvanometer driven by an electronic amplifier whose input is obtained from a pair of phototubes. The optical system shown in Fig. 19.42 is arranged so that when the mirrors attached to the measuring and recording galvanometers are parallel, the light beams reaching both phototubes are balanced and no action is required of

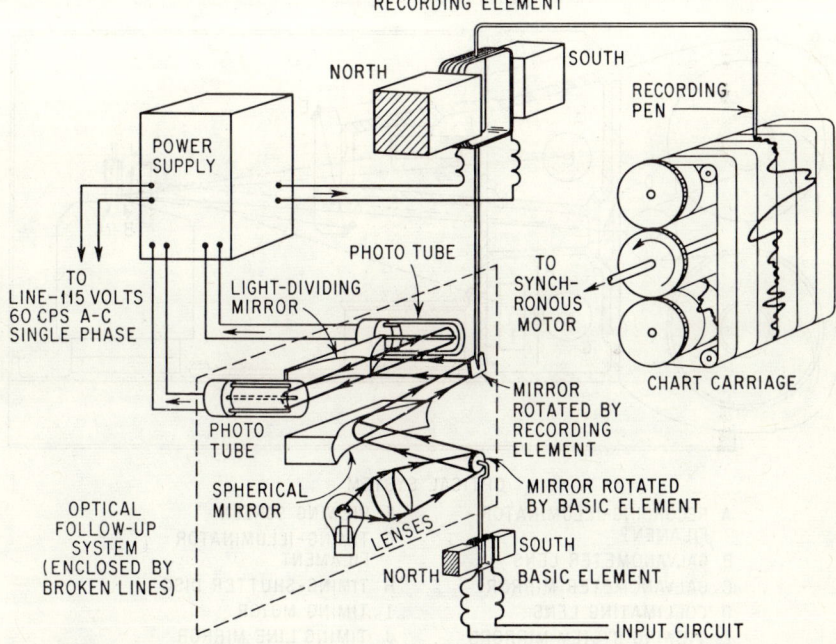

FIG. 19.42. Optical and electronic arrangements of a photoelectric recorder. (*Courtesy of General Electric Co.*)

the recording galvanometer. Changes of input current are accompanied by changes in angular position of the measuring galvanometer mirror, and result in one phototube being more intensely illuminated than the other. The amplifier input unbalance then is amplified and applied to the recording galvanometer as a current of the proper magnitude and phase to cause it to rotate in the direction which restores balance, namely, with the recording and measuring galvanometer mirrors again parallel.

Since the optical link imposes no load on the measuring galvanometer, this type of recorder has the sensitivity and response characteristics of optical galvanometers. The frequency range extends from d-c to about 5 cps. Sensitivities of 1 microampere full-scale are available with good stability and accuracy and with negligible errors from power-line disturbances or changes in amplifier characteristics.

MAGNETIC OSCILLOGRAPHS

Magnetic oscillographs photographically record the time-history of a variable by means of a beam of light reflected from a small mirror attached to the moving element of the recording galvanometer. Sometimes they are referred to as *string oscillographs*.

19-42 SHOCK AND VIBRATION INSTRUMENTATION

Substitution of a massless light beam for the mechanical stylus used by direct-writing oscillographs greatly reduces the torque requirements on the galvanometer. Galvanometers then may be made quite small and light. In combination with the long optical lever arm, this miniaturization permits high natural frequencies at reasonable deflection sensitivities.

Coil-type galvanometers are available which have a wide range of sensitivities and natural frequencies up to about 5,000 cps. When damped to the optimum value of 0.65 of critical (see *Pen Recorders, Damping*), they are capable of a flat frequency response within 2 per cent from d-c to about 3,000 cps. Bifilar galvanometers can be made with somewhat higher natural frequencies. Some low-frequency units, usable to several

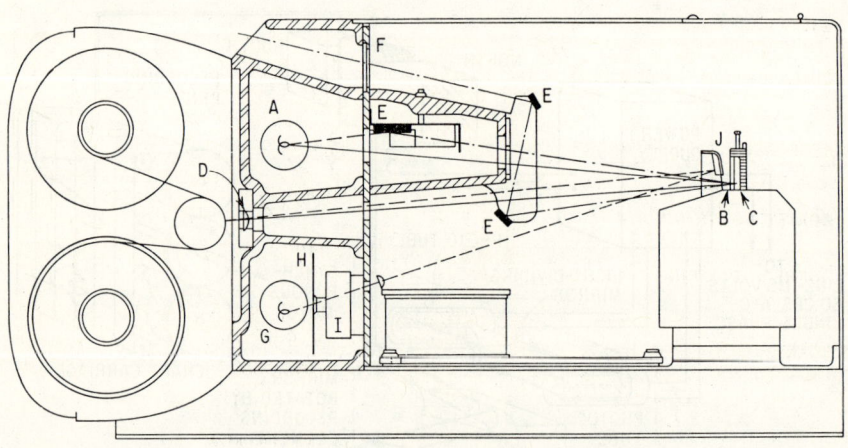

OPTICAL SYSTEM

A RECORDING-ILLUMINATOR FILAMENT
B GALVANOMETER LENS
C GALVANOMETER MIRROR
D COLLIMATING LENS
E VIEWING SYSTEM MIRRORS
F VIEWING SCREEN
G TIMING-ILLUMINATOR FILAMENT
H TIMING-SHUTTER DISC
I TIMING MOTOR
J TIMING LINE MIRROR

FIG. 19.43. Mechanical arrangement of a typical magnetic oscillograph. (*Courtesy of Consolidated Electrodynamics Corp.*)

hundred cycles per second, have been designed to operate directly from strain-gage bridge circuits. Electronic amplification is mandatory with higher-frequency galvanometers due to their greatly reduced sensitivity.

MECHANICAL ARRANGEMENTS. A schematic of the mechanical arrangement of a typical magnetic oscillograph is shown in Fig. 19.43. The main optical path consists of the direct beam from the recording lamp to the galvanometer mirrors and the reflected beam, modulated by the galvanometer motion, to a collimating lens and recording camera. Simultaneous viewing of the recorded signal can be provided by a beam-splitting mirror which projects part of the modulated beam onto a ground-glass viewing screen. Timing lines are recorded on the record by chopping the beam from a second light source.

In spite of their unimpressive frequency range, as compared to cathode-ray oscillographs and tape recorders, the basic simplicity and multichannel capabilities of magnetic oscillographs make them quite popular—particularly in field applications. Complete systems containing from 12 to 50 galvanometers are available.

GALVANOMETERS. Coil Type. Coil-type galvanometers are essentially miniature D'Arsonval movements, as shown in Fig. 19.44. The coil is air-wound of many turns of fine wire and made relatively long and narrow to minimize its inertia. It is positioned between magnetic pole-piece extensions with the plane of the coil parallel to the mag-

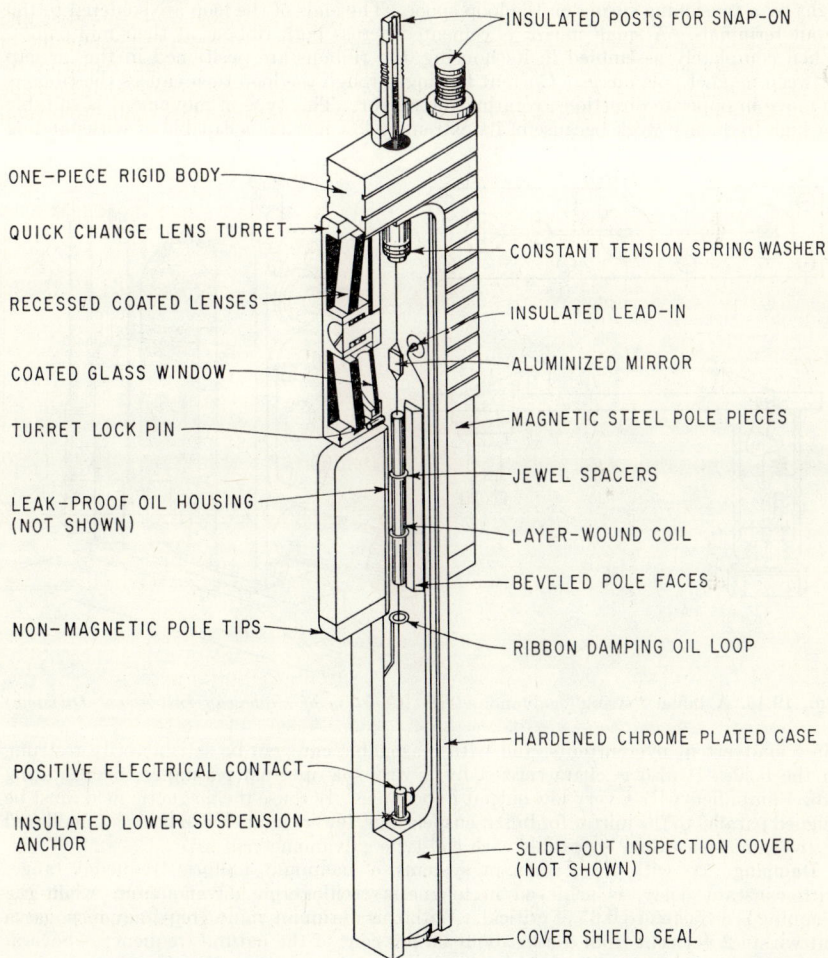

Fig. 19.44. Cutaway drawing of a coil-type galvanometer. (*Courtesy of Consolidated Electrodynamics Corp.*)

netic field. It is supported by tension ribbons which serve as the torsional restoring springs and the electrical connections to the coil. A mirror, cemented to the upper yoke, rotates with the coil. The entire system is sealed in a housing which may be filled with damping fluid.

Because of their small transverse dimensions, many coil-type galvanometers can be accommodated in a common magnet block, making this type of construction well suited for multichannel recording. In multiple assemblies, however, the removal of one galvanometer modifies the sensitivities of those adjacent to it. When these changes cannot be tolerated or compensated for by an over-all system calibration, dummy units of the same magnetic reluctance as the real galvanometer can be inserted in the vacancy.

Bifilar String. The bifilar string galvanometer, Fig. 19.45, is simpler in construction than the coil type, and is capable of much higher natural frequencies. The suspension is a single loop of conducting ribbon passed across two insulating bridges and stretched tight by a tensioning spring on the loop spool. The ends of the loop are soldered to the input terminals. A small mirror is cemented across both ribbons at their mid-points. When completely assembled in its housing, the ribbons are positioned in the air gap between magnet pole pieces. Current flowing through the loop then causes the ribbons to move in opposite directions, rotating the mirror. This type of movement is suitable for high-frequency work because of its extremely low inertia, is capable of withstanding

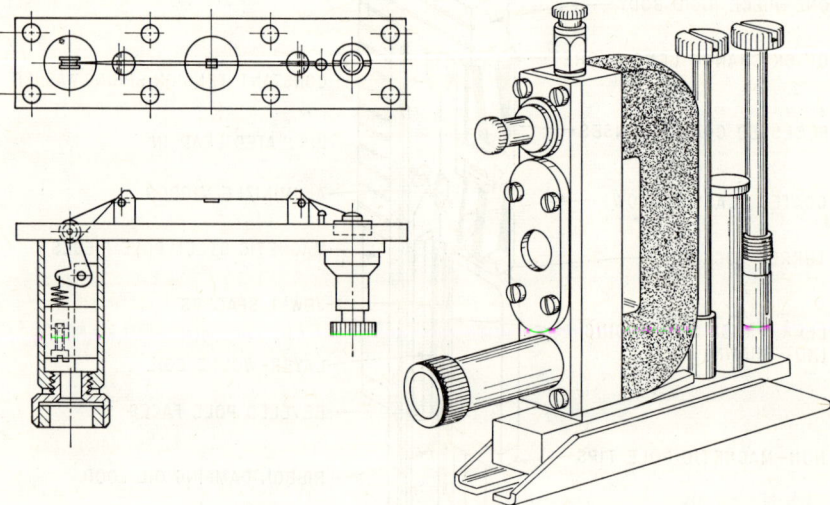

FIG. 19.45. A bifilar "string" galvanometer. (*Courtesy of Hathaway Instrument Division.*)

large inadvertent overcurrents, and with reasonable care can be satisfactorily restrung in the field. It also is characterized by a very low internal resistance, necessitating power amplifiers with a very low output impedance. Because the magnetic field must be aligned parallel to the mirror for bifilar suspensions, fewer channels can be accommodated in the same space than is possible with coil-type galvanometers.

Damping. As with other resonant systems, a maximum, uniform frequency range, with constant delay, is achieved in magnetic oscillograph galvanometers when the damping is adjusted to 0.65 of critical. With this optimum value, frequency response is flat within 2 per cent from d-c to about 60 per cent of the natural frequency. Several methods are used to provide damping. Air damping, which is a function of mechanical design, is negligible except for very low-frequency galvanometers, where it may reach 2 or 3 per cent of critical. Magnetic damping, resulting from the back-generated currents flowing in the driver impedance, is the most convenient for coil-type galvanometers. It is popular because the coupling network can readily be varied to suit special needs. Unfortunately the optimum damping resistance varies inversely with the natural frequency and a point is reached, around 600 cps in practical galvanometers, where the optimum damping resistance becomes smaller than the galvanometer's internal resistance. When this condition is obtained, galvanometers cannot be damped properly, even with their terminals short-circuited. For this reason, fluid damping is provided for high-frequency coil-type galvanometers and for all bifilar galvanometers which inherently have an extremely low internal resistance.

Fluids used for damping are especially compounded to the proper viscosity to provide optimum damping (see Fig. 16.8). The stability of silicone oils over long periods of time and temperature conditions, with a minimum change in viscosity, makes them particularly useful in this application. Thermostatically controlled, heated magnet blocks also minimize viscosity changes under adverse environmental conditions. Damping of a fluid-filled galvanometer is controlled almost entirely by the fluid viscosity, and is virtually independent of output circuit driving impedance unless it is extremely low. Fluid damping coefficients are sometimes purposely made several per cent lower than optimum with the intent that they will be trimmed to the correct value by the driving impedance.

COUPLING NETWORKS. Manufacturers' literature specifies the proper terminating resistance needed to properly damp a given galvanometer to the optimum value of 0.65 of critical. This is the total effective resistance that the coupling network and driving source should have, as seen from the galvanometer's terminals, and is in addition to the galvanometer's own internal resistance. For other degrees of damping the required network resistance can be calculated from

$$R_d = \frac{0.65(R_D + R_g)}{c/c_c} - R_g \qquad (19.54)$$

where c/c_c = fraction of critical damping
R_g = galvanometer's internal resistance
R_D = external resistance specified to give 0.65 per cent of critical damping
R_d = network resistance to produce a specified degree of damping

Resistive networks have properties which make them useful for coupling a galvanometer to a driving source. The most general form is composed of three resistance elements arranged in a T or π configuration. Conversion from a T to a π network, or vice versa, can be accomplished without in any way affecting their external characteristics. The three independent resistors permit three properties of the network to be specified arbitrarily. Usually the network input and output resistances are chosen, along with a term which describes voltage, current, or power transfer from input to output. These networks are passive, and can never exhibit any gain—only attenuation. Resistive networks are independent of frequency.

Table 19.2 gives design equations for several networks useful for coupling a galvanometer to its driver amplifier. It is assumed that the amplifier output resistance R_0 and the galvanometer's internal resistance R_g are known. The network parameters are then chosen so that (1) the network output resistance R_d [Eq. (19.54)] satisfies the galvanometer damping requirements; (2) the network input impedance R_1 is some suitable value, usually taken equal to R_0; and (3) the network transfer conductance $S = i_g/e$ gives the desired sensitivity. Although S is mathematically arbitrary, its range must be restricted in practice to those values which produce positive resistance values for R_1, R_2, and R_3. In normal configurations where R_0 is much larger than R_g, the series arm R_2 is the first to go negative when S becomes too large. The maximum value that S can have practically in a T network is too involved to express mathematically, but can be approximated as somewhat less than $1/(R_0 + R_1)$. The sensitivity factor S as defined here does not include amplifier gain or galvanometer deflection sensitivity. These should be accounted for to compute the over-all system capabilities.

Simpler L networks using two resistances also can be used, at the expense of losing one arbitrary network parameter. Network output resistance and transfer sensitivity are the more important and usually are specified, leaving the network input resistance uncontrolled. The two possible arrangements for the L network are shown in Table 19.2, columns B and C, along with their design equations. A consistently higher network input resistance is obtained for the configuration with the series leg pointing toward the drive source and is to be preferred. The choice of S also is restricted with this network to values which produce positive resistance elements. Its maximum value is relatively simple to evaluate, however.

These coupling networks can be degenerated one step further to the simple shunt and series resistance circuits of columns D and E. The one resistance element permits only the adjustment of R_d.

Table 19.2. Galvanometer Coupling Networks

	(A)	(B)	(C)	(D)	(E)
ADJUSTABLE PARAMETERS	R_L, R_d, S	R_d, S	R_d, S	$R_d (R_d < R_0)$	$R_d (R_d > R_0)$
DESIGN EQUATIONS	$R_1 = \dfrac{1}{\dfrac{1}{(R_d+R_g)} - S} - R_0$ $R_2 = \dfrac{1}{\dfrac{1}{(R_L+R_0)(R_d+R_g)} - S} - R_g$ $R_3 = \dfrac{S}{\dfrac{1}{(R_L+R_0)(R_d+R_g)} - S^2}$ $\dfrac{1}{(R_L-R_0)} - S$ $-S^2$	$R_1 = \dfrac{R_d}{S(R_d+R_g)} - R_0$ $R_3 = \dfrac{R_d}{1-S(R_d+R_g)}$ $R_L = R_1 + \dfrac{R_3 R_g}{(R_3+R_g)}$	$R_2 = R_d - SR_0(R_d+R_g)$ $R_3 = \dfrac{SR_0(R_d+R_g)}{1-S(R_d+R_g)}$ $R_L = \dfrac{R_3(R_2+R_g)}{R_2+R_3+R_g}$	$R_3 = \dfrac{R_0 R_d}{R_0 - R_d}$ $R_L = \dfrac{R_3 R_g}{R_3+R_g}$ $S = \dfrac{R_3}{(R_0+R_3)(R_3+R_g)}$	$R_1 = R_d - R_0$ $R_L = R_1 + R_g$ $S = \dfrac{1}{(R_0+R_1+R_g)}$
UNCONTROLLED PARAMETERS	NONE				

$S = i_g/e$. WHILE S MAY BE ARBITRARY, IT MUST BE RESTRICTED TO VALUES SUCH THAT R_1, R_2, AND R_3 ARE POSITIVE

PHOTOGRAPHIC RECORDING. Cameras. Magnetic oscillograph cameras are of two general types: the continuous-film magazine camera and the drum-type camera. In the magazine camera, the record is passed from a supply spool, over a drive roller where it is exposed to the galvanometer light beams, and then wound up on a take-up spool. The record is driven at constant speed by the drive roller with proper tension being maintained on the supply and take-up spools by friction clutches. Magazine capacity is usually sufficient to hold between 100 and 200 ft of record, depending on the thickness of the record base. Because this method involves the transfer of recording material from one spool to another, record speeds are limited to a maximum of about 10 ft/sec. Magazine cameras are useful where the test record is of long duration or where a series of short test records are to be made.

Drum cameras are capable of recording at high speed for a short duration. In this camera, a length of recording material is wrapped around the periphery of a drum contained within the camera housing and the drum rotated at high speed. The camera shutter is then opened in synchronism with the event to be recorded, and is closed again after one revolution of the drum. Record speeds up to 30 ft/sec are obtained easily. Synchronization problems and short record lengths make drum cameras less attractive than magazine types except where their high-speed capabilities are necessary to resolve trace details.*

Record Materials. A wide range of recording papers and films are available commercially which are especially made for magnetic oscillograph recording. The more common types, with their characteristics, are given in Table 19.3. Medium-speed recording papers, having an ASA speed range up to about 15, are to be preferred for all normal recording, since they give a higher-quality record with sharper contrast and are less likely to fog at low camera speeds or in the developing process. High-speed emulsions have a wider exposure latitude than normal emulsions, however, and are more suitable for simultaneously recording high- and low-velocity traces. Their contrast is generally not as good. Use of the newer thin-base papers will nearly double the magazine footage capacity, as compared to standard-weight papers.

Recording papers sensitive to ultraviolet light have been developed which require no chemical development to produce an immediately usable record. The paper is exposed by reflected light beams from an intense ultraviolet recording lamp, and is developed by exposure to normal room illumination. The record is stable for long periods of time under indoor conditions without further processing, and may be made permanent by chemical fixing.

Development. Short records, up to about 5 ft in length, can be hand-developed using normal darkroom techniques of development and fixing. Hand processing becomes extremely arduous for long or continuous film lengths and generally has been replaced by machine development using a stabilization process. These machines are capable of developing a continuous record several hundred feet long at a rate of several feet per minute. Records are processed by rapidly passing them through tanks containing developer, a stop bath, a stabilizing agent, and then through a drying drum. The principal difference of this method is that the stabilizers do not remove the undeveloped silver halide crystals as is done with a fixer (hypo), but render them essentially insensitive to light, heat, and humidity. A stabilized record is sufficiently permanent to last several years with normal care.

CATHODE-RAY OSCILLOGRAPHS

It has become accepted practice to use the terms oscilloscope and oscillograph interchangeably for cathode-ray-tube instruments, without regard to the etymological connotation of a visual or recorded presentation. In less formal fields it is simply a "scope." The popularity of cathode-ray oscillographs in both visual and recording applications is due mainly to the all-electronic nature of the cathode-ray tube. By using a virtually massless electron beam as the pointer, there is no mechanical natural frequency to limit its upper-frequency response. In fact, for most applications both the high- and the low-frequency cutoff points are set by the deflection amplifiers rather than by the cathode-ray tube.

* A single trace on an oscilloscope may be photographed conveniently with a Polaroid camera.

Table 19.3. Oscillograph Recording Paper Characteristics

Product	Speed, ASA	Base	Average thickness, in.	Spectral sensitivity	Wratten safelight recommendation	Processing
EASTMAN						
Linagraph 480	10	Alpha-cellulose paper	0.0051	Orthochromatic	Series 2, or Series 1 with caution	Linagraph Developer for manual processing
Linagraph 483	10	Alpha-cellulose paper	0.0030	Orthochromatic	Series 2, or Series 1 with caution	
Linagraph 809	10	Alpha-cellulose paper	0.0041	Orthochromatic	Series 2, or Series 1 with caution	
Linagraph Drift Survey Film	4	Acetate	0.0061	Orthochromatic	Series 2, or Series 1 with caution	
Verichrome Pan Film	125	Acetate	0.0042	Panchromatic	Series 3	D-19
Linagraph Ortho Film	160	Acetate	0.0061	Orthochromatic	Series 2	
Linagraph Pan Film	350	Acetate	0.0061	Panchromatic	Series 3 or total darkness	
Royal-X Pan Film	1,600	Acetate	......	Panchromatic	Total darkness	DK-50, DK-60a
DU PONT						
Lino-Writ 1 Type B	6	Rag and alpha-cellulose paper	0.0045	Orthochromatic	Series 2	Lino-Writ Developer for manual processing
Lino-Writ 1 Type W	5½	All rag paper	0.0030	Orthochromatic	Series 2	
Lino-Writ 2 Type B	11	Rag and alpha-cellulose paper	0.0045	Orthochromatic	Series 2	
Lino-Writ 2 Type W	10	All rag paper	0.0030	Orthochromatic	Series 2	
Lino-Writ 3 Type B	40	Rag and alpha-cellulose paper	0.0045	Orthochromatic	Series 2	
Lino-Writ 3 Type W	44	All rag paper	0.0030	Orthochromatic	Series 2	
Lino-Writ 4	50	All rag paper	0.0025	Orthochromatic	Series 2	
GRANT						
Linatrace 3 Thin	40	All rag paper	0.0039	Orthochromatic	Series 2	D-72 Developer for manual processing
Linatrace 3 Vellum	35	Translucent paper	0.0037	Orthochromatic	Series 2	
Linatrace 3 Waterproof	30	Waterproof paper	0.0037	Orthochromatic	Series 2	D-72 Developer for manual processing. Not suitable for hot drum drying
AGFA						
Isopan Record	640 / 2,000	Acetate	0.0058	Panchromatic	Total darkness	D-76, UFG, Agfa Atomal, Agfa Rodinal

19-48

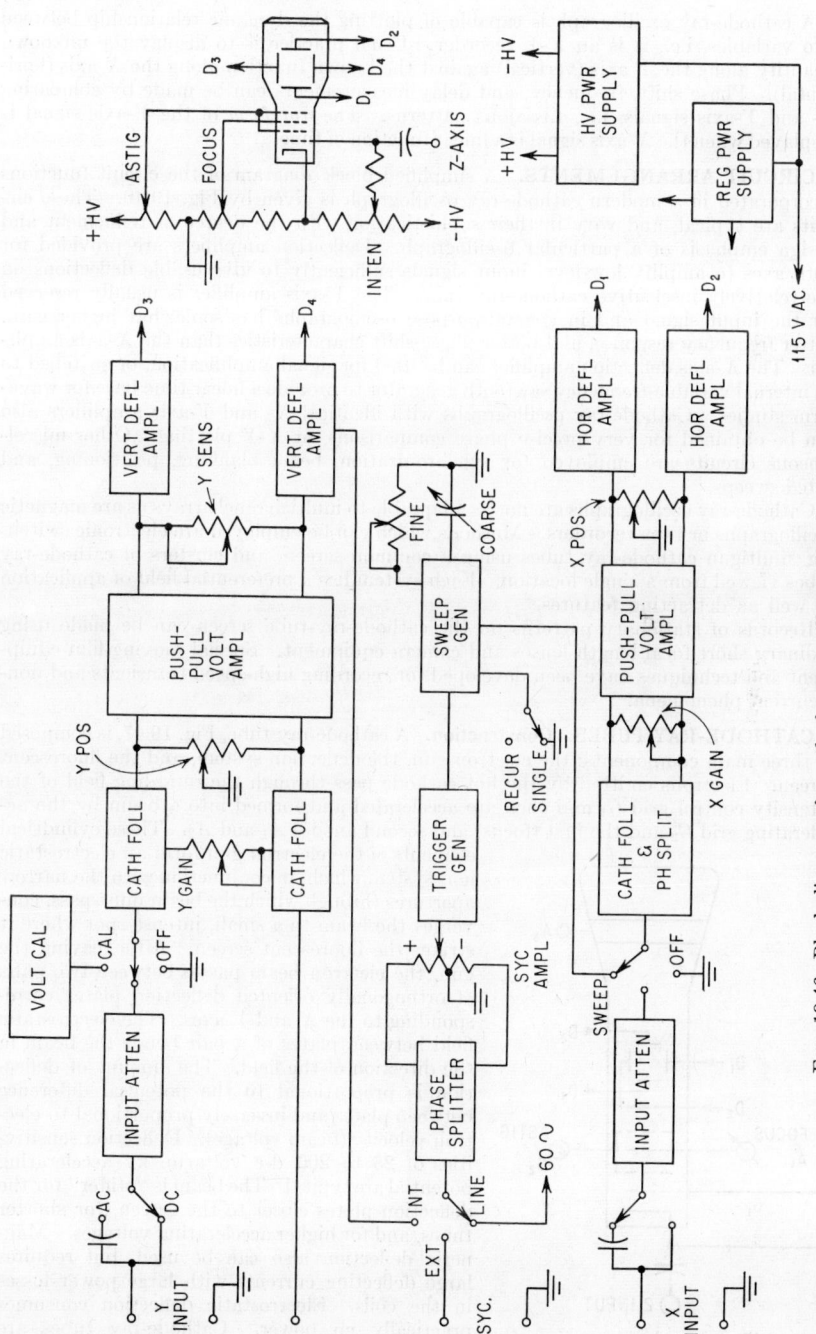

Fig. 19.46. Block diagram of a cathode-ray oscillograph indicating circuit functions.

A cathode-ray oscillograph is capable of plotting the dynamic relationship between two variables, i.e., it is an X-Y recorder. Usual practice is to display the unknown quantity along the Y axis (vertical) against the known function along the X axis (horizontal). Phase shift, frequency, and delay measurements can be made by comparing X- and Y-axis signals, i.e., Lissajous patterns. The *waveform* of the Y-axis signal is displayed when the X-axis signal is a linear function of time.

CIRCUIT ARRANGEMENTS. A simplified block diagram of the circuit functions incorporated in a modern cathode-ray oscillograph is given by Fig. 19.46. These circuits are typical, and vary in their sophistication with the degree of refinement and design emphasis of a particular oscillograph. Deflection amplifiers are provided for both axes to amplify low-level input signals sufficiently to give usable deflections on the relatively insensitive cathode-ray tube. The Y-axis amplifier is usually reserved for the input signal and in general-purpose oscillographs has somewhat higher gain, better frequency response, and better phase-shift characteristics than the X-axis amplifier. The X-axis deflection amplifier can be used for signal amplification, or switched to an internal variable-frequency sawtooth generator to provide a linear time base for waveform studies. Cathode-ray oscillographs with identical X- and Y-axis amplifiers also can be obtained for very precise phase comparisons, or X-Y plotting. Other miscellaneous circuits are employed for synchronization, beam blanking, positioning, and gated sweeps.

Cathode-ray oscillographs are not as adaptable to multichannel arrays as are magnetic oscillographs or tape recorders. Methods which can be employed are electronic switching, multigun cathode-ray tubes using a common screen, and clusters of cathode-ray tubes viewed from a single location. Each system has a preferential field of application as well as detracting features.

Records of stationary patterns on the cathode-ray-tube screen can be made using ordinary short focal length lenses and camera equipment. Special moving-film equipment and techniques have been developed for recording high-speed transients and non-recurrent phenomena.

CATHODE-RAY TUBES. Construction. A cathode-ray tube, Fig. 19.47, is composed of three main components: the electron gun, the deflection system, and the fluorescent screen. Electrons emitted by the hot cathode pass through the retarding field of the intensity control grid G_1 and then are accelerated and formed into a beam by the accelerating grid G_2 and the first (focus) and second anodes A_1 and A_2. These cylindrical elements of the electron gun form an electrostatic lens system which, in conjunction with the narrow apertures through which the beam must pass, converges the beam to a small, intense spot where it strikes the fluorescent screen. After leaving the gun, the electron beam passes between two pairs of orthogonally oriented deflection plates corresponding to the X and Y axes. The electrostatic field between plates of a pair bends the beam in the direction of the field. The amount of deflection is proportional to the potential difference between plates and inversely proportional to electron velocity (beam voltage). Deflection sensitivities of 25 to 200 d-c volts/in./kv accelerating potential are typical. The beam is "stiffer" for the deflection plates closer to the screen, for shorter tubes, and for higher accelerating voltages. Magnetic deflection also can be used, but requires large deflecting currents with large power losses in the coils. Electrostatic deflection consumes practically no power. Cathode-ray tubes are usable up to the frequency where the deflecting field begins to reverse before a particular electron

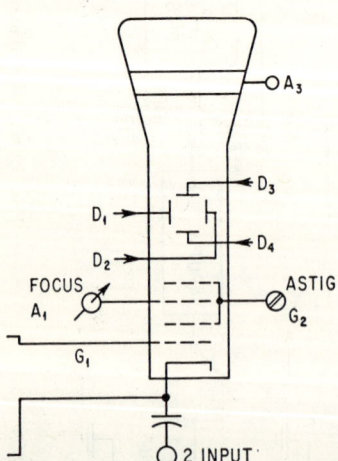

FIG. 19.47. Internal construction of a typical cathode-ray tube. (*Courtesy of A. B. Du Mont Laboratories, Inc.*)

in the beam has completed its transit. In typical cathode-ray tubes this point is seldom reached below 100 megacycles.

Intensity. The brightness of the trace on the cathode-ray-tube screen depends upon both the velocity of the electrons at impingement and the number of electrons in the beam. These parameters are controlled by the over-all accelerating potential and the intensity control grid G_1, respectively. Most modern cathode-ray tubes are "post-deflection acceleration" types, wherein the electron velocity through the deflecting fields is kept relatively low to provide good deflection sensitivity, and the principal accelerating field is applied by a conductive coating on the inside of the cone after the electron beam has been deflected. Post-deflection accelerator-type cathode-ray tubes have deflection factors several times larger than equivalent monoaccelerator types operated at the same over-all accelerating potential. Trace brightness can be adjusted by the intensity control grid G_1 to produce easily observable traces under normal indirect ambient light conditions. Beyond an optimum brightness, further advancement of the intensity control tends to defocus the beam, particularly at screen extremes, without any real increase in trace contrast. Even with a maximum trace brightness and stationary patterns damage to the screen is unlikely as long as the trace is spread out over a reasonable portion of the screen. However, an undeflected spot should not be made any brighter than necessary, nor allowed to remain on the screen for any time longer than required (as in moving-film recording), as it will rapidly cause screen burning with subsequent flaking of the screen at the affected area.

Astigmatism. Astigmatism in cathode-ray tubes is the defocusing and spot distortion which occurs when a previously focused spot is moved to a new position on the screen. Astigmatism can be minimized by adjusting the second anode potential to the average potential of the deflection plates. If the average potentials of the X- and Y-axis deflection plates are too different, optimum focus over the entire screen cannot be obtained for any setting of the astigmatism control, shown in Fig. 19.47.

Fluorescent Screens. The cathode-ray-tube screen is coated with a "phosphor" which emits light when bombarded by a beam of electrons. By varying the composition of the screen material, various combinations of fluorescence, phosphorescence, and persistence can be obtained to suit special applications. These characteristics have been standardized and are designated by a P number included in the cathode-ray-tube type number. Table 19.4 lists the characteristics of the phosphors most commonly used in oscillographic work. Cathode-ray tubes can be obtained with any of these screens. It is

Table 19.4. Cathode-ray-tube Screens

Phosphor type No.	Color		Persistence and decay time	Spectrum range (peak), angstroms	Normal service
	Fluorescence	Phosphorescence			
P1	Green	Green	Medium 20 milliseconds	4,900–5,800 (5,250)	General visual oscillographic use
P4	White	White	Medium 33 milliseconds	3,260–7,040 (4,100)(5,400)	TV receivers
P5	Blue	Blue	Very short 18 microseconds	3,480–5,750 (4,300)	Very-high-speed moving-film recording
P7	Blue-white	Yellow	Blue, short White, long 2-layer screen	3,900–6,500 (4,400)(5,580)	Low-speed recurrent or medium-speed nonrecurrent signals
P11	Blue	Blue	Short 2 milliseconds	4,000–5,500 (4,600)	High-speed moving-film records
P12	Orange	Orange	Medium–long	5,450–6,800 (5,900)	Low- and medium-speed recurrent data
P14	Purple	Orange	Purple, short Orange, long 2-layer screen	3,900–7,100 (4,400)(6,010)	Low- and medium-speed nonrecurrent transient

therefore possible to convert an oscillograph from one type of service to another merely by changing the cathode-ray tube.

The P1 screen produces an intense green fluorescence of medium persistence that is most satisfactory for general-service visual observation. For all normal recording work, the P11 screen is to be preferred. It produces a brilliant actinic blue spot which peaks in the spectral region where film emulsions are the most sensitive. Persistence of the P11 screen is sufficiently short to preclude smearing of the recorded image except at the highest writing speeds. It is also satisfactory for recurrent visual observations above 20 cps. Synchronized and stationary patterns can be photographed from any of the screen materials, provided their light output is sufficient to expose the emulsion.

DEFLECTION AMPLIFIERS. The voltage gain needed to provide usable trace deflection on the screen of the rather insensitive cathode-ray tube is provided by deflection amplifiers, one for each axis. Because the operating frequency capabilities of cathode-ray tubes are so extensive—d-c to 100 megacycles—it is technically difficult to design a single deflection-amplifier system which will perform optimumly, and with sufficient gain, over this entire region. Consequently oscillographs are usually designed for a particular range, and designated low- or high-frequency types, depending on the characteristics of their deflection amplifiers. Low-frequency types are utility oscillographs having deflection amplifiers flat from a very low frequency to several hundred kilocycles and a maximum sensitivity of approximately 1 to 10 millivolts (rms) per inch deflection. This frequency range is adequate to handle virtually all shock and vibration components, including higher-order harmonics in shock transients. Response to d-c is extremely desirable in many instances, although it tends to complicate the deflection-amplifier design and introduces some zero shifting over long periods of time. High-frequency oscillographs usually are a-c-coupled, have a flat frequency range from about 30 cps to the 10-megacycle region, sensitivities in the order of 50 millivolts (rms) per inch, and due to the extra complexity of video-type circuits are much more expensive.

Low-frequency Deflection Amplifiers. Circuit functions representative of a typical low-frequency vertical-deflection amplifier are indicated in the block diagram of Fig. 19.46. The arrangement is basically a high-gain d-c push-pull amplifier with sufficient gain and inverse feedback to give excellent short-term freedom from gain variations and drift, and good long-term over-all stability. Instabilities resulting from warm-up, voltage fluctuations, and other environmental influences, which affect both sections of an individual stage, are considerably attenuated in the output by the common-mode cancellation properties of the push-pull circuits.

The amplifier can be made to respond to only the a-c components of a mixed input signal by switching in an input series blocking capacitor. Low-frequency response of the deflection amplifier then depends on the time constant of the input network, and is usually good down to 3 or 4 cps. Its high-frequency response is not affected. Techniques for evaluating frequency response are considered in frequency, phase, and waveform determination, below.

Figure 19.48 shows a typical input stage, consisting of a frequency-compensated step input attenuator, the input amplifier stages (usually some form of cathode follower), and a second, vernier gain control in the stage output. If the amplifier gain is standardized to some predetermined value, attenuator settings then can be marked directly to indicate the voltage sensitivity per unit deflection. The input attenuator is often a decade voltage divider, used to prevent overloading the input stage, while the output attenuator may be a 1-2-5-10 vernier between the decade steps. In some oscillographs both attenuators are operated from a single panel control and the switching from one attenuator to the other is handled by internal switch connections. Because of the convenience of always having a calibrated deflection sensitivity, the continuous gain control so predominant on older oscillographs has been relegated to the minor task of standardizing the over-all amplifier gain.

Voltage division at high impedance by resistive networks, as by the input attenuator, suffers frequency and phase distortion unless the stray capacitances in shunt with the dividing resistances happen to be in exactly the inverse ratio as the dividing resistances. For no distortion the time constant for both sections should be the same. Frequency compensation is accomplished by adding small capacitances in shunt with the attenuator resistances in the proper ratio, one of which is variable to permit balancing the stray wiring capacities. The best procedure for exactly compensating the attenuator is to pass a square wave through it, and to adjust the padding condenser for the most nearly perfect attenuated square-wave output. The output attenuator in the circuit of Fig. 19.48 need not be compensated, since this gain adjustment is made by varying the stage loading rather than by voltage division.

The input stage of Fig. 19.48 is a push-pull cascade phase inverter, typical of the circuits used to derive a push-pull output signal from either single-ended or balanced input signals. For single-ended signals, namely, signals with one side grounded, the input to V_{1b} is shorted to ground by a jumper on its input terminal, and the signal is applied

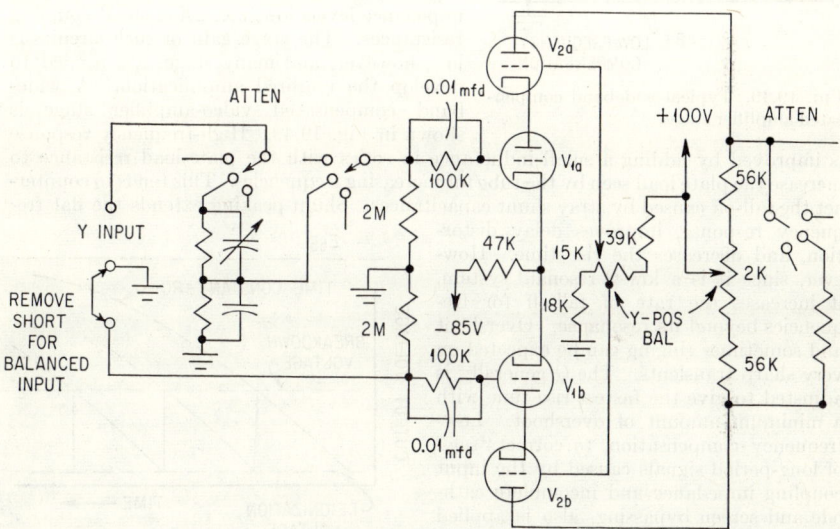

Fig. 19.48. Input circuit of a typical low-frequency cathode-ray oscillograph.

to V_{1a} through the input attenuator. Tubes V_{1a} and V_{1b} are coupled together by a common cathode resistor. A positive signal on the grid of V_{1a} makes its plate current increase, increases the voltage drop across the common cathode resistor, and raises the voltage on the two cathodes. This is normal cathode-follower action for V_{1a}. However, since the grid of V_{1b} is grounded at a fixed potential, raising its cathode potential increases the grid-to-cathode bias on this tube and makes its plate current decrease. The plate currents in V_{1a} and V_{1b} change in opposite directions, producing a push-pull output signal. The magnitudes of these opposing plate-current changes can never be quite equal. If they were, there would be no net voltage change across the common cathode resistor which is required to produce the push-pull action.

For balanced signals, the input grounding jumper is removed. Tubes V_{1a} and V_{1b} then operate as independent stages except that their common cathode connection helps reduce degeneration and loss of gain by keeping the developed bias voltage across the unbypassed cathode resistor constant. This connection also assists in rejecting common-mode signals, such as hum, from the output.

The cascade stages V_{2a} and V_{2b} are series-connected to their drivers and provide an extra pair of grids which may be used for amplifier balancing or trace positioning. They also provide voltage gain.

The remainder of the deflection amplifier consists of additional push-pull d-c amplifier stages with sufficient gain after feedback to produce the required voltage sensitivity. D-C amplifier principles are discussed in a previous section.

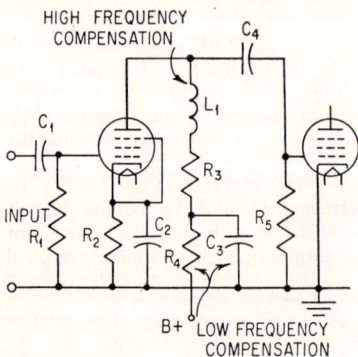

FIG. 19.49. Typical wideband compensated amplifier stage.

High-frequency Deflection Amplifiers. Vertical-deflection amplifiers intended for high-frequency oscillographs are wideband video amplifiers. They are designed with particular attention to a flat frequency response that extends well into the megacycle region, linear phase shift (constant delay), and good transient response.

The upper-frequency response of resistance-capacitance coupled amplifiers is limited by shunt loading from stray tube and wiring capacitances and can be extended by keeping impedance levels low, i.e., using small coupling resistances. The stage gain of such circuits is low, however, and many stages are needed to develop the required amplification. A wideband compensated video-amplifier stage is shown in Fig. 19.49. High-frequency response is improved by adding a small inductance in series with the plate-load resistance to increase the plate load seen by the tube for increasing frequencies. This tends to counteract the fall-off caused by stray shunt capacitances. Shunt peaking extends the flat frequency response, improves delay distortion, and decreases the rise time. However, since it is a low-Q resonant system, it increases the rate of fall-off for frequencies beyond its resonance. Overshoot and sometimes ringing can be expected on very sharp transients. The Q generally is adjusted to give the fastest rise time with a minimum amount of overshoot. Low-frequency compensation, to correct "sag" of long-period signals caused by the input coupling impedance and incomplete cathode and screen bypassing, also is applied in the plate circuit. It takes the form of a decoupling filter, sized to increase the plate load at the same rate as the low-frequency rolloff. Wideband amplifier compensation is discussed in the many textbooks on electronic engineering.[1,2,9] In shock and vibration work the wideband properties of high-frequency oscillographs are not needed and their lack of low-frequency response is an inconvenience, and so they are seldom used in this application.

SWEEP CIRCUITS. Sweep circuits are linear time-base generators which cause the beam to move at a uniform rate across the face of the cathode-ray tube to provide a time axis for the unknown signal. The ideal waveform of this generator is the sawtooth wave of Fig. 19.50A. The output voltage e_0 increases linearly with time until it reaches a predetermined level and then abruptly returns to its initial point to repeat the cycle. The sweep occurs as the sawtooth voltage varies from the deioni-

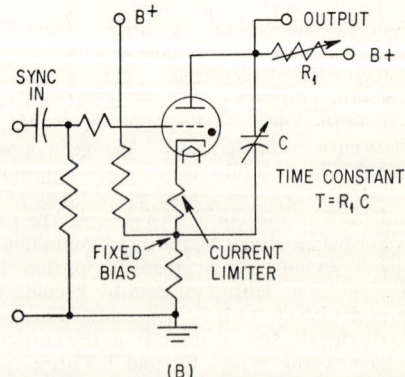

FIG. 19.50. Sweep voltage and circuit. (A) Sawtooth waveform produced by a charging R-C circuit is nearly linear in its lower region. (B) Simple thyratron circuit for generating sawtooth waveforms.

zation voltage to its maximum value. If this segment is a straight line, the sweep is linear with respect to time. The flyback time is kept to a minimum.

There are many electronic circuits capable of approximating this waveform with varying degrees of fidelity. The simplest is the relaxation oscillator of Fig. 19.50B, where the capacitor C is charged from the supply voltage through resistor R_1. The thyratron, held nonconducting by its negative bias voltage, is connected across the capacitor and will fire when the capacitor voltage reaches the breakdown potential of the thyratron. The capacitor is then quickly discharged through the low resistance of the conducting thyratron, the thyratron ceases to conduct, and the cycle repeats. The charging rate, and therefore the sweep rate, is controlled by the time constant R_1C. In practical circuits R_1 is a variable resistance which controls the fine-frequency adjustment. Coarse frequency control is provided by switching different capacitors in for C. This circuit is inherently nonlinear because of the exponential charging rate of an RC circuit. However, if breakdown is made to occur at a small fraction of the available charging volt-

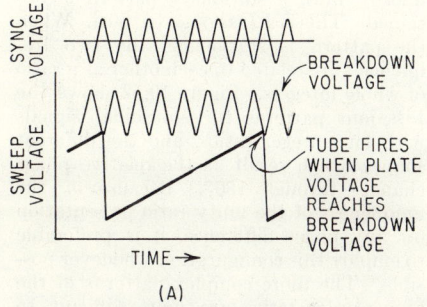 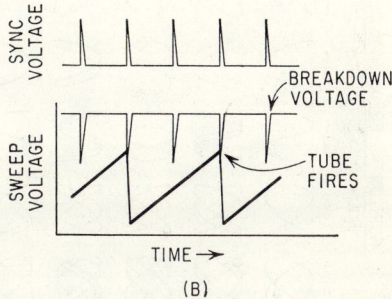

Fig. 19.51. Sweep synchronizing voltages. (A) Sine-wave synchronization. (B) Pulse synchronization.

age, the deviation from linearity is acceptably small. Much higher degrees of linearity are possible with special circuits, such as the phantastron,* bootstrap, and hard-tube oscillators, which include refinements especially designed to this application.

Stationary patterns of recurrent waveforms occur when the sweep frequency is synchronized to a submultiple of the input signal. Synchronization is accomplished by injecting a voltage of the proper phase, derived from the input signal, onto the grid of the thyratron. As Figs. 19.51A and B show, this momentarily lowers the breakdown potential and permits the thyratron to fire at the same point for successive sweeps, yielding a stationary pattern. Auxiliary circuits also can be added to the basic sweep generator to provide single-sweep operation, i.e., driven sweeps. This is useful for still recording of a nonrecurring transient, and can be set up so that the transient triggers the sweep.

Z-AXIS MODULATION. The intensity of the cathode-ray-tube trace can be modulated by momentarily varying the bias voltage on the intensity control grid to produce a distinguishing mark on the trace. This is a third arbitrary variable of a cathode-ray tube, and is appropriately termed the *Z axis*. It is regularly used as a timing marker for single-sweep or moving-film recordings. As a device for recording single events, it can be used as a reference marker to correlate the time scales on a multichannel record. Beam blanking pulses are easier to locate on the final record than intensifying pulses. They produce a sharp intensity discontinuity in the trace when the Z-axis signal pulse sides are steep, and its height (about −50 volts) is sufficient to produce beam current cutoff. Blanking has the disadvantage that important peaks in the trace may be missed completely if the pulse width is too great. However, intensifying pulses tend to defocus the beam—particularly if they are large enough to reduce the intensity grid bias to zero

* A type of one-tube oscillator employing feedback to generate a linear timing waveform.

or beyond. They are also much harder to locate on records where the trace brightens frequently because of periods of low velocity.

FREQUENCY, PHASE, AND WAVEFORM DETERMINATIONS. Some of the more important uses for cathode-ray oscillographs are in the determination of unknown frequencies, phase relationships, waveform analysis, and response characteristics of amplifiers and networks. These applications are made possible by the low loading and wide frequency range of the oscillograph system, and the availability of two separate deflection axes, one of which can be converted to a linear time scale.

Frequency Measurement. The simplest procedure for determining the frequency of an unknown cyclic signal is to compare it to a second signal whose frequency is variable and known. This can be done with a cathode-ray oscillograph by connecting the unknown signal to the vertical axis and the known signal to the horizontal axis (with the sweep turned off). Adjustments then are made to the comparison oscillator frequency until a stationary pattern is obtained. This is a *Lissajous* pattern. When the pattern is stationary, the two frequencies are related to each other as a ratio of whole integers. Figure 19.52 shows the Lissajous patterns for sinusoidal signals with low integer ratios, and the different shapes which result as the relative phase changes through 180°. Because of the uniqueness of the unity ratio presentation for any phase difference, it is preferable to employ this configuration whenever possible. The more complex patterns of the higher-order ratios are more difficult to evaluate, especially if the signals are derived from independent sources. Usual frequency instabilities tend to make the phase difference change continually, causing the pattern to squirm from one pattern to the next.

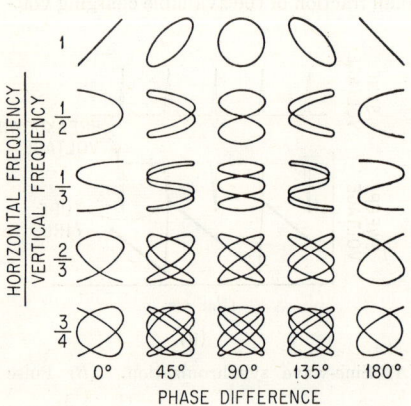

Fig. 19.52. Lissajous patterns for several common frequency ratios as a function of their relative phase.

A rule for determining the frequency ratio of the Lissajous frequencies is to count the number of loops which touch a horizontal line tangent to either the top or bottom of the raster,* and divide that number by the number of loops which touch a vertical line tangent to either side. Phase differences of 0 and 180° appear to violate this rule because some of the forward and return traces coincide and are not separately distinguishable. A more general rule is to trace out the complete pattern, counting loop crossings at both top and bottom (or both sides), until the original starting point is reached.

Input signals need not be sinusoidal. However, the Lissajous patterns become more complicated for other waveforms and may not be decipherable except for ratios less than 2:1. For example, two in-phase (or 180° out-of-phase) square-wave signals of the same frequency will simply produce dots at diagonal corners; at 90° phase difference, dots will appear at all four corners.

Phase Determination. Lissajous patterns can be employed to measure the phase difference between two signals of the same frequency, provided the pattern remains stationary. This requirement is automatically met when the two deflecting signals are related, e.g., the input and output voltages of an amplifier or network. Typical patterns are shown in Fig. 19.53 for sinusoidal voltages of varying relative phase and voltage. The relative phase angle θ between the two signals is given by

$$\theta = \sin^{-1}\left(\pm \frac{B}{A}\right) \qquad (19.55)$$

* Raster: the area upon which the image is produced on a cathode-ray tube.

where A and B are the maximum height and Y-axis intercept, respectively, of the pattern measured on the face of the cathode-ray tube or on a photographic record. Amplitudes of the two signals need not be equal, but the pattern must be centered at the origin of the calibrated scale on the cathode-ray tube for these measurements to be valid. When the pattern shifts spasmodically because of voltage fluctuations (as distinguished from phase variations) before measurements can be taken, photographic methods can be employed to record the trace during one of its quiescent periods. Photographic techniques are extremely useful under any condition, especially if Polaroid equipment is available, since centering is no longer important. Reference directions of the deflection axes must

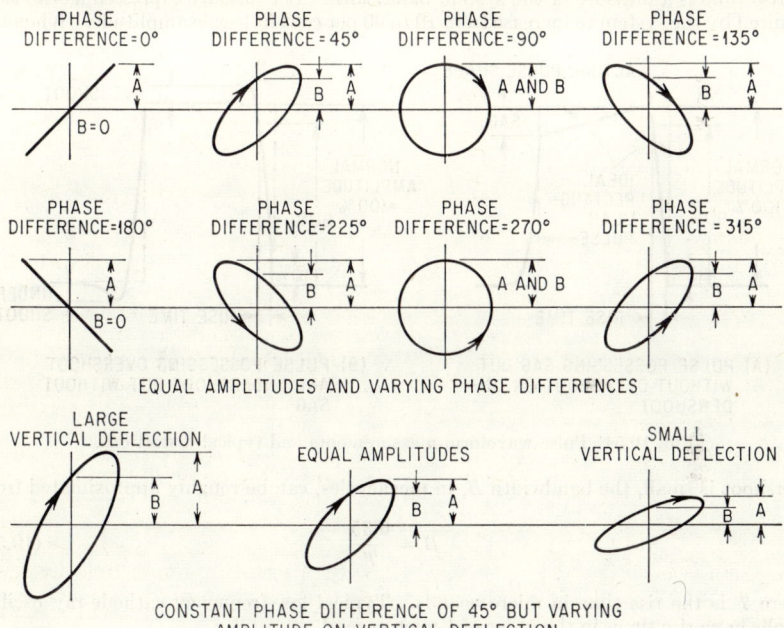

Fig. 19.53. Phase-shift patterns for various relative phase angles and amplitudes of two sinusoidal signals of the same frequency.

be indicated on the record and may be obtained by photographing the trace through an aligned, illuminated scale, or by double-exposing the record with one deflecting voltage removed. The origin of the recorded trace can then be located by boxing it in by a rectangle parallel to the reference deflection-axis marker.

Ambiguity exists as to the proper quadrant to assign to the calculated phase angle since it is impossible to determine the direction of rotation of the spot except at the very lowest frequencies. Uncertainty can be resolved if the exciting frequency can be lowered to a point where the phase characteristics are known, or until the rotation can be seen. The frequency then is increased back to the original frequency of interest, keeping track of the proper phase angles as the pattern passes through the distinguishing patterns of 0°, 90°, etc. This assumes that the phase shift is progressive with frequency. Accuracy of the elliptical-pattern measuring technique is not extremely high, especially in the vicinity of 0 and 90°. Other methods have been developed using variable phase-shift networks and circular time bases where higher degrees of accuracy are required.

Pulse Waveshapes. More information can be obtained about the performance of an amplifier or network from an examination of how it passes a square-wave signal or repetitive pulse than can be obtained from a whole series of frequency-response data using sine-wave point-to-point measurements. The cathode-ray oscillograph is well adapted

19-58 SHOCK AND VIBRATION INSTRUMENTATION

to this type of waveform analysis, and with proper interpretation of its visual pattern, can indicate what remedial steps should be employed to correct deficiencies.

An ideal square wave contains a fundamental sinusoid and an infinite number of odd harmonics, all precisely related in amplitude and phase. Any deficiency of an amplifier or network which discriminates nonuniformly against components in the square-wave structure will cause the output signal to be distorted and shaped differently from the original input pulse. The types of deficiencies present in a system show up characteristically by the manner in which they distort the output pulse. Important parameters are rise time, overshoot, undershoot, and sag—depicted graphically in Fig. 19.54.

Rise time is a measure of the system bandwidth. It is usually expressed as the time required by the system to increase from 10 to 90 per cent full-scale amplitude. When the

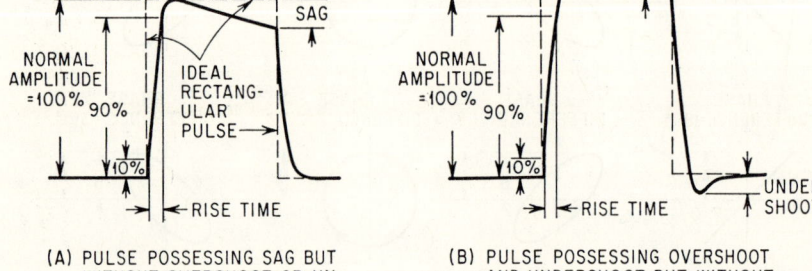

Fig. 19.54. Pulse waveform measurements and typical distortions.

overshoot is small, the bandwidth B, in megacycles, can be roughly approximated from

$$B = \frac{0.35}{T} \tag{19.56}$$

where T is the rise time in microseconds. Typical low-frequency cathode-ray oscillographs have rise times in the order of 1 microsecond.

Overshoot and *undershoot* are manifestations of the same deficiency and are normally symmetrical. They are caused by a nonuniformity in the frequency-response characteristic, usually from insufficient amounts of damping. High-Q, underdamped, compensating systems may even cause "ringing" to appear.

Systems with poor high-frequency response will have an excessively rounded nose and approach their final value somewhat gradually. *Sag*, sometimes termed *sawtooth*, is a deficiency in the low-frequency response and represents the system's inability to maintain a constant voltage. D-C amplifiers do not evidence sag. Overshoot, undershoot, and sag are all expressed as a percentage of the normal pulse height.

As a general rule, amplifiers which acceptably pass a square wave of a given frequency have response characteristics which are reasonably flat for sine-wave excitation out to frequencies ten times as great as the fundamental square-wave frequency. This rule is derived from the amplitude distribution levels of the square-wave harmonics. The ninth and eleventh harmonics (there is no tenth) are just large enough to make their presence (or absence) discernible in the output.

RECORDING METHODS. The following general rules apply to all cathode-ray-oscillograph recording techniques. They relate to the trace intensity, camera lens, and film.

The intensity of the trace on the cathode-ray-tube screen should be adjusted to produce the brightest image which can be sharply focused over its entire length. Increasing the intensity beyond this point will defocus the beam and not produce any real increase in contrast in the recorded image. If the intensity is more than adequate, a more sharply focused record will be obtained if the camera lens is "stopped down" several additional numbers before reducing the brightness.

The camera lens should be the best available. Recommended minima are: $f/4.5$ for stationary recording and $f/2$ or $f/1.5$ for moving films and single sweep. The lens should be stopped down at least one and preferably two stop numbers below its maximum aperture to decrease lens distortion. Excessive trace brightness is best adjusted by stopping down the lens further.

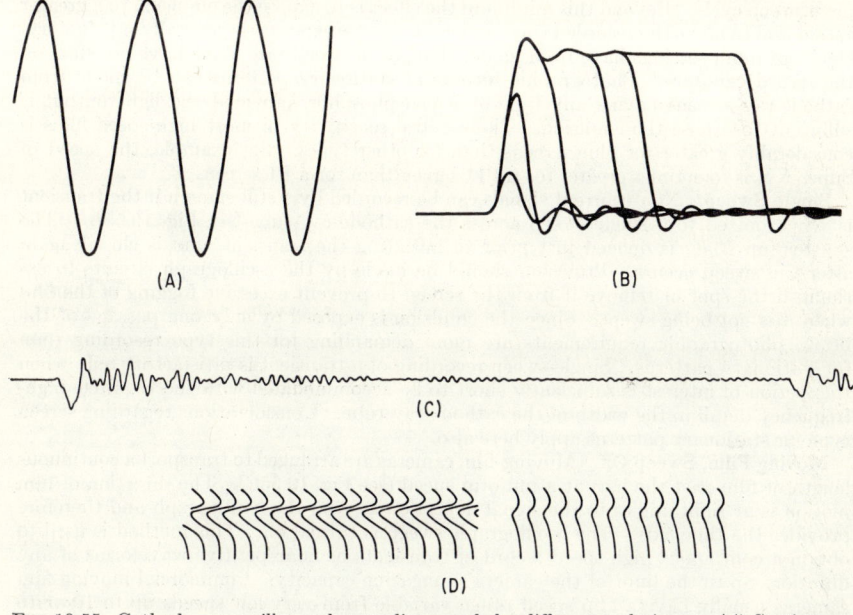

FIG. 19.55. Cathode-ray oscillograph recordings made by different techniques. (A) Stationary pattern recording of a synchronized and steady sine wave. (B) Single-sweep recording with multiple exposure. A standard timing and amplitude calibration sine wave may be superposed on the record. (C) Moving-film recording, sweep off. Z-axis timing can be indicated by blanking the trace at equal time intervals. (D) Moving-film recording, sweep on.

It is better to standardize on one fast film emulsion speed such as Super XX and stop down the lens than to use slower-speed emulsions. The somewhat larger grain of the higher-speed emulsions will have negligible influence on the quality of the record.

These rules are quite general; usable, but not necessarily good, records can be obtained over a wide latitude of adjustment. Many operators prefer to overexpose and overdevelop the emulsion to ensure that the trace will be "burned-in." Some experimentation is usually necessary to produce optimum results, but there is little danger of getting a completely unusable record once approximate settings have been established.

Four principal methods are currently used to record photographically the beam patterns from the face of a cathode-ray tube. These are (1) still recording of stationary patterns, (2) still recording of a single driven sweep, (3) moving-film recording with the oscillograph sweep turned off, and (4) moving-film recording with the sweep on. Typical records obtained by each of these methods are shown in Fig. 19.55. The first two methods employ the simplest of camera equipment, while the latter two require a shutterless camera especially developed for oscillographic work. Lens requirements for both types are similar. The lens should be capable of focusing down to about 12 to 15 in., and have an image reduction ratio no greater than necessary to cover the major portion of the film area. Lenses for still cameras should be at least as fast as $f/4.5$, and equipped with a shutter having speeds of approximately $\frac{1}{30}$ sec, *Time*, and *Bulb;* other speeds are superfluous. Moving-film camera lenses should be faster (e.g., $f/2$ or $f/1.5$) but need no shutter. Many cameras employ 35-mm film, but roll film and plate cameras are equally satisfactory although their writing speed is somewhat less because of the larger area covered by the image. Polaroid cameras are used extensively for still shots because they provide a finished print almost immediately, without the bother of darkroom development and printing.

Stationary Patterns. For photographing stationary patterns (see Fig. 19.55A), it is necessary only that the shutter be kept open at least long enough to record one complete sweep cycle. Beyond this minimum the effect is to expose the emulsion to a greater extent and to make the recorded trace wider. A $\frac{1}{30}$-sec exposure is generally satisfactory. Once optimum settings have been made, changes in sweep rate alone have no effect on the record exposure. Photographic records of stationary patterns can be made from cathode-ray screens having any type of screen phosphor, provided the light output is sufficient to expose the emulsion. The spectral sensitivity of most high-speed films is considerably greater for blue screens than for other types. For example, the speed of Super XX is four times greater for a P11 screen than for a P1 screen.

Single Sweep. Nonrecurrent signals can be recorded by a still camera if the transient is synchronized to a single sweep across the cathode-ray tube (see Fig. 19.55B). The shutter (on *Bulb*) is opened just prior to initiating the transient, and is closed again after the sweep occurs. Provision should be made by the oscillograph circuits to extinguish the spot or remove it from the screen to prevent excessive fogging of the film when it is not being swept. Since the emulsion is exposed by only one passage of the beam, photographic requirements are more demanding for this type recording than for stationary patterns. Single-sweep recording of a transient is satisfactory only when the section of interest is sufficiently short to be accommodated with the required high-frequency detail in the width of the cathode-ray tube. Considerations regarding screen type for stationary patterns apply here also.

Moving Film, Sweep Off. Moving-film cameras are arranged to transport a continuous length of film past the lens at a uniform speed (see Fig. 19.55C). The direction of film motion is at right angles to the signal deflection axis on the oscillograph and therefore provides the time axis. The oscillograph sweep is turned off. This method is used to obtain a continuous high-speed record of transients or nonrepetitive waveforms of any duration, up to the limit of the camera's magazine capacity. Commercial moving-film cameras usually have a film speed range variable from very low speeds up to 10 or 15 ft/sec.

The most suitable film speed depends upon the frequencies in the signal to be recorded. In shock and vibration work, frequencies above 5,000 cps are seldom considered important. If it is assumed that the camera resolution is such that it will just resolve 100 cycles/in., then a 5,000-cps signal can just be resolved when the film runs at 50 in./sec, or about 4 ft/sec. Principal frequencies in shipboard shock often range between several cps and 300 cps, permitting even lower film speeds. Low-frequency components can be "lost" by excessive stretching, just as effectively as high frequencies can be lost by insufficient resolution. The highest speeds of 10 to 15 ft/sec impose heavy demands on trace brightness, lens, aperture, and film emulsion, and are extremely wasteful of film. If these speeds are necessary, it might be better to use a lens with greater image reduction and lower the film speed in the same ratio. Both of these improve the focus and trace density on the emulsion. The record can subsequently be restored to usable size by additional photographic enlargement. Short-persistence screens, such as the P11, are required in this type recording to prevent smearing of the trace by the afterglow.

Moving Film, Sweep On (see Fig. 19.55D). It is sometimes necessary to record an extremely short-duration signal that is not regularly repetitive, not adaptable to synchronization, or not recordable with sufficient detail by the above method. This type of signal can be recorded by rotating the moving-film camera (or exchanging axes on the cathode-ray tube) so that the direction of film motion coincides with the signal deflection axis. With the sweep on, consecutive sweeps will appear across the width of the film separated by a distance depending upon the relative speeds of sweep and film. The transients then are expanded along the sweep line and permit a much more detailed study. This technique makes very efficient use of the film area. Short-persistence screens are also required here.

TIMING. Several methods are available for providing timing on oscillograph records. Where possible, it is preferable that the timing signal be displayed on the same record with the test signal, rather than as a separate record.

Stationary patterns do not need a timing signal. If the pattern can be synchronized, it is repetitive, and its frequency can be obtained directly from the exciting mechanism, or by Lissajous comparison to a standard oscillator.

Z-axis blanking (see *Z-axis Modulation*, above) can be used for either single-sweep or moving-film records. Z-axis modulation is shown on the record of Fig. 19.55C. To ensure clean breaks in the trace, the blanking pulse should have steep sides and sufficient amplitude to produce intensity cutoff. The shortest pulse duration acceptable depends on the resolving capabilities of the camera, film, and sweep speed. The approximate minimum pulse width in milliseconds can be computed from

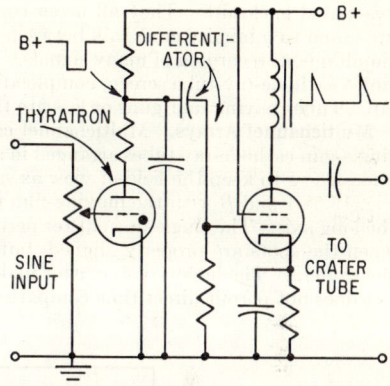

Fig. 19.56. Pulse-shaping circuit for crater tube timer.

$$\text{Pulse width (milliseconds)} = 5/\text{film speed (in./sec)} \qquad (19.57)$$

Single-sweep records are frequently calibrated for time and amplitude by double-exposing the record, first to the signal, and then to a sinusoidal voltage whose amplitude and frequency are both known (see Fig. 19.55B). No changes should be made in the amplifier or sweep controls between exposures. Less interference between the signal and calibration traces will result if the lens is stopped down more on the calibration exposure so that just the low-velocity peaks are recorded.

Some moving-film cameras have a built-in neon or argon bulb which can be flashed at line frequency or somewhat beyond to supply timing markers on the film edge. These are satisfactory for low-film speeds up to 2 or 3 ft/sec. For higher film speeds more frequent indications are desirable, especially if the film speed is changing. A facsimile reproduction tube (Sylvania Type R1130B), positioned in the camera's field of view, will produce an intense, small, short-duration spot of light at frequencies up to several thousand cycles per second. A suitable circuit is shown in Fig. 19.56.

MULTICHANNEL ARRAYS. Electronic Switch. It is often desirable to observe or record two related signals, as for example the input and output signals from a circuit. When these are repetitive signals, an electronic switch can be employed to display both waveforms on a single cathode-ray oscillograph. An electronic switch, Fig. 19.57, is arranged to accept simultaneously the two signals to be compared and to present alternately each of them for display on successive sweeps of the cathode-ray oscillograph. When the synchronized switching and sweep frequencies are high enough, the persistence of the screen and the eye make it appear that both signals are present simultaneously. Positioning circuits permit the two traces to be overlapped or separated vertically relative to each other.

In its normal application the switching frequency is low compared to frequencies in the input signals, and many cycles of both waveforms are displayed during each sweep. This technique is obviously unsatisfactory for multiplexing two nonrepetitive signals because of the large sections of information which would alternately be lost from each signal. However, if the frequencies in the two signals are low compared to the switching frequency, the electronic switch can be used as a two-pole commutator.

Multigun Cathode-ray Tubes. Cathode-ray tubes having as many as ten separate and distinct electron-gun and deflection-plate assemblies have been built within the same evacuated enclosure. They all use a common fluorescent screen. Special precautions are taken to minimize cross talk between channels by shielding. This technique permits simultaneous recording of many signals. However, on the comparatively small area of a single cathode-ray-tube screen, complications can arise if the traces are allowed to overlap. Tubes having four guns or less are the most popular.

Multichannel Arrays. Multichannel cathode-ray oscillographs can be built up from single-gun cathode-ray tubes arranged in a compact cluster. Tube diameters are usually 3 in. or less to keep the field of view as small as possible. Clusters may be as shown in Fig. 19.58A and B with the moving-film camera oriented in both cases at right angles to the long axis. The single-line cluster permits direct phase comparison between channels when the spots are properly aligned, but presents an abnormally wide field of view to the camera. The honeycomb cluster makes more efficient use of a smaller field of view, but does not permit direct time comparison between channels of a different row. Trace

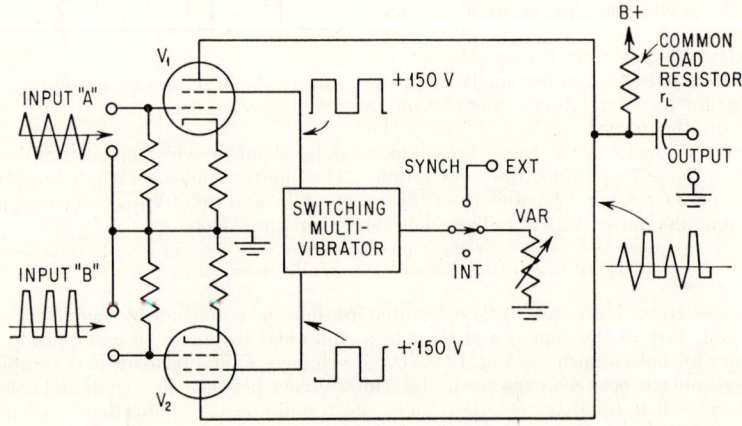

Fig. 19.57. Simplified schematic for an electronic switch.

overlapping can occur with this arrangement, but only to the extent of adjacent channels and between displaced time bases. Transients, which are usually large only in their initial phase, seldom interfere. Because of space economies, this latter method is preferred.

The time-displacement problem of the honeycomb cluster is not as serious as it might appear. First, it is difficult to get all the spots lined up on a common line which is perpendicular to the camera film axis and to have them stay there. Second, it is characteristic of moving-film recording that the spots soon burn holes in the screen at their normal rest positions, decreasing spot intensity. Therefore it is convenient to be able to displace them slightly into undamaged areas as the original positions become unsuitable. Because of these reasons, the spots seldom are accurately aligned, tending to nullify the sole advantage of the single-line cluster.

The individual traces of a cluster, or even of additional clusters, can be referenced to a common time scale by a distinctive marker superimposed simultaneously on all traces. Markers can take the form of Z-axis modulation or a single, short-duration pulse applied to the signal axis. When the Z axis is used for timing, the reference mark can be a pulse of the same polarity that is considerably longer in duration than the ordinary timing markers, or a pulse of opposite polarity. Z-axis timing on multichannel arrays need be applied to only one channel. However, some convenience will be derived, without appreciably loading the pulse generator, if all channels are modulated.

Referencing by a common deflection on the signal axis can take several forms. One of the most useful is to apply simultaneous calibrations to each channel, either as a d-c step change or as a sine wave. When these voltages are known, the channels are automatically calibrated in amplitude, as well as correlated in time. Another scheme is to apply an exponential pulse, derived from a capacitively coupled pulse generator, to some convenient point of the deflection amplifiers. This technique is useful where it is inconvenient to attach onto the input circuits, as with high-impedance transducers.

MAGNETIC-TAPE RECORDERS

Magnetic-tape recorders possess a combination of desirable features which cannot be duplicated by any other type of recorder. Some of the advantages, such as being able to play back (i.e., reproduce) a recording innumerable times without deterioration, being

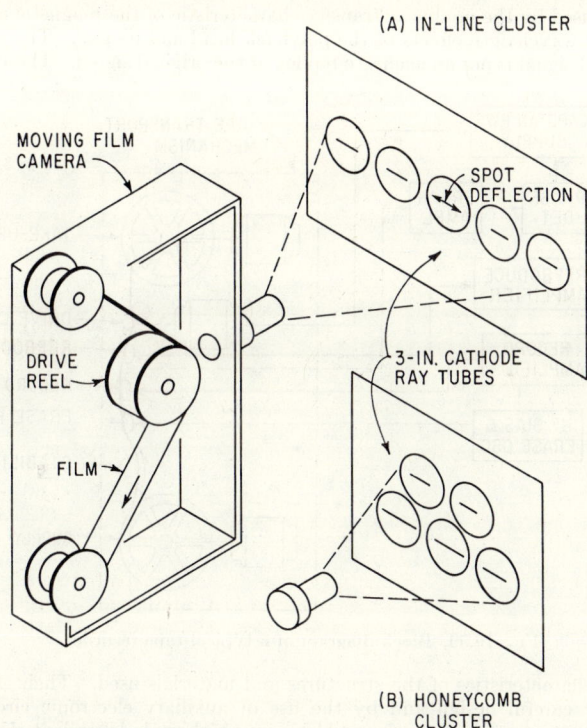

Fig. 19.58. Multichannel cathode-ray-tube arrays. (A) In-line cluster. (B) Honeycomb cluster.

able to read or store a record immediately without further processing, and being able to erase and rerecord on the same medium, are inherent advantages attributable to the magnetic medium. Other attractive properties of magnetic records, such as their extended frequency range, linearity, and accuracy, are not intrinsic qualities, but have been attained only as a result of extensive engineering development.

The basic components of all magnetic-tape recorders, Fig. 19.59, are (1) the magnetic tape, (2) the tape transport mechanism, (3) the record, reproduce, and erase heads, and (4) the electronic circuitry necessary to process the recording and playback head signals. Each of these has limitations which affect the over-all system accuracy. Depending on the intended function of the magnetic recorder, various degrees of refinement are incorporated in the individual components to make their contribution to the system inaccuracies as small as possible. An overriding requirement for magnetic-tape recorders intended for data acquisition is that they be as nearly perfect as the technological state of the art permits.

In the recording process, an initially neutral magnetic tape is magnetized as it is drawn past a gap in the recording head. The induction impressed on the tape is proportional to the magnetizing field and is a function of the input signal. In this manner variations of the input signal with respect to time are transformed into variations in remanent flux with respect to distance along the magnetic tape. The oxide coating on the tape is magnetically "hard" and will retain the induction impressed on it until acted upon by another external magnetic field. On playback, flux lines leaving the surface of the tape link with the magnetic circuit of the reproduce head and generate a voltage in its winding.

Between the recording and reproducing processes the signal undergoes a series of distortions caused by the nonlinear transfer characteristic of the magnetic medium, and frequency and wavelength effects of the playback head and its gap. The result is that the reproduced signal is not an accurate replica of the original signal. These phenomena

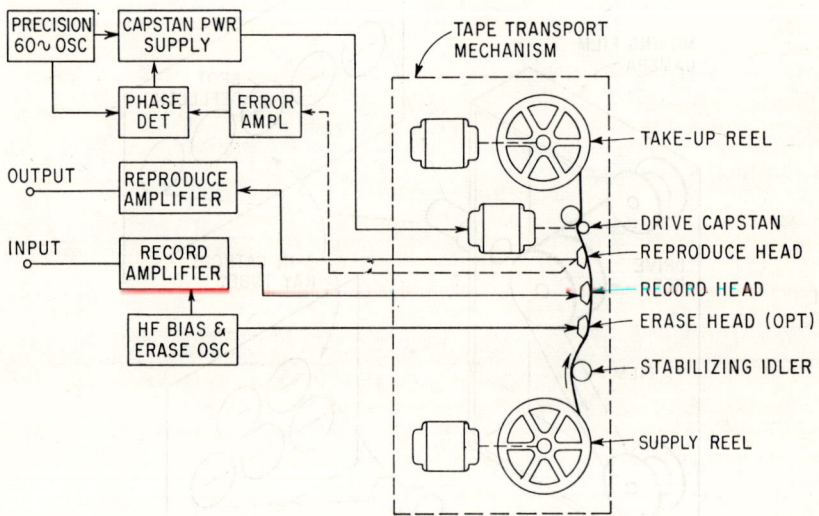

Fig. 19.59. Block diagram of a typical tape recorder.

are inherent characteristics of the structures and materials used. Their effects can be minimized by careful design and by the use of auxiliary electronic circuitry which linearizes the system or which employs techniques which are independent of these characteristics.

Three different techniques are used in analog magnetic-tape recording. These are amplitude (or direct) modulation (AM), frequency modulation (FM), or pulse-duration modulation (PDM). Each has a preferential field of application based on the characteristics of the data to be recorded and on their own limitations. In AM recording the signal amplitude is mixed with an a-c bias signal to set the operating point in a linear region and is recorded directly on the tape. It provides the widest bandwidth—100 to 100,000 cps at 60 in./sec tape speed—and is used predominantly where its high-frequency capabilities are required and its lack of low-frequency response is not important. In FM recording the amplitude of the data frequency-modulates a carrier which is recorded on the tape at saturation levels. This system is virtually independent of tape characteristics and will produce extremely accurate records over the frequency range from d-c to about 10,000 cps. Applications include all data handling where low-frequency components must be preserved. PDM is used for recording the variable-duration pulse train produced by a pulse-duration modulation telemetry system. Since only the pulse width carries information, the leading and trailing edges of the pulse are differentiated and recorded as spikes on the magnetic tape. Suitable demodulators permit reconstruction of the original signal on playback. This technique is also independent of the tape characteristics.

For all methods of recording, the upper frequency limit is dependent primarily on tape speed. Standard tape speeds for instrumentation tape recorders * are 30 and 60 in./sec. Most machines are also capable of additional tape speeds of 15, $7\frac{1}{2}$, $3\frac{3}{4}$, and $1\frac{7}{8}$ in./sec. The ability to record at one tape speed and play back at another can be used to expand or contract the data time scale, and permits many interesting and unusual processes to be developed.

The multichannel capabilities of magnetic-tape recorders have been one of the contributing factors to their enormous popularity where large quantities of data are to be handled. Seven channels of information can be recorded on a $\frac{1}{2}$-in.-wide tape, or 14 channels on a 1-in. tape. When used with time-multiplexed or frequency-multiplexed telemetry systems, tremendous amounts of data can be simultaneously recorded, stored, and reproduced.

HEAD CHARACTERISTICS. The ring head of Fig. 19.60 offers the best compromise between good magnetic performance and ease of manufacture, and has supplanted all other forms for recording and reproducing heads. It has a sharply defined longitudinal field that is well contained within the pole pieces, and a proportionally small vertical component. These features are desirable to contain the active portion of the gap to a very narrow region on the tape. The head is positioned relative to the tape so that its main field is parallel to the direction of tape motion, i.e., it produces longitudinal recording. Transverse and vertical head arrangements are possible, but seldom used. Longitudinal recording permits more efficient use of the available tape area and better stacking arrangements in multichannel recorders, and is used exclusively in modern tape recorders.

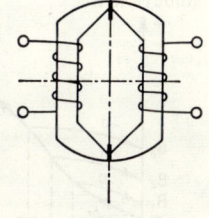

Fig. 19.60. Ring-type magnetic head assembly.

Recording and reproducing heads are quite similar in design. Their principal difference lies in the inductance and number of turns of their windings, and a somewhat more critical tolerance on the playback-head gap. A primary requirement of the magnetic-head circuit is that it be made of magnetically "soft" materials, i.e., that they have a high permeability to ensure an adequate field in the gap and a low coercivity to keep energy requirements low. Eddy-current losses are minimized by laminating the head structure. Mu-metal and Permalloy often are used in this application.

Head-gap width adjustment and alignment (for multihead stacks) are very critical operations in the manufacture of magnetic heads. Gap width should be as narrow as possible to extend the high-frequency range, while simultaneity of multiple data can be preserved only by restricting the scatter away from a common center line to very small values. With modern manufacturing techniques, gap widths of 0.25 mil have become standard on high-quality recorders. Gap scatter of multiple assemblies is contained with a band 100 microinches wide.

The advent of large data-gathering centers, such as are used in missile test ranges, has made it desirable to standardize the configuration of multichannel tape records to facilitate the interchange of these records between machines of different manufacture and between various test centers. A set of standards has been drawn up which defines the characteristics of a standard magnetic-tape record.

Recording Heads. The function of the recording head is to induce a state of magnetization in the magnetic tape proportional to the signal which was applied to the head when the tape was in contact with the head, at each position along the tape. Magnetic materials have a nonlinear magnetization curve. A plot, Fig. 19.61, of the remanent induction in an initially neutral magnetic material, as a function of magnetizing force, shows an extremely nonlinear transfer characteristic. Applied signals would be badly distorted in this recording process, and utterly useless in AM recording. Fortunately the transfer characteristic can be straightened by *magnetic biasing*. Both d-c and a-c biasing methods

* See Inter-Range Instrumentation Group, Magnetic Recorder/Reproducer Specifications No. 101–57.

are possible, but the former is seldom used because it is more critical of the recording gap size. In a-c biasing a high-frequency bias current is mixed with the input signal and recorded on a neutral magnetic tape. The bias frequency should be three or four times as high as the highest frequency to be recorded, and may be conveniently derived from the same oscillator used to excite the erase head. As long as the a-c bias current exceeds a minimum level, its adjustment is not too critical and linearity is very good. Optimum bias is approximately that which produces maximum output.

If the field in the recording head gap is reasonably uniform and terminates sharply, the width of the recording head gap is relatively unimportant when a-c bias techniques are employed. Then the remanent induction in a particular element of the magnetic tape is proportional to the recording field strength at the instant the element leaves the gap—irrespective of other variations it may have encountered while passing through the

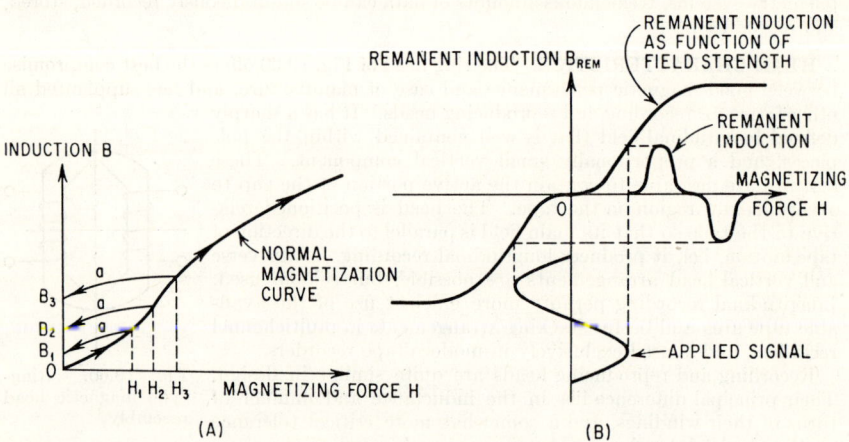

Fig. 19.61. Curves of remanent induction in a magnetic medium, as a function of magnetizing force.

gap. High-quality recorders nevertheless employ narrow recording gaps of 0.5 mil or less to contain their fields to sharp concentrated patterns. These small gap widths are important with frequency modulation (FM) and pulse-duration modulation (PDM) systems which operate at saturation levels and do not depend on tape transfer characteristics or employ biasing arrangements.

Tape Flux Distribution. Magnetic flux lines must form closed loops. Therefore, the remanent induction laid down within the magnetic medium during recording must emerge from the surface, pass through the surrounding air, and reenter the tape surface to complete its circuit. Since the internal flux density varies from point to point along the tape in accordance with the original magnetizing signal, it follows that the flux lines leaving or entering the surface at any point are proportional to the rate of change of internal flux at that point:

$$B_s = \frac{d\phi}{dx}\bigg|_{x=x_1} \tag{19.58}$$

where B_s is the flux density at the surface and ϕ is the internal flux in the medium. This is shown by Fig. 19.62, where it has been assumed that the magnetizing signal is sinusoidal and that a linear transfer characteristic exists:

$$\phi = \phi_{max} \sin 2\pi ft$$

If it is assumed that this signal was recorded at a tape speed v, then from the fundamental definitions

$$v = \frac{x}{t} \quad \text{and} \quad \lambda = \frac{v}{f} \tag{19.59}$$

It can be shown that

$$B_s = \frac{d\phi}{dx} = \frac{1}{v}\frac{d\phi}{dt}$$

from which

$$B_s = \frac{2\pi f \phi_{max}}{v} \cos 2\pi ft \qquad (19.60)$$

or in terms of recorded wavelength on the tape,

$$B_s = \frac{2\pi \phi_{max}}{\lambda} \cos 2\pi \frac{x}{\lambda} \qquad (19.61)$$

Equations (19.60) and (19.61) state that the magnitude of the surface flux density is a wavelength effect and that for a fixed tape speed the flux density increases linearly with frequency.

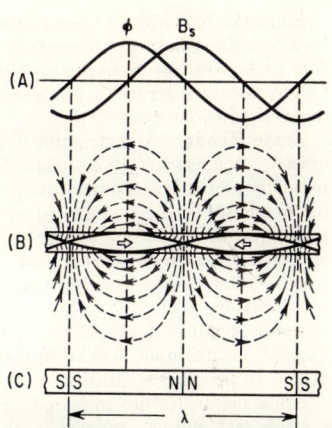

FIG. 19.62. Flux distribution ϕ and surface density B_s of a magnetic recording medium for a recorded sinusoidal signal.

Reproducing Heads. Ring heads used for reproducing are formed so that the pole pieces make contact with the tape for a considerable distance on both sides of the gap. When this distance is longer than a half wavelength of the recorded signal, most of the surface flux is shunted through the magnetic circuit of the head and is available for linking with the head winding. For a constant tape speed, the voltage induced in the winding under these conditions is proportional to the rate of change of the internal induction and (as has already been shown) is proportional to the surface flux density. The reproduced voltage from a sinusoidal input signal and constant playback tape speed then is

$$e_0 = kB_s = \frac{2\pi fk}{v} \cos 2\pi ft \qquad (19.62)$$

The voltage is also proportional to frequency, i.e., it increases 6 db/octave. The linearity of this response curve, Fig. 19.63, is disturbed at both extremities. On the low-frequency side the output drops off when the recorded half wavelength becomes appreciably greater than the contact length of the pole piece and only a portion of the flux lines can be shunted through the head. On the high-frequency side the output decreases when the recorded half wavelength becomes comparable to the gap width. Under these conditions, flux lines that emerge in one pole piece find that they must close through the air in the space between poles because their wavelength is too short for them to close their loop through the magnetic structure. These lines cannot then induce a voltage in the pickup coil. For these reasons the reproduce head gap is made as physically small as possible. Decreased output, particularly at short wavelengths, can also result if the reproducing head gap is not precisely normal to the direction of tape travel. This effect, called *gap tilt*, is minimized by careful manufacturing techniques which hold the gap azimuth to within 1 minute or less of the perpendicular.

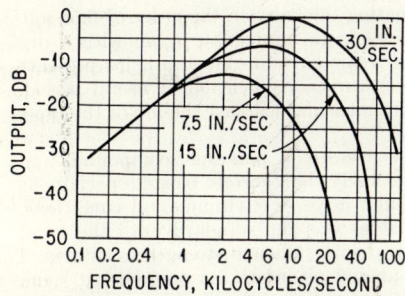

FIG. 19.63. Playback head output voltage for a variable-frequency signal recorded at saturation level, recorded and reproduced at the same speed. Typical head playback curves; saturation recording, no bias; head inductance, 50 millihenrys; head current, 1.5 milliamperes; ¼ mil gap.

Since the majority of the external flux lines lie close to the tape surface, the tape must be maintained in intimate contact with the magnetic structure of the head. Attenuation due to separation increases as the ratio of the separation distance to recorded wavelength becomes greater. For a signal which is 1.0 mil in length, a 0.5-mil separation reduces the head output 30 db.

Erase Heads. An attractive feature of tape recording is that a previous record can be erased. A necessary prerequisite for linearity in AM recording is that the tape be magnetically neutral before recording. The erasing head is placed upstream from the recording head and is dimensioned to cover the entire tape width. The normal demagnetizing technique of successively magnetizing the medium in opposite directions as the field strength is slowly reduced is used. In this case the tape motion provides the field reduction by removing an element from its influence. In order to complete the large number of alterations necessary to get a clear erasure at the maximum tape speed of 60 in./sec, the erase head is designed with a broad trailing field and operated at a very high frequency, for example, 300 kc to 400 kc. The magnetizing force is several times that required to produce saturation.

Erase heads are not included on most newer instrumentation tape recorders, because a more satisfactory, quieter, and quicker erasure can be obtained with bulk erasers. Erase heads are not required with FM or PDM systems, since these methods record at saturation levels and will completely obliterate any previous recording.

Head and Track Dimensions. Tape recorders are available with any number of heads—usually up to fourteen. Seven-track machines recording on tape ½ in. wide and fourteen-track machines using tape 1 in. wide are by far the most popular.

Track width and spacing are a compromise between good signal-to-noise characteristics and minimum cross talk. Standard dimensions established by the IRIG (see *Telemetry*) provide a track width of 0.050 in. per channel with a 0.020-in. guard band between tracks.

This spacing is too close to permit adequate shielding between adjacent magnetic heads. To provide better isolation, the heads are arranged in two stacks—odd-numbered heads in one stack, and even-numbered heads in the other. Stacks are separated 1.5 in. To preserve time correlation between data on different tracks, very stringent requirements are imposed on gap alignment. Scatter is maintained within a band 100 microinches wide.

TAPE CHARACTERISTICS. The coating on magnetic tapes is an especially prepared gamma or red ferric oxide having acicular particles uniformly smaller than 1 micron in length. Particles are embedded in a resinous vinyl binder applied to one side of the base material. Thickness of the coating varies from 0.35 mil for long-wear tapes up to 0.65 mil for high-output tapes. General-purpose tapes of the type most frequently used in instrumentation work have a coating thickness in the order of 0.55 mil. A recent innovation in the manufacture of tapes is to impregnate them with a dry silicone lubricant. This drastically reduces the wear on the heads and helps preserve their gap dimensions and high-frequency response over a much longer useful period. The lubricant occupies the space between the granules and will last the life of the tape, much in the manner of Oilite bearing materials.

Base materials commonly used are cellulose acetate, a polyester compound, and Mylar.* The latter two are more stable dimensionally than acetate bases, especially as regards humidity effects, and are preferable under adverse environmental conditions. Nominal thicknesses are 1 and 1.5 mils; the latter was the standard until the 1-mil polyester and Mylar-base materials were developed with sufficient strength to withstand the normal accelerating and braking loads of analog recorders. A standard 10½-in. NAB reel will accommodate 2,500 ft of 1.5-mil tape, sufficient to provide 8 min of continuous operation at 60 in./sec. A 50 per cent increase in capacity results from employing the thinner 1-mil tape; 14-in. NAB reels provide twice the capacity of 10½-in. reels.

Magnetic tapes intended for instrumentation use are subjected to rigorous quality controls during manufacture to minimize nonuniformities and irregularities which might cause signal errors or complete loss of information. Most of the common faults

* Mylar, Du Pont trademark.

have been recognized and corrective steps taken to minimize their occurrence. The principal defects are defined in the following paragraphs.

Dropouts are holes in the magnetic coating. When large enough, they can cause serious errors in AM or digital recordings if a peak or information bit should occur at a dropout.

Inclusions are foreign nonmagnetic particles imbedded in the binder along with the magnetic coating. Their effects are similar to those caused by dropouts.

Nodules are clumps of magnetic oxides which protrude predominantly above the surrounding surface level. Their effect is to push the tape out of intimate contact with the magnetic structure of the head and greatly reduce the output level. AM recordings are affected drastically, but FM and PDM records are not since they operate at saturation levels.

Magnetic tapes are very strong and durable for the thinness of their base materials. They can be preserved for long periods of time, including innumerable playings and recordings without appreciable deterioration. A few simple handling precautions will ensure their optimum performance.

Tape relaxation and uneven stretching over long periods of time can be minimized by reeling the tape under uniform tension. When a reel is to be stored, it should be rewound to the supply reel in the normal manner, and then passed completely through the machine to the take-up reel at a constant tape speed without stopping. With this method the successive layers are laid down uniformly directly on top of one another so that their edges are supported. Normal rewinding to the supply reel at extremely high speed, particularly if the process is stopped and started several times in the length of the tape, produces an extremely ragged winding. The uneven tensions cause the individual layers to slip laterally within the guide clearance of standard reels. Unsupported edges tend to curl and scallop and may be a contributing factor to poor contact on the edge channels. Precision reels with a minimum guide clearance are available. Reels prepared for storage by this reverse winding process should be marked "Rewind to Playback" as a reminder of their inside-out wrap.

For best protection, the tape should be stored in containers to protect it from mechanical damage, dust contamination, and marked humidity changes. Storage in their original plastic bags and boxes is generally satisfactory, or special storage cans may be obtained.

Tapes should be kept away from intense magnetic fields which could alter their remanent magnetism. Similarly, the record and reproduce heads should be periodically demagnetized as a precautionary measure to prevent inadvertent record contamination from this source. Occasionally, stored tapes experience a type of self-induced magnetization called "print-through." This is a transfer of the magnetic pattern from one layer to those immediately adjacent to it. Print-through levels are about 40 to 50 db down from normal recording levels and are bothersome only on AM recordings during periods of minimum signal levels.

TAPE TRANSPORT MECHANISMS. The tape transport mechanism includes all the elements necessary to control, guide, and store the tape during operation of the recorder. Reel-to-reel tape recorders are the most familiar and are used mostly for the permanent acquisition of data. They record on a long continuous length of tape as it passes over the magnetic heads during transfer from one reel to the other. Tape-loop transport mechanisms pass a short (4 to 75 ft) endless loop of tape over the magnetic heads; they are used to provide time delay, to provide repetitive cycling of selected data for analysis by wave analyzers, to provide a temporary record which is continually monitored by a playback head and erased if of no value, or to provide a temporary record which is transferred to a second reel-to-reel recorder for permanent retention if a preselected type of signal occurs. This latter technique allows important data from unpredictable random occurrences to be preserved, including the important signals just prior to the event, without accumulating large amounts of worthless data in between events.

The most basic function of the tape transport mechanism is to drive the tape past the recording and playback heads at constant speed. *Instantaneous* tape-speed variations cause "flutter" and "wow" in AM recordings and spurious amplitude components in

FM and PDM recordings. *Constant* tape-speed errors produce proportional frequency and amplitude errors in the respective operating modes. The basic problem of a constant mechanical drive speed is complicated by supply voltage and frequency fluctuations, and changes in load as the bulk of the tape transfers from the supply to take-up reel. Effective speed errors also can be caused by tape stretch. These errors are the result of over-all tape elongation from humidity or tension variations, and localized stretching as the tape alternately sticks and springs back to normal due to head friction. Dry lubrication of the tape during its manufacture greatly reduces this latter effect.

Tape speed is governed by a capstan and pinch-roller arrangement driven by a constant-torque, hysteresis-synchronous motor. A large flywheel helps to minimize further instantaneous speed variations. Tape speed is therefore as constant as the supply frequency. In many instances, particularly in the field where the main supply may be from auxiliary generators or inverters, the line frequency may vary enough to cause appreciable errors. Several techniques have been developed to circumvent this problem. One method is to drive the capstan motor from an internal, precision 60-cps oscillator which is heavily compensated against line and environmental variations. Oscillator accuracies as high as 20 parts per million can be obtained. This technique provides the same record and playback capstan speeds, but does not make allowance for over-all tape dimensional changes or localized stick flutter.

The precision 60-cps oscillator also can be used to compensate for dimensional changes of the tape by using it to modulate a reference signal on an extra channel. During playback the reference signal modulation is matched against the precision standard frequency in a phase comparator circuit. Differences in tape speed or dimensions produce an error signal which is used to drive a servo amplifier and modify the phase of the voltage supplied to the capstan motor. This servo speed-control technique is very effective for compensating long-term speed and dimensional changes. Because of the large inertia of the rotating parts, it cannot change speed sufficiently fast to compensate for frictional flutter.

If higher accuracies over a wider bandwidth are required an all-electronic compensation system must be employed. One system for accomplishing this is similar to the servo speed-control technique above except that the correction is done electronically in an FM reproduce amplifier detector circuit. The scheme is not applicable to AM recording. The high-frequency, unmodulated FM center-frequency signal is recorded as a reference on a spare channel. Speed variations during either record or playback will frequency-modulate the reference signal. On playback, the signal is demodulated by an FM reproduce amplifier for the flutter components. The error voltage is fed to the FM demodulators of other channels, in the proper phase and amount, to subtract exactly from their output voltage the contribution due to flutter. Servo speed control of the capstan motor is used in conjunction with electronic compensation. The former eliminates long-term speed errors, while the latter compensates for high-frequency flutter.

AMPLIFIER CHARACTERISTICS. Most modern instrumentation tape recorders are designed with modular, interchangeable, plug-in record and reproduce amplifiers, and compatible magnetic heads so that the type of modulation—AM, FM, or PDM—on any track may be chosen to suit particular test conditions. Proper selection depends upon the number of inputs, their frequency ranges, and the type of multiplexing (if any is employed), as well as characteristics of the amplifiers themselves.

Amplitude Modulation. Amplitude modulation often is referred to as *direct modulation* since the intensity of the magnetizing flux in the recording head is directly proportional to the signal amplitude, and the recorded frequencies are identical to those in the input. High-frequency bias is added in the AM record amplifier to straighten the magnetic transfer characteristics as discussed in *Head Characteristics*. Block diagrams of typical direct-record and -reproduce amplifiers are shown in Fig. 19.64A and B. The direct-record amplifier consists of a number of a-c coupled amplifier stages, with inverse feedback to minimize distortion and increase its flat frequency range and stability. A high-impedance output stage provides a constant-current source for the record head at all frequencies in the passband region. Bias voltage is supplied by a high-frequency (300 to 400 kc) oscillator and linearly mixed with the data signal after the final amplifier. In multitrack units provision is made to synchronize bias oscillators on all channels in order to minimize beat-note interference between channels.

The reproduce amplifier, Fig. 19.64B, consists of an a-c coupled preamplifier and a number of amplifier stages to boost the signal to its original proportions. Generous amounts of inverse feedback stabilize the gain and improve frequency-response characteristics at both ends of the frequency range. Equalization networks compensate for the

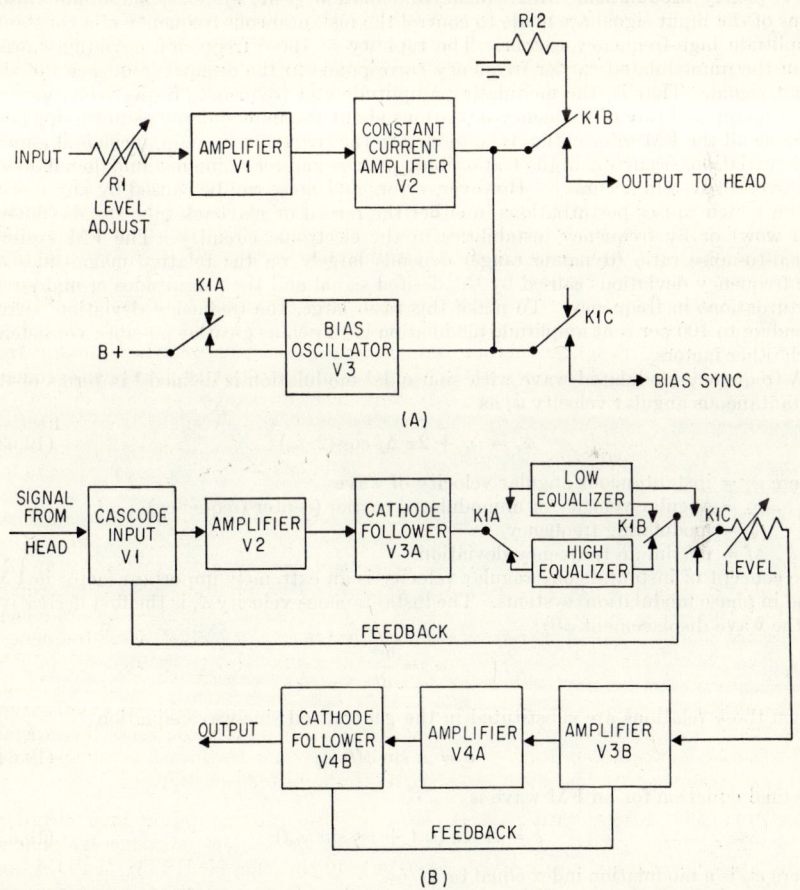

Fig. 19.64. Block diagrams of amplitude-modulated (direct) record and reproduce amplifiers. (A) AM record amplifier. (B) AM reproduce amplifier. (*Courtesy of Ampex Corp.*)

frequency-dependent characteristic of the reproduce heads to yield a flat over-all frequency response.

Typical direct-recording systems have a usable frequency range, flat within 3 db, from 100 to 100,000 cps at 60 in./sec. The upper frequency is limited by the size of the smallest wavelength that can be reproduced and is therefore a function of the tape speed, e.g., 50,000 cps at 30 in./sec. Low frequencies generally are limited by the signal-to-noise ratio of the system and the time constant of coupling networks. The low-frequency response seldom extends below 50 cps.

AM recordings are used predominantly for recording data signals containing high-frequency components which must be preserved, such as vibration and acoustic measurements in aircraft and missiles. Each data input requires one recording track. A second important application of direct recording is in the IRIG frequency-multiplexed FM/FM subcarrier telemetry system. Up to 18 subcarriers, proportionately spaced throughout the frequency range from 400 to 70,000 cps, can be added and made to modulate a single r-f transmitter. On the ground, the video complex from the receiver can be recorded on a single AM track of a magnetic-tape recorder.

Frequency Modulation. In frequency-modulation (FM) systems, amplitude variations of the input signal are made to control the instantaneous frequency of a constant-amplitude high-frequency carrier. The rapidity of these frequency deviations away from the unmodulated carrier frequency corresponds to the original frequencies of the input signal. That is, the modulating amplitude and frequency, respectively, govern the amount and rate of frequency deviations about the unmodulated center frequency.

Since all the FM information is contained in the frequency domain, incidental amplitude variations occurring in the transcribing process and recording medium do not affect the over-all system accuracy. However, errors and noise can be caused by any mechanism which causes perturbations in either the record or playback tape speeds (flutter and wow) or by frequency instabilities in the electronic circuitry. The FM system signal-to-noise ratio (dynamic range) depends largely on the relative magnitudes of the frequency deviations caused by the desired signal and the magnitudes of undesired perturbations in frequency. To make this ratio large, the frequency deviation corresponding to 100 per cent amplitude modulation is chosen as great as possible, consistent with other factors.

A frequency-modulated wave with sinusoidal modulation is defined [2] in terms of its instantaneous angular velocity ω_i as

$$\omega_i = \omega_c + 2\pi \, \Delta f \cos (2\pi f_m) \tag{19.63}$$

where ω_i = instantaneous angular velocity of wave
ω_c = angular velocity of unmodulated carrier (center frequency)
f_m = modulating frequency
Δf = maximum frequency deviation

The concept of instantaneous angular velocity is an extremely important factor in FM (and in phase modulation) systems. The instantaneous velocity ω_i is the first derivative of the wave displacement $\phi(t)$:

$$\omega_i = \frac{d\phi}{dt}$$

When these relations are substituted in the generalized sine-wave equation

$$e = A \sin \phi(t) \tag{19.64}$$

the final equation for an FM wave is

$$e = A \sin (\omega_c t + m_f \sin \omega_m t) \tag{19.65}$$

where m_f is a modulation index equal to $\Delta f / f_m$.

Energy in a frequency-modulated signal is contained in a set of sidebands on both sides of the carrier, spaced apart by an amount equal to the modulating frequency. The relative magnitude of these sidebands depends upon the value of the modulation index and varies in accordance with the magnitudes of the successive orders of a Bessel function with an argument m_f. In general, significant sidebands are contained within a frequency band on each side of the carrier equal to the sum of the maximum deviation and modulation frequencies.

A practical compromise of all factors, selected by tape recorder manufacturers and adapted by the IRIG,* makes maximum use of the tape recorder passband. The FM carrier is located at 54,000 cps (for 60 in./sec tape speed) and employs ±40 per cent frequency deviation, equivalent to a 100 per cent amplitude-modulation input level of 1.0 volt rms. If the maximum modulating frequency is restricted to 10,000 cps, the

---

* Inter-Range Instrumentation Group, Magnetic Recorder/Reproducer Specifications No. 101-57.

modulation index becomes approximately 2, which has only four significant sideband pairs (the fifth contains less than 1 per cent of the total energy). The sidebands are symmetrically spaced about the carrier at 10,000-cps intervals and cover the frequency range from 14,000 to 94,000 cps. These parameters afford sufficient clearance between the highest modulating frequency (10,000 cps) and the fourth-order low sideband harmonic (14,000 cps) to preclude intermodulation distortion. They also maintain the

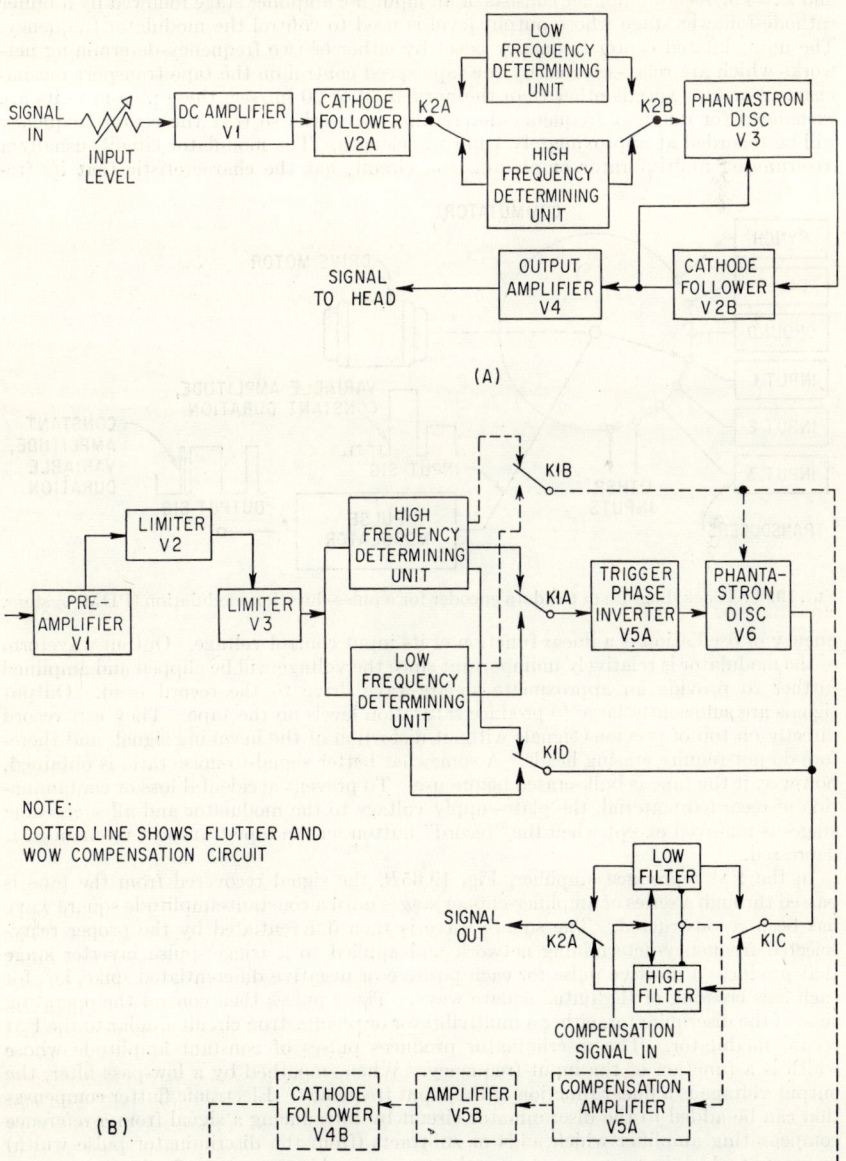

Fig. 19.65. Block diagrams of frequency-modulated (FM) record and reproduce amplifiers. (A) FM record amplifier. (B) FM reproduce amplifier. (*Courtesy of Ampex Corp.*)

entire frequency spectrum within the upper-frequency capabilities of the playback system. At lower tape speeds, the center frequency is lowered proportionally. Frequency response to twice that indicated above for each tape speed is possible if the per cent deviation is reduced by a factor of 2. That is, at 60 in./sec tape speed, response to 20,000 cps can be obtained by deviating the FM carrier only ±20 per cent.

Block diagrams of typical FM record and reproduce amplifiers are given in Fig. 19.65A and B. The record amplifier consists of an input d-c amplifier stage followed by a buffer cathode-follower stage whose output level is used to control the modulator frequency. The unmodulated center frequency is set by either of two frequency-determining networks which are relay-selected by the tape speed control on the tape transport mechanism. For tape speeds other than the normal 30 or 60 in./sec, these plug-in units are exchanged for different frequency-determining networks so that the center frequency will be recorded at approximately 1 mil wavelength. The modulator circuit, usually a free-running multivibrator or phantastron circuit, has the characteristic that its fre-

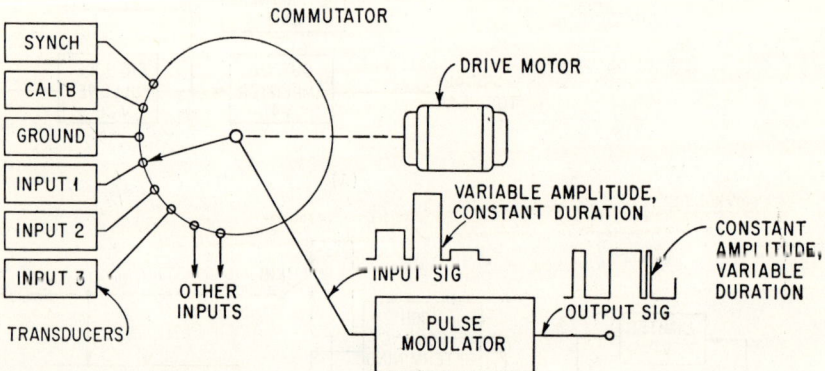

Fig. 19.66. Block diagram of the data encoder for a pulse-duration modulation (PDM) system.

quency of oscillation is a linear function of its input control voltage. Output waveform of the modulator is relatively unimportant since the voltage will be clipped and amplified further to provide an approximate square-wave drive to the record head. Output signals are sufficiently large to produce saturation levels on the tape. They can record directly on top of previous signals without distortion of the incoming signal, and therefore do not require erasing heads. A somewhat better signal-to-noise ratio is obtained, however, if the tape is bulk-erased before use. To prevent accidental loss or contamination of recorded material, the plate-supply voltage to the modulator and all succeeding stages is removed except when the "record" button on the tape transport mechanism is depressed.

In the FM reproduce amplifier, Fig. 19.65B, the signal recovered from the tape is passed through a series of amplifier-clipper stages until a constant-amplitude square wave has been reconstructed. The square wave is then differentiated by the proper relay-selected frequency-determining network and applied to a trigger-pulse inverter stage that produces a positive pulse for each positive or negative differentiated spike, i.e., for each axis crossing of the initial square wave. These pulses then control the operating time of the discriminator, either a multivibrator or phantastron circuit similar to the FM record modulator. The discriminator produces pulses of constant amplitude whose width is a function of the input frequency. When smoothed by a low-pass filter, the output voltage is a linear function of the input frequency. Electronic flutter compensation can be added to the discriminator circuit by introducing a signal from a reference compensating amplifier which adds or subtracts (from the discriminator pulse width) an amount which just compensates for the error caused by tape perturbations.

FM methods are unexcelled as a precision procedure for recording signals in the d-c to 10,000-cps range. The restricted upper frequency may be somewhat of a disadvantage in some applications, but is adequate for most shock and vibration work. Signal-to-noise ratios usually are about 50 db in uncompensated FM systems and about 10 db better with compensation.

Pulse-duration Modulation (PDM). In many applications it is desirable to record information from a very large number of data points, all of which vary rather slowly throughout the duration of a test. Temperatures, pressures, flow rates, position indicators, etc., are typical of these quasi-static types of data. If the output voltage from each of the desired data point transducers were assigned a separate recording track, the number of tape recorders necessary would be prohibitively bulky and expensive. In addition, the slowly varying nature of this type of data is very wasteful of the channel bandwidth.

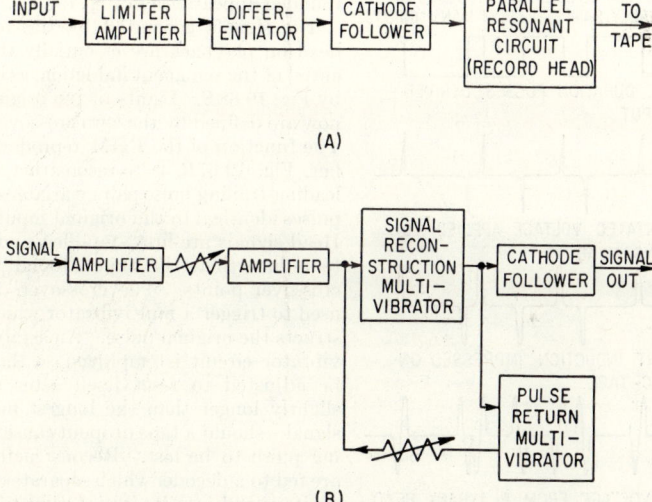

Fig. 19.67. Block diagrams of pulse-duration modulation (PDM) record and reproduce amplifiers. (A) PDM record amplifier. (B) PDM reproduce amplifier. (*Courtesy of Ampex Corp.*)

Pulse-duration modulation systems (PDM), sometimes termed *pulse-width modulation* (PWM), have been developed particularly for this type of data recording. Elements of the system are shown in the block diagram of Fig. 19.66. The numerous inputs are connected to a motor-driven or electronic commutator which is essentially a high-speed switch. It samples the output from each transducer in turn and commutates their voltages to a common amplifying system. Commutators may have as many as 90 contacts: 85 data inputs, zero, mid-scale, and full-scale calibration voltages and two for synchronizing. They usually operate at about 10 rps (600 rpm), providing a data bandwidth of d-c to about 2 cps.

The output from the commutator is used to control the time duration of a one-shot multivibrator. In this technique, variable-amplitude, constant-duration input signals are converted to constant-amplitude, variable-duration output pulses, time-multiplexed into a single channel of information. Typical waveforms as they appear at the input and output of the encoder are shown in Fig. 19.66. These sequential pulses can be recorded directly on one track of a tape recorder, or transmitted over a telemetering link and then recorded. Since the data are contained in the frequency domain, variations in amplitude levels of either the transmitting or recording medium do not affect the accuracy of the recorded data.

The commutator, pulse-duration encoder, and the decoder used to unscramble the pulses and reconstruct the individual original data signals are separate equipments and not part of the tape recorder circuitry. The tape recorder amplifiers used with PDM recordings are special circuits designed to generate sharp recording spikes to designate the beginning and ending of each pulse, and to reconstruct the original rectangular pulses on playback. Block diagrams are shown as Fig. 19.67A and B. Waveforms at various locations in a PDM system are shown in Fig. 19.68. The record amplifier includes an amplifier-clipper stage to limit all pulses to approximately the same amplitude. The rectangular pulses are then differentiated (Fig. 19.68C) to produce positive and negative pulses which mark the limits of the original input pulse. In the output amplifier the spikes are amplified sufficiently to produce tape saturation. Inductance of the recording head winding and diffusion effects from its gap cause the remanent flux induced on the magnetic tape to assume the more rounded appearance in Fig. 19.68D.

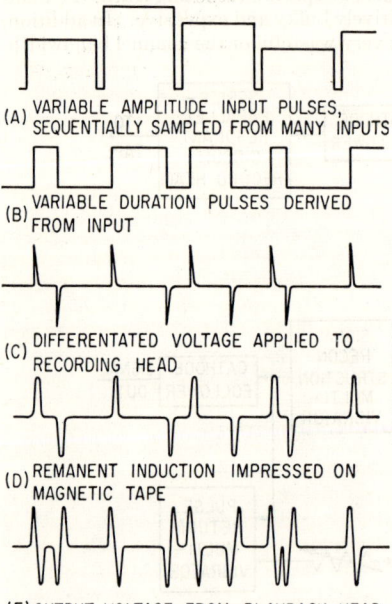

FIG. 19.68. Waveforms at various locations in a pulse-duration modulation (PDM) system. (A) Variable-amplitude input pulses, sequentially sampled from many inputs. (B) Variable-duration pulse derived from input by encoder. (C) Differentiated voltage applied to record head. (D) Remanent induction on magnetic tape. (E) Output voltage from reproduce head.

The signals generated by the reproduce head on playback are essentially the differential of the remanent induction, as indicated by Fig. 19.68E. Limits of the original pulse now are defined by the zero crossover points. The function of the PDM reproduce amplifier, Fig. 19.67B, is to reconstruct, from the leading-trailing pulse pairs, variable-duration pulses identical to the original input signals. Head signals are first amplified and clipped to reduce their rise time around the zero crossover points. The crossovers are then used to trigger a multivibrator which reconstructs the original pulse. An astable multivibrator circuit is employed so that it can be adjusted to reset itself after a period slightly longer than the longest modulated signal—should a tape dropout cause the trailing pulse to be lost. Reconstructed pulses are fed to a decoder which separates the composite signal into its initial components and presents each set of data to an individual pen recorder if a permanent record of the data is to be made.

Minimum pulse duration is limited by the rise time of reproduce head voltages and the diffusion of the remanent magnetic induction. Present tape recorders are typically just capable of distinguishing a leading-trailing pulse pair about 1.5 mils apart; e.g., a 50-microsecond pulse at 30 in./sec. Maximum pulse duration is generally variable over the range from about 300 to 6,000 microseconds to accommodate different sampling rates. Accuracy of the over-all system is of the order of ±2 microseconds.

TELEMETRY

Telemetry is defined as the indication, measurement, or integration of a quantity at a distance by electrical translating means. This definition implies a complete system: transducers to convert the measurand to an electrical equivalent, encoding and transmitting equipment, and receiving and decoding equipment capable of providing an indication or record of events as they occur at the remote location. The definition also implies that the actual distance between point of measurement and recording is immaterial, merely that the measurand is at a remote, untenable, or hazardous location. Remote control is different from telemetering in that it requires an additional function of control at the distant location.

TELEMETERING CONCEPTS

Telemetering was originally employed as a means of monitoring the load distribution at various points in an extensive power-line network. Transmission of this information was by direct-wired connections or by carrier currents on existing lines. This type of telemetry has been extended to convey temperatures, pressures, levels, flow rates, and other vital information to a central location in large manufacturing and industrial plants. Telemetry of pertinent data from hazardous areas, as in investigations or processes involving radioactive elements, represents another important application.

The first widespread use of radio transmission in telemetering was in connection with radiosondes for meteorological soundings. In this application, temperature and humidity data alternately tone-modulate the r-f carrier which is keyed off and on by the passage of a baroswitch control over grounding segments.

The most extensive and complex application of telemetry is in connection with aircraft, missile, and rocket development programs. Here the environment, space and weight limitations, and large quantities of data required dictate that a minimum amount of instrumentation be carried aloft and that the bulk of the recording equipment remain on the ground connected by a radio link. Many schemes increase the data-handling capabilities of each telemetering channel by multiplexing signals from numerous input transducers onto a single r-f carrier. Multiplex systems employ either time-sharing or frequency-sharing techniques. Proper choice depends on their relative advantages and disadvantages as related to the number, frequency range, bandwidth, and response characteristics of the input data, as well as space and weight limitations of the vehicle. Several of these methods are particularly good with respect to capacity, linearity, stability, ease of cyclic calibration, and ground terminal equipment requirements; they have been standardized to promote uniformity of equipment at ground receiving and data reduction centers. The standards, sponsored by an Inter-Range Instrumentation Group (IRIG), define characteristics of the telemetered signals rather than the characteristics of the equipment, and allow for and encourage continual improvement of equipment quality. At present, two distinct telemetering systems have been defined by standards. These are designated as FM/FM and PDM/FM, representing frequency-multiplexed and time-multiplexed systems, respectively. In addition, there are combinations of these basic types, as well as other systems using every method of modulation and multiplexing conceivable and practical. With little exception, these telemetering systems employ frequency modulation of the r-f carrier to minimize amplitude errors from signal-level variations due to vehicle height, distance, or aspect.

TELEMETRY SYSTEMS. FM/FM telemetering systems are of the frequency-shared multiplex type. The transmitter r-f carrier is simultaneously frequency-modulated by a group of subcarrier oscillators whose output signals are, in turn, frequency-modulated by the voltage amplitude from their respective input transducers. Up to eighteen data channels can be handled by a standard IRIG FM/FM system. Information from each input channel is continuous in time, but restricted in frequency so as not to interfere with adjacent subcarrier channels.

PDM/FM (pulse-duration modulation on a frequency-modulated carrier) telemetering systems are of the time-division multiplex type. Data channels are sequentially sampled by a commutator. Individual signals are encoded by a keyer to produce output pulses whose durations are proportional to the amplitude of their corresponding input amplitudes. The transmitter r-f signal is frequency-modulated by the keyer, the intelligence being contained in the length of time the instantaneous carrier frequency is deviated away from its quiescent frequency. A PDM/FM system based on IRIG specifications can handle up to 900 samples per second. Many different combinations of number of channels and frame rate can be selected.

PAM/FM (pulse-amplitude modulation on a frequency-modulated carrier) is similar to PDM/FM except that the amplitudes of the sequentially sampled data inputs modulate the transmitter r-f carrier directly, eliminating the keyer. Intelligence is contained in the amount of instantaneous frequency deviation of the r-f carrier. Because of time sharing, both PDM/FM and PAM/FM systems are restricted to data that change slowly in comparison to the sampling rate.

Combinations of these techniques also may be employed. For example, a commutator can be used in one or more of the FM/FM subcarrier bands to increase the data-handling capacity on that band. Modulation on the time-division multiplexed subcarrier band may be either pulse duration or pulse amplitude resulting in PDM-FM/FM or PAM-FM/FM designations. Restricted bandwidths imposed by the subcarrier frequency generally limit commutation rates to lower values than are allowable with unshared PDM or PAM systems.

PPM/AM (pulse-position modulation on an amplitude-modulated carrier) is a sequentially sampled, time-multiplexed system, similar in many respects to PDM/FM. In this system the amplitudes of the individual input signals control the position of a short-duration data pulse with respect to a reference pulse at the beginning of each sampling period. Often the reference pulse is suppressed and only the data pulse transmitted. Reference pulses are reconstructed at the ground station from a synchronizing pulse transmitted once each frame. Sampling rate is typically in the order of 5,000 samples per second, distributed among 10 to 15 channels. A PPM/AM system can produce considerably greater peak r-f power than a PDM/FM system, for the same average input power, because its duty cycle with only one short-duration pulse is so much less than that of a PDM/FM system where full-duration pulses are transmitted. Recording at the ground station is usually direct from cathode-ray-tube decoders because the data pulse is too short (2 to 4 microseconds) to be handled properly with the usual 100,000-cps bandwidth instrumentation tape recorders. Video tape recorders with at least 1 megacycle bandwidth and low jitter are required.

PPM telemetry systems have been incorporated in radar beacon transponders. Here the data pulse is transmitted after the beacon response pulse—the delay time between pulses being a measure of data amplitude. Signals can be multiplexed at the radar interrogation rate to yield data from numerous inputs.

The task of reducing all the data that modern telemetering systems are capable of conveying is enormous and time-consuming. It is not unusual for data reduction to lag many weeks behind the acquisition of the raw telemetry data. Because of this, other systems which are more adaptable to automatic data-reduction techniques are employed. One approach to this problem is pulse-code modulation, PCM. The system includes an analog-to-digital encoder following the commutator of time-division multiplex systems to pulse-modulate the transmitter r-f carrier in a binary code sequence. In addition to providing a higher packing density per channel, the digitized data are in a form suitable for automatic decoding and analysis.

DATA INPUT TRANSDUCERS. Transducers used with telemetering systems are similar to those discussed in earlier chapters. Special precautions may be included in their design to negate variations in their calibrations brought about by rapid and extreme changes in their environment. Transducers operating on the potentiometer or variable-reluctance principles are especially popular and compatible with telemetry systems because of their relatively high sensitivities and high outputs and because they are particularly adaptable to transducing a variety of mechanical motions, positions, and rates to an equivalent electrical signal. The remainder of this section is devoted to specifications and techniques of telemetering systems which are well developed and standardized.

FM/FM TELEMETERING SYSTEMS

FM/FM SPECIFICATIONS. The FM/FM IRIG frequency-multiplex system of radio telemetry employs an r-f transmitter simultaneously frequency-modulated by up to 18 subcarrier oscillators. The subcarriers, each on a different center frequency in the audio-to-ultrasonic frequency range, are individually frequency-modulated by separate data transducers. Subcarrier bands often are designated as *RDB bands*.* Current

---

* In reference to their original standardization by the now-extinct Research and Development Board.

standards are under the sponsorship of the Inter-Range Instrumentation Group (IRIG). Characteristics of standard IRIG FM/FM subcarrier bands are given in Table 19.5.

Subcarrier oscillator center frequencies range from 400 cps (band 1) to 70 kc (band 18). Ratios between adjacent subcarrier center frequencies approximate 1.3 to the closest round number. A standard frequency deviation of ± 7.5 per cent, equivalent to 100 per cent amplitude modulation of the input, is allowed on bands 1 through 18, giving the upper and lower frequency limits listed. By restricting the modulation index m_f to 5 ($m_f = \Delta f/f_m$; see *Frequency-modulated Tape Recorder Amplifiers*) the maximum signal frequency range on each subcarrier band is as given in the last column in Table 19.5. With this modulation index all significant sideband components are contained within a total bandwidth equal to twice the sum of the maximum frequency deviation and maximum modulating frequency. Limits given for each band provide ample clearance to minimize intermodulation distortion between adjacent channels.

The total frequency coverage of the FM/FM subcarrier system provides optimum use of the available spectrum and largely reflects the frequency-response capabilities of magnetic-tape recorders used for recording and temporary data storage. One complete IRIG FM/FM data system, containing all eighteen subcarrier channels, can be recorded on a single AM track of a magnetic-tape recorder.

Table 19.5. IRIG FM/FM Telemetry Subcarrier Oscillator Bands

Band	Center frequency, cps	± Deviation, cps	Lower limit, cps	Upper limit, cps	± Max. deviation, %	Frequency response,* cps
1	400	30	370	430	7.5	6.0
2	560	42	518	602	7.5	8.4
3	730	55	675	785	7.5	11.0
4	960	72	888	1,032	7.5	14.0
5	1,300	98	1,202	1,398	7.5	20.0
6	1,700	128	1,572	1,828	7.5	25.0
7	2,300	173	2,127	2,473	7.5	35.0
8	3,000	225	2,775	3,225	7.5	45.0
9	3,900	293	3,607	4,193	7.5	59.0
10	5,400	405	4,995	5,805	7.5	81.0
11	7,350	551	6,799	7,901	7.5	110.0
12	10,500	788	9,712	11,288	7.5	160.0
13	14,500	1,088	13,412	15,588	7.5	220.0
14	22,000	1,650	20,350	23,650	7.5	330.0
15	30,000	2,250	27,750	32,250	7.5	450.0
16	40,000	3,000	37,000	43,000	7.5	600.0
17	52,500	3,940	48,560	56,440	7.5	790.0
18	70,000	5,250	64,750	75,250	7.5	1,050.0
A †	22,000	3,300	18,700	25,300	15.0	660.0
B	30,000	4,500	25,500	34,500	15.0	900.0
C	40,000	6,000	34,000	46,000	15.0	1,200.0
D	52,500	7,880	44,620	60,380	15.0	1,600.0
E	70,000	10,500	59,500	80,500	15.0	2,100.0

* Frequency response given is based on maximum deviation and a modulation index of five.
† Bands A through E are optional and may be used by omitting adjacent bands as follows:

Band Used	Omit Bands
A	15 and B
B	14, 16, A and C
C	15, 17, B, and D
D	16, 18, C, and E
E	17 and D

Note: In the process of magnetic-tape recording of the above-listed subcarriers at a receiving station, provision also may be made to record tape speed control tone and tape speed error compensation signals. The speed control tone frequency is 17,000 cps. Standard tape flutter compensation signal frequencies are 50,000 cps and 100,000 cps for tape speeds of 30 and 60 in./sec, respectively.

Table 19.6. Typical R-F Transmitter Characteristics for FM/FM Telemetry

Carrier frequency	216 to 235 mc
Stability	0.1 per cent of carrier frequency
R-F modulation	Frequency modulation or phase modulation
R-F deviation	±75 kc minimum to ±125 kc maximum (equivalent deviation for phase modulation)
Power	100 watts maximum
Spurious radiation	Radiated power of all harmonics 60 db below fundamental

The frequency-response range may be doubled on the five highest subcarrier bands by doubling the maximum frequency deviation to ±15 per cent. When this option is exercised, adjacent bands (as noted in the footnote to Table 19.5) are omitted.

An extra-wide gap is provided between subcarrier bands 13 and 14 to allow insertion of a 17,000-cps 60-cps-modulated subcarrier speed-control tone by the ground station during recording with magnetic-tape recorders. This signal is a reference signal used to ensure that subsequent playback speeds are identical to those used during the initial recording.

Specifications covering the transmitter r-f carrier characteristics are given in Table 19.6. Most telemetering transmitters are capable of being tuned over the entire 216- to 235-megacycle telemetering band, and permit the carrier to be adjusted to any clear frequency. Bandwidth of an FM/FM telemetry system, when fully modulated on the highest subcarrier band, approaches 0.5 megacycle. At least 1 megacycle clearance should be maintained between adjacent telemetering channels to preclude intermodulation and cross talk.

FM/FM COMPONENTS. There are commercially available telemetering systems which utilize modular construction and which usually can be physically arranged to fit into available voids in a test vehicle. In extreme cases, the circuits of a proved system can be repackaged with little electronic redesign. Completely transistorized telemetering systems very greatly reduce space, weight, and power requirements.

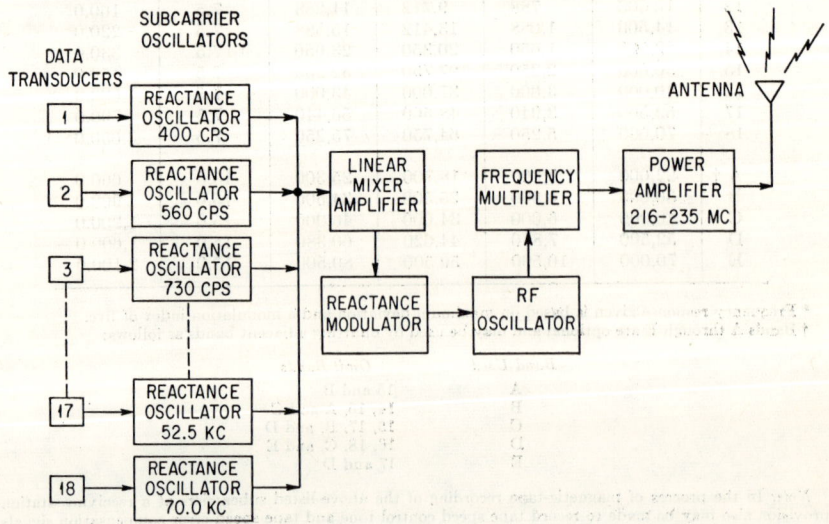

Fig. 19.69. Block diagram of a typical FM/FM telemetering system.

TELEMETRY

Although specific circuits and components vary between equipments developed by different organizations, the general electronic arrangement of an FM/FM telemetering transmitter is as shown in Fig. 19.69. Signals from the separate data transducers frequency modulate their individual subcarrier oscillators. Only those channels required to handle all the desired information are included. Output voltages from the subcarrier oscillators are linearly mixed and increased in level in a mixer amplifier and are used to frequency-modulate the r-f transmitter through some form of reactance modulator. Frequency multiplier stages increase the r-f signal to the proper frequency for transmission.

Subcarrier Oscillators. Subcarrier oscillators may be frequency-modulated by variations in inductance, voltage, resistance, or current. Inductance and voltage-controlled subcarrier oscillators are by far the most widely used.

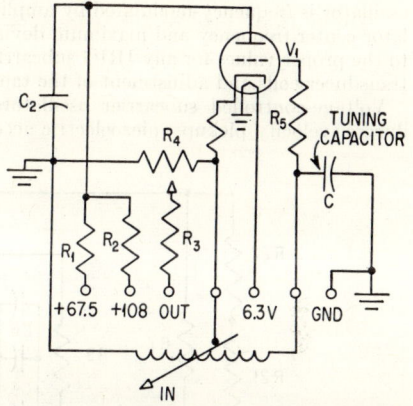

FIG. 19.70. Inductance-controlled subcarrier oscillator. Hartley circuit is tuned to one of the IRIG bands by adjustment of C or by changing the inductance of the transducer coil. (*Courtesy of Bendix-Pacific Division.*)

Variable-reluctance transducers (sensitive to acceleration, pressure, angular position etc.) can be used for this purpose which employ one or more coils wound around an E-shaped magnetic core. Spacing between the E core and a movable armature is a function of the variable being measured, causing the coil inductance to vary. When the transducer coil is used as the tank circuit of an oscillator, Fig. 19.70, the subcarrier

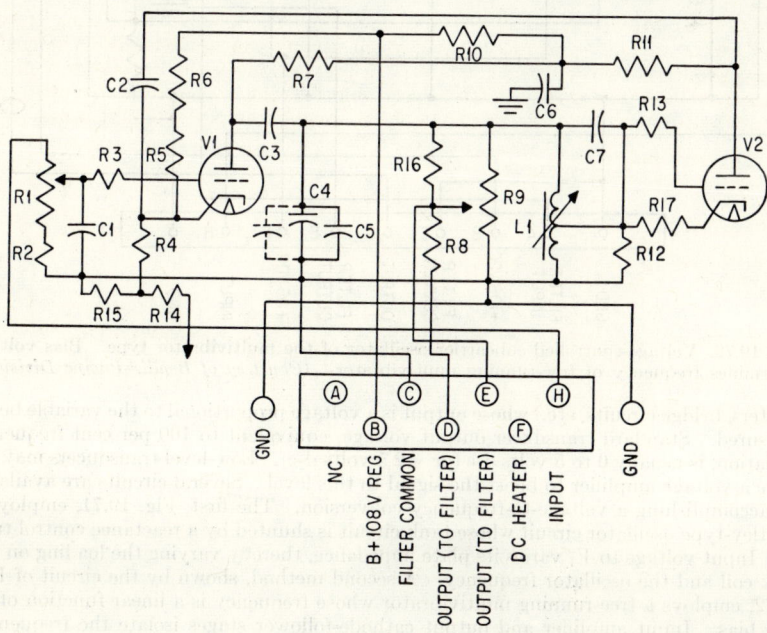

FIG. 19.71. Voltage-controlled subcarrier oscillator of the Hartley type. Input grid voltage controls plate impedance of reactance tube V_1 connected across tank circuit. (*Courtesy of Bendix-Pacific Division.*)

oscillator is frequency-modulated by amplitude variations of the input stimulus. Oscillator center frequency and maximum deviation for full-scale excitation can be adjusted to the proper values for any IRIG subcarrier band by a combination of interchangeable transducer coils and adjustment of the tank trimming condenser C.

Voltage-controlled subcarrier oscillators are used with self-generating-type transducers (velocity pickups, piezoelectric accelerometers, etc.) and other devices (potenti-

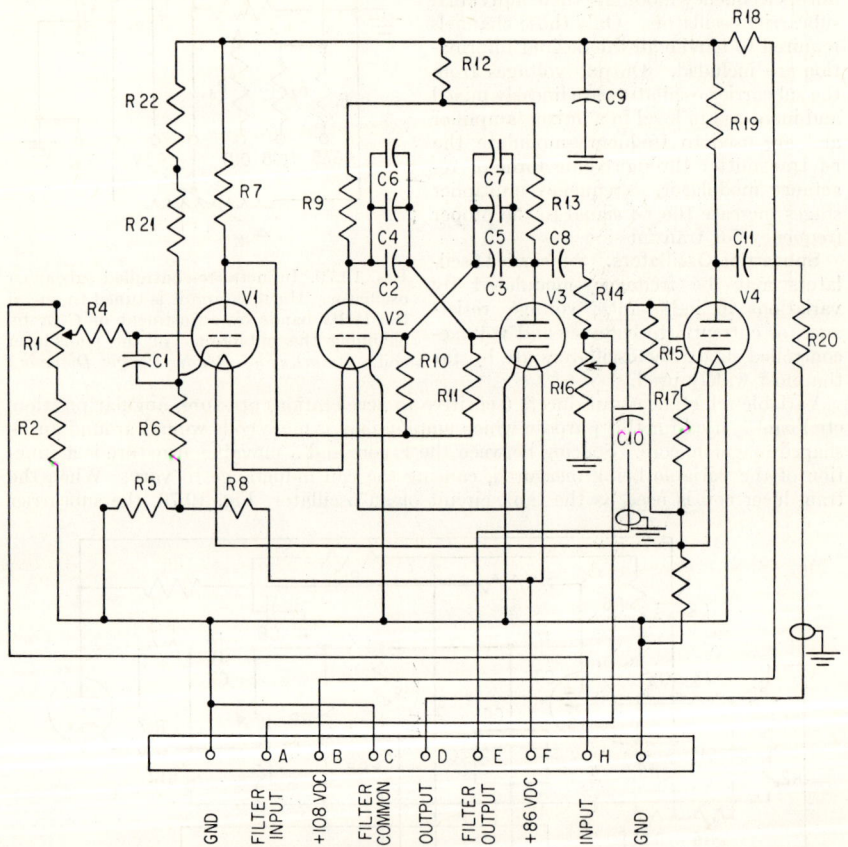

Fig. 19.72. Voltage-controlled subcarrier oscillator of the multivibrator type. Bias voltage determines frequency of free-running multivibrator. (*Courtesy of Bendix-Pacific Division.*)

ometers, bridge circuits, etc.) whose output is a voltage proportional to the variable being measured. Standard transducer output voltage, equivalent to 100 per cent frequency deviation, is usually 0 to 5 volts d-c (or ±2.5 volts d-c). Low-level transducers may require a voltage amplifier to boost the signal to this level. Several circuits are available for accomplishing a voltage-to-frequency conversion. The first, Fig. 19.71, employs a Hartley-type oscillator circuit whose tank circuit is shunted by a reactance control tube V_1. Input voltage to V_1 varies its plate impedance, thereby varying the loading on the tank coil and the oscillator frequency. A second method, shown by the circuit of Fig. 19.72, employs a free-running multivibrator whose frequency is a linear function of its grid bias. Input amplifier and output cathode-follower stages isolate the frequency-determining elements from the effects of external loading. This type of circuit requires a low-pass filter in its output to prevent harmonics of the square-wave output from causing interference on higher-frequency bands.

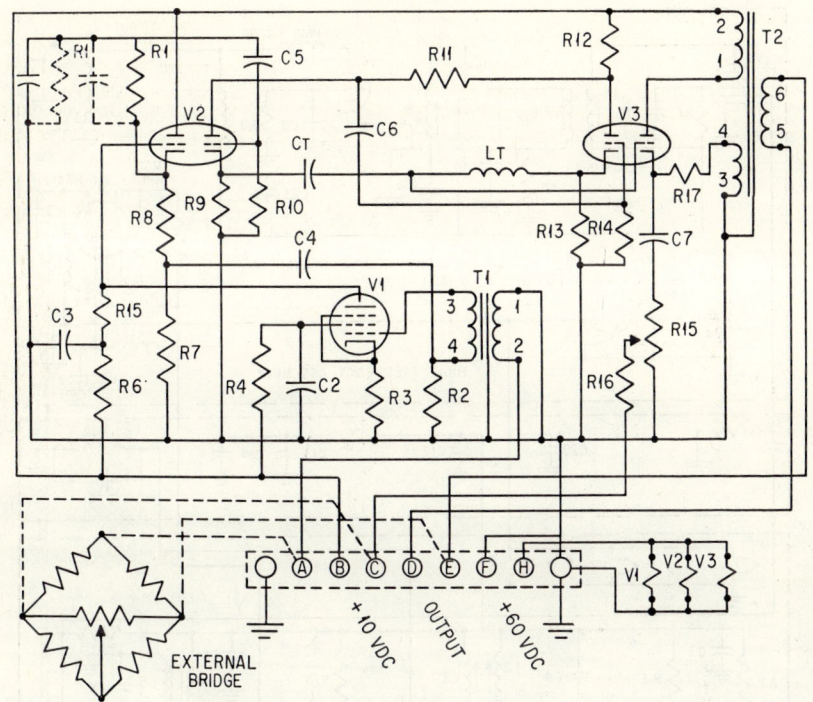

Fig. 19.73. Bridge-controlled subcarrier oscillator. (*Courtesy of Bendix-Pacific Division.*)

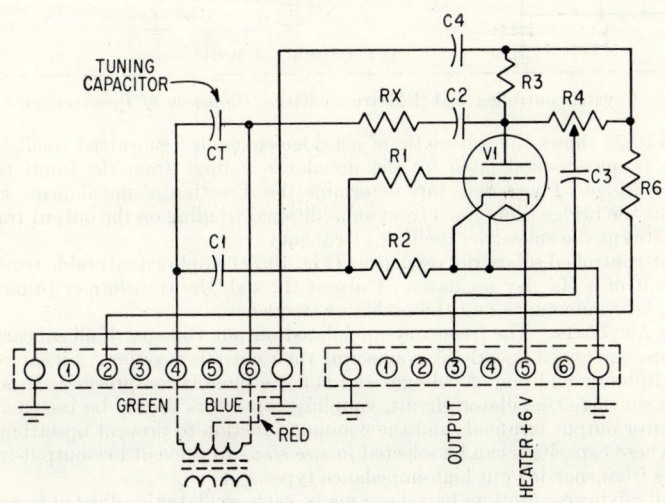

Fig. 19.74. Current-controlled subcarrier oscillator. Tank circuit of Hartley oscillator is a saturable reactor. (*Courtesy of Bendix-Pacific Division.*)

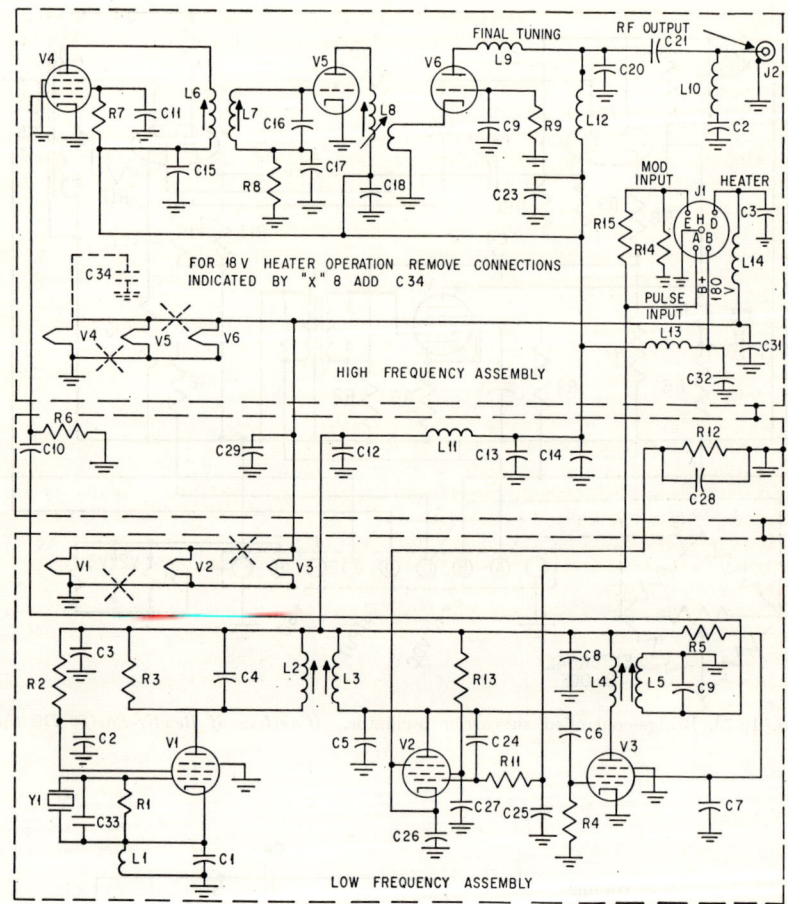

Fig. 19.75. Crystal-controlled FM/PM transmitter. (*Courtesy of Bendix-Pacific Division.*)

Figure 19.73 shows the schematic of a bridge-controlled subcarrier oscillator. This circuit is frequency-modulated by the unbalance voltage from the input transducer resistance bridge. Phase detectors determine the direction of unbalance. Excitation voltage for the bridge is obtained from an additional winding on the output transformer and operates at the subcarrier oscillator frequency.

Current-controlled subcarrier oscillators (Fig. 19.74) employ a saturable reactor as the tank circuit of a Hartley oscillator. Current through the transformer primary varies the secondary inductance and the oscillator frequency.

Mixing Amplifiers. The frequency-modulated output voltages of all subcarrier oscillators in use are mixed linearly by connecting their outputs together. Linear mixing is a simple arithmetic addition of voltages and is not a modulation process. If not part of the basic subcarrier oscillator circuit, coupling capacitors should be inserted between the oscillator output terminals and the common junction to prevent upsetting local d-c levels. These capacitors can be selected in size so as to prevent low-output-impedance oscillators from shorting out high-impedance types.

After all mixing connections have been made, each oscillator is adjusted to provide the proper output into the combined junction. During this adjustment, all other subcarrier oscillators remain connected to the junction but are temporarily rendered inoperative.

Typical output voltages are of the order of 0.2 volt (rms) or more per oscillator. A mixer amplifier may be employed to increase the combined level, as required by the deviation sensitivity of the transmitter reactance modulator.

R-F Transmitters. The circuit diagram of a typical low-power frequency-modulated telemetering transmitter is given by Fig. 19.75. The circuit includes a crystal-controlled oscillator, reactance-tube modulator, frequency multipliers, and final power amplifier. It produces an output of approximately 2 watts in the 216- to 235-megacycle telemetering band. Deviation sensitivities vary with individual units but usually average about ± 65 kc/volt input, and may range up to a maximum of ± 125 kc/volt input away from the center frequency. The several watts of power delivered to the antenna by this type of circuit is suitable for line-of-sight distances up to about 200 miles. For longer distances, or poorer propagation conditions, this transmitter can be used as a driver for a still higher-power final stage.

Ground-station Equipment. Figure 19.76 is a block diagram of typical ground-station equipment required for tracking, receiving, recording, storing, and processing FM/FM telemetry data. Quite often several radio links are necessary to transmit all the required data. Some of these additional channels may employ different telemetering systems, further complicating the receiving, separating, and recording processes.

The antenna system is usually made up of a high-gain helical antenna mounted on a rotary pedestal to permit complete coverage in azimuth and elevation. Circularly polarized helical antennas have the sensibly constant gain required to maintain reception, regardless of the angle of polarization that may result from gyrations of the missile. Antenna position during a test may be under the control of an operator observing signal strength meters tied in with the receivers, or in more elaborate arrangements may be entirely servo-controlled. When more than one radio telemetering link is to be received over the same antenna, a low-noise preamplifier and multicoupler should be employed to prevent intercoupling between receivers on adjacent frequencies.

A frequency-modulated receiver with good bandwidth, high signal-to-noise ratio, and low-distortion video system is required for each FM/FM telemetering system. After

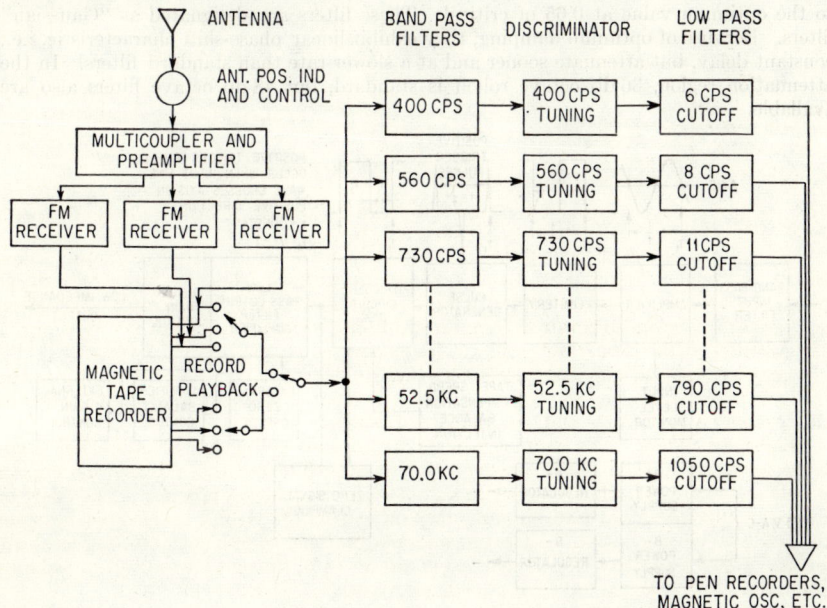

FIG. 19.76. Block diagram of typical FM/FM ground-station equipment. Separation equipment is sufficient for one radio link.

being demodulated, the data complex containing the combined frequency-modulated signals from all subcarrier oscillators is fed to an AM channel of a magnetic-tape recorder for permanent storage, or temporary retention until the data can be separated and recorded by pen recorders or magnetic oscillographs. Signals from other radio links may be recorded simultaneously on other tracks of the tape recorder.

If ground-station equipment facilities permit, the signal also can be fed to parallel-connected subcarrier discriminators to be separated and recorded immediately. This technique is useful where very rapid evaluations of vital information are required. When the number of simultaneous data signals exceeds the ground-station separating and recording equipments (as is most often the case), the tape recorded signal is played back repeatedly, stripping off the data a group at a time, until all channels have been recorded as permanent records. Without this temporary storage feature afforded by magnetic-tape recorders, immediate separation and recording of telemetry signals would impose a serious limit on the number of information channels that could be handled simultaneously.

An FM/FM subcarrier discriminator, Fig. 19.77, consists of a bandpass filter, amplifier, FM discriminator, low-pass filter, and output d-c amplifier. Center frequencies and passbands are available for each of the IRIG subcarrier bands in use. Attenuation characteristics of both bandpass and low-pass filters are standardized relative to IRIG band frequencies as shown by Fig. 19.78. Bandpass filters, Fig. 19.78A, for bands 1 through 18 are down 3 db at the ± 7.5 per cent frequency deviation points and have a 36 db/octave rolloff outside the passband. Bands A through E have similar characteristics except that the 3-db down points occur at ± 15 per cent frequency deviation. Both types of bandpass input filters are centered on the subcarrier band center frequency. Several types of low-pass filters, Fig. 19.78B, are available. The standard low-pass filter has a frequency response flat within ± 0.5 db in the passband region, and gives the sharpest transition near the cutoff frequency. Because of this discontinuity in its frequency-response characteristic, standard filters tend to "ring" on square waves, e.g., subcommutated data. Damping usually is approximately 30 per cent of critical. However, when ringing cannot be tolerated, more damping is added to the filter to increase it to the optimum value at 0.65 of critical. These filters are designated as "Gaussian" filters. Because of optimum damping, they exhibit linear phase-shift characteristic, i.e., constant delay, but attenuate sooner and at a slower rate than standard filters. In the attenuation region, 36 db/octave rolloff is standard, but 48 db/octave filters also are available.

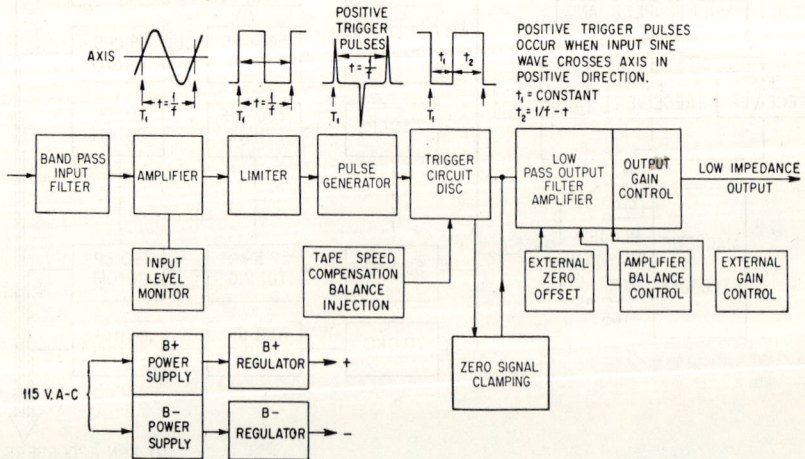

FIG. 19.77. Block diagram of an FM/FM subcarrier discriminator. (*Courtesy of Electro-Mechanical Research.*)

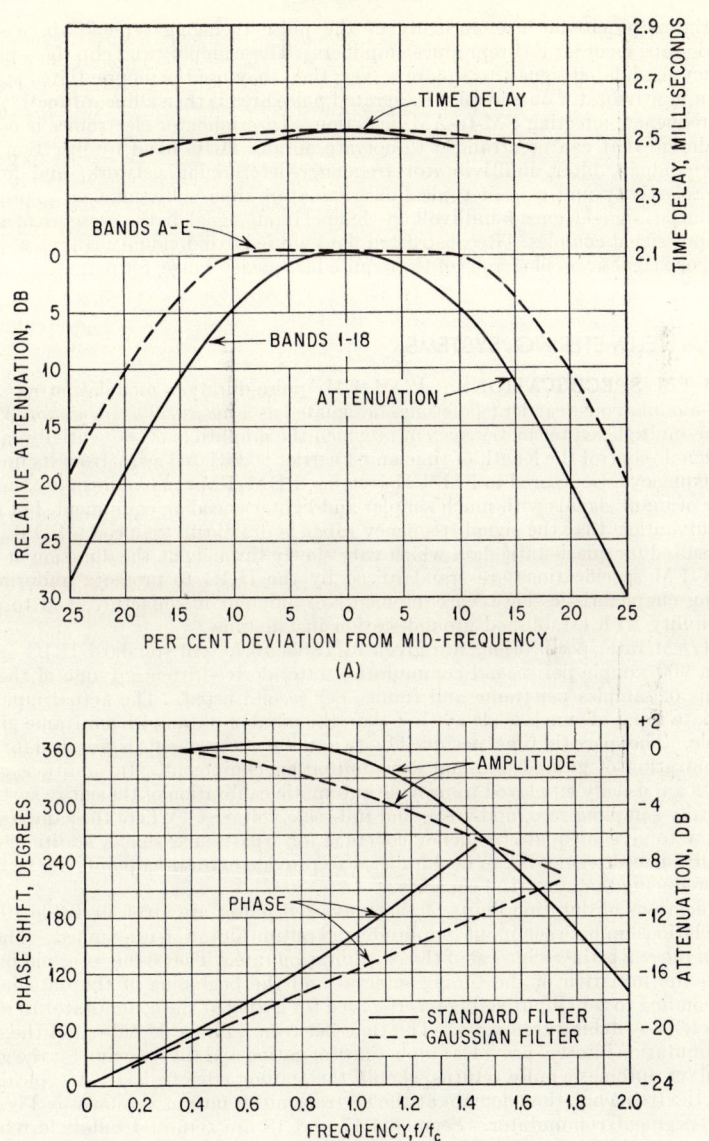

Fig. 19.78. Attenuation and phase-shift characteristics of bandpass and low-pass filters for FM/FM subcarrier discriminators. (A) Bandpass filter characteristics. (B) Standard and "Gaussian" low-pass output filters. (*Courtesy of Electro-Mechanical Research.*)

Discriminator circuits are normally of the pulse-averaging type as discussed for magnetic-tape recorder FM reproduce amplifiers. These amplify and clip the separated, frequency-modulated signals to square waves; then they use the differentiated spikes to trigger a multivibrator on and off. Integrated pulse area is then a linear function of the input frequency, affecting FM-to-AM detection. Discriminator electronics is of a universal design that can be arranged to operate on any IRIG band by insertion of the proper bandpass filter, multivibrator frequency-determining network, and low-pass filter. Selectivity and gain of typical discriminators are usually sufficient to produce a usable output signal from a 5-millivolt on-channel input signal, in the presence of a 5-volt total input-signal complex. Reconstructed data are fed to individual channels of pen recorders, or magnetic oscillographs of the required frequency range, for permanent recording.

PDM/FM TELEMETERING SYSTEMS

PDM/FM SPECIFICATIONS. PDM/FM, pulse-duration modulation of a frequency-modulated r-f carrier (sometimes designated as a *pulse-width modulation*, PWM), is a time-multiplexed telemetry system in which the amplitude of sequentially sampled input signals control the length of time an r-f carrier is deviated away from its unmodulated frequency. Compared to FM/FM systems, PDM/FM can handle many times the number of input signals with much simpler and lighter encoding equipment, but suffers the disadvantage that the signal frequency range is drastically reduced. PDM/FM is ideally suited for quasi-static data which vary slowly throughout the duration of a test.

PDM/FM specifications are standardized by the IRIG to promote uniformity of operating characteristics between equipments of different manufacture, and to ensure compatibility with established ground-station installations.

PDM/FM rate specifications are given in Table 19.7. All standard IRIG systems utilize a 900 sample per second commutation rate, derived from any one of the combinations of samples per frame and frames per second listed. The actual number of active data input channels is always less than the number of samples per frame given in this table. The space (in time) occupied by two successive segments is reserved for frame synchronization of ground-terminal decommutation equipment. In addition, several segments are usually employed to provide automatic calibration of the entire system by periodically sampling zero, mid-scale, and full-scale voltages. Where the sampling rate is too low to give adequate frequency coverage for a particular signal, additional symmetrically spaced segments can be paralleled to provide extra data points at the cost of further reducing the numerical capacity.

Specifications of duration limits of the converted pulses are given in Table 19.8. A linear relationship between input amplitude and output duration is assumed. The *commutation interval* is the reciprocal of the commutation rate. The 90-microsecond interval between the initiation of the timing interval and the beginning of the data interval (corresponding to zero input voltage) is reserved for gating at the commutator to remove the effects of switching transients. The time between a full-scale pulse and the end of the commutation interval provides ample time for gating and the insertion (at the ground terminal) of automatic pulse returns, should the trailing edge be lost. A typical pulse train in the frequency-time domain at the r-f transmitter output is shown by Fig. 19.79 for a 20-segment commutator. Segments 17 and 18 are zero and full-scale reference voltages; 19 and 20 are omitted for frame synchronization; and the remaining contacts contain active data signals.

On the basis that 10 samples per cycle are sufficient to define adequately the instantaneous amplitude of a varying signal, the maximum frequency response of a PDM/FM

Table 19.7. IRIG PDM/FM Telemetry Commutation Specifications

Samples per frame	30	45	60	90
Frame rate (frames/sec)	30	20	15	10
Commutation rate (samples/sec)	900	900	900	900

TELEMETRY

Table 19.8. IRIG PDM/FM Telemetry Pulse-duration Specifications

Zero signal duration (minimum duration).............	90 ± 30 microseconds
Maximum signal duration (maximum duration)......	700 ± 50 microseconds
Commutation interval.............................	$1,100 \pm 25$ microseconds
Pulse rise and decay time (between 10 and 90 per cent levels)	10 to 20 microseconds jitter ± 3 (microseconds)

system is not much greater than 1 to 3 cps for normal frame rates. Increased frequency range for a particular channel can be obtained by paralleling symmetrical (in space) contacts until the desired range is achieved.

R-F transmitter specifications for PDM/FM are similar to those for FM/FM telemetry except that the maximum frequency deviation is limited to a range between ± 25 kc and ± 45 kc. R-f bandwidth approaches 200 kc. The guard bands that are required depend somewhat on prevailing occupancy of the telemetering band, but a 0.5-megacycle separation between adjacent carriers is normally sufficient.

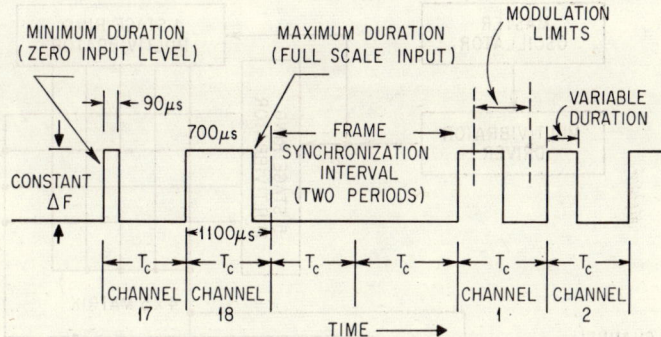

Fig. 19.79. Representative pulse-duration-modulated wave train.

PDM/FM COMPONENTS. Figure 19.80 is a block diagram of a typical PDM/FM airborne telemetering system. It comprises a commutator for sequentially sampling the output voltages from a large number of input channels; a keyer which converts the amplitude of each input pulse to a corresponding duration of an associated output pulse; and a frequency-modulated r-f transmitter. Because of the simplicity of the multiplexing equipment, PDM/FM systems generally impose a smaller size and weight penalty on their parent vehicles than do FM/FM systems.

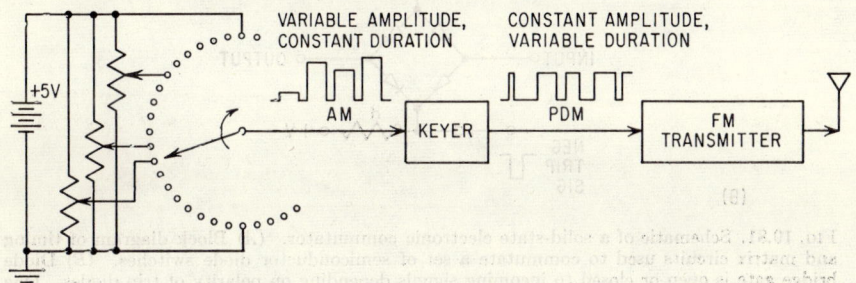

Fig. 19.80. Block diagram of a PDM/FM airborne system.

Commutators. PDM/FM techniques are predicted on time sharing by commutation. The high-speed mechanical switch is the most basic form of this device. It usually takes the form of a group of contact segments, circularly arranged on a contact plate to allow continuous rotary operation of the wiper assembly. Multiple-pole commutators may be obtained by including additional contact segment rings and wipers to a common contact plate or by ganging additional contact plates to the same rotor shaft. Synchronism between poles requires extreme precision during manufacture and assembly. Wiper rotation is provided by a small permanent-magnet motor with a small speed-reducer gearbox. Standard specifications suggest that the motor be governor-controlled between +5 and −15 per cent of its design speed for all environmental conditions.

Electrical performance and mechanical life of a mechanical switching commutator represent a design compromise. Too light a contact pressure introduces noise from contact bounce and a high and uncertain contact resistance; too heavy a contact pressure rapidly wears away and deteriorates the contact surfaces, causing resistance irregularities and depositing conducting particles over insulating surfaces. Contact resistances generally average in the range of a fraction of an ohm; contact life may extend to several hundred hours.

Noise in mechanical commutators arises from variations in contact resistance, contact bounce, and edge noises due to making and breaking electrical connection. When

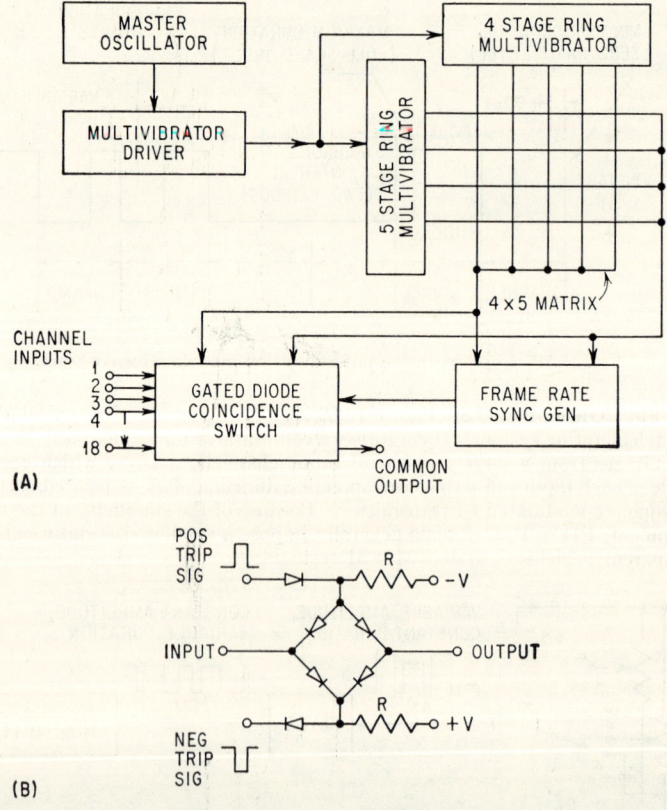

FIG. 19.81. Schematic of a solid-state electronic commutator. (A) Block diagram of timing and matrix circuits used to commutate a set of semiconductor diode switches. (B) Diode bridge gate is open or closed to incoming signals depending on polarity of trip diodes. One gate is required for each channel.

potentiometer-type transducers are used, having the normal 5-volt full-scale sensitivity, noise contributions due to switching have negligible effects on system accuracy, and can be tolerated. However, with low-level transducers such as strain-gage bridges, thermocouples, etc., and especially with high switching speeds and automatic data-reduction equipment, switching transients become bothersome. These disturbances can be eliminated by "gating" the commutated signal, i.e., grounding the wiper for a short interval during the transition between segments. In addition to eliminating noise, this technique produces a cleaner, more abrupt discontinuity at the beginning and end of the pulse, which is beneficial for syn-

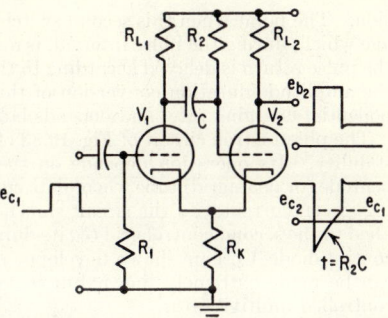

Fig. 19.82. One-shot multivibrator. Pulse duration is a function of input grid potential.

chronization at the ground station. Mechanical gating can be obtained by grounding alternate contact segments on a single-pole "make-before-break" commutator or by employing a separate pole to perform this function.

Because of difficulties experienced with mechanical commutators, other commutating techniques are sometimes employed, including electronic tubes and transistors, semiconductor diodes, photoconductors, magnetic cores, and nonlinear resistances. While these methods generally improve circuit noise and operating life they suffer from extra mechanical and electrical complexity, larger power consumption, size, and weight. Circuit details of a semiconductor commutator are shown by Fig. 19.81. The system employs a master oscillator to drive simultaneously a four- and a five-stage multivibrator ring. Output is taken in the form of a 4×5 matrix to control the individual opening of

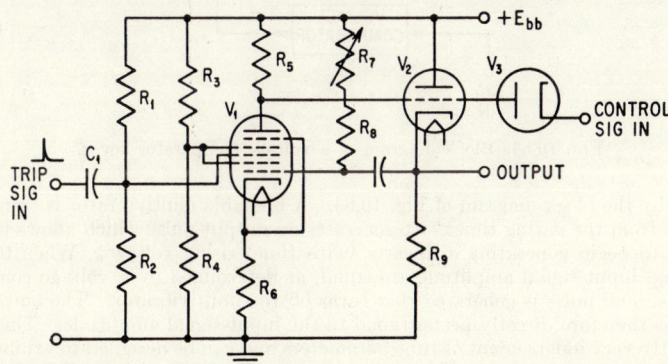

Fig. 19.83. A phantastron circuit in which pulse duration is a linear function of the voltage E_1 on the control diode.

a system of diode gates when coincidence pulses occur. This system is entirely transistorized, and is reasonably sized and stable.

Keyers. Conversion from pulse amplitude to pulse duration is accomplished in an encoder circuit termed a "keyer." Several circuits are capable of good performance, the simplest of which is the "one-shot" multivibrator shown in Fig. 19.82. In this circuit, vacuum tube V_2 is normally conducting because of its positive grid return; V_1 is cut off by virtue of the large bias voltage developed by the common cathode resistance. A positive pulse on the input initiates the switching action, cutting off V_2. The grid of V_2 recharges exponentially according to the time constant R_2C until it reaches a point that permits plate current to flow again. The circuit then returns to its original condition. The point when this second switching action occurs depends on the cathode voltage which, during the pulse interval, is a function of the input signal voltage. Therefore the pulse return is delayed according to the input voltage level. While relatively simple, the amplitude-duration conversion of this circuit is somewhat nonlinear due to the exponential charging rate, and depends largely on tube characteristics.

The phantastron circuit of Fig. 19.83 offers considerable improvement in linearity and stability. Its operation depends on the redivision of voltages that takes place in a pentode (or pentagrid) tube when plate current is initiated in a tube that is already drawing screen current. In the circuit shown the phantastron is triggered on by a pulse applied to the second control grid G_3; its duration is a linear function of the voltage E on the control diode V_3, here shown developed on potentiometer R_7. The phantastron circuit can be made extremely linear, but is somewhat more cumbersome than the voltage-controlled multivibrator.

The most precise technique for amplitude-to-duration conversion employs comparison between input-signal amplitude and an internally generated linearly rising voltage, as

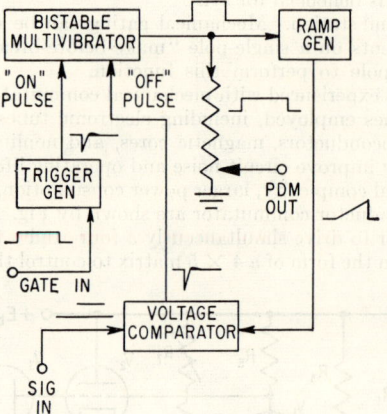

Fig. 19.84. Block diagram of a voltage comparator keyer.

indicated by the block diagram of Fig. 19.84. A bi-stable multivibrator is triggered on by signals from the gating timer. It generates an output pulse which allows the ramp generator to begin generating a linearly (with time) rising voltage. When the ramp voltage and input signal amplitude are equal, as determined by a voltage comparator circuit, a second pulse is generated that turns off the multivibrator. The output pulse duration is therefore directly proportional to the input-signal amplitude. This system is inherently very independent of tube parameters and can be designed to exhibit a very linear conversion characteristic.

R-F Transmitter. The frequency-modulated transmitter used in PDM/FM telemetering systems is identical in characteristics and design requirements to those previously discussed in connection with FM/FM systems, with the exception that the maximum deviation for PDM/FM is not as great.

PDM/FM Ground-station Equipment. The antenna and FM receiver system for PDM/FM telemetry signals are identical to those shown in Fig. 19.76 for FM/FM reception. From that point on the equipments are different. The amplifiers for the magnetic-tape recorder are special PDM amplifiers (see section on *Tape Recorder Amplifiers*) designed especially to record and reproduce pulse durations with a minimum amount of time distortion and jitter.

Because of the extremely large number of data input channels that can be telemetered concurrently by one PDM/FM link, it is extremely unlikely that the ground station will have sufficient reconversion equipment and permanent recording instrumentation to decommutate and record all these data as they are being received. The usual procedure, as is also followed with FM/FM, is to record the complete PDM video complex on magnetic tape, and to patch the available decommutation equipment onto those channels which contain information essential to the in-flight performance of the vehicle for a quick look in real time. After the test has been completed, the remaining channels can be chart-recorded a few at a time by repeatedly reproducing the tape-recorded signal.

Figure 19.85 is a block diagram of typical decommutation circuits used to separate a PDM/FM pulse train into its component channels. The signal, derived either directly from an FM receiver or from a tape recorder on playback, is passed through a series of

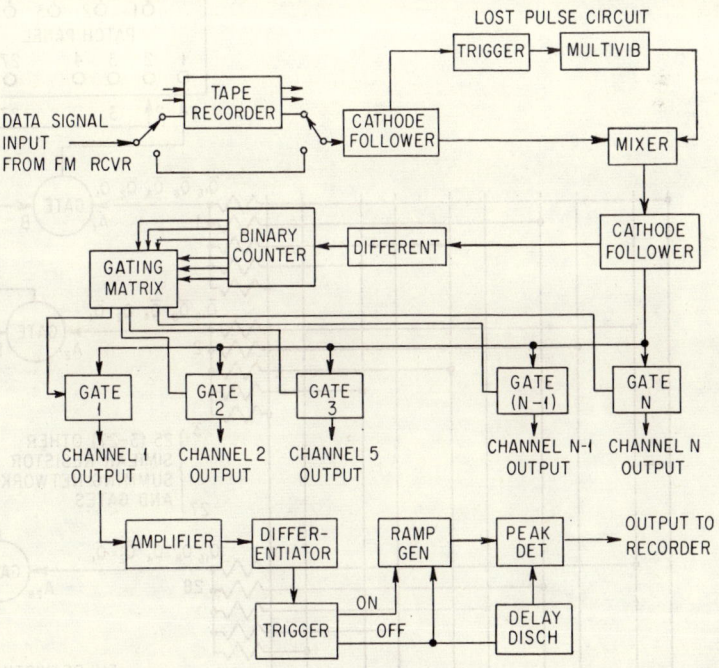

FIG. 19.85. Block diagram of a typical PDM/FM ground station.

buffer amplifiers that provide take-off points for auxiliary trigger circuits. It is applied simultaneously to one terminal of a set of gate circuits, the number of which corresponds to the number of data channels in the PDM complex. The first auxiliary circuit is a "free-running" multivibrator whose period is slightly longer than the maximum pulse duration and whose repetition rate is equal to the sampling rate. Its function is to insert artificial pulses into the pulse train in the proper position to maintain correct synchronism in the event signal fading or tape dropouts cause a pulse to be missed. The second auxiliary circuit includes some form of electronic commutator that steps one count for each pulse in the pulse train and resets itself during the synchronizing period. Figure 19.86 is a block diagram of a typical five-stage binary counter used to drive a twenty-eight-channel gating matrix. Differentiated pulses, derived from the pulse train, trigger the binary counter one step for each pulse. The matrix gates are connected as "AND" circuits, so arranged that a particular gate is open only when all its associated binary stages are in a similar state. The matrix opens one gate at a time in synchronism with the pulsed signal so that the mth pulse always appears on the signal bus when the mth gate is open.

The periodic signals appearing at the output of each gate are still in pulse-duration form and require reconversion to an amplitude variable for recording. The conversion circuit indicated in Fig. 19.85 is typical. The input pulse-duration signal is amplified, clipped, and differentiated to sharp pulses to define the beginning and end of the pulse. The first trigger starts a ramp generator whose voltage increases linearly with time until it is stopped by the second pulse. The maximum voltage attained by the ramp generator is proportional to the time between triggers, and is therefore proportional to the original

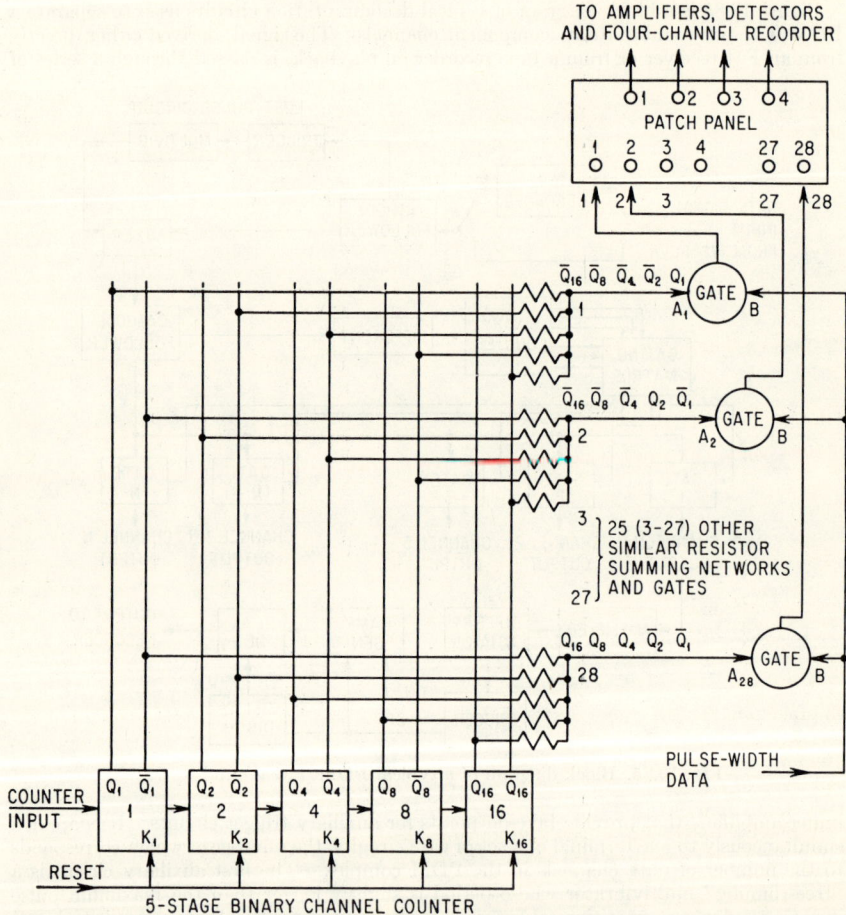

Fig. 19.86. PDM channel separator employing a five-stage binary counter and matrix gate.

input amplitude. This voltage is transferred by a peak detector circuit to the output for recording; then it is returned to zero in preparation for the next pulse by a delayed discharge from the "off" trigger.

COMBINATIONS OF TELEMETERING SYSTEMS

The two principal telemetering systems described above—FM/FM and PDM/FM—have preferential applications dependent mainly on the number and bandwidth of the input data. FM/FM is indicated where a moderate number of channels (up to eighteen) are sufficient and where the frequency range may extend to many hundreds of cycles per second. A second complete telemetering link could be employed if additional wideband

Table 19.9. IRIG Unseparated Data Commutation Rates for PDM-FM/FM or PAM-FM/FM Telemetry

Band	Center frequency, cps	Sample duration, milliseconds		Commutation rate,* samples per second	
		Conservative values	Minimum values	Conservative values	Maximum values
1	400	670.0	170.0	1.5	6.0
2	560	480.0	120.0	2.1	8.4
3	730	370.0	91.0	2.7	11.0
4	960	280.0	70.0	3.6	14.0
5	1,300	210.0	51.0	4.9	20.0
6	1,700	160.0	39.0	6.4	25.0
7	2,300	120.0	29.0	8.6	35.0
8	3,000	89.0	22.0	11.0	45.0
9	3,900	68.0	17.0	15.0	59.0
10	5,400	49.0	12.0	20.0	81.0
11	7,350	36.0	9.1	28.0	110.0
12	10,500	25.0	6.4	39.0	160.0
13	14,500	18.0	4.6	55.0	220.0
14	22,000	12.0	3.0	83.0	330.0
15	30,000	8.9	2.2	110.0	450.0
16	40,000	6.7	1.7	150.0	600.0
17	52,500	5.1	1.3	200.0	790.0
18	70,000	3.8	0.95	260.0	1,050.0
A	22,000	6.1	1.5	170.0	660.0
B	30,000	4.4	1.1	230.0	900.0
C	40,000	3.3	0.83	300.0	1,200.0
D	52,500	2.5	0.63	390.0	1,600.0
E	70,000	1.9	0.48	530.0	2,100.0

* Frame rate times the number of samples per frame. This assumes no loss time between samples. Multiply this value by the duty cycle for the actual values.

channels are needed to convey the desired amount of data. Quite often the input information is not cyclic, but rather a parameter that varies slowly throughout the measurement period. Bandwidths of a few cycles per second are more than sufficient. For this type of signal, a PDM/FM telemetering system provides adequate response for a very large number of input channels.

A normal telemetering application may include both narrow- and wide-frequency-range data, and therefore will not be exclusively compatible with either the FM/FM or PDM/FM systems. This difficulty can be resolved by employing a time-division multiplex on one of the RDB subcarrier bands to take care of the low-frequency information while the remaining FM/FM channels can be reserved for the higher-frequency information. Pulse-amplitude or pulse-duration modulation can be employed on the commutated subcarrier channel, leading to a compound designation of PAM-FM/FM or PDM-FM/FM for systems using these combined techniques.

Because of the large number of possible commutation schemes available, combining characteristics have been standardized by the IRIG to ensure compatibility with ground-handling equipment. These specifications are given in Tables 19.9 and 19.10 for unseparated and separated (automatic decommutation) data, respectively. The maximum commutation rates (samples per second), Table 19.9, are limited to the maximum frequency response allowed of each subcarrier band as previously given in Table 19.5 for a standard, uncommutated subcarrier band. These maximum rates are to be used only under extreme circumstances, only with manual analyses, and only with a subcarrier discriminator low-pass filter having four times the cutoff frequency normal

Table 19.10. IRIG Separated Data (Automatic Decommutation) Commutation Rates for PDM-FM/FM or PAM-FM/FM Telemetry

No. of samples per frame *	Frame rate, frames/sec	Commutation rate,† samples/sec	Lowest recommended subcarrier bands, cps
18	5	90	14,500 (± 7.5 per cent)
18	10	180	22,000 (± 15 per cent) or 30,000 (± 7.5 per cent)
18	25	450	30,000 (± 15 per cent) or 70,000 (± 7.5 per cent)
30	2.5	75	10,500 (± 7.5 per cent)
30	5	150	22,000 (± 7.5 per cent)
30	10	300	22,000 (± 15 per cent) or 40,000 (± 7.5 per cent)
30	20	600	40,000 (± 15 per cent)
30	30	900	70,000 (± 15 per cent)

* The number of samples per frame available to carry information is two less than the number indicated, because the equivalent of two samples is used in generating the frame synchronizing pulse.
† Frame rate times number of samples per frame.

for that particular subcarrier channel. This extra range helps preserve the pulse sharpness and reading accuracy. For all usual unseparated data, conservative values should be used. Conservative rates are one-fourth the maximum rates and provide sample durations sufficiently long to be free of switching transients which occur on the initial section of the pulse.

When automatic decommutation at the ground station is employed, commutation patterns should conform to one of the combinations listed in Table 19.10. The number of samples per frame corresponds to the number of commutation segments while the frame rate is the cyclic speed of the wiper; their product is the commutation rate. The lowest recommended subcarrier band to be used with a given commutation rate is based on its having a maximum allowable frequency response twice the commutation rate. Higher sampling rates for certain channels can be obtained by paralleling equally spaced segments.

It is important, especially with automatically decommutated data, that a uniform configuration be maintained in the subcarrier frequency-time domain. These patterns, and the equipment used to generate them, depend upon the mode in which signal intelligence is carried in the pulse train. PDM-FM/FM is similar to PDM/FM previously

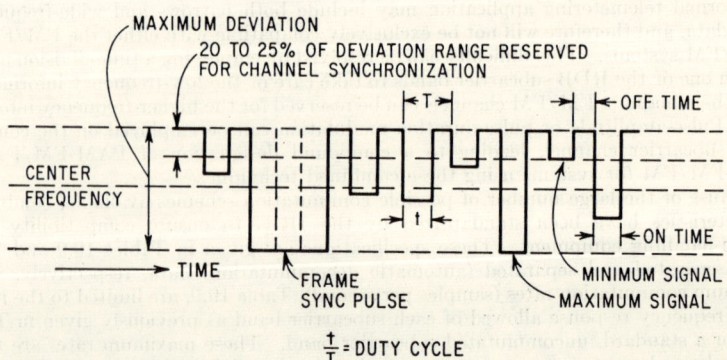

Fig. 19.87. Typical PAM-FM subcarrier oscillator waveform in the frequency-time domain.

discussed, except that the pulse duration is contained on a subcarrier band rather than directly on the r-f transmitter. Equipment is similar to that shown in Fig. 19.80 and includes a commutator and amplitude-to-duration keyer.

PAM-FM/FM is the more usual method of time multiplexing on an FM/FM subcarrier and is obtained by letting the amplitudes of the individual commutated inputs frequency-modulate the subcarrier oscillator directly. A representative PAM-FM pulse train in the subcarrier frequency-time domain is given by Fig. 19.87. Generally, the subcarrier is offset to its upper deviation limit in the absence of any input signal and modulated downward (in frequency) by the commutated intelligence. Zero signal input corresponds to 25 per cent of the deviation range, this amount being reserved for synchronization of ground receiving equipment. The remaining 75 per cent deviation is made linearly proportional to the amplitude of the individual input signals. The duty cycle of a PAM pulse is approximately 50 per cent.

REFERENCES

1. Landee, R. E., D. C. Davis, and A. P. Albrecht: "Electronic Designers' Handbook," McGraw-Hill Book Company, Inc., New York, 1957.
2. Terman, F. E.: "Electronic and Radio Engineering," 4th ed., McGraw-Hill Book Company, Inc., New York, 1955.
3. Skilling, H. H.: "Electric Transmission Lines," McGraw-Hill Book Company, Inc., New York, 1951.
4. Perls, T. A.: *J. Appl. Phys.*, **23**(6):674(1952).
5. RCA Tube Handbook, Series HB-3, Tube Department, Radio Corporation of America, Harrison, N.J.
6. Dewitt, D., and A. L. Rosoff: "Transistor Electronics," McGraw-Hill Book Company, Inc., New York, 1957.
7. Hunter, L. P.: "Handbook of Semiconductor Electronics," McGraw-Hill Book Company, Inc., New York, 1956.
8. Coblenz, A., and H. L. Owens: "Transistors: Theory and Applications," McGraw-Hill Book Company, Inc., 1955.
9. "Reference Data for Engineers," 4th ed., Federal Telephone and Radio Corp.
10. Shea, T. E.: "Transmission Networks and Wave Filters," D. Van Nostrand Company, Inc., Princeton, N.J., 1943.
11. Everitt, W. L., and G. E. Anner: "Communication Engineering," 3d ed., McGraw-Hill Book Company, Inc., New York, 1956.
12. Hurley, R. B.: *Electronic Equipment Eng.*, **6**(5):32(1958).
13. *Trans. IRE*, TRC-3, no. 2, December, 1957, p. 13.
14. National Telemetering Conference, yearly series.

20

MEASUREMENT TECHNIQUES

Richard D. Baxter
Convair, Division of General Dynamics

John J. Beckman
Convair, Division of General Dynamics

Harold A. Brown
Convair, Division of General Dynamics

INTRODUCTION

This chapter outlines many of the techniques employed in shock and vibration measurement. It includes a discussion of planning the test objectives, selecting types of measurements to implement them, selecting the measurement system best suited to these requirements, installing the components of the system, and testing and calibrating the system prior to field measurements. In addition, other factors must be considered, such as the method of data analysis to be employed. Many of the measurement techniques referred to here are treated in detail in preceding chapters.

DEFINING THE PROBLEM

The first step toward measurement is to define the nature of the test and what is to be measured. Careful pretest planning may save much time in making the measurements and obtaining the most useful information from the test data. Planning should start with a clear, written definition of the test objectives. The next step is to establish the various measurement requirements. Examples of such requirements are listed in Table 20.1. In the more simple vibration measurement problems, only a few of these factors need be considered. On the basis of this information, one can select specific instrumentation for the test.

SELECTION OF EQUIPMENT—PRETEST PLANNING

The selection of a measurement system for a particular test depends on many different requirements, for example, on the number and characteristics of variables to be measured. A close examination of these requirements may indicate that the desired data may be obtained from any of several different sets of measurements of variables. For example, a determination of the amount of damping in a single degree-of-freedom mass-spring sys-

Table 20.1. List of Measurement Requirements and Considerations

Measurement location
Measurement direction
Frequency range
Amplitude range
Required accuracy
Test condition for recording data (e.g., from 10 sec before to 10 sec after wheel touchdown on landing of an aircraft)
Total length of recording time required
Time correlation between channels and with test conditions
Method of structural excitation to obtain vibration
Method of record recovery
Space available for installation
Accessibility for service or calibration
Electrical power available
Total number of channels of recording
Number of vibration measurement channels of similar range
Redundancy requirement to ensure against loss of data
Specific automatic or manual data reduction operations
Type of playback equipment already available or needed
Calibration method and special fixtures needed (pretest, test, and posttest)
Percentage of the units under test that are required as spare units (or procedure to be followed in case of component failure)
Environment of operation
   Temperature
   Humidity
   Corrosiveness
   Magnetic and radio-frequency fields
   Acoustic fields
   Nuclear radiation
   Pressure (altitude)
Expected calendar duration of test program and total expected operating hours of equipment
Schedule considerations
Financial considerations

---

tem which is less than critically damped can be made from measurements of any of the following sets of variables:

1. Mass; spring constant; instantaneous values of driving force, displacement, velocity and acceleration of the mass.
2. Peak amplitudes of successive cycles (logarithmic decrement) and the frequency of oscillation when the mass is released at rest from a displaced position.
3. Mass, spring constant, steady-state values of driving force, displacement, and frequency.

Each possible measurement method should be examined to determine whether it satisfies the requirements which have been established and an appropriate choice should be made. Thus, in the above example, choice 3 probably is easiest to instrument and to employ.

In a proposed measurement system, the number of variables to be recorded is a most important consideration because the number may exceed the available number of recording channels. If the maximum available frequency response of the recorder is greater than that required for the data, multiplexing techniques may be used which increase the effective number of recording channels by restricting the frequency response of each channel. This may be accomplished by the frequency-sharing or time-division techniques described in Chap. 19.

In any multichannel recording system, it is desirable that the total number of recording channels include some spare channels to cover additional requirements for data not envisioned when the test run was planned and to anticipate breakdowns in the recording system.

In certain types of measurements, one must decide whether on-board recording or remote transmission is to be used. The flight testing of a long-range missile over a considerable distance, for example, requires a choice between radio telemetry or a recoverable on-board recorder. Short-range missiles can use either of these methods or a "trailing-wire" transmission system. The latter technique is generally useful during the period shortly after launch. On the other hand, radio telemetry usually is more satisfactory than wire transmission from any moving vehicle requiring ranges greater than a few hundred feet. This is because the volume of wire required is large for long distances and the handling of the wire becomes excessively difficult.

One advantage that the telemetering or trailing-wire transmission systems have in the testing of controlled vehicles, compared to the on-board recorder, is that instantaneous monitoring can be performed. For example, this allows the engineers to detect unsafe values or to redirect the test plans as the test progresses.

The choice of recorders usually is dependent on such factors as the magnitude of the data analysis task contemplated, the required frequency range, and the weight of the recorder. Magnetic-tape recordings are reproducible in a form suitable for high-speed automatic analyzers. In contrast, photographic records usually are more difficult to analyze, but they are relatively simple.

SELECTION OF A TRANSDUCER

The various characteristics of transducers are outlined in Chap. 12, and various types of transducers are described in Chaps. 13 to 17. The engineer must select the most appropriate transducer for the specific application from the many that are available. The selection may be determined by size, weight, electrical characteristics, environment in which the transducer is to operate, or limitations imposed by required auxiliary equipment. In addition, the selection is determined by the vibratory characteristics of the member to be studied. Some a priori knowledge of the motion is important in deciding whether to select a displacement, velocity, acceleration, or strain-measuring transducer. The following considerations often are of importance in determining the type of measurement to be made:

Displacement measurements may be useful:
1. Where the amplitude of displacement is particularly important (for example, in assemblies where vibrating parts must not touch).
2. Where the magnitude of the measured displacement may be an indication of stresses to be analyzed.
3. In studying low-frequency vibration where corresponding velocity and acceleration measurements may yield outputs which are too small for practical use.

Velocity measurements may be useful:
1. At intermediate frequencies when displacement amplitudes are too small to measure conveniently.
2. In correlating acoustic and vibration measurements, because a vibrating member may produce sound pressure in air which is proportional to velocity.

Acceleration measurements may be useful:
1. At high frequencies, where the highest signal output usually can be obtained from such measurements.
2. Where forces, loads, and stresses must be analyzed, since force is proportional to acceleration.
3. Where suitable displacement or velocity pickups would be too large because of clearance requirements.

Strain measurements may be useful:
1. Where a portion of the specimen that is being vibration tested has an appreciable variation in strain caused by vibration.

AUXILIARY EQUIPMENT *

When the full-scale output voltage of the transducer is insufficient to drive the recording equipment with satisfactory accuracy, a signal amplifier is required. The various

* See Chap. 19.

types of amplifiers are described in Chap. 19. Many commercially available units are miniaturized, and have excellent characteristics with respect to noise and sensitivity.

The amplified or direct transducer output may feed into a visual or a photographic recorder in simple systems, or into a commutator or a subcarrier oscillator for tape recording or radio telemetry, as described in Chap. 19.

In mobile installations serious limitations frequently are imposed on the size and weight of instruments and by the availability of power. All these factors are more critical on smaller vehicles. In experimental vehicles, it usually is important to select instruments that will operate satisfactorily when the electrical voltage or frequency is considerably outside the design tolerance. Battery power is advisable in many cases, but it carries the penalty of additional volume and weight. Sometimes vehicle power "backed up" by emergency batteries is the best choice. The final selection depends, to a large extent, on the balance between weight and reliability requirements. Volume limitations may be largely overcome by using modules specially shaped to fit into space which may otherwise be wasted. Such installations are more costly as well as more difficult to manufacture and maintain. Equipment maintenance should not be overlooked during system design or installation.

TIMING AND DATA CORRELATION METHODS

In tests involving observation of many variables, it is important to be able to compare the data on all recorded channels at a given instant of time and to correlate these data with external events. This usually is accomplished by providing a direct timing channel on the recording system and a visual indication of the timing channel for the system operator.

The choice of time coding depends on length of the test and ultimate use of the data. For very short tests, a continuous periodic pulse of known frequency will suffice. For longer tests, a coded pulse whose code indicates the elapsed time from a starting reference often proves more satisfactory for reasons given below. For tests which run continuously for days, or where correlation with time of day is important, a pulse-coded time of day in hours, minutes, and seconds may be useful.* The coded pulse has the following advantages over a continuous periodic pulse:

1. A specific value of time t on a long-duration record may be located much more rapidly, since it is unnecessary to count the elapsed time from zero to t.

2. Loss of a number of coded time signals does not affect the accuracy of time measurement of those remaining, while loss of a few of the continuous periodic pulses results in the wrong value of time from point of loss.

Code time signals are particularly useful when a correlation must be provided between different recorders and when the tests are so long that the recorders must be stopped for servicing or to save recording medium during the test. Coded time signals also may be used to provide a knowledge of the "off" time. "Off" time sometimes is related to instrumentation drifts; hence this technique can be used as an indirect measure of confidence one should place in data anomalies separated by periods of "no record."

It is desirable that all timing channels of the recording system be driven from a common timing mechanism, so that synchronization between channels and recorders is maintained even if absolute time is lost because of a malfunction of the timing mechanism. The ability to synchronize data from a number of channels at a given instant of time is limited by the accuracy with which time delays in the various channels of the recording or playback system are known.

ANALYSIS CONSIDERATIONS. The method of data analysis to be employed should be considered in selecting measurement instrumentation.

CATALOGING METHODS. Instrumentation selection in a large test organization may be aided considerably by a carefully indexed catalog (such as punched cards) of each test and test installation that has been made over a period of years. When a prob-

---

* Such a coded pulse may be synchronized with signals from radio station WWV when time power is first applied and periodically thereafter; this radio station, in Washington, transmits frequency standard signals and Greenwich Mean Time signals at frequent intervals.

lem is presented, a quick search can be made to see whether a similar measurement has been performed previously. Any such information may be very helpful in the new test, occasionally supplying the complete answer.

INSTRUMENTATION INSTALLATIONS

In studying the motion of a specimen or vehicle one must consider the effect of the added mass and the change in stiffness of the specimen which is introduced by the measuring instruments. It is important in testing to choose measurement equipment which will not affect the characteristics of the system under test. Thus, such considerations may dictate that test equipment weigh less than a specified value, or that special mountings be provided so that the test equipment does not increase the stiffness of the system under test.

TRANSDUCER MOUNTING TECHNIQUES

The basic problem in designing a mounting is to couple the transducer to the system under test so that the transducer accurately follows the motion of the surface to which it is attached. This requires that the effective stiffness of the transducer mounting be large in the frequency range of interest; otherwise the mounting will deflect under the inertia load of the transducer mass.

Many mounting fixtures or brackets have resonant frequencies which are below 2,000 cps and have little damping. The use of such a bracket may result in significant measurement errors as a result of resonant amplification or attenuation of vibration in the bracket. This is illustrated in Fig. 20.1, which shows the frequency response of a stand-

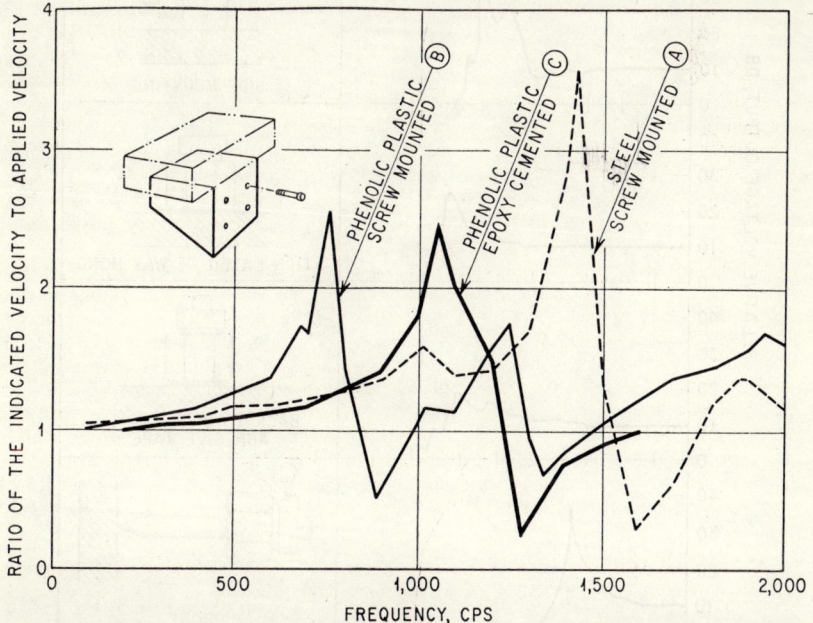

FIG. 20.1. Relative frequency response of a standard velocity transducer mounted on various brackets having identical geometry but fabricated of different materials. Curve A shows the response using a steel bracket attached to the specimen with four screws. Curve B shows the response using a bracket fabricated of a cloth-reinforced phenolic plastic and attached to the specimen with four screws. Curve C shows the response using a bracket fabricated of cloth-reinforced phenolic plastic and attached to the specimen with an epoxy resin adhesive.

ard velocity pickup mounted on brackets which are identical in geometry but which are fabricated from different materials. Note that a change in bracket material from steel to a cloth-reinforced plastic halves the resonant frequency of the mounting. A change in bracket attachment, from four screws to an epoxy resin adhesive bond, increases the frequency of the mounting resonance 60 or 70 per cent. Although these results are not of a general nature, they illustrate that minor variations in the transducer mounting bracket may produce significant changes in the output characteristics. Examples of the effects of various types of mountings on the frequency response of a transducer are shown in Fig. 20.2.

In order to design a transducer mounting properly, one must know the nature of its use, the frequency range and the maximum acceleration of the measurements, and the mechanical specifications of the test object. Specially designed mountings may be required for each test setup. Not only must mountings be designed carefully, but they should be tested under conditions closely approximating the service environment.

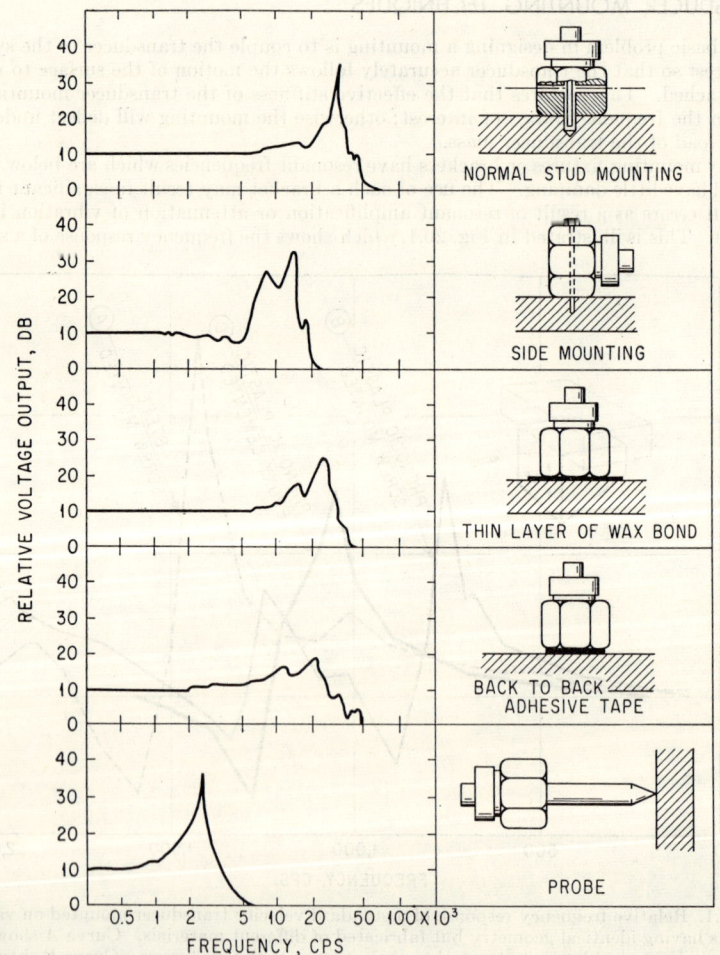

FIG. 20.2. Typical methods of coupling a crystal accelerometer to a test item and corresponding frequency-response curves. (*Courtesy of Brüel and Kjaer.*[7])

The effect of the transducer-mounting system can be estimated if it is assumed that it behaves as a simple spring-mass system driven at the end of the spring. Then, the acceleration of the transducer is:

$$\ddot{x} = \ddot{u}\frac{k}{k + m(2\pi f)^2}$$

where $\ddot{u}$ is the specimen acceleration, m is the transducer mass, k is the spring constant of the mounting, and f is the frequency of vibration. For $\ddot{x}$ to be within 1 per cent of $\ddot{u}$, $k > 100m(2\pi f)^2$. Since $f_n = (1/2\pi)(k/m)^{1/2}$ is the undamped natural frequency of the transducer-mounting system, then $f_n > 10f$. For example, suppose an accelerometer weighs 0.1 lb. In order that the data obtained with this transducer be accurate to 1 per cent at a frequency of 100 cps, the stiffness of the mounting must be such that $k > 10,000$ lb/in., i.e., the mounting must have a resonant frequency greater than 1,000 cps. Because the mass of the mounting is not negligible, the values of k calculated in this manner represent a lower limit of the required stiffness.

GENERAL RULES FOR TRANSDUCER MOUNTING DESIGN

Several types of mounting brackets are illustrated in Fig. 20.3. Some general rules to observe in the mechanical design of transducer mounts follow: *

1. The mounting must be rigid, but should be as light as possible.
2. The use of long, thin structural members and long bolts should be avoided. (Such members have relatively low spring constants and contribute to a low resonant frequency of the mounting.)
3. The major resonant frequency of the mounting must be well above the test frequency range.
4. In order to obtain maximum damping in the transducer mounting, cast mountings are preferred; next in order of preference are welded constructions and machined assemblies.
5. Methods of mechanical attachment of the transducer, in order of preference, are:
    a. Transducer bolted directly to the structure (subject to surface mounting conditions).
    b. Transducer bolted to a mounting which is itself attached to the structure.
    c. Transducer attached to a multiple mounting bracket (where a and b are not possible).
6. Employ flat machined mating surfaces between the transducer and the specimen or bracket. (This helps to avoid mechanical distortion of the transducer, when it is attached to the test article, with consequent effect on the response of the transducer.)
7. Avoid locating transducers on thin skins or delicate structural members in the test specimen. (This reduces the possibility of introducing a spurious low-frequency resonance due to the combination of the transducer mass and the stiffness of the skin.)

ADHESIVE TRANSDUCER MOUNTING. Adhesive bonding of transducers to test

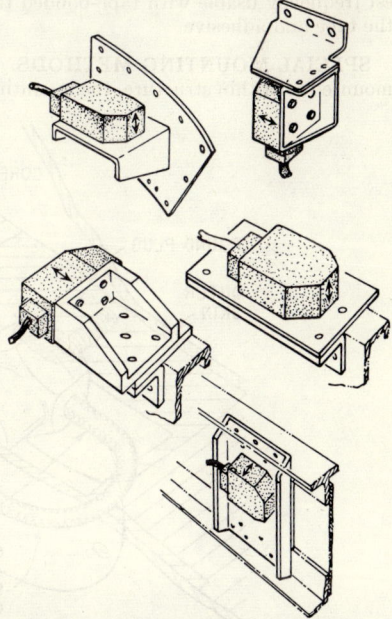

Fig. 20.3. Typical mounting brackets for velocity-type transducers. Arrows on the transducers indicate the direction of sensed motion.

* Also see *Effects of Mounting on Pickup Characteristics*, Chap. 16.

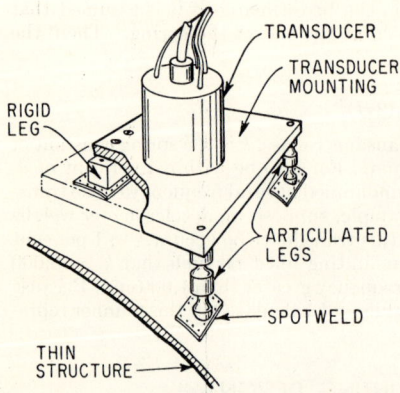

Fig. 20.4. A transducer mounting bracket for use on thin structure. This tripod arrangement is comprised of one rigid leg and two articulated legs, allowing the structure to flex without distorting the transducer.

specimens usually requires little preparation of the surface to which the transducer is to be attached. Often, a bracket is not required. However, the success of such adhesive bonds largely depends on the care with which the bond is made. Adequate cleaning of the mating surfaces with solvents to remove grease and wax is absolutely essential; furthermore, a thin layer of glue must be used to prevent decoupling resulting from elasticity of the cement. Mechanically hard, catalytic, or thermosetting cements are suitable for transducer mountings.* Solvent-drying cements are undesirable because metal surfaces prevent reasonably rapid escape of the solvent which leaves the cement in a plastic condition. Nonhardening cements are not suitable because they act like soft springs, thereby decoupling the transducer from the specimen.

A very fast method of transducer bonding (suitable for use with lightweight pickups which are used at low acceleration levels) employs double-backed pressure-sensitive tape as the bonding agent. Such tape consists of a thin plastic ribbon, coated with pressure-sensitive adhesive on both sides. It is available from several manufacturers. The highest frequency usable with tape-bonded transducers often is limited by the elasticity of the tape and adhesive.

SPECIAL MOUNTING METHODS. Thin Structures. When a transducer is to be mounted on a thin structure, the mounting design must avoid producing stress concen-

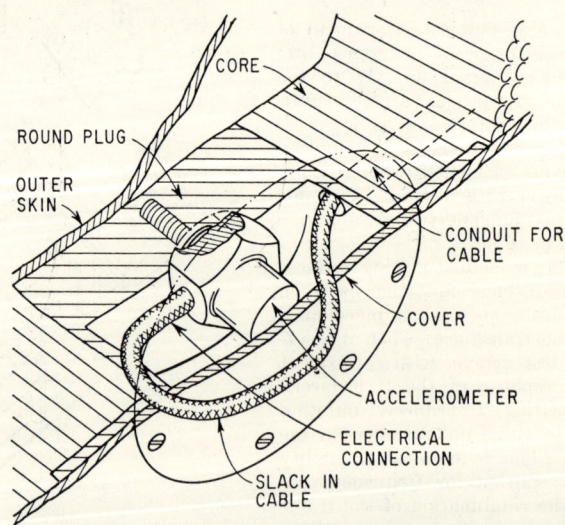

Fig. 20.5. Preferred installation of a transducer in a honeycomb material requires the incorporation of a suitable transducer mounting base in the core material prior to assembly of the honeycomb skin.

* Eastman No. 910 cement, Eastman Kodak Co., Kingsport, Tenn.; EC-1294, Minnesota Mining & Mfg. Co., Detroit, Mich.; EPON 828, Shell Chemical Corp., New York. (Also see information on cements in Chap. 17.)

trations in the structure and modifying the characteristics of the structure. The mass of the transducer bracket chosen for such an application should be as small as practical, to reduce loading on the specimen. A typical bracket design, shown in Fig. 20.4, employs a tripod arrangement with one rigid leg and two articulated legs connecting the pickup to the thin structure. The articulated legs allow the structure to flex without distorting the transducer.

Honeycomb Structures. The mounting of a transducer on a honeycomb structure can be difficult because honeycomb structures have very little local strength. The best solution is to install a mounting base for the transducer during manufacture of the honeycomb, as shown in Fig. 20.5. When circumstances prohibit a solution of this type, some other method of distributing the mounting loads over a considerable area of the honeycomb skin must be employed. For example, an adapter plate tapering in thickness to a feather edge can be bonded to the skin as shown in Fig. 20.6. To avoid possible delamination of the honeycomb skin, the thickened part of such an adapter should be reduced to the smallest practical size.

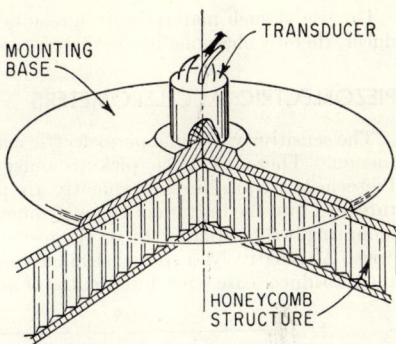

FIG. 20.6. A typical transducer base bonded to honeycomb skin. Such a base is required to distribute the mounting loads of the transducer to avoid producing stress concentrations and local failure. To avoid delamination of the honeycomb skin from the core, the thickened part of the transducer mounting base should be as small as practical.

SPECIAL TRANSDUCER CONSIDERATIONS

Various classes of transducers have characteristic features which must be considered in mounting design. Some of these are given in the following sections.

SOFT-SPRING TRANSDUCERS

In mounting a transducer that employs a mass-spring system having a low natural frequency (as found in some velocity and displacement transducers), the attitude of the transducer relative to gravity or other static acceleration fields should be considered. The effect of gravity will deflect the mass toward one of the internal stops in the transducer, thus limiting the maximum vibration amplitude which can be applied. Some commercially available transducers incorporate spring fixture adjustments which may be used to restore the mass of the mass-spring system to its normal position in the "installed" pickup attitude. In using other types of transducers, judicious orientation of the transducer and test article can reduce the effect of gravity.

In transducers employing a pivoted arm as the support, orientation of the pivoted arm parallel to the static acceleration vector, or in the opposite direction, respectively increases or decreases the effective restoring force applied to the mass of the mass-spring system. This effect is particularly important in velocity pickups having soft springs. For example, in one common pickup, the resonant frequency changes ± 12 per cent, depending on its orientation in the earth's gravitational field. This results in a sensitivity change of ± 7 per cent at 7.5 cps, decreasing to ± 2 per cent at 20 cps. Since this effect cannot be eliminated, it must be accounted for in transducer calibration.

INDUCTIVE-TYPE PICKUPS

When using a pickup which incorporates a magnet as part of the transducing element or for magnetic damping, avoid using magnetic materials in the transducer supports or mounting bolts.

PIEZOELECTRIC ACCELEROMETERS

The sensitivity of some piezoelectric accelerometers varies if the case is subject to distortion. Therefore, such pickups must be provided with distortion-free mountings. Piezoelectric transducers frequently are provided with a single mounting screw or stud threaded into the bottom of the instrument case. Then it is convenient to provide a flat mounting bed by spot facing on the test specimen a circle just larger than the instrument case. Alternatively, a special stud having a small diameter shoulder can be used to raise the transducer case a few hundredths of an inch above a specimen whose surface is rough,

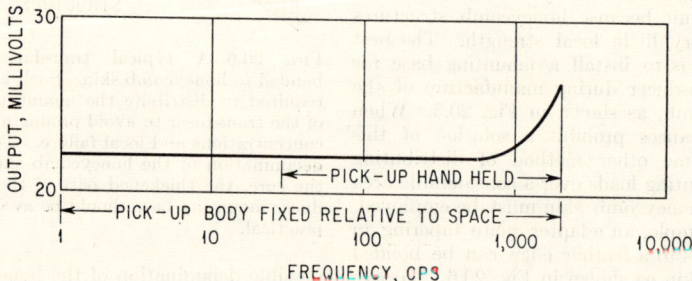

Fig. 20.7. Typical frequency response of a velocity pickup employing a probe. A pickup of this type will follow the typical curve within 10 per cent when hand-held or within 5 per cent when the case is rigidly mounted. The output voltage shown is for a sinusoidal excitation at 0.22 in./sec rms velocity.

irregular, or under flexure. Shoulder studs should not be used for very high levels of acceleration since they limit the area of contact between the specimen and the pickup.

The length of the screw or stud used in installing a piezoelectric pickup should be selected with care, since it is possible to bottom a long stud in the pickup, thereby causing the case to distort and possibly causing permanent damage. The torque applied to a mounting stud should follow the manufacturer's recommendations.

HAND-HELD VELOCITY PICKUPS

The application of seismic velocity pickups sometimes is limited by the amount of mass that can be added to the test specimen when the pickup is installed. Velocity pickups of the probe type are commercially available and largely overcome this limitation. The body of the transducer is hand-held and the motion of the test surface is sensed by a probe protruding from the transducer. In this case, the only mass added to the specimen is that of the probe itself. Such transducers find wide application in vibration survey work due to their low effective mass and the ease with which they can be moved about. Figure 20.7 shows a frequency response curve of a typical hand-held velocity pickup. Note that the response at low frequencies is limited if the transducer is hand-held. As indicated in Chap. 13, an extraneous signal is introduced by any motion of the hand.

INSTRUMENTATION WIRING CONSIDERATIONS

The use of electromechanical transducers requires the installation of a certain amount of wiring in the equipment under test. Where a number of pickups are required, the amount of wire needed may become quite large. The installation of this wiring should take account of the following facts.

EFFECTS OF THE MECHANICAL STIFFNESS AND MASS

The mechanical stiffness and mass of a bundle of wire may be considerable. For example, a well-laced bundle of 50 No. 22 copper conductors, weighing about ¼ lb/ft, has elastic properties similar to a 0.25-in.-diameter aluminum tube with 0.040-in. wall thickness. This wire has structural properties which, if neglected, can appreciably affect the natural frequencies of a test object. Mass loading effects of the wire on mode shapes can be minimized by distributing the wire over the specimen surface. This also reduces the cross-sectional moment of inertia of the wires, thereby reducing their bending stiffness. Routing of wires along the neutral axes of beams will further reduce specimen stiffening effects.

ELECTRICAL NOISE SUSCEPTIBILITY

Wiring installations should be designed to minimize self-generated and induced noise. The induced noise results when electrical energy from local electrical equipment is coupled into the measurement circuits by one or more of the following mechanisms:
1. Ground loops
2. Varying magnetic fields
3. Varying electric fields
4. Resistive leakage paths

Noise introduced by these effects can be limited significantly by judicious application of the following techniques:

1. Use separate signal and power current return leads to reduce ground-loop coupling between signal and power circuits.

2. Use common-current return leads of sufficiently low resistance where an inadequate number of current return leads prevent separate returns for each signal current.

3. Use twisted pair or coaxial cable to reduce inductive pickup in signal leads.

4. Reorient or relocate circuits to reduce inductive or capacitive coupling between circuits.

5. Use twisted-pair power-distribution leads to reduce the magnetic fields responsible for inductive pickup.

6. Use coaxial cable or shielding to reduce capacitive coupling between circuits.

7. Use single-point ground bonding to aid in location of unintentional ground loops and/or leakage paths.

The principles of these coupling mechanisms and the details of application of the remedies listed above appear in Chap. 19.

Electrical noise may be generated by motion of some part of the wiring because of variations in contact resistance in connectors, because of changes in geometry of the wire (particularly coaxial cable) or as a result of voltages induced by motion of the conductors through magnetic fields which may be present. In general, such electrical noise will be reduced if cable harnesses securely fasten the wire cable to a structure at frequent intervals, and if connectors are provided with mechanical locks and strain-relief loops in their cables. Cables are available commercially which are especially constructed to minimize the noise resulting from flexure. Such cable should be employed when the wiring will be subject to high accelerations or to high shock loads. Noise generated by wire motion in a magnetic field can best be eliminated by removing or reorienting the field. (See Chap. 19 for further details.)

WIRING INSTALLATION AND ELECTRICAL CHECKOUT PROVISIONS

The mechanical problems associated with wiring installation, especially when making a measurement in a complex vehicle such as an airplane, often result in the cutting of the cable into sections for ease of handling. It should be noted that the wires are much more prone to failure at connectors or terminals than at the intermediate points. When cables must be spliced, soldered connections provided with adequate stress reliefs are preferable to mechanical plugs or connectors. This is because the electrical resistance is lower and the possibility of the connectors being opened (thereby stressing the mechanical contacts) for trouble shooting is removed.

Electrical checkout considerations dictate that test points be available at various points in the circuits. To maintain high circuit reliability, such test points should not introduce any additional joints into the wiring system. Ordinarily, test point locations and installation splice locations can be chosen so that a single break in the cable will serve both purposes. Electrical test points should be of a parallel type, allowing connection of the test equipment to circuits without disturbing circuit continuity. Such connections allow observation of the circuit in operation if desired, and do not require the making and breaking of mechanical circuit connections.

FIELD CHECKOUT

Ordinarily a complete field calibration of a vibration measurement system is impractical. Calibration in the field usually is restricted to functional checkout of the system and of the auxiliary equipment such as the recorders and amplifier. Methods of calibration and channel identification are described in the next section. This section considers cross talk and electrical noise in the over-all measurement system, and the problem of spurious transducer-mounting resonances.

CROSS TALK

Cross talk, i.e., signal interference between channels in a measurement system, usually is due to coupling between measurement channels occurring in some common element, for example, a power supply having high internal impedance circuits. The source of coupling must be eliminated. Decoupling in common power supplies sometimes can be achieved by adding isolation networks such as those shown in Fig. 20.8. If other elements couple several measurement channels, often the simplest solution is to replace the common element by separate components in each channel.

ELECTRICAL NOISE

Significant electrical noise may be evident from the output of the measurement system even though no input signal is supplied. Such noise results: (1) from coupling between circuits in the measurement system and those of the power circuits, (2) from mechanical strain or vibration sensitive elements other than transducers, or (3) from improper equipment design. These sources and methods for their elimination are considered in detail in Chap. 19. Often it is most convenient to locate the source of noise by

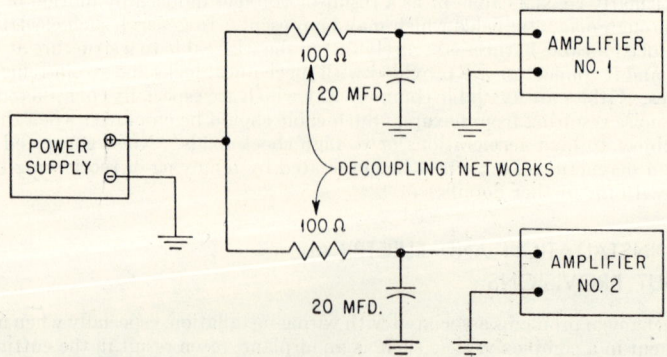

FIG. 20.8. Block diagram showing decoupling networks used to reduce cross talk between two amplifiers connected to a common high-voltage power supply having a high source impedance.

using an oscilloscope which is first connected to the transducer output with no vibration applied. Then the connection to the oscilloscope is moved component by component through the measurement system toward the recorder until the noise is observed. It is necessary that the system be electrically intact during this operation or the source of noise may be disconnected while the search for it is conducted. Another approach is to short-circuit the signal path at various points in the system, one at a time until the system noise disappears. Usually, this pin-points the source as the component next nearest the transducer from the last short circuit.

After locating the position of the noise source in the measurement system, the mechanism of coupling must be determined before corrective action can be taken. Often, elimination of the coupling mechanism eliminates the problem of cross talk.

Mechanical or acoustical sources must be eliminated or controlled if they result in noise in the measurement system. The component which is excited and which produces the noise must be decoupled from the driving source by isolating techniques or by physical separation. The latter is usually the more successful method.

SPURIOUS TRANSDUCER-MOUNTING RESONANCES

On initial installation, and whenever transducers are removed and remounted, it is advisable to check the transducer installation for spurious mounting resonances. These resonances may result from the selection of a transducer bracket which is not stiff enough. They also may result from faulty fasteners (such as loose rivets or bolts) or from improper seating of the transducer. Hence a careful visual check of the transducer mounting is helpful. It often is very useful to excite the transducer-mounting system by a blow while observing the transducer's output to detect resonances other than the resonant frequency of the transducer. The other resonant frequencies which appear may be due to (1) resonances in the test specimen or (2) resonances in the transducer-mounting system. Loose mountings usually produce "noisy" records and may produce audible buzzing sounds or clattering noises. Often it is difficult to determine the difference between resonances in the mounting and the resonances in the actual test specimen. If serious doubt exists, a different type of transducer mounting can be substituted and the test repeated. If the resonant frequencies are identical for the new mounting, the resonances are probably due to the test specimen, and original mount probably was satisfactory.

FIELD CALIBRATION TECHNIQUES

EARTH'S GRAVITATIONAL FIELD METHOD

This calibration technique is useful for calibrating accelerometers having useful sensitivity down to 0 cps. To employ this technique, the accelerometer is dismounted, but left electrically connected. Its output is observed for a $2g$ change in acceleration. Such a change is obtained by first orienting the accelerometer with the positive direction up (along the vertical) and then rotating the accelerometer through 180°. It is also possible to employ increments of acceleration within the $1g$ range [see Eq. (18.1)].

This technique is not applicable to crystal accelerometers or velocity pickups, or any instrument sensitive only to vibratory motion. It is not recommended for the calibration of accelerometers having a significant transverse sensitivity.

CHATTER METHOD

This calibration technique for accelerometers provides a $1g$ calibration point at any desired frequency. Such calibrations can be performed by mounting the accelerometer in a vertical position and placing a small mass on top of the transducer. The accelerometer is then placed on a small "shake table" which is vibrated. When the downward acceleration exceeds $1g$, the mass will become separated from the accelerometer. At this level, the test mass will "chatter" or bounce on the accelerometer case—providing

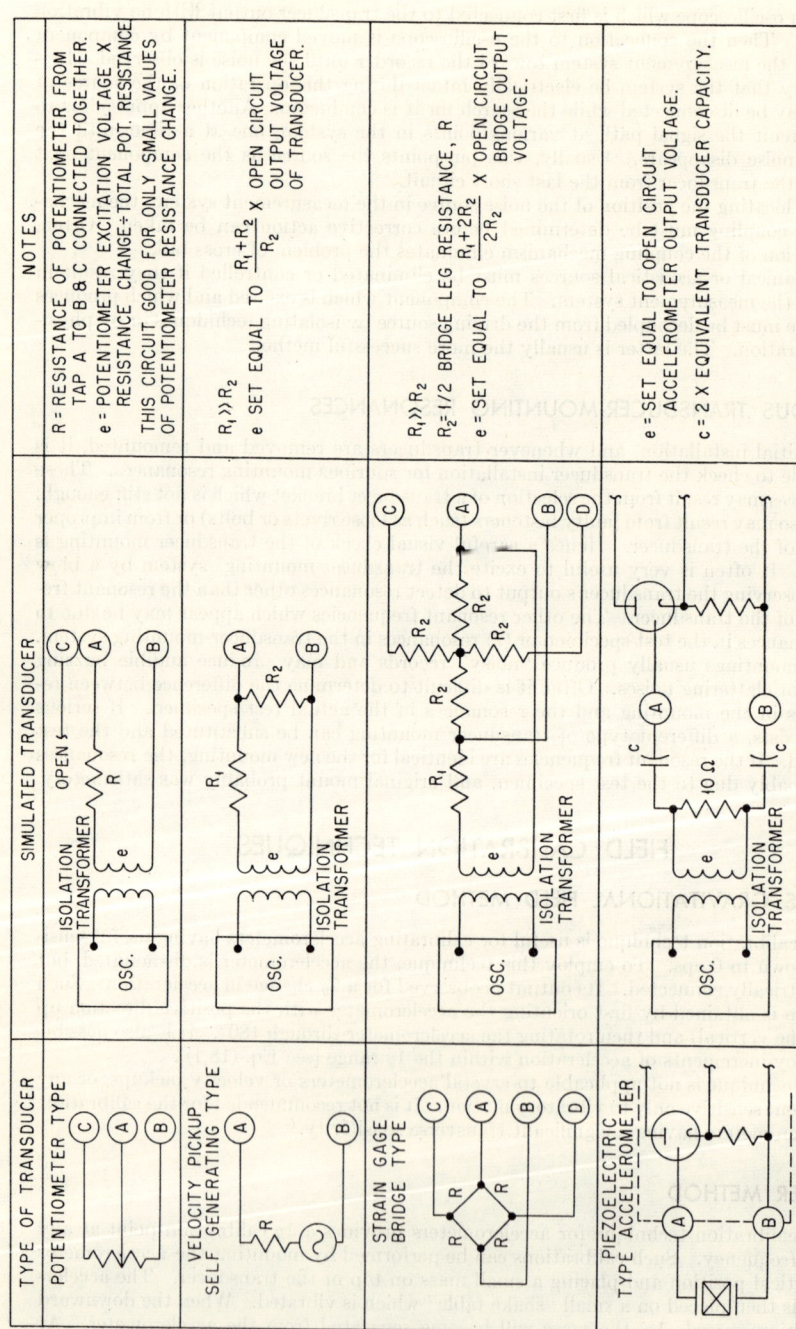

Fig. 20.9. Electrical schematic diagrams of some common types of transducers and typical circuits used to simulate them during field calibration. Terminals labeled A and B are the signal lead connections to which either the transducer or simulated transducer is connected.

an audible indication and producing a spike on the accelerometer output which may be observed on an oscilloscope (see Figs. 18.12 and 18.13).

SYSTEM CALIBRATION BY COMBINING THE CHARACTERISTICS OF THE SYSTEM COMPONENTS

An over-all system calibration can be determined by combining the measured electrical characteristics of all components in the system from transducer to recorder. Obtaining a system calibration in this way circumvents the difficulties of precise field calibration, but it requires that each element in the system be calibrated in the laboratory with extreme care and that the effects of the source and load impedances be completely accounted for. Thus, a system calibration is subject to the sum of the experimental errors introduced by the calibration of each element in addition to any errors resulting from improper simulation of, or accounting for, loading effects. In general, the calibration of each element is performed before the system is assembled, and so this method is subject to (1) error because of the possibility of undetected damage to components between calibration and use and (2) errors resulting from improper connections, misidentification, or confusion in polarity. While this type of calibration saves time and conserves trained personnel, an independent check (such as that provided by the method given below) usually is advisable.

VOLTAGE SUBSTITUTION METHOD OF SYSTEM CALIBRATION

A suitable "simulated transducer" for use in field checkout must duplicate the electrical outputs of the actual transducer for the various vibration conditions to be simulated. To do this, it must either (1) reproduce the electrical voltage or current-generating characteristics of the actual transducer and must have the same output impedance or (2) duplicate the electrical quantity generated by the actual transducer when connected to its load. Failure to meet these conditions will result in different electrical loading of the actual and simulated transducers, and will probably cause calibration errors. It also is important that the simulated transducer have the same electrical grounding configuration as the actual transducer; otherwise electric-circuit noise and cross-talk effects will not be represented accurately when the simulated transducer is in use. Typical examples of circuits which simulate transducers are shown in Fig. 20.9. The "simulated transducer" introduces an electrical signal into the measurement system, thereby simulating the response of the actual transducer. For checks of phase shift and polarity, it is necessary to observe the output of both the simulated transducer and the over-all system output of a signal measurement channel, as functions of time. This can often be done by temporarily connecting the simulated transducer reference output to some other recorder channel with known phase-shift characteristics.

If a simulated transducer output is connected to only one measurement channel and the output of the other measurement channels is observed, the amount of interference or cross talk between channels can be observed. Such cross talk may appear as a reproduction of the simulated transducer output or as some "noise" of unrecognized form. It can be distinguished from other electrical disturbances because it is a function of the output of the simulated transducer and disappears when the simulated transducer output is removed.

Small, self-contained, self-excited vibration calibrators that can generate a known vibration level are available for field use. Such devices can be used to compare transducers or to calibrate working transducers against a laboratory standard.*

REFERENCES

1. Anon.: *Instruments and Control Systems*, **32**:536 (1959), Instruments Publishing Co., Pittsburgh, Pa.

* For example, General Radio Co., Vibration Calibrator Type 1557-A.

2. Hernandez, J. S.: "Introduction to Transducers for Instrumentation," Statham Instruments, Inc., 1959.
3. Anon.: "How to Select an Electromagnetic Velocity Pickup," MB Mfg. Co., New Haven, Conn., 1959.
4. Dippel, R.: *C. E. C. Recordings*, **10**(5):8 (1956), Consolidated Electrodynamics Corp., Pasadena, Calif.
5. Bradley, W.: *Instr. Soc. Amer. J.*, **6**:55 (1959).
6. Jacobsen, L. S., and R. S. Ayre: "Engineering Vibrations," p. 517, McGraw-Hill Book Company, Inc., New York, 1958.
7. Brüel and Kjaer: *"Instructions and Applications, Accelerometer Sets 4308 and 4309,"* Naerum, Denmark, May, 1957.

Date Due